Molecular and Cellular Biology of the Vitamins

Molecular and Cellular Biology of the Vitamins is a key resource describing how vitamins function as physiologically active molecules at the cellular level. The contents of the book are divided into four sections: a thorough introduction; biological perspectives; fat-soluble vitamins; and water-soluble vitamins. Vitamin chapters cover information on chemical structures; intestinal absorption; plasma transport and metabolism; biochemical and physiological actions; regulations of gene expression; immunological properties; and deficiency-related diseases. The 'perspectives' chapters facilitate the understanding of vitamin biology, including the theory of biochemistry, physiology, endocrinology, molecular genetics, and immunology.

Features

- Facilitates learning and understanding through a logical flow of information.
- Discusses vitamin 'behavior' across a wide range of biological disciplines.
- Discusses immunological and deficiency-related diseases including coronary artery disease, diabetes, and cancer; and potential toxicity.

Molecular and Cellular Biology of the Vitamins appeals to those involved in vitamin research or teaching, postgraduate students studying nutrition or health-related topics, health practitioners, and scientists.

Molecular and Cellular Biology of the Vitamins

G.F.M. Ball

CRC Press
Taylor & Francis Group
Boca Raton London New York

CRC Press is an imprint of the
Taylor & Francis Group, an **informa** business

First edition published 2025
by CRC Press
2385 NW Executive Center Drive, Suite 320, Boca Raton FL 33431

and by CRC Press
4 Park Square, Milton Park, Abingdon, Oxon, OX14 4RN

CRC Press is an imprint of Taylor & Francis Group, LLC

Library of Congress Cataloging-in-Publication Data
Description: First edition. | Boca Raton, FL: CRC Press, 2024. | Includes bibliographical references and index. | Summary: "Molecular and Cellular Biology of the Vitamins is a key resource describing how vitamins function as physiologically active molecules at the cellular level. The contents of the book are divided into four sections including a thorough introduction; biological perspectives; fat-soluble vitamins; and water-soluble vitamins. Vitamin chapters cover information on chemical structures; intestinal absorption; plasma transport and metabolism; biochemical and physiological actions; regulations of gene expression; immunological properties; deficiency-related diseases. The "perspectives" chapters facilitate the understanding of vitamin biology; including the theory of biochemistry, physiology, endocrinology, molecular genetics, and immunology"-- Provided by publisher.
Identifiers: LCCN 2024008649 (print) | LCCN 2024008650 (ebook) | ISBN 9781032525808 (hardback) | ISBN 9781032531038 (paperback) | ISBN 9781003410300 (ebook)
Subjects: MESH: Vitamins--physiology | Vitamins--chemistry | Avitaminosis
Classification: LCC QP771 (print) | LCC QP771 (ebook) | NLM QU 160 | DDC 612.3/99--dc23/eng/20240703
LC record available at https://lccn.loc.gov/2024008649
LC ebook record available at https://lccn.loc.gov/2024008650

ISBN: 9781032525808 (hbk)
ISBN: 9781032531038 (pbk)
ISBN: 9781003410300 (ebk)

DOI: 10.1201/9781003410300

Typeset in Times
by Deanta Global Publishing Services, Chennai, India

Dedication

'A vitamin is a substance that makes you ill if you don't eat it.'

—Albert Szent-Györgyi

Nobel Laureate for physiology or medicine, 1937

Contents

PART I Introduction

PART II Biological Perspectives

PART IV The Water-Soluble Vitamins

Foreword

There has been a plethora of published books on vitamins including several by this author. What stands out about *Molecular and Cellular Biology of the Vitamins* is how the author connected the dots by providing insights into the biochemistry, physiology, and molecular biology of how fat-soluble and water-soluble vitamins interact with the body to improve health. Unlike many books published on vitamins which simply list the water-soluble and fat-soluble vitamins and explain their physiologic functions, G.F.M. Ball has taken a different approach. He wrote this book specifically for readers who have some background and knowledge of biochemistry and molecular biology. The first 14 chapters provide an in-depth understanding of the complexities associated with organ function including digestion and absorption which play a key role in the utilization of both water- and fat-soluble vitamins. Once these fundamental physiologic processes are understood he then delves into the specifics of the biochemistry and molecular biology associated with various physiologic processes, including bone metabolism, lipid metabolism, carbohydrate utilization, protein metabolism, cellular amino acid utilization, coagulation, cytokines signaling, and fatty acid metabolism. He relates how abnormalities in these biochemical processes can impact health. He then drills down further to explain how biologically active water-soluble and fat-soluble molecules including vitamins are recognized by organs, tissues, and cells through various signal transduction pathways. Larger molecules, such as proteins and peptides like insulin, can convey their physiologic messages to the cell by interacting with receptors in the cell membrane. This interaction can lead to a cascade of events or biochemical changes that amplify the initial signaling event at the plasma membrane. In some instances, small molecules such as cyclic AMP or changes in intracellular calcium concentrations are produced, which promote cascading events that ultimately lead to a biologic response. Lipid-soluble molecules like estrogen and some fat-soluble vitamins like vitamin D and vitamin A can enter the cell membrane and once inside the cell they are recognized by a receptor protein that ferries its precious cargo to either cellular organelles or to the nucleus to carry out their biologic function. G.M.F. Ball provides a guide for understanding the complexities of the immune system and how immune cells are capable of fighting infections not only with antibodies but also with antimicrobial peptides. He explains acute and chronic inflammation and furnishes insights into the biologic functions of cytokines and chemokines. Before embarking on explaining the biochemistry, physiology, and molecular biology of the water-soluble and fat-soluble vitamins, in Chapter 14 he discusses factors that help maintain cellular health including the all-important p53, a protein that plays a critical role in regulating genes involved in cell cycle arrest and apoptosis reducing risk for cellular malignancy.

With all the fundamental knowledge provided in the first 14 chapters, G.F.M. Ball focuses on each individual fat-soluble and water-soluble vitamin providing detailed information not only about its food source, bioavailability, and metabolism but also how at a biochemical and molecular level the vitamin is able to carry out its health-promoting biologic function(s). At the end of each vitamin chapter, he relates the health consequences of being deficient in the vitamin. Good examples include the influence of niacin deficiency on apoptosis and cancer, metabolic consequences of thiamine deficiency, and vitamin D deficiency associated with rickets and osteomalacia. A major theme of this book relates to immune function. G.F.M. Ball deftly provides detailed knowledge of how many of the vitamins play a critical role in our immune health. The author's goal for this book is to provide the reader with a true appreciation and understanding of the molecular dynamics of small vitamin molecules, many of which have been appreciated for over 100 years as critically important for preventing some of the most serious and devastating diseases, especially in children. This book is a must read for those who have a science background, especially in biochemistry, and who are interested in understanding and appreciating the health benefits of the vital small molecules, vitamins. Currently there is much sensationalism on social media and the Internet relating to health benefits or lack thereof of nutrients and vitamins for health. It is important to not just accept the latest news release or viral TikTok video as gospel but rather it is important to read the science associated with it. The knowledge in this book will most definitely help the reader in discerning whether there is adequate science to support such claims.

Michael F. Holick

Preface

This book is aimed at people with knowledge of biochemistry who wish to understand what vitamins, as molecules, actually do in the body. A true understanding means reading the laboratory reports published in journals to examine the evidence. I have done this to some extent and this book is based upon what I have learned. To facilitate understanding of complicated aspects such as cell signaling, molecular genetics, and immunology, I have written 13 chapters of perspectives. The perspectives chapters are dominated by 'A Guide to the Immune System' because immunology is a specialized subject that is relevant to vitamin biology.

The discerning reader will appreciate that knowledge of a physiological mechanism, such as the intestinal absorption of a vitamin, is based on evidence acquired piecemeal and may not be complete. Consequently, unexplainable inconsistencies may be apparent.

For dependable collaboration, I thank Lesley Haji-Gholam and Jeremy Nagle of the Science Reference Team at the British Library.

G.F.M. Ball

About the Author

George Frederick Mark Ball was born in Windsor, County of Berkshire, England on May 4, 1942. He spent most of his working life in various commercial and research laboratories concerned with pharmaceutical analysis, drug metabolism, clinical analysis, biochemical analysis, food analysis, and natural product chemistry (isolation of potential new antibiotics). He was also a technical writer preparing dossiers on potential new drugs for submission to the authorities for clinical trial and marketing permissions. He is the author of several books and book chapters on vitamins. He earned the B.Sc. honors degree in agricultural sciences from the University of Nottingham, School of Agriculture, Sutton Bonington, specializing in applied animal physiology.

Part I

Introduction

1 Nutritional Aspects of Vitamins

1.1 DEFINITION AND CLASSIFICATION OF VITAMINS

Vitamins are a group of chemically unrelated organic compounds that are essential in very small amounts for the normal functioning of the human body. They have widely varying biochemical and physiological functions and are broadly distributed in natural food sources. A vitamin in the true sense of the word is a dietary essential: insufficient amounts in the diet may cause deficiency diseases. It is thought that, during the course of evolution, the body's ability to synthesize most of these substances was lost. This presumably happened because obtaining vitamins from the diet is biologically more efficient than the complicated metabolic processes required to synthesize them from simple molecules. Plants have the ability to synthesize most of the vitamins and serve as primary sources of these dietary essentials.

Thirteen vitamins are recognized in human nutrition and these have been classified, according to their solubility, into two groups. The fat-soluble vitamins are represented by vitamins A, D. E, and K; also included are the 50 or so carotenoids that possess varying degrees of vitamin A activity. The water-soluble vitamins comprise vitamin C and the members of the vitamin B group, namely thiamin (vitamin B_1), riboflavin (vitamin B_2), niacin (vitamin B_6), pantothenic acid, biotin, folate and vitamin B_{12}. For several of the vitamins, biological activity is attributed to a number of structurally related compounds known as vitamers. The vitamers pertaining to a particular vitamin display, in most cases, similar qualitative biological properties to one another, but, because of subtle differences in their chemical structures, exhibit varying degrees of potency.

It is often stated that vitamins cannot be produced in the body. However, one must bear in mind the vitamin-synthesizing ability of the bacteria that normally inhabit regions of the gut. It is more accurate to say that vitamins cannot be synthesized by the body itself – yet there are exceptions. Vitamin D can be formed in the skin by solar radiation; vitamin A can be generated from β-carotene; and niacin can be synthesized from the amino acid tryptophan. Although individuals adequately exposed to sunlight can synthesize sufficient vitamin D to meet metabolic demands, vitamin D qualifies as a true vitamin because a deficiency in the diet causes disease if sunlight is inadequate.

1.2 NUTRITIONAL VITAMIN DEFICIENCY

In countries where diets are unbalanced and inadequate, or where there are particular dietary customs, certain typical disease patterns have been shown to be due to vitamin deficiency. Examples of the most commonly observed human diseases are xerophthalmia, rickets, beriberi, pellagra, and scurvy, which result from deficiencies of vitamin A, vitamin D, thiamin, niacin, and vitamin C, respectively. Deficiency of a single member of the vitamin B group is rare in humans because these vitamins are largely found together in nature, and foodstuffs lacking in one member of the complex are likely to be poor in the others. Specific vitamin deficiencies are more demonstrable in animals than in humans, and certain species show overt deficiency signs when deprived of a particular vitamin. Deficiency signs in animals provide visible evidence of biochemical disorders that may not manifest in humans and hence are worthy of mention.

Subclinical deficiency and marginal deficiency are synonymous terms used to describe conditions in individuals who are not clinically nutrient deficient, but who appear to be close to it. An alternative and perhaps better term, proposed by Victor Herbert in 1990, is 'early negative nutrient balance,' which is used when laboratory measurements indicate that an individual is losing more of a nutrient than is being absorbed.

By reference to the sequence of events in the development of vitamin deficiency, Pietrzik (1985) emphasized the importance of preventing functional metabolic disturbances that can evolve into overt clinical symptoms. This sequence can be subdivided into six stages as follows:

- **Stage 1.** Body stores of the vitamin are progressively depleted. A decreased vitamin excretion in the urine is often the first sign. Normal blood levels are maintained by homeostatic mechanisms in the very early stages of deficiency.
- **Stage 2.** The urinary excretion of the vitamin is further decreased and vitamin concentrations in the blood and other tissues are lowered. A diminished concentration of vitamin metabolites might also be observed.
- **Stage 3.** There are changes in biochemical parameters, such as low concentrations of the vitamin in blood, urine, and tissues, and low activity of vitamin-dependent enzymes or hormones. Immune response might also be reduced. Non-specific subclinical symptoms appear, such as general malaise, loss of appetite, or other mental changes.
- **Stage 4.** The biochemical changes become more severe and morphological or functional disturbances are observed. These disturbances might be corrected by vitamin dosing in therapeutic amounts within a relatively short time or vitamin supplementation in amounts of (or exceeding) the recommended dietary

DOI: 10.1201/9781003410300-2

allowances over a longer period. Malformation of cells is reversible at this stage.

- **Stage 5.** The classical clinical symptoms of vitamin deficiency appear. Anatomical alterations characterized by reversible damage of tissues might be cured in general by hospitalization of the patient. In most cases, there are deficiencies of several nutrients and a complicated dietetic and therapeutic regimen must be followed.
- **Stage 6.** The morphological and functional disturbances will become irreversible, finally leading to death in extreme cases.

From the health point of view, Pietrzik (1985) proposed that the borderline vitamin deficiency is represented by the transition from the third to the fourth stage.

The causes of nutritional vitamin deficiency are any one or a combination of the following: inadequate ingestion, poor absorption, inadequate utilization, increased requirement, increased excretion, and increased destruction in the body. The capacity to store vitamins in the body is another aspect to be considered: humans can store thiamin for only about two weeks, whereas vitamin B_{12} can be stored for several years. Other factors that contribute to nutritional vitamin deficiency are listed in Table 1.1.

1.3 BIOAVAILABILITY

The term 'bioavailability,' as applied to vitamins in human nutrition, refers to the proportion of the quantity of vitamin in the food ingested that undergoes intestinal absorption and utilization by the body. Utilization encompasses transport of the absorbed vitamin to the tissues, cellular uptake, and the subsequent fate of the vitamin. The vitamin may be converted to a form which can fulfill some biochemical function. Alternatively, the vitamin may be metabolized within the cell to a non-functional form for subsequent excretion or simply stored within the cell for future use. Bioavailability is an operational term defined by the method used to determine it.

Bioavailability should not be confused with nutrient stability. Whereas food processing can result in the loss of a labile vitamin, the bioavailability of the remaining amount of vitamin is not necessarily altered. Bioavailability is influenced by a diverse range of interacting parameters and therefore the amount of bioavailable vitamin in a diet or individual food can vary considerably.

Absorption of a vitamin depends on the chemical form and physical state in which the vitamin exists within the food matrix. These properties may be influenced by the effects of food processing and cooking, particularly in the case of niacin, vitamin B_6, and folate. In foods derived from animal and plant tissues, the B-group vitamins occur as their coenzyme derivatives, usually associated with their protein apoenzyme. In addition, niacin in cereals and vitamin B_6 in certain fruits and vegetables occur largely as bound storage forms. In milk and eggs, which are derived from animal secretions, the B-group vitamins occur, at least to some extent, in the underivatized form, a proportion of which is associated with specific binding proteins. Vitamins that exist naturally as chemically bound complexes with some other material in the food matrix exhibit lower efficiencies of digestion and absorption compared with the free (unbound) vitamin ingested, for example, in tablet form.

Certain dietary components can retard or enhance a vitamin's absorption, therefore the composition of the diet is an important consideration. For example, the presence of adequate amounts of dietary fat is essential for the absorption of the fat-soluble vitamins. Carotenoids exhibit low bioavailability relative to vitamin A due to the poor digestibility of fibrous plant material. Other ingested substances such as alcohol and drugs may interfere with the physiological mechanisms of absorption.

Biological factors can influence the absorption of a vitamin from a particular food or diet. For example, the absorption mechanism in intestinal epithelium can be adapted physiologically to meet changing metabolic requirements or food deprivation. Malabsorption may occur in the presence of gastrointestinal disorders or disease. Other general factors that influence absorption include the plane of nutrition, metabolic requirements, age, and state of health.

TABLE 1.1
Causes of Vitamin Deficiency

Due to	Caused by
Primary food deficiency	Crop failure
	Food storage losses
Diminished food intake	Poverty and ignorance
	Loss of appetite
	Apathy
	Food taboos and fads
	Pregnancy or sickness
	Dental problems
	Chronic disease
Diminished absorption	Absorption defect diseases
	Parasitic infections
	Malignant diseases
Increased requirements	Increased physical activity
	Infections
	Pregnancy and lactation
	Drug therapy
	Vitamin imbalance
	Rapid growth
Increased losses	Excessive sweating
	Diuresis
	Lactation

Source: from Marks (1975).

REFERENCES

Marks, J. (1975) *A Guide to The Vitamins. Their Role in Health and Disease*. MTP Press, Lancaster.

Pietrzik, K. (1985) Concept of borderline vitamin deficiencies. *International Journal for Vitamin and Nutrition Research*, **27**, 61–73.

Part II

Biological Perspectives

2 Membrane Physiology

2.1 EPITHELIAL CELL STRUCTURE

2.1.1 The Simple Epithelium

An epithelium is a sheet of closely apposed cells that forms a boundary at the interface of the body and its external environment or between fluid compartments within the body. The epithelium rests on a thin sheet of an extracellular matrix, the basal lamina, that is rich in mucopolysaccharides. We are concerned with epithelia that can translocate substances from one side of the epithelium to the other. Epithelia that perform this function are simple epithelia, consisting of a single layer of cells. The constituent cells are said to be polarized because distinct sets of functional surface components concerned with transepithelial transport are localized to separate plasma membrane domains – the apical and basolateral domains. These domains face the luminal and serosal compartments, respectively.

2.1.2 The Plasma Membrane

Individual animal cells of all types are enveloped by a plasma membrane (also called plasmalemma) composed of a lipid bilayer about 5 nm thick with globular proteins either partially embedded in it or attached to the outer or inner surface (Figure 2.1). The lipids are represented by phospholipids, glycolipids, and cholesterol, all of which are amphipathic, i.e. having both hydrophilic and hydrophobic portions. The phospholipids are predominantly phosphoglycerides, which have a glycerol backbone; sphingomyelin, a phospholipid which lacks a glycerol moiety, is a common constituent. The phospholipids are organized into a bilayer, only two molecules thick, with polar heads facing outward toward the aqueous environment and fatty acid tails forming a hydrophobic interior. Each phospholipid layer in this lamellar structure is called a leaflet. The molecular arrangement of the phospholipids represents the stable state of minimum free-energy expenditure and is due to a maximization of hydrophilic and hydrophobic interactions. Glycolipids are distinguished from phospholipids by the presence of one or more sugar residues at the polar head group. Cholesterol contributes to the membrane fluidity, whilst also providing mechanical stability. The cholesterol molecules are orientated in the bilayer with their hydroxyl group close to the polar heads of the phospholipids and glycolipids. This leaves the distal portions of the phospholipid/glycolipid fatty acid tails free and relatively flexible.

The plasma membrane regulates the movement of substances across the impervious lipid bilayer by virtue of its property of selective permeability. This property is attributed to the protein constituents of the membrane. Membrane proteins can be classified into two broad categories – integral and peripheral – according to how they interact with the lipid bilayer. Integral membrane proteins are amphipathic molecules whose hydrophobic regions are embedded within the lipid bilayer and hydrophilic regions are exposed at the membrane's surface. Almost all are transmembrane proteins that span the bilayer in a weaving fashion and account for most membrane-associated receptors and transporters and certain enzymes. Most of the integral membrane proteins are glycoproteins, whose chains of sugar residues invariably protrude to the outside of the cell. These chains, together with those of membrane glycolipids and other structures called proteoglycans, form a carbohydrate coat at the cell surface called the glycocalyx. Peripheral membrane proteins function almost entirely as enzymes. They do not interact with the hydrophobic interior of the lipid bilayer, but instead are bound ionically to the membrane's outer or inner surface. Binding is either indirectly to an integral membrane protein or directly to the polar head of a phospholipid.

2.1.3 The Junctional Complex

Each epithelial cell is adjoined with its neighbor by a junctional complex, which comprises three specialized attachment sites: the tight junction (zonula occludens), an intermediate junction (zonula adherens), and a desmosome (macula adherens) (Figure 2.2). The tight junction constitutes a gasket encircling the apical end of each epithelial cell and separating the luminal fluid from the fluid in the lateral intercellular space. If the epithelium is visualized as a six-pack of beer, with the cans representing the cells, the sheet of plastic holding the cans together corresponds to the tight junction. The intermediate junction is an adhesive band that also completely encircles the cell. Desmosomes are found all over the basolateral membrane and are sites of attachment to filaments of the cytoskeleton. In intermediate junctions and desmosomes, adhesion molecules of the cadherin family form Ca^{2+}-dependent homophilic interactions; that is, one cadherin on an epithelial cell binds to an identical molecule on an adjacent cell.

2.1.4 The Cytoskeleton

The cytoskeleton is an organized meshwork of cytoplasmic protein filaments that give the cytosol gel-like properties. As well as providing structural support for the cell, the cytoskeleton is involved in the movement of organelles and vesicles, chromosome segregation, and the positioning of messenger RNA in the cytoplasm. The three types

DOI: 10.1201/9781003410300-4

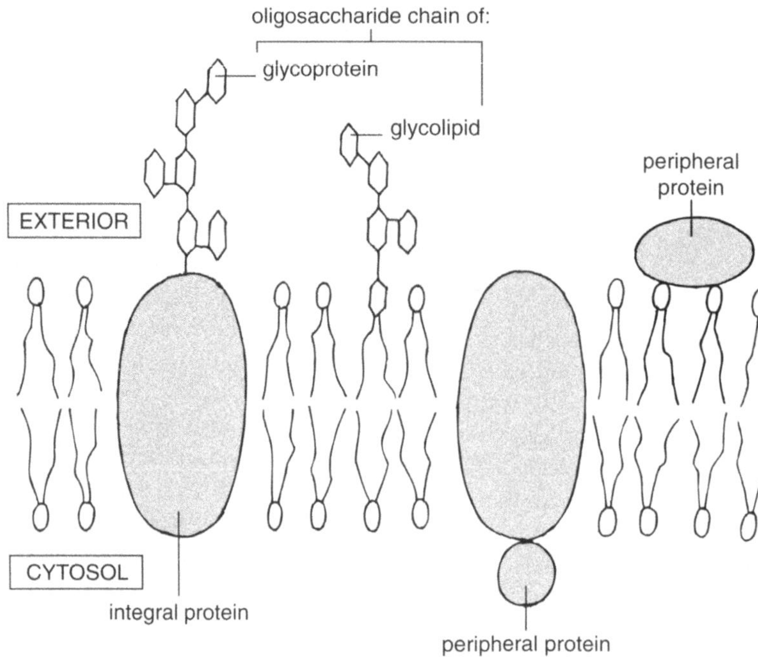

FIGURE 2.1 Schematic diagram of a typical plasma membrane. (Cholesterol molecules not shown.) The lipid bilayer consists of two leaflets of phospholipid molecules whose fatty acid tails form the hydrophobic interior of the bilayer; their polar hydrophilic head groups line both surfaces. Integral proteins mostly span the bilayer. Peripheral proteins are associated with the membrane by protein–protein and lipid–protein interactions. Chains of sugar residues (oligosaccharides) bind to membrane proteins and also to lipids, forming glycoproteins and glycolipids, respectively.

of filaments – intermediate filaments, actin filaments, and microtubules – are linear polymers of monomeric subunits. Intermediate filaments are tough, rope-like structures that provide mechanical stability to cells and tissues. Actin filaments are assembled from actin subunits and have a twisted two-stranded appearance. Microtubules are long, stiff, hollow cylinders composed of tubulin subunits. Actin filaments and microtubules have a distinct polarity (with designated plus and minus ends) and can interact with various motor proteins to produce directed intracellular movements. For example, cargos can be dragged along actin filaments by unconventional myosins, such as myosin-V, and along microtubules by kinesins and dyneins.

2.1.5 Tight Junctions

On freeze-fracture replica electron micrographs, tight junctions appear as a series of discrete sites of fusion involving the outer leaflet of the lateral plasma membrane of adjacent cells at the most apical point. Their function is to prevent the diffusion of macromolecules and impede, to varying degrees, the diffusion of small water-soluble molecules and ions across an epithelial sheet. They also prevent the diffusion of membrane proteins between the apical and basolateral regions of the plasma membrane, thereby maintaining the polarity of epithelial cells. Tight junctions are dynamic structures and their constituent proteins are subject to changes in gene expression, subcellular localization, post-translational modification, and protein–protein interactions. These processes are regulated by intracellular signaling pathways.

At the apparent fusion sites, paired tight-junctional strands (one from each cell) consist mainly of two transmembrane proteins, claudins and occludin. Claudins constitute the backbone of tight-junction strands. At least 24 claudins have been identified in mammals and distinct species of claudins are polymerized together in individual strands. Occludin molecules from adjacent cells interact to form the barrier portion of the tight junction. Another transmembrane protein, junctional adhesion molecule (JAM), associates laterally with the claudin-based backbone of tight-junction strands in epithelia and contributes to cell–cell adhesion. The transmembrane proteins are linked via accessory proteins to the actin cytoskeleton as a means of maintaining stability and function of the tight junction. Accessory proteins include three zonula occludens (ZO-1, ZO-2, and ZO-3), cingulin, AF-6, and 7H6. AF-6 is involved in Ras-mediated signal transduction and its interaction with ZO-1 may be critical in the modulation of the tight junction.

Tight junctions of intestinal epithelia are much more permeable to potassium ions (K^+) and sodium ions (Na^+) than to chloride ions (Cl^-). This cation selectivity has been explained by the tight junctions being lined with hydrated negative charges that discriminate against the passive movement of anions. The permeability of an epithelium to inorganic ions seems to be directly proportional to the transepithelial electrical resistance, measured by placing

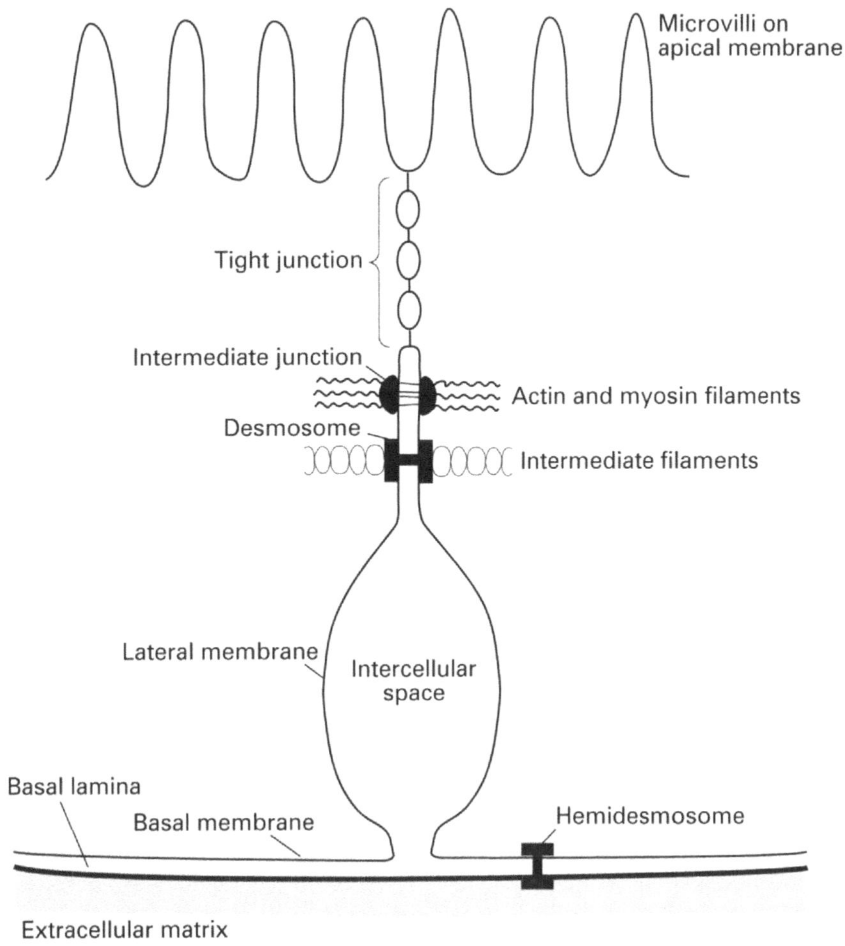

FIGURE 2.2 Diagram of the epithelial junctional complex. Tight junctions and intermediate junctions both form continuous zonular structures circumscribing the cell apex. Desmosomes and hemidesmosomes resemble spot welds between the lateral and basal plasma membranes, respectively.

microelectrodes on either side of the epithelium. Because plasma membrane resistances are, in most cases, relatively high, transepithelial resistance values reflect the resistance to current flow through the tight junction, and hence its ionic permeability. Different epithelia exhibit a wide range of resistance values, depending on the physiological requirements of the tissue. The transepithelial resistance correlates with the number of freeze-fracture strands in the tight junctions of various epithelia. For example, the low resistance (6 ohm cm^{-2}) epithelium of the mammalian renal proximal convoluted tubule has only a single continuous strand, whereas the high resistance (>2,000 ohm cm^{-2}) epithelium of the toad urinary bladder has more than eight strands lying in parallel. This correlation suggests that the strands represent the functionally important elements of the tight junction.

During various physiological processes, such as leukocyte transmigration across an epithelium, the tight junction permeability barrier is temporarily disrupted and then resealed. The resealing process is usually completed within 1 h, suggesting that resealing is accomplished by assembly of pre-existing elements rather than by resynthesis of new proteins. Phosphorylation of occludin may be a mechanism by which occludin localization and function at the tight junction are regulated (Wong, 1997).

It has been demonstrated that activation of Na^+-glucose co-transport in intestinal epithelial cells results in an increased flux of inert solutes through the tight junction (Turner & Madara, 1995) and coincides with myosin ATPase-mediated contraction of the perijunctional actinomyosin ring (Turner *et al.*, 1997). It is likely that ZO-1 plays a role in transducing cytoskeletal signals that regulate tight junction permeability. In this regard, Nusrat *et al.* (1995) reported that Rho, a guanosine triphosphate (GTP)-binding protein, regulated cytoskeletal organization, preferentially in the apical pole of columnar epithelial cells. Adenosine diphosphate (ADP)-ribosylation of Rho increased tight junction permeability and simultaneously displaced ZO-1 from the membrane at sites of tight junctions.

2.1.6 Gap Junctions

While individual cells are capable of maintaining their own functions, the co-ordination of cellular activities within a tissue depends on the exchange of information among constituent cells. Neural and endocrine co-ordination are ways to achieve this goal, but these involve long-range interactions. Intercellular communication can be achieved directly via the connecting channels of gap junctions. Gap junctions are plasma membrane structures formed at points of contact between two similar cells. As viewed by electron microscopy, they appear as clusters of particles (connexons) embedded in the plasma membrane. These clusters are present at very high density (~10^4 gap junctions μm^{-2}) and are referred to as gap junctional plaques. An individual connexon is formed from six subunits of transmembrane proteins of the connexin family. Each connexon is tightly joined to an identical connexon within the plasma membrane of the adjacent cell (Goodenough & Paul, 2009). The paired connexons form water-filled pipes (usually referred to as channels) which permit the ready diffusion of cytoplasmic ions and molecules from cell to cell. The pore size of the mammalian gap junction is such that it excludes molecules larger than ca. 1,000 daltons. Diffusible molecules include most sugars, amino acids, nucleotides, vitamins, and 'messenger' molecules such as cyclic adenosine monophosphate (cAMP).

One function of intercellular communication may be to maintain cellular homeostasis through the exchange of ions, metabolites, and water. In electrically active cells such as neurons, cardiocytes, and smooth muscle cells, gap junctions serve as electrical synapses for the rapid conduction of action potentials. In other cells, intercellular communication may co-ordinate tissue responses to hormones through the exchange of second messenger ions or molecules between hormone-stimulated cells and non-stimulated cells. Intercellular communication is also important in the regulation of mitotic cell proliferation. Many types of cancer cell have a decreased capacity for intercellular communication and a decreased number of gap junctions. This loss of intercellular communication has been correlated with the degree of malignancy of the neoplasm (Klaunig & Ruch, 1990).

Increasing concentrations of intercellular calcium ions (Ca^{2+}) cause a progressive decrease in gap junction permeability by a channel closure mechanism that resembles the shutter in a camera. This channel closure may be mediated by the protein calmodulin, which binds to gap junction proteins in the presence of Ca^{2+}. The Ca^{2+}-activated calmodulin may directly induce the conformational change in the gap junction protein(s) that results in channel occlusion. Calcium ions trigger channel closure whenever it is necessary to isolate a damaged cell.

2.2 MEMBRANE FLUIDITY

An artificial bilayer composed of a single type of phospholipid changes abruptly from a highly ordered gel-like state to a more mobile fluid when the temperature is raised. This change of state is called a phase transition and is due to increased motion about the C–C bonds of the fatty acid chains. The narrow temperature range over which the gel-like fluid transition takes place is known as the transition temperature. The lipid bilayer is in a more fluid state when the fatty acid chains of the phospholipids or glycolipids are not tightly packed. This condition is favored when the chains are unsaturated and of shorter length. *Cis* double bonds produce kinks in the fatty acid chains that make them more difficult to pack, and a shorter chain length reduces the tendency of the hydrocarbon chains to interact with one another. The addition of cholesterol restricts the C–C movements of the part of the fatty acid chains situated closest to the polar heads of the phospholipids, but it separates the flexible tails of the fatty acids and causes the inner regions of the bilayer to become slightly more fluid.

The fluidity of the lipid bilayer allows individual constituent phospholipids to diffuse laterally within their own monolayer, but they are always orientated with their polar heads facing outward. Integral membrane proteins also float quite freely, but tend to be restricted to a limited area of the membrane. Such restrictions give rise to a 'patchwork' of membrane domains, each exhibiting a different protein composition and function.

It is essential for many membrane functions that a precise fluidity of cell membranes be precisely maintained. Certain membrane transport processes and enzyme activities, for example, can be shown to cease when the viscosity of the plasma membrane is experimentally increased beyond a threshold level. Membrane fluidity provides the required motional freedom of the phospholipid molecules, allowing integral proteins to undergo conformational changes associated with their function. By synthesizing a diverse array of phospholipids intercalated with cholesterol, cells maintain an appropriate fluidity of their plasma membranes, as well as of all subcellular membranes.

2.3 THE MEMBRANE POTENTIAL

2.3.1 Some Basic Electrical Principles

The removal of electrons from a neutral body results in a positive charge, as when one electron is removed from the neutral sodium atom to form a sodium ion. The practical unit for measuring charge is the coulomb, which equals 6.2 × 10^{17} electrons. An electric field exists in the space around a charge and extends to infinity. In order to move a charge against the field, work must be performed. In experimental work, the difference in potential between two points is defined as the work necessary to move a unit charge from one point to the other. The practical unit of electrical potential difference is the volt, which is the potential difference against which 1 joule of work is done in the transfer of 1 coulomb.

A difference in potential is associated with a flow of current in a medium. The current can be charge-carrying ions in solutions of electrolytes. The current flow requires energy since it encounters resistance in the medium. In order to produce electrical currents, special sources of electrical energy are available, such as the battery and generator. Such sources of energy are said to produce an electromotive force. In solution, electrical energy is produced by chemical reactions which result in the separation of positive and negative charges at electrodes. Such separation of charge itself constitutes a form of stored energy, which manifests as a difference in electrical potential.

Movement of an ion in a solution in which a potential difference is present requires work to be done with or against two forces: the electric field and any concentration difference of the ion.

2.3.2 Establishment of the Membrane Potential

A difference in electrical charge – the membrane potential – exists across plasma membranes, due to a slight excess of positive ions over negative ions on one side of the membrane and a slight deficit on the other side. The magnitude of this membrane potential is influenced by the composition of the fluid bathing the membrane. The membrane potential of animal cell plasma membranes is typically within the range of –30 and –70 Mv. That is, the potential on the cytosolic (inner) face of the membrane is 30 to 70 mV more negative than the potential on the exoplasmic (outer) face. The membrane is therefore said to be 'polarized.' Taking the thickness of the plasma membrane to be 3.5 nm, a membrane potential of –70 mV works out at 200,000 volts per centimeter. To appreciate what this means, consider that high-voltage transmission lines for electricity utilize gradients of about 200,000 volts per kilometer. We will see that the membrane potential plays an essential role in the transport of substances across the membrane.

Figure 2.3 shows very simply how a membrane potential can be generated in a model system. Let us assume that the membrane separating the two aqueous compartments is permeable to K^+ but not to any other ions. The K^+ concentration of the intracellular compartment is ten times that of the extracellular compartment. Potassium ions tend to diffuse by random motion in both directions across the membrane. However, because the K^+ concentration is initially higher in the intracellular compartment, there is greater net movement of K^+ from inside to outside. This diffusion of K^+ down the concentration gradient, unaccompanied by anions, produces separation of charge and therefore the membrane becomes electrically charged, with the inner face negative with respect to the outer face. This build-up of electronegativity along the inner face of the membrane will tend to retard the outward diffusion of K^+. The membrane potential is the point at which the electric potential across the membrane becomes great enough to prevent further net diffusion of K^+ to the exterior, despite the high K^+ concentration gradient.

In an actual plasma membrane, the membrane potential is established in a similar manner as follows. An ATP-powered sodium-potassium pump in the membrane pumps three Na^+ ions out of the cell and two K^+ ions into the cell for each pumping cycle. This generates a high concentration

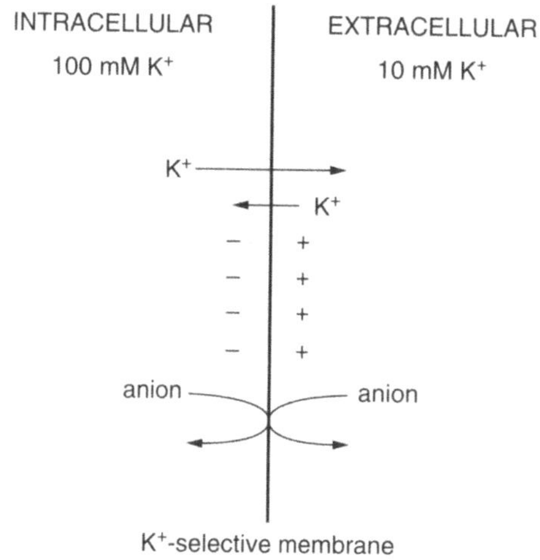

FIGURE 2.3 Diagram showing the principle of how a membrane potential is established. A membrane separates two compartments. The left (intracellular) compartment contains 100 mM K^+ and the right (extracellular) compartment contains 10 mM K^+. The membrane is freely permeable to K^+ but not to the accompanying anion. Net K^+ movement from left to right creates a membrane voltage, left side negative. This buildup of electronegativity will tend to retard the movement of K^+ from left to right until further net movement is prevented. The membrane voltage at this point defines the membrane potential.

of K^+ and a low concentration of Na^+ in the cytosol, relative to concentrations of these cations in the extracellular medium. The net outward diffusion of K^+ through permanently open potassium leak channels generates the inside-negative membrane potential aided by a small contribution by the sodium-potassium pump.

The separation of electric charges that creates the membrane potential occurs only in the immediate vicinity of the membrane, therefore the bulk of the cytosol and extracellular fluid remains electrically neutral. The charge separation involves only a minute fraction of the total number of positive and negative charges that exist in the cell, so ion concentrations are practically unaffected.

The membrane potential of most animal cells generally does not vary with time. In contrast, neurons and muscle cells are electrically active and undergo controlled changes in their membrane potential.

2.4 PROTEIN-MEDIATED MEMBRANE TRANSPORT SYSTEMS

2.4.1 Concept of Transport

The plasma membrane constitutes a selective barrier to the movement of molecules and ions between the extracellular and intracellular fluid compartments. Although physiological concentrations of fat-soluble molecules and small uncharged polar solutes such as urea can simply diffuse through the lipid bilayer, ions and water-soluble molecules having five or more carbon atoms cannot do so and require ion channels or carriers to mediate their transport. Most biologically important substances (e.g. glucose, essential amino acids, water-soluble vitamins, and certain inorganic ions) are translocated across the plasma membrane at rates that are sufficiently rapid to sustain essential metabolic processes.

In the context of this chapter, the term 'vectorial transport' refers to the directed movement of a solute from one side of a membrane to the other, mediated by a membrane-spanning transport protein (transporter). Most transporters are multisubunit protein complexes made up of identical or structurally similar polypeptides held together non-covalently. Each protein transports a particular type of molecule, or a group of closely related molecules. Specificity is imparted by the tertiary and quaternary structures of the transporter molecule – only if a solute's spatial configuration fits into the protein will the solute be transferred across the membrane. Transport proteins exert their effect through a change in their three-dimensional shape (conformational change), and it is this change that limits the rate of transport. The rapidity of protein-mediated transport is due to the fact that the transported molecules are prevented from entering the hydrophobic core of the membrane bilayer, and are therefore not slowed down.

If a transported substance carries a net electrical charge, it is responsive to the membrane potential as well as the concentration gradient. Together, these constitute the electrochemical gradient, which determines the energetically favorable direction of transport.

Solutes can be translocated across cell membranes either by passive movement or by active transport. Passive movement is diffusion down a concentration or electrochemical gradient and uses the energy of the normal kinetic motion of matter. Diffusion can be simple or facilitated, depending on whether a carrier is involved. Active transport is 'uphill' movement against a concentration or electrochemical gradient and utilizes ATP hydrolysis either directly or indirectly as a source of energy.

Membrane transport proteins fall into three main categories: carriers, ion channels, and ATP-powered pumps.

2.4.2 Carriers

Some physiologists prefer the term 'transporter' to 'carrier' because the latter implies that the protein ferries the transported solute from one side of the membrane to the other. This was once thought to be the case, but it is now known that it is the conformational change which brings about solute translocation. A carrier must undergo a conformational change in every transporting cycle and transports molecules at a rate of 10^2–10^4 per second.

Carriers transport specific molecules or ions. The interaction of a transportable solute with its carrier is characterized by saturation at high solute concentration, stereospecificity, and competition with structural analogs. These properties are shared by the interaction of a substrate and an enzyme, thus the carrier may be considered as a specialized membrane-bound enzyme. Two important parameters used to describe the kinetics of enzyme reactions are the Michaelis constant (K_m) and maximal velocity (V_{max}). These same parameters can be used to describe the kinetics of carrier-mediated transport. Figure 2.4 shows the relationship between solute concentration and the rate of carrier-mediated solute transport. The maximum rate of transport (V_{max}) is the point at which all of the available binding sites on the carrier are occupied by solute – a further increase in the solute concentration has practically no effect on the transport rate. V_{max} values are typically expressed in picomoles of solute per milligram protein per unit time (min). K_m describes the binding affinity of the carrier for the solute and is defined as the concentration of solute (expressed in units of molarity, mM) at which half of the available carrier sites are occupied. The value of K_m therefore represents the solute concentration at which the rate of transport is half-maximal ($V_{max}/2$). The lower the value of K_m, the greater the affinity of the carrier for the solute and the greater the transport rate. Each carrier protein has a characteristic K_m, which is independent of the amount of carrier.

When two solutes are transported simultaneously on the same carrier, the observed K_m for one of them will be an 'apparent K_m' and will depend on the concentration of the second solute. The concept of an apparent K_m in this situation is, in principle, the same as that in the context of an enzyme in the presence of a competitive inhibitor.

Some carrier proteins are uniporters, which simply transport a single solute from one side of the membrane to the other down a concentration gradient at a rate determined by V_{max} and K_m. Others, with more complex kinetics, are coupled carriers (also called co-transporters), which transport two different solutes simultaneously, either in the same direction (symport) or in opposite directions (antiport). In coupled transport, either symport or antiport, the free energy released during the movement of an inorganic ion (usually Na^+) down an electrochemical gradient is used to power the movement of the accompanying solute against its concentration gradient.

Carriers in sodium-coupled transport systems have sites for both the sodium ion and the accompanying substance. If

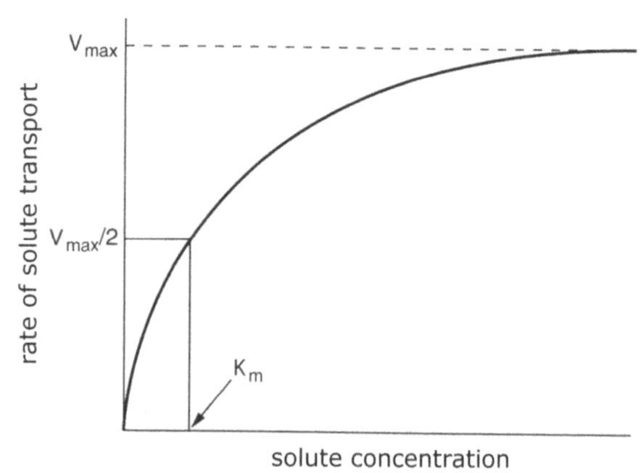

FIGURE 2.4 Relationship between solute concentration and the rate of carrier-mediated solute transport.

the accompanying substance is an anion (negative charge) and one anion is co-transported with one sodium ion (positive charge), the loaded carrier bears no net charge and responds solely to the sodium concentration gradient. If, however, the loaded carrier bears a net positive charge (e.g. a 2:1 ratio of Na^+ to anion$^-$), it is responsive to the negative membrane potential as well as the sodium concentration gradient; together, these constitute the electrochemical gradient for sodium.

2.4.3 Ion Channels

Ion channels are integral membrane proteins that allow small specific inorganic ions, mainly Na^+, K^+, Ca^{2+}, or Cl^-, to diffuse rapidly down their electrochemical gradients across the plasma membrane. The proteins form a narrow hydrophilic pore across the membrane through which ions can pass in single file at rates of 10^6–10^8 per second – very much faster rates than transport mediated by carrier proteins. The pore creates a direct link between the cytosol and the cell exterior, and therefore there is no need for solute binding. Ionic diffusion through channels may exhibit saturation, but usually only when concentrations are well beyond the physiological range.

Ion channels are highly selective and can distinguish, among ions of the same charge, those whose diameters differ by less than 0.1 nm. Selectivity results from the characteristics of the hydrophilic pore in the protein, namely its diameter, its shape, and the nature of the electrical charge along its inside surface. Some ion channels in nerve and muscle membranes are 100 times more permeable to K^+ than to Na^+, even though sodium ions are actually smaller than potassium ions.

Many types of ion channel open only in response to specific stimuli and close rapidly and spontaneously within milliseconds of having opened. Such channels are called gated channels, where the gate is the part of the channel protein that undergoes conformational change during opening and closing. Gated ion channels operate in an all-or-nothing fashion – the gate is either open or closed. They may open in response to changes in the membrane potential (voltage-gated channels), to mechanical stress (mechanically gated channels), or to a ligand binding to a cell-surface receptor (ligand-gated channels). The ligand can be either an extracellular mediator – specifically, a neurotransmitter – or an intracellular mediator, such as an ion or a nucleotide. Other types of ion channel are ungated and permanently open. Among these, K^+ leak channels play a crucial role in maintaining the membrane potential across plasma membranes.

2.4.4 ATP-powered Pumps

ATP-powered pumps are transmembrane proteins with one or more binding sites for ATP in their cytosolic domains. There are four classes of such pumps, designated P, V, and F class and the ABC (ATP-binding cassette) superfamily. P-class pumps transport only inorganic ions (H^+, Na^+, K^+, and Ca^{2+}), V- and F-class pumps transport only protons (H^+). Each of the several hundred ABC proteins transports a specific substrate or group of related substrates. ABCB11, for example, transports bile acids. The energy that drives these pumps is obtained directly from the hydrolysis of metabolically derived ATP by an ATPase that is inherent to the pump. The energy stored in the terminal anhydride bond of ATP is not dissipated, but is used to move ions (in the case of ion pumps) uphill against an electrochemical gradient.

The sodium-potassium pump of the P class (Na^+/K^+-ATPase) operates as an antiporter, actively transporting Na^+ out of the cell against its steep electrochemical gradient and transporting K^+ into the cell at a ratio of three Na^+ ions to two K^+ ions. The net outward movement of cations constitutes an electrical current, creating a potential difference across the membrane, with the inside negative relative to the outside. This electrogenic effect of the pump, however, seldom contributes more than 10% to the membrane potential. The remaining 90% is due to the net outward diffusion of K^+ through potassium leak channels, and so depends on the sodium-potassium pump only indirectly. The sodium-potassium pump is required to maintain osmotic balance and stabilize cell volume: without its function, most cells of the body would swell until they burst. In addition, the electrochemical gradient for Na^+ established by the sodium-potassium pump provides the energy for secondary active transport of solutes across membranes. The cardiac glycoside drug ouabain competes with K^+ for its binding site and specifically inhibits the sodium-potassium pump's action.

Proton pumps have been described in the membranes of various intracellular compartments concerned with endocytosis and potocytosis, such as clathrin-coated vesicles, plasmalemmal vesicles derived from caveolae, endosomes, and lysosomes, where their function is to acidify the lumen (Mellman *et al.*, 1986; Anderson & Orci, 1988). The pumping of protons across a vacuolar membrane from one compartment to another will generate an electrical potential across the membrane, and this will oppose further movement of protons. For proton movement to continue, there is an accompanying movement of an equal number of chloride anions in the same direction.

2.5 SIMPLE AND FACILITATED DIFFUSION

2.5.1 Simple Diffusion

The most straightforward mechanism for moving substances across cell membranes or within the cell itself is simple diffusion. Diffusion is the random movement of substances that ultimately results in their equal distribution. All molecules and ions in liquids or in gases undergo continual random jumping movements, driven by their inherent thermal energies, and are continually colliding with one another. These movements (collectively known as Brownian motion) and the impact of the collisions between individual molecules causes the displacement of molecules from one location to another. If two solutions of a particular solute at different concentration are separated by a permeable membrane, solute molecules will migrate unaided along the concentration gradient from the region of higher concentration to the region of lower concentration. Eventually, the molecules will be equally distributed throughout the area that encloses them – a state of equilibrium. At equilibrium, individual molecules will continue to migrate randomly back and forth across the membrane, but there will be no *net* migration of molecules across the membrane.

The net rate of diffusion of a solute across a cell membrane depends on the permeability of the membrane and the pressure difference across the membrane. In the case of an uncharged solute, the sole driving force is the difference in solute concentration between the two sides of the membrane. (Note that an uncharged solute refers to a molecule that bears no net electrical charge; this includes solutes that have an equal number of positive and negative charges). In the case of permeating ions, the driving force is the electrochemical gradient, which is a combination of the concentration difference and the membrane potential. Because the inside of the plasma membrane is negative with respect to the outside, the membrane potential favors the entry of cations into the cell, but opposes the entry of anions.

Lipid-soluble molecules readily pass through the plasma membrane by dissolving in the lipid matrix and diffusing through the phospholipid bilayer. Permeation of water and small non-electrolytes (e.g. urea and ammonia) through the bilayer is permitted by the formation of defects or kinks in the hydrocarbon chains, which disturb hydrophobic interactions between adjacent chains. The more fluid the lipid bilayer, the greater the permeability. Charged molecules, because of their shell of water molecules, are insoluble in the lipid bilayer and too big to pass through the narrow channels between adjacent hydrocarbon chains. However, specific inorganic ions can diffuse through membranes via ion channels.

2.5.2 Facilitated Diffusion

Facilitated diffusion refers to the carrier-mediated diffusion of molecules across the plasma membrane. Substances translocated in this way cannot usually pass through the membrane without the aid of a specific carrier protein. As in simple diffusion, the driving force is the concentration gradient or electrochemical gradient between the inside and the outside of the cell. Facilitated diffusion differs from simple diffusion in that it exhibits the characteristic properties of carrier-mediated processes, i.e. saturation kinetics, susceptibility to competitive inhibition, and solute specificity.

The classic example of facilitated diffusion is glucose transport across the plasma membrane of erythrocytes. The glucose concentrations in erythrocytes are lower than those in the extracellular fluid because the sugar is rapidly metabolized by these cells after gaining entry. The glucose transporter (a uniporter) contains a single glucose-binding site and alternates between two conformational states, one facing the exoplasmic (outside) surface of the membrane and the other facing the cytoplasmic (inside) surface. Only the thermal energy of the system is required for the conformational change to take place. A molecule of extracellular glucose binds to the outward-facing site, which then reorientates to face the inside of the cell. After release of the glucose into the cell interior, the transporter (without a bound glucose molecule) undergoes the reverse conformational change to recreate the outward-facing binding site.

The facilitated diffusion of glucose in erythrocytes occurs at a rate that is at least a hundred times faster than that predicted for simple diffusion. Specificity is high: for example, the K_m for the transport of the non-biological L-isomer of glucose is >3,000 mM (cf. 1.5 mM for D-glucose).

2.6 PRIMARY AND SECONDARY ACTIVE TRANSPORT

Primary active transport is carried out exclusively by ATP-powered pumps, which utilize the energy derived directly from ATP hydrolysis. In epithelial cells of small intestine and kidney, the sodium-potassium pump is situated in the basolateral membrane, and not in the apical membrane. The pumping of sodium ions out of the cell creates an inward downhill sodium concentration gradient across the cell. This gradient represents a storehouse of energy because the excess sodium outside the cell is always attempting to diffuse back into the cell. Transport systems driven by the energy of the sodium gradient are referred to as secondary active transport because the primary source of energy is the ATP-powered pump.

Secondary active transport is mediated by a coupled carrier – either a symporter or an antiporter. In sodium-coupled co-transport, a symporter mediates the movement of sodium ions and the accompanying solute in the same direction. The transport of glucose across the apical membranes of intestinal and renal epithelial cells is an example of sodium-coupled co-transport. In sodium-coupled countertransport, an antiporter mediates the movement of sodium and the accompanying solute in opposite directions. An example of an antiporter is the sodium–hydrogen exchanger, which mediates the influx of sodium ions and efflux of hydrogen ions as a means of regulating intracellular pH. The exchange takes place with a stoichiometry of 1 Na^+:1 H^+ and hence is electroneutral.

2.7 ENDOCYTOSIS AND POTOCYTOSIS

Endocytosis and potocytosis are mechanisms that cells use to internalize nutritive molecules from the extracellular environment. Both mechanisms are energy-dependent and tightly controlled. The principal function of endocytosis is to deliver macromolecules to lysosomes for hydrolytic processing or to transport them across polarized cells by transcytosis (Rodman *et al.*, 1990). Pinocytosis refers to the trapping of a portion of the extracellular fluid during endocytosis. For the internalization of certain vital macromolecules in a minimum of extracellular fluid, receptor-mediated endocytosis provides much greater rates than ordinary endocytosis. Potocytosis describes the sequestration and transport of molecules and macromolecules by caveolae.

2.7.1 Receptor-Mediated Endocytosis

A common feature among the various pathways of receptor-mediated endocytosis is that receptor–ligand complexes are conveyed in coated pits and coated vesicles. However, there are differences in the routes that ligands and receptors follow after entering the cell. In the endocytosis of plasma low-density lipoprotein (LDL), the macromolecular ligands bind to specific cell-surface receptors. This binding triggers clustering of receptor–ligand complexes at specialized internalization sites known as coated pits. The 'coat' is formed entirely of the protein clathrin, whose heavy and light chains are organized to form a network on the cytoplasmic face of the plasma membrane. The coated pits pinch off to form intracellular coated vesicles enclosing the receptor–ligand complexes. Within the cytosol, the vesicles are uncoated, after which they fuse with one another to form endosomes, whose contents are acidified by a proton pump. Within the endosomes, the ligands dissociate from their receptors, and the ligands are carried further to lysosomes, where they undergo enzymatic hydrolysis. The receptors leave the endosomes, apparently via incorporation into the membrane of vesicles that bud off from the endosome surface, and are recycled to the cell surface to bind new ligands.

2.7.2 Potocytosis

Potocytosis utilizes membrane proteins to concentrate small molecules before internalization. The membrane proteins are anchored by glycosylphosphatidylinositol (GPI) chains and are organized into clusters within flask-shaped invaginations of the plasma membrane known as caveolae. There are at least three distinct ways that GPI-anchored proteins can concentrate molecules: namely, receptor-mediated, enzyme-mediated, and carrier-mediated potocytosis (Anderson, 1993).

Molecules internalized by potocytosis follow one of four distinct intracellular routes. From the surface they travel either to (1) the cytoplasm, (2) the endoplasmic reticulum, (3) the opposite cell surface for transcytosis, or (4) a caveola-derived tubular/vesicular compartment (Anderson, 1998). Each pathway depends on the formation of a closed caveola compartment. This compartment can be a plasmalemmal vesicle that has detached from the membrane. Sometimes, however, caveolae can create compartments that are sealed off from the extracellular space, but not physically separated from the membrane (Mineo & Anderson, 2001).

Invagination of the plasma membrane to form caveolae is controlled by protein kinase Cα (PKCα), which is constitutively present in the membrane. The membrane also contains a 90-kDa protein PKCα substrate that, when phosphorylated, initiates invagination of the plasma membrane. Once the caveola has formed, the PKCα activity declines. The 90-kDa protein is then dephosphorylated by a resident protein phosphatase, which results in the return of the plasmalemmal vesicle to the cell surface. The kinase and phosphatase are targets for regulatory hormones such as histamine (Smart *et al.*, 1995). A 68-kDa protein identified as serum deprivation response protein (Sdr) is involved in targeting PKCα to caveolae (Mineo *et al.*, 1998).

2.8 TRANSEPITHELIAL MOVEMENT OF SOLUTES

At many places in the body, substances must be translocated all the way through an epithelium instead of simply through the cell membrane. Movement of this type occurs, for example, through the epithelia of the intestine, renal tubules, and choroid plexus. The vectorial nature of such movement is made possible by the polarity of the cell surface, whereby distinct sets of surface components (carriers, ion channels, and ion pumps) are localized to separate plasma membrane domains. Transepithelial movement may involve concentrative active transport through the apical membrane domain and facilitated diffusion for the downhill exit through the basolateral membrane domain.

Molecules crossing a simple epithelium (one consisting of a single layer of cells) can do so by (1) a transcellular route, via the apical cell membrane, cytoplasm, and basolateral cell membrane, or (2) by a paracellular route or shunt pathway, via the tight junction and underlying lateral intercellular space, bypassing the cells. By selectively controlling the passive diffusion of ions and small water-soluble molecules through the shunt pathway, the tight junction maintains any gradients created by the activity of pathways associated with the transcellular route. In leaky epithelia having electrical resistances of $<1,000$ ohm cm^{-2}, the shunt pathway is the major route for passive transepithelial flow. This is borne out by the work of Frizzell and Schultz (1972) who showed that at least 85% of passive ion flow across the fairly leaky epithelium of the small intestine takes the paracellular route.

REFERENCES

Anderson, R.G.W. (1993) Potocytosis of small molecules and ions by caveolae. *Trends in Cell Biology*, **3**, 69–72.

Anderson, R.G.W. (1998) The caveolae membrane system. *Annual Review of Biochemistry*, **67**, 199–225.

Anderson, R.G.W. & Orci, L. (1988) A view of acidic intracellular compartments. *Journal of Cell Biology*, **106**, 539–43.

Frizzell, R.A. & Schultz, S.G. (1972) Ionic conductances of extracellular shunt pathway in rabbit ileum. Influence of shunt on transmural sodium transport and electrical potential differences. *Journal of General Physiology*, **59**, 318–46.

Goodenough, D.A. & Paul, D.L. (2009) Gap junctions. *Cold Spring Harbor Perspectives in Biology*, **1**, a002576.

Klaunig, J.E. & Ruch, R.J. (1990) Role of inhibition of intercellular communication in carcinogenesis. *Laboratory Investigation*, **62**, 135–46.

Mellman, I., Fuchs, R. & Helenius, A. (1986) Acidification of the endocytic and exocytic pathways. *Annual Review of Biochemistry*, **55**, 663–700.

Mineo, C. & Anderson, R.G.W. (2001) Potocytosis. *Histochemistry and Cell Biology*, **116**, 109–118.

Mineo, C., Ying, Y.-S., Chapline, C., Jaken, S. & Anderson, R.G.W. (1998) Targeting of protein kinase Cα to caveolae. *Journal of Cell Biology*, **141**, 601–10.

Nusrat, A., Giry, M., Turner, J.R., Colgan, S.P., Parkos, C.A., Carnes, D., Lemichez, E., Boquet, P. & Madara, J.L. (1995) Rho protein regulates tight junctions and perijunctional actin organization in polarized epithelia. *Proceedings of the National Academy of Sciences of the United States of America*, **92**, 10629–33.

Rodman, J.S., Mercer, R.W. & Stahl, P.D. (1990) Endocytosis and transcytosis. *Current Opinion in Cell Biology*, **2**, 664–72.

Smart, E.J., Ying, Y.-S. & Anderson, R.G.W. (1995) Hormonal regulation of caveolae internalization. *Journal of Cell Biology*, **131**, 929–38.

Turner, J.R. & Madara, J.L. (1995) Physiological regulation of intestinal epithelial tight junctions as a consequence of Na^+-coupled nutrient transport. *Gastroenterology*, **109**, 1391–6.

Turner, J.R., Rill, B.K., Carlson, S.L., Carnes, D., Kerner, R., Mrsny, R.J. & Madara, J.L. (1997) Physiological regulation of epithelial tight junctions is associated with myosin light-chain phosphorylation. *American Journal of Physiology*, **273**, C1378–85.

Wong, V. (1997) Phosphorylation of occludin correlates with occludin localization and functions at the tight junction. *American Journal of Physiology*, **273**, C1859–67.

3 Functional Anatomy of the Small and Large Intestine, Liver, and Kidney

3.1 THE SMALL INTESTINE

3.1.1 Gross Anatomy

The small intestine (Figure 3.1) is about 500 cm in length and 3 cm in luminal diameter in the adult human. It starts with the duodenum, which is only about 25 cm long, and terminates at the ileocecal valve. The jejunum accounts for approximately two-fifths of the length of the small intestine and the ileum makes up the remaining three-fifths.

The intestinal wall comprises four principal layers as shown in Figure 3.2. The outermost serosal layer is covered with peritoneum and contains blood vessels, lymphatics, and nerve fibers which pass through it to the other layers. The next layer, the muscularis externa, consists of an inner band of circularly orientated and an outer band of longitudinally orientated smooth muscle fibers arranged in a spiral fashion. Contraction of these muscle fibers facilitates (1) segmentation, in which the gut contents are mixed by local contractions of small segments of gut wall, and (2) peristalsis, which refers to propulsive movements stimulated by distension of the gut wall. The third layer, the submucosa, comprises coarse areolar connective tissue and contains plexuses of blood vessels and lymphatics. The innermost layer, the mucosa, consists of an epithelium, lubricated by mucus, resting upon a basal lamina. This, in turn, is supported by a layer of loose, areolar connective tissue termed the lamina propria, which is attached to a thin sheet of smooth muscle, the muscularis mucosa.

To allow a maximum efficiency of absorption, the intestinal mucosa shows specialization that increases its surface area to approximately 300 m^2. Firstly, the mucosa, with a core of submucosa, is thrown into permanent concentric folds called plicae circulares which protrude into the intestinal lumen. The folds are up to 1 cm in height and any one fold may extend two-thirds or more around the circumference of the intestine. Rarely do the folds completely encircle the lumen. The plicae circulares are found most frequently in the distal duodenum and proximal jejunum. Thereafter, they become less frequent and are rare in the distal ileum. As well as increasing the absorptive area, the plicae circulares retard the passage of chyme, allowing more time for effective digestion and absorption.

The absorptive area is greatly increased by the presence of tightly packed projections called villi, each about 1 mm long. In the human, the villi are numerous in the duodenum and jejunum, but fewer in the ileum. In the proximal duodenum, they are broad, ridge-like structures, changing to tall, leaf-like villi in the distal duodenum/proximal jejunum. Thereafter, they gradually transform into short, finger-like processes in the distal jejunum and ileum.

3.1.2 The Villus

The villus (Figure 3.3) is the functional absorptive unit of the small intestine. Contained within the lamina propria core of each villus is a capillary network with a supplying arteriole and draining venule. A blind-ending lymphatic vessel (lacteal) in the center of each villus drains into a plexus of collecting vesicles in the submucosa. The lacteal is surrounded by smooth muscle elements that extend from the muscularis mucosa. During digestion, the muscularis mucosa contracts, causing the villi to sway and contract intermittently. These movements and contractions may have the effect of 'milking' the lacteals, forcing the lymph and absorbed nutrients (fats) into the lymphatic vesicles. Each villus is covered by an epithelium composed of a single layer of columnar absorptive cells (enterocytes) interspersed occasionally with mucus-secreting goblet cells. Undifferentiated cells in the epithelium of the intestinal glands (crypts) proliferate and migrate upward to replace enterocytes and other cells that are continuously being sloughed off from the tip of the villus into the intestinal lumen. During migration, the cells differentiate and increase their expression of genes involved in nutrient uptake. Complete cell renewal takes place every 5–6 days in the human small intestine.

3.1.3 The Enterocyte

The enterocyte constitutes the only anatomical barrier of physiological significance controlling the absorption of nutrients. As shown in Figure 3.4, the apical membrane of the enterocyte (i.e. the membrane facing the intestinal lumen) is covered with microvilli, which are minute projections of the plasma membrane. The brush-like appearance of the apical membrane under the microscope gives rise to the alternative term 'brush-border' membrane –for the apical membrane. The microvilli are packed at a density of 200,000 per mm^2 in the human jejunum. The basal membrane rests on a basal lamina composed of extracellular matrix proteins. The intercellular space between the lateral membranes only becomes dilated when nutrients and water are being absorbed. When absorption is not taking place, the intercellular space is collapsed and the lateral membranes are closely apposed.

DOI: 10.1201/9781003410300-5

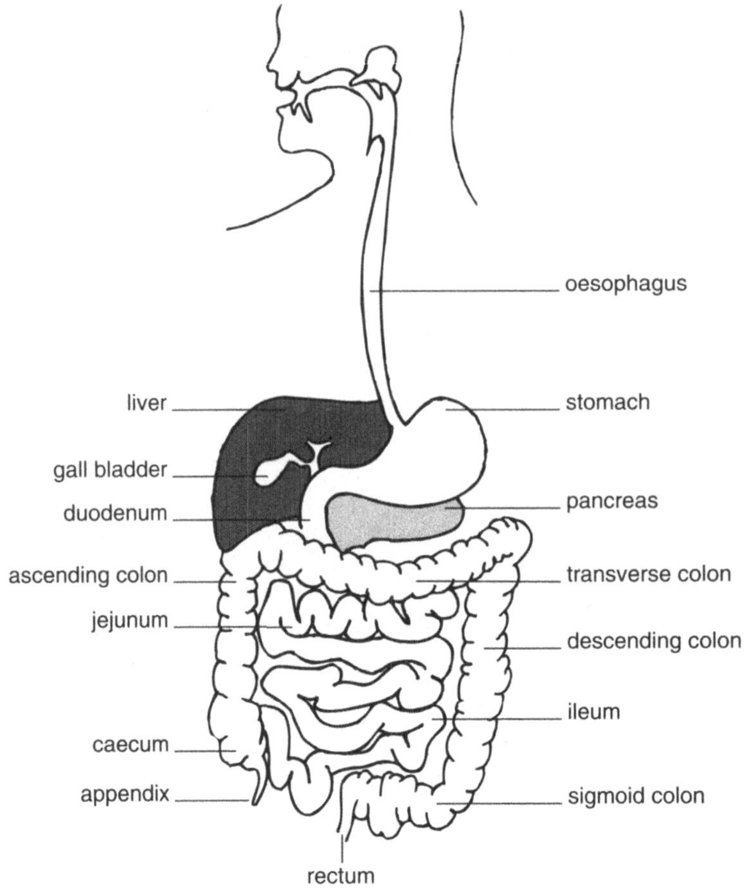

FIGURE 3.1 The gastrointestinal tract.

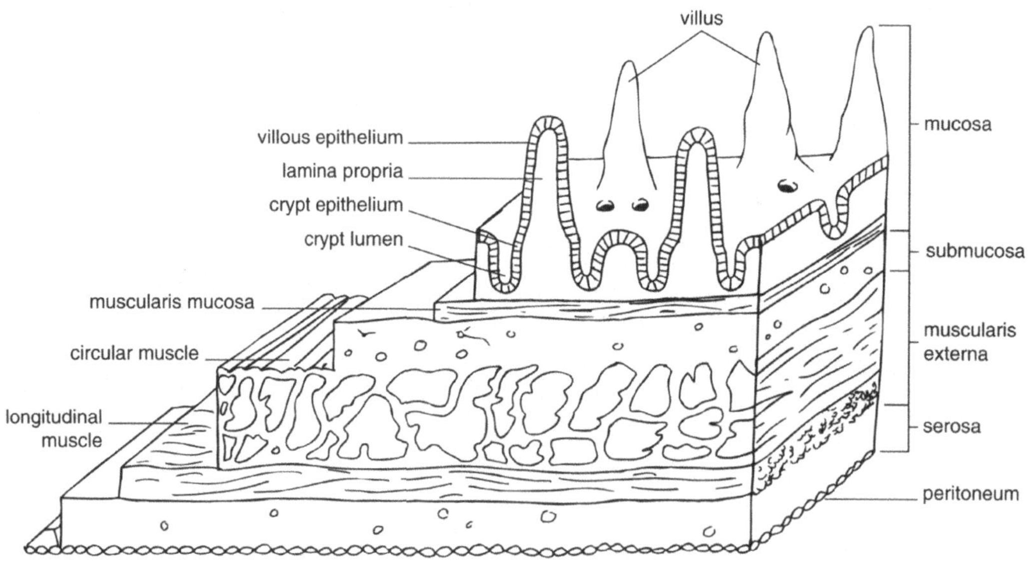

FIGURE 3.2 Structure of the small intestinal wall (sectional view) showing the four principal layers: serosa, muscularis externa, submucosa, and mucosa. The mucosa comprises the epithelium, lamina propria, and muscularis mucosa.

3.1.4 The Luminal Environment within the Small Intestine

The postprandial pH of the stomach contents is 2–3. The pH of the chyme in the small intestine gradually increases from 6 in the duodenum to about 7.4 in the terminal ileum.

Bulk contents of the intestinal lumen are mixed by segmentation and peristalsis, and water and solutes are brought to the surface of the mucosa by convection. However, the luminal environment immediately adjacent to the

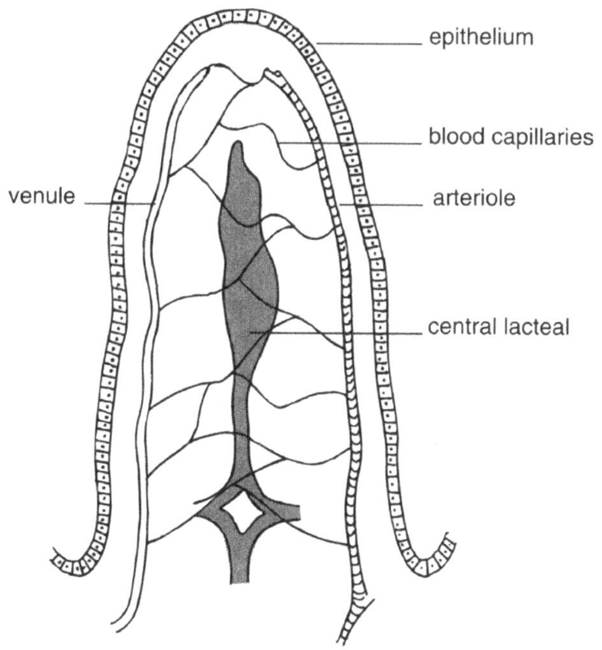

FIGURE 3.3 Blood and lymphatic vessels supplying the villus.

brush-border membrane is stationary and unaffected by gut motility. The lack of convective mixing in this region creates a series of thin layers, each progressively more stirred, extending from the surface of the enterocyte to the bulk phase of the lumen. Together, these are the so-called 'unstirred layer,' whose effective thickness in the human jejunum has been calculated to be 35 μm based on the rate of disaccharide hydrolysis at the brush border, rather than the 600 μm value derived from osmotic transient measurements (Levitt *et al.*, 1992). Solute movement within an unstirred layer takes place by diffusion, which is slow compared with the convective movement in the bulk luminal phase.

Said *et al.* (1986) confirmed earlier reports that the luminal surface of the small intestine is lined by a layer of hydrogen ions, creating an intestinal surface acid microclimate (ISAM). Using a pH microelectrode, the pH at the luminal surface was shown to be approximately two units lower than that of the luminal bulk phase and varied less than ±0.5 units despite large pH variations in the intestinal chyme. The ISAM was confined to the proximal half of the small intestine, where the digestion and absorption of nutrients is most active; the luminal surface of the distal small intestine and colon was found to be neutral. The existence of the ISAM creates an inwardly directed proton gradient across the apical membrane of the enterocyte. The proton-coupled absorption of folate uses the energy generated by the downward movement of protons.

Said *et al.* (1986) found that glucose and sodium were essential for the normal existence and maintenance of the ISAM, as their removal caused significant elevations in the luminal surface pH. The role of glucose appeared to be mediated through its metabolism in the enterocytes, because galactose, a hexose that shares the same transport carrier in the intestine with glucose but cannot be metabolized in the enterocytes, failed to support the ISAM. The Na^+/H^+ exchange mechanism that is known to exist in the intestine might be involved in providing the hydrogen ions of the ISAM. The effect of the metabolic inhibitors dinitrophenol and iodoacetate in elevating luminal surface pH further confirmed the dependence of the ISAM on intracellular metabolism

3.2 GROSS ANATOMY OF THE LARGE INTESTINE

Beginning at the ileocecal valve, the greater part of the large intestine (Figure 3.1) consists of the ascending, transverse,

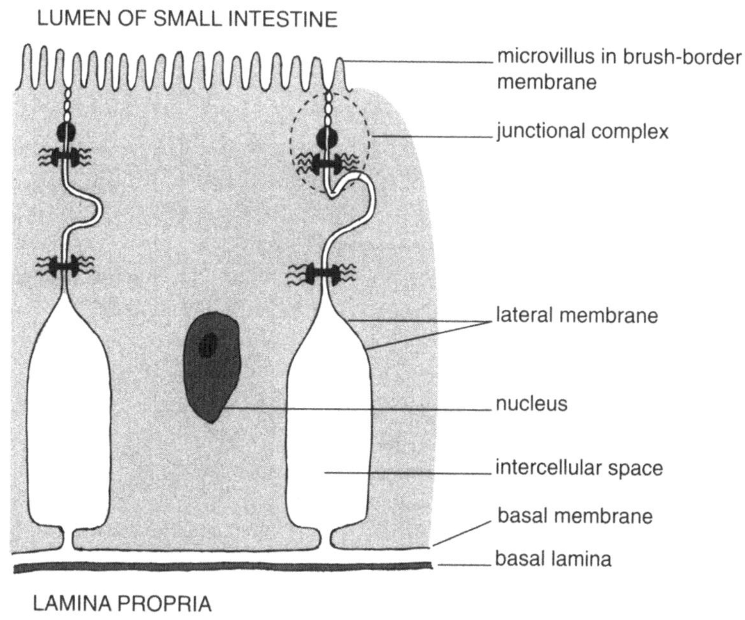

FIGURE 3.4 Longitudinal section of an enterocyte showing the membranes and junction points.

descending, and sigmoid colon. The cecum is a cul-de-sac that projects from the proximal part of the ascending colon and is the major site of bacterial vitamin synthesis in the large intestine. The vermiform appendix, which extends from the cecum, is vestigial in the human. The length of the large intestine in the adult human is around 160 cm. The luminal diameter is 7.5 cm at its commencement and diminishes gradually until it reaches a minimum of 2.5 cm at the rectosigmoid junction. The diameter then increases to form the rectal ampulla, which narrows abruptly at the tonically contracted anal canal.

The well-developed muscularis externa of the large intestine comprises an outer longitudinal and inner circular layer of smooth muscle. Most of the longitudinal muscle fibers aggregrate into three equidistantly placed bundles known as taeniae. The combined contractions of the circular and longitudinal muscles cause the unstimulated portion of the large intestine to bulge outward into baglike sacs called haustra. The mixing movements arising from haustral contractions gradually expose all the colonic contents to the intestinal epithelium. The slow but persistent haustral contractions gradually propel the fluid contents in the cecum and ascending colon analward. From the beginning of the transverse colon to the sigmoid, the haustral contractions are taken over by peristalsis-like mass movements. The submucosa and muscularis mucosa of the large intestine are similar to those of the small intestine. The mucosa of the entire large intestine is devoid of villi. The intestinal glands are larger and more numerous than those in the small intestine, and contain a greater abundance of mucus-producing goblet cells.

3.3 THE LIVER

The human liver is divided into right and left lobes of approximately equal size and is covered by fibrous connective tissue (the Glisson's capsule) from which thin septa enter the matrix. Approximately 75% of the blood entering the liver is venous blood from the portal vein; the remaining 25% is oxygen-enriched arterial blood delivered by the hepatic artery. The portal vein is formed by the union of the superior mesenteric vein and the splenic vein. The former brings nutrient-rich blood from a large part of the alimentary tract, and the latter conveys blood from the spleen, pancreas, gallbladder, and stomach. The liver is the first port of call for digested and absorbed material, including toxins, which the liver either detoxifies or excretes.

The liver matrix comprises 50,000 to 100,000 individual lobules, which take the shape of polygonal cylinders measuring several mm in length and 0.8 to 2 mm in diameter. Lobules are composed mainly of parenchymal cells (hepatocytes) which are organized into bilayered cell cords (also called plates) that radiate from a central vein like the spokes of a wheel. Lying between the hepatic cords are blood-filled spaces called sinusoids. Blood entering the liver through the portal vein and hepatic artery mixes together in the sinusoids and then drains out of the lobule through the central vein. The central vein joins with others to form larger sublobular veins. These unite to form collecting veins, which are tributaries of the right, middle, and left hepatic veins. The hepatic veins drain directly into the inferior vena cava, which leads to the right atrium of the heart.

Bile produced by hepatocytes is secreted into bile canaliculi, which are simply spaces, 1 to 2 μm across, between the two layers of cells. The canaliculi form a branching network which unites to form intralobular bile ductules. These join together in portal canals to form interlobular bile ductules. These ductules drain into the right and left hepatic ducts, which emerge from the liver and unite to form the common hepatic duct. The spatial relationship between the liver cell plates, sinusoids, and bile canaliculi is shown in Figure 3.5.

In cross section, the classical lobule is hexagonal, with a central vein at its center and portal canals at its peripheral corners. Portal canals are supported by a fibrous septum and contain a branch of the portal vein, a branch of the hepatic artery, a bile ductule, and usually a lymphatic vessel, which appears as a slit-like space lined by endothelium. Not all six corners of the hexagon have a portal canal; such carriers are supplied by branches of the portal vein and hepatic artery from an adjacent portal canal giving off smaller branches, which course along the border between adjacent lobules. Around portal canals, a sheet of smaller hepatocytes, one cell thick, is termed the limiting plate. The terminal branches of the portal vein, hepatic artery, and bile ducts penetrate the limiting plate and open into sinusoids.

Sinusoids are lined by a specialized endothelium that has no underlying basement membrane. The endothelial cells may overlap but do not have intercellular junctions. Hepatocytes are separated from the endothelium by a narrow perisinusoidal space called the space of Disse. Numerous long microvilli on the hepatocyte plasma membrane protrude into this space. Perforations in the endothelial cells, along with spaces between adjacent cells, allow plasma protein molecules of up to 0.1 μm in diameter to diffuse freely between the blood of the sinusoids and the interstitial fluid in the space of Disse. Thus, materials in the sinusoidal blood have direct access to the hepatocytes.

Interspersed among the endothelial cells lining the sinusoids are resident macrophages called Kupffer cells, which are capable of destroying certain pathogens when stimulated by an immune response. Wedged between many of the hepatocytes in the space of Disse, and comprising 5–15% of total liver cells, are ameba-like stellate cells (also called Ito cells) that store vitamin A (Figure 3.6).

Cirrhosis of the liver, common in chronic alcoholics, is caused by alcohol poisoning at such frequent intervals that the hepatocytes cannot recover fully between bouts of drinking. When this happens, fibroblasts grow in place of the hepatocytes and the liver becomes irreversibly clogged with connective tissue.

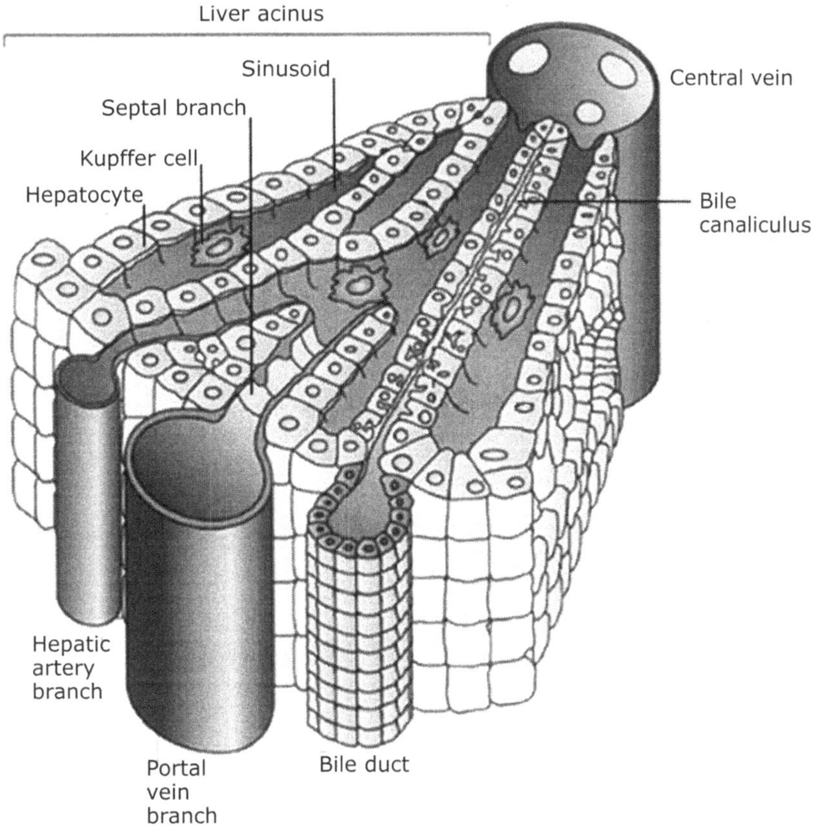

FIGURE 3.5 Diagram of a liver acinus showing the spatial relationship between the liver cell plates, sinusoids, and bile canaliculi. Blood entering the liver through branches of the portal vein and hepatic artery mixes together in the sinusoids and then drains out of the lobule through the central vein. Bile secreted into bile canaliculi by hepatocytes in the liver plate flows out of the liver via the common hepatic duct. Panwar, A.; Das, P.; Tan, L.P. 3D Hepatic organoid-based advancements in LIVER tissue engineering. *Bioengineering* 2021, **8**, 185. https://doi.org/10.3390/bioengineering8110185. Copyright: © 2021 by the authors. Licensee MDPI, Basel, Switzerland. This article is an open access article distributed under the terms and conditions of the Creative Commons Attribution (CC BY) license (https://creativecommons.org/licenses/by/4.0/). Figure 1 on page 3.

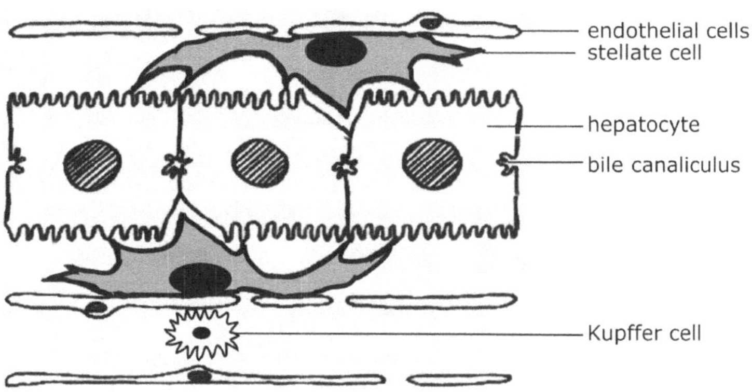

FIGURE 3.6 Stellate cells in the space of Disse. Also shown is a Kupffer cell in the sinusoid.

3.4 THE KIDNEY

The kidney facilitates the elimination of waste metabolic products in the urine, and also subserves functions of fluid and salt balance by excreting these materials in varying quantities dependent upon need. The kidney also plays a vital role in salvaging filtered water-soluble nutrients that would otherwise be wastefully excreted.

In histological section, the kidney consists of an enormous number of tortuous and intertwined uriniferous tubules. Each tubule consists of two functional units: the nephron of 30–40 mm length and the collecting tubules of about 20 mm length. The nephron is responsible for urine production and the collecting tubule is the excretory duct leading to the ureter.

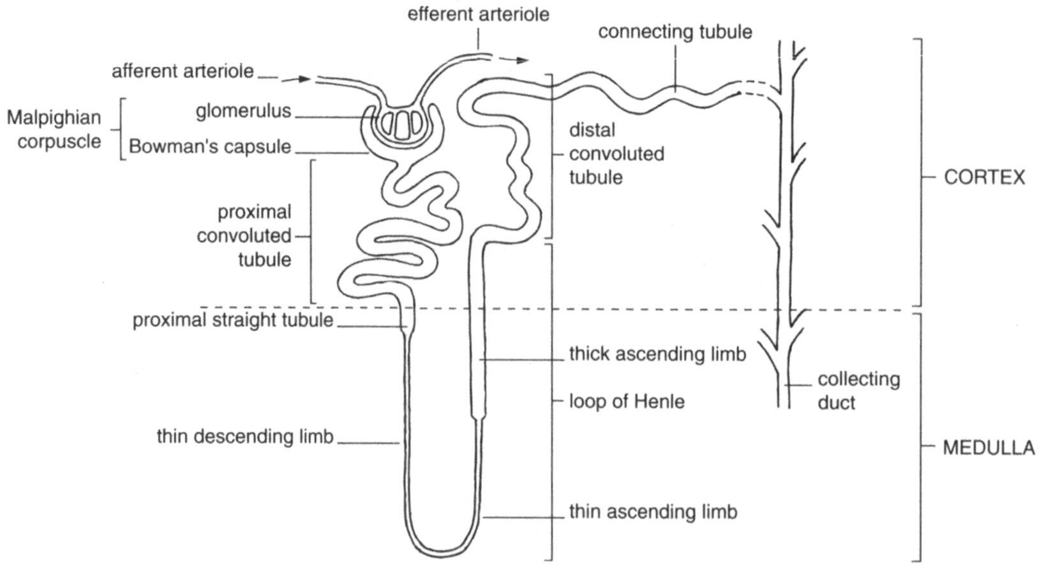

FIGURE 3.7 Diagrammatic structure of a nephron (not to scale).

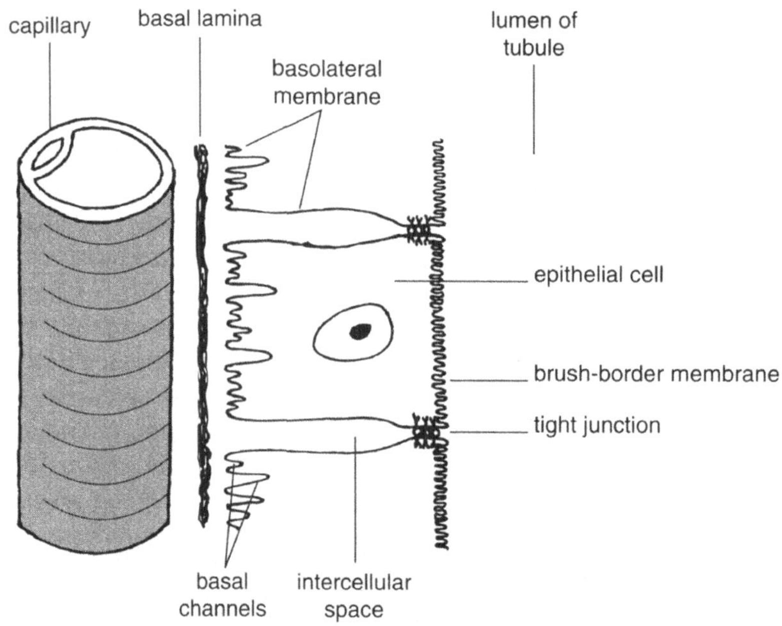

FIGURE 3.8 Longitudinal section of an epithelial cell of the proximal convoluted tubule.

As shown in Figure 3.7, the first part of the nephron is the cup-shaped Bowman's capsule containing a tuft of blood capillaries called the glomerulus. The entire structure is known as the Malpighian corpuscle wherein the process of filtration takes place. The glomerular pressure is sufficient to cause water and simple substances of low molecular weight to pass easily through the walls of the capillaries and Bowman's capsule into the lumen of the capsule. The glomerular filtration is discriminatory with regard to the size of macromolecules, and small proteins will traverse the glomerular barrier relatively easily. Filtration is not selective: both waste products and useful materials move from the blood into the nephron.

The selective reabsorption of essential constituents of the glomerular filtrate into the blood takes place in the proximal convoluted tubule – a region of the nephron which is composed of a single layer of epithelial cells with a brush border of microvilli at the luminal surface (Figure 3.8). The tubules are enveloped by blood capillaries whose walls are highly permeable. The outer membrane of the tubule epithelial cell rests on a basal lamina and is invaginated to form a labyrinth of basal channels. Adjacent tubule cells are separated by intercellular spaces and fluid circulates through these spaces and basal channels. This fluid bathes the cells of the proximal convoluted tubule and the surrounding capillary network.

With regard to the mechanism for transporting solutes across the tubule epithelial cell, an ATP-dependent sodium pump located in the basolateral membrane pumps sodium out of the epithelial cell (primary active transport). The removal of sodium from the cell creates an electrochemical gradient for sodium across the cell, which provides the energy for the carrier-mediated, sodium-coupled transport of solute across the brush-border membrane into the cell (secondary active transport). Transport of solutes across the basolateral membrane of the tubule cell takes place by sodium-independent facilitated diffusion.

REFERENCES

Levitt, M.D., Strocchi, A. & Levitt, D.G. (1992) Human jejunal unstirred layer: Evidence for extremely efficient luminal stirring. *American Journal of Physiology*, **262**, G593–6.

Said, H.M., Blair, J.A., Lucas, M.L. & Hilburn, M.E. (1986) Intestinal surface acid microclimate in vitro and in vivo in the rat. *Journal of Laboratory and Clinical Medicine*, **107**, 420–4.

4 General Physiology of Digestion and Absorption

4.1 GENERAL CONCEPTS

Digestion refers to the chemical and physical modifications which render ingested food constituents absorbable by the small intestine. Absorption is the process by which the products of digestion move from the lumen of the small intestine into the enterocytes and thence into the bloodstream or lymphatic system. Most absorption takes place and is completed in the small intestine. The principal functions of the colon are (1) absorption of water and inorganic salts (mainly sodium) from the chyme to form solid feces, and (2) storage of fecal matter until it can be expelled. A complex meal may be fully digested and absorbed in three hours. Many vital micronutrients, including folate and vitamin B_{12}, are conserved through their excretion in bile and subsequent reabsorption by the small intestine – a process known as enterohepatic circulation.

4.2 DIGESTION

Digestion commences as soon as the food enters the mouth. Chewing helps to break up large particles of food, whilst also mixing food with saliva, which acts as a lubricant and contains the enzymes salivary amylase and lingual lipase. Lingual lipase has only a minor role in digestion of dietary triglycerides; however, salivary amylase plays a major role in digestion of dietary starch. Most of the enzymatic activity of salivary amylase occurs in the stomach, where there is a much longer time for it to interact with the starch.

In the stomach, the churning action and the presence of hydrochloric acid and pepsin convert the food bolus into a liquid chyme. Dietary fat is converted into a coarse lipid emulsion. The stomach has three regions with regard to its glandular secretions: the fundus (in the proximal stomach), the body, and the antrum (in the distal stomach). The folds of the stomach lining contain microscopic gastric pits into each of which drain four or five gastric glands. The mucosa of the body and fundus contains oxyntic glands, whereas the mucosa of the antrum contains pyloric glands. Oxyntic glands are lined by parietal cells that secrete hydrochloric acid and intrinsic factor and by chief cells that secrete pepsinogen (the precursor of pepsin) and gastric lipase. The pyloric glands contain almost no parietal cells or chief cells but, rather, contain mucus-secreting cells and G cells, which produce the hormone gastrin.

The final stages of digestion take place in the duodenum. On arrival at the duodenum, the acidic chyme is buffered by the bicarbonate in pancreatic juice and bile. Brunner's glands in the duodenum produce an alkaline secretion containing mucus. Proteases, amylase, lipase, and other enzymes of pancreatic origin are also secreted into the duodenum. Bile produced by the liver is secreted continuously into the gallbladder, which discharges it intermittently into the duodenum. Bile emulsifies the coarse lipid emulsion into smaller globules, rendering the constituent triglycerides accessible to the hydrolytic activity of pancreatic lipase.

Secretions of the exocrine pancreas are stimulated by the hormones secretin and cholecystokinin. Secretin stimulates the ductile cells to secrete bicarbonate, whilst cholecystokinin stimulates the acinar cells to secrete enzymes. Cholecystokinin and (to a lesser extent) secretin also stimulate contraction of the gallbladder, releasing bile into the duodenum. The secretions of the exocrine pancreas and gallbladder emptying are also stimulated by the vagus nerve.

4.3 ABSORPTION

After the digested food has passed through the wall of the small intestine, nutrients other than fats or fat-soluble substances enter the venules within the lamina propria. The venules eventually lead to the portal vein, which conveys non-lipid products of digestion from the intestine to the liver. Most of the absorbed non-lipid material passes through the liver, whose cells extract nutrients for storage or metabolic processing. The nutrient-enriched blood that has passed through the liver is then sent to the heart, which pumps it to the lungs. The oxygenated blood returns to the heart for redistribution to all parts of the body. The nutrients pass from the blood capillaries into the extracellular fluid from which the cells take up the nutrients they need.

The triglycerides within the fat globules are hydrolyzed by pancreatic lipase and the 2-monoglycerides and free fatty acids combine with micelles composed of bile constituents to form mixed micelles. The mixed micelles dissociate in the acidic microclimate of the jejunum and release the fatty acid components. The fatty acids are absorbed, re-esterified within the enterocytes, and packaged into chylomicrons. The chylomicrons are discharged from the enterocytes and carried by the blood to all tissues. A lipase on capillary endothelia releases free fatty acids, which can be taken up by the tissue cells.

DOI: 10.1201/9781003410300-6

4.4 ADAPTIVE REGULATION OF INTESTINAL NUTRIENT TRANSPORT

Many patterns of adaptation fall into one or the other of two categories: (1) a non-specifically increased absorption of all nutrients, arising ultimately from an increase in the animal's overall nutrient requirements, and with an increase in absorptive surface area as the primary mechanism, and (2) phenomena involving the induction or repression of a specific transport mechanism, depending on the dietary availability or body store of the transported substrate.

4.4.1 Non-Specific Anatomical Adaptation to Changing Metabolic Requirements and Food Deprivation

Increases in metabolic requirements such as arise during pregnancy, lactation, growth, exercise, and cold stress are met by an increased absorption of all available nutrients, mediated at least in part by an induced increase in food intake (hyperphagia). The increased absorption is due to an increase in mucosal mass per unit length of intestine and a consequent increase in absorptive surface area. Not only is there an increase in the total number of cells (hyperplasia) but the villi become taller.

The mammalian intestine adapts to prolonged food deprivation by dramatically slowing the rate of epithelial cell production in the crypts in order to conserve proteins and biosynthetic energy. This effect on mitosis and enterocyte renewal leads to markedly shortened villi. Because cell migration along the crypt/villus axis is also slowed, more cells lining the villi are functionally mature. Therefore, food deprivation, by reducing mucosal mass and increasing the ratio of transporting to non-transporting cells, effectively increases solute transport per unit mass of intestine.

4.4.2 Dietary Regulation of Intestinal Nutrient Carriers

It is well established that certain intestinal nutrient carriers (e.g. those transporting glucose and amino acids) are adaptively regulated by their substrates. In response to a signal for regulation of transport, the number of carriers at both the apical and basolateral membranes of enterocytes is increased or decreased as appropriate. According to the adaptive modulation hypothesis of Karasov (1992), a carrier should be repressed when its biosynthetic and maintenance costs exceed the benefits it provides. The benefits can be provision of either metabolizable calories or an 'essential' nutrient, that is, a nutrient which cannot be synthesized in the body and must be obtained from the diet. Glucose carriers are up-regulated when the dietary supply of glucose is adequate or high because glucose provides valuable calories. The down-regulation of glucose carriers during a deficiency of glucose can be explained by the biosynthesis and maintenance costs outweighing the benefits of transporting this 'non-essential' nutrient.

One might expect carriers for water-soluble vitamins to be down-regulated by their substrates and up-regulated in deficiency of the vitamins. The rationale in this case is that carriers for these essential nutrients are most needed at low dietary substrate levels; at high levels the required amount of the vitamin could be extracted from the lumen by fewer carriers or even cross the enterocyte by passive movement. As vitamins do not provide metabolizable energy, there is nothing to gain from the cost of synthesizing and maintaining carriers when the vitamin supply is adequate or in excess.

The prediction of suppressed transport of vitamins at high dietary intakes has proved to be true for ascorbic acid, biotin, and thiamin, but not for pantothenic acid, for which carrier activity is independent of dietary levels (Ferraris & Diamond, 1989). It appears that intestinal carriers are regulated only if they make the dominant contribution to uptake, as is the case for the three regulated vitamins. It can also be reasoned that transporters for ascorbic acid, biotin, and thiamin would need to be regulated, because nutritional deficiencies of these vitamins can and do occur. In contrast, there is no need to regulate pantothenic acid transporters, because this vitamin is found naturally in almost all foods and cases of deficiency are rare.

4.5 ALCOHOLISM AND ITS EFFECT ON INTESTINAL TRANSPORT

Chronic alcoholism is a complex disorder often characterized by malnutrition, vitamin deficiencies (particularly thiamin, vitamin B_6, and folate), and diarrhea. Among the possible mechanisms for these disturbances are alterations in intestinal digestion, absorption, and secretion. The following is based on a review by Wilson and Hoyumpa (1979) of the possible mechanisms by which ethanol directly affects intestinal transport.

Chronic ingestion of ethanol (1.09 M) results in shortening of the villi in the jejunum and degenerative changes in cellular ultrastructure of both jejunal and ileal villi. However, disturbances in intestinal function have been observed in the absence of such changes and so the changes cannot fully account for the impaired transport of nutrients.

The ethanol molecules (CH_3CH_2–OH) insert into the lipid bilayer of the plasma membrane with their hydrocarbon portions positioned between hydrophobic regions of membrane lipids and proteins. Ethanol causes a molecular rearrangement of membranes, which loosens the membrane and makes it more fluid. This effect of ethanol may increase the permeability of the gut wall and allow passive movement of large antigenic molecules, toxins, and pathogenic organisms through the gut wall and into the bloodstream. It may also interfere with the conformational changes that carrier proteins need to make to effect solute transport. After chronic treatment with ethanol, cells adapt to this change in membrane fluidity by incorporating more rigid fatty acids into membrane phospholipids.

Ethanol inhibits the cell's sodium pump, thereby interfering with sodium-coupled active transport of glucose, amino acids, and certain water-soluble vitamins.

REFERENCES

Ferraris, R.P. & Diamond, J.M. (1989) Specific regulation of intestinal nutrient transporters by their dietary substrates. *Annual Review of Physiology*, **51**, 125–41.

Karasov, W.H. (1992) Tests of the adaptive modulation hypothesis for dietary control of intestinal nutrient transport. *American Journal of Physiology*, **263**, R496–502.

Wilson, F.A. & Hoyumpa, A.M. Jr. (1979) Ethanol and small intestinal transport. *Gastroenterology*, **76**, 388–403.

5 Glucose Transport

5.1 *IN VITRO* TECHNIQUES USED TO STUDY GLUCOSE TRANSPORT

5.1.1 Introduction

Glucose absorption has been well studied and the experimental techniques and postulated mechanisms help toward understanding absorption of water-soluble vitamins.

Caco-2 cells, a human-derived colonic carcinoma cell line, have been used extensively as an *in vitro* model system to study intestinal nutrient uptake. When grown as confluent monolayers, Caco-2 cells differentiate spontaneously into cells that exhibit the morphology of mature small intestinal cells and express many of the hydrolytic enzymes present in the small intestinal brush border (Ma *et al.*, 1994).

Discussed here are a few selected *in vitro* studies which have contributed to a model for the intestinal absorption of glucose. Of course, any conclusions derived from *in vitro* experiments should be confirmed and extended using *in vivo* techniques in which the normal vascular channels are intact. This is particularly true for studies of regulation of transport function.

5.1.2 Everted Intestinal Sacs

The everted sac consists of a short length of isolated rat intestine turned inside out so that the epithelial brush border is on the outer surface. The sac is filled with a physiological solution, sufficient in volume to distend the wall, and an oxygen bubble is introduced. The sac is tied at both ends and suspended in an oxygenated incubation medium maintained at 37°C. The radioactively labeled solute under investigation is placed in the large volume of incubation medium and the solute taken up is measured both in the tissue and in the physiological solution within the sac. A number of adjacent portions of intestine from the same animal can be studied simultaneously.

The everted sac preparation allows both net absorption and tissue uptake of solutes to be studied under physiological conditions. Net absorption includes not only events at both boundaries of the enterocytes, but also possible contributions from interstitial connective tissue and several layers of serosal muscle. Experiments using everted intestinal sacs led Crane in 1960 to propose his sodium-gradient hypothesis for the active transport of glucose (Crane, 1962). The hypothesis stated that the energy necessary for glucose accumulation in intestinal tissue is provided by an inwardly directed sodium concentration gradient. Crane envisioned membrane carriers which have binding sites for Na^+ as well as for glucose.

5.1.3 Short-Circuit Current Technique

In a typical experiment, a segment of rabbit ileum is cut open to produce a sheet of tissue and the external muscle layers and submucosa are surgically removed. The remaining mucosal strip, consisting of the epithelial cell layer, the underlying lamina propria, and the muscularis mucosa, is tightly mounted in the middle partition of an Ussing chamber, which separates solutions of identical composition and hydrostatic pressure. The bathing solutions are continuously oxygenated and circulated at the appropriate temperature during an experiment. In the open-circuited condition, in which no external current is applied to the system, the spontaneous electrical potential difference across the epithelium is continuously monitored using electrodes that are inserted into the chamber as close as possible to the surfaces of the mucosal strip.

In the short-circuited state, an external current just sufficient to abolish the transepithelial potential is passed across the tissue using another pair of electrodes connected to a variable electromotive force. In this so-called 'voltage-clamped' preparation, there is no driving force for the net diffusional flows of ions across the tissue. Thus, any net movements across the tissue must be the result of some form of active transport. Under these conditions, the current necessary to abolish the transepithelial potential difference (referred to as the short-circuit current) is equal to the sum of the net transepithelial flows of all ions that are actively transported across the intestine. The practical importance of this relationship is that, by using radioisotopes to measure unidirectional mucosal-to-serosal and serosal-to-mucosal fluxes of individual ions under short-circuit conditions, one can identify those ions that are actively transported across the intestine. It was demonstrated that the only such ion was Na^+.

Using a tracer quantity of the $^{24}Na^+$ isotope, Schultz and Zalusky (1964a, b) reported the following observations. The addition of glucose (11 mM) to the mucosal bathing solution results in a rapid increase in the transepithelial potential difference, the short-circuit current (I_{sc}), and the rate of active sodium transport from mucosa to serosa. These three effects are dependent upon the active transport of glucose and are independent of the metabolic fate of the sugar. Furthermore, they are inhibited by low concentrations of phloridzin (phlorizin) in the mucosa solution and by low concentrations of ouabain in the serosal solution. The increase in the short-circuit current (ΔI_{sc}) requires the presence of sodium in the perfusion medium and its magnitude is a linear function of the sodium concentration. On the other hand, ΔI_{sc} is a saturable function of the mucosal

DOI: 10.1201/9781003410300-7

glucose concentration, which is consistent with Michaelis-Menten kinetics. These data were important in establishing the concept that glucose entry across the apical membrane promotes a simultaneous entry of Na^+, which is partially extruded at the basolateral membrane via the sodium pump. The unidirectional serosa-to-mucosa sodium flux, on the other hand, is attributed to simple ionic diffusion, there being no evidence for the presence of a carrier or the influence of solvent drag.

5.1.4 Isolated Enterocytes

Isolated enterocytes allow determination of initial rates of solute uptake and provide information on the intracellular accumulation and metabolic fate of the solute. If normally energized cells are used, the membrane potentials interfere with an accurate determination of the coupling ratio. To circumvent this problem, the membrane potential is experimentally maintained at or near zero using biochemical agents to deplete the cells of ATP and to confer a high K^+ permeability on the membrane.

Freshly isolated chicken enterocytes were used to establish that the true Na^+:glucose transport coupling ratio is 2:1 rather than 1:1 (Kimmich & Randles, 1980). These authors also provided data consistent with a model of brush-border sodium-glucose co-transport in which Na^+ binding to the Na^+-dependent glucose carrier at the luminal surface of the apical membrane and de-binding at the cytoplasmic surface are both dependent on the membrane potential (Kimmich & Randles, 1988).

5.1.5 Membrane Vesicles

Brush-border membrane vesicles can be prepared by homogenizing isolated enterocytes or mucosal scrapings and precipitating non-brush-border particulate matter with 10 mM Ca^{2+} or Mg^{2+}. The brush borders, because of their greater density of negative surface charges, are not precipitated and remain in the supernatant fluid. Centrifugation of this fluid provides a pellet of membrane vesicles, 95% of which have the same membrane orientation as in the intact enterocyte (i.e. they are 'the right side out'). Basolateral membrane vesicles can be obtained from a source of isolated enterocytes by differential centrifugation followed by separation on a density gradient. Vesicles of brush-border and basolateral membranes can be prepared simultaneously using the technique of free-flow electrophoresis to isolate them (Murer & Kinne, 1980).

Membrane vesicles allow solute transport properties to be investigated independently in brush-border and basolateral membranes without the complications of cellular metabolism or intracellular compartmentalization. From such studies, the overall picture of intestinal glucose absorption has emerged. Three different membrane events have been identified: (1) entry of luminal glucose into the enterocyte by secondary active transport across the apical membrane via a sodium-glucose carrier; (2) facilitated diffusion of glucose out of the cell across the basolateral membrane via a sodium-independent carrier; and (3) primary active transport of sodium out of the cell via the sodium pump.

5.2 GLUCOSE ABSORPTION

Soon after eating a meal, the high concentration of glucose in the intestinal lumen may allow uptake to take place by simple diffusion. After a while, the luminal concentration of glucose becomes lower than the concentration in the enterocyte or in the intercellular space, and therefore the glucose must be absorbed against a concentration gradient.

Glucose and fructose transport have been well studied and the experimental techniques and postulated mechanisms help toward understanding the absorption of water-soluble vitamins. Figure 5.1 shows how physiological amounts of glucose and fructose are absorbed by the small intestine. Luminal glucose crosses the epithelial brush border and accumulates in the enterocyte by means of secondary active transport. Transport is mediated by a sodium-glucose co-transporting carrier (SGLT1), which binds the substrates at a stoichiometric ratio of two sodium ions to one glucose molecule. The immediate driving force for the sodium-coupled entry of glucose is the electrochemical gradient for sodium established by the constant extrusion of sodium out of the enterocyte by the action of the basolateral sodium pump. Fructose crosses the brush border by facilitated diffusion mediated by the glucose transporter GLUT5. Exit of both glucose and fructose from the enterocyte to the serosa takes place by facilitated diffusion at the basolateral membrane and is mediated by GLUT2.

The SGLT1 in the apical membrane mediates sodium-glucose co-transport through a sequence of conformational changes. At the extracellular surface of the co-transporter, the binding of two sodium ions results in a conformational change that permits the binding of a single glucose molecule. With both substrate species bound, the co-transporter undergoes a new conformational change that allows the release of Na^+ and glucose near the intracellular surface. On release of substrates to the cytoplasm, the co-transporter reverts to the conformation that exposes the sodium binding site to the extracellular surface (Stevens *et al.*, 1990).

The sodium-coupled entry of glucose causes a depolarization at the brush-border membrane, which markedly reduces the membrane potential. This component of the electrochemical driving force needs to be restored if the uptake process is to be perpetuated. The cellular model proposed by Schultz (1977) allows this to be achieved. This model takes into account the high conductivity of the intercellular tight junction, and the ease by which Na^+ equilibrates between the extracellular media on the basolateral and luminal sides of the enterocyte. By virtue of this rapid paracellular pathway for the flow of Na^+, the current generated at one of the limiting membranes can influence the potential difference across the other. This electrical coupling constitutes an electrochemical feedback between the co-transported entry of Na^+ and Na^+ extrusion at the

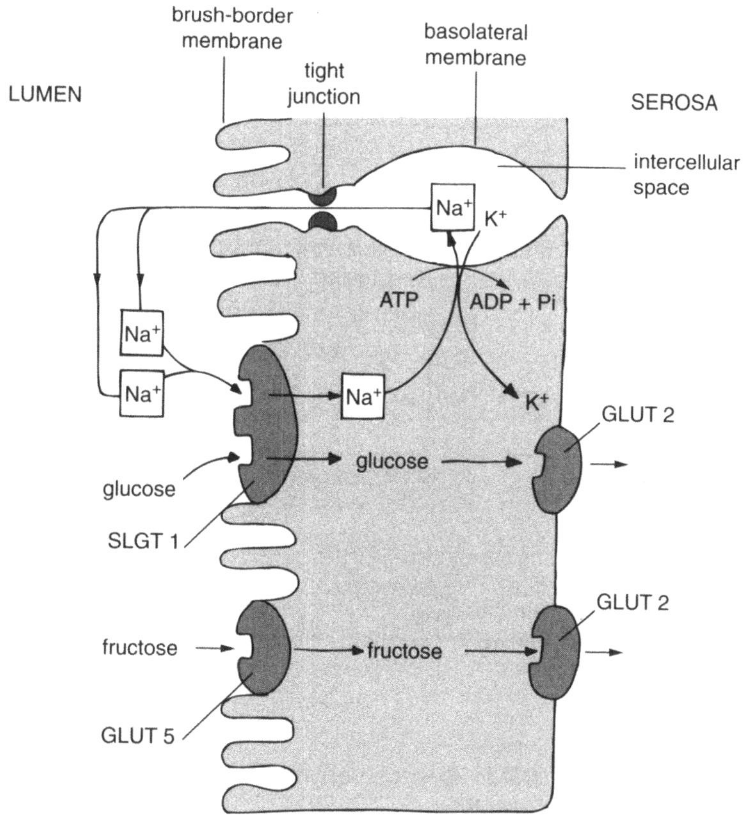

FIGURE 5.1 Diagram showing the carrier-mediated transport of D-glucose and D-fructose across the apical membrane and basolateral membrane of an enterocyte. Na^+ extruded into the intercellular space by the basolateral Na^+/K^+-ATPase (sodium pump) is able to equilibrate with Na^+ on the luminal side of the enterocyte by permeation through the tight junction. ATP, adenosine triphosphate; ADP, adenosine diphosphate; Pi, inorganic phosphate.

opposite membranes. The circular movement of Na^+ allows the co-transport of glucose to take place with minimal changes in the size of the intracellular sodium transport pool.

REFERENCES

Crane, R.K. (1962) Hypothesis for mechanism of intestinal transport of sugars. *Federation Proceedings*, **21**, 891–5.

Kimmich, G.A. & Randles, J. (1980) Evidence for an intestinal Na^+:sugar transport coupling stoichiometry of 2:0. *Biochimica et Biophysica Acta*, **596**, 439–44.

Kimmich, G.A. & Randles, J. (1988) Na^+-coupled sugar transport: Membrane potential-dependent K_m and K_i for Na^+. *American Journal of Physiology*, **255**, C486–94.

Ma, T.Y., Dyer, D.L. & Said, H.M. (1994) Human intestinal cell line Caco-2: A useful model for studying cellular and molecular regulation of biotin uptake. *Biochimica et Biophysica Acta*, **1189**, 81–8.

Murer, H. & Kinne, R. (1980) The use of isolated membrane vesicles to study epithelial transport processes. *Journal of Membrane Biology*, **55**, 81–95.

Schultz, S.G. (1977) Sodium-coupled solute transport by small intestine: A status report. *American Journal of Physiology*, **233**, E249–54.

Schultz, S.G. & Zalusky, R. (1964a) Ion transport in isolated rabbit ileum. I. Short-circuit current and Na fluxes. *Journal of General Physiology*, **47**, 567–84.

Schultz, S.G. & Zalusky, R. (1964b) Ion transport in isolated rabbit ileum. II. The interaction between active sodium and active sugar transport. *Journal of General Physiology*, **47**, 1043–59.

Stevens, B.R., Fernandez, A., Hirayama, B., Wright, E.M. & Kempner, E.S. (1990) Intestinal brush border membrane Na^+/glucose cotransporter functions *in situ* as a homotetramer. *Proceedings of the National Academy of Sciences of the United States of America*, **87**, 1456–60.

6 Metabolism and Transport of Lipids

6.1 DIGESTION AND ABSORPTION OF DIETARY LIPIDS

The fat content of a typical Western diet is composed mainly of triglycerides (triacylglycerols) accompanied by smaller amounts of phospholipids and sterols. Most of the dietary cholesterol is in the form of cholesteryl esters, which are combinations of free cholesterol and one molecule of fatty acid. The amount of dietary cholesterol is normally relatively low and the bulk of cholesterol in the intestinal lumen comes from the biliary cholesterol secreted by the liver. Dietary cholesterol is up to 65% esterified, while biliary cholesterol exists in the free form.

6.1.1 Digestion

Digestion of fat begins in the stomach. Lingual lipase, secreted by serous glands lying beneath the tongue, accounts for the limited digestion of fat in the stomach, made possible by the enzyme's particularly high stability at the low pH of the stomach contents. The stomach is the major site for the emulsification of dietary fat. The coarse lipid emulsion, on entering the duodenum, is emulsified into smaller globules by the detergent action of bile salts, aided by the churning action of the intestine. Pancreatic lipase, the principal enzyme of triglyceride digestion, hydrolyzes triglycerides at the 1 and 3 positions, producing 2-monoglycerides and free fatty acids (Figure 6.1). The pancreas also secretes cholesteryl ester hydrolase, which hydrolyzes cholesteryl esters to free cholesterol and a fatty acid, and phospholipase A_2, which hydrolyzes phospholipids to a fatty acid and the corresponding lysophospholipid. Secretion of bicarbonate from the pancreas and the biliary tract is needed to neutralize the pH to the optimal range for lipolysis. Bile salts inhibit lipase activity through displacement of the enzyme from its substrate at the surface of the lipid droplet. Colipase, a pancreatic protein, reduces this inhibition by promoting adhesion of lipase to the droplet.

During their detergent action, bile salts exist as individual molecules in solution. Above a critical concentration of bile salts, the bile constituents (bile salts, phospholipid, and cholesterol) form aggregates called micelles, in which the polar ends of the molecules are orientated toward the surface and the nonpolar portions form the interior. The 2-monoglycerides and long-chain free fatty acids are sufficiently polar to combine with the micelles to form mixed micelles. These are stable water-soluble structures that can dissolve fatty acids, cholesterol, and fat-soluble vitamins in their oily interior.

6.1.2 Absorption

Lipid absorption takes place mainly in the proximal jejunum. Mixed micelles do not cross the brush-border membrane of enterocytes as intact structures – the lipolytic products must dissociate from these structures before they can be absorbed. The low pH microclimate of the unstirred layer lining the luminal surface of the jejunum facilitates micellar dissociation. The bile salt constituents of the mixed micelles are reabsorbed in the ileum and recycled. The fatty acid components become protonated when the mixed micelles enter the unstirred layer. This protonation reduces fatty acid solubility in the mixed micelles, allowing release of the fatty acids together with other lipid constituents. Individual lipids, including cholesterol and fat-soluble vitamins, can then be passively absorbed across the brush-border-membrane. The absorption of cholesterol and plant-derived sterols is controlled by the ABC transporters ABCG5 and ABCG8, which combine into a single functional protein ABCG5/8 expressed in the brush-border membrane of enterocytes. This protein pumps excess or unwanted absorbed sterols out of the enterocytes back into the lumen to be eventually excreted in the feces.

After the lipolytic micellar products enter the enterocytes, a cytosolic fatty acid-binding protein (FABP) facilitates intracellular transport of fatty acids by directing them from the cell membrane to the smooth endoplasmic reticulum (ER), where triglyceride synthesis takes place. The triglycerides are packaged into chylomicrons, together with free and esterified cholesterol, phospholipids, apolipoproteins, fat-soluble vitamins, and carotenoids. The levels of lipids in chylomicrons are not strictly regulated, but depend on the dietary load. After further processing, the chylomicrons are discharged from the enterocyte by exocytosis across the basolateral membrane and enter the central lacteal of the villus. From there they pass sequentially into the larger lymphatic channels draining the intestine, into the thoracic duct, and ultimately into the circulating venous blood at the juncture of the left internal jugular vein and subclavian vein. It is the preponderance of triglyceride-laden chylomicrons that gives plasma its milky appearance after a fatty meal.

Medium-chain triglycerides, which contain fatty acids with a chain length of 6–12 carbon atoms, are not found in appreciable amounts in the normal diet. However, they deserve mention because they are included in specialized diets for patients who have fat malabsorption. Medium-chain triglycerides are absorbed in a more efficient manner than that described above for long-chain triglycerides.

DOI: 10.1201/9781003410300-8

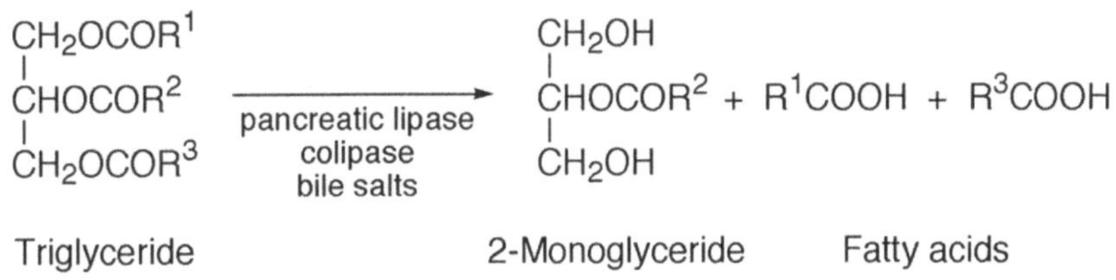

FIGURE 6.1 Lipolysis of triglycerides in the small intestine.

Being water-soluble, they can be absorbed directly as intact triglycerides in the absence of bile salts. Once inside the enterocyte, the triglycerides are hydrolyzed to medium-chain fatty acids by specific cellular lipases. The fatty acids are not re-esterified and are not packaged in chylomicrons. After leaving the enterocyte, the fatty acids enter the portal vein where they are bound to albumin and transported to the liver.

6.2 THE SYNTHESIS, MOVEMENT, AND RECYCLING OF BILE

6.2.1 Overview

Bile is an aqueous solution of bile salts and other solutes that emulsifies dietary lipids for digestion and absorption. Bile salts account for about one-half of the total solutes of the bile initially secreted by the liver. The other solutes are mainly bilirubin, cholesterol, phosphatidylcholine, and fatty acids. The electrolytes present are Na^+, K^+, Ca^{2+}, Cl^-, and HCO_3^-. Bile is synthesized continuously by hepatocytes and stored in concentrated form in the gallbladder between periods of digestion. Soon after a meal, the gallbladder begins to discharge bile into the upper duodenum. Greater than 90% of the bile salts secreted into the duodenum are reabsorbed in the ileum and returned via the portal vein to the liver – a cycle known as enterohepatic circulation.

The bile salts synthesized directly from cholesterol by the liver and released into the duodenum (cholic acid and chenodeoxycholic acid) are referred to as primary bile salts. The 10% or so of primary bile salts that are not reabsorbed in the ileum enter the colon, where they are metabolized by the resident bacteria. The bacterial enzymes catalyze hydrolysis and ring dehydroxylation of the primary bile salts to form the secondary bile salts, deoxycholate and lithocholate. Deoxycholate can be reabsorbed by the colon and returned to the liver, whereas lithocholate is excreted in the feces.

6.2.2 The Biliary Tract

The biliary tract depicted in Figure 6.2 is a system of ducts that directs secretions from the liver, gallbladder and pancreas into the duodenum. The gallbladder is a distensible thin-walled sac of approximately 50 mL capacity. It is composed mainly of longitudinal muscle and elastic tissue, bounded externally by peritoneum and lined internally by a layer of columnar epithelial cells. The cystic duct, which arises from the neck of the gallbladder, joins the common hepatic duct to form the common bile duct. The spiral valve of Heister in the wall of the cystic duct is formed by the folding of the mucous membrane and prevents sudden changes in capacity of the gallbladder following changes in pressure.

The pancreatic duct and the common bile duct usually unite to form the ampulla of Vater, but in about 10–15% of people, the bile and pancreatic ducts open separately into the duodenum. The ampullary opening into the duodenum is called the duodenal papilla. The smooth muscle, both longitudinal and circular, of the conjoined bile and pancreatic ducts is thickened over a length of about 1.5 cm to form the sphincter of Oddi. This sphincter controls the entry of bile and pancreatic secretions into the duodenum, as well as preventing duodenal contents from entering the ampulla.

6.2.3 Movement of Bile

Bile is secreted by hepatocytes into bile canaliculi, which empty into terminal bile ducts and then into progressively

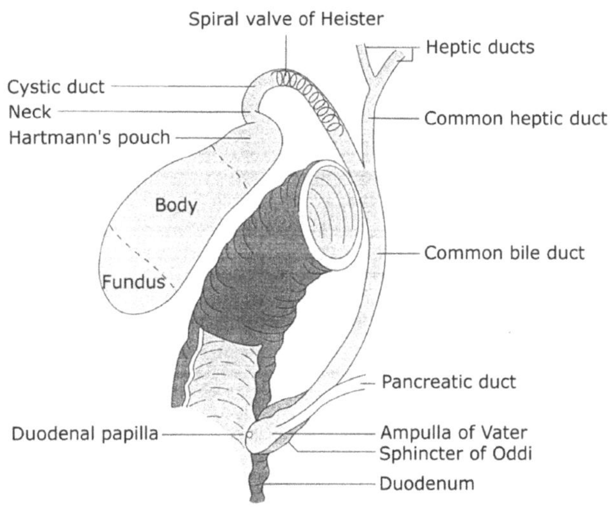

FIGURE 6.2 Gallbladder and biliary tract. From J.S. Dooley, A.S.F. Lok, A.K. Burroughs & E.J. Heathcote (eds) (2011). *Sherlock's Diseases of the Liver and Biliary* System, 12th ed. Figure 1.6 on page 5. Wiley-Blackwell Copyright 2011 by Blackwell Publishing Ltd.

larger ducts, finally reaching the common bile duct. A solution of sodium bicarbonate secreted by epithelial cells is added to the initial bile during its course through the bile ducts, doubling the volume of bile. This second secretion is stimulated by secretin, a peptide hormone released from the intestinal mucosa. From the common bile duct, the bile can enter the duodenum or be diverted via the cystic duct into the gallbladder. The gallbladder concentrates the bile by a factor of 5 to 20 as a result of water and electrolyte absorption by the mucosa. Gallbladder filling is regulated by the hormone FGF19, which is secreted from the ileum in response to bile acids.

During fasting, the sphincter of Oddi is contracted so that bile secreted by the liver flows into the gallbladder and not into the duodenum. About 30 minutes after a meal, the gallbladder begins to discharge bile into the upper duodenum. Gallbladder emptying is caused by a rhythmic contraction of the gallbladder wall. At the same time, the sphincter of Oddi relaxes, allowing bile and pancreatic juice to flow into the duodenum. Both gallbladder contraction and sphincter relaxation are mediated by vagal stimulation. However, the most important stimulus for gallbladder contraction is the hormone cholecystokinin, which is secreted by the duodenal and upper jejunal mucosa in response to fatty substances in the chyme. As digestion proceeds to its completion, the sphincter of Oddi contracts and bile is prevented from access to the intestine.

6.2.4 The Enterohepatic Circulation

In addition to gallbladder emptying, which provides an immediate supply of bile, recirculation of bile salts is necessary to ensure continuous secretion of bile during the digestive phase. Greater than 90% of the bile salts secreted into the duodenum are reabsorbed in the lower intestine and carried attached to plasma albumin back to the liver via the portal vein. This recycling of bile is called the enterohepatic circulation. The entire body pool of bile salts (approximately 2.5–5 g) is recirculated two to four times after each meal, providing 5–20 g of bile salts to the upper duodenum during the first hours after a meal.

The majority of bile salt reabsorption involves active transport in the distal ileum; the remainder is simple diffusion in the jejunum and colon. On reaching the liver, the bile salts are absorbed almost entirely into the hepatocytes and then used to make bile along with newly synthesized bile salts. New bile mixed with recirculated bile is sent via the cystic duct to the gallbladder. The small quantities of bile salts lost into the feces are replaced by similar amounts of newly synthesized bile salts.

The secretion and recycling of bile components are mediated by transport proteins in the membranes of hepatocytes and enterocytes. The expression of these transport proteins is stimulated by bile-activated FXR (liver X receptor). Three ABC proteins move bile components across the apical surface of hepatocytes into the canaliculi: ABCB4 exports phosphatidylcholine; ABCB11 exports bile salts; and ABCG5/8 exports cholesterol and other sterols.

In the ileum, the ileal bile acid transporter (IBAT), a Na^+-linked symporter, mediates the movement of about 95% of luminal bile salts across the brush-border membrane of enterocytes into the cytoplasm. The bile salts then move intracellularly, with the help of the ileal bile-acid-binding protein (I-BABP), to the opposite side of the enterocyte and cross the basolateral membrane aided by poorly characterized transport proteins. The gene encoding I-BABP is a target for FXR. The bile salts are carried in the portal blood to the liver, where another Na^+-linked symporter, the sodium-taurocholate co-transporting polypeptide (NTCP), mediates their uptake by hepatocytes.

6.3 GENERAL COMPOSITION OF LIPOPROTEINS

Following the digestion and absorption of dietary lipids, plasma lipoproteins are required to transport the lipids to their appropriate cellular destinations. The transport of cholesterol is particularly important, since this lipid is implicated in atherosclerosis. The lipoproteins in plasma are classified into chylomicrons, very low-density lipoproteins (VLDL), intermediate-density lipoproteins (IDL), low-density lipoproteins (LDL), and high-density lipoproteins (HDL). Lipid transport is carried out by three main pathways of lipoprotein metabolism: (1) the exogenous pathway involves the metabolism of chylomicrons that are assembled in the intestine during the absorption of dietary (exogenous) lipids; (2) the endogenous pathway involves the transport of endogenous lipids, starting with the secretion of VLDL from the liver; and (3) reverse cholesterol transport by which HDL takes up excess cholesterol from peripheral (extrahepatic) cells and returns it to the liver for excretion via the bile.

A typical lipoprotein particle has a hydrophobic core of neutral lipids consisting mostly of triglycerides and cholesteryl esters. Enclosing the core is a shell composed of a monomolecular layer of phospholipids with the polar head groups facing the outer aqueous environment. The phospholipids are interspersed with cholesterol molecules. The outer surface of the shell is hydrophilic, making the particle water-soluble and able to travel in the aqueous environment of the blood. Protruding from the surface are several different proteins, which are involved in the metabolism of the lipoproteins. These proteins are called apolipoproteins, or just apoproteins, where 'apo' designates the protein in its lipid-free form. Lipoprotein particles undergo continuous metabolic processing, so they have variable lipid composition (Table 6.1). The relative sizes of lipoproteins are shown in Figure 6.3. Lipoprotein densities increase with decreasing particle diameter because the density of their outer coating is greater than that of their inner core.

Note. In the published literature, lipoproteins are frequently grouped as follows:

TABLE 6.1
Characteristics of Human Plasma Lipoproteins

	Chylomicrons	VLDL	IDL	LDL	HDL
Density (g/mL)	<0.95	<1.006	1.006–1.019	1.019–1.063	1.063–1.210
Particle diameter (nm)	75–1200	30–80	25–35	18–25	5–12
Particle mass (kDa)	400,000	10,000–80,000	5,000–10,000	2,300	175–360
% Protein[a]	1.5–2.5	5–10	15–20	20–25	40–55
% Phospholipids[a]	7–9	15–20	22	15–20	20–35
% Free cholesterol[a]	1–3	5–10	8	7–10	3–4
% Triglycerides[b]	84–89	50–65	22	7–10	3–5
% Cholesteryl esters[b]	3–5	10–15	30	35–40	12
Major apoproteins important to lipid Metabolism	A-I, A-II B-48 C-I, II, III E	B-100 C-I, II, III E	B-100 E	B-100	A-I, A-II

[a]Surface components.
[b]Core lipids

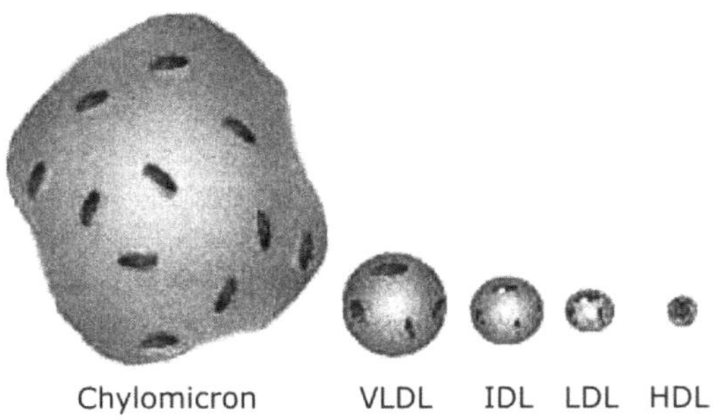

FIGURE 6.3 Relative sizes of lipoproteins.

- apoB-containing lipoproteins comprise VLDL, IDL, LDL, chylomicrons, and chylomicron remnants.
- triglyceride-rich lipoproteins (TRLs) comprise chylomicrons and VLDL.

6.4 APOLIPOPROTEINS

Each class of lipoproteins contains a distinct set of apolipoproteins (Table 6.2). Apolipoproteins play an important role in the structural and functional relationship among the lipoproteins. For example, they act as ligands for interaction with lipoprotein receptors in tissues. Polymorphisms in the genes encoding apolipoproteins alter their structure and function, and can determine an individual's susceptibility to atherosclerosis and cardiovascular disease.

6.4.1 ApoA-I and -II

ApoA-I is synthesized in both the intestine and liver, whereas apoA-II is synthesized only in the liver. Newly synthesized apoA-I and -II molecules acquire cholesterol and phospholipids and become associated with HDL particles. ApoA-I is loosely associated with the HDL particle, whereas apoA-II is tightly bound. In apoA-1, repeated amphipathic α helices provide the protein's lipid-binding regions. The α helices have their hydrophobic and hydrophilic residues on either side of the helical cylinders, so that they 'float' on phospholipid surfaces. ApoA-I exists as a dimer in the shape of a twisted elliptical ring that is ideal for wrapping around the HDL particle. Two such dimers can associate via hydrophobic interactions to form a tetramer on the surface of an HDL particle.

ApoA-I is the ligand for the scavenger receptor class B, type I (SR-BI) that mediates selective uptake of HDL cholesteryl esters by the liver in the process of reverse cholesterol transport. Overexpression of apoA-I increases plasma HDL cholesterol levels and suppresses atherosclerosis in the apoE-deficient mouse. In addition to reverse cholesterol transport, apoA-I can remove oxidized phospholipids from oxidized LDL and from cells. ApoA-I also enhances the esterification of cholesterol by activating lecithin: cholesterol acyltransferase (LCAT) present in the plasma; apoA-II enhances hepatic lipase activity.

6.4.2 ApoB

ApoB exists in two forms: apoB-100 and apoB-48. ApoB-100 is the fully translated protein and apoB-48 is the truncated form (48%) of the 513-kDa full-length apoB. ApoB-48 and apoB-100 are essential for the formation of chylomicrons and VLDL, respectively. ApoB-100 is the major apoprotein present in newly secreted VLDL and is the exclusive apoprotein of LDL; apoB-48 occurs exclusively in chylomicrons. In humans, apoB-100 is synthesized in the liver and apoB-48 in the intestine. In rodents, the liver synthesizes both apoB-100 and apoB-48. ApoB-100 wraps around the lipoprotein particle as a band of protein embedded into the surface phospholipid layer and is

TABLE 6.2
Characteristics of the Human Apolipoproteins Important to Lipid Metabolism

Apolipoprotein	Number of Amino Acids	Molecular Mass (kDa)[a]	Function
A-I	243	29	Activates LCAT[b]
A-II	77	17	Enhances hepatic lipase activity
B-48	2,152	241	Cholesterol clearance
B -100	4,536	513	Cholesterol clearance
C-I	56	6.6	Activates LCAT
C-II	79	8.9	Activates LPL[c]
C-III	79	8.8	Inhibits LPL and hepatic lipase
E	299	34	Cholesterol clearance

[a]All apolipoproteins are monomers except apoA-II, which is a disulfide-linked dimer.
[b]LCAT, lecithin: cholesterol acyltransferase.
[c]LPL, lipoprotein lipase.

not transferable. ApoB-100 is a ligand for the LDL receptor. ApoB-48 lacks the LDL receptor-binding domain of apoB-100.

6.4.3 ApoC-I, -II, and -III

ApoCs I, II, and III are synthesized in the liver and play important roles in the metabolism of chylomicrons and VLDL. They are located on the periphery of the lipoprotein particle and can be readily transferred from one lipoprotein type to another during circulation. ApoC-I enhances the esterification of cholesterol by activating LCAT. ApoC-II promotes the clearance of triglycerides from chylomicrons and VLDL by activating lipoprotein lipase attached to the luminal surface of capillary endothelia in extrahepatic tissues. ApoC-III inhibits lipoprotein lipase and hepatic lipase, thereby interfering with hepatic uptake and catabolism of apoB-containing lipoproteins.

6.4.4 ApoE

ApoE is a major protein component of triglyceride-rich lipoproteins (chylomicrons and VLDL). Unlike other apolipoproteins, which are synthesized exclusively in the liver or intestine, or both, apoE is synthesized (and perhaps secreted) by several tissues – liver, brain, spleen, kidney, lung, and adrenals. ApoE is located on the periphery of the lipoprotein particle and is readily transferable. It is a ligand for the LDL receptor and also binds avidly to the glycocalyx of hepatocytes, which facilitates LDL receptor-related protein (LPR)-mediated uptake of chylomicron remnants by the liver. ApoE is also abundant in atherosclerotic lesions and is made locally by intimal macrophages. This locally produced apoE mediates cholesterol efflux from macrophage foam cells in the intima. ApoE therefore plays a multifunctional protective role in atherosclerosis. ApoE-deficient mice fed a low-fat, low-cholesterol diet develop severe hypercholesterolemia and extensive atherosclerosis at ten weeks of age.

In humans, the single gene locus for apoE is polymorphic, giving rise to three common alleles which encode the three major isoforms of the apoE protein, namely E2, E3, and E4. The isoforms differ by cysteine-arginine interchanges at residues 112 and 158; the latter site is within the receptor binding region between residues 140 and 160. The most common allele gene product, apoE3, contains a cysteine at residue 112 and an arginine at residue 158. ApoE2 has a cysteine at both sites, while apoE4 has an arginine at both sites. These subtle molecular changes have marked effects on lipoprotein metabolism. Since humans have two copies of each chromosome (one from each parent) and both alleles are expressed, six phenotypes are possible. These are (with frequencies in parentheses) E3/E3 (60%), E3/E2 (15%), E3/E4 (20%), E4/E2 (2%), E4/E4 (2%), and E2/E2 (1%).

ApoE2 is defective in its ability to bind to lipoprotein receptors, owing to the loss of a positive charge at residue 158 that results from the amino acid interchange. The binding activity of apoE2 is <2% of normal apoE3. The reduced binding capacity of apoE2 impairs uptake of chylomicron remnants by the liver and results in an accumulation of remnants in the bloodstream. Individuals who are homozygous for the apoE2 genotype (E2/E2) can develop type III hyperlipoproteinemia, which is characterized by the presence of chylomicron remnants in the bloodstream after fasting. Individuals with the heterozygous E3/E2 genotype also exhibit delayed postprandial clearance of chylomicron remnants, but in a milder form than what is seen in E2/E2 individuals. ApoE4, although a metabolically abnormal isoform of apoE, has essentially the same binding capacity to the LDL receptor as apoE3. Despite this, individuals with the E3/E4 genotype clear chylomicron remnants into the liver more rapidly than E3/E3 individuals, and twice as fast as E3/E2 individuals. This greater uptake by the liver may be explained by the increased binding of apoE4 to the glycocalyx of hepatocytes.

6.5 CHYLOMICRONS

Chylomicrons are the largest and least dense of the lipoproteins. After synthesis in enterocytes, chylomicrons move through the lymphatic system and enter the bloodstream at the thoracic lymphatic duct. The major shell protein is apoB-48, which is exclusive to chylomicrons. The triglycerides in the interior make up about 90% of the weight of the particle. Also within the interior are cholesteryl esters, fat-soluble vitamins, and traces of other lipophilic molecules. The function of chylomicrons is to carry newly absorbed dietary triglycerides from the intestine to other tissues, where the triglycerides are hydrolyzed to free fatty acids and diglycerides that the tissue cells can assimilate. The much smaller triglyceride-depleted particles are called chylomicron remnants. These particles display surface apoE, which they acquire from other circulating lipoproteins. The remnants are carried in the blood to the liver and taken up. The apoE on the remnants serves as the ligand for receptors present on hepatocyte surfaces facing the space of Disse.

6.6 VERY LOW-DENSITY LIPOPROTEINS (VLDL)

VLDL are triglyceride-rich lipoproteins produced by the liver. Their primary function is to transport endogenous triglycerides to extrahepatic tissues, where the fatty acid products of triglyceride hydrolysis are stored or utilized as fuel.

6.6.1 Assembly of VLDL

The assembly of VLDL is achieved through multiple stages during which the various lipid constituents become associated with apoB-100 along the secretory pathway of hepatocytes. Assembly occurs as two lipidation steps. The first step gives rise to a partially lipidated apoB-100 particle, referred to as pre-VLDL; the second step involves the bulk addition of core lipid to the pre-VLDL particle, forming the mature VLDL.

The first step of VLDL assembly occurs in the rough endoplasmic reticulum during the translation of apoB-100 and extrusion of the polypeptide through the ER membrane channel (translocon) to the ER lumen. Microsomal triglyceride transfer protein (MTP) mediates the addition of small amounts of lipids to the apoB-100 polypeptide. Both neutral (triglycerides and cholesteryl esters) and polar lipids (phospholipids) are transferred, but neutral lipids are transferred more efficiently. This initial lipidation step is completed shortly after translocation of the lipid-poor apoB-100 to the ER lumen, where the polypeptide is folded into the protein structure. The fully formed pre-VLDL is a small, dense particle consisting of a neutral lipid core surrounded by a phospholipid monolayer in which the single molecule of apoB-100 is partially embedded. In the absence of sufficient lipids, nascent apoB-100 undergoes proteasomal degradation.

The second lipidation step occurs primarily by fusion between pre-VLDL particles and lipid droplets. Formation of the lipid droplets is dependent on MTP, which transfers lipids (mainly triglycerides) from cytosolic stores to the fusion site. Fusion takes place in the region of transition between the rough and smooth ER. Palmitoylation of apoB is necessary to target the protein to the fusion site. Ye *et al.* (2009) proposed that Cideb, a smooth ER- and lipid droplet-associated protein, mediates VLDL lipidation and maturation by interacting with apoB-100.

In hepatocytes, newly synthesized triglyceride enters the cytosolic pool and the stored triglycerides are hydrolyzed and re-esterified before being used in the core of the VLDL particle. Triacylglycerol hydrolase, an enzyme localized to the lumen of the ER, mobilizes stored triglycerides for VLDL assembly, but does not hydrolyze triglycerides once they have associated with VLDL. Phospholipids also contribute to VLDL triglyceride in a process which involves the activation of phospholipase D by ADP-ribosylation factor 1 (ARF-1), a small GTP-binding switch protein (Asp *et al.*, 2000). Phospholipase D hydrolyzes phosphatidylcholine to phosphatidic acid and choline. The phosphatidic acid produced is converted by phosphatidate phosphatase to diacylglycerol, which is the immediate precursor of triglycerides, phosphatidylcholine, and phosphatidylethanolamine. Bou Kahlil *et al.* (2009) showed that the level and compartmentalization of phosphatidic phosphatase controls the assembly and secretion of hepatic VLDL. Diacylglycerol acyltransferase (DGAT), the enzyme that catalyzes the final step in the pathway of *de novo* triglyceride synthesis, also influences VLDL particle size and secretion (Yamazaki *et al.*, 2005).

The cholesteryl ester that is packaged into the VLDL core is supplied through the action of the cholesterol-esterifying enzymes, acyl-coenzyme A: cholesterol acyltransferase-1 and -2 (ACAT1 and ACAT2). ACAT2 is the major enzyme responsible for hepatic cholesterol ester formation in mice and non-human primates, whereas ACAT1 is the predominant cholesterol esterification enzyme in human liver.

6.6.2 Intracellular Trafficking and Secretion of VLDL

Nascent VLDL are packaged into COPII-coated transport vesicles, which bud off from the smooth ER membrane. The vesicles traverse the ER–Golgi intracellular space and fuse with the membrane of the *cis*-Golgi network, delivering their VLDL cargo to the Golgi lumen. The docking and fusion of transport vesicles with their specific target membranes is facilitated by SNARE (SNAP REceptor) proteins. Vesicles that carry proteins such as albumin range in size from 55 to 70 nm diameter, limiting their ability to accommodate VLDL-sized particles. VLDL are carried in specialized VLDL transport vesicles (VTV), which, with a diameter of 100 to 120 nm, can comfortably contain VLDL-sized cargo. Upon delivery to the Golgi lumen, the apoB-100 component of VLDL undergoes modification, sorting, and packaging into vesicles. The mature VLD is transferred to later parts of the Golgi and is eventually secreted into the space of Disse. Nascent VLDL entering the blood contain a single molecule of apoB-100 and small amounts of E and C apoproteins. Additional amounts of E and C apoproteins are transferred to VLDL from circulating HDL.

6.6.3 Intermediate-Density Lipoproteins (IDL)

IDL are a short-lived transition stage between VLDL and LDL. They are actually VLDL remnants – the particles that result from removal of much of the triglycerides from VLDL by lipoprotein lipase, so that the concentrations of cholesterol and phospholipids are increased. A distinguishing feature of the IDL particle is its content of multiple copies of apoE in addition to a single copy of apoB-100. The presence of apoE enhances the binding of IDL to the LDL receptor.

6.7 LOW-DENSITY LIPOPROTEINS (LDL)

Most of the LDL are derived from IDL by removal of almost all the triglyceride, leaving an especially high concentration (35–40%) of cholesteryl ester in the core of the LDL particle. The protein component of LDL is a single molecule of apoB-100, which covers at least half of the particle surface and is non-transferable. LDL are the major carriers of cholesteryl ester in plasma and depositors of cholesteryl ester into peripheral cells. LDL cholesteryl ester is internalized via the LDL receptor pathway and hydrolyzed by a lysosomal acid cholesteryl ester hydrolase.

6.8 HIGH-DENSITY LIPOPROTEINS (HDL)

HDL participate in reverse cholesterol transport by removing excess cholesterol from peripheral tissues, esterifying it, and transferring the cholesteryl esters to the liver for excretion via the bile. The HDL fraction in human plasma comprises several subpopulations of particles of varying shape, size, density, electrophoretic mobility, and lipid and apolipoprotein composition. The distribution profile of the various subpopulations of HDL results from complex metabolic processes involving plasma factors that can continuously remodel circulating HDL particles; these factors include lipid-modifying enzymes and lipid transfer proteins.

6.8.1 Composition of HDL Particles

HDL particles consist of an outer amphipathic layer of apolipoproteins, phospholipids, and unesterified cholesterol and a core of cholesteryl esters and triglycerides. The apolipoproteins are represented by apoA-I and apoA-II. apoA-I is loosely bound to the particle surface and has a functional role, whereas apoA-II is tightly bound and has more of a structural role.

The HDL fraction in human plasma comprises several subpopulations of particles of varying size, density, electrophoretic mobility, and lipid and apolipoprotein composition. There are three subpopulations of apolipoprotein-specific HDL particles: (1) those (the majority) containing apoA-I but no apoA-II (LpA-I); (2) those containing both apoA-I and apoA-II (LpA-I:A-II); and (3) a minor subpopulation containing only apoA-II (LpA-II). LpA-I particles have been further separated into α_1, α_2, and α_3 subparticles with sizes of 11.2, 9.5, and 7.1 nm, respectively. LpA-I α_1, the most abundant LpA-I subparticle, is commonly referred to as α_1 HDL. Ultracentrifugation of the α_1 HDL fraction gives rise to larger and less dense HDL_2 particles and smaller and denser HDL_3 particles.

About 5% to 15% of apoA-I in human plasma is associated with particles that have electrophoretic pre-β mobility and which can be further distinguished into pre-β_i-LpA-I, pre-β_2-LpA-I, and pre-β_3-LpA-I. Pre-β_i-LpA-I, the smallest A-I HDL particle, is commonly referred to as preβ_1-HDL. These are disc-shaped particles, which contain a single molecule of apoA-1 complexed with a small amount of phospholipid and unesterified cholesterol. They are secreted from hepatocytes or enterocytes or shed during lipolysis of triglyceride-rich lipoproteins. As the preβ_1-HDL particles accumulate cholesterol, they are converted into larger discoidal complexes consisting of a phospholipid bilayer and a small amount of cholesterol surrounded by two molecules of apoA-I. Esterification of the cholesterol and sequestration of the cholesteryl esters into the core of the particle converts the preβ_1-HDL discs into small, spherical α-migrating HDL with a surface monolayer of phospholipids. Remodeling of the small, spherical HDL forms large, spherical particles surrounded by three or four molecules of apoA-I. These particles represent the plasma HDL cholesterol that is measured in routine lipid testing.

6.8.2 Assembly of HDL Particles

The assembly of HDL particles begins with the synthesis and secretion of lipid-free apoA-I particles by the liver and intestine. These particles acquire cholesterol effluxed from the liver, a process mediated by ABCA1. Phospholipids released during lipolysis of chylomicrons and VLDL are also transferred to the particles, which then become nascent preβ_1-HDL particles containing ≈10% by mass of lipid. The transfer of phospholipids is facilitated by phospholipid transfer protein (PLTP). Preβ_1-HDL particles acquire additional cholesterol from cells in large extrahepatic tissues (skeletal muscle, adipose tissue, and skin) via ABCAI-mediated efflux. The enzyme lecithin: cholesterol acyltransferase (LCAT), carried on HDL particles, then esterifies the cholesterol and the cholesteryl esters that are formed migrate to the core of the HDL particle. The nascent discoidal HDL continue to acquire cholesterol and phospholipids from extrahepatic tissues and lipid-rich lipoproteins until they eventually become large, spherical, mature HDL. These mature particles can acquire additional cholesterol from extrahepatic tissues and deliver it to the liver – the process of reverse cholesterol transport.

6.9 LDL RECEPTOR AND LDL RECEPTOR-RELATED PROTEIN

6.9.1 LDL Receptor

The LDL receptor is a transmembrane glycoprotein present in almost all cell types, with the greatest abundance in hepatocytes. It binds to the apoB-100 on the surface of LDL and mediates endocytosis of LDL and their cargo of cholesterol by the liver and non-hepatic tissues. The LDL receptor also recognizes apoE, which is present on chylomicron remnants. *In vivo*, the LDL receptor probably does not mediate endocytosis of chylomicron remnants by the liver to any significant extent, because the more numerous LDL particles win the competition for binding to the LDL receptor.

6.9.2 LDL Receptor-Related Protein

The LDL receptor-related protein (LRP) occurs only in the liver, on the surface of hepatocytes. Its best-known role is to mediate endocytosis of partially catabolized chylomicron remnants by the liver. ApoE-enriched chylomicron remnants bind avidly to the glycocalyx, specifically to heparan sulfate proteoglycans, on hepatocyte surfaces. This interaction is required for the uptake of remnants by the LRP, as uptake can be inhibited by removal of heparan sulfate proteoglycans with heparinase. Ji *et al.* (1994) proposed that the remnants interacting with heparan sulfate proteoglycans may be transferred to the LPR, or the heparan sulfate proteoglycans may form a complex with the LPR and together mediate internalization.

6.10 LIPID TRANSFER PROTEINS

Two distinct lipid transfer proteins, cholesteryl ester transfer protein (CETP) and phospholipid transfer protein (PLTP), both hydrophobic plasma glycoproteins, can markedly alter the distribution of plasma lipoproteins by shuttling lipid components from one lipoprotein particle to another.

6.11 CHOLESTERYL ESTER TRANSFER PROTEIN

CETP is present in humans, monkeys, rabbits, and hamsters, but not in mice and rats. It is secreted mainly from the liver and circulates in plasma bound to HDL. CETP mediates the transfer of cholesteryl esters from the HDL core to apoB-containing lipoproteins (VLDL, IDL, and LDL) in exchange for triglycerides. The triglycerides transferred to HDL are hydrolyzed by hepatic lipase in the liver space of Disse. The concerted action of CETP and hepatic lipase converts larger HDL_2 into smaller HDL_3 and releases lipid-free apoA-I and/or preβ_1-HDL. The smaller HDL particles are the preferred binding partners for SR-BI, the major HDL receptor on hepatocytes. CETP also exchanges cholesteryl esters between LDL and VLDL. The physiological importance of CETP in HDL metabolism was conclusively demonstrated by the discovery of CETP-deficient patients who have an increased proportion of large HDL particles enriched in cholesteryl esters and apoE (Inazu *et al.*, 1990).

The presence of two natural inhibitors of CETP has been reported in human plasma. Lipid transfer inhibitor protein (LTIP), found within the LDL fraction, has been identified as apoF. The extent of inhibition by LTIP between lipoprotein pairs is very different. Lipid transfer from LDL to VLDL was reduced by 40%, transfer from HDL to LDL by 19%, and transfer from HDL to VLDL by only 6% (Morton & Greene, 1994). The second inhibitor, found within the HDL fraction, is apoC-I, which accounts for most of the inhibition of transfers involving HDL.

Most experimental evidence in animals favors a pro-atherogenic role for CETP. Rabbits have a naturally high level of CETP and are highly susceptible to the development of diet-induced atherosclerosis. Inhibiting CETP in rabbits results in a marked reduction in atherosclerosis. Human subjects with a genetic deficiency of CETP have abnormally high levels of HDL cholesterol, which suggests a pro-atherogenic role for CETP. However, there are circumstances in which CETP exhibits anti-atherogenic properties, such as polymorphism of the CETP gene. Whether inhibition of CETP prevents or enhances atherosclerosis appears to depend on which dyslipidemia is present. CETP has been suggested to be pro-atherogenic in hypercholesterolemia and anti-atherogenic in hypertriglyceridemia and hyperalphalipoproteinemia (Inazu *et al.*, 2000).

6.12 PHOSPHOLIPID TRANSFER PROTEIN (PLTP)

Phospholipid transfer protein (PLTP) facilitates the rapid mass transfer of phospholipids, and to a lesser extent cholesterol, from VLDL into HDL during VLDL lipolysis. Liver and adipose tissue are major sites of PLTP biosynthesis because of their large size. The mouse is an ideal model to study the role of PLTP *in vivo*, because this species is naturally deficient in CETP. PLTP knockout mice exhibited a marked 60–70% reduction in plasma HDL levels, with major reductions in HDL phospholipid, free cholesterol, cholesteryl ester, and apoA-I (Jiang *et al.*, 1999). The low HDL levels in these animals were due primarily to a marked increase in the catabolism of HDL protein and cholesterol ester (Qin *et al.*, 2000). These *in vivo* results demonstrate that PLTP-mediated transfer of phospholipid from VLDL to HDL retards the catabolism of HDL and is crucial for maintaining plasma HDL levels, at least in the mouse.

A second function of PLTP is its participation in HDL conversion, a remodeling process in which PLTP mediates the fusion of two molecules of HDL_3, forming the larger HDL_2 particle and releasing preβ_1-HDL. Huuskonen *et al.* (2000) demonstrated that phospholipid transfer is a prerequisite for efficient PLTP-mediated HDL conversion. The physiological significance of HDL conversion is its role in reverse cholesterol transport, whereby the preβ_1-HDL released mediates the efflux of excess cholesterol from peripheral cells. The preβ_1-HDL particles rapidly transfer the cholesterol to HDL, which transport it to the liver. von Eckardstein *et al.* (1996) verified that PLTP promotes the generation of preβ_1-HDL and enhances the ability of plasma to stimulate cholesterol efflux from cells. The *in vivo* relevance of PLTP-mediated HDL conversion was demonstrated in transgenic mice that overexpress human PLTP (van Haperen *et al.*, 2000). Plasma from these mice had an increased capacity to generate preβ_1-HDL and a higher capability to prevent cholesterol accumulation in macrophages than plasma from wild-type mice. Taken together, these results show that PLTP plays an important role in the prevention of atherosclerosis.

6.13 THE EXOGENOUS PATHWAY OF LIPID TRANSPORT

6.13.1 Chylomicron Lipolysis and the Formation of Chylomicron Remnants

The chylomicrons are cleared from the plasma in a matter of minutes. Circulating chylomicrons pick up apoC and apoE from HDL. Many tissues, as well as the liver, synthesize apoE, so the chylomicrons may acquire additional apoE as they circulate through the tissues. Attached to the luminal surface of capillary endothelia in specific peripheral tissues (adipose tissue, heart, skeletal muscle, and lactating mammary tissue) is lipoprotein lipase, which is activated by the apoC-II that the chylomicrons have acquired. Lipoprotein lipase hydrolyzes triglycerides in the chylomicrons, releasing free fatty acids and diglycerides, which are absorbed by the tissue cells. The remaining chylomicron remnants are carried in the blood to the liver,

6.13.2 Uptake of Chylomicron Remnants by the Liver

The liver has the capacity to rapidly remove chylomicron remnants from the circulation. The particles are reduced in size by the lipolytic action of hepatic lipase to allow passage through the fenestrations of hepatic capillary endothelial cells and accumulation in the space of Disse. Endocytosis of the particles by the liver is mediated by the LDL receptor-related protein (LRP). This receptor is present solely in the liver and is located on hepatocyte sinusoidal surfaces facing the space of Disse. The apoE-enriched remnants bind avidly to the glycocalyx on hepatocyte surfaces and this interaction has been shown to be required for the uptake of remnants by the LRP.

It is only after chylomicrons are partially degraded that the apoE on the surface of the resulting remnant particle binds with high affinity to the LRP. Borensztajn *et al.* (1988) found that it is changes in the phospholipid composition on the surface of chylomicrons during their transformation into remnants that determine how the particles are discriminated by the receptor. It is conceivable that in the intact chylomicrons, with a full phospholipid complement, the receptor binding domain of apoE is not exposed. As phospholipids are progressively removed, steric or conformational changes may occur, which allow access of the apoprotein to its receptor with increasing binding affinity.

Theoretically, chylomicron remnants could be cleared by the liver through interaction with LDL receptors. However, at the LDL receptor, the remnants must compete for binding with LDL. As the concentration of LDL is much higher than that of the chylomicron remnants (even more in the tissue fluid than in the plasma), there is little chance of significant uptake of the remnants by the LDL receptor.

6.14 THE ENDOGENOUS PATHWAY OF LIPID TRANSPORT

6.14.1 Source of Free Fatty Acids Used by the Liver for Triglyceride Synthesis

The liver receives free fatty acids from three main sources: (1) circulating free fatty acids derived from the hydrolysis of triglycerides stored in adipose tissue; (2) circulating chylomicron remnants; and (3) dietary medium-chain free fatty acids delivered via the portal vein. The liver is also capable of *de novo* synthesis of fatty acids from non-hepatic sources. The fatty acids from these sources are converted to triglycerides in the smooth endoplasmic reticulum of hepatocytes. A major portion of the triglyceride produced is used for the formation of VLDL.

6.14.2 Delivery of Triglycerides to the Tissues by VLDL

The main role of the endogenous pathway is to guarantee the delivery of triglycerides and cholesterol to tissues in both the fed and fasting states. The pathway begins in the liver with the packaging of endogenous lipids into triglyceride-rich nascent VLDL. The surface protein complement of nascent VLDL is a single molecule of apoB-100 and small amounts of apoC-I, apoC-II, apoC-III, and apoE. The VLDL particles are released into the circulation and delivered to the peripheral tissues (primarily muscle and adipose tissue). The secreted VLDL particles acquire additional amounts of E and C apoproteins from circulating

HDL. The VLDL come into contact with lipoprotein lipase attached to the blood-facing surfaces of small blood vessels and capillaries within adipose and muscle tissue. The lipase, activated by apoC-II on the VLDL surface, hydrolyzes the core triglycerides in VLDL and the free fatty acids produced are taken up by the tissues for storage or energy production. The remaining glycerol is available as a substrate for gluconeogenesis.

6.14.3 Formation of IDL and LDL

VLDL are cleared from the plasma within a few hours of their appearance in the bloodstream. The progressive removal of triglycerides from VLDL increases the concentrations of cholesterol and phospholipids and reduces the particle size. During lipolysis, the VLDL particles lose apoCs by transfer to HDL and become IDL. The presence of multiple copies of apoE on IDL, in addition to the single copy of apoB-100, enables some of these particles to bind with high affinity to LDL receptors on hepatocytes and non-hepatic tissues and be taken up.

The cholesteryl ester transfer protein mediates the exchange of triglycerides in VLDL and IDL for cholesteryl esters from HDL. The loss of triglycerides and gain of cholesteryl esters, accompanied by the transfer of apoE to HDL, transforms the VLDL/IDL to LDL. The LDL can be removed from the circulation by endocytosis mediated by LDL receptors on hepatocytes and extrahepatic cells.

6.14.4 LDL as a Depositor of Cholesterol into Peripheral Cells

All growing cells need cholesterol for membrane synthesis. This is provided by the esterified cholesterol within the core of LDL. Circulating LDL particles are small enough to cross the vascular endothelium and come into contact with virtually every cell of the body. Cells acquire cholesterol by LDL receptor-mediated endocytosis of LDL particles. The apoB-100 on the LDL particle binds to the LDL receptor, but the absence of apoE makes the binding affinity much lower than that of IDL. The receptor–ligand complexes are concentrated in clathrin-coated pits, internalized, and enclosed within endosomes, where the ligands and receptors dissociate. The LDL are then delivered to lysosomes, where the apoproteins are degraded and the cholesteryl esters are hydrolyzed to free cholesterol. The cholesterol is incorporated into the plasma membranes as necessary. Excess intracellular cholesterol is re-esterified by sterol *O*-acyltransferase 2 (SOAT2) for intracellular storage.

The intracellular concentration of cholesterol is subject to feedback control mediated by the LDL-derived cholesterol itself. The incoming cholesterol (1) suppresses the activity of 3-hydroxy-3-methylglutaryl coenzyme A reductase (HMC CoA reductase), which is the rate-limiting enzyme in cholesterol biosynthesis; (2) activates the cholesterol esterifying enzyme acyl coenzyme A: cholesterol acyltransferase (ACAT), thereby sequestering excess cholesterol as cholesteryl esters; and (3) turns off the synthesis of the LDL receptor, thereby preventing further entry of LDL-cholesterol into the cell.

LDL particles can also enter cells by simply passing through the plasma membrane. The absence of a receptor means that the binding is of low affinity, and thus at low plasma concentrations LDL entry by this route is insignificant. However, non-receptor-mediated LDL uptake is not saturable or subject to feedback control, but continues to increase with increasing extracellular LDL concentration. When an individual's plasma LDL levels are high, entry of cholesterol into cells by this route becomes significant and can lead ultimately to atherosclerosis.

6.15 REVERSE CHOLESTEROL TRANSPORT

Reverse cholesterol transport (RCT) describes the process by which excess cholesterol is discharged from peripheral tissues and delivered by HDL particles to the liver for secretion into the bile and eventual elimination from the body in the feces. An outstanding cell type that accumulates excess cholesterol is the macrophage. Macrophages that occupy the subendothelial space (intima) of artery walls can become so bloated with cholesterol that they are redefined as foam cells. Even so, the mass of cholesterol derived from macrophages is only a tiny fraction of that derived from all the peripheral tissues. Excess unesterified cholesterol is toxic to cells and can lead ultimately to apoptosis. Non-hepatic cells acquire cholesterol through uptake of lipoproteins and *de novo* synthesis; yet, with the exception of steroidogenic tissues that convert cholesterol to steroid hormones, are unable to catabolize it. RCT is therefore an important mechanism for maintaining cholesterol homeostasis.

The efflux of cholesterol from macrophages is primarily facilitated by ABCA1 and to a lesser extent by ABCG1 and the scavenger receptor B1 (SR-B1) (Yue *et al.*, 2010). The cholesterol taken up by the HDL is esterified and the cholesteryl esters move to the core of the particle. A proportion of the HDL cholesteryl esters is delivered directly to the liver. Another proportion is transferred by the CETP to VLDL and LDL, which deliver the cholesterol esters to the liver. A third proportion of HDL cholesterol esters is delivered to steroidogenic tissues (adrenal gland, testes, and ovary) for storage and hormone synthesis.

Hepatic uptake of cholesteryl esters takes place by three mechanisms: (1) selective uptake of HDL cholesteryl esters mediated by SR-BI; (2) uptake of the cholesteryl ester-laden VLDL and LDL by endocytosis mediated by the LDL receptor, which recognizes its ligand, apoB-100, on the lipoproteins; and (3) endocytosis of the entire HDL particle (holo-HDL) mediated by ecto-F_1 ATPase in conjunction with the $P2Y_{13}$ receptor (Fabre *et al.*, 2006). In the liver, the cholesteryl esters are de-esterified and the cholesterol is secreted into the bile, used for bile acid synthesis, or packaged and secreted in newly synthesized lipoproteins. The final step of the RCT pathway is excretion of the cholesterol in the feces.

The major route for delivery of HDL cholesteryl ester to the liver is selective uptake mediated by SR-BI. Selective uptake is a process by which the core cholesteryl ester is taken into the cell without the endocytic uptake and degradation of the entire HDL particle. The process occurs in two steps: (1) the apoA-I on the surface of HDL binds to SR-BI on the hepatocyte; and (2) the core cholesteryl ester is selectively transferred to the plasma membrane. The cholesteryl esters are efficiently delivered to a metabolically active membrane compartment, either in the plasma membrane or in an internal membrane, where they are hydrolyzed extra-lysosomally by a cell-type-specific neutral cholesteryl ester hydrolase. The lipid-depleted HDL is released into the extracellular fluid. Ji *et al.* (1999) showed that SR-BI promotes the hepatic uptake of HDL cholesterol *in vivo* and facilitates the secretion of cholesterol into bile.

Mice that lack the $P2Y_{13}$ receptor exhibit a substantial reduction in holo-HDL uptake into the liver, and as a result the fecal excretion of cholesterol originating from macrophages is reduced (Fabre *et al.*, 2010).

6.16 ABSORPTION OF THE FAT-SOLUBLE VITAMINS

Absorption of the fat-soluble vitamins takes place mainly in the proximal jejunum. along with lipids present in the meal. The efficiency of vitamin absorption parallels that of fat absorption and is affected by the nature of the lipid component of the diet. In the absence of fat in the meal, the fat-soluble vitamins are hardly absorbed at all.

In the jejunum, the fat-soluble vitamins are solubilized in mixed micelles to facilitate uptake by the enterocytes. Inside the enterocytes, the vitamins are incorporated into chylomicrons, which are released and transferred via the lymphatic system to the blood. The circulating chylomicrons are attacked by lipoprotein lipase, and the chylomicron remnants containing the vitamins are taken up by the liver. The liver synthesizes specific carrier proteins to mobilize and deliver vitamins A and D to target tissues, therefore transport of these vitamins is not dependent on lipoproteins. No specific carrier proteins for vitamins E and K have been identified and the liver secretes these newly absorbed vitamins within lipoproteins.

REFERENCES

Asp, L., Claesson, C., Borén, J. & Olofsson, S.-O. (2000) ADP-ribosylation factor 1 and its activation of phospholipase D are important for the assembly of very low density lipoproteins. *Journal of Biological Chemistry*, **275**(34), 26285–92.

Borensztajn, J., Getz, G.S. & Kotlar, T.J. (1988) Uptake of chylomicron remnants by the liver: Further evidence for the modulating role of phospholipids. *Journal of Lipid Research*, **29**, 1087–96.

Bou Khalil, M., Sundaram, M., Zhang, H.-Y., Links, P.H., Raven, J.F., Manmontri, B., Sariahmetoglu, M., Tran, K., Reue, K., Brindley, D.N. & Yao, Z. (2009) The level and compartmentalization of phophatidate phosphatase-1 (lipin-1) control the assembly and secretion of hepatic VLDL. *Journal of Lipid Research*, **50**, 47–58.

Fabre, A.C.S., Vasntourout, P., Champagne, E., Tercé, F., Rolland, C., Perret, B., Collet, X., Barbaras, R. & Martinez, L.O. (2006) Cell surface adenylate kinase activity regulates the F_1-ATPase/$P2Y_{13}$-mediated HDL endocytosis pathway on human hepatocytes. *Cellular and Molecular Life Sciences*, **63**, 2829–37.

Fabre, A.C., Malaval, C., Ben Addi, A., Verdia, C., Pons, V., Serhan, N., Lichtenstein, L., Combes, G., Huby, T., Briand, F., Collett, X., Nijstad, N., Tietge, U.J.F., Robaye, B., Perret, B., Boeynaems, J.-M. & Martinez, L.O. (2010) P2Y13 receptor is critical for reverse cholesterol transport. *Hepatology*, **52**, 1477–83.

Huuskonen, J., Olkkonen, V.M., Ehnholm, C., Metso, J., Julkunen, I. & Jauhiainen, M. (2000) Phospholipid transfer is a prerequisite for PLTP-mediated HDL conversion. *Biochemistry*, **39**, 16092–8.

Inazu, A., Brown, M.L., Hesler, C.B., Agellon, L.B., Koizumi, J., Takata, K., Maruhama, Y., Mabuchi, H. & Tall, A.R. (1990) Increased high-density lipoprotein levels caused by a common cholesteryl-ester transfer protein gene mutation. *New England Journal of Medicine*, **323**, 1234–8.

Inazu, A., Koizumi, J. & Mabuchi, H. (2000) Cholesteryl ester transfer protein and atherosclerosis. *Current Opinion in Lipidology*, **11**, 389–96.

Ji, Y., Wang, N., Ramakrishnan, R., Sehayek, E., Huszar, D., Breslow, J.L. & Tall, A.R. (1999) Hepatic scavenger receptor BI piomotes rapid clearance of high density lipoprotein free cholesterol and its transport into bile. *Journal of Biological Chemistry*, **274**(47), 33398–33402.

Ji, Z.-S., Fazio, S., Lee, Y.-L. & Mahley, R.W. (1994) Secretion-capture role for apolipoprotein E in remnant lipoprotein metabolism involving cell surface heparan sulfate proteoglycans. *Journal of Biological Chemistry*, **269**(4), 2764–72.

Jiang, X.-C., Bruce, C., Mar, J., Lin, M., Ji, Y., Francone, O.L. & Tall, A.R. (1999) Targeted mutation of plasma phospholipid transfer protein gene markedly reduces high-density lipoprotein levels. *Journal of Clinical Investigation*, **103**, 907–14.

Morton, R.E. & Greene, D.J. (1994) Regulation of lipid transfer between lipoproteins by an endogenous plasma protein: Selective inhibition among lipoprotein classes. *Journal of Lipid Research*, **35**, 836–47.

Qin, S., Kawano, K., Bruce, C., Lin, M., Bisgaier, C., Tall, A.R. & Jiang, X.-C. (2000) Phospholipid transfer protein gene knock-out mice have low high density lipoprotein levels, due to hypercatabolism, and accumulate apoA-IV-rich lamellar lipoproteins. *Journal of Lipid Research*, **41**, 269–76.

van Haperen, R., van Tol, A., Vermeulen, P., Jauhiainen, M., van Gent, T., van den Berg, P., Ehnholm, S., Grosveld, F., van der Kamp, A. & de Crom, R. (2000) Human plasma phospholipid transfer protein increases the antiatherogenic potential of high density lipoproteins in transgenic mice. *Arteriosclerosis, Thrombosis, and Vascular Biology*, **20**, 1082–8.

von Eckardstein, A., Jauhiainen, M., Huang, Y., Metso, J., Langer, C., Pussinen, P., Wu, S., Ehnholm, C. & Assmann, G. (1996) Phospholipid transfer protein-mediated conversion of high density lipoproteins generates preβ_1-HDL. *Biochimica et Biophysica Acta*, **1301**, 255–62.

Yamazaki, T., Sasaki, E., Kakinuma, C., Yano, T., Miura, S. & Ezaki, O. (2005) Increased very low density lipoprotein secretion and gonadal fat mass in mice overexpressing liver DGAT1. *Journal of Biological Chemistry*, **280**(22), 21506–14.

Ye, J., Li, J.Z., Liu, Y., Li, X., Yang, T., Ma, X., Li, Q., Yao, Z. & Li, P. (2009) Cideb, an ER- and lipid droplet-associated protein, mediates VLDL lipidation and maturation by interacting with apolipoprotein B. *Cell Metabolism*, **9**, 177–90.

Yue, P., Chen, Z., Nassir, F., Bernal-Mizrachi, C., Finck, B., *et al.* (2010) Enhanced hepatic apoA-I secretion and peripheral efflux of cholesterol and phospholipid in CD36 null mice. *PLoS ONE*, **5**(3), e9906. doi: 10.1371/journal.pone.0009906.

7 Topics of Biochemistry

7.1 CATABOLISM OF CARBOHYDRATES, LIPIDS, AND PROTEINS – AN OVERVIEW

Many of the enzymes involved in cellular metabolism consist of a protein (apoenzyme) and an organic coenzyme, which is either a derivative of a B-group vitamin or the vitamin itself. The entire enzyme (apoenzyme plus coenzyme) is referred to as the holoenzyme. Coenzymes of these metabolic enzymes and their parent vitamins are listed in Table 7.1. The coenzymes participate in the overall reaction by acting as a donor of a particular chemical group (e.g. acyl, amino, one-carbon fragments) or electrons.

The cell derives its energy from chemical reactions arising from the catabolism of dietary carbohydrates, fats, and proteins. After digestion of these energy-containing nutrients, organic fuel molecules (glucose, fatty acids, and some amino acids) are systematically oxidized through a pathway of intermediates, ultimately to carbon dioxide and water. All the oxidative reactions are enzyme-catalyzed and each exhibits a free energy change. Within a given pathway, some reactions may be energy-consuming. However, energy-producing reactions will prevail, so there is a net gain in energy.

The energy produced through the catabolism of nutrient molecules is used to form the high-energy anhydride bonds connecting the β- and γ-phosphates (terminal phosphates) of adenosine triphosphate (ATP) and these bonds can in turn be hydrolyzed to release the energy when needed. The released energy is free energy (G), which can be used to drive the various energy-requiring processes and anabolic reactions. ATP, as the reservoir of cellular energy, is the major link between energy-releasing and energy-demanding reactions. In nearly all cases, the energy is released by enzymatic hydrolysis of the terminal anhydride bond. The products of this hydrolysis are adenosine diphosphate (ADP) and a free phosphate group. The latter can be transferred to various phosphate acceptors, thereby activating the acceptors to a higher energy level.

The dietary macronutrients – polysaccharides, lipids, and proteins – are degraded in a three-stage process (Figure 7.1).

Stage 1 is digestion, during which the macronutrients are hydrolyzed to their component substituents: polysaccharides to glucose and other monosaccharides, lipids to glycerol and fatty acids, and proteins to amino acids. The energy released from these hydrolytic processes is made available to the body as heat.

In Stage 2, glucose is oxidized anaerobically to pyruvate in the cytosol by the process of glycolysis. Glycolysis produces a net gain of two molecules of ATP for each molecule of glucose degraded. This ATP is used for energy-consuming processes by the cells carrying out the anaerobic degradation of glucose. Skeletal muscle, with its poor oxygen supply, is ideally designed for carrying out glycolysis. If the oxygen supply to the tissue is adequate, pyruvate is oxidized to acetyl coenzyme A (acetyl-CoA) after entering the mitochondria. Long-chain fatty acids are broken down to acetyl-CoA by mitochondrial enzymes of the β-oxidation system. The glycerol released from hydrolyzed lipids is utilized in the glycolytic pathway. The carbon skeletons of degraded amino acids yield either an intermediate of the tricarboxylic acid (TCA) cycle or acetyl-CoA.

In Stage 3, acetyl-CoA is oxidized by means of the TCA cycle, which takes place in the mitochondria. Three steps in the TCA cycle produce reduced nicotinamide adenine dinucleotide (NADH), while one step produces reduced flavin adenine dinucleotide ($FADH_2$).

The NADH and $FADH_2$ produced in the TCA cycle are reoxidized in the mitochondria by the electron transport chain, in which a sequence of cytochromes is alternately reduced and oxidized, and molecular oxygen is the terminal electron acceptor. The reduced coenzymes reduce ubiquinone, which, when subsequently reoxidized, transfers electrons to the first cytochrome (cytochrome *b*) of the chain. The downhill flow of electrons generates sufficient energy to power the synthesis of ATP at three specific sites along the chain – the process of oxidative phosphorylation.

7.2 OXIDATION AND REDUCTION OF NICOTINAMIDE NUCLEOTIDE COENZYMES AND FLAVIN COENZYMES

The ubiquitous nicotinamide nucleotide coenzymes (NAD and NADP) and the flavin coenzymes (FMN and FAD) undergo reduction and oxidation reactions involving both protons and electrons. The redox-active moieties of NAD(P) and the flavins are nicotinamide and riboflavin, respectively.

7.2.1 NAD

The oxidized form of NAD is depicted as NAD^+, the plus sign indicating that the nitrogen atom in the pyridine ring carries a positive charge. The oxidation of a substrate catalyzed by an NAD-dependent dehydrogenase is shown in Figure 7.2. Two hydrogen atoms, comprising a total of two protons ($2H^+$) and two electrons ($2e^-$), are removed from the substrate to be oxidized (AH_2). The pyridine ring of

DOI: 10.1201/9781003410300-9

TABLE 7.1
Coenzymes Derived from B Group Vitamins

Coenzyme	Vitamin Source	Major Metabolic Roles in Mammals	Mechanistic Role
Thiamin pyrophosphate (TPP)	Thiamin	Oxidative decarboxylation of α-keto groups; transketolation	Prosthetic group
Flavin mononucleotide (FMN) and flavin adenine dinucleotide (FAD)	Riboflavin	Oxidation-reduction reactions involving one- and two-electron transfers	Prosthetic group
Nicotinamide adenine dinucleotide (NAD) and nicotinamide adenine dinucleotide phosphate (NADP)	Niacin	Oxidation-reduction reactions involving two-electron transfers	Cosubstrate
Pyridoxal phosphate (PLP)	Vitamin B_6	Reactions involved in amino acid metabolism, including transamination and decarboxylation	Prosthetic group
Coenzyme A (CoA)	Pantothenic acid	Transfer of acyl groups	Cosubstrate
Biotin–lysine complexes (biocytin)	Biotin	ATP-dependent carboxylation	Prosthetic group
Tetrahydrofolate (THF)	Folate	Transfer of one-carbon substituents	Cosubstrate
Adenosylcobalamin	Vitamin B_{12}	Conversion of L-methylmalonyl-CoA to succinyl-CoA	Prosthetic group
Methylcobalamin	Vitamin B_{12}	Methylation of homocysteine to methionine	Prosthetic group

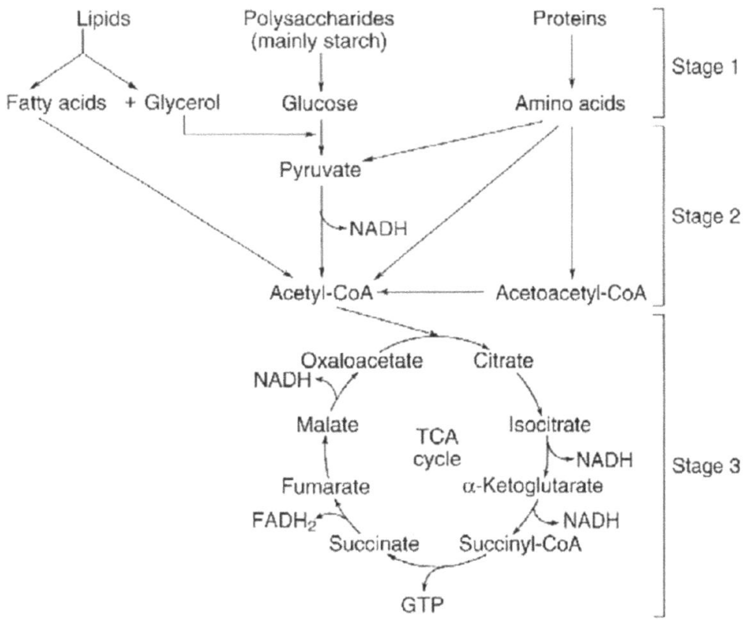

FIGURE 7.1 Main stages in the catabolism of lipids, carbohydrates, and proteins. Stage 4 is the electron transport chain, which generates the proton-motive force that powers ATP synthesis.

NAD^+ accepts a hydride ion (H^-), consisting of the two electrons and one of the protons. One of these electrons and its associated proton create a second hydrogen atom at carbon-4 and the other electron neutralizes the positive charge on the nitrogen atom. The proton (H^+) that always accompanies NADH in biochemical equations is the remaining proton donated by the substrate. This proton is released into the aqueous medium, where it remains associated with the coenzyme. The two hydrogens at position 4 in NADH project out from the planar pyridine ring in opposite directions. One hydrogen (solid bond in Figure 7.2) projects to the front of the ring (toward the reader) and the other (dotted bond) projects to the back of the ring (away from the reader). NAD-dependent dehydrogenases have stereospecificity with regard to hydride ion transfer, some enzymes for the front of the ring and others for the back. Note that oxidation of a substrate by NAD^+ or reduction by NADH always involves the transfer of two electrons at a time.

NAD^+ is loosely associated with the apoenzyme of a dehydrogenase. After being reduced in an oxidative reaction, the coenzyme (NADH) can readily diffuse to another dehydrogenase, for which it acts as a hydrogen donor. This mobility enables limited amounts of NAD to participate in oxidations and reductions catalyzed by different enzymes.

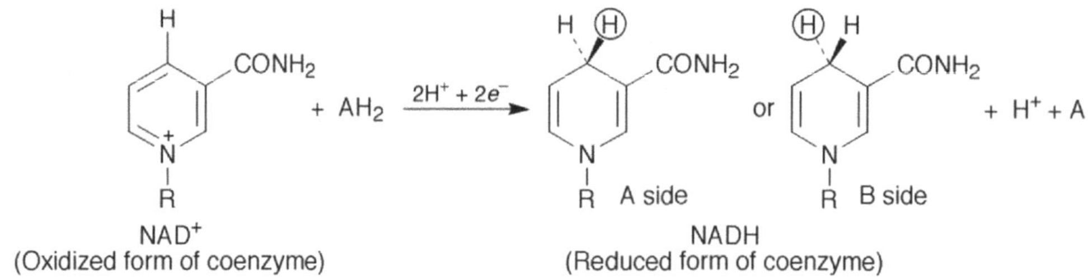

FIGURE 7.2 The oxidation of a substrate (AH_2) by NAD^+. A given dehydrogenase catalyzes transfer of a hydride ion exclusively to one face of the pyridine ring – the A side or the B side. The encircled H denotes the hydrogen atom transferred from the substrate. R, nucleotide moiety.

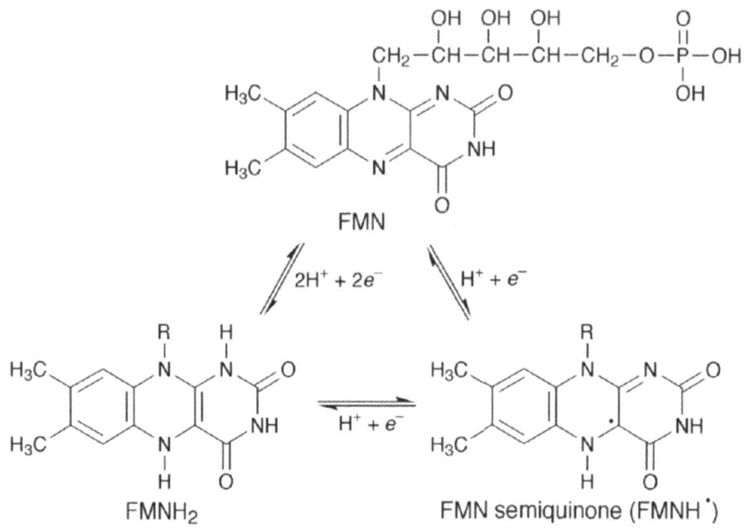

FIGURE 7.3 Reversible reduction of FMN (riboflavin-5′-phosphate). The reduction of FMN to $FMNH_2$ may proceed as either a single two-electron reaction or as two one-electron reactions with intermediate formation of the FMN semiquinone radical.

Another nicotinamide coenzyme, nicotinamide adenine dinucleotide phosphate (NADP), undergoes reversible reduction in the same way as NAD. However, these two coenzymes in reduced form have entirely different functions in the cell. The major role of NADH is to transfer its electrons gained from metabolic intermediates to the electron transport chain, thereby producing ATP. NADPH, in contrast, acts as a reducing agent in many biosynthetic pathways.

7.2.2 FMN and FAD

FMN and FAD are tightly bound prosthetic groups in the flavin-containing enzymes known as flavoproteins. The two flavin nucleotides are functionally equivalent, and whichever one is involved with a given enzyme appears to be a matter of enzymatic binding affinity. FMN and FAD can be reduced to $FMNH_2$ and $FADH_2$, respectively, either in a single step or in two steps. Figure 7.3 shows the reversible reduction of FMN (riboflavin-5′-phosphate). In the one-step reaction, FMN or FAD accepts two hydrogen atoms (each atom a proton plus an electron) from a reduced substrate, forming $FMNH_2$ or $FADH_2$. In the two-step reaction, FMN or FAD accepts one hydrogen atom to form the semiquinone radical (FMNH• or FADH•), which is stabilized by protein binding in enzyme complexes. Acceptance of the second hydrogen atom yields the fully reduced nucleotide. Because flavoproteins can be reduced in two one-electron reactions, they can act at the junction between a two-electron donor and a one-electron acceptor.

7.3 HORMONAL REGULATION OF FUEL METABOLISM

7.3.1 Overview

Blood glucose concentrations of about 4.5 mM between meals are maintained by the balance between two peptide hormones, insulin and glucagon, secreted by the pancreatic beta and alpha cells, respectively. After a meal, the pancreas responds to the rise in blood glucose and amino acids by secreting insulin and suppressing glucagon secretion. The opposing effects of insulin and glucagon on blood glucose are summarized in Table 7.2. The overall effect of insulin is to favor the conversion of excess blood glucose to two storage forms: glycogen (in the liver and muscle) and triacylglycerols (in adipose tissue). Glucagon counters the effects of insulin by breaking down liver glycogen to release

TABLE 7.2
Effects of Insulin and Glucagon on Fuel Metabolism and Their Target Enzymes

Metabolic Effect	Target Enzyme
Insulin	
↑ Glucose uptake (muscle, adipose tissue)	↑ Glucose transporter (GLUT4)
↑ Glucose uptake (liver)	↑ Glucokinase (increased expression)
↑ Glycogen synthesis (liver, muscle)	↑ Glycogen synthase
↓ Glycogen breakdown (liver, muscle)	↓ Glycogen phosphorylase
↑ Glycolysis, acetyl-CoA production (liver, muscle)	↑ Phosphofructokinase-1 (by ↑ PFK-2)
	↑ Pyruvate dehydrogenase complex
↑ Fatty acid synthesis (liver)	↑ Acetyl-CoA carboxylase
↑ Fatty acid uptake (adipose tissue)	↑ Lipoprotein lipase
Glucagon	
↑ Glycogen breakdown (liver)	↑ Glycogen phosphorylase
↓ Glycogen synthesis (liver)	↓ Glycogen synthase
↓ Glycolysis (liver)	↓ PFK-1
↑ Gluconeogenesis (liver)	↑ FBPase-2
	↓ Pyruvate kinase
	↑ PEP carboxylase
↑ Fatty acid mobilization (adipose tissue)	↑ Hormone-sensitive triacylglycerol lipase
↓ Fatty acid synthesis (liver)	↓ Acetyl-CoA carboxylase

glucose and liberating free fatty acids from adipose tissue. Adrenaline, secreted by the adrenal medulla in response to stress signals from the brain, stimulates glucagon secretion and inhibits insulin secretion, reinforcing its effect of mobilizing fuels and inhibiting fuel storage.

7.3.2 Insulin and Glucagon

Although the liver is responsible for glucose homeostasis, this organ cannot by itself sense the blood glucose level – it relies on hormonal signals transmitted by the pancreas. When the blood glucose level becomes high after a meal, the beta cells of the pancreatic islets of Langerhans secrete insulin. This hormone causes an increase in the number of glucose transporters in the plasma membrane of liver cells, thereby stimulating glucose uptake. The liver then converts the glucose to glycogen. When the blood glucose level becomes low, the alpha cells of the pancreatic islets secrete glucagon. This hormone stimulates both glycogenolysis and gluconeogenesis, leading to hepatic secretion of glucose into the bloodstream. Glucagon and insulin regulate one another's secretion from the islet alpha and beta cells. The effects of both hormones are promptly reversed when the need for them has ended.

Insulin is secreted into the blood and unbound insulin is cleared from the circulation by the action of insulinase within 10 to 15 minutes. To initiate its effects on target cells, insulin first binds with and activates the insulin receptor. This is a tyrosine-specific protein kinase that undergoes autophosphorylation upon insulin binding. Autophosphorylation opens up the active site so that the kinase can phosphorylate tyrosine residues of several intracellular proteins. The net effect is to activate some of these proteins while inactivating others. In this way, insulin directs the intracellular metabolic machinery to produce the desired effects on carbohydrate, fat, and protein metabolism.

Glucagon binds to a specific receptor on hepatocytes and signals through 3′,5′-cyclic adenosine monophosphate (cAMP) to activate protein kinase A (PKA), which in turn initiates a cascade of various protein kinases. Signaling is inhibited by cAMP phosphodiesterase, which hydrolyzes cAMP to 5′-AMP.

7.3.3 Cellular Uptake of Glucose

Insulin causes rapid uptake of glucose by the muscles and adipose tissue, thereby lowering blood glucose. The hormone circulating in the blood binds to insulin receptors on myocytes (skeletal muscle cells) and adipocytes (fat-storing cells). The stimulated insulin receptor can transduce signals through a phosphoinositol pathway leading to the activation of protein kinase B (PKB). By an unknown mechanism, PKB triggers the fusion of small intracellular vesicles containing the GLUT4 glucose transporter with the plasma membrane. The resulting increased number of GLUT4 transporters on the cell surface of the myocytes and adipocytes increases glucose influx. When the insulin level drops, the GLUT4 transporters are internalized by endocytosis, thereby lowering glucose influx. The internalized vesicles fuse with endosomes, and patches of endosomes enriched with GLUT4 bud off to become vesicles, ready to be translocated to the cell surface.

Glucose rapidly traverses the blood–brain barrier by means of facilitated diffusion and is transported into brain neurons by GLUT3, independently of insulin. The brain preferentially uses glucose as fuel, but can adapt to the use of acetoacetate or D-β-hydroxybutyrate under starvation conditions, when glucose is unavailable. If the blood glucose concentration is allowed to fall too low, symptoms of hypoglycemic shock develop, characterized by progressive nervous irritability that leads to fainting, seizures, and even coma.

Most of the glucose absorbed after a meal is taken up by the liver and immediately converted into glucose-6-phosphate. Glucose transport into hepatocytes is mediated by GLUT2, which is always present in the plasma membrane, and is independent of insulin. Insulin enhances glucose uptake by increasing the expression of glucokinase (hexokinase IV), the enzyme that catalyzes the initial phosphorylation of glucose after it diffuses into liver cells. Once phosphorylated, the glucose is temporarily trapped inside the liver cells, because phosphorylated glucose cannot diffuse back through the cell membrane.

7.3.4 Breakdown and Synthesis of Glycogen

Glucagon stimulates the phosphorylation of phosphorylase kinase *b*, the enzyme which catalyzes the conversion of low-activity glycogen phosphorylase *b* to the catalytically active glycogen phosphorylase *a*. Glucagon also stimulates the phosphorylation of glycogen synthase, which decreases the enzyme's activity. Glucagon therefore stimulates glycogen breakdown and inhibits glycogen synthesis. This results in liver cells breaking down glycogen and secreting glucose into the bloodstream within minutes of the initial signal from the pancreas. Glucagon does not stimulate glycogenolysis in muscle, because muscle cells do not possess receptors for this hormone. However, adrenaline can stimulate glycogenolysis in both liver and muscle. In muscle, the glucose-1-phosphate produced in response to adrenaline provides fuel for glycolysis to sustain muscle contraction for the fight-or-flight response.

Insulin maintains the activity of glycogen synthase through PKB, which inactivates glycogen synthase kinase-3 (GSK3) by phosphorylation. Inactive GSK3 cannot convert active glycogen synthase *a* to the inactive *b* form by phosphorylation, so glycogen synthase remains active. At the same time, insulin stops glycogenolysis by stimulating phosphoprotein phosphatase I (PP1), the enzyme that inactivates glycogen phosphorylase. Insulin-stimulated protein kinase activates PP1 by phosphorylating site 1 of the glycogen-binding G_M subset, thereby promoting binding of PP1 to glycogen. The net effect of all of these actions is to increase the amount of glycogen stored in the liver.

7.3.5 Regulation of Glycolysis and Gluconeogenesis

The hormonal regulation of glycolysis and gluconeogenesis is mediated by fructose 2,-6-biphosphate, an allosteric effector which activates the glycolytic enzyme

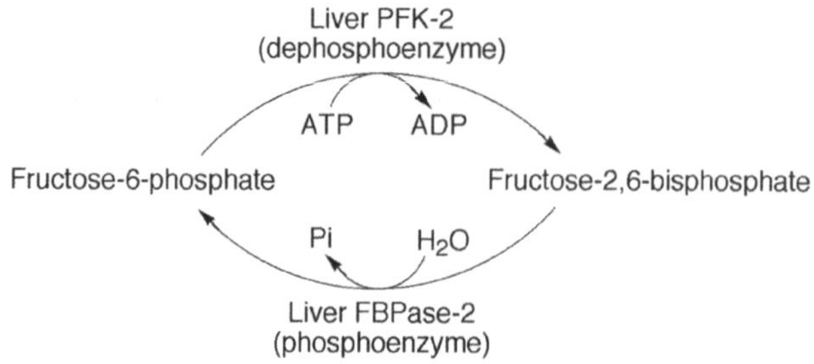

FIGURE 7.4 The formation and breakdown of fructose-2,6-biphosphate by the bifunctional enzyme phosphofructokinase-2/fructose-2,6-biphosphatase (PFK-2/FBPase-2).

phosphofructokinase-1 (PFK-1) and, at the same time, inhibits the gluconeogenic enzyme fructose 1,6-biphosphatase (FBPase-1). The cellular concentration of fructose 2,6-biphosphate is set by the relative rates of its formation and breakdown. It is formed by phosphorylation of fructose 6-phosphate, catalyzed by phosphofructokinase-2 (PFK-2) and is broken down by fructose 2,6-biphosphatase (FBPase-2) (Figure 7.4). These two enzymes exist together as a single, bifunctional protein. The balance of these two activities in the liver is regulated by the opposing effects of glucagon and insulin. Glucagon, acting through cAMP, causes the phosphorylation of PFK-2/FBPase-2, enhancing its FBPase-2 activity and inhibiting its PFK-2 activity. This lowers the cellular level of fructose 2,6-biphosphate, thereby inhibiting glycolysis and stimulating gluconeogenesis. The resulting production of more glucose enables the liver to replenish blood glucose in response to glucagon. Insulin, secreted in the fed state, stimulates the removal of the phosphoryl group from PFK-2/FBPase-2, thereby enhancing its PFK-2 activity. The increased level of fructose-2,6-biphosphate stimulates glycolysis and inhibits gluconeogenesis.

In addition, glucagon signaling phosphorylates and inactivates the glycolytic enzyme pyruvate kinase. The reaction catalyzed by pyruvate kinase (step 12 of glycolysis) is essentially irreversible, owing to the large overall free energy change. The resulting accumulation of phosphoenolpyruvate (PEP) favors gluconeogenesis, an effect augmented by glucagon's stimulation of the synthesis of the gluconeogenic enzyme PEP carboxykinase.

7.3.6 Regulation of Fatty Acid Metabolism

When more carbohydrates are ingested than can be used for immediate energy, the excess glucose can be used for fat synthesis. Almost all of this synthesis occurs in the liver cells. Insulin stimulates the oxidation of pyruvate (the end-product of glycolysis) to acetyl-CoA by activating the pyruvate dehydrogenase complex. Acetyl-CoA carboxylase, the enzyme that converts acetyl-CoA to malonyl-CoA in the first step in fatty acid synthesis, is regulated by hormonally controlled reversible phosphorylation. Glucagon, adrenaline, and noradrenaline trigger the enzyme's cAMP-dependent increase in phosphorylation, which inactivates the enzyme. Insulin signaling, on the other hand, dephosphorylates and thus activates the enzyme. Most of the fatty acids produced are used to form triacylglycerols, which are released from the liver cells to the blood in the very low-density lipoproteins (VLDLs) and carried to the adipose tissues. Insulin activates lipoprotein lipase attached to the luminal side of capillary vessels in the adipose tissue. The activated lipase splits the triacylglycerols again into fatty acids, which are absorbed into the adipose cells and again converted to triacylglycerols for storage.

When there is a requirement for energy, glucagon and adrenaline stimulate hydrolysis of triacylglycerols in the adipose tissue. The long-chain free fatty acids produced are oxidized to acetyl-CoA, which starts a turn of the TCA cycle by donating its acetyl group to oxaloacetate. Neutral lipids are stored in adipocytes in the form of lipid droplets, which have a core of sterol esters and triacylglycerols surrounded by a monolayer of phospholipids. A surface coating of proteins known as perilipins restricts access to the lipid droplets. The binding of glucagon to its receptor in the adipocyte membrane leads to activation of PKA, which phosphorylates hormone-sensitive triacylglycerol lipase, doubling or tripling its activity. More importantly, PKA also phosphorylates the perilipin molecules, which causes the hormone-sensitive lipase to move from the cytosol to the lipid droplet surface, where it can begin hydrolyzing triacylglycerols to free fatty acids and glycerol. Adrenaline mobilizes fatty acids in a similar manner to glucagon, whereas insulin decreases the activity of hormone-sensitive lipase.

Glucagon and adrenaline also slow fatty acid synthesis by inhibiting acetyl-CoA carboxylase. Inhibition is mediated by cAMP-activated PKA, which phosphorylates the carboxylase. Insulin relieves this inhibition of the carboxylase.

7.3.7 Stimulation of Cellular Amino Acid Uptake

Insulin stimulates the transport of amino acids into cells in two distinct ways: (1) directly by increasing the affinity of the membrane carrier for amino acids and (2) indirectly by initiating the synthesis of a specific protein(s) which enhances transport (Elsas *et al.*, 1968). Insulin stimulation of amino acid uptake in rat diaphragm muscle is influenced by age. Muscle from 10-day-old rats responds only to the direct stimulatory mechanism; muscle from 25-day-old rats responds to both direct and indirect stimulation; and muscle from 50-day-old rats has lost responsivity to direct stimulation and relies completely on indirect stimulation through new protein synthesis (Elsas *et al.*, 1971).

7.3.8 Regulation of Protein Metabolism

Insulin signaling via PKB rapidly activates protein synthesis by activating components of the translational machinery, including eIFs (eukaryotic initiation factors) and eEFs (eukaryotic elongation factors (Proud, 2006). In the absence of insulin, the ribosomes simply stop working. For example,

- The insulin-induced inactivation of glycogen synthase kinase-3 (GSK3) leads to the dephosphorylation and activation of eIF2B, the factor required for recycling eIF2.
- Insulin elicits phosphorylation of 4E-BP1 (eIF4E-binding protein 1), thereby releasing it from eIF4E and allowing eIF4E to form initiation factor complexes.
- PKB phosphorylates the TSC1–TSC2 complex to relieve its inhibitory action on mTOR (the mechanistic target of rapamycin) signaling that impinges upon the translational machinery.
- Insulin accelerates elongation by activating eEF2 by dephosphorylation and inactivating eEF2 kinase by phosphorylation.
- In the long term, insulin stimulates synthesis of ribosomal proteins to augment the capacity for protein synthesis.

Insulin promotes protein formation and prevents the degradation of proteins by the following processes.

- Insulin promotes storage of proteins in the tissues during the few hours after a meal by stimulating transport of many of the amino acids into cells.
- Insulin conserves the amino acids in the protein stores by suppressing gluconeogenesis.
- Insulin inhibits protein catabolism.

Virtually all protein storage comes to a halt when insulin is not available. The catabolism of proteins increases, protein synthesis stops, and large quantities of amino acids are dumped into the plasma. Most of the excess amino acids in the plasma are used either directly for energy or as substrates for gluconeogenesis. The resulting protein wasting is one of the most serious of all the effects of severe diabetes mellitus.

7.4 THE FASTING AND STARVATION STATES

The fuel reserves of a healthy adult human are of three types: glycogen stored in the liver and muscles, triacylglycerols in adipose tissues, and proteins in various tissues. In the first few hours after a meal, the blood glucose level falls slightly and tissues receive glucose released from liver glycogen. The post-absorptive state evolves into a fasting state after 48 hours of no food intake. A fast longer than two to three days exhausts the liver's glycogen reserve and about half the glycogen in muscle. Triacylglycerols now become the primary fuel for muscles and liver, but not for the brain, which cannot utilize free fatty acids. The brain is constantly dependent on incoming glucose from the blood. During fasting, when liver glycogen stores have been depleted, the brain relies on gluconeogenesis in the liver as a means of supplying glucose. The precursors of gluconeogenesis are glucogenic amino acids derived from the degradation of easily mobilizable muscle proteins. Gluconeogenesis is signaled by the secretion of glucagon in response to low levels of blood glucose. Glucocorticoid hormones, especially cortisol, accelerate proteolysis in muscle tissue to provide the glucogenic amino acids. Gluconeogenesis is responsible for 64% of total glucose production over the first 22 hours of the fast and accounts for almost all the glucose production by 46 hours.

The body is in a state of starvation when glucose levels are reduced to the physiologically allowable limit of 70% of the normal fasting level of about 90 mg 100 ml^{-1} blood. The clinical manifestations of chronic starvation are due to protein and energy deficiency, hence the term protein-energy malnutrition (PEM). Dietary protein and energy deficiencies usually occur together, but sometimes one predominates and, if severe enough, may lead to the clinical syndrome of kwashiorkor (predominantly protein deficiency) or marasmus (mainly energy deficiency). PEM develops gradually over weeks or months. This allows a series of metabolic and behavioral adjustments that result in decreased nutrient demand and a nutritional status compatible with a lower level of cellular nutrient availability. If the supply of nutrients becomes progressively lower, the patient can no longer adapt and will eventually die.

During the early phase of starvation, the pool of readily mobilizable protein used for glucogenesis becomes depleted and the remaining protein is not so easily removed. At this time, the rate of glucogenesis decreases to between one-third and one-fifth of its previous rate, and the rate of protein depletion becomes greatly decreased. The lessened availability of glucose then initiates a series of events that leads to excessive fat utilization. The glycerol released from hydrolysis of stored triacylglycerols takes over from amino acids as the main precursor of gluconeogenesis.

By now, the use of TCA cycle intermediates for gluconeogenesis has depleted the supply of oxaloacetate, preventing entry of acetyl-CoA into the cycle. Acetyl-CoA therefore accumulates and is converted into ketone bodies in the liver by the pathway shown in Figure 7.5. Ketone bodies – acetone, acetoacetate, and D-β-hydroxybutyrate – are water-soluble, energy-rich compounds that are tolerated at high concentrations. The useful ketone bodies, acetoacetate and D-β-hydroxybutyrate, are carried by the blood to highly energetic tissues such as skeletal muscle, the heart, and the renal cortex, where they are converted into acetyl-CoA by the pathway shown in Figure 7.6. The acetyl-CoA produced is oxidized in the TCA cycle, providing much of the energy required by these tissues. Importantly, ketone bodies can cross the blood–brain barrier and gain access to the brain cells. During prolonged fasting, ketone bodies provide about two-thirds of the brain's energy requirement.

Because adipose tissue is the primary source of energy during starvation, the survival time of a starving person is dependent on that person's fat reserves before food deprivation and not on their muscle mass. A normal-weight adult has enough stored fat to maintain a basal rate of metabolism for about three months without food. When fat reserves are exhausted, the last resort is the degradation of the remaining protein that is essential for maintenance of cellular function: death is now inevitable in the absence of nourishment.

Although a person's energy requirements can be met for several months by utilization of stored fat, starvation will lead to vitamin deficiency over a shorter period. Except for vitamin B_{12}, intakes of water-soluble vitamins in excess of the body's immediate metabolic needs are excreted in the urine; therefore, these vitamins are stored to only a slight extent in cells and are depleted within weeks of starvation. Dramatically, a lack of vitamin C can cause death from scurvy in 20 to 30 weeks. Vitamin B_{12} is stored efficiently in the liver and, because its requirement does not normally exceed 1 μg per day, stores can last for several years during starvation. Fat-soluble vitamins, by their very nature, are stored to a significant extent in body tissues. Vitamins A and K are stored mainly in the liver, vitamin D in the adipose tissue and skeletal muscle, and vitamin E mainly in adipose tissue. The hepatic stores of vitamin A can last for five to ten months during starvation.

7.5 DIABETES MELLITUS

Diabetes mellitus is a syndrome of impaired carbohydrate, fat, and protein metabolism. There are two distinct types of diabetes mellitus. Type 1, insulin-dependent diabetes mellitus (IDDM) or juvenile-onset diabetes mellitus, strikes suddenly in childhood, often at the age of 14. Type 2, non-insulin-dependent diabetes mellitus (NIDDM) or maturity-onset diabetes mellitus, develops rather gradually after the age of 40. Type 2 diabetes accounts for over 90% of the diagnosed cases of diabetes and affects 18% of the population over 65 years of age in the United States.

In both types of diabetes mellitus, metabolism of all the main foodstuffs is altered. The basic effect of insulin lack on glucose metabolism is to prevent the efficient uptake and utilization of glucose by most cells of the body, except those of the brain. As a result, blood glucose concentration increases, cell utilization of glucose becomes progressively lower, and utilization of fats and proteins for energy increases.

7.5.1 Type 1 Diabetes

Type 1 diabetes in genetically susceptible individuals is caused by lack of insulin secretion due to the selective destruction of the pancreatic beta cells by an aberrant autoimmune response. The classic symptoms of diabetes emerge suddenly when about 80% of the beta cells have been destroyed over a period of years. The lack of insulin decreases the efficiency of peripheral glucose utilization and augments glucose production, raising the plasma glucose concentration to 300 to 1200 mg 100 ml^{-1}. When the plasma glucose concentration exceeds the renal threshold

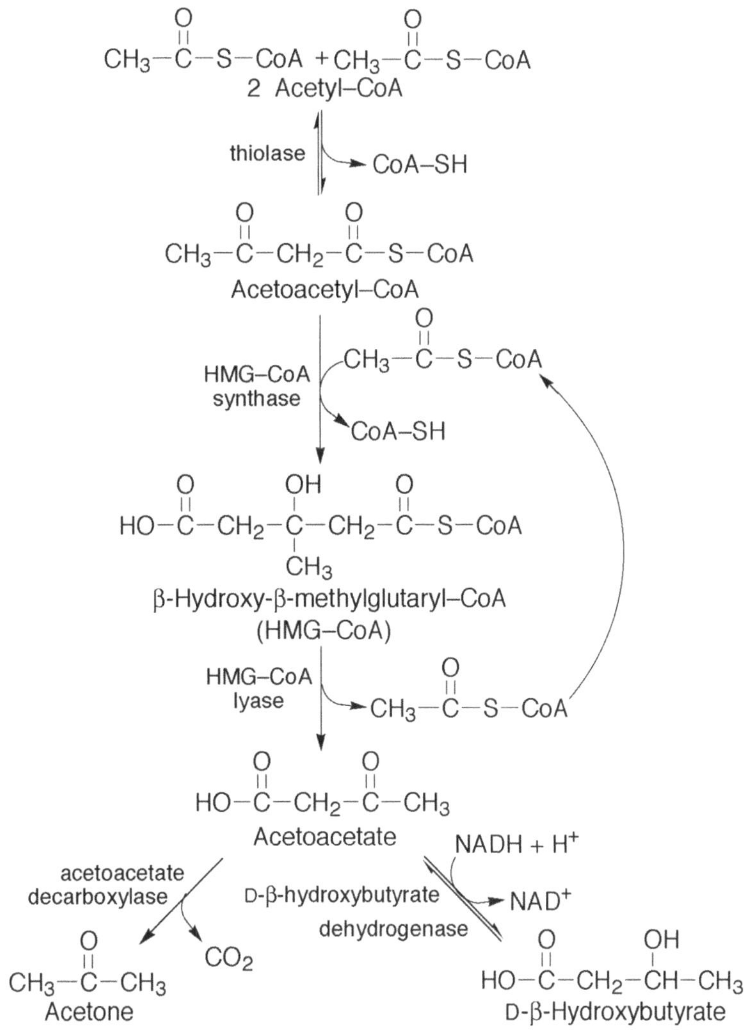

FIGURE 7.5 The synthesis of ketone bodies (acetoacetate, D-β-hydroxybutyrate, and acetone) in the mitochondrial matrix of liver cells. The first step, in which two molecules of acetyl-CoA condense to form acetoacetyl-CoA, is a reversal of the last step in fatty acid oxidation. The equilibrium of the thiolase-catalyzed reaction changes to favor acetoacetyl-CoA formation when levels of acetyl-CoA rise. A third molecule of acetyl-CoA condenses with acetoacetyl-CoA to form β-hydroxy-β-methylglutaryl-CoA (HMG-CoA), which is cleaved to form acetoacetate and acetyl-CoA. The acetoacetate is reversibly reduced to D-β-hydroxybutyrate by a dehydrogenase that is specific for the D-stereoisomer. In healthy people, acetone is formed in very small amounts from the decarboxylation of acetoacetate, either spontaneously or enzymatically. Most of the acetone thus produced is wastefully excreted in urine and in exhaled air.

level of 180 mg 100 ml^{-1}, the excess glucose that cannot be reabsorbed is excreted in the urine.

The very high levels of blood glucose can cause severe cell dehydration throughout the body due to the increased osmotic pressure in the extracellular fluids. In addition, the osmotic effect of glucose in the renal tubules greatly decreases tubular reabsorption of fluid, leading to a massive loss of fluid in the urine. This osmotic diuresis causes dehydration of the extracellular fluid, which in turn causes compensatory dehydration of the intracellular fluid. Thus, polyuria (excessive urine excretion) and polydipsia (increased thirst) are classic symptoms of diabetes.

The shift from carbohydrate to fat metabolism in diabetes causes ketone bodies to be released into the plasma more rapidly than they can be taken up and oxidized by the tissue cells. The high levels of ketone bodies in plasma, in association with the dehydration of the extracellular fluid, can cause severe acidosis. This leads rapidly to diabetic coma unless the condition is treated immediately with large amounts of insulin.

Failure to use glucose for energy leads to increased utilization and decreased storage of proteins as well as fat. Therefore, a person with severe untreated diabetes mellitus suffers rapid weight loss and asthenia (lack of energy) despite eating large amounts of food (polyphagia). Without insulin treatment, these metabolic abnormalities can cause severe wasting of the body tissues and death within a few weeks.

Treatment of type 1 diabetes is to administer insulin in doses such that the patient will have carbohydrate, fat, and protein metabolism that is as normal as possible. In

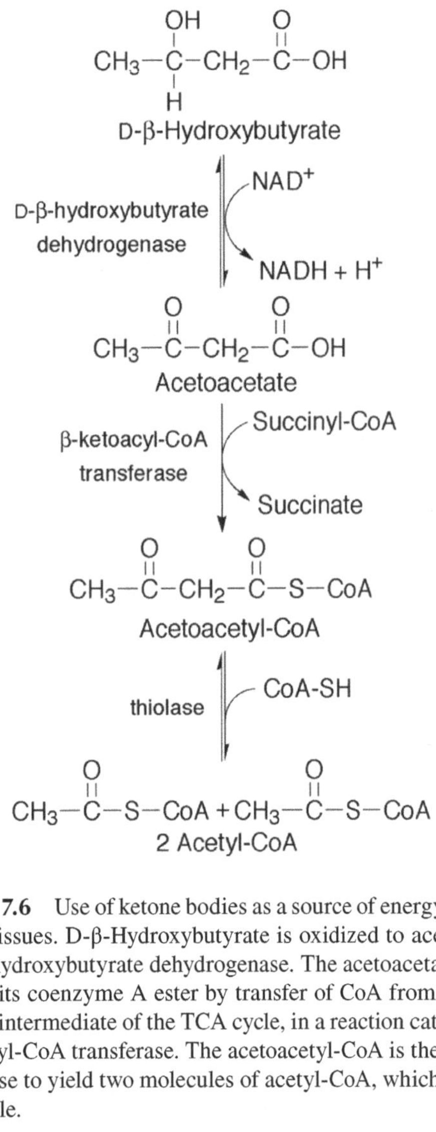

FIGURE 7.6 Use of ketone bodies as a source of energy in extrahepatic tissues. D-β-Hydroxybutyrate is oxidized to acetoacetate by D-β-hydroxybutyrate dehydrogenase. The acetoacetate is activated to its coenzyme A ester by transfer of CoA from succinyl-CoA, an intermediate of the TCA cycle, in a reaction catalyzed by β-ketoacyl-CoA transferase. The acetoacetyl-CoA is then cleaved by thiolase to yield two molecules of acetyl-CoA, which enter the TCA cycle.

practice, each patient is provided with an individual pattern of treatment.

7.5.2 Type 2 Diabetes

Type 2 diabetes is caused by decreased sensitivity of target tissues to the metabolic effect of insulin, a condition referred to as insulin resistance. The condition develops gradually in genetically susceptible individuals aged over 40 and has a strong association with obesity. These individuals have normal or even greatly elevated blood insulin levels. This is a compensatory response by the pancreatic beta cells for the decrease in carbohydrate utilization and storage, and the consequent increase in blood glucose. However, even the increased levels of insulin are not sufficient to maintain normal glucose regulation and increasing severity of hyperglycemia will occur, especially after ingestion of a carbohydrate-rich meal.

Insulin resistance is thought to be caused by an interruption in the insulin signaling pathway that leads to impaired synthesis or mobilization of the cell's glucose transporter, GLUT4. In skeletal muscle cells, the insulin signal that initiates the translocation of GLUT4-containing vesicles from the cell interior to the surface is interrupted. The result is a reduced number of transporters at the cell surface and a consequent reduction in the rate of glucose uptake. In adipocytes, insulin resistance is due mainly to suppression of mRNA encoding GLUT4, resulting in depleted intracellular stores of the protein.

In many cases, type 2 diabetes can be effectively treated, at least in the early stages, with caloric restriction and weight reduction, and no exogenous insulin administration is required. Drugs that increase insulin sensitivity or drugs that cause additional pancreatic secretion of insulin may also be used. However, in the later stages, insulin administration is often required to control plasma glucose concentrations.

7.6 BIOSYNTHESIS OF EICOSANOIDS

Eicosanoids are a series of oxygenated lipid metabolites that include prostaglandins, thromboxanes, and leukotrienes. Their precursors are polyunsaturated fatty acids with chain lengths of 20 carbon atoms. Eicosanoids are very potent signaling molecules; they do not exist free in tissues, but have to be synthesized and released in response to an appropriate stimulus. Prostaglandins act in many tissues by regulating the synthesis of the 'second messenger' cAMP. Because cAMP mediates the action of diverse hormones, the prostaglandins have an amazing variety of biological effects. Thromboxanes, produced by platelets, induce constriction of blood vessels and platelet aggregation, which are early steps in blood clotting. Leukotrienes are formed in leukocytes and mast cells and, like prostaglandins, are mediators of inflammation.

Many, but not all, eicosanoids are synthesized from arachidonic acid. The intracellular concentration of free arachidonic acid is very low, but it can be released from cell membrane phospholipids through the hydrolytic action of phospholipase A_2: indirect pathways involve phospholipases C and D (Smith, 1992).

The biosynthesis of eicosanoids is shown in Figure 7.7. In response to hormonal or other stimuli, phospholipase A_2 attacks membrane phospholipids, releasing arachidonate from the middle carbon (C-2) of the glycerol moiety. Enzymes of the smooth endoplasmic reticulum then convert arachidonate into prostaglandins, beginning with the formation of prostaglandin H_2 (PGH_2), the immediate precursor of many other prostaglandins and of thromboxanes. The arachidonate undergoes a cyclooxidation to form a peroxide, PGG_2, and this in turn is reduced to a transient hydroxyendoperoxide, PGH_2, by peroxidation. These two reactions are catalyzed by dual cyclooxygenase (COX) and peroxidase functions of prostaglandin H synthase (PGHS). PGH_2 is converted to one of several possible prostaglandins and thromboxanes.

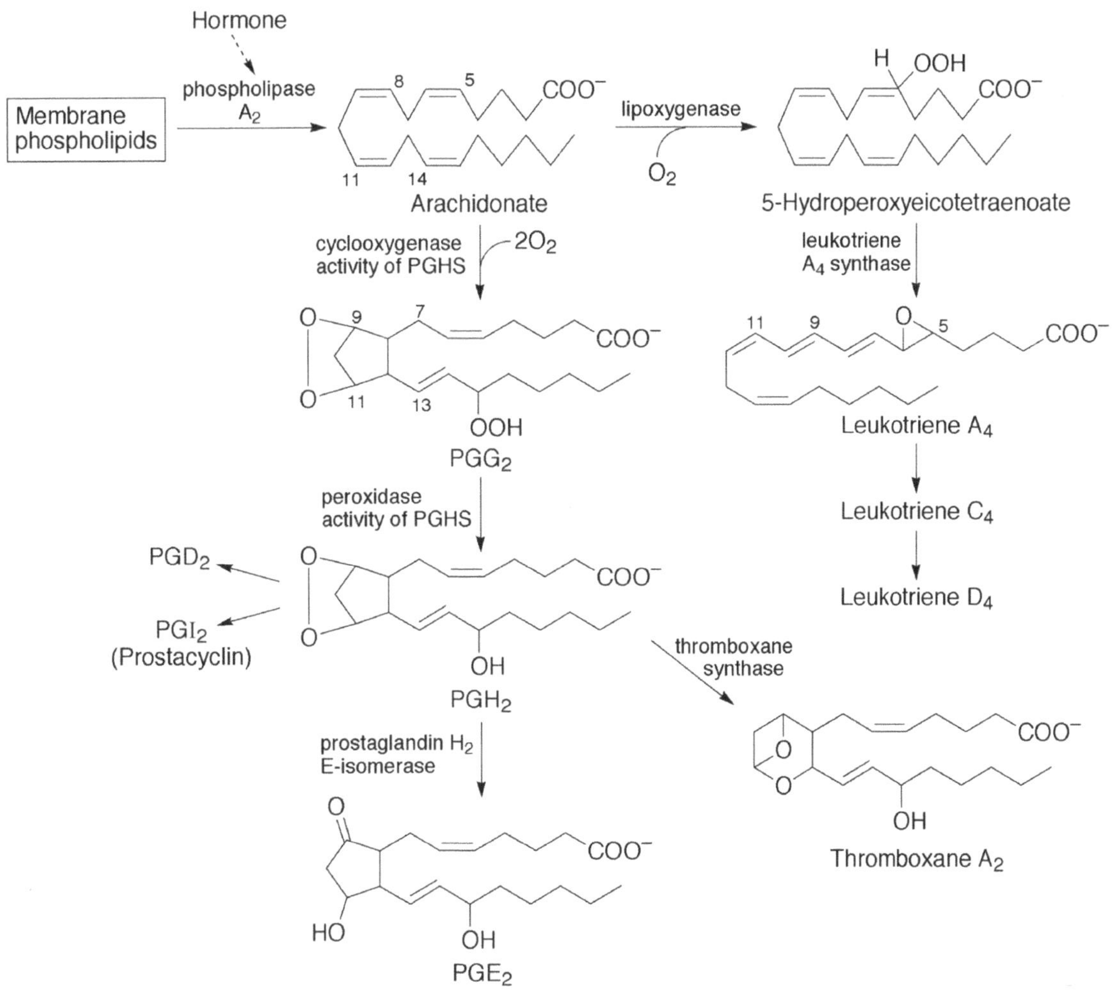

FIGURE 7.7 Biosynthesis of eicosanoids. PGHS, prostaglandin H synthase.

Leukotriene synthesis begins with the action of several lipoxygenases that catalyze the incorporation of molecular oxygen into arachidonate. The various leukotrienes differ in the position of the peroxide group introduced by the lipoxygenases.

7.7 UBIQUITIN-PROTEASOME PATHWAY OF SELECTIVE PROTEIN DEGRADATION

Differential rates of protein degradation are an important aspect of cell regulation. For example, the rapid degradation of cyclins in response to specific signals is necessary to ensure the correct sequence of events in the cell cycle. In addition, misfolded or otherwise abnormal proteins are recognized and destroyed.

The major pathway of selective protein degradation in eukaryotic cells uses ubiquitin, a 76-amino-acid polypeptide, to tag proteins for rapid proteolysis. Ubiquitin attaches covalently to lysine residues of proteins and additional ubiquitin molecules are added to form a polyubiquitin chain. Polyubiquitinated proteins are recognized by and degraded within a large multisubunit protein complex called the proteasome, which contains numerous proteases. Each proteasome consists of a hollow cylinder formed from protein subunits that assemble as a stack of four heptameric rings. Individual ubiquitin molecules are released intact in the degradation process, so they can be reused in another cycle. Both the attachment of ubiquitin and the degradation of the tagged proteins depend on ATP hydrolysis to provide energy.

Polyubiquitination is a three-step process (Figure 7.8). First, ubiquitin is activated by being attached via a high-energy thioester bond to ubiquitin-activating enzyme (E1). The ubiquitin is then transferred to ubiquitin-conjugated enzyme (E2), which combines with an accessory protein (E3) to form ubiquitin ligase. In the third step, the E3 component of ubiquitin ligase binds to a specific recognition sequence in the protein substrate, and the ubiquitin is transferred from the E2 component to the protein. Many repetitions of these three steps result in the attachment of a polyubiquitin chain to the protein. The polyubiquitin tag directs the protein to a proteasome within which the protein is degraded into small peptide fragments.

There are about 30 structurally similar but distinct E2 enzymes in mammalian cells and hundreds of different E3

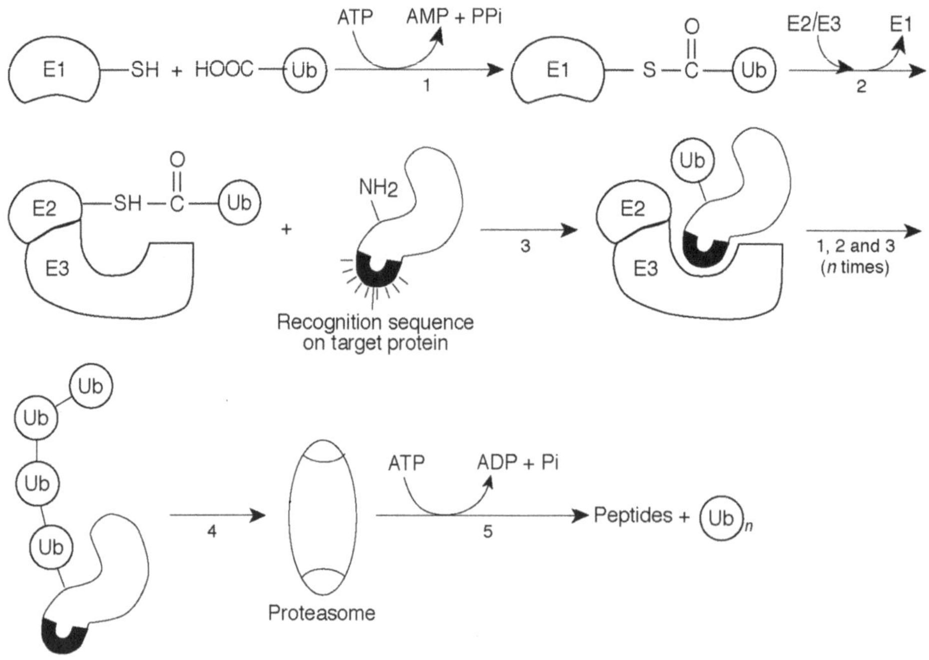

FIGURE 7.8 Ubiquitin-proteasome pathway of selective protein degradation. Steps: (1) Ubiquitin is activated at its carboxyl-terminus by formation of a high-energy thioester bond with a cysteine residue of ubiquitin-activating enzyme, E1. (2) The activated ubiquitin is transferred to a cysteine residue of a ubiquitin-conjugating enzyme, E2, which exists as a complex with an accessory protein, E3. The functional E2–E3 complex is known as ubiquitin ligase. (3) The protein substrate binds to the E3 component of ubiquitin ligase and the ubiquitin molecule is transferred to a lysine residue of the protein. Steps 1, 2, and 3 are repeated several times, resulting in the attachment of a polyubiquitin chain to the protein. (4) The polyubiquitin tag directs the selected protein to a proteasome. (5) Within the proteasome, the protein is degraded into small peptide fragments and the individual ubiquitin molecules are released.

proteins, but there is only one E1 enzyme. Distinct ubiquitin ligases recognize different amino acid sequences, which allows selective targeting of intracellular proteins for degradation. Whereas polyubiquitination requires all three enzymes, monoubiquitination requires E1 and E2 activities only.

Misfolded proteins are instantly recognized because hydrophobic amino acid sequences that are normally buried within the folded protein are exposed at the surface. The degradation of regulatory proteins is controlled by exposing recognition sequences at the appropriate time. Recognition can be achieved by causing a conformational change through phosphorylation or binding to a specific protein. Alternatively, a recognition sequence can be unmasked by the regulated dissociation of a protein subunit.

7.8 DEFENSES AGAINST REACTIVE OXYGEN SPECIES

7.8.1 Reactive Oxygen Species

Reactive oxygen species (ROS) comprise oxygen-centered free radicals and non-radical derivatives of oxygen. The superoxide anion radical ($O_2{\bullet}^-$), the hydroxyl radical ($\bullet OH$), and hydrogen peroxide (H_2O_2) are primary forms that can give rise to secondary reactive products such as the lipid peroxyl radical (LOO•).

Reactive oxygen species can be generated within the body by a variety of processes and, if not adequately removed, can subject the body to oxidative stress. Severe oxidative stress produces harmful effects such as (1) physical damage to biological membranes and lipoproteins through lipid peroxidation; (2) modification of proteins, manifested by an inactivation of certain enzymes; and (3) damage to DNA with the potential for causing mutations. The accumulated damage may form the basis of age-dependent diseases such as atherosclerosis, arthritis, neurodegenerative disorders, and cancer.

Animals have evolved mechanisms to deplete reactive oxygen species, to limit their formation or to repair the damage caused by them. Many of these mechanisms arise through the direct activation of oxidative stress-inducible genes and the resultant synthesis of antioxidant enzymes, DNA repair enzymes, and a host of other beneficial proteins (Storz *et al.*, 1990).

7.8.1.1 Free Radicals

The electrons in an atom revolve in orbitals around the nucleus. Each orbital can hold a maximum of two electrons paired with antiparallel spin, resulting in no net spin. A free radical can be defined as an atom or molecule capable of independent existence that contains one or more unpaired electron(s). The unpaired electron is found alone in the outer orbital and is denoted by a superscript dot next to the element. The simplest free radical is an atom of hydrogen (H•), with one proton and a single electron. Reduction of oxygen by the transfer to it of a single electron will produce the superoxide anion radical (O_2•$^-$), usually referred to simply as superoxide. The oxygen molecule itself is a free radical because it has two unpaired electrons in different orbitals. Owing to their unstable electronic configuration, free radicals are much more reactive than non-radicals. They readily extract electrons from other molecules with which they collide, and these molecules in turn become free radicals. Thus a chain reaction is propagated.

Free radicals are generally produced by electron transfer reactions catalyzed either enzymatically or non-enzymatically through the redox chemistry of transition metal ions. There are many ways of generating free radicals. Examples include phagocytosis during infection; metabolic processing of foreign compounds (e.g. constituents of tobacco smoke, drugs, pesticides, solvents, and pollutants); ultraviolet irradiation of the skin; and ionizing radiation. A major source of superoxide is electron 'leakage' from the mitochondrial electron transport chain to molecular oxygen. Another source of superoxide is the metal-catalyzed autoxidation of certain compounds, including catecholamines, ascorbic acid, thiols (e.g. glutathione, cysteine), tetrahydrofolate, and reduced flavins. Physical exercise leads to a greater production of free radicals relative to the greater oxygen consumption.

Free radicals range from very oxidizing to very reducing and their reduction potentials can be used to predict a hierarchy for free radical reactions (Buettner, 1993). The most oxidizing free radical is the hydroxyl radical (•OH) which, with a half-life of only 1 nanosecond at 37°C, is so reactive that there is no time for any molecule or radical to remove it before damage is done. Damage to DNA and proteins is therefore unavoidable and must be dealt with by repair enzymes.

7.8.1.2 Singlet Oxygen

Singlet oxygen ($^1O_2$) is an exceptionally reactive and destructive form of oxygen, which has been implicated in the inactivation of enzymes, the peroxidation of lipids in membranes and lipoproteins, and in DNA strand breakage. It is produced when a molecule of ground-state triplet oxygen is excited by absorbing energy from a photochemical reaction. Excitation can be achieved when biological pigments such as retinaldehyde, flavins, and porphyrins are illuminated in the presence of oxygen. The pigment absorbs light, enters a higher electronic state, and transfers energy to the oxygen molecule. The result is a movement of one of the unpaired electrons in a way that alleviates the spin restriction. As there are no unpaired electrons, singlet oxygen does not qualify as a radical. Certain biological antioxidants, particularly carotenoids, are able to inactivate singlet oxygen by quenching.

7.8.1.3 Hydrogen Peroxide

Hydrogen peroxide is produced by several oxidase enzymes and can also be formed enzymatically from superoxide. Hydrogen peroxide is important in free radical biochemistry because it is a potential source of the highly reactive hydroxyl radical in the presence of free transition metal ions. It is relatively stable and its high diffusibility, both within and between cells, extends the range of free radical reactions.

7.8.2 Defense Systems

Transition metal ions in the free state promote free radical reactions and so the body ensures as far as possible that iron and copper ions, at least, are safely bound in storage or transport proteins. There is three times as much transferrin iron-binding capacity in plasma as iron needing to be transported. Iron ions bound to transferrin cannot stimulate lipid peroxidation or formation of hydroxyl free radicals. The same is true of copper ions bound to the plasma protein ceruloplasmin or albumin. The binding of iron to proteins depends on the plasma or tissue pH values being normal. A slight displacement of pH towards acidosis (as occurs in cases of ischemia, trauma, inflammatory states, and tumors) is sufficient to allow detachment of the metal and the initiation of radical-forming reactions.

The body's next line of defense is the removal of reactive oxygen species or their precursors by certain enzymes. For example, most of the superoxide generated in the body undergoes a dismutation reaction, forming hydrogen peroxide and oxygen (reaction 7.1). This reaction can take place spontaneously (albeit rather slowly) or can be catalyzed by either of two superoxide dismutases, one a cytosolic copper-zinc enzyme and the other a mitochondrial manganese-centered enzyme.

$$2O_2{\bullet}^- + 2H^+ \longrightarrow H_2O_2 + O_2 \qquad (7.1)$$

Two enzymes remove hydrogen peroxide in mammalian cells: catalase (reaction 7.2) and selenium-dependent glutathione peroxidase (reaction 7.3). The latter enzyme is the more important of the two: catalase is probably limited to destroying hydrogen peroxide generated by oxidase enzymes located within peroxisomes.

$$2H_2O_2 \longrightarrow 2H_2O + O_2 \qquad (7.2)$$

$$H_2O_2 + 2GSH \longrightarrow GSSG + 2H_2O \qquad (7.3)$$

Oxidized glutathione (GSSG) is then converted back to glutathione (GSH) by an NADPH-dependent enzyme, glutathione reductase.

Enzymes do not completely prevent the formation of free radicals, and the body must now rely on antioxidants that 'scavenge' free radicals and convert them to harmless stable derivatives. Among naturally occurring antioxidants, vitamin E, vitamin C, and carotenoids are the most important. Vitamin E is the major chain-breaking antioxidant in membrane lipids and lipoproteins. Vitamin C is an effective scavenger of several reactive oxygen species in the aqueous environment of the cytosol and extracellular fluids. Carotenoids have an important role in inactivating singlet oxygen. The ways in which each of these antioxidants function are discussed in the relevant chapters.

7.9 LIPID PEROXIDATION

The polyunsaturated fatty acids (PUFAs) incorporated into the phospholipids of cell plasma membranes, the membranes of subcellular organelles (e.g. mitochondria), and plasma lipoproteins are susceptible to attack by certain free radicals such as •OH, LOO•, and LO•. These free radicals initiate chain reactions that, if not stopped by chain-breaking antioxidants, result in the accumulation of toxic products. The process of lipid peroxidation can be conveniently divided into three stages: initiation, propagation, and termination.

Initiation

$$LH + \bullet OH \longrightarrow L\bullet + H_2O \qquad (7.4)$$

Propagation

$$L^\bullet + O_2 \longrightarrow LOO^\bullet \qquad (7.5)$$

$$LOO^\bullet + L'H \longrightarrow LOOH + L'^\bullet \qquad (7.6)$$

Termination

$$LOO\bullet + LOO\bullet \longrightarrow \text{molecular products} \qquad (7.7)$$

Transition metals can decompose lipid hydroperoxides into peroxyl and alkoxyl radicals that bring about a dramatic intensification of the propagative stage.

$$LOOH + Fe^{2+}\ (\text{or } Cu^{+}) \longrightarrow LO\bullet + OH^{-} + Fe^{3+}\ (\text{or } Cu^{2+}) \qquad (7.8)$$

$$LOOH + Fe^{3+}\ (\text{or } Cu^{2+}) \longrightarrow LOO\bullet + H^{+} + Fe^{2+}\ (\text{or } Cu^{+}) \qquad (7.9)$$

In reactions 7.4 to 7.9, LH represents a PUFA side chain of a phospholipid; •OH, a hydroxyl radical; L•, a carbon-centered lipid radical; LOO•, a lipid peroxyl radical; LO•, a lipid alkoxyl radical; and LOOH, a lipid hydroperoxide. Reactions 7.4 to 7.6 are illustrated in Figure 7.9.

Initiation involves removal of a hydrogen atom (H•) from a methylene ($-CH_2-$) carbon in the PUFA molecule by any chemical species that is sufficiently reactive. Reaction 7.4 shows initiation by a hydroxyl radical. Removal of the hydrogen atom leaves behind an unpaired electron on the methylene carbon atom (–C•H–). The presence of a double bond in the fatty acid weakens the C–H bonds on the carbon atom adjacent to the double bond and so makes H• removal easier. Thus the greater the number of double bonds in a PUFA, the more susceptible is the PUFA to peroxidation.

The lipid radical L• undergoes a molecular rearrangement to form a conjugated diene that reacts with molecular oxygen to give LOO•. The peroxyl radical then attacks an adjacent PUFA side chain of a phospholipid (L′H), removing a hydrogen atom and converting itself into a lipid hydroperoxide. The new L′• reacts with oxygen to yield another peroxyl radical and thus a chain reaction is propagated. In this manner, a single initiation can result in hundreds of

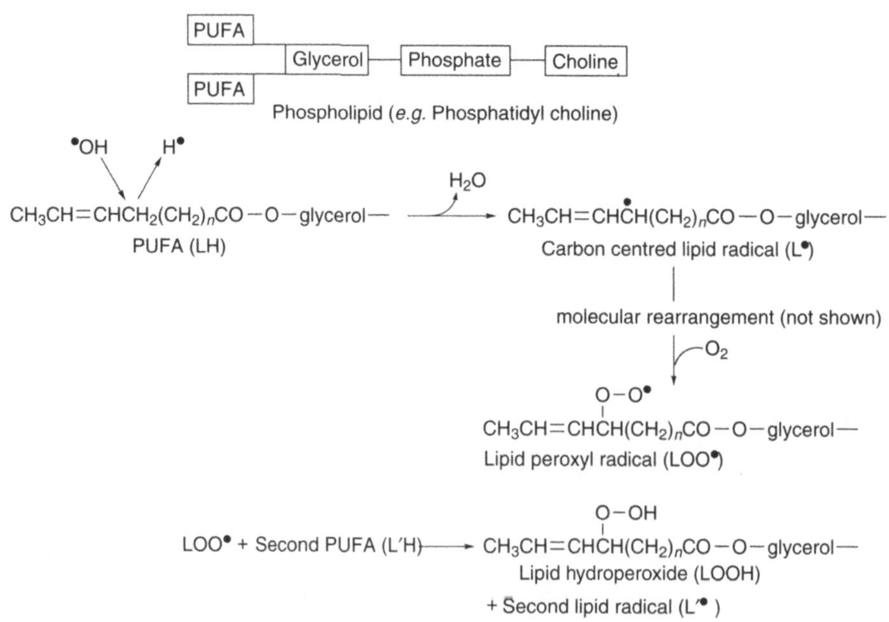

FIGURE 7.9 Initiation and propagation stages of lipid peroxidation.

PUFA side chains being converted into lipid hydroperoxides. The length of the propagation chain depends on the lipid-protein ratio in a membrane, the fatty acid composition, the oxygen concentration, and the presence or absence within the membrane of chain-breaking antioxidants. In the absence of a chain-breaking antioxidant, the propagation chain (reactions 7.5 and 7.6) is eventually broken when two peroxyl radicals react together to give molecular products (reaction 7.7).

In the body, lipid hydroperoxides are cleaved by phospholipase A_2, releasing the peroxidized fatty acid from the phospholipid. The peroxidized free fatty acid can now be reduced to a harmless hydroxy fatty acid by the action of glutathione peroxidase (reaction 7.10). The reducing equivalents for this reaction are provided by the tripeptide glutathione (reduced form, GSH; oxidized form, GSSG).

$$\text{PUFA-OOH} + 2\text{GSH} \longrightarrow \text{PUFA-OH} + H_2O + \text{GSSG} \qquad (7.10)$$

Alternatively, a phospholipid hydroperoxide glutathione peroxidase can convert lipid hydroperoxides directly to a hydroxy acid (Buettner, 1993). Both of these peroxidase enzymes uniquely require the trace element selenium as a co-factor.

The PUFAs in LDL are protected from peroxidation by the presence of several indigenous lipid-soluble antioxidants. By far the major antioxidant is α-tocopherol (vitamin E): on average about six molecules of this antioxidant are present in each LDL molecule. Vitamin E is a very efficient free radical scavenger and terminates chains almost as soon as they are initiated. In doing so, vitamin E and other antioxidants are gradually used up. When the LDL is finally depleted of its antioxidants, the chain reaction will proceed unchallenged until all LDL PUFAs are converted into lipid hydroperoxides. At termination, the lipid hydroperoxides break down to a wide range of products, including aldehydes, hydrocarbons, epoxides, and alcohols. Some of these breakdown products are diffusible toxins which irritate endothelial cells and may contribute to disruption of endothelial integrity.

7.10 HEMOSTASIS

7.10.1 Overview

Immediately after a blood vessel has been severed or ruptured, bleeding is stemmed by local vasoconstriction resulting from nervous reflexes, vascular spasm, and humoral factors released from the traumatized tissues and blood platelets. The damaged vessel is then sealed temporarily with a plug of aggregated platelets. Localized blood coagulation leads to subsequent formation of a blood clot composed of a three-dimensional meshwork of fibrin fibers entrapping blood cells, platelets, and plasma. A large amount of plasminogen is also trapped in the clot along with other plasma proteins. The clot begins to develop in 15–20 s if the trauma to the vascular wall has been severe and in 1–2 min if the trauma has been minor. Within 3–6 min of rupture of a vessel (if the opening is not too large), the entire opening or broken end of the vessel is filled with clot. After 20 min to one hour, the clot retracts, closing the vessel still further. Clot retraction is effected by entrapped platelets which activate themselves to contract, thereby pulling attached fibrin fibers towards them and compressing the fibrin meshwork into a smaller mass. Within a few hours of clot formation, the clot is invaded by fibroblasts and the deposition of fibrous connective tissue throughout the clot seals the vessel permanently. A day or so after the clot has stopped the bleeding, the injured tissues and vascular endothelium very slowly release tissue plasminogen activator. This activator converts the plasminogen within the clot to plasmin, which dissolves the clot. After one or two weeks, the clot has been replaced by connective tissue.

7.10.2 Characteristics of Platelets

Platelets are colorless, round or oval discs of diameter 1–4 μm (cf. erythrocytes with a mean diameter of 7.8 μm). They arise by fragmentation of megakaryocytes within the bone marrow or soon after the megakaryocytes enter the blood. Although platelets have no nuclei and cannot reproduce, they have many functional characteristics of whole cells. They have the capacity to synthesize and store various factors involved in adhesion and aggregation; they generate ADP and ATP; and they store large quantities of calcium ions. Contractile proteins in the platelets' cytoplasm enable them to contract. On the platelet surface is a coat of glycoproteins (the glycocalyx) that prevents adherence to normal endothelium and yet causes adherence to injured areas of the vessel wall.

7.10.3 Platelet Activation

When platelets come in contact with collagen exposed at the damaged vascular surface, they undergo an immediate and drastic change, an event known as activation. The platelets bind directly to collagen by means of collagen-specific glycoprotein (GP) Ia-IIa surface receptors. This adhesion is strengthened further by von Willebrand factor (vWF), which is released from the endothelium and from platelets. vWF forms additional links between the platelets' GP Ib-IX-V receptor complex and the exposed collagen fibrils. This localization of platelets to the extracellular matrix promotes collagen interaction with platelet GP VI. This triggers a signaling cascade that results in activation of platelet integrins. Activated integrins mediate tight binding of platelets to the extracellular matrix at the site of injury.

Activated platelets swell to a spherical form and extend long pseudopodia. Their contractile proteins contract forcefully and cause the release of granules that contain multiple active factors, including ADP, serotonin, platelet-activating factor, vWF, and thromboxane A_2. The secreted ADP acts on nearby platelets to activate them as well, and the stickiness of these additional platelets causes them to adhere to the originally activated platelets. Thromboxane A_2 is both a vasoconstrictor and an aggregating agent. The granular contents activate a G_q-linked protein receptor cascade in the platelets that stimulates the release of Ca^{2+} from the endoplasmic reticulum into the cytosol. After stimulation, the cytosolic Ca^{2+} concentration is restored to the resting level by sequestration of the ion into the endoplasmic reticulum and extrusion out of the cell. The calcium activates protein kinase C, which, in turn, activates phospholipase A_2 (PLA_2). PLA_2 then modifies the platelets' integrin receptor GP IIb-IIIa, increasing its affinity to bind fibrinogen present in the plasma. The activated platelets change shape from spherical to stellate, and the fibrinogen cross-linking with GP IIb-IIIa aids adhesion of adjacent platelets. The aggregation of platelets at the damage site leads to the formation of a platelet plug, which physically blocks the breach in the vessel wall.

7.10.4 Inhibition of Platelet Aggregation

Platelet aggregation can be inhibited by prostacyclin and prostaglandin E_1 (PGE_1) acting through IP receptors. Prostacyclin is the main product of arachidonic acid metabolism in the endothelial cells of blood vessels and is not synthesized in platelets. PGE_1 can be synthesized in platelets from dihomo-γ-linolenic acid.

Inhibition of platelet aggregation by potential drugs is studied *in vitro* by a turbidometric method using an aggregometer. Platelet-rich plasma is placed in a cuvette and incubated at 37°C (with stirring) with the molecule of interest. A threshold aggregation concentration of the aggregating agent, such as ADP, collagen, or thrombin, is then added to the cuvette, and the degree of aggregation inhibition is measured as the percentage of light transmitted through the cuvette.

7.10.5 Initiation of Coagulation: Formation of Prothrombin Activator

The complexity of the blood coagulation system ensures that it only becomes active when necessary. There are two pathways leading to the formation of prothrombin activator: the extrinsic pathway (Figure 7.10) and the intrinsic pathway (Figure 7.11). In both of these pathways, a series of plasma proteins, the blood-clotting factors, create a cascade which enormously amplifies the original stimulus for clotting. Most of the blood clotting factors circulate as inactive precursors (zymogens) of proteolytic enzymes. Each factor is activated by partial proteolysis, and then in turn activates the next factor. Most of the clotting factors (Table 7.3) are designated by Roman numerals. In the pathways described below, a small letter 'a' after the Roman numeral denotes the activated form of the factor.

7.10.5.1 Extrinsic Pathway

This pathway is triggered by injury to the tissues and is initiated by a glycoprotein known as tissue factor (also called tissue thromboplastin). Tissue factor is an integral membrane protein found on the surface of nonvascular cells in close association with phospholipids. When injury occurs, tissue factor is exposed to blood and forms a one-to-one complex with circulating Factor VII in the presence

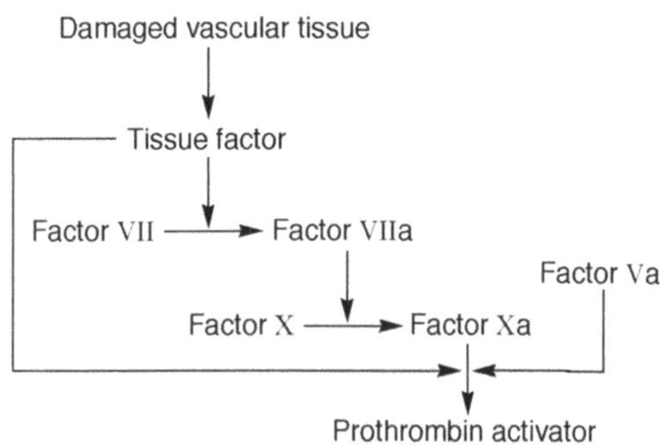

FIGURE 7.10 Extrinsic pathway for initiating blood coagulation.

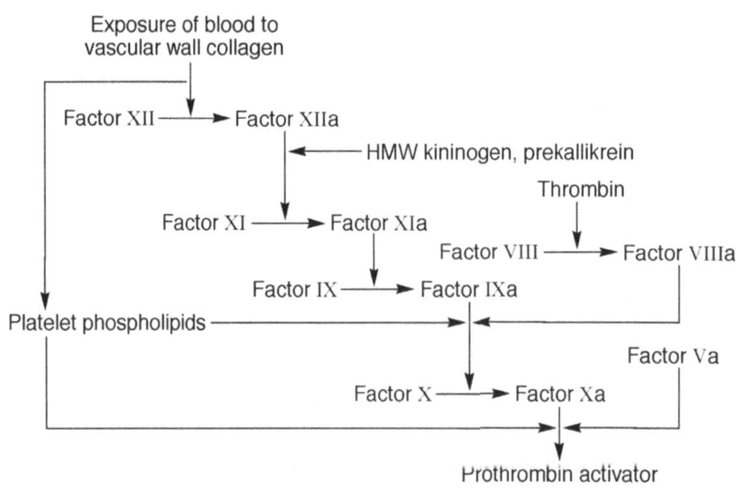

FIGURE 7.11 Intrinsic pathway for initiating blood coagulation.

TABLE 7.3
Components of the Blood Coagulation Cascade

Factor	Identity	Function
Factor I	Fibrinogen	Structural
Factor II	Prothrombin	Protease zymogen
Factor III	Tissue factor (thromboplastin)	Co-factor/initiator
Factor IV	Calcium	Co-factor
Factor V	See text	Co-factor
Factor VII	See text	Protease zymogen
Factor VIII	See text	Co-factor
Factor IX	See text	Protease zymogen
Factor X	See text	Protease zymogen
Factor XI	See text	Protease zymogen
Factor XII	See text	Protease zymogen
Factor XIII	See text	Protease zymogen
Protein C	See text	Protease zymogen
Protein S	See text	Co-factor
Protein Z	See text	Co-factor
von Willebrand factor	See text	Adhesion

of calcium ions. Formation of this complex facilitates the conversion of Factor VII to its active protease form (Factor VIIa) by minor proteolysis. The Factor VIIa–tissue factor complex then converts Factor X to another protease (Factor Xa) by the cleavage of a single peptide bond. Factor Xa combines immediately with the tissue factor, as well as with Factor Va and phospholipids provided by activated platelets, to form the complex prothrombin activator.

7.10.5.2 Intrinsic Pathway

This pathway begins in the blood itself. Exposure of the blood to vascular wall collagen following damage to blood vessels causes circulating Factor XII to change its molecular configuration, converting it into a proteolytic enzyme (Factor XIIa). Simultaneously, platelets are stimulated to release phospholipids. Factor XIIa activates Factor XI, a reaction which also requires HMW (high-molecular-weight) kininogen and is accelerated by prekallikrein. Activated Factor XI then activates Factor IX. Activated Factor IX, acting in concert with activated Factor VIII and with the platelet phospholipids, activates Factor X. The last step in the intrinsic pathway is the same as that in the extrinsic pathway. That is, activated Factor X combines with Factor Va and platelet or tissue phospholipids to form prothrombin activator.

Factor VIII is the factor that is missing in a person who has classic hemophilia. The bleeding disease called thrombocytopenia is caused by a deficiency of platelets.

7.10.6 Conversion of Prothrombin to Thrombin and Clot Formation

A scheme of these events is given in Figure 7.12. Prothrombin activator, in the presence of sufficient amounts of calcium ions and phospholipids provided by activated platelets,

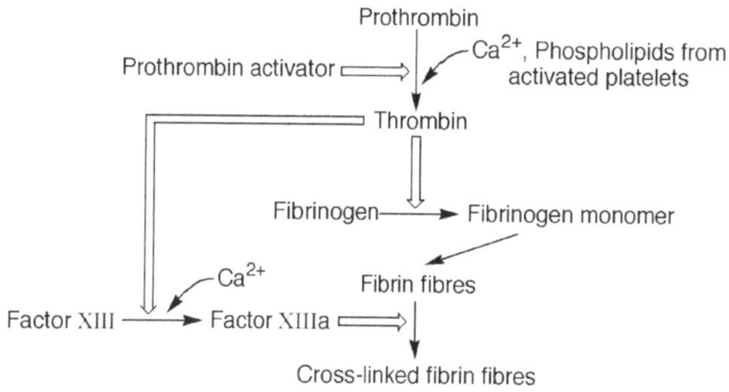

FIGURE 7.12 Scheme for conversion of prothrombin to thrombin and conversion of fibrinogen to cross-linked fibrin fibers.

causes conversion of prothrombin into thrombin. Once generated, thrombin exerts positive feedback upon its formation by activating Factor V. Prothrombin is a plasma protein that is formed continually in the liver; it splits easily into smaller compounds, one of which is thrombin. Much of the prothrombin first formed attaches to prothrombin receptors on the platelets that have already stuck to the damaged tissue. This binding stimulates the formation of still more thrombin in the localized area of damage.

Fibrinogen is a soluble plasma protein of high molecular weight (MW = 340,000) which is formed in the liver. Thrombin, a weakly proteolytic enzyme, selectively removes four peptides from each molecule of fibrinogen to form a molecule of fibrin monomer. Many fibrin monomer molecules then spontaneously polymerize into long fibers of insoluble fibrin that constitute the reticulum of the clot. The formation of fibrin accelerates the conversion of fibrin-stabilizing factor (Factor XIII) to an enzyme (Factor XIIIa) by thrombin in the presence of calcium ions. Fibrin-stabilizing factor is a plasma protein, but is also released from platelets entrapped in the clot. Factor XIIIa causes covalent bonding between fibrin monomer molecules as well as multiple cross-linking between adjacent fibrin fibers; it also causes cross-linking of other plasma proteins to fibrin, so incorporating them into the clot. The result is a greatly strengthened reticulum.

7.10.7 Regulation of the Coagulation Cascade

Many plasma proteins are involved in the regulation of coagulation. Among them, antithrombin III is a plasma protease inhibitor that scavenges any of the blood-clotting enzymes that move away from the growing clot. Protein C inactivates the active co-factor forms of Factors V and VIII, thus rapidly slowing blood clotting. Protein C is converted to its active enzyme form on endothelial cell membranes by a complex of two proteins, thrombomodulin and thrombin. Activated protein C, anchored to the membrane through binding to membrane-bound protein S, converts the Factors Va and VIIIa to their inactive forms by limited proteolysis. Protein Z binds to Factor Xa, and this complex in turn binds to the tissue-factor–Factor VII complex, thus inactivating the extrinsic pathway.

7.10.8 Lysis of Blood Clots

Circulating in the bloodstream is a protein called plasminogen that, when activated, becomes a proteolytic enzyme called plasmin. Whenever plasmin is formed, it causes lysis of the clot by destroying fibrin fibers, as well as fibrinogen, prothrombin, and Factors V, VIII, and XII. During clot formation, plasminogen is trapped within the clot along with other plasma proteins. The injured tissues and vascular endothelium very slowly release a potent substance called tissue plasminogen activator that, after the clot has stopped the bleeding, eventually converts plasminogen to plasmin. Plasmin plays an important role in removing minute clots from millions of tiny peripheral vessels that otherwise would become occluded.

FURTHER READING

Halliwell, B., Gutteridge, J.M.C. & Cross, C.E. (1992) Free radicals, antioxidants, and human disease: Where are we now? *Journal of Laboratory and Clinical Medicine*, **119**, 598–620.

REFERENCES

Buettner, G.R. (1993) The pecking order of free radicals and antioxidants: Lipid peroxidation, α-tocopherol, and ascorbate. *Archives of Biochemistry and Biophysics*, **300**, 535–43.

Elsas, L.J., Albrecht, I. & Rosenberg, L.E. (1968) Insulin stimulation of amino acid uptake in rat diaphragm. *Journal of Biological Chemistry*, **243**(8), 1846–53.

Elsas, L.J. II, MacDonell, R.C. Jr. & Rosenberg, L.E. (1971) Influence of age on insulin stimulation of amino acid uptake in rat diaphragm. *Journal of Biological Chemistry*, **246**(21), 6452–9.

Proud, C.G. (2006) Regulation of protein synthesis by insulin. *Biochemical Society Transactions*, **34**, 213–16.

Smith, W.L. (1992) Prostanoid biosynthesis and mechanisms of action. *American Journal of Physiology*, **263**, F181–9.

Storz, G., Tartaglia, L.A. & Ames, B.N. (1990) Transcriptional regulator of oxidative stress-inducible genes: Direct activation by oxidation. *Science*, **248**, 189–94.

8 Transduction of Intracellular Signals

8.1 HORMONES

Hormones are chemical messengers, which are synthesized by the cells of endocrine glands and secreted from these glands directly into the bloodstream. They travel in the systemic circulation to the cells of distal target tissues where they initiate their biological response. There are three general classes of hormones: (1) peptide and polypeptide hormones include thyroid-stimulating hormone (thyrotrophin) secreted by the anterior pituitary gland and parathyroid hormone; (2) steroid hormones such as estrogens, testosterone, and cortisol; and (3) derivatives of the amino acid tyrosine, e.g. the thyroid hormones thyroxine and triiodothyronine, and adrenaline (epinephrine). Peptide and amine hormones are stored in secretory vesicles until needed. Steroid hormones are not stored: they are synthesized from intracellular stores of cholesteryl esters after a stimulus.

Hormone action is initiated by the binding of hormone to receptors in the cells of target tissues. Hormonal receptors are large proteins which have both high affinity and high specificity for their ligands. Each cell within a target tissue contains some 2,000 to 100,000 receptors. The target tissues for a particular hormone are those whose cells contain receptors for that hormone: cells that lack such receptors do not respond. Thus receptors provide the first level of specificity for hormone action. Formation of the hormone–receptor complex triggers a cascade of reactions in the target cell, with each stage becoming more powerfully activated, so that very low concentrations of circulating hormone can elicit a robust biological response.

Whereas the receptors for the steroid hormones reside within the cell, the receptors for peptide hormones and amine hormones are located in or on the surface of the cell plasma membrane. These hormones, being hydrophilic, cannot cross the lipid bilayer of the plasma membrane and so many of them recruit a 'second messenger' to mediate their intracellular action (the hormone itself is the 'first messenger'). The best-known second messenger is cyclic AMP (cAMP); others include diacylglycerol and inositol triphosphate.

Positive or negative feedback to the endocrine gland may regulate any step during the synthesis, processing, or release of the hormone. Positive feedback results in additional secretion of the hormone, while negative feedback has the opposite effect. Most hormones are controlled through negative feedback in order to prevent overactivity of the target tissues and to allow responses to subsequent signals. The controlled variable is often the degree of activity of the target tissue. Above a certain level of activity, negative feedback signals to the endocrine gland become powerful enough to shut down hormone production or prevent secretion. Some hormones can cause up-regulation of receptors by inducing their synthesis. When this occurs, the target tissue becomes progressively more sensitive to the stimulating effects of the hormone.

8.2 INTRACELLULAR SIGNAL TRANSDUCTION

Signal molecules synthesized and released by signaling cells are able to pass their message to distant target cells where they control transcription of specific genes by modifying the activity of activators and repressors. This section discusses signaling pathways initiated by hydrophilic signal molecules that cannot simply diffuse across the lipid bilayer of the plasma membrane.

Hydrophilic signal molecules such as peptide hormones, growth factors, and cytokines pass their message to target cells that have specific receptors on their surface. The signaling molecules act as ligands which bind to structurally complementary sites on the receptor. Ligand binding initiates a chain of intracellular biochemical reactions that distributes the signal to appropriate parts of the cell and ultimately induces a specific cellular response. Amplifier proteins, which are usually either enzymes or ion channels, greatly increase the signal they receive, either by producing large amounts of small intracellular mediators or by activating large numbers of downstream intracellular signaling proteins. When there are multiple amplification steps in a relay chain, the chain is often referred to as a signaling cascade. The overall process of converting transient cell-surface stimuli into long-term cellular responses is called signal transduction.

Signal transduction pathways employ a variety of sequentially acting protein kinases, which catalyze the covalent attachment of a phosphate group to specific proteins. Two general classes of kinase exist in eukaryotes: those phosphorylating serine and threonine residues and those phosphorylating tyrosine residues. Interaction among signal transduction pathways (cross-talk) allows signals to be fine-tuned by additive, synergistic, or antagonistic effects. Positive signals are normally followed by immediate negative feedback control to prevent overshoot, and to allow responses to subsequent signals. The stimulus can also be diminished by decreasing the number of receptors at the cell surface. This can be achieved by decreased production of the receptors by the protein-manufacturing mechanism of the cell. Alternatively, the receptor proteins

DOI: 10.1201/9781003410300-10

themselves are inactivated or destroyed during the course of their function. In any event, this down-regulation of the receptor decreases the responsiveness of the target tissue to the stimulus.

In many signaling pathways, ligand binding to a surface receptor leads to activation of transcription factors in the cytosol, permitting them to translocate into the nucleus and stimulate (or occasionally repress) transcription of their target genes. Alternatively, receptor stimulation may lead to activation of cytosolic protein kinases that then translocate into the nucleus and regulate the activity of nuclear transcription factors. Activation of transcription factors generally involves signal-induced phosphorylation. The activation of some transcription factors, such as nuclear factor-kappa B (NF-κB), involves phosphorylation and subsequent ubiquitin-mediated degradation of an inhibitory protein.

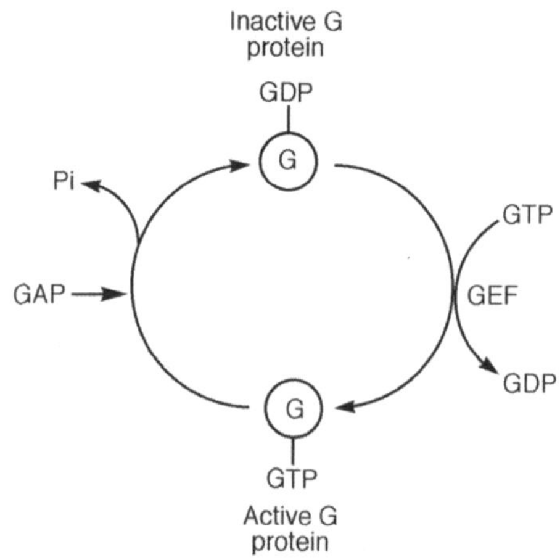

FIGURE 8.1 Regulation of G protein activity. In the course of the exchange reaction, the GEF displaces the bound nucleotide (GDP) and, subsequently, a new nucleotide (GTP) displaces the GEF. GAP, GTPase-activating protein; GEF, guanine nucleotide-exchange factor; Pi, inorganic phosphate.

8.3 SIGNALING THROUGH G PROTEIN-COUPLED RECEPTORS

8.3.1 G Proteins and G Protein-Coupled Receptors

The term 'G protein' refers to three classes of mammalian guanine nucleotide-binding proteins: (1) heterotrimeric G proteins in which GTP or GDP is attached to the α subunit; (2) the Ras superfamily of small G proteins, whose members include Rab, Ran, and Ras; and (3) intermediate-sized G proteins, such as the translation elongation factors eEF1A and eEF2. G proteins are GTPases, but intrinsic hydrolysis of bound GTP is very slow. All G proteins function as molecular switches in a variety of biological processes, being turned 'on' when bound to GTP and turned 'off' when bound to GDP. GTP or GDP binding is stabilized by interaction of the phosphates with the G protein's P loop and by a Mg^{2+} ion that is co-ordinated by elements of switch I and switch II domains (Vetter & Wittinghofer, 2001). The conformational change induced by GTP binding enables the G protein to interact with an effector protein, which initiates downstream signaling events. GTP hydrolysis deactivates the G protein, thereby terminating the transmitted signal.

The shuttling of G proteins between these active and inactive states requires the participation of two classes of regulatory proteins: a guanine nucleotide-exchange factor (GEF) promotes the exchange of bound GDP for GTP, and a GTPase-activating protein (GAP) accelerates hydrolysis of bound GTP (Figure 8.1). GEFs catalyze the release of GDP and the rebinding of GTP, thereby regenerating the active G protein. GEFs drastically decrease the affinity of nucleotide binding by disturbing the structure of the P loop and pushing the Mg^{2+} ion out of its position (Vetter & Wittinghofer, 2001). GEFs actually have no preference for the two nucleotides and the replacement of GDP by GTP is due to the approximately ten times higher cellular concentration of GTP over GDP. Different GAPs increase the G protein's intrinsic GTPase activity by supplying a particular amino acid to stabilize the orientation of the catalytic machinery. GEFs and GAPs are multidomain proteins that are regulated by extracellular signals and localized cues that control cellular events in time and space (Bos *et al.*, 2007).

The human genome encodes several thousand G protein-coupled receptors (GPCRs). These include receptors for many neurotransmitters and most of the receptors for hormones that control carbohydrate, amino acid, and fat metabolism. The light-activated protein, rhodopsin, in the eye is a GPCR and so are a great number of olfactory (smell) and gustatory (taste) receptors. A particular ligand can activate many different GPCR family members; for example, at least nine distinct receptors are activated by adrenaline, another five or more by acetylcholine, and at least 15 by serotonin.

The binding of ligands to the extracellular domain of GPCRs results in the binding of the receptor to a heterotrimeric G protein attached to the inner leaflet of the plasma membrane. The protein's three subunits are designated α, β, and γ, in order of decreasing molecular weight. Heterotrimeric G proteins relay signals from ligand-activated GPCRs to appropriate effector proteins. Effector proteins are either membrane-bound ion channels or enzymes that catalyze formation of intracellular second messengers, such as cAMP.

There are various types of mammalian heterotrimeric G proteins, each type being specific for a particular set of receptors. Each protein is defined by its specific α-subunit, which binds to a guanine nucleotide. For example, G_s transmits signals from a wide variety of GPCRs to adenylyl cyclase (formerly called adenylate cyclase), an enzyme which catalyzes the formation of cAMP from ATP within cells. Other examples of heterotrimeric G proteins and their effectors are given in Table 8.1.

Heterotrimeric G proteins fulfill several purposes in cell signaling. (1) They provide an extra step for signal

TABLE 8.1
Examples of Mammalian Heterotrimeric G proteins and Their Effectors

Gα Class	Associated Effector	2nd Messenger	Receptor Examples
$G_s\alpha$	Adenylyl cyclase	cAMP (increased)	β_1- and β_2-adrenergic receptor
$G_i\alpha$	Adenylyl cyclase	cAMP (decreased)	α_1-adrenergic receptor
	K^+ channel ($G_{\beta\gamma}$ activates effector)	Change in membrane potential	Muscarinic acetylcholine receptor
$G_q\alpha$	Phospholipase C	IP_3, DAG (increased)	α_2-adrenergic receptor
$G_o\alpha$	Phospholipase C	IP_3, DAG (increased)	Acetylcholine receptor in endothelial cells

amplification, because an activated receptor can stimulate multiple G proteins, which, in turn, regulate multiple effector proteins. (2) They integrate signals from several stimuli. For example, the G_s protein triggers the synthesis of cAMP in response to receptor activation by several different hormones. (3) They act as a point of signal divergence by regulating more than one effector in response to a stimulus. Divergence can also occur through the action of a receptor on different G proteins. This allows the same hormone to either stimulate or inhibit a particular action. For example, adrenaline can bind to both α- and β-adrenergic receptors. The β_1- and β_2-adrenergic receptors are coupled to G_s, and therefore the binding of adrenaline leads to an increase in cAMP levels. The α_1-adrenergic receptor is coupled to an inhibitory G protein, G_i, so that adrenaline binding leads to a decrease in cAMP levels.

In many cases, GTP hydrolysis to GDP is enhanced by a protein termed RGS (regulator of G protein signaling), whose activity also may be controlled by extracellular signals. An RGS protein does not, however, appear to exist for G_s (Weinstein *et al.*, 2001).

An additional mechanism of regulating signal transduction, known as desensitization, allows a receptor to be rapidly inactivated and then reactivated despite the continued presence of a stimulus of constant intensity. Persistent or excessive stimulation of a GPCR leads to activation of a G protein-coupled receptor kinase (GRK) that phosphorylates the receptor. The receptor then binds an arrestin protein which sterically prevents signaling to the G protein. The family of GRKs includes at least six members (GRKs 1–6), of which the most thoroughly investigated are rhodopsin kinase (GRK1) and β-adrenergic receptor kinase (GRK2). GRKs recognize only the liganded or stimulus-modified forms of the receptors that they phosphorylate. Desensitization of rhodopsin, the photoreceptor in the rod cells of the retina, occurs in less than one second following light stimulation, thereby preventing a brief flash of light from being perceived as continuous illumination.

8.3.2 The Adenylyl Cyclase–Cyclic AMP Second Messenger System

The main components of the adenylyl cyclase–cAMP second messenger system are shown in Figure 8.2. The

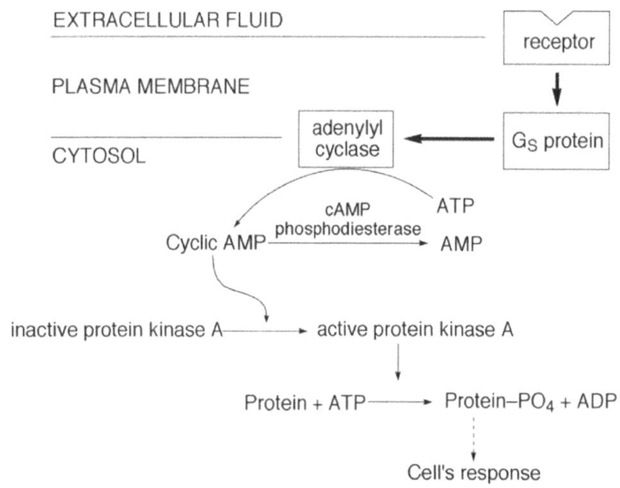

FIGURE 8.2 The adenylyl cyclase–cyclic AMP second messenger system.

synthesis and degradation of cAMP are shown in Figure 8.3. Amplification of the hormonal signal is achieved as follows. The binding of a single molecule of hormone to one receptor molecule activates many G protein molecules, each of which activates a molecule of adenylyl cyclase. Each molecule of adenylyl cyclase catalyzes the formation of multiple cAMP molecules, each of which activates one molecule of cAMP-dependent protein kinase. This kinase, known as protein kinase A (PKA), phosphorylates serine and threonine residues in target proteins. PKA initiates a cascade of various protein kinases by activating a second kinase, which activates a third kinase, and so forth. The rapid hydrolysis of cAMP to AMP by cAMP phosphodiesterase turns off the signal.

The specific action that occurs in response to increases or decreases of cAMP in each type of target cell depends on the nature of the intracellular machinery – some cells having one set of enzymes, and other cells having other enzymes. Therefore, different functions are elicited in different target cells – such functions as alterations in cell membrane permeability, muscle contraction or relaxation, secretion of substances by the cell, and synthesis of intracellular chemicals. The cAMP signaling system is also able to control transcription of specific genes. PKA-mediated phosphorylation of the cAMP response element-binding protein (CREB) enhances the binding of this protein to a

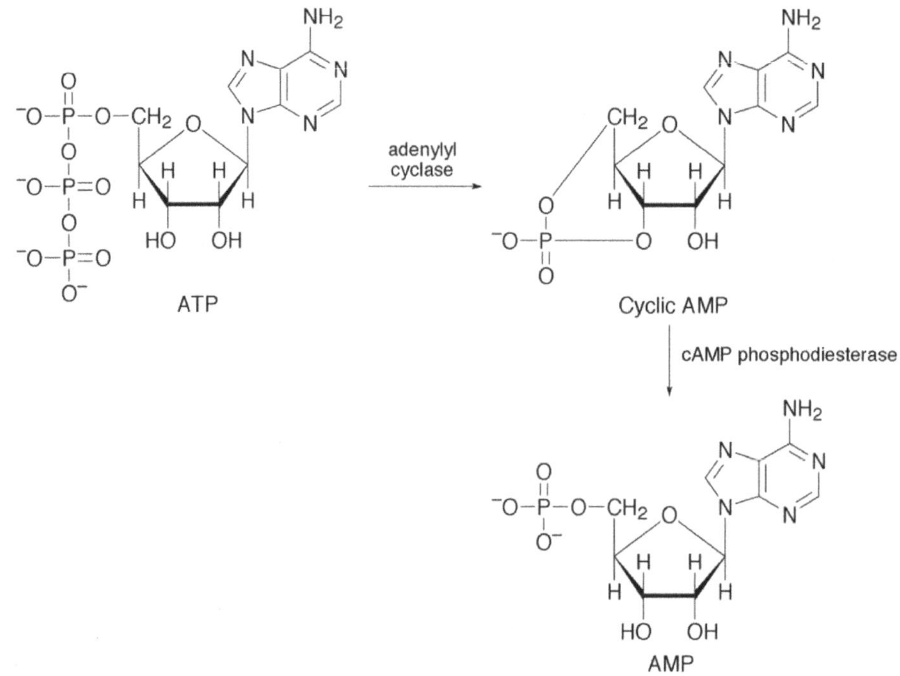

FIGURE 8.3 Synthesis and degradation of cyclic AMP.

specific site on the gene termed the *cAMP response element* (*CRE*). This protein–DNA interaction initiates transcription, leading to synthesis of new protein.

A similar alternative second messenger, 3′,5′-cyclic guanosine monophosphate (cGMP), can be formed from GTP by guanylyl cyclase in response to a few peptide hormones. cGMP specifically activates protein kinase G (PKG), which is located predominantly in smooth muscle and the cerebellar region of the brain. Whereas PKA is ubiquitous and phosphorylates many intracellular proteins, PKG is likely to have a more restricted role.

8.3.3 Phospholipase-Mediated Cell Signaling Systems

Receptor-mediated hydrolysis of cellular phospholipids is of central importance in cell signal transduction initiated by a variety of extracellular agents, including peptide hormones, neurotransmitters, growth factors, and cytokines. The phospholipid hydrolysis is catalyzed by three different phospholipases, namely phospholipases A_2, C, and D. The main pathways are shown in Figure 8.4. Phosphatidylinositol, a phospholipid constituent of the plasma membrane, can be doubly phosphorylated to form phosphatidylinositol 4,5-biphosphate (PIP_2). Activation of phospholipase C in response to appropriate ligand binding of the cell-surface receptor results in the hydrolysis of PIP_2. The hydrolysis products are both second messengers: inositol 1,4,5-triphosphate causes the release of Ca^{2+} from storage compartments in mitochondria and the endoplasmic reticulum, while diacylglycerol activates protein kinase C (PKC). The mobilized Ca^{2+} and activated PKC act synergistically to cause a variety of cellular responses. Many receptors that are coupled to phospholipase C also stimulate phospholipase D, which hydrolyzes phosphatidylcholine to generate phosphatidic acid and choline. Phosphatidic acid can subsequently be dephosphorylated by phosphatidic acid phosphorylhydrolase to produce diacylglycerol. Phosphatidic acid may also activate PKC directly. In some cell types, phosphatidic acid can also be hydrolyzed by phospholipase A_2 to form the multifunctional signaling molecule, lysophosphatidic acid (Moolenaar, 1995).

The diacylglycerol produced by receptor-mediated hydrolysis of PIP_2 is short-lived, either being re-incorporated into phosphoinositides or hydrolyzed to arachidonic acid, the precursor of prostaglandins and thromboxane. Long-term cellular responses, such as cell proliferation, require sustained activation of PKC through the indirect formation of diacylglycerol via phosphatidylcholine and phosphatidic acid.

In intact cells, a variety of agonists acting through G protein-coupled receptors and receptor tyrosine kinases stimulate phospholipase D. In many systems, receptor-mediated activation of phospholipase D appears to depend on PKC, since down-regulation of PKC impairs the effect of agonists. Only the α, βI, and βII isozymes of PKC are effective in activating phospholipase D. Pachter *et al.* (1992) reported that in PKCβI-over-expressing cells, PKC down-regulated agonist-stimulated phospholipase C activity and up-regulated phospholipase D activity. This reciprocal effect appeared to occur, at least in part, at the level of G protein coupling to the effector. Ohguchi *et al.* (1996) confirmed several previous reports that phosphatidylinositol biphosphate is required for PKC-mediated phospholipase D activity. Singer *et al.* (1996) obtained partially purified phospholipase D from membranes of porcine brain and

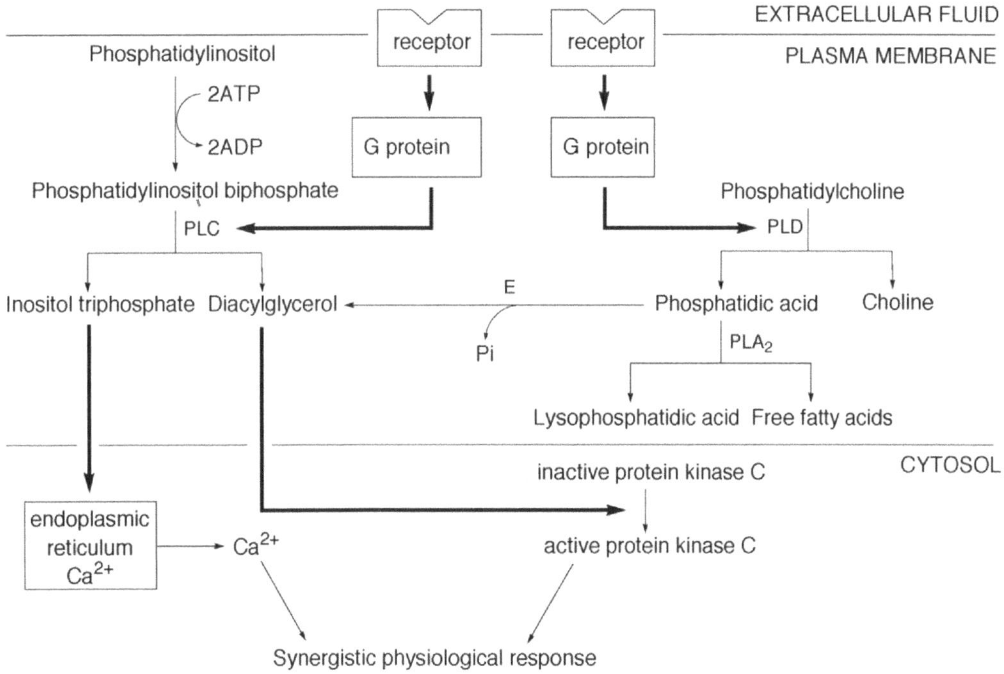

FIGURE 8.4 Phospholipase-mediated cell signaling systems. PLC, phospholipase C; PLD, phospholipase D; PLA_2, phospholipase A_2; E, phosphatidic acid phosphatase; Pi, inorganic phosphate.

studied its regulation by purified PKCα. They found that PKCα is capable of stimulating phospholipase D activity in the presence of activators such as Ca^{2+} or phorbol ester. This action by PKCα was synergistic with stimulation evoked by either Arf or RhoA, which are monomeric GTP-binding proteins. Stimulation of phospholipase D by PKCα occurred in the absence of added nucleotides and without involving the enzyme's kinase activity, i.e. stimulation was independent of phosphorylation. Presumably, stimulation takes place by an allosteric mechanism involving direct interaction between the two proteins or indirect interaction via an intermediate protein. Hammond *et al.* (1997) established that two alternately spliced forms of phospholipase D1 are targets for direct regulation by PKCα.

Signaling is desensitized in response to chronic activation by dephosphorylation of the PKC, effected by a membrane-associated heterotrimeric type 2A phosphatase (Hansra *et al.*, 1996). This event is followed by increased proteolysis of the kinase. These two independent desensitization processes are triggered by the membrane-associated activation of PKC itself.

8.3.4 Functional Properties of the Protein Kinase C Family

PKC catalyzes the covalent addition of a phosphate group to serine and threonine residues of various proteins, thereby altering their activity. In addition to its role in signal transduction, PKC is involved in a multitude of cellular processes, including proliferation and cell cycle progression, differentiation, apoptosis, cytoskeletal remodeling, modulation of ion channels, release of certain hormones and neurotransmitters, secretion of certain enzymes, smooth muscle contraction and relaxation, activation of T and B lymphocytes, and inhibition of gap junctions. When intracellular Ca^{2+} concentrations become too high, PKC activates the Ca^{2+}-transport ATPase (calcium pump) and the Na^+/Ca^{2+} exchange protein, both of which remove Ca^{2+} from the cell.

PKC comprises a large family of isozymes, ten of which are the products of separate genes. The isozymes are identified by a Greek letter suffix, PKCα being the most ubiquitous. Isoforms of PKCβ (βI and βII) are generated post-translationally by the alternative splicing of the primary mRNA transcript. All PKC isozymes are single polypeptides consisting of an N-terminal regulatory region and a C-terminal catalytic region. The catalytic regions are very similar and the individual isozymes exhibit little difference in substrate specificity *in vitro*. The N-terminal substrate-binding site is immediately preceded by an autoinhibitory pseudosubstrate sequence. In the absence of activators, the basic pseudosubstrate is protected from proteases by a cluster of acidic residues.

The isozymes of PKC so far identified have been divided into three subfamilies based on their co-factor requirements for activation (Table 8.2). All known members of the PKC family require phosphatidylserine for their activation. Inositol triphosphate, by mediating Ca^{2+} mobilization from intracellular stores, acts synergistically with diacylglycerol to activate classical PKCs. The novel PKCs lack Ca^{2+}-binding sites and can be activated by diacylglycerol independently of Ca^{2+}. The sequence of PKCμ is markedly different from its fellow novel members and the autoinhibitory pseudosubstrate domain is absent. The atypical PKCs

TABLE 8.2
PKC Isozymes in Mammalian Cells

Subfamilies	Isozymes	Activated by Diacylglycerol?	Activated by Phorbol Esters?	Dependent on Calcium for Activation?
cPKCs (classical or conventional)	α, βI, βII, γ	Yes	Yes	Yes
nPKCs (novel)	δ, ε, η, θ, μ	Yes	Yes	No
aPKCs (atypical)	ζ, ι	No	No	No

Abbreviations: α, alpha; βI, beta I; βII, beta II; γ, gamma; δ, delta; ε, epsilon; ζ, zeta; η, eta; θ, theta; ι, iota; μ, mu.

TABLE 8.3
Tissue Distribution of PKC Isoenzymes

Isoenzyme	Tissue Distribution
PKC-α	Ubiquitous (e.g. brain, lung, spleen, thymus, kidney, skin), but not in liver.
PKC-βI/II	Ubiquitous, but not in kidney. Major isoenzyme in liver.
PKC-γ	Exclusively in the central nervous system.
PKC-δ	Ubiquitous, including liver and kidney.
PKC-ε	Ubiquitous, but not in liver.
PKC-ζ	Ubiquitous, including liver and kidney.
PKC-η	Strongly expressed in skin and lung; slightly expressed in brain and spleen.
PKC-θ	Predominantly expressed in skeletal muscle; less expressed in lung, spleen, skin, and brain.

lack binding sites for diacylglycerol and phorbol ester, but are activated by other lipid mediators independently of Ca^{2+} (Musashi *et al.*, 2000). PKC isozymes exhibit distinct patterns of tissue distribution (Table 8.3), which supports the hypothesis that they have unique functions in different cell types.

In the absence of diacylglycerol, PKC resides in the cytosol. When diacylglycerol is generated, it recruits the kinase from the cytosol to new subcellular sites wherein are found specific anchoring proteins known as RACKs (Receptors for Activated C Kinase). These sites include the plasma membrane, cytoskeletal elements, and nuclei, depending on which particular isozyme is activated. At these binding sites, the PKC adopts a conformation in which its pseudosubstrate is removed. In this open conformation, mature PKC binds substrates and effects downstream signaling. Diacylglycerol, aided by *cis*-unsaturated fatty acids, also increases the affinity of PKC for Ca^{2+}, thereby causing full enzyme activity when Ca^{2+} concentrations are basal. Isozymes translocated to the nucleus phosphorylate nuclear regulatory proteins (transcription factors), providing a mechanism for the direct regulation of gene expression that leads to changes in cell proliferation and differentiation. The differential localization of isozyme-specific RACKs provides some of the specificity of the physiological responses mediated by the various PKC isozymes. Inhibitors of the interaction between the kinases and their anchoring proteins have potential as therapeutic agents (Mochly-Rosen & Gordon, 1998).

PKC is a target for phorbol esters, a number of which are well-known as potent tumor promoters. Extremely low concentrations of phorbol esters are able to substitute for diacylglycerol and activate PKC in the presence of phospholipid. Whereas diacylglycerol is metabolized quickly and activates PKC only transiently, phorbol ester is metabolically stable and activates PKC in a sustained manner.

PKC exerts negative feedback control over various steps in signal transduction pathways, operating in both short-term and long-term responses of the cell. In the short-term, for example, PKC often inhibits the receptor-mediated hydrolysis of inositol phospholipids, thereby blocking the activation of the Ca^{2+}-signaling pathway. In the case of cell proliferation, a long-term response, PKC phosphorylates the receptor for epidermal growth factor. This decreases the affinity of ligand binding and inhibits ligand-induced tyrosine phosphorylation to give a functionally down-regulated receptor.

8.4 THE PHOSPHATIDYLINOSITOL 3-KINASE PATHWAY

Some activated receptor tyrosine kinases and cytokine receptors can initiate the phosphatidylinositol 3-kinase (PI3K) pathway, which is not shared with G protein-coupled receptors. These activated receptors recruit PI3K to the plasma membrane, where the enzyme catalyzes the phosphorylation of phosphatidylinositol 4-phosphate and phosphatidylinositol 4,5-biphosphate to PI 3,4-biphosphate and PI 3,4,5-triphosphate, respectively. The 3-phosphate added to these substrates by PI3K is a binding site for various signal transduction proteins. One such protein is protein kinase B (PKB), which binds to the 3-phosphate of PI-3,4-biphosphate and PI 3,4,5-triphosphate through its PH domain. PKB (also known as Akt) is a serine/threonine kinase that is implicated in regulating multiple biological processes, including gene expression, apoptosis, and cellular proliferation.

In unstimulated cells, PKB is in the cytosol with its PH domain masking its catalytic domain. Following receptor activation and the resulting rise in PI-3,4-biphosphate and PI 3,4,5-triphosphate, PKB binds to these membrane-bound phosphoinositides. The binding of PKB to the phosphoinositides through its PH domain relieves the inhibition of its catalytic domain, rendering PKB partially active. Complete activation of PKB depends on recruitment of another kinase, PDK1, to the plasma membrane via binding of its PH domain to PI-3,4-biphosphate and PI 3,4,5-triphosphate. The binding of PDK1 to the phosphoinositides induces a conformational change that permits PDK1 to phosphorylate and hence activate PKB. PDK1 phosphorylates PKB at two sites: threonine 308 located in PKB's activation loop, and serine 473 within a hydrophobic region close to the carboxyl terminus. Once fully activated, PKB can dissociate from the plasma membrane and phosphorylate its many target proteins.

8.5 RECEPTOR TYROSINE KINASES

Receptor tyrosine kinases (RTKs) are cell-surface receptors that contain intrinsic protein tyrosine kinase activity in their cytosolic domains. They directly initiate signal transduction without the involvement of second messengers. More than 50 RTKs have been identified, including the receptors for several polypeptide growth factors (Table 8.4) and the hormone insulin. All RTKs consist of an extracellular ligand-binding domain, a single membrane-spanning α helix, and a cytosolic domain containing catalytic and regulatory regions. In the absence of ligand, most RTKs are monomers, but some, such as the insulin receptor, form disulfide-linked dimers.

Ligand binding to a RTK leads to receptor dimerization. Most RTK ligands, including epidermal growth factor (EGF), are monomers and the binding of one ligand molecule to a monomeric receptor induces dimerization, primarily through interactions between activated loop segments. Some other ligands, such as platelet-derived growth factor (PDGF) and nerve growth factor (NGF), are themselves dimers, and so their binding brings two receptor monomers together directly. Dimerization promotes autophosphorylation, whereby each member of the pair *trans*-phosphorylates its partner on several tyrosine residues. Autophosphorylation increases the tyrosine kinase activity of the receptor, resulting in the phosphorylation of regulatory regions and the creation of specific binding sites for proteins involved in downstream signal transduction.

The binding of a ligand to an RTK leads to activation of a Ras protein. Ras proteins regulate cell signaling events that lead to cell growth and differentiation, and are the prototype subfamily of the ras superfamily of small G proteins. There are multiple Ras proteins, and different ones act in different cell types. Ras can also be activated by extracellular signals that activate G protein-coupled receptors.

The signal transduction pathway leads ultimately to activation of a mitogen-activated protein (MAP) kinase. Nearly 20 MAP kinases are known in mammals. These fall into three major groups: the extracellular signal-regulated protein kinases (ERK), the p38 MAP kinases, and the c-Jun NH_2-terminal kinases (JNK). Two upstream protein kinases activated in series lead to activation of a MAP kinase. In all currently known MAP kinase cascades, the kinase immediately upstream of the MAP kinase is a member of the MAP kinase/ERK kinase (MEK) family. These are dual-specificity enzymes that can phosphorylate hydroxyl side chains of serine/threonine and tyrosine residues in their MAP kinase substrates. In the ERK1 and ERK2 pathways depicted in Figure 8.5, the binding of a monomeric ligand to the extracellular domain of the RTK induces dimerization. Autophosphorylation of the receptor's cytosolic domain promotes the binding of an adaptor protein (Grb2) to a specific phosphotyrosine on the receptor and to a GEF known as Sos. The resultant binding of Sos to the inactive Ras promotes the exchange of GDP for GTP, thereby activating Ras, which then dissociates from Sos. Active Ras recruits, binds, and activates a serine/threonine kinase (Raf). Subsequent hydrolysis of Ras-GTP to Ras-GDP releases active Raf, which phosphorylates

TABLE 8.4
Some Growth Factors That Signal via Receptor Tyrosine Kinases

Growth Factor	Cellular Responses
Epidermal growth factor (EGF)	Stimulates proliferation of various cell types
Insulin-like growth factors (IGF-1 and -2)	Stimulate cell growth and survival
Nerve growth factor (NGF)	Stimulates survival and growth of some neurons
Platelet-derived growth factors (PDGF AA, BB, AB)	Stimulate survival, growth, and proliferation of various cell types
Macrophage colony-stimulating factor (M-CSF)	Stimulates monocyte/macrophage proliferation and differentiation
Fibroblast growth factors (FGF-1 and -24)	Stimulate proliferation of various cell types; inhibit differentiation of some precursor cells; inductive signals in development

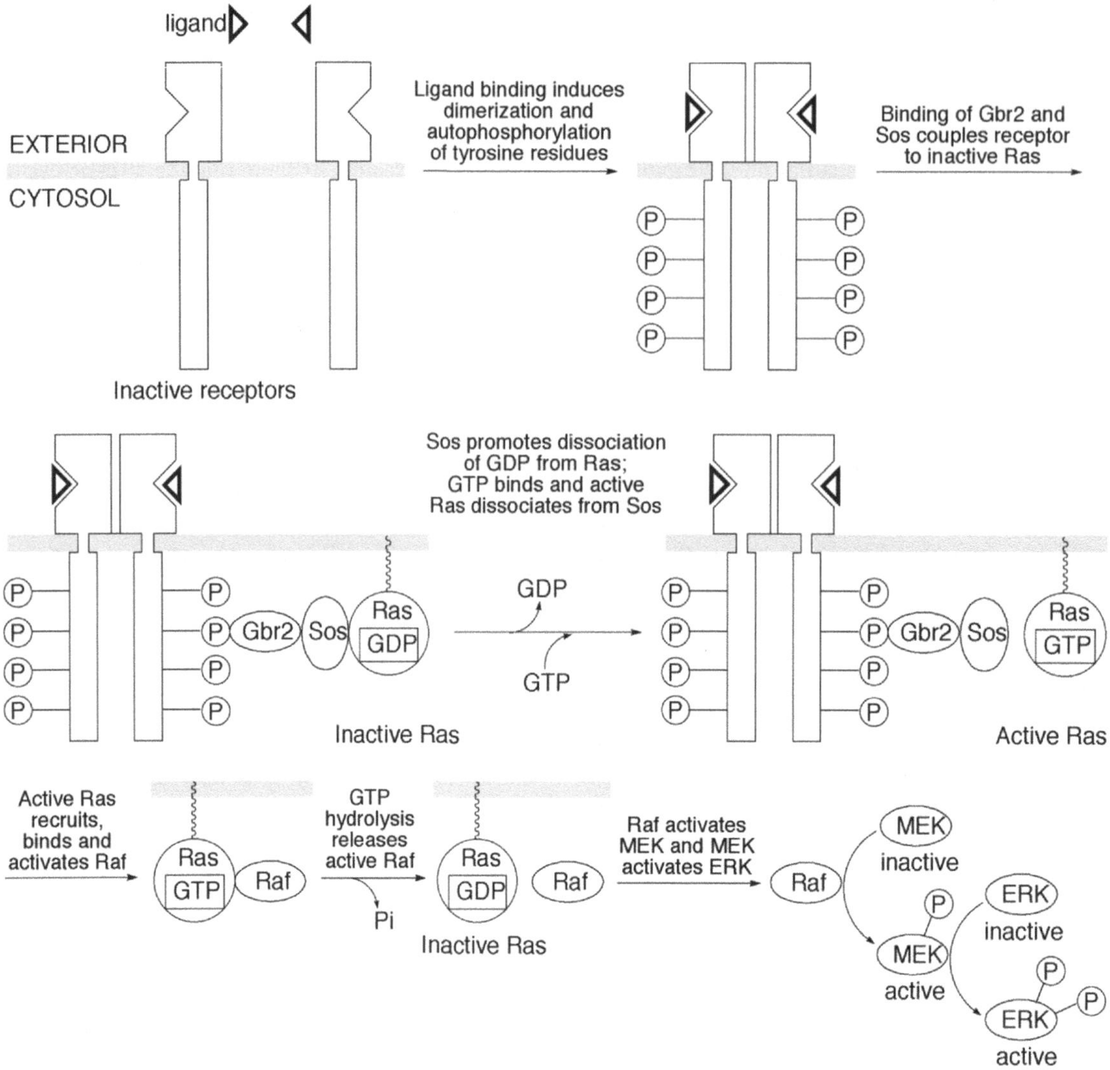

FIGURE 8.5 Activation of ERK1 and ERK2 via Ras signaling. See text for details.

and thereby activates MEK1 and MEK2. MEK1 and MEK2 then phosphorylate ERK1 and ERK2, respectively. Activated ERK1 and ERK2 in dimeric form translocate to the nucleus where they phosphorylate many different proteins, including transcription factors that regulate expression of proteins involved in cell proliferation and differentiation.

MAP kinases can be inactivated at the appropriate time and locations by dual-specificity MAP kinase phosphatases (MKPs) (Pearson *et al.*, 2001).

8.6 CYTOKINE SIGNALING VIA THE JAK–STAT PATHWAY

Broadly speaking, cytokines form a family of secreted proteins that control immune responses by mediating gene activation or repression. Most cytokine receptors do not have intrinsic kinase activity, but are permanently associated with a cytosolic protein tyrosine kinase known as Janus kinase (Jak). There are four known members of the Jak family: Jak1, Jak2, Jak3, and tyrosine kinase 2 (Tyk2). Jaks can activate several downstream signaling pathways. The Jak–STAT pathway refers to the activation of cytosolic STAT transcription factors. There are seven mammalian STATs: STAT1, STAT2, STAT3, STAT4, STAT5A, STAT5B, and STAT6.

The Jak–STAT signaling pathway depicted in Figure 8.6 provides a fast track to the nucleus. Ligand binding induces receptor dimerization, which allows the adjacent Jaks to phosphorylate one another to increase their kinase activity. The activated Jaks then phosphorylate several tyrosine residues on the receptor's cytosolic domain. A monomeric STAT protein binds to a phosphotyrosine in the receptor and is phosphorylated by active Jak. Phosphorylated STATs spontaneously dissociate from the receptor and dimerize, thereby exposing nuclear localization signals. The STAT dimer migrates to the nucleus where it binds to promoter sequences on the DNA and activates transcription of target genes.

Jak–STAT signaling can be regulated over the short term (minutes) by SHP-1 and over a longer period by SOCS. Cross-talk between different Jak–STAT pathways and other

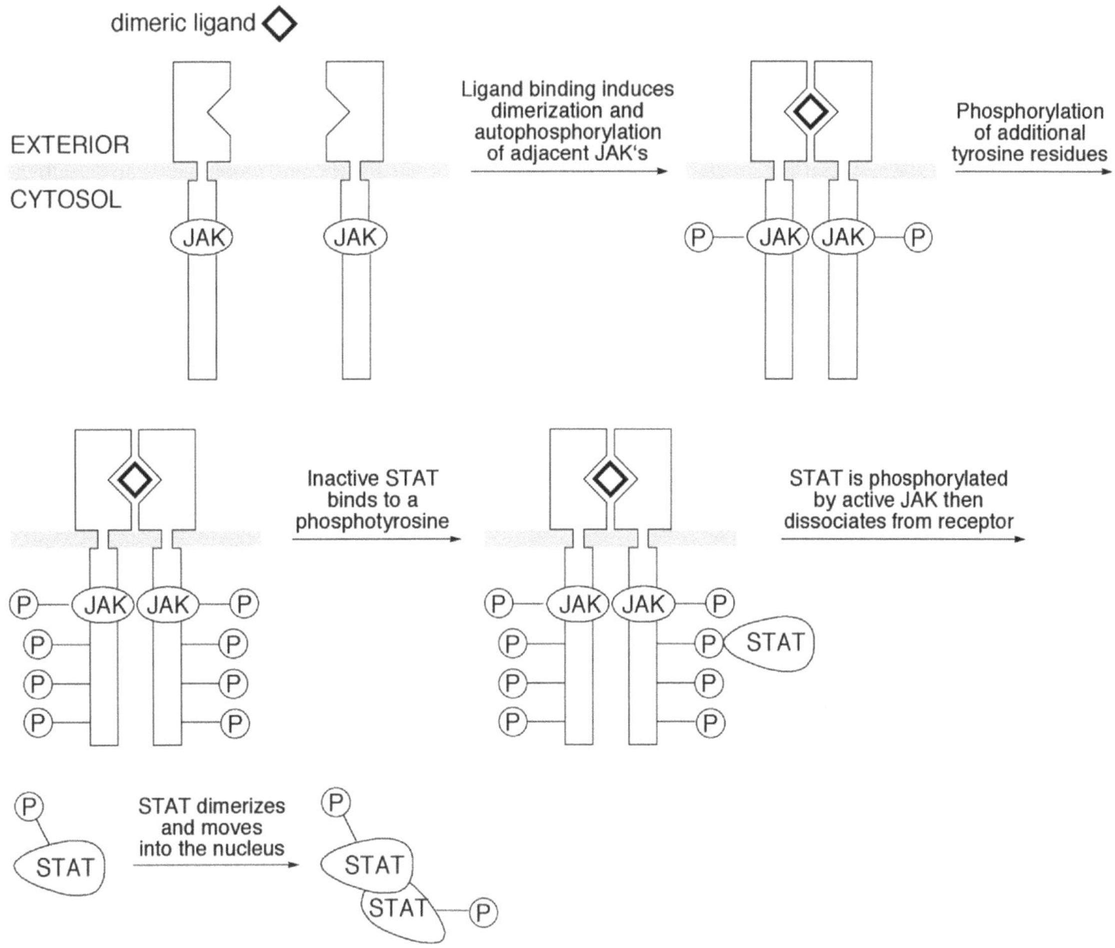

FIGURE 8.6 The Jak–STAT signaling pathway.

signaling pathways provides additional levels of regulation (Shuai & Liu, 2003).

SHP-1 is a tyrosine phosphatase that negatively regulates signaling from several types of cytokine receptors. In addition to a phosphatase catalytic domain, SHP-1 has two SH2 domains. When cells are not stimulated by a cytokine, one of the SH2 domains physically binds to and inactivates the catalytic site. When cells are stimulated, the blocking SH2 domain binds to a specific phosphotyrosine residue in the activated receptor, causing a conformational change that unmasks the SHP-1 catalytic site and brings it near to the phosphorylated tyrosine in the activation lip of Jak associated with the receptor. Removal of the phosphate from this tyrosine by SHP-1 inactivates the Jak, thereby blocking the Jak–STAT signaling pathway.

SOCS proteins, also known as CIS proteins, negatively regulate the Jak–STAT signaling pathway in two ways. First, several SOCS proteins bind via their one SH2 domain to multiple phosphotyrosines on an activated cytokine receptor, preventing binding of other signaling proteins and thus blocking receptor signaling. One SOCS protein, SOCS-1, also binds to the critical phosphotyrosine in the activation lip of Jak-2, thereby inhibiting its kinase activity. Second, all SOCS proteins contain a domain, called the SOCS box, that targets Jak-2 for degradation by the ubiquitin–proteosome pathway.

8.7 CALCIUM SIGNALING

Ionized calcium (Ca^{2+}) functions as a ubiquitous intracellular second messenger controlling many diverse processes, including muscle contraction, neuronal excitability, pancreatic secretion, transcription factor activation, fertilization, embryonic pattern formation, cell proliferation, cell differentiation, and apoptosis (Berridge *et al.*, 2000). In order to fulfill this function, cytosolic concentrations of free Ca^{2+} must be kept extremely low (ca. 0.1 μM in quiescent cells) – 10,000-fold lower than the Ca^{2+} concentration in the extracellular fluid. Cells maintain this low level of cytosolic Ca^{2+} by ion pumps that move Ca^{2+} either out of the cell or into intracellular storage compartments. When functional proteins need to be activated, the concentration of Ca^{2+} in the cytosol is raised to ca. 0.5 μM.

The endoplasmic reticulum is an extensive three-dimensional network of intracellular membranes, extending from the nuclear membrane throughout the cytoplasm. This structure actively sequesters Ca^{2+} within its intraorganellar space. When the second messenger inositol triphosphate

(IP_3) is released into the cytosol, it activates IP_3-gated Ca^{2+} channels in the endoplasmic reticulum membranes, causing a rise in the cytosolic Ca^{2+} level. This rise is only transient because (1) IP_3 is rapidly hydrolyzed to the biphosphate (IP_2), which does not stimulate Ca^{2+} release from the endoplasmic reticulum, and (2) Ca^{2+} that enters the cytosol is rapidly pumped out of the cell. The increased cytosolic Ca^{2+} binds with high affinity to the protein calmodulin, which has four Ca^{2+}-binding sites. When three or four of these sites have bound with Ca^{2+}, the calmodulin molecule undergoes a conformational change that allows it to activate a number of enzymes and other proteins.

Calcium/calmodulin-dependent protein kinases (CaM-kinases) phosphorylate serine and threonine residues in proteins. CaM kinase II, frequently referred to as the multifunctional or multiprotein CaM kinase, phosphorylates a variety of proteins, including enzymes involved in metabolism, ion channels, and transcription factors. One of the transcription factors phosphorylated by CaM II is CREB, which (as mentioned in Section 8.3.2) is phosphorylated at the same site by PKA. Two other multifunctional CaM kinases (CaM I and IV) have been identified. Other CaM kinases are classed as specific kinases; these include myosin light-chain kinase, phosphorylase kinase and elongation factor 2 kinase that regulate muscle contraction, glycogenolysis, and protein synthesis, respectively.

Other specific enzymes activated by the Ca^{2+}/calmodulin complex include (1) cAMP phosphodiesterase, which hydrolyzes cAMP to AMP, terminating its effects as a second messenger; (2) Ca^{2+}-ATPase, which pumps Ca^{2+} out of the cell; (3) nitric oxide synthase, which catalyzes the formation of nitric oxide by the oxidative deamination of L-arginine; (4) protein phosphatase type 2B (also termed calcineurin); and (5) adenylyl cyclase. The Ca^{2+}/calmodulin complex also activates phosphatases that, in turn, dephosphorylate transcription factors, thereby regulating gene expression.

In many cell types, the transient increase in cytosolic Ca^{2+} mediated by IP_3 is followed by a more sustained increase resulting from extracellular Ca^{2+} entry via ion channels in the plasma membrane. The entry of extracellular Ca^{2+} is particularly important in the electrically excitable cells of nerve and muscle, in which voltage-gated channels are opened by membrane depolarization. The resulting increase in cytosolic Ca^{2+} stimulates a further release of Ca^{2+} from intracellular stores by activating distinct Ca^{2+} channels known as ryanodine receptors. In neurons, one effect of increased cytosolic Ca^{2+} is to trigger the release of neurotransmitters. In muscle cells, the release of additional Ca^{2+} from the sarcoplasmic reticulum triggers muscle contraction.

8.8 THE NUCLEAR FACTOR-KAPPA B PATHWAY

8.8.1 The NF-κB Family of Rel Proteins

Nuclear factor-kappa B (NF-κB) is a collective name for a set of dimeric transcription factors composed of various combinations of members of the Rel family of DNA-binding proteins. The NF-κB family consists of five members: p50 (NF-κB1), p52 (NF-κB2), RelA (p65), RelB, and c-Rel. These proteins form various homo- and heterodimers that regulate target gene transcription differentially. Most of the transcriptionally active dimers are inducible, that is, activatable by extracellular stimuli via a signal transduction pathway. One exception is the c-Rel–p50 heterodimer, which is largely responsible for constitutive NF-κB activity in B cells. The predominant and most thoroughly studied inducible NF-κB dimer is the RelA–p50 heterodimer; others include RelA–c-Rel, RelA–RelA, RelB–p50, and RelB–p52. The NF-κB proteins share a Rel homology domain (RHD), which contains sequences responsible for dimerization, nuclear localization, and DNA binding. RelA, RelB, and c-Rel contain a C-terminal transactivation domain and therefore dimers containing these proteins have transactivation potential. p50 and p52 lack the transactivation domain and homodimers of these proteins are transcriptionally repressive.

NF-κB, in common with many (perhaps all) transcription factors, requires interaction with co-activators in order to exert its *trans*-activation effect. Co-activators are adaptor molecules that bridge enhancer-bound activators to the general transcription machinery. Sheppard *et al.* (1999) demonstrated that transcription mediated by the p50–RelA heterodimeric form of NF-κB requires interaction with CBP/p300, SRC-1, and PCAF, all of which are histone acetyltransferases (HATs) that form functional complexes *in vivo*. The transcriptionally repressive p50 homodimer fails to recruit co-activators.

RelA, RelB, and c-Rel are synthesized as their mature forms, whereas p50 and p52 are formed post-translationally by ubiquitin-dependent proteolysis of larger precursor proteins – p105 and p100, respectively. Whereas p50 and RelA are expressed widely in various cell types, the expression of RelB is restricted to specific regions of the thymus, lymph nodes, and Peyer's patches. The expression of c-Rel is confined to hematopoietic cells and lymphocytes.

NF-κB exists in the cytoplasm in an inactive form associated with an inhibitory protein termed IκB (inhibitor of NF-κB). There is a small family of IκBs of which the best-studied and major member is IκBα. The binding of IκB to an NF-κB dimer sterically blocks the latter's nuclear localization sequence, preventing it from getting to the nuclear site of action. To relieve this inhibition, the NF-κB–IκB complex needs to be disrupted.

8.8.2 NF-κB Activating Stimuli and the Target Genes Transcribed

Pahl (1999) compiled a list of all stimuli known to activate NF-κB and another list of its 150 or so target gene products. NF-κB is activated by a wide variety of different stimuli, including pro-inflammatory cytokines (e.g. IL-1 and TNF-α), pathogenic bacteria, viruses, protozoan parasites such as *Trypanosoma cruzi*, bacterial products (e.g. lipopolysaccharide) and viral products (e.g. double-stranded RNA).

NF-κB is also activated during physiological stress conditions (e.g. hemorrhagic shock), physical stress (e.g. radiation), and environmental stress (e.g. cigarette smoke). The cell responds to immunogenic stimuli by increasing the expression of genes encoding a wide range of molecules involved in the immune system; these include cytokines/chemokines, immunoreceptors, adhesion molecules, acute phase proteins, growth factors, antimicrobial peptides, and transcription factors. Stress stimuli promote the increased production of enzymes that generate secondary mediators of inflammation; such enzymes include inducible nitric oxide synthase (iNOS), cyclooxygenase-2 (COX-2), and phospholipase A_2 (PLA_2). Because NF-κB can induce the synthesis of other transcription factors (e.g. c-*myc* and p53), an initial NF-κB activation may indirectly induce the synthesis of many more proteins than those identified. The selectivity of NF-κB activation resides mainly in the cell type: not all cell types possess the required receptor or signal transduction molecules for NF-κB to be activated.

Schreck *et al.* (1991) reported that treatment of T cells with micromolar concentrations of hydrogen peroxide activated NF-κB. Bear in mind that after its passive diffusion through the cell plasma membrane, the relatively innocuous hydrogen peroxide can be converted into the highly reactive hydroxyl radical. The activation of NF-κB by hydrogen peroxide was inhibited by the antioxidant and radical scavenger *N*-acetyl-L-cysteine (NAC). These findings led to the suggestion that reactive oxygen species might activate NF-κB by acting as second messengers. However, Hayakawa *et al.* (2003) showed that NAC inhibited NF-κB activation independently of antioxidative function. NAC selectively blocked TNF-stimulated NF-κB activation by lowering the affinity of the TNF receptor to TNF, and had no effect on IL-1- or TPA-stimulated NF-κB activation.

8.8.3 Overview of NF-κB Activation

NF-κB exists in the cytoplasm in an inactive form associated with an inhibitory IκB protein. Activation of NF-κB is achieved through the signal-induced proteolytic degradation of IκB. The pro-inflammatory cytokines TNF-α and IL-1β initiate a signaling cascade leading to activation of two IκB kinases, IKKα and IKKβ, which phosphorylate IκB at specific amino-terminal serine residues. Phosphorylated IκB is then selectively ubiquitinated by an E3 ubiquitin ligase. In the last step of this signaling cascade, phosphorylated and ubiquitinated IκB, which is still associated with NF-κB in the cytoplasm, is selectively degraded by the 26S proteasome. This process exposes the nuclear localization sequence, thereby freeing NF-κB to interact with the nuclear import machinery and translocate to the nucleus, where it binds its target genes to initiate transcription.

8.8.4 The IκB Kinases

The IκB kinases IKKα and IKKβ (or IKK-1 and IKK-2) exist as functional kinases within a multiprotein complex called the IKK signalsome. Another component of this complex, IKKγ (also known as NEMO [NF-κB essential modulator]), serves as its regulatory subunit and is essential for IKK activation. The C-terminal region of IKKγ appears to be necessary for recruitment of upstream activators. IKKα and IKKβ contain a kinase domain at their N termini and a leucine zipper and helix–loop–helix (HLH) motif in their C-terminal portions. Within the kinase domain is an activation loop. The leucine zipper facilitates dimerization and the HLH motif functions as an intrinsic activator of the kinase. IKKγ does not contain a recognizable catalytic domain but is composed mostly of three large α-helical regions, including a leucine zipper. The predominant form of IKK is an IKKα–IKKβ heterodimer associated with either a dimer or trimer of IKKγ. Yang *et al.* (2000) provided genetic evidence that mitogen-activated protein kinase kinase kinase 3 (MEKK3) is required for IKK activation, and functions downstream of receptor-interacting protein (RIP) and TNF receptor-associated factor 2 (TRAF2) in the TNF-induced NF-κB activation pathway.

IKK activity is regulated by phosphorylation of serine residues at two sites on the IKKα and IKKβ subunits. The two catalytic subunits of IKK (IKKα and IKKβ) associate into a heterodimer via their leucine zippers.. In non-stimulated cells, IKK is not phosphorylated and both subunits are configured with their HLH in contact with the kinase domain. The activation loop within the kinase domain folds back onto the kinase domain, thus interfering with entry of peptide substrates and ATP into the catalytic pocket. Upon cell stimulation with TNF-α or IL-1, two critical serines in the activation loop of IKKβ are phosphorylated. IKKα is not a target for upstream activators, but undergoes *trans*-autophosphorylation at the same two sites. Phosphorylation moves the activation loop away from the catalytic pocket, allowing the entry and binding of substrates. Once activated, IKK phosphorylates IκBs. In addition, IKK undergoes progressive autophosphorylation at its C-terminal serine cluster. When at least nine of the C-terminal serines are phosphorylated, electrostatic repulsion weakens the interaction between the HLH and the kinase domain, and IKK activity decreases. This conformational change may facilitate the recruitment of a phosphatase that shuts off IKK activity by dephosphorylation of the activation loop. This mode of regulation explains why IKK is usually activated in a highly transient fashion.

8.8.5 Pathways of NF-κB Activation

NF-κB activation represents the terminal step in a signal transduction pathway. Two major pathways, the classical and the alternative NF-κB pathways, lead to the activation of NF-κB (Bonizzi & Karin, 2004).

8.8.5.1 The Classical Pathway

The classical pathway is activated by microbial and viral infections and various stress conditions. These stimuli induce the transcription of genes encoding many of the proteins that participate in the innate immune response or stress response.

In the absence of a stimulus, NF-κB dimers bind to an inhibitory IκB protein. Activation of NF-κB requires it to be freed from its association with IκB. In the cytoplasm, upstream kinases phosphorylate the IKK-β catalytic subunit of the IKK complex, which then phosphorylates the IκB component of the RelA–p50–IκB complex. Phosphorylation targets IκB for ubiquitination and subsequent proteasome-mediated degradation. This process frees the RelA–p50 dimer, which then translocates to the nucleus and initiates transcription. TNF-α and IL-1β generated through the activation of NF-κB are stimulators of this pathway and therefore an amplifying loop is established.

Activated NF-κB induces the synthesis of the IκBα protein, which, like all gene products, is assembled in the cytoplasm by the process of translation and subsequent modification. The presence of nuclear localization and nuclear export motifs on the IκBα molecule allows this inhibitor to terminate the transcriptional activity of NF-κB. Newly synthesized IκBα enters the nucleus, displaces NF-κB from its DNA-binding site, and conveys the transcription factor back to the cytoplasm. This process of post-induction repression restores the pre-induction state of the NF-κB–IκBα complex in readiness for a new activation signal. A feedback mechanism is thus established to ensure a transient activation of transcription.

8.8.5.2 The Alternative Pathway

The alternative pathway plays a central role in the expression of genes involved in development and maintenance of secondary lymphoid organs. Only a few stimuli such as lymphotoxin β (LT-β) and B lymphocyte stimulator (Blys, also known as BAFF) trigger this pathway, which is dependent on the α subunit and not the β subunit of IKK.

Upon stimulation by Blys, for example, IKKα homodimers are phosphorylated by NF-κB-inducing kinase (NIK) and the activated homodimer then phosphorylates the p100 associated with RelB. This targets p100 for ubiquitination and subsequent partial degradation to produce p52. The RelB–p52 dimer then translocates to the nucleus and initiates transcription.

8.8.6 Regulation of Transcriptional Activity of NF-κB

Nuclear NF-κB can undergo reversible acetylation, which acts as a molecular switch to control the duration of the transcriptional response. When RelA is acetylated by CBP/p300 (but not by PCAF), it interacts only weakly, if at all, with newly synthesized IκBα, allowing binding of dimeric NF-κB to the DNA and activation of transcription. Deacetylation of RelA by HDAC3 (but not by HDAC1) increases its affinity for IκBα, resulting in the nuclear export of the RelA–p50–IκBα complex and termination of transcription. The nuclear export of this complex also serves to replenish the depleted cytoplasmic pool of latent NF-κB–IκBα complexes needed for the next inductive NF-κB response.

8.9 APOPTOSIS SIGNALING PATHWAYS

8.9.1 Introduction

Apoptosis is programed cell death and is a physiological mechanism to eliminate unwanted or potentially dangerous cells without inducing an inflammatory response. Most cell types are programed to undergo apoptosis unless cell death is actively suppressed by survival signals from other cells. These signals initiate pathways that lead to the prevention of apoptosis. Survival pathways are typically initiated by the binding of growth factors to receptor tyrosine kinases.

Apoptosis is initiated via two central signaling pathways: (1) an intrinsic pathway, in which mitochondria control the mobilization of pro-apoptotic proteins in response to a variety of intracellular stresses, and (2) an extrinsic pathway, in which external signaling proteins initiate apoptosis by binding to death receptors on the cell surface. An overview of these pathways is shown in Figure 8.7.

8.9.2 The Caspases

Caspases (cysteinyl *asp*artate-specific prote*ases*) are a family of specialized proteases that have cysteine residues at their active sites and cleave substrates at specific Asp-Xxx bonds (i.e. after aspartic acid residues). The preferred tetrapeptide recognition motif differs among caspases and explains their different substrate specificities. Caspases cleave a restricted set of target proteins, usually at a single site, causing a loss of function. In some cases, a functionally dormant protein can be activated by cleaving an inhibitory subunit. Blocking caspase activity with pharmacological inhibitors nearly always prevents the occurrence of apoptosis.

Fourteen mammalian caspases have been characterized and numbered 1–14 according to the chronology of their identification. Based on their main biological function, caspases are divided into two subfamilies. The CED-3 subfamily is directly involved in apoptosis and comprises caspases-2, -3, -6, -7, -8, -9 and -10. The ICE subfamily processes pro-inflammatory cytokines and comprises caspases-1, -4, -5, -11, -12, -13 and -14. The following discussion is confined to the caspases of the CED-3 subfamily.

Caspases exist in the cell as inactive precursors, referred to as procaspases, and a specific apoptotic stimulus is required to activate them. This strategy minimizes the risk of cells being inadvertently killed. The procaspases are single-chain polypeptides that typically comprise four domains: a long or short amino-terminal prodomain, a large α-chain subunit, a linker segment, and a small β-chain subunit. Activation entails proteolytic cleavage of specific Asp-Xxx bonds between domains followed by association of the large and small subunits to

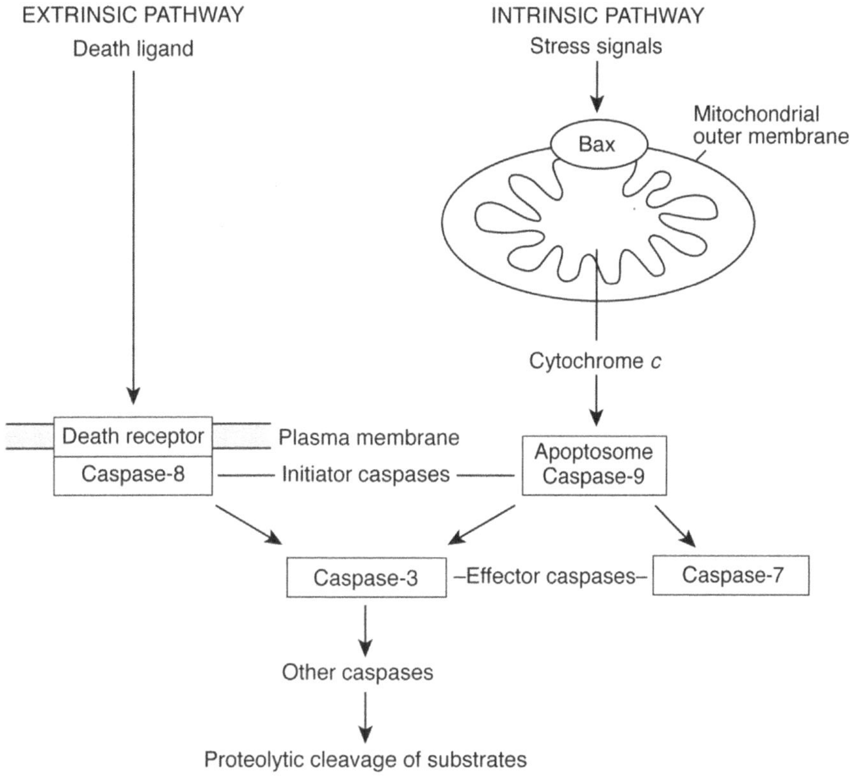

FIGURE 8.7 Overview of apoptosis. Cell death may be signaled by the binding of an extracellular death ligand to a specific receptor on the cell surface, which leads to the activation of caspase-8. This initiator caspase then directly activates the effector caspase-3. Alternatively, intracellular stress signals converge on mitochondria, freeing the membrane-bound pro-apoptotic protein Bax from its inhibitors. The subsequent release of cytochrome *c* from the mitochondrion triggers the formation of an apoptosome that contains caspase-9. This initiator caspase then simultaneously activates caspases-3 and -7.

form a heterodimer. This bond cleavage is catalyzed either by the caspase itself (autocatalysis) or by other caspases. Two heterodimers then align to form a tetramer in which the two chains of each dimer are folded to form the catalytic site. The prodomain and linker segment are released. The subunit structures of a generalized procaspase with a long prodomain and the corresponding active caspase are shown in Figure 8.8.

Caspases-8 and -9 are involved in upstream regulatory events and are referred to as initiator caspases. Initiator procaspases are activated by forming aggregates with the aid of adaptor proteins, which bind to the long prodomains. Activated initiator caspases cleave downstream effector procaspases, thereby creating a proteolytic cascade that leads ultimately to cell death (Figure 8.9).

Caspase-8, activated in the extrinsic pathway, can transduce the apoptotic signal in two ways: (1) caspase-8 activates the effector caspase-3, thereby initiating a caspase cascade that results in dismantling of the cell; or (2) caspase-8 cleaves the inactive Bid to generate the active truncated form (tBid). The latter protein translocates from the cytosol to the mitochondrion where, in concert with Bax and Bak, it alters the permeability of the outer membrane to allow the release of pro-apoptotic proteins like cytochrome *c* and Smac/DIABLO. The resultant activation of caspase-9

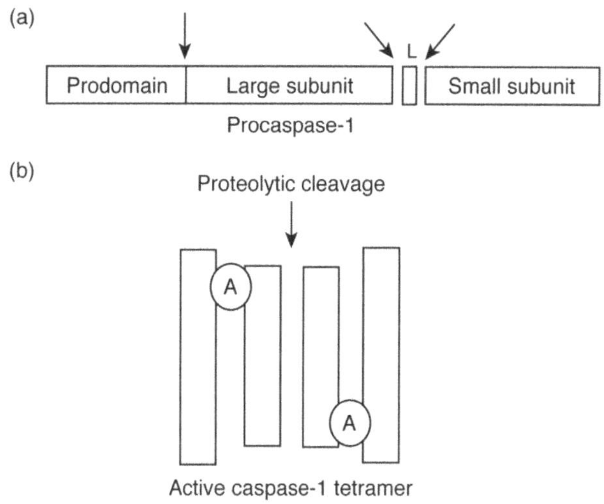

FIGURE 8.8 Subunit structures of (a) procaspase-1 (arrows indicate Asp-Xxx cleavage sites) and (b) active caspase-1. L, linker segment; A, active site.

further amplifies the caspase cascade initiated by direct activation of caspase-8 at death receptors.

Caspase-9, activated in the intrinsic pathway, initiates the cascade by simultaneously activating procaspases-3 and -7. Caspase-3 then activates procaspases-2 and -6.

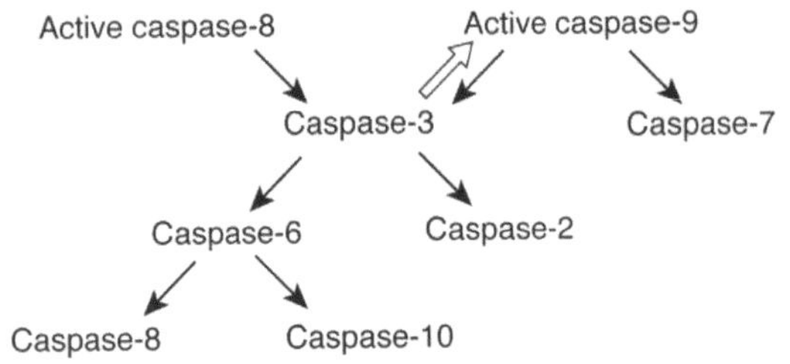

FIGURE 8.9 The caspase proteolytic cascade. The open arrow shows how caspase-3 provides a caspase-9-activating amplification loop.

Caspase-3 also promotes further processing of caspase-9 to create a positive feedback amplification loop. Caspase-6 activates procaspases-8 and -10. Caspases-2, -3, -6, -7, -8 and -10 are directly responsible for cell dismantling and are referred to as effector or executioner caspases. Caspases are not exclusively initiator or effector – some of them may play dual roles, with others being more dominant in either role.

The membrane blebbing observed in apoptotic cells seems to be initiated by caspase-mediated cleavage of a kinase (PAK-2) between a negative regulatory subunit and the catalytic subunit, resulting in a constitutive activation of the enzyme. Cleavage of the inhibitory subunit of DNA fragmentation factor (DFF) by caspase-3 and caspase-7 results in the release and activation of the DNA-fragmenting catalytic subunit. Caspase-3 also activates ACINUS (apoptotic chromatin condensation inducer in the nucleus), leading to chromatin condensation (Saelens *et al.*, 2004). Caspases degrade the cellular infrastructure by cleaving actin and fodrin (cytoskeletal proteins) and lamins (structural proteins of the nuclear envelope). Other proteins cleaved by caspases include gelsolin (cytoskeletal regulator), poly(ADP-ribose) polymerase, DNA-dependent protein kinase (enzymes involved in DNA break repair), retinoblastoma-associated protein (a regulator of cell cycle progression), and protein kinase Cδ.

8.9.3 DNA Fragmentation Factor

The nuclease mainly responsible for apoptotic nuclear DNA fragmentation is DNA fragmentation factor (DFF). DFF is a heterodimeric protein comprising 40-kDa and 45-kDa subunits termed DFF40 and DFF45, respectively. DFF40 (also known as CAD [caspase-activated DNase] or CPAN [caspase-activated nuclease]) contains an intrinsic endonuclease activity, whereas DFF45 (also known as ICAD [inhibitor of caspase-activated DNase]) serves as both a chaperone and an inhibitor of DFF40. DFF resides in the nucleus of normal cells. Upon activation of apoptosis, DFF45 is cleaved at two sites by caspase-3, caspase-7, or granzyme B and the cleavage products dissociate from DFF40. This results in the activation of DFF40, which then cleaves DNA into oligonucleosome-sized fragments. Chromatin-associated proteins, including high-mobility group (HMG) proteins, histone H1, and topoisomerase II, can potentiate the endonuclease activity of DFF40 *in vitro*. In addition to DNA fragmentation, DFF induces chromatin condensation *in vitro*.

8.9.4 Bcl-2 Proteins

The Bcl-2 family of proteins are the key regulators that control mitochondria-mediated apoptosis. The founder member is Bcl-2 itself. The 22 known members of the Bcl-2 family are characterized by the presence of up to four highly conserved Bcl-2 homology (BH) domains, designated BH1–4. The Bcl-2 family can be divided into two main groups according to whether they promote or inhibit apoptosis. The pro-apoptotic family members are further classified according to whether they contain multiple BH domains or only BH3 (Antonsson, 2004; Table 8.5).

Many members of the Bcl-2 family can homodimerize. More importantly, heterodimers can be formed by the pairing of a pro- and an anti-apoptotic member, in which case the dimer has no net effect on apoptosis. The propensity of the cell to undergo mitochondrial-mediated apoptosis is determined by the ratio of pro- to anti-apoptotic Bcl-2 proteins.

8.9.5 The Mitochondrial Pathway of Apoptosis

The mitochondrial pathway (intrinsic pathway) starts with the release of mitochondrial proteins in response to intracellular stresses, such as DNA damage, heat shock, and oxidative stress. The release of mitochondrial proteins is controlled by members of the Bcl-2 family in the following manner. In normal healthy cells, Bax is monomeric and resides in the cytosol. After apoptotic stimulation, Bax undergoes a conformational change that induces its oligomerization and translocation to the mitochondria, where it inserts into the outer membrane. In the absence of a pro-apoptotic signal, Bax is inhibited by its association with Bcl-2 and Bcl-x_L, which are also membrane-bound. In the presence of a pro-apoptotic signal, Bad migrates from the cytosol to the

TABLE 8.5
The Bcl-2 Protein Family (from Antonsson, 2004)

	Pro-Apoptotic	
Anti-Apoptotic	**Multidomain**	**'BH3 Domain Only'**
Bcl-2	Bax	Bid
Bcl-x_L	Bak	Bad
Bcl-w	Bok	Bim/Bod
A1	Bcl-x_5	Bik/Nbk
DIVA/BOO	Bcl-rambo	Hrk
NR-13		Blk
		Bmf
		Nix
		PUMA
		Noxa
		BNip3

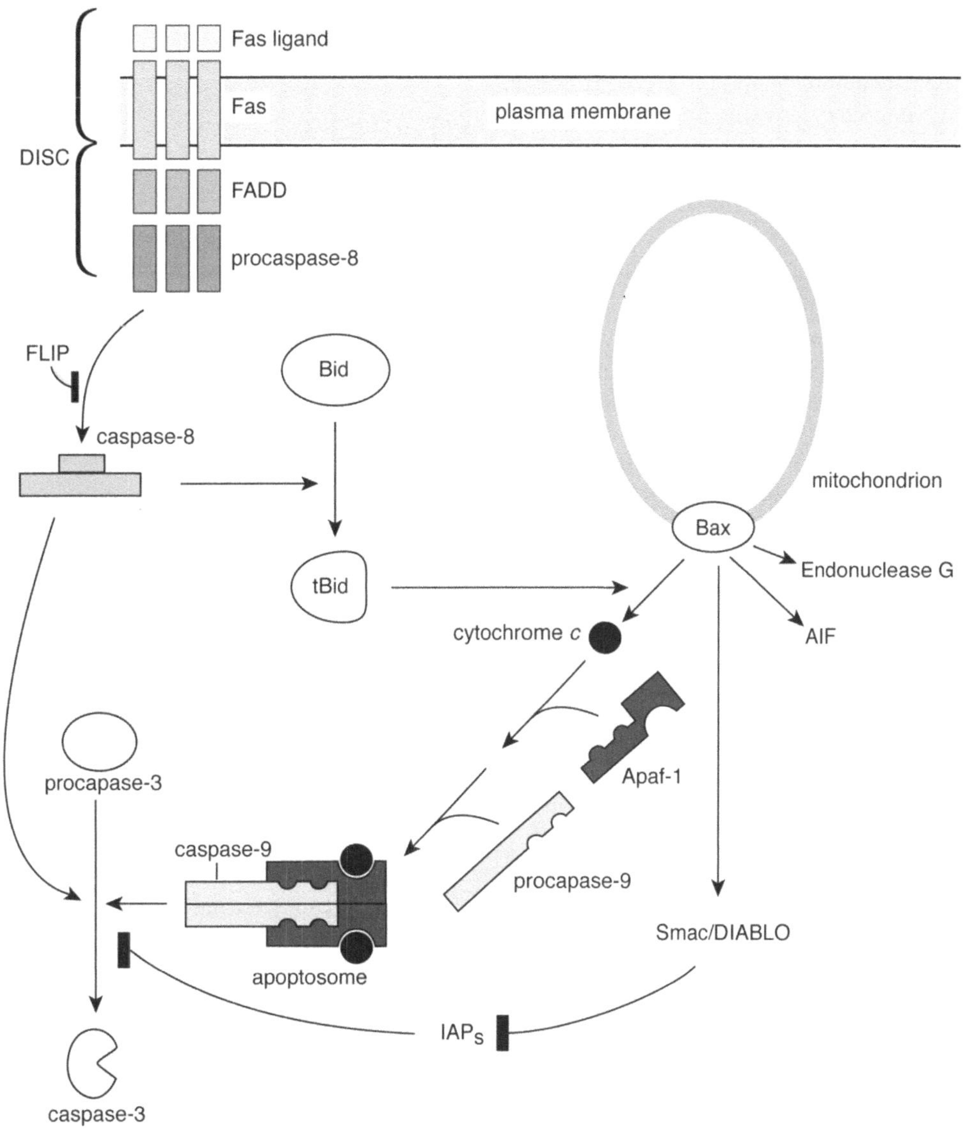

FIGURE 8.10 Integration of the mitochondrial and death-receptor pathways into the caspase cascade.

mitochondrion where it binds to Bcl-2 and Bcl-x_L, thereby freeing them from their association with Bax. Bax and Bak, together with truncated Bid (tBid) produced in the cytosol by the action of caspase 9, alter the permeability of the mitochondrial outer membrane to allow the release of proteins from the intermembrane space into the cytosol. The mitochondrial proteins released can be divided into two groups. One group, cytochrome *c*, Smac/DIABLO, and HtrA2/Omi, activates the caspase-dependent mitochondrial pathway. The second group, endonuclease G and AIF, translocates to the nucleus and induces DNA degradation (chromatinolysis) in a caspase-independent manner. Figure 8.10 shows the release of protein factors from stimulated mitochondria.

8.9.5.1 Activation of the Caspase-Dependent Mitochondrial Pathway

The release of cytochrome *c* triggers the deoxyadenosine triphosphate (dATP)-dependent assembly of a multiprotein complex, the apoptosome, which is composed of seven trimers each of cytochrome *c*, the adaptor protein Apaf-1, and procaspase-9. The procaspase-9 is activated in the apoptosome through a conformational change. The initiator caspase-9 then triggers the caspase cascade by simultaneously activating procaspases-3 and -7.

Human Smac (Second mitochondria-derived activator of caspase) and its murine ortholog DIABLO (Direct IAP-Binding protein with Low PI) are known collectively as Smac/DIABLO. This protein, together with HtrA2/Omi (high-temperature requirement protein A2), facilitates caspase activation. When released into the cytosol, the two proteins bind to IAPs (inhibitor of apoptosis proteins). IAPs are a group of proteins that binds to both procaspases and activated caspases, thereby inhibiting their activation and activity, respectively. The binding of Smac/DIABLO or HtrA2/Omi to IAPs prevents the latter from inhibiting caspases.

The presence of IAPs in the cytosol ensures that the apoptotic signaling cascade is not accidentally triggered, for example by damaged mitochondria in a cell. However, the release of IAP inhibitory proteins from the mitochondria during apoptosis ensures a fast propagation of the caspase cascade and execution of the cytochrome *c*-mediated apoptosis death program.

8.9.5.2 Activation of Caspase-Independent Mitochondrial Pathways

Upon apoptosis induction, apoptosis-inducing factor (AIF) and endonuclease G translocate from mitochondria to the cytosol and then to the nucleus where they induce DNA fragmentation and chromatin condensation, two hallmarks of apoptosis. AIF does not exhibit any detectable DNase activity and its intrinsic NADH oxidase activity is not required for its apoptotic function. The possibility of interaction with another protein was explored by Candé *et al.* (2004), who discovered that AIF and cyclophilin A co-operate to induce DNA degradation both *in vitro* and *in vivo*. Widlak *et al.* (2001) demonstrated that exonuclease and DNase 1 each can co-operate with endonuclease G to ensure efficient DNA fragmentation.

8.9.6 The Death-Receptor Pathway of Apoptosis

The death-receptor pathway (extrinsic pathway) is triggered by the binding of extracellular signal molecules to so-called death receptors that span the plasma membrane of target cells. These receptors belong to the tumor necrosis factor (TNF) receptor superfamily that have a structurally and functionally homologous cytoplamic death domain within their cytoplasmic-facing tails. Death receptors are represented by the Fas receptor (FasR/APO-1/CD95), TNF receptor 1 (TNF-R1), and TNF-related apoptosis-inducing ligand receptors 1 and 2 (TRAIL-R1 and TRAIL-R2). These receptors are expressed constitutively on most cell types, whereas expression of their ligands is tightly regulated. Viral infection often leads to sensitivity of the infected cell to apoptosis mediated by FasR or TNF-R1. The TRAIL receptors mediate apoptosis in a wide variety of tumor cell lines, but do not seem to be cytotoxic to normal cells *in vitro* (Walczak *et al.*, 1999).

8.9.6.1 Signaling by Fas

Fas ligand (FasL), produced by activated natural killer cells and cytotoxic T lymphocytes, can trigger death of virus-infected cells, certain types of cancer cells, foreign graft cells, and excess lymphocytes. The Fas signaling pathway is depicted in Figure 8.10. In the absence of FasL, FasR exists as an inactive trimeric complex. Binding of FasL induces receptor clustering and formation of a death-inducing signaling complex (DISC) that comprises FasL, FasR, Fas-associated death domain (FADD, an adaptor protein), and procaspase-8. Conformational changes within the DISC lead to activation of procaspase-8 and subsequent release of large amounts of mature caspase-8. Fas-mediated apoptosis is regulated by FLIP, a degenerate caspase homolog which blocks caspase-8 activation and release.

REFERENCES

Antonsson, B. (2004) Mitochondria and the Bcl-2 family proteins in apoptosis signaling pathways. *Molecular and Cellular Biochemistry*, **256/257**, 1412–55.

Berridge, M.J., Lipp, P. & Bootman, M.D. (2000) The versatility and universality of calcium signalling. *Nature Reviews Molecular Cell Biology*, **1**, 11–21.

Bonizzi, G. & Karin, M. (2004) The two NF-κB activation pathways and their role in innate and adaptive immunity. *Trends in Immunology*, **25**, 280–8.

Bos, J.L., Rehmann, H. & Wittinghofer, A. (2007) GEFs and GAPs: Critical elements in the control of small G proteins. *Cell*, **129**, 865–77.

Candé, C., Vahsen, N., Kouranti, I., Schmitt, E., Daugas, E., Spahr, C., Luban, J., Kroemer, R.T., Giordanetto, F., Garrido, C., Penninger, J.M. & Kroemer, G. (2004) AIF and cyclophilin A cooperate in apoptosis-associated chromatinolysis. *Oncogene*, **23**, 1514–21.

Hammond, S.M., Jenco, J.M., Nakashima, S., Cadwallader, K., Gu, Q., Cook, S., Nozawa, Y., Prestwich, G.D., Frohman, M.A. & Morris, A.J. (1997) Characterization of two alternatively spliced forms of phospholipase D1. *Journal of Biological Chemistry*, **272**(6), 3860–8.

Hansra, G., Bornancin, F., Whelan, R., Hemmings, B.A. & Parker, P.J. (1996) 12-*O*-Tetradecanoylphorbol-13-acetate-induced dephosphorylation of protein kinase Cα correlates with the presence of a membrane-associated protein phosphatase 2A heterotrimer. *Journal of Biological Chemistry*, **271**(51), 32785–8.

Hayakawa, M., Miyashita, H., Sakamoto, I., Kitagawa, M., Tanaka, H., Yasuda, H., Karin, M. & Kikugawa, K. (2003) Evidence that reactive oxygen species do not mediate NF-κB activation. *EMBO Journal*, **22**, 3356–66.

Mochly-Rosen, D. & Gordon, A.S. (1998) Anchoring proteins for protein kinase C: A means for isozyme selectivity. *FASEB Journal*, **12**(1), 35–42.

Moolenaar, W.H. (1995) Lysophosphatidic acid, a multifunctional phospholipid messenger. *Journal of Biological Chemistry*, **270**(22), 12949–52.

Musashi, M., Ota, S. & Shiroshita, N. (2000) The role of protein kinase C isoforms in cell roliferation and apoptosis. *International Journal of Hematology*, **72**, 12–19.

Ohguchi, K., Banno, Y., Nakashima, S. & Nozawa, Y. (1996) Regulation of membrane-bound phospholipase D by protein kinase C in HL60 cells. *Journal of Biological Chemistry*, **271**(8), 4366–72.

Pachter, J.A., Pai, J.-K., Mayer-Ezell, R., Petrin, J.M., Dobek, E. & Bishop, W.R. (1992) Differential regulation of phosphoinositide and phosphatidylcholine hydrolysis by protein kinase C-β1 overexpression. *Journal of Biological Chemistry*, **267**(14), 9826–30.

Pahl, H.L. (1999) Activators and target genes of Rel/NF-κB transcription factors. *Oncogene*, **18**, 6853–66.

Pearson, G., Robinson, F., Gibson, T.B., Xu, B.-E., Karandikar, M., Berman, K. & Cobb, M.H. (2001) Mitogen-activated protein (MAP) kinase pathways: Regulation and physiological functions. *Endocrine Reviews*, **22**, 153–83.

Saelens, X., Festjens, N., Vande Walle, L., van Gurp, M., van Loo, G. & Vandenabeele, P. (2004) Toxic proteins released from mitochondria in cell death. *Oncogene*, **23**, 2861–74.

Schreck, R., Rieber, P. & Baeuerle, P.A. (1991) Reactive oxygen intermediates as apparently widely used messengers in the activation of the NF-κB transcription factor and HIV-1. *EMBO Journal*, **10**, 2247–58.

Sheppard, K.-A., Rose, D.W., Haque, Z.K., Kurokawa, R., McInerney, E., Westin, S., Thanos, D., Rosenfeld, M.G., Glass, C.K. & Collins, T. (1999) Transcriptional activation by NF-κB requires multiple coactivators. *Molecular and Cellular Biology*, **19**, 6367–78.

Shuai, K. & Liu, B. (2003) Regulation of JAK–STAT signalling in the immune system. *Nature Reviews Immunology*, **3**, 900–911.

Singer, W.D., Brown, H.E., Jiang, X. & Sternweis, P.C. (1996) Regulation of phospholipase D by protein kinase C is synergistic with ADP-ribosylation factor and independent of protein kinase activity. *Journal of Biological Chemistry*, **271**(8), 4504–10.

Vetter, I.R. & Wittinghofer, A. (2001) The guanine nucleotide-binding switch in three dimensions. *Science*, **294**, 1299–1304.

Walczak, H., Miller, R.E., Ariail, K., Gliniak, B., Griffith, T.S., Kubin, M., Chin, W., Jones, J., Woodward, A., Le, T., Smith, C., Smolak, P., Goodwin, R.G., Rauch, C.T., Schuh, J.C.L. & Lynch, D.H. (1999) Tumoricidal activity of tumor necrosis factor-related apoptosis-inducing ligand *in vivo*. *Nature Medicine*, **5**, 157–63.

Weinstein, L.S., Yu, S., Warner, D.R. & Liu, J. (2001) Endocrine manifestations of stimulatory G protein α-subunit mutations and the role of genomic imprinting. *Endocrine Reviews*, **22**, 675–705.

Widlak, P., Li, L.Y., Wang, X. & Garrard, W.T. (2001) Action of recombinant human apoptotic endonuclease G on naked DNA and chromatin substrates. Cooperation with exonuclease and DNase I. *Journal of Biological Chemistry*, **276**(51), 48404–9.

Yang, J., Lin, Y., Guo, Z., Cheng, J., Huang, J., Deng, L., Liao, W., Chen, Z., Liu, Z. & Su, B. (2000) The essential role of MEKK3 in TNF-induced NF-κB activation. *Nature Immunology*, **2**, 620–4.

9 Type II Nuclear Receptor-Mediated Transcription

9.1 TYPE II NUCLEAR RECEPTORS

Vitamin A (as retinoic acid) and vitamin D (as 1α,25-dihydroxyvitamin D_3) induce the synthesis of specific proteins through receptor-mediated regulation of gene transcription. Nuclear receptors for retinoic acid (RAR), vitamin D_3 (VDR), thyroid hormone (TR), and prostanoids (PPAR) are categorized as type II nuclear receptors. (Type I nuclear receptors comprise receptors for the classic steroid hormones.) Type II nuclear receptors are able to form stable heterodimers with the retinoid X receptor (RXR) *in vitro*. This constitutive dimerization permits DNA binding in the absence of ligand.

9.2 HORMONE RESPONSE ELEMENTS

The hormone response elements on the DNA are binding sites for nuclear receptors and act as enhancers to increase the transcriptional activity of an adjacent promoter. The response elements are composed of two half-sites or core recognition motifs, each with six base pairs, separated by nucleotide spacers. The dyad symmetry of these elements permits receptor–ligand complexes to bind as dimers, with each monomer recognizing one of the half-sites.

Three main features determine the specificity of receptor binding to hormone response elements: (1) the precise nucleotide sequence of the half-site, (2) the orientation of half-sites with respect to each other, and (3) the spacing between half-sites. The orientation of half-sites can be as inverted repeats, everted repeats, or direct repeats, as illustrated in Figure 9.1. Table 9.1 shows the features of response elements preferentially recognized by the principal nuclear receptors. The half-site sequence recognized by type II nuclear receptors is 5′-AGGTCA with a direct repeat orientation; that is, 5′-AGGTCAnAGGTCA-3′, where n is a nucleotide spacer. The preferred spacing between the direct repeats (DR) is three nucleotides (DR3) for VDR/RXR, four nucleotides (DR4) for TR/RXR, and five nucleotides (DR5) for RAR/RXR. RAR/RXR is also capable of binding to DR1 and DR2 elements, but with somewhat lower affinity than to the DR5 element. DR1 is also a response element for an RXR homodimer and for a PPAR/RXR heterodimer. Thyroid hormone response elements (TREs) are typically composed of two or more sequences of a consensus AGGTCA half-site arranged as an inverted repeat, everted repeat, or direct repeat.

9.3 RECEPTOR DOMAINS

Based on amino-acid sequence similarity, the protein structures of nuclear receptors can be divided into six domains, designated A to F (Figure 9.2). The A/B domain is at the N (amino) terminus of the molecule and the F domain is at the C (carboxy) terminus. Significant sequence homology is found among the members of the nuclear receptor superfamily; those domains exhibiting strong homology have been conserved during evolution.

The most highly conserved domain is the C domain, which contains the DNA-binding site and regions that contribute to receptor dimerization and nuclear localization. The functionally complex E domain encompasses about 250 amino acids and contains subdomains responsible for ligand binding, dimerization, nuclear localization, and transcriptional activation. A region essential for transcriptional activation, AF-2, is one of two activation functions. The A/B domain is highly variable in size, ranging from 25 (VDR) to 603 amino acids (mineralocorticoid receptor, MR), and exhibits negligible sequence conservation. It contains the second activation function, AF-1, whose length and sequence are highly variable and unique to each specific receptor isoform. AF-1 functions independently of ligand, as opposed to AF-2, which is ligand-inducible. Although only weakly active by itself, AF-1 synergizes with and modulates the activity of AF-2. The D domain is a flexible 'hinge' region, allowing a degree of rotation within the protein molecule. The F domain is a variable region for which no specific function has been identified.

9.4 CHROMATIN REMODELING

In the nucleus, the DNA is compacted into chromatin by winding around specific DNA-binding proteins called histones. A histone is a barrel-shaped octamer comprising two sets of core histones (H2A, H2B, H3, and H4) and linker histones (H1). Each of the core histones has an amino-terminal tail containing a number of positively charged amino acid residues. The fundamental repeating unit of chromatin is the nucleosome, which comprises a histone octamer around which 146 base pairs of DNA are wrapped in two nearly complete turns. The linker histones act as a clamp, preventing the unwinding of DNA from the octameric complex. The histone tails lie on the outside of the nucleosome, where they can interact ionically with the negatively charged phosphates of the DNA backbone.

The organization of chromatin into nucleosomes is an essential feature in the regulation of transcription. Nucleosomes impede access of the basal transcription machinery to the promoter on the DNA. When it becomes necessary for a gene to be expressed, changes in cell

DOI: 10.1201/9781003410300-11

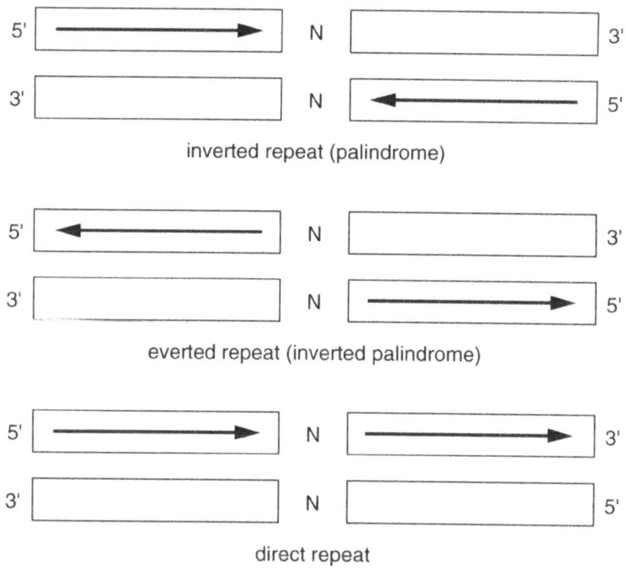

FIGURE 9.1 Half-site orientations of hormone response elements. Arrows indicate the 5′ to 3′ direction. N, nucleotide spacer(s).

TABLE 9.1
Features of Hormone Response Elements Preferentially Recognized by Nuclear Receptors

Hexanucleotide Sequences	Type of Response Element	Receptor Dimers
Inverted repeat sequences		
5′–AGAACAnnnTGTTCT–3′	IR3	GR/GR, MR/MR, AR/AR, PR/PR
3′–TCTTGTnnnACAAGA–5′		
5′–AGGTCAnnnTGACCT–3′	IR3	ER/ER
3′–TCCAGTnnnACTGGA–5′		
Direct repeat sequences		
5′–AGGTCAnAGGTCA–3′	DR1	RXR/RXR, RAR/RXR,
3′–TCCAGTnTCCAGT–5′		
5′–AGGTCAnnAGGTCA–3′	DR2	RAR/RXR
3′–TCCAGTnnTCCAGT–5′		
5′–AGGTCAnnnAGGTCA–3′	DR3	VDR/RXR
3′–TCCAGTnnnTCCAGT–5′		
5′–AGGTCAnnnnAGGTCA–3′	DR4	TR/RXR
3′–TCCAGTnnnnTCCAGT–5′		
5′–AGGTCAnnnnnAGGTCA–3′	DR5	RAR/RXR
3′–TCCAGTnnnnnTCCAGT5′–		

Base pairs: adenine (A); thymine (T); cytosine (C); guanine (G).
IR3, inverted repeat with three nucleotide spacers (n). DR1, DR2, DR3, etc., direct repeats with one, two, three, etc. nucleotide spacers.
Abbreviations: Receptors (R) for: GR, glucocorticoid; MR, mineralocorticoid; AR, androgen; PR, progesterone; ER, estrogen; RXR, retinoid X; RAR, retinoic acid; VDR, vitamin D_3; TR, thyroid hormone.

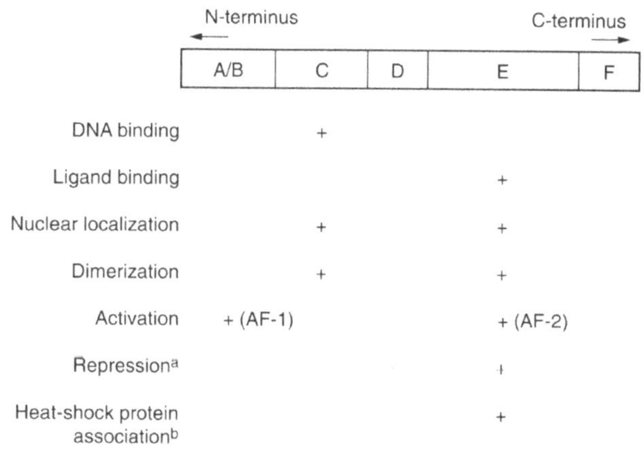

FIGURE 9.2 General functional organization of type II nuclear receptors into six domains, A–F.

physiology elicit a remodeling of chromatin, which allows binding of transcription factors at promoters. Two important chromatin remodeling systems have been studied. In one system, the changes in chromatin structure are mediated by ATP-driven remodeling complexes; the other system involves the post-translational modification of core histones by acetylation of their tails. Both systems disrupt histone–DNA interactions in a transient and reversible manner.

Three families of ATP-dependent chromatin remodeling complexes have been characterized, namely, SWI/SNF, ISWI, and Mi-2/CHD, each containing several members. SWI/SNF mediates chromatin remodeling by creating a loop of under-twisted DNA that propagates around the nucleosome in a wave-like manner. This loop propagation mechanism disrupts only one or two histone–DNA contacts at a time (Saha *et al.*, 2006).

Nuclear histone acetyltransferases (HATs) catalyze the transfer of an acetyl group from acetyl coenzyme A onto the ε-amino group of specific lysine residues present exclusively in the amino-terminal tails of each of the core histones. Neutralization of the positively charged lysines reduces the net positive charge of the histone tails and weakens their association with DNA. The displacement of the flexible tails permits subtle changes in nucleosomal structure and a partial unwinding or loosening of the core DNA. The overall result is an increase in accessibility of transcription factors to their DNA binding sites. The opposing activity of histone deacetylases (HDACs) maintains the nucleosomes in the transcriptionally repressive state in which the histone tails are in close association with the DNA.

9.5 TYPE II NUCLEAR RECEPTOR ACTIVATION AND REPRESSION

Liganded type II nuclear receptors bound to hormone response elements on the DNA act as transcriptional activators. Various primary and secondary co-activators remodel chromatin and accelerate the assembly of the basal transcription machinery at the promoter. Primary co-activators bind directly to the activated receptor and recruit secondary co-activators.

PCAF is a nuclear receptor co-activator that first associates with the DNA-bound RAR/RXR heterodimer and then interacts with p300/CBP, which exhibits HAT activity (Blanco *et al.*, 1998). The arginine-specific protein methyltransferase CARM1 binds to the AD2 domain of p160 co-activators and potentiates receptor activation by modifying chromatin structure through histone methylation (Chen *et al.*, 1999). CBP and p300 also interact with p160 co-activators but, in this case, via the p160's AD1 domain (Ma *et al.*, 1999). In these interactions, CARM1 and p300/CBP serve as secondary co-activators for nuclear receptors, as they are recruited to the promoter by binding to the primary p160 co-activators.

Type II nuclear receptors repress basal transcription in the absence of their ligand. RAR possesses a silencing function, the CoR (co-repressor) box, which, in the absence of ligand, interacts with a co-repressor. Neither of the SMRT or N-CoR co-repressors possesses intrinsic repressive activity, but they interact with large complexes such as SIN3 and NuRD, which contain HDACs (Privalsky, 2004). The binding of ligand (all-*trans* retinoic acid) to the RAR partner of the DNA-bound heterodimer causes the release of the co-repressor complex and the recruitment of a co-activator complex.

FURTHER READING

Collingwood, T.N., Urnov, F.D. & Wolffe, A.P. (1999) Nuclear receptors: coactivators, corepressors and chromatin remodeling in the control of transcription. *Journal of Molecular Endocrinology*, **23**, 255–75.

REFERENCES

Blanco, J.C.G., Minucci, S., Lu, J., Yang, X.-J., Walker, K.K., Chen, H., Evans, R.M., Nakatini, Y. & Ozato, K. (1998) The histone acetylase PCAF is a nuclear receptor coactivator. *Genes & Development*, **12**, 1638–51.

Chen, D., Ma, H., Hong, H., Koh, S.S., Huang, S.-M., Schurter, B.T., Aswad, D.W. & Stallcup, M.R. (1999) Regulation of transcription by a protein methyltransferase. *Science*, **284**, 2174–7.

Ma, H., Hong, H., Huang, S.-M., Irvine, R.A., Webb, P., Kushner, P.J., Coetzee, G.A. & Stallcup, M.R. (1999) Multiple signal input and output domains of the 160-kilodalton nuclear receptor coactivator proteins. *Molecular and Cellular Biology*, **19**, 6164–73.

Privalsky, M.L. (2004) The role of corepressors in translational regulation by nuclear hormone receptors. *Annual Review of Physiology*, **66**, 315–60.

Saha, A., Wittmeyer, J. & Cairns, B.R. (2006) Chromatin remodelling: The industrial revolution of DNA around histones. *Nature Reviews Molecular Cell Biology*, **7**, 437–47.

10 A Guide to the Immune System

10.1 OVERVIEW OF THE IMMUNE RESPONSE

When pathogenic microorganisms breach the barriers imposed by the skin and mucous membranes, the innate immune system comes into play. Cells of the innate immune system comprise (1) phagocytes, represented by neutrophils, monocytes, macrophages, and immature dendritic cells (DCs); (2) cells that secrete toxic granules, namely basophils, eosinophils, mast cells, and natural killer (NK) cells; and (3) innate lymphoid cells that secrete cytokines. The innate immune system elicits a local inflammatory response, which is modulated by cytokines secreted by macrophages and innate lymphoid cells resident in epithelial barrier tissues. The reactions that comprise the inflammatory response bring fluid, proteins, and phagocytic leukocytes from the blood to the site of infection, where they effect host defense and tissue repair. The inflammatory response may lead to the destruction of the pathogen, in which case the response will be acute and confined to the infected area. The production of antimicrobial peptides by neutrophils, intestinal Paneth cells, epithelial cells of the skin, and the trachea-bronchial lining can be regarded as an integral component of the inflammatory response. Innate immune defense also includes complement – a system of circulating proteins that, when activated, generates a cascade of proteolytic reactions that ultimately destroys the pathogen by lysis.

If the pathogen is not destroyed by the innate immune system, the continued production of cytokines will set the adaptive immune response in motion. The chief protagonists of adaptive immunity are the T and B lymphocytes and mature DCs, whose actions and numbers are controlled by cytokines. Previously unstimulated (naïve) $CD4^+$ and $CD8^+$ T cells need to be activated (primed), which triggers their proliferation and differentiation into effector T cells. Cells of the adaptive immune system are directed to locations where their specialized functions are most suitably employed.

Immature DCs capture foreign antigens in the peripheral tissues and process them into peptides, which are displayed on the cell surface in association with a membrane-bound major histocompatibility (MHC) class II molecule. The DCs then mature and migrate to the lymph nodes, where they present the antigenic peptide to the rare naïve $CD4^+$ T cells whose receptors specifically recognize the antigen displayed. Engagement of the naïve $CD4^+$ T cell receptor (priming) triggers cell proliferation and differentiation. Most of the expanded $CD4^+$ T cells differentiate into three subsets of helper T (Th) cells, namely Th1, Th2, and Th17. These cells leave the lymph nodes and disperse to peripheral tissues, where they produce cytokines that dictate the behavior of other cells of the immune system. Th1 cells drive type 1 immune responses that protect against intracellular infections with viruses, bacteria, fungi, and protozoan parasites. Th2 cells contribute to type 2 immune responses that provide help to B cells and defend against extracellular parasites such as helminths. Th17 cells are effective against extracellular bacteria and fungi. Some of the expanded $CD4^+$ T cells differentiate into follicular helper T cells (T_{FH}). These cells remain in the lymph nodes, where they activate naïve B cells to proliferate and differentiate into antibody-producing plasma cells.

Host cells that have been infected with a virus or DCs that have ingested virus-infected cells process the viral antigens and display the peptide fragments on the cell surface in association with a MHC class I molecule. The peptide–MHC class I complex can be recognized by a naïve $CD8^+$ T cell, which differentiates into a cytotoxic T cell capable of killing the infected cell.

B cells differentiate into antibody-producing plasma cells when activated by antigen. Activation can be T-independent or T-dependent, depending on the requirement for T cell help. Marginal zone B cells resident in the spleen make early T-independent antibody responses to blood-borne antigens. The most effective antibody response results from the T-dependent germinal center reaction in which follicular B cells differentiate into plasma cells and memory B cells, and high-affinity antibodies of different classes are produced. Antibody responses are described as primary and secondary. Primary responses result from the activation of naïve B cells, whereas secondary responses are due to stimulation of memory B cells produced after the primary response. The secondary response develops more rapidly than the primary response and produces larger amounts of antibodies.

Upon disappearance of the source of antigen that initiated their production, the vast majority of effector T cells and B cells are eliminated via apoptosis, leaving a small proportion of antigen-reactive T and B cells to survive as long-lived memory cells. Memory cells can generate more rapid and effective responses if a re-infection occurs at some time in the future.

Immune responses to pathogens must be tightly controlled, otherwise they can result in severe inflammation and collateral tissue damage. Infection-induced immunopathology can be controlled through the production of anti-inflammatory cytokines and through the generation of regulatory T cells that release soluble inhibitors of the immune system. Natural killer T (NKT) cells have the potential to produce cytokines with opposing activities

DOI: 10.1201/9781003410300-12

and so provide a delicate means of regulating the immune response. Unless prevented, the immune system can respond to self antigens, leading to autoimmune diseases. Tolerance to self antigens is achieved by preventing the development of self-reactive lymphocytes (central tolerance) and by mechanisms that render mature self-reactive lymphocytes anergic (peripheral tolerance).

The immune system must be able to discriminate between harmful and innocuous materials that are ingested in order to prevent hypersensitivity reactions to harmless dietary antigens. Oral tolerance is the state of local and systemic unresponsiveness that is initiated in the small intestine by oral administration of innocuous antigen.

10.2 GENERAL FEATURES OF THE IMMUNE SYSTEM

10.2.1 Introduction

Immunity refers to the body's resistance to invasive pathogens (viruses, bacteria, fungi, protozoa, and multicellular parasites) or their toxic products, and to allergens (pollen, animal hair, chemicals, etc.). The immune system comprises (1) cellular defense mechanisms mediated by several types of leukocytes (white blood cells) and (2) humoral defense mechanisms mediated by soluble antibodies produced by B lymphocytes. Any immune response involves, firstly, recognition of the pathogen or other foreign material and, secondly, destruction of these invaders.

Because the immune system is primarily destructive, it is vital that it can discriminate between what is normally present in the body ('self') and what should not be there ('non-self'). The immune system has evolved a number of different mechanisms to limit the production of self-reactive B and T lymphocytes, a process known as immmunological tolerance. Any breakdown in the self versus non-self discrimination could result in immune responses being directed against the individual's own body tissues (autoimmunity) and the manifestation of an autoimmune disease.

10.2.2 Pathogens

The mucosal surfaces of the respiratory, intestinal, and genital tracts, as well as skin lesions, are vulnerable sites where pathogens could gain access to the body's interior. Pathogens are capable of colonizing most of the body tissues, where they proliferate in the cytosol, the intracellular vesicular system, extracellular spaces, epithelial surfaces, and body fluids (blood and lymph). Viruses are obligate intracellular parasites that use the biosynthetic machinery of the host cell to propagate. Some pathogenic bacteria (e.g. *Listeria*) are specialized to enter the cytoplasm of epithelial cells. Mycobacteria, such as *Mycobacterium tuberculosis* and *Mycobacterium leprae*, grow only in intracellular vesicles of macrophages. *Leishmania* species, such as *Leishmania major*, are parasitic protozoa that also infect macrophages. Other bacteria and larger parasites replicate in extracellular spaces and body fluids. The immune response depends on the site of infection and the nature of the pathogen. To clear a viral infection, it is necessary to destroy the infected host cells. Extracellular pathogens are selectively destroyed and their toxic products neutralized.

Certain pathogens, such as human immunodeficiency virus (HIV), hepatitis C virus, and many parasites, have evolved cunning strategies to subvert the host immune response and cause chronic infections. One strategy to evade immune recognition is to alter antigenic components such as surface proteins of parasites. Other strategies include induction of host immunosuppressive cytokines and T cells with suppressor activity.

10.2.3 Antigens

Originally, the term antigen was used for any molecule that induced B lymphocytes to produce a specific antibody (*anti*body *gen*erator). Nowadays, however, the term is much more widely used to indicate any molecule that lymphocytes and soluble antibodies can bind to. Antigens are nearly always macromolecules, such as proteins, glycoproteins, and lipoproteins. These molecules may be on the surface of a pathogen or they may be soluble products, such as toxins secreted by pathogens or heat shock proteins leaked from damaged bacterial cell walls. The part of an antigen molecule that is recognized by an antigen receptor is known as an antigenic determinant, which consists of one or more epitopes. Antigens usually function as immunogens; that is, they generate a positive immune response. However, antigens can also function as tolerogens, which induce a specific immunological unresponsiveness (tolerance). The degree of foreignness of an antigen is an important factor: molecules exhibiting the greatest differences from constituents of the host's body are usually the most immunogenic.

Foreign antigens are conventionally categorized as exogenous or endogenous, depending on whether they are derived from outside the cell or synthesized within the cell. Thus an exogenous antigen could be a protein from a bacterium that has been captured by a phagocyte, and an endogenous antigen could be a viral protein that is synthesized within a cell after it has been infected with a virus.

10.2.4 Innate and Adaptive Immunity

The immune system is divided into innate immunity and adaptive immunity. In addition, a group of specialized lymphocytes, acting within a 'natural memory' compartment, responds to infection within a time-frame that is intermediate between innate and adaptive immunity.

Innate immunity, being genetically determined, is present from birth and is in full readiness for an attack by invading pathogens. The innate defense mechanisms act non-specifically against a wide range of microorganisms and foreign material and can be mobilized at the site of infection within minutes to hours. The innate response to an invading pathogen involves the recognition of general molecular patterns, such as carbohydrates, present on the pathogens themselves or in the infected cell. The main characteristics and components of innate immunity occur in the inflammatory response. There is no immunological memory, that is, the response is not dependent upon prior exposure to a particular infectious agent. Innate immune defense begins in the skin and the mucous membranes of the respiratory, intestinal, and urogenital tracts. The outer keratin layer of skin is impenetrable to microorganisms, although hair follicles are vulnerable to infection by *Staphylococcus aureus*. Fluids secreted by the skin and mucous membranes contain bactericidal and fungicidal molecules. Innate immunity in the blood and tissues is provided by phagocytic and secretory cells, and also by soluble mediators, such as proteins of the complement system.

Adaptive immunity is initiated when an innate response fails to eradicate a new infection. A unique feature of adaptive immunity is the ability to recognize individual pathogens specifically. This feature is based on clonal selection of lymphocytes bearing antigen-specific receptors. It takes several weeks to develop an effective adaptive immune response to invasion by a novel pathogen; however, that particular pathogen will be recognized and rapidly dealt with the next time it is encountered. Moreover, the second response will be much more vigorous and effective than the initial response.

The delayed appearance (up to 5 days) of effector cells of the adaptive immune system is due to the needs for antigen processing, DC maturation and migration, and T cell clonal expansion and migration to sites of infection.

Pathogens that survive the innate immune response may be present outside host cells (e.g. bacteria and parasites in extracellular spaces or body fluids) or inside host cells (e.g. viruses in the cytosolic compartment and bacteria in the vesicular system). The B and T lymphocytes (B and T cells) of the adaptive immune system are able to cope with both situations. Soluble antibodies secreted by terminally differentiated B cells (termed plasma cells) enable extracellular pathogens to be recognized and ingested by phagocytic cells of the innate immune system. T cells recognize intracellular pathogens through MHC-restricted recognition. This involves the binding of an antigen fragment to a cell-surface MHC molecule on an antigen-presenting cell, such as a mature DC.

10.2.5 Origin of Cells of the Immune System

The majority of cells in the immune system were formerly (but not nowadays) classified as white blood cells or leukocytes. Cell types of the immune system, except for the lately discovered innate lymphoid cells, are listed in Table 10.1. Immune cells are all blood cells and have in common a limited life span and an inability to replicate; they therefore must be replaced continuously throughout life. In adults, blood cells are formed in the bone marrow by the process of hematopoiesis, which is sustained by a variety of cytokines and other growth factors produced by bone marrow stromal cells. The $CD34^+$ bone marrow fraction contains pluripotent hematopoietic stem cells that give rise to all blood cell types through a series of mitotic cell divisions called a cell lineage. A cell lineage traces the development of the blood cells through a gradual process of commitment and differentiation. In all lineages, extensive proliferation occurs during differentiation.

Pluripotent stem cells can undergo self-renewal or they can differentiate. Self-renewal involves a symmetrical

TABLE 10.1
Cells of the Immune System[a]

Cell Type	Description	Type of Immunity
Circulating cells		
Neutrophil	Polymorphonuclear leukocyte	Innate
Eosinophil	Polymorphonuclear leukocyte	Innate
Basophil	Polymorphonuclear leukocyte	Innate
Monocyte	Mononuclear leukocyte	Innate
Platelet	Fragment of megakaryocyte	Innate
Natural killer cell (NK cell)	Large granular leukocyte	Innate
T lymphocyte (T cell)	Mononuclear leukocyte	Adaptive
Natural killer T cell (NKT cell)	Mononuclear leukocyte	Bridges innate and adaptive immune systems
B lymphocyte (B cell)	Mononuclear leukocyte	Adaptive
Plasma cell	Differentiated B cell	Adaptive
Tissue cells		
Macrophage	Differentiated monocyte	Innate
Mast cell	Similar to basophil	Innate
Immature dendritic cell	Stellate	Innate
Mature dendritic cell	Numerous cytoplasmic processes	Adaptive

[a]Excludes innate lymphoid cells.

division to yield two daughter stem cells identical to the mother cell. Differentiation involves an asymmetrical division to generate an identical stem cell and a multipotential progenitor cell. Self-renewal and differentiation of stem cells and commitment of progenitors appear to be stochastic (random) events, occurring with a given probability for the cell population as a whole.

Hematopoietic stem cells and progenitors express Flt3, a receptor tyrosine kinase which is important for their normal development. (Receptor tyrosine kinases are cell-surface receptors with intrinsic kinase activity that directly initiate signal transduction without the involvement of second messengers.) Flt3 ligand is expressed by bone marrow stromal cells and synergizes with other growth factors to stimulate proliferation of stem cells, progenitors, DCs, and NK cells.

In the dual lymphoid–myeloid model of hematopoiesis (Figure 10.1), hematopoietic stem cells differentiate into multipotential progenitor cells, which give rise to common myeloid progenitors (CMPs) or common lymphoid progenitors (CLPs). These progenitors give rise to precursors for various lineages. Hematopoiesis produces DCs of two subsets, conventional DCs (also known as myeloid DCs) and plasmacytoid DCs, both of which can be generated, directly or indirectly, from Flt3-expressing CMPs or CLPs.

The specialized precursors of the myeloid lineage differentiate in the bone marrow and the differentiated cells are subsequently released into the blood. B cell maturation takes place in the spleen where newly arrived cells are referred to as transitional B cells. Precursors of T cells migrate from the bone marrow to the thymus, where, as thymocytes, they undergo Notch 1 receptor-induced commitment to the T cell lineage. During early development in the thymus, thymocytes express either α β-type or γ δ-type antigen-binding receptors. Ultimately, about 90% of these cells mature into the α β T cell lineage. NKT cells derive from thymocytes that randomly express a CD1d-specific T cell receptor and undergo positive selection by CD1d expressed by CD4$^+$CD8$^+$ (double-positive) cortical thymocytes. NK cells and innate lymphoid cells (ILCs) develop from the CLP under the direction of dedicated transcription factors. ILC precursors migrate from their primary site of production into infected and injured tissues, where they complete their maturation.

10.2.6 The CD Classification of Cell-Surface Molecules

The surfaces of immune-system cells are covered with receptors and other functional proteins that indicate the differentiation of the cell. The expression level of cell-surface molecules often changes as cells differentiate or become activated. The different cell-surface molecules have been classified by the use of monoclonal antibodies and assigned a CD number. CD stands for 'cluster of differentiation,' referring to the cluster of monoclonal antibodies that define a particular cell-surface molecule. Approximately 340 CD molecules have been assigned.

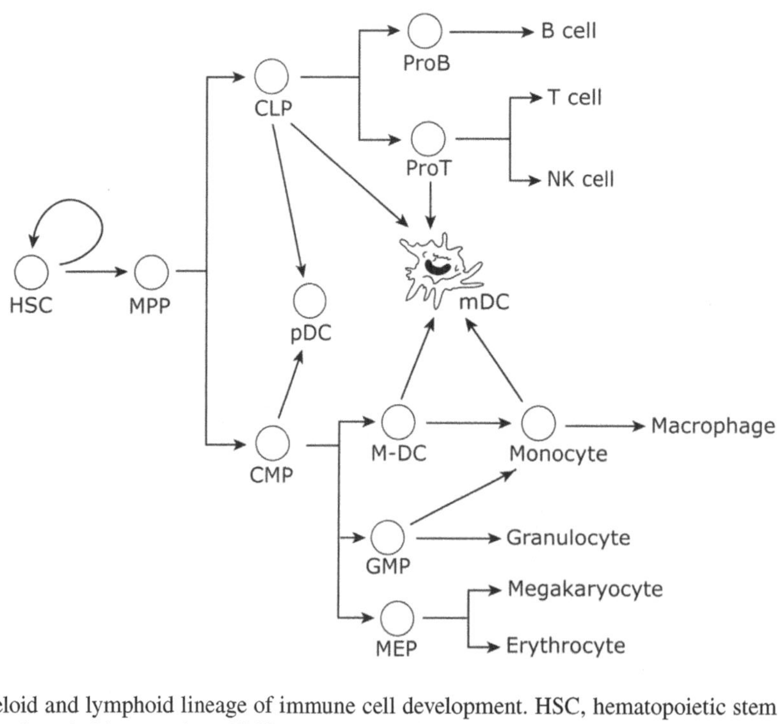

FIGURE 10.1 The myeloid and lymphoid lineage of immune cell development. HSC, hematopoietic stem cell; MPP, multipotential progenitor; CLP, common lymphoid progenitor; CMP, common myeloid progenitor; ProB, B cell precursor; ProT, T cell precursor; M DC, monocyte and dendritic cell precursor; GMP, granulocyte and monocyte precursor; MEP, megakaryocyte and erythrocyte precursor; pDC, plasmacytoid dendritic cell; mDC, myeloid dendritic cell; NK cell, natural killer cell. Granulocytes comprise neutrophils, basophils, eosinophils, and mast cells.

10.2.7 Bacterial Cell Walls

The structure of bacterial cell walls is important in the response of the body's immune system to pathogenic bacteria. The gram stain (named after Christian Gram) is a crystal violet–iodine reagent used to classify bacteria according to the nature of the cell wall. Bacteria that take up the stain are designated gram-positive (e.g. *Staphylococcus aureus*) and those that do not are designated gram-negative (e.g. *Escherichia coli*).

10.2.8 General Mechanisms of Immune Suppression

Immune responses can be suppressed by a variety of mechanisms, and some examples follow:

- Suppression through the action of regulatory T cells.
- Histamine, after first mediating the immune response, helps to suppress it by activating regulatory T cells.
- High concentrations of soluble antibody block the antigen-recognizing receptors (membrane-bound antibody) on B lymphocytes. The consequent inability of the B cells to recognize their specific antigen renders them unable to differentiate and produce soluble antibody.
- Antibodies may be regulated by idiotype–anti-idiotype interactions.
- Macrophages can produce prostaglandins, which non-specifically inhibit lymphocyte responses.

10.2.9 Immune Privileged Sites

Immune responses and associated inflammation in certain parts of the body, including the brain, anterior chamber of the eye, testis, and placenta, carry a high risk of organ dysfunction or reproductive failure. Such tissues possess various means of protection against the damaging effects of an immune response. and hence are called immune privileged sites. Blood–tissue barriers, such as the blood–brain barrier, are anatomical features of these tissues. Other protective mechanisms include the local production of immunosuppressive cytokines and expression of cell surface molecules that inactivate or kill lymphocytes.

10.3 CELLS OF THE INNATE IMMUNE SYSTEM

10.3.1 Introduction

The cell types that function as effector cells in the innate immune system are neutrophils, basophils, eosinophils, and mast cells (collectively known as the granulocytes), monocytes and macrophages (mononuclear phagocytes), and immature DCs. NK cells are included in the most recently identified immune cell types known collectively as innate lymphoid cells. The cells of this family are considered to be innate counterparts of certain T cell subsets. All of these cell types derive from a common hematopoietic stem cell in the bone marrow. Various tissue cells also play a part by secreting cytokines; these include keratinocytes (which also produce antimicrobial peptides), mucosal epithelial cells, endothelial cells, fibroblasts, osteoblasts, chondrocytes, bone stromal cells, hepatocytes, and Kupffer cells. The bone marrow–derived effector cells can be broadly categorized as phagocytic and secretory. Neutrophils and macrophages are dedicated 'professional' phagocytes, which internalize and kill bacteria, fungi, protozoa, viruses, virus-infected cells, and tumor cells by generating highly reactive and toxic molecules. Immature conventional DCs are also phagocytic, but do not destroy the particles. Instead, these cells differentiate into mature non-phagocytic DCs after a period of days to weeks. The mature DCs migrate from the peripheral tissues into the lymphoid organs where, as cells of the adaptive immune system, they display antigen fragments on their surface for presentation to T lymphocytes. The highly specialized DCs are discussed in Section 10.13, with no further mention in this section.

The secretory cells of the innate immune system are mast cells, basophils, and eosinophils. These cells function mainly in immune defense against parasites that are too large to be internalized by phagocytes. They release the toxic contents of cytoplasmic granules to the exterior on activation, thereby creating an environment hostile to invading organisms or (in the case of eosinophils) directly killing the parasites.

10.3.2 Neutrophils

Neutrophils are the most abundant type of leukocyte and the first inflammatory cells to respond to infections. Circulating neutrophils are targeted by chemotaxis to the site of infection, where they phagocytose, kill, and digest the invading microorganisms. Neutrophils kill the microorganisms intracellularly by releasing a cocktail of microcidal compounds and enzymes from membrane-bound vesicles called granules. Degranulation of neutrophils is induced by IL-8. There are two distinct types of granule, primary (azurophilic) and secondary (specific), whose contents are listed in Table 10.2. The combined action of the granular contents is to disrupt the integrity of microbial outer membranes, making them permeable. Neutrophils are short-lived and die after two or three days. In animals with neutropenia, infections which normally are mild can be life-threatening.

Defensins and two peptides derived from cleavage of a single human cathelicidin are antimicrobial peptides that insert into microbial membranes and oligomerize to form aqueous pores. Bactericidal permeability-increasing protein (BPI) binds to lipopolysaccharide and makes the outer envelope of gram-negative bacteria permeable. Cathepsin G has several other functions in addition to its proteolytic activity: it is required for the release of elastase from neutrophils; it can activate neutrophil progelatinase B and procollagenase; and it acts as an antimicrobial peptide.

TABLE 10.2
Contents of Neutrophil Granules

Constituent	Antimicrobial Effect
Azurophilic granules	
Defensins	Cationic peptides which form aqueous pores in microbial membranes
Bacterial permeability-increasing protein (BPI)	Increases permeability of outer envelope of gram-negative bacteria
Myeloperoxidase	Catalyzes formation of hypochlorite from hydrogen peroxide and chloride ion
Elastase, cathepsin G, proteinase 3	Serine proteinases
Lysozyme	Degrades peptidoglycan component of bacterial cell wall
Specific granules	
Cathelicidins	Cationic peptides which form aqueous pores in microbial membranes
Progelatinase B, procollagenase	Matrix metalloproteinases
Plasminogen activator	Activates plasminogen to become plasmin, a proteolytic enzyme
Lysozyme	Degrades peptidoglycan component of bacterial cell wall
NADPH oxidase	Creates reactive oxygen intermediates upon induced assembly of enzyme components

TABLE 10.3
Mediators Present in Mast Cells

Mediator	Main Biological Effect
Released from granules	
Histamine	Vasodilation, increased vascular permeability
Heparin	Anticoagulant
Proteolytic enzymes	Degradation of blood vessel endothelial membrane
Platelet-activating factor	Release of histamine from platelets
Neutrophil chemotactic factor	Attracts neutrophils from blood
Eosinophil chemotactic factor	Attracts eosinophils from blood
Tumour necrosis factor-α	Promotes inflammation
Cathelicidin (LL-37 in human cells)	Antimicrobial peptide
Newly synthesized	
Prostaglandin D_2	Vasodilation and bronchoconstriction, neutrophil chemotaxis
Leukotriene C_4	Bronchoconstriction

Lysozyme splits sugars off the peptidoglycan component of bacterial cell walls. Gram-positive bacteria are easily lysed by lysozyme because their cell wall is composed largely of peptidoglycan. In gram-negative bacteria, lysozyme can gain access to the underlying peptidoglycan after pore-forming molecules have disrupted the integrity of the outer envelope.

10.3.3 Mast Cells

Mast cells originate from hematopoietic bone marrow prercursors, circulate in the blood as immature progenitors, and complete their differentiation in the mucosal and connective tissues. Mature mast cells are sessile and present in tissues throughout the body, but chiefly in perivascular connective tissue, epithelia, and lymph nodes. Their cytoplasm is full of membrane-bound granules containing a variety of preformed mediators of allergic and inflammatory responses, including TNF-α (Table 10.3). Mast cell granules also contain antimicrobial peptides that can kill bacteria by disrupting their membranes. The binding of antigen-bound IgE antibodies to FcεRI receptors on mast cells stimulates release of the preformed mediators from the granules. A variety of other microbial coatings (opsonins), including the complement fragments C3a and C5a, can trigger mast cell degranulation independently of antibody binding. Mast cells also possess pattern recognition receptors, such as Toll-like receptors, enabling these cells to recognize pathogenic bacteria and kill them by releasing the antimicrobial peptides from the granules. In addition to degranulation, triggering FcεRI receptors results in the rapid synthesis of arachidonic acid metabolites such as leukotrienes and prostaglandin D_2, which have more prolonged inflammatory effects than histamine. Finally, stimulation of mast cells induces the synthesis and secretion of inflammatory cytokines, such as TGF-β.

10.3.4 Basophils

Basophils circulating in the bloodstream are recruited to epithelia where they perform a similar function to mast cells. They release preformed mediators of allergic and inflammatory responses from granules when bound by IgE–antigen, and they also release newly synthesized leukotrienes when stimulated by complement.

10.3.5 Eosinophils

Eosinophils constitute 1–2% of circulating leukocytes, but their numbers increase in cases of allergy and invasion by parasites. They are highly efficient at killing extracellular parasites that are too large to be phagocytosed. The granules of eosinophils consist of a crystalloid core containing a so-called major basic protein and a matrix containing other basic (cationic) proteins, a neurotoxin, a peroxidase, and various other enzymes. When suitably stimulated, eosinophils exocytose these molecules, which kill and digest the parasites.

10.3.6 Monocytes and Macrophages

Monocytes and macrophages are collectively known as mononuclear phagocytes. Monocytes are circulating phagocytes that represent about 5–10% of peripheral blood leukocytes in humans and mice. Their half-life in blood is relatively short, about one day in mice and three days in humans. Circulating mouse monocytes comprise two main subsets: Gr-1^{high} 'classical' monocytes, which home preferentially to sites of inflammation, and Gr-1^{low} 'non-classical' monocytes, which seed non-inflamed tissues. Human counterparts of the two murine subsets have also been identified. In the absence of infection, Gr-1^{high} blood monocytes return to the bone marrow and are converted to Gr-1^{-} monocytes, thus conserving the monocyte population.

During inflammation and infection, monocytes are recruited to the affected sites, where they differentiate into either macrophages or DCs. Monocytes can also migrate into tissues spontaneously, without a defined stimulus, in order to replenish macrophage or DC populations. The monocyte-derived macrophages within a tissue acquire a characteristic morphology and specialized function, depending on the local environment. In the absence of a stimulus, macrophages remain relatively quiescent at a particular site and in this state they are known as *resident* macrophages. Resident macrophages are found at sites that are vulnerable to invasion by pathogens; thus serosal macrophages are found in capillaries and alveolar macrophages are found in the lung. Resident macrophages are also strategically placed in the spleen, lymph nodes, and various other sites. Examples of resident macrophages are given in Table 10.4.

TABLE 10.4
Examples of Resident Macrophages

Tissue	Type of Cell
Lining of peritoneal cavity	Peritoneal macrophage
Lung	Alveolar macrophage
Spleen (sinus)	Splenic sinus macrophage
Capillary	Serosal macrophage
Liver	Kupffer cell
Kidney	Mesangial cell
Brain	Microglial cell

Resident macrophages have low microbicidal activity, but they possess a variety of cell-surface receptors which allow them to respond to specific stimuli very rapidly. A variety of inflammatory agents stimulate the resident macrophages to develop into *inflammatory* macrophages, which express a greater number of receptors and secrete a wide variety of inflammatory mediators in copious amounts. Resident and/or inflammatory macrophages can be triggered by interferon-γ during immune responses to become *activated* macrophages, now with a greatly enhanced ability to kill microbes and to release cytokines. The activated macrophages degrade material that they engulf and are important for the local clearance of dead cells in inflamed and non-inflamed tissues. Unlike neutrophils, macrophages are long-lived.

10.4 PHAGOCYTOSIS

Phagocytes engulf foreign particles, internalize them, and destroy them using an impressive array of killing mechanisms.

10.4.1 Target Recognition

Circulating phagocytes are directed to the site of infection by chemotaxis and attach to the surface of the particle to be engulfed. They recognize a particle as being foreign either directly by means of pattern receptors or because the particle has been coated with an opsonin. Target recognition is greatly enhanced when the particle surface is opsonized. A classic example of an opsonin is a complement protein called C3b, which binds to a complement receptor on the phagocyte. Antibodies can act directly as opsonins by binding to an antigen on the surface of the pathogen and to Fc receptors on the phagocyte. A pathogen can be coated with both C3b and antibody, in which case the binding and activation of the phagocyte are increased synergistically (Figure 10.2).

10.4.2 Internalization

The edges of the phagocyte's plasma membrane around the points of attachment extend outward to engulf the particle, finally enclosing it within a vacuole called the phagosome. Actin and other contractile fibrils in the cytoplasm surround the phagosome around its outer edge, pushing it to the interior. The phagosome then fuses with cytoplasmic storage granules and lysosomes to form an enlarged vacuole called a phagolysosome. Inside the phagolysosome,

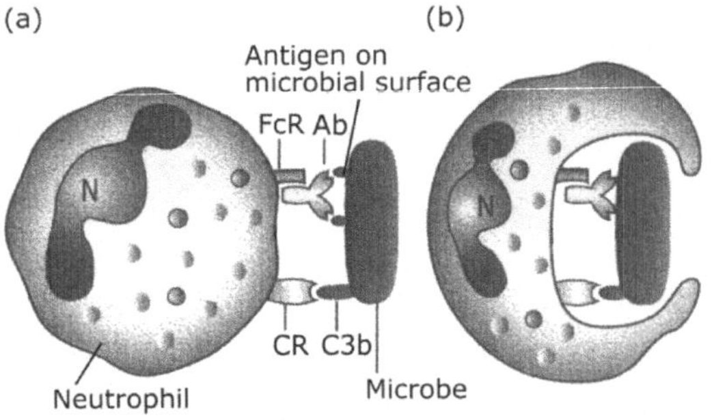

FIGURE 10.2 Phagocytosis of a microbe by a neutrophil. (a) Recognition of a target opsonized by antibody and complement; (b) engulfment of the target.

microorganisms are killed by a variety of mechanisms and subsequently digested by lysozyme and proteases. Finally, the digested contents of the phagolysosome are expelled from the phagocyte by exocytosis.

10.4.3 Killing Mechanisms

Having internalized a pathogenic bacterium, for example, the phagocyte destroys it in a two-stage process. Firstly, the phagocyte is stimulated to generate lethal reactive oxidants; secondly, a large number of enzymes and other bioactive molecules are released from the granules and lysosomes into the phagolysosomes. The lysosomal products are responsible for digesting the bacterium which has been killed by the reactive oxidants.

10.4.4 Self-Protection of Phagocytes

The phagocytes themselves are susceptible to attack by the reactive oxidants that they produce. However, they are able to defend themselves against self-destruction, at least to a limited extent. The antioxidant systems of the phagocytes include superoxide dismutase, which converts superoxide to oxygen and hydrogen peroxide; catalase and glutathione peroxidase, both of which reduce hydrogen peroxide to water; and other antioxidants, such as α-tocopherol and ascorbic acid. These systems protect neutrophils long enough for them to deliver their lethal cocktail of toxins to extracellular and phagocytized pathogens, but after phagocytosis the neutrophils die. Macrophages live much longer than neutrophils and can phagocytose during their whole life span by resynthesis of lysosomal enzymes. The content of ascorbic acid in macrophages is about twice as high as that of neutrophils and monocytes, which may account for their superior protection against self-destruction.

10.5 RECEPTORS AND DETECTORS IN INNATE IMMUNITY

10.5.1 Introduction

The cells of the innate immune system carry an array of receptors that enable them to recognize microbial targets either directly or indirectly. The binding of these receptors by the appropriate ligand activates or inhibits the immune cells. The regulation of immune responses by inhibitory receptors is necessary to avoid damage to host cells or uncontrolled proliferation of activated immune cells.

Phagocytic receptors present at the surfaces of neutrophils, macrophages, and immature DCs mediate the internalization of particles by phagocytosis. Some phagocytic receptors, such as the mannose receptor, recognize and bind specific types of molecules uniquely present on microbial cell surfaces. The term 'pathogen-associated molecular patterns' (PAMPs) can be applied to surface patterns of carbohydrate molecules, and the term 'pattern-recognition receptors' (PRRs) can be applied to the innate immune receptors that recognize them. Recognition of molecular patterns common to entire classes of microorganisms allows a small number of PRRs to detect a multitude of potential microbial pathogens. Other phagocytic receptors recognize and bind protein-coated (opsonized) particles. These so-called opsonin receptors fall into two categories: complement receptors and Fc receptors. Complement receptors bind to particles opsonized by complement components in the innate immune system; Fc receptors bind to particles opsonized by antibodies in the adaptive immune system. A group of structurally unrelated phagocytic receptors known as scavenger receptors can mediate the uptake of components of apoptotic host cells as well as components of pathogens by macrophages.

Toll-like receptors (TLRs) are a family of signaling molecules in diverse pathways that lead to the transcription of genes involved in immunity. Cytosolic bacterial detectors, such as NOD1, NOD2, and NALP3, detect pathogenic bacteria that have invaded the cytosol of host cells and that have escaped detection by TLRs. RIG-like receptors (RLRs) are cytosolic sensors of viral RNA that induce the production of the antiviral type I interferons (IFN-α/β). A large number of immunoglobulin-like receptors with a tyrosine-based inhibitory motif (known collectively as immunoglobulin-like transcripts [ILTs]) inhibit immune cell functions such as phagocytosis by macrophages and killing by NK cells.

10.5.2 The Mannose Receptor

The mannose receptor is a pattern-recognition C-type lectin receptor. It is present on resident macrophages, DCs, and various types of endothelial cells and can recognize a wide range of gram-negative and gram-positive bacteria, yeasts, parasites, and mycobacteria.

In immature DCs, the mannose receptor functions to rapidly internalize glycosylated antigens and deliver them efficiently to endocytic compartments. Here, they can be processed into antigenic peptides that are transported to the cell surface as peptide–MHC class II complexes. In immature DCs, the antigen-bound mannose receptor is endocytosed in clathrin-coated vesicles, whereas nonmannosylated antigens are internalized by macropinocytosis. Mannose receptor-mediated uptake of antigens by immature DCs results in a ~100-fold more efficient presentation to T cells as compared with antigens internalized by macropinocytosis. Unlike Fc receptors, the mannose receptor is not targeted to lysosomes for destruction, but recycles between the plasma membrane and an intracellular early endocytic compartment while releasing its cargo. This recycling allows continuous capture of antigens by a finite number of receptors, leading to early clearance of many pathogens as part of the innate immune response. In addition to its antigen-capturing ability, the mannose receptor is a signal transducer, triggering a variety of responses including secretion of mediators, lysosomal enzyme secretion, cytokine production, and modulation of other cell surface receptors.

The mannose receptor binds ligands carrying terminal mannose, fucose, *N*-acetylglucosamine, and glucose residues in a calcium-dependent manner. This specificity is based on these hexose sugars having equatorially placed hydroxyl groups at carbons C3 and C4. These sugars are commonly present in terminal positions on glycoproteins that decorate the surfaces of microoganisms, but do not usually occur on mammalian glycoproteins. Therefore, the mannose receptor can distinguish between self and non-self.

10.5.3 Toll-Like Receptors

The Toll-like receptors (TLRs) are a family of pattern-recognition receptors that initiate signaling in diverse pathways that lead to inflammatory and immune defense responses. There are at least ten mammalian TLRs, of which TLR1–9 are conserved between the human and mouse. TLRs are expressed by various immune cell types and also by non-immune cells such as epithelial and endothelial cells. Expression is regulated in response to pathogens, a variety of cytokines, and environmental stresses. A variety of microbial-specific proteins, lipoproteins, polysaccharides, and nucleic acids are TLR ligands (see summary in Table 10.5).

TLRs that are involved in the recognition of bacterial and fungal ligands (TLRs 1, 2, 4, 5, and 6) are localized on the cell surface. In contrast, TLRs that recognize bacterial and viral nucleic acids (TLRs 3, 7, 8, and 9) congregate in the endoplasmic reticulum. When bacteria are degraded in phagolysosomes, TLRs are recruited to these intracellular compartments where they are exposed to their ligands. Viruses that enter cells by receptor-mediated endocytosis and are degraded in endosomes liberate the nucleic acids that are ligands for TLRs 3, 7, 8, and 9.

10.6 CYTOKINES AND CHEMOKINES

10.6.1 General

Cytokines are a diverse group of short-term acting signaling molecules that provide a communication network among the various cell types of the immune system. They are proteins whose actions are mediated by their binding to specific receptors on target cells. The various cytokine families include interleukins (ILs), tumor necrosis factors (TNFs), interferons (IFNs), colony-stimulating factors (CSFs), and transforming growth factor-β (TGF-β). Cytokines play a crucial role in the amplification of the immune response because their release from just a few stimulated cells results in the activation of multiple cell types, which are not necessarily located in the immediate area. Cytokines are not

TABLE 10.5
Human Toll-Like Receptors (TLRs), Their Ligands and Immune Cells That Express Them

TLRs	Expression	Ligands
TLR2	Cell surface of monocytes, neutrophils, immature cDCs, epithelial cells	Gram-positive bacterial peptidoglycan, mycobacterial lipoarabinomannan, fungal phospholipomannan
TLR2/TLR1 dimer	Cell surface of most immune cell types	Bacterial triacylated lipoproteins
TLR2/TLR6 dimer	Cell surface of most immune cell types	Bacterial diacylated lipoproteins and lipoteichoic acid, fungal zymosan
TLR3	Intracellular compartments of immature cDCs	Viral dsRNA
TLR4	Cell surface of monocytes, macrophages, gram-negative bacterial lipopolysaccharide,	Immature cDCs, epithelial cells, some viral proteins
TLR5	Immature cDCs, NK cells, T cells, epithelial cells	Bacterial flagellins
TLR7	Intracellular compartments of monocytes, immature cDCs and pDCs, B cells	Viral ssRNA
TLR8	Intracellular compartments of monocytes, immature cDCs, NK cells, B cells, T cells	Viral ssRNA
TLR9	Intracellular compartments of neutrophils, macrophages, immature pDCs, NK cells, B cells, epithelial cells	Unmethylated viral and bacterial CpG DNA

Abbreviations: dsRNA, double stranded RNA; ssRNA, single stranded RNA; cDC, conventional dendritic cell; pDC, plasmacytoid dendritic cell; NK natural killer cell.

stored pre-formed within their cells of origin; rather, they are synthesized by a cell in response to a specific stimulus and immediately secreted. Cytokines usually act locally in a paracrine fashion, but they can also act in an autocrine fashion or systemically in an endocrine fashion. A given cytokine can act on several different cell types (pleiotropy) and several different cytokines can act on the same cell type to elicit a particular response. This latter feature (redundancy) is due to several cytokines sharing a common multi-chain receptor. Each cytokine can affect the synthesis and activity of another in a stimulatory or inhibitory manner. An immune response can be fine-tuned not only by varying the type and amount of cytokine secreted, but also by regulating the density and affinity of cytokine receptors.

Cytokines exhibit a great diversity of stimulatory or inhibitory effects upon cells, affecting development, survival or death, proliferation, differentiation, activation, and chemotaxis. Table 10.6 lists selected effects of cytokines.

- Interleukins are produced largely by helper T cells and many are involved in directing cells of the immune system to proliferate and differentiate. Approximately 30 interleukins have been characterized (IL-1, IL-2, IL-3, etc. up to at least IL-33). IL-1 exists in two forms, IL-1α and IL-1β. Both bind with high affinity to the same receptors and exert parallel pro-inflammatory responses. They differ, however, in their synthesis pathways and their distribution. Whereas newly synthesized IL-1α is active, IL-1β is synthesized as an inactive precursor that becomes active upon proteolytic cleavage. In humans, IL-1α remains within the cells or anchored to the surface of cells such as macrophages, whereas IL-1β is preferentially released as a free hormone-like mediator. As a product of macrophages, IL-1 plays an important role in the activation of T cells. It both augments the production of IL-2 and induces expression of the IL-2 receptor on T cells activated by antigen via the T cell receptor. These effects permit antigen-stimulated T cells to rapidly expand in numbers. IL-9 acts indirectly via promotion of IL-5 and IL-13 production. IL-18 synergizes with IL-12 to stimulate the secretion of IFN-γ.
- The TNF family comprises at least 18 members, which exhibit 25–30% sequence homology and individually participate in a wide variety of immune responses. TNF family members are synthesized as membrane proteins. Some members, including TNF itself, are cleaved from the membrane and released as soluble regulators. Other members are retained on the plasma membrane and engage with TNF receptors to mediate cell–cell contact signaling. There are two subsets of TNF, namely TNF-α, which is frequently referred to as TNF (no Greek letter suffix), and TNF-β, which is also known as lymphotoxin-α.
- Two functionally overlapping but distinct types of interferon, designated type I and type II, are well known. Type I interferons comprise multiple variants of IFN-α, a single IFN-β, and a lesser-known IFN-ω (omega); type II is represented solely by IFN-γ. There is also a family of type III interferons with similar functions to the type I family. All interferons function primarily in defense against viruses. When a host cell is infected with a virus, it secretes interferon, which then binds to receptors on uninfected neighboring cells. The signaling cascade induces IFN-stimulated genes (ISGs) whose protein products inhibit viral replication by targeting multiple points in the viral life cycle, including the degradation of viral RNA. In addition, IFN-γ activates macrophages and influences naïve CD4$^+$ T cell differentiation toward a Th1 phenotype.
- Colony-stimulating factors direct the maturation of hematopoietic stem cells in the bone marrow.
- TGF-β was discovered as a tumor product that promoted the survival of tumor cells *in vitro*. Mammals express three isoforms of TGF-β, namely TGF-β1, TGF-β2, and TGF-β3, encoded by unique genes located on different chromosomes. TGF-β1, the main isoform, is a homodimeric protein that is synthesized and secreted by CD4$^+$ regulatory T cells, activated macrophages, and several other cell types. TGF-β has anti-inflammatory properties and can repress the production of pro-inflammatory cytokines by many cell types. In the presence of pro-inflammatory cytokines, such as IL-1 and IL-6, TGF mediates pro-inflammatory actions. TGF-β is also an immunosuppressive cytokine: it inhibits proliferation of B cells and T cells, and it blocks differentiation of naïve T cells into effector helper T cells.

10.6.2 Pro-Inflammatory Cytokines

Several cytokines produced mainly by activated macrophages mediate inflammation. The three major pro-inflammatory cytokines, IL-1, IL-6, and TNF-α, induce the synthesis of hepatic acute-phase proteins. IL-1 and TNF-α induce the synthesis of IL-6, activate endothelial cells, stimulate chemokine production, and increase neutrophil production by the bone marrow. IL-12 and IL-18 stimulate production of the macrophage-activating cytokine IFN-γ by NK cells and T cells. The production of pro-inflammatory cytokines is inhibited by IL-4, IL-10, and TGF-β, which makes them anti-inflammatory cytokines.

10.6.3 Cytokines and Cell Signaling

Many cytokines are ligands for receptors that have no intrinsic kinase activity, but are permanently associated with cytosolic tyrosine kinases known as Jaks. There are

TABLE 10.6
Principal Immunological Effects of Selected Cytokines

Cytokine	Cellular Source	Effects
Interleukins		
IL-1α and -β	Mono, Macro, cDC, Endo, Fibro, keratinocytes, NK, B	**Pro-inflammatory**: Induces the synthesis of adhesion molecules on vascular endothelium.
		Induces the synthesis of hepatic acute-phase proteins.
		Augments the production of IL-2 and induces the expression of the IL-2 receptor on activated T cells
IL-2	Th1, cDC	Stimulates the proliferation of newly activated T cells *in vitro*.
		Promotes the survival of effector T cells *in vivo*.
		Facilitates differentiation of $CD4^+$ T cells to Th1 and Th2 cells.
		Stimulates activation of $CD8^+$ T cells.
		Expands $CD8^+$ memory T cells and NK cells.
		Stimulates B cell proliferation and differentiation into plasma cells.
		Maintains Treg cells and expands them at high doses.
		Facilitates TGF-β-dependent differentiation of naïve T cells to Treg cells.
		Inhibits TGF-β/IL-6-dependent differentiation of naïve T cells to Th17 cells.
		Sensitizes activated T cells to undergo apoptosis or activation-induced cell death
IL-3	BM stroma, MC, NK, Th1, Th2	Stimulates mast cell proliferation.
IL-4	MC, Baso, Eosin, NK, NKT, Th2, Tc2, ILC2, B	**Anti-inflammatory**: Inhibits the production of pro-inflammatory cytokines and adhesion molecules.
		Induces differentiation of naïve $CD4^+$ T cells into Th2 cells.
		Promotes the proliferation of T cells and mast cells.
		Inhibits IL-2-dependent proliferation and differentiation of B cells.
		Acts with IL-13 to direct antibody class switching to IgE and IgG1.
		Acts with IL-13 to recruit eosinophils.
		Acts with IL-13 to induce an alternative pathway of macrophage activation that enhances tissue repair.
IL-5	MC, Th2, ILC2	Induces proliferation of eosinophils and activated B cells.
		Directs antibody class switching to IgA.
IL-6	Mono, Macro, cDC, pDC, Fibro, Th2, ILC2, BM stroma, MC, Endo	**Pro-inflammatory**: Induces the synthesis of hepatic acute-phase proteins.
		Polarizes naïve $CD4^+$ T cells to effector Th2 cells by inducing the initial production of endogenous IL-4.
		Antagonizes the IL-12-mediated differentiation of Th1 cells.
		Stimulates B cell proliferation and differentiation of B cells into plasma cells.
		Stimulates T cell proliferation by activating IL-2 synthesis.
IL-7	BM and thymic stromal cells	Essential cytokine for survival of naïve and memory T cells.
IL-8	Mono, Macro, Endo	Induces chemotaxis and degranulation of neutrophils.
IL-9	Th2, ILC2	Promotes IL-5 and IL-13 production.
IL-10	Mono, Macro, cDC, Th2, Tc2, B	**Anti-inflammatory**: Inhibits the production of pro-inflammatory cytokines.
		Immunosuppressive: Down-regulates MHC class II and co-stimulatory molecules in immature dendritic cells and reduces release of IL-1β, IL-6, IL-12, and TNF-α.
		Prevents secretion of IL-2 by Th1 cells.

(Continued)

TABLE 10.6 (CONTINUED)
Principal Immunological Effects of Selected Cytokines

Cytokine	Cellular Source	Effects
		Negatively regulates IFN-γ production from T cells and NK cells.
		Reduces the Th1 response by inhibiting the expression of IL-12 by antigen-presenting cells.
IL-12	Macro, cDC, pDC, B	**Pro-inflammatory**: Stimulates IFN-γ production by NK cells and T cells.
		Enhances cytotoxic activity of NK cells and CD8⁺ T cell. Induces differentiation of Th1 cells and inhibits differentiation of Th2 cells.
IL-13	MC, Baso, Eosin, NKT, Th2, ILC2	**Anti-inflammatory**: Inhibits the release of pro-inflammatory cytokines and the synthesis of adhesion molecules.
		Promotes the differentiation of naïve CD4⁺ T cells into Th2 cells.
		Acts with IL-4 to direct antibody class switching to IgE and IgGf1.
		Acts with IL-4 to recruit eosinophils.
		Acts with IL-4 to induce an alternative pathway of macrophage activation that enhances tissue repair.
		Stimulates mucus production by goblet cells.
IL-15	Mono, Macro, cDC	Binds to a common IL-2Rβ chain and has multiple IL-2-like activities.
		Acts with IL-7 to increase the homeostatic proliferation and survival of resting memory T cells.
		Stimulates proliferation and cytotoxicity of NK cells and CD8⁺ T cells.
IL-17	Th17 cell, ILC3	**Pro-inflammatory**: Stimulates production of IL-1β, IL-6, IL-8, and TNF-α by many cell types.
		Stimulates the production of antimicrobial peptides by epithelial cells.
		Stimulates endothelial cells to produce chemokines that recruit neutrophils.
IL-18	Mono, Macro, cDC, keratinocytes, Kupffer cells, chondrocytes and other cell types	**Pro-inflammatory:** Enhances the functions of NK cells.
		Induces both Th1 and Th2 cytokines depending on the stimulus.
IL-21	Th2, Th17, T_{FH}	Induces proliferation of B cells and their differentiation into plasma cells
		Stimulates development of Th17 and T_{FH} cells.
IL-22	Th17, ILC3	**Pro-inflammatory**: Induces the production of IL-1α and inhibits the production of IL-4.
		Induces the synthesis of hepatic acute phase proteins.
		Induces the production of antimicrobial peptides by epithelial cells.
		Promotes tissue repair at barrier surfaces.
IL-23	Macro, cDC	**Pro-inflammatory**: Promotes proliferation and maintenance of Th17 cells.
		Stimulates proliferation of CD4⁺ memory T cells.
		Induces IL-22.
Tumour necrosis factors		
TNF-α	Mono, Macro, MC, keratinocytes, cDC, pDC, NK, B, Th1, Th2, Tc	**Pro-inflammatory**: Induces the synthesis of hepatic acute-phase proteins.
		Stimulates the phagocytic activity of macrophages.
		Signals caspase-dependent apoptosis.
TNF-β (lymphotoxin-α)	Th1, Tc1, B	Enhances phagocytosis by neutrophils and macrophages.
Interferons		
IFN-α and –β (Type I IFN)	pDC (IFN-α/β), Macro (IFN-α), Fibro, epithelial cells (IFN-β)	**Anti-inflammatory (IFN-α)**: Inhibits the production of pro-inflammatory cytokines.
		Anti-viral activity: Inhibits viral replication.
		Induces the differentiation of monocytes to cDC.
IFN-γ	NK, NKT, Th1, Tc1, ILC1	**Pro-inflammatory**: Up-regulates expression of inflammatory cytokines on macrophages.
		Anti-viral activity: Inhibits viral replication.

(Continued)

TABLE 10.6 (CONTINUED)
Principal Immunological Effects of Selected Cytokines

Cytokine	Cellular Source	Effects
		Inhibits the generation of Th2 cells while simultaneously enhancing the development of Th1 cells.
Colony stimulating factors		
Granulocyte-CSF (G-CSF)	Macro, Fibro, Endo	Stimulates proliferation and maturation of neutrophil precursors in bone marrow.
Macrophage-CSF (M-CSF)	Macro, Fibro, Endo, bone marrow cells	Stimulates proliferation and maturation of monocyte/macrophage precursors in bone marrow
Granulocyte-macrophage-CSF (GM-CSF)	Macro, Fibro, Endo MC, NK, Th1, Th2	Stimulates development of granulocytes, monocytes/macrphages and conventinal dendritic cells in bone marrow.
		Activates macrophages.
Transforming growth factor		
TGF-β	Macro, MC, B, Tc, NK, epithelial cells, stromal cells, cancerous cells	**Anti-inflammatory**: Inhibits the production of pro-inflammatory cytokines.
		Pro-inflammatory: Potent chemoattractant for monocytes, macrophages and neutrophils.
		Induces hepatocytes to produce acute-phase proteins..
		Immunosuppressive: Inhibits macrophage activation.
		Inhibits the expression and production of IL-2 by T cells.
		Inhibits IL-2-driven proliferation of B and T cells.
		Inhibits the expression of MHC class II molecules.
		Inhibits the T cell stimulatory function of antigen-presenting cells.
		Blocks differentiation of naïve $CD4^+$ T cells into either the Th1 or Th2 subset.
		Blocks differentiation of naïve $CD8^+$ T cells into cytotoxic T cells.
		Promotes the development of Th3 regulatory T cells.
		Antibody class switching: Promotes switching to IgA.

Abbreviations: BM, bone marrow; Endo, endothelial cell; Fibro, fibroblast; MC, mast cell; Baso, basophil; Eosin, eosinophil; Mono, monocyte; Macro, macrophage; cDC, conventional dendritic cell; pDC, plasmacytoid dendritic cell; B, B cell; Th1, Th2, and Th17, helper T cells; T_{FH}, follicular helper T cell; Treg cells, $Foxp3^+$ regulatory T cells; Tc, cytotoxic T cell; NK, natural killer cell; NKT, natural killer T cell; ILC1/2/3, groups 1, 2, and 3 innate lymphoid cells.

four known members of the Jak family: Jak1, Jak2, Jak3, and Tyk2. Each cytokine activates a defined set of Jaks by virtue of specific interactions between a particular receptor subunit and a Jak molecule. Jaks in turn can activate a family of cytosolic transcription factors known as STATs, giving rise to the Jak–STAT pathway of signal transduction. Cytokines that activate the Jak–STAT pathway include many interleukins, such as IL-2, -3, -4, -5, -6, -7, -9, -10, -11, -12, -13, -15, and -21, as well as the interferons IFN-α, -β, and -γ and the colony-stimulating factor GM-CSF.

Receptors for a few cytokines are composed of a single type of chain that dimerizes on ligand binding. Most receptors, however, are composed of two different chains, one of which uniquely binds a single cytokine, while the other is common to multiple receptors.

10.6.3.1 Interleukin Signaling

The interleukins IL-2, IL-4, IL-7, IL-9, IL-15, and IL-21 are known as the γ_c family of cytokines because they bind to receptors having a common γ chain (γ_c chain). The Jaks and STATS activated by the γ_c cytokines are shown in Table 10.7. The γ_c chain (CD132) associates exclusively with Jak3, while the cytokine-specific receptor chain associates with Jak1. The IL-2 and IL-15 receptors are unique in that they share the IL-2Rβ chain (CD122), specificity being provided by the IL-2Rα chain or IL-15Rα chain. Unlike other γ_c cytokines, IL-15 is a cell surface-associated cytokine, being tightly bound to the IL-15Rα chain. The γ_c cytokines function primarily in T cell development, proliferation, and survival.

IL-8 transmits signals through receptors coupled to heterotrimeric G proteins. There are two distinct types of IL-8 receptor, designated α and β. The IL-8 receptors can interact with different members of the Gα subunits, including $G_i\alpha$ and $G_q\alpha$, thereby activating distinct signaling pathways in a cell-type-specific manner. M-CSF signals via a receptor with intrinsic tyrosine kinase activity. The binding of TNF-α to TNF-α receptor 1 (TNFR-1) triggers the death-receptor pathway that leads to apoptotic cell death.

TABLE 10.7
Jaks and STATs Activated by γ_c Cytokines

Cytokine	Receptor Chains	Jak	STAT
IL-2	IL-2Rα, IL-2Rβ, γ_c	Jak1, Jak3	**STAT5a-b**, STAT1, STAT3
IL-4	IL-4R, γ_c	Jak1, Jak3	**STAT6**, STAT5a-b
IL-7	IL-7R, γ_c	Jak1, Jak3	**STAT5a-b**, STAT1, STAT3
IL-9	IL-9R, γ_c	Jak1, Jak3	**STAT5a-b**
IL-15	IL-15Rα, IL-2Rβ, γ_c	Jak1, Jak3	**STAT5a-b**
IL-21	IL-21R, γ_c	Jak1, Jak3	**STAT3**, STAT1, STAT5a-b

The main signal transducer and activator of transcription (STAT) proteins that are activated by these cytokine receptors are shown in bold.

10.6.3.2 TGF-β Signaling

TGF-β transmits signals through a unique pathway. A functional receptor complex for TGF-β is composed of two essential subunits, type I (TβRI) and type II (TβRII), both of which are dimeric transmembrane proteins with serine/threonine kinases as part of their cytosolic domains. TβRII is a constitutively active kinase that phosphorylates itself in the absence of TGF-β, whereas TβRI is kept inactive by a wedge-shaped GS region pressed against the kinase domain. The immediate downstream transducers of TGF-β signaling are certain members of the SMAD family of proteins. SMADs 2 and 3 are collectively referred to as receptor-phosphorylated SMAD (R-SMAD); SMAD4 is a co-SMAD; and SMADs 6 and 7 are inhibitory SMADs. In the basal state, R-SMAD is retained in the cytoplasm by proteins such as SARA. SARA not only tethers SMAD2, but also occludes a nuclear-localization signal within SMAD2.

In the absence of TGF-β, TβRI and TβRII dimers are dissociated. TGF-β binds directly to TβRII and this stimulates the recruitment of TβRI to form a functional receptor complex. In some cell types, TGF-β binds first to a membrane-anchored proteoglycan known as β-glycan, which enhances receptor formation. TβRII phosphorylates the inhibitory GS domain within TβRI, causing a conformational change that exposes the kinase domain. Activated TβRI then phosphorylates R-SMAD, which allows its release from cytoplasmic anchoring proteins and unmasks its nuclear-localization signal. Two molecules of phosphorylated R-SMAD interact with one molecule of unphosphorylated SMAD4 and with an importin-β molecule. The resulting SMAD complex is now free to enter the nucleus by transport through the nuclear pore complex and exert its transcriptional activation function. The SMADs are linked to the general transcriptional machinery on the DNA through direct contact with the co-activator CBP/p300. Nucleocytoplasmic shuttling of the SMAD proteins in the unphosphorylated state also occurs. One mechanism for switching off the TGF-β signal involves SMAD ubiquitination regulatory factors (SMURFS), which initiate proteasome-mediated degradation of the activated TGF-β receptor and SMADs that have entered the nucleus. SMAD6 and SMAD7 inhibit TGF-β signaling by blocking R-SMAD binding to TβRI or R-SMAD oligomerization with SMAD4. TGF-β signaling induces expression of SMAD7, which provides a TGF-β-induced negative feedback loop.

10.6.4 Cytokines and Hematopoiesis

During the innate immune response to an infection, the number of circulating leukocytes can increase more than 10-fold. This is due in part to the stimulatory action of certain cytokines on hematopoietic stem cells and progenitor cells. These particular cytokines are also necessary to maintain the viability of progenitor cells, which would otherwise die by apoptosis. Cytokines involved in hematopoiesis can be divided into two categories, according to their target cell specificity. One group comprises the factors whose activity is restricted to the differentiation of a particular cell type. Examples are macrophage-CSF (M-CSF) for monocytes/macrophages, granulocyte-CSF (G-CSF) for neutrophils, and IL-5 (or Eo-CSF) for eosinophils. The second group has a wider spectrum of activities, and comprises IL-3 and granulocyte-macrophage-CSF (GM-CSF). The chief function of IL-3 is to stimulate early hematopoietic progenitor cells, rendering them responsive to more specific cytokines for differentiation. GM-CSF promotes the development of granulocytes and monocytes/macrophages. A combination of GM-CSF and IL-4 promotes the differentiation of monocytes into functionally immature DCs *in vitro*. Hematopoiesis can be regulated negatively by a heterogeneous set of cytokines, such as interferons, TNF-α, and TGF-β.

10.6.5 Chemokines

Chemokines are chemoattractant molecules that direct the migration of immune-system cells to target sites by chemotaxis. The cells migrate up the concentration gradient established by the decreasing concentration of chemokine with increasing distance from its source. Different chemokines attract different types of cells. Chemokines are secreted proteins that bind to G-protein-coupled receptors on the surface of cells. They can be divided into two categories: inflammatory and homeostatic (or lymphoid). Inflammatory chemokines attract circulating immune-system cells to sites of inflammation and are strongly up-regulated by inflammatory stimuli. Constitutively produced homeostatic chemokines are involved in cell migration to and within secondary lymphoid tissues and serve to maintain the normal population of immune-system cells in these tissues.

There are two main families of chemokines, CC and CXC, based on the relative positions of two highly conserved cysteine residues in the amino-terminal region (Zlotnik & Yoshie, 2000). Each of these cysteines forms a disulfide bond with a cysteine located toward the carboxyl terminus. In the CC family, the two conserved cysteines are adjacent, whereas in the CXC family they have an amino acid between them. Chemokines are ligands for chemokine receptors, therefore they include an L in their nomenclature (e.g. CCL1). Around 40 CCL/CXL chemokines have been identified. In addition, one chemokine, CX3CL1, has three residues between the two conserved cysteines, and two chemokines (XCL1 and XCL2) lack the first conserved cysteine. Most chemokine receptors are capable of binding one of several related chemokines; for example, CXC receptor 2 (CXCR2) is capable of binding seven different ligands within the CXCL family. Also, a few chemokines can bind to more than one receptor; for example, the inflammatory chemokine CCL5 can bind to CCR1, CCR3, and CCR5. A homeostatic chemokine that attracts T cells to lymphoid tissue follicles is CXCL13, whose receptor is CXCR5. Chemokines probably act in an overlapping, sequential fashion to control lymphocyte navigation.

10.7 INFLAMMATION

10.7.1 Introduction

Inflammation is the body's physiological response to tissue damage caused by infection or trauma, manifested as redness, heat, swelling, and pain. The local inflammatory response is the recruitment of immune cells from the blood to the site of infection or injury. There is also a systemic inflammatory response in which the liver is stimulated to synthesize and secrete acute-phase proteins into the bloodstream. The intensity of inflammation depends on the balance between the activities of pro-inflammatory cytokines and anti-inflammatory cytokines. IL-1 (α and β forms), IL-6, and TNF-α are pro-inflammatory and their release can be amplified by IL-2, IL-3, IFN-γ, and GM-CSF. The anti-inflammatory cytokines, IL-4, IL-10, IL-13, TGF-β, and IFN-α modulate the inflammatory response. Inflammation is normally acute, taking minutes to hours to develop and lasting for several days. Acute inflammation is self-limiting, because the production of pro-inflammatory cytokines is followed closely by anti-inflammatory cytokines. If the infection is not eliminated or the tissue injury is prolonged, chronic inflammation will take over and the ensuing tissue damage will manifest as a disease of some kind.

10.7.2 Summary of the Acute Inflammatory Response

The reactions that comprise the acute inflammatory response are designed to destroy the infective agent, prevent ongoing tissue damage, and activate the repair processes. The main events that facilitate the response are as follows:

- Dilation of the arterioles (vasodilation) to increase blood flow to the affected area.
- Localized production of chemokines that attract circulating leukocytes.
- Expression of adhesion molecules on the surfaces of vascular endothelial cells, rendering them able to capture leukocytes.
- Increased vascular permeability, which allows cells and proteins to pass through the endothelium and enter the tissue.
- Elimination of the foreign agent by phagocytosis.
- A rapid programed clearance of recruited inflammatory cells by apoptosis and phagocytosis.
- Tissue repair.

Vasoactive molecules secreted by macrophages and mast cells or present in plasma are listed in Table 10.8.

10.7.3 Migration of Leukocytes to Inflammatory Sites

The microbial invaders at the site of infection activate tissue cells to secrete cytokines that induce migration of circulating leukocytes to the infection site. Tissue macrophages and mast cells are the major sources of these cytokines, although endothelial and epithelial cells can also produce IL-1 and IL-6. Neutrophils are the first cells to arrive, followed by monocytes, which differentiate into macrophages. Once activated, macrophages are the main source of growth factors and cytokines, which profoundly affect endothelial, epithelial, and mesenchymal cells in the local

TABLE 10.8
Vasoactive Molecules Involved in the Inflammatory Response

Molecule	Main Source	Function
Histamine	Mast cells, platelets	Causes vasodilation and increased permeability; up-regulates endothelial adhesion molecules.
Bradykinin	Plasma	Causes vasodilation and increased permeability.
Prostaglandin E_2	Neutrophils, monocytes, macrophages	Causes vasodilation; potentiates other mediators in increasing vascular permeability.
Leukotriene B_4	Macrophages	Increases vascular permeability; up-regulates adhesion molecules on neutrophils; chemotactic for neutrophils; causes mast cells to release mediators.
Complement component C5a	Plasma	Causes vasodilation and increased permeability; causes mast cells to release mediators; chemotactic for monocytes and neutrophils.

microenvironment. Mast cells are also important in acute inflammation, owing to their release of stored and newly synthesized inflammatory mediators.

In order to reach the inflammatory site, the leukocytes must cross the vascular endothelium that forms the interface between tissue and blood. This migration of leukocytes from the bloodstream into tissue is called extravasation, which is depicted in Figure 10.3.

10.7.4 Cell–Cell Adhesion

Extravasation depends upon interactions between adhesion molecules on leukocytes and those on endothelial cells, which allow the leukocytes to stick to the endothelium. Some adhesion molecules are expressed permanently on the cell surface (constitutive expression), while others are induced by signaling molecules such as cytokines. By altering the expression or activity of adhesion molecules on leukocytes and endothelial cells, it is possible to control the entry of the leukocytes into the tissue. The leukocytes are attracted toward the specific site in the tissue by traveling along the concentration gradient established by a chemokine. Cell movement in this manner is termed chemotaxis.

Broadly speaking, there are three main families of cell adhesion molecules concerned with immunity: selectins, integrins, and the immunoglobulin superfamily. The designations, patterns of expression, and ligands of representative adhesion molecules are summarized in Table 10.9.

Selectins interact mainly with carbohydrate moieties of glycoproteins through their N-terminal lectin domain. There are three well-described selectins: L-selectin, E-selectin, and P-selectin, the letters standing for leukocyte, endothelium, and platelet, respectively, to indicate the predominant cell type expressing each selectin. E-selectin and P-selectin bind PSGL-1, a dimeric sialomucin, which is constitutively expressed by neutrophils. L-selectin binds GlyCAM-1, MadCAM-1, and CD34 on endothelial cells.

Integrins are a family of 24 different heterodimers composed of α and β polypeptide chains linked covalently and are classified according to the structure of the β chain. For example, the β_1 family includes the VLA integrins and the β_2 family includes LFA-1 and Mac-1. A distinctive feature of integrins is their ability to reversibly and rapidly modulate their adhesiveness in response to activating signals. This is achieved through changes in the conformation of the α β heterodimers that affect their affinity for ligand.

Cell adhesion molecules of the immunoglobulin superfamily comprise a wide number of molecules that interact among themselves or with integrins and selectins. Included in this superfamily are the intercellular adhesion molecules-1, -2, and -3 (ICAM-1, -2, and -3) and the vascular cell adhesion molecule-1 (VCAM-1). ICAM-1 is expressed on various immune cells as well as on endothelial cells, mainly after their activation. ICAM-2 is expressed by resting leukocytes and non-activated endothelial cells, and its expression is not increased by cell activation. ICAM-3 has a more restricted pattern of expression and is detected solely on cells of bone marrow origin. ICAM-3 plays a key role in leukocyte–leukocyte interactions, whereas ICAM-1 and ICAM-2 are involved in these interactions as well as in leukocyte–endothelial cell binding. VCAM-1 is mainly expressed by endothelial cells activated with different stimuli; it is also expressed by some bone marrow–derived cells, including macrophages and DCs.

10.7.4.1 Neutrophil–Endothelial Cell Adhesion

Neutrophils constitutively express L-selectin and endothelial cells constitutively express P-selectin, whilst endothelial E-selectin expression requires induction and protein synthesis. The affinity of the P-selectin–ligand interaction is high and the association/dissociation rates are fast, even under static conditions. Neutrophil activation by numerous signaling molecules results in up-regulation of the integrins LFA-1 and Mac-1 already on the neutrophil surface. On the endothelium, ICAM-2 is constitutively expressed, whereas ICAM-1 is induced by various activating molecules. The strong binding of integrins on the neutrophil to ICAMs on the endothelial cell is a key factor in the migration of neutrophils to inflammatory sites.

10.7.4.2 Eosinophil–Endothelial Cell Adhesion

Eosinophils alone appear to express VLA-1, which binds VCAM-1 on the endothelium. This interaction may explain the selective migration of eosinophils over neutrophils in certain pathological conditions.

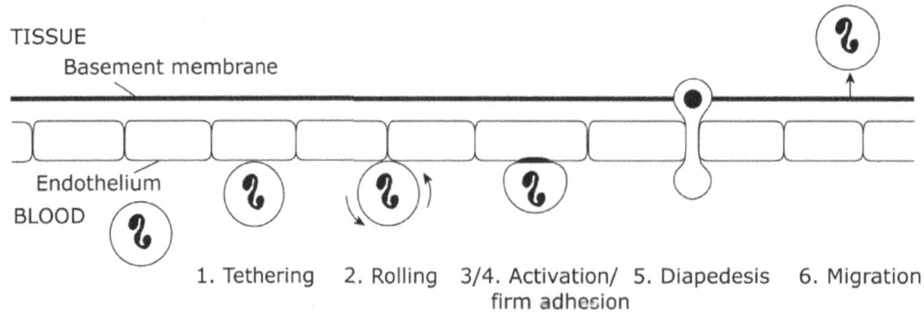

FIGURE 10.3 Steps in the migration of circulating leukocytes to inflammatory sites.

TABLE 10.9
Designations and Characteristics of Major Cell Adhesion Molecules

Usual Designation	CD Designation	Pattern of Expression	Main Ligands and their Sites
Selectins			
L-selectin	CD62L	All types of leukocyte	GlyCAM-1, MadCAM-1, and CD34 on endothelial cells
E-selectin (ELAM-1)	CD62E	Activated endothelial cells	ESL-1 and PSGL-1 on neutrophil cells
P-selectin	CD62P	Activated platelets and	PSGL-1 on neutrophils, endothelial cells
Integrins			
$\alpha_L\beta_2$ (LFA-1)	CD11a/CD18	All types of leukocyte	ICAM-1 and -2
$\alpha_M\beta_2$ (Mac-1, CR-3)	CD11b/CD18	Monocytes, macrophages, neutrophils	ICAM-1
$\alpha_E\beta_7$	CD103/–	T cells	E-cadherin on epithelial cells
$\alpha_1\beta_1$ (VLA-1)	CD49a/CD29	Eosinophils	VCAM-1
$\alpha_2\beta_1$ (VLA-2)	CD49b/CD29	T cells	Collagen type I
$\alpha_4\beta_1$ (VLA-4)	CD49d/CD29	T cells, monocytes	VCAM-1
$\alpha_4\beta_7$ (LPAM-1)	CD49d/–	B cells, T cells	MadCAM-1, VCAM-1
Immunoglobulin Superfamily			
ICAM-1	CD54	Many cell types, including endothelial cells, macrophages, and lymphocytes, mainly after their activation	LFA-1, Mac-1
ICAM-2	CD102	Resting leukocytes, non-activated endothelial cells	LFA-1
ICAM-3	CD50	Leukocytes	LFA-1
VCAM-1	CD106	Activated endothelial cells, macrophages, dendritic cells	VLA-1, VLA-4, LPAM-1

–, no CD designation yet assigned.

10.7.5 Activation of Endothelial Cells

Inflammatory mediators activate resting endothelial cells to move P-selectin from intracellular storage vesicles to the cell surface where it interacts with the PSGL-1 ligand on neutrophils. Cytokines induce the endothelial synthesis of E-selectin, which interacts with its ligands, ESL-1 and PSGL-1, also on neutrophils. These interactions are responsible for the tethering of circulating neutrophils, which then roll along the blood vessel wall. The tethering and rolling of other types of leukocyte, such as eosinophils, monocytes, and lymphocytes, are mediated by the VLA-1, VLA-4, and LPAM-1 integrins through their interaction with VCAM-1 and MadCAM-1. L-selectin may also participate in leukocyte tethering under inflammatory conditions. Cytokines stimulate the endothelial cells to synthesize ICAM-1 and VCAM-1, both expressed on the cell surface along with ICAM-2, which is expressed constitutively. Integrins (mainly LFA-1, Mac-1, and VLA-4) on the surface of leukocytes are activated to their high-affinity forms. The increased affinity/avidity of the integrins for their ICAM-1, ICAM-2, and VCAM-1 ligands on activated endothelial cells leads to cessation of rolling and to the firm adhesion of leukocytes to the luminal surface of the endothelium. Platelet-activating factor (PAF) and other local activation signals triggered through rolling lead to activation of the leukocytes, thereby increasing their responsiveness to chemotactic agents such as C5a and leukotriene B_4. The leukocytes squeeze between adjacent endothelial cells (the process of diapedesis), secrete enzymes to digest the vascular basement membrane, and follow the chemotactic gradient to the inflammatory site in the underlying tissue. Cadherins (adhesion molecules) in endothelial intermediate junctions and desmosomes participate in the diapedesis.

10.7.6 The Systemic Inflammatory Response

In a vigorous immune response, levels of pro-inflammatory cytokines rise in the blood to the point where they act upon the liver and brain to induce the systemic inflammatory response. The lipopolysaccharide (LPS) on gram-negative bacteria is a potent inducer of cytokines, as are the peptidoglycan and lipoteichoic acid on gram-positive bacteria. In addition, exotoxins behave as superantigens and trigger the release of cytokines from T cells.

IL-1, IL-6, and TNF-α stimulate hepatocytes to synthesize and secrete acute-phase proteins that join in the fight against infection and help restore the inflamed tissues to their normal state after the infection has been cleared. The combined action of these proteins is known as the acute-phase response. The acute-phase proteins can be divided into two categories according to whether their increase is dramatic (100- to 1,000-fold) or moderate (1.5- to fivefold). Those that increase dramatically have the shortest response time (e.g. 6–10 h for C-reactive protein, cf. 48–72 h for C3).

Acute-phase proteins showing a dramatic increase in concentration include:

- **C-reactive protein** activates the classical pathway of complement on binding to phosphorylcholine-containing membranes of certain microorganisms.

- **Mannose-binding protein** activates the lectin pathway of complement on binding to mannose and certain other sugar residues on a wide variety of microorganisms.
- **Serum amyloid A protein** displaces apolipoprotein A-I from high-density lipoprotein (HDL), thereby enhancing the binding of HDL to macrophages concomitant with a decrease in the binding capacity of HDL to hepatocytes. This suggests that the cholesterol within HDL can be redistributed by macrophages and used for tissue repair, rather than being transferred to the liver for excretion.

Acute-phase proteins showing a more modest rise in concentration include:

- **C3**, a complement component, is cleaved to C3a, which activates mast cells, and C3b, which is an opsonin.
- **Factor B** is a factor in the alternative pathway of complement activation.
- **Fibrinogen**, a major blood coagulation protein, is involved in the generation of fibrinopeptides, which are chemotactic for phagocytes.
- **Haptoglobin** and **hemopexin**, as antioxidants, protect against reactive oxygen species.
- **α_1-Protease inhibitor**, **α_1-antitrypsin**, and **α_1-antichymotrypsin** antagonize the activity of proteolytic enzymes; α_1-antichymotrypsin also inhibits the generation of the superoxide anion.
- **Ceruloplasmin** (ferroxidase) is an enzyme which oxidizes iron(II) transferrin to iron(III) transferrin. It removes iron from inflammatory sites and may act as a scavenger of the superoxide anion radical.

IL-1, IL-6, and TNF-α cause fever by raising the temperature set-point in the brain via production of PGE_2. The resultant rise in body temperature enhances various cellular and biochemical processes concerned with immunity. Other effects of cytokines on the brain are drowsiness and anorexia which, by reducing physical activity, conserve energy.

10.7.7 Antimicrobial Peptides

Antimicrobial peptides are produced constitutively by keratinocytes (the major cell type of the epidermis) and epithelial cells lining the respiratory, intestinal, and urinogenital tracts. During inflammation and infection at mucosal surfaces, the peptides are released from neutrophil granules upon stimulation of neutrophils by IL-8. The peptides provide innate immune defense through their microbicidal activity against a wide range of bacteria, fungi, enveloped viruses, and protozoa. Several additional functions of antimicrobial peptides associated with inflammation have been reported, including tissue repair through promotion of fibroblast chemotaxis and inhibition of tissue injury by inhibiting certain proteases.

Antimicrobial peptides carry a net positive charge and have a folded amphipathic tertiary structure, which enables these molecules to bind to microbial surfaces, insert into the membranes, and form ion channels or aqueous pores. The peptides are attracted to bacterial surfaces by electrostatic bonding to the negative charges that exist on the outer envelope of gram-negative bacteria – for example, phosphate groups on LPS – and to the teichoic and lipoteichoic acids on the surface of gram-positive bacteria. The peptides have to traverse the peptidoglycan cell wall before they can interact with the plasma membrane of gram-positive bacteria. The polysaccharide coating of capsulated bacteria is another barrier to overcome. Once peptides have gained access to the plasma membrane, they aggregate and insert into the lipid bilayer so as to form transmembrane pores with the hydrophilic regions of the peptide forming the pore interior. The permeabilization of the plasma membrane destroys the transmembrane potential, affects osmotic regulation, and inhibits respiration before membrane rupture eventually leads to lysis of the bacterial cells.

In humans, the two major families of antimicrobial peptides are defensins and cathelicidins, and these are described in the following.

10.7.7.1 Defensins

Defensins are divided into two subfamilies, α- and β-defensins, which differ structurally in their disulfide bond connections. In humans, α-defensins are produced by neutrophils, NK cells, and intestinal Paneth cells, whereas β-defensins are produced by blood monocytes, alveolar macrophages, and immature DCs, as well as by keratinocytes and epithelial cells lining the trachea-bronchial tree. Defensins are absent in mouse neutrophils. A defensin called cryptdin is expressed in the small intestine of newborn mice before the developmental appearance of Paneth cells. This suggests that defensins contribute to intestinal immunity during postnatal development of the intestinal mucosa.

The α-defensins produced by neutrophils are termed human neutrophil peptides (HNPs). Four HNPs, designated HNP1–4, have been characterized and these accumulate in the azurophilic granules of neutrophils at remarkably high concentrations (>10 mg ml^{-1}). NK cells release α-defensins upon recognition of bacteria through Toll-like receptors. α-Defensins, packaged in granules within Paneth cells, are secreted into the crypts of Lieberkühn by degranulation that is both constitutive and inducible by bacterial DNA (CpG oligonucleotides).

Expression of four characterized human β-defensins, hBD-1–hBD-4, is inducible at the transcriptional level in many tissues. For example, expression of hBD-2 is stimulated in several cell types by 1,2,5-dihydroxyvitamin D_3 (Wang *et al.*, 2004) and hBD-1 and hBD-2 expression by monocytes is increased upon exposure to LPS and/or IFN-γ (Duits *et al.*, 2002). Chertov *et al.* (1996) identified HNP-1 and HNP-2 as T cell chemoattractants released from IL-8-stimulated neutrophils; HNP-3 lacked T cell chemotactic activity. Human neutrophil-derived α-defensins up-regulate antigen-specific immunoglobulin production *in vivo* (Lillard *et al.*, 1999; Tani *et al.*, 2000). These antibody responses are associated with the induction of antigen-specific $CD4^+$ T cell proliferation and secretion of type I (IFN-γ) and type II (IL-5, IL-6, and IL-10) cytokines by Th1 and Th2 helper T cells, respectively. These findings of chemotactic activity and immunoglobulin up-regulation demonstrate that defensins not only engage in innate host defense, but also contribute to adaptive immune responses.

10.7.7.2 Cathelicidins

The human genome contains only one member of the cathelin gene family, namely *FALL39***.** This gene is expressed in squamous epithelia throughout the body and in neutrophils, NK cells, and mast cells. The one human gene product is an 18-kDa polypeptide called human cationic antimicrobial peptide-18 (hCAP-18). This polypeptide is stored in the specific granules of neutrophils and is cleaved extracellularly to produce a C-terminal 5-kDa peptide called LL-37, which is the active sequence.

LL-37 is chemotactic for neutrophils, monocytes, and $CD4^+$, but not $CD8^+$, T cells. LL-37 is also chemotactic for mast cells and stimulates these cells to release histamine and generate prostaglandin D_2. A positive feedback loop is established because recruited neutrophils release additional defensins and cathelicidin to amplify innate immune responses.

10.7.7.3 Bacterial Resistance to Antimicrobial Peptides

In the human gut, pathogenic bacteria from the environment must compete with the commensal bacteria to establish a viable colony. The commensal bacteria are relatively resistant to the action of endogenous antimicrobial peptides and can chronically stimulate epithelial surfaces to express such peptides at levels that kill pathogens. *Shigella* species that cause dysentery can invade the gut mucosa by down-regulating the expression of hBD-1 and LL-37: bacterial DNA is a potential effector of this down-regulation.

10.7.8 Chronic Inflammation

In contrast to the largely vascular changes of acute inflammation, chronic inflammation is characterized by infiltration of damaged tissues by macrophages and lymphocytes, together with tissue destruction and attempts at repair. The affected area is walled off from the surrounding healthy tissue by a ball of cells known as a granuloma. A granuloma is composed largely of macrophages with varying morphology, some forming what are termed 'epithelioid' cells and others fusing to form giant cells. The macrophages phagocytose microbes and cell debris, whilst continuing to secrete cytokines that stimulate the recruitment of monocytes and lymphocytes. Persistent macrophage activation results in continued tissue damage, which underlies a variety of disease processes such as inflammatory bowel diseases.

10.8 ANTIBODIES

10.8.1 Introduction

Antibodies are glycoproteins of the immunoglobulin family synthesized and secreted by plasma cells in response to antigenic stimulation of naïve B cells. Occurring freely in the blood and lymph, antibodies act as flexible adaptors, allowing various mediators of the immune system to recognize specific pathogens and their products and to exert their action. The body can make several million different antibodies, each one able to recognize a specific infectious agent. There may be several different antibodies for a given pathogen, each binding to a different antigen on that pathogen's surface.

10.8.2 Structure

The basic unit of all antibodies is the immunoglobulin monomer. This is depicted in diagrams as a Y-shaped structure consisting of four polypeptide chains – two identical heavy chains and two identical light chains – held together by disulfide bridges (Figure 10.4). A linear stretch of amino acids rich in proline, the hinge region, imparts flexibility to the molecule. The light chains (molecular mass 25 kDa) are one of two types designated κ (kappa) and λ (lambda). These are structurally similar but are coded by different genes; only one type is found in a given antibody molecule. Each of the arms of the Y-shaped molecule constitutes a Fab (fragment with antigen binding) region, which is responsible for binding the antigen. Thus there are two binding sites with the same specificity for antigen. The stem constitutes the Fc (fragment crystalline) region, which mediates the effector functions of the antibody, such as binding complement components or Fc receptors on a variety of cell types.

The domains of the amino termini of the light and heavy chains are called the variable domains (V_L and V_H, respectively), because their amino acid sequences differ markedly from one immunoglobulin to another. Within the

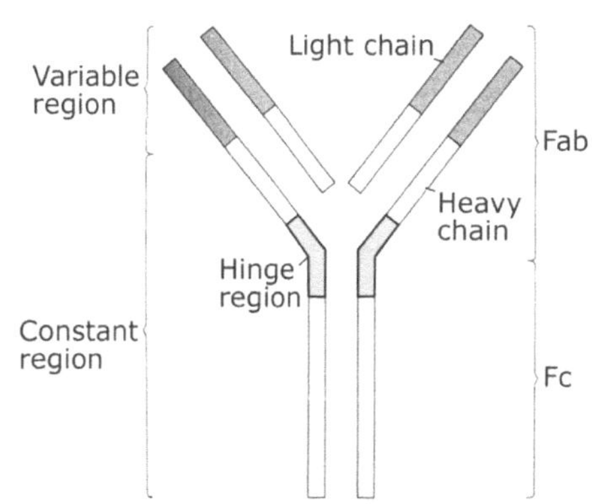

FIGURE 10.4 Basic structure of an immunoglobulin monomer.

variable domain of each chain, the variability of amino acid sequence is restricted to three short segments termed hypervariable regions, each 5–15 amino acids in length. The variable domains of the heavy and light chains are configured in a way that brings the six hypervariable regions together to create the antigen-binding site. The remaining segments of the variable domains are called the framework regions and are essential for maintaining the integrity of the antigen-binding site. The domains that occupy the remainder of the light and heavy chains are termed constant domains (C_L and C_H, respectively), because they vary very little among immunoglobulins. The antigen specificity of an antibody is determined by the variable regions of the immunoglobulin heavy and light chains, V_H and V_L. In contrast, the effector action of the antibody is determined by the constant region of its heavy chain, C_H.

10.8.3 Classification

Antibody molecules have been placed into five classes (isotypes) of immunoglobulin (Ig) based on differences in the structure of their heavy chain regions. These five isotypes are IgA, IgG, IgD, IgE, and IgM, whose heavy chains are designated α, γ, δ. E, and μ, respectively. In their soluble, secreted form, IgD, IgE, and IgG exist as monomers, IgA exists as a dimer, and IgM exists as a pentamer or less commonly as a hexamer. The structures of secretory IgA and IgM are shown in Figure 10.5. In humans, there are four subclasses (subtypes) of IgG (IgG1, IgG2, IgG3 and IgG4), which have different effector functions, and two of IgA (IgA1 and IgA2), which appear to have identical functions. Mice also have four subclasses of IgG (IgG1, IgG2a, IgG2b and IgG3), but these do not correspond functionally to the human IgG subclasses. Certain strains of mice lack the gene for IgG2a, but synthesize a related subclass called IgG2c. The structural differences between antibody classes are most pronounced in the Fc regions and this leads to the triggering of different effector functions on binding to antigen (Table 10.10). For example, IgM recognition of antigen might lead to complement activation, whereas IgE recognition (possibly of the same antigen) might lead to mast cell degranulation. The specificity of antibody for antigen is not related to its class and different classes of antibody can have the same antigen specificity.

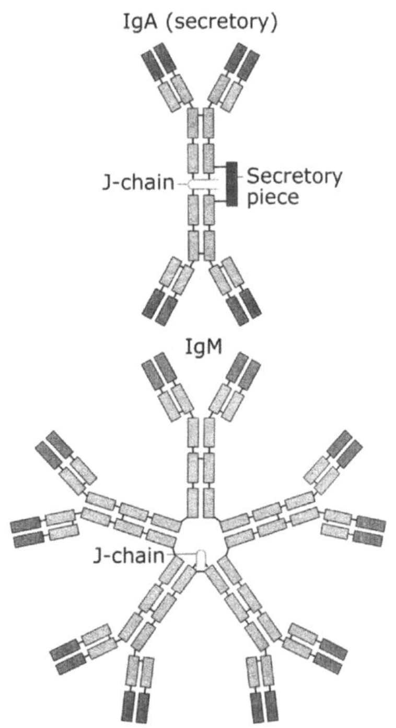

FIGURE 10.5 Structures of secretory forms of IgA and IgM. The J chain in pentameric IgM is an additional peptide that holds the monomeric subunits together. The secretory piece in the IgA dimer may protect IgA from proteolytic cleavage.

10.8.4 Antibody Isotype Switching

Naïve B cells (B cells that have never encountered foreign antigen) continuously produce IgM without any evidence of infection, so are ready to provide immediate defense against a variety of pathogens. IgM has a relatively low affinity for antigen, and if an infection develops, the appropriate higher-affinity IgG, IgA, or IgE isotype takes over. This transition is achieved in part by some of the progeny of activated IgM-expressing B cells undergoing isotype switching. This process allows the IgM C_H (Cμ) exon to be exchanged for an alternative C_H exon such as that for Cε, which is characteristic of IgE. Isotype switching is controlled by the cytokines to which the activated B cells are exposed. IL-4 preferentially induces IgM switching to IgG1 and IgE; IFN-γ preferentially induces switching to IgG2a and IgG3; and TGF-β preferentially induces switching to IgG2b and IgA.

10.8.5 Antibody–Antigen Binding

An antibody molecule does not bind to the whole of an antigen – rather, it binds specifically to a part of the antigen that is called an antigenic determinant or epitope. A particular antigen can have several different epitopes or repeated epitopes. Antibodies recognize the unique surface topography of epitopes. The difference between just one amino acid in an epitope can determine whether or not it will bind antibody. Binding is a combination of various individually weak interactions such as hydrogen bonds, electrostatic forces, van der Waals forces, and hydrophobic forces. The *affinity* of binding refers to the strength of the interaction between an epitope and an individual antibody combining site. The better the fit of the epitope to a particular antibody combining site, the higher the affinity of that site for antigen. The *avidity* of antibody for the antigen refers to the overall strength of the interaction between antibody and antigen. The avidity will increase, for example, if an antibody uses two of its combining sites to bind two epitopes on the same antigen rather than binding only one epitope.

TABLE 10.10
Structural Features and Effector Functions of the Different Classes of Antibody

Class	Structure and Function
IgM	Pentamer containing 10 potential antigen-binding sites. Because of steric hindrance, only up to six sites are actually able to bind antigen. Mainly confined to the bloodstream. The first antibody to be produced during a primary immune response. Activates the classical complement pathway. Its production is down-regulated as soon as sufficient IgG has been generated.
IgG	Monomer. Four subclasses: IgG1, IgG2, IgG3, and IgG4. IgG1 comprises ~ 70% of total IgG. The most abundant antibody in serum; also found in fluids of non-mucosal tissues. Major antibody of the secondary response. The only class able to cross the placenta and confer immunity to the fetus. Activates the classical complement pathway (IgG3>IgG1>>IgG2I>>>gG4) and also the alternative complement pathway. Recruits and activates a pool of destructive cell types, including NK cells, macrophages, and granulocytes.
IgA	Monomer in serum. Two subclasses: IgA1 and IgA2, with apparently the same function. Dimer in various body secretions such as saliva, tears, sweat, milk, colostrum, and secretions of the respiratory, gastrointestinal, and urogenital tracts. Crucial in protecting the mucosal surfaces of the body against attack by pathogenic microorganisms. Epithelial cells synthesize a protein called the secretory piece, which is incorporated into the IgA dimer and facilitates the transport of IgA from the blood to mucosal sites. The secretion of IgA in breast milk provides protection for the gastrointestinal tract of newborns.
IgE	Monomer. Scarce in serum. Most IgE is bound to Fc receptors on mast cells and basophils. When complexed with specific antigen, IgE can trigger the release of inflammatory mediators from these cells. The triggering of mast cell degranulation by IgE leads to the recruitment of eosinophils which kill parasitic worms.
IgD	Monomer. Low concentration in serum and other body fluids. Does not mediate any of the usual effector functions attributed to other antibodies. Found mainly on the surface of early B cells as an antigen receptor together with monomeric IgM.

10.8.6 Antibodies as Antigens: Idiotopes and Idiotypes

The variable regions of antibodies contain unique topographical features called idiotopes that are analogous to the epitopes of antigens. The sum of all the idiotopes expressed by a particular antibody is called its idiotype. Idiotopes can be recognized by other antibodies, creating an idiotype network of interacting antibodies. An immune response elicited by the idiotype of an antibody is called an anti-idiotypic response and the antibodies generated are termed anti-idiotypic antibodies. An immune response that is produced by an individual against idiotypes of their own antibody molecules is called an auto-anti-idiotypic response and the antibodies that are generated are termed auto-anti-idiotypic antibodies.

10.8.7 Functions of Antibodies

Antibodies deal with pathogens and their toxic products in a number of ways. Apart from neutralization, the functions listed below result from the formation of immune complexes between multivalent antigens and divalent antibody.

- **Neutralization**. The direct binding of antibody to the surfaces of bacteria or viruses so as to block binding to host cells and prevent infection.
- **Agglutination.** Antibody binding to antigens on bacteria causes the bacteria to clump or agglutinate. This makes bacteria more susceptible to phagocytosis.
- **Opsonization**. Antibodies can act as opsonins by binding to antigen on microbial surfaces via their Fab antigen-binding site. Phagocytes have Fc receptors that bind to the Fc part of the antibody molecule and the antibody–microbial complex is phagocytosed. There are a number of different Fc receptors that bind different classes of antibody. No Fc receptor for IgM has been characterized in humans, although in mice there is a receptor for IgM that is shared with IgA. When antibody binds to antigen, the Fc region undergoes a conformational change that enables it to bind to the Fc receptor on phagocytes. The binding of antibody to the Fc receptor triggers the phagocyte to phagocytose the opsonized bacterium. Opsonization in this manner is highly effective in the removal of encapsulated bacteria (e.g. *Streptococcus pneumoniae*), which are otherwise resistant to phagocytosis.
- **Activation of complement**. When antibody binds to antigen, the Fc portion of the antibody alters conformation so that it can bind the first component of complement, C1, and thereby activate the classical complement cascade. A single molecule of antibody can induce a multitude of opsonizing complement fragments which facilitate phagocytosis.
- **Antibody-dependent cellular cytotoxicity (ADCC)**. ADCC is mediated by cytotoxic effector cells whose Fc receptors recognize antibodies on the surface of host cells that have been infected by viruses, bacteria, or parasites. Toxic molecules secreted by the effector cells kill the infected host cells. Effector cells include neutrophils, monocytes, macrophages, and NK cells. The latter are frequently referred to as 'killer cells' when they mediate ADCC, the term 'natural killer cell' being reserved for occasions when they function in the

absence of antibody. Eosinophils are particularly efficient at killing parasitic worms whose surfaces are coated with antibody.

10.9 COMPLEMENT

10.9.1 Introduction

Complement is a system of plasma proteins that produces a rapid and amplified response to an immune stimulus. Complement proteins, produced in the liver, circulate in the blood and extracellular fluid as inactive precursor enzymes, called zymogens. Complement activation involves proteolytic cascades in which a zymogen is altered to become an active protease that cleaves and thereby induces the proteolytic activity of the next complement protein in the cascade. Proteolytic cascades allow tremendous amplification because each enzyme molecule activated at one step can generate multiple enzyme molecules at the next step. A number of the proteins involved in the cascade are split into two fragments, designated a and b fragments, by the products of the previous step. Precursor molecules are each assigned a capital letter followed by a number (e.g. C4) and the cleavage products are denoted by an a or b suffix (e.g. C4a and C4b).

The products of complement activation become covalently attached to microbial surfaces or to antibodies bound to microbes. This ensures that the biological functions of the complement system are limited to microbial cell surfaces or to antigen-bound antibodies and do not occur in the blood. The overall system is inherently controlled by the lability of the active components. More specific regulation is provided by a number of circulating or membrane-bound proteins. Protection of surrounding host cells from membrane attack is largely ensured by membrane-bound negative regulatory proteins.

There are three major pathways of complement activation: the classical, the alternative, and the lectin pathways. These pathways differ in the ways in which they are initiated, but converge into a final common pathway. The terminal product, called the membrane attack complex, causes death of pathogenic microbes by lysis (bursting of the cell).

10.9.2 The Classical Pathway

The protein components of the classical pathway are each assigned a number and react in the order: C1, C4, C2, C3, C5, C6, C7, C8, and C9. C1 is a complex consisting of C1q, C1r, and C1s. C1q binds to antibodies and C1r and C1s are proteases. Binding of antigen to antibody unmasks a specific reactive site on the antibody, which binds directly with C1q. The binding of C1q to antibody activates C1r and C1s, which cleave C4 and C2 to generate C4b2a. This protease, termed the classical pathway C3 convertase, cleaves C3 to produce C3a and C3b fragments. Some of the C3b molecules bind to the C3 convertase to form a C4b2a3b complex. This complex functions as the classical pathway C5 convertase, which will cleave C5 and initiate the late steps of complement activation.

10.9.3 The Lectin Pathway

The lectin pathway is initiated in the absence of antibody by the binding of mannose-binding lectin (MBL) to mannose residues (and certain other sugar residues) on the surface of a wide variety of pathogens, including gram-positive and gram-negative bacteria, yeasts, viruses, and parasites. This association activates two MBL-associated serine proteases (MASP-1 and MASP-2) which, like C1r and C1s, cleave C4 and C2 to generate the C3 convertase C4b2a. Subsequent events are identical to those that occur in the classical pathway.

10.9.4 The Alternative Pathway

The alternative pathway commences when C3b, produced by cleavage of C3 in the classical/lectin pathway, binds covalently to the microbial cell surface. Once C3b is deposited, it can bind a plasma protein called factor B. Another plasma protein called factor D cleaves factor B to release a small Ba fragment, leaving Bb attached to C3b. The C3bBb dissociates rapidly unless it is stabilized by the binding of a protein called properdin. The stabilized C3bBb complex is another C3-cleaving enzyme (termed the alternative pathway C3 convertase), which, by generating more molecules of C3b, amplifies the complement cascade. Some of the C3b molecules bind to the C3 convertase to form a C3bBb3b complex, which functions as the alternative pathway C5 convertase.

10.9.5 The Final Common Pathway

The cleavage of C5 to C5a and C5b by either of the C5 convertases (C4b2a3b or C3bBb3b) is the last proteolytic event in the complement system. Once C5b has been generated, the classical/lectin and alternative pathways converge into the final common pathway that leads to the formation of the membrane attack complex. The C5a fragment diffuses away and C5b binds in turn to C6 and C7 to form a complex that binds transiently to the microbial cell surface. The C7 component of the C5b,6,7 complex inserts into the lipid bilayer of the cell membrane, where it becomes a high-affinity receptor for recruitment of the C8 molecule. The inserted C7 and C8 molecules anchor the C5b,6,7,8 complex to the cell. Finally, the membrane attack complex is formed by the stepwise addition of up to 16 molecules of C9, which assemble to form a transmembrane pore. The high osmotic pressure within the cell allows a net influx of water through the pore, causing cell lysis. The membrane attack complex is effective against gram-negative bacteria, enveloped viruses, and thin-walled parasites, but the thick peptidoglycan layer of gram-positive bacteria may protect these organisms.

10.9.6 Additional Functions of Complement

Additional functions of complement are mediated by the interaction of complement components with receptors on a variety of host cell types. For example:

- **Opsonization**. The C3b fragment is an opsonin by virtue of its binding to receptors on macrophages and neutrophils, which promotes phagocytosis of the opsonized particle.
- **Chemotaxis**. The C5a fragment causes chemotactic migration of monocytes and neutrophils toward the site of complement activation. The C3a fragment is chemotactic for eosinophils.
- **Activation of mast cells and basophils**. Fragments C3a and C5a bind to their respective receptors on mast cells and basophils and induce the release of preformed inflammatory mediators from the cytoplasmic granules. C5a is up to 20 times more potent than C3a in this activity.
- **Stimulation of the respiratory burst**. C3a and C5a stimulate the generation of reactive oxygen species in phagocytes, especially neutrophils.

10.10 SECONDARY LYMPHOID ORGANS AND TISSUES

10.10.1 Overview

Adaptive immune responses are initiated in secondary lymphoid organs and tissues, which comprise the spleen, lymph nodes, and non-encapsulated mucosa-associated lymphoid tissues (MALT). Blood-borne particulate antigens are drained into the spleen; antigens from skin and other epithelia are drained into lymph nodes; and ingested and inhaled antigens are drained to MALT. Secondary lymphoid tissues are perfused by recirculating naïve T and B lymphocytes, and recognition of antigen localized at discrete T cell and B cell zones causes selective sequestration of antigen-reactive lymphocytes.

10.10.2 The Lymphatics

The lymphatics comprise a network of lymphatic vessels and associated lymph nodes that drain the interstitial fluid from body tissues and return it to the blood. Most of the fluid filtering from the blood capillaries flows among the cells of the body as interstitial fluid and finally is reabsorbed back into the venous ends of the blood capillaries. However, about 10% of the fluid, now called lymph, enters the lymphatic capillaries, rather than the venous capillaries. The lymphatic capillaries join afferent lymphatic vessels which eventually drain into lymph nodes. Efferent lymphatic vessels leaving lymph nodes may drain into other lymph nodes. Eventually, efferent lymphatic vessels join to become larger vessels that drain into the bloodstream via two major lymphatic ducts, the thoracic duct and right lymphatic duct.

10.10.3 Lymph Nodes

Lymph nodes are bean-shaped structures, 1–25 mm in diameter, that are clustered at strategic sites such as the groin, armpits, and neck and along the small intestine. Those lining the small intestine are known as mesenteric lymph nodes, as they are located between the membranous layers of the mesentery. Each lymph node has its own arterial and venous blood supply. The circulation of lymph through a lymph node involves afferent lymphatic vessels, a system of lymph sinuses within the node, and efferent lymphatic vessels. The interior of the lymph node consists of a network of reticular cells and fibers in which immune cells are segregated into specific zones. The function of the lymph nodes is to filter off and deal with foreign material draining from the tissues.

The lymph nodes are enveloped by a fibrous capsule with trabeculae extending radially to the interior. Beneath the capsule is a space called the subcapsular sinus. Two main regions can be distinguished histologically, the cortex and the medulla. The cortex is further divided into the outer cortex and the paracortex. The outer cortex is occupied by aggregates of B cells called B cell follicles. In the unstimulated lymph node, the B cell follicles are composed of naïve B cells and are referred to as primary follicles. After antigenic challenge, the primary follicles develop into secondary follicles, each with a germinal center. Germinal centers are sites where B cells, activated by helper T cells, proliferate, then undergo affinity maturation and immunoglobulin isotype switching before differentiating into either antibody-producing plasma cells or memory B cells. The paracortex contains densely packed T cells together with resident immature DCs. Blood-borne lymphocytes enter the T cell area by passing between the large cuboidal cells of post-capillary high endothelial venules. The medulla contains some T and B cells, but most of the cells in this region are macrophages and plasma cells arranged along strands of lymphoid tissue called medullary cords.

10.10.4 The Spleen

The human spleen is an oval-shaped organ of about 12 cm length and 5 cm diameter located on the left side of the abdomen. Like the lymph node, the spleen is enclosed by a capsule and divided by trabeculae into communicating compartments, but there is no division into cortex and medulla. Blood vessels enter and leave the spleen at one point, the hilum, and the larger arteries and veins run together in the trabeculae. High endothelial venules, which mediate the extravasation of recirculating cells into lymph nodes, are absent in the spleen. Blood-borne antigens are delivered to the spleen via the splenic artery; there are no afferent lymphatic vessels. The splenic artery branches like a tree until the finest branches become arterioles, called central arterioles. Each central arteriole divides into two or more terminal arterial capillaries. Venous sinuses empty into pulp veins, which unite to form trabecular veins and finally the splenic vein. Lymph collects into efferent lymphatic vessels present in the capsule and in the larger trabeculae.

The tissue inside the spleen consists mostly of red pulp within which are many discrete areas of white pulp. Red pulp consists of a network of cellular cords containing a large proportion of monocytes and macrophages supported by a stroma of reticular cells and fibers. Within this network are the venous sinuses, whose walls are composed of elongated endothelial cells lying parallel with the long axis of the sinus and connected by stress fibers to circularly placed bands composed of extracellular matrix components. Contractility of the stress fibers allows the formation of slits between the endothelial cells.

The function of the red pulp is to filter the blood, removing aged or damaged erythrocytes and all manner of cellular debris. Terminal arterial capillaries open into the red pulp spaces from which blood is forced into the venous sinuses through the slits between the endothelial cells. This passage is difficult for aging erythrocytes, which have stiffening membranes, giving the opportunity for phagocytosis by the macrophages located in the cords.

The white pulp fulfills the immune function of the spleen, which is to deal with blood-borne particulate antigens. It is organized as sheaths of lymphoid tissue around a central arteriole. A T cell zone rich in T cells and DCs immediately surrounds a central arteriole and is known as the periarteriolar lymphoid sheath (PALS). Adjacent to the PALS is the B cell area containing primary and secondary follicles. As in the lymph nodes, the secondary follicles contain a germinal center composed of rapidly proliferating B cells.

Situated between the white pulp areas and the red pulp is a specialized region called the marginal zone, which is based upon a framework of reticular fibers. Much of the blood entering the spleen is released from terminal arterioles into sinuses between the white pulp and the surrounding marginal zone. The slowing of the blood flow in these sinuses allows time for intimate contact between blood-borne antigen and immune cells. The monolayer of sinus-lining cells bordering the marginal zone is leaky, allowing blood to percolate through the marginal zone before reaching the red pulp and returning to the circulation.

10.10.5 Mucosa-Associated Lymphoid Tissue

Mucosa-associated lymphoid tissue (MALT) distributed throughout the mucosa of the respiratory, gastrointestinal, and urogenital tracts provides protection from pathogenic microorganisms at these vulnerable sites. Gut-associated lymphoid tissues (GALT) are the gastrointestinal version of MALT. The lymphoid cells may be present as dispersed aggregates of non-encapsulated lymphoid tissue, such as those occurring throughout the lung and the lamina propria of the intestinal wall. In other tissues, such as the lamina propria of the tonsils and appendix, and Peyer's patches in the small intestine, the lymphoid cells are organized into follicles. Most of the follicles contain germinal centers where plasma cells and memory B cells are generated. The interfollicular region surrounding these secondary follicles is rich in T cells, which, when activated by antigen-presenting cells, will provide the necessary helper T cells for antibody production.

The most prominent GALT structures are Peyer's patches, found mainly in the distal ileum. Peyer's patches are macroscopic aggregates of lymphoid follicles surrounded by T cell-rich interfollicular regions situated in the lamina propria of the small intestine. They are separated from the intestinal lumen by a single layer of columnar epithelial cells, known as the follicle-associated epithelium, and a more diffuse area immediately below the epithelium, called the subepithelial dome. The follicle-associated epithelium is interspersed with microfold (M) cells, which lack microvilli and the normal thick layer of mucus. M cells capture invasive pathogens in the lumen and transfer them to mucosal DCs for presentation to T cells in the interfollicular region.

10.10.6 Lymphocyte Recirculation

The astronomical number (hundreds of millions) of different varieties of B and T cells, each specific for a particular epitope on an antigen, implies that the chances of a B or T cell encountering its specific antigen must be very slim. This problem is overcome by the unique ability of naïve lymphocytes to travel from blood to lymph and thence back to the blood. This lymphocyte recirculation brings cells and antigen together in secondary lymphoid organs whose architecture is designed to allow effective interactions to occur. Lymphocyte recirculation also enables the activated lymphocytes to return to the infected tissues,

Circulating lymphocytes are conveyed by the splenic artery to the spleen and enter the white pulp through venule walls. The transit time in the spleen is about five hours, after which the lymphocytes return to the bloodstream via the splenic vein. Blood-borne lymphocytes enter the specialized areas of lymph nodes and MALT via high endothelial venules. They leave through the efferent lymphatics and pass through other lymph nodes, finally re-entering the bloodstream via the thoracic duct. In most non-lymphoid tissues, lymphocytes can also migrate from the vascular capillaries into the lymphatic capillaries, providing a route to the draining lymph node via the afferent lymphatics.

10.11 LYMPHOCYTES: T CELLS AND B CELLS

10.11.1 Introduction

Lymphocytes refer to T cells and B cells – non-phagocytic cells that exert their adaptive immune functions in secondary lymphoid organs. Mature T and B cells that have not previously been exposed to antigen are described as naïve. Naïve T cells and B cells circulate from blood to secondary lymphoid organs, to efferent lymph, and then back again to the blood. A group of specialized lymphocytes, which do not circulate through secondary lymphoid organs, respond to infection within a time-frame (days) that is intermediate between innate immunity (minutes to hours) and adaptive immunity (several weeks). These cell types include $\gamma\,\delta$ T cells, B-1 cells, and marginal zone B cells.

Unlike cells of the innate immune system, which have non-specific recognition systems, lymphocytes have unique surface receptors that can recognize a specific antigen or a limited set of structurally related antigens, referred to as cognate antigens. Therefore, before exposure to antigen, there exists in an individual a vast repertoire of lymphocytes that can recognize virtually any antigen in existence. Lymphocytes specific for individual antigens are not sufficient in number to fight an infection and therefore naïve lymphocytes with an appropriate receptor specificity must be activated by antigen to proliferate before the daughter cells differentiate into effector cells.

The activation of T and B cells starts with a process of transformation known as blastogenesis, which produces enlarged cells called lymphoblasts. These cells undergo clonal expansion, duplicating themselves two to four times every 24 hours for three to five days so that one naïve lymphocyte gives rise to a clone of around 1,000 daughter cells of identical specificity. The lymphoblasts then differentiate into effector lymphocytes. Most effector cells undergo apoptosis when the invading pathogen has been eliminated, but a small fraction remains as long-lived memory T and B cells. Memory cells are poised to respond rapidly and effectively to subsequent infection by a given pathogen.

10.11.2 Types of T Cells

10.11.2.1 CD4+ and CD8+ Effector T Cells

T cells are classified as $CD4^+$ and $CD8^+$ T cells on the basis of CD4 and CD8 cell surface molecules that were first recognized as markers of helper and cytotoxic T cells, respectively, before they were found to be co-receptors.

Effector T cells can be divided into three main categories based on their function –helper T cells, cytotoxic T cells, and regulatory T cells (formerly called suppressor T cells). Helper T cells produce cytokines that act upon other immune cell types to regulate their function. Cytotoxic T cells bind to host tissue cells infected by viruses and other intracellular pathogens, and then destroy these cells by releasing cytotoxins and inducing apoptosis. Regulatory T cells control the activation of other T cell types by releasing soluble inhibitors and are necessary to maintain peripheral tolerance to self antigens. T cells of all types express pro-apoptotic members of the Bcl-2 family of apoptotic regulators and are thus programed to die unless actively protected by survival signals.

T cells exist as two types, $\alpha\,\beta$ and $\gamma\,\delta$, according to whether their antigen-binding receptors (or T cell receptors, TCRs) are composed of α and β chains or γ and δ chains. Over 90% of peripheral T cells are $\alpha\,\beta$ cells, which undertake the established functions of T cells. $\gamma\,\delta$ T cells selectively populate epithelial boundaries between the host and the external environment. At these boundaries, the $\gamma\,\delta$ T cells act as a first line of defense by providing inflammatory and cytotoxic responses to antigens produced by invading microbes or damaged tissues.

10.11.2.2 Differentiation of Effector T Cells

Naïve T cells need to be activated in order for them to proliferate and differentiate into effector T cells. The activation of naïve T cells, known as priming, takes place in secondary lymphoid tissues. $\alpha\,\beta$T cells are primed when their TCRs recognize their cognate antigens presented as peptide–MHC class II complexes by DCs. Most of the expanded $CD4^+$ T cells differentiate into three subsets of helper T cells, namely Th1, Th2, and Th17. These subsets leave the lymphoid tissues and disperse to peripheral tissues, where they secrete cytokines that act upon other types of immune cell. Some of the expanded $CD4^+$ T cells differentiate into follicular helper T (T_{FH}) cells. These cells remain in the lymphoid tissues, where they activate naïve B cells to proliferate and transform into antibody-producing plasma cells.

The expanded $CD8^+$ T cells differentiate into cytotoxic T cells, which migrate to sites of infection and kill virus-infected host cells. Cloning of human $CD8^+$ cytotoxic T cells produces two subsets, Tc1 and Tc2, which characteristically secrete particular cytokines. Thus, Tc1 cells secrete IFN-γ but not IL-4, while Tc2 cells secrete IL-4 but not IFN-γ. These two subsets show no differences in their cytotoxic function and therefore the Tc1/Tc2 nomenclature is rarely used.

10.11.3 Types of B Cells

B cells express immunoglobulins (Igs) on their surface, which act as receptors, and produce soluble antibodies after differentiation into plasma cells. Two distinct populations of B cells develop from hematopoietic precursors: B-1 cells from precursors in the fetus and neonate, and B-2 cells from precursors in the adult. Marginal zone B cells occupy the marginal zone of the spleen and defend against blood-borne pathogens

10.11.3.1 Conventional B-2 Cells

B-2 cells are the conventional B cells that differentiate into long-lived plasma cells and memory B cells after a one- to three-week development period. They express low-surface IgM and high-surface IgD and produce high-affinity soluble antibodies in an adaptive response. B-2 cells are called follicular B cells because they encounter their specific antigen in the follicles of secondary lymphoid organs. When circulating naïve B cells enter a lymph node, they home to the primary follicles, attracted by the follicular chemokine CXCL13, which recognizes its receptor, CXCR5, on the B cell surface. Activation by antigen causes clonal expansion and migration of the B cells to the boundary of the T cell zone within the lymphoid tissue. In the germinal center reaction, signals from specialized follicular helper T cells (T_{FH} cells) stimulate the differentiation of the follicular B cells into plasma cells.

10.11.3.2 B-1 Cells

B-1 cells represent a minor population that predominates in early life. They express high-surface IgM and low-surface IgD and are a major source of natural IgM antibody, together with natural IgA and IgG. The natural antibodies are produced continuously, providing an immediate and non-adaptive defense against a restricted range of microbial antigens. Unlike B-2 cells, B-1 cells do not circulate through lymph nodes. The B-1 population is divided into B-1a cells, which express CD5 on their surface, and B-1b cells that do not. CD5 negatively regulates Ig-receptor-mediated signaling in B-1a cells by recruiting the SHP-1 protein tyrosine phosphatase into the receptor complex.

B-1 cells develop from hematopoietic B-0 cell precursors during fetal and early postnatal life. The uncommitted B-0 cells have the general features of B-2 cells. Those B-0 cells that express weakly autoreactive surface immunoglobulin acquire the typical B-1 phenotype and are positively selected by self antigens, whereas those cells that cannot bind autoantigens retain the B-2 cell phenotype. The B-1 cell population is maintained throughout life by a feedback mechanism that controls self-renewal of mature cells.

B-1 cells occupy peritoneal and pleural cavities, which, in the healthy body, are not actual spaces, but refer to the thin film of fluid that separates the membranes enveloping the viscera and lungs. The cells are rare in the spleen and extremely few, if any, are found in lymph nodes. Migration of peritoneal B-1 cells into the gut-associated lymphoid tissue is essential for their differentiation into antibody-producing plasma cells. Watanabe *et al.* (2000) found that $\gamma\,\delta$ T cells provide non-cognate help for B-1 cell migration and differentiation, and that other cell types, probably macrophages, are required to produce the necessary chemokines and cytokines.

10.11.3.3 Marginal Zone B Cells

Marginal zone B cells occupy the marginal zone of the spleen and produce IgM antibody as a defense against blood-borne bacteria and viruses. Their response fills the gap between the fast natural antibody response by B-1 cells and the adaptive response of B-2 cells. Marginal zone B cells are continuously replenished by lymphopoiesis within the bone marrow. The cells do not recirculate through lymph nodes, but migrate directly to the sites of their effector action as soon as they are formed.

Infections with blood-borne pathogens constitute a major health risk because of their rapid access to all parts of the body. The strategic location of marginal zone B cells in the spleen and their low activation threshold enables them to respond early to blood-borne antigens. In rodents, marginal zone B cells are restricted to the spleen, whereas in humans extra-splenic marginal zone B cells are found in the subcapsular sinus of lymph nodes, in and beneath the epithelium of tonsils, and in the dome region of Peyer's patches. The localization of marginal zone B cells in the marginal zone depends on the cells' expression of the sphingosine-1-phosphate receptor, SIPR1. The cells can resist the shear forces associated with local blood flow because of their high expression of the $\alpha_L\beta_2$ and $\alpha_4\beta_1$ integrins. The binding of marginal zone B cells to lipopolysaccharide or cognate antigen down-regulates SIPR1, allowing the cells to relocate to the white pulp and interact with helper T cells at the B cell–T cell boundary.

Marginal zone B cells have a unique surface phenotype, with high expression of the complement receptors CR1 (CD35) and CR2 (CD21) and the non-classical MHC class I molecule CD1d. The high expression of CR1/2 facilitates the binding of marginal zone B cells to complement-coated IgM-containing immune complexes present in thc blood. This binding triggers migration of the marginal zone B cells to the splenic follicles, where the IgM-containing immune complexes are transferred to follicular DCs. The localization of antigen on follicular DCs is implicated in the germinal center reaction, which results in the generation of long-lived plasma cells and memory B cells. The high expression of CD1d enables the marginal zone B cells to present lipid antigens to NKT cells, which elicit a rapid immune response. The activated NKT cells stimulate marginal zone B cell proliferation and antibody production via CD40L–CD40 interactions.

Marginal zone B cells make early T cell-independent antibody responses to blood-borne pathogens. The cells may contact antigen in its free form through recognition of repetitive bacterial cell wall polysaccharides and viral coat proteins. Alternatively, a subset of DCs binds blood-borne bacteria and delivers them intact to the marginal zone of the spleen. The DCs stimulate the marginal zone B cells by presenting the bacterium and secreting the signaling molecules BAFF (B cell-activating factor of the TNF family) and/or APRIL (another TNF family member) (Figure 10.6). The stimulated marginal zone B cells differentiate into short-lived plasma cells, which secrete large amounts of IgM whose principal effector action is the activation of complement.

Besides their ability to make antibody responses, marginal zone B cells participate directly in the priming of T cells, as demonstrated by Attanavanich and Kearney (2004). After immunization with protein antigen, both marginal zone B cells and follicular B cells captured antigen and migrated

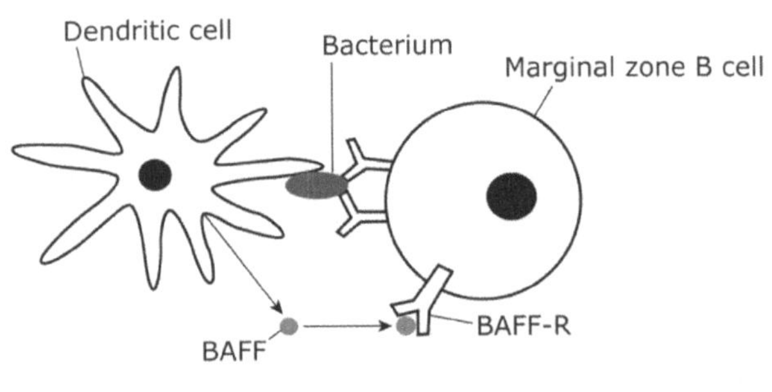

FIGURE 10.6 Activation of marginal zone B cells by immature blood dendritic cells.

to T cell areas of the spleen. Marginal zone B cells were far superior to follicular B cells in activating naïve CD4$^+$ T cells by antigen presentation. The activated T cells reciprocated by secreting cytokines that stimulate the marginal zone B cells to proliferate and differentiate into plasma cells.

Cinamon *et al.* (2008) demonstrated that in the spleen, marginal zone B cells continuously shuttle between the marginal zone and follicular areas, and that many of the cells visit a follicle every few hours. This constitutive shuttling of marginal zone B cells provides an efficient mechanism for systemic antigen capture and delivery to follicular DCs. The authors speculated that the marginal zone B cells are attracted to the S1P-high environment of the blood-rich marginal zone, where they can adhere because of their high expression of the $\alpha_L\beta_2$ and $\alpha_4\beta_1$ integrins. After a while, the cells are exposed to a sufficient amount of S1P to down-regulate their surface expression of SIPR1 and are no longer retained efficiently in the marginal zone. This situation gives CXCL13 the opportunity to attract the cells to the follicles. After a short period of migration, the cells regain their expression of SIPR1 and migrate back to the marginal zone. Exposure of marginal zone B cells to lipopolysaccharide or cognate antigen down-regulates SIPR1 and SIPR3, allowing migration to the follicles in response to CXCL13 (Cinamon *et al.*, 2004).

10.12 PROCESSING AND PRESENTATION OF PROTEIN ANTIGENS

10.12.1 Introduction

The major histocompatibility complex (MHC) is a collection of genes whose protein products are involved in the presentation of protein-derived antigens to T cells. Use of the term MHC refers to this collection of genes in general, irrespective of species. MHC genes have been divided into three classes, I, II, and III. The class I and class II MHC gene products (MHC-I and MHC-II) are involved in T cell recognition of peptide antigen. Class III MHC gene products have various other immune-related functions not directly concerned with antigen presentation and will not be mentioned further.

MHC-I molecules present peptides derived from intracellular proteins to CD8$^+$ cytotoxic T cells. The intracellular protein (endogenous antigen) could be a viral protein synthesized within a host cell. MHC-II molecules present peptides derived from extracellular proteins to CD4$^+$ helper T cells. The extracellular protein (exogenous antigen) could be a protein from a bacterium that has been captured by an antigen-presenting cell (APC). The two classes of MHC molecules acquire their associated peptides in different pathways of intracellular transport.

10.12.2 Nomenclature of MHC Molecules

The term HLA (human leukocyte antigen) refers specifically to the human MHC. There are three class I HLA genes (*HLA-A*, *HLA-B*, and *HLA-C*) and three class II HLA genes (*HLA-DP*, *HLA-DQ*, and *HLA-DR*). There are three mouse class I MHC genes (*H-2K*, *H2-D*, and *H2-L*) and two mouse class II MHC genes (*I-A* and *I-E*). All of these genes (human and mouse) encode MHC molecules with the same names as the respective genes.

10.12.3 Polymorphism of MHC Genes

Genes encoding MHC-I and MHC-II molecules are highly polymorphic; that is, each genetic locus within a population of human individuals can be occupied by one of many alternative forms (variants) of a gene known as alleles. These multiple alleles have arisen during evolution from mutations. For some of these genes, over 600 allelic variants have been identified in a population. Polymorphism of the MHC genes means that the protein products will vary in their amino acid sequence. The polymorphism maps mainly to the α_1 and α_2 domains of MHC-I molecules and to the α_1 and β_1 domains of MHC-II molecules. These are the domains that form the margins of the peptide-binding groove.

Polymorphism of MHC genes has evolved to provide a means for increasing the variety of antigenic peptides that can be presented to T cells. This ensures that the population as a whole is equipped to deal with any possible pathogen. Because allelic MHC-I and MHC-II molecules have differences in their binding grooves, their ability to interact with a particular peptide will vary from one extreme to the other. An individual may be unable to make an adaptive immune response to a particular pathogen because the antigenic peptide does not bind to that individual's MHC molecule. By contrast, other individuals may mount good responses to the same pathogen because the peptide binds well to their MHC molecules. Thus unrelated individuals have a diverse susceptibility to disease. The polymorphism decreases the chances of a population being exterminated by a potentially lethal pathogen, because at least some individuals will probably be able to mount a robust immune response.

A population can develop increased resistance to a pathogen by increasing the frequency of certain MHC alleles through natural selection. For example, there is a higher prevalence of the HLA-B53 allele in regions of Africa where malaria is endemic than in other parts of the world where malaria is absent. Possession of this allele is associated with protection against a lethal form of malaria.

10.12.4 Structure and Cellular Distribution of MHC Molecules

10.12.4.1 Structure

The MHC-I molecule is composed of two chains – a glycoprotein α chain and a smaller chain called β_2-microglobulin. The α chain is organized into three globular domains (α_1, α_2, and α_3), which protrude from the cell surface. A hydrophobic section of the α chain spans the cell plasma membrane and a C-terminal tail penetrates the cytoplasm. The β_2-microglobulin chain is non-covalently associated with the α_3 domain and does not span the membrane. The MHC-II molecule consists of two non-covalently associated polypeptides, α and β. Each of these chains has two globular domains, a membrane-spanning hydrophobic region and a C-terminal cytoplasmic tail. Both class I and class II MHC molecules have a single peptide-binding groove that binds one peptide at a time.

10.12.4.2 Cellular Distribution

MHC-I molecules are expressed constitutively on all nucleated cells of the body, enabling virus-infected cells to be recognized and killed by CD8+ cytotoxic T cells. The level of expression (i.e. the number of MHC-I molecules on the cell surface) can vary enormously between different cell types. For example, the molecules are abundantly expressed on lymphocytes, less so on the cells of liver, lung, and kidney, and only sparsely on neurons and skeletal muscle cells. Expression is increased during an immune response, thereby increasing the chances of a virus-infected cell being recognized and killed by a cytotoxic T cell.

MHC-II molecules are expressed constitutively on the so-called professional antigen-presenting cells (APCs), namely macrophages, mature DCs, and certain B cells, and their levels increase significantly when the APCs are activated. Their function is to bind peptides generated by the cellular processing of internalized particles and display the peptides on the APC surface for recognition by CD4+ helper T cells. Non-professional APCs include fibroblasts (skin), thymic epithelial cells, thyroid epithelial cells, glial cells (brain), pancreatic beta cells, and vascular endothelial cells. These APCs do not constitutively express MHC-II molecules; these molecules are expressed only upon stimulation of the APCs by certain cytokines, such as IFN-γ.

10.12.5 The MHC Class I Antigen Presentation Pathways

There are two pathways leading to MHC class I presentation of antigens to CD8+ cytotoxic T cells, direct presentation and cross-presentation. In direct presentation, antigens synthesized by pathogens such as viruses within host cells (endogenous antigens) are processed by the host cell for entry into the MHC class I pathway. The more virulent viruses have devised mechanisms that impair or block the direct presentation pathway, and cross-presentation outsmarts these viruses. Cross-presentation is the process by which cells infected with intracellular microbes, such as viruses, are ingested by DCs, and the antigens of the infectious microbes are processed and presented in association with MHC-I molecules to CD8+ cytotoxic T cells. This process is called cross-presentation (or cross-priming) to indicate that one cell type (e.g. the DC) can present antigens from another cell type (the virus-infected cell) and prime CD8+ T cells specific for these antigens.

10.12.5.1 Direct Presentation

Antigenic viral proteins present in the cytosol of the infected host cell are tagged by ubiquitin and degraded in proteosomes. The peptides thus generated are transported by TAP-1 and -2 proteins to the lumen of the endoplasmic reticulum where stable peptide—MHC-I complexes are produced. The complexes move through the Golgi apparatus and are transported in Golgi vesicles to the cell surface where they can be recognized by specific CD8+ cytotoxic T cells.

10.12.5.2 Cross-Presentation

Immature DCs internalize virus-infected cells or cellular fragments by receptor-mediated phagocytosis or endocytosis. The cell-derived antigens that are phagocytosed are contained within phagosomes. The phagosomes fuse with endoplasmic reticulum (ER)-derived vesicles, creating a compartment that contains the antigens and newly synthesized MHC-I molecules, plus the components required for peptide loading onto the MHC molecules. The antigens escape from this compartment and gain entry into the cytosol via the Sec 61 multimolecular channel. The antigens are then degraded by proteasomes and the resulting peptides are transferred back into the phagosome by TAP proteins for selective loading onto MHC-I molecules. The antigens that are endocytosed by the immature DCs are degraded by the endosomal protease cathepsin S. The peptides generated are loaded directly onto recycling MHC-I molecules within the endosome without the need for prior processing in the cytosol. The peptide–MHC-I complexes are transported to the surface of the DC and presented to CD8+ cytotoxic T cells, which are said to be cross-primed. Of all the DC subsets tested, only the lymphoid splenic CD8α + cells are able to cross-present cell-derived antigens on MHC-I molecules to CD8+ T cells *in vivo*.

The DC is the major cell type known for its capacity to cross-present exogenous antigens. B cells can cross-present exogenous particulate antigens that they internalize by receptor-mediated endocytosis via their antigen-specific Ig receptors. Monocytes can also participate in cross-presentation, although they themselves are not APCs. In experimental settings, antigens carried to the bone marrow by B cells and neutrophils can be transferred to monocytes, which migrate out of the bone marrow and, in the presence of GM-CSF, differentiate into DCs. The monocyte-derived DCs migrate to lymph nodes and cross-present the previously engulfed antigen.

10.12.6 The MHC Class II Antigen Presentation Pathway

Exogenous antigenic proteins are captured by APCs and internalized into early endosomes. The proteins are degraded enzymatically in late endosomes and lysosomes to generate antigenic peptides. In the endoplasmic reticulum, newly synthesized MHC-II binds to a trimeric protein called the invariant chain in a way that blocks the peptide-binding cleft. The MHC-II–invariant chain complexes pass through the Golgi apparatus and emerge within Golgi vesicles, which convey the complexes toward the cell surface. During this passage, the Golgi vesicles fuse with endosomal vesicles containing the antigenic peptides to form a transport vesicle. Within this vesicle, the invariant chain is displaced from the MHC groove by an antigenic peptide. Finally, the stable peptide–MHC-II complexes are delivered to the surface of the APC where they are displayed for recognition by $CD4^+$ helper T cells.

10.13 DENDRITIC CELLS

This section deals with immunogenic dendritic cells (DCs) that promote the expansion and differentiation of effector T cells to provide protection against pathogens and tumor antigens. Tolerogenic DCs (tolDCs), which promote the development of regulatory T cells, are dealt with in Section 10.28.2.

10.13.1 Introduction

Dendritic cells (DCs) obtained from bone marrow precursors occur as two morphologically and functionally distinct types: $CD11c^{high}$ conventional dendritic cells (cDCs, previously called myeloid DCs) and $CD11c^{low}$ plasmacytoid DCs (pDCs). DCs have the potential to transform from immature to mature states. Mature DCs are characterized by their membranous processes, which resemble the dendrites of nerve cells. cDCs are classed, along with B cells and macrophages, as 'professional' APCs because, in addition to presenting antigen to naïve $CD4^+$ T cells, they express adhesive and co-stimulatory molecules that enable T cell priming. cDCs are by far the most efficient APCs and are unique in their ability to trigger a primary T cell response.

Immature cDCs capture foreign antigens in the peripheral (non-lymphoid) tissues and process them into peptides, which are displayed on the cell surface in association with a membrane-bound MHC class II molecule. Inflammatory signals induce the cDCs to mature and migrate to secondary lymphoid organs, where they prime naïve $CD4^+$ T cells through antigen presentation. Immature pDCs circulate in the blood and internalize virus particles (virions), which activates the pDCs to secrete large amounts of antiviral type I interferon (IFN-α/β). The pDCs migrate to secondary lymphoid organs and differentiate into functional APCs, referred to as pDC-derived DCs.

In the steady state (when the immune system is not being challenged), DC homeostasis is maintained by means of a dynamic balance of three processes: (1) constant replenishment by precursors from blood, (2) cell division, and (3) cell death after a limited number of divisions.

Antigen-presenting DCs that are present in T cell areas of secondary lymphoid tissues are referred to as interdigitating DCs. Follicular DCs are restricted to the germinal centers of B cell follicles and do not qualify as interdigitating DCs.

Throughout this book, the term 'dendritic cell' or 'DC' (with no c or p prefix) refers to conventional dendritic cells (cDCs).

10.13.2 Development of Dendritic Cells from Bone Marrow Precursors

DCs can be generated in bone marrow from precursors downstream of the common myeloid and lymphoid progenitors (CMPs and CLPs) that express the Flt3 receptor (Figure 10.1). The ligand for this receptor (Flt3L) is essential for the development of steady-state DC populations. Which of the two progenitors will actually generate DCs *in vivo* will depend on the availability of the progenitor and the local environment. CMPs are the main source of DCs in most circumstances. Diao *et al.* (2004) isolated a population of DC precursors in mouse bone marrow that could be separated into two subpopulations, which they designated $B220^-$ DCp and $B220^+$ DCp. $B220^-$ DCp generated exclusively cDCs, whereas $B220^+$ DCp was a precursor of pDCs with some capacity to generate cDCs. The $CD8\alpha^+$ and $CD8\alpha^-$ subclasses of DCs appear to be derived from lymphoid and myeloid lineages, respectively (Shortman & Caux, 1997).

Newly formed cDCs leave the bone marrow and travel in the blood to various non-lymphoid peripheral tissues, where they take up residence as interstitial DCs with an antigen-capturing role. These immature DCs occur in the epidermal layers of the skin and in the epithelia of intestinal, respiratory, and urogenital tracts. Non-epithelial populations of interstitial cells include dermal DCs and DCs in the heart, kidney, lung, and liver.

10.13.3 Differentiation of Monocytes into Conventional Dendritic Cells

Monocytes are established circulating precursors for tissue macrophages and cDCs. Two major human monocyte subsets have been identified based on the differential expression of CD14 (part of the receptor for lipopolysaccharide) and CD16 (the FcγRIII receptor). The so-called 'classical' $CD14^+CD16^-$ subset represents up to 95% of blood monocytes in a healthy individual, and the 'non-classical' $CD14^{low}CD16^+$ subset comprises the remainder. These subsets differ in many respects, including adhesion molecules and chemokine receptor (CCR) expression. The equivalent classical and non-classical mouse monocytes, based on Gr-1 (Ly6C/G) expression, are Gr-1^{high} and Gr-1^{low}, respectively.

Mouse Gr-1^{high} blood monocytes are rapidly recruited to sites of inflammation, where they give rise to immature cDCs. Immature human and mouse cDCs become mature cDCs after stimulation with CD40 ligand (CD154). Recruitment of monocytes to inflamed skin is critically dependent on their expression of CCR2 and on the secretion of CCR2-binding chemokines by the inflamed skin. CCR2 is not expressed on Gr-1^{low} monocytes and these cells migrate scarcely or not at all to inflamed tissue in mice. In the absence of inflammation, Gr-1^{low} monocytes migrate to non-lymphoid organs, such as the intestinal lamina propria and the lung, where they replenish macrophages or DCs.

10.13.4 Influence of Cytokines on the Differentiation of Blood Monocytes

Human monocytes can be induced, without proliferation, to differentiate into either macrophages or immature cDCs when cultured with the appropriate cytokines. M-CSF induces the formation of macrophages from monocytes. GM-CSF is essential, but not sufficient on its own to induce cDC formation from human peripheral blood – additional cytokines are required. In cultured monocytes, a combination of GM-CSF and IL-4 generates functionally immature cDCs. IL-4 suppresses the macrophage differentiation potential of monocytes and allows extensive generation of DCs to ensue. DCs thus generated can be matured by the addition of TNF-α to the GM-CSF/IL-4 combination. Monocytes cultured in the presence of type 1 interferon (IFN-α/β) and GM-CSF differentiate into mature cDCs. Whereas GM-CSF/IL-4-treated monocytes required at least six days to fully acquire the immature cDC phenotype, the GM-CSF/IFN-α/β combination induced mature cDCs within three days.

Tbe generation of DCs from cultured monocytes is useful for experimental purposes, but *in vivo* evidence is required to prove that monocytes can differentiate into DCs under physiological conditions. Such evidence is provided by two observations. (1) Monocytes differentiate into DCs as they traverse endothelial monolayers without exogenous cytokines. The monocytes migrate in the basal-to-apical direction (reverse transmigration), as would occur during entry into lymphatic vessels. Monocytes that remain in the subendothelial matrix become macrophages (Randolph *et al.*, 1998). (2) About 25% of monocytes that phagocytose subcutaneously injected latex microspheres migrate to the T cell areas of draining lymph nodes, where they differentiate into DCs. The remaining 75% of monocytes are retained in subcutaneous tissue and become macrophages (Randolph *et al.*, 1999). Thus, blood-borne monocytes represent an available source of precursors that can polarize towards macrophages or DCs, depending on which cytokines are prevalent at the time.

10.13.5 Conventional Dendritic Cell Subsets

10.13.5.1 Langerhans Cells

Langerhans cells refer to DCs that are present in the epidermal layer of the skin and in the epithelia of intestinal, respiratory and reproductive tracts. These immature DCs capture foreign antigens, then migrate via the lymphatics to the skin-draining lymph nodes. Langerhans cells act as transporters and donate their antigenic cargo to lymph node-resident DCs for presentation and T cell priming.

Langerhans cells have long residence times in skin, but turn over rapidly on arrival at the lymph nodes. Under steady-state conditions, Langerhans cells in the skin are gradually depleted owing to a slow base rate of migration to the lymph nodes. The population of Langerhans cells in non-inflamed skin is maintained not by circulating precursors, but by local clonal expansion of hematopoietic precursors that probably seeded the skin during embryonic development. Once an inflammatory threshold is reached, Langerhans cells rapidly disappear from the affected skin area and are replaced by blood-borne precursors within two weeks. The recruitment of precursors is dependent on their expression of CCR2 and on the secretion of CCR2-binding chemokines (CCL2 and CCL7) by the inflamed epithelial cells. Ginhoux *et al.* (2006) demonstrated that the circulating Langerhans cell precusor is the classical Gr-1^{high} monocyte subset and that M-CSF is critical for these monocytes to proliferate and differentiate into Langerhans cells in inflamed skin.

10.13.5.2 Gut-Associated Dendritic Cells

Various subsets of DCs are found throughout the intestine, including the lamina propria of the small and large intestine, the GALT, and the mesenteric lymph nodes. A minor population of CD103^{+} DCs in mesenteric lymph nodes (MLNs) presents orally derived soluble antigen to T cells. In addition, these cells play crucial roles in generating gut-homing T cells and inducing regulatory T cell development. The CD103 marker is α_E, the alpha chain of the $\alpha_E\beta_7$ integrin, which is the receptor for the epithelial cell adhesion molecule E-cadherin. The MLN CD103^{+} DCs are thought to be derived from circulating DC precursors that continually seed the small intestinal lamina propria and migrate from there to the MLNs. In contrast, MLN CD103^{-} DCs represent a resident population derived from blood precursors. These cells present systemic soluble antigen to T cells.

A subset of mouse lamina propria DCs expresses Toll-like receptor 5. When stimulated by the TLR5 ligand (flagellin), these DCs induced the differentiation of naïve B cells into IgA-producing plasma cells by a mechanism independent of GALT. In the process, the TLR5^{+} DCs up-regulated expression of the CCR9 homing receptor on the B cells. The TLR5^{+} DCs also promoted the differentiation of antigen-specific Th17 and Th1 cells. Retinoic acid produced by the TLR5^{+} DCs supported the generation and retention of the plasma cells in the lamina propria and positively regulated the differentiation of Th17 cells.

Two distinct DC subsets have been identified in murine Peyer's patches. CD11b^{+} DCs are present in the subepithelial dome region, whereas CD8α^{+} DCs are present in the T cell-rich interfollicular region. The strategically positioned CD11b^{+} DCs capture and process luminal antigens transported by overlying M cells. The antigen-loaded DCs migrate to the interfollicular region, where they transfer the antigen to resident Th1-inducing CD8α^{+} DCs for subsequent presentation to naïve T cells by a process known as cross-priming.

10.13.5.3 Follicular Dendritic Cells

A unique subset of $CD4^{+}CD3^{-}CD11c^{+}$ DCs is capable of stimulating T cells in the germinal centers of secondary lymphoid tissues. These cells, termed follicular DCs (FDCs), differ from the other types of DC in that they are of mesenchymal rather than bone marrow origin, non-phagocytic, and lack lysosomes. The cells are restricted to the germinal centers, where their surface receptors bind intact antigen complexed with antibody. These FDC-bound antibody–antigen complexes bind to post-mitotic germinal center B cells (centrocytes) and play a part in the selection of high-affinity B cells.

Follicular DCs release vesicular fragments of plasma membrane called iccosomes that have intact antigen, antibody Fc, and complement receptor CR2 (CD21) ligand attached. Iccosomes bind tightly to B cells and are rapidly endocytosed. The B cells process this FDC-derived antigen and present it as a peptide–MHC II complex to T cells, thereby obtaining the necessary T cell help.

10.13.5.4 Splenic Dendritic Cells

The $B220^{-}$ DC precursors originally found in mouse bone marrow are also widely distributed in spleen and other secondary lymphoid tissues. In the mouse spleen, mature DCs have a short life span and are entirely replaced within 10–14 days. Mature DCs are capable of a degree of homeostatic proliferation and can divide when seeded onto monolayers of splenic stromal cells. Blood monocytes do not have the capacity to fully replenish splenic DCs in the steady state.

Murine splenic DCs can be divided into two subsets based on expression of lineage-specific surface markers, myeloid CD11b and lymphoid CD8α, and the C-type lectin receptor DEC-205. These are referred to as the myeloid ($CD11b^{+}CD8\alpha^{-}DEC\text{-}205^{-}$) and the lymphoid ($CD11b^{-}CD8\alpha^{+}DEC\text{-}205^{+}$) subsets. The myeloid subset resides in the red pulp and marginal zone and the lymphoid subset resides in the T cell-rich area of the periarteriolar lymphoid sheath. Myeloid splenic $CD8\alpha^{-}$ DCs are specialized to present captured antigens (ignoring dying cells) on MHC class II molecules to $CD4^{+}$ helper T cells. Lymphoid splenic $CD8\alpha^{+}$ DCs are specialized to capture and ingest virus-infected cells, dying cells, and tumor cells and cross-present the endogenous antigens on MHC class I molecules to $CD8^{+}$ cytotoxic T cells.

10.13.5.5 Blood Dendritic Cells

A subset of blood-derived immature DCs, referred to as blood DCs, binds blood-borne particulate antigens and delivers them intact to the spleen. In a systemic infection, these DCs provide critical survival signals to antigen-specific marginal zone B cells and initiate the T-independent antibody responses (see Section 10.21.5). Blood DCs are singularly responsible for initiating these responses and are part of the first line of innate defense mechanisms involved in the capture and delivery of particulate antigens to marginal zone B cells.

10.13.6 Conventional Dendritic Cells in the Initiation of Adaptive Immunity

Immature cDCs internalize microorganisms and cellular debris by phagocytosis and receptor-mediated endocytosis. In addition, large amounts of extracellular fluid containing soluble antigens can be taken up by macropinocytosis. At the site of infection, danger signals arising from the recognition of microbial components by specific receptors and the local secretion of pro-inflammatory cytokines trigger a process of cDC differentiation called maturation, which transforms the immature antigen-capturing cells into mature APCs.

Ashany *et al.* (1999) showed that DCs express the Fas death receptor upon maturation, yet the mature cells are highly resistant to Fas-mediated apoptosis. The authors observed that expression of the apoptosis inhibitor FLIP was significantly higher in Fas-resistant DCs than in Fas-sensitive macrophages, which suggested that FLIP protects DCs from Fas-mediated apoptosis. This resistance to Fas-mediated apoptosis likely contributes to the efficiency of DCs as professional APCs.

Maturation takes place concomitantly with the migration of the cDCs from their niches in the peripheral tissues to the lymph nodes via the afferent lymphatic vessels. The maturing cDCs up-regulate MHCs, co-stimulatory molecules, and the chemokine receptor CCR7. During maturation, the captured antigen is processed into peptides, which bind to MHC molecules in the endosomal compartments of the DCs. The peptide–MHC complex then moves to the cell surface. The migratory cDCs (formerly known as 'veiled' cells) are guided to the T cell areas of the lymph nodes by CCR7, which recognizes CCL19 and CCL21 chemokines expressed by stromal cells.

On arrival at the lymph nodes, the matured cDCs present the antigen to naïve antigen-specific T cells. The 'dendrites' of mature cDCs enable these cells to interact with several T cells simultaneously. The primed T cells undergo clonal expansion and differentiate into effector T cells. The antigen sampling and migratory capacities of cDCs allow naïve T cells to come into contact with peripheral antigens that they would otherwise not have encountered. By contrast, lymphoid tissue-resident cDCs capture, process, and present antigen in the lymphoid organ itself.

10.13.7 Plasmacytoid Dendritic Cells

A subset of immature bone marrow–derived dendritic cells, called plasmacytoid dendritic cells (pDCs) because of their plasma cell-like morphology, corresponds to the previously known natural interferon-producing cells (IPCs) that produce large amounts of type I interferon (IFN-α/β) upon viral stimulation. The pDC is defined as $CD11c^{-}CD45RA^{+}CD123^{high}$ cells in humans and $CD11c^{+}B220^{+}NK1.1^{-}CD19^{-}CD11b^{-}$ cells in mice.

pDCs at the immature state can only make weak and unstable interactions with naïve antigen-specific T cells and, without co-stimulation, the T cells die by neglect. Virus-activated pDCs differentiate into mature cells with a characteristic DC morphology; these cells have been called pDC-derived DCs. To clarify the situation, pDCs are the counterparts of immature cDCs, and pDC-derived DCs are the counterparts of mature cDCs. pDCs are crucial in innate immunity as they stimulate the antiviral activity of NK cells. In adaptive immunity, pDC-derived DCs function as APCs and induce cytotoxic T cell responses. Virus-activated pDCs also induce the differentiation of CD40-activated B cells into plasma cells that secrete virus-specific antibodies.

Unlike immature cDCs, which are phagocytic and express cell-surface Toll-like receptors (TLRs), pDCs are non-phagocytic and selectively express intracellular TLR7 and TLR9. TLR7 recognizes single-stranded viral RNA, whereas TLR9 recognizes CpG-containing viral DNA. After binding and/or fusing to pDCs, viruses are internalized and delivered to the lysosomal compartment, where nucleic acids are released to interact with the TLRs. Engagement of the TLRs triggers MyD88-dependent signaling pathways that activate transcription of the IFN-α/β genes. In addition to its well-known pro-inflammatory properties, the IFN-α/β secreted is anti-viral by virtue of its many effects on immune cells.

Besides IFN-α/β, activated pDCs secrete pro-inflammatory TNF-α. They also secrete IL-12, which contributes to Th1 polarization and induces IFN-γ production in NK cells, and IL-6, which facilitates differentiation of B cells into plasma cells.

After secreting IFN-α/β in response to viral infection, pDCs rapidly differentiate into the mature pDC-derived DCs through an autocrine mechanism mediated by IFN-α/β and TNF-α. These differentiated cells continue to produce lower amounts of IFN-α/β and express high levels of MHC classes I and II and the co-stimulatory molecules CD80 and CD86. Up-regulation of the chemokine receptor CCR7 facilitates migration of the differentiated pDCs to the T cell areas of secondary lymphoid tissues, attracted by CCL19 and CCL21. At these sites, the pDCs play an important part in the activation and survival of T and B cells.

A distinct population of $CD11c^+B220^+Ly6C^+$ pDCs in the small intestinal mucosa (mostly in the lamina propria) expresses high amounts of the chemokine receptor CCR9, which targets the cells to the small intestinal epithelium through binding to CCL25 expressed by epithelial cells. The numbers of pDCs increase when the intestine becomes inflamed, suggesting that their rapid recruitment strengthens the first line of defense, particularly against viruses.

10.14 T CELL ACTIVATION

10.14.1 Introduction

The T cell population of a given person has a repertoire of many millions of different T cell receptors (TCRs). An individual T cell will express receptor specificity for only one antigen. Naïve $CD4^+$ and $CD8^+$ T cells circulate through secondary lymphoid tissues, where they search for their cognate antigen in the form of a peptide–MHC complex displayed on the surface of an APC or naïve $CD8^+$ T cell. $CD4^+$ T cells recognize antigen bound to MHC-II molecules, and $CD8^+$ T cells recognize antigen bound to MHC-I molecules. A specialized cell–cell contact termed the immunological synapse concentrates the peptide–MHC complexes by greater than 100-fold in the center of the synapse. The interaction of the TCR and the peptide–MHC complex is stabilized by the T cell co-receptor molecules, CD4 or CD8. The initial activation of naïve T cells by binding of antigen to the TCR is known as priming, which informs the T cell that antigen has been encountered. The signal initiated by ligation of the TCR instructs the naïve T cell to proliferate and differentiate to become an effector T cell. Antigenic stimulation of naïve T cells must be sustained for at least 20 hours to commit to proliferation.

The information presented in this section is based on data pertaining to the activation of the $\alpha\beta$ type of T cell in response to an acute infection. Chronic antigen exposure confuses the situation, as discussed by Bevan (2004).

10.14.2 Recognition of Antigen by Naïve T Cells

Antigen recognition by the TCR is the necessary first signal for the priming of naïve T cells, ensuring that the resultant immune response remains specific for the antigen. The TCR does not bind peptide or MHC molecules alone: it only binds effectively to peptides that are immobilized within the groove of a MHC molecule. This dual recognition system, MHC-restricted recognition, creates a diversity of binding specificity, which extends to alleles of a MHC molecule.

Dendritic cells have an essential role as APCs in T cell activation. When immature DCs encounter pathogenic microbes in the tissues, they internalize the protein antigens and begin to mature as they migrate to the T cell zones of secondary lymphoid tissues. During this migration, the DCs process the antigens into peptides, which they present in association with MHC-II molecules to naïve $CD4^+$ T cells specific for the antigen in question. Lymph nodes also contain resident immature DCs that have migrated to these sites directly from the blood. In the lymph nodes, these cells capture, process, and present antigen to naïve T cells before the arrival of migrant mature DCs.

Protein antigens that are produced in the cytoplasm of host cells as a result of viral infection are processed into peptides and presented to naïve $CD8^+$ T cells in association with a MHC-I molecule. Viral antigens internalized by DCs can also be processed and cross-presented on MHC I molecules. Antigens that enter the circulation may be captured and processed by resident DCs in the spleen.

10.14.3 Priming of Naïve T Cells

The priming of naïve $CD4^+$ T cells takes place in the T cell areas of lymph nodes and spleen. Dendritic cells present the antigenic peptide to the rare naïve $CD4^+$ T cells whose receptors recognize the specific extracellular antigen that was captured. The small number of peptide–MHC-II complexes can engage sufficient TCRs to transduce an activation signal. This is possible because the low affinity of the TCR enables a single peptide–MHC-II complex to serially engage and trigger up to ~200 TCRs. Antigens derived from intracellular pathogens can be processed in the infected cell and the peptide–MHC-I complex presented directly to $CD8^+$ T cells.

Soluble antigens are delivered to lymph nodes via the afferent lymphatic ducts within minutes of infection. The antigens leak through the lymph node conduits connecting the subcapsular sinus with the T cell area, and are captured and processed by resident immature DCs that cluster round the conduits. The stimulated DCs mature and present the antigen as peptide–MHC-II complexes to the nearby T cells that have arrived by extravasation across the high endothelial venules. Interactions of soluble antigens with resident dendritic cells are transient, and production of fully competent effector T cells must await the arrival of migratory DCs some 8–12 hours later. Migratory dendritic cells that have captured antigen in infected peripheral tissues display higher levels of antigen on their surface than resident DCs, and are able to make sustained contacts with several T cells at a time. Blood-borne antigens are delivered directly to the spleen, where they are captured and processed by resident DCs for presentation to T cells.

10.14.4 Activation of Primed T Cells

Since each T cell expresses receptors of a single antigen specificity, only a small subset of T cells is activated by any particular antigen. During activation, the naïve T cell grows to double its size and then undergoes rapid clonal expansion, while also differentiating into an effector T cell. Clonal expansion comprises two distinct cellular processes: proliferation (cell cycling) and enhanced survival. The activated T cells also acquire a polarized morphology that facilitates migration into inflamed peripheral tissues. Most effector T cells die by about 15 days after activation, and normally only a few memory cells remain. Cell death occurs whether or not the infection is cleared: this is passive cell death, not apoptosis, and is due to lack of sufficient survival signals.

The priming of naïve T cells through the TCR and CD4 or CD8 co-receptor provides antigen specificity. Full T cell activation requires complementary signals from ligated co-stimulatory receptors constitutively expressed on naïve T cells. The CD80 (B7.1) and CD86 (B7.2) ligands of the CD28 co-stimulatory receptor, and the CD70 ligand of the CD27 co-stimulatory receptor are important contributors to T cell activation. Co-stimulation lowers the threshold for T cell activation and supports proliferation and survival by (1) down-regulating cell cycle inhibitors; (2) up-regulating anti-apoptotic molecules; and (3) inducing the production of IL2 and expression of the IL-2 receptor α chain (IL-2Rα). The IL-2–IL-2R interaction is responsible for the proliferation of antigen-activated T cells. The proliferative effects of IL-2 are opposed by TGF-β, so the balance between these two cytokines is essential to maintain immune homeostasis. In the absence of CD28-mediated co-stimulation, T cells that encounter antigens either fail to respond and die by apoptosis, or enter a state of anergy in which they are refractory to subsequent stimulation.

The activation signals induce the T cells to proliferate vigorously, with one naïve cell giving rise to more than 10,000 daughter cells over the course of 5–8 days. The expanded $CD8^+$ T cell population differentiates into cytotoxic T cells, and the expanded $CD4^+$ T cell population differentiates into the Th1, Th2, and Th17 subsets of helper T cells. These effector T cells migrate to discrete sites throughout the body where antigen may be found. Whether or not the T cell response totally eliminates the pathogen, most of the antigen-specific effector T cells undergo passive cell death because of loss of survival signals. Cells that survive circulate as long-lived memory T cells, which migrate to infected tissues and effect an enhanced secondary immune response upon re-encounter with the pathogen.

10.14.5 Survival of Naïve T Cells

The thymus starts to atrophy after puberty and, consequently, there is a reduced output of new T cells. The problem of maintaining the existing pool of antigen-specific T cells in the periphery is resolved in part by prolonging cell survival. Naïve T cells in the peripheral pool continue to receive survival signals from self peptide–MHC presented by dendritic cells, and from IL-7 on fibroblastic reticular cells in the T cell zone of secondary lymphoid tissues. Survival also partly depends on ligation of IL-15Rα with IL-15 expressed on the surface of dendritic cells (Surh & Sprent, 2008).

10.14.6 Alternative Pathway of T Cell Activation

The non-integrin adhesion molecule CD2 and a subfamily member called SLAM (signaling lymphocytic activation molecule) function as co-stimulatory receptors in the so-called 'alternative pathway' of T cell activation. The principal ligand of CD2 in humans is leukocyte function-associated antigen 3 (LFA-3, or CD58), which is expressed on a wide variety of cells. In mice, the principal ligand for CD2 is CD48, which is structurally similar to LFA-3. CD2-mediated signaling involves a heterophilic reaction between CD2, expressed on a T cell, and CD58, expressed on an APC. SLAM-mediated signaling involves a homophilic interaction between SLAM expressed on a T cell and SLAM expressed on an APC. Knockout mice lacking both CD28 and CD2 have more profound defects in T cell responses than do mice lacking either molecule alone. This indicates that CD28 and CD2 may compensate for each other, an example of the redundancy of co-stimulatory receptors of T cells.

10.14.7 Clonal Expansion of T Cells

The frequency of naïve T cells specific for a given antigen is 1 in 10^5 to 10^6 of naïve T cells. After antigen recognition, the frequency of antigen-specific T cells increases from 1 in 3 to 1 in 10 for $CD8^+$ cells (a >50,000-fold expansion) and 1 in 100 to about 1 in 1,000 for $CD4^+$ cells. This tremendous expansion of the antigen-specific population of T cells occurs within one week after an acute viral infection in mice. The major cytokine responsible for T cell clonal expansion is IL-2, which stimulates cell cycle progression through the synthesis of cyclins and promotes cell survival by inducing the anti-apoptotic protein Bcl-2. IL-2 is synthesized and secreted by $CD4^+$ T cells within two to four hours of recognition of the antigen. The fact that the T cells that recognize antigen and produce IL-2 also preferentially respond to IL-2 ensures that only these antigen-specific T cells will proliferate.

T cell proliferation is greatly increased by a positive feedback loop that is established in the lymph node. Activation of the T cell through the TCR and CD28 up-regulates the expression of CD40 ligand (CD40L) on the T cell surface. CD40, the receptor for CD40L, is expressed on activated DCs. The binding of CD40L to CD40 induces up-regulation of CD80/86 on the DCs, which, by amplifying the CD28-mediated signal, up-regulates CD40L and further amplifies the reciprocal signaling between the two cells. This positive feedback loop is broken by CTLA-4 (CD152), which, when expressed on activated T cells, competes with CD28 for binding to CD80/86.

IL-4 directly enhances proliferation of highly purified, mitogen-activated human T cells by promoting cell cycle entry and subsequent DNA synthesis. In addition, IL-4 enhances production of IL-2 and increased responsiveness to IL-2, in part by up-regulating IL-2R expression. IL-2 and IL-4 act synergistically to promote T cell proliferation via the alternative CD2 pathway of T cell activation. This synergy can be explained by IL-4-induced production of IL-2 and IL-2-induced enhancement of IL-4 responsiveness. The proliferative response to IL-4 is enhanced by monocyte-produced IL-1 and IL-6.

10.14.8 Protection of T Cells from Apoptosis

During the expansion phase of the immune response, effector T cells are protected from apoptosis (mitochondrial pathway) by anti-apoptotic factors such as Bcl-2, which are produced in amounts sufficient to overwhelm the activity of pro-apoptotic factors such as Bim. Toward the end of the immune response, co-stimulatory molecules that induce Bcl-2 are down-regulated, allowing Bim to exert its function. Certain cytokines protect T cells from apoptotic death through their ability to induce expression of Bcl-2.

Resting T cells undergo spontaneous apoptosis when cultured *in vitro*. When resting T cells are cultured in the presence of IL-6, Bcl-2 is not down-regulated and the cells survive. This suggests that IL-6 rescues T cells from apoptosis by mediating maintenance or induction of Bcl-2 expression. IL-4 and IL-7 have also been shown to prevent the death of resting T cells *in vitro*. The effect of IL-4 is independent of STAT6, demonstrating that the IL-4 receptor on resting T cells uses a signaling pathway other than the Jak–STAT pathway.

T cells activated with antigen will undergo apoptosis unless rescued by cytokines. Vella *et al.* (1998) showed that four cytokines, IL-2, IL-4, IL-7, and IL-15, prevent the deaths of activated T cells *in vitro*. These cytokines share a component of their receptors, the common γ chain, therefore it is likely that they rescue T cells via a common pathway. Two of these cytokines, IL-2 and IL-4, were tested *in vivo* for their rescuing activities; both were found to be active, IL-4 particularly so. This result suggested that all four cytokines can promote T cell survival *in vivo* as well as *in vitro*.

10.14.9 Activation-Induced Cell Death

Previously activated T cells die by apoptosis when exposed to antigen. This phenomenon, known as activation-induced cell death (AICD), involves the interaction between the death receptor Fas and its ligand FasL. Naïve T cells do not express Fas and are resistant to AICD. Mice carrying a mutated Fas gene or FasL gene develop a profound autoimmune disease characterized in part by the accumulation of large numbers of abnormal T cells.

Brunner *et al.* (2003) addressed the possible role of the Fas–FasL interaction in AICD in the A1.1 line of T cell hybridomas, which rapidly undergoes apoptosis when activated. Activation of the A1.1 cells through the TCR-induced expression of Fas and FasL on the cell surface (both membrane-bound) and interference with the ensuing interaction inhibited AICD. TCR ligation could induce apoptosis in a single cell, suggesting that the activation-induced apoptosis is suicide rather than fratricide, in which an activated cell bearing FasL kills neighboring activated cells bearing Fas.

10.14.10 Dissociation of the T Cell from the Antigen-Presenting Cell

Most encounters of T cells with APCs do not result in the recognition of an antigen. In this case, the T cell must be able to separate efficiently from the APC so that it can continue to migrate through the lymph node, eventually leaving via the efferent lymphatic vessels to re-enter the blood and continue circulating. Dissociation, like stable binding, may also involve signaling between the T cell and the APC, but little is known of its mechanism.

10.14.11 T Cell Migration

The migration patterns of T cells are determined by interactions between chemosensory receptors and their ligands and between adhesion molecules. The binding of the T cell to an APC induces the rapid reorientation of the T cell cytoskeleton. As a result, the T cells acquire a polarized morphology similar to that of moving amebae: a leading edge followed by the cell body, and a trailing edge or uropod. Chemokine receptors, which bind chemokines and provide direction of movement, are redistributed to the leading edge. Adhesion molecules accumulate at the uropod and bind adhesion molecules on the extracellular matrix. During migration, the uropod anchors the cell to the extracellular matrix, while acto-myosin-based contraction of the actin network pulls the cell body forward (Sánchez-Madrid & del Pozo, 1999).

Blood-borne naïve T cells migrate through the walls of lymph node post-capillary high endothelial venules to enter the T cell-rich regions of the cortex. The T cells are attracted to these regions by the concentration gradients of CCR7 ligands, and through the binding of T cell CD62L (L-selectin) to carbohydrate ligands on lymph node endothelial cells. In the cortex, the T cells scan interdigitating DCs for the presence of pathogen-derived peptide–MHC complexes that they bind with high avidity. If TCR ligation does not occur, desensitization of CCR7 and recognition of efferent lymph sphingosine 1-phosphate (S1P) eventually drive naïve T cells out of the lymph node and back into the blood. By excluding naïve T cells from non-lymphoid peripheral tissues, potentially autoreactive T cells will not gain access to tissue-resident cells with high expression of self antigens.

Effector T cells down-regulate homing molecules for lymph node entry and up-regulate molecules involved in migration to particular non-lymphoid tissues. The directed migration of effector T cells to sites of infection would maximize the efficiency of the response and minimize immunopathological damage, but might lack the flexibility to clear an infection that is spreading to other tissues. To test the migratory abilities of $CD8^+$ T cells responding to a local infection, Masopust *et al.* (2004) orally infected mice with the EDIM strain of rotavirus, a natural mouse pathogen that preferentially infects intestinal epithelial cells. The results indicated that despite a gut-restricted infection, activated $CD8^+$ T cells were able to migrate to uninfected non-lymphoid sites beyond intestinal tissue. Similar results were obtained in response to intranasal infection with Sendai virus, which preferentially infects the lung. These experiments demonstrated that effector $CD8^+$ T cells migrate pervasively to all non-lymphoid organs, regardless of the site of initial encounter with antigen. This action may be essential to ensure complete elimination of the pathogen in cases where the infection has spread to other tissues.

10.14.12 Bystander T Cell Activation

Bystander T cell activation is the phenomenon whereby bystander cells receive signals or microbial-derived products from infected cells and produce cytokines that activate T cells. Thus the infection results in the activation of T cells that are not specific for the infecting pathogen. A bystander cell can be any type of cell that is in the vicinity of an infected cell and is non-antigen-specific. Bystander cells can be activated in the following manner: (1) infected cells transfer cell signaling molecules via gap junctions to bystander cells that are in direct contact; (2) bacterial outer membrane vesicles containing pathogen-associated molecular patterns (PAMPs) or immune signaling complexes (inflammazones) are released from infected cells and taken up by bystander cells; and (3) infected cells release soluble molecules, such as cytokines, that are detected by bystander cells (Holmgren *et al.*, 2017).

10.15 CONTROL OF T CELL ACTIVATION BY CTLA-4 AND PD-1

T cell activation is inhibited by CTLA-4 and PD-1 receptors through distinct and potentially synergistic mechanisms (Parry *et al.*, 2005). The control of T cell activation is achieved by balancing activating and inhibitory signals mediated by co-stimulatory CD28 and inhibitory CTLA-4 and PD-1.

10.15.1 Inhibition by CTLA-4

10.15.1.1 Expression and Intracellular Trafficking of CD28 and CTLA-4

CD28 and CTLA-4 display markedly different patterns of expression and localization within T cells. CD28 is constitutively expressed on the T cell surface. CTLA-4 is undetectable on naïve T cells, but is rapidly up-regulated upon T cell priming. Differential control of CD28 and CTLA-4 expression is provided by differences in their stability. CTLA-4 undergoes fairly rapid degradation in activated T cells due to lysosomal targeting. Its short half-life (approximately two hours) allows the levels of CTLA-4 to be rapidly adjusted by TCR-mediated signals that affect its rate of synthesis. Altering the expression levels of CTLA-4 could then impact its availability at the immunological synapse. In contrast, CD28 is quite stable and can exert its function even after its synthesis is reduced or terminated.

CD28 and CTLA-4 exhibit completely different trafficking patterns in T cells. Prior to TCR stimulation, CD28 is uniformly distributed around the plasma membrane. However, upon T cell priming, cytoskeletal reorganization moves CD28 to the immunological synapse in less than one minute. In contrast, CTLA-4 resides in intracellular vesicles at a site facing away from the leading edge of the T cell. Upon priming, cytoskeletal reorganization moves these vesicles towards the synapse where the CTLA-4 accumulates. This translocation process is completed within one hour and is facilitated by cytoskeletal rearrangements that are initiated by TCR signaling. The amount of CTLA-4 that translocates to the synapse correlates with the strength of the TCR signal, suggesting that T cells receiving stronger stimuli are more susceptible to CTLA-4-mediated inhibition. This may represent a novel feedback control mechanism in which a stimulatory signal regulates the recruitment of an inhibitory receptor to a functionally relevant site on the T cell surface.

Ligand binding is important for the localization of CD28 and CTLA-4 to the immunological synapse. CD28 shares its ligands, CD80 and CD86, with CTLA-4. Due to affinity differences, CD86 is preferred to CD80 for CD28 localization, whereas CD80 is the preferred ligand for CTLA-4 localization. Both B7 molecules stabilize CD28 and CTLA-4 by preventing their diffusion away from the synapse.

10.15.1.2 Signaling through CTLA-4

In one proposed mechanism for T cell inactivation by CTLA-4, the interaction of recruited CTLA-4 with the TCR zeta chain (TCRζ) recruits the tyrosine phosphatase, SHP-2, which promotes TCRζ dephosphorylation and thereby inhibits TCR signal transduction (Lee *et al.*, 1998).

The observation that CTLA-4 expression is maximal two to three days after T cell activation; Walunas *et al.* (1994) suggests that CTLA-4 functions late in the immune response. It was proposed that the initial binding of CD80/86 to CD28 would provide co-stimulation to T cells activated by antigen presentation. At maximum expression later on, CTLA-4 interactions with CD80/86 would be dominant and the inhibitory effects of CTLA-4 would terminate the immune response.

CTLA-4 signaling inhibits production of the primary T cell proliferation factor, IL-2, and down-regulates expression of the IL-2 receptor alpha chain (IL-2Rα) by T cells. Inhibition of T cell proliferation manifests as a blocking of the cell cycle; there is no evidence that CTLA-4 ligation promotes apoptosis. It appears that CTLA-4 does not terminate ongoing responses, but rather restricts the progression of T cell activation. The effects of CTLA-4 signaling are mostly apparent in the presence of optimal CD28-mediated co-stimulation. This is consistent with the finding that CTLA-4 expression on naïve T cells is regulated by co-stimulation through CD28–CD86 interactions (Walunas *et al.*, 1996).

During the activation of normal T cells, the CD28 and CTLA-4 co-receptors on the T cell surface independently bind to CD80 and CD86 expressed on the antigen-presenting DC. CTLA-4 has a 20- to 50-fold higher affinity than CD28 for CD80/86 molecules and thus normally prevents CD80/86 on DCs from engaging CD28. In addition, CTLA-4 delivers inhibitory signals that regulate the activating signals triggered by the TCR. T cell activation is controlled by a balance between CD28–CD80/86 and CTLA-4–CD80/86 interactions. It has been postulated that the level of CD80/86 expression on the DC influences the relative engagement of CD28 or CTLA-4, and this in turn determines if the response is initiated or terminated. Invading microbes activate DCs to increase the expression of CD80/86, so that CD28 can be engaged despite its relatively low affinity. By contrast, DCss that are presenting self antigens normally express low levels of CD80/86, but these levels are sufficient to engage the high-affinity CTLA-4. Once engaged, CTLA-4 may competitively inhibit binding of CD28 to CD80/86 co-stimulatory molecules.

10.15.2 Inhibition by PD-1

10.15.2.1 Expression and Intracellular Trafficking of PD-1

PD-1 can be expressed on T cells, B cells, natural killer T cells, activated monocytes, and DCs. PD-1 is not expressed on resting T cells, but is inducibly expressed after activation. PD-1 binds either of two ligands, PD-L1 (B-H1) or PD-L2 (B7-DC), which are B7 family members. PD-L1 is constitutively expressed on T and B cells, macrophages, and DCs, and further up-regulated following activation of these cells. PD-L1 expression is also detected on various non-hematopoietic cell types, such as pancreatic beta cells. Both type I and type II interferons (IFN-α/β and IFN-γ) up-regulate PD-L1. By contrast, PD-L2 expression is regulated more tightly: it is induced on macrophages and DCs by IFN-γ, and also on monocytes and macrophages by IL-4 and GM-CSF. Both PD-L1 and PD-L2 are expressed on tumor cells from various organs.

In activated T cells, PD-1 is localized to an intracellular compartment near the Golgi apparatus and to the plasma membrane. During a T cell–APC interaction, a cytoskeletal reorganization moves PD-1 to the immunological synapse, where it engages with its ligand expressed on the APC. PD-L2 is the preferred ligand for PD-1 recruitment to the synapse, consistent with the higher affinity of PD-1 for PD-L2 relative to PD-L1. PD-L2 can therefore mediate stronger T cell inhibition in diseases such as asthma and conjunctivitis that involve high surface levels of PD-L2 on professional APCs. In diseases such as diabetes and chronic viral infections that involve low PD-L2 expression, PD-L1 is the more important ligand mediating T cell inhibition.

10.15.2.2 Signaling through PD-1

Upon PD-1 ligation at the immunological synapse, two tyrosine phosphatases, SHP-1 and SHP-2, bind to the immunoreceptor tyrosine-based switch motif (ITSM) in PD-1. The cross-linking of PD-1 and CD3 in the synapse increases the amount of SHP-2, but not SHP-1, associated with PD-1. The recruited SHP phosphatase activity inhibits downstream signal transduction through direct dephosphorylation of signaling intermediates. PD-1 engagement inhibits a TCR signal transduction pathway involving PKC-θ (theta), which is required for T cell IL-2 production.

Engagement of PD-1 by PD-L1 or PD-L2 inhibits T cell activation when CD28-mediated co-stimulation is suboptimal. Proliferation and cytokine production are inhibited in both $CD4^+$ and $CD8^+$ T cells, with proliferation being inhibited as a result of compromised IL-2 production. $CD8^+$

T cells may be more sensitive to the PD-1–PD-L inhibitory pathway than CD4+ T cells because of their intrinsic inability to produce significant levels of IL-2. PD-1 typically has greater effects on cytokine production than on cellular proliferation, with significant effects on IFN-γ, TNF-α, and IL-2. PD-1 signaling blocks cell cycle progression, but does not induce apoptosis; inhibits the induction of the cell survival factor Bcl-x_L; and inhibits the expression of transcription factors associated with T cell function. PD-1 signaling also maintains previously activated T cells in an anergic state, in part by inhibiting stable interactions with DCs and preventing further TCR engagement.

The inhibitory PD-L–PD-1 signal overcomes the stimulatory CD80/86-CD28 signal when antigenic stimulation of the TCR is weak or limiting. At high levels of TCR stimulation, the CD80/86-CD28 signal is dominant, allowing T cell activation to take place. The threshold for T cell activation is a balance between these stimulatory and inhibitory signals. Signaling through CD28 can overcome PD-1-mediated inhibition of proliferation by inducing IL-2 production.

PD-1 is not only a ligand for PD-1, but also interacts with CD80 (B7-1) to inhibit T cell proliferation and cytokine production. CD86 (B7-2) does not bind to PD-L1 or PD-L2, and PD-L2 does not bind to CD80. The PD-L1:CD80 interface overlaps with the CD80:CTLA-4 and PD-L1:PD-1 interfaces, creating additional interactions between T cells and APCs.

10.16 DIFFERENTIATION OF NAÏVE CD4+ T CELLS INTO DISTINCT HELPER T CELLS AND NAÏVE CD8+ T CELLS INTO CYTOTOXIC T CELLS

10.16.1 Cytokines Produced by Th1 and Th2 Cells

Naïve CD4+ helper T cells can differentiate into either Th1 or Th2 effectors, depending on the cytokine milieu present at the time of initial activation. The Th1 and Th2 subsets are defined by the cytokines that they produce, referred to as type 1 and type 2 cytokines, respectively. The corresponding immune responses are referred to as types 1 and 2. Representative cytokines produced by Th1 and Th2 cells are listed in Table 10.11. The signature cytokine for Th1 cells is IFN-γ, while that for Th2 cells is IL-4. Th1 cells support cell-mediated immunity and, as a consequence, promote inflammation, cytotoxicity, and delayed-type hypersensitivity. Th2 cells support humoral immunity and down-regulate the inflammatory actions of Th1 cells.

The development of each Th cell subset is induced by the types of microbes which that subset is best able to combat. Thus, Th1 cells, which act against intracellular microbes, develop in response to infection by such microbes. Similarly, Th2 cells, which defend against extracellular parasites, develop in response to such parasites. Th17 cell development is induced by the extracellular bacteria and fungi that these cells are effective against.

TABLE 10.11
Cytokines Secreted by Human CD4+ Th1 and Th2 Cells

	Th1	Th2
IFN-γ	++	–
IL-2	++	–
TNF-β	++	–
TNF-α	++	+
GM-CSF	++	+
IL-3	++	++
IL-4	–	++
IL-5	–	++
IL-6	–	++
IL-9	–	++
IL-10[a]	++	++
IL-13	–	++

[a]IL-10 is classed as a Th2 cytokine in the mouse, but is produced by both Th1 and Th2 cells in the human.

A normal immune response includes substantial numbers of naïve CD4+ T cells that have not differentiated into the Th1 or Th2 subsets. These proliferating, undifferentiated T cells (called Th0 cells) produce mainly IL-2 (the major cytokine responsible for T cell proliferation), but not IFN-γ, IL-4, or IL-5. They constitute an expanded pool of antigen-specific T cells with the potential to differentiate into the appropriate helper cell and thus provide extra flexibility for immune responses.

10.16.2 Contribution of Cytokines to the Th1/Th2/Th17 Balance

The differentiation of naïve CD4+ T cells into the Th1, Th2, and Th17 subsets of helper T cells is initiated by cytokines produced by immature DCs and other immune cells early during an immune response. The cytokines bind to receptors on the T cells and activate transcription factors that stimulate production of the cytokines that characterize each subset. Each subset expresses its own characteristic set of transcription factors. With continued activation, chromatin remodeling occurs so that genes encoding that subset's signature cytokines are accessible to transcriptional regulators, and genes that encode cytokines not produced by that subset are rendered inaccessible. These epigenetic changes ensure that each subset can produce only its characteristic set of cytokines.

The cytokines produced by a particular Th subset promote further differentiation of that subset and suppress the development of the other subsets. This mutual antagonism ensures the predominance of a particular immune response. For instance, IFN-γ secreted by Th1 cells enhances Th1 cell differentiation and inhibits the generation of Th2 and Th17 cells. Similarly, IL-4 produced by Th2 cells enhances

Th2 cell development and inhibits the generation of Th1 and Th17 cells.

During the initiation of a Th1 response, IL-12 is produced particularly by macrophages in response to certain microbial antigens, while NK cells are the main sources of IFN-γ in response to IL-12. In the case of a Th2 response, the initial source of IL-4 is less clear, since none of the professional APCs make this cytokine. Some non-APCs, such as mast cells and basophils, can produce IL-4, but these cell types are not abundant in the lymphoid organs where T cell priming occurs. Rincón *et al.* (1997) showed that IL-6, probably secreted by activated APCs, is able to induce the initial production of IL-4 by the CD4$^+$ T cells. The low levels of secreted IL-4 are sufficient to induce the production of endogenous IL-4 in an autocrine manner, while inhibiting the production of IFN-γ. The boosted levels of IL-4 can then polarize the differentiation of naïve CD4$^+$ T cells to effector Th2 cells.

10.16.3 Expression of IL-12 Receptors

The expression of functional IL-12 receptors is important for early regulation of Th phenotype development. Antigen activation of naïve CD4$^+$ T cells is required for induction of the two IL-12R subunits, IL-12Rβ1 and IL-12Rβ2, but receptor expression is also regulated by cytokines and other factors. Developing Th1 cells express both β1 and β2 subunits and are IL-12 responsive, whereas developing Th2 cells do not express the β2 subunit and become unresponsive to IL-12. The IL-12 signaling defect in Th2 cells is due to the inhibition of IL-12Rβ2 expression by IL-4 during the differentiation process. The Th2 cells' lack of responsiveness to the Th1-inducing IL-12 provides a potential mechanism for stabilizing the Th2 cell lineage. If IFN-γ is added to the developing Th2 cells *in vitro*, IL-12Rβ2 expression is maintained and the responsiveness to IL-12 is restored. IFN-γ therefore overrides the inhibitory effect of IL-4 and prevents early Th cells from premature commitment to the Th2 pathway.

10.16.4 Flexibility of Dendritic Cells in Directing Helper T Cell Development

Because helper T cell development requires the presence of different cytokines produced by immature DCs early during the immune response, several research groups have investigated whether distinct DC subsets may produce different cytokines and directly induce Th1 or Th2 development. Human monocyte-derived DCs matured *in vitro* with CD40 ligand were found to induce Th1 differentiation, whereas similarly matured pDC-derived DCs induced Th2 differentiation (Rissoan *et al.*, 1999). Mouse antigen-pulsed splenic CD8α$^+$ DCs matured *in vitro* with GM-CSF (Maldonado-López *et al.*, 1999) or *in vivo* by Flt3 ligand administration (Pulendran *et al.*, 1999) induced the development of Th1 cells when injected into mice, whereas similarly matured CD8α$^-$ DCs induced the development of Th2 cells. These findings suggest that distinct subsets of DCs are pre-programed to direct the differentiation of CD4$^+$ T cells into either Th1 or Th2 cells. However, Boonstra *et al.* (2003) showed that both mouse cDCs and pDCs generated from bone marrow, as well as splenic CD8α$^+$ and CD8α$^-$ DCs, can induce the development of both Th1 and Th2 effector cells, depending on the dose of antigen presented. In general, high antigen doses induced Th1 cell development, whereas low antigen doses induced Th2 cell development.

10.16.5 Th1 Cell Development

Th1 cell development is driven mainly by IL-12 and IFN-γ produced by DCs, macrophages, and NK cells in response to activating microbes. The master regulator of Th1 cell development is the transcription factor T-bet. Antigen-stimulated NK cells secrete IFN-γ, which binds to its receptor on the T cell and activates the transcription factor STAT1. STAT1 then stimulates expression of T-bet, which activates transcription of the IFN-γ and IL-12Rβ2 genes, as well as a second transcription factor, Hlx. Hlx interacts with T-bet to establish an open chromatin state at the IFN-γ locus. The ability of IFN-γ to stimulate T-bet expression and the ability of T-bet to enhance IFN-γ transcription sets up a positive amplification loop that drives differentiation of T cells toward the Th1 effector type. TLR4 ligation by microbial LPS induces DCs and macrophages to secrete IL-12. T cells further enhance IL-12 secretion by DCs and macrophages by virtue of CD40L on T cells engaging CD40 on these APCs. IL-12 binds to its up-regulated receptor on the T cell and further enhances IFN-γ production through activation of STAT4. IL-18, produced by macrophages and DCs, synergizes with IL-12 to stimulate IFN-γ production by the T cells and NK cells.

The events that drive differentiation of naïve CD4$^+$ T cells to the Th1 subset are coupled with transcriptional events that block differentiation into the Th2 and Th17 subsets. IFN-γ induces expression of the transcription factors IRF-1 and IRF-2, which repress IL-4 expression. (IL-4 is the major cytokine inducing Th2 cell development). Furthermore, Hlx represses IL-4Rα expression, rendering Th1 cells unresponsive to IL-4. Activated T-bet also interferes with the activity of GATA-3 (a transcription factor that promotes Th2 cell differentiation), while other signals repress and silence the GATA-3 gene.

10.16.6 Th2 Cell Development

Th2 cell development is stimulated by IL-4 and occurs in response to helminths that reside in the extracellular spaces. IL-4 is produced early in an immune response by mast cells, basophils, and eosinophils, and later by the differentiated Th2 cells themselves. The master regulator of Th2 cell differentiation is the transcription factor GATA-3. GATA-3 is low in naïve CD4$^+$ T cells in part because it is actively degraded. Stimulation of the TCR during antigen presentation triggers signals that lead to impaired GATA-3 degradation, allowing GATA-3 to accumulate and bind to regulatory sites across a cluster of genes that includes IL-4, IL-5, and IL-13. GATA-3 can directly activate these genes, as well as mediate chromatin remodeling at their loci. GATA-3 enhances its own expression through a positive feedback loop, thereby committing differentiating cells toward the Th2 effector type. Furthermore, GATA-3 blocks Th1 cell differentiation by down-regulating the expression of IL-12Rβ2. The T-bet and IFN-γ genes that are involved in Th1 cell development become silenced by mechanisms that remain to be elucidated. The IL-4 produced by the Th2 cells binds the constitutively expressed IL-4R on the undifferentiated T cells and activates STAT6, which further augments the expression of GATA-3 and thereby amplifies Th2 cell development.

10.16.7 Effect of TGF-β on Differentiation of Naïve CD4+ T Cells into Effector Th1 and Th2 Cells

Regulatory T cells produce TGF-β and stimulate TGF-β production by DCs. TGF-β blocks Th1 cell differentiation by inhibiting the expression of T-bet (Gorelik *et al.* (2002) and blocks Th2 cell differentiation through the inhibition of GATA-3 expression (Gorelik *et al.*, 2000).

10.16.8 Th17 Cell Development

Th17 cell development is stimulated by pro-inflammatory cytokines produced in response to bacteria and fungi. Th17 cells develop via a lineage distinct from the Th1 and Th2 lineages. IL-1 and IL-6 produced by antigen-stimulated DCs and TGF-β produced by various cells engage their receptors on the naïve CD4+ T cells and co-operate to activate expression of the orphan nuclear receptor RORγt, which acts as a transcription factor. IL-6 also activates the transcription factor STAT3, which functions with RORγt to regulate Th17 cell development. Activated macrophages and DCs secrete IL-23, which promotes proliferation and maintains survival of Th17 cells. TGF-β acts to up-regulate the IL-23 receptor (IL-23R) on Th17 cells, thereby conferring responsiveness to IL-23. RORγt also up-regulates the expression of IL-23R, and engagement of IL-23R by IL-23 further induces RORγt expression. In addition to RORγt, IFN regulatory factor 4 (IRF4), a transcription factor induced by TCR/CD3 signaling, is required for Th17 cell development. Th17 cells produce high levels of IL-21, which functions in a positive feedback autocrine loop to enhance Th17 cell development. IL-27 negatively regulates Th17 cell development.

Although IL-23 is required for the accumulation and function of Th17 cells in mice infected with pathogenic organisms, it is dispensable for the *in vitro* polarization of naïve CD4+ T cells to the Th17 lineage. Naïve CD4+ T cells need to be pre-conditioned by IL-6 and IL-21 in order to be responsive to IL-23 and differentiate into Th17 cells. Accordingly, IL-6 induces IL-21, which induces expression of IL-23R on Th17 cells. IL-23 produced by activated APCs can then act in synergy with TGF-β to induce expression of IL-17 and the other cytokines that characterize the Th-17 lineage.

The *in vitro* differentiation of naïve CD4+ T cells to Th17 cells is induced directly by a combination of TGF-β and IL-6, and amplified by IL-1 and TNF-α. Neither TGF-β nor IL-6 alone induces the generation of Th17 cells. The TGF-β/IL-6-dependent Th17 cell induction is inhibited by IL-2. TGF-β also acts indirectly by suppressing Th1 and Th2 cell development. The action of TGF-β on naïve CD4+ T cells is antagonized by IFN-γ and IL-4. TGF-β deficiency has been shown to impair Th17 cell development *in vivo* (Mangan *et al.*, 2006).

IL-27 limits inflammation by suppressing the development of Th17 cells. Evidence presented by Ivanov *et al.* (2007) suggested that this suppression is not simply a diversion of naïve T cells to the Th1 pathway. IL-27 was shown to prevent expression of RORγt in response to TGF-β plus IL-6, and IL-27 inhibition of the Th17 lineage was relieved by forced expression of RORγt. Thus, one mechanism by which IL-27 inhibits Th17 cell development is by inhibiting induction of RORγt.

10.16.9 Differentiation of Naïve CD8+ T Cells into Cytotoxic T Cells

The full activation of naïve CD8+ T cells and their differentiation into functional cytotoxic T cells and memory cells may require the co-operation of CD4+ helper T cells, depending on the strength of antigenic stimulation. Infectious agents can activate APCs directly, bypassing the need for CD4+ T cell help. CD4+ T cell help may be required for CD8+ T cell responses to latent viral infections, organ transplants, and tumors, all of which tend to elicit relatively weak innate immune reactions. In such cases, the CD4+ T cells may secrete cytokines that stimulate the differentiation of CD8+ T cells. The CD4+ T cells may also enhance the ability of APCs to stimulate CD8+ T cell differentiation in the following manner. Immature DCs capture antigen and process it for MHC class II-restricted presentation to CD4+ T cells. The latter cells up-regulate expression of CD40L, which interacts with CD40 expressed on the antigen-specific DCs. This interaction endows DCs with the capacity to prime naïve CD8+ T cells that recognize MHC class I-restricted antigen on the DCs. In this interaction, the DCs are said to be 'licensed.' The CD40-dependent priming of CD8+ T cells is critically dependent on CD70 expression on the licensed DCs and its specific binding to the CD27 receptor on the CD8+ T cells.

10.17 EFFECTOR FUNCTIONS OF T CELLS

10.17.1 Introduction

After leaving the thymus, naïve T cells circulate continually through secondary lymphoid organs in search of their cognate peptide–MHC complex on the surface of APCs. Following their activation in secondary lymphoid tissues, the naïve T cells differentiate into effector T cells, the functions of which are determined, at least in part, by the environment in which they are initially activated. Some CD4+ effector T cells migrate towards the B cell follicle in the secondary lymphoid organ in which they were activated, and provide help to B cells. Others, together with CD8+ effector T cells, enter the circulation via the thoracic duct and localize to extra-lymphoid tissues, where they help to co-ordinate immune responses in the periphery.

10.17.2 Helper Functions of Th1 Cells

The cytokines secreted by Th1 cells (IFN-γ, IL-2, TNF-β, etc.) activate cell-mediated immunity, i.e. cytotoxic and inflammatory responses mediated by CD8+ T cells, NK cells, and macrophages. Thus, Th1 cells provide immunity against intracellular infections with viruses, bacteria, fungi, and protozoan parasites.

A major function of Th1 cells is to activate macrophages to ingest and destroy microbes. Macrophages are relatively inefficient at killing internalized microbes; indeed, some intracellular pathogens, notably *Mycobacterium* and *Leishmania* species, find refuge and proliferate in the internal vesicles of these cells. Fragments of bacteria ingested by macrophages are degraded in lysosomes and the peptide degradation products are loaded onto MHC-II molecules. The peptide–MHC-II complexes move to the cell surface where they are recognized by Th1 cells (Figure 10.7). The Th1 cells then secrete cytokines that activate the microbicidal machinery of the macrophage.

Macrophages are activated by IFN-γ and CD40L working synergistically. CD40L expressed on the surface of the Th1 cell interacts with CD40 on the surface of the macrophage, and IFN-γ secreted by the Th1 cell binds to the IFN-γ receptor on the macrophage. IFN-γ signaling increases the phagocytic activity of macrophages by stimulating the synthesis of several enzymes within these cells. Among these enzymes, NADPH oxidase in the phagolysosome produces lethal reactive oxygen species, and iNOS in the cytoplasm produces toxic nitric oxide, which diffuses into the phagolysosome. Lysosomal enzymes digest the organisms that have been killed. In addition to enhanced killing of phagocytosed bacteria, activated macrophages secrete inflammatory cytokines (IL-1, TNF-α), and increase the expression of accessory molecules required for T cell activation (CD80/CD86 and MHC molecules).

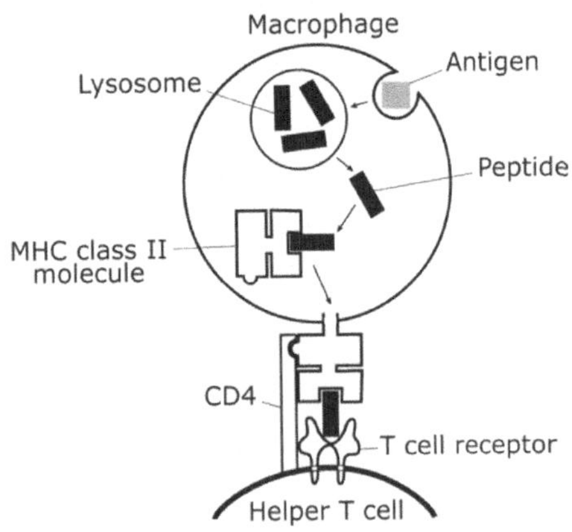

FIGURE 10.7 Antigen presentation to Th1 helper T cells by phagocytic macrophages.

10.17.3 Helper Functions of Th2 Cells

The cytokines secreted by Th2 cells activate humoral immunity, i.e. antibody production by B cells, and defend against extracellular parasites, including helminths and nematodes that are too large to be phagocytosed The type 2 cytokines co-operate to enhance the proliferation and function of eosinophils, basophils, and mast cells. In particular, IL-5 activates eosinophils, which are efficient killers of helminths. IL-13 stimulates mucus secretion from airways and the intestinal mucosa, which helps to expel parasites. Both IL-4 and IL-13 can activate B cells to switch antibody production to IgE, which triggers mast cell degranulation. IL-5, IL-9, and IL-13 facilitate the repair of tissue that is physically damaged by the migrating helminths.

10.17.4 Helper Functions of Th17 Cells

Th17 cells are abundant in mucosal tissues, particularly those of the gastrointestinal tract where they combat intestinal infections. They are also present in the skin where they maintain epithelial integrity. The signature cytokine secreted by Th17 cells is IL-17, which has pleiotropic activities. The IL-17 receptors are expressed on stromal, epithelial, and endothelial cells. IL-17 co-ordinates tissue inflammation by stimulating the local production of pro-inflammatory cytokines, chemokines, and matrix metalloproteases. IL-17 also stimulates local endothelial cells to produce chemokines that recruit neutrophils to inflammatory sites. The recruited neutrophils ingest and kill extracellular bacteria and fungi. IL-17 stimulates the production of defensins and other antimicrobial peptides from numerous cell types. Another cytokine produced by Th17 cells is IL-22, which stimulates repair reactions in epithelial tissues. A third cytokine, IL-21, promotes the differentiation of Th17 cells, providing an autocrine pathway for amplifying Th17 responses.

10.17.5 Cytotoxic T Cells

The term cytotoxic T lymphocyte (CTL) originally applied exclusively to CD8+ T cells, but since the discovery that CD4+ T cells can be cytotoxic as well, the term now applies to both CD8+ T cells and cytotoxic CD4+ T cells. To avoid ambiguity, CD8+ T cells and cytotoxic CD4+ T cells will be designated CD8+ CTL and CD4+ CTL in this book, respectively. When describing killing mechanisms, CTLs and NK cells are collectively called cytotoxic lymphocytes (CLs).

10.17.5.1 CD8+ CTLs

CD8+ CTLs are specialized to deal with viruses and bacteria that have found their way into the interior of host cells. Such intracellular pathogens are inaccessible to antibodies and phagocytes and the only way to eliminate them is to destroy the infected cell. Viruses are unable to replicate when deprived of the replicative machinery provided by the host cell. CD8+ CTLs recognize infected host cells through specific binding of their TCR to an antigenic peptide–MHC class I complex displayed on the surface of the infected cell (Figure 10.8).

The major mechanism of CD8+ CTL cytotoxicity is the release of cytolytic molecules from cytoplasmic granules by exocytosis – a process termed degranulation. The granules are specialized secretory organelles with a morphology similar to lysosomes. In response to a signal generated upon specific recognition of the target by the CD8+ CTL, the granules re-orient within the cell and move towards the point of contact where they fuse with the plasma membrane and release their contents into the intercellular space. After destroying the target cell, the CD8+ CTL disengages and seeks another target cell.

Target cell recognition triggers the *de novo* synthesis of lytic proteins, a proportion of which acquires a granule targeting signal and re-fills granules. The constant replenishment of granule contents enables a single CD8+ CTL to kill multiple targets. A larger proportion of the newly synthesized proteins fails to acquire the granule targeting signal and is secreted directly via a non-granule route. These directly secreted proteins mediate killing of non-antigen-bearing bystander cells.

Two types of cytolytic molecules released from the granules are perforin and granzymes. Perforin is a pore-forming protein and granzymes are serine proteases. The targeted delivery of newly synthesized perforin from the endoplasmic reticulum to the secretory granules is rapid and efficient and does not allow this toxic protein to damage the CTL itself during transit. Perforin is inactive in the acidic environment within the secretory granules. It is released as a monomer, but in the presence of Ca^{2+} the monomers polymerize to form tubular structures which insert into the target cell plasma membrane. The transmembrane pores thus formed may cause death of the target cell by osmotic lysis. However, many target cells are not killed by perforin alone and granzymes are necessary to kill them. The perforin pores permit the entry of granzymes into the target cell. Of the many granzymes, the most abundant and best understood are granzymes A and B. Granzyme A can promote the activation of a nuclease through proteolysis of its inhibitor, and this results in multiple breaks of single-stranded DNA within the target cell. Granzyme B induces apoptosis by triggering the activation of multiple caspases. In addition, granzyme B induces DNA fragmentation without the requirement for caspases. Newly synthesized granzymes A and B are targeted to the lytic granules by the mannose-6-phosphate receptor, which is the signal for the targeting of lysosomal enzymes to lysosomes. The granzymes are recognized by the mannose-6-phosphate receptor in the Golgi apparatus and shuttled to the granules by this receptor. A less efficient mannose-6-phosphate receptor-independent pathway to the lytic granules also exists for these granzymes. In addition to perforin and granzymes, the granules contain a full complement of lysosomal enzymes.

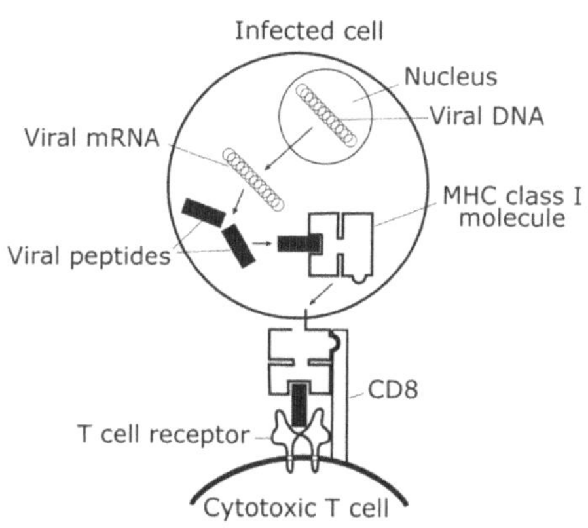

FIGURE 10.8 Recognition of a virus-infected cell by a CD8+ cytotoxic T cell.

CD8+ CTLs can also kill bystander cells by inducing apoptosis. Human activated CD8+ CTLs express Fas ligand (FasL) and TNF-α on their surface and also release soluble TNF-α. The induced expression of these ligands allows apoptotic signaling through FasR and TNF-R1 death receptors expressed on neighboring target cells. In addition to specific target cell killing by perforin, both FasR- and TNF-R1-mediated apoptosis are used by hepatitis C virus (HCV)-specific human CD8+ CTLs to kill virally infected hepatocytes and uninfected bystander hepatocytes. FasL-mediated bystander lysis requires a close cell–cell contact between activated CD8+ CTLs and bystander target cells, but TNF-α-mediated cytolysis does not.

10.17.5.2 CD4+ CTLs

Activated macrophages are susceptible to Fas-mediated apoptosis induced by CD4+ Th1 cells (Ashany *et al.*, 1995). Th1 cells can also mediate Fas-dependent apoptosis of bystander cells. In the direct killing mechanism, the macrophage presents antigen to the Th1 cell. The activated Th1 cell up-regulates FasL, which can interact with FasR expressed by the macrophage. Fas signaling leads to death of the macrophage by apoptosis. In bystander killing, the FasL induced by antigen presentation can interact with FasR expressed by any nearby cell, leading to apoptosis of the bystander target as well as the macrophage (Thilenius *et al.*, 1999).

The death receptor TRAIL is constitutively expressed on human CD4+ T cell clones and is responsible for spontaneous bystander cytotoxicity against certain tumor target cells. TRAIL does not induce apoptosis in normal cells, possibly because of preferential expression of the decoy receptor TRAIL-R3.

10.17.6 How Do Cytotoxic T Cells Kill Target Cells without Killing Themselves and Causing Collateral Damage?

A single $CD8^+$ CTL can kill multiple target cells within a few hours without itself being killed, indicating that these cells somehow resist self-induced or fraternally-induced cytolysis. $CD8^+$ CTLs are resistant to lysis induced by CTL granule extracts and purified perforin. Balaji *et al.* (2002) showed that CTLs and NK cells die within a few hours via perforin-dependent suicide when induced to degranulate in the presence of nontoxic, membrane-impermeable cathepsin B-specific protease inhibitors. This finding implicates cathepsin B in the protection of these killer cells from self destruction after degranulation. Cathepsin B is a lysosomal protease also found in lytic granules. Balaji and colleagues showed that resting $CD8^+$ CTLs express low to undetectable levels of cathepsin B on their plasma membrane, but substantial levels soon after degranulation. They demonstrated that cathepsin B readily cleaves membrane-associated perforin before it has a chance to form a pore. These results provide evidence that surface cathepsin B protects $CD8^+$ CTLs from self-destruction after degranulation.

A serine protease inhibitor (serpin) called proteinase inhibitor 9 (PI-9) in the cytosol of $CD8^+$ and $CD4^+$ CTLs specifically inhibits granzyme B. Human PI-9 has seven murine homologs, one of which is serine protease inhibitor 6 (SPI-6). PI-9 is a non-secretory protein and occupies a separate subcellular compartment to granzyme B. PI-9 protects CTLs against cytolysis by misdirected granzyme B without interfering with their capacity to kill target cells by the Fas pathway of apoptosis. PI-9 is up-regulated in CTLs in response to granzyme B production and degranulation. Because $CD8^+$ CTLs express both Fas and FasL, they can kill each other (fratricide) at the end of an immune response to control the size of the memory T cell pool.

In addition to its expression in CTLs., PI-9 is expressed in specific DC subsets of both myeloid and lymphoid lineage, and is up-regulated during their maturation. The presence of PI-9 in APCs (DCs, macrophages, and B cells) affords protection from collateral damage mediated by granzyme B during their association with the CTLs. In mice, Th1 cells induce SPI-6 in DCs and subsequent DC resistance to granzyme B, whereas Th2 cells neither induce SPI-6 nor convey protection. PI-9 is also expressed in endothelial and mesothelial cells likely to be exposed to CTLs and is up-regulated by inflammatory stimuli. Furthermore, PI-9 protects cells present in immune privileged sites, such as Sertoli cells in the testis, granuloma cells in the ovary, intermediate trophoblastic cells in placental tissue, and lens cells in the eye.

10.17.7 γ δ T Cells

γ δ T cells make up less than 5% of circulating T cells, but selectively populate epithelial-rich tissues such as the skin, intestine, reproductive tract, and lungs, where they make up about 50% of total T cells. The high frequency of pre-existing γ δ T cells enables responses to occur rapidly without the need for prior expansion. γ δ T cells do not recognize peptide antigen presented by classical MHC class I or class II molecules and do not express either CD4 or CD8 co-receptors. The TCR of γ δ T cells can bind directly to soluble antigens released from microbial pathogens and damaged host cells. Antigens recognized by γ δ T cells include lipids, small phosphorylated molecules, and nucleotide conjugates. Damage to epithelia induces the surface expression of non-classical MHC class I-like molecules, such as MIC-A and MIC-B, which are ligands for the γ δ T cell-activating receptor, natural killer group 2, member D (NKG2D). The engagement of inhibitory receptors on γ δ T cells maintains a high activation threshold necessary to avoid unprovoked tissue injury. Animal studies have indicated a biphasic action of γ δ T cells. The early secretion of pro-inflammatory cytokines and cytotoxic molecules provides a rapid host defense in advance of the α β T cell response. After the arrival of α β T cells, recruited blood-borne γ δ T cells secrete epithelial cell growth factors that contribute to tissue repair.

Bonneville *et al.* (2010) reviewed the role of γ δ T cells in sensing cellular stress and infection. γ δ T cells can kill infected or transformed cells through pathways that involve the engagement of death-inducing receptors, such as Fas and TRAIL receptor. Activated γ δ T cells secrete cytotoxic perforin and granzymes, as well as bacteriostatic molecules such as granulysin. γ δ T cells can produce cytokines that are involved in protective immunity against viruses and intracellular pathogens (TNF-α and IFN-γ), extracellular bacteria and fungi (IL-17), and extracellular parasites (IL-4, IL-5, and IL-13). γ δ T cells can also produce immunosuppressive cytokines (IL-10 and TGF-β).

10.18 MEMORY T CELLS

10.18.1 Overview

Once a pathogen has been effectively dealt with in a primary immune response, most antigen-specific T cells die through loss of survival signals, thus restoring their initial number. A minor fraction of the antigen-specific T cells, about 5% of the peak frequency, survives and differentiates into memory T cells. In contrast to naïve T cells, which have a lifetime of weeks or months, memory T cells may persist for years in the absence of cognate antigen. Long-lasting immunity in the steady state is attributable to homeostatic proliferation of memory T cells held temporarily in the bone marrow and a slow turnover of tissue-resident memory T (T_{RM}) cells. Recirculating memory T cells provide systemic immunity, while T_{RM} cells provide protection at the site of pathogen invasion.

Upon re-infection with the same pathogen, the memory T cells rapidly proliferate and differentiate into secondary effector T cells. This secondary (or recall) response is faster (within hours of re-infection) and more efficient than the primary response. This is because T cells specific for a given antigen have undergone clonal expansion during the primary response and the derived memory cells are typically 10- to 100-fold more numerous than the pool of naive T cells before antigen encounter. Furthermore, the memory cells have a lower activation threshold than their naïve counterparts and they secrete cytokines faster and at a higher quantity.

Both $CD4^+$ and $CD8^+$ T cells can become memory cells with helper and cytotoxic effector functions, respectively. Some $CD4^+$ memory T cells are derived from fully differentiated Th1, Th2, or Th17 effectors and retain their respective cytokine secretion profiles upon re-activation. Other $CD4^+$ memory T cells are derived from precursors before commitment to these effectors and display a random selection of cytokines upon re-activation. This section concentrates mainly on memory $CD8^+$ T cells, since these have been more intensively studied than memory $CD4^+$ T cells.

10.18.2 Memory $CD8^+$ T Cell Differentiation during Acute Viral Infection

During a $CD8^+$ T cell primary response to an acute viral infection, there are three characteristic phases: (1) initial activation of naïve $CD8^+$ T cells and a 10^4- to 10^5-fold expansion of effector $CD8^+$ T cells lasting for seven to ten days; (2) apoptotic death of 90–95% of the activated effector $CD8^+$ T cells following pathogen clearance; and (3) the terminal differentiation of surviving effector cells into end-stage effectors and memory cells along a continuum. The end-stage effectors die soon after pathogen clearance, leaving a stable population of antigen-experienced memory cells.

Prlic and Bevan (2008) proposed that the fate of a $CD8^+$ effector T cell is predetermined before the onset of contraction. Death of the end-stage effectors could be the default pathway and a combination of epigenetic modifications and transcriptional changes could allow a subset of cells to survive and progress to the memory stage.

10.18.3 Memory T Cell Recirculation

Mackay *et al.* (1990) compared the preferred recirculation pathways of naïve and memory T cells. Naïve T cells traffic from blood to lymph nodes, whereas memory T cells traffic from blood to peripheral tissues. These different pathways may be of significance in immunological surveillance of the body for foreign antigens. Recirculation through lymph nodes is the most effective route for naïve T cells to make contact with a new antigen. Conversely, it makes sense that memory T cells should migrate through peripheral tissues such as skin, where antigens are first encountered. Gebhardt *et al.* (2011) demonstrated different patterns of peripheral migration by memory $CD4^+$ and $CD8^+$ T cells.

The bone marrow is a preferred site for memory T cells undergoing homeostatic proliferation in the absence of antigen. *In vivo* experiments have shown that the bone marrow represents a temporary stopping point for recirculating memory T cells. Bone marrow memory T cells exhibit some phenotypical differences when compared with corresponding cells from lymphoid periphery and blood. For example, a high proportion of bone marrow memory $CD4^+$ and $CD8^+$ T cells express CD69, which is essential for the retention and persistence of the memory cells in the bone marrow (Shinoda *et al.*, 2012).

When circulating memory T cells reach the bone marrow, some pass straight through the parenchymal tissue and out into the blood, while others are held for a while within survival niches composed of IL-7-producing stromal cells. The secreted IL-7 provides survival signals for resting $CD4^+$ memory T cells. A minor proportion of the retained cells self-renew, which counterbalances losses due to cell death. This dynamic process of homeostatic proliferation contributes to maintaining constant numbers of recirculating memory T cells over time. Recirculating memory T cells provide immunity against systemic pathogens such as the measles virus.

10.18.4 Central and Effector Memory T Cells

Two subsets of memory $CD8^+$ T cells have been characterized – central memory T (T_{CM}) cells and effector memory T (T_{EM}) cells. These cells circulate throughout the body and enter selected tissues by means of specific homing molecules. Immediately after pathogen clearance, the antigen-specific memory T cell pool is composed mainly of T_{EM} cells, which gradually transform into the T_{CM} phenotype over time. T_{CM} cells express the lymphoid homing receptors CCR7 and CD62L (L-selectin), which enable them to enter lymph nodes and the spleen. T_{CM} cells lack effector function, but they proliferate rapidly and differentiate into T_{EM} cells when activated by antigen-bearing DCs in the lymph nodes. T_{CM} cells therefore constitute a pool of already clonally expanded cells that can generate large numbers of antigen-specific T_{EM} cells.

T_{EM} cells do not express CCR7 or CD62L; instead, they express tissue-specific homing receptors, such as CD103 and CLA, which direct the cells to peripheral non-lymphoid tissues. Entry of T_{EM} cells into these tissues triggers the induction of granzyme B and cytolytic activity. T_{EM} cells also express receptors for inflammatory chemokines, such as CCR1, CCR3, and CCR5, which direct the cells to inflammatory sites in the tissues. T_{EM} cells have a lower proliferative capacity than T_{CM} cells and decline more rapidly with time, but they exert effector functions immediately upon antigen challenge.

Commitment to a particular memory $CD8^+$ T cell lineage is dictated by the number of naïve T cell precursors. In physiological conditions of low precursor frequency, distinct T_{CM} and T_{EM} lineages develop without the ability to interconvert. This supports a model in which antigenic stimulation of naïve T cells gives rise to effector T cells that then develop into T_{CM} and T_{EM} memory cells as separate lineages (Lefrancois & Marzo, 2006). These memory T cell types do not interconvert and both are self-renewing, but at different rates.

10.18.5 Tissue-Resident Memory T Cells

Populations of non-circulating $CD4^+$ and $CD8^+$ memory T cells reside permanently in secondary lymphoid organs, skin, and the surface linings of many peripheral tissues, including lung, intestine, liver, brain, and female reproductive tract. The highest frequencies of these T_{RM} cells are found in previously infected tissues. T_{RM} cells are distinct from circulating T_{CM} and T_{EM} cells in that they express CD69. In addition, human T_{RM} cells have an enhanced capacity to rapidly mobilize an immune response at the preferred sites of re-infection with specific pathogens, whilst limiting inflammation-induced tissue damage. The quiescent state of T_{RM} cells promotes longevity and prevents inappropriate activation by non-pathogenic antigens to which many human tissues are continually exposed. Thus, T_{RM} cells provide a rapid localized first line of defense that is reinforced by the delayed arrival of recirculating memory T cells.

10.18.6 Primary Versus Secondary Memory $CD8^+$ T Cell Responses

Most studies of memory $CD8^+$ T cell behavior have focused on the primary memory response that follows an initial infection or immunization. Many vaccination regimes require booster immunizations to engender the most robust protective memory T cell response. The $CD8^+$ T cell response to initial and subsequent antigen challenge shows an increase in secondary memory $CD8^+$T cell numbers compared with primary stimulation (Harty & Badovinac, 2008). Secondary responses generate cells that retain a T_{EM} phenotype and accumulate in non-lymphoid tissues. Secondary memory cells display higher and more sustained levels of cytotoxic molecules than their primary counterparts, and provide better protection against challenge by a virulent pathogen.

It has been proposed that the number of memory $CD8^+$ T cells in the host is inflexible, and that individual cells are constantly competing for limited space. Consequently, vaccines that introduce over-abundant quantities of memory $CD8^+$ T cells specific for a particular antigen could displace memory $CD8^+$ T cells specific for all previous infections. Vezys *et al.* (2009) tested this proposal by introducing into mice as many long-lived memory $CD8^+$ T cells specific for a single vaccine as there were memory $CD8^+$ T cells in the host before vaccination. Contrary to expectations, the size of the $CD8^+$ T cell compartment doubled to accommodate these new cells, and pre-existing memory $CD8^+$ T cells specific for a previously encountered infection were largely preserved. It therefore appears that the size of the memory $CD8^+$ T cell compartment increases with immunological experience.

10.18.7 Migratory Behavior of Memory $CD8^+$ T Cells

Circulating memory $CD8^+$ T cells migrate into the surrounding tissue at common entry sites (bone marrow, lymph nodes, spleen, liver, lung parenchyma) and at restricted entry sites (brain, lung airways, peritoneal cavity, intestinal lamina propria, skin). Migration at common entry sites requires that memory T cells express homing molecules that are not tissue-specific, such as $\alpha_L\beta_2$ and $\alpha_4\beta_1$ integrins and P-selectin ligand. By contrast, migration at restricted entry sites requires that memory T cells express tissue-specific adhesion molecules and/or chemokine receptors. For example, gut-homing memory T cells express $\alpha_4\beta_7$ integrin, which binds to the mucosal adhesion molecule MadCAM-1, and the chemokine receptor CCR9, which interacts with the gut-associated chemokine CCL25. The migratory behavior of memory T cells can be modified during a secondary pathogen challenge, depending on the route of infection. This flexibility allows T cells to respond efficiently to secondary infections that are localized at different sites from the primary infection.

10.18.8 Requirement for $CD4^+$ T Cell Help in Generating Memory $CD8^+$ T Cells

Helper $CD4^+$ T cells are absolutely required for the differentiation of $CD8^+$ T cells into long-lived, functional memory cells. The helper cells imprint a differentiation program during the priming of naïve $CD8^+$ T cells by first activating DCs via CD40–CD40L interactions. During infection with influenza virus, CD27, OX40, and 4-1BB collectively determine formation of memory $CD8^+$ T cells and their capacity for secondary expansion, with CD27 making the greatest contribution. CD27 transmits the programed $CD4^+$ T cell help signal from DC to $CD8^+$ T cell. Once primed in the presence of $CD4^+$ T cells, the $CD8^+$ T cells acquire the ability to undergo secondary expansion without the need for T cell help. Memory $CD8^+$ T cells primed in the absence of help proliferate poorly and their cytokine secretion is low and similar to that of naïve cells, even if $CD4^+$ T cells are present. The priming of $CD8^+$ T cells by CD40-licensed DCs is critically dependent on the expression of CD70 (the ligand for the CD27 receptor) on the DCs. Blockade of CD70 during priming does not affect the primary $CD8^+$ T cell response, but inhibits the secondary response by reducing the number of memory $CD8^+$ T cells generated. The prevention of Fas-dependent apoptosis of primary $CD8^+$ T cells by late signals from CD27 in response to viral infection is required for optimal quality of memory $CD8^+$ T cells.

10.19 NATURAL KILLER T CELLS

10.19.1 Introduction

Natural killer T (NKT) cells are unique lymphocytes that interact specifically with lipid antigens presented by CD1d.

In mice, NKT cells occur at the highest frequency in liver (10–40% of liver lymphocytes), but they are present at lower frequencies (typically <1%) in thymus, bone marrow, spleen, lymph nodes, and blood. Murine NKT cells can be divided into the subgroups of classical and non-classical. Classical NKT cells express a TCR consisting of an invariant Vα14-Jα18 α chain paired with a Vβ8.2, Vβ2, or Vβ7 β chain. Non-classical NKT cells do not express the Vα14-Jα18 α chain and utilize a diverse TCR repertoire. Classical NKT cells constitute the majority of NKT cells in the mouse and are the most studied, therefore the generic term 'NKT cells' will be used from now on to represent solely these cells.

10.19.2 NKT Cell Cytokine Secretion Pattern

Activated NKT cells rapidly secrete large amounts of cytokines of both Th1 (primarily IFN-γ) and Th2 (primarily IL-4) types. NKT cells seem to store or pre-form some mRNA for cytokines even before activation with exogenous antigens. The cytokines secreted can stimulate DCs, macrophages, NK cells, T cells, and B cells, therefore NKT cells have an impact on both the innate and adaptive arms of the immune system. Individual NKT cells are able to make both Th1- and Th-2-type cytokines simultaneously, described as a Th0 cytokine secretion pattern. Th1 cytokines often antagonize the action of Th2 cytokines and vice versa. The potential for NKT cells to produce a broad spectrum of cytokines with opposing activities provides a delicate means of regulating the immune response.

10.19.3 Activation of NKT Cells

NKT cells are activated indirectly by a limited repertoire of glycolipid antigens that are ignored by T cells. Activation occurs through interaction with an APC. The APC (typically an immature DC) internalizes the antigen, processes it, and presents it bound specifically to a CD1d molecule expressed on the APC. The internalized antigens are delivered to late endocytic and lysosomal compartments within which they are loaded onto CD1d molecules. The antigen-loaded CD1d molecules then travel to the cell surface where they can be recognized by the TCR of an NKT cell.

The major natural ligand of mouse CD1d is a glycosylphosphatidylinositol. Of the lipid antigens tested *in vitro*, the most efficient is α-galactosylceramide (α-GalCer) (Figure 10.9), a synthetic compound that was originally isolated from a marine sponge. α-GalCer is a selective NKT cell ligand that binds to CD1d and strongly stimulates NKT cell function. CD1d is expressed by hepatocytes, macrophages, DCs, and B cells, therefore each of these cell types has the potential to present α-GalCer.

Dendritic cells present α-GalCer in association with CD1d to NKT cells, which respond by secreting type 1 (IFN-γ) and type 2 (IL-4 and IL-13) cytokines. IFN-γ activates NK cells and T cells, whereas IL-4 and IL-13 activate B cells. α-GalCer recognition also induces the NKT cell

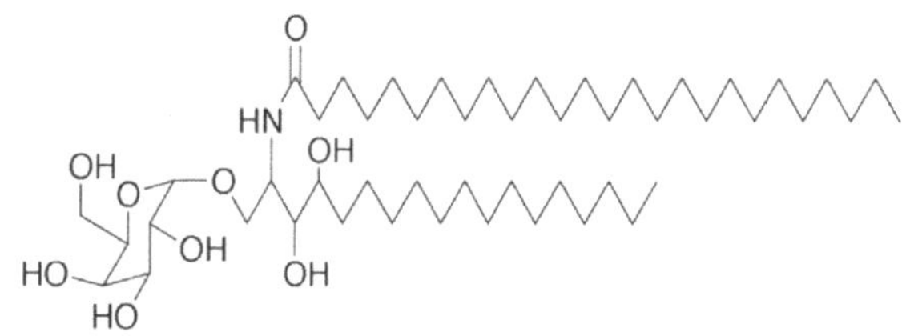

FIGURE 10.9 Structure of α-galactosylceramide.

to up-regulate CD40L, which engages CD40 on DCs and stimulates them to produce IL-12. IL-12 activates NK cells and T cells and also functions to bias cytokine production by the NKT cells toward IFN-γ rather than IL-4 (Vincent *et al.*, 2003).

10.19.4 NKT Cell Help for B Cells

Human NKT cells activated with α-GalCer can provide CD1d-restricted help to B cells, inducing proliferation of naïve and memory B cells and antibody production. Engagement of CD40 and CD80/86 co-stimulatory molecules is important for B cell proliferation and antibody responses. A small fraction of NKT cells express the CCR7 and CXCR5 homing chemokine receptors for lymphoid tissues, indicating that interactions between NKT cells and B cells can occur to some extent in these tissues. The great majority of NKT cells is equipped with the CXCR3 and CCR5 homing receptors for inflamed tissues, which suggests that NKT cells could activate B cells that have migrated to these sites.

10.19.5 Anti-Tumor Activity of NKT Cells

Both exogenous α-GalCer and IL-12 promote tumor rejection through activation of NKT cells. NK cells and cytotoxic T cells have been implicated as downstream effector cells in the α-GalCer-induced anti-tumor response. NKT cells activated by IL-12 preferentially produce large amounts of IFN-γ. Suboptimal doses of α-GalCer and IL-12 act synergistically *in vivo* to activate NKT cells and the IFN-γ produced induces the expression of IL-12 receptors on these cells. The selective activation of NKT cells by α-GalCer requires IL-12 produced by DCs and necessitates direct contact between the two cell types through CD40–CD40L interactions. This implies that IL-12 produced by activated DCs stimulates NKT cells to produce IFN-γ that in turn activates NK cells and cytotoxic T cells to mediate tumor rejection.

10.20 INNATE LYMPHOID CELLS AND NATURAL KILLER CELLS

10.20.1 Nomenclature

Innate lymphoid cells (ILCs) are a family of lymphocyte-like immune cells that are resident at mucosal sites such as the lung and small intestine, where they contribute to barrier immunity. They have also been found in bone marrow, secondary lymphoid organs, and many nonlymphoid tissues. ILCs are maintained at their anatomical sites by self-renewal. This sedentary existence is consistent with the proposed roles of ILCs as sentinels and local keepers of tissue function.

Initial studies of ILCs were confused by the number of different names used to characterize these cells. In 2013, an international group of experts proposed that ILCs should be categorized into three groups – group 1 (ILC1), group 2 (ILC2), and group 3 (ILC3) – based on the similar cytokine secretion patterns of CD4$^+$ Th1, Th2, and Th17 cells, respectively (Spits *et al.*, 2013). In this nomenclature, NK cells are placed in group I ILCs because of their shared capacity to produce IFN-γ. However, because of the different developmental requirements between NK cells and ILC1s, NK cells are no longer classified as group 1 ILCs and are regarded as innate counterparts of CD8$^+$ cytotoxic T cells. An ILC/NK precursor derived from a common lymphoid progenitor (CLP) in either the fetal liver or adult bone marrow can develop into all ILC subsets and NK cells (Artis & Spits, 2015).

10.20.2 Common Properties of Innate Lymphoid Cells

Although few studies have demonstrated the expression of Toll-like receptors (TLRs) in mouse ILCs, expression of TLR2 and TLR9 has been reported in human peripheral blood NK cells and also expression of TLR2 in group 3 ILCs (Sonnenberg & Artis, 2012), indicating that NK cells and ILCs can directly sense bacterial components. In contrast to T cells and B cells, ILCs do not express antigen-specific receptors or undergo clonal expansion and differentiation when stimulated. This allows the ILCs to act early against pathogens before the initiation of an adaptive immune response. Signaling molecules produced in infected tissues regulate the maturation and activation of ILCs, which in turn secrete cytokines that mediate the appropriate immune response. Strikingly, the different ILC populations have distinct patterns of cytokine production that mirror the cytokine-secreting profiles of T helper (Th) cell subsets. Cytokine signaling through their receptors promotes antimicrobial immunity, inflammation, and tissue repair at barrier surfaces. Inappropriate activation of ILCs has been linked to the pathogenesis of multiple chronic diseases, including inflammatory bowel disease and asthma.

10.20.3 Group 1 ILCs

Group 1 ILCs promote type 1 cytokine-dependent immunity to viruses and intracellular bacteria and protozoa. Klose *et al.* (2014) identified an ILC1 subset in mouse intestine whose lineage specification was distinct from that of NK cells. The cells produced copious amounts of IFN-γ in response to IL-12. A similar ILC1 subset was identified in the lamina propria of human intestine (Bernink *et al.*, 2013). This subset expressed high amount of CD127, which is the α chain of the IL-7 receptor (IL-7Rα), so the cells are referred to as CD127$^+$ ILC1s. At least a portion of these cells differentiated from ILC3 cells under the influence of IL-12; these differentiated cells are referred to as ex-ILC3s. Another ILC1 subset found in the intraepithelial space in the intestine of both mice and humans lacked CD127 expression (Fuchs *et al.*, 2013). This subset expressed CD103, which is the α_E subunit of the $\alpha_E\beta_7$ integrin: this adhesion molecule interacts with E-cadherin to promote interaction with epithelial cells. The cells of this subset, referred to as CD103$^+$ ILC1s, also expressed the NK cell-activating receptors NKp44 (human cells) and NKp46 and NK1.1 (mouse cells). Approximately 50% of the human CD103$^+$ ILC1s contained intracellular perforin and granzymes, consistent with potential cytotoxic activity; therefore, this population could be classified as a subset of NK cells. Both CD127$^+$ and CD103$^+$ ILC1 cell numbers were increased in Crohn's disease patients. Thus, ILC1s may initiate IFN-γ-mediated immune responses against pathogens, but contribute to inflammatory bowel disease (IBD) when dysregulated.

10.20.4 Group 2 ILCs

ILC2s promote type 2 cytokine-dependent immunity to extracellular parasitic worms (helminths and nematodes). The cells are found at many sites, including spleen. liver, lung, intestinal lamina propria, skin, bone marrow, and adipose tissue. Extracellular parasitic worm infection results in the release of IL-25, IL-33, and thymic stromal lymphopoietin (TSLP) from epithelial cells. These cytokines induce the proliferation of ILC2s and their secretion of type 2 cytokines, e.g. IL-4, IL-5, IL-9, and IL-13. The combined action of these cytokines mediates worm expulsion. ILC2s also secrete amphiregulin, a member of the epidermal growth factor family that can help repair epithelia damaged by infection.

ILC2 cells and CD4$^+$ T cells co-operate to mediate type 2 immune responses. ILC2s influence the differentiation of naïve CD4$^+$ T cells to a Th2 phenotype *in vitro*. The differentiation effect requires cell–cell contact via the TCR and is apparently independent of the cytokine milieu. ILC2s express the MHC class II and so are able to present peptide antigen to CD4$^+$ T cells, thereby augmenting dendritic cell-induced T cell activation. The activated T cells produce IL-2, which induces the ILC2s to proliferate and secrete IL-13, thereby amplifying type 2 immune responses. The cross-talk between ILC2s and CD4$^+$ T cells contributes to their mutual maintenance, expansion, and cytokine production. Th2 cell responses are impaired in the absence of ILC2s and deletion of MHC class II renders ILC2s incapable of efficiently inducing parasitic helminth expulsion. Thus ILC2s function not only as a first line of defense via their early production of type 2 cytokoines, but also are important for the induction of the adaptive immune response.

10.20.5 Group 3 ILCs

Group 3 ILCs are defined by their capacity to produce the Th17 cell-associated cytokines IL-17A and/or IL-22 when stimulated by IL-23 and IL-1β. The cytokines produced protect against extracellular bacteria and fungi. In addition, IL-22 promotes anatomical containment of symbiotic bacteria resident in gut-associated lymphoid tissue (Sonnenberg *et al.*, 2012).

ILC3s can be divided into two subgroups: one that expresses the natural cytotoxicity triggering receptor (NCR^+) and one that does not (NCR^-) (Spits *et al.*, 2013). (*Note.* The NCR comprises NKp46 in mice and NKp44 in humans.) NCR^+ ILC3s produce IL-22 (but not IL-17A), while NCR^- ILC3s secrete IFN-γ in addition to IL-17A and IL-22. IL-22 acts on IL-22R$^+$ intestinal epithelial cells and Paneth cells to increase production of the Reg3 family of antimicrobial peptides.

ILC3s located in the spleen integrate stromal and immunological signals to enhance antibody production by splenic marginal zone B cells (Magri *et al.*, 2014). von Burg *et al.* (2014) showed that splenic but not intestinal NCR^- ILC3s express MHC class II and the co-stimulatory molecules CD80 and CD86 act upon IL-1β stimulation and become *bona fide* APCs capable of promoting ovalbumin-specific CD4$^+$ T cell proliferation in mice. The splenic ILC3s are strategically located at the interface between B and T cell zones in the lymphoid microenvironment. Antigen-driven T and B cell responses were impaired *in vivo* when this cellular interaction was disrupted, hence splenic ILC3s are fully capable of priming T cell responses.

10.20.6 Natural Killer Cells

Natural killer (NK) cells are innate cytotoxic cells that produce the signature type 1 cytokine IFN-γ. NK cells are dispersed in secondary lymphoid organs, blood, and peripheral organs. Their targets are virally or bacterially infected host cells, some tumor cells, and cells that suffer stress or DNA damage. NK cells express an array of activating and inhibitory receptors on their surface. In general, the activating receptors recognize ligands on infected and abnormal cells, and the inhibitory receptors recognize normal healthy cells. Almost all healthy cells in the body express surface MHC class I molecules, which are recognized by inhibitory receptors on NK cells. Expression of MHC class I therefore protects healthy cells from attack by NK cells. Virus infection or cell damage causes a loss of MHC class I expression, thus depriving the NK cell inhibitory receptors of their ligands. At the same time, the NK cells will receive activation signals from the infected cells through activating receptors. The net result will be activation of the NK cells to kill the infected or damaged cells.

Binding of surface components on the target cells to activate receptors on NK cells triggers the release of cytotoxic granules, which contain perforins and granzymes in addition to proteases and other destructive enzymes. Release of the granules into the extracellular space results in the formation of perforin pores in the target cell membrane. This facilitates the entry into the cell of enzymes that cause lysis and granzymes that induce apoptosis – a lethal combination. NK cells express the granzyme B inhibitor, PI-9, which protects them from self-destruction by misdirected granzyme B.

NK cells also produce IFN-γ in response to IL-12 and IL-18. IFN-γ promotes an antiviral state in the local site, activates macrophages to kill phagocytosed bacteria, and promotes Th1 cell differentiation in secondary lymphoid organs. Other cytokines produced by NK cells include TNF-α (pro-inflammatory), IL-10 (immunosuppressive), and the growth factors IL-3, GM-CSF, and G-CSF. At a later stage of the immune response, NK cells bind to antibody-coated cells by Fc receptors and destroy these cells by the process of antibody-dependent cellular cytotoxicity (ADCC). Finally, NK cells play a role in cancer prevention by expressing FasL and TRAIL, which effect apoptosis of tumor cells.

Lucas *et al.* (2007) demonstrated a unique *in vivo* role of DCs for the priming of naïve NK cells. After TLR stimulation by viruses and bacteria, NK cells are recruited to local lymph nodes where priming takes place. Priming is mediated by IL-15, which is presented by DCs on the IL-15 receptor α chain (IL-15Rα). IL-15 is produced by DCs in response to microbial stimulation and expression of the IL-15Rα is up-regulated by TLR triggering in a type I IFN-dependent manner. A likely source of type I IFN for NK cell activation is plasmacytoid DCs, which secrete this cytokine in response to viral stimulation. Once primed, effector NK cells leave the lymph nodes and preferentially migrate to sites of infection or cell damage.

10.21 B CELLS AND HUMORAL IMMUNITY

10.21.1 Introduction

The human body makes millions of different types of B cells each day that circulate in the blood and lymphatic system. Each clone of B cells has a unique surface receptor protein, the B cell receptor (BCR), that will bind to one particular antigen, its cognate antigen. Cellular interactions that lead to differentiation of B cells into antibody-producing plasma cells and memory B cells take place in distinct areas of secondary lymphoid tissues. Antigen-induced antibody responses can be either T-independent (TI) or T-dependent (TD), depending on the requirement for T cell help in activating the B cells. Antigens that induce TI and TD responses are referred to as TI and TD antigens, respectively. Natural antibodies, primarily IgM, are produced continuously without any evidence of infection and provide immediate non-adaptive defense against a variety of pathogens.

10.21.2 Primary and Secondary Antibody Responses

Activation of naïve B cells during an animal's first exposure to a particular TD antigen elicits a primary antibody response and the formation of memory B cells that are specific for that antigen. In mice, the primary TD response is detected 5–10 days after antigen challenge, rises over the next 10–20 days, and then declines to a sustained low level. IgM of relatively low affinity is produced first, but as the response progresses, higher-affinity antibody, mostly IgG, takes over. Re-exposure to the same TD antigen several weeks later stimulates memory B cells to elicit a secondary response in which IgM production resembles that seen in the primary response, whereas IgG is produced in greater abundance together with some IgA and IgE. The lag phase of the secondary response is shorter (two to three days) and high-affinity antibody production is greater and more persistent compared with the primary response. By contrast, the primary TI response produces mainly IgM over a longer period than that produced in the TD response, and there is a short, weak IgG response. There is no significant production of memory B cells, so the secondary response is a weaker profile of the primary response.

10.21.3 Natural Antibodies

B-1 cells continuously synthesize and secrete natural IgM antibody in the absence of infection or immunization, probably stimulated by commensal microorganisms. IgM circulates in the blood and provides an immediate non-adaptive defense against a variety of blood-borne pathogens. The pentameric structure of IgM provides high cross-reactivity for a wide range of antigens. Mice deficient in secretory IgM exhibit a delay in the maturation of their response to TD antigens, as measured by diminished titers of specific IgG compared with control mice. This observation implies that development of the primary response to a TD antigen is accelerated either by the natural IgM that is present in serum before antigen encounter or by the first wave of specific IgM production that is elicited immediately after antigen challenge, or by both.

10.21.4 The B Cell Receptor (BCR)

The BCR recognizes and binds free (soluble) protein antigen in the blood or lymph. The BCR serves two roles in B cell activation. First, antigen-bound receptors trigger intracellular signals that initiate the process of activation. Second, the BCR binds and internalizes the antigen, which is subsequently processed and presented as a peptide–MHC II complex to helper T cells.

The BCR consists of two interacting subunits, an antigen-binding surface immunoglobulin (sIg), and a signaling subunit. The sIg is a Y-shaped antibody of the IgM or IgD class consisting of two immunoglobulin heavy chains (IgHs) and two light chains (IgLs). The signaling subunit is a disulfide-linked heterodimer of Ig-α and Ig-β proteins. Cell-surface expression of the BCR is essential for B cell progenitors to develop into mature B cells. The binding of a multivalent antigen to the BCR generates intracellular signals that are required for the initiation of antibody responses to foreign antigens and for the establishment of tolerance to self antigens. Because different BCR signaling pathways have different thresholds for activation and different degrees of feedback regulation, the extent and duration of BCR engagement by antigen determines which pathways are activated and to what degree each one is activated. The differential activation of the pathways determines whether BCR engagement leads to activation, inactivation (anergy), or apoptosis of mature and immature B cells.

The BCR is accompanied by a co-receptor complex composed of three cell-surface proteins, CD21, CD19, and CD81 (Figure 10.10). CD21 (also known as complement receptor 2, CR2), is a receptor for the complement fragments C3d and C3dg, which are deposited on the surface of pathogens. Signaling through the CD19 component of the co-receptor augments the signaling from the BCR, thereby lowering the threshold for B cell activation (Matsuuchi & Gold, 2001).

10.21.5 T-Independent Antibody Responses

In TI antibody responses, marginal zone B cells in the spleen and B-1 cells in peritoneal and pleural cavities produce mainly IgM within 48 hours of infection. The most important TI antigens are polysaccharides, glycolipids, and nucleic acids. These antigens cannot be processed and presented in association with MHC molecules and therefore they cannot be recognized by CD4$^+$ helper T cells. TI antigens are categorized as either type 1 or type 2 (TI-1 or TI-2) based on different mechanisms of B cell activation. Many TI-1 antigens, notably lipopolysaccharide (LPS), are components of bacterial cell walls that can stimulate the BCR and a Toll-like receptor (TLR) together and induce a strong B cell response. TI-2 antigens comprise repetitive polysaccharides, such as those found in bacterial capsules. The extensive cross-linking of the BCR by the repeating epitopes is sufficient to activate the B cell. In addition, many polysaccharides activate the complement cascade via the alternative pathway, generating C3d, which binds to the antigen. The complement-coated antigen can simultaneously ligate the BCR and complement receptor 2 (CR2), thereby enhancing BCR signaling. Several viruses, including influenza,

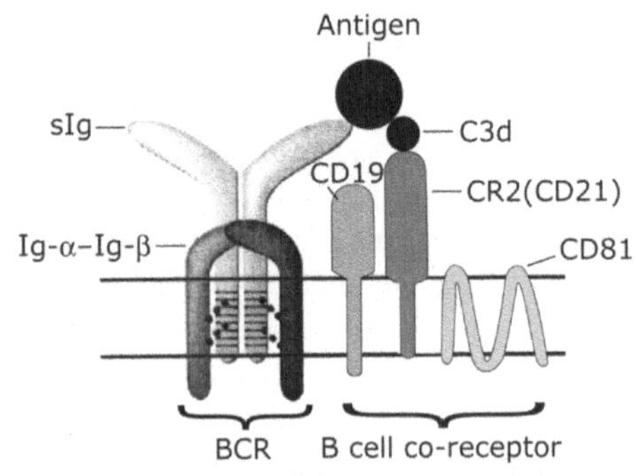

FIGURE 10.10 Antigen binding to the B cell receptor (BCR) and B cell co-receptor.

rabies, polio, and vesicular stomatitis virus (VSV), have been shown to induce TI responses. Apart from VSV, which acts as a TI-1 antigen, there are no data as to whether these viruses represent T1-1 or TI-2 antigens.

The TI antibody response is initiated by CD11c^{low} immature blood DCs, which bind blood-borne particulate antigens and deliver them intact to the spleen. In a systemic infection, these DCs provide critical survival signals to antigen-specific marginal zone B cells and promote their differentiation into IgM-secreting plasma cells. DC-derived soluble ligands (BLyS and APRIL) bind to their B cell receptors (TACI, BCMA, and BAFF-R) and provide a crucial survival signal required for plasma cell generation. By providing constant high levels of TACI/BCMA/BAFF-R ligands, blood DCs sustain B cell activation and survival. In the absence of soluble TACI ligands, marginal zone- and B-1-derived plasma cells cannot survive and undergo apoptosis, resulting in severely impaired antibody responses.

10.21.6 T-Dependent Antibody Responses

TD antigens are proteins in the form of soluble proteins, whole cells, viruses, and parasites. The chances of a B cell randomly encountering its cognate T cell – that is, a helper T cell primed by the same antigen – are estimated to be between 1 in 10^8 and 1 in 10^{12}. Also, T cells and B cells occupy two distinct zones in lymph nodes and spleen. Therefore, TD antibody responses depend upon the regulated migration of activated T and B cells into distinct locations within the lymphoid tissues.

TD antibody responses occur in two phases – an early phase and a late phase. In the early phase, B cells in extrafollicular foci proliferate and differentiate into plasmablasts that produce mainly IgM by three to four days after infection. In the late phase, germinal centers are established in secondary follicles. These are sites for the germinal center reaction, in which follicular helper T cells (T_{FH} cells) specialize in providing cognate help to antigen-primed B cells. The germinal center reaction takes one to three weeks to develop fully. Follicular B cells in germinal centers undergo somatic hypermutation, affinity maturation, and immunoglobulin class switch recombination. B cells selected for high antigen-binding affinity can either become memory B cells with the ability to survive for long periods without antigenic stimulation, or can differentiate into long-lived plasma cells, producing high-affinity IgG, IgA, and IgE.

10.21.7 Sequence of Events Leading up to the Germinal Center Reaction

- Naïve CD4$^+$ T cells are primed by antigen-presenting resident and migrant DCs in the T cell zone of secondary lymphoid organs. The activated cells proliferate and some differentiate into the Th1, Th2, and Th17 subsets, which leave the lymphoid organs and disperse to peripheral tissues. Other activated CD4$^+$ T cells differentiate into T_{FH} cells, which have up-regulated expression of CXCR5 and migrate toward the nearest primary follicle along a concentration gradient of CXCL13.
- Soluble antigen filtering through the lymph node percolates through the primary follicles where it can be recognized and bound by follicular B cells. In addition, the resident follicular DCs can capture antibody–antigen complexes, display the antigen intact on their surface, then transfer the antigen to the follicular B cells. The antigen bound to the BCR is internalized and returned to the B cell surface as a peptide bound to a MHC class II molecule. If the follicular B cells do not encounter their specific antigen, they leave the lymph node after about 16 hours to sample another location.
- Antigen-primed B cells increase their expression of CCR7 and, attracted by the T-zone chemokines CCL19 and CCL21, migrate toward the T cell zone. The B cells also express CD69, which blocks surface expression of $S1P_1$, thereby retaining the cells in the lymphoid tissues.
- The T_{FH} cells recognize their specific antigen displayed as peptide–MHC class II on the B cell surface. A cell–cell interaction described as linked recognition takes place at the interface between the T cell zone and primary follicle. During this interaction, the B cell serves as an antigen-presenting cell to activate the T_{FH} cell (Figure 10.11). The T_{FH} cells reciprocate by activating the B cells to proliferate and differentiate into plasma cells secreting high-affinity antibodies against the original antigen. Antigen-primed B cells that fail to interact with their cognate T cells die within 24 hours.
- B cell activation requires prolonged contact with T_{FH} cells. To meet this requirement, stimulation of

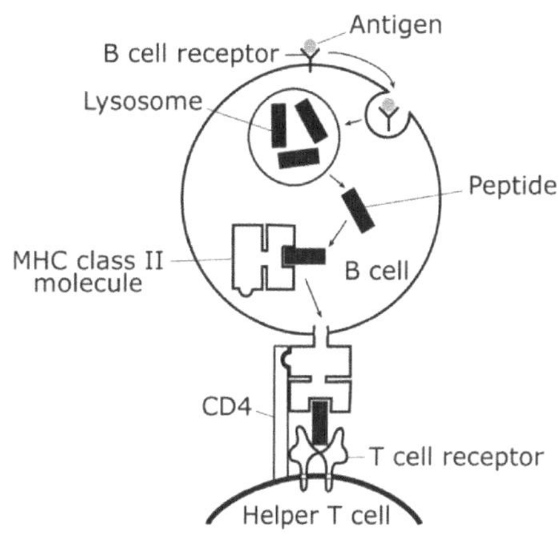

FIGURE 10.11 Antigen presentation to helper T cells by B cells. Antigens bound to the B cell receptor are internalized by receptor-mediated endocytosis and degraded in the endosomal–lysosomal system. The peptide degradation products are loaded onto MHC-II molecules and carried to the cell surface. There the peptideMHC-II complexes are recognized by helper T cells, which activate the B cells.

the BCR and/or TLR up-regulates the expression of CD86 on the B cell and this molecule engages with CD28 on the T cell surface. Also, ICOS ligand (ICOSL), which is constitutively expressed on B cells, engages with ICOS on the T cell surface

- The activation of T_{FH} cells induces the expression of CD40L, which recognizes its receptor, CD40, constitutively expressed on B cells. The ligation of CD40 provides an essential signal for T-dependent B cell activation. When activated T_{FH} cells interact with antigen-primed B cells, a conformational change in the ligated CD40 induces the association of cytosolic proteins called TRAFs (TNF receptor-associated factors) with the cytoplasmic domain of CD40. The recruited TRAFs initiate enzyme cascades that lead to the activation and nuclear translocation of transcription factors, which collectively stimulate B cell proliferation and differentiation to plasma cells. The CD40L–CD40 interaction is also crucial for subsequent germinal center formation. The activated T_{FH} cells also synthesize and secrete cytokines, among which IL-2, IL-6, and IL-21 stimulate B cell proliferation and differentiation.
- B cells that have been activated by T_{FH} cells in lymph nodes migrate from the follicle border to the medullary cords, where they form collections of proliferating cells called extracellular foci. In the spleen, the activated B cells establish similar foci, called PALS foci, between the periarteriolar lymphoid sheath and the red pulp. These foci are the sites of the early phase of a TD antibody response in which B cells proliferate for several days and some of them differentiate into plasmablasts. Plasmablasts are cells that have begun to secrete antibody, yet are still dividing. They secrete only IgM, but, as no class switching is involved, the antibody response is rapid. After several days, the plasmablasts stop dividing and either die or undergo terminal differentiation into short-lived plasma cells (life span of about three days) that remain in the lymph nodes or spleen.
- Those proliferating B cells that do not differentiate into plasmablasts migrate back into a primary follicle along with some specific helper T cells. Within the follicle, the B cells continue to proliferate and ultimately form a central region of vigorous cell division called a germinal center. Follicles in which germinal centers have formed are called secondary follicles. Each fully formed germinal center contains cells derived from only one or a few antigen-specific B cell clones.

10.21.8 The Germinal Center Reaction

The proliferating B cells, now called centroblasts, displace the unstimulated B cells toward the periphery of the secondary follicle, forming the follicular mantle. The centroblasts localize at one pole of the germinal center called the dark zone, where they undergo somatic hypermutation. Somatic hypermutation introduces randomized point mutations into the gene segments encoding the variable regions of the B cell's surface immunoglobulin, thereby changing the conformation of the antigen-binding site. The mutation rate is about one base pair change per 10^3 base pairs per cell division; in contrast, the mutation rate in the rest of the DNA is about one base pair change per 10^{10} base pairs per cell division. Most mutations have a negative impact on the ability of the BCR to bind the original antigen; cells that harbor such mutations are eliminated by apoptosis. Less frequently, mutations improve the affinity of a BCR for antigen; cells that harbor these mutations are efficiently selected and expanded. Expansion seems to be due to the prevention of apoptosis rather than an increased rate of proliferation.

After a while, centroblasts slow their proliferation rate and become centrocytes. These cells stop expressing CXCR4, allowing them to migrate to an adjacent light zone where there are abundant follicular DCs bearing intact antigen complexed with antibody. There, the centrocyte's ability to bind follicular DC-bound antigen is tested in competition with the other clonally related centrocytes harbouring different mutations. Since the mutations are random, the conformational change in the antigen-binding site could either increase or decrease the affinity of the binding site for antigen, or have no effect on the affinity. At high antigen concentrations both high- or low-affinity cells will be stimulated. As the immune response progresses, the amount of antigen available for continued stimulation of centrocytes will fall and the centrocytes will begin to compete with each other for the limiting amount of antigen. Only those centrocytes that have mutated their surface immunoglobulin into a higher affinity will win this competition and be the only ones that will continue to be stimulated and proliferate. Thus, centrocytes with high affinity for antigen will be selected and low-affinity centrocytes will die by apoptosis. The positively selected centrocytes, on binding antigen, are stimulated to undergo additional rounds of division and repeated cycles of mutation and selection. This leads to an increase in the average antigen-binding affinity of the centrocyte population over time, the process of affinity maturation. Antigen-stimulated centrocytes then undergo class switch recombination at the immunoglobulin heavy-chain gene locus so that the antibody isotypes that will be secreted are the high-affinity IgG, IgA, and IgE antibodies, rather than the low-affinity IgM. Finally, the selected, class-switched centrocytes differentiate into either plasmablasts or memory B cells. The plasmablasts are precursors of long-lived plasma cells that secrete high-affinity antibodies.

The germinal center reaction takes one to three weeks to develop fully and can last for six or seven weeks. T_{FH} cells are susceptible to apoptotic death and express a unique profile of genes that control apoptosis and cell cycle (Marinova *et al.*, 2006). Thus T_{FH} cells are programed to be short-lived, which limits the germinal center reaction as appropriate.

10.21.9 Fas-Mediated Apoptosis of B Cells

The need to control the magnitude and duration of humoral immune responses after antigenic stimulation suggests that the pathways of B cell activation and apoptosis may be closely linked. Productive B cell activation requires the delivery of at least two signals: the binding of antigen to surface Ig receptors (signal 1), followed by cognate interaction with an antigen-specific T cell (signal 2). In the absence of sIg signaling, B cells up-regulate their expression of FasR in response to CD40 ligation and are destroyed by Fas-dependent, Th1 cell-mediated apoptosis. B cells stimulated through engagement of their sIg are resistant to Fas-mediated apoptosis. Protection against Fas-dependent apoptosis afforded by sIg engagement constitutes a fail-safe mechanism that ensures survival of antigen-specific B cells, but eliminates bystander B cells activated by CD40L-expressing T cells.

10.22 B CELL MEMORY

B cell memory is mediated by long-lived plasma cells and memory B cells that are derived from centrocytes in the germinal center reaction

10.22.1 Long-Lived Plasma Cells

Plasma cells are specialized in secreting vast quantities of antibody – about 2,000 molecules per cell per second. They are situated in highly vascularized tissues that facilitate antibody secretion directly into the bloodstream. The secreted antibody has the same molecular structure as the immunoglobulin that was expressed on the cell surface of the precursor B cell, except for the extreme carboxy-terminal part of the heavy chains. Thus the antibody secreted by a particular plasma cell has the same antigen specificity as the previously expressed surface immunoglobulin. Plasma cells have very low levels of surface immunoglobulin and lack MHC class II molecules, therefore they can no longer interact directly with helper T cells. Nevertheless, helper T cells provide vital signals for plasma cell survival.

There are two types of plasma cells – short-lived and long-lived. Short-lived plasma cells reach peak numbers at days 8–10 post-immunization and secrete IgM or IgG. They are generated during TI antibody responses and early during TD antibody responses in extrafollicular foci in the spleen and lymph nodes, and remain at these sites until death. Long-lived plasma cells can first be detected 7–10 days after exposure to antigen and accumulate over the next several weeks, reaching a plateau level by 6–7 weeks. They are derived from plasmablasts generated in germinal centers during late TD antibody responses. The plasmablasts are released into the circulation and home to the bone marrow. Individual plasmablasts occupy survival niches composed of bone marrow stromal cells. The stromal cells secrete the chemokine CXCL12 which attracts the CXCR4 receptor expressed on the plasmablasts and retains the cells in the niche. Plasmablasts that succeed in the acquisition of a survival niche differentiate into long-lived plasma cells that are non-dividing and immobile. The long-lived plasma cells in the bone marrow continue to secrete IgG, IgA, and IgE antibodies for months or even years after the antigen is no longer present. These high-affinity antibodies can provide immediate protection if the antigen is encountered later.

The control of germinal center B cell differentiation to plasma cells is an intricate process that centers on two transcriptional repressors called Blimp-1 and Bcl-6. Blimp-1 promotes terminal differentiation by switching off genes required for B cell proliferation, somatic hypermutation, and isotype switching. During the germinal center reaction, Blimp-1 is inhibited by Bcl-6, whose expression is up-regulated via signaling through CD40 on the B cell surface. Differentiation is initiated when Bcl-6 expression is turned off. Blimp-1 is now relieved of inhibition and can exert its repressive effect.

10.22.2 Memory B Cells

Memory B cells are generated mainly in germinal centers. Some memory B cells may remain in the secondary lymphoid organ in which they were generated, whereas others exit germinal centers and recirculate between the blood and lymphoid organs. The memory cells circulate in a resting state and can remain viable for long periods without antigenic stimulation. Their long life span is due in part to their expression of the anti-apoptotic protein Bcl-2.

Memory B cells express surface immunoglobulin, but do not secrete antibody. If a re-infection overwhelms the levels of circulating antibody produced by the long-lived plasma cells in the bone marrow, excess antigen will bind to memory B cell receptors. The memory B cells are now able to present the antigen to their cognate helper T cells. This elicits a secondary immune response in which the memory B cells undergo accelerated clonal expansion and rapid differentiation to short-lived plasma cells producing high-affinity antibodies. The secondary response is more rapid than the primary response because the population of antigen-specific memory cells has already undergone clonal expansion. The secondary response is also more vigorous because the production of isotype-switched, high-affinity antibodies is greatly accelerated after antigen recall.

10.23 VIRUSES

10.23.1 Introduction

Viruses are the simplest form of life, being composed only of a nucleic acid (DNA or RNA) encased by a coat of protein. Since they lack metabolic machinery, viruses do not qualify as cells. Viruses are obligate intracellular parasites that use host cell components to replicate. Some viruses possess genes coding for their own DNA polymerase and RNA polymerase enzymes, but many depend on the host polymerases for genome replication. All viruses make use of the host's protein synthesis machinery to synthesize the protein coats of their progeny. Viral infections cause many diseases in humans, including the common cold, influenza, and life-threatening diseases such as polio, measles, smallpox, and rabies.

10.23.2 Structure of a Typical Virus Particle

The virus particle (virion) consists essentially of a nucleic acid genome encased by a protein coat (capsid). The nucleic acid is either DNA or RNA, both of which may be single-stranded (ss) or double-stranded (ds). The RNA of ssRNA viruses may be either the positive strand (+ssRNA) denoting the mRNA or the negative strand (−ssRNA) denoting the strand complementary to the mRNA. The entire genome may occupy either one nucleic acid molecule (monopartite or linear genome) or several nucleic acid molecules (multipartite or segmented genome). A capsid plus the enclosed nucleic acid is called a nucleocapsid. The capsid is composed of multiple copies of a particular protein arranged in a symmetrical fashion. The capsid protein units in icosahedral viruses aggregate as closed polyhedral shells that are based on the icosahedron, an approximately spherical body with 20 triangular faces.

The human immunodeficiency virus (HIV) is a retrovirus, so-called because it transcribes its genomic RNA to DNA, which is then inserted into the host cell's chromosomal DNA. The general features of the mature HIV virion are shown in Figure 10.12. The virus is enveloped by a lipid bilayer that is derived from the membrane of the host cell. Embedded in the lipid bilayer are several cellular membrane proteins derived from the host cell. Exposed surface glycoproteins (gp120) are anchored to the virus via interactions with the transmembrane protein (gp41). Lining the inner surface of the lipid envelope is a matrix shell made up of the matrix protein (p17). A conical capsid composed of the capsid protein (p24) surrounds the core, which contains two copies of the viral RNA genome stabilized as a ribonucleoprotein complex with the nucleocapsid protein (p7). The core also contains the virus-encoded

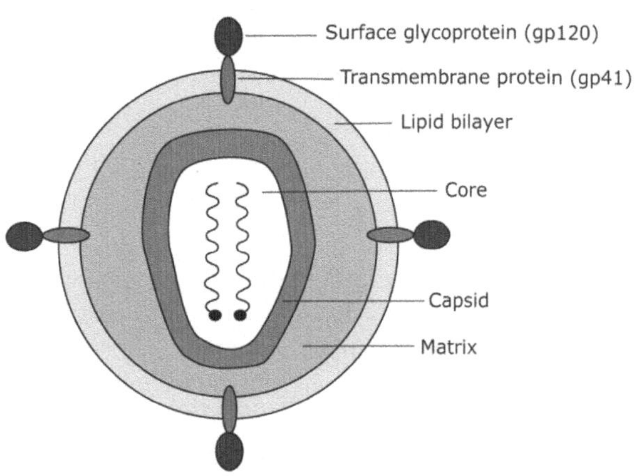

FIGURE 10.12 Morphology of the mature HIV virion.

enzymes, protease, reverse transcriptase, and integrase, together with a number of accessory proteins that function in virus replication.

10.23.3 General Plan of Viral Replication

The replication of a typical virus proceeds through the following general steps: (1) attachment to the host cell; (2) internalization; (3) uncoating; (4) replication; (5) assembly of progeny virions; (6) maturation; and (7) release of the progeny from the host cell.

Attachment. The surface of a particular virus binds specifically to multiple copies of a receptor protein on a target host cell. Most viruses display tissue tropism, meaning that they bind receptors with restricted tissue expression patterns and hence can only infect certain cell types; for example, the hepatitis B virus only infects liver cells.

Internalization. Enveloped viruses enter host cells by either of two mechanisms: (1) fusion between the viral and cellular membranes, mediated by a viral fusion protein; or (2) engulfment by receptor-mediated endocytosis or macropinocytosis. In the fusion mechanism, the nucleocapsid is released into the cell interior, leaving the viral envelope as a patch on the cellular plasma membrane. In receptor-mediated endocytosis, the entire virus is internalized in an endocytotic vesicle, which becomes internally acidified after fusion with a lysosome. The acidic environment within the vesicle promotes conformational changes in the viral envelope proteins, which facilitate fusion between the viral and endocytic membranes and the subsequent release of the nucleocapsid into the cytoplasm. Macropinocytosis is a form of endocytosis which is not receptor-mediated.

Uncoating. In some viruses, the capsid is degraded, thereby releasing the genome into the host cell. In other viruses there is no degradation of the capsid because the capsid proteins play a role in viral replication; for these viruses, uncoating refers to changes in the nucleocapsid.

Replication. This is the step in which viral genomes are replicated and viral structural proteins and enzymes are synthesized. The different types of genome necessitate different replication strategies. The genome replication of most RNA viruses occurs in the cytoplasm of the host cell, catalyzed by a viral polymerase that the host cell cannot provide. In contrast, most DNA viruses replicate their genomes in the nucleus and utilize the host's DNA- and RNA-synthesizing machinery, along with the host's RNA-processing machinery. This means that the viral genome must traverse the nuclear membrane in order to utilize the transription machinery. The enzymes involved in RNA and DNA synthesis are shown in Figure 10.13.

Assembly. The formation of stable immature virions occurs when sufficient concentrations of viral components localize at specific sites within the infected cell.

Maturation. Newly assembled virions need to mature in order to become infectious. This often involves proteolytic cleavage of a capsid protein by a viral protease to make the particle more stable.

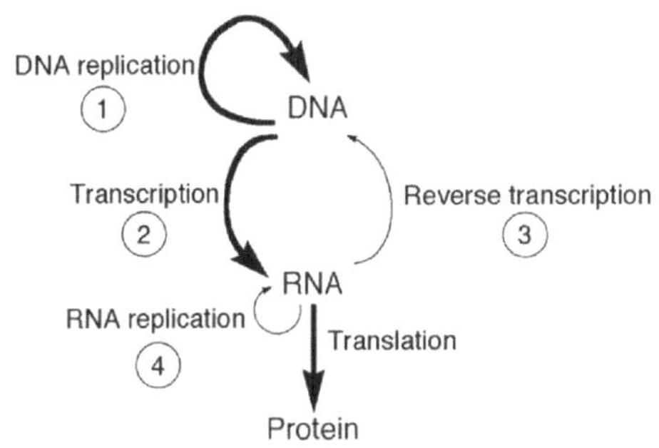

FIGURE 10.13 The enzymes involved in RNA and DNA synthesis. 1, DNA polymerase; 2, DNA-dependent RNA polymerase II; 3, RNA-dependent DNA polymerase (reverse transcriptase); 4, RNA-dependent RNA polymerase (RNA replicase).

Release. Lytic viruses (e.g. polio) are released upon lysis of the host cell, whereas other viruses (e.g. influenza) are released by budding. Most enveloped viruses bud through the plasma membrane, but a few, such as rotavirus, bud from the endoplasmic reticulum membrane and are secreted from the cell. Viruses that are released via budding may damage the cell or they may not (as is the case with retroviruses).

10.23.4 Immune Responses to Viral Infection

10.23.4.1 Introduction

Immune responses to viruses are aimed at preventing infection and eliminating infected host cells once infection has been established. The principal mechanisms of innate immunity against viruses are inhibition of viral replication by type I interferons (IFN-α/β) produced in infected cells and NK cell-mediated killing of infected cells. Adaptive immunity against viral infections is mediated by antibodies, which block virus binding and entry into host cells, and by CTLs, which eliminate the infection by killing infected host cells. NK cells can respond to virus infection within hours, well before CTLs, which take up to five days to respond.

10.23.4.2 Antiviral Effects of Type I Interferons

Plasmacytoid dendritic cells (pDCs) produce by far the largest amounts of type I interferons in response to viral infection. These cells are believed to take up virions by macropinocytosis or receptor-mediated endocytosis. They express TLR7 and TLR9 and make large amounts of IFN-α in response to internalized virus particles.

In addition to its well known pro-inflammatory properties, IFN-α/β is anti-viral by virtue of the following effects.

- Activates transcription of several genes whose protein products collectively inhibit viral replication and assembly. One such protein is dsRNA-activated serine/threonine protein kinase (PKR), which, on binding dsRNA, becomes activated to phosphorylate and thereby deactivate the translation initiation factor eIF2α. Another induced protein is RNase L, which promotes viral RNA degradation. The actions of IFN-α/β-induced proteins confers on host cells a resistance to viral infection, called an antiviral state.
- Increases the cytotoxicity of NK cells.
- Directly inhibits replication of viral RNA and DNA in infected cells.
- Stimulates cross-priming of CTLs by cDCs during virus infection.
- Increases the effectiveness of antigen presentation to $CD8^+$ T cells by up-regulating expression of MHC class I molecules.
- Promotes the differentiation of naïve T cells to the Th1 subset of $CD4^+$ helper T cells.
- Induces a molecule called CD69 on the T cell surface that binds to and reduces surface expression of $S1P_1$, the receptor that mediates egress of T cells from lymphoid organs. The consequent retention of cytotoxic T cells in lymphoid organs leads to increased proliferation, which maximizes the chances of the T cells encountering their specific antigen.
- Promotes the migration and maturation of cDCs from peripheral tissues to draining lymph nodes, thus augmenting the adaptive immune response.

10.23.4.3 NK Cell-Mediated Killing of Infected Cells

The inhibitory receptors on the surface of NK cells recognize normal healthy cells that express MHC class I molecules on their surface, ensuring that these healthy cells are not attacked by NK cells. Virus infection causes a loss of MHC class I expression, thus depriving the NK cell inhibitory receptors of their ligands. The activating receptors on NK cells, which recognize ligands on infected cells, will be stimulated to a greater extent than the inhbitory receptors, allowing the NK cells to kill the infected cells.

10.23.4.4 Antibody-Mediated Viral Immunity

Antibodies can block the initial infection of a cell by a virus and help prevent the spread of viruses released from cells. Neutralizing antibodies bind to proteins on the surface of viruses, thereby preventing attachment and entry of the virus into host cells. The simultaneous binding of antibody to the virus and phagocytes (opsonization) stimulates phagocytosis of the virus. NK cells recognize antibody bound to virally infected cells and mediate antibody-dependent cellular cytotoxicity (ADCC). Complement activation also participates in antibody-mediated viral immunity by promoting phagocytosis and direct lysis of viruses.

Antibodies are effective against viruses only during the extracellular stage of their lives: once viruses enter cells and begin to replicate intracellularly, they are inaccessible to antibodies. Therefore, humoral immunity induced by previous infection or vaccination is able to protect individuals from viral infection but cannot eradicate established infection.

10.23.4.5 Cytotoxic T Cell-Mediated Killing of Infected Cells

Activation of naïve $CD8^+$ T cells in response to viral infection results in massive clonal expansion of antigen-specific cells. This expansion is followed by differentiation into effector $CD8^+$ CTLs. Upon viral clearance, the vast majority of antigen-specific CTLs undergoes apoptosis, with a small fraction remaining as long-lived memory cells.

In the direct antigen presentation pathway, viral proteins within the host cell are converted by proteolysis to peptides, which are transported to the cell surface in a complex with a MHC class I molecule. The peptide–MHC-I complex is recognized by $CD8^+$ CTLs and the cell–cell contact activates the CTLs to destroy the infected host cell. In the cross-presentation pathway, host cells infected with viruses are internalized by DCs. Antigenic peptides produced by proteolysis of viral proteins are transported to the DC surface in a complex with a MHC class I molecule and the complex is recognized by CTLs.

10.23.5 Viral Strategies to Evade Immune Responses

Viruses have developed various strategies to evade host's antiviral responses to ensure their own replication and survival. Examples of viral immune evasion strategies are as follows.

- Viruses can alter their antigens by antigenetic variation, such as point mutations, and are thus no longer targets of immune responses.
- Some viruses produce molecules that inhibit the immune response. For example, poxviruses encode molecules that are secreted by infected cells and bind to several cytokines and chemokines. The secreted cytokine-binding proteins may function as competitive antagonists of the cytokines.
- Some chronic viral infections are associated with failure of cytotoxic T cell responses, which allows viral persistence. This type of immune deficit may result from activation of inhibitory T cell pathways, such as the PD-1 pathway, which normally functions to maintain T cell tolerance to self antigens. This phenomenon has been called exhaustion, implying that immune responses against the virus are initiated but then shut off prematurely.
- Some viruses can avoid detection by the host by entering a dormant state known as latency. In proviral latency, the retrovirus genome is integrated into the DNA of a host cell and replicates along with normal host cell division. Host infection by a different pathogen activates transcription of the proviral DNA.
- Viruses interfere with antigen presentation by decreasing the transcription of MHC class I genes, inhibiting proteasome-mediated generation of peptides, inhibiting the binding of peptides onto MHC class I molecules, and down-regulating the expression of co-stimulatory molecules on host cells.
- Viruses have many strategies to evade NK cell responses.
- Viruses may infect and either kill or inactivate the host's immune cells. For example, the HIV specifically infects and eliminates $CD4^+$ helper T cells, which are important in the generation of virus-specific CTLs and antibodies.
- Viruses that induce death receptor ligands have evolved several distinct counter strategies that specifically suppress apoptosis mediated through death receptors, including antagonism of ligand binding, blockade of adaptor protein recruitment, and inhibition of the apoptotic cascade, which encompass virtually every step in the signaling pathway. For example, some viruses produce v-FLIPs, which interfere with Fas-mediated apoptosis by inhibiting the recruitment and activation of the protease FLICE (procaspase-8) by the Fas death receptor.
- Vaccinia virus encodes a potent complement inhibitor, vaccinia complement-control protein (VCP), which disrupts both the classical and alternative complement pathways.
- Some viruses avoid recognition by the complement system by removing antibody–antigen complexes from the surface of the infected cell.
- By expressing an Fc receptor, the herpes simplex virus avoids recognition by the complement system via two mechanisms: (1) sterically hindered access of virus-specific immune IgG or Fc-dependent effector cells to the virus or virus-infected cell by bound non-immune IgG; (2) antibody bipolar bridging whereby immune IgG binds with its hypervariable region end to a viral envelope or cell surface protein and with its Fc end to the virus's Fc receptor.
- Some viruses encode and express complement-interfering proteins, while others can incorporate host complement control proteins in their viral envelope and/or up-regulate expression of these proteins in infected cells.
- Viruses counter the responses of several cytokines by encoding inhibitors, decoy receptors, or modified viral versions of these soluble mediators.
- Different viruses antagonize type I interferon-mediated defense responses by inhibiting distinct steps in IFN-activated signaling and antiviral pathways.
- By encoding and/or inducing the expression of superantigens, certain viruses may evade host's

antiviral cellular immunity via activating non-specific T cells and thus shifting the focus of the immune response away from the virus.
- Glycoprotein C protects herpes simplex virus types 1 and 2 from complement-mediated neutralization.

10.24 IMMUNE SUPPRESSION

Regulation of the immune system is of paramount importance for the suppression of autoimmune reactions and for the regulation of the immune response against foreign pathogens.

10.24.1 Regulatory T Cells as Mediators of Immune Suppression

Regulatory T cells (Tregs) function to control immune responses and maintain self-tolerance. Treg-mediated suppressive activity can contribute to the immune escape of pathogens or tumors. Indeed, elimination of Tregs in tumor-bearing mice can improve anti-tumor immune responses and survival of the animals (Zou, 2006). Manipulation of Treg function is an important goal of immunotherapy, since this T cell subset affects the balance between immune tolerance and immune responsiveness toward pathogens.

An obvious question is what factors decide whether effector or regulatory responses will be dominant in a given situation? Baecher-Allan *et al.* (2002) suggested that different strengths or modes of stimulation may alter the mechanism by which Tregs regulate immune responses. With a weak TCR stimulus, Tregs were suppressive and dominant over effector T cells. Wth a strong TCR stimulus, such as would occur in response to a microbial antigen, the Tregs became refractory to suppression, allowing effector T cells to be dominant.

Tregs offer considerable promise for adoptive immunotherapy of autoimmune diseases and allograft rejection. To be acceptable for clinical use, these cells must exhibit antigen specificity and not compromise the host's ability to respond to environmental pathogens.

10.24.2 The Natural Subset of Regulatory T Cells

Various subsets of $CD4^+$ and $CD8^+$ Tregs occur in the immune system (Zou, 2006). The most studied is $CD4^+CD25^+Foxp3^+$, commonly referred to as the natural (or constitutive) Treg (nTreg). The specific surface marker CD25 is the IL-2 receptor α chain (IL-2Rα) and the Foxp3 marker is the transcription factor forkhead box p3.

nTregs develop in the thymus and, once generated, migrate to peripheral tissues. The roles of nTregs include suppression of DCs that support the generation of effector T cells; inhibition of function and migration of effector Th1, Th2 and Th17 cells; and suppression of mast cells, basophils and eosinophils.

10.24.3 Generation of Natural Regulatory T Cells

nTregs are positively selected in the thymus through relatively high avidity interactions between the TCR of self-reactive thymocytes and a self peptide associated with a MHC class II molecule expressed on a thymic cortical epithelial cell. This selection process is controlled by Foxp3, which acts as a transcription factor. Lower avidity interactions would promote the development of effector $CD4^+CD25^-$ T cells, whereas higher avidity interactions would lead to clonal deletion. The narrow avidity window ensures that only a small proportion of positively selected thymocytes develop into $CD4^+CD25^+$ nTregs. The mature nTregs migrate to peripheral tissues, where they survive as a long-lived population of self antigen–specific cells poised to prevent pathological autoimmune reactions. IL-2 functions during nTreg development by up-regulating Foxp3 and CD25 and by increasing the number of thymic T cells. Without a functional IL-2R, only immature $CD4^+Foxp3^{low}CD25^-$ T cells develop.

10.24.4 Dependence of Natural Regulatory T Cells on Paracrine IL-2

IL-2 is essential for the development, maintenance, and suppressive function of nTregs. Mice deficient in IL-2 or IL-2R develop lethal autoimmune diseases, which can be prevented by transfer of small numbers of wild-type nTregs into neonatal IL-2Rβ-deficient mice. Unlike effector T cells, nTregs do not produce IL-2 after TCR stimulation and have to rely upon paracrine sources of this cytokine. The main sources of IL-2 are other T cell types, especially activated $CD4^+$ T cells. In a negative feedback loop, IL-2 produced by activated non-regulatory T cells contributes to the maintenance, expansion, and activation of nTregs, which in turn limit the expansion of non-regulatory T cells. Disruption of this IL-2-mediated feedback loop promotes the development of autoimmune/inflammatory disease. Further, manipulation of this feedback loop is instrumental in tuning the intensity of nTreg-mediated suppression and hence the strength of a variety of immune responses (Sakaguchi *et al.*, 2008).

10.24.5 Reciprocal Developmental Pathways for the Generation of Th17 Cells and Natural Regulatory T Cells

Reciprocal developmental pathways have been demonstrated for the generation of Th17 effector cells and nTregs. IL-2 facilitates the differentiation of naïve $CD4^+$ T cells into nTregs, but inhibits their differentiation into Th17 cells. IL-2 therefore regulates the balance between the two contrasting cell types. During the steady state, when the immune system is not activated, the production of TGF-β will suppress the generation of effector T cells and induce nTregs, and thereby maintain self-tolerance. However, upon infection or inflammation, IL-6 produced by the activated innate immune system will suppress the generation of TGF-β-induced nTregs and induce a pro-inflammatory T cell response predominated by Th17 cells.

10.24.6 CTLA-4 as a Mediator of Natural Regulatory T Cell Function

CTLA-4 is an important mediator of the suppressive function of nTregs. Whereas effector T cells only express CTLA-4 after they have been activated, nTregs express CTLA-4 constitutively. Moreover, CTLA-4 interacts more strongly with the CD80/CD86 (B7.1/B7.2) co-stimulatory molecules on antigen-presenting DCs than does the alternative co-receptor CD28. nTregs can therefore be activated by a small amount of self antigen presented by immature DCs despite their low expression of B7 molecules.

CTLA-4 knockout mice develop uncontrolled lymphocyte activation manifested by massively enlarged lymph nodes and spleen and fatal multi-organ lymphocytic infiltrates. A specific deficiency of CTLA-4 in nTregs results in fatal T cell-mediated autoimmune disease and enhanced tumor immunity in mice. Blocking CTLA-4 with a specific monoclonal antibody (mAb) elicits organ-specific autoimmune disease and colitis in otherwise healthy mice. and exacerbates diabetes in the non-obese diabetic (NOD) mouse. Polymorphisms in the CTLA-4 gene are associated with several autoimmune diseases in humans, including type 1 diabetes and Graves' disease (hyperthyroidism).

Read *et al.* (2006) investigated the effects of CTLA-4 deficiency and anti-CTLA-4 mAb treatment on nTreg cell function *in vivo*. Anti-CTLA-4 mAb treatment of mice completely inhibited nTreg function, but did not reduce nTreg numbers or their homing to the gut-associated lymphoid tissue (GALT). This suggested that the antibody mediates its function by blockade of a signal required for nTreg activity. In contrast to the striking effect of the antibody, CTLA-4-deficient mice were able to produce functional nTregs, suggesting that, under some circumstances, other immune regulatory mechanisms are able to compensate for the loss of the CTLA-4-mediated pathway.

10.24.7 Mechanisms of Bystander Suppression by Natural Regulatory T Cells

Once activated through their TCR, Tregs suppress by various mechanisms that are independent of the antigen specificity of the responding effector T cells. This antigen-non-specific suppression is called bystander suppression. It allows a Treg with one antigen specificity to suppress effector T cells with many other distinct antigen specificities, provided that the antigens involved are present in the same vicinity. It is likely that several mechanisms operate synergistically. Selected mechanisms of suppression by nTregs are described below.

10.24.7.1 Cell Contact-Dependent Suppression

nTregs produce surface-bound TGF-β as well as soluble TGF-β when stimulated by APCs. nTregs mediate immunosuppression via the binding of surface-bound TGF-β to TGF-β receptor on the responder cell. nTregs can suppress activation of effector T cells *in vitro* by a cell-contact-dependent mechanism that comprises two steps. The first step is LFA-1-dependent and CTLA-4-independent; and the second step is both LFA-1- and CTLA-4-dependent. (1) Antigen-stimulated nTregs out-compete naïve effector T cells to form aggregates around DCs, thereby impeding access of the effector T cells to the DCs. (2) After forming aggregates, nTregs directly and specifically down-regulate the expression of B7 molecules on DCs. Both steps are required to prevent naïve effector T cells from being activated by antigen-presenting DCs, resulting in specific immune suppression and tolerance (Onishi *et al.*, 2008).

10.24.7.2 Killing of Immune Cells by the Perforin/Granzyme Pathway

Several *in vitro* studies have indicated that activated nTregs may use the perforin/granzyme mechanism of cell lysis to kill immune cells, including $CD4^+$ and $CD8^+$ T cells, B cells, monocytes, and both immature and mature DCs. Cao *et al.* (2007) used tumor-bearing mice to investigate whether nTregs activated *in vivo* express granzymes and, if so, whether these molecules are important for nTreg-mediated suppression of anti-tumor immune responses. They found that tumor cells induced granzyme B (but not granzyme A) expression in 5–30% of Tregs in the tumor environment. These nTregs used the perforin/granzyme B system to kill NK cells and $CD8^+$ T cells, thereby suppressing tumor clearance mediated by these cytotoxic cells. Although tumor cells gain a survival advantage by activating granzyme B, this advantage is lost in the setting of granzyme B deficiency, because NK cells and $CD8^+$ T cells can utilize granzyme A and destructive enzymes to kill tumor cells.

10.24.7.3 Suppression of Cytotoxic T Cell Function

Mempel *et al.* (2006) showed that nTregs suppress the cytotoxic activity of $CD8^+$ T cells by attenuating the antigen-triggered release of cytolytic granules in a reversible and TGF-dependent manner. This inability to degranulate was the only detectable difference beween regulated and non-regulated $CD8^+$ T cells. Regulated $CD8^+$ T cells exhibited no deficiency in granule content and no defects in proliferation or ability to form antigen-dependent conjugates with target cells. Thus, nTregs reversibly suppress $CD8^+$ T cell-mediated immunity by allowing acquisition of full effector potential, but withholding the license to kill.

10.24.7.4 Suppression by Inhibitory Cytokines

Natural Tregs promote bystander suppression through the production of suppressive cytokines. Tregs regulate the expansion of peripheral activated/memory $CD4^+$ T cells through a mechanism involving the production of IL-10

(Annacker *et al.*, 2001). IL-10 inhibits the antigen presentation function of macrophages and DCs by down-regulating their expression of MHC class II and co-stimulatory molecules in combination with reduced release of IL-1β, IL-6, IL-12, and TNF-α. In this way, IL-10 indirectly inhibits T cell activation and terminates cell-mediated immune reactions. IL-10 directly inhibits T cell and B cell proliferation and differentiation by preventing secretion of IL-2 by Th1 cells. IL-10 is also an important negative regulator of IFN-γ production from T cells and NK cells. In the case of NK cells, IL-10 prevents macrophage secretion of TNF-α and IL-12, thereby preventing NK cell activation and IFN-γ production (Taylor *et al.*, 2006).

TGF-β secreted by nTregs controls immune and inflammatory responses by suppressing the activation of macrophages and promoting the development of Tregs. TGF-β can also inhibit the development of Th1 and Th2 helper T cell subsets. TGF-β knockout mice develop uncontrolled inflammatory lesions and lymphoproliferation.

10.24.7.5 Suppression by Cytokine Deprivation

Pandiyan *et al.* (2007) showed that nTregs induced apoptosis of responder T cells *in vitro* and *in vivo*. The apoptosis was caused by the nTregs sequestering cytokines that signal through the common γ-chain (IL-2, IL-4, IL-7, IL-9, IL-15, and IL-21) and depriving the responder T cells of the cytokines required for their survival.

10.24.7.6 Inhibition of Dendritic Cell Activation by LAG-3

LAG-3 is a transmembrane protein selectively expressed on chronically stimulated nTregs that binds MHC class II molecules on immature DCs with higher affinity than CD4. Liang *et al.* (2008) demonstrated that engagement of LAG-3 with MHC class II during interaction between nTregs and antigen-presenting DCs inhibited DC activation. Their data suggested that MHC II cross-linking on DCs triggers an ITAM-mediated signaling pathway in which ERK recruits the inhibitory phosphatase SHP-1 that then suppresses DC maturation and immunostimulatory capacity.

10.24.7.7 CTLA-4-Dependent Downmodulation of Dendritic Cell Antigen Presentation Ability

nTregs induce down-regulation of co-stimulatory CD80 and CD86 on DCs in a CTLA-4-dependent way, thereby inhibiting antigen presentation to effector T cells (Oderup *et al.*, 2006).

10.24.8 Blockade of Natural Regulatory T Cell-Mediated Suppresion by Dendritic Cells

It is obviously important that effector T cells are not suppressed during infection. Pasare and Medzhitov (2003) showed that induction of Toll-like receptor (TLR) signaling blocks the suppressive effect of nTregs, allowing activation of pathogen-specific adaptive immune responses. Specifically, ligation of TLR4 and TLR 9 with LPS and CpG, respectively, stimulates maturation of DCs, which then produce IL-6. The binding of IL-6 to IL-6 receptors on activated effector T cells makes these cells insensitive to the action of the nTregs. IL-6 needs to act in synergy with other unspecified factors to exert this effect.

10.24.9 Direct Suppression of B Cells by Natural Regulatory T Cells

Lim *et al.* (2005) reported that nTregs accumulate at the interface beteween the B cell and T cell zones and within germinal centers of secondary lymphoid tissues – sites where Th cell–B cell interactions and immunoglobulin (Ig) production occur. nTregs were shown to directly suppress B cell Ig production and to inhibit Ig class switch recombination without having to first suppress Th cells.

10.24.10 Linked Suppression by Anergic T Cells

One of the features of immune suppression observed in some *in vivo* systems is that the induction of non-responsiveness to one antigenic determinant can lead to the regulation of T cell responses to other determinants, provided that the determinants are displayed by the same APC. Similarly, tolerance to one alloantigen can regulate the response to other alloantigens, provided that they are expressed by the same tissue or cells. This phenomenon, known as linked suppression, has potential benefits in the context of immunotherapy (Frasca *et al.*, 1997).

Lombardi *et al.* (1994) reported that cultured anergic human T cells caused inhibition of antigen-specific and allospecific T cell proliferation. In order for the inhibition to occur, the anergic T cells had to be specific for the same APCs as the T cells that were suppressed. This report was an *in vitro* demonstration of linked suppression. A more detailed analysis (Frasca *et al.*, 1997) revealed that the regulatory effect of anergic T cells was dependent on cell contact and was maintained during *in vitro* culture. The mechanisms of suppression appeared to be competition for the APC surface and for locally produced IL-2.

When anergic T cells were cultured with bone-marrow-derived DCs, the DCs exhibited reduced expression of MHC class II, CD80, and CD86 molecules, and lost their ability to stimulate responding antigen-specific T cells. This inhibition of antigen presentation by DCs only affected immature DCs. Inhibition was independent of inhibitory cytokines. The data suggested that anergic T cells deliver a negative cell-contact-dependent signal to the DCs that arrests their maturation into functional APCs (Vendetti *et al.*, 2000).

10.25 IMMUNITY IN THE GUT

10.25.1 Introduction

The term 'gut' in the present context refers to the small intestine and colon. Immunity in the gut aims to destroy pathogenic microbes in the lumen and hinder the passage of any survivors across the intestinal epithelium. The intestinal epithelium constitutes an efficient barrier against the commensal flora and pathogens. This is due to the tightly bound epithelial cells and the presence of the glycocalyx on the microvilli, the mucus layer, and an array of secreted antibacterial factors, such as lactoferrin. Lymphocytes, DCs, and macrophages interspersed in various sites in the intestine and gut-associated lymphoid tissues (GALT) mediate innate and adaptive immunity.

In the intestine, M cells in Peyer's patches take up foreign matter from the lumen and transfer it to DCs in the subepithelial dome. Whole bacteria, viruses, and soluble microbial products can be taken up in various ways. The DCs migrate into the interfollicular T cell zone of the Peyer's patch or mesenteric lymph node where they activate naïve T cells to become helper T cells. The helper T cells enable B cells in the B cell zone to become IgA-secreting plasma cells (IgA-PCs). Most of the plasma cells find their way via the thoracic duct to the lamina propria.

Selective homing of lymphocytes to the gut is an essential feature of gut immunity. Effector lymphocytes that are generated in the gut-associated lymphoid tissues (GALT) and mesenteric lymph nodes acquire gut-homing receptors and migrate in the blood to the intestinal mucosa. In this location, T cells can respond to invading pathogens, and IgA-PCs can secrete antibodies that are transported into the lumen and neutralize pathogens before they get the chance to invade.

10.25.2 The Intestinal Microbiota

In healthy humans, mucosal surfaces at the interfaces with the external environment are colonized by populations of commensal microorganisms. This microflora (or microbiota) comprises mostly anaerobic bacteria and relatively small numbers of fungi and protozoa. The overwhelming majority and diversity of the bacterial flora resides at the distal alimentary tract, most notably at the colon. At least 1,000 different species of bacteria are estimated to be present in the gut, amounting to over 10^{13} cells in an individual human. The acidity of the stomach contents prevents continuous access of most bacterial species from the external environment to the intestine. Certain pathogenic bacteria can survive the stomach acidity, and these species may become part of the gut microbiota if they are ingested in contaminated food or water. However, many intestinal bacterial pathogens are poorly adapted to compete with the gut microflora for dietary nutrients, restricting their luminal colonization.

The relationship between host and commensal bacteria is symbiotic, with the host providing a protected, nutrient-rich environment for the bacteria. The metabolic activity of the commensals benefits the host by promoting the development of organized lymphoid structures and the function of immune cells. Thus, in germ-free mice, isolated lymphoid follicles in the small intestine fail to develop and B cells in mucosal sites fail to differentiate into IgA-PCs. The microbiota also enhances host digestive efficiency by degrading indigestible plant polysaccharides.

Microbial colonization of the intestine starts soon after birth and increases rapidly after weaning. Intestinal commensal bacteria are anatomically restricted to either the intestinal lumen, the luminal face of the intestinal epithelium, or within the underlying GALT. The anatomical barrier to commensal bacteria in the intestine is the single layer of epithelial cells that separates the gut microflora from the lamina propria. If commensal bacteria manage to breach this barrier, the systemic immune response to the commensals will be dysregulated, leading to inflammatory bowel disease (IBD). The most common forms of IBD in humans are Crohn's disease, which can affect any part of the gastrointestinal tract, and ulcerative colitis, in which pathology is restricted to the colonic mucosa.

10.25.3 Strategies for Pathogen Invasion of the Intestinal Epithelium

Entry of enteroinvasive bacteria into the enterocyte is key to a successful invasive process. There are essentially two major mechanisms of bacterial internalization. The 'zippering' process involves a surface-bound bacterial protein binding an adherence molecule of the mammalian cell surface with high affinity. The 'trigger' process corresponds to bacteria inducing massive cytoskeletal changes in the mammalian cell underneath the site of contact, thereby causing a ruffling process that internalizes the bacterial cell in a macropinocytic vacuole. A variety of bacterial species, viruses, and protozoa translocate through the intestinal epithelium via M cells. These pathogens take advantage of a physiological route of mucosal sampling of antigens to cross the epithelial barrier.

10.25.4 Distribution of $\gamma\ \delta$ T Cells in the Human Intestine

$\gamma\ \delta$ T cells are relatively rare in the lymphoid organs, but are well represented in epithelia. The epithelium of human small and large intestine contains more $\gamma\ \delta$ T cells (10–30% of total T cells) than any of the other epithelia investigated. The $\gamma\ \delta$ T cells are concentrated in the columnar epithelium of the intestinal crypts. The relative number of $\gamma\ \delta$ T cells within the epithelium varies greatly, not only between different sites in the intestinal tract, but also between sections of the same site originating from different subjects.

10.25.5 Tissue Tropism

Effector/ memory T cells and plasma cells are programed to migrate to a particular type of non-lymphoid tissue – a feature called tropism. Of particular importance are tissues at body surfaces that are exposed to the external environment, such as the intestinal mucosa and inflamed skin.

Tropism is determined by the induced expression of selective tissue-homing receptors on T cells and plasma cells. The corresponding ligands for these receptors are displayed on the surface of epithelial and endothelial cells in the target tissues.

The homing potential of activated lymphocytes depends on the lymphoid microenvironment in which the naïve lymphocytes are activated. Thus, T cells that are activated in the GALT and mesenteric lymph nodes acquire gut-homing receptors and migrate to the intestinal mucosa, whereas those activated in skin-draining lymph nodes acquire skin-homing receptors and migrate to the skin. Evolution has cleverly adapted food-derived vitamin A to imprint homing of T cells and IgA-PCs cells to the small intestinal lamina propria (Sigmundsdóttir & Butcher, 2008).

10.25.6 Homing of T Cells to the Small Intestine

Effector T cells leave the lymphoid tissues, enter the blood, and home to the small intestine, ready to respond to invasive pathogens that may be present in the intestinal lumen. Gut homing of T cells requires their expression of the integrin $\alpha_4\beta_7$ and the chemokine receptor CCR9, which are attracted by their respective ligands, MadCAM-1 and CCL25. MadCAM-1 is expressed constitutively on postcapillary venular endothelial cells in the lamina propria and CCL25 is expressed by enterocytes, especially those in the crypt region. Gut-homing T cells also express the $\alpha_E\beta_7$ integrin that can bind to epithelial (E-) cadherin, allowing retention of the T cells in the epithelium. Both $\alpha\beta$ and $\gamma\delta$ T cell subsets are found in the gut mucosa. Localization and expansion of $\alpha\beta$ T cells increases in response to increased colonization of normal intestinal flora. In contrast, $\gamma\delta$ T cell localization and expansion is not influenced by normal microbial colonization.

The gut homing process is, in principle, the same for effector/memory $CD4^+/CD8^+$ T cells and $CD4^+$ nTreg cells. Imprinting of activated $CD8^+$ T cells with the gut-homing receptors $\alpha_4\beta_7$ and CCR9 is mediated by DCs from Peyer's patches (PP-DCs) and mesenteric lymph nodes (MLN-DCs) and not by DCs from peripheral lymph nodes (PLN-DCs) or spleen. All PP-DC subsets have the ability to generate gut-tropic T cells, but among MLN-DCs only the minority population of lamina propria-derived $CD103^+$ DCs have this ability. MLN-DCs down-regulate L-selectin (CD62L) on responding T cells, thereby preventing T cell re-entry into secondary lymphoid organs.

Iwata *et al.* (2004) made the ground-breaking discovery that the vitamin A metabolite, retinoic acid, enhances the expression of $\alpha_4\beta_7$ and CCR9 on T cells upon activation and imprints them with the gut tropism. DCs in Peyer's patches and mesenteric lymph nodes were shown to express retinaldehyde dehydrogenases (RALDH), which catalyze the oxidation of retinaldehyde (derived from dietary vitamin A) to retinoic acid. Vitamin A deficiency caused a depletion of T cells from the intestinal lamina propria. The Iwata group also found that retinoic acid suppresses the expression of the skin-homing receptors CCR4 and the ligands for E- and P-selectin on T cells upon activation. Thus, when naïve T cells are activated by antigen presented by DCs in the GALT or mesenteric lymph nodes, they are exposed to the retinoic acid produced by the DCs, and this induces the up-regulation of $\alpha_4\beta_7$ and the *de novo* induction of CCR9 on the differentiated effector cells

Migration of T cells to the colon is relatively poor, supporting the notion that the small and large intestine have distinct homing requirements. The integrin $\alpha_4\beta_7$ is involved in T cell migration into the colon, but CCR9 plays no part as its ligand CCL25 is not expressed by colonic epithelial cells. Retinoic acid does not appear to be necessary for T cell homing to the colon

10.25.7 Homing of IgA-Secreting Plasma Cells to the Small Intestine and Colon

The predominant class of antibody providing humoral immunity in the gut is IgA and it is estimated that around 80% of IgA-PCs reside in the gut mucosa. The secreted IgA antibody is transported through the intestinal epithelium into the lumen by the polymeric Ig receptor. IgA-PCs acquire gut-homing properties in the GALT and mesenteric lymph nodes where they have undergone differentiation from naïve B cells. The major population of mouse intestinal IgA-PCs has a half life of 4.7 days under physiological conditions of stimulation. This short half-life means that IgA-PCs need to be constantly replenished in the gut by recently activated naïve B cells or memory B cells.

IgA PCs in Peyer's patches, mesenteric lymph nodes, small intestine, and colon mostly co-express CCR9, CCR10, and CXCR4 on their surface and are capable of responding to their respective ligands (CCL25, CCL28, and CXCL12). IgA PCs homing into the small intestine are guided by CCL25 and CCL28, whereas those homing into the colon are guided by CCL28 and CXCL12. Both CCL25 and CCL28 enhance $\alpha_4\beta_7$ integrin-dependent adhesion of IgA PCs to MadCAM-1, which is critical for homing of IgA-PCs to the gut. Retinoic acid produced by PP-DCs confers small intestine tropism on IgA-PCs in the manner previously described for T cells.

10.25.8 Innate Immunity

Innate immune protection in the gut is mediated in part by the non-specific physical and chemical barrier provided by the mucosal epithelial cells and their secreted mucus. α-Defensins produced by intestinal Paneth cells insert into the outer membrane of bacteria and form transmembrane pores. The consequent loss of membrane integrity results in lysis of the bacterial cell. Those Toll-like receptors (TLRs) that are expressed by intestinal epithelial cells recognize specific molecules on the surface of microbes and initiate signaling pathways that lead to inflammatory and protective immune responses. The cytosolic bacterial detector NOD1 detects pathogenic bacteria that have escaped detection by TLRs.

10.25.9 Adaptive Immunity

Resident CX3CR1+ macrophages in the intestinal mucosa sample luminal antigens by extending cytoplasmic processes through gaps between adjacent epithelial cells and transfer the antigen to CD103+ DCs in the lamina propria. The DCs migrate to mesenteric lymph nodes where they present processed antigen to naïve T cells and induce the differentiation of these T cells into IFN-γ- or IL-17-producing effector cells. The M cells that overlie the domes of Peyer's patches capture invasive pathogens in the lumen and transport them across the epithelial barrier to underlying DCs for processing and presentation to naïve T cells in the interfollicular region.

Humoral immunity in the gut is mediated mainly by IgA antibodies that plasma cells secrete into the lumen. In fact, IgA is produced in larger amounts than any other antibody isotype. The dominance of IgA production is partly due to the selective induction of IgA class switching in B cells in GALT and mesenteric lymph nodes. TGF-β and other molecules that promote IgA class switching are expressed by intestinal epithelial cells and GALT DCs in response to TLR signaling. The dominance of IgA in the gut is enhanced by the imprinting of gut-homing on IgA-producing plasma cells (IgA-PCs) in GALT and mesenteric lymph nodes. Activated B cells that have undergone class switching enter the blood and home to the intestinal lamina propria, where they reside as plasma cells. Within the gut lumen, secreted IgA binds to microbes and toxins and neutralizes them by preventing their binding to receptors on host cells. Antigens in the lumen that have been opsonized by antibodies can be delivered by Fc receptor-mediated pathways to DCs in the GALT, leading to T cell responses to the antigens.

T cells are found within the gut epithelium, scattered throughout the lamina propria and submucosa, and within Peyer's patches and other organized collections of follicles. Intraepithelial lymphocytes (IELs) are specialized T cells found sporadically between adjacent epithelial cells of the gut epithelium. In humans, most of the IELs are CD8+ cells comprising two major populations. The first represents conventional α β T cells that are activated by antigen in association with classical MHC molecules in draining lymph nodes and migrate into intestinal epithelia to await a second exposure to antigen, akin to memory T cells in other tissues. They are prevalent in the colon, where the antigen load is particularly high. The second IEL population is made up of α β and γ δ T cells. In mice, about 50% of IELs are γ δ T cells, but in humans these cells represent only about 10% of IELs. Of the α β cells, one-third possess the conventional heterodimeric CD8α β co-receptor and are restricted by MHC class I molecules for antigen recognition; two-thirds express a homodimeric CD8α α co-receptor and are not MHC class I restricted. Of the IEL γ δ T cells, most express CD8α α, with the remainder expressing neither CD4 nor CD8 co-receptors (CD4CD8 double-negative). The TCR of γ δ T cells can bind directly to soluble antigens released from microbial pathogens and damaged host cells. γ δ T cells secrete toxic and bacteriostatic molecules and produce cytokines that are involved in protective immunity against a variety of intracellular and extracellular pathogens. Thus, γ δ T cells act early in the defense of the mucosal barrier. The repertoire of IELs in an individual is restricted to those that can recognize antigens likely to be encountered in the gut lumen, as opposed to the vast repertoire of most T cells that can recognize virtually any antigen in existence.

Th17 cells are abundant in the gastrointestinal tract, where they combat intestinal pathogens. The signature cytokine secreted by Th17 cells is IL-17, whose receptors are expressed on stromal, epithelial, and endothelial cells. IL-17 has pleiotropic activities: it co-ordinates tissue inflammation, stimulates the production of chemokines that recruit neutrophils to inflammatory sites, and stimulates the production of antimicrobial peptides. Th17 cells also produce IL-22, which stimulates epithelial repair, and IL-21, which promotes the differentiation of Th17 cells.

10.25.10 Maintenance of Intestinal Homeostasis

A homeostatic relationship between the microbiota and the intestinal immune system has evolved to minimize the adverse health effects of intestinal microorganisms, even during overt pathogenic challenge. This intestinal homeostasis ensures that the resident bacteria breach the epithelial barrier as rarely as possible; most of those that do are killed rapidly and do not penetrate to systemic sites. Homeostasis is maintained by a hierarchy of three immunological barriers. First, the likelihood of tissue invasion across the epithelium is decreased through the production of mucin, antimicrobial proteins, and IgA. Second, bacteria that manage to penetrate intestinal tissues are killed by resident macrophages and NK cells. In addition, phagocytosed microorganisms activate intestinal DCs whose secreted cytokines stimulate intestinal CD4+ effector T cells. Acute inflammation is normally avoided through the induction of nTreg cells. Third, bacteria that have crossed the epithelial barrier or have been taken up by intestinal DCs do not penetrate further than the mesenteric lymph nodes and so are not exposed to the systemic immune system. Gut commensal bacteria contribute to the maintenance of intestinal homeostasis, partly through their ability to control the differentiation of effector T cells in the mucosa and to modulate inflammatory responses through the induction of Treg cells and IL-10 production.

10.25.11 Innate Lymphoid Cell Interactions with the Intestinal Microbiota

Signals derived from commensal bacteria can be directly recognized by a number of immune cell receptors. Human LTi cells express TLR2 and ligation of this receptor induces IL-2, which stimulates the secretion of not only IL-22, but also IL-13 and IL-5 by the LTi cells. Commensal bacteria promote steady-state expression of IL-1β in intestinal macrophages and the IL-1β produced enhances IL-22 production from ILC3s. The gut microflora and IFN-γ promote steady-state production of IL-7, which is a critical factor for ILC2 and ILC3 development and maintenance.

Signals originating from the microbiota impact the function of ILCs. For example, mice devoid of commensal bacteria exhibit impaired antiviral immune responses, indicating that commensal bacteria may be essential for promoting optimal NK cell responses. In contrast, the microbiota repressed the activity of ILC3s in healthy mice by inducing the epithelial expression of IL-25, which then suppressed the constitutive production of IL-22 by ILC3s. ILC3s did not express the IL-25 receptor and the signal from IL-25 was relayed by DCs, which express the IL-25 receptor. IL-23, a chief inducer of IL-22, was not involved in this regulatory process. The fact that both Th17 cells and ILC3s produce IL-22 suggests that these two cell types might compete for regulating factors and therefore regulate each other's activity, This was shown to be the case as the production of IL-22 by ILC3s was much greater in RAG-2-deficient mice (mice lacking mature B and T cells). As the microbiota induces the generation of Th17 cells, the data indicated that the microbiota represses IL-22 production by ILC3s through both IL-25 and adaptive immunity, Of note, adaptive immunity was not required for the IL-25-mediated repression of IL-22 production by ILC3s, as the administration of IL-25 efficiently repressed this activity in RAG-2-deficient mice. When intestinal inflammation was induced in mice by oral administration of dextran sodium sulfate (DSS), expression of both IL-22 and IL-23 increased in the intestine, concomitant with much lower IL-25 expression by epithelial cells and a three- to fourfold population expansion of ILC3. Thus, whereas the microbiota and adaptive immunity repress the activity of ILC3s in the steady state, epithelial damage that induces intestinal inflammation fully de-represses this activity and forces a return to intestinal immune homeostasis.

In addition to direct and indirect effects of commensal bacteria on ILC populations, ILCs can reciprocally regulate the composition or anatomical location of commensal bacteria through the production of cytokines. ILC3 are critical in promoting the anatomical containment of commensal bacteria. Depletion of ILC3s resulted in peripheral dissemination of commensal bacteria and systemic inflammation, which was prevented by administration of IL-22. IL-22 was shown to act directly on the intestinal epithelium and induce expression of tissue-protective mucin genes and antimicrobial proteins. REG-3γ, a secreted antibacterial lectin, is essential for maintaining a ~50-μm zone that physically separates the microbiota from the small intestinal epithelial surface.

ILC3s maintain intestinal immune homeostasis through MHC class II-dependent interactions with $CD4^+$ T cells that limit potentially damaging T cell responses to commensal bacteria. ILC3s express MHC class II and are able to process and present antigens to T cells; however, intestinal ILC3s lack expression of the co-stimulatory molecules required to activate T cell responses. Antigen presentation by intestinal ILC3s inhibited $CD4^+$ T cell responses to symbiotic bacteria independently of IL-17A, IL-22, or IL-23. Selective deletion of MHC class II in mice resulted in exaggerated commensal bacteria-dependent $CD4^+$ T cell responses that promoted spontaneous intestinal inflammation. The regulation of $CD4^+$ T cells by ILC3s in the intestine is countered by the regulation of ILC3 population size and function by $CD4^+$ T cells, demonstrating dynamic cross-regulation between these cell types that contributes to intestinal homeostasis.

A bacterial component of the intestinal microbiota (a species of the *Clostridium* genus) promoted the accumulation of Treg cells in the mouse colon. Mortha *et al.* (2014) demonstrated how ILC3s can relay signals from the intestinal microbiota that lead to Treg accumulation. Murine ILC3s lack TLRs and cannot directly sense microbial signals in the gut. However, bacterial components of the microbiota are recognized by TLRs expressed on macrophages, and MyD88-dependent TLR signaling elicits the production of IL-1β by the macrophages. IL-1β activates the IL-1R–MyD88 pathway in ILC3s to produce GM-CSF, which in turn primes DCs and macrophages to produce retinoic acid and IL-10. These two regulatory molecules promote the differentiation of naïve $CD4^+$ T cells to Treg cells and expansion of the differentiated cells, thereby promoting a tolerogenic state in both the small and large intestine. Cell-type-specific ablation of IL-1R-dependent signaling in ILC3s abrogated oral tolerance to dietary antigens and compromised intestinal homeostasis *in vivo.* These results demonstrated that microbiota-dependent cross-talk between macrophages and ILC3s contributes to intestinal homeostasis.

10.25.12 Protective Effect of Toll-Like Receptor Ligation by Commensal Bacterial Products against Intestinal Injury

It has been widely held that disruption of the mucosal barrier upon injury to intestinal epithelial cells exposes the multitude of Toll-like receptor (TLR) ligands expressed by commensal bacteria to TLR-expressing macrophages resident in the lamina propria, and this results in a potent inflammatory response and corresponding injury. Rakoff-Nahoum *et al.* (2004) showed that commensal bacterial products interact with host TLRs under normal steady-state conditions, and this interaction is critical for the protection against gut injury and associated mortality in mice. This unexpected TLR-mediated protection was postulated to operate through two mechanisms that are not mutually exclusive. The first is that protective factors are induced when TLRs expressed on colonic epithelium interact with TLR ligands expressed on commensal bacteria. The second possible mechanism is that commensal-derived TLR ligands may induce the production of protective factors upon epithelial damage. The detection of such ligands by TLRs expressed by resident macrophages may signal that disruption of the epithelial barrier has occurred. The authors proposed that mammalian TLRs have at least two distinct functions – protection from infection and control of intestinal epithelial homeostasis, which depend on the recognition of ligands expressed on pathogens and commensals, respectively.

The above findings demonstrate that a crucial balance exists between the protective effect of TLR activation by commensals, and the detrimental effect of this interaction when it becomes dysregulated. This balance is an important consideration in clinical regimens associated with intestinal tissue damage, such as colonic surgery, which are accompanied by treatment with antibiotics to prevent opportunistic infections. The experimental findings suggest that complete depletion of microflora may be detrimental and that controlled stimulation of TLRs by their ligand (natural or synthetic) may be beneficial in such situations.

10.26 CENTRAL AND PERIPHERAL TOLERANCE

10.26.1 Introduction

Immunological tolerance provides the means of rendering the immune system unresponsive to self antigens and preventing autoimmune diseases. Tolerance can also be directed toward transplanted tissue, infectious agents, therapeutic proteins, and tumor antigens. This section discusses self-tolerance, where the immune system does not attack our own bodies.

The random rearrangement of the genes encoding antigen-specific receptors during the early development of T and B cells creates a vast diversity of receptors. Inevitably, lymphocytes whose receptors are specific for self antigens will be generated. The mechanisms involved in self-tolerance limit the production of self-reactive T and B cells, and also prevent those that are produced from actually generating immune responses against self antigens.

Natural tolerance to self is an active process that depends on the activity of auto-reactive Tregs. Direct evidence for the role of Treg cells in self-tolerance is that autoimmune diseases can be produced in athymic (T cell-deficient) mice by inoculating Treg-depleted spleen cell suspensions, and that reconstitution of Tregs prevents these diseases in a dose-dependent fashion. The removal of Tregs also breaks immunological unresponsiveness to autologous tumors *in vivo* and *in vitro*, indicating that tumor immunity and autoimmunity share a common regulatory basis.

Self-tolerance mechanisms have been divided into two broad categories. Central tolerance is due to the death of developing T cells and B cells when they encounter self antigens in the thymus and bone marrow, respectively. Peripheral tolerance is maintained by mechanisms that act on mature lymphocytes when they encounter self antigens in peripheral tissues. Central and peripheral tolerance have been well discussed by Xing and Hogquist (2012).

10.26.2 Tolerogenic Dendritic Cells

Tolerogenic dendritic cells (tolDCs) dampen chronic inflammation and autoimmunity by promoting the development of regulatory T cells (Tregs) and are essential for the induction and maintenance of central and peripheral T cell tolerance. Thymic tolDCs contribute to central tolerance induction by participating in the negative selection of autoreactive T cells in the thymus. Immature tolDCs contribute to peripheral tolerance by inducing the differentiation of Tr1 and $CD8^+$ Treg cells. In the periphery, tolDCs exhibit reduced secretion of pro-inflammatory IL-12, but high production of anti-inflammatory cytokines like IL-10. In the absence of infection, immature tolDCs capture and process self antigens from apoptotic tissue cells, but are not stimulated to express CD80/CD86 or other co-stimulatory molecules. Tbe partially matured tolDCs migrate to secondary lymphoid tissues where they present the self antigen to specific naïve T cells. Without the necessary co-stimulation, the majority of the T cells become anergic (unresponsive to the antigen) or are deleted by apoptosis. Some naïve $CD4^+$T cells recognized by self antigen in the absence of co-stimulation develop into Tregs. Although immaturity is considered to be a good indicator of DC tolerogenicity, phenotypically mature DCs can retain their tolerogenic function if they have been previously exposed to certain differentiation signals. Adoptive transfer of tolDCs in animals can limit the incidence of transplant rejection and curb the progression of certain autoimmune diseases.

Probst *et al.* (2003) developed a system that allowed direct *in vivo* comparison of activated DCs and resting DCs for their respective capacity to prime, to be ignored by, or to tolerize naïve $CD8^+$ T cells in the periphery. They found that antigen presentation by activated DCs primed endogenous $CD8^+$ T cells to proliferate and to develop protective effector function, whereas antigen presentation by resting DCs resulted in antigen-specific tolerance. Probst *et al.* (2005) reported that tolerance induced by resting DCs occurs only if the responding $CD8^+$ T cells can receive signals through PD-1 and CTLA-4. If either of these inhibitory signals were missing, the T cells are primed. It can be surmised that the opposing functions of DCs (tolerance and priming) are controlled by the balance between positve and negative signals, which changes when DCs undergo activation. Resting DCs express greater levels of PD-L than other co-stimulatory ligands, so the inhibitory PD-1–PD-L interaction is dominant and tolerance ensues. By contrast, activated DCs express lower levels of PD-1 than other co-stimulatory ligands, so that the negative effect of PD-1 is overcome by positive signals and priming ensues.

Conventional DCs produce the immunosuppressive cytokines IL-10, IL-27, and TGF-β. IL-10 plays an indispensable role in the induction of tolerance and the limitation of excessive immune responses. In general, IL-10 acts on DCs by down-regulation of MHC II and co-stimulatory molecules in combination with reduced release of IL-1β, IL-6, TNF-α, and, above all, IL-12. Additionally, DCs that are cultured in the presence of IL-10 up-regulate the expression of inhibitory molecules and produce more IL-10, resulting in a positive feedback loop. IL-27 signaling in DCs promotes tolerogenic function by inducing expression of the immunoregulatory molecule CD39. IL-10 and TGF-β drive T cell differentiation toward Tregs.

The tolerogenic potential has been found to be highest for plasmacytoid dendritic cells (pDCs), followed by $CD103^+$ DCs, with $CD11b^+$ DCs having low activity in most *in vivo* assays. pDCs are the major cellular producers of IFN-α/β and are involved in the induction of central and peripheral tolerance; they also mediate oral tolerance. Recirculating pDCs present self antigens in the thymus and contribute to the inactivation, or deletion, of autoreactive T cells. In the periphery, immature pDCs promote T cell anergy, T cell deletion, and Treg development.

10.26.3 Central Tolerance

10.26.3.1 T Cells

Central tolerance of T cells takes place during their development in the thymus and allows only those thymocytes that recognize a ligand complex comprising peptide fragments of foreign antigen bound to self MHC molecules to develop into mature T cells. The process has two components, positive selection and negative selection. In positive selection, complexes of self peptides and MHC molecules are displayed on the surface of thymic cortical epithelial cells. The weak binding of thymocyte T cell receptors (TCRs) to these ligands triggers survival signals, whereas non-binding leads to rapid cell death by a default pathway of apoptosis. The surviving thymocytes receive additional signals that inhibit RAG gene expression (thereby closing off the option of editing), increase cell-surface TCR expression, and induce the expression of homing receptors for chemokines found in the thymic medulla. This process of positive selection also determines whether a T cell ultimately becomes a $CD4^+$ cell or a $CD8^+$ cell. Before positive selection, all thymocytes are double-positive ($CD4^+CD8^+$) i.e. bear both CD4 and CD8 co-receptors. During positive selection, they become either $CD4^+CD8^-$ or $CD8^+CD4^-$ T cells, depending on whether they recognize MHC class II or class I molecules, respectively.

Negative selection (also known as clonal deletion) is the process in which thymocytes whose TCRs bind strongly to self peptides in association with self MHC molecules undergo apoptosis by the mitochondrial pathway or differentiate into Tregs. The thymic APCs that mediate negative selection are primarily bone marrow–derived DCs and macrophages, both abundant in the medulla. Thymic medullary epithelial cells express a nuclear protein called AIRE (autoimmune regulator) which induces the expression of a number of genes that are otherwise expressed only in specific tissues outside the thymus. Expression of these genes makes many tissue-specific peptides available for presentation to thymocytes. Thymocytes that do not react with self peptide–MHC are allowed to finish the maturation process and emerge from the thymus as mature naïve $CD4^+$ or $CD8^+$ T cells with specificity for foreign antigen–MHC. The naïve cells circulate continually through secondary lymphoid organs in search of their cognate peptide–MHC complex.

10.26.3.2 B Cells

If immature B cells in the bone marrow bind to multivalent self antigens with high avidity, B cell receptor (BCR) signaling will initiate a new round of V(D)J recombination in which secondary rearrangements substitute another *V* gene onto an already rearranged *V(D)J* segment. This process, known as receptor editing, can also occur by wholesale replacement of an entire light chain. Receptor editing is mediated by two proteins, RAG-1 and RAG-2, which are expressed exclusively in developing lymphocytes. If the new BCR is no longer self-reactive, the receptor signaling will stop and the immature B cells (now called transitional B cells) will migrate to the spleen. In the spleen, further maturation and selective events result in the transitional B cells becoming one of the three types of mature B cells (B1 cells, marginal zone B cells, and recirculating follicular B cells). If the initial editing is non-productive, light chain rearrangement can be repeated at other gene loci. Failure to eradicate self-reactivity after receptor editing results in apoptotic death of the B cell within 1–2 days. Immature B cells that bind to soluble self antigens with relatively weak avidity become anergic and die relatively quickly after migration to the spleen. Anergy is due to down-regulation of antigen receptor expression as well as a block in antigen receptor signaling. Finally, some immature B cells with self-reactive receptors are unaffected by the presence of self antigens and remain in a state of immunological ignorance. Ignorant self-reactive B cells should not be considered inert as they can be activated and cause an autoimmune disease under certain conditions.

10.26.4 Peripheral Tolerance

10.26.4.1 T Cells

Self-reactive T cells with low binding avidity to self peptides escape from the thymus and must undergo peripheral tolerance. In addition, T cells must remain tolerant to harmless environmental antigens present in the respiratory tract and intestine. Tolerogenic DCs are essential for the induction and maintenance of peripheral T cell tolerance and their antigen presentation in the absence of co-stimulatory molecules leads to T cell anergy or apoptosis.

Among splenic DCs, the myeloid ($CD11b^+CD8\alpha^-$-$DEC205^-$) subset and not the lymphoid ($CD11b^-CD8\alpha^+$-$DEC205^+$) subset was identified as the tolerogenic population in peripheral T cell tolerance. The $CD8\alpha^-$ DCs produced immunosuppressive IL-10 and supported reversal of experimental allergic encephalomyelitis in mice. The IL-10 produced by the $CD8\alpha^-$ DCs had the capacity to inhibit the function of T cells directly by alteration of

CD28 expression, and/or indirectly by down-regulation of IL-12 from the CD8α⁺ DCs. Thus CD8α⁻ DCs mediated both tolerance and bystander suppression, which effectively regulated autoreactive T cells.

PD-1 and CTLA-4 have been shown to be negative regulators of T cell activation. Disruption of the mouse PD-1 gene results in the spontaneous development of lupus-like autoimmune diseases, demonstrating that PD-1 is essential for maintaining peripheral T cell tolerance. The PD-L2 ligand is required for induction of T cell tolerance to oral antigens. Mice deficient in CTLA-4 develop massive lymphoproliferation and fatal widespread tissue destruction. Antibody-mediated blockade of CTLA-4 prevents development of tolerance.

Keller *et al.* (2008) used CD70 transgenic mice to confirm previous reports that expression of CD70 on immature DCs breaks $CD8^+$ T cell tolerance and allows the naïve T cells to spontaneously differentiate into effector and memory cells. Their data strongly suggested that CD27–CD70 signaling at the interface between $CD8^+$ T cells and DCs can override instructions for anergy mediated by PD-1 and CTLA-4. Although reversal of the tolerant state is generally unwanted, there are instances in which it can be beneficial, in particular in cancer patients.

10.26.4.2 B Cells

Peripheral tolerance mechanisms prevent activation of self-reactive B cells that have not been eliminated by central tolerance. These mechanisms also prevent activation of self-reactive B cell clones that may be generated by somatic hypermutation in germinal centers.

The survival of peripheral B cells depends on competition for limiting quantities of BAFF. Constant stimulation of self-reactive B cells by self antigens elevates their survival requirement for BAFF, so they fail to receive enough BAFF and are competitively deleted.

When a recirculating self-reactive B cell encounters self antigen in some peripheral tissue, such as the thyroid, it can process the antigen and present self peptide on an MHC class II molecule. In the absence of activating signals, the B cell will enter the unresponsive state of anergy. Until they die, anergic B cells are located in the region of secondary lymphoid organs where activated helper T cells are present. Antigen recognition induces expression of CD40 ligand and Fas ligand on the T cell surface. CD40L stimulation of B cells induces expression of Fas receptor on their surface, making them susceptible to apoptosis induced by FasL. Thus an anergic B cell presenting antigen to a helper T cell will be induced to undergo apoptosis by the death receptor pathway. In contrast, a non-anergic B cell will have active signaling through the antigen receptor, which prevents Fas-mediated apoptosis.

Mature B cells that recognize self antigens with low affinity may be prevented from responding by the engagement of various inhibitory receptors. These receptors set a threshold for B cell activation: they allow responses to foreign antigens with T cell help, but do not allow responses to self antigens. This mechanism of peripheral tolerance became evident when mice with defects in the SHP-1 tyrosine phosphatase or the CD22 inhibitory receptor were shown to develop autoimmunity. Phosphorylation of ITIM motifs in the cytoplasmic tail of CD22 stimulates the recruitment of SHP-1, which attenuates BCR signaling by removing the activating phosphate groups in the signaling pathway.

10.27 ORAL TOLERANCE TO FOOD PROTEIN

10.27.1 Introduction

The immune system must be able to discriminate between innocuous and harmful materials that are ingested and make the right decision as to how to deal with them. Oral tolerance is the state of local and systemic immune unresponsiveness to an innocuous antigen that is induced by prior administration of the antigen by the oral route. The fed antigen can be a protein or a small chemical molecule known as a hapten. Numerous investigators have found that animals fed proteins such as ovalbumin or sheep red blood cells are unresponsive to these antigens when subsequently immunized, but do respond normally to other antigens. Oral tolerance has also been observed in humans fed and immunized with keyhole limpet hemocyanin. The effects of oral tolerance are measured typically as reductions in systemic delayed-type hypersensitivity and T cell proliferation. A failure of oral tolerance can result in food allergies exemplified by celiac disease (gluten-sensitive enteropathy or non-topical sprue).

Oral tolerance induction begins with antigen uptake in GALT and lamina propria. Immune mechanisms underlying oral tolerance include anergy or deletion by apoptosis of antigen-specific effector T cells and bystander suppression of effector T cells by Tregs. A process analogous to oral tolerance occurs when environmental allergens enter the respiratory mucosa via the nasal passages. The more general term ‘mucosa tolerance’ is used to describe tolerance induced by either oral or nasal antigen administration.

Intestinally derived antigen can disseminate via the blood and lymph and generate widespread systemic tolerance. Antigen taken up into GALT or lamina propria enters the bloodstream via the portal vein, first reaching the liver before it becomes distributed into the wider circulation. Immature DCs in the lamina propria capture intestinal antigen and transport it to the mesenteric lymph nodes via afferent lymph. Systemic tolerance occurs because antigen presentation by resident APCs in the liver, peripheral lymph nodes, and spleen leads to T cell anergy or apoptosis in the absence of CD28-mediated co-stimulation.

10.27.2 Requirement of Natural Regulatory T Cells for Oral Tolerance

Oral tolerance is evidently the consequence of an active regulatory mechanism, because mice fed high doses of a protein antigen become refractory to subsequent parenteral adminstration of the same antigen. Feeding low doses of antigen expands the numbers of Th3 cells and nTregs.

Dubois *et al.* (2003) provided evidence that $CD4^+CD25^+$ nTregs are critical for oral tolerance induction using the model of contact hypersensitivity (CHS) to the hapten 2,4-dinitrofluorobenzene (DNFB), which is mediated by $CD8^+$ CTLs independently of $CD4^+$ T cell help. They found that (1) transfer of naïve $CD4^+CD25^+$, but not $CD4^+CD25^-$ T cells, into MHC class II-associated invariant chain knock-out mice, which are deficient in $CD4^+$ T cells, completely prevents the CHS response and skin infiltration by $CD8^+$ T cells by blocking development of hapten-specific $CD8^+$ T cells; (2) *in vivo* depletion of nTregs by antibody treatment in normal mice impairs oral tolerance; and (3) nTregs inhibit hapten-specific $CD8^+$ T cell proliferation and IFN-γ production *in vitro*. Thus, both nTregs and oral hapten are required to achieve oral tolerance.

10.27.3 Role of the Intestine in Oral Tolerance

A population of $CD103^+$ DCs that makes up 2–3%of total leukocytes in the intestinal lamina propria of normal mice plays a critical role in the induction of oral tolerance. The CD103 marker is α_E, the α chain of the $\alpha_E\beta_7$ integrin whose best-known ligand, E-cadherin, is expressed by intestinal epithelial cells. The interaction between CD103 on DCs and E-cadherin on epithelial cells tethers the DCs close to the intestinal epithelium. This may allow the DCs to be conditioned, perhaps by epithelial cell-derived factors. $CD103^+$ DCs undergo a rapid turnover and are replenished continually by blood-borne precursors. They acquire antigens from the gut lumen and migrate to gut-draining mesenteric lymph nodes. It is not the $CD103^+$ DCs, but rather a subset of resident $CX3CR1^+$ macrophages that samples luminal antigen by extending processes through gaps between adjacent intestinal epithelial cells. Presumably, the $CX3CR1^+$ macrophages transfer antigen to the $CD103^+$ DCs in the lamina propria. In the mesenteric lymph nodes, the pre-conditioned $CD103^+$ DCs metabolize dietary vitamin A and the retinoic acid produced promotes TGF-β-mediated conversion of naïve $CD4^+$ T cells into Treg cells. $CD103^+$ DCs also produce IDO, depleting tryptophan from the microenvironment and generating toxic catabolites (kynurenines). The combined effect of these occurrences is to inhibit the generation of effector T cells and promote the induction of Treg cells.

The $Foxp3^+$ Tregs that are involved in oral tolerance differentiate from naïve $Foxp3^-$ precursors and are not simply expanded nTregs that have migrated from the thymus. The generation of iTregs occurs in the gut-draining mesenteric lymph nodes and requires presentation of intestinal antigens by migrating $CD103^+$ DCs. In order for oral tolerance to be fully established, the iTregs need to home to the intestinal mucosa and undergo local expansion. In the lamina propria, resident $CX3CR1^+$ macrophages act as local APCs. By presenting antigenic epitopes in a tolerogenic context, the macrophages deliver the necessary signal to trigger Treg proliferation. In CX3CR1-deficient mice, reduced production of IL-10 by intestinal macrophages correlated with impaired local Treg expansion, implicating IL-10 in the induction of oral tolerance.

Mortha *et al.* (2014) established that the gut microflora (microbiota) stimulates intestinal macrophages to release IL-1β, which then stimulates group 3 innate lymphoid cells (ILC3s) to release GM-CSF. This cytokine stimulates macrophages and DCs to release retinoic acid and IL-10, which promote the differentiation of naïve T cells to Tregs and expansion of the differentiated cells. When mice deficient in GM-CSF were fed ovalbumin (OVA), differentiation and expansion of OVA-specific Treg cells in the small and large intestine were impaired. These findings demonstrate the contribution of ILC3-derived GM-CSF to the control of oral tolerance to dietary antigens.

10.27.4 Role of the Liver in Oral Tolerance

10.27.4.1 Introduction

The liver is an immune-privileged organ with exquisite mechanisms of immune regulation that ensure upkeep of local and systemic immune tolerance to self and foreign antigens, but that is also able to mount effective immune responses against pathogens. Blood from the intestine carries bacterial products and mainly harmless food-derived antigens to the liver via the portal vein. Circulating innocuous antigens derived from aged or damaged cells also pass through the liver and neo-antigens arise in the liver from the degradation of toxic or waste products. All of these antigens are trapped by specialized APCs resident in the liver and presented to naïve T cells passing through the sinusoids. Thomson and Knolle (2010) have described the liver microanatomy and the function of liver cells as APCs in the tolerogenic liver environment.

10.27.4.2 Modulation of Liver Tolerance by Liver-Resident Antigen-Presenting Cells and Induction of Regulatory T Cells

Antigen presentation by immature liver DCs and non-professional APCs like liver sinusoidal endothelial cells (LSECs), Kupffer cells, hepatic stellate cells (HSCs), and even hepatocytes induces tolerance rather than immunity. These hepatic APCs promote tolerance by induction of T cell anergy or apoptosis (Horst *et al.*, 2016).

Unstimulated pDCs in the liver can also function as APCs and induce tolerance. Goubier *et al.* (2008) showed that feeding mice with a protein or a hapten induced anergy and/or deletion of nearly 70–80% of antigen-specific T cells. *In vivo* depletion of pDCs abrogated oral tolerance and suppression of delayed-type hypersensitivity (DTH) responses. Conversely, adoptive transfer of pDCs from antigen-fed mice prevented T cell priming and induction of DTH responses in an antigen-specific way.

The tolerogenic potential of the liver includes conversion of effector T cells to iTregs. In the gut this conversion is mediated by $CD103^+$ DCs, but this subset has not been identified in the liver. However, the factors necessary for iTreg induction are abundantly expressed in the liver. TGF-β is produced by Kupffer cells, HSCs, and to a lesser extent by LSECs. Moreover, vitamin A conversion into retinoic acid is prominantly controlled by HSCs and LSECs, with LSECs as the major inducers of $CD4^+CD25^+Foxp3^+$ iTregs (Horst *et al.*, 2016).

10.27.4.3 Generation of CD4+ Regulatory T Cells Expressing the Fas Ligand in the Liver

Watanabe *et al.* (2002) described a new subset of hepatic Tregs that may confer a tolerogenic property on the liver. Oral administration of a high dose of ovalbumin (OVA) to mice led to the selective generation of hepatic $CD4^+$ Tregs expressing the ligand of the Fas death receptor, FasL, and secreting suppressive Th2-type cytokines IL-4, IL-10, and TGF-β. The emergence of these $CD4^+FasL^+$ Tregs occurred concomitantly with massive cell death of OVA-specific $CD4^+$ T cells by Fas–FasL-mediated apoptosis in the liver. The authors postulated that a death signal provided by the FasL expressed on the Tregs preferentially delete antigen-specific T cells that constitutively express Fas and lack IL-4 production. Adoptive transfer of the $CD4^+FasL^+$ Tregs to naïve mice and subsequent antigenic challenge led to suppression of T cell proliferation as well as as IgG antibody responses to OVA; this suppressive effect was mostly abrogated by a blocking antibody to FasL. Watanabe *et al.* (2003) demonstrated that hepatic $CD4^+FasL^+$ Tregs can be generated *in situ* by antigenic activation of naïve $CD4^+$ T cells by hepatic $CD11c^+$ professional APCs. A part of OVA ingested is captured by the APCs and presented to OVA-specific $CD4^+$ T cells. Activation of the T cells up-regulates FasL expression, thereby creating the regulatory phenotype.

10.27.4.3 Tolerogenic Properties of the Liver

The liver stroma programs the differentiation of hematopoietic $CD117^+$ progenitors into tolerogenic DCs that suppress T cell proliferation and induce apoptosis of activated T cells. These liver-resident DCs produce prostaglandin E_2 (PGE_2), which inhibits T cell proliferation by enhancing the production of immunosuppressive IL-10.

Watanabe *et al.* (2003) reported that stimulation of hepatic $CD11c^+$ cells from mice fed OVA reduced the production of IL-12 and increased that of IL-18. An increased secretion of IL-18 promotes Fas-mediated apoptosis, because IL-18 up-regulates FasL expression in T cells. A decrease of IL-12 secretion enhances cell death signal mediated by Fas and also decreases survival signal by down-regulating IL-2R expression of OVA-specific $CD4^+$ T cells. A decrease of IL-12 secretion, together with an increase of IL-18 secretion, also drives $CD4^+$ T cells to Th2 differentiation. Overall, effects of decreased IL-12 secretion by hepatic $CD11c^+$ cells can result in enhanced cell death of antigen-specific Th1 cells by Fas-mediated apoptosis.

10.28 AUTOIMMUNITY

Autoimmunity refers to the state of adaptive immune responsiveness to self antigens that occurs when the mechanisms of self-tolerance in T or B cells fail. Failure of self-tolerance may be due to (1) defects in deletion of T or B cells or receptor editing in B cells during the maturation of these cells; (2) defective apoptosis of mature self-reactive lymphocytes; (3) insufficient numbers or defective functions of regulatory T cells; (4) inadequate function of inhibitory receptors; or (5) activation of APCs, which overcome regulatory mechanisms and result in excessive T cell activation. Once an autoimmune reaction develops, many amplification mechanisms are activated that perpetuate the response and lead to an autoimmune disease.

Autoimmune diseases are described as either organ-specific or systemic. Organ-specific autoimmune diseases include type 1 diabetes and experimental autoimmune encephalomyelitis (EAE), which is an animal model of multiple sclerosis. In systemic lupus erythematosus, antibodies form complexes with ubiquitous self antigens and these immune complexes activate effector mechanisms to cause widespead damage to tissues and organs. A characteristic feature of autoimmune diseases is that they are chronic, progressive, and self-perpetuating.

Autoimmune diseases have a strong genetic component. In most cases, affected individuals inherit multiple genetic polymorphisms that make them susceptible to disease. Susceptibility genes disrupt self-tolerance mechanisms, leading to the survival of self-reactive lymphocytes. Autoimmune reactions may be triggered by environmental stimuli, such as infections or tissue injury, in genetically susceptible individuals. Infection or necrosis in tissues promote the influx of self-reactive T cells and activation of these cells, resulting in tissue injury.

An autoantigen-directed immune response could be activated in the following manner. An infectious microbe may contain an antigen with an epitope that cross-reacts with a structurally similar epitope of a self protein, so that the immune response to the microbial epitope is also directed to uninfected host tissue cells displaying the similar epitope. This cross-reaction of microbial and self antigens that initiates an autoimmune attack is called molecular mimicry. Once an autoimmune reaction has developed, it may become self-perpetuating due to epitope spreading, whereby the release of multiple self antigens from damaged tissues leads to a broadening of the immune response to self antigens.

10.29 TYPE 1 (INSULIN-DEPENDENT) DIABETES MELLITUS

10.29.1 Introduction

Type 1 (insulin-dependent) diabetes mellitus, hereafter referred to simply as diabetes, is an organ-specific autoimmune disease in which the insulin-producing β cells in the pancreatic islets of Langerhans are destroyed by the body's own immune system. The disease is characterized by an initial inflammatory response manifested as a cellular infiltration in and around the islets of Langerhans. When more than 80% of β cells have been destroyed, insulin production becomes insufficient, causing a loss of glucose homeostasis. A number of β cell autoantigens have been identified, including insulin, insulinoma-associated protein 2, islet-specific glucose-6-phosphatase catalytic subunit-related protein, and glutamic acid decarboxylase.

Diabetes results from T cell-mediated autoimmunity against the pancreatic β cells and may be due to a thymic failure to delete autoreactive T cells. Cytokine imbalance plays a key role in disease emergence, with overexpression of IFN-γ-secreting diabetogenic Th1 cells relative to IL-4-secreting regulatory Th2 cells. Systemic administration of IL-4 or anti-IFN-γ monoclonal antibody (mAb) prevents diabetes onset. This section focuses on the cytokines involved in diabetes and the protective role of NKT cells.

10.29.2 The Non-Obese Diabetic (NOD) Mouse as a Model of Human Type 1 Diabetes

The non-obese diabetic (NOD) mouse is an animal model of human diabetes which develops the disease spontaneously. The NOD strain was developed during a breeding program aimed at establishing a cataract-prone subline. The subline did not manifest cataract susceptibility, but after 26 generations of inbreeding it showed a high incidence of diabetes. Diabetes pathogenesis in NOD mice begins at three to four weeks of age with the infiltration of immune cells into the perivascular duct and peri-islet regions of the pancreas, a process called insulitis. DCs and macrophages infiltrate first, followed by T cells ($CD4^+$ and $CD8^+$) and B cells. This stage is followed by the slow, progressive, and selective T cell-mediated destruction of islet β cells. By four to six months of age, the β cells are totally destroyed and overt diabetes is evident. NOD female mice develop a more destructive insulitis and incur a higher incidence (80–90%) of diabetes than males (10–40%). This pronounced female gender bias is not observed in humans.

Considerable evidence suggests that diabetes in NOD mice is mediated by autoreactive T cells. Diabetes is prevented by neonatal thymectomy, by immunosuppressive agents that target T cells, and by anti-CD4 and anti-CD8 mAb treatments. Both $CD8^+$ and the Th1 subset of $CD4^+$ cells are able to kill β cells. $CD8^+$ T cells recognize β cells through class I MHC molecules expressed on the β cells. Th1 cells do not interact directly with β cells and may recruit other effector cell types such as macrophages, NK cells, or $CD8^+$ T cells.

10.29.3 Role of Cytokines in the Pathogenesis of Diabetes

Cytokines released by infiltrating $CD4^+$ and $CD8^+$ T cells, monocytes, and macrophages have been implicated in diabetes by exerting direct effects on β cell function and also by recruiting lymphocytes and macrophages to the site of inflammation. Considerable evidence indicates that the Th1 subset of $CD4^+$ helper T cells is diabetogenic and the Th2 subset is protective. Correlations were observed between β cell destructive insulitis and expression of pro-inflammatory IL-1, TNF-α, and type 1 IFN-γ, and other type 1 cytokines, IL-2 and TNF-β. In contrast, expression of anti-inflammatory cytokines, TGF-β, and type 2 IL-4 and IL-10, correlated with benign insulitis (Rabinovitch, 1998). IL-4 production by Th2 cells was strongly impaired in patients with new onset diabetes (Berman *et al.*, 1996).

A thymic defect in NOD mice results in the loss of thymic and peripheral Th2 cells that are crucial for the immune regulation of potentially diabetogenic Th1 cells (Zipris *et al.* 1991a, b). Cameron *et al.* (1997) showed that systemic administration of IL-4 to NOD mice prevents insulitis and diabetes. This prevention could be explained by at least two mechanisms. First, IL-4 treatment interferes with the migration of autoreactive effector Th1 cells to the pancreatic islets during the inductive phase of diabetes. Second, IL-4 treatment generates Th2 regulatory cells with suppressive activity over diabetogenic Th1 effector cells in the thymus, spleen, and pancreatic islets. These findings are consistent with the idea that diabetes occurs as a result of highly skewed Th1 responses, whereas protection and recovery from this autoimmune disease is a result of prevailing Th2 responses.

10.29.4 Protective Effect of Natural Killer T Cells

NKT cells are abundant in pancreatic lymph nodes and are also present in islets. The number of NKT cells relative to total T cells in the pancreatic islets of female NOD mice decreases during the transition from peri-insulitis to invasive insulitis and diabetes. This deficiency of NKT cells promotes the development of diabetes, whereas activating NKT cells *in vivo* by injection of the synthetic ligand α-galactosylceramide (α-GalCer) suppresses disease in the NOD mouse. NKT cells induce DC maturation and accumulation in pancreatic lymph nodes, where they subsequently recruit and tolerize diabetogenic T cells (Chen *et al.*, 2005). A lower frequency of NKT cells and an extreme bias toward IFN-γ secretion has been demonstrated in human diabetic patients.

NKT cells are a significant source of IL-4 in response to priming. Young prediabetic NOD mice exhibit a clear deficit in the number of NKT cells and those cells that are present lack the ability to produce IL-4. Injection of NKT cells into NOD mice exerts a protective effect against diabetes, which is mediated by the Th2-associated IL-4 and/or IL-10.

10.30 TRANSPLANTATION AND GRAFT-VERSUS-HOST DISEASE

Definitions Used in Transplantations

The following transplantation terms describe the genetic relationship between the donor and recipient.

- **Allogeneic**: Genetically different individuals of the same species. **Syngeneic**: Genetically identical individuals, such as identical twins or members of an inbred strain of mice. **Xenogeneic**: Individuals of different species.
- **Alloreactive T cells:** T cells from one individual that will recognize antigens on cells or tissues of another genetically non-identical individual.
- **Alloantigen:** A cell or tissue antigen that is present in some members of a species but not in others and that is recognized as foreign on an allograft.
- **Alloantigen recognition (allorecognition)**: Pathways by which T cells may recognize alloantigens and lead to transplant rejection after an organ transplant

Bone marrow transplantation is used to treat genetic diseases of bone marrow cells, such as severe combined immunodeficiency, and in situations where the bone marrow is affected by cancer. In this procedure, donor bone marrow containing hematopoietic stem cells is infused intravenously into the recipient. The stem cells are capable of regenerating all cellular elements of the blood and immune system. In allogeneic transplantation, the transplanted organ/tissue is called an allograft. A major problem in allogeneic bone marrow transplantation is that mature T cells present in the donor bone marrow are activated by MHC molecules on host DCs, leading to clonal expansion and differentiation into effector cells. Cytokines secreted by the effector cells signal cytotoxic T cells and NK cells to attack host tissues. This graft-versus-host reaction is called allograft rejection and is manifested clinically as graft-versus-host disease (GVHD). Clinical GVHD has both an acute and a chronic form. Acute GVHD occurs within the first 100 days of transplantation and is characterized by damage to the skin, liver, and gastrointestinal tract. Chronic GVHD occurs after 100 days post-transplantation and manifests as a systemic autoimmune disease

The clinical manifestations of acute GVHD are largely due to dysregulated production of the pro-inflammatory cytokines IL-1 and TNF-α. Activated donor Th1 cells initiate GVHD by secreting IL-2, which stimulates host tissues to release IL-1, TNF-α, and IFN-γ. The resulting inflammatory response causes additional release of cytokines and amplifies local organ injury. Activated Tc1 donor cells propagate GVHD by secreting high levels of IFN-γ, mediating perforin- and granzyme-induced cytolysis, and inducing Fas-based apoptosis. Intestinal lesions of GVHD permit increased translocation of bacterial lipopolysaccharide (LPS) endotoxin into the systemic circulation. LPS triggers release of TNF-α from macrophages primed by IFN-γ during the graft-versus-host reaction.

Strategies for preventing acute GVHD have been directed largely against donor T cells. The use of cytokines to induce Th2 polarization of donor helper T cells reduces morbidity and mortality in experimental acute GVHD, while preserving cytotoxic cellular activity against tumors and infectious agents. Krenger *et al.* (1996) reported that infusion of *in vitro*-polarized Th2 cells increased GVHD-free survival in mice after allogeneic bone marrow transplantation. Increased survival correlated with diminished production of IFN-γ and TNF-α and increased production of IL-4 two weeks after transplantation.

10.31 NEUROENDOCRINE MODULATION OF IMMUNE RESPONSES

It has long been known that stressful conditions can suppress immune functions, reducing the ability of an individual to recover from infection. This can be explained by the fact that the neuroendocrine and immune systems are interconnected, forming an integrated system with common mediators and receptors. Many, but not all, of the regulatory effects of the neuroendocrine stress response on immune-mediated diseases occur through the actions of the glucocorticoid hormones on immune cell functions. Cells of the immune system have receptors for many hormones, hormone-releasing factors, and neurotransmitters – evidence that these molecules regulate immune responses. Moreover, immune cells themselves produce hormones and endorphins. Both excess or inadequate stress hormone responses are associated with disease: excess responses with enhanced susceptibility to infection, and inadequate responses with enhanced susceptibility to inflammatory, autoimmune, and allergic diseases.

The interplay between the neuroendocrine and immune systems is bi-directional, as exemplified by the control of cortisol release from the adrenal glands during stress. Cortisol has well-known anti-inflammatory properties and it also suppresses the immune system. The signal for the release of cortisol originates in the brain. Electrical signals generated in the brain, and also IL-1 and IL-6 synthesized in brain cells, stimulate the hypothalamus to produce corticotrophin-releasing hormone (CRH), which induces the anterior lobe of the pituitary gland to release adrenocorticotrophic hormone (ACTH). CRH can also induce lymphocytes to produce their own ACTH. The ACTH is released into the bloodstream and, on reaching the target organ (adrenal glands), stimulates the adrenal cells to secrete cortisol. At excessive concentrations, ACTH inhibits the release of CRH and ACTH by negative feedback to the hypothalamus and anterior pituitary gland.

10.32 EFFECT OF ADVANCING AGE ON IMMUNE RESPONSIVENESS

It is well established that human immune responsiveness declines with age and this decline is associated with a decrease in absolute numbers of T and B cells. Macrophages interact with T and B cells and therefore intervene in immune responses. Macrophages have a high concentration of arachidonic acid in their membrane phospholipids and, upon stimulation, up to 50% of this fatty acid can be released for conversion to prostaglandins and other eicosanoids. These compounds suppress lymphocyte proliferation and cytokine synthesis. Prostaglandin E_2 (PGE_2) production by macrophages increases with increasing age in mice. This is due to increased cyclooxygenase (COX) activity resulting from higher expression of COX mRNA and protein. At low concentrations, PGE_2 is believed to be necessary for certain aspects of cellular immunity, while at higher concentrations it suppresses T cell-mediated immunity, as demonstrated by decreases in mitogen-induced T cell proliferation and IL-2 production. The inhibitory effects of PGE_2 result from the specific binding of PGE_2 to a receptor on the T cell surface. Receptor binding stimulates the production of cAMP, which mediates the PGE_2 effect by activating protein kinase A. Goodwin and Messner (1979) showed that proliferation of cultured lymphocytes from human subjects aged over 70 is more sensitive to inhibition by PGE_2 compared to cells from younger subjects. This decreased responsiveness to PGE_2 was partially reversed by blocking endogenous production of PGE_2 with indomethacin, an inhibitor of prostaglandin synthesis. The above observations suggest that the decline in immunoresponsiveness with advancing age could be at least partly explained by an increased production of immunosuppressive PGE_2.

FURTHER READING

Abbas, A.K., Lichtman, A.H. & Pillai, S. (2012) *Cellular and Molecular Immunology*, 7th edn. Saunders (imprint of Elsevier Inc.), Philadelphia, PA.

DeFranco, A.L., Locksley, R.M. & Robertson, M. (2007) *Immunity*. New Science Press Ltd., London.

Delves, P.J., Martin, S.J., Burton, D.R. & Roitt, I.M. (2006) *Roitt's Essential Immunology*, 11th edn. Blackwell Publishing Ltd., Oxford.

Murphy, K. (2012) *Janeway's Immunobiology*, 8th edn. Garland Science, London and New York.

REFERENCES

Annacker, O., Pimenta-Araujo, R., Burlen-Defranoux, O., Barbosa, T.C., Cumano, A. & Bandeira, A. (2001) $CD25^+CD4^+$ T cells regulate the expansion of peripheral CD4 T cells through the production of IL-10. *Journal of Immunology*, **166**, 3008–18.

Artis, D. & Spits, H. (2015) The biology of innate lymphoid cells. *Nature*, **517**, 293–301.

Ashany, D., Song, X., Lacy, E., Nikolic-Zugic, J., Friedman, S.M. & Elkon, K.B. (1995) Th1 CD4+ lymphocytes delete activated macrophages through the Fas/APO-1 antigen pathway. *Proceedings of the National Academy of Sciences of the U.S.A.*, **92**, 11225–9.

Ashany, D., Savir, A., Bhardwaj, N. & Elkon, K.B. (1999) Dendritic cells are resistant to apoptosis through the Fas (CD95/APO-1) pathway. *Journal of Immunology*, **163**, 5303–11.

Attanavanich, K. & Kearney, J.F. (2004) Marginal zone, but not follicular B cells, are potent activators of naive CD4 T cells. *Journal of Immunology*, **172**, 803–11.

Baecher-Allan, C., Viglietta, V. & Hafler, D.A. (2002) Inhibition of human $CD4^+CD25^{high}$ regulatory T cell function. *Journal of Immunology*, **169**, 6210–17.

Balaji, K.N., Schaschke, N., Machleidt, W., Catalfamo, M. & Henkart, P.A. (2002) Surface cathepsin B protects cytotoxic lymphocytes from self-destruction after degranulation. *Journal of Experimental Medicine*, **196**, 493–503.

Berman, M.A., Sandborg, C.J., Wang, Z., Imfeld, K.L., Zaldivar, F. Jr., Dadufalza, V. & Buckingham, B.A. (1996) Decreased IL-4 production in new onset type I insulin-dependent diabetes mellitus. *Journal of Immunology*, **157**, 4690–6.

Bernink, J.H., Peters, C.P., Munneke, M., te Velde, A.A., Meijer, S.L., Weijer, K.W., Hreggvidsdottir, H.S., Heinsbroek, S.E., Legrand, N., Buskens, C.J., Bemelman, W.A., Mjösberg, J.M. & Spits, H. (2013) Human type 1 innate lymphoid cells accumulate in inflamed mucosal tissues. *Nature Immunology*, **14**, 221–9.

Bevan, M.J. (2004) Helping the $CD8^+$ T-cell response. *Nature Reviews Immunology*, **4**, 595–602.

Bonneville, M., O'Brien, R.L. & Born, W.K. (2010) $\gamma\ \delta$ T cell effector functions: A blend of innate programming and acquired plasticity. *Nature Reviews Immunology*, **10**, 467–78.

Boonstra, A., Asselin-Paturel, C., Gilliet, M., Crain, C., Trinchieri, G., Liu, Y.-J. & O'Garra, A. (2003) Flexibility of mouse classical and plasmacytoid-derived dendritic cells in directing T helper type 1 and 2 cell development: Dependency on antigen dose and differential Toll-like receptor ligation. *Journal of Experimental Medicine*, **197**, 101–9.

Brunner, T., Wasem, C., Torgler, R., Cima, I., Jakob, S. & Corazza, N. (2003) Fas (CD95/Apo-1) ligand regulation in T cell homeostasis, cell-mediated cytotoxicity and immune pathology. *Seminars in Immunology*, **15**, 167–76.

Cameron, M.J., Arreaza, G.A., Zucker, P., Chensue, S.W., Strieter, R.M., Chakrabarti, S. & Delovitch, T.L. (1997) IL-4 prevents insulitis and insulin-dependent diabetes mellitus in nonobese diabetic mice by potentiation of regulatory T helper-2 cell function. *Journal of Immunology*, **159**, 4686–92.

Cao, X., Cai, S.F., Fehniger, T.A., Song, J., Collins, L.I., Piwnica-Wormf, D.R. & Ley, T.J. (2007) Granzyme B and Perforin are important for regulatory T cell-mediated suppression of tumor clearance. *Immunity*, **27**, 635–46.

Chen, Y.-G., Choisy-Rossi, C.-M., Holl, T.M., Chapman, H.D., Besra, G.S., Porcelli, S.A., Shaffer, D.J., Roopenian, D., Wilson, S.B. & Serrez, D.V. (2005) Activated NKT cells inhibit autoimmune diabetes through tolerogenic recruitment of dendritic cells to pancreatic lymph nodes. *Journal of Immunology*, **174**, 1196–204.

Chertov, O., Michiel, D.F., Xu, L., Wang, J.M., Tani, K., Murphy, W.J., Longo, D.L., Taub, D.D. & Oppenheim, J.J. (1996) Identification of defensin-1, defensin-2, and CAP37/azurocidin as T-cell chemoattractant proteins released from interleukin-8-stimulated neutrophils. *Journal of Biological Chemistry*, **271**(6), 2935–40.

Cinamon, G., Matloubian, M., Lesneski, M.J., Xu, Y., Low, C., Lu, T., Proia, R.L. & Cyster, J.G. (2004) Sphingosine 1-phosphate receptor 1 promotes B cell localization in the splenic marginal zone. *Nature Immunology*, **5**, 713–20.

Cinamon, G., Zachariah, M.A., Lam, O.M., Foss, F.W. Jr. & Cyster, J.G. (2008) Follicular shuttling of marginal zone B cells facilitates antigen transport. *Nature Immunology*, **9**, 54–62.

Diao, J., Winter, E., Chen, W., Cantin, C. & Cattral, M.S. (2004) Characterization of distinct conventional and plasmacytoid dendritic cell-committed precursors in murine bone marrow. *Journal of Immunology*, **173**, 1826–33.

Dubois, B., Chapat, L., Gouibier, A., Papiernik, M., Nicolas, J.-F. & Kaiserlian, D. (2003) Innate $CD4^+CD25^+$ regulatory T cells are required for oral tolerance and inhibition of $CD8^+$T cells mediating skin inflammation. *Blood*, **102**, 3295–3301.

Duits, L.A., Ravensbergen, B., Rademaker, M., Hiemstra, P.S. & Nibbering, P.H. (2002) Expression of β-defensin 1 and 2 RNA by human monocytes, macrophages and dendritic cells. *Immunology*, **106**, 517–25.

Frasca, L., Carmichael, P., Lechler, R. & Lombardi, G. (1997) Anergic T cells effect linked suppression. *European Journal of Immunology*, **27**, 3191–7.

Fuchs, A., Vermi, W., Lee, J.S., Lonardi, S., Gilfillan, S., Newberry, R.D., Cella, M. & Colonna, M. (2013) Intraepithelial type 1 innate lymphoid cells are a unique subset of IL-12- and IL-15-responsive IFN-γ-producing cells. *Immunity*, **38**, 769–81.

Gebhardt, T., Whitney, P.G., Zaid, A., Mackay, L.K., Brooks, A.G., Heath, W.R., Carbone, F.R. & Mueller, S.N. (2011) Different patterns of peripheral migration by memory $CD4^+$ and $CD8^+$ T cells. *Nature*, **477**, 216–9.

Ginhoux, F., Tacke, F., Angeli, V., Bogunovic, M., Loubeau, M., Dai, X.-M., Stanley, E.R., Randolph, G.J. & Merad, M. (2006) Langerhans cells arise from monocytes *in vivo*. *Nature Immunology*, **7**, 265–73.

Goodwin, J.S. & Messner, R.P. (1979) Sensitivity of lymphocytes to Prostaglandin E_2 increases in subjects over age 70. *Journal of Clinical Investigation*, **64**, 434–9.

Gorelik, L., Fields, P.E. & Flavell, R.A. (2000) Cutting edge: TGF-β inhibits Th type 2 development through inhibition of GATA-3 expression. *Journal of Immunology*, **165**, 4773–7.

Gorelik, L., Constant, S. & Flavell, R.A. (2002) Mechanism of transforming growth factor β-induced inhibition of T helper type 1 differentiation. *Journal of Experimental Medicine*, **195**, 1499–1505.

Goubier, A., Dubois, B., Gheit, H., Joubert, G., Villard-Truc, F., Asselin-Paturel, C., Trinchieri, G. & Kaiserlian, D. (2008) Plasmacytoid dendritic cells mediate oral tolerance. *Immunity*, **29**, 464–75.

Harty, J.T. & Badovinac, V.P. (2008) Shaping and reshaping $CD8^+$ T cell memory. *Nature Reviews Immunology*, **8**, 107–19.

Holmgren, A.M., McConkey, C.A. & Shin, S. (2017) Outrunning the Red Queen: Bystander activation as a means of outpacing innate immune subversion by intracellular pathogens. *Cellular & Molecular Immunology*, **14**, 14–21.

Horst, A.K., Neumann, K., Diehl, L. & Tiegs, G. (2016) Modulation of liver tolerance by conventional and nonconventional antigen-presenting cells and regulatory immune cells. *Cellular & Molecular Immunology*, **13**, 277–92.

Ivanov, I.I., Zhou, L. & Littman, D.R. (2007) Transcriptional regulation of Th17 cell differentiation. *Seminars in Immunology*, **19**, 409–17.

Iwata, M., Hirakiyama, A., Eshima, Y., Kagechika, H., Kato, C. & Song, S.-Y. (2004) Retinoic acid imprints gut-homing specificity on T cells. *Immunity*, **21**, 527–38.

Keller, A.M., Schildknecht, A., Xiao, Y., van den Broek, M. & Borst, J. (2008) Expression of costimulatory ligand CD70 on steady-state dendritic cells breaks $CD8^+$ T cell tolerance and permits effective immunity. *Immunity*, **29**, 934–46.

Klose, C.S.N., Flach, M., Möhle, L., Rogell, L., Hoyler, T., Ebert, K., Fabiunke, C., Pfeifer, D., Sexl, V., Fonseca-Pereira, D., Dominguez, R.G., Veiga-Fernandes, H., Arnold, S.J., Busslinger, M., Dunay, I.R., Tanriver, Y. & Diefenbach, A. (2014) Differentiation of type 1 ILCs from a common progenitor to all helper-like innate lymphoid cell lineages. *Cell*, **157**, 340–56.

Krenger, W., Cooke, K.R., Crawford, J.M., Sonis, S.T., Simmons, R., Pan, L., Delmonte, J. Jr., Karandikar, M. & Ferrara, J.L.M. (1996) Transplantation of polarized type 2 donor T cells reduces mortality caused by experimental graft-versus-host disease. *Transplantation*, **62**, 1278–85.

Lee, K.-M., Chuang, E., Griffin, M., Khattri, R., Hong, D.K., Zhang, W., Straus, D., Samelson, L.E., Thompson, C.B. & Bluestone, J.A. (1998) Molecular basis of T cell inactivation by CTLA-4. *Science*, **282**, 2263–6.

Lefrancois, L. & Marzo, A. (2006) The descent of memory T-cell subsets. *Nature Reviews Immunology*, **6**, 618–23.

Liang, B., Workman, C., Lee, J., Chew, C., Dale, B.M., Colonna, L., Flores, M., Li, N., Schweighoffer, E., Greenberg, S., Tybulewicz, V., Vignali, D. & Clynes, R. (2008) Regulatory T cells inhibit dendritic cells by lymphocyte activation gene-3 engagement of MHC class II. *Journal of Immunology*, **180**, 5916–26.

Lillard, J.W. Jr., Boyaka, P.N., Chertov, O., Oppenheim, J.J. & McGhee, J.R. (1999) Mechanisms for induction of acquired host immunity by neutrophil peptide defensins. *Proceedings of the National Academy of Sciences of the U.S.A.*, **96**, 651–6.

Lim, H.W., Hillsamer, P., Banham, A.H. & Kim, C.H. (2005) Cutting edge: Direct suppression of B cells by $CD4^+CD25^+$ regulatory T cells. *Journal of Immunology*, **175**, 4180–3.

Lombardi, G., Sidhu, S., Batchelor, R. & Lechler, R. (1994) Anergic T cells as suppressor cells in vitro. *Science*, **264**, 1587–9.

Lucas, M., Schachterle, W., Oberle, K., Aichele, P. & Diefenbach, A. (2007) Dendritic cells prime natural killer cells by trans-presenting interleukin 15. *Immunity*, **26**, 503–17.

Mackay, C.R., Marston, W.L. & Dudler, L. (1990) Naive and memory T cells show distinct pathways of lymphocyte recirculation. *Journal of Experimental Medicine*, **171**, 801–17.

Magri, G., Miyajima, M., Bascones, S., Mortha, A., Puga, I., Cassis, L., Barra, C.M., Comerma, L., Chudnovskiy, A., Gentile, M., Llige, D., Cols, M., Serrano, S., Aróstegui, J.I., Juan, M., Yagüe, J., Merad, M., Fagarasan, S. & Cerutti, A. (2014) Innate lymphoid cells integrate stromal and immunological signals to enhance antibody production by splenic marginal zone B cells. *Nature Immunology*, **15**, 354–64.

Maldonado-López, R., De Smedt, T., Michel, P., Godfroid, J., Pajak, B., Heirman, C., Thielemans, K., Leo, O, Urbain, J. & Moser, M. (1999) $CD8\alpha^+$ and $CD8\alpha^-$ subclasses of dendritic cells direct the development of distinct T helper cells in vivo. *Journal of Experimental Medicine*, **189**, 587–92.

Mangan, P.R., Harrington, L.E., O'Quinn, D.B., Helms, W.S., Bullard, D.C., Elson, C.O., Hatton, R.D., Wahl, S.W., Schoeb, T.R. & Weaver, C.T. (2006) Transforming growth factor-β induces development of the T_H17 lineage. *Nature*, **441**, 231–4.

Marinova, E., Han, S. & Zheng, B. (2006) Human germinal center T cells are unique T_h cells with high propensity for apoptosis induction. *International Immunology*, **18**, 1337–45.

Masopust, D., Vezys, V., Usherwood, E.J., Cauley, L.S., Olson, S., Marzo, A.L., Ward, R.L., Woodland, D.L. & Lefrancois, L. (2004) Activated primary and memory CD8 T cells migrate to nonlymphoid tissues regardless of site of activation or tissue of origin. *Journal of Immunology*, **172**, 4875–82.

Matsuuchi, L. & Gold, M.R. (2001) New views of BCR structure and organization. *Current Opinion in Immunology*, **13**, 270–7.

Mempel, T.R., Pittet, M.J., Khazaie, K., Weninger, W., Weissleder, R., von Boehmer, H. & von Andrian, U.H. (2006) Regulatory T cells reversibly suppress cytotoxic T cell function independent of effector differentiation. *Immunity*, **25**, 129–41.

Mortha, A., Chudnovskiy, A., Hashimoto, D., Bogunovic, M., Spencer, S.P., Belkaid, Y. & Merad, M. (2014) Microbiota-dependent crosstalk between macrophages and ILC3 promotes intestinal homeostasis. *Science*, **343**, 1249288. DOI:10.1126/science.1249288

Oderup, C., Cederbom, L., Makowska, A., Cilio, C.M. & Ivars, F. (2006) Cytotoxic T lymphocyte antigen-4-dependent down-modulation of costimulatory molecules on dendrtic cells in $CD4^{+}CD25^{+}$ regulatory T-cell-mediated suppression. *Immunology*, **118**, 240–9.

Onishi, Y., Fehervari, Z., Yamaguchi, T. & Sakaguchi, S. (2008) $Foxp3^{+}$ natural regulatory T cells preferentially form aggregates on dendritic cells *in vitro* and actively inhibit their maturation. *Proceedings of the National Academy of Sciences of the U.S.A.*, **105**(29), 10113–8.

Pandiyan, P., Zheng, L., Ishihara, S., Reed, J. & Lenardo, M.J. (2007) $CD4^{+}CD25^{+}Foxp3^{+}$ regulatory T cells induce cytokine deprivation-mediated apoptosis of effector $CD4^{+}$ T cells. *Nature Immunology*, **8**, 1353–62.

Parry, R.V., Chemnitz, J.M., Frauwirth, K.A., Lanfranco, A.R., Braunstein, I., Kobayashi, S.V., Linsley, P.S., Thompson, C.B. & Riley, J.L. (2005) CTLA-4 and PD-1 receptors inhibit T-cell activation by distinct mechanisms. *Molecular and Cellular Biology*, **25**, 9543–53.

Pasare, C. & Medzhitov, R. (2003) Toll pathway-dependent blockade of $CD4^{+}CD25^{+}$ T cell-mediated suppression by dendritic cells. *Science*, **299**, 1033–6.

Prlic, M. & Bevan, M.J. (2008) Exploring regulatory mechanisms of $CD8^{+}$ T cell contraction. *Proceedings of the National Academy of Sciences of the U.S.A.*, **105**(43), 16689–94.

Probst, H.C., Lagnel, J., Kollias, G. & van den Broek, M. (2003) Inducible transgenic mice reveal resting dendritic cells as potent inducers of $CD8^{+}$ T cell tolerancer. *Immunity*, **18**, 713–20.

Probst, H.C., McCoy, K., Okazaki, T., Honjo, T. & van den Broek, M. (2005) Resting dendritic cells induce peripheral $CD8^{+}$ T cell tolerance through PD-1 and CTLA-4. *Nature Immunology*, **6**, 280–6.

Pulendran, B., Smith, J.L., Caspary, G., Brasel, K., Pettit, D., Maraskovsky, E. & Maliszewski, C.R. (1999) Distinct dendritic cell subsets differentially regulate the class of immune response *in vivo*. *Proceedings of the National Academy of Sciences of the U.S.A.*, **96**, 1036–41.

Rabinovitch, A. (1998) An update on cytokines in the pathogenesis of insulin-dependent diabetes mellitus. *Diabetes/Metabolism Reviews*, **14**, 129–51.

Rakoff-Nahoum, S., Paglino, J., Eslami-Varzaneh, F., Edberg, S. & Medzhitov, R. (2004) Recognition of commensal microflora by Toll-like receptors is required for intestinal homeostasis. *Cell*, **118**, 229–41.

Randolph, G.J., Beaulieu, S., Lebecque, S., Steinman, R.M. & Muller, W.A. (1998) Differentiation of monocytes into dendritic cells in a model of transendothelial trafficking. *Science*, **282**, 480–3.

Randolph, G.J., Inaba, K., Robbiani, D.F., Steinman, R.M. & Muller, W.A. (1999) Differentiation of phagocytic monocytes into lymph node dendritic cells in vivo. *Immunity*, **11**, 753–61.

Read, S., Greenwald, R., Izcue, A., Robinson, N., Mabdelbrot, D., Francisco, L., Sharpe, A.H. & Powrie, F. (2006) Blockade of CTLA-4 on $CD4^{+}CD25^{+}$ regulatory T cells abrogates their function in vivo. *Journal of Immunology*, **177**, 4376–83.

Rincón, M., Anguita, J., Nakamura, T., Fikrig, E. & Flavell, R.A. (1997) Interleukin (IL)-6 directs the differentiation of IL-4-producing $CD4^{+}$ T cells. *Journal of Experimental Medicine*, **185**, 461–9.

Rissoan, M.-C., Soumelis, V., Kadowski, N., Grouard, G., Briere, F., de Waal Malefyt, R. & Liu, Y.-J. (1999) Reciprocal control of T helper cell and dendritic cell differentiation. *Science*, **283**, 1183–6.

Sakaguchi, S., Yamaguchi, T., Nomura, T. & Ono, M. (2008) Regulatory T cells and immune tolerance. *Cell*, **133**, 775–87.

Sánchez-Madrid, F. & del Pozo, M.A. (1999) Leukocyte polarization in cell migration and immune interactions. *EMBO Journal*, **18**, 501–11.

Shinoda, K., Tokoyoda, K., Hanazawa, A., Hayashizaki, K., Zehentmeier, S., Hosokawa, H., Iwamura, C., Koseki, H., Tumes, D.J., Radbruch, A. & Nakayama, T. (2012) Type II membrane protein CD69 regulates the formation of resting T-helper memory. *Proceedings of the National Academy of Sciences of the U.S.A.*, **109**(19), 7409–14.

Shortman, K. & Caux, C. (1997) Dendritic cell development; multiple pathways to nature's adjuvants. *Stem Cells*, **15**, 409–19.

Sigmundsdóttir, H. & Butcher, E.C. (2008) Environmental cues, dendritic cells and the programming of tissue-selective lymphocyte trafficking. *Nature Immunology*, **9**, 981–7.

Sonnenberg, G.F. & Artis, D. (2012) Innate lymphoid cell interactions with microbiota: Implications for intestinal health and disease. *Immunity*, **37**, 601–10.

Sonnenberg, G.F., Monticelli, L.A., Alenghat, T., Fung, T.C., Hutnick, N.A., Kunisawa, J., Shibata, N., Grunberg, S., Sinha, R., Zahm, A.M., Tardif, M.A., Sathaliyawala, T., Kubota, M., Farber, D.L., Collman, R.G., Shaked, A., Fouser, L.A., Weiner, D.B., Tessier, P.A., Friedman, J.R., Kiyono, H., Bushman, F.D., Chang, K.-M. & Artis, D. (2012) Innate lymphoid cells promote anatomical containment of lymphoid-resident commensal bacteria. *Science*, **336**, 1321–5.

Spits, H., Artis, D., Colonna, M., Diefenbach, A., DiSanto, J.P., Eberl, G., Koyasu, S., Locksley, R.M., McKenzie, A.N.J., Mebius, R.E., Powrie, F. & Vivier, E. (2013) Innate lymphoid cells – A proposal for uniform nomenclature. *Nature Reviews Immunology*, **13**, 145–9.

Surh, C.D. & Sprent, J. (2008) Homeostasis of naive and memory T cells. *Immunity*, **29**, 848–62.

Tani, K., Murphy, W.J., Chertov, O., Salcedo, R., Koh, C.Y., Utsunomiya, I., Funakoshi, S., Asai, O., Herrmann, S.H., Wang, J.M., Kwak, L.W. & Oppenheim, J.J. (2000) Defensins act as potent adjuvants that promote cellular and humoral immune responses in mice to a lymphoma idiotype and carrier antigens. *International Immunology*, **12**, 691–700.

Taylor, A., Verhagen, J., Blaser, K., Akdis, M. & Akdis, C.A. (2006) Mechanisms of immune suppression by interleukin-10 and transforming growth factor-β: The role of T regulatory cells. *Immunology*, **117**, 433–42.

Thilenius, A.R.B., Sabelko-Downes, K.A. & Russell, J.H. (1999) The role of the antigen-presenting cell in Fas-mediated direct and bystander killing: Potential in vivo function of Fas in experimental allergic encephalomyelitis. *Journal of Immunology*, **162**, 643–50.

Thomson, A.W. & Knolle, P.A. (2010) Antigen-presenting cell function in the tolerogenic liver environment. *Nature Reviews Immunology*, **10**, 753–66.

Vella, A.T., Dow, S., Potter, T.A., Kappler, J. & Marrack, P. (1998) Cytokine-induced survival of activated T cells *in vitro* and *in vivo*. *Proceedings of the National Academy of Sciences of the U.S.A.*, **95**, 3810–15.

Vendetti, S., Chai, J.-G., Dyson, J., Simpson, E., Lombardi, G. & Lechler, R. (2000) Anergic T cells inhibit the antigen-presenting function of dendritic cells. *The Journal of Immunology*, **165**, 1175–81.

Vezys, V., Yates, A., Casey, K.A., Lanier, G., Ahmed, R., Antia, R. & Masopust, D. (2009) Memory CD8 T-cell compartment grows in size with immunological experience. *Nature*, **457**, 196–9.

Vincent, M.S., Gumpers, J.E. & Brenner, M.B. (2003) Understanding the function of CD1-restricted T cells. *Nature Immunology*, **4**, 517–23.

von Burg, N., Chappaz, S., Baerenwaldt, A., Horvath, E., Dasgupta, S.B., Ashok, D., Pieters, J., Tacchini-Cottier, F., Rolink, A., Acha-Orbea, H. & Finke, D. (2014) Activated group 3 innate lymphoid cells promote T-cell-mediated immune responses. *Proceedings of the National Academy of Sciences of the U.S.A.*, **111**(35), 12835–40.

Walunas, T.L., Lenschow, D.J., Bakker, C.Y., Linsley, P.S., Freeman, G.J., Green, J.M., Thompson, C.B. & Bluestone, J.A. (1994) CTLA-4 can function as a negative regulator of T cell activation. *Immunity*, **1**, 405–13.

Walunas, T.L., Bakker, C.Y. & Bluestone, J.A. (1996) CTLA-4 ligation blocks CD28-dependent T cell activation. *Journal of Experimental Medicine*, **183**, 2541–50.

Wang, T.-T., Nestel, F.P., Bourdeau, V., Nagai, Y., Wang, Q., Liao, J., Tavera-Mendoza, L., Lin, R., Hanrahan, J.H., Mader, S. & White, J.H. (2004) Cutting edge: 1,25-dihydroxyvitamin D_3 is a direct inducer of antimicrobial peptide gene expression. *Journal of Immunology*, **173**, 2909–12.

Watanabe, N., Ikuta, K., Fagarasan, S., Yazumi, S., Chiba, T. & Honjo, T. (2000) Migration and differentiation of autoreactive B-1 cells induced by activated γ/δ T cells in antierythrocyte immunoglobulin transgenic mice. *Journal of Experimental Medicine*, **192**, 1577–86.

Watanabe, T., Yoshida, M., Shirai, Y., Yamori, M., Yagita, H., Itoh, T., Chiba, T., Kita, T. & Wakatsuki, Y. (2002) Administration of an antigen at a high dose generates regulatory $CD4^+$ T cells expressing CD95 ligand and secreting IL-4 in the liver. *Journal of Immunology*, **168**, 2188–99.

Watanabe, T., Katsukura, H., Shirai, Y., Yamori, M., Nishi, T., Chiba, T., Kita, T. & Wakatsuki, Y. (2003) A liver tolerates a portal antigen by generating $CD11c^+$ cells, which select *Fas* $ligand^+$ Th2 cells via apoptosis. *Hepatology*, **38**, 403–12.

Xing, Y. & Hogquist, K.A. (2012) T-cell tolerance: Central and peripheral. *Cold Spring Harbor Perspectives in Biology*, **4**, a006957.

Zipris, D., Crow, A.R. & Delovitch, T.L. (1991a) Altered thymic and peripheral T-lymphocyte repertoire preceding onset of diabetes in NOD mice. *Diabetes*, **40**, 429–35.

Zipris, D., Lazarus, A.H., Crow, A.R., Hadzija, M. & Delovitch, T.L. (1991b) Defective thymic T cell activation by concanavalin A and anti-CD3 in autoimmune nonobese diabetic mice. Evidence for thymic T cell anergy that correlates with the onset of insulitis. *Journal of Immunology*, **146**, 3763–71.

Zou, W. (2006) Regulatory T cells, tumor immunity and immunotherapy. *Nature Reviews Immunology*, **6**, 295–307.

11 Physiology of the Vascular System

11.1 VASOMOTOR CONTROL

The blood flow to a particular tissue is regulated at the minimal level that will supply the tissue's requirements. This ensures that the tissues are always adequately supplied whilst keeping the workload on the heart to a minimum. Acute control of local blood flow is achieved within seconds to minutes by changes in local vasodilation/vasoconstriction of the arterioles and small arteries effected through relaxation/contraction of the vascular smooth muscle. The state of contraction of the vascular smooth muscle is referred to as vascular tone. The arteries are normally maintained in a partial state of contraction (vasomotor tone) by the continual firing of nervous impulses from the vasomotor center.

The vascular endothelium lining the lumen can modulate vascular tone by the synthesis and release of potent vasorelaxant and vasoconstrictor substances. Release of these substances is usually balanced in favor of vasodilators, notably epithelium-derived relaxing factor, prostacyclin, and endothelium-derived hyperpolarizing factor. Epithelium-derived relaxing factor has been identified as nitric oxide (NO), which is formed within endothelial cells by the oxidative deamination of L-arginine catalyzed by endothelial nitric oxide synthase (eNOS). The synthesis/release of NO is triggered by the shear stress on the endothelial cells caused by the viscous drag that results from an increase in blood flow. A constant basal release of NO is stimulated by pulsatile blood flow. NO is also synthesized and released in response to large numbers of agents that activate endothelial surface receptors; these agents include local mediators such as acetylcholine and bradykinin, and the hormone noradrenaline. Shear stress and activation of the endothelial receptors triggers an influx of Ca^{2+}, which stimulates the activity of eNOS. The NO generated in the endothelium diffuses to the underlying smooth muscle, where it activates an intracellular NO receptor with guanylyl cyclase activity. The 3′,5′-cyclic guanosine monophosphate (cGMP) produced activates protein kinase G (PKG), whose phosphorylation of myosin light chains leads to muscle relaxation.

11.2 INTERACTIONS BETWEEN PLATELETS AND THE VASCULAR ENDOTHELIUM

The vascular endothelium plays an important role in maintaining the normal fluidity of the blood. Among other mechanisms, this is achieved by the synthesis and release of anticoagulants (e.g. thrombomodulin), fibrinolytic (e.g. tissue plasminogen activator), and platelet inhibitory (e.g. NO, prostacyclin) substances.

Prostacyclin (PGI_2) is a potent stimulator of adenylyl cyclase, the enzyme that converts adenosine triphosphate (ATP) to 3′,5′-cyclic adenosine monophosphate (cAMP). Because platelet aggregation is inhibited by cAMP, prostacyclin acts as a platelet anti-aggregating agent. This effect upon platelets is opposed by another product of arachidonic acid metabolism, thromboxane A_2, which inhibits adenylyl cyclase. In experimental atherosclerosis, the capacity of arterial cells to generate prostacyclin at the site of plaque formation is considerably impaired (Gryglewski *et al.*, 1978).

Platelet aggregation is an essential process in preventing blood loss in the event of vascular injury, but it is equally important for platelets to circulate freely at other times. Platelets are easily activated so there is a potential danger of inappropriate aggregation. This is prevented by the action of NO and prostacyclin continuously released by the endothelial cells into the lumen of the blood vessel. The NO diffuses into the platelets and increases the concentration of cGMP through activation of guanylyl cyclase. The cGMP initiates a number of protein phosphorylation steps that finally lead to suppression of Ca^{2+} release from intracellular storage sites (Bassenge, 1991). The consequent inhibition of Ca^{2+} signaling limits platelet activity. Prostacyclin acts in a parallel manner to NO, except that the second messenger is cAMP rather than cGMP. Because their actions are mediated by different second messengers, there is synergism between NO and prostacyclin in inhibiting platelet aggregation. Thus, even when the concentrations of each are subthreshold, they are effective in combination. Unlike prostacyclin, NO also inhibits platelet adhesion.

At sites of injury, ADP released by aggregating platelets and thrombin formed during blood clotting stimulate the release of endothelial NO and prostacyclin, which act synergistically to cause relaxation of the vascular smooth muscle. Aggregating platelets also release vasoconstrictors such as serotonin, thromboxane A_2, and prostaglandin E_2. However, the endothelial NO produced in response to aggregating platelets normally overrides the direct vasoconstrictor effects of the platelet products and so the net effect is vasodilation. This effect of platelet aggregation and blood clotting upon the vascular endothelium constitutes a protective mechanism against vasospasm, which is associated with angina. The release of endothelial inhibitors of platelet aggregation (NO and prostacyclin) in response to platelet aggregation controls adhesion and aggregation in the manner of negative feedback.

Endothelium-dependent relaxation in response to aggregating platelets is severely impaired in atherosclerosis and

DOI: 10.1201/9781003410300-13

moderately impaired in hypercholesterolemia. Thus endothelial damage or dysfunction shift the balance of vasoactive compounds toward vasoconstriction, a state conducive to vasospasm. Under clinical conditions, the spasms can be complicated by thrombosis, leading to vessel occlusion and myocardial infarction.

11.3 REGULATION OF PLASMA CHOLESTEROL LEVELS

Cholesterol is essential for the synthesis of steroid hormones, vitamin D, and bile acids, as well as being a component of cell membranes. The level of plasma cholesterol is determined by cholesterol absorption, synthesis, storage, and excretion. Excess cholesterol can contribute to disease processes such as atherosclerosis and gallstone formation, therefore cholesterol biosynthesis and catabolism must be co-ordinately regulated to maintain homeostasis. Although much cholesterol is obtained from the diet, the body can synthesize all the cholesterol it requires. Cholesterol is synthesized in all nucleated cells. About one-third of the total daily synthesis (about 800 mg) is formed in the liver, and much of it is exported in lipoproteins for uptake by extrahepatic tissues.

Rats and mice cope with high-cholesterol diets by stimulating cholesterol catabolism to bile acids, which are excreted from the body in the bile. African green monkeys, humans, and other species living on a lower cholesterol diet have evolved an alternative strategy for maintaining cholesterol homeostasis in which the absorption of cholesterol in the intestine is reduced through decreases in bile acid production. The latter species are prone to hypercholesterolemia if their diets are cholesterol-rich.

As the cellular cholesterol level rises, the need to import cholesterol through the low-density lipoprotein (LDL) receptor or to synthesize additional cholesterol decreases. An accumulation of cholesterol modulates the following three processes. (1) The cell's ability to make its own cholesterol is stopped by turning off the synthesis of 3-hydroxy-3-methylglutaryl-coenzyme A (HMG-CoA) reductase, the rate-controlling enzyme in cholesterol biosynthesis. Suppression of this enzyme leaves the cell dependent on external cholesterol derived from the receptor-mediated uptake of LDL. (2) The incoming LDL-derived cholesterol activates acyl-CoA:cholesterol acyl transferase (ACAT), which converts cholesterol into the esterified storage form. (3) The cell regulates the amount of incoming cholesterol by suppressing the synthesis of the LDL receptor. When the cellular cholesterol level begins to fall as cells use more cholesterol, synthesis of the LDL receptor and HMG-CoA reductase increases and the activity of ACAT decreases.

11.4 THE LDL AND HDL CHOLESTEROL TRANSPORT PROTEINS

LDL is popularly known as "bad cholesterol" because its core of cholesterol can be taken up by macrophages at vulnerable sites in artery walls, transforming the macrophages into foam cells. Accumulation of foam cells leads to the formation of an atheromatous plaque. High-density lipoprotein (HDL) is called "good cholesterol" because it can remove excess cholesterol from macrophages and transport it to the liver for disposal. A person with a high blood level of LDL and a low HDL level is at risk for atherosclerotic disease.

In clinical laboratories, attention has moved away from monitoring changes in total cholesterol concentration toward examining the distribution of cholesterol between LDL and HDL. Assayed cholesterol associated with the LDL fraction is designated as LDLC and cholesterol associated with the HDL fraction as HDLC. It is desirable for a patient to have a relatively low LDLC and a relatively high HDLC.

Values for plasma LDLC and HDLC concentrations used by practitioners for the risk assessment are as follows (Banerjee & Captain, 2014).

- **Total cholesterol**. Desirable level: 140–199 mg/dL; borderline risk: 200–239 mg/dL; high risk >250 mg/dL.
- **LDL cholesterol**. Desirable level: <130 mg/dL; borderline risk: 140–150 mg/dL; high risk: >160 mg/dL.
- **HDL cholesterol**. Normal level for men: 35–65 mg/dL; normal level for women: 35–80 mg/dL. Levels <25 mg/dL increase the risk of atherosclerosis.

Plasma cholesterol values are strongly influenced by genetic, environmental, and lifestyle factors, hence it is not easy to define normal values. In order to overcome this problem, 'target' values for the different lipid fractions have been tailored to accommodate the treatment requirements of different risk categories in the United States National Cholesterol Education Program (National Institute of Health, 2002). The target values of various risk categories are shown in Table 11.1.

The causal relationship between high LDL cholesterol levels and atherosclerosis is demonstrated unequivocally by the genetic disorder familial hypercholesterolemia (FH). This disorder is an example of an inborn error of metabolism and is due to impaired production of the LDL receptor. There are two forms of FH: a heterozygous form and a more severe homozygous form. FH heterozygotes inherit one mutant gene and number about one in 500 people in most ethnic groups. The cells of these individuals produce approximately half the normal number of LDL receptors. As a result, LDL is removed from the circulation at half the normal rate, the lipoprotein accumulates in blood to levels twofold above normal, and heart attacks occur typically in the fourth and fifth decades of life. If two heterozygotes of the opposite sex have a child, that child has a one in four chance of inheriting two copies of the mutant gene, one from each parent. Such FH homozygotes (about one in

TABLE 11.1
ATP III Classification of Risk Categories for LDL, Total, and HDL Cholesterol

	mmol/L	mg/dL	Category
LDL cholesterol	<2.6	<100	Optimal
	2.6–3.3	100–129	Near or above optimal
	3.4–4.1	130–159	Borderline high
	4.1–4.9	160–189	High
	≥4.9	≥190	Very high
Total cholesterol	<5.2	<200	Desirable
	5.2–6.2	200–239	Borderline high
	>6.2	≥240	High
HDL cholesterol	<1.0	<40	Low
	≥1.6	≥60	High

Source: Extracted from National Cholesterol Education Program. National Institute of Health (2002).

a million people) have plasma LDL levels six- to tenfold above normal. Heart attacks can occur at the age of two and are almost inevitable by the age of 20 (Goldstein & Brown, 1987).

11.5 OXIDATIVE MODIFICATION OF LDL

The modification of LDL to oxidized LDL (oxLDL) involves the peroxidation of constituent polyunsaturated fatty acids. During modification, there is extensive degradation of lecithin (phosphatidylcholine) to lysolecithin catalyzed by phospholipase A_2 present in the apoB-100 moiety of LDL (Parthasarathy & Barnett, 1990). The removal of lysolecithin from the surface of the particle aids the peroxidation of core lipids. There is also fragmentation of apoB-100 by direct oxidative scission. Extensive oxidation of LDL leads to aggregation of the LDL particles.

Immunological evidence exists for a role of oxLDL in the development of atherosclerosis. Human and rabbit atherosclerotic lesions contain an immunoglobulin (IgG) that recognizes epitopes present in oxLDL but not in native LDL (Ylä-Herttuala *et al.*, 1994). The titers of antibodies that recognize epitopes of oxLDL increase with the progression of atherosclerosis in mice (Palinski *et al.*, 1995) and humans (Salonen *et al.*, 1992).

It seems likely that oxidative modification of LDL is confined to microdomains within the arterial intima where LDL would be isolated from the many antioxidants present in extracellular fluid. It has been shown *in vitro* that LDL can be oxidatively modified to cytotoxic oxLDL by all the major cell types in the arterial wall, namely endothelial cells, smooth muscle cells, and macrophages. These cells generate reactive substances which can be transferred to the LDL.

The aldehydes arising from lipid hydroperoxide decomposition are particularly important in atherogenesis. Malondialdehyde, 4-hydroxynonenal, and probably other aldehydes bind covalently to the ε amino group of lysine residues on the apoB protein moiety of LDL particles, causing a progressive decrease in the positive charge and a conformational change. This modification results in a loss of recognition by the LDL receptor and the creation of new epitopes of apoB that are recognized by the macrophage scavenger receptor. Evidence for the involvement of aldehydes is the observation by Picard *et al.* (1992) that aminoguanidine, a compound with high binding affinity for aldehydes, prevents apoB lysine modification, thereby inhibiting the oxidatively induced uptake of LDL by macrophages.

Several biochemical systems have been shown to promote LDL oxidation *in vitro*; these systems involve thiols, lipoxygenase, myeloperoxidase, superoxide, and peroxynitrite radical (Berliner & Heinecke, 1996). Physiologically relevant mechanisms underlying LDL oxidation *in vivo* are largely a matter of speculation.

11.6 ATHEROGENIC PROPERTIES OF OXIDIZED LDL

11.6.1 Mechanisms

OxLDL may be atherogenic by numerous mechanisms, some of which are listed below.

- The lysolecithin present in oxLDL increases vascular superoxide production, resulting in a self-perpetuating cycle of LDL oxidation.
- OxLDL, by virtue of its lysolecithin content, exhibits chemotactic activity for circulating monocytes and T lymphocytes.
- OxLDL induces monocyte chemotactic protein-1 (MCP-1) synthesis in endothelial cells and smooth muscle cells.
- Both oxLDL and lysolecithin can induce the expression of intracellular adhesion molecule-1 (ICAM-1) and vascular cell adhesion molecule-1 (VCAM-1).
- Monocytes/macrophages exposed to oxLDL release cytokines which stimulate endothelial cells to synthesize cell adhesion molecules.
- OxLDL inhibits the motility of resident macrophages in the subendothelial space, thereby favoring macrophage accumulation in developing lesions.
- OxLDL induces endothelial cells to synthesize macrophage colony-stimulating factor (M-CSF), which can induce differentiation of monocytes into macrophages, including an increased expression of scavenger receptor.
- OxLDL stimulates proliferation of macrophages and aortic smooth muscle cells in culture.
- Lysolecithin present in oxLDL induces migration of smooth muscle cells in human coronary arteries.

- OxLDL enhances lipopolysaccharide-induced tissue factor expression in human adherent monocytes. Normally, tissue factor is not expressed within the vasculature; however, its expression can be induced in monocytes and endothelial cells by a variety of agonists, including lipopolysaccharide. Activation of the blood coagulation cascade by aberrant expression of tissue factor on the surface of monocytes or endothelial cells could lead to thrombus formation in ruptured plaques.
- OxLDL stimulates platelet aggregation.
- Endothelial cells chronically exposed to oxLDL have a severely impaired capacity to release prostacyclin, a potent cytoprotective agent, which inhibits cell proliferation, platelet adhesion and aggregation, and monocyte adhesion.
- Lysolecithin in oxLDL decreases the responsiveness of isolated rabbit aorta to endothelium-dependent relaxants.
- OxLDL decreases the expression of eNOS, the enzyme responsible for producing the epithelium-derived relaxing factor, NO.
- OxLDL enhances agonist-induced vasoconstrictions by a direct effect on the vascular smooth muscle.

11.6.2 Cytotoxicity

Oxidative modifications of LDL by lipid peroxidation renders the oxLDL toxic to cells *in vitro*. Cytotoxicity of oxLDL to endothelial cells, smooth muscle cells, fibroblasts, monocytes, neutrophils, and macrophages in culture has been demonstrated. The capacity of oxLDL to injure or kill cells may contribute to endothelial damage and may account for the necrotic debris found in the core regions of atherosclerotic plaques. Electron microscopic examination of mouse peritoneal macrophages exposed to oxLDL revealed the following ultrastructural changes: chromatin condensation, reduction in cell volume, cell break-up into membrane-bound bodies, and blebbing of the plasma membrane. These changes are more characteristic of apoptosis than necrosis. In contrast, cells exposed to native LDL exhibited no such changes.

The presence during lipoprotein preparation of general free radical scavengers (vitamin E, butylated hydroxytoluene) or the divalent cation chelator ethylenediaminetetraacetic acid (EDTA) prevented the formation of cytotoxic oxLDL. However, the toxic action of oxLDL could not be prevented by the addition of any of these agents during incubation of the oxLDL with cells (Morel *et al.*, 1983). These observations indicate that an oxidized lipid is responsible for cytotoxicity rather than free radicals generated in culture by the action of oxLDL. Hughes *et al.* (1994) extracted the toxic components from lipid extracts of oxLDL and identified them as the oxysterols 7-ketocholesterol and 7-hydroxycholesterol. When the two toxins were added to native LDL at concentrations equivalent to those present in oxLDL, the resultant LDL was not rendered cytotoxic. This suggests that the oxysterols are released more rapidly from oxLDL than from native LDL because of the severe disruption of the lipoprotein's lipid environment during oxidative modification. The mechanism(s) by which oxysterols exert their cytotoxicity has not been established.

11.7 PROTECTIVE EFFECTS OF HDL

Epidemiological studies have generally shown a strong inverse correlation between plasma HDL concentration and incidence of coronary heart disease. The role of HDL in reverse-cholesterol transport is well established and many other protective effects against development of atherosclerosis have been reported. Some of these effects are listed below.

- HDL inhibits the oxidative modification of LDL.
- HDL inhibits the transmigration of monocytes induced by oxLDL.
- HDL blocks the induction by oxLDL of monocyte adhesion to the endothelium.
- HDL prevents oxLDL-induced cytotoxicity in vascular smooth muscle cells and endothelial cells.
- HDL removes cholesterol from cells in the arterial intima, thereby preventing the formation of foam cells.
- HDL ameliorates the abnormal vasoconstriction that is a feature of early atherosclerosis.
- HDL stimulates prostacyclin synthesis and prolongs its half-life by binding to it.
- HDL is protected from oxidative damage by vitamin E present in its core.

11.8 ATHEROSCLEROSIS

11.8.1 Introduction

Atherosclerosis is a disease of large- and medium-sized arteries with a predisposition for critical arterial beds such as the coronaries, which supply heart muscle with nutritive, oxygenated blood. It is the principal cause of cardiovascular and cerebrovascular disease, leading to heart attacks and strokes, respectively. Atherosclerosis is characterized by the progressive formation of an atheromatous plaque upon the innermost layer of the arterial wall. This build-up of fatty material reduces the bore of the artery, whilst also causing hardening and loss of elasticity of the arterial wall. Although there may be severe narrowing of arteries, atherosclerosis alone rarely results in complete blockage. However, rheological forces may cause the plaque to rupture or fissure, resulting in hemorrhage into the plaque, thrombosis, and occlusion. This event is a major cause of myocardial infarction. There is irrefutable evidence linking atherosclerosis with chronically elevated levels of plasma cholesterol (hypercholesterolemia). At one time, atherosclerosis was thought to be a degenerative disease that was an inevitable consequence of aging. The disease is now regarded as a chronic inflammatory response that has become excessive (Ross, 1993).

11.8.2 Anatomy of the Arterial Wall

The wall of a typical artery is composed of three coats. The innermost coat, the tunica intima, consists of an inner endothelium, a subendothelial layer of delicate fibroelastic connective tissue, and an external band of elastic fibers, which may be absent in many vessels. The middle coat, the tunica media, consists chiefly of circularly arranged smooth muscle cells. The outer coat, the tunica adventitia, is composed principally of connective tissue, most of the elements of which run parallel to the long axis of the vessel. The structure and relative thickness of each coat vary according to the type and size of the vessel.

11.8.3 Development of the Atherosclerotic Plaque

The lesions of atherosclerosis represent a continuum from the early fatty streak to the intermediate fibrofatty lesion to the mature fibrous plaque. Early lesions are commonly found at intimal sites where changes in blood flow occur at branches and curves in the arterial system. Changes in the mature lesion involve all three coats of the arterial wall. Not all fatty streaks progress to fibrous plaques; some remain stable or regress and disappear.

At the destined site of plaque formation, cholesterol-rich LDLs enter the subendothelial space of the tunica intima where their core lipids are minimally oxidized. Local vascular cells produce MCP-1 (monocyte chemotactic protein 1), which attracts circulating monocytes, and also M-CSF (macrophage colony-stimulating factor), which stimulates monocyte differentiation to macrophages. The macrophages internalize the LDLs and become so bloated with cholesteryl ester in lipid droplets that they are redefined as foam cells. The mode of oxLDL uptake by macrophages changes with increasing degrees of LDL oxidation from one involving endocytosis by scavenger receptors to an alternative one in which aggregates of oxLDL particles are phagocytosed. The accumulation of foam cells forms a fatty streak within the intima. The overlying endothelium is frequently thinned and distorted by the subjacent foam cells. Local smooth muscle cells are mitogen-stimulated to proliferate and the new cells migrate from the tunica media into the intima. Some smooth muscle cells internalize LDL cholesterol to also become foam cells. Endothelial cells and smooth muscle cells synthesize and secrete collagen and elastin, and the deposition of connective tissue advances the fatty streak to the intermediate fibrofatty lesion. The fibrofatty lesion is characterized by necrosis of the foam cells and the release of foam cell lipids to the interstitium. Monocyte recruitment continues and the endothelium begins to fragment. The fibrofatty lesion progresses to an advanced plaque, which is covered by a dense cap of fibrous connective tissue containing multiple layers of smooth muscle cells that occupy slit-like spaces. Foam cells accumulate at the shoulders of the plaque. Beneath the fibrous cap is a region rich in macrophages, T lymphocytes, and smooth muscle cells. The plaque has a core of extracellular lipid, cholesterol crystals, and the remnants of dead foam cells. Calcification of the atherosclerotic plaque begins at the fatty streak stage as small scattered deposits of hydroxyapatite crystals. Over time, the crystals coalesce to produce larger solid crystals and eventually a histology characteristic of bone. Microvascular channels ramify throughout the thickened lesion, providing nutrients and oxygen to the cells of the plaque.

FURTHER READING

Glass, C.K. & Witztum, J.L. (2001) Atherosclerosis: The road ahead. *Cell*, **104**, 503–16.

Negre-Salvayre, A., Dousset, N., Ferretti, G., Bacchetti, T., Curatola, G. & Salvayre, R. (2006) Antioxidant and cytoprotective properties of high-density lipoproteins in vascular cells. *Free Radical Biology and Medicine*, **41**, 1031–40.

Reaven, P.D. & Witztum, J.L. (1996) Oxidized low density lipoproteins in atherogenesis: Role of dietary modification. *Annual Review of Nutrition*, **16**, 51–71.

REFERENCES

Banerjee, P. & Captain, C. (2014) *Common Laboratory Tests Used by TCM Practitioners*. Singing Dragon, London and Philadelphia.

Bassenge, E. (1991) Antiplatelet effects of endothelium-derived relaxing factor and nitric oxide donors. *European Heart Journal*, **12**(Supplement E), 12–15.

Berliner, J.A. & Heinecke, J.W. (1996) The role of oxidized lipoproteins in atherogenesis. *Free Radical Biology & Medicine*, **20**, 707–27.

Goldstein, J.L. & Brown, M.S. (1987) Regulation of low-density lipoprotein receptors: Implications for pathogenesis and therapy of hypercholesterolemia and atherosclerosis. *Circulation*, **76**, 504–7.

Gryglewski, R.J., Dembińska-Kięć, A., Żmuda, A. & Gryglewska, T. (1978) Prostacyclin and thromboxane A_2 biosynthesis capacities of heart, arteries and platelets at various stages of experimental atherosclerosis in rabbits. *Atherosclerosis*, **31**, 385–94.

Hughes, H., Mathews, B., Lenz, M.L. & Guyton, J.R. (1994) Cytotoxicity of oxidized LDL to porcine aortic smooth muscle cells is associated with the oxysterols 7-ketocholesterol and 7-hydroxycholesterol. *Arteriosclerosis and Thrombosis*, **14**, 1177–85.

Morel, D.W., Hessler, J.R. & Chisolm, G.M. (1983) Low density lipoprotein cytotoxicity induced by free radical peroxidation of lipid. *Journal of Lipid Research*, **24**, 1070–6.

National Institute of Health (2002) Third report of the National Cholesterol Education Program (NCEP) expert panel on detection, evaluation, and treatment of high blood cholesterol in adults (Adult Treatment Panel III). Final Report. *Circulation*, **106**(25), 3143–421.

Palinski, W., Tangirala, R.K., Miller, E., Young, S.G. & Witztum, J.L. (1995) Increased autoantibody titers against epitopes of oxidized LDL in LDL receptor-deficient mice with increased atherosclerosis. *Arteriosclerosis, Thrombosis, and Vascular Biology*, **15**, 1569–76.

Parthasarathy, S. & Barnett, J. (1990) Phospholipase A_2 activity of low density lipoprotein: Evidence for an intrinsic phospholipase A_2 activity of apoprotein B-100. *Proceedings of the National Academy of Sciences of the U.S.A.*, **87**, 9741–5.

Picard, S., Parthasarathy, S., Fruebis, J. & Witztum, J.L. (1992) Aminoguanidine inhibits oxidative modification of low density lipoprotein and the subsequent increase in uptake by macrophage scavenger receptors. *Proceedings of the National Academy of Sciences of the U.S.A.*, **89**, 6876–80.

Ross, R. (1993) Rous-Whipple award lecture. Atherosclerosis: A defense mechanism gone awry. *American Journal of Pathology*, **143**, 987–1002.

Salonen, J.T., Ylä-Herttuala, S., Yamamoto, R., Butler, S., Korpela, H., Salonen, R., Nyyssönen, K., Palinski, W. & Witztum, J.L. (1992) Autoantibody against oxidized LDL and progression of carotid atherosclerosis. *Lancet*, **339**, 883–7.

Ylä-Herttuala, S., Palinski, W., Butler, S.W., Picard, S., Steinberg, D. & Witztum, J.L. (1994) Rabbit and human atherosclerotic lesions contain IgG that recognizes epitopes of oxidized LDL. *Arteriosclerosis and Thrombosis*, **14**, 32–40.

12 Structure of Bone and Its Growth and Development

12.1 COMPOSITION OF BONE MATRIX

Bone is a rigid form of connective tissue that consists of cells and an intercellular matrix. The matrix is a well-ordered structure with an organic portion comprising about 35% of the bone weight and a mineral portion comprising 65%. The organic component is composed mainly of type I collagen and the mineral component is mainly calcium phosphate in the form of hydroxyapatite [$Ca_{10}(PO_4)_6(OH)_2$] crystals. The collagen fibers and hydroxyapatite crystals are embedded in an amorphous ground substance consisting largely of mucopolysaccharides. Individual collagen molecules become interconnected by the formation of pyridinoline cross-links, which are unique to bone. The hydroxyapatite crystals are regularly distributed along the length of the collagen fibers and take the form of needles, thin plates, or leaves.

12.2 BONE CELLS

There are three types of functional cells in bone tissue: osteoblasts, osteocytes, and osteoclasts (Figure 12.1).

12.2.1 Osteoblasts

The primary function of osteoblasts is to form a mineralized bone matrix. Osteoblasts arise from mesenchymal stem cells present in the stroma (framework) of the bone marrow. The stem cells proliferate and then commit themselves to a particular pathway of lineage progression, differentiation, and maturation. The commitment event and lineage progression involve the action of locally generated cytokines and growth factors, many of which are controlled by circulating hormones. Stem cell-derived cells that differentiate into osteoblasts are referred to as osteoprogenitor cells or simply as osteoblast precursors.

Osteoblasts synthesize and secrete type I collagen and the following noncollagen proteins:

- Fibronectin – a multi-adhesive matrix protein.
- Bone morphogenetic protein-2 – promotes osteoblast differentiation and mineralization.
- Bone sialoprotein – stimulates hydroxyapatite formation.
- Osteocalcin – required for the alignment of apatite crystallites parallel to the collagen fibers.
- Osteonectin – plays a role in the survival of bone cells.
- Osteopontin – modulates osteoclast differentiation and bone resorption.
- Osteoprotegerin – inhibits osteoclast activation and bone resorption.

Hydroxyapatite crystals fail to precipitate in soft tissues because of the presence of inhibitors: one powerful inhibitor is inorganic pyrophosphate. To overcome this inhibition in bone, osteoblasts secrete alkaline phosphatase into the osteoid to split the pyrophosphate.

12.2.2 Osteocytes

Calcification induces morphological and metabolic changes in osteoblasts, converting them into osteocytes, which become entrapped in the calcified matrix. Osteocytes are mature bone cells and have no mitotic potential or secretory activity. In developing bone, their cytoplasmic processes extend for considerable distances in narrow channels called canaliculi and make contact with processes of neighboring osteocytes via gap junctions. The network thus formed allows molecules to be passed from cell to cell. In mature bone the processes are almost completely withdrawn, but the canaliculi remain to provide an avenue for the exchange of nutrients and waste products between the blood and the imprisoned osteocytes.

12.2.3 Osteoclasts

Osteoclasts are multinucleated giant cells that are solely responsible for the resorption (destruction) of bone matrix. Osteoclastogenesis starts from hematopoietic stem cells, which differentiate into macrophage colony-forming cells. These cells proliferate and make cell-to-cell contact with osteoblasts/stromal cells. This interaction induces the cells to differentiate into mononuclear prefusion osteoclasts and then into multinucleated osteoclasts, which finally fuse together to form mature osteoclasts. Macrophage colony-stimulating factor (M-CSF) produced by osteoblasts/stromal cells is required for both proliferation and differentiation of osteoclast precursors (Suda *et al.*, 1999).

Resorption is important in the development, growth, maintenance, and repair of bone. Osteoclasts are found in or near cavities called Howship's lacunae on bone surfaces. They express many surface receptors for calcitonin, which is a potent hormonal inhibitor of the cell's resorptive activity. The surface of the osteoclast facing the bone matrix is

DOI: 10.1201/9781003410300-14

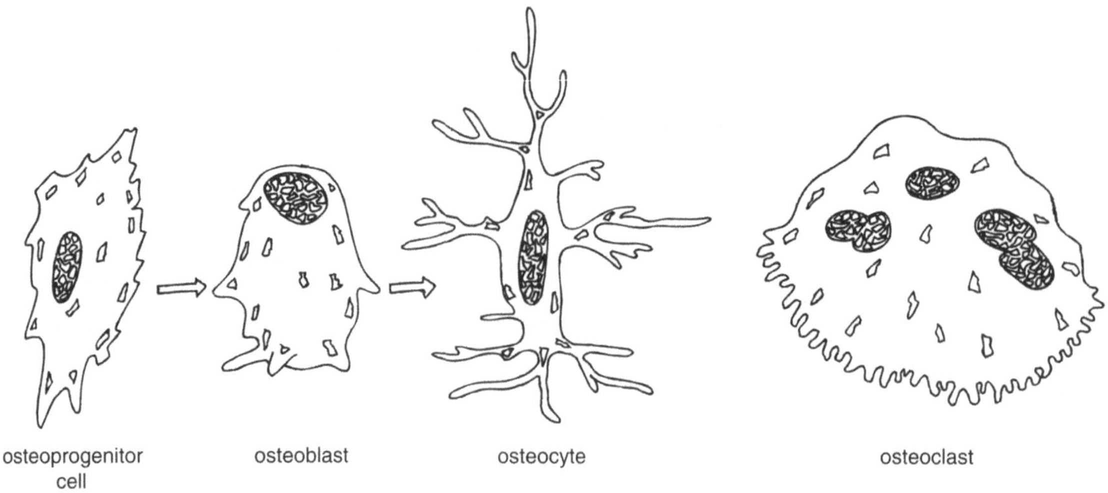

FIGURE 12.1 Types of cells in bone tissue.

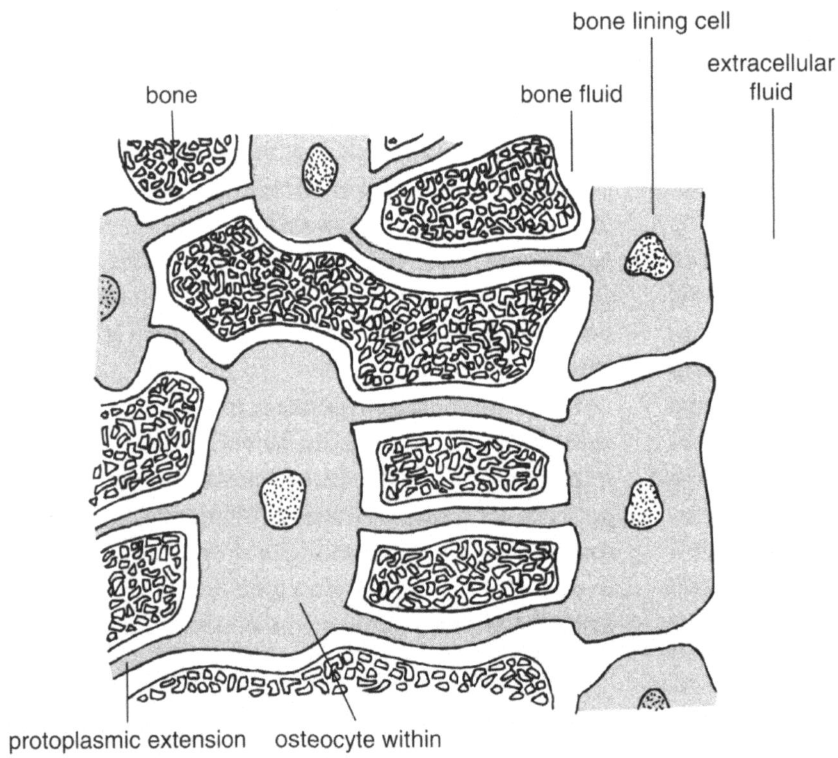

FIGURE 12.2 The bone fluid compartment.

termed the 'ruffled border,' owing to extensive infoldings. Adjacent to the ruffled border in the cell's interior is a so-called clear zone that is rich in actin filaments but devoid of organelles. The clear zone is a site of adhesion of the osteoclast to the bone matrix and creates a microenvironment of low pH and lysosomal enzymes for bone resorption. Osteoclasts secrete a bone-dissolving mixture of hydrochloric acid and proteolytic enzymes such as collagenase. The ruffled border is essential for the activity of the osteoclast. In osteopetrosis, a genetic disease characterized by dense heavy bone, the osteocytes lack ruffled borders and bone resorption is defective.

12.3 THE BONE FLUID COMPARTMENT

Bone contains a fluid compartment that is separated from the extracellular fluid by the lining cells on the bone surface. As shown in Figure 12.2, the bone fluid compartment comprises the fluid-filled space between the lining cells and bone matrix, around the protoplasmic extensions in the canaliculi, and around osteocytes in their lacunae. The bone fluid is therefore in direct contact with bone crystals or amorphous calcium phosphate deposits.

12.4 ARCHITECTURE OF BONE

There are two types of bone tissue: woven or primary bone and lamellar or mature bone. All of the bone in the newborn baby is woven and this bone type occurs also in localized regions of the growing skeleton and in repairing fractures. Woven bone is characterized by a random orientation of its collagen fibers and a low mineral content relative to that of lamellar bone. Except in a very few locations, such as tooth sockets, woven bone is replaced in infants over one year old by lamellar bone, so called because the collagen fibers are orientated in parallel arrays. The bone tissue described in the following is lamellar bone.

There are two types of lamellar bone: cortical bone and cancellous or trabecular bone. Cortical bone is a dense solid mass which forms the external layer of all bones of the body and predominates in the shafts of long bones. Cancellous bone takes the form of an irregular latticework of rods, plates, and arches individually known as trabeculae. Cancellous bone makes up most of the bone tissue of short, flat, and irregularly shaped bones such as the vertebrae, pelvis, and shoulder blade; it is also found in the bulbous ends of long bones. In early life, the spaces between the trabeculae of all cancellous bones are filled with hematopoietic (blood-producing) red bone marrow, but later the red marrow is confined to the cancellous bones of the trunk. The spaces in other cancellous bones are filled with yellow bone marrow, which consists mostly of fat.

In a typical long bone (Figure 12.3) the shaft (diaphysis) is composed almost entirely of cortical bone with a small component of cancellous bone surrounding the medullary cavity. In the adult, the medullary cavity contains yellow bone marrow. The bulbous ends (epiphyses) and the metaphyses consist of cancellous bone covered by a shell of cortical bone. In the growing animal, the epiphysis is separated from the metaphysis by a thick plate of hyaline cartilage known as the epiphyseal plate; it is in this region that bone elongation occurs. In the adult, the cartilaginous plate has been replaced by cancellous bone.

The outer surfaces of most bones, but not the articular surfaces of joints, are covered by a bilayered sheath called the periosteum. The outer layer of the periosteum consists of dense fibrous connective tissue and contains a network of blood vessels. The inner layer is composed of more loosely arranged connective tissue and numerous osteoprogenitor cells. Some component collagenous fibers of the periosteum penetrate the bone matrix as Sharpey's fibers. Inside the bone, the medullary cavity and the cavities of cancellous bone are lined by the endosteum. This membrane is composed of a single layer of osteoprogenitor cells and a very small amount of connective tissue. The principal functions of the periosteum and endosteum are nutrition of the osseous tissue and the provision of new osteoblasts for bone repair.

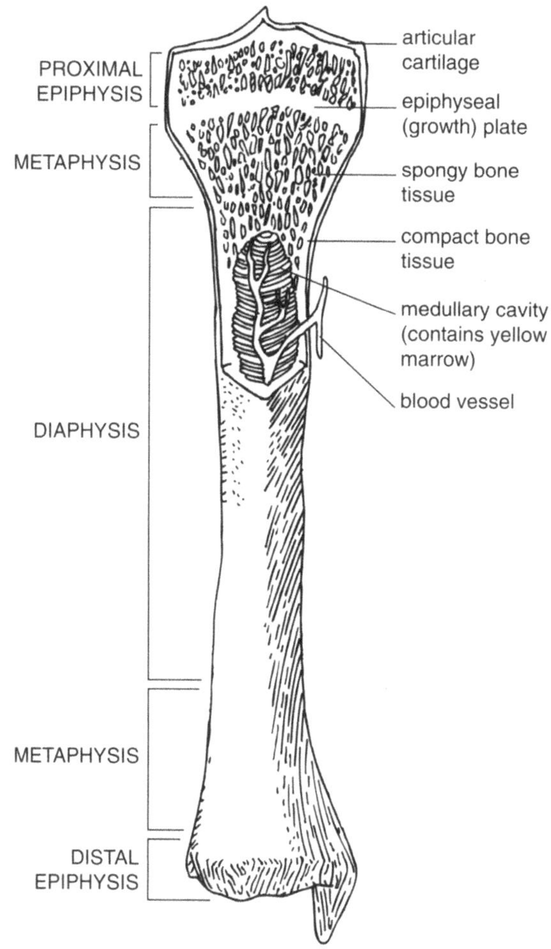

FIGURE 12.3 Gross anatomy of a typical long bone with interior partially exposed.

12.4.1 Cortical Bone

In cortical bone, the calcified bone matrix is formed of layers called lamellae arranged in a manner determined by the distribution of blood vessels. The lamellae are organized into osteons (Haversian systems) in which four to twenty lamellae are concentrically arranged around a central Haversian canal containing the blood vessels and nerves. A segment of a Haversian system is shown diagrammatically in Figure 12.4. Between, and occasionally within, the lamellae there are small cavities called lacunae which contain osteocytes. Radiating from each lacuna are the canaliculi. These channels penetrate adjacent lamellae to unite with canaliculi of neighboring lacunae and eventually open to extracellular fluid at bone surfaces.

Each osteon is a long often bifurcated cylinder which in long bones is orientated mainly in the long axis of the diaphysis. The Haversian canals communicate with the periosteum bone marrow cavity and with each other through transverse or oblique Volkmann's canals, which

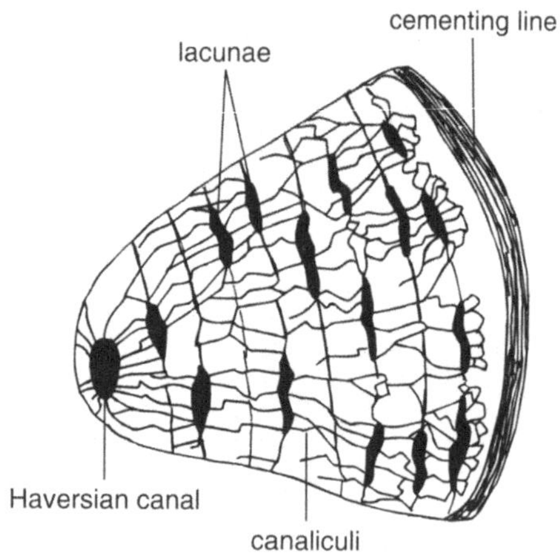

FIGURE 12.4 Diagram of a segment of an osteon (Haversian system) in cross section.

perforate the lamellae. The intervals between osteons are occupied by interstitial lamellae, which are the remnants of osteons partly destroyed during the internal reconstruction of the bone.

12.4.2 Cancellous Bone

Each trabecula consists of a mosaic of angular segments formed by parallel sheets of lamellae. These segments of lamellar bone are called trabecular packets and are functionally analogous to the osteons of cortical bone. The latticework of trabeculae is orientated along lines of stress.

Within the trabeculae are osteocytes that lie in lacunae. The entrapped osteocytes are nourished by diffusion through canaliculi extending to the surface. Most trabeculae are less than 0.2 mm in thickness. A few trabeculae are thicker than 0.2 mm and their central portion generally contains concentric lamellae surrounding a blood vessel.

12.5 BONE FORMATION: OSSIFICATION

Ossification begins around the sixth or seventh week of embryonic life and continues throughout adulthood. There are two types of ossification, which differ by the initial material on which the bone is formed. Intramembranous ossification refers to the formation of bone directly within mesenchymal connective tissue. Endochondral ossification refers to the replacement of cartilage by bone in a miniature cartilage 'model.' In both types of ossification, woven bone is formed initially.

12.5.1 Intramembranous Ossification

The frontal and parietal bones of the skull are examples of bones that develop by intramembranous ossification. This process also contributes to the growth of short bones and the thickening of long bones.

During the eighth week of life in the human fetus, an area of mesenchyme becomes richly vascularized. This prompts mesenchymal cells to differentiate into a variety of cell types, including osteoblasts. Osteoblasts secrete ground substance and collagen monomers, which polymerize rapidly to form collagen fibers. The resulting tissue is called the osteoid in which the entrapped osteoblasts become osteocytes. Within a few days, minute crystals of calcium salts are deposited on the surfaces of the collagen fibers, eventually forming hydroxyapatite crystals. Lacunae and canaliculi are formed around the osteocytes and their cytoplasmic processes. As the bone matrix forms, it develops into trabeculae that fuse with one another to create the open latticework appearance of cancellous bone. The spaces between trabeculae fill with vascularized connective tissue, which differentiates into red bone marrow. The mesenchyme located near the endosteal surface transforms into the endosteum and the remaining envelope of connective tissue at the periosteal surface develops into the periosteum. Eventually, most surface layers of the cancellous bone are replaced by cortical bone, but cancellous bone remains in the center of the bone.

12.5.2 Endochondral Ossification

Most bones of the body are formed by endochondral ossification. This is a slow process which is not achieved until the bone has reached its full size and growth has ceased. The process is best observed in a long bone.

At the site where the bone is going to form, mesenchymal cells cluster together in the shape of the future bone. The mesenchymal cells differentiate into chondroblasts that produce the cartilage model. In addition, a membrane called the perichondrium develops around the cartilage model. The cartilage model grows in length by continual cell division of chondrocytes accompanied by further secretion of cartilage matrix by the daughter cells. This pattern of growth from within is called interstitial growth. The increase in thickness of the cartilage model is due mainly to appositional growth. This refers to the addition of more matrix to its periphery by new chondroblasts that develop from the perichondrium.

As the cartilage model continues to grow, chondrocytes in its midregion hypertrophy, probably because they accumulate glycogen for adenosine triphosphate production. Some hypertrophied cells burst, releasing their contents, and changing the pH of the matrix. The resultant chemical changes trigger calcification. Once the cartilage becomes calcified, other chondrocytes die because nutrients no longer diffuse quickly enough through the matrix. The lacunae of the cells that have died are now empty, and the thin partitions between them break down forming small cavities.

In the meantime, a nutrient artery penetrates the perichondrium and then the developing bone through a hole in the midregion of the model. This stimulates osteoprogenitor cells in the perichondrium to differentiate into osteoblasts. The cells lay down a thin sheet of cortical bone under the perichondrium called the periosteal bone collar.

Once the perichondrium starts to form bone, it is known as the periosteum. Near the midregion of the model, the periosteum sends out osteogenic buds into the disintegrating calcified cartilage through holes made by osteoclasts in the bone collar. The buds, containing blood capillaries and osteoprogenitor cells, enter the spaces left by the dead and degenerating chondrocytes. The invading osteoprogenitor cells proliferate and develop into osteoblasts, which begin to deposit bone matrix over the remnants of calcified cartilage, forming cancellous bone. This region of bone deposition is called the primary ossification center, in which ossification proceeds inward from the external surface of the model. As the ossification center expands toward the ends of the model, osteoclasts break down the newly formed cancellous bone. This bone-resorbing activity leaves a hollow cavity, the medullary cavity, in the core of the diaphysis along its length. The cavity then fills with red bone marrow. The longitudinal expansion of the primary ossification center is accompanied by a widening of the periosteal bone collar in the same direction. The bone collar also thickens, providing support to the central zone of resorbing cartilage prior to its replacement by bone.

At about the time of birth, epiphyseal arteries enter the epiphyses and secondary ossification centers develop. Osteoprogenitor cells invade the area via vascular osteogenic buds originating from the diaphysis. Cartilage removal and bone matrix deposition follow. Bone formation is similar to that in the primary ossification center. One difference, however, is that no medullary cavities are formed in the epiphyses. Secondary ossification proceeds outward in all directions from the center of the epiphysis until there is almost complete replacement of cartilage by cancellous bone. Two regions at each epiphysis, the articular surface and the epiphyseal plate, do not undergo secondary ossification. Articular cartilage persists throughout adult life and, in the absence of a perichondrium, no equivalent of a bone collar is formed here. The epiphyseal plate is converted to cancellous bone much later in life, when growth of the long bone is complete.

12.6 BONE GROWTH

As a child grows, long bones lengthen by addition of bone material at the cartilaginous epiphyseal plate. In this process, chondrocytes are produced mitotically on the epiphyseal side of the plate. The chondrocytes then die and the cartilage is replaced by bone on the diaphyseal side. In this way, the thickness of the plate remains almost constant, but the bone on the diaphyseal side increases in length. Eventually, chondrocytes stop dividing and bone replaces the cartilage in the epiphyseal plate. The newly formed bony structure is called the epiphyseal line and, with its appearance, bone stops growing in length but continues to thicken. In the thickening of long bones, bone lining the medullary cavity is first destroyed by osteoclasts in the endosteum so that the cavity increases in diameter. At the same time, osteoblasts from the periosteum add new bone tissue to the outer surface. During the growth spurt, there is a rapid increase in bone mineral density, followed by a slower increase until peak bone mass is achieved sometime in young adulthood. Exercise and sports activity in children and young adults increases peak bone mass, particularly if the sport is weight-bearing.

Before puberty, bone growth is stimulated mainly by growth hormone. Estrogens and testosterone, the sex hormones produced at puberty, are responsible for the acceleration in growth of the long bones during the teenage years. The sex hormones also promote morphological changes in the skeleton that are typical of males and females.

12.7 BONE REMODELING

Bone tissue is continuously being resorbed and reformed. This remodeling of bone is a natural process of renewal and repair. Remodeling occurs at discrete foci called basic multicellular units (BMUs) and involves a sequence of highly co-ordinated cellular events.

Cancellous bone remodeling starts on the bone surface, which is covered by a single layer of flat lining cells. These cells respond to stimuli such as parathyroid hormone (PTH) and mechanical stress, which initiate the remodeling cycle. The lining cells retract and secrete matrix metalloproteinase, which removes the underlying surface layer of bone. The exposed osteoid serves as a chemoattractant for migration of osteoclast precursors to the site. The bone lining cells express RANKL (a ligand belonging to the tumor necrosis factor-α [TNF-α] superfamily of trimeric signaling molecules) and stimulate expression of its receptor, RANK, on the osteoclast precursors. (RANK stands for receptor activator of nuclear factor [NF]-κB and was formerly known as osteoclast differentiation factor [ODF] and stromal osteoclast-forming activity [SOFA].) Interaction of RANK with RANKL initiates multiple intracellular signaling pathways that induce osteoclast differentiation (Boyce & Xing, 2008). Osteoblasts produce M-CSF, which is required for survival of osteoclast precursors. The multinucleate osteoclasts then begin to erode the bone.

The transition phase is a period when osteoclastic bone resorption is inhibited and osteoblast precursors differentiate into bone-forming osteoblasts. The coupling of osteoclastic and osteoblastic activities ensures that the amount of bone removed by resorption is quantitatively replaced with new bone.

During the termination phase of cancellous bone remodeling, osteoblastic bone formation is suppressed by sclerostin, which is produced by osteocytes and secreted through osteocyte canaliculi. Bone resorption is suppressed by osteoprotegerin, a decoy RANKL receptor secreted by osteoblasts. Binding of RANKL to osteoprotegerin prevents the RANKL–RANK interaction, thereby inhibiting osteoclast differentiation. Osteoblasts remaining at the surface become quiescent and flatten to form a layer of lining cells over new bone. Osteoblasts imprisoned within the bone become osteocytes.

An imbalance in bone remodeling leads to a progressive decrease in bone density (osteopenia) and a breakdown of bone architecture, leading ultimately to osteoporosis. Remodeling imbalance may be due to increased osteoclastic activity, creating resorption pits which are too deep for normal osteoblasts to fill. Alternatively, or in addition, osteoclastic activity may be normal, but the ability of osteoblasts to fill the resorption pits is impaired.

12.8 OSTEOPOROSIS

Osteoporosis is a disease which causes reduced bone mass and deterioration of bony microarchitecture. The bones become so fragile that they fracture after only minor trauma, such as falling from a standing height. Increased activity of individual osteoclasts or an increase in the overall number of osteoclasts can cause osteoporosis. Elderly women during their lifetime lose about a third of cortical bone mass and half of cancellous bone mass from the skeleton. This section discusses involutional osteoporosis of which there are two types, I and II.

12.8.1 Type I Osteoporosis

This syndrome manifests in women typically between 50 and 75 years of age, and results from an acceleration of cancellous bone loss after the menopause. About one in five postmenopausal women will develop type I osteoporosis unless treated. Within the first 1 to 5 years after the onset of menopause, the rate of cancellous bone loss is two to six times the premenopausal rate of about 1% per year, but it gradually returns to the premenopausal rate about the tenth year after onset of menopause (Krall & Dawson-Hughes, 1999). There is only a slight corresponding increase in cortical bone loss. Fractures occur most commonly in the distal radius (forearm) and the spinal vertebrae. The vertebral fractures are of the crush or collapse type, causing a >25% reduction of vertebral height, and may be acutely painful.

Osteoporosis in postmenopausal women is due to the dramatic decrease in estrogen production that accompanies menopause; this is evident by the well-established efficacy of hormone replacement therapy. (*Note.* Estrogens are a group of female sex hormones that include estradiol and estrone.) The accelerated phase of bone loss is associated with increased osteoclastic resorption and estrogens exert their protective effect against bone loss mainly by inhibiting resorption. Inhibition of resorption is due to both decreased osteoclastogenesis and diminished resorptive activity of mature osteoclasts. Estrogen receptors have been found in both osteoblasts and osteoclasts. Estrogens could therefore directly modulate the secretion of local regulatory factors by these cells or modulate the cells' response to regulatory factors. Estrogens could also act indirectly by modulating the production of factors involved in bone resorption. For example, the cytokines interleukin-1 (IL-1) and TNF-α, which stimulate bone resorption, activate osteoclasts indirectly via a primary effect on osteoblasts. In addition, these two cytokines stimulate osteoclast progenitor cell proliferation and fusion. Furthermore, they act synergistically to increase the secretion of other bone-resorbing cytokines such as M-CSF and IL-6. The finding of increased expression of IL-1 and TNF-α in the bone cells of women with postmenopausal osteoporosis is consistent with the suppressive effect of estrogens (when present) upon the production and activity of these cytokines. Estrogens can also act positively by inducing secretion of osteoprotegerin, which inhibits bone resorption, and increasing the production of transforming growth factor-β (TGF-β), a cytokine which decreases both recruitment and activity of osteoclasts. Finally, estrogens prevent apoptosis of bone-forming osteoblasts.

12.8.2 Type II Osteoporosis

Type II osteoporosis is a purely age-related syndrome, which affects men and women over 70 and is twice as common in women as in men. Bone loss increases gradually with aging, unlike the accelerated loss seen in type I osteoporosis. In both sexes, cancellous bone loss begins at about age 40 and continues into old age. Cortical bone loss commences five to ten years later but slows or ceases later in life. In elderly women, type I and type II osteoporosis overlap and result in a disproportionate loss of cancellous bone. Type II osteoporotic fractures occur most commonly in the proximal femur (hip) and the spinal vertebrae. The vertebral fractures are of the compression type, causing the normally cube-shaped vertebrae to become wedge-shaped. A severe consequence of vertebral compression fractures is kyphosis, known as 'dowager's hump.'

Age-induced factors that contribute to type II osteoporosis are (1) decreased osteoblast activity, (2) decreased renal synthesis of 1,25-dihydroxyvitamin D_3, (3) decreased intestinal calcium absorption, and (4) elevated blood phosphate concentration. When combined, these factors cause an increase in blood PTH concentration. High blood PTH stimulates bone absorption and a consequent net increase in bone loss.

REFERENCES

Boyce, B.F. & Xing, L. (2008) Functions of RANKL/RANK/OPG in bone modeling and remodeling. *Archives of Biochemistry and Biophysics*, **473**, 139–46.

Krall, E.A. & Dawson-Hughes, B. (1999) Osteoporosis. In: *Modern Nutrition in Health and Disease*, 9th edn. (Ed. by M.E. Shils, J.A. Olson, M. Shike & A.C. Ross), pp. 1353–64. Lippincott Williams & Wilkins, Philadelphia.

Suda, T., Takahashi, N., Udagawa, N., Jimi, E., Gillespie, M.T. & Martin, T.J. (1999) Modulation of osteoclast differentiation and function by the new members of the tumor necrosis factor receptor and ligand families. *Endocrine Reviews*, **20**, 345–57.

13 The Blood–Brain, Blood–Cerebrospinal Fluid, and Placental Barriers

13.1 THE BLOOD–BRAIN AND BLOOD–CEREBROSPINAL FLUID BARRIERS

13.1.1 Introduction

The brain must be protected from abrupt changes in the concentrations of circulating adrenaline and other neurally active substances which would interfere with synaptic communication, as well as from potentially toxic molecules that may find their way into the bloodstream. On the other hand, essential molecules such as glucose, amino acids, and water-soluble vitamins must have access to the brain cells (neurons and glial cells). The blood–brain barrier serves to control the chemical composition of the extracellular fluid compartment of the brain, i.e. the interstitial fluid surrounding the brain cells, which comprise the intracellular compartment. The cerebrospinal fluid bathes the exterior of the brain and spinal cord in the subarachnoid space and also fills the ventricles of the brain. The interstitial fluid and cerebrospinal fluid are separated by only a single layer of ependymal cells, which form the lining of the ventricles. These cells are loosely linked by gap junctions so that the ependymal layer allows interchange of solutes by simple diffusion between the two fluid compartments. Because of the leaky ependymal layer, a blood–cerebrospinal fluid barrier is also necessary to maintain a constant chemical environment for the brain cells. The surface area of this barrier is only about 0.02% of the surface area of the blood–brain barrier. The relationship between the blood–brain and blood–cerebrospinal fluid barriers is shown in Figure 13.1.

For a typical water-soluble vitamin to enter the extracellular space of brain or the cerebrospinal fluid, the circulating molecule must pass through the blood–brain and/or blood–cerebrospinal fluid barriers. Specialized transport systems are required for water-soluble vitamins to negotiate these barriers. The blood–brain barrier is due to the presence of high resistance tight junctions between adjacent endothelial cells of cerebral capillaries. The blood–cerebrospinal fluid barrier is due to the presence of tight junctions between epithelial cells of the choroid plexus. There is no anatomical barrier to the movement of substances between the cerebrospinal fluid and the extracellular space of the brain. Once inside the extracellular space, the molecules must pass through the plasma membrane of the brain cell itself before finally entering the cell.

An important fact to be considered in brain nutrition is that the cerebrospinal fluid is constantly leaving the central nervous system and being replaced by new fluid; the total turnover time is about 6 hours. Thus there is a clear need for continuing transfer of essential nutrients from blood into cerebrospinal fluid and the extracellular space of brain.

13.1.2 The Blood–Brain Barrier

The properties of the cerebral capillary endothelial cells account for the blood–brain barrier. The wall of a cerebral capillary is composed of endothelial cells surrounded by a basal lamina, similar to the walls of capillaries in peripheral tissues. Here, however, the similarities end. Whereas spaces between adjacent peripheral endothelial cells allow small molecules and ions to diffuse freely through the capillary wall, adjacent endothelial cells in cerebral capillaries abut against each other and the cells are completely sealed together by continuous tight junctions of high electrical resistance (1,000 ohm cm^{-2} or more). These high resistance junctions present an effective barrier, even to ions. This means that all substances entering or leaving the brain must pass through two plasma membranes and cytoplasm of the endothelial cells rather than between the cells. In addition, cerebral capillaries lack the fenestrations and vesicular transport systems found in peripheral capillaries, thus excluding the passage of small molecules and macromolecules via these routes.

Certain regions of the brain, accounting for less than 1% of the brain volume, lack a blood–brain barrier. In these brain regions, neurons secrete hormones and other factors that require rapid and uninhibited access to the systemic circulation. These leaky regions are isolated from the rest of the brain by specialized cells called tanycytes.

Neurons in the brain do not abut directly upon capillaries except in certain special structures, e.g. parts of the hypothalamus. Usually they are separated from capillaries by the protoplasmic processes of astrocytes, a type of glial cell. The intervening astrocytes are therefore implicated in the exchange of materials between the blood and the nervous tissue. Around the capillary lies a virtual perivascular space occupied by another cell type, the pericyte. The functions of pericytes include the secretion of basal lamina components, the regulation of revascularization and repair, and the regulation of vascular tone in capillaries. The association between endothelial cells, pericytes, and astrocytes is illustrated in Figure 13.2.

The blood–brain barrier does not exclude the passage of small lipophilic molecules into the extracellular fluid compartment, as such molecules dissolve readily in the lipid components of plasma membranes. Water can also enter

DOI: 10.1201/9781003410300-15

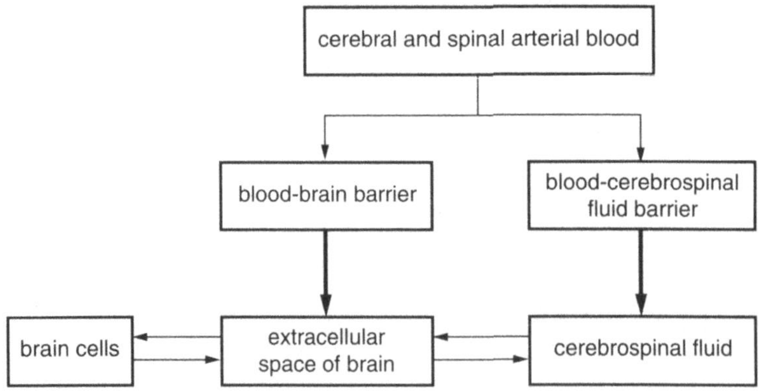

FIGURE 13.1 Relationship between the blood–brain and blood–cerebrospinal fluid barriers. Before water-soluble molecules in the blood can enter the brain cells, the molecules must enter the extracellular space of brain (the interstitial fluid surrounding the brain cells) or the cerebrospinal fluid which bathes the brain's surfaces. To enter the extracellular space from blood, the molecule must pass through the blood–brain barrier, while to enter the cerebrospinal fluid, the molecule must pass through the blood–cerebrospinal fluid barrier. There is no barrier to the movement of molecules between the extracellular space and the cerebrospinal fluid.

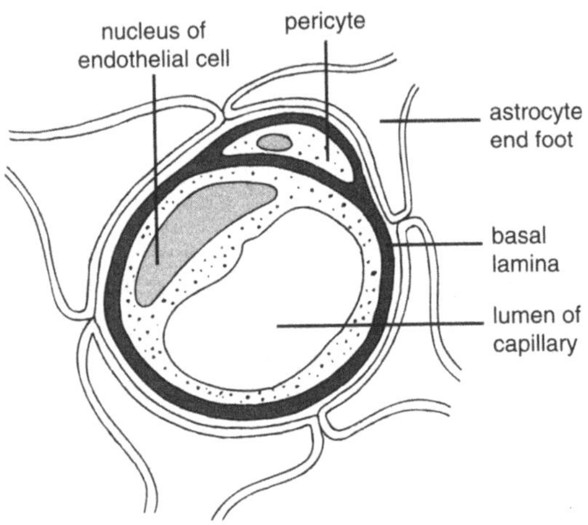

FIGURE 13.2 The blood–brain barrier is imposed by the specialized endothelial cells of the cerebral capillaries. The diagram shows a cerebral capillary and associated cells in cross-section. A single endothelial cell can completely surround the lumen of a capillary. The capillary endothelium is encased within a basal lamina, which also houses pericytes. Almost completely enclosing the basal lamina are terminal processes (end feet) of astrocytes.

the brain by simple diffusion. Essential hydrophilic molecules that the brain consumes rapidly and in large quantities, such as glucose and amino acids, rapidly traverse the blood–brain barrier by means of facilitated diffusion. Similar carrier-mediated transport systems also move surplus substances out of the brain and into the blood.

13.1.3 The Blood–Cerebrospinal Fluid Barrier

The inner surfaces of the cranium and vertebral column are lined by three membranes called the meninges; from the outside inwards these are termed the 'dura mater,' the 'arachnoid membrane,' and the 'pia mater.' The dura mater consists of two layers – an outer layer of dense connective tissue containing numerous blood vessels and nerves, and an inner layer of dense fibrous tissue. The arachnoid membrane and pia mater are delicate nonvascular membranes covering the brain surface. These two membranes are separated by the subarachnoid space, which contains some of the cerebrospinal fluid. The arachnoid membrane forms an impermeable barrier between the blood supply of the dura mater and the cerebrospinal fluid in the subarachnoid space.

Projecting into the cavities of the brain ventricles are fronds of blood capillary networks collectively called the choroid plexus. The capillaries are surrounded by a single layer of epithelial cells continuous with the ventricular lining of ependymal cells. Like the epithelial cells that line the intestine and other organs, the choroid epithelial cells are structurally polarized. The basolateral surfaces of the choroid epithelial cells are in contact with the blood plasma that filters through the leaky walls of the capillaries. The apical surfaces contain villi that extend into the ventricular cerebrospinal fluid. Adjacent choroid epithelial cells are sealed together by tight junctions that impede the passage of even small water-soluble molecules. These tight junctions constitute the blood–cerebrospinal fluid barrier of the choroid plexus (Spector & Johanson, 1989).

The choroid plexus controls the one-way transfer of essential micronutrients that the brain needs in relatively small amounts over extended periods. Such substances include vitamin B_6, folate, and ascorbic acid. Active transport mechanisms, mediated by carriers on the basolateral surface of choroid epithelial cells, extract micronutrients out of the blood and into the cells' cytoplasm. Once concentrated therein, the micronutrients are released into the cerebrospinal fluid through the apical cell surface, probably by facilitated diffusion.

13.1.4 Transport Regulation

In many cases of carrier-mediated micronutrient transport from blood to brain, the K_m (the concentration of micronutrient at which the carrier system is half-saturated) is

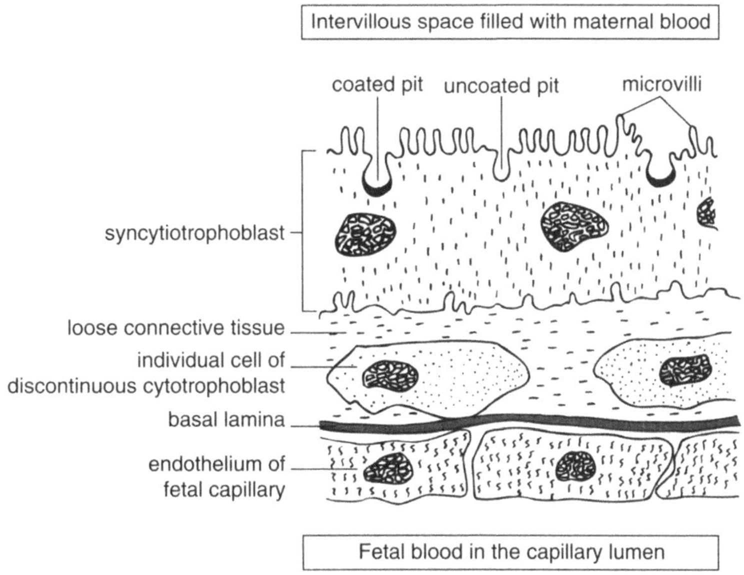

FIGURE 13.3 Diagram of part of the chorionic villus of the human placenta near term showing the cellular barrier between maternal and fetal blood.

approximately equal to the normal concentration of micronutrient in the plasma. In other words, the system is already half-saturated when plasma concentrations are normal. In such cases, typified by thiamin, folate, and ascorbic acid, an increase in plasma concentration will progressively decrease brain uptake as the carrier system approaches saturation. Nicotinamide, pantothenic acid, and biotin are exceptions: they have K_m values that are much higher than their normal plasma concentration, creating the potential for high uptake by the brain (Spector, 1989).

13.2 THE PLACENTAL BARRIER

The functional unit of the placenta concerned with exchange of materials between mother and fetus is the chorionic villus depicted in Figure 13.3. Passage of dissolved substances occurs between the maternal blood in the intervillous spaces and the fetal blood in the capillaries of the chorionic villi. The trophoblast comprises an outer syncytiotrophoblast and an inner cytotrophoblast. The syncytiotrophoblast is a multinuclear epithelial layer which is devoid of lateral cell boundaries. Its microvillous surface is bathed by maternal blood circulating in the intervillous space, while its basal surface faces the fetal circulation. The microvillous surface of the syncytiotrophoblast presents a large area for nutrient absorption. Also present are coated and uncoated pits that facilitate endocytosis of macromolecules. The cytotrophoblast consists of discrete cells that become progressively fewer in number.

REFERENCES

Spector, R. (1989) Micronutrient homeostasis in mammalian brain and cerebrospinal fluid. *Journal of Neurochemistry*, **53**, 1667–74.

Spector, R. & Johanson, C.E. (1989) The mammalian choroid plexus. *Scientific American*, **261**(5), 48–53.

14 The Cell Cycle, Apoptosis, and Cancer

14.1 THE CELL CYCLE

14.1.1 TERMINOLOGY

It is important to consider the relationship between cell proliferation and cell growth in the context of the cell cycle. Proliferation is an increase in cell number resulting from mitotic cell division: growth is an increase in the volume and mass of individual cells, with no increase in cell number. The cell must grow before it can proliferate, otherwise it will divide into ever smaller cells. Except for the G_1 and G_2 growth phases of the cell cycle, the commonly used term 'cell growth' actually means cell proliferation.

The terms 'hyperplasia' and 'hypertrophy' apply to aggravated cell proliferation and growth, respectively.

- **Hyperplasia**. An increase in the number of cells in an organ or tissue in response to a specific stimulus, such as an excess of growth hormone; hyperproliferation will cease if the stimulus is removed. A clinical example is benign prostate hyperplasia (prostate enlargement). Hyperplasia may be a normal physiological response, as in the case of milk-secreting glandular cells in mammary tissue in response to pregnancy.
- **Hypertrophy**. An increase in the size of cells, as exemplified by hypertrophy of skeletal muscle cells during steroid use.

14.1.2 GENERAL FEATURES OF THE CELL CYCLE

The cell cycle is the ordered sequence of events which allows somatic cells to grow and divide only when conditions are favorable. The cell cycle is organized, for simplicity, into a sequence of four phases (Figure 14.1). During a post-mitotic gap 1 phase (G_1 phase), the newly formed daughter cell increases in volume and mass while synthesizing mRNAs and proteins. The cell then doubles its complement of chromosomes during a DNA synthesis phase (S phase), increases further in volume and mass during a post-synthetic gap 2 phase (G_2 phase), and then divides by mitosis and cytokinesis (M phase). All phases except M phase can be grouped together and called interphase, which is the intervening period between mitoses. Most post-mitotic cells exit the cell cycle in G_1 and enter a quiescent phase, G_0, in which they can survive for hours, days, or weeks, depending on cell type. Re-entry is regulated, thereby providing control of cell proliferation. Some cell types (e.g. nerve cells, skeletal muscle cells, and cells of the eye lens) are in a terminally differentiated G_0 state and never divide. Almost all the variation in cell cycle duration in the adult body occurs during the time the cell spends in G_1 or G_0. The time taken for a cell to progress from the beginning of S phase through mitosis is typically 12–24 hours and relatively constant, regardless of the interval from one cell division to the next.

The cell possesses biochemical monitoring systems designed to make specific cell cycle events dependent on the successful completion of preceding events. At a 'restriction point' in late G_1 the integration of growth regulatory information originating from various external signals determines whether the cell will proceed to S phase and complete the cycle. After negotiating the restriction point, the cells are committed to complete a cycle, barring major defects such as DNA damage or metabolic disturbance. The committed cells become refractory to mitogen-induced signals and come to rely upon intrinsic factors for their continued progression. In addition, cells are rendered insensitive to antimitogenic signals after the restriction point.

Exit from the cell cycle is necessary to allow cells to differentiate. Differentiation-inducing signals activate mechanisms of cell cycle arrest that lead to cells in the G_1 phase withdrawing into G_0. Quiescent cells re-enter the cell cycle in response to mitogens such as platelet-derived growth factor (PDGF) and epidermal growth factor (EGF). Addition of such a mitogen to cultured cells in G_0 causes a rapid increase in the expression of as many as 100 different genes. These genes, which include c-*jun*, c-*fos*, c-*myc*, and c-*myb*, are called primary response genes because they are induced well before cells enter the S phase and replicate their DNA. Primary response genes encode transcription factors (e.g. c-Jun and c-Fos), which induce the expression of many secondary response genes encoding proteins necessary for cells to progress through the cell cycle.

If a problem in DNA replication, DNA repair, or chromosome replication is detected, a checkpoint is activated and signals are generated that arrest the cell cycle. This allows the opportunity for cell repair or elimination by apoptosis.

14.1.3 REGULATION OF THE CELL CYCLE BY CYCLIN-DEPENDENT PROTEIN KINASES

The cell cycle is regulated by cyclin-dependent protein kinases (CDKs) in association with regulatory subunits called cyclins. The cyclin–CDK complexes regulate the activities of multiple proteins involved in DNA replication and mitosis by phosphorylation, activating some and inhibiting others to co-ordinate their activities. Mammalian cells express four CDKs, named CDK1, CDK2, CDK4, and CDK6, and four major classes of cyclins, designated A, B, D, and E. For each cyclin class, multiple family members

DOI: 10.1201/9781003410300-16

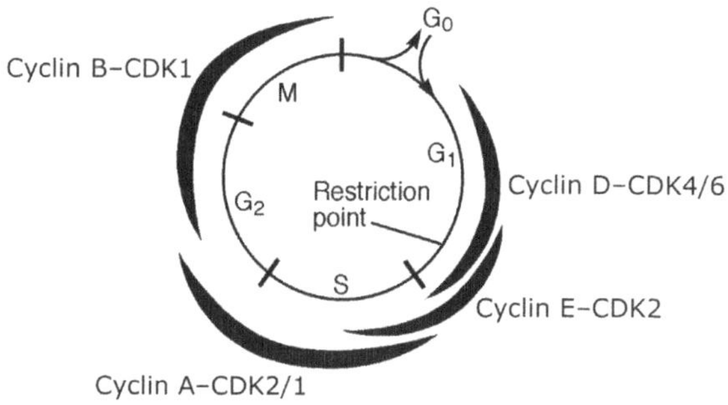

FIGURE 14.1 Waves of specific CDK activities during the cell cycle. The activities of CDKs associated with specific cyclins are depicted by arrows. Growth factor-dependent signaling pathways control the accumulation of cyclin D/CDK complexes. The activities of cyclin E-, A-, and B-associated kinases are in part determined by the regulated accumulation of the cyclins via E2F-dependent control of transcription. The abrupt loss of cyclins E, A, and B kinase activities at specific points in the cell cycle is largely the result of regulated degradation of the cyclin subunits. The restriction point is the point beyond which the cell is committed to completing the cycle, barring major defects.

can contribute to the activity; for example, cyclin D has three family members, cyclins D1, D2, and D3. Progression through the different phases of the cell cycle requires the successive activation of a particular cyclin–CDK combination. Figure 14.1 shows the waves of specific cyclin–CDK activities during the cell cycle. The programed synthesis and destruction of individual cyclins maintains an appropriate order of cell cycle events.

14.1.4 Cell Cycle Checkpoints

Cell cycle checkpoints serve as surveillance mechanisms, ensuring that critical transitions occur in the correct order and are accurate. There are two checkpoints, one at the G_1 to S transition and the other at the G_2 to M transition. The checkpoint in late G_2 ensures that cells do not initiate mitosis until they have repaired damaged DNA after replication. In the event of DNA damage, signals detected by 'sensor' proteins are relayed by 'transducer' proteins to 'effector' proteins, whose actions result in cell cycle arrest. This allows time for DNA repair enzymes to repair the damage before the cycle is allowed to resume. Many of the transducers are protein kinases that will phosphorylate effectors to regulate the response.

14.1.5 Mitogenic Stimulation of the Cell Cycle

The cells of a multicellular organism will divide only if stimulated by a mitogen. In the absence of a mitogenic signal, the cell cycle arrests and the cell may exit into G_0. Mitogens act in the G_1 phase by binding to cell-surface receptors to initiate a complex array of intracellular signals. The ultimate result is the activation of G_1- and G_1/S-phase cyclin–CDK complexes, which overcome the inhibitory barriers that otherwise block progression into S phase. In addition to mitogens, some members of the transforming growth factor-β (TGF-β) family act on some cell types to stimulate proliferation and on others to inhibit proliferation. Other extracellular factors stimulate proliferation at one concentration and inhibit at another.

14.1.6 p53 as a Transcriptional Activator of Specific Genes Involved in Cell Cycle Arrest and Apoptosis

The tumor-suppressor protein, p53, acts as a transcription factor to increase synthesis of p21, which arrests damaged cells in G_1 and G_2 stages of the cell cycle until the damage is repaired. The C-terminal domain of p53 binds to the exposed ends of DNA at the points of damage. Under normal conditions, p53 is highly unstable and generally does not accumulate to high enough levels to stimulate transcription. This instability is due to polyubiquitination by a ligase called Mdm2 and subsequent proteasomal degradation. DNA damage provokes activation of ATM, a serine kinase that inhibits p53 polyubiquitination by phosphorylating the protein at the binding site for Mdm2. The resultant stabilization of p53 leads to a marked increase in its concentration. If the DNA repair is successful, the levels of p53 and p21 will fall and the cell cycle can proceed. When DNA damage is extensive, p53 also activates expression of genes that lead to apoptosis. Inhibition of poly(ADP-ribose)polymerase (PARP) results in decreased basal p53 protein levels and impairs p53 stabilization after DNA damage, demonstrating that PARP is critical for p53 function.

14.1.7 Replicative Senescence

The biology of replicative senescence and its relevance to aging and tumorigenesis has been explained by Campisi (1997). A summary of this paper, focusing on the process of replicative senescence, is given here.

Normal somatic mammalian cells, with the possible exception of certain stem cells, cannot divide indefinitely and their proliferative capacity declines exponentially with each division. Upon completing a finite number of divisions, cell proliferation is arrested in the G_1 phase of the cell cycle. Once arrested, cells cannot be stimulated to enter the S phase by any known combination of physiological mitogens. The process that limits the proliferative potential of cells has been termed 'replicative senescence,' meaning an irreversible proliferation arrest that can be triggered by a number of factors, including DNA damage and oxidative stress. Somatic cells acquire this finite replicative life span sometime during embryonic development. The number of divisions that normal cells complete before they senesce can be 60–80 for fetal or neonatal human cells. Importantly, senescent cells acquire resistance to apoptotic stimuli, emphasizing that replicative senescence and apoptosis (programed cell death) are two distinct processes.

The means by which cells sense the number of divisions they have completed appears to depend, at least in part, on the length of their telomeres, but the mechanism is unknown. Telomeres are nucleoprotein complexes located at the ends of chromosomes in mammalian cells. They are not replicated during the S phase of the cell cycle and, hence, are shortened after each cell division. When telomeres reach a critical length, cell proliferation is arrested.

Senescent cells do not fail to proliferate because of a general breakdown of growth factor signal transduction. Rather, replicative senescence causes the selective repression of a few positive-acting growth regulatory genes whose expression has been shown to be important for G_1 progression and DNA synthesis. In addition, replicative senescence causes overexpression of negative growth regulators, in particular p21 and p16 inhibitors of CDKs.

Replicative senescence is genetically dominant and replicative immortality is recessive. This conclusion is based on somatic cell fusion studies in which proliferating normal cells were fused to immortal tumor-derived cells. In the majority of cases, the hybrid cells proliferated for some time, but eventually they senesced. In conclusion, replicative senescence prevents cells from dividing indefinitely, and appears to be a powerful, albeit imperfect, tumor suppressive mechanism.

14.2 APOPTOSIS

14.2.1 Introduction

Apoptosis (Greek: 'falling off,' as leaves from a tree) is programed cell death. Its purpose is to eliminate senescent cells and activated lymphocytes at the end of an immune response, and to rid the body of defective cells, such as virus-infected cells and oncogenically transformed cells. Apoptosis is essential to a variety of physiological processes, such as differentiation during embryogenesis, establishment of immune self-tolerance, and killing of virally infected host cells by cytotoxic immune cells. The cells respond to a variety of stimuli, including DNA damage, growth factor withdrawal, and viral infection. Signal transduction pathways culminating in cell death have been described in Section 8.9. A single cell within a tissue can be targeted and, once sentenced to death, may completely disappear within 4 hours. Apoptosis is controlled through the expression of specific genes, which are turned on by transcription factors such as NF-κB and p53. Dysregulation of apoptosis may represent significant factors in pathogenesis of various diseases, including cancer.

14.2.2 Some Essential Functions of Apoptosis

Apoptosis eliminates unwanted cells in the developing embryo, such as neurons that fail to make the correct connections. Cells with damaged DNA are removed before they get the chance to mutate and possibly transform into a tumor cell. In the immune system, autoreactive and redundant T lymphocytes are eliminated by apoptosis during the selection processes in the thymus. Cytotoxic T lymphocytes kill virus-infected host cells by releasing a cytotoxic enzyme (granzyme B) that activates apoptosis.

14.2.3 Morphological Features of Apoptosis

The course of apoptosis is accompanied by a characteristic sequence of changes in cell morphology. In the initial phase, the cell becomes rounded and loses contact with its neighbors. The chromatin condenses, the DNA fragments, and the nucleus begins to break down into discrete chromatin bodies. Condensation of cytoplasmic proteins leads to cell shrinkage, but most intracellular organelles remain intact at this stage. In the next phase, membrane blebbing leads to disintegration of the cell into membrane-enclosed vesicles called apoptotic bodies. Finally, neighboring cells phagocytose the apoptotic bodies and completely degrade them in phagosomes. If the apoptotic cells line vessels, ducts, or body cavities, the apoptotic bodies shed into lumens.

The important difference between apoptosis and necrotic cell death is that apoptosis does not incur release of cell contents into the extracellular milieu and therefore does not elicit an inflammatory response.

14.2.4 Regulation and Dysregulation of Apoptosis

14.2.4.1 Regulation

Many cellular stimuli result in the induction of both p53 and NF-κB transcription factors. Stimulation of p53 induces apoptosis, whereas stimulation of NF-κB inhibits apoptosis. A regulatory mechanism must exist to integrate these opposing outcomes. Webster and Perkins (1999) showed that p53 and NF-κB inhibit each other's ability to transcribe genes and that this process is controlled by their relative levels of expression.

14.2.3.2 Dysregulation

Dysregulation of apoptosis, causing unwarranted cell death or cell survival, can have pathological consequences. For example, the premature death of particular subsets of neurons is associated with degenerative neurological diseases, such as Alzheimer's disease and Parkinson's disease. An abnormal increase in the death of T lymphocytes contributes to AIDS. Conversely, impaired apoptosis allows the survival of self-reactive T and B lymphocytes during their development, thus promoting autoimmune disease. Failure to kill cells with damaged DNA may result in mutations that lead to cancer.

14.3 CANCER

The purpose of this section is to familiarize the reader with some of the terminology used in the field of cancer and to implicate some factors in the etiology of this disease.

14.3.1 Introduction

Cancer is the accumulation of cells caused by a loss of cellular regulation. Cell accumulation can result from either increased cell proliferation or the failure of cells to undergo apoptosis. In most or all cases of cancer, the losses of cellular regulation are due to genetic damage of some kind. Cancer kills because the relentlessly proliferating cancer cells take up more and more of the available nutrients, so that normal tissues eventually die from lack of nutrition. The ultimate goal of cancer therapy is to find drugs that will specifically and efficiently kill cancer cells, while leaving normal cells unharmed.

14.3.2 The Genetic Basis of Cancer

Cancer is caused by an accumulation of genetic mutations, either acquired or inherited. The genes that are altered are those involved in normal cellular processes, such as cell cycle regulation, signaling, and differentiation. Environmental agents that may cause mutations include viruses, chemicals, and radiation. Cells whose DNA has been altered by such agents can only become cancerous after they have been exposed to a second stimulus, most often a chronic irritant or inflammatory agent (Coussens & Werb, 2002).

Cancer can result from the activation of proto-oncogenes to oncogenes (cancer genes), whose altered protein products cannot be properly controlled by the cell. Activation can be induced by point mutations (altered triplet codon) and frameshift mutations (insertions or deletions of a single nucleotide or a segment of DNA); a proto-oncogene can fuse with another cellular gene; and a normal proto-oncogene product can be overexpressed. One common feature of oncogenes is their dominance: only one of the two copies needs to be altered to induce tumor formation. Furthermore, the biochemical activity of the oncoprotein is usually greater than that of the normal gene product.

Examples of proto-oncogenes that are important in cancer development are the three *ras* genes, H-*ras*, K-*ras*, and N-*ras*. Their encoded Ras proteins, when activated by incoming signals, transduce the signal to other proteins, which ultimately turn on genes involved in cell proliferation, differentiation, and survival. Qiu *et al.* (1995) showed that the Rho family GTPase Rac1 acts downstream of oncogenic Ras and drives a signaling cascade that co-operates with the Ras-driven MAP kinase pathway to cause transformation. Ras proteins are molecular switches, which can be turned on or off by binding to guanosine triphosphate (GTP) and guanosine diphosphate (GDP), respectively. A guanine nucleotide-exchange factor (GEF) promotes the exchange of bound GDP for GTP, and a GTPase-activating protein (GAP) increases the rate of hydrolysis of bound GTP to GDP. Point mutations in codons 12, 13, or 61 convert the normal *ras* gene into an oncogene. In the oncogenic Ras protein, the binding of GAPs is prohibited, preventing hydrolysis of GTP and locking Ras into its active conformation. The constitutive activity of the mutated Ras proteins results in hyperproliferation and cancer. Some 20% of human tumors have activating point mutations in *ras*, most frequently in K-*ras* (about 85% of total), then N-*ras* (about 15%), then H-*ras* (less than 1%) (Downward, 2003). The incidence of mutated *ras* genes varies greatly among the different tumor types. For example, more than 80% of pancreatic tumors and nearly 50% of colonic tumors harbor a mutated K-*ras* gene. In several tumor types, a mutant gene is found only occasionally and in breast tumors no mutated gene has been identified at all (Bos, 1989).

Cancer can also result from inactivation of tumor-suppressor genes, allowing the unconstrained growth of the cancer cell. The protein products of these genes serve as transducers of anti-proliferative signals. Tumor-suppressor genes are recessive and both copies must become defective for tumor formation to occur. *p53* is classified as a tumor-suppressor gene because its encoded protein activates a number of genes that are responsible for cell cycle arrest and apoptosis. *p53* also represses expression of other genes that must be switched off to facilitate these processes.

14.3.3 Tumorigenesis

A tumor can be benign (non-cancerous) or malignant (cancerous). Benign tumors, which include skin moles, do not spread beyond the local tissue and therefore are not considered to be cancerous. However, large benign tumors can put pressure on organs and be fatal, as is the case with brain tumors, which occupy a limited space within the skull. Malignant tumors have the ability to invade surrounding tissues and the potential to spread to areas outside the local tissues, a process called metastasis. Metastatic tumors account for the majority of cancer deaths. In addition to these defined types of tumors, there are tumors of uncertain or unknown behavior.

Malignant tumors can be solid or hematological. Solid malignant tumors comprise the proliferating cancer cells and a supportive stroma of connective tissue and blood vessels. They are classified as carcinomas if they derive from endoderm (gut epithelium) or ectoderm (skin and neural epithelia) and sarcomas if they derive from mesoderm (muscle, blood, and connective tissue precursors). Hematological malignant tumors involve over-production of bone marrow-derived white blood cells and are classified as leukemias if they develop in blood itself and lymphomas if they develop in the lymphatic system.

In order for most oncogenic mutations to induce cancer, they must occur in dividing cells so that the mutation is passed on to many progeny cells. When such mutations occur in non-dividing cells (e.g. muscle cells and neurons), they generally do not induce cancer. Nonetheless, cancer can occur in tissues composed mainly of non-dividing differentiated cells, such as the absorptive cells (enterocytes) that line the small intestine, keratinized cells that form the skin, and red and white blood cells. The cells that initiate the tumors are not the differentiated cells, but rather their proliferating precursor cells.

Tumor formation entails a sequence of morphological changes described by the following terms.

- **Metaplasia**. The reversible transformation of a cell into another cell type in response to an environmental stimulus.
- **Dysplasia**. Irreversible morphological changes in cells caused by mutations in the genome or by epigenetic changes. Graded into low-, intermediate-, and high-grade dysplasia according to the thickness of the epithelium affected. High-grade dysplasia includes the histological changes previously called 'carcinoma *in situ*.'
- **Anaplasia**. Irreversible loss of structural and functional differentiation of normal cells throughout almost the entire thickness of the epithelium. Anaplastic cells fail to recognize patterns of orientation to one another and may grow in sheets. There is a total loss of communal structures such as gland formation or stratified squamous architecture. This lack of differentiation is considered a hallmark of malignant tumors.
- **Neoplasia**. Uncontrolled and irreversible proliferation of metaplastic or dysplastic cells that results in a neoplasm. Neoplastic growths include both benign and malignant tumors.

14.3.4 Induction of Cancer by Chemical Carcinogens

The induction of cancer by chemical carcinogens can be divided into three stages: initiation, promotion, and progression. Initiation refers to the damage to DNA by a chemical or its metabolite. Promotion induces the proliferation of initiated cells, thereby passing the DNA damage to the daughter cells. The third stage, progression, is the least defined and involves the irreversible transition from preneoplastic to neoplastic cells. The combined sequence of events is referred to as neoplastic transformation.

14.3.5 Acquired Capabilities of Cancer

Hanahan and Weinberg (2000) have described the following six acquired capabilities of cancer.

- **Self-sufficiency in proliferation signals**. Tumor cells generate many of their own proliferation signals, thereby reducing their dependence on stimulation from their normal tissue microenvironment.
- **Insensitivity to anti-proliferative signals**. Anti-proliferative signals are funneled through the retinoblastoma-associated protein (Rb) and its two relatives, p107 and p130. The Rb signaling pathway can be disrupted in a variety of ways in different types of tumors, rendering cells insensitive to anti-proliferative factors that normally operate along this pathway.
- **Evasion of apoptosis**. The p53 transcription factor can elicit apoptosis by up-regulating expression of proapoptotic Bax in response to sensing DNA damage. Resistance to apoptosis can be acquired by cancer cells through mutation of *p53*.
- **Limitless replicative potential**. In most normal cells, mechanisms that extend the length of telomeres are strongly suppressed to deny them unlimited replicative potential. In malignant cells, telomeres are maintained at a length above a critical threshold and this permits unlimited multiplication of descendant cells.
- **Sustained angiogenesis**. The ability to induce and sustain angiogenesis (the formation of new blood vessels). The oxygen and nutrients supplied by the vasculature are crucial for cell function and survival, obligating all cells in a tissue to reside within 100 μm of a capillary blood vessel. Besides providing oxygen and nutrients for tumor growth, angiogenesis provides an expanding endothelial surface for detached tumor cells to enter the circulation during metastasis.
- **Tissue invasion and metastasis**. The capability for invasion and metastasis enables cancer cells to escape the primary tumor mass and colonize new terrain in the body where, at least initially, nutrients and space are not limiting.

14.3.6 Angiogenesis

The angiogenic process is characterized by a series of stages.

- **Stage 1**. Degradation of the vascular basement membrane.
- **Stage 2**. Sprouting, elongation, migration, and proliferation of endothelial cells.
- **Stage 3**. Association of endothelial cells into new tubular channels.
- **Stage 4**. Synthesis of new basement membrane, recruitment of perivascular cells, and vessel maturation.

Tumors appear to activate angiogenesis by changing the balance of positive and negative regulatory factors (Pepper, 1997).

14.3.7 Metastasis

Metastasis is the spread of malignant cells from the tumor of origin to another part of the body. From a clinical perspective, metastasis is the most critical aspect of tumorigenesis. Although surgery and radiation therapy effectively control many cancers at the primary site, the development of metastatic disease signals a poor prognosis. Chemotherapy, hormonal therapy, and radiation serve palliative purposes in the metastatic setting, and some offer a modest but statistically significant extension of survival (Steeg, 2006).

In the process of metastasis, cancer cells break away from the original (primary) tumor, travel through the blood or lymph system, and form a new, metastatic (secondary) tumor in another organ or tissue. Oncogenic transformation is not sufficient for metastatic competence: transformed cells must acquire additional abilities to surmount the natural barriers against metastasis. It has been estimated that only about 0.01% of cancer cells that enter the circulation will eventually survive and give rise to micrometastases (Joyce & Pollard, 2009). The principal sites of metastasis for a particular primary tumor are characteristic for that tumor (Nguyen *et al.*, 2009; Table 14.1). Although established in a completely different organ, the metastatic tumor is the same type of cancer as the primary tumor.

The process of metastasis has been described by Steeg (2006) and can be divided into four steps. In this sequence, steps 1 and 2 leading to intravasation of tumor cells into blood vessels and lymphatics are together referred to as invasion.

- **Step 1**. Individual tumor cells (historically known as 'seeds') break out of their confining basement membranes, enter the extracellular matrix, and reach efferent blood vessels or lymphatics. This step involves changes in tumor cell adherence to cells and to the extracellular matrix, proteolytic degradation of surrounding tissue, and motility to physically propel the seed through tissue.
- **Step 2**. The tumor cells intravasate into blood vessels and lymphatics, and circulate to target tissues. Invasion signaling components confer tumor cell resistance to detachment-induced cell death (anoikis) during circulation.
- **Step 3**. Tumor cells arrest in capillary beds and move into tissues by extravasation. A proportion of cells non-specifically arrest by binding coagulation factors and by size restriction. Endothelial cell adhesion molecules and tumor-derived integrins also contribute to tumor cell arrest.
- **Step 4**. This final step is metastatic colonization – the process by which cancer cells escape the primary tumor and colonize another tissue to form a secondary metastatic tumor.

TABLE 14.1
Principal Sites of Metastasis for Different Tumor Types

Tumor Type	Principal Sites of Metastasis
Breast	Bones, lungs, liver, and brain
Lung adenocarcinoma	Brain, bones, adrenal gland, and liver
Skin melanoma	Lungs, brain, skin, and liver
Colorectal	Liver and lungs
Pancreatic	Liver and lungs
Prostate	Bones
Sarcoma	Lungs
Uveal melanoma	Liver

Source: Nguyen *et al.* (2009).

14.3.8 Association between the Development of Cancer and Chronic Inflammation

It is well established that chronic inflammation (defined as unresolved, persistent inflammation) is associated with an increased risk of cancer (Lu *et al.*, 2006). Almost 20% of human cancers worldwide are related to chronic inflammation caused by infection, exposure to irritants, or autoimmune disease (Balkwill & Mantovani, 2001). Common associations between chronic inflammation and cancer include *Helicobacter pylori* infection and stomach cancer; hepatitis B and C virus infections and liver cancer; and inflammatory bowel disease and colorectal cancer (Shacter & Weitzman, 2002).

A strong tumor-associated inflammatory response can also be initiated by cancer therapy. Radiation and chemotherapy cause massive necrotic death of cancer cells and surrounding tissues. Proteins released by necrosis trigger an inflammatory reaction, which can have two opposing outcomes. First, activation of the immune system by products of necrotic cells can enhance the efficiency of tumor antigen presentation, resulting in anti-tumor immunity that helps the host eradicate the remaining tumor. Second, inflammatory mediators released by dying cancer cells can lead to activation of tumor-infiltrating macrophages. The macrophages produce cytokines that activate oncogenic transcription factors in remaining cancer cells and ultimately stimulate tumor growth. The balance between therapy-induced tumor eradication and regrowth is very delicate and is likely to depend on the extent of therapy-induced cell death, the type of cancer being treated, and the inflammatory microenvironment associated with the tumor (Grivennikov & Karin, 2010).

Mantovani *et al.* (2008) viewed the connection between inflammation and cancer as consisting of two pathways: the intrinsic pathway and the extrinsic pathway. The intrinsic pathway is activated by genetic events that cause neoplasia. These events include the activation of various types of oncogene by mutation, chromosomal rearrangement, or amplification, and the inactivation of tumor-suppressor genes. Cells that are transformed in this manner produce inflammatory mediators, thereby generating an inflammatory microenvironment in tumors for which there is no underlying inflammatory condition (for example, breast tumors). By contrast, in the extrinsic pathway, inflammatory or infectious conditions augment the risk of developing cancer at certain anatomical sites (for example, the colon, prostate, and pancreas). The two pathways converge, resulting in the activation of transcription factors, most notably NF-κB, in tumor cells. These transcription factors co-ordinate the production of inflammatory mediators, including cytokines and chemokines, as well as the production of cyclooxygenase 2 (Cox-2), which, in turn, results in the production of prostaglandins. The factors produced recruit and activate various leukocytes, most notably cells of the myeloid lineage (e.g. macrophages, neutrophils, eosinophils, and mast cells). The cytokines activate the same key transcription factors in inflammatory cells, stromal cells, and tumor cells, resulting in more inflammatory mediators being produced and a cancer-related inflammatory microenvironment being generated.

Lipopolysaccharide (LPS), a cell wall component of gram-negative bacteria, provides a strong pro-inflammatory stimulus that can increase tumor growth. In a model of metastatic colon adenocarcinoma, the major inflammatory mediator responsible for LPS-induced tumor growth was tumor necrosis factor-α (TNF-α). In this model, the growth-promoting activity of TNF-α was dependent on NF-κB activation (Luo *et al.*, 2004). Although high doses of TNF-α administered in close proximity to solid tumors can kill both cancer cells and the tumor neovasculature, endogenous TNF-α promotes tumor development and growth (Wilson & Balkwill, 2002).

14.3.9 Tumor Immunology

Dranoff (2004) summarized the potential for immune destruction of tumors. Innate immune cells recognize cancer cells through pattern-recognition receptors and other cell-surface molecules. For example, NKG2D receptors on natural killer (NK) cells recognize the non-classical MHC class I molecules MICA and MICB expressed on cancer cells. Dendritic cells phagocytose apoptotic tumors, and macrophages ingest heat-shock proteins complexed with tumor-derived peptides that are released from necrotic cancer cells.

Adaptive immunity toward cancer cells is initiated by dendritic cells, which capture dying tumor cells or debris, migrate to regional lymph nodes, and present the tumor-derived material onto CD1d for presentation to natural killer T (NKT) cells, and onto MHC class I and class II molecules for presentation to helper ($CD4^+$) and cytotoxic ($CD8^+$) T cells. Dendritic cells secrete IL-12, which induces $CD4^+$ cells to differentiate into Th1 cells. Th1 cells (and also NK cells) secrete IFNγ, which inhibits tumor cell proliferation, enhances tumor cell apoptosis, improves tumor antigen presentation, and inhibits angiogenesis. Mast cells secrete IL-4, which induces $CD4^+$ cells to differentiate into Th2 cell. Th2 cells secrete IL-4, IL-5, IL-6, IL-10, and IL-13, which enhance eosinophil function and increase antibody production by B cells. Antibodies to cancer cell surface molecules can inhibit oncogenic signaling and/or stimulate tumor destruction through engaging Fc receptors on macrophages, granulocytes, and NK cells. Dendritic cells also secrete IL-18, which up-regulates the cytotoxic activity of NK cells and $CD8^+$ cells. The latter two cell types can lyse tumors through the perforin/granzyme pathway or death ligands such as the soluble form of TNF-related apoptosis-inducing ligand (TRAIL), which signals through death receptors on the cancer cell. Macrophages can lyse tumors through the production of nitric oxide and reactive oxygen species. Tumor blood vessels can be attacked by lymphocytes and granulocytes.

Several properties of tumor cells enable them to escape host defenses.

- Failure of effector T cells to recognize tumor antigens due to mutations or deletions in genes encoding the antigens.
- Masking of antigens due to excessive production of glycocalyx molecules by tumor cells.
- Failure of tumors to induce strong effector T cell responses due to lack of expression of co-stimulators or class II MHC molecules by tumor cells.
- Secretion of immunosuppressive cytokines by tumor cells.
- Creation of immune tolerance to tumors by tolerogenic dendritic cells.

The mixture of cytokines produced in the tumor microenvironment has an important role in cancer pathogenesis. Cytokines that are released in response to infection and inflammation can function to inhibit tumor development and progression. Alternatively, cancer cells can respond to host-derived cytokines that promote growth, attenuate apoptosis, and facilitate metastasis. As discussed by Dranoff (2004), manipulation of the cytokine balance in the tumor microenvironment can be exploited for cancer therapy. For example, modulation of this environment could block the tumorigenic function of tumor-associated macrophages and allow the macrophages to present antigens to infiltrating T cells with the consequent destruction of the tumor.

14.3.10 Role of TGF-β in Cancer Development

Many human malignant tumors secrete abnormally high amounts of the immunosuppressive cytokine TGF-β and this is correlated with the severity of tumor grade. Systemic administration of TGF-β promotes tumor progression and metastasis in mice. The TGF-β signaling pathway stimulates the activation of Smad proteins that accumulate in the nucleus and control transcription of a large number of target genes. Studies with transgenic mice genetically engineered so that all of their T cells were insensitive to TGF-β signaling showed that T cell-specific blockade of TGF-β signaling leads to the enhancement of anti-tumor immunity (Gorelik & Flavell, 2001). When challenged with live tumor cells, the transgenic mice were able to mount an immune response, eradicate the tumor, and survive. Abrogation of TGF-β signaling was shown to be necessary for the generation of tumor-eradicating $CD8^+$ T cells, but not for the generation or effector function of helper $CD4^+$ T cells. Ablating TGF-β with monoclonal antibodies or soluble TGF-βRII also suppressed tumorigenesis in mice (Won *et al.*, 1999).

Tumor-derived TGF-β can affect several cell types that are in proximity to the growing tumor – such as stromal fibroblasts, endothelial cells, and immune cells – to produce an environment that is conducive to tumor growth, invasion, and metastasis. In addition, TGF-β can exert tumorigenic effects by acting on the tumor cells themselves.

It is now clear that TGF-β can act as both a tumor suppressor and a stimulator of tumor progression, invasion, and metastasis. At early stages of tumorigenesis, when the tumor is still benign, TGF-β acts directly on the cancer cell to suppress tumor outgrowth. As the tumor progresses, genetic and/or biochemical changes allow TGF-β to stimulate tumor progression by its pleiotropic activities on both the cancer cell itself and on non-malignant stromal cell types of the tumor (Akhurst & Derynck, 2001).

In summary, cancer cells defend themselves against attack from cytotoxic $CD8^+$ T cells by producing abnormally high amounts of TGF-β, which represses the transcription of genes encoding cytotoxic molecules and thereby renders the $CD8^+$ T cells ineffective. The above results indicate that T cell-specific blockade of TGF-β signaling has strong potential to shift the balance of the immune response in favor of anti-tumor immunity.

14.3.11 Role of Infiltrated Leukocytes in Tumor Growth and Spread

The microenvironment of solid tumors contains the cancer cells and surrounding stromal cells as well as leukocytes that have infiltrated the tumor. The leukocytes are recruited by continuous signals sent out by the unregulated epithelial cells. They comprise innate immune cells (including macrophages, dendritic cells, neutrophils, mast cells, and NK cells) and adaptive immune cells (T and B lymphocytes). Under normal physiological circumstances, leukocytes are recruited in response to injury or infection and act as immune cells to heal the injury or combat the infection. In contrast, the leukocytes recruited by tumors are acted upon by an array of growth factors, proteases, and angiogenic mediators that promote tumor growth, angiogenesis, and metastasis. Thus, the tumor itself regulates the infiltration and activity of the leukocytes to promote its survival and progression.

14.3.12 Intercellular Communication through Gap Junction Channels

Certain cells seem to produce signal molecules which regulate the growth of cell populations. These putative molecules are transmitted to neighboring cells by diffusion through gap junction channels and never leave the intercellular space. Many studies have shown that intercellular communication is necessary for normal growth regulation and that cell proliferation occurs when intercellular communication is decreased. Inhibiting the gap-junctional flow of growth-controlling factors leads to neoplastic transformation of cells and the formation of cancerous tumors. The specific implication of intercellular communication in growth control has been demonstrated by experiments in which an exogenous gene for the gap junctional protein connexin43 was incorporated into communication-deficient cancer cells in culture. Expression of that gene led to the formation of functional gap junctions and concomitant increase in intercellular communication, and to inhibition of growth of these cancer cells (Mehta *et al.*, 1991). Furthermore, injection of exogenous connexin-expressing cells into mice suppressed tumor formation (Rose *et al.*, 1993).

REFERENCES

Akhurst, R.J. & Derynck, R. (2001) TGF-β signaling in cancer – A double-edged sword. *Trends in Cell Biology*, **11**, S44–51.

Balkwill, F. & Mantovani, A. (2001) Inflammation and cancer: Back to Virchow? *The Lancet*, **357**, 539–45.

Bos, J.L. (1989) *ras* oncogenes in human cancer: A review. *Cancer Research*, **49**, 4682–9.

Campisi, J. (1997) The biology of replicative senescence. *European Journal of Cancer*, **33**, 703–9.

Coussens, L.M. & Werb, Z. (2002) Inflammation and cancer. *Nature*, **420**, 860–7.

Downward, J. (2003) Targeting RAS signalling pathways in cancer therapy. *Nature Reviews Cancer*, **3**, 11–22.

Dranoff, G. (2004) Cytokines in cancer pathogenesis and cancer therapy. *Nature Reviews Cancer*, **4**, 11–22.

Gorelik, L. & Flavell, R.A. (2001) Immune-mediated eradication of tumors through the blockade of transforming growth factor-β signaling in T cells. *Nature Medicine*, **7**, 1118–22.

Grivennikov, S.I. & Karin, M. (2010) Inflammation and oncogenesis: A vicious connection. *Current Opinion in Genetics and Development*, **20**, 65–71.

Hanahan, D. & Weinberg, R.A. (2000) The hallmarks of cancer. *Cell*, **100**, 57–70.

Joyce, J.A. & Pollard, J.W. (2009) Microenvironmental regulation of metastasis. *Nature Reviews Cancer*, **9**, 239–52.

Lu, H., Ouyang, W. & Huang, C. (2006) Inflammation, a key event in cancer development. *Molecular Cancer Research*, **4**, 221–33.

Luo, J.-L., Maeda, S., Hsu, L.-C., Yagita, H. & Karin, M. (2004) Inhibition of NF-κB in cancer cells converts inflammation-induced tumor growth mediated by TNFα to TRAIL-mediated tumor regression. *Cancer Cell*, **6**, 297–305.

Mantovani, A., Allavena, P., Sica, A. & Balkwill, F. (2008) Cancer-related inflammation. *Nature*, **454**, 436–44.

Mehta, P.P., Hotz-Wagenblatt, A., Rose, B., Shalloway, D. & Loewenstein, W.R. (1991) Incorporation of the gene for a cell-to-cell channel protein into transformed cells leads to normalization of growth. *Journal of Membrane Biology*, **124**, 207–25.

Nguyen, D.X., Bos, P.D. & Massague, J. (2009) Metastasis: from dissemination to organ-specific colonization. *Nature Reviews Cancer*, **9**, 274–84.

Pepper, M.S. (1997) Manipulating angiogenesis: from basic science to the bedside. *Arteriosclerosis, Thrombosis and Vascular Biology*, **17**, 605–19.

Qiu, R.-G., Chen, J., Kirn, D., McCormick, F. & Symons, M. (1995) An essential role for Rac in Ras transformation. *Nature*, **374**, 457–9.

Rose, B., Mehta, P.P. & Loewenstein, W.R. (1993) Gap-junction protein gene suppresses tumorigenicity. *Carcinogenesis*, **14**, 1073–5.

Shacter, E. & Weitzman, S.A. (2002) Chronic inflammation and cancer. *Oncology*, **16**, 217–26.

Steeg, P.S. (2006) Tumor metastasis: Mechanistic insights and clinical challenges. *Nature Medicine*, **12**, 895–904.

Webster, G.A. & Perkins, N.D. (1999) Transcriptional cross talk between NF-κB and p53. *Molecular and Cellular Biology*, **19**, 3485–95.

Wilson, J. & Balkwill, F. (2002) The role of cytokines in the epithelial cancer microenvironment. *Seminars in Cancer Biology*, **12**, 113–20.

Won, J., Kim, H., Park, E.J., Hong, Y., Kim, S.J. & Yun, Y. (1999) Tumorigenicity of mouse thymoma is suppressed by soluble type II transforming growth factor β receptor therapy. *Cancer Research*, **59**, 1273–7.

Part III

The Fat-Soluble Vitamins

15 Vitamin A
Retinoids and the Provitamin A Carotenoids

15.1 CHEMISTRY AND BIOLOGICAL FUNCTIONS

Vitamin A-active compounds, defined as compounds having qualitatively the biological activity of retinol, are represented by retinoids and provitamin A carotenoids. The retinoids comprise retinol, retinaldehyde, and retinoic acid, together with their naturally occurring and synthetic analogs. The naturally occurring retinoids are sometimes referred to as preformed vitamin A because they do not require metabolic conversion, as do carotenoids, in order to become biologically active. Carotenoids are represented by β-carotene and chemically related pigments that are responsible for the color of many vegetables and fruits. From a nutritional viewpoint, these pigments are classified as provitamin A carotenoids and inactive carotenoids. In nature, carotenoids are synthesized exclusively by higher plants and photosynthetic microorganisms, in which they function as accessory light-harvesting pigments to chlorophyll. Although animals are unable to synthesize carotenoids, they can assimilate them through their diet.

The structures of some physiologically important retinoids are shown in Figure 15.1. The parent vitamin A compound, retinol, comprises a cyclohexenyl (β-ionone) ring attached at the carbon-6 position to a polyene side chain. The four double bonds in the side chain give rise to *cis/trans* isomerization. The predominant isomer, all-*trans* retinol, possesses maximal (100%) vitamin A activity and is frequently accompanied in foodstuffs by smaller amounts of 13-*cis* retinol, which has 75% relative activity. An aldehyde form, 11-*cis* retinaldehyde, is the chromophore in the retina of the eye, while all-*trans* and 9-*cis* retinoic acid are active metabolites of retinol found in most if not all tissues. Changes in the molecular state of oxidation and *cis/trans* isomerization are of physiological importance in modifying the biological activity of retinoids.

Carotenoids are classified chemically as carotenes, which are hydrocarbons, and xanthophylls, which have one or more oxygen-containing groups (e.g. hydroxyl) either on the ring or in the chain. Most naturally occurring carotenoids contain 40 carbon atoms. In some instances, C_{40}-carotenoids undergo partial oxidative cleavage in the plant tissues to give shortened molecules known as apocarotenoids. The majority of xanthophylls in plant tissues occur as mono or bis esters of saturated long-chain fatty acids (e.g. palmitic acid).

For a carotenoid to have vitamin A activity, its structure must incorporate a molecule of retinol, i.e. an unsubstituted β-ionone ring with an 11-carbon polyene chain. The most ubiquitous and nutritionally most important carotenoid, β-carotene (Figure 15.2), is composed of two molecules of retinol joined tail to tail, thus β-carotene possesses maximal provitamin A activity. The structures of all other provitamin carotenoids incorporate only one molecule of retinol, hence theoretically contribute 50% of the biological value of β-carotene. Over 500 naturally occurring carotenoids have been isolated and characterized; of these, about 50 possess provitamin A activity in varying degrees.

In plant and animal tissues the carotenoids are usually found associated with lipid fractions in non-covalent association with membranes and lipoproteins, and they accumulate in the chloroplasts of green leaves. In nature, carotenoids exist mainly as their all-*trans* forms. Food processing and preservation methods, especially canning, induce *cis-trans* isomerization, leading to a reduced vitamin A potency.

Vitamin A is required for several essential life processes, including hematopoiesis, bone development, pattern formation during embryogenesis, the maintenance of differentiated epithelia, and immunocompetence. These processes can be supported by all forms of vitamin A, including the provitamin A carotenoids. The other vitamin A-dependent processes, namely vision and reproduction, specifically require either retinol or retinaldehyde. Retinoic acid cannot support these functions because it cannot be reduced back metabolically to retinaldehyde. Hence animals maintained on retinoic acid as the only source of vitamin A become both blind and sterile, but are otherwise in good general health.

Retinoids promote the differentiation of a variety of cell lines in culture, including epithelial cells and chondrocytes. That this function occurs *in vivo* is demonstrated by the replacement of secretory cells in epithelia by keratin-producing cells when an animal is deprived of vitamin A. The effects of vitamin A on cellular differentiation are due to the control of gene expression by retinoic acid in selected tissues, the protein products being responsible for the effects. Not all vitamin A-responsive genes are up-regulated by retinoic acid; some are down-regulated.

In this text the precise terms 'retinoids' and 'carotenoids' are used where the specificity is known. The non-specific term 'vitamin A' is used in the sense of vitamin A activity

DOI: 10.1201/9781003410300-18

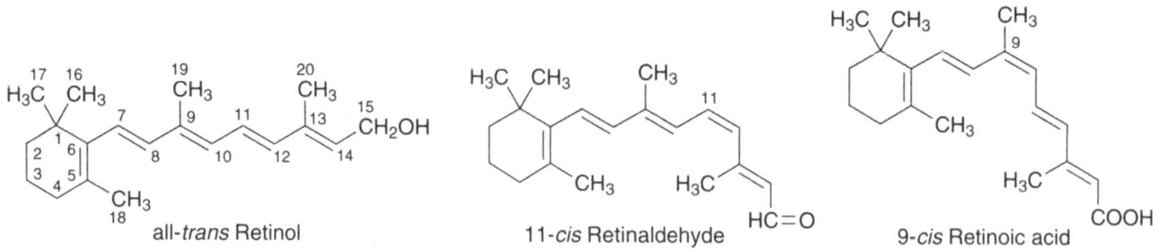

FIGURE 15.1 Structures of some physiologically important retinoids.

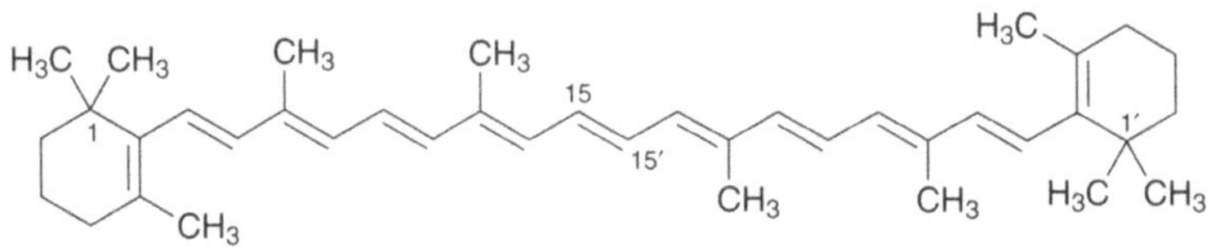

FIGURE 15.2 Structure of β-carotene.

and therefore accounts for provitamin A carotenoids as well as retinoids.

15.2 FOOD SOURCES

All natural sources of vitamin A in the diet are derived ultimately from provitamin A carotenoids. For much of the world's human population, and particularly in parts of the developing world, vegetables and fruits provide the main dietary sources of vitamin A in the form of β-carotene and other provitamin carotenoids. In other parts of the world, milk, butter, cheese, and eggs are important dietary sources of vitamin A. The liver of meat animals is a particularly rich source as this organ stores the vitamin for the body's use. Preformed vitamin A is present in animal tissues and in milk as a consequence of the enzymatic conversion of ingested provitamin carotenoids in the intestinal wall of the animal. Dietary preformed vitamin A consists mainly of retinol esterified with long-chain fatty acids, particularly palmitic acid. Retinyl esters are also found in processed foods supplemented with vitamin A.

15.3 OVERVIEW OF VITAMIN A METABOLISM

O'Byrne and Blaner (2013) have reviewed the various processes involved in vitamin A metabolism.

Retinol in the free state can enter the lipid bilayer of biological membranes and disrupt membrane structure and function. In order to permit transport within an aqueous environment and to limit its level in membranes, retinol (and other retinoids) are bound to proteins, both extracellularly and intracellularly. The intracellular retinol–protein complex is the metabolically active form of retinol. Alternatively, retinol is esterified with long-chain fatty acids for transport in lipoproteins or storage in cytoplasmic lipid droplets. Protein binding and esterification prevent the disruptive action of free retinol on membrane structure and function, whilst also protecting retinoids from unwanted metabolic processing and decomposition.

In the human, the main events that take place from absorption of dietary vitamin A to cellular metabolism in the target tissues are as follows. Ingested retinyl esters and carotenoids are incorporated into mixed micelles in the intestinal lumen and the retinyl esters are hydrolyzed. The micelles dissociate and the retinol and carotenoids are absorbed. Within the enterocytes, varying proportions of the provitamin carotenoids are converted first to retinaldehyde and then to retinol, which is subsequently esterified. Some of the retinaldehyde is oxidized irreversibly to retinoic acid. Retinoic acid can also be produced from β-carotene via apo-carotenals of varying chain length without involving retinaldehyde as an intermediate. The absorbed retinol is also esterified and the combined retinyl esters, accompanied by varying small amounts of unchanged carotenoids, are released into the bloodstream via the lymphatic system as components of chylomicrons. The absorption of β-carotene and retinyl ester and principal metabolic events within the enterocyte are shown diagrammatically in Figure 15.3.

The circulating chylomicrons undergo lipolysis and the resultant chylomicron remnants are taken up by the liver and to a lesser extent by extrahepatic tissues. Within the liver, the chylomicron remnant retinyl esters are hydrolyzed and most of the retinol is transferred from hepatocytes to stellate cells where it is esterified and stored. Carotenoids do not accumulate in liver cells: they are released into the circulation as components of very low-density lipoproteins (VLDLs), which are subsequently delipidated to low-density lipoproteins (LDLs). The entire LDL particle is taken up by the cells of peripheral tissues, particularly adipose tissue, where the carotenoids accumulate.

Upon demand, the retinyl esters in the liver are hydrolyzed and the retinol is released into the bloodstream bound to retinol-binding protein (RBP), which is synthesized in the liver. In the plasma the retinol–RBP complex forms

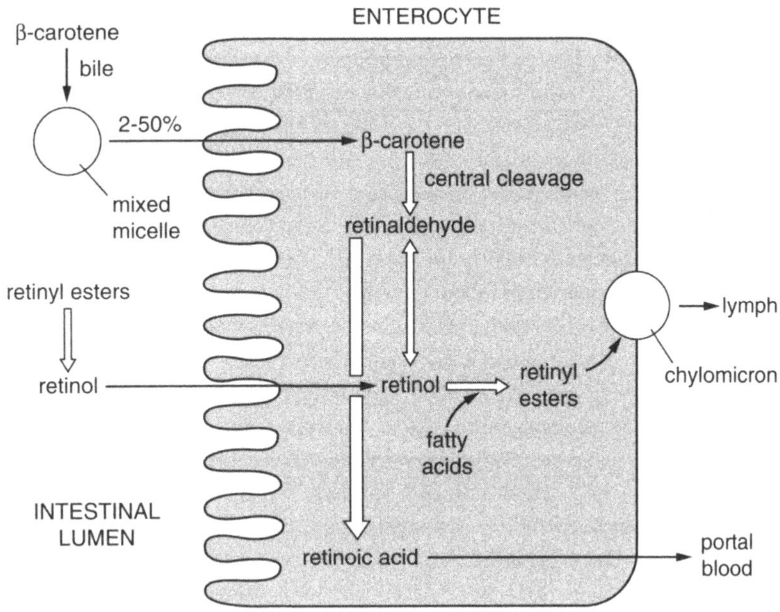

FIGURE 15.3 Intestinal absorption and metabolism of β-carotene and retinyl esters. See text for additional metabolic reactions.

a larger complex with a protein called transthyretin (formerly known as prealbumin), which also transports thyroid hormone. The retinol–RBP–transthyretin complex delivers retinol to a cell surface receptor expressed in vitamin A-requiring cells. RBP is recognized by the receptor and, after negotiating the lipid bilayer of the plasma membrane, the retinol interacts with a specific cellular-binding protein. Cytoplasmic retinol is subject to a variety of metabolic fates, including oxidation to retinoic acid. Cellular-binding proteins play major roles in controlling retinol metabolism and may regulate the movement of retinoic acid to the nucleus where it acts as a hormone to affect gene expression.

15.4 RETINOID-BINDING PROTEINS

A number of retinoid-binding proteins (Table 15.1) play an essential role in transporting the hydrophobic retinoids within aqueous extracellular, cytosolic, and nuclear compartments. The binding proteins in their free uncomplexed state are referred to as apo-proteins (e.g. apo-RBP); when complexed with their ligand they are referred to as holo-proteins (e.g. holo-RBP).

RBP is a single polypeptide of molecular weight 21 kDa, which specifically binds retinol for transport in the plasma. The protein is synthesized in the liver as a larger molecular weight pre-RBP, which is rapidly processed to the 21-kDa molecule by the removal of a 3.5-kDa peptide. RBP is also synthesized in a large number of extrahepatic tissues. The protein molecule has a barrel core that completely encapsulates a single retinol molecule, with the β-ionone ring lying deep in the hydrophobic pocket and the polyene tail extending almost to the surface. This arrangement protects the retinol from oxidation and exposure to enzymes during transport. The ligand is tightly bound and can only be removed *in vitro* by extraction with organic solvents or by denaturing the protein.

RBP possesses two additional binding sites: one for attachment to transthyretin and the second for linking to receptors on the surface of target cells. Transthyretin, which is also synthesized in the liver, is a tetrameric protein composed of apparently identical subunits, each with a molecular weight of 14 kDa, and possesses a single thyroid hormone-binding site. Although transthyretin would seem to be capable of binding four molecules of RBP, it only binds one. Apparently, the binding of one molecule of RBP to one transthyretin subunit causes the remaining subunits to lose their affinity for RBP. Hence, under normal physiological conditions, with transthyretin in molar excess, holo-RBP–transthyretin exists as a 1:1 molar complex. In blood, approximately 95% of the retinol–RBP is complexed with transthyretin – the rest is free. As will become evident, retinol nutritional status strikingly influences the metabolism and tissue levels of RBP.

The cellular retinoid-binding proteins (CRBPs) facilitate normal vitamin A metabolism by delivering retinoids to the appropriate enzyme reaction sites within the cell. These cytoplasmic proteins are distinct from plasma RBP. The four cellular proteins for binding retinol and retinoic acid share a high degree of homology and have molecular weights in the range of 15.0 to 15.7 kDa. CRBP-I and cellular retinoic acid-binding protein (CRABP)-I have been found in all rat tissues and organs tested, suggesting that they are involved in essential cellular processes throughout the body. The levels vary widely among different tissues, the highest concentrations being found in tissues of the reproductive tract. They have an additional role in delivering their respective ligands to specific binding sites within the nucleus. CRABP-I also sequesters the retinoic

TABLE 15.1
Principal Retinoid-Binding Proteins Involved in Vitamin A Metabolism

Binding Protein	Major Ligand	Location
Intercellular transport		
Retinol-binding protein, RBP	all-*trans* retinol	Synthesized in hepatocytes Circulates in plasma
Interphotoreceptor or interstitial retinol-binding protein, IRBP cells	all-*trans* and 11-*cis* retinol	Matrix between pigment epithelium and photoreceptor of the retina
Intracellular transport		
Cellular retinol-binding protein epididymis, liver, type I, CRBP-I	all-*trans* retinol	Proximal kidney (1) Bone (2)
Cellular retinol-binding protein (enterocytes) type II, CRBP-II	all-*trans* retinol and retinaldehyde	Intestine Liver and intestine of developing embryo
Cellular retinoic acid-binding protein type I, CRABP-I	all-*trans* retinoic acid	Developing embryo Seminal vesicle, vas deferens, skin (1)
Cellular retinoic acid-binding protein type II, CRABP-II	all-*trans* retinoic acid	Developing embryo Adult skin cells
Cellular retinaldehyde-binding protein, CRALBP	11-*cis* retinaldehyde	Retina

(1) Kato *et al.* (1985). CRBP-I and CRABP-I were found in every rat tissue tested (21 tissues in males, 18 in females). Those listed are the top three locations.
(2) Harada *et al.* (1995).

acid that is in excess of cellular needs, and facilitates its catabolism into a variety of polar metabolites. Tissue distribution of CRBP-II and CRABP-II is much more restricted. The former is found mainly in the enterocytes of the small intestine, where it is involved in the metabolism of newly absorbed or newly formed retinol. The unique proteins, cellular retinaldehyde-binding protein (CRALBP) and interphotoreceptor retinol-binding protein (IRBP), are found only in visual tissue; their respective molecular weights are 36.0 and 135.0 kDa.

There is a problem in studies of retinoid metabolism of promiscuity of retinoids *in vitro*. For example, a plethora of non-specific enzymes, including alcohol dehydrogenases, convert 'free' retinol into retinoic acid *in vitro*, but have no access to the protein-bound retinol *in vivo*. To ensure that only the physiologically relevant enzymes catalyze retinoid metabolism in the body, the substrate is not the retinoid *per se*, but the holo-cellular retinoid-binding protein, and the enzyme recognizes the substrate through protein–protein interaction. The enzyme forms a transient complex with the holo-binding protein and the retinoid is transferred directly from the binding protein to the active site of the enzyme without diffusing into the aqueous medium (Napoli, 1993).

With regard to CRBP-I and CRABP-I, the amount and availability of their respective ligands, retinol, and retinoic acid does not influence the tissue levels of these cellular-binding proteins. Thus a low dietary supply of vitamin A is accompanied by a concomitant increase in the level of apo-CRBP-I relative to holo-CRBP-I – this is an important concept in retinol metabolism. It has been demonstrated that tissue levels of both CRBP-I and CRABP-I are maintained constant in the face of considerable differences in retinoid nutritional status (Kato *et al.*, 1985; Blaner *et al.*, 1986). However, tissue levels of CRBP-I (but not CRABP-I) are decreased if the diet is totally devoid of retinoid.

15.5 INTESTINAL ABSORPTION

Harrison (2012) has reviewed mechanisms involved in the intestinal absorption of dietary vitamin A and provitamin A carotenoids.

Ingested retinyl esters and provitamin A carotenoids are liberated from their association with membranes and lipoproteins by the action of pepsin in the stomach and proteolytic enzymes in the small intestine. In the stomach the free carotenoids and retinyl esters congregate in fatty globules, which then pass into the duodenum. In the presence of bile salts, the globules are broken up into smaller globules, which renders them more easily digestible by a variety of pancreatic lipases and results in the formation of mixed micelles. Extensive hydrolysis of retinyl esters takes place within the duodenum, catalyzed mainly by a non-specific pancreatic hydrolase that can act on a wide variety of esters as substrates. Retinyl ester hydrolysis is completed by brush-border hydrolases. With the exception of esterified xanthophylls, which are hydrolyzed by esterases, carotenoids are absorbed without prior metabolic conversion.

The retinol and carotenoids contained within the mixed micelles cross the unstirred layer of the intestinal lumen and are released as a result of micelle dissociation in the brush-border region. Physiological concentrations of retinol derived from natural food sources are absorbed by facilitated diffusion; at higher concentrations a process of simple diffusion takes over. The carrier-mediated absorption of retinol shows specificity toward all-*trans* retinol and 3-dehydroretinol; uptake of 9-*cis* and 13-*cis* retinol and retinaldehyde takes place by simple diffusion (Dew & Ong, 1994). *In vitro* studies in Caco-2 cells have suggested that carotenoid uptake by enterocytes is a facilitated process (During *et al.*, 2002); however, no transporter proteins have as yet been identified for carotenoids.

Preformed vitamin A and provitamin A carotenoids (provitamins) of dietary origin differ in their efficiency of absorption. When foods containing normal physiological amounts of these compounds are ingested, retinol is absorbed with an efficiency of 70–90% compared with 20–50% for the provitamins. The absorption efficiency of retinol remains high as the amount ingested increases beyond physiological levels, whereas that of the provitamins falls markedly with increased ingestion to less than 10%. Whereas crystalline β-carotene in antioxidant-stabilized commercial form is absorbed with an efficiency of about 50%, the absorption of carotenoids from raw carrot can be as low as 1%. The cooking of vegetables increases absorption, probably as a result of dissociation of carotenoids from plant cell membranes and lipoproteins.

In a human study of β-carotene absorption, Blomstrand and Werner (1967) fed single small doses of radioactive β-carotene dissolved in vegetable oil to hospitalized patients in whom cannulae had been inserted into the thoracic duct. The percentages of administered radioactivity recovered in the lymph of three patients were 14.6, 8.7, and 16.8%. In these patients most of the radioactivity was found in the retinyl ester fraction which contained 68 to 88% of the total radioactivity in the lymph lipids. Smaller amounts of radioactivity (1.7, 11.3, and 27.9%) were recovered in the β-carotene fractions. These results demonstrate that only a part of β-carotene consumed is absorbed in the intestine. Most of that absorbed is converted in the enterocytes to retinaldehyde and further to retinol and retinyl esters; some remains as intact β-carotene.

There is a large variability, three- to fourfold among healthy male humans, in efficiency of carotenoid absorption (Brown *et al.*, 1989). In some people there is no plasma response to a single oral dose of β-carotene. Johnson and Russell (1992) measured β-carotene concentrations in plasma and various lipoproteins in healthy males for ten days after a single oral dose (120 mg) of β-carotene in capsule form. Seven of the 11 subjects were non-responders, showing little or no increase in plasma β-carotene and only a small response in chylomicrons. This lack of response may be caused by inefficient uptake of luminal β-carotene by enterocytes, inefficient incorporation of β-carotene into chylomicrons, or extensive conversion of β-carotene to retinyl esters. Of interest with the four responders was that surges of chylomicron β-carotene occurred every few days following the single dose, suggesting delayed release of β-carotene from enterocytes. It is possible that these surges were the result of re-uptake of β-carotene from the intestinal lumen following sloughing off of the epithelial cells.

15.6 METABOLIC EVENTS IN THE INTESTINE

15.6.1 Esterification of Retinol

Within enterocytes retinol becomes bound in a 1:1 molar ratio to CRBP-II, which is present exclusively and abundantly in these cells. CRBP-II binds all-*trans* and 13-*cis* retinol with high affinity; it also binds retinaldehyde, but not retinoic acid. The protein-bound retinol is esterified with saturated long-chain fatty acids, preferentially palmitic acid. The esterification uses a different pool of fatty acids and hence different enzymes than are used for the synthesis of triglycerides. Two microsomal enzymes are involved in the esterification of retinol, namely acyl coenzyme A (CoA):retinol acyltransferase (ARAT) and lecithin:retinol acyltransferase (LRAT). The substrates for the ARAT-catalyzed reaction are free retinol and acyl-CoAs; retinol bound to CRBP-II is not a substrate for ARAT. LRAT is an unusual enzyme in that it utilizes a membrane phospholipid, phosphatidylcholine (lecithin), as an endogenous donor of fatty acids for esterification. The enzyme activity shows positional selectivity as only the fatty acid from position 1 of the phospholipid is transferred to retinol. Position 1 is usually occupied by a saturated fatty acid (palmitic or stearic acid), which explains the predominance of saturated fatty acids in the retinyl esters. Unlike ARAT, LRAT can utilize CRBP-II-bound retinol as well as free retinol. However, because of the abundance of CRBP-II in enterocytes, the majority of retinol will be bound and thus restricted to esterification by LRAT.

Whether ARAT or LRAT is involved in retinol esterification depends on the amount of available retinol. LRAT is responsible for the esterification of retinol–CRBP-II when normal loads of vitamin A are ingested. In contrast, ARAT activity is important if the amount of retinol absorbed exceeds the saturation level of CRBP-II. It seems that CRBP-II functions both to direct retinol to the microsomes for esterification by LRAT and to prevent retinol from participating in the ARAT reaction.

15.6.2 Conversion of Provitamin A Carotenoids to Retinoids

A scheme of the intestinal metabolism of β-carotene is shown in Figure 15.4 (Sharma *et al.*, 1977). Both symmetric and asymmetric oxidative cleavage of carotenoids take place within the enterocyte. Symmetric cleavage at the central 15,15′ double bond of β-carotene produces two molecules of retinaldehyde; other provitamin A carotenoids produce one molecule of retinaldehyde. The enzyme responsible for the central cleavage of β-carotene was cloned independently by von Lintig and Vogt (2000) and Wyss *et al.* (2000) and shown to be a dioxygenase (dela Seña *et al.*,

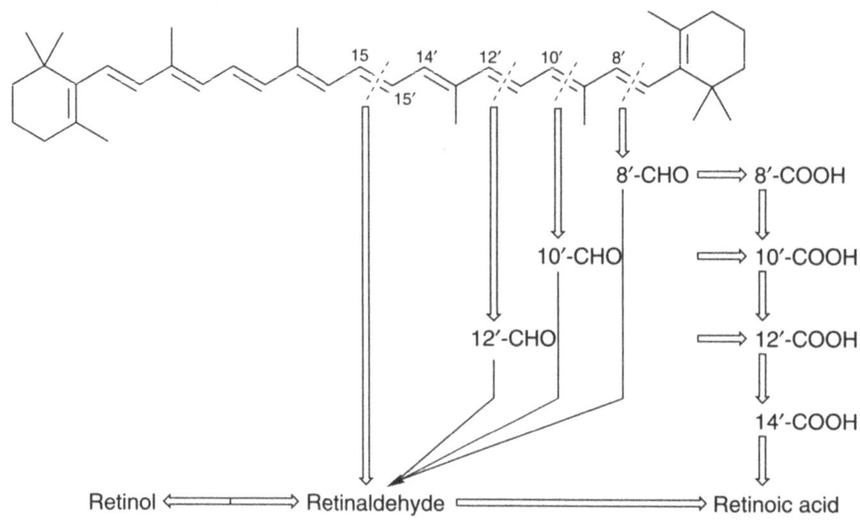

FIGURE 15.4 Intestinal metabolism of β-carotene. The enzyme β-carotenoid-15,15′-dioxygenase forms retinaldehyde directly. Cleavage at other double bonds forms β-apocarotenals (e.g. 8′-CHO), which can be shortened to retinaldehyde. β-Apocarotenals may be oxidized to β-apocarotenoic acids (e.g. 8′-COOH), which can form retinoic acid. Retinol is esterified, incorporated in chylomicrons together with some intact β-carotene, and secreted into lymph. Retinoic acid enters portal blood accompanied by other polar metabolites. Reprinted from Sharma *et al.*, Studies on the metabolism of β-carotene and apo-β-carotenoids in rats and chickens, *Biochimica et Biophysica Acta*, Vol. 486, pp. 183–94, © 1977, with permission from Elsevier. Scheme 1 on p. 193.

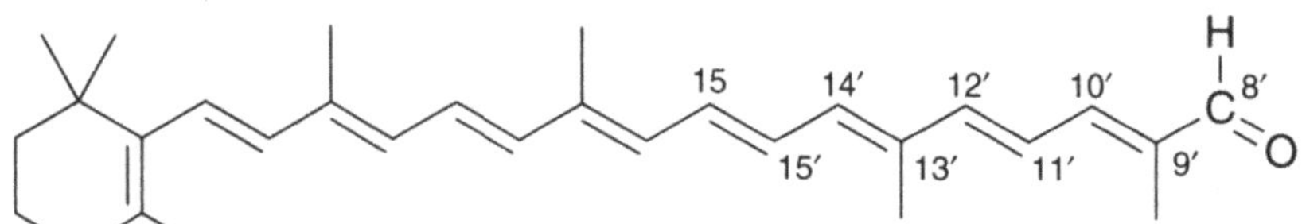

FIGURE 15.5 Structure of β-apo-8′-carotenal.

2014). Its accepted name (IUBMB enzyme nomenclature) is β-carotene 15,15′-dioxygenase (EC 1.13.11.63).

In the asymmetric cleavage of β-carotene shown in Figure 15.4, one molecule of β-carotene produces β-apocarotenal molecules with different chain lengths, all of which can be shortened to retinaldehyde (Glover, 1960). The initial reaction is enzymatic cleavage of the terminal 7′-8′ double bond of β-carotene to produce β-apo-8′-carotenal (Figure 15.5). The β-apocarotenals are oxidized to their corresponding β-apocarotenoic acids and these in turn undergo a stepwise oxidation to retinoic acid (Wang *et al.*, 1996). Studies using whole intestinal homogenates showed that only trace amounts of β-apocarotenals were detected, and the conversion of β-apo-8′-carotenal to retinaldehyde was very slow (Nagao *et al.*, 1996). It is generally agreed that the enzymatic central cleavage of β-carotene plays the major role in β-carotene metabolism under normal health conditions.

Most of the retinaldehyde formed from carotenoids becomes bound to CRBP-II and reversibly reduced to retinol by retinaldehyde reductase – a relatively non-specific aldehyde reductase which does not appear to be zinc-dependent (Fidge & Goodman, 1968). The resulting retinol–CRBP-II complex is then used as a substrate for esterification by LRAT.

The bioconversion of provitamins to retinoids may be regulated both up and down at the level of the intestinal cleavage enzyme. Using a dioxygenase assay, van Vliet *et al.* (1992) found a 130% higher cleavage activity in hamsters fed a low vitamin A diet compared with normally fed controls. This up-regulation was confirmed in rats by van Vliet *et al.* (1996), who also found that a high intake of either retinyl ester or β-carotene down-regulated cleavage activity.

15.7 METABOLIC EVENTS IN THE LIVER

15.7.1 Hepatic Uptake of Chylomicron Remnants

The chylomicrons, containing retinyl esters and small amounts of intact carotenoids, are released from enterocytes by exocytosis into the lymph. They enter the bloodstream where lipase activity and apolipoprotein exchange result in their conversion to chylomicron remnants, which are taken up primarily by the parenchymal cells (hepatocytes) of the liver. Uptake appears to involve binding of the chylomicron remnants to lipoprotein receptors in the space of Disse, followed by lipolytic processing and receptor-mediated endocytosis. After uptake of chylomicron remnants, the constituent carotenoids are incorporated into VLDLs, which are released into the circulation without delay. The retinyl esters are hydrolyzed, probably by a hydrolase located in the plasma membrane of hepatocytes and/or in early endosomes. In contrast to many other ligands that are transferred to lysosomes after endosomal processing, retinol is transferred from endosomes to the endoplasmic reticulum, where apo-RBP is found in high concentration. Whether this transfer occurs via vesicular transport or via CRBP-I in the cytosol is not known.

Although chylomicron remnants are mainly cleared by the liver, uptake of remnants also takes place in the bone marrow and spleen, and to a lesser extent in adipose tissue, skeletal muscle, testes, lungs, and kidneys. Considering the role of retinoids in cell differentiation, chylomicron remnants may be important for delivering retinyl esters to tissues with intensive cell proliferation and differentiation – tissues such as bone marrow and spleen.

15.7.2 Esterification of Retinol

Within hepatic stellate cells, retinol binds to CRBP-I which directs retinol to LRAT for esterification when the vitamin is present in normal amounts. When retinol is present at high levels and the pool of CRBP-I becomes saturated, ARAT may esterify the excess. Thus, as in the intestine, both enzymes may be involved in hepatic retinol esterification, depending on the amount of retinol present and on whether it is bound to CRBP-I.

15.7.3 Storage of Retinyl Esters

Vitamin A is unique among vitamins in that it is massively stored by the liver. The level of stored retinyl ester fluctuates in accordance with dietary intake. Normally, most of the newly absorbed retinol is transferred within 2 to 4 hours from hepatocytes to stellate cells for storage. However, during periods of vitamin A insufficiency, newly absorbed retinol is secreted from hepatocytes, as holo-RBP, directly into the blood to satisfy the immediate needs of the tissues.

The transfer of newly absorbed retinol from hepatocytes to stellate cells within the liver is mediated by RBP; thus other components of chylomicron remnants, such as cholesterol and vitamin D, are not transferred. Binding of retinol to RBP in hepatocytes initiates a translocation of holo-RBP from the endoplasmic reticulum to the Golgi apparatus, followed by secretion of the retinol complex from the cell. Within stellate cells, retinol is esterified and the esters are stored in cytoplasmic lipid droplets. The storage capacity of hepatic stellate cells is high and accounts for about 50 to 80% of the total body pool of vitamin A. The normal reserve of vitamin A in stellate cells is adequate to last for several months in humans. Stellate cells are also found in the intestine, kidney, heart, large blood vessels, ovaries, and testes; these cells store retinyl esters when large amounts of vitamin A are consumed. When the stellate cells contain so much retinol that they can accept no more, hypervitaminosis A occurs.

The amount of vitamin A stored in the liver influences retinol utilization by extrahepatic tissues, and therefore hepatic liver reserves are a true indication of vitamin A status. Green *et al.* (1987) determined the retinol utilization rate in rats provided with different intakes of vitamin A, such that the rats had low, marginal, or high liver vitamin A reserves. Vitamin A-depleted rats exhibited a lower utilization rate, which was positively correlated with the size of the plasma retinol pool; i.e. the lower the plasma retinol concentration, the lower the vitamin A utilization rate. The increased rate of utilization observed in rats of higher vitamin A status was reflected in an increased rate of retinol catabolism. It appeared that some minimal utilization rate is maintained as long as dietary supply and/or liver stores of vitamin A can maintain normal plasma retinol concentrations. The decreased utilization rate in depleted states could be a way of conserving vitamin A for its most critical functions, whereas in vitamin A sufficiency increased catabolism prevents excessive accumulation of retinol. Accelerated catabolism as a function of increase in liver vitamin A stores was also reported in rats fed excessive amounts of vitamin A (Leo *et al.*, 1989a).

15.7.4 Mobilization of Retinol and Carotenoids

The vitamin A stored in stellate cells can be readily mobilized for use in a highly regulated process. Thus an individual's plasma vitamin A levels remain quite constant over a wide range of dietary intakes and liver stores. Only when liver reserves of vitamin A are nearly depleted do plasma concentrations of retinol decrease significantly. A kinetic model of retinol dynamics in rats (Green *et al.*, 1993) predicted that the stellate cell retinol pool responsible for the secretion is small and rapidly turning over. This is compatible with the relatively small amounts of apo-RBP observed in stellate cells (more than 90% of the apo-RBP in liver is found in hepatocytes). Upon demand, the retinyl esters are hydrolyzed to retinol, which then combines with RBP to form holo-RBP. This complex is secreted into the bloodstream where it becomes reversibly complexed in a 1:1 molar ratio with transthyretin. The formation of the larger retinol–RBP–transthyretin complex minimizes the loss of holo-RBP in the urine during its passage through the kidney.

One factor that specifically regulates the release of holo-RBP from the liver is the dietary intake of vitamin A. Studies in the rat (Muto *et al.*, 1972; Smith *et al.*, 1973) showed that in the retinol-depleted state, the secretion of holo-RBP from the liver was blocked, resulting in the accumulation of an enlarged pool of apo-RBP in the liver and a concomitant decline in plasma RBP levels (as holo-RBP–transthyretin). Oral administration of retinyl acetate to the depleted rats stimulated the rapid secretion of holo-RBP from the expanded liver pool into the plasma and within 5 hours after administration the plasma RBP levels had returned to normal. This effect of retinol upon RBP secretion took place without affecting RBP synthesis (Soprano *et al.*, 1982); this is unusual as the synthesis of other binding proteins (e.g. transferrin, ferritin, and zinc metallothionein) is controlled by their specific ligands. These experiments demonstrated that the livers of vitamin A-deficient animals contain a pool of previously formed RBP, which can be released rapidly into the plasma, as holo-RBP, as soon as vitamin A becomes available. The delayed (5-hour) response to oral administration of vitamin A is due to the processes of intestinal absorption, hepatic uptake of chylomicron remnant retinyl ester, and hydrolysis of ester to provide retinol.

The rapid release of holo-RBP from the liver in response to vitamin A dosing during vitamin A deficiency provides the basis of the relative dose response (RDR) test for assessing vitamin A nutritional status. The test measures the amount by which plasma retinol levels are below normal due to inadequate liver stores (Loerch *et al.*, 1979). In subjects with low hepatic vitamin A reserves, a test dose of vitamin A will provoke a prominent increase in plasma retinol because of accumulated apo-RBP in the liver.

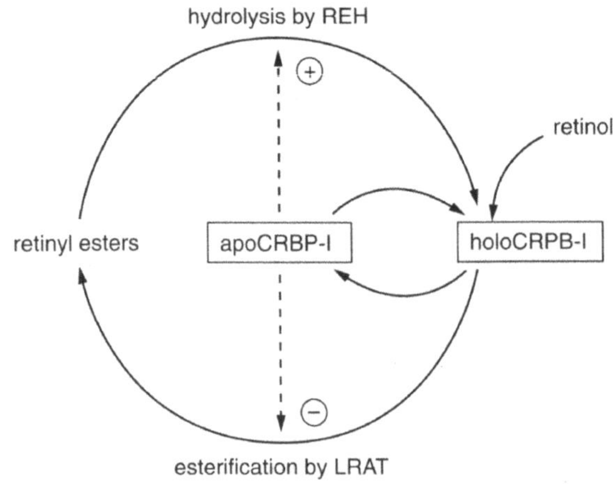

FIGURE 15.6 Effect of the apo-/holo-CRBP-I ratio on retinol esterification and hydrolysis in the liver. Holo-CRBP-I, which is formed as a result of an influx of exogenous retinol, serves as substrate for LRAT to facilitate liver storage of retinyl esters. Apo-CRBP-I, resulting from depletion of holo-CRBP-I, stimulates retinyl ester hydrolase (REH) and inhibits LRAT, leading to mobilization of vitamin A.

15.8 REGULATION OF RETINOL METABOLISM IN THE LIVER AND INTESTINE

The activity of hepatic LRAT is regulated by vitamin A nutritional status. LRAT activity decreased to undetectable levels in the livers of vitamin A-depleted rats and returned to normal levels on repletion with an oral dose of retinyl ester (Randolph & Ross, 1991). This decrease would inhibit liver storage of vitamin A in the chronically deficient state. Retinoic acid was shown to be more potent than an equimolar dose of retinol in restoring hepatic LRAT activity (Matsuura & Ross, 1993). The dose of retinoic acid required to restore the activity (2–20 µg) is consistent with physiological concentrations of this retinoid in tissues. The increase in LRAT activity in response to retinoic acid dosing was blocked completely by both actinomycin D and cycloheximide, which are inhibitors of RNA and protein synthesis, respectively. It can be inferred from these results that the vitamin A metabolite, retinoic acid, regulates the esterification of newly available retinol in the liver through an induction of LRAT, which requires synthesis of both new RNA and new protein. In contrast, neither the activity of intestinal LRAT nor that of ARAT in either liver or intestine was affected by vitamin A deficiency (Randolph & Ross, 1991).

In vitro experiments using liver microsomes have shown that when the ratio of apo-CRBP-I to retinol exceeds 1, apo-CRBP-I inhibits the esterification of retinol by LRAT (Herr & Ong, 1992). Working in concert with the inhibition of hepatic LRAT is the ability of apo-CRBP-I to stimulate retinyl ester hydrolysis in liver microsomes (Boerman & Napoli, 1991). The changing ratio of apo- to holo-CRBP-I during variable vitamin A status could regulate vitamin A storage and mobilization in the liver as shown in Figure 15.6. During times of low vitamin A intake or deficiency, when apo-CRBP-I will predominate over holo-CRBP-I, hydrolysis of stored retinyl esters will be stimulated and esterification of retinol will be inhibited, leading to mobilization of vitamin A. Any retinol arriving at the liver during such times would be immediately released, bound to RBP, for distribution to other organs, rather than stored in the liver as retinyl esters. When the supply of exogenous retinol becomes plentiful, holo-CRBP-I will predominate over apo-CRBP-I, hydrolysis of esters will not be stimulated, and synthesis of esters will not be inhibited, leading to net storage. The inhibitory and stimulatory effects of apo-CRBP-I on retinol esterification and ester hydrolysis respectively provide a mechanism for the animal to respond more rapidly to changes in intracellular retinol levels.

In the intestine, esterification of retinol is essential for completion of the absorptive process because the esters are then incorporated into chylomicrons for release and distribution to the rest of the body. Not surprisingly, therefore, apo-CRBP-II has relatively little effect on the rate of esterification. The difference between hepatic and intestinal LRAT in their response to vitamin A deficiency is attributable to the different ways in which the holo forms of CRBP-I and CRBP-II interact with LRAT. This difference fulfills the physiological needs of the animal during vitamin A deficiency by allowing efficient intestinal processing of the low vitamin A intake, while minimizing the storage of retinyl esters in the liver.

15.9 RECYCLING OF PLASMA RETINOL

For many years it was assumed that, once retinol left the plasma, it was taken up by tissues and irreversibly utilized. It has now been established that retinol recycles extensively among liver, plasma, interstitial fluid, and extrahepatic tissues. Recycling is a means of conserving vitamin A by the body and, indeed, excretory loss of the vitamin in vitamin A-depleted animals is much reduced relative to the loss in vitamin A-sufficient animals. The rate of plasma retinol turnover is nine to ten times the rate of irreversible utilization of retinol, and this ratio is not affected by vitamin A status. This means that a relatively small and constant fraction (approximately 10%) of retinol molecules that leaves the plasma is irreversibly utilized, and approximately 90% is recycled to plasma for distribution to liver and to other target organs or tissues. More than half of the retinol secreted by the liver as holo-RBP is provided by recycled plasma retinol.

Retinol recycling involves cellular uptake of circulating retinol, interaction with CRBP, possible esterification and hydrolysis, and then a transfer to apo-RBP and secretion of holo-RBP. Changes in the dynamics of retinol recycling allow for rapid adjustment in vitamin A distribution in response to changes in nutritional, metabolic, or physiological conditions. The vehicle for retinol recycling is RBP, which is synthesized in adequate amounts in a wide variety of extrahepatic tissues, including the kidneys and adipose tissue, as well as in the liver. Little of the RBP associated with retinol recycling to the liver is degraded in the liver; rather it too is recycled to the plasma. Thus there are adequate amounts of apo-RBP in plasma to complex with retinol that has been reabsorbed by the kidney.

15.10 HOMEOSTASIS OF CIRCULATING RETINOL

Circulating retinol concentrations are homeostatically regulated to remain constant, despite great variations in the dietary supply and liver stores of vitamin A. The controlled release of vitamin A from liver stores is necessary to provide tissues with optimal amounts of retinol, without releasing excessive amounts which would lead to toxicity. Although various schemes for the homeostasis of circulating retinol have been proposed, an unequivocal mechanism has not yet been established.

The concentration of retinol bound to plasma RBP is maintained within a normal range of concentrations, referred to as its homeostatic set point, as long as there is some minimal concentration of vitamin A in the liver and in extrahepatic tissues. Thus rats fed a vitamin A-deficient diet maintained a relatively stable plasma retinol level, which did not drop below 30 μg/dl until liver reserves fell below 10 μg/g tissue (Underwood *et al.*, 1979). Control of plasma retinol concentration is mediated by factors that affect the balance between retinol input to plasma and retinol output from plasma. Controlling factors include the enzymes that esterify retinol and hydrolyze retinyl esters in the tissues. The activity of hydrolytic enzymes is enhanced during vitamin A deprivation, releasing holo-RBP into the bloodstream. The activity of esterifying enzymes, on the other hand, is enhanced when vitamin A intake is plentiful, allowing surplus vitamin A to be stored. Ultimately, the set point for plasma retinol depends on the rate of release of holo-RBP from the liver. Underwood *et al.* (1979) proposed that hepatic secretion of retinol is controlled by a signal generated in proportion to the uptake or utilization of retinol in extrahepatic target tissues.

The plasma retinol homeostatic set point is influenced by several dietary and hormonal factors; these include protein, calorie and zinc nutriture, and fluctuating steroid hormone levels that occur during the estrous cycle or as a result of stress. It is likely that steroid hormones act by influencing the synthesis of RBP (Borek *et al.*, 1981). Ahluwalia *et al.* (1980) used hypophysectomized rats to demonstrate that without growth there was no vitamin A utilization. They showed that, in addition to dietary protein, growth hormone was required for mobilization of liver vitamin A stores. The data suggested that growth hormone may play an important role in vitamin A homeostasis by regulating retinol entry at the tissue level.

It is well documented that humans with chronic renal failure have elevated plasma levels of retinol. Gerlach and Zile (1990), using rats with surgically induced acute renal failure, established that the rise in plasma retinol is almost entirely due to retinol in the retinol–RBP–transthyretin complex. Gerlach and Zile (1991a) provided evidence that this retinol is derived from the hepatic pool of retinol newly acquired from the diet. When hepatic release of retinol–RBP was blocked by colchicine, the expected rise in plasma retinol was much reduced. Further, neither peripheral tissue uptake nor hepatic cycling of retinol were altered by the short-term absence of renal function. These findings suggest that the kidney has a physiological role in regulating the homeostatic set point for circulating retinol concentrations.

Gerlach and Zile (1991b) postulated the following regulatory mechanisms. (1) The intact kidney provides a specific regulatory substance (negative feedback signal) which prevents the release of hepatic retinol–RBP; in renal failure, the decreased signal will allow the release of retinol–RBP. (2) A regulatory substance originating in the peripheral tissues (positive feedback signal) is normally removed by the kidney and therefore hepatic retinol–RBP will not be released; in renal failure the substance will accumulate in the circulation and elicit the release of retinol–RBP. Investigations by Gerlach and Zile (1991b) led them to favor the second regulatory mechanism, with apo-RBP acting as the positive feedback signal molecule. They added apo-RBP to the circulation of rats with renal failure and observed an increase in plasma retinol concentration above that already caused by the renal failure. This was evidence that apo-RBP is a positive physiological feedback signal from the periphery for the regulation of release of hepatic retinol–RBP into circulation.

15.11 UPTAKE OF CIRCULATING RETINOL AND CAROTENOIDS BY PERIPHERAL TISSUE CELLS

Because of the disruptive effects of free (uncomplexed) retinol on membrane structure and function (Roels *et al.*, 1969), it is desirable to discourage non-specific uptake of retinol by cells that have no particular requirement for it.

This might be achieved by means of specific RBP receptors on the plasma membrane of vitamin A-requiring cells, which recognize holo-RBP, bind it, and facilitate the transfer of holo-RBP or free retinol across the membrane and into the cytoplasm.

RBP binding to cell membranes has been reported in the following specific locations: the choroidal surface of retinal pigment epithelial cells, but not in rod outer segments (Heller, 1975); interstitial cells of testes, but not on epithelial cells of seminiferous tubules (McGuire *et al.*, 1981); human placental cells (Sivaprasadarao & Findlay, 1988a); and epithelial cells of the choroid plexus, signifying movement of retinol across the blood–brain barrier (MacDonald *et al.*, 1990). Evidence of RBP receptor-mediated uptake of retinol has also been reported in enterocytes of the small intestine (Rask & Peterson, 1976) and testicular Sertoli cells (Shingleton *et al.*, 1989). Among these locations, cellular uptake of retinol without a concomitant uptake of the RBP has been reported to take place in retinal pigment epithelium (Chen & Heller, 1977), enterocytes (Rask & Peterson, 1976), and Sertoli cells (Shingleton *et al.*, 1989). The latter authors found that the amount of retinol accumulated by cultured Sertoli cells from holo-RBP was approximately equal to the cellular content of CRBP-I. This implied that the ligand saturation of CRBP-I may be the factor regulating the cellular uptake of retinol.

There appear to be two entirely different mechanisms for receptor-mediated uptake of circulating retinol by the cells of target tissues. The first mechanism applies to the tissues mentioned above, at least some of which have been shown to take up retinol unaccompanied by RBP. Upon recognition and binding by a cell-surface receptor, holo-RBP releases its retinol molecule and, as a result, undergoes a conformational change which reduces its affinity for both the receptor and transthyretin. The transthyretin portion of the holo-RBP– transthyretin complex is not involved in the recognition process. Indeed, transthyretin has been found to inhibit the binding of RBP to plasma membranes *in vitro* (Sivaprasadarao & Findlay, 1988a). Retinol is internalized into the cytoplasm where it interacts with its specific cytoplasmic-binding protein, CRBP-I. The low-affinity form of apo-RBP is circulated to the kidneys where it is filtered at the glomerulus, endocytosed in the cells of the proximal convoluted tubules, and degraded in lysosomes within the tubular cells.

Sivaprasadarao and Findlay (1988b) suggested that the ratio of apo- to holo-RBP levels in the plasma can regulate retinol distribution among various tissues. Vitamin A deficiency can effect a change in this ratio: holo-RBP levels decrease to almost zero, while apo-RBP levels remain unchanged. Since apo-RBP has poor affinity for transthyretin, most of it will exist in the free state. As both holo- and apo-RBP bind to the cell surface receptor, the high ratio of apo- to holo-RBP in the plasma would not only result in the decreased uptake of retinol by most tissues, but actually stimulate the secretion of retinol by extrahepatic tissues. The retinol so liberated might then be used for critically dependent tissues such as the eye and gonads.

The second mechanism of retinol uptake, exemplified in the liver and kidney, also involves the interaction of holo-RBP with a specific cell-surface receptor, but the entire retinol–RBP complex is internalized by receptor-mediated endocytosis. Evidence for this uptake mechanism was provided by Senoo *et al.* (1990), who studied the *in vivo* uptake of RBP in rat liver cells by immunocytochemistry at the electron microscopic level. Rats were injected intravenously with human RBP and simultaneously, for comparison, with asialo-orosomucoid, a protein known to be taken up by hepatocytes by receptor-mediated endocytosis. The native, unmodified RBP was subsequently identified in the liver sections by a sheep anti-human RBP antibody, which does not recognize endogenous rat RBP. Ten minutes after injection, RBP was found in close contact with the plasma membrane of parenchymal and stellate cells. Ten minutes later, RBP was also found attached to the membranes of small vesicles near the cell surface. At two hours after injection, RBP was detected in larger vesicles deeper in the cytoplasm, remote from the vesicle membranes as though dissociated from its putative receptor. Asialo-orosomucoid was also localized in these larger vesicles.

Blomhoff's research group (Gjøen *et al.*, 1987) studied the uptake of RBP in various organs at different times. RBP was labeled with ^{125}I–tyramine cellobiose (^{125}I–TC–RBP) and injected intravenously into rats. (The advantage of using ^{125}I–TC–RBP is that the radioactive degradation products do not escape from the cells in which the protein is degraded, as is the case for proteins radioiodinated directly.) Of all the organs tested (liver, kidneys, intestine, spleen, heart, lungs, etc.) the liver contained the most radioactivity, followed by the kidneys. After 1 hour, approximately 20% and 10% of the injected dose was recovered in the liver and kidneys, respectively; other organs contained less than 3%. Of the liver cell types, parenchymal and stellate cells took up about equal amounts of ^{125}I–TC–RBP when calculated per gram of liver; Kupffer cells and endothelial cells accumulated insignificant amounts. The liver was shown to be the main organ for tissue catabolism of plasma RBP. The same laboratory (Senoo *et al.*, 1990) labeled RBP and several other proteins with ^{125}I–TC and compared their *in vivo* uptake in different rat liver cell types. RBP was the only protein that was taken up selectively by parenchymal and stellate cells. The data from these two studies suggest that hepatic parenchymal and stellate cells and cells in kidneys contain receptors for RBP, and that RBP is internalized in these cells, probably by receptor-mediated endocytosis.

Proof of an RBP receptor-mediated uptake of circulating retinol requires isolation and characterization of an RBP receptor. An abundant 63-kDa terminally glycosylated membrane protein which specifically binds RBP has been identified as the RBP receptor in microsomal membranes of retinal pigment epithelial cells (Båvik *et al.*,

1991) and further characterized (Båvik *et al.*, 1992; Båvik *et al.*, 1993). The receptor cannot discriminate between the apo- and the holo-form of RBP. Particularly large numbers of receptor-binding sites were also found in microsome fractions of liver and kidney, whereas lung and muscle contained few, if any. The abundance of RBP receptors in retinal pigment epithelium and in liver and kidney cells is in accordance with the responsibility of these cells for the uptake and transport of large amounts of retinol for functional purposes. In contrast, it is expected that only relatively small amounts of retinol are needed in cells where accumulated retinol is used mainly for the synthesis of retinoic acid.

Carotenoids are exported from the liver as components of VLDLs, which are converted to LDLs by lipoprotein lipase on the surface of blood vessels. Plasma membrane-associated receptors of peripheral tissue cells bind apolipoprotein B-100 on the surface of LDLs, initiating receptor-mediated uptake of the entire LDL particle. Carotenoids released from lipoproteins are taken up by many tissues, particularly adipose tissue, where they accumulate.

15.12 BIOSYNTHESIS OF RETINOIC ACID

The *in situ* synthesis of retinoic acid has been demonstrated in a wide array of tissues and in a variety of species. Despite its physiological importance, retinoic acid is a quantitatively minor metabolite of retinol. The irreversible nature of the terminal oxidation step (Step 3) explains the observation that retinoic acid is unable to support those processes, namely vision and reproduction, that specifically require either retinol or retinaldehyde.

The reversible interconversion of retinol and retinaldehyde is the rate-limiting step in retinoic acid synthesis. The class IV alcohol dehydrogenase catalyzes the oxidation of retinol to retinaldehyde in the presence of nicotinamide adenine dinucleotide (NAD). In the presence of reduced nicotinamide adenine dinucleotide phosphate (NADPH), a microsomal retinol dehydrogenase catalyzes the reduction of retinaldehyde to retinol to promote retinoid storage. The irreversible oxidation of retinaldehyde to retinoic acid is catalyzed by some members of the aldehyde dehydrogenase and cytochrome P450 enzyme families (Duester, 1996).

In vitro studies using ferret liver led Wang *et al.* (1993) to suggest that retinoic acid regulates its own synthesis via feedback inhibition of retinol oxidation to retinaldehyde and stimulation of retinol esterification. This suggestion is supported by the observation that addition of retinoic acid to a vitamin A-deficient rat diet resulted in an increase in hepatic retinyl ester concentration and a concomitant decrease in plasma retinol concentration (Bhat & Lacroix, 1991).

Retinoic acid synthesized in the cytoplasm moves to the nucleus where it binds to a nuclear retinoid receptor. The binding protein CRABP-I may be important in establishing equilibrium between the cytoplasmic and nuclear concentrations of retinoic acid.

15.13 FATE OF UNMETABOLIZED CAROTENOIDS

Intact carotenoids taken up by the liver as components of chylomicron remnants are secreted from hepatocytes into the bloodstream as components of VLDLs. Triglyceride stripping of VLDLs results in a transfer of carotenoids to LDLs, which are the major vehicles for transport of hydrocarbon carotenoids (carotenes) in plasma. The intact carotenoids are ultimately stored in adipose tissues, liver, and other organs without a toxic effect. Cellular uptake of plasma carotenes might be mediated by LDL receptors. Recycling of carotenoids from tissues back to plasma takes place among VLDL, LDL, and high-density lipoproteins (HDLs). Carotenoid accumulation in tissues does not seem to be homeostatically regulated, but is related to plasma levels.

Rats absorb β-carotene in a different manner to humans. When rats are fed radioactive β-carotene, the β-carotene is converted almost entirely to retinyl esters, and there is no labeled β-carotene in the lymph (Huang & Goodman, 1965). This means that accumulation of β-carotene in rat tissues does not occur and neither does it occur in the tissues of mice, hamsters, guinea pigs, rabbits, chickens, pigs, or sheep. The low tissue concentrations of β-carotene in rats, mice, and hamsters confounds experimental studies on the putative anti-cancer activity of β-carotene. In the search for a more suitable animal model, the ferret has been found to resemble the human in its ability to absorb and accumulate intact carotenoids (Ribaya-Mercado *et al.*, 1989).

15.14 CATABOLISM

Retinoic acid is a key metabolite in the classic cytosolic pathway of retinol metabolism in the liver and extrahepatic tissues. Retinoic acid undergoes a variety of metabolic reactions, including conjugation with glucuronic acid, decarboxylation, oxidation at the 4 position of the cyclohexenyl ring, epoxidation, isomerization, and esterification (Frolik, 1984). The major pathway of retinoic acid catabolism is oxidation to 4-hydroxyretinoic acid, which is further oxidized to 4-oxoretinoic acid. The latter compound may be progressively metabolized to chain-shortened oxidation products that are excreted in the urine. Alternatively, 4-oxoretinoic acid may form its β-glucuronide conjugate, which is excreted via the bile into the feces.

Leo and Lieber (1985) discovered a retinol oxidase in rat liver microsomes that can convert retinol to polar metabolites without producing retinaldehyde. This microsomal enzyme is dependent on cytochrome P450, NADPH, and molecular oxygen. Leo *et al.* (1989c) established that human liver microsomes can metabolize retinol and retinoic acid to polar products. The reaction was specifically dependent on P450IIC8, as neither P450IIC9 nor P450IIE1 displayed significant activity. Leo *et al.* (1987) also discovered an oxygen-independent retinol dehydrogenase in rat liver microsomes that can convert retinol to retinaldehyde using NAD as a coenzyme. Finally, Leo *et al.* (1989b) discovered a retinaldehyde dehydrogenase in rat liver cytosol that catalyzes the oxidation of retinaldehyde to retinoic acid. The physiological substrate for the enzyme could be the retinaldehyde derived from β-carotene or formed from retinol by the cytosolic or microsomal retinol dehydrogenases.

Leo *et al.* (1989a) showed that degradation of vitamin A through the above-mentioned microsomal pathways of retinol catabolism in the liver may provide a homeostatic mechanism which offsets in part excessive vitamin A accumulation in the liver. They fed rats increasing levels of vitamin A (including very high levels) and found that, as intake increased, progressively smaller fractions of the administered vitamin were recovered in tissues. This decrease in fractional recovery could be accounted for in part by enhanced rates of vitamin A degradation secondary to the increased concentration of retinol in the liver.

15.15 EXCRETION

Approximately 5–20% of ingested vitamin A and a larger percentage of carotenoids are not absorbed from the intestinal tract and are eliminated from the body in the feces. Some 10–40% of the absorbed vitamin A is oxidized and/or conjugated in the liver and then is secreted into the bile. Although some of these biliary metabolites, such as retinoyl β-glucuronide, are reabsorbed to some extent and returned to the liver in an enterohepatic cycle, most of the biliary metabolites are eliminated in the feces.

15.16 NUTRITIONAL FACTORS THAT INFLUENCE VITAMIN A STATUS

Among various nutrients which influence vitamin A nutritional status, fat and protein are the most important.

15.16.1 Fat Deficiency

The presence of adequate amounts of dietary fat is essential in forming micelles and providing a lipid vehicle for vitamin A absorption and transport. The absorption of retinol and carotenoids is markedly lower than normal when diets contain very little fat (less than 5 g per day). A study conducted in a Rwandan village area in Africa showed that supplementation of the carotenoid-sufficient but low-fat diet with 18 g per day of olive oil increased the absorption of vegetable carotenoids from 5 to 25% in boys that were clearly vitamin A-deficient (Roels *et al.*, 1958).

15.16.2 Protein Deficiency

In cases of severe protein–energy malnutrition, the lack of dietary protein leads to an impairment of all stages of vitamin A metabolism due to a depressed synthesis of enzymes, retinoid-binding proteins, and receptors. Even where dietary carotenoids are abundant, a reduced enzymatic cleavage of carotenoids will result in vitamin A deficiency. Furthermore, liver stores of vitamin (even if plentiful) will not be released into the bloodstream due to the depressed synthesis of RBP.

15.16.3 Zinc Deficiency

Zinc is an essential co-factor for many enzymes, some of which are directly critical to vitamin A metabolism; for example, zinc deficiency significantly reduces the enzymatic oxidation of retinol to retinaldehyde in the retina (Huber & Gershoff, 1975). Other zinc-dependent enzymes are involved in the synthesis of proteins, including perhaps retinoid-binding proteins and receptors. The synthesis of one particular protein, opsin in the rod cells of the eye, is depressed in zinc-deficient rats (Dorea & Olson, 1986). The consequent depression in rhodopsin levels, rather than a depressed enzymatic oxidation of retinol, probably explains the impaired dark adaptation found in zinc-deficient humans.

Zinc-deficient animals exhibit low plasma vitamin A levels even if large supplements of vitamin A are given. Zinc deficiency also reduces food intake, with a consequent impairment of growth, and these parameters confound experiments on zinc–vitamin A interactions. Smith *et al.* (1976) showed that a severely food-restricted group of rats fed a zinc-adequate diet had plasma vitamin A levels similar to zinc-deficient rats. They concluded that either zinc deficiency or growth depression decreases plasma vitamin A levels, and postulated that zinc deficiency *per se* may impair hepatic RBP synthesis. It was not possible from their data to establish if the reduction in plasma vitamin A in the zinc-deficient animals was a result of zinc deficiency *per se* or a secondary effect of reduced food intake accompanying zinc deficiency.

15.16.4 Vitamin E Deficiency

It is generally recognized that vitamin E is necessary for optimal utilization of vitamin A. More vitamin A is stored in the liver when the diet contains vitamin E than when the diet is deficient in vitamin E (Ames, 1969).

15.17 THE ROLE OF VITAMIN A IN VISION

15.17.1 Structure of the Retina

The retina is composed of nervous tissue forming a photosensitive inner lining within the posterior half of the eyeball. The first layer of the retina is a monolayer of cells called the pigment epithelium, which forms the protective blood–retina barrier. Beneath the pigment epithelium is a layer or rods and cones, which make synaptic connections with dendrites of bipolar neurons. These interconnecting neurons integrate the electrical signals and conduct them to ganglion neurons at the innermost surface of the retina. The ganglion neurons lead to the optic nerve, which conducts the signals to the visual cortex of the brain. The image on the retina is actually inverted, but the brain learns to turn it the right way up.

15.17.2 Rods and Cones

The photoreceptor cells – the rods and cones – are modified neurons. The human retina contains about 3 million cones and 100 million rods. Rod cells make it possible to form black-and-white images in dim light, while cones are responsible for color vision in bright light. The rod cell is a long, thin structure divided into two parts – the outer segment and the inner segment (Figure 15.7). The outer segment consists of a narrow tube filled with a stack of some 2,000 tiny discs that contain the light-absorbing pigment rhodopsin, also known as visual purple. The discs originate near the bottom of the tube as invaginations of the plasma membrane. They gradually migrate toward the top, replacing those that are shed and phagocytosed. The inner segment contains the nucleus and mitochondria that generate the required energy.

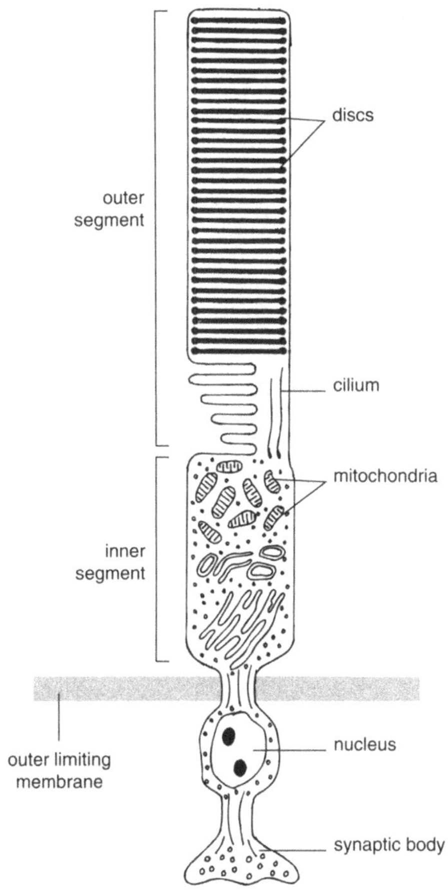

FIGURE 15.7 Diagram of a rod cell. The visual pigment rhodopsin is contained in the discs that are stacked in the outer segment.

15.17.3 Rhodopsin

Rhodopsin is a stereospecific combination of a chromophore (11-*cis* retinaldehyde) and a protein (opsin), joined covalently in a Schiff base linkage by the condensation of the aldehyde group of retinaldehyde with the epsilon (ε)-amino group of a lysine residue (Figure 15.8). The bent shape of the 11-*cis* isomer of retinaldehyde allows it to fit snugly into the opsin molecule and also holds the protein in the conformation specific to rhodopsin. No other stereoisomer of retinaldehyde can combine with opsin in this manner. Whereas free 11-*cis* retinaldehyde in solution absorbs radiation in the ultraviolet (UV) region of the spectrum (maximal absorbance at a wavelength of 380 nm), the chromophore in rhodopsin absorbs the much more plentiful radiation in the visible (green) region (maximal absorbance, 498 nm).

15.17.4 The Visual Cycle

The function of vitamin A in vision is based upon isomerization reactions occurring during the visual cycle. The essential isomerization reactions of the visual cycle are shown in Figure 15.9. Cells of the pigment epithelium take up RBP-bound all-*trans* retinol from the blood by receptor-mediated endocytosis, esterify it, and store it in lipid globules. When needed, the retinyl ester is hydrolyzed and the retinol is isomerized to 11-*cis* retinol. This isomer is then oxidized to 11-*cis* retinaldehyde, which is transported to the rod outer segment and combined with opsin to form rhodopsin.

When light strikes the dark-adapted retina, the rhodopsin undergoes progressive decomposition through a series of unstable intermediates (Figure 15.10). The decomposition process is known as bleaching, because it causes rhodopsin to lose its color. The photoactivation of electrons in the 11-*cis* retinaldehyde moiety of rhodopsin leads to its instantaneous conversion to a strained form of all-*trans* retinaldehyde, which begins to pull away from the opsin moiety. This initial photochemical reaction occurs within picoseconds (10–12 s) of illumination and is the only

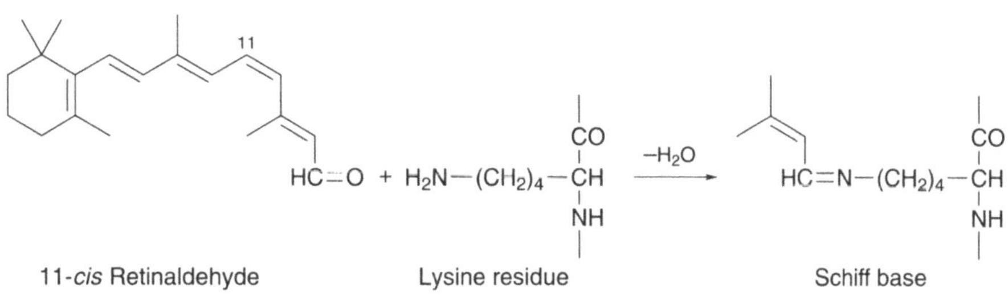

FIGURE 15.8 Schiff base formation between 11-*cis* retinaldehyde and a lysine residue in opsin.

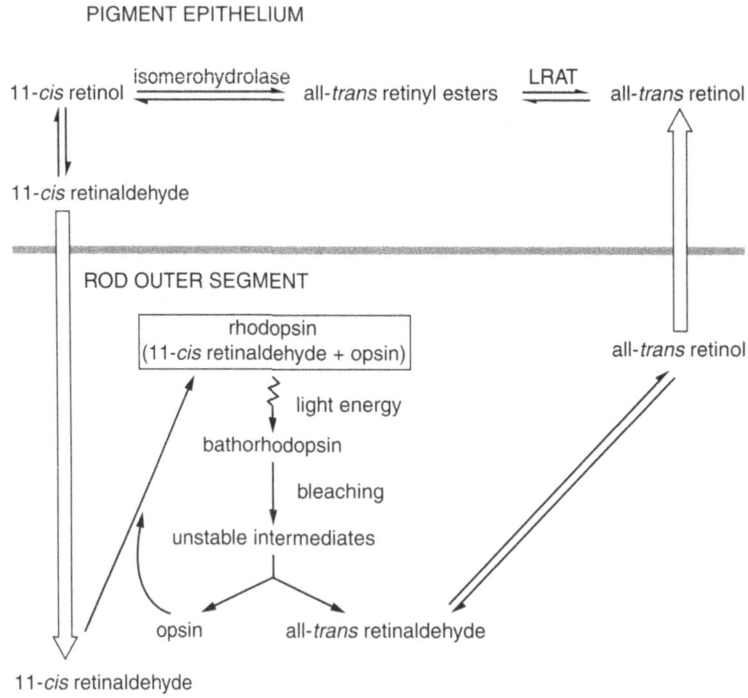

FIGURE 15.9 The visual cycle. Two-way arrows indicate reversible enzymatic reactions; solid arrows represent transport mediated by interphotoreceptor retinoid binding protein (IRBP); LRAT, lecithin:retinol acyltransferase.

light-dependent step in the visual cycle; all subsequent steps are thermal reactions proceeding in the dark. The immediate decomposition product is bathorhodopsin, which is a partially split combination of all-*trans* retinaldehyde and opsin. The pathway of rhodopsin decomposition ends in the separation of all-*trans* retinaldehyde from opsin.

After its dissociation from opsin, all-*trans* retinaldehyde is reduced to all-*trans* retinol by a membrane-bound, NADP-dependent retinol dehydrogenase in the rod outer segment. The all-*trans* retinol is then transported by IRBP to the pigment epithelium, where it is esterified by LRAT. The all-*trans* retinyl ester (mainly retinyl palmitate) is then converted to 11-*cis* retinol by an isomerohydrolase and the 11-*cis* retinol is oxidized to 11-*cis* retinaldehyde. This isomer is transported by IRBP to the rod outer segment, where it spontaneously combines with opsin to form rhodopsin. Additional small amounts of 11-*cis* retinaldehyde are available from the degradation of phagocytosed discs shed from the distal end of the rod outer segment.

By regulating the amount of rhodopsin present in the rods, the visual system is able to alter its sensitivity according to the amount of available light. The increase in sensitivity which occurs while the eyes are in darkness or near darkness is called dark-adaptation, and the decrease in sensitivity caused by exposure to bright light is called light-adaptation.

Color vision in cone cells involves a path of sensory transduction essentially the same as that described above for rod cells, but using a photoreceptor that is slightly different to rhodopsin.

15.17.5 The Excitatory Cascade

A rod cell can be excited by a single photon. Excitation initiates an amplification cascade of molecular events, which lead ultimately to the conductance of a nerve impulse from the eye to the brain (Figure 15.10). It is the metarhodopsin II (also called activated rhodopsin) in the rhodopsin decomposition pathway that excites electrical changes in the rods. Metarhodopsin II functions as an enzyme to activate many molecules of transducin – a protein with three subunits, α (alpha), β (beta), and γ (gamma), that is present in an inactive form in the rod cell membrane and membranes of the discs. In response, the α-subunit of transducin binds guanosine triphosphate (GTP) in place of endogenously associated guanosine diphosphate (GDP). The production of hundreds of α-transducin–GTP complexes by a single molecule of metarhodopsin II is the first stage of amplification in vision. α-Transducin with its GTP then triggers the activity of a phosphodiesterase enzyme which hydrolyzes 3′,5′-cyclic guanosine monophosphate (cyclic GMP) to the noncyclic form (GMP). Activation of the phosphodiesterase takes place because it is subject to an inhibitory constraint. When the α-subunit of transducin with its attached GTP binds to the phosphodiesterase, it carries away an inhibitory γ-subunit of the enzyme. The departure of this subunit unleashes the catalytic activity of the phosphodiesterase. That activity is powerful: each activated enzyme molecule can hydrolyze 4,200 molecules of cyclic GMP per second. Because cyclic GMP specifically maintains the sodium channels in the plasma membrane of the rod cell in an open

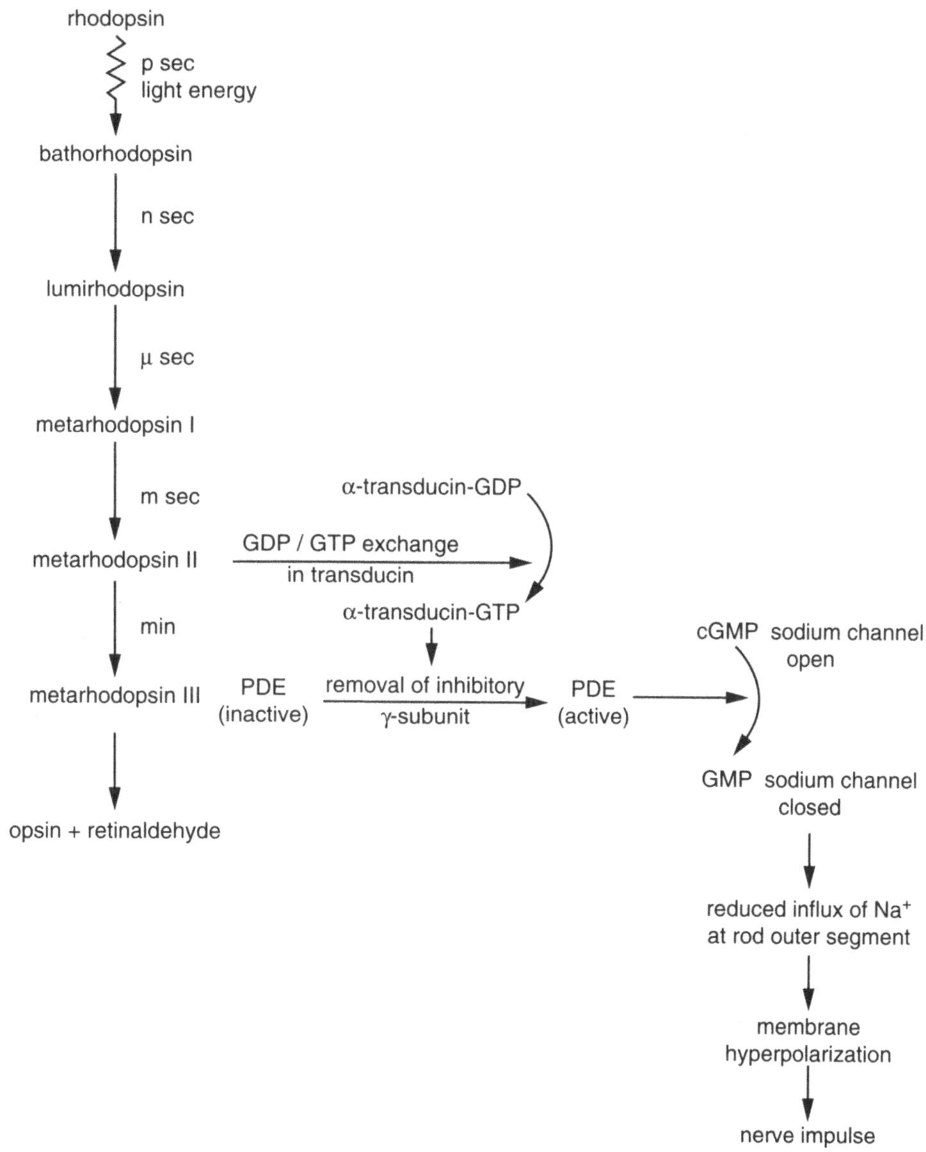

FIGURE 15.10 The amplification cascade of molecular events in the retina. GDP, guanosine diphosphate; GTP, guanosine triphosphate; cGMP, 3′,5′-cyclic guanosine monophosphate (cyclic GMP); PDE, phosphodiesterase.

state, a decrease in its concentration closes the channels, leading to a marked reduction in the influx of sodium ions into the rod outer segment. The membrane consequently becomes hyperpolarized and a nerve impulse is transmitted.

The biochemical system must be turned off and reset after nerve propagation. A GTPase intrinsic to the α-subunit of transducin terminates the activated state by converting the bound GTP to GDP. During this hydrolysis step the α-subunit of transducin releases the inhibitory γ-unit of phosphodiesterase. The γ-unit returns to the phosphodiesterase, binds to it, and restores the enzyme to the quiescent state. Transducin is then restored to its preactivation form by the recombination of the α-subunit and the joined β-γ unit.

A mechanism of regulating signal transduction, known as desensitization, allows a receptor to be rapidly inactivated and then reactivated despite the continued presence of a stimulus of constant intensity. The light-stimulated rhodopsin is desensitized by the binding of arrestin I following phosphorylation catalyzed by rhodopsin kinase. Arrestin I is an inhibitor that prevents the activation of transducin. Desensitization of rhodopsin occurs in less than one second following light stimulation, thereby preventing a flash of light from being perceived as continuous illumination.

The vitamin A stored in the eye can exchange fairly rapidly with the general metabolic pool of vitamin A as long as the pool is adequate. If a rat is maintained on a low vitamin A diet, its eyes take up retinol much more avidly than other tissues which also have a need for it. Moreover, the eyes can hold tenaciously to their stores of esterified vitamin long after other tissues are depleted. The eyes take priority because rats are nocturnal and night vision is important for searching for food.

15.18 RETINOIDS AS REGULATORS OF GENE EXPRESSION

15.18.1 Nuclear Retinoid Receptors

Nuclear retinoid receptors are implicated in various aspects of cell growth, development, and homeostasis by controlling expression of specific genes through their hormonal ligands, which act as signals. The two classes of receptor – retinoic acid receptors (RARs) and retinoid X receptors (RXRs) – differ in their primary structure, their binding affinity to retinoids, and their ability to regulate expression of different target genes. The retinoid receptors belong to a subfamily of the superfamily of receptors that function as ligand-activated transcription factors; the subfamily includes receptors for steroid hormones, thyroid hormone, and vitamin D.

In contrast to steroid hormone receptors, which are encoded by single-copy genes located on separate chromosomes, the RARs and RXRs are each encoded by three distinct genes, giving rise to receptor subtypes α, β, and γ. The receptors show characteristic patterns of distribution in adult and embryonic tissues (Table 15.2). The three RAR subtypes within a given animal species have very similar amino acid sequences in their DNA-binding C domains and ligand-binding E domains. The same is true for the three RXR subtypes. The lack of similarity among domains A/B, D, and F suggests that functional differences between subtypes may be conferred by these regions. Between species (mouse and human) the sequence of a particular subtype is nearly 100% conserved through domains A–F. The sequence homology between RARs and RXRs is not strong, the most conserved region lying in the C domain. Somewhat surprisingly, there is greater homology between RARs and thyroid hormone receptors than between RARs and RXRs.

The diversity of the retinoid signaling pathways is further increased by the presence of two promoters for each of the six genes that encode the α, β, and γ subtypes of RAR and RXR, giving rise to two alternative primary mRNA transcripts for each gene. These transcripts generate receptor isoforms, which are identified by numbers behind the Greek letter. Isoforms for a given receptor subtype differ only in the amino acid sequences of their amino-terminal A regions. Further isoforms are generated post-translationally by the alternative splicing of the primary transcripts. At least seven isoforms of RARγ, for example, have been identified in the mouse, but RARγ3 to γ7 are minor and lack the A region altogether (Zelent *et al.*, 1991). The major isoforms of the RARs and RXRs are listed in Table 15.3. Some target genes are activated preferentially by one RAR isoform over another and antagonism between various isoforms has been demonstrated (Husmann *et al.*, 1991).

The RARs bind both all-*trans* retinoic acid and 9-*cis* retinoic acid with approximately the same high affinity and activate transcription as a consequence of this binding. The two ligands compete with each other for binding to distinct but overlapping binding sites within AF-2 of RARα (Tate *et al.*, 1994). The RXRs are incapable of high-affinity binding to all-*trans* retinoic acid, but they bind and are activated by 9-*cis* retinoic acid. In addition, the 9-*cis* isomer of 3,4-didehydro-retinoic acid, but not the all-*trans* isomer, is an activating ligand for RXRs (Allenby *et al.*, 1993).

RARs are unable to form functional homodimers, and require heterodimerization with RXRs for efficient DNA

TABLE 15.3
Major Isoforms of RAR and RXR α, β, and γ Subtypes

RAR		
Type 1	**Type 2**	**RXR**
α1	α2	α1, α2
β1, β3	β2, β4	β1, β2
γ1	γ2	γ1, γ2

Type 1 and type 2 RAR isoforms are transcribed from an upstream promoter (P1) and a downstream promoter (P2), respectively. The two pairs of RARβ isoforms (β1/β3 and β2/β4) are generated by alternative splicing. The isoforms of RXRα, RXRβ, and RXRγ are also transcribed from different promoters.
Source: Chambon (1996).

TABLE 15.2
Distribution of RAR and RXR Receptors in Mammalian Adult and Embryonic Tissues

Receptor	Adult	Developing Embryo
RARα	Ubiquitous	Ubiquitous
RARβ	Kidney, adrenal glands, pituitary gland, prostate gland, spinal cord, brain, muscle, bone	Neural crest cells
		Developing peripheral nervous system
		Epithelia of lungs, intestine, and genital tract
RARγ	Skin, bone	Neural crest craniofacial domains
		Differentiating squamous and mucous epithelia
RXRα	Skin, liver, kidney, lung, spleen, muscle, bone	Epithelia of digestive tract, liver, skin
RXRβ	Ubiquitous	Ubiquitous
RXRγ	Liver, kidney, adrenal glands, lung, brain, muscle	Discrete areas of developing central nervous system, pituitary gland

binding and transcriptional activity. In contrast, the RXRs can function both as heterodimers and homodimers. The ability of the RXRs to homodimerize is attributable to a unique third helix in the C/D domain (Lee *et al.*, 1993) and is dependent upon the binding of its ligand, 9-*cis* retinoic acid.

15.18.2 Role of Retinoid Receptors in Transcriptional Regulation

RAR, in dimeric combination with RXR, binds in a ligand-independent manner to a retinoic acid response element (RARE) on the DNA. In the absence of its ligand, the RAR partner of the DNA-bound heterodimer interacts via its repressor function (the CoR box) with either or both of the co-repressors N-CoR and SMRT, which in turn interact with a Sin3–histone deacetylase (HDAC) complex. The RXR partner does not bind co-repressors. Deacetylation of histone tails creates a tight nucleosomal structure, which represses transcription. The binding of ligand to RAR causes a conformational change in the receptor's ligand-binding domain, which triggers release of the co-repressor–HDAC complex and exposes the activation surface of the receptor's AF-2 motif. This allows the recruitment of a co-activator complex containing combinations of CBP/p300, PCAF, SRC-1, and ACTR, all of which exhibit histone acetyltransferase (HAT) activity. Acetylation of histone tails loosens the nucleosomal structure and, together with the chromatin remodeling activity of Swi/Snf, allows recruitment of the transcription machinery to the promoter and the initiation of transcription. Thus RAR functions as a transcriptional activator in the presence of its ligand and as a repressor in the absence of ligand (Figure 15.11). Binding of the RAR ligand allows the subsequent binding of the RXR ligand that then enhances the transcriptional response to the RAR ligand (Minucci *et al.*, 1997).

15.18.3 Specifying the Retinoic Acid Response

The retinoic acid responses can be limited to certain cell types because the retinoid receptors can influence the magnitude of response. For instance, the RARγ1 isoform can inhibit the activation of the βRARE by other RARs (Husmann *et al.*, 1991). βRARE is a response element found in the promoter of the *RARβ* gene. The RARγ1-mediated inhibition is isoform-specific, as RARγ2 activates the βRARE. The inhibition is also dose-dependent and may involve competition among receptors for the response element. Thus in cells that produce RARγ1, retinoic acid-responsive genes that contain βRARE-like response elements may show a reduced or complete lack of response to retinoic acid.

A given RAR–RXR heterodimer can have different ligand affinities, and thus different levels of activation, depending on the relative polarity of the two receptors and the exact structure of the DNA binding site. It is therefore the receptor–response element complex and not the receptor alone that determines the hormonal response. The diversity of RAREs, together with tissue- and cell type-specific expression patterns of the various RAR and RXR subtypes and isoforms, allows for a pleiotropic and fine-tuned hormonal response.

La Vista-Picard *et al.* (1996) further showed that synthetic retinoids with restricted molecular conformations can be highly selective for binding to certain receptor–RARE combinations, and could serve as gene-selective activators. Because of their limited biological activity, such compounds could have therapeutic value since they would

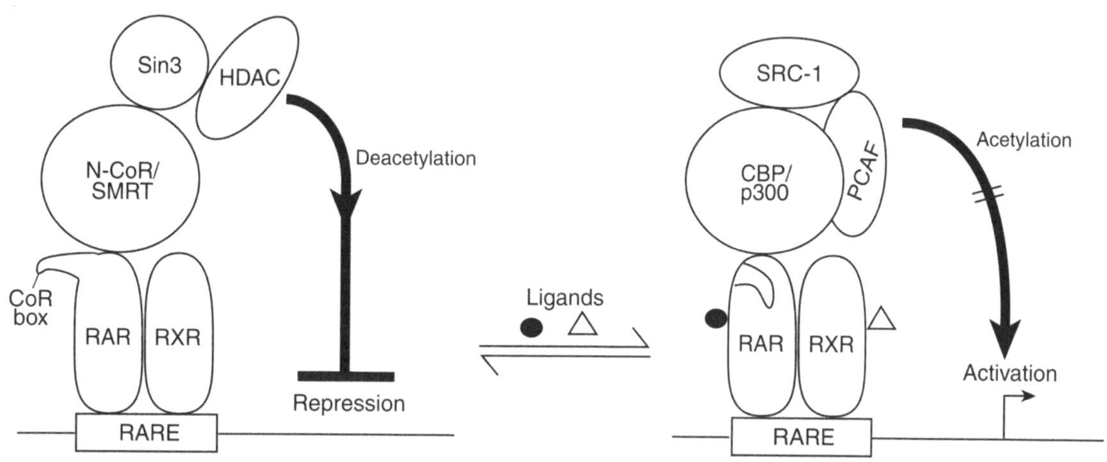

FIGURE 15.11 Function of RAR in a DNA-bound RXR–RAR heterodimer as a transcriptional activator in the presence of ligand or as a repressor in the absence of ligand. The N-CoR/ SMRT–Sin3 combination recruited by unliganded RAR is the co-repressor complex to which is linked a histone deacetylase (HDAC). Deacetylation of the histone tails leads to transcriptional repression. Ligation of RAR causes the release of the co-repressor–HDAC complex and recruitment of a co-activator complex (CBP/p300, PCAF, SRC-1) whose components possess HAT activity. Acetylation of the histone tails leads to transcriptional activation. Ligation of RAR allows the subsequent ligation of RXR, which then enhances the transcriptional response to RAR ligation.

have fewer side effects than the flexible natural compounds with their broad-spectrum activities.

15.18.4 Interaction between the Thyroid Hormone Receptor and the Retinoic Acid Receptor

The thyroid hormone receptors (TRs) are encoded by two different genes, giving rise to subtypes α and β. Thyroid hormone response elements (TREs) are typically composed of two or more sequences of a consensus AGGTCA half-site arranged as an inverted repeat (palindrome), everted repeat, or direct repeat (Ribeiro *et al.*, 1995), thereby allowing a great flexibility in half-site arrangement and orientation. Palindromic TREs are responsive to both thyroid hormone (tri-iodothyronine) and retinoic acid. The two hormones interact co-operatively to stimulate transcription of the growth hormone gene in the rat pituitary gland (Bedo *et al*, 1989; Morita *et al.*, 1989, 1990). The presence or absence of thyroid hormone dictates how these dual-responsive elements are regulated. In the presence of thyroid hormone, TRs activate the TRE and also allow some degree of positive co-operativity with RARs. In the absence of thyroid hormone, the unliganded TR functions as a repressor, preventing either TR and RAR from activating the TRE. The unliganded RAR does not show a negative effect upon TR (Brent *et al.*, 1989; Graupner *et al.*, 1989). The repressive effect of unliganded TR on retinoic acid-induced gene expression through a common response element could account for well-characterized developmental abnormalities associated with hypothyroidism.

15.18.5 Experimental Use of Embryonal Carcinoma Cells

The establishment of retinoic acid-responsive genes has been facilitated by the use of cultured mouse teratocarcinoma stem cells, also known as embryonal carcinoma (EC) cells. These cells exhibit many features of normal cell differentiation pathways and cell–cell interactions. EC cell lines can be isolated from tumors (teratocarcinomas) that arise when normal preimplantation (1–7.5-day-old) inbred mouse embryos (blastocysts) are transplanted to an extra-uterine site such as the kidney or testis of an adult mouse (Martin, 1981) or by injecting cultured embryonic stem cells taken from late blastocysts into mice (Evans & Kaufman, 1981). A typical mouse teratocarcinoma contains undifferentiated stem cells (the EC cells) interspersed in a disorganized mixture of differentiated cell types. The continued proliferation of the EC cell population is responsible for the malignant properties of teratocarcinomas. EC cells share morphological, biochemical, and genetic properties with pluripotent stem cells from the inner cell mass of mouse blastocysts. (Pluripotent stem cells are capable of forming derivatives of all three primary germ layers, namely, endoderm, mesoderm, and ectoderm.) Many murine EC cell lines are capable of undergoing a controlled differentiation in culture in a manner that closely parallels the normal behavior of embryonic cells. The EC cells therefore provide an *in vitro* model for biochemical studies of cellular differentiation and early embryonic development.

Cultures of mouse EC cells can either proliferate exponentially in an undifferentiated state or, if stimulated, differentiate into cells similar to definitive embryonic cells. Retinoic acid is a potent inducer of differentiation in a number of EC cell lines, including F9 and P19. Differentiation of these stem cells is accompanied by a decrease in the rate of proliferation and also by the induction of expression of several genes.

The particular type of differentiated cell depends on the EC cell line, the manner in which the cells are cultured, and the concentration of retinoic acid added. Treatment of low-density monolayer cultures of F9 stem cells with retinoic acid converts the cells to primitive endoderm cells, and subsequent treatment with dibutyryl cAMP induces further differentiation to parietal endoderm (Strickland, 1981). If F9 cells are grown as suspension aggregates or as high-density monolayer cultures, treatment with retinoic acid converts the outer surface cells to visceral rather than parietal endoderm (Hogan *et al.*, 1981). Retinoic acid treatment of P19 stem cells in monolayer cultures results in fibroblasts or, in the case of aggregate cultures, in neurons and glial cells (Jones-Villeneuve *et al.*, 1982). The effect of retinoic acid concentration was demonstrated by Edwards and McBurney (1983) using a subclone of the P19 cell line in aggregate culture. A concentration of 10^{-9} M retinoic acid induced an abundance of cardiac muscle; at a higher concentration (10^{-8} M), skeletal muscle became abundant, and at 10^{-7} to 10^{-5} M neurons and glial cells appeared.

15.18.6 Retinoic Acid-Responsive Genes

Retinoic acid-responsive genes in which a RARE has been identified fall into the category of primary response genes or, in some cases, delayed primary response genes. These genes, by definition, bind the hormone-receptor complex – in this case, the RAR–RXR heterodimer – and produce the hormone responsive protein with no requirement for ongoing protein synthesis. Primary response genes produce mRNA within minutes to a few hours after addition of retinoic acid to cell cultures, whereas delayed primary response genes produce mRNA only after several hours. Vitamin A-responsive genes, which apparently have no RARE and which produce mRNA only after several hours, are categorized as secondary response genes. These genes, by definition, do not bind the hormone-receptor complex. They may be regulated by a transcription factor produced by a primary response gene or they may be regulated at the post-transcriptional level. In the latter case, for example, retinol acts by stabilizing newly synthesized alkaline phosphatase transcripts (Zhou *et al.*, 1994), regulating the processing of interleukin-1 precursor transcripts into mature mRNA (Jarrous & Kaempfer, 1994) and increasing the half-life of connexin43 by inducing adhesion molecules (Bex *et al*., 1995).

The many retinoic acid-responsive genes encode proteins having very diverse functions. The proteins include nuclear retinoid receptors, cellular retinoid-binding proteins, transcription factors, enzymes, extracellular matrix proteins, and growth factors (Table 15.4). The RAREs that are present in primary response genes frequently consist of direct repeats of two core recognition motifs with consensus sequence AGGTCA separated by 5, 2, or 1 base pairs and designated DR5, DR2, and DR1 elements, respectively. Sequence differences from the consensus do occur and generally result in a reduced transcriptional response to retinoic acid owing to a lower receptor binding affinity. RAREs are not exclusively of the direct repeat type: everted repeat elements were identified in the genes encoding γF-crystallin

TABLE 15.4
A Selection of Retinoic Acid-Responsive Genes

Gene or Gene Product	Up- or Down-Regulated by Retinoic Acid	Type of Gene (If Known)	Type of RARE* (If Identified) or AP-1 Binding Site	References
Nuclear receptors				
RARα2 (M)	Up	Primary response	DR5*	Leroy *et al.* (1991a, b)
RARβ2 (H)	Up	Primary response	DR5*	de Thé *et al.* (1989, 1990)
RARγ2 (H)	Up	Primary response	DR5*	Lehmann *et al.* (1992)
Cellular retinoid-binding proteins				
CRBP-I (R)	Up	Primary response	DR2*	Husmann *et al.* (1992)
CRBP-II (M)	Up	Primary response	DR1* and DR2*	Nakshatri and Chambon (1994)
CRABP-I (M)	Up	Secondary response	Not identified	Wei *et al.* (1989, 1995)
CRABP-II (H)	Up	Primary response	DR5*	Åström *et al.* (1994)
Homeobox genes				
Hoxa-1 (M)	Up	Primary response	DR5*	Langston and Gudas (1992)
HOXB1 (H)	Up	Primary response	Two DR2* elements	Ogura and Evans (1995)
Hoxd-4 (M)	Up	Primary response	DR5*	Pöpperl and Featherstone (1993)
Enzymes				
PEPCK (R)	Up	Primary response	DR1* and DR5*	Scott *et al.* (1996)
MCAD (H)	Up	Primary response	Complex*	Raisher *et al.* (1992)
Alcohol dehydrogenase (H)	Up	Primary response	DR5*	Duester *et al.* (1991)
Alkaline phosphatase (M)	Up	Secondary response	Not identified	Gianni *et al.* (1991)
Collagenase (H)	Down	Primary response	AP-1	Lafyatis *et al.* (1990)
Stromelysin (R)	Down	Primary response	AP-1	Nicholson *et al.* (1990)
Transglutaminase (M)	Up	Primary response	Not identified	Chiocca *et al.* (1988)
Extracellular matrix proteins				
Laminin B1 (M)	Up	Delayed primary response	Complex*	Vasios *et al.* (1989, 1991)
Collagen type IV (M)	Up	Secondary response	Not identified	Gudas *et al.* (1990)
Matrix Gla protein (H)	Down	Primary response	Novel*[a]	Kirfel *et al.* (1997)
Growth factors and their receptors				
IGF-I (R)	Down	Secondary response	Not identified	Lowe *et al.* (1992)
Interleukin-2 (H)	Down	Secondary response	Not identified	Felli *et al.* (1991)
PDGF-Rα (M)	Up	Primary response	Not identified	Wang *et al.* (1990)
Gap-junctional protein				
Connexin43 (M)	Up	Secondary response	Not identified	Guo *et al.* (1992)
Other proteins				
Complement factor H (M)	Up	Primary response	DR5*	Muñoz-Cánoves *et al.* (1990)
Apolipoprotein A1 (H)	Up	Primary response	DR2*	Rottmann *et al.* (1991)
Lactoferrin (H)	Up	Primary response	DR1*	Lee *et al.* (1995)
γF-crystallin (M)	Up	Primary response	ER8*	Tini *et al.* (1993)

(M), mouse; (R), rat; (H), human.

DR, direct repeat motif with 1, 2, or 5 base pair spacers between half-sites; ER, everted repeat motif.

Abbreviations: PEPCK, phosphoenolpyruvate carboxykinase; MCAD, medium-chain acyl-coenzyme A dehydrogenase; IGF-I, insulin-like growth factor-I; PDGF-Rα, platelet-derived growth factor receptor-α.

(Tini *et al.*, 1993) and medium-chain acyl-coenzyme A dehydrogenase (Raisher *et al.*, 1992). In the latter gene, the response element was activated by RXRα in preference to RARα or RARβ and thus could be described as an RXRE.

The promoters of some genes contain composite response elements which can selectively bind and initiate transcription from multiple nuclear hormone receptors, thereby allowing cross-talk between different hormonal pathways. For example, the rat oxytocin promoter contains a RARE that is predominantly responsive to retinoic acid, but also permits binding of TR and ER (estrogen receptor) to mediate transcription in the presence of thyroid hormone and estradiol (Adan *et al.*, 1993). Kato *et al.* (1995) described a novel class of response elements which are widely spaced (10 to 200 base pairs) direct repeats of the consensus AGGTCA motif (DR10 to DR200) and which act as promiscuous transactivation sites for ER, RAR-RXR, and vitamin D receptor (VDR)-RXR, but not TR-RXR.

In the following discussion of selected retinoic acid-responsive genes, special attention is given to the genes encoding the retinoid receptors and the cellular retinoid-binding proteins, as these proteins are physiologically important in vitamin A function.

15.18.6.1 The Retinoid Receptors

All three genes coding for the RAR proteins (subtypes α, β, and γ) contain a RARE in one of their two promoters and have been shown to be transcriptionally up-regulated in response to all-*trans* and 9-*cis* retinoic acid. This form of autoregulation may amplify a retinoic acid signal by increasing the number of available receptors in target tissues. Target gene responsiveness is cell-type specific (Davis & Lazar, 1993). Autoregulation of RARs has not only been demonstrated in tissue culture cell lines (Zelent *et al.*, 1991), but also in the adult rat (Haq *et al.*, 1991). Each RAR subtype is known to contain at least one isoform that can control its own synthesis. The RARE in the *RARβ2* gene promoter confers the greatest sensitivity to transcriptional activation. This element is located only about one helical turn from the TATA box in comparison with RAREs from other genes which are located further upstream. The induction of *RARβ* transcription is a direct response to retinoic acid treatment in that it does not require new protein synthesis. In this respect the *RARβ* gene is a primary response gene whose induction occurs within 3 hours of retinoic acid addition. The βRARE is a target for all three RAR subtypes, but not for TR or VDR (Sucov *et al.*, 1990). However, RARγ1 can repress the activation of the βRARE by other RARs, thereby limiting the autoregulated stimulation of *RARβ* expression (Husmann *et al.*, 1991). The induction of RARs by their retinoid ligands provides another possible mechanism by which retinoic acid can control the availability of RXR for other nuclear receptors. The induced increase in RAR molecules is likely to encourage RAR/RXR heterodimerization and, under conditions where RXR molecules are limited, this could impair signal responses mediated by other nuclear receptors that require RXR for DNA binding.

15.18.6.2 CRBP-I

Several whole animal studies have examined the effects of oral administration of retinyl acetate or retinoic acid on CRBP-I mRNA levels in various tissues. Rajan *et al.* (1990) reported that the levels of CRBP-I mRNA in certain tissues (lung, testis, spleen, and small intestine) of vitamin A-deficient rats were lower than those in normally fed rats. Oral repletion with retinyl acetate restored the levels to normal. Retinoid deficiency did not affect the levels of CRBP-I mRNA in the three tissues with the highest content of CRBP-I protein and CRBP-I mRNA (proximal epididymis, liver, and kidney). Haq and Chytil (1988) showed that administration of retinoic acid to vitamin A-deficient rats elicited a more rapid response than observed with retinyl acetate, with a two- to threefold increase in CRBP-I mRNA in lung tissue within 1 hour. Increased expression of CRBP-I was also demonstrated in the lung tissue of normally fed, vitamin A-replete rats after oral administration of retinoic acid (Rush *et al.*, 1991). Consistent with these studies was the demonstration that topical application of retinol or retinoic acid to adult human skin led to increased levels of CRBP-I mRNA and protein (Fisher *et al.*, 1995). These experiments indicate that, in certain tissues at least, dietary retinol (or more directly retinoic acid) directly induces the expression of the *CRBP-I* gene. Thus organs in which CRBP-I expression is depressed during vitamin A deficiency are able to respond rapidly to retinoids as soon as they become available.

The expression of CRBP-I can be modulated in the vitamin A-replete animal by glucocorticoid hormones. The administration of dexamethasone to vitamin A-replete rats decreased the levels of CRBP-I mRNA levels in lung and liver tissues. In addition, the increase in CRBP-I mRNA induced by oral administration of retinoic acid was blocked when dexamethasone was administered at the same time (Rush *et al.*, 1991). This effect may be maintained by putative glucocorticoid-response elements identified in the promoter of the *CRBP-I* gene.

The hormonal changes that occur during pregnancy and lactation also affect the expression of CRBP-I. Toward the end of pregnancy in the rat, the level of CRBP-I mRNA in the maternal liver rises markedly then drops abruptly at term (Levin *et al.*, 1987). This event coincides with a significant mobilization of hepatic vitamin A stores so that the milk of the lactating rat can meet the vitamin A needs of the suckling pups.

Wei *et al.* (1989) demonstrated that treatment of tissue cultures with retinoic acid induced an increase in CRBP-I mRNA followed by the protein itself. The 3-hour lag period and the nondependence on concomitant protein synthesis identified the *CRBP-I* gene as a primary response gene. CRBP-I induction occurred both at 10^{-7} and 10^{-9} M retinoic acid in EC cells that differentiate into neurons and extra-embryonic endoderm and also in a non-EC cell that differentiates into glial cells.

The rapid synthesis of the CRBP-I protein in response to retinoic acid, observed for the whole animal as well as in cultured cells, suggested that the *CRBP-I* gene would contain a RARE. Such a response element has been demonstrated in the mouse (Smith *et al.*, 1991) and the rat (Husmann *et al.*, 1992). In both species, the RARE is located 1 kilobase upstream of the transcription start site. The RARE in the rat *CRBP-I* gene is a DR2-type element and is activated *in vitro* by the RAR/RXR heterodimer. The presence of a RARE in the *CRBP-I* gene indicates that CRBP-I protein levels are regulated primarily at the level of transcription.

15.18.6.3 CRBP-II

In the adult, expression of the *CRBP-II* gene is essentially confined to the small intestinal enterocytes. In contrast to CRBP-I, whose expression is decreased in various organs during vitamin A deficiency, expression of intestinal CRBP-II is actually increased. Rajan *et al.* (1990) reported an increase in the CRBP-II mRNA level of 42% in the small intestine of the vitamin A-deficient rat. The contrasting expression profiles of these two similar proteins reflect the physiological needs of the animal. During vitamin A deficiency, the animal conserves its precious supply of retinoids to maintain only essential functions, allowing the animal to hunt for food. In addition, the intestine is kept at the ready to receive dietary vitamin A as soon as the animal finds it.

During the last five days of pregnancy in the rat, mRNA for CRBP-II in the small intestine shows a fourfold increase, accompanying the increase observed for CRBP-I in the maternal liver. Intestinal CRBP-II levels remain elevated during the suckling period and return to normal by one-week post-weaning (Quick & Ong, 1989). This increase in production of intestinal CRBP-II, like that of hepatic CRBP-I, is an adaptive response, ensuring that sufficient vitamin A reaches the mammary gland during lactation.

Nakshatri and Chambon (1994) showed that the mouse *CRBP-II* gene contains three response elements, which they designated RE1, RE2, and RE3. The elements RE2 and RE3, which have DR2 and DR1 sequence motifs, respectively, are putative RAREs and are conserved in the rat and mouse *CRBP-II* gene 5′-flanking regions. The more distal RE1 element, which also has a DR1 motif, is a truncated form of an element previously found in the rat and designated an RXRE by Mangelsdorf *et al.* (1991). Nakshatri and Chambon (1994) showed that RE2 and RE3 were required for maximal RAR–RXR-mediated retinoic acid inducibility to the mouse *CRBP-II* promoter in transfection experiments. RE2 had no effect on its own, but cooperated with RE3. RE1 was not involved in the induction; if anything, it exerted a slight inhibitory effect. Thus RE3 appeared to be the only element within the mouse *CRBP-II* promoter that responded on its own to RAR and RXR. The RE3 and RE1 response elements have a much higher affinity for HNF-4 and ARP-1 than for RAR-RXR heterodimers and RXR homodimers. HNF-4 and ARP-1 are members of the nuclear receptor superfamily and are expressed in liver and intestinal cells. Nakshatri and Chambon (1994) showed that HNF-4 constitutively activates the mouse *CRBP-II* promoter, whereas ARP-1 competitively represses its activation by RAR, RXR, and HNF-4. Although the data reported by Nakshatri and Chambon (1994) and also by Mangelsdorf *et al.* (1991) show that the mouse and rat *CRBP-II* promoters can respond to overexpressed RXR and RAR in co-transfection experiments, for these results to be physiologically relevant, it must be demonstrated that expression of the *CRBP-II* gene can be stimulated by retinoic acid treatment *in vivo*. There are no available data concerning a possible retinoic acid induction of CRBP-II mRNA or protein in adult intestine and in prenatal liver or intestine, which are also known to express the *CRBP-II* gene (Nakshatri & Chambon, 1994). It appears that HNF-4 is the major transcriptional activator of the *CRBP-II* gene *in vivo*. Inducement of the *CRBP-II* gene by retinoic acid may only take place in tissues where HNF-4 and ARP-1 are lacking or present in low amounts.

15.18.6.4 CRABP-I

CRABP-I is ubiquitously expressed in adult animal tissues at a low basal level, except in several retinoic acid-sensitive tissues such as the eye and the testis where the expression is highly elevated. In adult rats, neither CRABP-I protein nor mRNA levels were affected by dietary vitamin A deficiency (Rajan *et al.*, 1991), suggesting that the binding protein is not regulated by overall vitamin A status. CRABP-I is also constitutively expressed at a required level in certain regions of the developing embryo and at certain times of development. In the developing mouse, CRABP-I mRNA has been detected in the neural crest, the eye region, and the craniofacial region (Perez-Castro *et al.*, 1989) and in rudimentary limbs (Dollé *et al.*, 1989).

Wei *et al.* (1989), using tissue cultures, showed that expression of the *CRABP-I* gene was induced by 10^{-7} M (but not 10^{-9} M) retinoic acid in a manner that required the synthesis of new protein. A detectable increase in CRABP-I mRNA was not observed until 12 hours after retinoic acid treatment. These observations, and the absence of a RARE with any known consensus sequence, place the *CRABP-I* gene in the category of secondary response genes. Induction occurred in cells that differentiate into neurons, but not in cells that differentiate into extra-embryonic endoderm or glial cells.

15.18.6.5 CRABP-II

Transcription of the human *CRABP-II* gene was markedly induced by retinoic acid in skin *in vivo* and in skin fibroblasts *in vitro*. Retinoic acid had no such effect on cultured lung fibroblasts, demonstrating cell-specific regulation of the *CRABP-II* gene (Åström *et al.*, 1991). In the cultured skin fibroblasts, the increase in CRABP-II mRNA was detected 1 hour after addition of retinoic acid. The level peaked at 2 hours and returned to basal levels within 6 hours. Ongoing protein synthesis was required for this transient increase of transcription (Åström *et al.*, 1992). The human *CRABP-II* gene contains one far upstream RARE of the DR5 type that binds the RAR-RXR heterodimer (Åström *et al.*, 1994). The binding of receptor to the response element and the early response are consistent with *CRABP-II* being a primary response gene, but the ongoing protein synthesis is inconsistent with this definition. Interestingly, the mouse *CRABP-II* gene contains two co-operating response elements, a DR2 RARE and a DR1 RXRE, mediating differential induction of transcription in response to both all-*trans* and 9-*cis* retinoic acid (Durand *et al.*, 1992).

15.18.6.6 Gene Encoding a Gluconeogenesis Enzyme

Phosphoenolpyruvate carboxykinase (PEPCK) catalyzes a rate-controlling step (oxaloacetate to phosphoenolpyruvate) in gluconeogenesis. The *PEPCK* gene contains two RAREs; RARE1 is a DR1 and RARE2 is a DR5 type of element. RARE1 is a binding site for TRs as well as for RARs. This allows thyroid hormone to regulate *PEPCK* gene transcription through RARE1 in the following manner (Lucas *et al.*, 1991). During hyperthyroidism, intracellular levels of retinoic acid are elevated owing to a stimulatory effect of thyroid hormone on activities of liver enzymes responsible for retinoic acid synthesis. This leads to the binding of a functional RAR–RAR homodimer to the RARE1 and stimulation of *PEPCK* gene transcription. Expression of TRs is reduced during hyperthyroidism owing to negative feedback regulation of TR synthesis by thyroid hormone, thus there is little antagonism of the RAR-mediated effect on transcription. In contrast, during hypothyroidism, TR levels are increased, leading to the binding of a non-functional TR–RAR heterodimer to the RARE1. This form of repression is not acutely relieved by readministration of thyroid hormone to a hypothyroid animal because the ligand-bound TR is still capable of repressing RAR activation of *PEPCK* gene transcription through RARE1.

15.18.7 Regulation of Two Retinoid Signaling Pathways by RXR

The regulation of two retinoid signaling pathways by RXR is shown in Figure 15.12. Which pathway is favored depends on ligand availability and receptor concentration. At high intracellular concentrations, 9-*cis* retinoic acid promotes the formation of a stable RXR homodimer on DR1 RXREs, thus favoring the transcriptional activation of RXRE target genes. Under conditions of low 9-*cis* retinoic acid and/or high RAR concentrations, the RAR –RXR heterodimer is stabilized and ligation by 9-*cis* retinoic acid and all-*trans* retinoic acid favors activation of DR2 and DR5 RAREs. The blocking of RXR homodimer function by the RAR–RXR heterodimer can be explained by the heterodimer having a greater affinity for DNA binding than the homodimer. The heterodimer outcompetes the homodimer for binding to the DR1 RXRE and forms a transcriptionally inactive complex

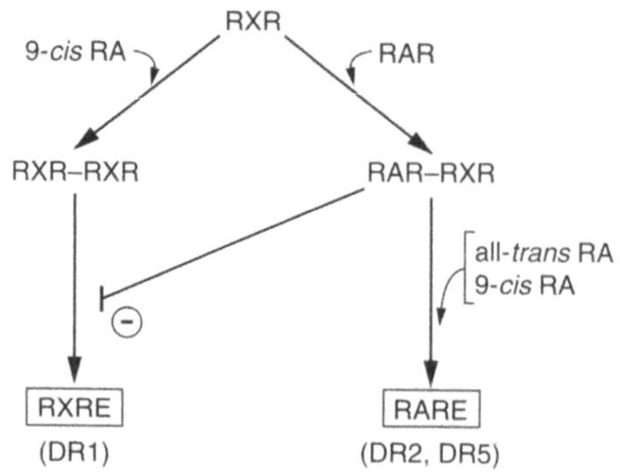

FIGURE 15.12 Regulation of two retinoid signaling pathways by RXR. See text for explanation.

with this response element. By regulating the production of all-*trans* and 9-*cis* retinoic acids, the cell is able to influence which receptor pathway will be activated (Mangelsdorf *et al.*, 1994).

15.18.8 Heterodimerization between RXR and Orphan Receptors

A large number of orphan receptors dimerize with RXR and allow ligand binding to RXR, thereby switching RXR from a silent to an active heterodimerization partner. The ability of orphan receptors to allow RXR to bind and be activated by ligand (9-*cis* retinoic acid) increases the diversity of 9-*cis* retinoic acid signaling through RXR. The many different orphan receptors confer DNA-specific binding of the heterodimer to a variety of response elements, including DR1, DR4, and DR5 types.

15.18.8.1 COUP-TF–RXR

The chicken ovalbumin upstream promoter transcription factor (COUP-TF) inhibits retinoid hormone-mediated transcription on several response elements that are activated by RAR–RXR heterodimers or RXR homodimers (Tran *et al.*, 1992). DR1 response elements have been the most extensively studied. COUP-TF is expressed only in selected tissues or cell types, such as liver and enterocytes. In the developing embryo, it is expressed in a spatially and temporally regulated manner. Furthermore, COUP-TF antagonizes only certain subsets of retinoid receptors; for instance, it does not repress RARβ2-mediated transcription (Tran *et al.*, 1992). Overall, the ability of COUP-TF to block transcription from certain RAREs and RXREs provides an important mechanism of restricting the broad spectrum of responses to retinoid hormones.

A member of the apolipoprotein (apo) gene family, *apoAI*, is an example of a gene containing a DR1 RXRE. The encoded protein, apoAI, is the major protein constituent of plasma HDL, which is involved in the transport of cholesterol and other lipids in the plasma. Widom *et al.* (1992) studied the effects of ARP-1 (apoAI regulatory protein, a homolog of COUP-TFII) on the responsiveness of the *apoAI* gene in human liver cells to RXRα and its ligand

9-*cis* retinoic acid. ARP-1 and RXRα were found to bind individually to the RXRE with similar affinity. These receptors also bound to the same site as heterodimers with an affinity approximately ten times greater than that of either ARP-1 or RXRα alone. Repression of gene transcription by ARP-1 was overcome by overexpression of RXRα in the presence of 9-*cis* retinoic acid. These observations indicate that regulation of *apoAI* gene expression is controlled by the balance of the intracellular levels of ARP-1 and liganded RXRα.

Retinoic acid (either all-*trans* or 9-*cis*) induces high levels of inhibitory COUP-TFI and II during differentiation of P19 EC cells (Jonk *et al.*, 1994), thus creating a negative feedback system involved in the down-regulation of retinoic acid-activated genes.

15.18.8.2 LXR–RXR

The orphan receptor LXR in heterodimeric complex with RXR binds to a specific DR-4-type response element, the LXRE. The complex is activated by the RXR ligand 9-*cis* retinoic acid, there being no observable repression in the absence of ligand. Ligand binding induces a conformational change in the entire heterodimer complex, exposing the AF-2 of LXR, and allowing transcription through interaction with co-activator proteins (Willy & Mangelsdorf, 1997).

15.18.8.3 OR1–RXR

The orphan receptor OR1 is also known as ubiquitous receptor (UR), RIP15, and NER. Heterodimeric interaction between OR1 and RXRα on a DR4 response element activates transcription in both the absence and presence of 9-*cis* retinoic acid. The activation potential is mainly dependent on the AF-2 of OR1, which is unmasked by a dimerization-induced conformational change (Wiebel & Gustafsson, 1997). The DR4-type response element is shared by the TR–RXR heterodimer and therefore OR1 has the potential to modulate the thyroid hormone signal pathway.

15.18.8.4 RLD-1–RXR

The particular response element recognized by the RLD-1–RXR heterodimer is a unique DR4, which is also capable of binding RAR–RXR and TR–RXR heterodimers. Retinoids and thyroid hormone can induce transcription through the RLD-1 response element in the presence of RAR–RXR or TR–RXR heterodimers, but only in the absence of TRs or RARs can the RLD-1–RXR heterodimer constitutively mediate transcription. In contrast to OR1, the presence of 9-*cis* retinoic acid inhibits the activity of RLD-1–RXR (Apfel *et al.*, 1994). These observations suggest a highly specific role for RLD-1 within the network of gene regulation by the TR/RAR subfamily.

15.18.8.5 NGFI-B–RXR

The orphan receptors NGFI-B (also called nur77) and the closely related nurr-1 can initiate transcription on binding both as monomers and heterodimers to DNA. NGFI-B is found in a variety of tissues, including the muscle, brain, heart, ovary, and testis, while nurr-1 expression is restricted to the brain. NGFI-B binds as a monomer to a specific response element that contains a single extended half-site, 5′-AAAGGTCA-3′. The receptor's zinc fingers interact with the consensus half-site (AGGTCA), while an unusual amino acid motif (the A box) interacts with the two A/T base pairs immediately 5′ of the consensus half-site (Wilson *et al.*, 1993). NGFI-B and nurr-1, as monomers, are constitutively active, but in heterodimeric complex with RXR they require 9-*cis* retinoic acid in order to activate transcription (Perlmann & Jansson, 1995). The heterodimers bind selectively to DR-5 response elements, with RXR occupying the 5′ half-site. NGFI-B and nurr-1 are encoded by primary response genes, which are ultimately induced by various growth factors. From a physiological viewpoint, the heterodimeric interaction of these receptors with RXR indicates a novel mechanism for convergence between retinoid and growth factor signaling pathways.

15.18.8.6 MB67–RXR

Another orphan receptor, MB67, which occurs predominantly in liver, can heterodimerize with RXR and activate transcription from DR-5 response elements in the absence of a ligand. Baes *et al.* (1994) demonstrated this constitutive activity in RAREs that control expression of the *RARβ2* and alcohol dehydrogenase 3 genes. Activation of the βRARE by the MB67– RXR heterodimer is weaker than that conferred by RAR–RXR heterodimers, possibly because MB67 essentially lacks the A/B domain, which contains amino acid sequences necessary for full transcriptional activation. The lack of dependence upon a retinoid ligand suggests that MB67 acts to stimulate expression of retinoid-responsive genes (i.e. those containing a DR-5 element) when retinoic acid levels are low. In this respect, MB67 activation of the βRARE could contribute to the basal level of expression of RARβ2 mRNA observed in the livers of retinol-deficient rats (Haq *et al.*, 1991).

15.18.8.7 FXR–RXR

Some nuclear receptors, in heterodimeric association with RXR, are activated by nonhormonal intracellular metabolites, thereby transducing metabolic cues into genomic responses. Although referred to as orphan receptors, the metabolites may possibly act as their ligands. One such receptor, FXR, is activated by farnesol and related metabolites, the transcriptional response being determined by the FXR component of the FXR–RXR complex (Forman *et al.*, 1995). Farnesyl pyrophosphate is a key intermediate in the mevalonate pathway leading to the biosynthesis of cholesterol.

15.18.9 Transcriptional Cross-Modulation between the Retinoid Receptors and the AP-1 Signaling Pathways

Cross-modulation can take place between the retinoid receptors and the AP-1 signaling pathways, allowing one type of transcription factor to regulate the function of the other and thereby increase the level of complexity. Since, in the main, retinoids induce cell differentiation and AP-1 induces cell proliferation, cross-modulation provides a molecular switch, allowing a finely tuned balance between these cellular activities.

Experimental results have led to various mechanisms being suggested for transrepression of AP-1 by liganded RAR. (1) RARα competes with Jun protein for binding to target DNA, forming a non-productive RARα/c-Jun complex (Schüle *et al.*, 1991). (2) RARα disrupts c-Jun–c-Fos dimerization, thereby preventing the formation of an AP-1 complex capable of DNA binding. Inhibition of dimerization is cell-specific, occurring only in those cells that exhibit retinoic acid-induced repression of AP-1 activity (Zhou *et al.*, 1999). (3) RAR and Jun or Fos proteins compete for binding to limiting amounts of transcriptional co-activators (Kamei *et al.*, 1996). (4) RAR inhibits the activity of Jun amino-terminal kinase (JNK), thereby interfering with c-Jun activation (Caelles *et al.*, 1997).

The inhibitory effect of RAR on AP-1 activity is thought to be responsible for the clinical effects of retinoids as anti-neoplastic, anti-arthritic, and immunosuppressive agents. In the case of the stromelysin gene, inhibition can occur in the absence of protein synthesis, indicating a primary transcriptional response (Nicholson *et al.*, 1990). All three subtypes of RAR (α, β, and γ) can repress transcriptional induction of the human collagenase gene (Schüle *et al.*, 1991). RXR in the presence of its ligand, 9-*cis* retinoic acid, has also been shown to inhibit AP-1 activation (Salbert *et al.*, 1993). Multiple regions of RAR, including the DNA-binding domain, are required for inhibitory activity and both c-Jun homodimers and c-Jun–c-Fos heterodimers can be repressed. The inhibitory effect of retinoic acid can be removed by treatment of cells with 12-*O*-tetradecanoylphorbol-13-acetate (TPA) and also by treatment with both c-Jun and c-Fos (Yang-Yeng *et al.*, 1991). At least *in vitro*, inhibition of AP-1 activity requires an excess of ligand-activated RAR, whereas removal of inhibition requires an excess of AP-1 protein. This reciprocal effect is due to an interaction between RAR and AP-1 proteins that results in mutual loss of DNA-binding activity. The interaction appears to be weak in that a stable complex between c-Jun and RARβ could not be detected in the absence of a chemical cross-linking agent (Yang-Yen *et al.*, 1991). In spite of the apparently weak interaction, however, endogenous RAR levels appear sufficient to allow inhibition of AP-1 activity.

15.19 ROLE OF VITAMIN A IN BONE METABOLISM AND EMBRYONIC DEVELOPMENT

15.19.1 Effects of Vitamin A on Bone Metabolism

The gene expression of the nuclear retinoid receptors (RARα, RARβ, RARγ, RXRα, RXRβ, but not RXRγ) and one of the cellular retinoid-binding proteins (CRBP-I) has been demonstrated in adult rat tibia (Harada *et al.*, 1995). Retinoic acid (either all-*trans* or 9-*cis*) increased the mRNA levels of the *RARβ* and *CRBP-I* genes. These observations indicate that, in bone, the actions of vitamin A are exerted through these nuclear receptors by regulating target gene expression, and through CRBP-1 by modulating the intracellular transport of vitamin A.

The regulatory effects of vitamin A on bone metabolism appear somewhat contradictory. Oreffo *et al.* (1998) reported that retinol and all-*trans* retinoic acid stimulate bone resorption through direct effects on osteoclasts. On the other hand, retinoic acid acts as an inducer of osteocalcin (Oliva *et al.*, 1993), matrix Gla protein (Cancela & Price, 1992), and bone morphogenetic proteins (Rogers *et al.*, 1992).

15.19.2 Consequences of Retinoid Imbalance during Embryogenesis

All-*trans* retinoic acid, and also 9-*cis* retinoic acid, 4-*oxo*-all-*trans* retinoic acid, and 3,4-didehydro-all-*trans* retinoic acid, are implicated in pattern formation, i.e. the development of the embryo in a spatially organized fashion. In particular, endogenous retinoids play a role in the anterior-posterior development of the central body axis and in limb development. The concentrations of retinoic acid and nuclear retinoid receptors in the developing embryo are highly regulated, both spatially and temporally. Treatment of embryos with exogenous retinoids produces teratogenic effects, while maternal insufficiency of vitamin A during pregnancy results in either fetal death or severe congenital malformations.

After the formation of the primary tissue layers in vertebrate embryogenesis (gastrulation), the anterior-posterior axis is established with the formation of the neural tube. Cranial neural crest cells arising from the developing hindbrain migrate and differentiate to form parts of the face, both jaws, periocular tissues, bones of the ear, the thymus, and the septa of the heart and its major arteries. Many of the malformations resulting from retinoic acid excess occur in tissues whose origins can be traced back to the early development of the neural tube. Malformations may arise through the premature induction of gene transcription by retinoic acid or through death of many neural crest cells by apoptosis. Abnormalities are frequently found in craniofacial structures, the central nervous system, the heart, and the thymus.

The developing eye is the most sensitive organ to vitamin A deprivation. The eye may be completely absent; otherwise, the lens is reduced in diameter and the lens cells fail to elongate and form the necessary crystalline fibers. Cardiovascular defects include failure of closure of the interventricular septum in the heart. In the developing central nervous system, vitamin A deficiency leads to incomplete formation of the myelencephalon (part of the posterior hindbrain) and a failure of the neural tube to extend neurites into the periphery. Development of the lungs, diaphragm, and urinogenital tract is also impaired. The above effects have been reported in animals. Vitamin A deficiency sufficient to cause morphological abnormalities in human embryonic development is rare, although functional defects, especially of the lungs, are quite common.

Boylan and Gudas (1992) showed that the level of CRABP-I expression influences the metabolism of retinoic acid in the cytoplasm of differentiating embryonic stem cells: the higher the CRABP-I level, the faster the metabolism of retinoic acid to inactive metabolites. Retinoic acid not immediately metabolized presumably moves to the nuclear receptors and initiates the activation of responsive genes, leading to differentiation. The level of CRABP-I expression also influences the sensitivity of the stem cells to the addition of retinoic acid: the higher the level of CRABP-I, the greater the concentration of retinoic acid required to initiate differentiation (Boylan & Gudas, 1991). It appears, therefore, that the function of CRABP-I in embryonic stem cells is to regulate the concentration of intracellular retinoic acid. This function is very important in the developing embryo where excessive retinoic acid could lead to teratogenesis.

The distributions of the nuclear retinoid receptors (RARs and RXRs) in the mammalian embryo are shown in Table 15.2. The characteristic spatial and temporal expression of the different receptor subtypes correlates with pattern formation, suggesting that they play a role in morphogenesis and other developmental events (Underhill *et al.*, 1995).

15.19.3 Anterior-Posterior Development of the Central Body Axis

Pattern formation along the central anterior-posterior axis is determined by the expression patterns of *Hox* genes in the developing embryo. Within the Hox locus of the chromosome, *Hox* genes are arranged in decreasing order of sensitivity to retinoic acid in the 3′ to 5′ direction. Similarly, *Hox* genes at the 3′ ends of the cluster are expressed earlier in embryonic development and in more anterior regions, while *Hox* genes closer to the 5′ ends of the cluster are expressed at later times in development and in posterior regions of the embryo. The order of induction will correspond to the gene expression patterns (anterior to posterior) and will be influenced by the concentration of retinoic acid (Gudas, 1994).

15.19.4 Limb Development

Three regions of the developing limb bud have been identified that profoundly influence limb development, namely the zone of polarizing activity (ZPA), the apical ectodermal ridge (AER), and the progress zone. The ZPA regulates the anterior-posterior axis, whereas the AER is required for proper formation of structures along the anterior-posterior axis. The progress zone receives signals from both the ZPA and the AER and translates this information into distinct cell identities (Johnson *et al.*, 1994). Several members of the fibroblast growth factor (FGF) family are expressed in the AER and these may initiate the proliferation of the mesenchyme in normal limb development.

Transplantation of mesenchyme from the ZPA results in mirror-image duplication of the limb along the anterior-posterior axis, a property called polarizing activity. A secreted protein called Sonic hedgehog has been identified as the likely mediator of polarizing activity within the limb bud. The synthesis of Sonic hedgehog is regulated by *Hoxb-8*, a primary response gene whose expression is induced by retinoic acid (Tabin, 1995). When a bead containing retinoic acid is placed on the anterior of a limb bud, it causes polarizing activity by inducing an ectopic ZPA adjacent to the bead. This effect is due to the retinoic acid-induced secretion of Sonic hedgehog, which is localized and does not extend around the bead.

15.20 EFFECTS OF VITAMIN A ON THE IMMUNE SYSTEM

15.20.1 Introduction

Vitamin A deficiency is strongly associated with depressed immunity (Smith *et al.*, 1987) and therefore a combination of deficiency and infectious disease is potentially fatal, especially among infants and children. Randomized trials in Sumatran (Tarwotjo *et al.*, 1987) and Indian (Rahmathullah *et al.*, 1990) preschool-age children showed that vitamin A supplementation decreased mortality due to infection by 96 and 54%, respectively.

15.20.2 *In situ* Synthesis of Retinoic Acid

The active metabolite of vitamin A, retinoic acid, modulates a broad range of immune responses. The capacity of cells to synthesize retinoic acid depends on the expression of retinaldehyde dehydrogenase (RALDH), the enzyme that catalyzes the irreversible conversion of retinaldehyde to retinoic acid. RALDH expression is restricted largely to cell types at barrier interfaces, such as the lung, skin, and gut. In adult mammals, three RALDH isoforms (RALDH1, RALDH2, and RALDH3) have been described in gut-associated cells, including small- and large-intestinal epithelial cells, mesenteric lymph node stromal cells, and dendritic cells (DCs) in Peyer's patches, small-intestinal lamina propria, and mesenteric lymph nodes (Cassani *et al.*, 2012). Macrophages in the lamina propria also produce retinoic acid (Denning *et al.*, 2007).

15.20.3 Effects of Vitamin A on the Formation and Size of Peripheral Lymphoid Organs

Vitamin A deficiency usually involves a decrease in the size of the major peripheral lymphoid organs – the spleen and the thymus. Thymus involution is probably one of the major causes for immune deficiency associated with hypovitaminosis A. The response of spleen cells to mitogens is reduced in vitamin A-deficient animals. Vitamin A repletion has a thymotropic effect: the thymus is repopulated and the mitogenic response of spleen cells is restored to normal (Lotan, 1986).

Lymphoid tissue inducer (LTi) cells, a subset of group 3 innate lymphoid cells (ILC3 cells), are essential for the formation of lymph nodes during embryogenesis (Hazenberg & Spits, 2014). Van de Pavert *et al.* (2014) demonstrated that mouse embryonic secondary lymphoid organs contain pre-ILC cells and that local differentiation of these progenitors into mature LTi cells is controlled by maternal vitamin A intake and cell-autonomous retinoic acid signaling in the uterus. They also established that maternal levels of dietary vitamin A control the size of secondary lymphoid organs.

15.20.4 Effects of Vitamin A on Cells of the Innate Immune System

Cells of the innate immune system are targets of active vitamin A metabolites and synthetic retinoids in innate immunity.

15.20.4.1 Neutrophils

Retinoic acid in conjunction with hematopoietic growth factors plays a crucial role in the terminal differentiation of promyelocytes to neutrophils in the bone marrow (Tsai & Collins, 1993). In vitamin A-deficient rats, when plasma and liver retinol concentrations are 1% of normal, the bone marrow retinol concentrations are increased fourfold due to sequestration of retinol mobilized from liver stores (Twining *et al.*, 1996a). Therefore, the bone marrow is able to supply neutrophils during dietary deprivation of vitamin A. However, the neutrophils of vitamin A-deficient rats show impaired chemotaxis, a lower rate of adhesion to pathogenic organisms, and reduced respiratory burst activity (Twining *et al.*, 1997). Neutrophil azurophilic granules from vitamin A-deficient rats contain decreased concentrations of cathepsin G, but the concentrations of other neutrophil proteases are not altered (Twining *et al.*, 1996b). In addition to its proteolytic activity, cathepsin G activates neutrophil progelatinase B and procollagenase and acts as an antimicrobial peptide. This impairment of neutrophil function probably contributes to the depressed immunity observed under vitamin A deficiency.

15.20.4.2 Monocytes/Macrophages

Retinoic acid increased the proliferation of granulocyte/macrophage precursor cells even in the presence of maximally stimulating concentrations of colony-stimulating factor (Tobler *et al.*, 1986). Retinoids can modulate macrophage function by enhancing phagocytosis and IL-1 production (Dillehay *et al.*, 1988). An increase in transglutaminase activity during macrophage activation is thought to contribute to the enhanced functional capacity of activated macrophages. Moore *et al.* (1984) reported that physiological concentrations of retinoic acid specifically induced the synthesis of transglutaminase by cultured mouse macrophages. Retinoic acid stimulated the capacity of macrophages to phagocytose and kill the unicellular parasite *Trypanosoma cruzi* (Wirth & Kierszenbaum, 1986). Transglutaminase activity appeared to be involved in the retinoid effect, because inhibition of the enzyme activity canceled the effect. Retinoic acid at physiological concentrations does not act alone in inducing IL-1β production in human monocytes; rather it strongly enhances phorbol ester-induced IL-1β production at the level of transcription (Matikainen *et al.*, 1991). This effect of phorbol ester is mediated through the activation of its cellular receptor, protein kinase C. Macrophages from normal healthy donors are usually not cytotoxic to tumor cells *in vitro*, but they can be rendered tumoricidal by interaction *in vitro* with bacterial products, lymphokines, or vitamin A (retinyl palmitate) (Tachibana *et al.*, 1984). Macrophages from rats or mice receiving very high intakes of retinyl palmitate also exhibit an increased ability to kill tumor cells (Tachibana *et al.*, 1984; Moriguchi *et al.*, 1985).

15.20.4.3 Dendritic Cells

Retinoic acid signaling regulates the final differentiation step of pre-cDCs between two functionally distinct splenic and intestinal DC subsets. Either acute or chronic deprivation of retinoic acid signaling in mice causes a selective loss of the splenic subset (Klebanoff *et al.*, 2013).

Geissmann *et al.* (2003) investigated the effects of retinoids on cultured immature Langerhans cell-type DCs and dissected the nuclear pathways that underlie these effects. In the absence of inflammatory signals, retinoids induced DC apoptosis through binding to RARα–RXR heterodimers. In the presence of inflammatory signals, retinoids activated the DCs through binding to RARα–RXR heterodimers and also to RXR–RXR homodimers or RXR–orphan receptor heterodimers. The apoptotic signaling pathway transduced by RARα –RXR was distinct from the activation pathway induced by RARα–RXR. Activation of immature DCs via cross-talk with pro-inflammatory cytokines ensured their survival and induced their differentiation into mature antigen-presenting DCs. On the basis of these results, the opposite effects of vitamin A on the immune response that have been reported could be due to the presence or absence of inflammatory signals and to the dose of retinoids.

15.20.4.4 Natural Killer Cells

Vitamin A deficiency in rats reduced the absolute number of natural killer (NK) cells in blood and to a lesser extent in spleen. The activity per NK cell was normal in blood and only slightly, but significantly, reduced in the spleen (Zhao *et al.*, 1994). When vitamin A-deficient rats were fed a retinoic acid supplement, the number of NK cells equaled those of control rats and NK cell cytotoxicity was significantly elevated (Zhao & Ross, 1995). NK cell cytotoxicity was increased in vitamin A-deficient rats when their production of cytokines was stimulated by polyinosinic acid:cytidylic acid (Ross, 1996).

15.20.4.5 Innate Lymphoid Cell Subsets ILC1, ILC2, and ILC3

ILCs provide early innate immunity against microbes in a variety of human tissues (Hazenberg & Spits, 2014). Goverse *et al.* (2016) studied the effects of vitamin A deficiency on RORγt$^+$ ILC3 cells and the formation of enteric lymphoid tissue in the mouse. Vitamin A deficiency resulted in a specific reduction of ILC3s, including LTi cells, and a diminished production of IL-17 and IL-22. Also, the numbers of postnatally developed solitary intestinal lymphoid tissues (SILTs) were reduced, but the numbers of prenatally developed Peyer's patches were not altered. Paradoxically, only the NCR$^-$ subset was enhanced when the vitamin A-deficient mice received dietary retinoic acid. This suggested that NCR$^+$ cells, which may arise from NCR$^-$ cells, need a different trigger than retinoic acid for the induction of NCR expression. Blockade of retinoic acid signaling shortly after birth reduced the numbers of NCR$^-$ and NCR$^+$ ILC3s without affecting LTi cells or the postnatal development of enteric lymphoid tissue. This demonstrated that the onset of retinoic acid signaling might take place before RORγt is expressed on LTi cells.

15.20.5 Homing of Innate Lymphoid Cells to the Gut

The ILC1, ILC2, and ILC3 subsets can actively migrate to the gut and gut-associated mesenteric lymph nodes by means of homing receptors expressed on these cells. There are two gut-homing receptors, $\alpha_4\beta_7$ and CCR9. $\alpha_4\beta_7$ acts as an adhesion molecule to bind MadCAM-1, which is widely expressed on intestinal endothelial cells in the small intestine and colon. CCR9 mediates chemotaxis when activated by CCL25, a chemokine that is specifically expressed in the small intestine (Zabel *et al.*, 1999). The homing receptor required for migration of ILCs to mesenteric lymph nodes is CCR7. ILC migration to the spleen does not involve specific homing receptors. ILC1 and ILC3 cells can migrate from lymphoid tissues to the gut by switching expression of homing receptors from CCR7 to $\alpha_4\beta_7$ and CCR9 (Kim *et al.*, 2015). DCs in mesenteric lymph nodes produce retinoic acid, which is a key factor in up-regulating the expression of both $\alpha_4\beta_7$ and CCR9 on ILC1 and ILC3 cells. Retinoic acid activates RARα, which induces transcription of the genes encoding the two receptors. Only ILC1 and ILC3 cells undergo the retinoic acid-dependent homing receptor switch. In contrast, ILC2 cells acquire gut-homing receptors in a largely retinoic acid-independent manner during their development in the bone marrow and can migrate directly to the intestine. ILC2 cells do not highly express $\alpha_4\beta_7$ and rely on CCR9 for migration to the small intestine. Without retinoic acid, gut-homing receptors on ILC1 and ILC3 cells would not be expressed and defective mucosal immunity would arise.

Ruiter *et al.* (2015) observed that retinoic acid acts in synergy with IL-2 and other ILC-activating cytokines to induce production of effector cytokines in human ILCs. IL-5 and IL-13 were induced in ILC2 cells, and IFN-γ was induced in ILC1 and ILC3 cells. In addition, retinoic acid in combination with activating cytokines enhanced the expression of the gut-homing integrin $\alpha_4\beta_7$ in all three types of ILCs. In contrast, retinoic acid alone completely inhibited the IL-2-induced expression of the skin-homing receptor CLA in ILCs.

15.20.6 Effects of Vitamin A on Antibody Production

Immunoglobulin (Ig)A is produced in larger amounts than any other antibody isotype and is the predominant antibody in the gut mucosa. Activated B cells that undergo IgA isotype switching in gut-associated lymphoid tissue (GALT) and mesenteric lymph nodes gain entry to the systemic circulation and home to the intestinal lamina propria where they reside as plasma cells. Retinoic acid has a specific activity for IgA isotype switching and this activity is more selective than that of TGF-β1, a well-known switching factor for mouse IgA and IgG2b. Seo *et al.* (2013) showed that retinoic acid and TGF-β synergize to increase IgA production, but not production of IgG2b; rather, retinoic acid diminished the TGF-β-mediated IgG2b response.

Retinol is essential for proliferation of activated human B cells (Buck *et al.*, 1990) and, moreover, is a normal modulator of B cell function (Blomhoff *et al.*, 1992). In a study of the effects of retinoids on B cell immune function in newborn infants (Ballow *et al.*, 1996), retinoic acid enhanced the synthesis of IgM in antigen-stimulated umbilical cord blood mononuclear cells, whereas in adult peripheral blood mononuclear cells it enhanced the synthesis of IgG. IgM is the first antibody produced during a primary immune response and is 15 times more effective than IgG in activating the complement cascade. Highly purified T cells preincubated with retinoic acid released a factor (probably a cytokine) which acted on cord blood B cells to augment IgM synthesis.

Infantile tetanus arising from infection with *Clostridium tetanii* remains a significant cause of morbidity and mortality in parts of the developing world (Henderson *et al.*, 1988), even though it is preventable by vaccination with tetanus toxoid (TT). Kinoshita *et al.* (1991) reported that vitamin A deficiency in the rat led to a decreased production of TT-specific IgM and IgG antibodies in both the primary and secondary responses. However, levels of total plasma IgG (as opposed to anti-TT IgG) were abnormally elevated. When the rats were repleted with retinol 1 day after primary immunization with tetanus toxoid, anti-TT IgM and IgG concentrations in both the primary and secondary responses reached vitamin A-sufficient control levels. When the deficient rats were repleted 2 days before the booster immunization, the secondary response was normal in terms of antibody concentrations and class switching from anti-TT IgM to IgG. These experiments showed that immunological memory to TT could be established and maintained in the vitamin A-deficient rat. In the early retinol-repleted rats, the increase in anti-TT IgG production during the secondary response persisted for 26 days without any additional retinol in the interim. This strong secondary response was specific for the IgG2a and IgG2b subclasses of IgG and reflected the previous development of subclass-specific memory B cells (Kinoshita & Ross, 1993). It appeared that the only vitamin A-dependent process was that of B cell differentiation, either from the naive B cell to the antibody-secreting plasma cell during the primary response or from the quiescent memory B cell to the plasma cell during the secondary response. The observation that the total plasma IgG of vitamin A-deficient rats is elevated above normal shows that the B cells are at least capable of producing antibodies, albeit not the *specific* antibodies required to deal with a particular antigen. This would explain reports of decreased host defense against certain types of infection in human vitamin A deficiency.

Pasatiempo *et al.* (1990) studied the effects of vitamin A deficiency on the antibody response to T-independent type 1 and type 2 bacterial antigens and to protein T-dependent antigens (see Table 15.5 for summary). The type 2 antigens were capsular polysaccharides from *Streptococcus pneumoniae*, type III and *Neisseria meningitidis*, serotype C. The former causes meningitis and otitis media as well as pneumonia in humans, while the latter causes meningitis. The type 1 antigens were lipopolysaccharides from *Pseudomonas aeruginosa*, an opportunistic pathogen associated with respiratory and ocular infections, and from *Serratia marcescens.* The T-dependent protein antigens were tetanus toxoid and sheep red blood cells. Young growing male rats raised on vitamin A-deficient or sufficient diets were immunized either near 40 days of age, before signs of vitamin A deficiency were apparent, or near 47 days when symptoms of deficiency were beginning to manifest. The results showed that asymptomatic vitamin A-deprived rats, immunized when they were still growing, had a markedly reduced specific antibody response to the bacterial polysaccharides (type 2 antigens) and to T-dependent protein antigens. Symptomatic vitamin A-deprived rats, immunized later, were essentially unable to produce specific antibodies to these antigens. Repletion of vitamin A-deficient rats with retinol near the time of immunization restored antibody production to control quantities. In striking contrast, vitamin A-deprived rats immunized with bacterial lipopolysaccharides (type 1 antigens) produced an antibody response indistinguishable from vitamin A-sufficient animals. In summary, specific antibody production in response to T-independent, type 2 antigens and T-dependent antigens depends on adequate vitamin A status. Antibody responses to type 1 antigens (at least those tested) are not affected by vitamin A status.

Lipopolysaccharide (LPS), a component of the outer surface of gram-negative bacteria, is a mitogen, which can stimulate B cell proliferation. Many of the immunogenic actions of LPS are mediated by macrophages through the release of cytokines, principally TNF-α and IL-1β, that in turn stimulate release of additional soluble factors. Arora and Ross (1994) showed that co-immunization with tetanus toxoid and LPS significantly elevated the IgG concentration in the plasma of vitamin A-sufficient and deficient rats as compared with identical rats immunized with tetanus toxoid alone. In a similar set of experiments, Pasatiempo *et al.* (1994) showed that even symptomatic vitamin A-deficient rats were able to mount a strong antibody response to SSS-III after co-immunization with SSS-III and LPS. These observations were not unexpected in view of the normal antibody response to LPS (Table 15.6) and its known adjuvant properties (McGhee *et al.*, 1979). Arora and Ross (1994) further showed that the adjuvant effect of LPS is mediated through the release of TNF-α. TNF-α is known to augment the capacity of macrophages to release IL-1 and prostaglandin E_2, among many other effects. These results demonstrate that vitamin A-deficient rats fail to produce specific antibodies to protein antigens and type 2

TABLE 15.5
Antibody Production in Vitamin A-Depleted Rats after Immunization with Bacterial or Protein Antigens

	Antigen		AntibodyVitaminA-
Bacterium Production in Depleted Rats	**Composition**	**Type**	
Streptococcus pneumoniae, type III	Polysaccharide (SSS-III)	TI, type 2	Impaired
Neisseria meningitidis, serotype C	Polysaccharide	TI, type 2	Impaired
Pseudomonas aeruginosa, FD-type 1	Lipopolysaccharide	TI, type 1	Normal
Serratia marcescens	Lipopolysaccharide	TI, type 1	Normal
—	Tetanus toxoid (protein)	TD	Impaired
—	Sheep red blood cells (protein)	TD	Impaired

Abbreviations: TI, T-independent; TD, T-dependent.

TABLE 15.6
Relations among Various Biological Properties of Carotenoids and α-Tocopherol

Compound	Provitamin A Status[a]	Chemopreventive Activity in 10T½ cells	Induction of Gap Junctional Communication	Relative Inhibition of Lipid Peroxidation
β-Carotene	++++	++++	++++	++
Canthaxanthin	–	++++	++++	+++
Lutein	–	+	+++	+++
α-Carotene[b]	++	++	+	+
Lycopene	–	+	++	++
Methyl-bixin	–	–	–	+++
α-Tocopherol	NA	+	±[c]	++++

[a]Assessed by a classical method involving growth rate determination of rodents deficient in vitamin A.
[b]α-Carotene is much less stable in cell culture than β-carotene.
[c]Tendency to enhance junctional communication, but does not reach statistical significance.
NA, not applicable.
Source: Zhang *et al.* (1991).

antigens because TNF-α and probably other soluble factors are missing or non-functional. Stimulation of cytokines by co-immunization with LPS induces specific antibody responses in vitamin A-deficient rats.

Wiedermann *et al.* (1993) studied the *in vivo* and *in vitro* immune responses in vitamin A-deficient rats to parenterally administered antigens. The serum IgG and IgM antibody responses to a T cell-dependent antigen (β-lactoglobulin) were significantly lower in the deficient rats than in the pair-fed controls, but there were no such differences in the IgG and IgM responses to a type 2, T cell-independent antigen (Ficoll conjugated to picrylsulfonic acid). This observation indicates that B cell function is not directly affected by vitamin A deficiency and the decreased antibody response to the T cell-dependent antigen might be due to cytokine production by T helper cells. However, the biliary IgA and the serum IgE antibodies against both antigens were decreased in the deficient animals. In the *in vitro* experiments, IL-2 and IFN-γ levels in supernatants from mitogen-stimulated lymphocytes from vitamin A-deficient rats were higher than control levels, whereas IL-6 levels were lower. Il-2 and IFN-γ are produced by Th1 cells, while IL-6 is produced by Th2 cells. IgE production is dependent on the presence of IL-4, which is also produced by Th2 cells. These results indicate an up-regulation of Th1 cell function and a down-regulation of Th2 cell function. Despite an increase in T cell proliferation *in vitro*, vitamin A-deficient rats had a lower delayed-type hypersensitivity than the control rats as well as suppressed antibody production.

Smith and Hayes (1987) reported that in vitamin A-deficient mice, the IgG1 and IgG3 responses to a protein antigen (hemocyanin) were impaired earlier and to a greater extent than were IgM responses. The impaired responses were attributable to fewer antibody-secreting cells (plasma cells) rather than to a decreased secretion rate per cell. The reduction in plasma cell number resulted from a defect in the ability of T helper cells to provide stimulatory signals to B cells (Carman *et al.*, 1989). The IgG responses were restored by adding retinyl acetate to the murine diet (Chun *et al.*, 1992). Carman and Hayes (1991) reported that the T cells of vitamin A-deficient mice produced excess interferon-γ (IFN-γ).

Mora *et al.* (2006) showed that retinoic acid produced by DCs resident in GALT could not by itself induce mouse B cells to secrete IgA. However, GALT DC-derived retinoic acid potently synergized with GALT DC-derived IL-5 or IL-6 to induce IgA secretion. Interestingly, retinoic acid plus IL-5/IL-6 did not induce IgA secretion in the absence of DCs, which suggests that DCs contribute additional essential factor(s). Vitamin A-depleted mice had decreased numbers of IgA-secreting B cells in the small intestinal lamina propria, but had normal serum IgA levels.

Morikawa and Nonaka (2005) reported that retinoic acid accelerates the differentiation of human B cells into plasma cells. As the effect of retinoic acid was accompanied by an increase of CD38 expression, CD38-mediated signaling may operate in this process. Uematsu *et al.* (2008) identified a subset of lamina propria DCs in the small intestine that expressed Toll-like receptor 5 (TLR5). When stimulated by the TLR5 ligand flagellin, these DCs produced retinoic acid, which induced the differentiation of naïve B cells into IgA-producing plasma cells by a mechanism independent of GALT. In the process, the $TLR5^+$ DCs also induced the up-regulation of CCR9 on the B cells, which promoted retention in the lamina propria.

15.20.7 Effect of Retinoic Acid on Production of Pro-inflammatory Cytokines

During infection, the pro-inflammatory cytokines TNF-α, IL-1, and IL-6 are produced by tissue cells at the site of inflammation. The retinoic acid produced metabolically by the intestinal $CD103^+$ DCs in response to infection signals via RARα the production of additional pro-inflammatory cytokines IFN-γ, IL-17, and IL-22 by recruited Th1 and Th17 cells (Hall *et al.*, 2011a).

15.20.8 Dependency of CD4+ T Cell-Mediated Immunity on Retinoic Acid Signaling

Pino-Lagos *et al.* (2011) established that retinoic acid signaling via RARα is critical as an early mediator in the development of CD4+ T cell-mediated immunity. In response to infection, retinoic acid signaling in CD4+ T cells is enhanced at the site of inflammation. CD4+ T cell helper function and migration to the site of inflammation are inhibited when retinoic acid signaling in these cells was disrupted. The absence of retinoic acid signaling in CD4+ T cells also alters T cell polarity, resulting in a shift from a Th1/Th17 to a Th2 phenotype. Similar findings were reported by Hall *et al.* (2011a) who proposed that the retinoic acid–RARα axis mediates both regulatory and inflammatory circuits of the immune response.

15.20.9 Differential Effects of Vitamin A on Th1 and Th2 Cell Responses

Carman *et al.* (1992) investigated the immunological effects of vitamin A deficiency in mice infected with the parasitic helminth *Trichinella spiralis.* They showed that vitamin A deficiency shifts the immune response toward an early Th1-type response and delays the development of a Th2-type response. The normal Th2 response to *T. spiralis* infection is characterized by high levels of parasite-specific IgG antibodies and increased bone marrow synthesis of eosinophils, the cells that specialize in killing extracellular parasites. The inappropriate Th1 response elicits high levels of IFN-γ, low Th2 and IgG-secreting cell frequencies, and impaired eosinophil production. Th2 cells secrete IL-5, which induces growth and differentiation of eosinophils; therefore, by suppressing Th2 cell growth through excess production of IFN-γ, one would expect decreased eosinophil production in vitamin A deficiency.

Cytokines influence the differentiation of naïve CD4+ T cells into Th1 or Th2 helper T cells. Cantorna *et al.* (1994, 1995) confirmed that the imbalance between Th1 and Th2 responses in vitamin A deficiency was caused by a distorted pattern of cytokine gene expression, in particular an excessive production of IFN-γ and IL-12. IFN-γ, by antagonizing IL-4, can inhibit proliferation of Th2 cells. IL-12 stimulates the differentiation of Th1 cells. Excessive production of IFN-γ was due to dysregulation of IFN-γ gene expression and excess antigen-presenting capacity for IFN-γ stimulation. Vitamin A repletion decreased the synthesis of IFN-γ and IL-12, limited the antigen presentation capacity to stimulate IFN-γ, and increased the Th2 cell frequency, thereby restoring the Th1/Th2 balance to normal. Retinoic acid was 100-fold more active than retinol for restoring IgG responses in vitamin A-deficient cell cultures (Chun *et al.*, 1992). Retinoic acid was shown to inhibit IFN-γ gene transcription through its role as a ligand for nuclear RAR (Cantorna *et al.*, 1996).

Hoag *et al.* (2002) found that the inhibitory effect of retinoic acid on IFN-γ synthesis was not sufficient to enhance Th2 cell frequency. Furthermore, retinoic acid had no effect on fully differentiated Th2 cells, ruling out a direct enhancing effect on IL-4 gene expression. Retinoic acid did, however, act on antigen-presenting cells, which responded by somehow maximizing Th2 cell development in an IL-4-dependent manner. Hoag and colleagues speculated that retinoic acid binding to one of the RAR–RXR receptor dimers enhances transcription of one or more genes encoding co-stimulatory molecules on the antigen-presenting cells. Iwata *et al.* (2003) found that all-*trans* retinoic acid and 9-*cis* retinoic acid, acting through the RAR, suppress Th1 development and enhance Th2 development *in vitro.* All-*trans* retinoic was shown to regulate the expression of genes essential for Th1 and Th2 development. RAR antagonists, but not a RXR antagonist, canceled the effect of retinoic acid on Th1 and Th2 development, showing that RXR is not involved in the effect. These results show a direct effect of retinoic acid on Th1/Th2 development, but the molecular mechanism remains to be elucidated.

15.20.10 Retinoic Acid-Induced Generation of Th17 Cells

Contrary to reports (mentioned later) of negative regulation of Th17 cell development by retinoic acid, there is evidence that retinoic acid is critical for the *in vivo* development and/or survival of Th17 cells (Hall *et al.*, 2011b). Mice reared on a vitamin A-deficient diet exhibited decreased generation of Th17 cells in the small intestine during steady state (Cha *et al.*, 2010). The ability of lamina propria DCs to produce IL-6, which promotes Th17 cell polarization, was reduced in vitamin A-deficient mice (Hall *et al.*, 2011a). Retinoic acid produced by stimulated TLR5+ lamina propria DCs positively regulated Th17 cell differentiation (Uematsu *et al.*, 2008). These findings suggest that a synergy between retinoic acid signaling and microbial-driven signals promotes Th17 cell development *in vivo.* Importantly, the ability of retinoic acid to support Th17 cell differentiation probably results from combined actions on DCs and T cells, because addition of retinoic acid to Th17 cell-polarizing conditions in antigen-presenting cells (APC)-less cultures did not enhance Th17 cell differentiation (Wang *et al.*, 2010).

The vitamin A-deficient diet also elicited hyperplasia of mucin-producing goblet cells and consequent reduction of gut microflora. Among the gut microflora, segmented filamentous bacteria have been shown to induce generation of Th17 cells (Ivanov *et al.*, 2009), demonstrating that disruption of the gut microflora can inhibit the differentiation of CD4+ T cells into Th17 cells. A similar decrease of intestinal Th17 cells in vitamin A deficiency was observed in a mouse model of chronic intestinal inflammation (Kang *et al.*, 2009).

The relevance of retinoic acid-induced generation of Th17 cells was demonstrated by the discovery that distinct Th17 cell subsets with specific gut-homing tropisms are generated depending on the available concentrations of retinoic acid (Wang *et al.*, 2010). At physiological concentrations of retinoic acid, Th17 cells migrated to the small intestine and induced inflammation, whereas at suboptimal concentrations of retinoic acid, Th17 cells migrated mainly to the large intestine and induced less severe inflammation. The gut tropism of Th17 cells is determined by differential expression of the gut-homing receptors $\alpha_4\beta_7$ and CCR9, regulated by the retinoic acid signal. $\alpha_4\beta_7$ directs the migration of Th17 cells to the small intestinal lamina propria and GALT, whereas CCR9 directs the Th17 cells only to the lamina propria. Although induction of Th17 cells can be somewhat suppressed by retinoic acid produced in the small intestine, the increase of Th17 cells expressing gut-homing receptors would result in an overall increase of these inflammatory cells.

15.20.11 Induction of Regulatory T Cells by Retinoic Acid during the Steady State

Yokota *et al.* (2009) identified the CD103$^+$ DC subset as one of several intestinal DC subsets that produce retinoic acid. They showed that the expression of a retinaldehyde dehydrogenase, which catalyzes the conversion of retinaldehyde to retinoic acid, is synergistically induced by GM-CSF and IL-4. The retinoic acid produced synergizes with TGF-β to promote conversion of naïve T cells into Foxp3$^+$ iTreg cells (Coombes *et al.*, 2007; Sun *et al.*, 2007). This role of retinoic acid contributes to the maintenance of gut mucosal homeostasis during the steady state. The presence of retinoic acid attenuates the inhibitory impact of co-stimulation on iTreg cell induction (Benson *et al.*, 2007). The induction of iTreg cells by retinoic acid occurs with reciprocal inhibition of Th17 cell generation (Kang *et al.*, 2007). This regulatory effect of retinoic acid is mediated at least in part by RARα, which itself is a transcription factor (Schambach *et al.*, 2007). Xiao *et al.* (2008) showed that retinoic acid inhibits expression of the cytokine receptors IL-6Rα and IL-23R, and also IFN regulatory factor 4 (IRF4), a transcription factor required for Th17 development (Brüstle *et al.*, 2007). Xiao *et al.* (2008) also showed that retinoic acid enhances TGF-β signaling in naïve CD4$^+$ T cells by increasing the expression and phosphorylation of the signaling molecule Smad3, and this results in increased *Foxp3* gene expression even in the presence of Th17-inducing IL-6 and IL-21. However, despite a large percentage decrease in Foxp3 protein production, retinoic acid could still strongly enhance the conversion of Smad3-deficient naïve CD4$^+$ T cells into iTreg cells (Nolting *et al.*, 2009). Retinoic acid, acting through IL-15-stimulated DCs, can block Treg cell generation and induce Th1 cell responses *in vitro* (DePaolo *et al.*, 2011).

Mucida *et al.* (2007) demonstrated the regulatory effect of retinoic acid on iTreg and Th17 cell development *in vivo*. Mice were orally infected with *Listeria monocytogenes* and treated with retinoic acid or an RAR inhibitor (LE540). Infected mice treated with retinoic acid exhibited a decreased frequency of mucosal Th17 cells compared to the controls (no treatment), whereas LE540-treated mice showed no apparent difference in frequency. Mucida *et al.* (2007) also showed that culturing CD4$^+$ T cells in the presence of TGF-β and IL-6 induced high levels of RORγt, the orphan receptor that is implicated in the gene transcription of Th17 cells. The addition of retinoic acid greatly reduced the levels of RORγt, thereby opposing Th17 cell development. Furthermore, retinoic acid promoted the expression of CTLA-4, a Treg cell-surface receptor, on most TGF-β-generated Treg cells. This retinoic acid-induced differentiation of CD4$^+$ T cells into Treg cells at the expense of Th17 cells, together with the induction of Treg cells in synergy with TGF-β, demonstrates a regulatory role of retinoic acid in inducing immune tolerance in the intestine (Kim, 2008).

IL-2 is required for the production of TGF-β-dependent iTreg cells and exogenously added IL-2 also suppresses Th17 cell development (Mucida *et al.*, 2007). Kim (2008) reviewed the evidence that IL-2 signaling activates the transcription factor STAT5, which in turn is involved in the expression of *Foxp3*. RARα can interact with STAT5 and co-regulate gene expression. This implies that IL-2 signaling and STAT5 are required for the retinoic acid-induced *Foxp3* expression and generation of functional iTreg cells. This, however, proved not to be the case: STAT5-deficient naïve T cells, although less efficient than wild-type T cells, could still be converted to iTreg cells in response to retinoic acid and TGF-B1 (Elias *et al.*, 2008). Therefore, the retinoic acid and TGF-β-mediated induction of *Fox3p* expression can be enhanced by the IL-2 signaling pathway, but does not absolutely require the IL-2/STAT5 signaling pathway.

In contrast to its enhancement of TGF-β-mediated induction of *Foxp3* and generation of iTregs, retinoic acid potently inhibits the TGF-β-mediated induction of IL-10 in naïve CD4$^+$ T cells, thus limiting their differentiation into Tr1 regulatory T cells. Accordingly, mice with vitamin A deficiency have increased numbers of Tr1 cells producing immunosuppressive IL-10 (Maynard *et al.*, 2009).

Several mechanisms may contribute to the enhancement of naïve CD4$^+$ T cell conversion into iTreg cells by retinoic acid. Xiao *et al.* (2008) showed that retinoic acid elicits this effect by enhancing TGF-β signaling directly in naïve CD4$^+$ T cells and inhibiting the expression of IL-6Rα, IL-23R, and IRF-4. Hill *et al.* (2008) proposed an indirect effect of retinoic acid involving CD4$^+$CD44^{high} memory T cells. These cells actively inhibit the TGF-β-induced conversion of naïve CD4$^+$ T cells by secreting IL-4, IL-21, and IFN-γ. Retinoic acid acts by preventing the secretion of these cytokines. Takaki *et al.* (2008) found that cultured iTreg cells were rapidly overwhelmed by Th2 cells because IL-4 secreted by the Th2 cells suppressed *Foxp3* expression by recruiting STAT6 to an IL-4-responsive silencer region on the *Foxp3* promoter. STAT6 binding to this site reduced TGF-β1-mediated *Foxp3* promoter activation and prevented chromatin remodeling. Retinoic acid induced the dissociation of STAT6 from the *Foxp3* promoter and enhanced TGF-β-mediated chromatin remodeling (histone acetylation) in the silencer region, thereby reverting the suppressive effect of IL-4. The dimeric retinoic acid receptor, RARα/RXRα, was shown to bind to a RARE located 300 bp upstream of the STAT6 binding site. This effect of retinoic acid does not apply to nTreg cells, because IL-4 does not suppress *Foxp3* expression in these cells.

15.20.12 Inhibition of Interleukin-2 Gene Expression and T Cell Activation by Retinoic Acid in Mice

H.K. Blomhoff and colleagues showed that retinoic acid enhances the secretion of IL-2 from human peripheral blood T cells *in vitro*, followed by increased proliferation and inhibition of activation-induced death of the T cells (Ertesvag *et al.*, 2002; Engedal *et al.*, 2004). They then examined the situation *in vivo* using a transgenic mouse strain carrying 10.7 kilobases of the murine IL-2 gene promoter/enhancer region (Ertesvag *et al.*, 2009). Surprisingly, IL-2 gene expression and T cell activation were inhibited in the transgenic mice treated with retinoic acid. Accordingly, vitamin A-deficient mice exhibited increased basal IL-2 gene expression. The inhibitory effect of retinoic acid on murine T cells was confirmed *in vitro*, indicating a different effect of retinoic acid on IL-2 gene expression in murine and human T cells. The authors concluded that vitamin A modulates IL-2 gene expression and hence T cell activation *in vivo*.

15.20.13 Effect of Retinoic Acid on Liver-Associated T Cell Tolerance

The liver is a tolerogenic organ in which food-derived antigens and antigens arising from the degradation of toxic or waste products are trapped by resident APCs and presented to naïve T cells passing through the sinusoids. Antigen presentation by immature liver DCs and non-professional APCs like liver sinusoidal endothelial cells, Kupffer cells, and hepatic stellate cells induces tolerance rather than immunity.

Dunham *et al.* (2013) found that HSCs alone were unable to directly induce proliferation and cytokine production in naïve CD4+ T cells. In the presence of DCs and TGF-β1, HSCs preferentially induced the expression of *Foxp3* in naïve CD4+ T cells: none of the other hepatic cell types were capable of this. Importantly, this Treg development was dependent on HSC production of retinoic acid, as antagonism of RARα signaling inhibited the up-regulation of *Foxp3* in naïve CD4+ T cells. Thus, HSCs preferentially induce Treg differentiation during priming in a retinoic acid-dependent manner. This finding may provide a mechanism for the tolerogenic nature of the liver and the expansion of Tregs during hepatic B virus (HBV) and hepatic C virus (HCV) infections.

15.20.14 Effect of Retinoic Acid on Effector and Memory CD8+ T Cell Differentiation

Allie *et al.* (2013) investigated the role of retinoic acid in CD8+ T cell effector and memory formation in the context of vaccinia virus infection. Using a mouse model expressing a dominant negative RARα in the T cell compartment, the absence of RAR signaling resulted in a lack of effector CD8+ T cells accompanied by an increase in the proportion of central memory CD8+ T cells. Thus, retinoic acid enhances the effector CD8+ T cell response, but at the detriment of the long-term central memory response. This property of retinoic acid has significant implications for optimal vaccine design in which vitamin A is used as an adjuvant to enhance the CD8+ T cell response (Tan *et al.*, 2011).

15.20.15 Regulation of T Cell and B Cell Gut Tropism

Dendritic cells from GALT, but not DCs from other lymphoid organs, confer gut-homing on effector T cells and memory T cells (Mora *et al.*, 2003; Johansson-Lindbom *et al.*, 2003). T cells activated in GALT migrate to the mucosa of the small intestine by way of efferent lymph, the thoracic duct, and the bloodstream. Lung DCs also induce T cell migration to the gut (Ruane *et al.*, 2013).

Iwata *et al.* (2004) found that physiological concentrations of retinoic acid enhanced the expression of the gut-homing molecules, integrin $\alpha_4\beta_7$ and chemokine receptor CCR9, in CD4+ T cells during TCR-mediated activation, whilst suppressing the expression of the skin-homing receptor CCR4. They also found that retinoic acid produced by DCs in mesenteric lymph nodes and Peyer's patches during antigen presentation is at least partly responsible for imprinting T cells with gut-homing specificity. The enhanced expression of $\alpha_4\beta_7$ on T cells by antigenic stimulation with these DCs was suppressed by the RALDH inhibitor citral and the RAR antagonist LE135, supporting the involvement of retinoic acid and the RAR. Accordingly, vitamin A deficiency caused a reduction in $\alpha_4\beta_7^+$ memory/activated T cells in GALT and a depletion of T cells from the intestinal lamina propria. Thus, vitamin A is physiologically critical for the homing of antigen-experienced T cells to the small intestine. Svensson *et al.* (2008) demonstrated a central and competing role for DC subset-induced RAR signaling and antigen dose-dependent inhibitory signals in regulating gut-homing receptor expression on CD8+ T cells.

The large majority of IgA-secreting plasma cells reside in the gut mucosa, where they provide protection against intestinal pathogens. After activation of naïve B cells in the

GALT, newly generated IgA-secreting plasma cells leave the GALT, enter the blood, and home to the lamina propria. GALT-DC-derived retinoic acid confers gut tropism on the plasma cells in the manner described for T cells (Mora & von Andrian, 2008).

Given that retinoic acid confers gut tropism to effector T and B cells, Hammerschmidt *et al.* (2011) tested their hypothesis that mucosal protection can be induced by antigen application in the presence of retinoic acid. Mice were injected subcutaneously (s.c.) with antigen *per se* or with antigen and retinoic acid. The results showed that retinoic acid administered s.c. induced increased RALDH activity in skin-draining lymph node DCs, inducing autonomous retinoic acid production. These DCs subsequently triggered expression of gut-homing molecules on activated T and B cells. Thus, retinoic acid-supplemented vaccination redirects immune responses initiated in skin-draining lymph nodes toward the small intestine and establishes a pronounced IgA response. Importantly, this immunization technique protected mice from cholera toxin-induced diarrhea and diminished bacterial loads in Peyer's patches after oral infection with *Salmonella*.

15.20.16 Stabilizing Effect of Retinoic Acid on Natural Regulatory T Cells

Treg cells are inherently unstable, as shown by their ready conversion to other cell types in the presence of certain cytokines *in vitro*. For example, IL-6 induces the conversion of nTregs (but not iTregs) to Th17 cells in a TGF-β-dependent manner (Zheng *et al.*, 2008). Zhou *et al.* (2010) found that retinoic acid not only maintains the phenotypical stability of nTregs, but also sustains their functional activities in the presence of IL-6. Pretreatment of nTreg cells with retinoic acid down-regulated IL-6R expression and JL-6R signaling, thereby restraining conversion to Th17 cells and sustaining Fox3p expression of Tregs. Adoptive transfer treated with retinoic acid to mice with established collagen-induced arthritis markedly suppressed the progression and ameliorated the severity of arthritis. Furthermore, retinoic acid was shown to alter the stability and function of Tregs from autoimmune arthritic mice, implicating that this strategy may have an important clinical value.

15.20.17 Protection against Colonic Inflammation

IL-22 regulates the expression of antimicrobial peptides and can promote both pathological inflammation and tissue repair at barrier surfaces, depending on the context (Sonnenberg *et al.*, 2011). Mielke *et al.* (2013) showed that retinoic acid protects against colonic inflammation by inducing IL-22 production by ILC3 cells and γ δ T cells. Colitis was induced in mice by dextran sodium sulfate treatment or *Citrobacter rodentium* infection. In addition, retinoic acid treatment enhanced production of the antimicrobial peptides Reg3β and Reg3γ in the colon. The attenuating effects of retinoic acid on colitis were reversed by treatment with an anti-IL-22 neutralizing antibody, demonstrating that retinoic acid mediates protection by enhancing IL-22 production. Retinoic acid is also able to induce the production of IL-22 binding protein by DCs (Martin *et al.*, 2014). This binding protein is a soluble receptor for IL-22; it has a higher affinity compared with the membrane-bound receptor and functions as a decoy receptor that controls the activity of IL-22 (Jones *et al.*, 2008).

15.20.18 Impact of Vitamin A Supplementation on the Immune System

Vitamin A deficiency is strongly associated with depressed immunity (Smith *et al.*, 1987) and a combination of deficiency and infectious disease is potentially fatal, especially among infants and children (Sommer, 1993). For example, human immunodeficiency virus (HIV)-infected Ugandan infants with low plasma carotenoid concentrations had a significantly increased risk of death from infection compared with those with high carotenoid concentrations (Melikian *et al.*, 2001). Large-dose vitamin A supplementation significantly reduced the incidence of childhood diarrhea and respiratory disease (Lie *et al.*, 1993). Meta-analyses of community-based studies in Indonesia, India, Nepal, Sudan, and Ghana have shown that vitamin A supplementation reduces childhood mortality by ~30% (Semba, 1994).

Clinical trials have suggested that vitamin A supplementation reduces morbidity and mortality in different infectious diseases, such as measles, diarrheal disease, HIV infection, and malaria (Semba, 1999). The effect of vitamin A on immune function depends on the type of infection and the immune responses involved (Semba, 1999). Puzzlingly, vitamin A supplementation has no worldwide effect on death rates from childhood pneumonia (Vitamin A and Pneumonia Working Group, 1995).

Griffiths (2000) discussed some important issues regarding vitamin A supplementation as a strategy for combating childhood mortality. Vitamin A supplementation has been shown to reduce overall mortality in children 6 months and older in the great majority of community studies (Fawzi *et al.*, 1993). As a result, public health and child health authorities have recommended that all children at risk for vitamin A deficiency receive vitamin A supplements, regardless of their nutritional or vitamin status. Vitamin A supplements are often given at the same time as immunizations during mass vaccination campaigns. Griffiths (2000) cited global studies suggesting that vitamin A supplementation may be detrimental in adequately nourished children, actually increasing the incidence of respiratory tract infections. If the results of these studies are confirmed, vitamin A supplementation may only be appropriate for targeted populations, minimizing harm while maximizing benefit.

15.21 VITAMIN A AND CANCER

The anti-cancer properties of retinoids administered as drugs are attributed to their ability to inhibit cell proliferation, induce differentiation, or trigger apoptosis of cancer cells. The actual response *in vitro* depends on the given retinoid, the type of cell, and the growth conditions (Grunt *et al.*, 1998). Chronic pharmacological administration of classic retinoids is limited by their toxicity and the search is on for synthetic retinoids that have increased anti-cancer activity and reduced *in vivo* toxicity.

15.21.1 Anti-Cancer Properties of Retinoids

Animal studies have shown that vitamin A deficiency increases the susceptibility to neoplasia due to polycyclic hydrocarbons (Cone & Nettesheim, 1973; Nettesheim & Williams, 1976; Dogra *et al.*, 1985). Retinoids suppress malignant transformation of cultured cells in which the transformation is induced by ionizing radiation or chemical carcinogens. A majority of primary human cancers arise in epithelial tissues that depend upon retinoic acid for normal cell proliferation and differentiation. Anti-tumor activity is at least partially due to induction of cellular differentiation and/or inhibition of cell proliferation (Bollag & Holdener, 1992). Retinoids are most effective in the promotion stage of carcinogenesis; they have been postulated to delay the clonal expansion of premalignant initiated cells (Lippman *et al.*, 1987).

15.21.2 Retinoids and Apoptosis

Apoptosis is a tightly regulated process of programed cell death, which involves changes in the expression of distinct genes. The physiological role of apoptosis in cancer prevention is to eliminate DNA-damaged cells whose replication could lead to mutations and possibly cancer. Retinoids induce apoptosis in many tumor cell types and therefore have potential use for cancer chemotherapy and prevention.

The induction of apoptosis by retinoids is related to cell proliferation and differentiation in various ways, depending on the cell model and on the different retinoid molecules involved. Terminal differentiation and apoptosis may be related in one of three ways: (1) retinoids first induce differentiation, and then the differentiated cells undergo apoptosis; (2) retinoids induce differentiation and apoptosis concurrently; (3) retinoids induce apoptosis in a differentiation-independent manner (Lotan, 1995).

Toma *et al.* (1997) examined the *in vitro* effects of all-*trans* retinoic acid and 13-*cis* retinoic acid on three human breast cancer cell lines in terms of inhibition of proliferation and induction of apoptosis. Two of the cell lines, MCF-7 and ZR-75.1, were estrogen receptor-positive (ER^+) and the other, MDA-MB-231, was estrogen receptor-negative (ER^-). Both retinoids were potent growth inhibitors in ER^+ cells, but high retinoid concentrations for a long period were needed to produce growth inhibition in ER^- cells. The anti-proliferative activity of all-*trans* retinoic acid was more pronounced than that of 13-*cis* retinoic acid on all three cell lines. The two retinoids induced apoptosis in MCF-7 and MDA-MB-231 cells, with the all-*trans* isomer being the more effective; however, neither retinoid had any apoptotic effect on ZR-75.1 cells. After retinoid treatment, the levels of ER mRNA in the ER^+ cells decreased in ZR-75.1 but not in MCF-7. Levels of ER expression are known to be lower in normal mammary epithelial cells than in malignant tumors, and during the tumorigenic process there is an increase of the ER level (Guilbaud *et al.*, 1990). Taking ER status as a marker of differentiation, the observed decrease of ER expression may represent a differentiation criterion in the ZR-75.1 cells. Thus, ZR-75.1 cells treated with retinoic acids may undergo a differentiation process and die in a way that is distinct from apoptosis. The same group (Raffo *et al.*, 2000) investigated the activity of four synthetic retinoids with marked and high selectivity for RARα, RARβ, RARγ, and RXRα in the same three cell lines. RAR-selective retinoids, and especially the RARα-selective compound (Ro 40-6055), were found to be the most effective growth inhibitors and apoptosis inducers in ER^+ cells, while ER^- cells were more sensitive to an RXR-selective activity.

Treatment of breast cancer cells with all-*trans* retinoic acid (Wang *et al.*, 2000) or synthetic RARα/β-selective retinoids (Vivat-Hannah *et al.*, 2001) lowers the threshold for killing by the chemotherapy agents paclitaxel (Taxol) and Adriamycin. This indicates that combination regimens of RARα/β-selective retinoids with chemotherapy agents may have potential use for the treatment of breast cancer.

Pratt *et al.* (2003) compared the regulation of the anti-apoptotic protein Bcl-2, the cell cycle proteins cyclin D1 and CDK2, and survivin (a member of the inhibitor of apoptosis family of proteins) in ER^+ MCF-7 and ZR-75 breast cancer cells by natural and synthetic retinoids selective for the α, β, and γ RARs. In both cell lines, all-*trans* retinoic acid and 9-*cis* retinoic acid most strongly down-regulated all of these proteins and also reduced percentages of S-phase cells. Inhibition of Bcl-2 and survivin expression by individual retinoids correlated well with their effects on Taxol sensitization. The synthetic RARα agonist Ro 40-6055 also decreased expression of Bcl-2 and survivin, predominantly in MCF-7 cells, although it did not produce an equivalent decrease in S-phase percentages. In agreement with this result, Schneider *et al.* (2000) used the synthetic RARα-selective agonists AM80 and AM580 to demonstrate that the RARα was the pivotal receptor mediating growth inhibition and apoptosis in ER^- SK-BR-3 and ER^+ T47-D breast cancer cells. In contrast, the RARβ-selective agonist CD417 failed to inhibit growth and apoptosis in these cells. This could be explained by the finding by F. Lin *et al.* (2000) that the RARβ receptor mediates its anti-tumor effect by inhibition of AP-1 activity in a ligand-independent manner.

15.21.3 Retinoic Acid as a Stimulator of TGF-Mediated Anti-Proliferation in Acute Promyelocytic Leukemia

Acute promyelocytic leukemia (APL) is a blood cancer caused by an abnormal build-up of promyelocytes in the bone marrow and the abnormal appearance of these cells in the blood. Promyelocytes are intermediates in the myeloid lineage of hematopoiesis in the bone marrow. Myoblasts arising from hematopoietic stem cells mature into promyelocytes, which develop into myelocytes. Three types of myelocyte, called neutrophilic, basophilic, and eosinophilic myelocytes, differentiate into neutrophils, basophils, and eosinophils, collectively called granulocytes. The uncontrolled proliferation of promyelocytes leads to a depletion of normal white and red blood cells and platelets in the body. APL typically presents as a bleeding disorder, attributable in some cases to the progressive depletion of agents that would inhibit fibrinolysis, such as α2-anti-plasmin and plasminogen-activator inhibitor (Warrell *et al.*, 1993).

The HL-60 cell line, derived from a patient with APL, proliferates continuously in suspension culture and consists predominantly of promyelocytes. HL-60 cells have the capacity to differentiate *in vitro* to a variety of different cell types of the myeloid lineage (Collins, 1987). Retinoic acid treatment of patients with APL induces remission that is associated with the terminal differentiation of leukemic cells into mature granulocytes, followed by the emergence of normal hematopoietic cells as the patient achieves remission (Warrell *et al.*, 1993). Treatment of HL-60 cells with retinoic acid also results in granulocytic differentiation and cessation of growth (Breitman *et al.*, 1980), making HL-60 cells a good tissue culture model for studying granulocytic leukemic cell differentiation.

The cytokine transforming growth factor-β (TGF-β) is a disulfide-linked homodimer that is secreted by cells as small or large latent TGF-β complexes. The small latent TGF-β complex is derived from a single gene product. In this complex, each partner of the homodimer is linked by electrostatic forces to a peptide called latency-associated peptide (LAP), which prevents association of TGF-β with its receptor. The large latent TGF-β complex is composed of two distinct gene products – the small latent TGF-β complex and an additional protein called latent TGF-β binding protein (LTBP) that is disulfide-linked to one of the LAP dimers (Figure 15.13). Activation of the small or large latent TGF-β complex involves the release of the mature TGF-β through disruption of the electrostatic interactions with LAP (Harpel *et al.*, 1992).

Kamijo *et al.* (1990) showed that TGF-β1 dose-dependently induced differentiation of human monocytic leukemia U-937 cells into cells with macrophage characteristics, but had little effect on proliferation. However, combinations of TGF-β1 and tumor necrosis factor-α (TNF-α) were synergistic in inducing differentiation and inhibiting proliferation of U-937 cells. In contrast to U-937 cells, TGF-β1 only slightly induced differentiation of promyelocytic leukemic HL-60 cells, either alone or in combination with TNF-α. Falk *et al.* (1991) showed that addition of TGF-β1 to HL-60 cells treated with a suboptimum concentration (1 nM) of retinoic acid resulted in a marked decrease in proliferation with no effect on granulocytic differentiation. The retinoic acid treatment increased the levels of TGF-β receptors and active TGF-β protein in a dose-dependent manner. Retinoic acid treatment of HL60 cells had no effect on the level of TGF-β mRNA expression. These results show that suboptimal concentrations of retinoic acid sensitize HL-60 cells to exogenous TGF-β by up-regulating the expression of type I TGF-β receptors. Nunes *et al.* (1996) used neutralizing anti-TGF-β IgG to demonstrate that the anti-proliferative activity of retinoic acid in HL-60 cells is TGF-β-dependent, but that differentiation is not. They further showed that a TGF-β-dependent effect requires cells to activate the latent form of TGF-β. These observations provide evidence that retinoic acid may stimulate a TGF-β-mediated autocrine anti-proliferative loop during differentiation of HL-60 cells.

15.21.4 Anti-Cancer Effect of Retinoic Acid via Repression of AP-1 Activity

Regulation of AP-1 activity is thought to be a key mechanism by which retinoids inhibit proliferation of breast cancer cells (Pfahl, 1993). To this end, new retinoids have

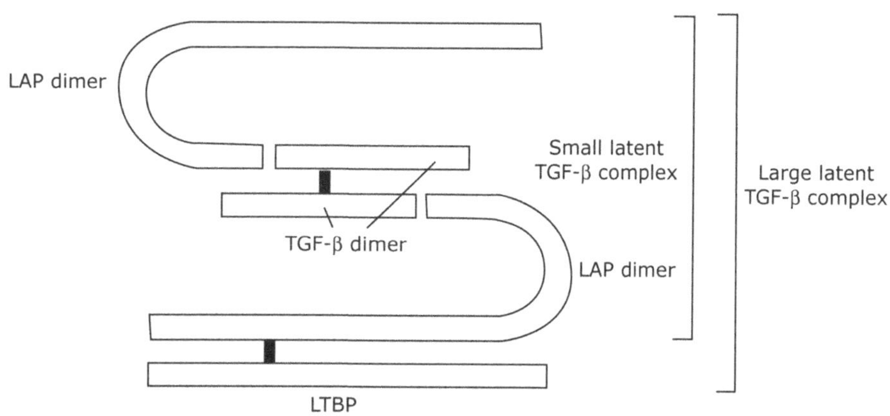

FIGURE 15.13 Schematic representation of the small and large TGF-β complexes. The short solid lines indicate disulfide bonds, one linking the two TGF-β dimers and the other linking one of the LAP dimers with LTBP. The juxtapositions of LAP and TGF-β dimers are maintained by electrostatic interactions between them.

been developed with the ability to block AP-1 activity but not induce transcription. Huang *et al.* (1997) investigated whether inhibition of AP-1 or activation of RARE is responsible for the anti-tumor promotion effect of retinoic acid. The results showed that SR11302, a retinoid with AP-1 inhibiting activity, but lacking RARE transactivating activity, markedly inhibited tumor formation. In contrast, SR11235, a retinoid with RARE transactivating activity, but lacking AP-1 inhibiting activity, did not inhibit tumor formation. These results provided the first *in vivo* evidence that inhibition of tumor promotion by retinoids is mediated by blocking AP-1 activation, not by activating RARE-dependent transcription. Pratt *et al.* (2003) found that all retinoids tested, except the RARβ-specific retinoid CD-417, were almost equally capable of inhibiting AP-1 activity.

Lin *et al.* (2002) demonstrated that interaction between the orphan receptor COUP-TF and AP-1 is involved in regulating inhibition of AP-1 activity by retinoic acid in cancer cells. They first showed that COUP-TF expression in various cancer cell lines correlated with the repressive effect of retinoic acid upon AP-1 transactivation. COUP-TF effectively suppressed AP-1 transcriptional activity by physically interacting with c-Jun to cause loss of c-Jun binding to the DNA. This interaction required the COUP-TF DNA binding domain and the c-Jun leucine-zipper domain. The authors speculated that the inhibitory effect of COUP-TF could also involve the recruitment of a transcriptional co-repressor, such as N-CoR or SMRT. Expression of both RARα/β and COUP-TF was required for optimal inhibition of AP-1 capable of inhibiting AP-1 activity by retinoic acid.

15.21.5 Induction and Regulation of the RARβ Gene

RARβ plays a critical role in mediating anti-cancer effects of retinoids. The RARβ gene maps to a region on the short arm of chromosome 3, which is frequently deleted in lung cancer (Kok *et al.*, 1987). Abnormalities in structure and expression of the RARβ gene occur with high frequency in human lung cancer (Gebert *et al.*, 1991). A progressive decrease in RARβ mRNA levels has been observed during breast carcinogenesis (Xu *et al.*, 1997).

The RARβ gene is unique among RAR genes in that its expression is highly induced by retinoic acid through a response element (βRARE) present in its promoter that is activated by RAR–RXR heterodimers. The βRARE consists of two direct repeats with a five-nucleotide spacer and is separated from the TATA box also by five nucleotides (de Thé *et al.*, 1990; Sucov *et al.*, 1990). The close apposition between the βRARE and the TATA box is striking in comparison with other RAREs that are located further upstream of the transcription start site. The induction of *RARβ* transcription is a direct response to retinoic acid treatment in that it does not require new protein synthesis. In this respect, the RARβ gene is a primary response gene whose induction occurs within 3 hours of retinoic acid addition. The autoinduction of *RARβ* transcription amplifies the effects of retinoic acid by raising the level of the receptor.

Berkenstam *et al.* (1992) proposed a model for the transcriptional regulation of *RARβ* in which the ligand-activated RAR–RXR heterodimer functionally interacts with the adjacent TFIID via an adenovirus E1A-like protein present in the cell. Kruyt *et al.* (1993) also found that a cellular E1A functions as a coregulator of retinoic acid-dependent *RARβ* promoter activation.

15.21.6 Role of COUP-TF in the Induction of RARβ Expression by Retinoic Acid

Zhang *et al.* (1994) reported that the *RARβ* promoter in a number of human lung cancer cell lines failed to respond to retinoic acid, despite high expression of RARs and RXRs. Similarly, Kim *et al.* (1995) found that retinoid refractoriness occurs during lung carcinogenesis despite the presence of functional retinoid receptors. Thus, expression of RAR and RXR in cancer cells is not sufficient for retinoic acid to induce *RARβ* transcription. B. Lin *et al.* (2000) demonstrated that COUP-TF is required for retinoic acid to induce RARβ expression, growth inhibition, and apoptosis in cancer cells. They further showed that COUP-TF could strongly induce transcriptional activity of the *RARβ* promoter in a retinoic acid- and RARα-dependent manner through its binding to a DR-8 response element in the promoter. COUP-TF, through its direct interaction with RARα, strongly enhanced the interaction of RARα with its co-activator CBP. It is likely that COUP-TF induces a RARα conformational change that is more favorable for its interaction with CBP. The results of this study demonstrated that COUP-TF induces *RARβ* promoter transcriptional activity by acting as an accessory protein for RARα to recruit its co-activator. The finding in this study that COUP-TF is underexpressed in many retinoic acid-resistant cancer cell lines suggests that a loss of COUP-TF may represent one of the important mechanisms for retinoid resistance in cancer cells.

15.21.7 Anti-AP-1 Activity of RARβ

Expression of RARβ in certain, but not all, breast cancer and lung cancer cells induces apoptosis (Liu *et al.*, 1996; Li *et al.*, 1999), indicating that apoptosis induction is not the sole mechanism mediating the anti-cancer effects of RARβ. F. Lin *et al.* (2000) evaluated anti-AP-1 activity of RARβ in cancer cells. Unlike RARα and RARγ, which repressed AP-1 activity in a retinoic acid-dependent manner, RARβ could repress phorbol ester-induced AP-1 activity in a retinoic acid-independent manner. By uniquely inhibiting AP-1 transcriptional activity, RARβ may inhibit cell proliferation often associated with cancer development.

15.21.8 Anti-Cancer Properties of Carotenoids

In humans, a significant body of epidemiological evidence correlates the intake of carotenoid-rich fruits and vegetables with protection from some forms of cancer (Ziegler, 1989). The potential use of carotenoids for the chemoprevention of cancer in humans would appear to be of great significance, as β-carotene is essentially nontoxic. However, data obtained from experimental studies using rats or mice (the well-defined cancer models) cannot be extrapolated to humans on account of the rodent's limited ability to absorb carotenoids as intact molecules and by the rapid metabolism of any absorbed carotenoid.

Studies of carotenoids in cell culture have been hindered in the past by difficulties in supplying the highly lipophilic carotenoids in a bioavailable form. Bertram *et al.* (1991) overcame this problem by the use of tetrahydrofuran as solvent, allowing the delivery of diverse carotenoids to cultured cells at high concentration in a bioavailable, micelle-like form. Using this delivery system and the 10T½ line of transformable mouse fibroblasts, Bertram's group demonstrated that many dietary carotenoids can inhibit neoplastic transformation in the post-initiation phase of carcinogenesis.

15.21.9 Carotenoids as Biological Antioxidants

The anti-carcinogenic effect of carotenoids may be due, at least partly, to their antioxidant activity. This is attributable to two main properties of carotenoids: (1) the ability to trap peroxyl free radicals implicated in lipid peroxidation (Burton & Ingold, 1984) and (2) the ability to deactivate singlet oxygen by physical quenching (Foote & Denny, 1968). The extensive conjugated double-bond system of carotenoids is important for both of these properties, but provitamin A activity is irrelevant.

15.21.9.1 Trapping of Peroxyl Radicals

β-Carotene acts as a chain-breaking antioxidant in lipid peroxidation by trapping the chain-propagating lipid peroxyl radical, LOO•, in an addition reaction (reaction 15.1). The chain-breaking activity of β-carotene is weak compared with that of α-tocopherol, the major lipid-soluble antioxidant.

$$\text{LOO}\bullet + \beta\text{-carotene} \rightarrow \text{LOO} - \beta\text{-carotene}\bullet \qquad (15.1)$$

The carbon-centered β-carotenyl radical thus formed is resonance-stabilized because of the delocalization of the unpaired electron in the conjugated polyene system. The β-carotenyl radical can react with another peroxyl radical, leading to a termination reaction (reaction 15.2).

$$\text{LOO} - \beta\text{-carotene}\bullet + \text{LOO}\bullet \rightarrow \text{LOO} - \beta\text{-carotene} - \text{OOR (non-radical)} \qquad (15.2)$$

Alternatively, the β-carotenyl radical can react with molecular oxygen to form a β-carotene-peroxyl radical (reaction 15.3).

$$\text{LOO} - \beta\text{-carotene}\bullet + O_2 \leftarrow\rightarrow \text{LOO} - \beta\text{-carotene} - \text{OO}\bullet \qquad (15.3)$$

Reaction 15.3 is reversible and dependent on the partial pressure of oxygen (pO_2) in the system. If pO_2 is sufficiently low, the equilibrium shifts to the left, reducing the amount of β-carotene-peroxyl radical. On the other hand, if pO_2 is high, the equilibrium shifts to the right, increasing the amount of β-carotene-peroxyl radical, which is capable of propagating the lipid peroxidation chain reaction. The physiological pO_2 in most tissues is below 20 torr, but in lung it is much higher – around 160 torr. Kennedy and Liebler (1992) investigated the effect of pO_2 on the ability of β-carotene to inhibit peroxidation in phospholipid liposomes (a model membrane bilayer). They found β-carotene to be an effective antioxidant up to 160 torr oxygen, with no difference in effectiveness between 160 and 15 torr; at 760 torr inhibition of peroxidation markedly declined. These data imply that β-carotene could be as effective an antioxidant in lung as in other tissues. Canthaxanthin and astaxanthin, which possess oxo groups at the 4 and 4′-positions in the β-ionone ring, are purported to be more effective than β-carotene in trapping peroxyl radicals (Terao, 1989). It may be no coincidence that the length of the β-carotene molecule approximates to the width of a typical cell membrane. With an orientation perpendicular to the plane of the membrane, the conjugated double-bond system of the carotene molecule could trap radicals at any depth in the membrane.

15.21.9.2 Quenching of Singlet Oxygen

Carotenoids deactivate singlet oxygen ($^1O_2$) by physical quenching, which involves the transfer of excitation energy from $^1O_2$ to the carotenoid, forming ground state triplet oxygen ($^3O_2$) and triplet excitated carotenoid (reaction 15.4).

$$^1O_2 + \text{carotenoid} \rightarrow {}^3O_2 + {}^3\text{carotenoid} \qquad (15.4)$$

The energy is harmlessly dissipated through rotational and vibrational interactions between the triplet carotenoid and the surrounding medium to regenerate ground state carotenoid (reaction 15.5).

$$^3\text{Carotenoid} \rightarrow \text{carotenoid} + \text{thermal energy (heat)} \qquad (15.5)$$

The rate of quenching is a function of the length of the conjugated polyene chain and parallels the protective action of carotenoids (Foote *et al.*, 1970). Lycopene, an open chain carotenoid, is approximately twice as effective as β-carotene in quenching $^1O_2$ (Di Mascio *et al.*, 1989).

15.21.9.3 Prooxidant Actions of Carotenoids

The antioxidant activity of carotenoids may shift into prooxidant activity, depending on the redox potential of the carotenoid molecules as well as on the biological environment in which they act. If an inappropriate prooxidant activity were to develop in normal cells, the reactive oxygen metabolites generated would lead to damage of cellular lipids, proteins, and DNA, and possibly induce neoplastic transformation. In contrast, if carotenoids were to act as prooxidants in already transformed cells, the production of cytotoxic products would be beneficial (Palozza, 1998).

15.21.10 Enhancement of Gap Junctional Communication by Retinoids and Carotenoids

There is increasing evidence that one biochemical mechanism underlying the chemopreventive action of retinoids is enhanced gap junctional communication of growth controlling signals. Hossain *et al.* (1989) showed good correlation between retinoid inhibition of neoplastic transformation and enhanced gap junctional communication in carcinogen-initiated 10T½ cells. Retinoid-enhanced junctional communication is achieved in mouse 10T½ fibroblasts (Rogers *et al.*, 1990) and in human skin (Guo *et al.*, 1992) through the synthesis of the gap junction protein, connexin43, and its mRNA.

Zhang *et al.* (1991) demonstrated excellent correlation between the inhibition of neoplastic transformation of 10T½ cells by carotenoids and their ability to enhance gap junctional communication. The magnitude of induced junctional communication was similar to that induced by retinoids, but required higher concentrations of carotenoids (up to 1,000-fold) and longer treatment times (3–4 days vs. 1 day for retinoids). The order of potency was β-carotene/canthaxanthin (approximately equipotent), lutein, lycopene, and α-carotene. The antioxidant activity of the carotenoids was found not to correlate with their ability to inhibit neoplastic transformation or their ability to enhance junctional communication. Furthermore, the antioxidant α-tocopherol exhibited only limited activity in the transformation assay, and methyl-bixin was inactive (Table 15.6). These data suggest that the antioxidant properties of carotenoids are not significant in their action as inhibitors of neoplastic transformation of 10T½ cells. Three of the carotenoids tested (canthaxanthin, lutein, and lycopene) are reported to be without provitamin A activity in mammals, thus their activity in the 10T½ cells is presumably not attributable to conversion to retinoids. The carotenoids enhanced gap junctional communication by up-regulating *connexin43* gene expression (Zhang *et al.*, 1992).

15.21.11 Potential of Commercial Retinoids as Anti-Cancer Agents

Altucci and Gronemeyer (2001) have discussed the therapeutic potential of commercial retinoids as anti-cancer agents. As previously mentioned, the retinoid receptors mediate two distinct types of action: (1) cell differentiation by transcriptional activation through specific response elements on the DNA and (2) anti-proliferative activity by inhibition of AP-1. Because the two types of action are mechanistically distinct, various laboratories have investigated whether novel retinoids can be synthesized that inhibit AP-1 but do not activate transcription. The potential advantage of such 'dissociated' retinoids as receptor ligands is that they could be therapeutically useful in the treatment of hyperproliferative and inflammatory diseases without the side effects resulting from the activation of retinoid-responsive genes. Synthetic retinoids having this dissociation ability have been described (Fanjul *et al.*, 1994; Chen *et al.*, 1995; Nagpal *et al.*, 1995; Li *et al.*, 1999), but they display distinct receptor selectivities and their differential action is limited to only some of the RAR subtypes or isoforms. The aim is to discover retinoids that are completely specific for AP-1 antagonism through all the retinoid receptors. Chronic pharmacological administration of retinoids is limited by their potential toxicity, and the physician and patient have to evaluate the risk/benefit ratio.

15.22 VITAMIN A DEFICIENCY

15.22.1 Animals

The first sign of vitamin A deficiency in animals is loss of appetite. In young animals, the decreased food intake results in lack of growth. Controlled experiments using laboratory animals have shown that loss of appetite is not due to poor palatability of the vitamin A-deficient diet, nor is it due to impaired taste function (Anzano *et al.*, 1979). Metabolic disturbances caused by vitamin A deficiency may be sufficiently detrimental to the animal that it reduces its food intake in order to minimize these disturbances; however, the neurophysiological control systems are unknown.

When rats are deprived of vitamin A, mucus-secreting epithelia such as found in the trachea, certain parts of the urinogenital tract, and the conjunctiva are replaced by a keratin-producing squamous epithelium that is not secretory. This results in a drying-up (xerosis) of the mucous membranes with loss of function and a greater susceptibility to infection. Mucus-secreting goblet cells in the intestinal crypts are also reduced in number. Reproductive changes in the male rat include shrinkage of the testes and atrophy of the accessory sex organs. The effects on the seminiferous tubules include degeneration of the germinal epithelium, decrease in tubule size, and a halting spermatogenesis. These histological and reproductive changes can be prevented or corrected by restoring vitamin A to the animal's diet. Female rats deprived of vitamin A fail to maintain pregnancy and may resorb their fetuses. This is more likely to be a direct effect on the fetus than a reduction in placental transport of nutrients (Anon, 1977).

In adult cattle, a mild deficiency of vitamin A is associated with roughened hair and scaly skin. Prolonged deficiency affects the cornea, which becomes dry, soft, and cloudy. Bulls continue to produce viable spermatozoa even when blindness through xerophthalmia has developed. Continued vitamin A deprivation leads to degeneration of the seminiferous tubules with a consequent reduction in semen volume and sperm count, and an increased production of abnormal spermatozoa. The lack of protective mucus in the alimentary and respiratory tracts of cattle leads to scours and pneumonia, resulting often in death. In pigs, compression of the brain, due to improper modeling of bone, gives rise to nervous disorders such as unco-ordinated movements and convulsions. Vitamin A-deficient poultry suffer a high mortality rate. Early signs of deficiency include retarded growth, weakness, ruffled plumage, and a staggering gait. The keratinization of intestinal epithelia leads to parasitic infestations and the impaired production of antibodies reduces the bird's resistance to infections such as coccidiosis. In mature birds suffering from severe vitamin A deficiency, egg production and hatchability are reduced.

15.22.2 Humans

Sommer (2008) has presented an historical overview of vitamin A deficiency and clinical disease.

The clinical effects of vitamin A deficiency in adults are usually seen only in people whose diet has been deficient for a long time in both dairy produce and vegetables. An early sign of vitamin A deficiency is night blindness, which is caused by an insufficient amount of visual purple in the retina. In more advanced deficiency, the epithelial cells of the skin and mucous membranes lining the respiratory, gastrointestinal, and urinogenital tracts cease to differentiate, and lose their secretory function. The undifferentiated cells are flattened and multiply at an increased rate, so that the cells pile up on one another and the surface becomes keratinized. This condition promotes dry skin and loss of hair sheen. The lack of protective mucus in the affected mucosae leads to an increased susceptibility to infections.

Xerophthalmia, a disease which mainly affects very young children, refers to keratinization of the conjunctiva, which later spreads to the cornea causing ulceration. The ultimate condition is keratomalacia which, if not treated, leads to permanent blindness.

15.23 VITAMIN A TOXICITY

15.23.1 Acute and Chronic Toxicity of Vitamin A

An excessive intake of preformed vitamin A produces symptoms of toxicity, either acute or chronic. In either case, toxicity results from the indiscriminate use of pharmaceutical supplements, and not from the consumption of usual diets. The only naturally occurring products that contain sufficient vitamin A to induce toxicity in humans are the livers of animals at the top of long food chains, such as large marine fish and carnivores (e.g. bear and dog).

Acute toxicity results from the ingestion of a single or several closely spaced very large doses of vitamin A, usually more than 100 times the recommended intake. Such doses produce a variety of toxic signs that include vomiting, severe headache, dizziness, blurred vision, muscular incoordination, and malaise. These signs are usually transient and disappear within a few days. In acute hypervitaminosis A, the excess retinol circulating in the bloodstream is not subject to the normal regulation of RBP binding, and the unbound retinol disrupts the integrity of the cell membranes.

Chronic toxicity results from the recurrent ingestion over a period of weeks to years of excessive doses of vitamin A that are usually more than ten times the recommended intake. Toxic signs commonly include headache, bone and joint pain, hair loss, nose bleed, bleeding lips, and cracking and peeling of the skin. After terminating dosing, most patients recover fully from toxicity. Chronic megadosing with vitamin A may eventually result in cirrhosis of the liver. Once hypervitaminosis A is established, no effective treatment is available, save for low vitamin A diets (Leo & Lieber, 1988).

The most serious consequences of an excessive intake of vitamin A are its teratogenicity. Fetal resorption, abortion, and malformed fetuses or infants are the most serious teratogenic effects. The acidic retinoids, both synthetic and natural, are more powerful teratogens than are retinol and its esters.

15.23.2 Toxicity of b-Carotene

Carotenoids in foods are not thought to be toxic, even when ingested in large amounts. Hypercarotenosis, a benign condition characterized by a jaundice-like yellowing of the skin, can result when large amounts of carotene-rich foods (e.g. carrot juice) are ingested regularly.

FURTHER READING

D'Ambrosio, D.N., Clugston, R.D. & Blaner, W.S. (2011) Vitamin A metabolism: An update. *Nutrients*, **3**, 63–103.

Harrison, E.H. (2005) Mechanisms of digestion and absorption of dietary vitamin A. *Annual Review of Nutrition*, **25**, 87–103.

Lefebvre, P., Martin, P.J., Flajollet, S., Dedieu, S., Billaut, X. & Lefebvre, B. (2005) Transcriptional activities of retinoic acid receptors. *Vitamins and Hormones*, **70**, 199–264.

Raverdeau, M. & Mills, K.H.G. (2014) Modulation of T cell and innate immune responses by retinoic acid. *Journal of Immunology*, **192**, 2953–8.

Ross, A.C. (1992) Vitamin A status; relationship to immunity and the antibody response. *Proceedings of the Society for Experimental and Biological Medicine*, **200**, 303–20.

Semba, R.D. (1998) The role of vitamin A and related retinoids in immune function. *Nutrition Reviews*, **56**, S38–48.

REFERENCES

Adan, R.A.H., Cox, J.J., Beischlag, T.V. & Burbach, J.P.H. (1993) A composite hormone response element mediates the transactivation of the rat oxytocin gene by different classes of nuclear hormone receptors. *Molecular Endocrinology*, **7**, 47–57.

Ahluwalia, G.S., Kaul, L. & Ahluwalia, B.S. (1980) Evidence of facilitory effect of growth hormone on tissue vitamin A uptake in rats. *Journal of Nutrition*, **110**, 1185–93.

Allenby, G., Bocquel, M.-T., Saunders, M., Kazmer, S., Speck, J., Rosenberger, M., Lovey, A., Kastner, P., Grippo, J.F., Chambon, P. & Levin, A.A. (1993) Retinoic acid receptors and retinoid X receptors: Interactions with endogenous retinoic acids. *Proceedings of the National Academy of Sciences of the U.S.A.*, **90**, 30–34.

Allie, S.R., Zhang, W., Tsai, C.-Y., Noelle, R.J. & Usherwood, E.J. (2013) Critical role for all-trans retinoic acid for optimal effector and effector memory CD8 T cell differentiation. *Journal of Immunology*, **190**, 2178–87.

Altucci, L. & Gronemeyer, H. (2001) The promise of retinoids to fight against cancer. *Nature Reviews Cancer*, **1**, 181–93.

Ames, S.R. (1969) Factors affecting absorption, transport, and storage of vitamin A. *American Journal of Clinical Nutrition*, **22**, 934–5.

Anon. (1977) Vitamin A and retinol-binding protein in fetal growth and development of the rat. *Nutrition Reviews*, **35**, 305–9.

Anzano, M.A., Lamb, A.J. & Olson, J.A. (1979) Growth, appetite, sequence of pathological signs and survival following the induction of rapid, synchronous vitamin A deficiency in the rat. *Journal of Nutrition*, **109**, 1419–31.

Apfel, R., Benbrook, D., Lernhardt, E., Ortiz, M.A., Salbert, G. & Pfahl, M. (1994) A novel orphan receptor specific for a subset of thyroid hormone-responsive elements and its interaction with the retinoid/thyroid hormone receptor subfamily. *Molecular and Cellular Biology*, **14**, 7025–35.

Arora, D. & Ross, A.C. (1994) Antibody response against tetanus toxoid is enhanced by lipopolysaccharide or tumor necrosis factor-alpha in vitamin A-sufficient and deficient rats. *American Journal of Clinical Nutrition*, **59**, 922–8.

Åström, A., Tavakkol, A., Pettersson, U., Cromie, M., Elder, J.T. & Voorhees, J.J. (1991) Molecular cloning of two human cellular retinoic acid-binding proteins (CRABP). Retinoic acid-induced expression of CRABP-II but not CRABP-I in adult human skin *in vivo* and in skin fibroblasts *in vitro*. *Journal of Biological Chemistry*, **266**(26), 17662–6.

Åström, A., Pettersson, U. & Voorhees, J.J. (1992) Structure of the human cellular retinoic acid-binding protein II gene. Early transcriptional regulation by retinoic acid. *Journal of Biological Chemistry*, **267**(35), 25251–5.

Åström, A., Pettersson, U., Chambon, P. & Voorhees, J.J. (1994) Retinoic acid induction of human cellular retinoic acid-binding protein-II gene transcription is mediated by retinoic acid receptor-retinoid X receptor heterodimers bound to one far upstream retinoic acid-responsive element with 5-base pair spacing. *Journal of Biological Chemistry*, **269**(35), 22334–9.

Baes, M., Gulick, T., Choi, H.-S., Martinoli, M.G., Simha, D. & Moore, D.D. (1994) Anew orphan member of the nuclear hormone receptor superfamily that interacts with a subset of retinoic acid response elements. *Molecular and Cellular Biology*, **14**, 1544–52.

Ballow, M., Wang, W. & Xiang, S. (1996) Modulation of B-cell immunoglobulin synthesis by retinoic acid. *Clinical Immunology and Immunopathology*, **80**, S73–S81.

Båvik, C.-O., Eriksson, U., Allen, R.A. & Peterson, P.A. (1991) Identification and partial characterization of a retinal pigment epithelial membrane receptor for plasma retinol-binding protein. *Journal of Biological Chemistry*, **266**(23), 14978–85.

Båvik, C.-O., Busch, C. & Eriksson, U. (1992) Characterization of a plasma retinol-binding protein membrane receptor expressed in the retinal pigment epithelium. *Journal of Biological Chemistry*, **267**(32), 23035–42.

Båvik, C.-O., Lévy, F., Hellman, U., Wernstedt, C. & Eriksson, U. (1993) The retinal pigment epithelial membrane receptor for plasma retinol-binding protein. Isolation and cDNA cloning of the 63-kDa protein. *Journal of Biological Chemistry*, **268**(27), 20540–6.

Bedo, G., Santisteban, P. & Aranda, A. (1989) Retinoic acid regulates growth hormone gene expression. *Nature*, **339**, 231–4.

Benson, M.J., Pino-Lagos, K., Rosemblatt, M. & Noelle, R.J. (2007) All-trans retinoic acid mediates enhanced T reg cell growth, differentiation, and gut homing in the face of high levels of co-stimulation. *Journal of Experimental Medicine*, **204**, 1765–74.

Berkenstam, A., del Mar Vivanco Ruiz, M., Barettino, D., Horikoshi, M. & Stunnenberg, H.G. (1992) Cooperativity in transactivation between retinoic acid receptor and TFIID requires an activity analogous to E1A. *Cell*, **69**, 401–12.

Bertram, J.S., Pung, A., Churley, M., Kappock, T.J., IV, Wilkins, L.R. & Cooney, R.V. (1991) Diverse carotenoids protect against chemically induced neoplastic transformation. *Carcinogenesis*, **12**, 671–8.

Bex, V., Mercier, T., Chaumontet, C., Gaillard-Sanchez, I., Flechon, B., Mazet, F., Traub, O. & Martel, P. (1995) Retinoic acid enhances connexin43 expression at the post-transcriptional level in rat liver epithelial cells. *Cell Biochemistry and Function*, **13**, 69–77.

Bhat, P.V. & Lacroix, A. (1991) Effects of retinoic acid on the concentrations of radioactive metabolites of retinol in tissues of rats maintained on a retinol-deficient diet. *Canadian Journal of Physiology and Pharmacology*, **69**, 826–30.

Blaner, W.S., Das, K., Mertz, J.R., Das, S.R. & Goodman, D.S. (1986) Effects of dietary retinoic acid on cellular retinol- and retinoic acid-binding protein levels in various rat tissues. *Journal of Lipid Research*, **27**, 1084–8.

Blomhoff, H.K., Smeland, E.B., Erikstein, B., Rasmussen, A.M., Skrede, B., Skjønsberg, C. & Blomhoff, R. (1992) Vitamin A is a key regulator for cell growth, cytokine production, and differentiation in normal B cells. *Journal of Biological Chemistry*, **267**(33), 23988–92.

Blomstrand, R. & Werner, B. (1967) Studies on the intestinal absorption of radioactive β- carotene and vitamin Ain man. *Scandinavian Journal of Clinical and Laboratory Investigation*, **19**, 339–45.

Boerman, M.H. & Napoli, J.L. (1991) Cholate-independent retinyl ester hydrolysis. Stimulation by apo-cellular retinol-binding protein. *Journal of Biological Chemistry*, **266**(33), 22273–8.

Bollag, W. & Holdener, E.E. (1992) Retinoids in cancer prevention and therapy. *Annals of Oncology*, **3**, 513–26.

Borek, C., Smith, J.E., Soprano, D.R. & Goodman, D.S. (1981) Regulation of retinol-binding protein metabolism by glucocorticoid hormones in cultured H4IIEC3 liver cells. *Endocrinology*, **109**, 386–91.

Boylan, J.F. & Gudas, L.J. (1991) Overexpression of the cellular retinoic acid binding protein-I (CRABP-I) results in a reduction in differentiation-specific gene expression in F9 teratocarcinoma cells. *Journal of Cell Biology*, **112**, 965–79.

Boylan, J.F. & Gudas, L.J. (1992) The level of CRABP-I expression influences the amounts and types of all-trans retinoic acid metabolites in F9 teratocarcinoma stem cells. *Journal of Biological Chemistry*, **267**(30), 21486–91.

Breitman, T.R., Selonick, S.E. & Collins, S.J. (1980) Induction of differentiation of the human promyelocytic leukemia cell line (HL-60) by retinoic acid. *Proceedings of the National Academy of Sciences of the U.S.A.*, **77**(5), 2936–40.

Brent, G.A., Dunn, M.K., Harney, J.W., Gulik, T., Larsen, P.R. & Moore, D.D. (1989) Thyroid hormone aporeceptor represses T3-inducible promoters and blocks activity of the retinoic acid receptor. *New Biologist*, **1**, 329–36.

Brown, E.D., Micozzi, M.S., Craft, N.E., Bieri, J.G., Beecher, G., Edwards, B.K., Rose, A., Taylor, P.R. & Smith, J.C. Jr. (1989) Plasma carotenoids in normal men after a single ingestion of vegetables or purified β-carotene. *American Journal of Clinical Nutrition*, **49**, 1258–65.

Brüstle, A., Heink, S., Huber, M., Rosenplänter, C., Stadelmann, C., Yu, P., Arpaia, E., Mak, T.W., Kamradt, T. & Lohoff, M. (2007) The development of inflammatory TH-17 cells requires interferon-regulatory factor 4. *Nature Immunology*, **8**, 958–66.

Buck, J., Ritter, G., Dannecker, L., Katta, V., Cohen, S.L., Chait, B.T. & Hämmerling, U. (1990) Retinol is essential for growth of activated human B cells. *Journal of Experimental Medicine*, **171**, 1613–24.

Burton, G.W. & Ingold, K.U. (1984) β-Carotene: An unusual type of lipid antioxidant. *Science*, **224**, 569–73.

Caelles, C., González-Sancho, J.M. & Muñoz, A. (1997) Nuclear hormone receptor antagonism with AP-1 by inhibition of the JNK pathway. *Genes & Development*, **11**, 3351–64.

Cancela, M.L. & Price, P.A. (1992) Retinoic acid induces matrix Gla protein gene expression in human cells. *Endocrinology*, **130**, 102–8.

Cantorna, M.T., Nashold, F.E. & Hayes, C.E. (1994) In vitamin A deficiency multiple mechanisms establish a regulatory T helper cell imbalance with excess Th1 and insufficient Th2 function. *Journal of Immunology*, **152**, 1515–22.

Cantorna, M.T., Nashold, F.E. & Hayes, C.E. (1995) Vitamin A deficiency results in a priming environment conducive for Th1 cell development. *European Journal of Immunology*, **25**, 1673–9.

Cantorna, M.T., Nashold, F.E., Chun, T.Y. & Hayes, C.E. (1996) Vitamin A down-regulation of IFN-γ synthesis in cloned mouse Th1 lymphocytes depends on the CD28 costimulatory pathway. *Journal of Immunology*, **156**, 2674–9.

Carman, J.A. & Hayes, C.E. (1991) Abnormal regulation of IFN-γ secretion in vitamin A deficiency. *Journal of Immunology*, **147**, 1247–52.

Carman, J.A., Smith, S.M. & Hayes, C.E. (1989) Characterization of a helper T lymphocyte defect in vitamin A-deficent mice. *Journal of Immunology*, **142**, 388–93.

Carman, J.A., Pond, L., Nashold, F., Wassom, D.L. & Hayes, C.E. (1992) Immunity to *Trichinella spiralis* infection in vitamin A-deficient mice. *Journal of Experimental Medicine*, **175**, 111–20.

Cassani, B., Villablanca, E.J., De Calisto, J., Wang, S. & Mora, J.R. (2012) Vitamin A and immune regulation: Role of retinoic acid in gut-associated dendritic cell education, immune protection and tolerance. *Molecular Aspects of Medicine*, **33**, 63–76.

Cha, H.-R., Chang, S.-Y., Chang, J.-H., Kim, J.-O., Yang, J.-Y., Kim, C.-H. & Kweon, M.-N. (2010) Downregulation of Th17 cells in the small intestine by disruption of gut flora in the absence of retinoic acid. *Journal of Immunology*, **184**, 6799–6806.

Chambon, P. (1996) A decade of molecular biology of retinoic acid receptors. *FASEB Journal*, **10**, 940–54.

Chen, C.-C. & Heller, J. (1977) Uptake of retinol and retinoic acid from serum retinol-binding protein by retinal pigment epithelial cells. *Journal of Biological Chemistry*, **252**(15), 5216–21.

Chen, J.-Y., Penco, S., Ostrowski, J., Balaguer, P., Pons, M., Starrett, J.E., Reczek, P., Chambon, P. & Gronemeyer, H. (1995) RAR-specific agonist/antagonists which dissociate transactivation and AP1 transrepression inhibit anchorage-independent cell proliferation. *EMBO Journal*, **14**, 1187–97.

Chiocca, E.A., Davies, P.J.A. & Stein, J.P. (1988) The molecular basis of retinoic acid action. Transcriptional regulation of tissue transglutaminase gene expression in macrophages. *Journal of Biological Chemistry*, **263**(23), 11584–9.

Chun, T.Y., Carman, J.A. & Hayes, C.E. (1992) Retinoid repletion of vitamin A-deficient mice restores IgG responses. *Journal of Nutrition*, **122**, 1062–9.

Collins, S.J. (1987) The HL-60 promyelocytic leukemia cell line: Proliferation, differentiation, and cellular oncogene expression. *Blood*, 1233–44.

Cone, M.V. & Nettesheim, P. (1973) Effects of vitamin A on 3-methylcholanthrene-induced squamous metaplasias and early tumors in the respiratory tract of rats. *Journal of the National Cancer Institute*, **50**, 1559–1606.

Coombes, J.L., Siddiqui, K.R.R., Arancibia-Cárcamo, C.Y., Hall, J., Sun, C.-M., Belkaid, Y. & Powrie, F. (2007) A functionally specialized population of mucosal $CD103^+$ DCs induces $Foxp3^+$ regulatory T cells via a TGF-β- and retinoic acid-dependent mechanism. *Journal of Experimental Medicine*, **204**, 1757–64.

Davis, K.D. & Lazar, M.A. (1993) Induction of retinoic acid receptor-β by retinoic acid is cell specific. *Endocrinology*, **132**, 1469–74.

dela Seña, C., Riedl, K.M., Narayanasamy, S., Curley, R.W. Jr., Schwartz, S.J. & Harrison, E.H. (2014) The human enzyme that converts dietary provitamin A carotenoids to vitamin A is a dioxygenase. *Journal of Biological Chemistry*, **289**(19), 13661–6.

Denning, T.L., Wang, Y.C., Patel, S.R., Williams, I.R. & Pulendran, B. (2007) Lamina propria macrophages and dendritic cells differentially induce regulatory and interleukin 17-producing T cell esponses. *Nature Immunology*, **8**, 1086–94.

DePaolo, R.W., Abadie, V., Tang, F., Fehlner-Peach, H., Hall, J.A., Wang, W., Matietta, E.V., Kasarda, D.D., Waldmann, T.A., Murray, J.A., Semrad, C., Kupfer, S.S., Belkaid, Y., Guandalini, S. & Jabri, B. (2011) Co-adjuvant effects of retinoic acid and IL-15 induce inflammatory immunity to dietary antigens. *Nature*, **471**, 220–4.

de Thé, H., Marchio, A., Tiollais, P. & Dejean, A. (1989) Differential expression and ligand regulation of the retinoic acid receptor α and β genes. *EMBO Journal*, **8**, 429–33.

de Thé, H., Vivanco-Ruiz, M.D.M., Tiollais, P., Stunnenberg, H. & Dejean, A. (1990) Identification of a retinoic acid responsive element in the retinoic acid receptor β gene. *Nature*, **343**, 177–80.

Dew, S.E. & Ong, D.E. (1994) Specificity of the retinol transporter of the rat small intestine brush border. *Biochemistry*, **33**, 12340–5.

Dillehay, D.L., Walia, A.S. & Lamon, E.W. (1988) Effects of retinoids on macrophage function and IL-1 activity. *Journal of Leukocyte Biology*, **44**, 353–60.

Di Mascio, P., Kaiser, S. & Sies, H. (1989) Lycopene as the most efficient biological carotenoid singlet oxygen quencher. *Archives of Biochemistry and Biophysics*, **274**, 532–8.

Dogra, S.C., Khanduja, K.L. & Gupta, M.P. (1985) The effect of vitamin A deficiency on the initiation and post-initiation phases of benzo(a)pyrene-induced lung tumorigenesis in rats. *Britixh Journal of Cancer*, **52**, 931–5.

Dollé, P., Ruberte, E., Kastner, P., Petkovich, M., Stoner, C.M., Gudas, L.J. & Chambon, P. (1989) Differential expression of genes encoding α, β, and γ retinoic acid receptors and

CRABP in the developing limbs of the mouse. *Nature*, **342**, 702–5.

Dorea, J.G. & Olson, J.A. (1986) The rate of rhodopsin regeneration in the bleached eyes of zinc-deficient rats in the dark. *Journal of Nutrition*, **116**, 121–7.

Duester, G. (1996) Involvement of alcohol dehydrogenase, short-chain dehydrogenase/reductase, aldehyde dehydrogenase, and cytochrome P450 in the control of retinoid signaling by activation of retinoic acid synthesis. *Biochemistry*, **35**, 12221–7.

Duester, G., Shean, M.L., McBride, M.S. & Stewart, M.J. (1991) Retinoic acid response element in the human alcohol dehydrogenase gene *ADH3*: Implications for regulation of retinoic acid synthesis. *Molecular and Cellular Biology*, **11**, 1638–46.

Dunham, R.M., Thapa, M., Velazquez, V.M., Elrod, E.J., Denning, T.L., Pulendran, B. & Grakoui, A. (2013) Hepatic stellate cells preferentially induce Foxp3$^+$ regulatory T cells by production of retinoic acid. *Journal of Immunology*, **190**, 2009–16.

Durand, B., Saunders, M., Leroy, P., Leid, M. & Chambon, P. (1992) All-trans and 9-cis retinoic acid induction of CRABPII transcription is mediated by RAR–RXR heterodimers bound to DR1 and DR2 repeated motifs. *Cell*, **71**, 73–85.

During, A., Hussain, M.M., Morel, D.W. & Harrison, E.H. (2002) Carotenoid uptake and secretion by Caco-2 cells: β-carotene isomer selectivity and carotenoid interactions. *Journal of Lipid Research*, **43**, 1086–95.

Edwards, M.K.S. & McBurney, M.W. (1983) The concentration of retinoic acid determines the differentiated cell types formed by a teratocarcinoma cell line. *Developmental Biology*, **98**, 187–91.

Elias, K.M., Laurence, A., Davidson, T.S., Stephens, G., Kanno, Y., Shevach, E.M. & O'Shea, J.J. (2008) Retinoic acid inhibits Th17 polarization and enhances FoxP3 expression through a Stat-3/Stat-5 independent signaling pathway. *Blood*, **111**, 1013–20.

Engedal, N., Ertesvag, A. & Blomhoff, H.K. (2004) Survival of activated human T lymphocytes is promoted by retinoic acid via induction of IL-2. *International Immunology*, **16**, 43–53.

Ertesvag, A., Engedal, N., Naderiu, S. & Blomhoff, H.K. (2002) Retinoic acid stimulates the cell cycle machinery in normal T cells: Involvement of retinoic acid receptor-mediated IL-2 secretion. *Journal of Immunology*, **169**, 5555–63.

Ertesvag, A., Austenaa, L.M., Carlsen, H., Blomhoff, R. & Blomhoff, H.K. (2009) Retinoic acid inhibits in vivo interleukin-2 gene expression and T-cell activation in mice. *Immunology*, **126**, 514–22.

Evans, M.J. & Kaufman, M.H. (1981) Establishment in culture of pluripotent cells from mouse embryos. *Nature*, **292**, 154–6.

Falk, L.A., De Benedetti, F., Lohrey, N., Birchenall-Roberts, M.C., Ellingsworth, L.W., Faltynek, C.R. & Ruscetti, F.W. (1991) Induction of transforming growth factor-β1 receptorv expression and TGF-β1 protein production in retinoic acid-treated HL-60 cells: Possible TGF-β1-mediated autocrine inhibition. *Blood*, **77**, 1248–55.

Fanjul, A., Dawson, M.I., Hobbs, P.D., Jong, L., Cameron, J.F., Harlev, E., Graupner, G., Lu, X.-P. & Pfahl, M. (1994) Anew class of retinoids with selective inhibition of AP-1 inhibits proliferation. *Nature*, **372**, 107–11.

Fawzi, W.W., Chalmers, T.C., Herrera, M.G. & Mosteller, F. (1993) Vitamin A supplementation and child mortality. *Journal of the American Medical Association*, **269**, 898–903.

Felli, M.P., Vacca, A., Meco, D., Screpanti, I., Farina, A.R., Maroder, M., Martinotti, S., Petrangeli, E., Frati, L. & Gulino, A. (1991) Retinoic acid-induced down-regulation of the interleukin-2 promoter via *cis*-regulatory sequences containing an octamer motif. *Molecular and Cellular Biology*, **11**, 4771–8.

Fidge, N.H. & Goodman, D.S. (1968) The enzymatic reduction of retinal to retinol in rat intestine. *Journal of Biological Chemistry*, **243**(16), 4372–9.

Fisher, G.J., Reddy, A.P., Datta, S.C., Kang, S., Yi, J.Y. & Chambon, P. (1995) All-trans retinoic acid induces cellular retinol-binding protein in human skin *in vivo*. *Journal of Investigative Dermatology*, **105**, 80–6.

Foote, C.S. & Denny, R.W. (1968) Chemistry of singlet oxygen. VII. Quenching by β- carotene. *Journal of the American Chemical Society*, **90**, 6233–5.

Foote, C.S., Chang, Y.C. & Denny, R.W. (1970) Chemistry of singlet oxygen. X. Carotenoid quenching parallels biological protection. *Journal of the American Chemical Society*, **92**, 5216–18.

Forman, B.M., Goode, E., Chen, J., Oro, A.E., Bradley, D.J., Perlmann, T., Noonan, D.J., Burka, L.T., McMorris, T., Lamph, W.W., Evans, R.M. & Weinberger, C. (1995) Identification of a nuclear receptor that is activated by farnesol metabolites. *Cell*, **81**, 687–93.

Frolik, C.A. (1984) Metabolism of retinoids. In: *The Retinoids*, Vol. 2 (Ed. by M.B. Sporn, A.B. Roberts & D.S. Goodman), pp. 177–208. Academic Press, Inc., Orlando, FL.

Gebert, J.F., Moghal, N., Frangioni, J.V., Sugarbaker, D.J. & Neel, B.G. (1991) High frequency of retinoic acid receptor β abnormalities in human lung cancer. *Oncogene*, **6**, 1859–68.

Geissmann, F., Revy, P., Brousse, N., Lepelletier, Y., Folli, C., Durandy, A., Chambon, P. & Dy, M. (2003) Retinoids regulate survival and antigen presentation by immature dendritic cells. *Journal of Experimental Medicine*, **198**, 623–34.

Gerlach, T.H. & Zile, M.H. (1990) Upregulation of serum retinol in experimental acute renal failure. *FASEB Journal*, **4**, 2511–7.

Gerlach, T.H. & Zile, M.H. (1991a) Metabolism and secretion of retinol transport complex in acute renal failure. *Journal of Lipid Research*, **32**, 515–20.

Gerlach, T.H. & Zile, M.H. (1991b) Effect of retinoic acid and apo-RBP on serum retinol concentration in acute renal failure. *FASEB Journal*, **5**, 86–92.

Gianni, M., Studer, M., Carpani, G., Terao, M. & Garattini, E. (1991) Retinoic acid induces liver/bone/kidney-type alkaline phosphatase gene expression in F9 teratocarcinoma cells. *Biochemical Journal*, **274**, 673–8.

Gjøen, T., Bjerkelund, T., Blomhoff, H.K., Norum, K.R., Berg, T. & Blomhoff, R. (1987) Liver takes up retinol-binding protein from plasma. *Journal of Biological Chemistry*, **262**(23), 10926–30.

Glover, J. (1960) The conversion of β-carotene into vitamin A. *Vitamins and Hormones*, **18**, 371–86.

Goverse, G., Labao-Almeida, C., Ferreira, M., Molenaar, R., Wahlen, S., Konijn, T., Koning, J., Veiga-Fernandes, H. & Mebius, R.E. (2016) Vitamin A controls the presence of RORγ+ innate lymphoid cells and lymphoid tissue in the small intestine. *Journal of Immunology*, **196**, 5148–55.

Graupner, G., Wills, K.N., Tzukerman, M., Zhang, X.-K. & Pfahl, M. (1989) Dual regulatory role for thyroid-hormone receptors allows control of retinoic-acid receptor activity. *Nature*, **340**, 653–6.

Green, M.H., Green, J.B. & Lewis, K.C. (1987) Variation in retinol utilization rate with vitamin A status in the rat. *Journal of Nutrition*, **117**, 694–703.

Green, M.H., Green, J.B., Berg, T., Norum, K.R. & Blomhoff, R. (1993) Vitamin A metabolism in rat liver: A kinetic model. *American Journal of Physiology*, **264**, G509–21.

Griffiths, J.K. (2000) The vitamin A paradox. *Journal of Pediatrics*, **137**, 604–6.

Grunt, ThW., Dittrich, E., Offterdinger, M., Schneider, S.M., Dittrich, Ch. & Huber, H. (1998) Effects of retinoic acid and fenretinide on the c-erbB-2 expression, growth and cisplatin sensitivity of breast cancer cells. *British Journal of Cancer*, **78**, 79–87.

Gudas, L.J. (1994) Retinoids and vertebrate development. *Journal of Biological Chemistry*, **269**(22), 15399–402.

Gudas, L.J., Grippo, J.F., Kim, K.-W., LaRosa, G.J. & Stoner, C.M. (1990) The regulation of the expression of genes encoding basement membrane proteins during the retinoic acid-associated differentiation of murine teratocarcinoma cells. *Annals of the New York Academy of Sciences*, **580**, 245–51.

Guilbaud, N.F., Gas, N., Dupont, M.A. & Valette, A. (1990) Effects of differentiation-inducing agents on maturation of human MCF-7 breast-cancer cells. *Journal of Cellular Physiology*, **145**, 162–72.

Guo, H., Acevedo, P., Parsa, F.D. & Bertram, J.S. (1992) Gap-junctional protein connexin 43 is expressed in dermis and epidermis of human skin: Differential modulation by retinoids. *Journal of Investigative Dermatology*, **99**, 460–7.

Hall, J.A., Cannons, J.L., Grainger, J.R., Dos Santos, L.M., Hand, T.W., Naik, S., Wohlfert, E.A., Chou, D.B., Oldenhove, G., Robinson, M., Grigg, M.E., Kastenmayer, R., Schwartzberg, P.L. & Belkaid, Y. (2011a) Essential role for retinoic acid in the promotion of $CD4^+$ T cell effector responses via retinoic acid receptor alpha. *Immunity*, **34**, 435–47.

Hall, J.A., Grainger, J.R., Spencer, S.P. & Belkaid, Y. (2011b) The role of retinoic acid in tolerance and immunity. *Immunity*, **35**, 13–22.

Hammerschmidt, S.I., Friedrichsen, M., Boelter, J., Lyszkiewicz, M., Kremmer, E., Pabst, O. & Förster, R. (2011) Retinoic acid induces homing of protective T and B cells to the gut after subcutaneous immunization in mice. *Journal of Clinical Investigation*, **121**, 3051–61.

Haq, R. & Chytil, F. (1988) Retinoic acid rapidly induces lung cellular retinol-binding protein mRNA levels in retinol deficient rats. *Biochemical and Biophysical Research Communications*, **156**, 712–6.

Haq, R., Pfahl, M. & Chytil, F. (1991) Retinoic acid affects the expression of nuclear retinoic acid receptors in tissues of retinol-deficient rats. *Proceedings of the National Academy of Sciences of the U.S.A.*, **88**, 8272–6.

Harada, H., Miki, R., Masushige, S. & Kato, S. (1995) Gene expression of retinoic acid receptors, retinoid X receptors, and cellular retinol-binding protein I in bone and its regulation by vitamin A. *Endocrinology*, **136**, 5329–35.

Harpel, J.G., Metz, C.N., Kojima, S. & Rifkin, D.B. (1992) Control of transforming growth factor-β activity: Latency vs. activation. *Progress in Growth Factor Research*, **4**, 321–35.

Harrison, E.H. (2012) Mechanisms involved in the intestinal absorption of dietary vitamin A and provitamin A carotenoids. *Biochimica et Biophysica Acta*, **1821**, 70–7.

Hazenberg, M.D. & Spits, H. (2014) Human innate lymphoid cells. *Blood*, **124**, 700–709.

Heller, J. (1975) Interactions of plasma retinol-binding protein with its receptor. Specific binding of bovine and human retinol-binding protein to pigment epithelium cells from bovine eyes. *Journal of Biological Chemistry*, **250**(10), 3613–9.

Henderson, R.H., Keja, J., Hayden, G., Galazka, A., Clements, J. & Chan, C. (1988) Immunizing the children of the world: Progress and prospects. *Bulletin of the World Health Organization*, **66**, 535–43.

Herr, F.M. & Ong, D.E. (1992) Differential interaction of lecithin-retinol acyltransferase with cellular retinol binding proteins. *Biochemistry*, **31**, 6748–55.

Hill, J.A., Hall, J.A., Sun, C.-M., Cai, Q., Ghyselinck, N., Chambon, P., Belkaid, Y., Mathis, D. & Benoist, C. (2008) Retinoic acid enhances Foxp3 induction indirectly by relieving inhibition from $CD4^+CD44^{hi}$ cells. *Immunity*, **29**, 758–70.

Hoag, K.A., Nashold, F.E., Goverman, J. & Hayes, C.E. (2002) Retinoic acid enhances the T helper 2 cell development that is essential for robust antibody responses through its action on antigen-presenting cells. *Journal of Nutrition*, **132**, 3736–9.

Hogan, B.L.M., Taylor, A. & Adamson, E. (1981) Cell interactions modulate embryonal carcinoma cell differentiation into parietal or visceral endoderm. *Nature*, **291**, 235–7.

Hossain, M.Z., Wilkens, L.R., Mehta, P.P., Loewenstein, W. & Bertram, J.S. (1989) Enhancement of gap junctional communication by retinoids correlates with their ability to inhibit neoplastic transformation. *Carcinogenesis*, **10**, 1743–8.

Huang, C., Ma, W.-Y., Dawson, M.I., Rincon, M., Flavell, R.A. & Dong, Z. (1997) Blocking activator protein-1 activity, but not activating retinoic acid response element, is required for the antitumor promotion effect of retinoic acid. *Proceedings of the National Academy of Sciences of the U.S.A.*, **94**, 5826–30.

Huang, H.S. & Goodman, D.S. (1965) Vitamin Aand carotenoids. 1. Intestinal absorption and metabolism of ^{14}C-labeled vitamin A alcohol and β-carotene in the rat. *Journal of Biological Chemistry*, **240**(7), 2839–44.

Huber, A.M. & Gershoff, S.N. (1975) Effects of zinc deficiency on the oxidation of retinol and ethanol in rats. *Journal of Nutrition*, **105**, 1486–90.

Husmann, M., Hoffmann, B., Stump, D.G., Chytil, F. & Pfahl, M. (1992) A retinoic acid response element from the rat CRBPI promoter is activated by an RAR/RXR heterodimer. *Biochemical and Biophysical Research Communications*, **187**, 1558–64.

Husmann, M., Lehmann, J., Hoffmann, B., Hermann, T., Tzukerman, M. & Pfahl, M. (1991) Antagonism between retinoic acid receptors. *Molecular and Cellular Biology*, **11**, 4097–103.

Ivanov, I.I., Atarashi, K., Manel, N., Brodie, E.L., Shima, T., Karaoz, U., Wei, D., Goldfarb, K.C., Santee, C.A., Lynch, S.V., Tanoue, T., Imaoka, A., Itoh, K., Takeda, K., Umesaki, Y., Honda, K. & Littman, D.R. (2009) Induction of intestinal Th17 cells by segmented filamentous bacteria. *Cell*, **139**, 485–98.

Iwata, M., Eshima, Y. & Kagechika, H. (2003) Retinoic acids exert direct effects on T cells to suppress Th1 development and enhance Th2 development via retinoic acid receptors. *International Immunology*, **15**, 1017–25.

Iwata, M., Hirakiyama, A., Eshima, Y., Kagechika, H., Kato, C. & Song, S.-Y. (2004) Retinoic acid imprints gut-homing specificity on T cells. *Immunity*, **21**, 527–38.

Jarrous, N. & Kaempfer, R. (1994) Induction of human interleukin-1 gene expression by retinoic acid and its regulation at processing of precursor transcripts. *Journal of Biological Chemistry*, **269**(37), 23141–9.

Johansson-Lindbom, B., Svensson, M., Wurbel, M.-A., Malissen, B., Márquez, G. & Agace, W. (2003) Selective generation of gut tropic T cells in gut-associated lymphoid tissue (GALT): Requirement for GALT dendritic cells and adjuvant. *Journal of Experimental Medicine*, **198**, 963–9.

Johnson, E.J. & Russell, R.M. (1992) Distribution of orally administered β-carotene among lipoproteins in healthy men. *American Journal of Clinical Nutrition*, **56**, 128–35.

Johnson, R.L., Riddle, R.D. & Tabin, C.J. (1994) Mechanisms of limb patterning. *Current Opinion in Genetics and Development*, **4**, 535–42.

Jones, B.C., Logsdon, N.J. & Walter, M.R. (2008) Structure of IL-22 bound to its high-affinity IL-22R1 chain. *Structure*, **16**, 1333–44.

Jones-Villeneuve, E.M.V., McBurney, M.W., Rogers, K.A. & Kalnins, V.I. (1982) Retinoic acid induces embryonal carcinoma cells to differentiate into neurons and glial cells. *Journal of Cell Biology*, **94**, 253–62.

Jonk, L.J.C., de Jonge, M.E.J., Vervaart, J.M.A., Wissink, S. & Kruijer, W. (1994) Isolation and developmental expression of retinoic-acid-induced genes. *Developmental Biology*, **161**, 604–14.

Kamei, Y., Xu, L., Heinzel, T., Torchia, T., Kurokawa, R., Gloss, B., Lin, S.-C., Heyman, R.A., Rose, D.W., Glass, C.K. & Rosenfeld, M.G. (1996) ACBP integrator complex mediates transcriptional activation and AP-1 inhibition by nuclear receptors. *Cell*, **85**, 403–14.

Kamijo, R., Takeda, K., Nagumo, M. & Konno, K. (1990) Effects of combinations of transforming growth factor-β1 and tumor necrosis factor on induction of differentiation of human myelogenous leukemic cell lines. *Journal of Immunology*, **144**, 1311–16.

Kang, S.G., Lim, H.W., Andrisani, O.M., Broxmeyer, H.E. & Kim, C.H. (2007) Vitamin A metabolites induce gut-homing FoxP3$^+$ regulatory T cells. *Journal of Immunology*, **179**, 3724–33.

Kang, S.G., Wang, C., Matsumoto, S. & Kim, C.H. (2009) High and low vitamin A therapies induce distinct FoxP3$^+$ T-cell subsets and effectively control intestinal inflammation. *Gastroenterology*, **137**, 1391–402.

Kato, M., Blaner, W.S., Mertz, J.R., Das, K., Kato, K. & Goodman, D.S. (1985) Influence of retinoid nutritional status on cellular retinol- and cellular retinoic acid-binding protein concentrations in various rat tissues. *Journal of Biological Chemistry*, **260**(8), 4832–6.

Kato, S., Sasaki, H., Suzawa, M., Masushige, S., Tora, L., Chambon, P. & Gronemeyer, H. (1995) Widely spaced, directly repeated PuGGTCA elements act as promiscuous enhancers for different classes of nuclear receptors. *Molecular and Cellular Biology*, **15**, 5858–67.

Kennedy, T.A. & Liebler, D.C. (1992) Peroxyl radical scavenging by β-carotene in lipid bilayers. Effect of oxygen partial pressure. *Journal of Biological Chemistry*, **267**(7), 4658–63.

Kim, C.H. (2008) Regulation of FoxP3$^+$ regulatory T cells and Th17 cells by retinoids. *Clinical and Developmental Immunology*, **2008**, Article ID 416910, 12 pages.

Kim, M.H., Taparowsky, E.J. & Kim, C.H. (2015) Retinoic acid differentially regulates the migration of innate lymphoid cell subsets to the gut. *Immunity*, **43**, 107–19.

Kim, Y.-H., Dohi, D.F., Han, G.R., Zou, C.-P., Oridate, N., Walsh, G.L., Nesbitt, J.C., Xu, X.-C., Hong, W.K., Lotan, R. & Kurie, J.M. (1995) Retinoid refractoriness occurs during lung carcinogenesis despite functional retinoid receptors. *Cancer Research*, **55**, 5603–10.

Kinoshita, M. & Ross, A.C. (1993) Vitamin A status and immunoglobulin G subclasses in rats immunized with tetanus toxoid. *FASEB Journal*, **7**, 1277–82.

Kinoshita, M., Pasatiempo, A.M.G., Taylor, C.E. & Ross, A.C. (1991) Immunological memory to tetanus toxoid is established and maintained in the vitamin A-depleted rat. *FASEB Journal*, **5**, 2473–81.

Kirfel, J., Kelter, M., Cancela, L.M., Price, P.A. & Schüle, R. (1997) Identification of a novel negative retinoic acid responsive element in the promoter of the human matrix Gla protein gene. *Proceedings of the National Academy of Sciences of the U.S.A.*, **94**, 2227–32.

Klebanoff, C.A., Spencer, S.P., Torabi-Parizi, P., *et al.* (2013) Retinoic acid controls the homeostasis of pre-cDC-derived splenic and intestinal dendritic cells. *Journal of Experimental Medicine*, **210**, 1961–76.

Kok, K., Osinga, J., Carritt, B., Davis, M.B., van der Hout, A.H., van der Veen, A.Y., Landsvater, R.M., de Leij, L.F.M.H., Berendsen, H.H., Postmus, P.E., Poppema, S. & Buys, C.H.C.M. (1987) Deletion of a DNA sequence at the chromosomal region 3p21 in all major types of lung cancer. *Nature*, **330**, 578–81.

Kruyt, F.A.E., Folkers, G.E., Walhout, A.J.M., van der Leede, B.M. & van der Saag, P.T. (1993) E1Afunctions as a coactivator of retinoic acid-dependent retinoic acid receptor-β2 promoter activation. *Molecular Endocrinology*, **7**, 604–15.

Lafyatis, R., Kim, S.-J., Angel, P., Roberts, A.B., Sporn, M.B., Karin, M. & Wilder, R.L. (1990) Interleukin-1 stimulates and *all-trans*-retinoic acid inhibits collagenase gene expression through its 5' activator protein-1-binding site. *Molecular Endocrinology*, **4**, 973–80.

Langston, A.W. & Gudas, L.J. (1992) Identification of a retinoic acid responsive enhancer 3' of the murine homeobox gene Hox-1.6. *Mechanisms of Development*, **38**, 217–27.

La Vista-Picard, N., Hobbs, P.D., Pfahl, M., Dawson, M.I. & Pfahl, M. (1996) The receptor-DNA complex determines the retinoid response: A mechanism for the diversification of the ligand signal. *Molecular and Cellular Biology*, **16**, 4137–46.

Lee, M.-O., Liu, Y. & Zhang, X.-K. (1995) A retinoic acid response element that overlaps an estrogen response element mediates multihormonal sensitivity in transcriptional activation of the lactoferrin gene. *Molecular and Cellular Biology*, **15**, 4194–207.

Lee, M.S., Kliewer, S.A., Provencal, J., Wright, P.E. & Evans, R.M. (1993) Structure of the retinoid X receptor α DNA binding domain: A helix required for homodimeric DNA binding. *Science*, **260**, 1117–21.

Lehmann, J.M., Zhang, X.-K. & Pfahl, M. (1992) RARγ2 expression is regulated through a retinoic acid response element embedded in Sp1 sites. *Molecular and Cellular Biology*, **12**, 2976–85.

Leo, M.A. & Lieber, C.S. (1985) New pathway for retinol metabolism in liver microsomes. *Journal of Biological Chemistry*, **260**(9), 5228–31.

Leo, M.A. & Lieber, C.S. (1988) Hypervitaminosis A: A liver lover's lament. *Hepatology*, **8**, 412–7.

Leo, M.A., Kim, C.-I. & Lieber, C.S. (1987) NAD+-dependent retinol dehydrogenase in liver microsomes. *Archives of Biochemistry and Biophysics*, **259**, 241–9.

Leo, M.A., Kim, C.-I. & Lieber, C.S. (1989a) Role of vitamin A degradation in the control of hepatic levels in the rat. *Journal of Nutrition*, **119**, 993–1000.

Leo, M.A., Kim, C.-I., Lowe, N. & Lieber, C.S. (1989b) Increased hepatic retinal dehydrogenase activity after phenobarbital and ethanol administration. *Biochemical Pharmacology*, **38**, 97–103.

Leo, M.A., Lasker, J.M., Raucy, J.L., Kim, C.-I., Black, M. & Lieber, C.S. (1989c) Metabolism of retinol and retinoic acid by human liver cytochrome P450IIC8. *Archives of Biochemistry and Biophysics*, **269**, 305–12.

Leroy, P., Krust, A., Zelent, A., Mendelsohn, C., Garnier, J.-M., Kastner, P., Dierich, A. & Chambon, P. (1991a) Multiple isoforms of the mouse retinoic acid receptor α are generated by alternative splicing and differential induction by retinoic acid. *EMBO Journal*, **10**, 59–69.

Leroy, P., Nakshatri, H. & Chambon, P. (1991b) Mouse retinoic acid receptor α2 isoform is transcribed from a promoter that contains a retinoic acid response element. *Proceedings of the National Academy of Sciences of the U.S.A.*, **88**, 10138–42.

Levin, M.S., Li, E., Ong, D.E. & Gordon, J.I. (1987) Comparison of the tissue-specific expression and developmental regulation of two closely linked rodent genes encoding cytosolic retinol-binding proteins. *Journal of Biological Chemistry*, **262**(15), 711–24.

Li, Y., Hashimoto, Y., Agadir, A., Kagechika, H. & Zhang, X.-K. (1999) Identification of a novel class of retinoic acid receptor β-selective retinoid antagonists and their inhibitory effects on AP-1 activity and retinoic acid-induced apoptosis in human breast cancer cells. *Journal of Biological Chemistry*, **274**(22), 15360–6.

Lie, C., Ying, C., Wang, E.L., Brun, T. & Geissler, C. (1993) Impact of large-dose vitamin A supplementation on childhood diarrhoea, respiratory disease and growth. *European Journal of Clinical Nutrition*, **47**, 88–96.

Lin, B., Chen, G.-Q., Xiao, D., Kolluri, S.K., Cao, X., Su, H. & Zhang, X.-K. (2000) Orphan receptor COUP-TF is required for induction of retinoic acid receptor β, growth inhibition, and apoptosis by retinoic acid in cancer cells. *Molecular and Cellular Biology*, **20**, 957–70.

Lin, F., Xiao, D., Kolluri, S.K. & Zhang, X.-K. (2000) Unique anti-activator protein-1 activity of retinoic acid receptor β. *Cancer Research*, **60**, 3271–80.

Lin, F., Kolluri, S.K., Chen, G.-Q. & Zhang, X.-K. (2002) Regulation of retinoic acid- induced inhibition of AP-1 activity by orphan receptor chicken ovalbumin upstream promoter-transcription factor. *Journal of Biological Chemistry*, **277**(24), 21414–22.

Lippman, S.M., Kessler, J.F. & Meyskens, F.L. Jr. (1987) Retinoids as preventive and therapeutic agents (part II). *Cancer Treatment Reports*, **71**, 493–515.

Liu, Y., Lee, M.O., Wang, H.G., Li, Y., Hashimoto, Y., Klaus, M., Reed, J.C. & Zhang, X. (1996) Retinoic acid receptor beta mediates the growth-inhibitory effect of retinoic acid by promoting apoptosis in human breast cancer cells. *Molecular and Cellular Biology*, **16**, 1138–49.

Loerch, J.D., Underwood, B.A. & Lewis, K.C. (1979) Response of plasma levels of vitamin A to a dose of vitamin A as an indicator of hepatic vitamin A reserves in rats. *Journal of Nutrition*, **109**, 778–86.

Lotan, R. (1986) Immunomodulatory effects of retinoids. *Journal of Nutrition, Growth and Cancer*, **3**, 57–65.

Lotan, R. (1995) Retinoids and apoptosis: Implications for cancer chemoprevention and therapy. *Journal of the National Cancer Institute*, **87**, 1655–7.

Lowe, W.L. Jr., Meyer, T., Karpen, C.W. & Lorentzen, L.R. (1992) Regulation of insulin-like growth factor I production in rat C6 glioma cells: Possible role as an autocrine/paracrine growth factor. *Endocrinology*, **130**, 2683–91.

Lucas, P.C., Forman, B.M., Samuels, H.H. & Granner, D.K. (1991) Specificity of a retinoic acid response element in the phosphoenolpyruvate carboxykinase gene promoter: Consequences of both retinoic acid and thyroid hormone receptor binding. *Molecular and Cellular Biology*, **11**, 5164–70.

MacDonald, P.N., Bok, D. & Ong, D.E. (1990) Localization of cellular retinol-binding protein and retinol-binding protein in cells comprising the blood-brain barrier of rat and human. *Proceedings of the National Academy of Sciences of the U.S.A.*, **87**, 4265–9.

Mangelsdorf, D.J., Umesono, K., Kliewer, S.A., Borgmeyer, U., Ong, E.S. & Evans, R.M. (1991) A direct repeat in the cellular retinol-binding protein type II gene confers differential regulation by RXR and RAR. *Cell*, **66**, 555–61.

Mangelsdorf, D.J., Umesono, K. & Evans, R.M. (1994) The retinoid receptors. In: *The Retinoids: Biology, Chemistry, and Medicine*. 2nd edn. (Ed. by M.B. Sporn, A.B. Roberts & D.S. Goodman), pp. 319–48. Raven Press, Ltd, New York.

Martin, G.R. (1981) Isolation of a pluripotent cell line from early mouse embryos cultured in medium conditioned by teratocarcinoma stem cells. *Proceedings of the National Academy of Sciences of the U.S.A.*, **78**, 7634–8.

Martin, J.C.J., Bériou, G., Hesian, M., Shauvin, C., Utriainen, L., Aumeunier, A., Scott, C.L., Mowat, A., Cerovic, V., Houston, S.A., Leboeuf, M., Hubert, F.X., Hémont, C., Meradi, M., Milling, S. & Josien, R. (2014) Interleukin-22 binding protein (IL-22 BP) is constitutively expressed by a subset of conventional dendritic cells and is strongly induced by retinoic acid. *Mucosal Immunology*, **7**, 101–11.

Matikainen, S., Serkkola, E. & Hurme, M. (1991) Retinoic acid enhances IL-1β expression in myeloid leukemia cells and in human monocytes. *Journal of Immunology*, **147**, 162–7.

Matsuura, T. & Ross, A.C. (1993) Regulation of hepatic lecithin:retinol acyltransferase activity by retinoic acid. *Archives of Biochemistry and Biophysics*, **301**, 221–7.

Maynard, C.L., Hatton, R.D., Helms, W.S., Oliver, J.R., Stephensen, C.B. & Weaver, C.T. (2009) Contrasting roles for all-trans retinoic acid in TGF-β-mediated induction of *Fox3p* and *Il10* genes in developing regulatory T cells. *Journal of Experimental Medicine*, **206**, 343–57.

McGhee, J.R., Farrar, J.J., Michalek, S.M., Mergenhagen, S.E. & Rosenstreich, D.L. (1979) Cellular requirements for lipopolysaccharide adjuvanticity. Arole for both T lymphocytes and macrophages for in vitro responses to particulate antigens. *Journal of Experimental Medicine*, **149**, 793–807.

McGuire, B.W., Orgebin-Christ, M.-C. & Chytil, F. (1981) Autoradiographic localization of serum retinol-binding protein in rat testis. *Endocrinology*, **108**, 658–67.

Melikian, G., Mmiro, F., Ndugwa, C., Perry, R., Jackson, J.B., Garrett, E., Tielsch, J. & Semba, R.D. (2001) Relation of vitamin A and carotenoid status to growth failure and mortality among Ugandan infants with human immunodeficiency virus. *Nutrition*, **17**, 567–72.

Mielke, L.A., Jones, S.A., Raverdeau, M., Higgs, R., Stefanska, A., Groom, J.R., Misiak, A., Dungan, L.S., Sutton, C.E., Streubel, G., Bracken, A.P. & Mills, K.H.G. (2013) Retinoic acid expression associates with enhanced IL-22 production by γ δ T cells and innate lymphoid cells and attenuation of intestinal inflammation. *Journal of Experimental Medicine*, **210**, 1117–24.

Minucci, S., Leid, M., Toyama, R., Saint-Jeannet, J.-P., Peterson, V.J., Horn, V., Ishmael, J.E., Bhattacharyya, N., Dey, A.,

Dawid, I.B. & Ozato, K. (1997) Retinoid X receptor (RXR) within the RXR-retinoic acid receptor heterodimer binds its ligand and enhances retinoid- dependent gene expression. *Molecular and Cellular Biology*, **17**, 644–55.

Moore, W.T., Murtaugh, M.P. & Davies, P.J.A. (1984) Retinoic acid-induced expression of tissue transglutaminase in mouse peritoneal macrophages. *Journal of Biological Chemistry*, **259**(20), 12794–802.

Mora, J.R. & von Andrian, U.H. (2008) Differentiation and homing of IgA-secreting cells. *Mucosal Immunology*, **1**, 96–109.

Mora, J.R., Bono, M.R., Manjunath, N., Weninger, W., Cavanagh, L.L., Rosemblatt, M. & von Andrian, U.H. (2003) Selective imprinting of gut-homing T cells by Peyer's patch dendritic cells. *Nature*, **424**, 88–93.

Mora, J.R., Iwata, M., Eksteen, B., Song, S.-Y., Junt, T., Senman, B., Otipoby, K.L., Yokota, A., Takeuchi, H., Ricciardi-Castagnoli, P., Rajewski, K., Adams, D.H. & von Andrian, U.H. (2006) Generation of gut-homing IgA-secreting B cells by intestinal dendritic cells. *Science*, **314**, 1157–60.

Moriguchi, S., Werner, L. & Watson, R.R. (1985) High dietary vitamin A (retinyl palmitate) and cellular immune functions in mice. *Immunology*, **56**, 169–77.

Morikawa, K. & Nonaka, M. (2005) All-*trans*-retinoic acid accelerates the differentiation of human B lymphocytes maturing into plasma cells. *International Immunopharmacology*, **5**, 1830–8.

Morita, S., Fernandez-Mejia, C. & Melmed, S. (1989) Retinoic acid selectively stimulates growth hormone secretion and messenger ribonucleic acid levels in rat pituitary cells. *Endocrinology*, **124**, 2052–6.

Morita, S., Matsuo, K., Tsuruta, M., Leng, S., Yamashita, S., Izumi, M. & Nagataki, S. (1990) Stimulatory effect of retinoic acid on GH gene expression: The interaction of retinoic acid and tri-iodothyronine in rat pituitary cells. *Journal of Endocrinology*, **125**, 251–6.

Mucida, D., Park, Y., Kim, G., Turovskaya, O., Scott, I., Kronenberg, M. & Cheroutre, H. (2007) Reciprocal TH17 and regulatory T cell differentiation mediated by retinoic acid. *Science*, **317**, 256–60.

Mūnoz-Cánoves, P., Vik, D.P. & Tack, B.F. (1990) Mapping of a retinoic acid-responsive element in the promoter region of the complement factor H gene. *Journal of Biological Chemistry*, **265**(33), 20065–8.

Muto, Y., Smith, J.E., Milch, P.O. & Goodman, D.S. (1972) Regulation of retinol-binding protein metabolism by vitamin A status in the rat. *Journal of Biological Chemistry*, **247**(8), 2542–50.

Nagao, A., During, A., Hoshino, C., Terao, J. & Olson, J.A. (1996) Stoichiometric conversion of all-trans-β-carotene to retinal by pig intestinal extract. *Archives of Biochemistry and Biophysics*, **328**, 57–63.

Nagpal, S., Athanikar, J. & Chandraratna, R.A.S. (1995) Separation of transactivation and AP1 antagonism functions of retinoic acid receptor α. *Journal of Biological Chemistry*, **270**(2), 923–7.

Nakshatri, H. & Chambon, P. (1994) The directly repeated RG(G/T)TCA motifs of the rat and mouse cellular retinol-binding protein II genes are promiscuous binding sites for RAR, RXR, HNF-4, and ARP-1 homo- and heterodimers. *Journal of Biological Chemistry*, **269**(2), 890–902.

Napoli, J.L. (1993) Biosynthesis and metabolism of retinoic acid: Roles of CRBP and CRABP in retinoic acid homeostasis. *Journal of Nutrition*, **123**, 362–6.

Napoli, J.L., Posch, K.P., Fiorella, P.D. & Boerman, M.H. (1991) Physiological occurrence, biosynthesis and metabolism of retinoic acid: Evidence for roles of cellular retinol-binding protein (CRBP) and cellular retinoic acid-binding protein (CRABP) in the pathway of retinoic acid homeostasis. *Biomedicine and Pharmacotherapy*, **45**, 131–43.

Nettesheim, P. & Williams, M.L. (1976) The influence of vitamin A on the susceptibility of the rat lung to 3-methylcholanthrene. *International Journal of Cancer*, **17**, 351–7.

Nicholson, R.C., Mader, S., Nagpal, S., Leid, M., Rochette-Egly, C. & Chambon, P. (1990) Negative regulation of the rat stromelysin gene promoter by retinoic acid is mediated by an AP1 binding site. *EMBO Journal*, **9**, 4443–54.

Nolting, J., Daniel, C., Reuter, S., Stuelten, C., Li, P., Sucov, H., Kim, B.-G., Letterio, J.J., Kretschmer, K., Kim, H.-J. & von Boehmer, H. (2009) Retinoic acid can enhance conversion of naive into regulatory T cells independently of secreted cytokines. *Journal of Experimental Medicine*, **206**, 2131–9.

Nunes, I., Kojima, S. & Rifkin, D.B. (1996) Effects of endogenously activated transforming growth factor-β on growth and differentiation of retinoic acid-treated HL-60 cells. *Cancer Research*, **56**, 495–9.

O'Byrne, S.M. & Blaner, W.S. (2013) Retinol and retinyl esters: Biochemistry and physiology. *Journal of Lipid Research*, **54**, 1731–43.

Ogura, T. & Evans, R.M. (1995) Evidence for two distinct retinoic acid response pathways for *HOXB1* gene regulation. *Proceedings of the National Academy of Sciences of the U.S.A.*, **92**, 392–6.

Oliva, A., Ragione, F.D., Fratta, M., Marrone, G., Palumbo, R. & Zappia, V. (1993) Effect of retinoic acid on osteocalcin gene expression in human osteoblasts. *Biochemical and Biophysical Research Communications*, **191**, 908–14.

Oreffo, R.O.C., Teti, A., Triffitt, J.T., Francis, M.J.O., Carano, A. & Zallone, A.Z. (1998) Effect of vitamin A on bone resorption: Evidence for direct stimulation of isolated chicken osteoclasts by retinol and retinoic acid. *Journal of Bone and Mineral Research*, **3**, 203–10.

Palozza, P. (1998) Prooxidant actions of carotenoids in biologic systems. *Nutrition Reviews*, **56**, 257–65.

Pasatiempo, A.M.G., Kinoshita, M., Taylor, C.E. & Ross, A.C. (1990) Antibody production in vitamin A-depleted rats is impaired after immunization with bacterial polysaccharide or protein antigens. *FASEB Journal*, **4**, 2518–27.

Pasatiempo, A.M.G., Kinoshita, M., Foulke, D.T. & Ross, A.C. (1994) The antibody response of vitamin A-deficient rats to pneumococcal polysaccharide is enhanced through coimmunization with lipopolysaccharide. *Journal of Infectious Diseases*, **169**, 441–4.

Perez-Castro, A.V., Toth-Rogler, L.E., Wei, L.-N. & Nguyen-Huu, M.C. (1989) Spatial and temporal pattern of expression of the cellular retinoic acid-binding protein and the cellular retinol-binding protein during mouse embryogenesis. *Proceedings of the National Academy of Sciences of the U.S.A.*, **86**, 8813–7.

Perlmann, T. & Jansson, L. (1995) A novel pathway for vitamin A signaling mediated by RXR heterodimerization with NGFI-B and NURR1. *Genes & Development*, **9**, 769–82.

Pfahl, M. (1993) Nuclear receptor/AP-1 interaction. *Endocrine Reviews*, **14**, 651–8.

Pino-Lagos, K., Guo, Y., Brown, C., Alexander, M.P., Elgueta, R., Bennett, K.A., De Vries, V., Nowak, E., Blomhoff, R., Sockanathan, S., Chandraratna, R.A., Dmitrovsky, E. & Noelle, R.J. (2011) A retinoic acid-dependent checkpoint in the development of $CD4^+$ T cell-mediated immunity. *Journal of Experimental Medicine*, **208**, 1767–75.

Pöpperl, H. & Featherstone, M.S. (1993) Identification of a retinoic acid response element upstream of the murine *Hox-4.2* gene. *Molecular and Cellular Biology*, **13**, 257–65.

Pratt, M.A.C., Niu, M. & White, D. (2003) Differential regulation of protein expression, growth and apoptosis by natural and synthetic retinoids. *Journal of Cellular Biochemistry*, **90**, 692–708.

Quick, T.C. & Ong, D.E. (1989) Levels of cellular retinol-binding proteins in the small intestine of rats during pregnancy and lactation. *Journal of Lipid Research*, **30**, 1049–54.

Rahmathullah, L., Underwood, B.A., Thulasiraj, R.D., Milton, R.C., Ramaswamy, K., Rahmathullah, R. & Babu, G. (1990) Reduced mortality among children in southern India receiving a small weekly dose of vitamin A. *New England Journal of Medicine*, **323**, 929–35.

Raisher, B.D., Gulick, T., Zhang, Z., Strauss, A.W., Moore, D.D. & Kelly, D.P. (1992) Identification of a novel retinoid-responsive element in the promoter region of the medium chain acyl-coenzyme A dehydrogenase gene. *Journal of Biological Chemistry*, **267**(28), 20264–9.

Raffo, P., Emionite, L., Colucci, L., Belmondo, F., Grazia Moro, M., Bollag, W. & Toma, S. (2000) Retinoid receptors: Pathways of proliferation inhibition and apoptosis induction in breast cancer cell lines. *Anticancer Research*, **20**, 1535–44.

Rajan, N., Blaner, W.S., Soprano, D.R., Suhara, A. & Goodman, D.S. (1990) Cellular retinol-binding protein messenger RNA levels in normal and retinoid-deficient rats. *Journal of Lipid Research*, **31**, 821–9.

Rajan, N., Kidd, G.L., Talmage, D.A., Blaner, W.S., Suhara, A. & Goodman, D.S. (1991) Cellular retinoic acid-binding protein messenger RNA: Levels in rat tissues and localization in rat testis. *Journal of Lipid Research*, **32**, 1195–204.

Randolph, R.K. & Ross, A.C. (1991) Vitamin A status regulates hepatic lecithin:retinol acyltransferase activity in rats. *Journal of Biological Chemistry*, **266**(25), 16453–7.

Rask, L. & Peterson, P.A. (1976) *In vitro* uptake of vitamin A from the retinol-binding Plasma protein to mucosal epithelial cells from the monkey's small intestine. *Journal of Biological Chemistry*, **251**(20), 6360–6.

Ribaya-Mercado, J.D., Holmgren, S.C., Fox, J.G. & Russell, R.M. (1989) Dietary β-carotene absorption and metabolism in ferrets and rats. *Journal of Nutrition*, **119**, 665–8.

Ribeiro, R.C.J., Apriletti, J.W., West, B.L., Wagner, R.L., Fletterick, R.J., Schaufele, F. & Baxter, J.D. (1995) The molecular biology of thyroid hormone action. *Annals of the New York Academy of Sciences*, **758**, 366–89.

Roels, O.A., Trout, M. & Dujacquier, R. (1958) Carotene balances on boys in Ruanda where vitamin A deficiency is prevalent. *Journal of Nutrition*, **65**, 115–27.

Roels, O.A., Anderson, O.R., Lui, N.S.T., Shah, D.O. & Trout, M.E. (1969) Vitamin A and membranes. *American Journal of Clinical Nutrition*, **22**, 1020–32.

Rogers, M., Berestecky, J.M., Hossain, M.Z., Guo, H., Kadle, R., Nicholson, B.J. & Bertram, J.S. (1990) Retinoid-enhanced gap junctional communication is achieved by increased levels of connexin 43 mRNA and protein. *Molecular Carcinogenesis*, **3**, 335–43.

Rogers, M.B., Rosen, V., Wozney, J.M. & Gudas, L.J. (1992) Bone morphogenetic proteins-2 and −4 are involved in the retinoic acid-induced differentiation of embryonal carcinoma cells. *Molecular Biology of the Cell*, **3**, 189–96.

Ross, A.C. (1996) Vitamin A deficiency and retinoid repletion regulate the antibody response to bacterial antigens and the maintenance of natural killer cells. *Clinical Immunology and Immunopathology*, **80**, S63–72.

Rottman, J.N., Widom, R.L., Nadal-Ginard, B., Mahdavi, V. & Karathanasis, S.K. (1991) A retinoic acid-responsive element in the apolipoprotein AI gene distinguishes between two different retinoic acid response pathways. *Molecular and Cellular Biology*, **11**, 3814–20.

Ruane, D., Brane, L., Reis, B.S., Cheong, C., Poles, J., Do, Y., Zhu, H., Velinzon, K., Choi, J.H., Studt, N., Mayer, L., Lavelle, E.C., Steinman, R.M., Mucida, D. & Mehandru, S. (2013) Lung dendritic cells induce migration of protective T cells to the gastrointestinal tract. *Journal of Experimental Medicine*, **210**, 1871–88.

Ruiter, B., Patil, S.U. & Shreffler, W.G. (2015) Vitamins A and D have antagonistic effects on expression of effector cytokines and gut-homing integrin in human innate lymphoid cells. *Clinical & Experimental Allergy*, **45**, 1214–25.

Rush, M.G., Haq, R. & Chytil, F. (1991) Opposing effects of retinoic acid and dexamethasone on cellular retinol-binding protein ribonucleic acid levels in the rat. *Endocrinology*, **129**, 705–9.

Salbert, G., Fanjul, A., Piedrafita, F.J., Lu, X.P., Kim, S.-J., Tran, P. & Pfahl, M. (1993) Retinoic acid receptors and retinoid X receptor-α down-regulate the transforming growth factor-β1 promoter by antagonizing AP-1 activity. *Molecular Endocrinology*, **7**, 1347–56.

Schambach, F., Schupp, M., Lazar, M.A. & Reiner, S.L. (2007) Activation of retinoic acid receptor-α favours regulatory T cell induction at the expense of IL-17-secreting T helper cell differentiation. *European Journal of Immunology*, **37**, 2396–9.

Schneider, S.M., Offterdinger, M., Huber, H. & Grunt, T.W. (2000) Activation of retinoic Acid receptor α is sufficient for full induction of retinoid responses in SK-BR-3 and T47D human breast cancer cells. *Cancer Research*, **60**, 5479–87.

Schüle, R., Rangarajan, P., Yang, N., Kliewer, S., Ransone, L.J., Bolado, J., Verma, I.M. & Evans, R.M. (1991) Retinoic acid is a negative regulator of AP-1-responsive genes. *Proceedings of the National Academy of Sciences of the U.S.A.*, **88**, 6092–6.

Scott, D.K., Mitchell J.A. & Granner, D.K. (1996) Identification and characterization of a second retinoic acid response element in the phosphoenolypyruvate carboxykinase gene promoter. *Journal of Biological Chemistry*, **271**(11), 6260–4.

Semba, R.D. (1994) Vitamin A, immunity, and infection. *Clinical Infectious Diseases*, **19**, 489–99.

Semba, R.D. (1999) Vitamin A and immunity to viral, bacterial and protozoan infections. *Proceedings of the Nutrition Society*, **58**, 719–27.

Senoo, H., Stang, E., Nilsson, A., Kindberg, G.M., Berg, T., Roos, N., Norum, K.R. & Blomhoff, R. (1990) Internalization of retinol-binding protein in parenchymal and stellate cells of rat liver. *Journal of Lipid Research*, **31**, 1229–39.

Seo, G.-Y, Jang, Y.-S., Kim, H.-A., Lee, M.-R., Park, M.-H., Park, S.-R., Lee, J.-M., Choe, J. & Kim, P.-H. (2013) Retinoic acid acting as a highly specific IgA isotype switch factor, cooperates with TGF-β1 to enhance the overall IgA response. *Journal of Leukocyte Biology*, **94**, 325–35.

Sharma, R.V., Mathur, S.N., Dmitrovskii, A.A., Das, R.C. & Ganguly, J. (1977) Studies on the metabolism of β-carotene and apo-β-carotenoids in rats and chickens. *Biochimica et Biophysica Acta*, **486**, 183–94.

Shingleton, J.L., Skinner, M.K. & Ong, D.E. (1989) Characteristics of retinol accumulation from serum retinol-binding protein by cultured Sertoli cells. *Biochemistry*, **28**, 9641–7.

Sivaprasadarao, A. & Findlay, J.B.C. (1988a) The interaction of retinol-binding protein with its plasma-membrane receptor. *Biochemical Journal*, **255**, 561–9.

Sivaprasadarao, A. & Findlay, J.B.C. (1988b) The mechanism of uptake of retinol by plasma-membrane vesicles. *Biochemical Journal*, **255**, 571–9.

Smith, J.C. Jr., Brown, E.D., McDaniel, E.G. & Chan, W. (1976) Alterations in vitamin A metabolism during zinc deficiency and food and growth restriction. *Journal of Nutrition*, **106**, 569–74.

Smith, J.E., Muto, Y., Milch, P.O. & Goodman, D.S. (1973) The effects of chylomicron vitamin A on the metabolism of retinol-binding protein in the rat. *Journal of Biological Chemistry*, **248**(5), 1544–9.

Smith, S.M. & Hayes, C.E. (1987) Contrasting impairments in IgM and IgG responses of vitamin A-deficient mice. *Proceedings of the National Academy of Sciences of the U.S.A.*, **84**, 5878–82.

Smith, S.M., Levy, N.S. & Hayes, C.E. (1987) Impaired immunity in vitamin A-deficient mice. *Journal of Nutrition*, **117**, 857–65.

Smith, W.C., Nakshatri, H., Leroy, P., Rees, J. & Chambon, P. (1991) A retinoic acid Response element is present in the mouse cellular retinol binding protein I (mCRBPI) promoter. *EMBO Journal*, **10**, 2223–30.

Sommer, A. (1993) Vitamin A, infectious disease, and childhood mortality: A 2 cent solution. *Journal of Infectious Diseases*, **167**, 1003–7.

Sommer, A. (2008) Vitamin A deficiency and clinical disease: An historical overview. *Journal of Nutrition*, **138**, 1835–9.

Sonnenberg, G.F., Fouser, L.A. & Artis, D. (2011) Border patrol: Regulation of immunity, inflammation and tissue homeostasis at barrier surfaces by IL-22. *Nature Immunology*, **12**, 383–90.

Soprano, D.R., Smith, J.E. & Goodman, D.S. (1982) Effect of retinol status on retinol-binding protein biosynthesis rate and translatable messenger RNA level in rat liver. *Journal of Biological Chemistry*, **257**(13), 7693–7.

Strickland, S. (1981) Mouse teratocarcinoma cells: Prospects for the study of embryogenesis and neoplasia. *Cell*, **24**, 277–8.

Sucov, H.M., Murakami, K.K. & Evans, R.M. (1990) Characterization of an autoregulated response element in the mouse retinoic acid receptor type β gene. *Proceedings of the National Academy of Sciences of the U.S.A.*, **87**, 5392–6.

Sun, C.-M., Hall, J.A., Blank, R.B., Bouladoux, N., Oukka, M., Mora, J.R. & Belkaid, Y. (2007) Small intestine lamina propria dendritic cells promote de novo generation of Foxp3 T reg cells via retinoic acid. *Journal of Experimental Medicine*, **204**, 1775–85.

Svensson, M., Johansson-Lindbom, B., Zapata, F., Jaensson, E., Austenaa, L.M., Blomhoff, R. & Agace, W.W. (2008) Retinoic acid receptor signaling levels and antigen dose regulate gut homing receptor expression on CD8+ T cells. *Nature Immunology*, **1**, 38–48.

Tabin, C. (1995) The initiation of the limb bud: Growth factors, *Hox* genes, and retinoids. *Cell*, **80**, 671–4.

Tachibana, K., Stone, S., Tsubura, E. & Kishino, Y. (1984) Stimulatory effect of vitamin A on tumoricidal activity of rat alveolar macrophages. *British Journal of Cancer*, **49**, 343–8.

Takaki, H., Ichiyama, K., Koga, K., Chinen, T., Takaesu, G., Sugiyama, Y., Kato, S., Yoshimura, A. & Kobayashi, T. (2008) STAT6 inhibits TGF-β1-mediated Foxp3 induction through direct binding to the *Foxp3* promoter, which is reverted by retinoic acid receptor. *Journal of Biological Chemistry*, **283**(22), 14955–62.

Tan, X., Sande, J.L., Pufnock, J.S., Blattman, J.N. & Greenberg, P.D. (2011) Retinoic acid as a vaccine adjuvant enhances CD8+ T cell response and mucosal protection from viral challenge. *Journal of Virology*, **85**, 8316–27.

Tarwotjo, I., Sommer, A., West, K.P. Jr., Djunaedi, E., Mele, L., Hawkins, B. & the Aceh Study Group (1987) Influence of participation on mortality in a randomized trial of vitamin A prophylaxis. *American Journal of Clinical Nutrition*, **45**, 1466–71.

Tate, B.F., Allenby, G., Janocha, R., Kazmer, S., Speck, J., Sturzenbecker, L.J., Abarzúa, P., Levin, A.A. & Grippo, J.F. (1994) Distinct binding determinants for 9-*cis* retinoic acid are located within AF-2 of retinoic acid receptor α. *Molecular and Cellular Biology*, **14**, 2323–30.

Terao, J. (1989) Antioxidant activity of β-carotene-related carotenoids in solution. *Lipids*, **24**, 659–61.

Tini, M., Otulakowski, G., Breitman, M.L., Tsui, L.-C. & Giguère, V. (1993) An everted repeat mediates retinoic acid induction of the γF-crystallin gene: Evidence of a direct role for retinoids in lens development. *Genes & Development*, **7**, 295–307.

Tobler, A., Dawson, M.I. & Koeffler, H.P. (1986) Retinoids. Structure—Function relationship in normal and leukemic hematopoiesis in vitro. *Journal of Clinical Investigation*, **78**, 303–9.

Toma, S., Isnardi, L., Raffo, P., Dastoli, G., Dre Francisci, E., Riccardi, L., Palumbo, R. & Bollag, W. (1997) Effects of all-trans-retinoic acid and 13-*cis*-retinoic acid on breast-cancer cell lines: Growth inhibition and apoptosis induction. *International Journal of Cancer*, **70**, 619–27.

Tran, P., Zhang, X.-K., Salbert, G., Hermann, T., Lehmann, J.M. & Pfahl, M. (1992) COUP orphan receptors are negative regulators of retinoic acid response pathways. *Molecular and Cellular Biology*, **12**, 4666–76.

Tsai, S. & Collins, S.J. (1993) A dominant negative retinoic acid receptor blocks neutrophil differentiation at the promyelocyte stage. *Proceedings of the National Academy of Sciences of the U.S.A.*, **90**, 7153–7.

Twining, S.S., Schulte, D.P., Wilson, P.M., Fish, B.L. & Moulder, J.E. (1996a) Retinol is sequestered in the bone marrow of vitamin A-deficient rats. *Journal of Nutrition*, **126**, 1618–26.

Twining, S.S., Schulte, D.P., Wilson, P.M., Zhou, X., Fish, B.L. & Moulder, J.E. (1996b) Neutrophil cathepsin G is specifically decreased under vitamin A deficiency. *Biochimica et Biophysica Acta*, **1317**, 112–8.

Twining, S.S., Schulte, D.P., Wilson, P.M., Fish, B.L. & Moulder, J.E. (1997) Vitamin A deficiency alters rat neutrophil function. *Journal of Nutrition*, **127**, 558–65.

Uematsu, S., Fujimoto, K., Jang, M.H., Yang, B.-G., Jung, Y.-J,, Nishiyama, M., Sayo, S., Tsujimura, T., Yamamoto, M., Yokoto, Y., Kiyono, H., Miyasaka, M., Ishii, K.J. & Akhira, S. (2008) Regulation of humoral and cellular gut immunity by lamina propria dendritic cells expressing Toll-like receptor 5. *Nature Immunology*, **9**, 769–76.

Underhill, T.M., Kotch, L.E. & Linney, E. (1995) Retinoids and mouse embryonic development. *Vitamins and Hormones*, **51**, 403–57.

Underwood, B.A., Loerch, J.D. & Lewis, K.C. (1979) Effects of dietary vitamin A deficiency, retinoic acid and protein quantity and quality on serially obtained plasma and liver levels of vitamin A in rats. *Journal of Nutrition*, **109**, 796–806.

van de Pavert, S.A., Ferreira, M., Dominguez, R.G., Ribeiro, H., Molenaar, R., Moreira-Santos, L., Almeida, F.F., Ibiza, S., Barbosa, I., Goverse, G., Labão-Almeida, C., Godinho-Silva, C., Konijn, T., Schooneman, D., O'Toole, T., Mizee, M.R., Habani, Y., Haak, E., Santori, F.R., Littman, D.R., Schulte-Merker, S., Dzierzak, E., Simas, J.P., Mebius, R.E. & Veiga-Fernandes, H. (2014) Maternal retinoids control type 3 innate lymphoid cells and set the offspring immunity. *Nature*, **508**, 123–7.

van Vliet, T., van Schaik, F. & van den Berg, H. (1992) Beta-carotene metabolism: The enzymatic cleavage to retinal. *Voeding*, **53**, 186–90.

van Vliet, T., van Vlissingen, M.F., van Schaik, F. & van den Berg, H. (1996) β-Carotene absorption and cleavage in rats is affected by the vitamin A concentration of the diet. *Journal of Nutrition*, **126**, 499–508.

Vasios, G.W., Gold, J.D., Petkovich, M., Chambon, P. & Gudas, L.J. (1989) A retinoic acid-responsive element is present in the 5′-flanking region of the laminin B1 gene. *Proceedings of the National Academy of Sciences of the U.S.A.*, **86**, 9099–103.

Vasios, G., Mader, S., Gold, J.D., Leid, M., Lutz, Y., Gaub, M.-P., Chambon, P. & Gudas, L. (1991) The late retinoic acid induction of laminin B1 gene transcription involves RAR binding to the responsive element. *EMBO Journal*, **10**, 1149–58.

Vitamin A and Pneumonia Working Group. (1995) Potential interventions of the prevention of childhood pneumonia in developing countries: A meta-analysis of data from field trials to assess the impact of vitamin A supplementation on pneumonia morbidity and mortality. *Bulletin of the World Health Organization*, **73**, 609–19.

Vivat-Hannah, V., You, D., Rizzo, C., Daris, J.-P., Lapointe, P., Zusi, F.C., Marinier, A., Lorenzi, M.V. & Gottardis, M.M. (2001) Synergistic cytotoxicity exhibited by combination treatment of selective retinoid ligands with Taxol (paclitaxel). *Cancer Research*, **61**, 8703–11.

von Lintig, J. & Vogt, K. (2000) Filling the gap in vitamin Aresearch. Molecular identification of an enzyme cleaving β-carotene to retinal. *Journal of Biological Chemistry*, **275**(16), 11915–20.

Wang, C., Kelly, J., Bowen-Pope, D.F. & Stiles, C.D. (1990) Retinoic acid promotes transcription of the platelet-derived growth factor α-receptor gene. *Molecular and Cellular Biology*, **10**, 6781–4.

Wang, C., Kang, S.G., Hogenesch, H., Love, P.E. & Kim, C.H. (2010) Retinoic acid determines the precise tissue tropism of inflammatory Th17 cells in the intestine. *Journal of Immunology*, **184**, 5519–26.

Wang, Q., Yang, W., Uytingco, M.S., Christakos, S. & Wieder, R. (2000) 1,25-Dihydroxyvitamin D_3 and all-*trans*-retinoic acid sensitize breast cancer cells to chemotherapy-induced cell death. *Cancer Research*, **60**, 2040–8.

Wang, X.-D., Krinsky, N.I. & Russell, R.M. (1993) Retinoic acid regulates retinol metabolism via feedback inhibition of retinol oxidation and stimulation of retinol esterification in ferret liver. *Journal of Nutrition*, **123**, 1277–85.

Wang, X.-D., Russell, R.M., Liu, C., Stickel, F., Smith, D.E. & Krinsky, N.I. (1996) β-Oxidation in rabbit liver *in vitro* and in the perfused ferret liver contributes to retinoic acid biosynthesis from β-apocarotenoic acids. *Journal of Biological Chemistry*, **271**(43), 26490–8.

Warrell, R.P. Jr., de Thé, H., Wang, Z.-Y. & Degos, L. (1993) Acute promyelocytic Leukemia. *New England Journal of Medicine*, **329**, 177–89.

Wei, L.-N., Blaner, W.S., Goodman, D.S. & Nguyen-Huu, M.C. (1989) Regulation of the cellular retinoid-binding proteins and their messenger ribonucleic acids during P19 embryonal carcinoma cell differentiation induced by retinoic acid. *Molecular Endocrinology*, **3**, 454–63.

Wei, L.-N., Lee, C.-H. & Chang, L. (1995) Retinoic acid induction of mouse cellular retinoic acid-binding protein-I gene expression is enhanced by sphinganine. *Molecular and Cellular Endocrinology*, **111**, 207–11.

Widom, R.L., Rhee, M. & Karathanasis, S.K. (1992) Repression by ARP-1 sensitizes apolipoprotein AI gene responsiveness to RXRα and retinoic acid. *Molecular and Cellular Biology*, **12**, 3380–9.

Wiebel, F.F. & Gustafsson, J.-Å. (1997) Heterodimeric interaction between retinoid X receptor α and orphan nuclear receptor OR1 reveals dimerization-induced activation as a novel mechanism of nuclear receptor activation. *Molecular and Cellular Biology*, **17**, 3977–86.

Wiedermann, U., Hanson, L.Å., Kahu, H. & Dahlgren, U.I. (1993) Aberrant T-cell function *in vitro* and impaired T-cell dependent antibody response *in vivo* in vitamin A-deficient rats. *Immunology*, **80**, 581–6.

Willy, P.J. & Mangelsdorf, D.J. (1997) Unique requirements for retinoid-dependent transcriptional activation by the orphan receptor LXR. *Genes & Development*, **11**, 289–98.

Wilson, T.E., Fahrner, T.J. & Milbrandt, J. (1993) The orphan receptors NGFI-B and steroidogenic factor 1 establish monomer binding as a third paradigm of nuclear receptor-DNAinteraction. *Molecular and Cellular Biology*, **13**, 5794–804.

Wirth, J.J. & Kierszenbaum, F. (1986) Stimulatory effects of retinoic acid on macrophage interaction with blood forms of *Trypanosoma cruzi*: Involvement of transglutaminase activity. *Journal of Immunology*, **137**, 3326–31.

Wyss, A., Wirtz, G., Woggon, W.-D., Brugger, R., Wyss, M., Friedline, A., Bachmann, H. & Hunziker, W. (2000) Cloning and expression of β,β-carotene 15,15′-dioxygenase. *Biochemical and Biophysical Research Communications*, **271**, 334–6.

Xiao, S., Jin, H., Korn, T., Liu, S.M., Oukka, M., Lim, B. & Kuchroo, V.K. (2008) Retinoic acid increases Foxp3$^+$ regulatory T cells and inhibits development of Th17 cells by enhancing TGF-β-driven Smad3 signaling and inhibiting IL-6 and IL-23 receptor expression. *Journal of Immunology*, **181**, 2277–84.

Xu, X.-C., Sneige, N., Liu, X., Nandagiri, R., Lee, J.J., Lukmanji, F., Hortoobagyi, G., Lippman, S.M., Dhingra, K. & Lotan, R. (1997) A progressive decrease in nuclear retinoic acid receptor acid receptor β message RNAlevels during breast carcinogenesis. *Cancer Research*, **57**, 4992–6.

Yang-Yen, H.-F., Zhang, X.-K., Graupner, G., Tzukerman, M., Sakamoto, B., Karin, M. & Pfahl, M. (1991) Antagonism between retinoic acid receptors and AP-1: Implications for tumor promotion and inflammation. *New Biologist*, **3**, 1206–19.

Yokota, A., Takeuchi, H., Maeda, N., Ohoka, Y., Kato, C., Song, S.-Y. & Iwata, M. (2009) GM-CSF and IL-4 synergistically trigger dendritic cells to acquire retinoic acid-producing capacity. *International Immunology*, **21**, 361–77.

Zabel, B.A., Agace, W.W., Campbell, J.J., Heath, H.M., Parent, D., Roberts, A.I., Ebert, E.C., Kassam, N., Qin, S., Zovko, M., LaRosa, G.J., Yang, L.-L., Soler, D., Butcher, E.C., Ponath, P.D., Parker, C.M. & Andrew, D.P. (1999) Human G protein-coupled receptor GPR-9-6/CC chemokine receptor 9 is selectively expressed on intestinal homing T lymphocytes, mucosal lymphocytes, and thymocytes and is required for thymus-expressed chemokine-mediated chemotaxis. *Journal of Experimental Medicine*, **190**, 1241–55.

Zelent, A., Mendelsohn, C., Kastner, P., Krust, A., Garnier, J.-M., Ruffenach, F., Leroy, P. & Chambon, P. (1991) Differentially expressed isoforms of the mouse retinoic acid receptor β are generated by usage of two promoters and alternative splicing. *EMBO Journal*, **10**, 71–81.

Zhang, L.-X., Cooney, R.V. & Bertram, J.S. (1991) Carotenoids enhance gap junctional communication and inhibit lipid peroxidation in C3H/10T1/2 cells: Relationship to their cancer chemopreventive action. *Carcinogenesis*, **12**, 2109–14.

Zhang, L.-X., Cooney, R.V. & Bertram, J.S. (1992) Carotenoids up-regulate *connexin43* gene expression independent of their provitamin A or antioxidant properties. *Cancer Research*, **52**, 5707–12.

Zhang, X.-K., Liu, Y., Lee, M.-O. & Pfahl, M. (1994) A specific defect in the retinoic acid response associated with human lung cancer cell lines. *Cancer Research*, **54**, 5663–9.

Zhao, Z. & Ross, A.C. (1995) Retinoic acid repletion restores the number of leukocytes and their subsets and stimulates natural cytotoxicity in vitamin A-deficient rats. *Journal of Nutrition*, **125**, 2064–73.

Zhao, Z., Murasko, D.M. & Ross, A.C. (1994) The role of vitamin A in natural killer cell cytotoxicity, number and activation in the rat. *Natural Immunity*, **13**, 29–41.

Zheng, S.G., Wang, J. & Horwitz, D.A. (2008) Cutting edge: $Foxp3^+CD4^+$ $CD25^+$ regulatory T cells induced by IL-2 and TGF-β are resistant to Th17 conversion by IL-6. *Journal of Immunology*, **180**, 7112–16.

Zhou, H., Manji, S.S., Findlay, D.M., Martin, T.J., Heath, J.K. & Ng, K.W. (1994) Novel action of retinoic acid. Stabilization of newly synthesized alkaline phosphatase transcripts. *Journal of Biological Chemistry*, **269**(35), 22433–39.

Zhou, X., Kong, N., Wang, J., Fan, H., Zou, H., Horwitz, D., Brand, D., Liu, Z. & Zheng, S.G. (2010) Cutting edge: All-trans retinoic acid sustains the stability and function of natural regulatory T cells in an inflammatory milieu. *Journal of Immunology*, **185**, 2675–9.

Zhou, X.-F., Shen, X.-Q. & Shemshedini, L. (1999) Ligand-activated retinoic acid receptor inhibits AP-1 transactivation by disrupting c-Jun/c-Fos dimerization. *Molecular Endocrinology*, **13**, 276–85.

Ziegler, R.G. (1989) A review of epidemiologic evidence that carotenoids reduce the risk of cancer. *Journal of Nutrition*, **119**, 116–22.

16 Vitamin D

16.1 CHEMISTRY AND BIOLOGICAL FUNCTIONS

Vitamin D is represented by cholecalciferol (vitamin D_3) and ergocalciferol (vitamin D_2), which are structurally similar secosteroids derived from the UV irradiation of provitamin D sterols. (Secosteroids are steroids in which one of the rings has broken.) Vitamin D_3 is produced endogenously by the photochemical action of sunlight on 7-dehydrocholesterol in the skin. Vitamin D_2 is produced in plants, fungi, and yeasts by the solar irradiation of ergosterol.

Vitamin D_3 and vitamin D_2 differ structurally only in the C-17 side chain, which in vitamin D_2 has a double bond and an additional methyl group (Figure 16.1).

Irradiation of the parent steroid results in breakage of the B ring at the 9,10 carbon–carbon bond, resulting in the conjugated triene system of double bonds. The vitamin D structures retain their numbering from the parent compound, cholesterol. Asymmetric carbon atoms are designated by using the *R*,*S* notation.

The vitamin D_2 and D_3 molecules are biologically inactive and must be converted to their active, hormonal forms by successive hydroxylation reactions (Figure 16.2).

Vitamin D_3 of endogenous or dietary origin arrives at the liver where it is hydroxylated at carbon 25 to yield 25-hydroxyvitamin D_3 [25(OH)D_3]. This metabolite, also known as calcidiol, is the major circulating form of vitamin D and is a good indicator of the combined effects of exposure to sunlight and dietary intake of the vitamin. Calcidiol, in passing through the kidney, is hydroxylated at the α-position of carbon 1 to generate 1α,25-dihydroxyvitamin D_3 [1α,25$(OH)_2D_3$], also known as calcitriol. Calcitriol circulates at low concentrations and acts in the manner of a steroid hormone.

The dihydroxylated metabolites of vitamins D_2 and D_3 show the same physiological activities as one another in mammals, including the maintenance of calcium and phosphate homeostasis, regulation of bone remodeling, and modulation of cell proliferation and differentiation. The results of a randomized, placebo-controlled, double-blinded study of healthy adults demonstrated that a 1,000 IU dose of vitamin D_2 daily is as effective as vitamin D_3 in maintaining circulating concentrations of 25(OH)D, indicating equivalent potency (Holick *et al.*, 2008).

The D_3 and D_2 designations are used in this chapter when experimental data specify these forms. The distinguishing subscript is not used where the situation applies to either vitamin D_2 or D_3.

16.2 FOOD SOURCES

The vitamin D activity in the human diet is contributed mainly by vitamin D itself and its immediate metabolite, 25(OH)D. The proportion of vitamin D obtained from the diet is normally very small compared with that synthesized in skin in response to sunlight. The richest natural sources of vitamin D_3 are fish-liver oils, especially halibut-liver oil. Fatty fish, such as herring, sardines, pilchards, and tuna, are rich natural food sources; smaller amounts of the vitamin are found in mammalian liver, eggs, and dairy products. Cereals, vegetables, and fruit contain no vitamin D, whilst meat, poultry, and white fish contribute insignificant amounts. Foodstuffs commonly enriched with vitamin D include margarine, skimmed milk powder, evaporated milk, milk-based beverages, breakfast cereals, dietetic products of all kinds, baby foods, and soup powders.

16.3 CUTANEOUS SYNTHESIS OF VITAMIN D_3

16.3.1 Structural Organization of the Epidermis

The skin is composed of two layers: the epidermis and the underlying dermis. The bulk of the epidermal cells (keratinocytes) undergoes a process of keratinization resulting in the formation of the dead superficial layers of skin. Other cells, melanocytes, which produce the melanin pigment, do not undergo keratinization. The superficial keratinized cells are continuously sloughed from the surface and must be replaced by cells that proliferate in the basal layers of the epidermis. The newly formed cells are displaced to higher levels and as they move upward their cytoplasm is replaced by keratin. Thus the structural organization of the epidermis into layers reflects stages in the dynamic processes of cell proliferation and differentiation.

The epidermis consists of five layers as shown in Figure 16.3.

Melanin is present mainly in the stratum basale and in the deeper layer of the stratum spinosum. Exposure of the skin to UV radiation increases the enzymatic activity of melanocytes leading to tanning. The Malpighian layer is responsible for cell proliferation and for initiation of the keratinization process.

16.3.2 Photosynthesis

Vitamin D_3 is synthesized in the skin from 7-dehydrocholesterol during exposure to sunlight by strictly chemical (i.e. non-enzymatic) reactions (Figure 16.4).

DOI: 10.1201/9781003410300-19

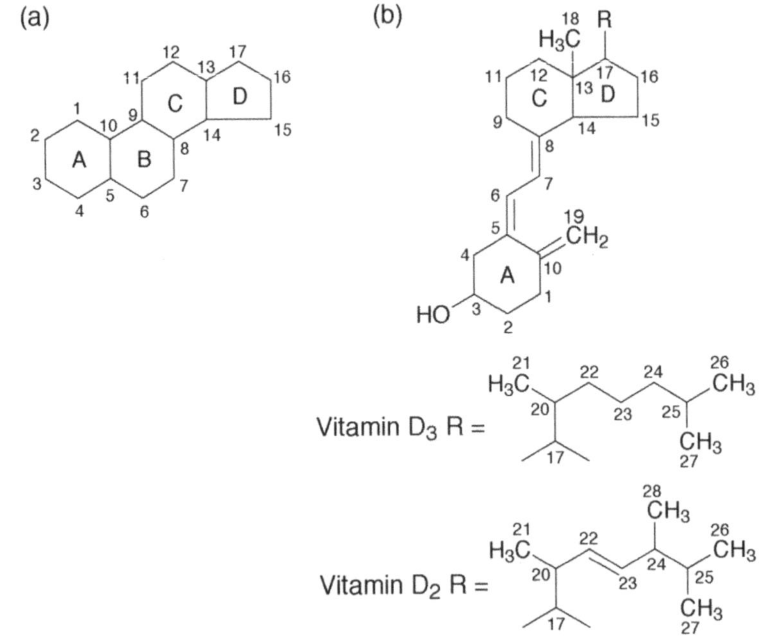

FIGURE 16.1 Structural relationship of (a) the parent steroid nucleus to (b) vitamin D.

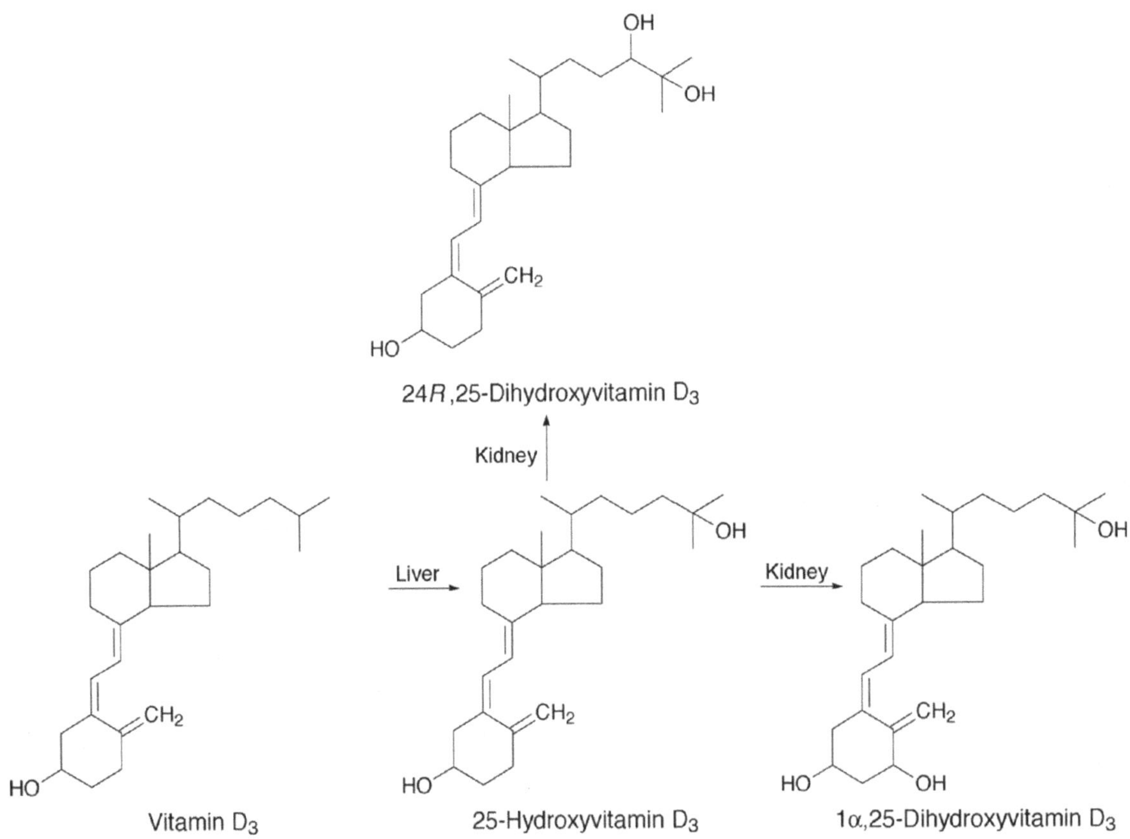

FIGURE 16.2 Conversion of vitamin D3 to hormonal metabolites.

Exposure to increasing doses of solar radiation results in a corresponding increase in serum vitamin D_3 concentration within 24 to 48 hours and a return to baseline concentration by seven days (Adams *et al.*, 1982). In young white adults, at least 30 μg of vitamin D_3 is made in the skin and released into the circulation from each square meter of body surface after exposure to one minimal erythema dose (MED), i.e. a dose causing minimal sunburn (Holick, 1987). Vitamin D_3

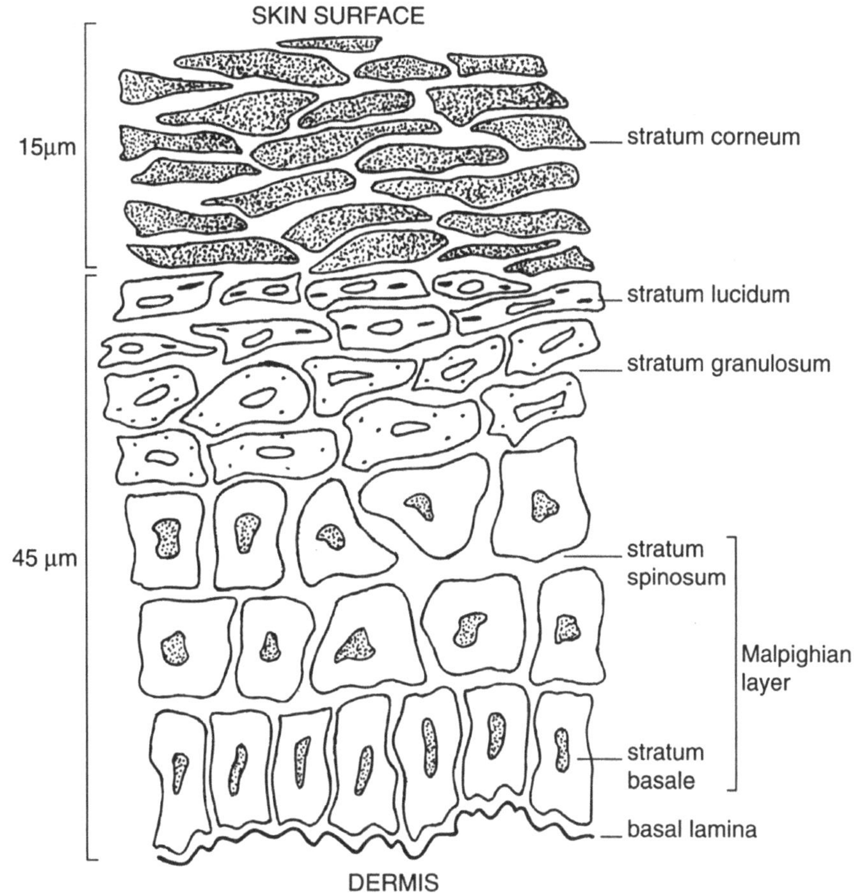

FIGURE 16.3 Diagrammatic representation of the epidermis.

that is synthesized in the skin during seasonal exposure to the sun can be stored in the body fat for use in the winter (Mawer *et al.*, 1972).

7-Dehydrocholesterol is converted photochemically to previtamin D_3, which is then converted to vitamin D_3 by a temperature-dependent process. 7-Dehydrocholesterol is equally distributed between the epidermis and the dermis, but because the penetrating radiation reaches the epidermis first, approximately 90% of the total cutaneous previtamin D_3 is made in the epidermis, mostly in the Malpighian layer. The waveband of solar radiation responsible for the conversion of 7-dehydrocholesterol to previtamin D_3 is that between 290 and 315 nm, known as the UV-B band. The optimum radiation was experimentally determined to be between 295 and 300 nm with an apparent maximum near 297 nm (MacLaughlin *et al.*, 1982). Solar radiation below 290 nm is prevented from reaching the Earth's surface by the ozone layer in the stratosphere. Complete cloud cover reduces UV energy by only 50% (Lawson, 1985).

The high-energy photons of the UV-B radiation are absorbed by the conjugated 6,7-diene in the B-ring of 7-dehydrocholesterol, resulting in ring opening at C9–C10 to produce previtamin D_3. Previtamin D_3, which is biologically inactive and thermodynamically unstable, undergoes temperature-dependent isomerization to vitamin D_3. In human skin *in vitro* at 37°C, it took only 24 hours for the conversion of previtamin D_3 to vitamin D_3 to reach equilibrium, compared to 7 to 10 days in hexane (Tian *et al.*, 1993).

During a single prolonged exposure to the sun, the synthesis of previtamin D_3 reaches a plateau at about 10 to 15% of the original cutaneous 7-dehydrocholesterol concentration. At this point, the previtamin, instead of converting to vitamin D_3, undergoes photoisomerization to two biologically inactive products, lumisterol and tachysterol. Tachysterol formation also plateaus (at about 5% of the original 7-dehydrocholesterol concentration), whereas lumisterol formation continues to increase. These events take place regardless of the amount of skin pigmentation. Lumisterol and tachysterol remain in the skin and are eventually discarded during the sloughing off of old epidermal cells. As soon as previtamin D_3 stores are depleted, exposure of cutaneous lumisterol and tachysterol to solar UV radiation may promote their conversion back to previtamin D_3 (Holick *et al.*, 1981). The photochemical conversion of previtamin D_3 to lumisterol and tachysterol is the first line of defense against the production of toxic levels of vitamin D after excessive exposure to sunlight. A second line of defense is the conversion of vitamin D_3 itself to 5,6-*trans*-vitamin D_3 and biologically inactive suprasterols (Webb *et al.*, 1989). These photochemical mechanisms for regulating the total cutaneous production of vitamin D_3 presumably

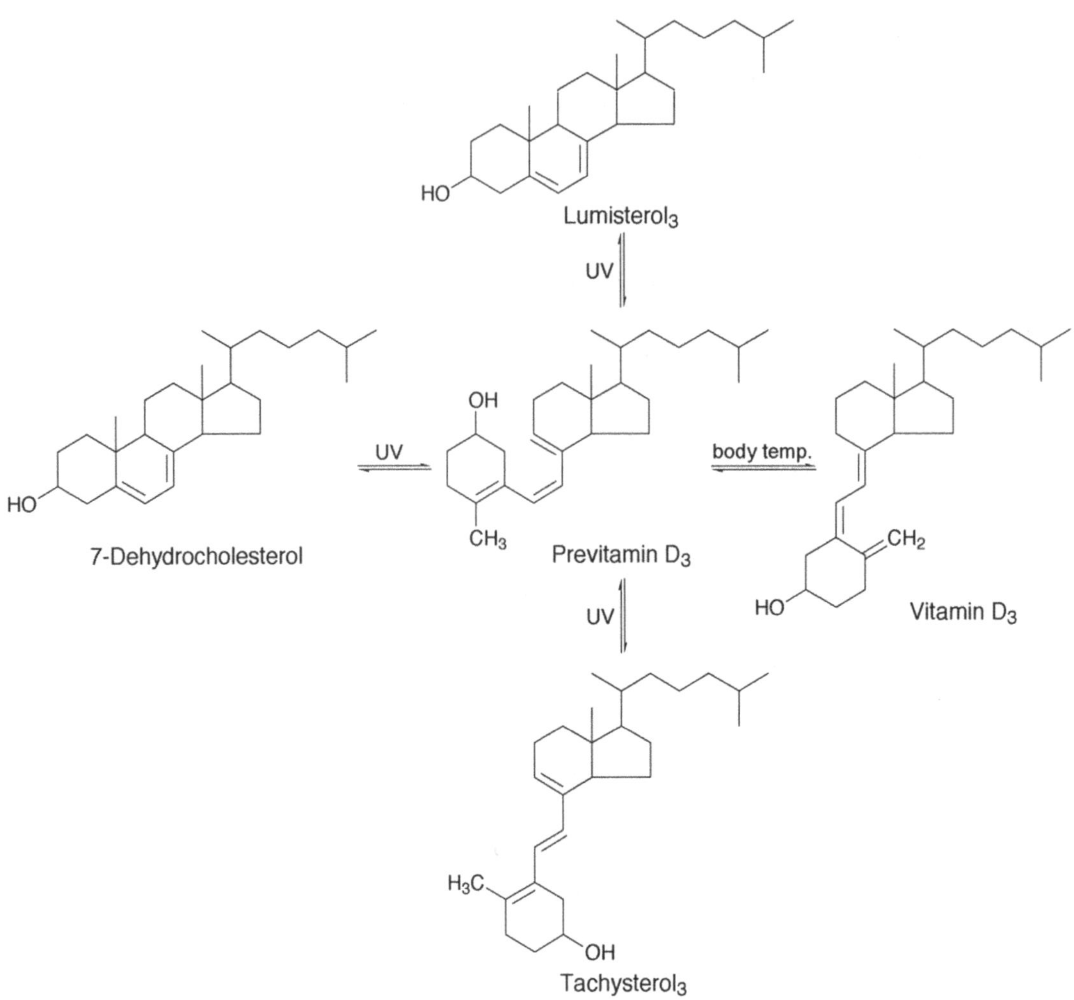

FIGURE 16.4 The cutaneous synthesis of vitamin D3.

account for the absence of reported cases of solar-induced vitamin D toxicity.

It is evident that only short exposures to bright sunlight produce sufficient vitamin D_3 in white people. After the initial beneficial exposure, use of a sunscreen is recommended to prevent the harmful effects of the sun. Individuals who have limited or no access to direct sunlight throughout the year are at risk from vitamin D deficiency unless they have an adequate dietary supply of the vitamin. Window glass does not transmit radiation of wavelengths below 334 nm, therefore natural sunlight that is transmitted indoors is of no benefit for cutaneous vitamin D_3 production. Fluorescent tubes emit short wavelength radiation (λ 280 nm) but this is filtered out by the glass or plastic fixtures in which they are mounted.

16.3.3 Factors Affecting Vitamin D_3 Production

Various biological and environmental factors affect the cutaneous photosynthesis of vitamin D_3, including aging, degree of skin pigmentation, occlusive clothing, duration of exposure to sunlight, the use of topical sunscreens, cloud cover, air pollution, season of year, and latitude.

16.3.3.1 Aging

After about 20 years of age, age-related changes in skin take place (Montagna & Carlisle, 1979) and the skin becomes progressively thinner. The concentration of epidermal 7-dehydrocholesterol also decreases with age and young adults can produce two to three times more previtamin D_3 in their skin than can elderly individuals (MacLaughlin & Holick, 1985). When healthy young and elderly volunteers were exposed to the same amount of simulated sunlight, serum vitamin D concentrations in the young subjects increased to a maximum of 30 ng/ml within 24 hours after exposure, whereas the elderly subjects achieved a maximum of only 8 ng/ml (Holick *et al.*, 1989). The results of a study of 433 postmenopausal women (Need *et al.*, 1993) showed that the tendency for serum 25(OH)D to fall with age is due in part to the age-related decline in skin thickness.

16.3.3.2 Degree of Skin Pigmentation

At relatively low levels of solar radiation, skin pigmentation is a limiting factor for previtamin D_3 synthesis because melanin competes with 7-dehydrocholesterol in absorbing UV-B radiation (Holick *et al.*, 1981). When surgical specimens of white skin and a range of black skins were exposed to the same dose of simulated sunlight (1.5 MED), the exposure times necessary to maximize previtamin D_3 formation increased with increasing skin pigmentation (Holick *et al.*, 1981). The 1.5 MED induced a 30-fold increase in serum vitamin D in Caucasian subjects, but did not alter serum vitamin D in heavily pigmented Black subjects. Re-exposure of one Black subject to 9 MED increased serum vitamin D to a concentration similar to those recorded in the Caucasian subjects after exposure to 1.5 MED (Clemens *et al.*, 1982). Hence Blacks have the same capacity as Caucasians to produce vitamin D_3, but require a longer exposure to, or higher intensity of, solar radiation to do so.

16.3.3.3 Use of Sunscreens

Sunscreen lotions play an important role in protecting against sunburn, photoaging, and skin cancer. The problem with sunscreens is that they filter out the same radiation that is responsible for vitamin D synthesis in the skin. When young Caucasian adults were entirely covered with a protection factor 8 sunscreen, they were unable to elevate their serum vitamin D_3 levels after a whole-body exposure to a dose of simulated sunlight that caused marked elevations of serum vitamin D_3 in unprotected subjects (Matsuoka *et al.*, 1987).

16.3.3.4 Season and Latitude

Webb *et al.* (1988) investigated the effect of season and latitude on the cutaneous production of vitamin D_3 in Boston, USA (42°N), and Edmonton, Canada (52°N). In Boston, the conversion of 7-dehydrocholesterol to previtamin D_3 is greatest in June and July and gradually declines after August. Between November and February, exposure to sunlight on cloudless days for about 5 hours does not result in any significant conversion. Previtamin D_3 synthesis resumes in the middle of March. In Edmonton (10° north of Boston), the photosynthesis of previtamin D_3 ceases in October and is not resumed until April. These observations are explained by an increase in the zenith angle of the sun during the winter months and with latitude. Filtration of sunlight through the ozone layer takes place through an increased path length, decreasing the UV-B radiation that reaches the Earth's surface.

16.4 INTESTINAL ABSORPTION OF DIETARY VITAMIN D

The vitamin D activity in the human diet is contributed mainly by free vitamin D, with 25(OH)D making a quantitatively small but nutritionally significant contribution. Animal tissues contain a small proportion of vitamin D esterified with both saturated and unsaturated fatty acids. Ingested vitamin D is solubilized within mixed micelles in the duodenum and esters of vitamin D, if present, are hydrolyzed during solubilization. Solubilization is highly bile salt-dependent and patients with intraluminal bile salt deficiency have inefficient absorption of vitamin D. The micelles dissociate in the acidic microclimate of the unstirred layer and the liberated vitamin D is absorbed in the jejunum by simple diffusion, along with other lipids. Vitamin D is incorporated into chylomicrons within the enterocytes and, when released, the chylomicrons convey the vitamin in the mesenteric lymph to the systemic circulation. During transport in the blood, lipolysis of the chylomicrons converts them into chylomicron remnants.

16.5 TRANSPORT OF SYNTHESIZED AND DIETARY VITAMIN D

Cutaneously synthesized vitamin D_3 slowly diffuses from the skin into the blood where it is picked up for transport by the vitamin D-binding protein (DBP), which has a single binding site for vitamin D and its metabolites. The affinity of DBP for binding the vitamin D metabolites is in the order 25(OH)D=24,25$(OH)_2$D > 1α,25$(OH)_2$D > vitamin D. At normal circulating concentrations of vitamin D metabolites, less than 5% of the total DBP binding sites are occupied. Some of the vitamin D of dietary origin is transferred from the chylomicrons to DBP. Both chylomicron remnants and DBP deliver vitamin D to the liver, but uptake from the remnants is more rapid than that from DBP. Much of the vitamin D bound to DBP is deposited in adipose tissue and skeletal muscle prior to hepatic uptake.

16.6 METABOLIC EVENTS IN THE LIVER

In the liver, the vitamin D of both cutaneous and dietary origin is hydroxylated to 25(OH)D or is catabolized to inactive compounds for excretion in the bile. The physiologically relevant 25-hydroxylase appears to be CYP2RI, a microsomal enzyme (Prosser & Jones, 2004). In contrast to the rapid hepatic uptake of dietary vitamin D and inactivation of surplus amounts, the supply of cutaneous vitamin D to the liver is gradual, allowing continuous, prolonged production of 25(OH)D. Thus, the plasma concentration of this metabolite is maintained even when exposure of the skin to sunlight occurs only intermittently. Hydroxylation of vitamin D is only loosely regulated and circulating concentrations of 25(OH)D continue to rise in response to ingestion of pharmacological doses of vitamin D. Levels can rise to more than 400 ng/ml, leading to vitamin D toxicity. In contrast, extensive UV irradiation of the skin does not cause hypervitaminosis D and raises 25(OH)D levels in plasma to no more than 80 ng/ml (Fraser, 1983).

The mammalian liver does not store vitamin D to any significant extent and any vitamin D that is not metabolized in the liver is stored in the adipose tissue and skeletal muscle (Mawer *et al.*, 1972). Almost all of the 25(OH)D produced in the liver is released without delay into the bloodstream, where it becomes tightly bound to DBP. The normal concentration range of plasma 25(OH)D is 10–40 ng/ml. The blood constitutes the largest single body pool of 25(OH)D, since extrahepatic tissues take up only small amounts. During vitamin D deprivation, the 25(OH)D blood pool is maintained through the prolonged release of vitamin D from its skin reservoir and from storage sites in muscle and adipose tissue.

16.7 METABOLIC EVENTS IN THE KIDNEY

On arrival at the kidney, the 25(OH)D–DBP complex is filtered through the glomerulus and reabsorbed by the epithelial cells of the proximal tubule via endocytosis mediated by two synergistic DBP receptors, cubilin and megalin (Christensen & Birn, 2001). Cubilin sequesters DBP on the cell surface and megalin facilitates the internalization. DBP is degraded in lysosomes, releasing 25(OH)D, which is conveyed to the mitochondria by at least two intracellular vitamin D binding proteins, IDBP-1 and IDBP-2.

The metabolic fate of 25(OH)D in the kidney is dependent on the calcium and phosphorus requirements of the animal. An urgent need for either element results in the formation of 1α,25$(OH)_2$D, whereas during states of adequate calcium or phosphate balance, 25(OH)D is converted predominantly to 24,25$(OH)_2$D, which circulates in the bloodstream bound to DBP. Two mitochondrial enzymes catalyze these conversions – 25(OH)D-1α-hydroxylase (CYP27B1) and 25(OH)D-24-hydroxylase (CYP24A1), respectively.

16.8 REGULATION OF VITAMIN D METABOLISM

Plasma concentrations of the highly potent 1α,25$(OH)_2$D are kept within the range of 25 to 70 pg/ml by reciprocal changes in the rates of synthesis and degradation at the cellular level. 1α,25$(OH)_2$D elicits its own degradation by strongly inducing expression of the CYP24 24-hydroxylase, while simultaneously down-regulating the CYP27B1 1α-hydroxylase (Zierold *et al.*, 2003). CYP24 converts 1α,25$(OH)_2$D to 1α,24,25$(OH)_3$D, which is ten times less biologically active than 1α,25$(OH)_2$D. (Note: CYP24 acts on both 25(OH)D and 1α,25$(OH)_2$D.) Circulating calcitriol levels decrease exponentially with age, a decrease which correlates with reduced CYP27B1 and increased CYP24 expression in the kidney. 1α,25$(OH)_2$D also induces skeletal endocrine cells to produce fibroblast growth factor 23 (FGF-23), which then feedback represses renal CYP27B1 to curtail 1α,25$(OH)_2$D biosynthesis (Kolek *et al.*, 2005). Parathyroid hormone (PTH) raises plasma levels of 1α,25$(OH)_2$D by inducing CYP27B1 while down-regulating CYP24 (Zierold *et al.*, 2003).

The formation of 1α,24,25$(OH)_3D_3$ is the first step in the C-24 oxidation pathway (Figure 16.5), which represents a self-induced mechanism for limiting the action of 1α,25$(OH)_2$D once the initial wave of gene expression has been initiated. Each oxidation step leads to a progressive loss of biological activity, culminating in the inactive water-soluble product, calcitroic acid, which is excreted in the urine. The C-24 oxidation pathway is expressed constitutively only by the kidney. However, after pretreatment with 1α,25$(OH)_2$D, this catabolic pathway is inducible in intestine and bone as well as kidney.

16.9 EXCRETION

The liver has a large capacity for catabolizing vitamin D into highly polar inactivation products and excreting these catabolites in bile. Many of these excretion products are conjugated with glucuronic acid. Only a negligible proportion of injected radioactively labeled vitamin D appears in the bile as unchanged vitamin D or 25(OH)D (Clements *et al.*, 1984). An enterohepatic circulation of vitamin D catabolites exists (Arnaud *et al.*, 1975) but, as the catabolites have only trivial biological activity, it does not constitute a functional mechanism for conserving vitamin D (Fraser, 1983). In conclusion, the majority of vitamin D degradation products is excreted in the feces via the bile, with only a few percent appearing in the urine.

16.10 GENOMIC ACTIONS OF 1A,25-DIHYDROXYVITAMIN D_3 (CALCITRIOL)

16.10.1 Introduction

The genomic action of calcitriol is mediated by the vitamin D receptor (VDR), which functions as a ligand-activated transcription factor in the nucleus of target cells. The sequence of events involved in the control of gene transcription by the VDR is (1) binding of calcitriol to the VDR in the cytosol; (2) translocation of the hormone–receptor complex to the nucleus; (3) binding of VDR–RXR heterodimers (RXR, retinoid X receptor) or, less commonly, VDR homodimers to the vitamin D response element (VDRE) in the promoter of primary vitamin D-responding genes; and (4) recruitment of other nuclear proteins into the transcriptional pre-initiation complex. The VDRE functions as a transcriptional enhancer.

One must always bear in mind that an increase in the level of mRNA is not necessarily due to increased transcription: an increased mRNA stability will also lead to its accumulation.

16.10.2 Expression and Regulation of the VDR

The VDR is a protein of 53 kDa which selectively binds calcitriol$_3$ with high affinity. The protein is a type II member of the nuclear hormone receptor superfamily and possesses the characteristic two zinc finger motifs in the DNA binding site. Relatively high levels of the VDR are present in the cells of bone, kidney, and small intestine – tissues that are important for calcium homeostasis – but the receptor is also present in cell types not involved in calcium homeostasis, including lymphocytes, colonocytes, hepatocytes, and cardiac myocytes. There are no isoforms of the VDR. Many allelic variants (polymorphisms) of the VDR gene occur naturally in the human population, with substantial differences between races and ethnic groups. Their expression associates with decreased bone density, propensity to hyperparathyroidism, resistance to vitamin D therapy, and susceptibility to infections, autoimmune diseases, and cancer (Dusso *et al.*, 2005).

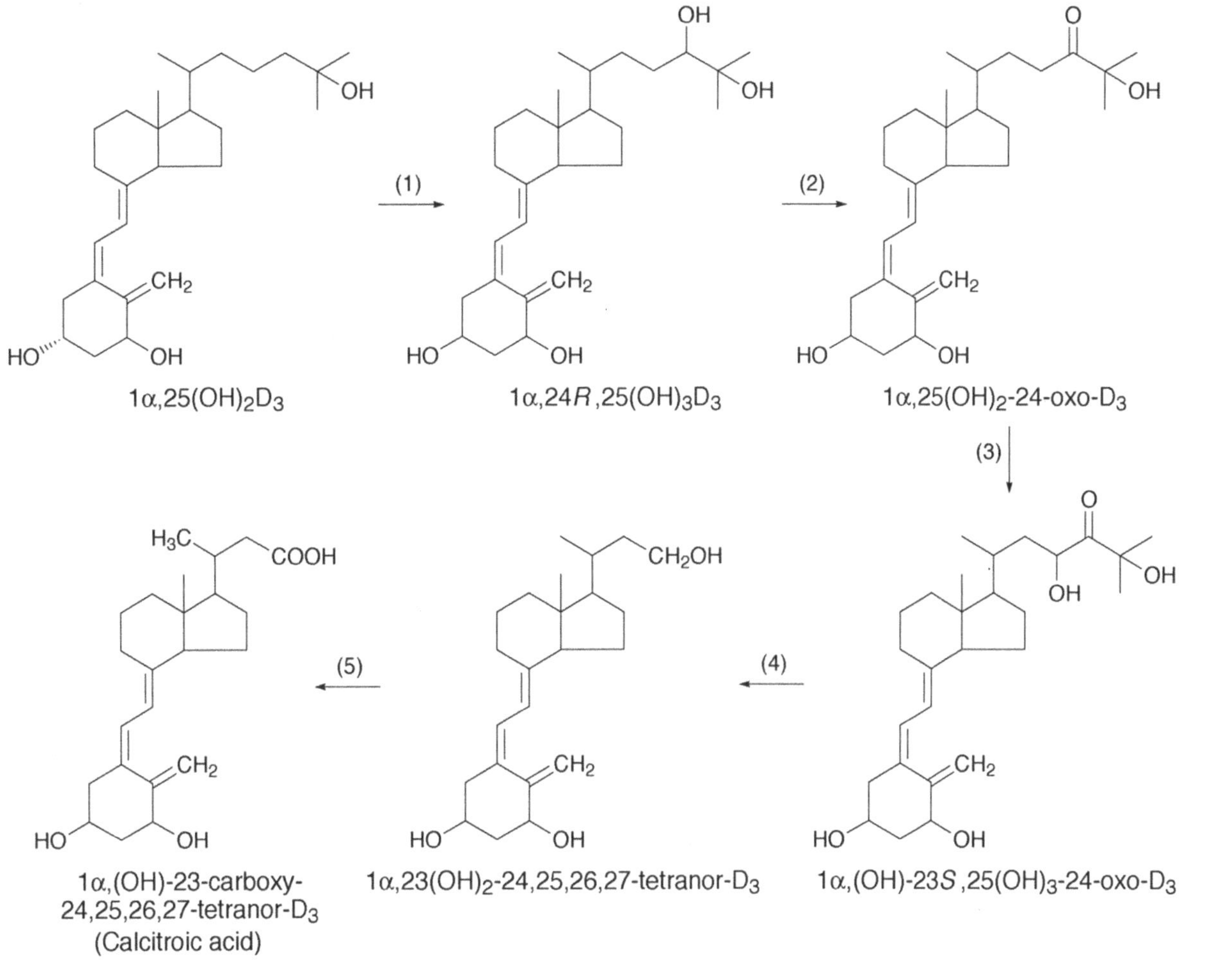

FIGURE 16.5 The carbon-24 (C-24) oxidation pathway of 1α,25-dihydroxyvitamin D_3 degradation. Steps: (1) Hydroxylation at C-24, (2) oxidation of the C-24 hydroxyl group to a ketone, (3) hydroxylation at C-23, (4) cleavage of the C-23–C-24 bond, and (5) oxidation of the C-23 hydroxyl group to a carboxyl group. Steps 1 to 5 are catalyzed by 24*R*-hydroxylase.

Desai *et al.* (1995) used staurosporine, an inhibitor of protein kinase C (PKC) and related protein kinases, and okadaic acid, an inhibitor of protein phosphatases, to investigate the contribution of VDR phosphorylation/dephosphorylation to vitamin D-stimulated transcription of the rat osteocalcin gene. The results suggested the presence of at least two functionally distinct phosphorylation sites on the VDR. At one site, staurosporine inhibits a phosphorylation event that is specifically required for the intrinsic transactivation function of VDR–RXR heterodimers. At the other site, okadaic acid inhibits a post-translational dephosphorylation event that is required for VDRE binding of VDR-containing transcription factor complexes.

The level of VDR expression in a cell determines the magnitude of the response evoked by calcitriol. Within each target tissue, the level of VDR is not fixed but rather it is dynamically regulated by multiple factors. These include calcitriol, which up-regulates the amount of receptor (homologous regulation), and other hormones and growth factors which may cause up- or down-regulation of receptor abundance (heterologous regulation). The mechanisms underlying the regulation of VDR abundance include alterations in the rate of transcription of the VDR gene, the stability of the VDR mRNA, and post-translational events. The protein kinase A (PKA) and PKC signal transduction pathways interact at the level of VDR regulation. Activation of the PKA pathway by forskolin up-regulates VDR gene expression, whereas activation of the PKC pathway by phorbol ester down-regulates VDR gene expression (Krishnan & Feldman, 1991, 1992). There are profound tissue- and cell-specific variations in VDR regulation.

16.10.3 Recruitment of Co-Activators by the VDR

The ligand-activated VDR may function to recruit co-activators that remodel chromatin structure and permit greater accessibility of the transcriptional machinery to DNA. Co-activators shown to interact with liganded VDR include the histone acetyltransferase (HAT) protein SRC-1 (Onate *et al.*, 1995) and the multiprotein complex known as DRIP (Rachez *et al.*, 1998). There have been no reports as yet of co-repressor proteins which bind to the VDR, so it is presently unclear whether the VDR interacts with co-repressors to bring about down-regulation of target genes. At least one co-repressor, N-CoR, fails to interact with the VDR (Hörlein *et al.*, 1995). Determination of whether vitamin D hormone action is stimulatory or inhibitory may be tissue-specific or dependent on the state of cellular differentiation. Most vitamin D-responsive genes are up-regulated by calcitriol, the human parathyroid hormone (PTH) gene being one of the few that is down-regulated.

16.10.4 Post-Translational Modification of the VDR by Phosphorylation

The binding of calcitriol to the VDR results in a substantial increase in phosphorylation of the receptor (Brown & DeLuca, 1990). Two sites of serine (Ser) phosphorylation on the human VDR have been identified, each with a different function. Ser-51 in the zinc finger region is phosphorylated by PKC (Hsieh *et al.*, 1991), a post-translational modification that inhibits the receptor's ability to interact with the VDRE (Hsieh *et al.*, 1993). Phosphorylation of Ser-208 in the ligand-binding domain by casein kinase II (CKII) potentiates transcriptional activation by modulating the interaction with transcriptional co-activators (Arriagada *et al.*, 2007).

Replacement of Ser-51 or Ser 208 with amino acids that are incapable of being phosphorylated does not affect DNA binding or attenuate calcitriol-mediated transcriptional activation. This demonstrates that phosphorylation of the two serine residues is not an obligatory switch for VDR function; rather, these modifications represent both positive (casein kinase II) and negative (PKC) modulatory mechanisms that apparently govern receptor activity under appropriate cellular conditions. Jurutka *et al.* (1996) envisaged two populations of liganded VDR: one that is hypophosphorylated at Ser-208, yet still active in transcriptional enhancement, and a superactive, hyperphosphorylated form that is even more effective at co-operatively recruiting co-activators and/or basal transcription factors.

16.10.5 Nuclear Localization of the Liganded VDR

The unoccupied VDR exists in equilibrium between the cytosolic and nuclear compartments of the target cell. Receptor–hormone binding shifts this equilibrium to favor nuclear localization (Walters *et al.*, 1986). The hormone–receptor complex rapidly translocates to the nucleus along microtubules facilitated by specialized motor proteins. Disruption of microtubular integrity impairs the genomic response to calcitriol in human monocytes, which clearly underlines the physiological importance of the intracellular tubulin transport system (Kamimura *et al.*, 1995).

16.10.6 Role of Ligands in VDR–RXR Heterodimer Binding to DNA

Thompson *et al.* (1998) investigated the molecular function of calcitriol and 9-*cis* retinoic acid ligands in the binding of the VDR and RXR to mouse osteopontin and rat osteocalcin VDREs. Efficient binding to either VDRE occurred as a VDR–RXR heterodimer, not as a VDR homodimer. Calcitriol dramatically enhanced heterodimer–VDRE interaction, whereas somewhat higher concentrations of 9-*cis* retinoic acid inhibited this association. A possible explanation for this inhibition is that the binding of 9-*cis* retinoic acid to the RXR partner destabilizes the heterodimer and induces RXR homodimer formation. MacDonald *et al.* (1993) offered this explanation when they demonstrated that 9-*cis* retinoic acid repressed vitamin D_3-induced transactivation of the osteocalcin gene. Thus liganded RXR is diverted from vitamin D-activated transcription toward expression of vitamin A-dependent genes.

Thompson *et al.* (1998) showed that the transcriptional response to hormone depends on the sequential order in which the components assemble. They proposed the existence of two alternative allosteric pathways for VDR–RXR association and response to ligand. (1) An unliganded VDR associates with RXR to form an apo-heterodimer in solution. Subsequent binding of calcitriol induces a conformational change in the heterodimer, which results in enhanced binding to the VDRE. The RXR partner can readily be dissociated from the DNA-bound heterodimer by the addition of 9-*cis* retinoic acid, leading to the formation of RXR homodimers that mediate retinoid-responsive pathways. This 9-*cis* retinoic acid-receptive conformation is proposed to exist also in monomeric RXR and in the apo-VDR–RXR. (2) When VDR binds calcitriol before heterodimer formation, it is postulated to acquire a conformation distinct from that in the first pathway. After heterodimerization, the liganded VDR allosterically alters the ligand-binding domain of its RXR partner, rendering the RXR unable to bind its own ligand and thus making the heterodimer resistant to 9-*cis* retinoic acid-elicited dissociation.

16.10.7 Interaction of the VDR with Basal Transcription Factor TFIIB

In the formation of the pre-initiation complex on a gene promoter, the binding of the basal transcription factor TFIIB to the TATA-binding protein is a requisite for the recruitment of RNA polymerase II. VDR interacts with TFIIB through a highly specific, ligand-independent, direct protein–protein contact and enhances transcription in the manner of an activator (Blanco *et al.*, 1995; MacDonald *et al.*, 1995).

16.10.8 Vitamin D-Responsive Genes Containing VDREs

Although over 50 genes have been reported to be regulated by calcitriol, only a few have been shown to contain VDREs. More than half of the known natural VDREs have a direct repeat structure of two six-nucleotide half-sites with a specific three-nucleotide spacer (DR3). Other natural VDREs have DR4, DR6, or IR9 (inverted palindrome) structures. In most vitamin D-responsive genes, RXR binds to the 5′

half-site and its heterodimeric partner, VDR, binds to the 3′ half-site. A selection of primary vitamin D-responding gene products is listed in Table 16.1.

Genes having a simple DR3-type VDRE include those encoding p21 protein, atrial natriuretic factor, integrin β3 subunit, and osteopontin. p21 protein is a potent inhibitor of cyclin-dependent kinases, which are regulators of the cell-cycle (Xiong *et al.*, 1993). Atrial natriuretic factor is a peptide hormone released by the atrial walls of the heart when they become stretched and which acts directly on the kidney to greatly increase salt and water excretion. Integrin $\alpha_v\beta_3$ is a receptor that is expressed on the osteoclast plasma membrane and recognizes the bone-residing proteins osteopontin and bone sialoprotein (Medhora *et al.*, 1993). The binding of osteopontin to cell-surface integrin $\alpha_v\beta_3$ initiates signals that elicit osteoclastic actions (Noda & Denhardt, 2002). The carbonic anhydrase isoform II is highly expressed in cells where acid-base regulation is a primary function, i.e. gastric parietal cells, salivary glands, renal tubular cells, and osteoclasts. The carbonic anhydrase II DR3-type VDRE is unusual in that the receptor complex binds in 5′-VDR–RXR-3′ polarity, the reverse of that normally found in VDREs (Quélo *et al.*, 1994).

Two functional VDREs have been identified in the promoters of the rat and human 24*R*-hydroxylase genes – a unique feature among vitamin D-responsive genes. The two response elements function synergistically due to co-ordinated interaction (Kerry *et al.*, 1996) and the combination represents the most powerful of the VDREs reported to date. Phorbol ester, acting via PKC, can potentiate the effect of calcitriol on induction of 24*R*-hydroxylase in both the kidney and intestine, but not in bone (Armbrecht *et al.*, 1997). There are several possible explanations for this potentiation. The phorbol ester could increase the phosphorylation of VDR by PKC and thus enhance its genomic effect as demonstrated by Hsieh *et al.* (1991). Calcitriol may also up-regulate phorbol ester receptors in renal tubular cells as it does in HL-60 leukemia cells (Martell *et al.*, 1987). This, in turn, could facilitate the effect of phorbol ester on VDR phosphorylation. Phorbol ester can induce transcription of the primary response gene c-*fos* in renal tubular cells (Chen *et al.*, 1993). The c-fos protein would bind to AP-1 binding sites in the 24-hydroxylase promoter and potentiate the genomic effect of VDR binding to the two VDREs.

Osteocalcin, synthesized exclusively by osteoblasts, is a negative regulator of bone formation (Ducy *et al.*, 1996). The VDRE of the rat osteocalcin gene has been characterized as an imperfect DR3 with the structure GGGTGAATGAGGACA (Demay *et al.*, 1992a). Sneddon *et al.* (1997) identified a sequence present from −420 to −414 in this gene that markedly enhances VDR-mediated transactivation and is absolutely required for calcitriol responsiveness. This sequence is unable to bind the VDR or confer calcitriol responsiveness independently of the

TABLE 16.1
A Selection of Protein Products Encoded by Vitamin D-Responsive Genes

Gene Product	Species	Up- or Down-Regulated by Hormone–VDR	Type of VDRE	Reference
24*R*-Hydroxylase	Rat	Up	Two DR3s	Zierold *et al.* (1995)
24*R*-Hydroxylase	Human	Up	Two DR3s	Chen and DeLuca (1995)
Osteocalcin	Rat	Up	DR3[a]	Demay *et al.* (1992a)
Osteocalcin	Human	Up	DR3[a]	Ozono *et al.* (1990)
Osteopontin	Mouse	Up	DR3	Noda *et al.* (1990)
Osteopontin	Pig	Up	DR3	Zhang *et al.* (1992)
Fibronectin	Mouse, rat, human	Up	DR6	Polly *et al.* (1996)
Carbonic anhydrase II	Chicken	Up	DR3	Quélo *et al.* (1994)
Calbindin-D_{9k}	Rat	Up	DR3	Darwish and DeLuca (1992)
Calbindin-D_{9k}	Human	Up	IP9	Schräder *et al.* (1995)
Calbindin-D_{28k}	Mouse	Up	DR4	Gill and Christakos (1993)
Na^+/phosphate co-transporter	Human	Up	DR3	Taketani *et al.* (1997)
Phospholipase C-γ1	Human	Up	DR6	Xie and Bikle (1997)
Pit-1	Mouse	Up	DR4	Rhodes *et al.* (1993)
Integrin β_3 subunit	Chicken	Up	DR3	Cao *et al.* (1993)
Atrial natriuretic factor	Rat	Up	DR3	Kahlen and Carlberg (1996)
p21	Human	Up	DR3	M. Liu *et al.* (1996)
c-fos	Mouse	Up	IP9	Schräder *et al.* (1997)
Parathyroid hormone	Chicken	Down	DR3	S. M. Liu *et al.* (1996)
Parathyroid hormone	Human	Down	DR3	Demay *et al.* (1992b)
PTH-related peptide	Rat	Down	DR3	Kremer *et al.* (1996)
Bone sialoprotein	Rat	Down	DR3	Kim *et al.* (1996)

[a] A DNA sequence which does not bind VDR is essential for the $1\alpha,25(OH)_2D_3$ responsiveness of the rat and human osteocalcin genes (Sneddon *et al.*, 1997).

VDRE. The sequence is also capable of augmenting transcription in concert with either an estrogen or cyclic AMP response element as shown by an enhanced transcriptional response to 17β-estradiol and forskolin. The human osteocalcin gene contains the same augmenting sequence in a similar position relative to the VDRE.

Pit-1, a transcriptional activator of the prolactin and growth hormone genes (Mangalam *et al.*, 1989), has an atypical DR4 VDRE, as does the mouse calbindin-D28 gene. The gene that encodes fibronectin, an important noncollagenous protein in the extracellular matrix, has a DR6-type VDRE that specifically binds VDR homodimers (Polly *et al.*, 1996). A DR6-type VDRE has also been identified in the human phospholipase C-γ1 gene.

The product of the c-*fos* gene, c-Fos, pairs with c-Jun to form AP-1, which is the terminal acceptor responsible for converting transient stimuli at the cell surface into long-term transcriptional responses. The mouse c-*fos* gene has an IP9-type VDRE which is bound by VDR–RXR heterodimers and is as functional as the more common DR3-type elements (Schräder *et al.*, 1997). The human calbindin-D9k gene VDRE is also of the IP9 type.

Repression of the human PTH gene appears to be attributable to a negative VDRE that preferentially binds a heterodimer consisting of VDR and a tissue-specific silencing partner rather than VDR–RXR (Haussler *et al.*, 1998). Simply changing two base pairs in the 3′ half-site of the avian PTH VDRE converted a negative element to a positive one. Binding of the VDR–nonRXR heterodimer to the negative VDRE may change the conformation of the heterodimer, preventing it from recruiting a co-activator, whereas binding to the positive VDRE permits co-activator recruitment. PTH-related peptide, which shares a common receptor with PTH, is also down-regulated by $1\alpha,25(OH)_2D_3$. Falzon (1996) identified a region in the gene encoding this protein which resembled the negative VDRE from the PTH gene.

Bone sialoprotein (BSP) initiates hydroxyapatite formation in developing bone. The down-regulation of the rat BSP gene appears to be attributable to a VDRE that overlaps a unique inverted TATA box and involves competition between the VDR and the TATA-binding protein. In comparison with activating VDREs, the rat BSP VDRE binds VDR homodimers more avidly than it does VDR–RXR heterodimers (Kim *et al.*, 1996).

16.10.9 Stimulation of Activator Protein-1-Dependent Caco-2 Cell Differentiation

The Caco-2 human colon cancer cell line provides a useful model to investigate the effects of calcitriol on proliferation and several parameters of enterocyte differentiation. VDR mRNA was detected in both pre- and post-confluent cells, and its abundance was unchanged with time and in response to calcitriol (Halline *et al.*, 1994). These authors reported that calcitriol causes a dose-dependent reduction in proliferation of Caco-2 cells and enhances the normal rise in alkaline phosphatase activity during differentiation. Increasing the expression of PKC-α in Caco-2 cells led to a less proliferative, more differentiated phenotype, whereas reducing PKC-α expression had the opposite effects (Scaglione-Sewell *et al.*, 1998).

The transcription factor, activator protein-1 (AP-1), activates many genes involved in cell proliferation. The functional forms of AP-1 are Jun–Fos heterodimers and Jun homodimers. Among the Jun and Fos families, c-Jun and c-Fos are the respective protein products of the genes c-*jun* and c-*fos*. Chen *et al.* (1999) showed that calcitriol causes a time-dependent increase in AP-1 DNA binding activity and a specific increase in AP-1 gene-transactivating activity. Calcitriol significantly enhanced the abundance of both the c-Jun protein and the c-*jun* mRNA; there was no change in the protein abundance of JunB, JunD, or c-Fos.

Receptor protein kinases (RTKs) are cell-surface receptors that directly initiate signal transduction without the involvement of second messengers. The binding of a ligand to a RTK leads to activation of a Ras protein that triggers signal transduction pathways leading ultimately to activation of a mitogen-activated protein (MAP) kinase. Signaling through MAP kinases, including ERKs and JNKs, influences AP-1 transcriptional activity by increasing both the abundance of AP-1 components and altering the phosphorylation of their subunits (Whitmarsh & Davis, 1996). Chen *et al.* (1999) showed that calcitriol induces a rapid but transient activation of ERK2 and a more persistent activation of JNK1 in Caco-2 cells.

Transfection experiments indicated that calcitriol-induced AP-1 transactivating ability is mediated by both ERK and JNK cascades and likely involves c-Jun phosphorylation. Inhibition of JNK activation or c-Jun expression significantly reduced calcitriol-induced alkaline phosphatase activity in calcitriol-treated cells.

In separate experiments, pretreatment of pre-confluent Caco-2 cells with a specific inhibitor of Ca^{2+}-dependent PKC isoforms blocked the activation of AP-1 and of ERK2 and JNK1 by calcitriol, indicating that stimulation of these proteins by calcitriol could be mediated by any one of the Ca^{2+}-dependent PKC isoforms. However, Caco-2 cells contain predominantly the alpha (α) and zeta (ζ) isoforms of PKC, and only PKC-α is activated by calcitriol in these cells (Bissonnette *et al.*, 1994). Chen *et al.* (1999) used antisense PKC-α cDNA to confirm that the AP-1 activation induced by calcitriol is PKC-α-dependent. A model was proposed in which treatment of Caco-2 cells with calcitriol rapidly stimulates PKC-α, which, in turn, activates ERK2 and JNK1. This leads to enhanced AP-1 transactivating activity and thus to alterations in genes involved in the control of differentiation.

16.10.10 Inhibition of Nuclear Factor κB Activation

Genetic deletion of the VDR causes elevated NF κB transcriptional activity, demonstrating that VDR signaling suppresses NF-κB activation (Sun *et al.*, 2006). Chen *et al.* (2013) showed that the VDR protein is able to directly interact with IKKβ protein. This interaction blocks the formation of the IKK complex, thereby reducing IKKβ phosphorylation. As a result, the IKKβ is unable to phosphorylate the inhibitory IκBα protein, consequently diminishing IκBα ubiquitination and degradation. The direct consequence of reduced IκBα degradation is the retention of the RelA(p65)–p50 heterodimer in the cytoplasm, leading to reduced NF-κB transcriptional activity.

16.11 REGULATION OF VITAMIN D-INDUCED GENE TRANSCRIPTION

16.11.1 Interaction of Runx2 with the VDR to Up-Regulate Osteocalcin Gene Expression

Osteocalcin is a bone-specific protein that is induced in osteoblasts with the onset of mineralization of the extracellular matrix (Owen *et al.*, 1990). Expression of the osteocalcin gene is regulated principally by the transcription factor Runx2, which has been proven to be essential for bone formation (Komori *et al.*, 1997). The rat osteocalcin gene promoter contains three recognition sites for Runx2 interactions, sites A, B, and C. Osteocalcin gene expression is also controlled by the VDRE, which recognizes the VDR upon ligand activation. The VDRE is located in the distal region of the osteocalcin promoter and functions as an enhancer to increase osteocalcin transcription. A tight functional relationship exists between Runx2 and the VDR in the regulation of the osteocalcin gene expression (Paredes *et al.*, 2004). In osteoblast-like cells expressing osteocalcin, the VDR interacts directly with Runx2 bound to Runx2 site B. This protein–protein interaction involves a domain located C terminally to the DNA binding region of Runx2 and the N-terminal end of the VDR. The activation factor AF-1 present in the N-terminal A/B domain of most steroid receptors is not involved in this interaction, as this factor is absent in the VDR. The role of Runx2 in the 1α,25(OH)$_2$D$_3$-dependent stimulation of the osteocalcin gene promoter in osteoblasts is to directly stabilize the binding of the VDR to the VDRE. Runx2 also recruits the co-activator p300 to the osteocalcin promoter, where it up-regulates both basal and calcitriol-enhanced osteocalcin expression (Sierra *et al.*, 2003). p300 can bind simultaneously to several transcription factors, forming large multimolecular protein complexes that regulate transcription.

16.11.2 Repression of Vitamin D-Induced Transactivation of the Osteocalcin Gene by YY1

Overlapping the recognition sequences for VDR–RXR within the VDRE of the osteocalcin gene are two binding motifs for the multifunctional transcription factor YY1. This protein can either activate, repress, or initiate gene transcription (Shrivastava & Calame, 1994). In the context of the osteocalcin promoter, YY1 represses calcitriol-induced transcription by competing with VDR–RXR for binding at the VDRE. YYI also interacts directly with TFIIB thereby interfering with the transactivation function of the VDR (Guo *et al.*, 1997). This repressive function of YY1 would prevent the precocious induction of osteocalcin gene transcription by VDR-mediated mechanisms under physiological conditions in which expression of the gene is inappropriate.

16.11.3 Suppression of Osteocalcin Gene Expression during Osteoblast Proliferation

The target cell for the hormonal action of vitamin D on bone is the osteoblast. Within this cell, 1α,25(OH)$_2$D$_3$ enhances osteocalcin gene expression at the three principal levels – transcription, mRNA accumulation, and protein synthesis. The enhanced transcription is dependent upon basal levels of gene expression (Owen *et al.*, 1991), suggesting that there is a co-ordinate transactivation involving the contribution of activities at the VDRE and basal elements. Two essential basal elements in the proximal promoter are the TATA box and the osteocalcin box (OC box), a 24-nucleotide sequence that contains a central CCAAT motif (Lian & Stein, 1992). The VDRE is located further upstream in the human osteocalcin gene promoter (Ozono *et al.*, 1990).

The sequential expression of vitamin D-dependent genes associated with bone tissue development has been studied in cultures of normal diploid rat osteoblasts. The first 10 to 12 days constitute the cell proliferation period, characterized by the expression of genes encoding AP-1 proteins (Jun and Fos) and extracellular matrix proteins. The *jun* and *fos* genes are transcribed in response to growth factors and other cell-surface stimuli, and the protein products are responsible for converting the transient stimuli into a long-term transcriptional response. During the following stages, the proliferative activity declines and the genes encoding the AP-1 proteins and extracellular matrix proteins are gradually down-regulated. During the next stage of matrix maturation (days 12 through 18), alkaline phosphatase mRNA and enzyme activity increases more than tenfold to peak levels, and matrix Gla protein is expressed maximally. Days 16 through 20 are characterized by the progressive mineralization of the extracellular matrix. As the cellular levels of alkaline phosphatase mRNA decline, the accumulation of calcium in the matrix increases co-ordinately with the up-regulated expression of genes encoding the calcium-binding proteins, osteocalcin and osteopontin (Stein & Lian, 1993).

The modular organization of the human osteocalcin gene promoter explains how gene expression might be suppressed during cell proliferation. Overlapping the VDRE and also within the OC box are AP-1 binding sites which bind Jun–Fos heterodimers and Jun homodimers. During the proliferation of osteoblasts, the Jun and Fos proteins are present at high levels. Occupancy of the AP-1 binding site overlapping the VDRE by Jun and Fos dimers blocks the binding of liganded VDR to its specific site and prevents transcription of the osteocalcin gene. At the end of the proliferation period, the levels of Fos and Jun decline, allowing the VDR to bind to the VDRE and the osteocalcin gene to be expressed. Thus the competition of the VDR and the AP-1 proteins for binding to the composite DNA element determines the activation or suppression of the osteocalcin gene (Lian & Stein, 1992). High levels of Jun and Fos also suppress basal transcription of the osteocalcin gene (Schüle *et al.*, 1990).

16.11.4 Repression of VDR-Mediated Transcription of the Osteocalcin and Osteopontin Genes by the Thyroid Hormone Receptor

Thompson *et al.* (1999) demonstrated two distinct repressive actions of the TR on VDR-mediated transcription of the rat osteocalcin and mouse osteopontin genes: (1) a thyroid hormone-independent action, perhaps due to TR–RXR outcompeting VDR–RXR for binding to the VDREs; and (2) a thyroid hormone-dependent repression, likely by diversion of limiting RXR from VDR–RXR toward the formation of TR–RXR heterodimers. The reverse of this phenomenon was indicated by a relatively weak binding of VDR–RXR to the TRE of the myosin heavy chain gene and modest repression by liganded VDR of thyroid hormone-mediated transactivation. This reciprocal inhibition of transactivation by VDR and TR is permitted by the half-site homology between the rat osteocalcin VDRE and the rat myosin heavy chain TRE. The formation of a TR–VDR heterodimer reported by Schräder *et al.* (1994) was not observed by Thompson *et al.* (1999) who concluded that this heterodimer is not biochemically or physiologically relevant. As far as is known, the only relevant heterodimerization among nuclear receptors (excluding RXR) is between the type I mineralocorticoid and glucocorticoid receptors.

16.11.5 Inhibitory Effect of the VDR on Transactivation of the Growth Hormone Gene by Thyroid Hormone and Retinoic Acid

The rat growth hormone gene, located exclusively in the somatotropic pituitary cells, contains a hormone response element that functions both as a retinoic acid response element (RARE) and a thyroid hormone response element (TRE) (García-Villalba *et al.*, 1993). This allows thyroid hormone and retinoic acid to interact co-operatively to stimulate transcription of the growth hormone gene in pituitary cells (Bedo *et al.*, 1989). García-Villalba *et al.* (1996) reported that incubation of rat pituitary cells with nanomolar concentrations of vitamin D_3 inhibits thyroid hormone and retinoic acid transactivation of the growth hormone gene by interference on the common response element. The results suggested that the liganded VDR can directly affect the pituitary response to other nuclear receptors, thereby contributing to the growth arrest that occurs in hypervitaminosis D. This negative effect on the growth hormone gene is apparently paradoxical, since a deficiency of vitamin D produces a rachitic state associated with a defect in growth.

16.12 NONGENOMIC ACTIONS OF CALCITRIOL

16.12.1 Introduction

A large number of responses occur within seconds to minutes following addition of vitamin D metabolites to the *in vitro* system – too rapid to involve changes in gene expression controlled by the VDR. Moreover, such responses are not blocked by inhibitors of transcription (actinomycin D) or protein synthesis (cycloheximide). For these reasons, the rapid responses are described as nongenomic. Nongenomic actions of calcitriol are mediated by a plasma membrane vitamin D receptor that is distinct from the nuclear VDR. The binding of calcitriol to the membrane VDR triggers signal transduction pathways, which involve second messengers, and which modulate the genomic actions of the hormone.

Rapid nongenomic responses to calcitriol have been observed in many cell types, including enterocytes, colonocytes, chondrocytes, osteoblasts, hepatocytes, skeletal and cardiac muscle cells, keratinocytes, mammary gland epithelial cells, parathyroid cells, and pituitary cells (Norman, 1998). The responses have been most clearly delineated in osteoblast-like osteosarcoma cells. At a molecular level, these include effects on membrane phospholipid metabolism (Grosse *et al.*, 1993), activation of voltage-sensitive calcium channels (Caffrey & Farach-Carson, 1989), and elevation of cytosolic (Baran *et al.*, 1991) and nuclear (Sorensen *et al.*, 1993) Ca^{2+} concentrations.

A number of nongenomic effects of calcitriol are concerned with the regulation of intracellular calcium, which is an important second messenger involved in the activation of many target enzymes. Additional effects include stimulation of prostaglandin production (Boyan *et al.*, 1994), increased intracellular pH (Jenis *et al.*, 1993), and increased membrane fluidity (Brasitus *et al.*, 1986).

16.12.2 Evidence for a Distinct Membrane Receptor for Calcitriol

There is strong evidence that the nongenomic responses to calcitriol are mediated by a membrane receptor that is biochemically different from the functional nuclear VDR. Ligand specificity for the rapid actions is different from that for genomic responses (Farach-Carson *et al.*, 1991). Using osteoblasts from mice in which the VDR gene had been genetically ablated, Wali *et al.* (2003) showed that the calcitriol-induced rapid increases in intracellular calcium and PKC activity are neither mediated nor dependent upon a functional VDR.

A 66-kDa protein that binds vitamin D analogs has been isolated from basolateral membranes of chick intestinal epithelium (Nemere *et al.*, 1994) and from both plasma membranes and matrix vesicles of rat chondrocytes (Nemere *et al.*, 1998). Antibody (Ab99) generated to a [^{3}H]calcitriol binding protein isolated from the basolateral membrane of chick intestinal epithelium blocked the rapid activation of PKC by calcitriol in chondrocytes (Nemere *et al.*, 1998) and enterocytes (Nemere *et al.*, 2000). Baran *et al.* (2000) reported that annexin II, a 36-kDa protein, can serve as a cell-surface receptor mediating calcitriol-induced increase in intracellular calcium in osteoblast-like ROS 24/1 cells that lack the VDR. The rapid effects of calcitriol to increase intracellular calcium were not observed in cells pretreated with anti-annexin II antibodies. It is likely that several membrane receptors mediate the rapid actions of calcitriol (Brown *et al.*, 1999).

16.12.3 Activation of Protein Kinase C

PKC occupies a central position in signal transduction and controls diverse cellular processes, including proliferation and differentiation (Nishizuka, 1992). The binding of calcitriol to a membrane receptor in a wide range of cell types stimulates the activity of phospholipase C, whose hydrolytic action on membrane-derived phosphatidylinositol biphosphate results in the generation of the two second messengers, diacylglycerol and inositol triphosphate. Diacylglycerol is an activator of PKC. Cell types examined include osteosarcoma cells (Civitelli *et al.*, 1990), myoblasts (Morelli *et al.*, 1993), enterocytes (Lieberherr *et al.*, 1989), colonocytes (Wali *et al.*, 1990), hepatocytes (Baran *et al.*, 1988), parathyroid cells (Bourdeau *et al.*, 1990), and keratinocytes (MacLaughlin *et al.*, 1990).

16.12.4 Activation of Phospholipase D

Calcitriol activates phospholipase D (PLD) in a pathway that requires the Ca^{2+}-dependent PKC isoform, PKC-α (Khare *et al.*, 1999). PLD hydrolyzes phosphatidylcholine to generate phosphatidic acid and choline. Phosphatidic acid can subsequently be hydrolyzed by phospholipase A_2 to form the second messenger, lysophosphatidic acid (LPA) in some cell types or it can be dephosphorylated to produce diacylglycerol (DAG). DAG, together with inositol triphosphate, is also produced by hydrolysis of phosphatidylinositol biphosphate, which is catalyzed by phospholipase C (PLC). DAG activates PKC through a conformational change, whilst also promoting the translocation of the kinase from the cytosol to the plasma membrane. Inositol triphosphate causes the release of Ca^{2+} from intracellular storage compartments. The observation that calcitriol-induced activation of PLD is preceded by the activation of PLC suggests that the PLC-dependent source of DAG could provide the active PKC-α.

16.12.5 Regulation of Intracellular Calcium

Transient increases in intracellular calcium can be initiated in two major ways.

1. Calcium can be released from intracellular storage sites associated with mitochondria and the endoplasmic reticulum. The binding of calcitriol to a membrane receptor stimulates phospholipase C to hydrolyse membrane phosphoinositides, thereby generating inositol triphosphate and releasing Ca^{2+} from the intracellular storage sites (Wali *et al.*, 1990).
2. Extracellular calcium can enter the cell down its electrochemical gradient by the opening of voltage-gated calcium channels. The presence of L-type (dihydropyridine-sensitive) voltage-activated calcium channels has been demonstrated in basolateral membranes of rabbit ileal enterocytes (Homaidan *et al.*, 1989) and in osteosarcoma cells (Guggino *et al.*, 1989). The opening of such channels is the fastest known response of osteoblasts to treatment with nanomolar concentrations of calcitriol. The hormone enhances the activity of voltage-gated channels in rat pituitary cells (Tornquist & Tashjian, 1989) and rat osteosarcoma cells (Caffrey & Farach-Carson, 1989).

Addition of calcitriol to the vascularly perfused duodenal loop prepared from vitamin D-replete chicks enhanced calcium transport from the lumen to the vascular effluent within 14 minutes. No effect of calcitriol on calcium translocation could be demonstrated in perfused duodena obtained from vitamin D-deficient chicks (Nemere *et al.*, 1984). This rapid effect of calcitriol, now termed 'transcaltachia,' was suppressed by 24*R*,25$(OH)_2D_3$ (Nemere, 1999), implying a means of regulating calcium transport. Calcitriol acted at the basolateral surface (where voltage-gated calcium channels are located) and not at the brush border. The involvement of calcitriol$_3$ in the activation of voltage-gated calcium channels appears to be an early effect in transcaltachia. Introduction of a calcium channel antagonist completely abolished the movement of calcium, while a calcium channel agonist mimicked the stimulatory response to calcitriol (de Boland *et al.*, 1990). De Boland and Norman (1990) provided evidence for the involvement of PKA and PKC in transcaltachia. Forskolin and phorbol ester, activators of PKA and PKC respectively, stimulated transcaltachia analogously to calcitriol. In addition, the transcaltachial response to calcitriol was respectively suppressed or abolished by inhibitors of PKA and PKC. Collectively, these observations suggest that calcitriol binds to a membrane receptor located in the basolateral membrane of the enterocyte. Signal transduction via G proteins, effector proteins, and second messengers then leads to activation of PKA and PKC. These kinases might stimulate calcium influx via phosphylation-dependent activation of voltage-gated calcium channels at the basolateral membrane. Studies with analogs of calcitriol have provided definite correlations between binding to the putative basolateral membrane receptor (66-kDa protein) and the ability to initiate transcaltachia (Nemere, 1995).

The $24,25(OH)_2D_3$, that is produced in the kidney during states of adequate calcium balance, appears to be a natural antagonist of the membrane actions of calcitriol. Because calcitriol and $24,25(OH)_2D_3$ do not compete with each other for binding to isolated basolateral membranes of chick duodenal epithelium, it is likely that $24,25(OH)_2D_3$ acts through a separate, non-nuclear receptor (Nemere & Farach-Carson, 1998).

16.12.6 Regulation of Phosphate Transport

Perfusion of chick duodenal loops with physiological concentrations of calcitriol stimulated phosphate transport within 2–8 minutes. This effect was observed at the basolateral surface and not at the brush border. High concentrations of calcitriol inhibited phosphate transport. The biphasic dose-response curve is characteristic of a receptor-mediated response. The stimulatory action of calcitriol was attenuated by $24,25(OH)_2D_3$ and abolished by $25(OH)D_3$ (Nemere, 1996a).

16.12.7 Modulation of the Genomic Actions of Calcitriol by Nongenomic Mechanisms

The 1β epimer, 1β,25-dihydroxyvitamin D_3, does not interact with the nuclear VDR and thus has no effect on basal gene transcription. It does, however, inhibit nongenomic effects by competing with $1\alpha,25(OH)_2D_3$ for the membrane VDR. Baran *et al.* (1992) demonstrated that, in osteosarcoma cells, the 1β epimer inhibits the $1\alpha,25(OH)_2D_3$-induced rapid rise in intracellular calcium and accompanying increase in osteocalcin gene transcription. The inhibition of transcription occurred without interfering with the binding of the $1\alpha,25(OH)_2D_3$ –nuclear VDR complex to the VDRE. Since $1\beta,25(OH)_2D_3$ has no genomic effects on osteocalcin, yet can block the transcription that accompanies the nongenomic effect of $1\alpha,25(OH)_2D_3$, this study suggests that the nongenomic effects of $1\alpha,25(OH)_2D_3$ can modulate the genomic actions of this hormone in some way.

Khoury *et al.* (1994) showed that the calcitriol-induced transcription of osteocalcin and osteopontin genes in osteosarcoma cells is independent of Ca^{2+} influx, suggesting that the nongenomic stimulation of calcium channels by calcitriol is not required for target gene activation. Jenis *et al.* (1993) discovered that the nongenomic calcitriol-induced increase in intracellular pH is necessary for osteocalcin and osteopontin expression in the osteoblast. The effect upon pH appears to be regulated by the Na^+/H^+ antiporter, since the incubation of cells in a Na^+-free medium eliminated the pH effect and blocked the hormone-induced increase in osteocalcin and osteopontin mRNA steady-state levels.

The genomic response may be modulated by second messengers generated at both the plasma and nuclear membranes in response to the binding of calcitriol to membrane receptors (Baran & Sorensen, 1994). Calcitriol increases PKC activity through stimulation of plasma membrane phosphoinositides and formation of diacyclglycerol (Wali *et al.*, 1990). In renal epithelial cells, calcitriol specifically induces translocation of PKCβ (but not PKCα) from the plasma membrane to the nuclear membrane, an event which enhances phosphorylation of nuclear proteins (Simboli-Campbell *et al.*, 1994). Phosphorylation of the nuclear VDR by PKC has been shown to down-regulate transcription (Hsieh *et al.*, 1993). Sorensen and Baran (1993) reported that calcitriol rapidly enhances phospholipase C activity in the nuclear membrane of osteosarcoma cells, resulting in an increased level of inositol triphosphate, which in turn releases calcium from intranuclear storage sites in Caco-2 cells. Calcitriol activates PKC-α, which in turn limits the further rise in intracellular Ca^{2+} in response to calcitriol (Bissonnette *et al.*, 1994). This feedback provides a potential mechanism limiting Ca^{2+}-dependent events stimulated by calcitriol, including activation of PKC-α.

16.13 CALCIUM AND PHOSPHATE TRANSPORT IN INTESTINE, KIDNEY, AND BONE

16.13.1 Calcium-Binding Proteins

Calcitriol regulates the synthesis of two classes of calcium-binding proteins known as calbindins. These proteins are named calbindin-D_{9k} and calbindin-D_{28k} according to their molecular weights of 9 and 28 kDA and are found in mammalian intestine and kidney, respectively. Calbindin-D_{9k} binds two calcium ions per molecule, and calbindin-D_{28k} binds five to six calcium ions per molecule (Lowe *et al.*, 1992). In birds, calbindin-D_{28k} is present in both intestine and kidney. Gross and Kumar (1990) reviewed extensive evidence that the calbindins are involved in transcellular calcium transport.

Present within the microvillar region of the enterocyte is a relatively high concentration of the ubiquitous calcium-binding protein calmodulin, much of which is combined with a 110-kDa protein in the form of brush-border myosin I. Treatment of Caco-2 cells with the calmodulin antagonist trifluoroperazine reduced calcitriol-stimulated apical-to-basolateral Ca^{2+} transport by 56% (Fleet & Wood, 1999), implicating calmodulin in the transport process.

16.13.2 Voltage-Independent Calcium Channels

Hoenderop *et al.* (1999a) cloned from rabbit kidney cells a novel epithelial calcium channel, named ECaC, that was distinct from voltage-activated calcium channels. This channel was exclusively present in the proximal small intestine, the distal nephron, and the placenta. Expression of ECaC was shown to coincide with that of calbindin-D_{9k} in intestine and placenta, and calbindin-D_{28k} in kidney. Immunofluorescence staining revealed that, in kidney, ECaC is abundantly present along the apical membrane of cells lining the distal convoluted tubule, connecting tubule and cortical collecting duct, where it colocalizes with calbindin-D_{28k}. This part of the nephron is the site of PTH- and calcitriol-regulated transcellular Ca^{2+} reabsorption. The presence of several potential phosphorylation sites for PKC and PKA indicated that ECaC is subject to hormonal regulation. Further studies revealed that ECaC is a Ca^{2+} selective channel, which is activated by hyperpolarization and regulated by cytosolic calcium (Hoenderop *et al.*, 1999b). Activation of ECaC is constitutive at physiological membrane potentials. Repetitive stimulation of ECaC-expressing cells with voltage ramps induced a decay of the current response in the presence of 30 mM Ca^{2+}, but when the Ca^{2+} concentration was decreased to 1 nM, the decay process was virtually abolished (Vennekens *et al.*, 2000).

Calcium transporters CaT1 and CaT2 have been cloned and shown to be expressed specifically in the intestine and distal nephron, respectively, of the rat (Peng *et al.*, 1999, 2000). CaT1 and CaT2 shared structural similarity and functional properties to ECaC, including autoregulation and insensitivity to L-type voltage-activated calcium channel blockers. The multiple PKA and PKC sites in the CaT2 protein may provide the basis for multiple levels of regulation by PTH and other factors regulating renal calcium reabsorption.

16.13.3 Intestinal Calcium Absorption

Calcium is present in foods and dietary supplements as relatively insoluble salts. Because calcium is absorbed only in its ionized form, it must first be released from the salts. Solubilization of most calcium salts takes place in the acidic medium of the stomach but, on reaching the alkaline environment of the small intestine, some of the Ca^{2+} may complex with minerals or other specific dietary constituents, thereby limiting calcium bioavailability

Calcium absorption takes place by the translocation of luminal Ca^{2+} through the enterocytes (transcellular route) and between adjacent enterocytes via the tight junctions (paracellular route). Transcellular movement is a saturable, energy-dependent process that is subject to regulation by vitamin D and is confined almost entirely to the duodenum and upper jejunum. Paracellular movement is passive, independent of vitamin D status, and exists all along the small intestine (Pansu *et al.*, 1983).

Two current models which describe the mechanism of transcellular absorption are the facilitated diffusion/active transport model and the vesicular transport model. These vitamin D-dependent mechanisms ensure the absorption of appropriate amounts of calcium in the face of changing needs and varying dietary calcium intakes.

16.13.3.1 The Facilitated Diffusion/ Active Transport Model

This transcellular pathway is a complex process involving three steps: (1) passive movement of Ca^{2+} from lumen through the brush-border membrane of the enterocyte via a calcium channel; (2) cytosolic diffusion via the ferrying action of vitamin D-dependent calbindin-D_{9k}; and (3) active extrusion across the opposing basolateral membrane by a high-affinity Ca^{2+}–ATPase (calcium pump) and Na^+–Ca^{2+} exchanger. In the vitamin D-deficient rat (Sampson *et al.*, 1970) and chick (Chandra *et al.*, 1990) luminal Ca^{2+} readily crosses the brush-border membrane, but remains bound in the microvillus. Vitamin D repletion of the deficient animal results in the mobilization of Ca^{2+} from the microvillar region into the body of the cell. Thus the regulatory action of vitamin D is on the steps beyond brush-border entry, as also shown by Roche *et al.* (1986). The concentration of cytosolic Ca^{2+} within the enterocyte is closely controlled at about 10^{-7} M, and this is ultimately a function of the rate of entry and the rate of exit of Ca^{2+} across the cell boundaries. Intracellular organelles, including mitochondria, microsomes, and lysosomes, play a major role in controlling cytosolic Ca^{2+} by storing and releasing the cation as appropriate.

16.13.3.2 Apical Entry

This step involves the movement of luminal Ca^{2+} across the brush-border membrane of the enterocyte into the cytosol. The downhill electrochemical gradient permits the entry of luminal Ca^{2+} without the input of metabolic energy (Stein, 1992). The observation that Ca^{2+} uptake at the intestinal brush border has a saturable component (Wilson *et al.*, 1989) suggests an interaction with a low-affinity binding site, which might be associated with a calcium channel or a carrier of some sort. The results of *in vitro* and *in vivo* studies using blockers of voltage-activated calcium channels suggest that such channels are unlikely to be present in brush border of enterocytes (Favus & Tembe, 1992). It seems likely that a voltage-independent calcium channel, such as CaT1, mediates the saturable Ca^{2+} uptake at the intestinal brush-border membrane. The duodenal mRNA level of CaT1 was not responsive to $1\alpha,25(OH)_2D_3$ or to a calcium-deficient diet *in vivo* (Peng *et al.*, 1999), which supports previous studies demonstrating that apical entry of Ca^{2+} into the enterocyte is independent of vitamin D.

16.13.3.3 Cytosolic Diffusion

This step involves vectorial movement of the transported Ca^{2+} through the cytosol from the apical to the basolateral pole. The rate of unaided Ca^{2+} diffusion through the cytosol of an enterocyte has been calculated to be nearly two orders of magnitude too slow to satisfy the experimentally determined value of V_{max} for transcellular duodenal Ca^{2+} transport (Bronner *et al.*, 1986). This implies that intracellular diffusion will have to be facilitated. A candidate protein for such facilitated diffusion is the vitamin D-inducible protein calbindin-D_{9k}. Active calcium transport rate in everted sacs was proportional to the intestinal concentration of calbindin-D_{9k}; transport was not detectable in vitamin D-deficient rats in which the intestinal calbindin-D_{9k} content is virtually zero (Roche *et al.*, 1986). In a model proposed by Wasserman and Fullmer (1995), Ca^{2+} in the intestinal lumen binds initially to brush-border calmodulin. It is then transferred to cytoplasmic calbindin and thence to the calcium-binding site of the basolateral calcium pump. The dissociation constants for the brush-border complex, calbindin, and the calcium pump favor the apical-to-basolateral movement of Ca^{2+}. The sequestration of Ca^{2+} by calmodulin contributes to protection against calcium-mediated cytotoxicity.

16.13.3.4 Basolateral Extrusion

The movement of Ca^{2+} out of the cell across the basolateral membrane into the vascular supply of the lamina propria is uphill in terms of both concentration and the prevailing electropotential. An active transport system requiring a source of metabolic energy is therefore needed to pump the Ca^{2+} out of the cell. A basolateral Ca^{2+}-ATPase (calcium pump) identified in enterocytes by Ghijsen *et al.* (1982) has highest activity in the duodenum with decreasing activity toward the ileum (Ghijsen & van Os, 1982). The calcium pump in the enterocyte basolateral membrane is stimulated by calbindin-D_{9k} (Walters, 1989). It seems reasonable to propose that the binding of Ca^{2+} to calbindin triggers an allosteric activation of the ATPase. Astimulatory effect of calcitriol on the ATP-dependent uptake of Ca^{2+} by duodenal basolateral membrane vesicles has been demonstrated in vitamin D-deficient rats (Ghijsen & van Os, 1982; Walters & Weiser, 1987; Favus *et al.*, 1989) and chicks (Takito *et al.*, 1990). Immunological studies have shown that feeding vitamin D to chicks previously deprived of the vitamin induces the synthesis of the calcium pump protein (Wasserman *et al.*, 1992). A vitamin D-induced increase in calcium pump mRNA confirmed the effect of vitamin D on gene expression (Zelinski *et al.*, 1991; Armbrecht *et al.*, 1992; Cai *et al.*, 1993). It is noteworthy that calcitriol does not stimulate calcium pump activity in vitamin D-replete animals (Favus *et al.*, 1989).

A $3Na^{+}:1Ca^{2+}$ antiporter plays a secondary role in transporting cytosolic Ca^{2+} across the basolateral membrane and is not regulated by vitamin D. The transport capacity of this exchanger is about 20% of that of the Ca^{2+}-ATPase. However, this transport capacity is constant along the length of the small intestine, and in the ileum both systems are of equal effect (Ghijsen *et al.*, 1983).

16.13.3.5 The Vesicular Transport Model

Nemere *et al.* (1986) proposed an alternative model for calcitriol-mediated calcium absorption in which Ca^{2+} is specifically bound at the brush-border membrane and internalized by endocytosis. The endocytotic vesicles are conveyed along microtubules to lysosomes where fusion of the membrane-bounded organelles occurs. The lysosomes, in turn, move along microtubules to the basolateral membrane where exocytosis of Ca^{2+} completes the transport process.

Vesicular transport of calcium requires intact microtubules to provide intracellular 'tracks.' Calcitriol induces the synthesis of tubulin in the intestine of vitamin D-deficient chicks (Nemere *et al.*, 1987). Disruption of the microtubules by colchicine completely inhibited the rapid effect of calcitriol on calcium transport from lumen to blood supply; there was no such inhibition in response to lumicolchicine, a light-inactivated analog of colchicine (Nemere *et al.*, 1984).

de Boland and Norman (1990) proposed a model that takes into account both transcaltachia and vesicular transport. Conceivably, receptor-mediated activation of PKA and PKC by calcitriol might stimulate basolateral entry of Ca^{2+} into the enterocyte (transcaltachia) through activation of voltage-gated calcium channels. The transient increase of intracellular Ca^{2+}, in turn, may activate exocytosis of calcium-containing vesicles as well as Ca^{2+} efflux by the calcium pump and the $Na^{+}:Ca^{2+}$ antiporter, resulting in a net increase of intestinal calcium transport.

16.13.3.6 Paracellular Route

The paracellular movement of calcium occurs largely by diffusion and is independent of vitamin D status and age. Even in total vitamin D deficiency, Ca^{2+} absorption is ensured by the paracellular route (Bronner, 1990). The rate of movement is essentially the same along the entire length of the small intestine. In the rat, some 16% of the luminal content of soluble calcium is absorbed by the paracellular route in 1 hour, regardless of the calcium concentration of the chyme (Bronner *et al.*, 1986). Paracellular calcium movement is dominant under conditions when transcellular transport is down-regulated, as in old age (Armbrecht *et al.*, 1979) or when the food content of calcium is abundant (Pansu *et al.*, 1981).

16.13.4 Effect of Age on Intestinal Calcium Absorption

Intestinal calcium absorption in humans declines with aging in both sexes, particularly after 70 years of age (Bullamore *et al.*, 1970; Alevizaki *et al.*, 1973). Active transport of calcium by the duodenum also rapidly declines in rats aged between 1 and 6 months, irrespective of the level of dietary calcium (Armbrecht *et al.*, 1979). Furthermore, the capacity of calcitriol to stimulate calcium absorption in rats declines with age (Armbrecht *et al.*, 1980; Wood *et al.*, 1998). Armbrecht *et al.* (1999) reported a 90% decrease in basolateral calcium pump protein in the duodena of rats aged between 2 and 12 months. There was little further decline between 12 and 24 months of age. The expression of intestinal calbindin tended to parallel that of the calcium pump. It appeared that the age-related decline in plasma calcitriol concentrations leads to a decrease in calcium pump and calbindin mRNA levels, which in turn results in a decrease of the corresponding proteins. If the age-related decline in calcitriol were the sole contributor to decreased calcium absorption, then administration of calcitriol would be expected to fully reverse the decline in the adult and old animal. However, even multiple doses of calcitriol over 48 hours to the adult rat did not increase calcium transport, the calcium pump, or calbindin to levels seen in the young, despite equivalent mRNA levels. This suggests that, with age, the intestine becomes refractory to calcitriol.

16.13.5 Renal Calcium Reabsorption

The kidney plays a crucial role in calcium homeostasis. To maintain a net calcium balance, more than 98% of the filtered load of calcium must be reabsorbed along the nephron. Tubular reabsorption of calcium can take place via the transcellular and paracellular routes.

Paracellular, diffusional movement predominates in the proximal convoluted tubule and thick ascending limb of Henle's loop, in which the epithelium is highly permeable. The lumen-positive transepithelial potential difference generated by Cl^- reabsorption provides a driving force for passive Ca^{2+} reabsorption via the paracellular route. Transcellular Ca^{2+} transport takes place in the distal convoluted tubule and connecting tubule where the epithelia are less permeable. The contribution of the collecting duct to overall calcium reabsorption is small. Although the active, transcellular reabsorption of calcium in the distal nephron amounts to only about 10% of the filtered load, it plays an important role in the fine-tuning of whole body calcium homeostasis.

Transcellular Ca^{2+} transport in the distal nephron involves passive Ca^{2+} influx at the apical membrane, diffusion through the cytosol, and active extrusion across the opposing basolateral membrane. The apical Ca^{2+} influx is mediated by voltage-dependent calcium channels, which are stimulated by PTH and diuretics. The PTH-dependent transport of calcium in distal convoluted tubule cells is accelerated by calcitriol (Friedman & Gesek, 1993). The extrusion of Ca^{2+} is mediated by the basolateral Na^+/Ca^{2+} exchanger, which is stimulated by calbindin-D_{28k}.

16.13.6 Na^+/Pi Co-Transporters

Three families of vertebrate Na^+/Pi co-transporter (Pi = inorganic phosphate) have been identified: type I, type II (IIa and IIb isoforms), and type III. Type I proteins have been found in the apical membrane of renal proximal tubules, but their function is not yet clearly established. Type II proteins are also located in apical membranes – type IIa in renal proximal tubules and type IIb in small intestinal enterocytes. Type III co-transporters are found in many tissues and appear to be located at the basolateral membrane (Murer *et al.*, 2000). Type II (IIa and IIb) Na^+/Pi co-transporters mediate secondary active phosphate transport in which the immediate energy source is the downhill concentration gradient for Na^+ maintained by the action of the sodium pump at the basolateral membrane. The parallel operation of other sodium-coupled transport systems will indirectly affect the Na^+/Pi co-transport rate due to competition for driving forces (Danisi & Murer, 1991). Type II co-transporters operate with a $3Na^+$ to 1Pi stoichiometry (Murer *et al.*, 2001). In the presence of divalent phosphate, these transporters interact with the substrate (Pi) followed by the loading of two Na^+ ions. Thus the translocation of the fully loaded transporter is an electroneutral process. The observed negative charge transfer within the transport cycle is the result of the reorientation of the unloaded transporter (Murer *et al.*, 2002).

16.13.7 Intestinal Phosphate Absorption

Dietary phosphorus is a mixture of inorganic phosphate (Pi) and organic phosphorus. The phosphorus in meat and fish exists largely in the form of phosphoproteins and phospholipids; over 80% of the phosphorus in grains such as wheat, rice, and maize is found as phytic acid (hexaphosphoinositol); about 33% of the phosphorus in milk exists as inorganic phosphate; milk protein (casein) is particularly highly phosphorylated. Regardless of its dietary form, most phosphorus is absorbed in inorganic form. Organically bound phosphorus is hydrolyzed enzymatically in the lumen of the small intestine. The phosphorus within phytic acid has poor bioavailability owing to incomplete hydrolysis.

Phosphate absorption takes place mainly in the jejunum by an energy-dependent transcellular route that is regulatable and a passive paracellular route that is not regulatable. The transcellular pathway involves secondary active transport at the brush-border membrane mediated by the type IIb Na^+/Pi co-transporter. Both the monovalent and divalent forms of phosphate are transported (Quamme, 1985). The capacity (V_{max}) of the transport system is significantly greater at pH 6.1 than at 7.4; thus the acidic environment of the unstirred layer favors Pi uptake (Borowitz & Ghishan, 1989). The basolateral membrane also contains a Na^+/Pi co-transporter, which has a lower capacity and a higher affinity compared to the brush-border co-transporter (Kikuchi & Ghishan, 1987).

A low-phosphorus diet leads to a rapid decrease of plasma Pi concentration and activation of the renal 1α-hydroxylase. The resultant increase in circulating calcitriol induces an increased capacity of the type IIb Na^+/Pi co-transporter in the brush border of the intestinal epithelium. The hormonal response requires several hours and involves protein synthesis. The rapid adaptive response observed in the kidney (see Section 16.13.8) does not appear to take place in the intestine (Hattenhauer *et al.*, 1999). However, studies using the perfused chick duodenal loop suggest that calcitriol interacts with a basolateral membrane receptor to initiate rapid vesicular transport of phosphate (Nemere, 1996b).

16.13.8 Renal Phosphate Reabsorption

Renal reabsorption of phosphate takes place in the proximal convoluted tubule and, under normal physiological conditions, ~80% of phosphate contained in the glomerular filtrate is reabsorbed. The brush-border uptake, which is mediated by the type IIa Na^+/Pi co-transporter, is rate-limiting in most situations and the target of physiological control mechanisms. Basolateral exit is ill-defined, but may involve another Na^+/Pi co-transporter (Schwab *et al.*, 1984). Transport of phosphate in the opposite direction (from the peritubular interstitium into the tubular cell) can take place across the basolateral membrane if apical entry is insufficient to satisfy the cell's metabolic requirements.

The existence of multiple systems for transporting phosphate across the brush-border membrane of the tubular epithelium has been recognized since 1977. Walker *et al.* (1987) demonstrated two sodium-dependent systems in the early segments of the proximal tubule: a high-capacity, low-affinity system and a low-capacity, high-affinity system. A third transport system in the late proximal segments was found by the same group to be independent of sodium and mediated by a hydroxyl ion/Pi exchanger or a proton/Pi co-transport system (Yan *et al.*, 1988). The renal control of acid-base balance requires secretion of hydrogen ions into the tubular lumen by the tubular epithelial cells, causing a progressive decrease in luminal pH from 7.4 to 6.8 along the length of the proximal tubule. Such a decline in pH results in a fall of the divalent to monovalent phosphate ratio from 4:1 to 1:1 at any given Pi concentration (Quamme & Wong, 1984). It has been demonstrated that the two sodium-dependent systems mentioned above transport the divalent form of phosphate, whereas the sodium-independent system located in the late proximal tubule has a preference for the monovalent form (Yan *et al.*, 1988). Accordingly, the multiple transport systems act in concert to reclaim filtered phosphate along the proximal tubule.

The rate of renal phosphate reabsorption is adjusted in response to deviations in plasma Pi concentration to achieve a correct phosphate homeostasis. A low-phosphorus diet induces a rapid (2–4 hours) nongenomic response followed by a long-term genomic response if phosphorus deprivation persists. Both types of response involve stimulation of sodium-dependent phosphate transport by calcitrol. The rapid response may be mediated by a microtubule-dependent translocation of the Na^+/Pi co-transporter protein from intracellular compartments to the brush-border membrane (Lötscher *et al.*, 1996). Rapid down-regulation of the co-transporter in response to acute administration of a large dose of phosphate is probably mediated by endocytosis of the protein (Lötscher *et al.*, 1996). The long-term response involves transactivation of the Na^+/Pi co-transporter gene (Taketani *et al.*, 1997).

16.13.9 Calcium and Phosphate Movement in Bone

The Ca^{2+} concentration in the bone fluid compartment is normally about one-third that in the extracellular fluid. The lining cells exert the primary control on extracellular calcium homeostasis. These cells have open channels between them, permitting paracellular entry of Ca^{2+} into the bone fluid compartment down the concentration gradient. The uphill movement of Ca^{2+} from the bone fluid compartment to the extracellular fluid involves active pump-driven transport through the lining cells.

In the presence of vitamin D, PTH increases the calcium permeability of the lining cell plasma membrane facing the bone fluid, allowing Ca^{2+} to diffuse into the cells from the bone fluid. The increase in intracellular Ca^{2+} then activates the calcium pump on the opposite membrane. This action of PTH results in the rapid removal of Ca^{2+}, accompanied by Pi, from amorphous calcium phosphate deposits in the vicinity of the lining cells and transference of these ions to the extracellular fluid.

16.14 CALCIUM AND PHOSPHATE HOMEOSTASIS

16.14.1 Introduction

Calcium is the most abundant cation in the human body, but less than 1% is present in ionic form in the extracellular compartment. Ionic calcium (Ca^{2+}) is crucial for many cellular functions including neurotransmitter release and nerve impulse propagation; contraction of skeletal, cardiac, and smooth muscle; blood clotting; exocrine and endocrine secretory processes; and cell proliferation. Ca^{2+} also acts as a 'second messenger,' connecting some stimuli (certain hormones, growth factors, and neurotransmitters) with physiological responses. Bones and teeth contain about 99% of the body's calcium; the other 1% is distributed in both intra- and extracellular fluids.

The blood level of Ca^{2+} is very closely regulated: even small changes in Ca^{2+} concentration can be fatal. Acute hypocalcemia in the human causes essentially no other significant effects besides tetany, because tetany kills the patient before other effects can develop. In tetany, the low concentration of Ca^{2+} in the extracellular fluid causes the nervous system to become progressively more excitable because of increased neuronal membrane permeability. Eventually, peripheral nerve fibers become so excitable that they begin to discharge spontaneously, initiating nerve impulses that pass to the skeletal muscles, where they elicit the muscular spasms of tetany. In hypercalcemia, the nervous system is depressed, and reflex actions of the central nervous system become sluggish. There is also constipation and lack of appetite, probably because of depressed contractability of the muscular walls of the gastrointestinal tract. Above a certain high level of blood Ca^{2+}, calcium phosphate is likely to precipitate throughout the blood and the soft tissues. Deposition of calcium in the kidney or heart causes death due to renal failure or cardiac arrest.

Phosphorus is a component of hydroxyapatite in the skeleton, of phospholipids in cell membranes, and of the nucleic acids, DNA and RNA. In cells, phosphorus participates in energy metabolism and acid-base regulation. Many signaling molecules depend on the phosphorylation of enzymes to elicit an hormonal response. Approximately 85% of the body's phosphorus is in the skeleton, 14% is associated with soft tissue such as muscle, and 1% is found in the blood and body fluids. Within the physiological range of pH values, inorganic phosphate (Pi) is present in two different ionic species, $H_2PO_4^-$ (monovalent) and HPO_4^{2-} (divalent). The relative concentration of each is dependent on the ambient pH. Thus variation in the pH value may have marked effects on the transport of phosphate by altering the concentration ratio of these phosphate species.

Calcitriol restores low plasma concentrations of Ca^{2+} and Pi to normal by action at the three major targets, namely intestine, bone, and kidney. The hormone (1) stimulates the intestinal absorption of Ca^{2+} and Pi by independent mechanisms; (2) stimulates the transport of Ca^{2+} (accompanied by Pi) from the bone fluid compartment to the extracellular fluid compartment; and (3) facilitates the renal reabsorption of Ca^{2+}. These three mechanisms provide calcium for bone mineralization and prevent tetany.

16.14.2 Parathyroid Hormone

Parathyroid hormone (PTH) is a peptide hormone produced by the parathyroid glands. PTH, acting co-operatively with calcitriol, controls extracellular calcium and phosphate concentrations within a narrow physiological range. Calcitriol dramatically decreases PTH gene transcription (Silver *et al.*, 2002).

Phosphate reabsorption in the renal proximal tubule is carried out by types I, IIa, IIb, and III sodium-phosphate co-transporters, with ~70% of reabsorption mediated by the NaPi-IIa co-transporter located in the apical (brush-border) membrane. PTH regulates tubular phosphate reabsorption by reducing the amount of NaPi-IIa co-transporters in the microvilli. This occurs through activation of several signal transduction pathways. In particular, stimulation of PTH receptors leads to activation of both adenylyl cyclase and phospholipase C, with subsequent activation of protein kinases A and C, respectively (Murer *et al.*, 1999). PTH regulation of NaPi-IIa co-transporters is dependent on a A kinase anchoring protein that binds to the regulatory subunit of PKA. This association brings the PTH receptor in close proximity to PKA, thus ensuring the specific activation of PKA by PTH (Khundmiri *et al.*, 2003).

PTH has two effects on bone in causing mobilization of Ca^{2+} and phosphate. One is the rapid removal of calcium phosphate salts from the bone fluid compartment into the bloodstream, an effect called osteolysis. This is initiated by PTH increasing the Ca^{2+} permeability of the bone lining cell plasma membrane. The second effect is activation of the osteoclasts, causing bone resorption. The rapid loss of phosphate in the kidney is due to PTH diminishing proximal tubular reabsorption of phosphate ions.

16.14.3 Calcium Homeostasis

The control of calcium homeostasis involves three major sites: bone, the kidneys, and the intestine. Bone is the major reservoir of calcium in the body, storing around 99% of the total. The role of bone in calcium homeostasis is to 'buffer' blood calcium level, releasing Ca^{2+} to the blood when the blood level decreases and taking Ca^{2+} back when the level rises. In the kidney, PTH positively regulates 1α-hydroxylase gene expression via a PKA-dependent pathway (Murayama *et al.*, 1999). Simultaneously, PTH negatively regulates renal 24*R*-hydroxylase gene expression (Shinki *et al.*, 1992). This reciprocal regulation by PTH in the kidney allows for calcitriol production with minimal concomitant catabolism. The effect of calcitriol is to restore plasma calcium levels to normal.

In osteoblasts, PTH acts synergistically with calcitriol to increase transcription of the 24*R*-hydroxylase gene (Armbrecht *et al.*, 1998). Thus PTH exerts opposite effects in kidney and bone with respect to 24*R*-hydroxylase regulation. Activation of the 24*R*-hydroxylase in bone is a means of regulating the resorptive action of calcitriol. Cycloheximide has no effect on the capacity of PTH to increase 24*R*-hydroxylase mRNA levels in osteoblasts, showing that the action of PTH does not require new protein synthesis. One possible mechanism by which PTH could increase 24*R*-hydroxylase promoter activity in conjunction with calcitriol is phosphorylation of the cAMP-response element binding protein (CREB) by PKA and consequent binding of the CREB to a cAMP-response element (CRE). Interaction between CREB/CRE and VDR/VDRE complexes would synergistially increase 24*R*-hydroxylase promoter activity.

PTH and calcitriol mutually regulate each other's synthesis and/or secretion. PTH down-regulates VDR abundance in the kidney (Reinhardt & Horst, 1990), while calcitriol lowers preproPTH mRNA levels in the parathyroids (Silver *et al.*, 1985, 1986) with a subsequent reduction in PTH secretion (Cantley *et al.*, 1985). In osteoblastic cells, calcitriol reduces PTH-stimulated cAMP production (Pols *et al.*, 1986), while PTH up-regulates VDR abundance (Krishnan *et al.*, 1995).

The homeostatic control of calcium is represented schematically in Figure 16.6. In response to a lowered concentration of serum Ca^{2+}, the parathyroid glands are stimulated, within minutes, to secrete PTH. This hormone stimulates, within hours, the activity of renal 1α-hydroxylase. The calcitriol thus produced acts by itself to initiate active calcium transport in the intestine. In bone, calcitriol acts synergistically with PTH to mobilize Ca^{2+} (accompanied by Pi) from the bone fluid compartment into the bloodstream. The presence of both PTH and the vitamin D hormone are required for this system to operate *in vivo* (Jones *et al.*, 1998). In the kidney, PTH and calcitriol act in concert to cause the reabsorption of the last 1% of the filtered load of calcium into the plasma compartment. The resultant increase in the concentration of circulating Ca^{2+} provides a powerful negative feedback signal to the parathyroids, suppressing the secretion of PTH and ultimately stopping the renal production of calcitriol.

The inverse relationship between serum Ca^{2+} and PTH secretion is dependent on the sensing of serum Ca^{2+} by the

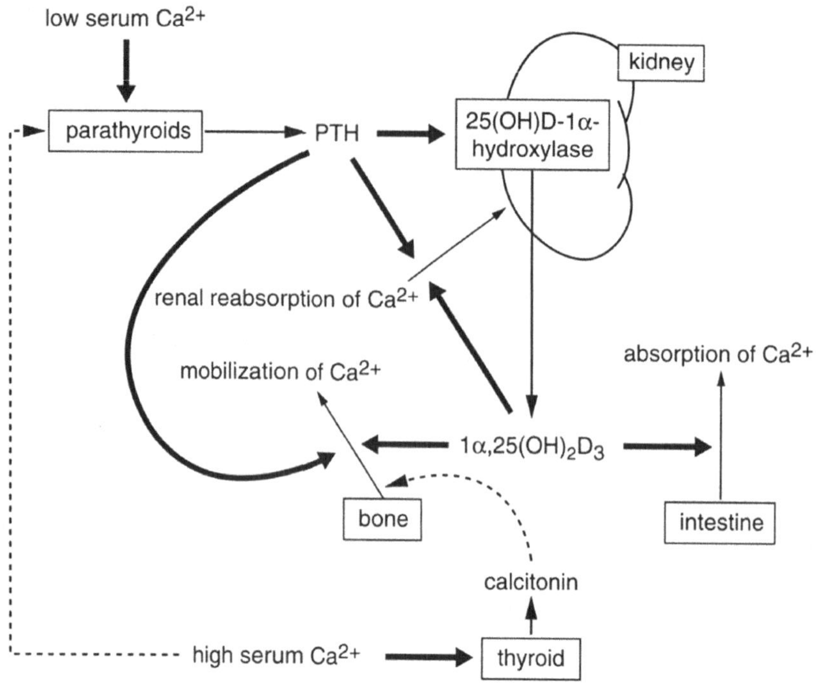

FIGURE 16.6 Simplified diagram of calcium homeostasis. The heavy solid lines represent stimulatory effects and the dotted lines represent inhibitory effects.

calcium-sensing receptor (CaR) on the parathyroid cell. The CaR is a G protein-coupled, seven-transmembrane receptor, which uses acidic amino acids to bind the millimolar concentrations of Ca^{2+} in the extracellular fluid. In the situation of hypocalcemia, the serpentine CaR is relaxed and PTH is secreted. The PTH activates its specific receptor on bone and kidney to release Ca^{2+}. The elevated serum Ca^{2+} activates the CaR, which transduces an inhibitory signal to the PTH secretory mechanism to complete the feedback loop. The parathyroid cell is therefore programed to synthesize and secrete PTH continuously and the CaR activates a brake on this process. In addition to preventing the secretion of PTH, the Ca^{2+}-activated CaR signals the degradation of preformed PTH (Silver *et al.*, 2002).

Vitamin A has been shown to antagonize the serum calcium response to vitamin D in rats at low intake levels of vitamin D and calcium (Rohde *et al.*, 1999). Similarly, in healthy human subjects, an intake of vitamin A corresponding to about one serving of liver antagonized the serum calcium response to physiological levels of vitamin D (Johansson & Melhus, 2001). The data from these two studies suggested that this antagonism was exerted at the level of intestinal calcium absorption. The antagonistic effect of vitamin A may explain the remarkably high incidence of osteoporosis found in Scandinavia, where the sunlight exposure is limited and the vitamin A intake is high.

Calcitonin, a peptide hormone and known inhibitor of osteoclastic bone resorption, is secreted by parafollicular cells of the thyroid gland in response to elevated serum Ca^{2+}. Total thyroidectomy without calcitonin replacement does not result in any demonstrable derangement of calcium or phosphate homeostasis, therefore the hormone does not seem to be important in 'normal' bone metabolism. The effect of calcitonin is to decrease the formation of new osteoclasts. Because osteoclastic resorption of bone leads to osteoblastic activity, decreased numbers of osteoclasts are followed some time later by decreased amounts of osteoblasts. Therefore, the effect of calcitonin on plasma Ca^{2+} concentration is transient, lasting for a few hours to a few days at most. It has been suggested that calcitonin may be important during periods of heightened calcemic challenge, such as growth, pregnancy, and lactation.

16.14.4 Phosphate Homeostasis

Plasma phosphate levels are maintained within a concentration range of 1.12–1.45 mM. Unlike calcium, dietary phosphate usually exceeds the body's nutritional requirement. A diet that is low in phosphorus is likely to be low also in calcium, which complicates the picture of phosphate homeostasis.

The restoration of the normal plasma calcium level in response to hypocalcemia is not accompanied by a rise in plasma phosphate, because PTH independently increases urinary phosphate excretion, and this effect is usually great enough to override increased phosphate mobilization from the bone.

Let us consider a hypothetical situation of a normal plasma calcium level during hypophosphatemia. A lowering of plasma phosphate will stimulate the kidney to release calcitriol. leading to increased renal reabsorption of phosphate. The calcitriol will also increase the intestinal absorption of phosphate and calcium. The parathyroids will not be stimulated to produce PTH. In the absence of PTH, mobilization of phosphate from the bone will be retarded and there will be no phosphate diuresis. The net effect will be an elevation of plasma phosphate.

Hyperphosphatemia is countered through phosphate excretion, governed by the fact that the blood phosphate concentration is maintained at or near the renal transport maximum for the ion.

16.14.5 Fibroblast Growth Factor 23 as a Hormone Regulator of Phosphate Metabolism

An elevation of plasma phosphate will stimulate osteocytes to secrete fibroblast growth factor 23 (FGF-23), which acts on the kidney to cause phosphaturia and a lowering of plasma phosphate.

FGF-23 is a glycoprotein comprising 251 amino acids. Osteocytes situated deep inside calcified bone synthesize and secrete FGF-23 in response to raised blood phosphate. The secreted FGF-23 is carried by the blood to target sites in the kidneys and parathyroid glands. This behavior qualifies FGF-23 as a hormone, with bone acting as the endocrine organ. FGF-23 binds to an FGF receptor–Klotho protein complex (FGFR–KL) formed by the binding of Klotho to the c-splice isoforms of FGF receptors 1, 3, and 4: it is actually the Klotho co-receptor that binds FGF-23 (Liu *et al.*, 2007). In the kidney, the Klotho co-receptor is expressed specifically on the cells of the distal tubule, but its site of action is in the proximal tubule. This suggests that either FGF-23 uses a signal transduction pathway to the proximal tubule or it induces the secretion of an intermediary phosphatonin at the distal tubule, which acts in a paracrine fashion on the proximal tubule (Bergwitz & Jüppner, 2010).

Shimada *et al.* (2004) investigated the short-term effects of a single intravenous injection of recombinant FGF-23 in normal and parathyroidectomized animals. The injection caused a reduction in serum phosphate and calcitriol levels without affecting serum calcium levels. FGF-23 inhibited the expression of the NaPi-IIa co-transporter in the renal proximal tubule. The impaired phosphate reabsorption would cause an increase in urinary excretion of phosphate (phosphaturia) leading to hypophosphatemia. This action of FGF-23 was maintained in parathyroidectomized animals, confirming it to be independent of PTH. In addition, FGF-23 directly suppressed the expression of 1α-hydroxylase that mediates the production of calcitriol and enhanced the expression of 24*R*-hydroxylase that converts calcitriol into less active metabolites. These effects on the renal enzymes started as early as 1 hour after injection and lasted more than 9 hours. A decrease in serum phosphate was first observed 9 hours after the injection and was accompanied with a reduction in renal mRNA and protein levels for the NaPi-IIa co-transporter. Because calcitriol enhances intestinal phosphate absorption, the decrease in serum calcitriol induced by FGF-23 would reduce phosphate absorption and lead to hypophosphatemia. Galitzer *et al.* (2008) showed that FGF-23 activates the MAPK signal transduction pathway to decrease PTH secretion by the parathyroid. Because PTH stimulates 1α-hydroxylase activity in the kidney, the decreased synthesis of calcitriol contributes to the development of hypophosphatemia. Finally, an injection of calcitriol into normal mice increased the serum FGF-23 level within 4 hours (Shimada *et al.*, 2004). In conclusion, FGF-23 is a potent regulator of vitamin D metabolism and phosphate homeostasis.

Mutations in the FGF-23 gene that render the protein resistant to proteolytic cleavage are responsible for the rare hereditary disease autosomal dominant hypophosphatemic rickets (ADHR) in which excessive excretion of phosphate over a long period leads to diminished calcification of the bones. Unlike the usual type of rickets, ADHR does not respond to vitamin D therapy (ADHR Consortium, 2000).

16.14.6 Secondary Hyperparathyroidism

Secondary hyperparathyroidism refers to the excessive secretion of PTH by the parathyroid glands in response to hypocalcemia and associated hyperplasia of the glands. Classically, this disorder is caused by chronic renal failure in which there is a loss of 1α-hydroxylase activity and a consequent decline in calcitriol production. Vitamin D deficiency is another cause. The progressive decline in levels of circulating calcitriol is a major factor in secondary hyperparathyroidism, because it normally suppresses PTH by several mechanisms. Calcitriol (1) directly inhibits transcription of the PTH gene (Silver *et al.*, 1986); (2) up-regulates expression of the CaR on the parathyroid cells (Canaff & Hendy, 2002); and (3) indirectly reduces PTH secretion by increasing intestinal calcium absorption. Calcitriol and its analogs are the mainstays for the prevention and treatment of secondary hyperparathyroidism in patients with renal failure (Silver *et al.*, 2002).

Low calcium levels also cause secondary hyperparathyroidism by stimulating PTH secretion through relaxation of the CaR. With persistent hypocalcemia, the expression of the PTH gene is enhanced due to the improved stability of PTH mRNA. Increased levels of serum phosphate also stabilize PTH mRNA. Prolonged hypocalcemia and hyperphosphatemia reduce p21 and lead to parathyroid gland hypertrophy (Ott & Elder, 2013).

16.15 VITAMIN D ACTION ON BONE

16.15.1 Introduction

It is well established that calcitriol is required for normal development and mineralization of bone, and for bone remodeling. Calcitriol has an indirect effect on bone through its control of calcium absorption in intestine and reabsorption in the kidney. This is shown in vitamin D-deficient rats where rickets can be cured by increasing the calcium and phosphorus content of the diet. A direct role for vitamin D in bone is indicated by the increased formation and decreased resorption of bone in the osteoblast-specific VDR overexpression mouse model (Gardiner *et al.*, 2000).

Anderson *et al.* (2005) demonstrated a strong positive relationship between 1α-hydroxylase (CYP27B1) and 24-hydroxylase (CYP24) mRNA levels in bone, unlike the inverse relationship between these enzymes in kidney. They further showed that locally synthesized calcitriol in bone tissue regulates bone CYP24 expression independently of circulating calcitriol levels. Thus, autocrine or paracrine vitamin D metabolism in the bone neither contributes to circulating calcitriol levels nor is it regulated by circulating vitamin D metabolites.

Calcitriol stimulates bone resorption by increasing the expression of proteins essential to the resorptive process. One such protein is carbonic anhydrase II, which facilitates proton (H^+) production and thus enables the drastic acidification of the resorptive lacuna (Lehenkari *et al.*, 1998). Calcitriol inhibits osteoblastic bone formation by decreasing alkaline phosphatase activity (Manolagas *et al.*, 1981) and collagen synthesis (Rowe & Kream, 1982), and increasing the synthesis of osteocalcin, a potent inhibitor of mineralization (Price, 1985).

16.15.2 Effect of Calcitriol on the Differentiation of Osteoblasts

Maehata *et al.* (2006) reported that calcitriol stimulates gene expression of Runx2/Cbfa1 and osterix in human osteoblastic cell cultures. These two proteins are transcription factors critical for osteoblast differentiation. In addition, calcitriol up-regulated alkaline phosphatase, osteocalcin, and the gene for the α1 chain of type I collagen (COL1A1), which are markers for osteoblast differentiation. The cell growth rate was decreased by the presence of calcitriol and ascorbic acid 2-phosphate in the culture medium. The data suggested that type I collagen regulates the growth properties and differentiation of human osteoblasts by causing the cells to adhere to the bone matrix, thus triggering the relevant cellular responses (Franceschi, 1999). Thus, both direct and collagen-mediated signals are required for active calcitriol-elicited differentiation of human osteoblasts.

16.15.3 Regulation of Osteoblast Growth by Interactions between Transforming Growth Factor-β and Calcitriol

In many studies, complex interactions have been demonstrated between transforming growth factor-β (TGF-β) and calcitriol, modulating their receptor expression, synthesis, and effects on osteoblast-specific gene expression. The co-operative interactions of TGF-β and calcitriol can be synergistic or antagonistic (Gurlek & Kumar, 2001).

Both calcitriol and TGF-β have diverse effects on osteoblasts, depending on whether the cells are at the proliferative or differentiation stage. Short-term treatment of osteoblast cultures with calcitriol during the proliferative stage inhibits proliferation, down-regulates collagen synthesis, and stimulates expression of osteoblast-specific markers, such as osteopontin. In contrast, TGF-β stimulates proliferation and collagen production, and inhibits expression of osteoblast-specific markers, such as osteocalcin. Thus, TGF-β retards development of the mature osteoblast phenotype, whereas calcitriol advances this development (Gurlek & Kumar, 2001).

Ingram *et al.* (1994) examined the effects of calcitriol and TGF-β separately and in combination on primary human osteoblastic cell cultures. TGF-β reduced the stimulatory effect of calcitriol on osteocalcin expression and secretion. TGF-β and calcitriol each marginally increased alkaline phosphatase activity, but the two together synergistically increased this activity. Procollagen type I synthesis and secretion were increased in a dose-dependent manner in the presence of TGF-β and these effects were not suppressible by calcitriol.

The function of ligand-inducible VDR as a transcription factor is modulated by regulatory and inhibitory Smad proteins of the TGF-β signaling pathway described in Section 10.6.3. Smad3 was shown to enhance the VDR-mediated action of calcitriol by forming a nuclear complex with the VDR co-activator SRC-1, which stabilized the interaction of VDR with Smad3 (Yanagisawa *et al.*, 1999). SRC-1 binds to the AF-2 activation subdomain located in the E domain of VDR. The N-terminal Mad homology 1 region (MH1) region of Smad3 mediated the interaction of Smad3 with the VDR. The ligand-dependent interaction of Smad3 with RXR/VDR heterodimer was evident. The enhanced function of VDR by Smad3 was also observed in VDREs derived from mouse osteopontin and human osteocalcin gene promoters. Subramaniam *et al.* (2001) demonstrated that the VDRE-containing human osteocalcin gene promoter contains two functional Smad3 binding elements through which TGF-β and vitamin D signaling pathways converge. Thus, Smad3 mediates cross-talk between vitamin D and TGF-β signaling pathways. Yanagi *et al.* (1999) showed that Smad7, an inhibitory Smad, also localizes in the nucleus, where it abrogates the Smad3-mediated potentiation of VDR signaling through unknown molecular mechanisms.

16.15.4 Effect of Circulating 25-Hydroxyvitamin D Levels on Mineralization

Anderson *et al.* (2007) developed a rodent model for studying the effects of frank vitamin D deficiency as well as graded depletion of circulating 25(OH)D levels on bone health. Ten-week-old male rats were fed a defined diet containing 0.4% calcium and various levels of dietary vitamin D_3 (0, 50, 100, 200, 300, 500 ng/rat/day) for 4 months to achieve stable mean serum 25(OH)D levels ranging between 10 and

115 nM. Dietary vitamin D was strongly related to serum 25(OH)D levels. No significant changes in serum calcium, phosphate, or PTH occurred in any of the dietary treatment groups and concentrations of these parameters were within normal ranges. Levels of serum calcitriol increased as dietary vitamin D increased between 0 and 200 ng/day, at which point a maximum level was achieved and retained with higher vitamin D intakes. Increasing dietary calcium from 0.4% to 1% suppressed serum calcitriol regardless of whether rats received 0 or 500 ng of vitamin D/day.

Anderson *et al.* (2008) used their rodent model of vitamin D depletion to quantify the 25(OH)D levels required for normal mineralization of cancellous bone. The histomorphology of distal and proximal femora and lumbar vertebrae was analyzed. Results showed a strong positive association between circulating 25(OH)D levels and trabecular bone mineral volume in the distal femoral metaphysis, but no significant changes in the epiphyses. Osteoclast surface (the proportion of bone surface occupied by osteoclasts) was reduced with increasing serum 25(OH)D levels. A marked rise in osteoclast surface at low serum 25(OH)D levels was mainly responsible for the low trabecular bone mineral volume observed in the distal femoral metaphysis. Importantly, there was no correlation between bone mineral volume or osteoclast surface and either serum levels of calcitriol, calcium, or PTH, demonstrating that the serum level of 25(OH)D is the major determinant of bone mineral volume and bone resorption. This role of 25(OH)D is evident from the observation that 25(OH)D administration improves bone mineralization in patients with chronic renal failure, who have impaired CYP27B1 activity (Eastwood *et al.*, 1977). The animals fed zero vitamin D, which had a mean serum 25(OH)D level of 10.9 nM, were the only group to show signs of osteomalacia. These animals had normal calcium and phosphate levels, and their serum PTH levels were only marginally increased. Thus, severe vitamin D deficiency can lead to osteomalacia when dietary calcium and phosphate intakes are adequate. Animals with serum 25(OH)D levels below 80 nM had marked osteopenia without accompanying hypocalcemia or marked induction of PTH. Only in animals with serum 25(OH)D levels greater than 80 nM was there no reduction in bone mineral volume, consistent with the level of 25(OH)D recommended to reduce the risk of hip fracture in elderly humans (Bischoff-Ferrari *et al.*, 2006).

16.15.5 Metabolism of 25-Hydroxyvitamin D in Human Osteoblasts

Atkins *et al.* (2007) demonstrated that human osteoblasts constitutively express CYP27B1 mRNA and secrete *de novo* synthesized calcitriol in response to 25(OH)D. Incubation of osteoblasts with physiological concentrations of 25(OH)D up-regulated transcription of CYP24 and the calcitriol-responsive genes, osteocalcin, osteopontin, and RANKL. Exposure to 25(OH)D also promoted the recognized calcitriol effect of inhibition of osteoblast proliferation and promotion of *in vitro* mineralization. Osteoblasts were shown to express the DBP receptors, cubilin and megalin, suggesting that osteoblasts are able to internalize DBP–25(OH)D complexes *in vivo*. Specific knockdown of CYP27B1 greatly reduced cellular responses to 25(OH)D, strongly implying that the 25(OH)D effect in osteoblasts is preceded by conversion to calcitriol. This study, together with another study (van Driel *et al.*, 2006), provides evidence for autocrine and paracrine activities of calcitriol that regulate key osteoblast functions independently of circulating, kidney-derived calcitriol.

16.15.6 Metabolism of 25-Hydroxyvitamin D in Human Osteoclasts

Kogawa *et al.* (2010a) demonstrated that human osteoclasts, like osteoblasts, can convert 25(OH)D to calcitriol. Kogawa *et al.* (2010b) provided evidence that 25(OH)D metabolism in human osteoclasts increases the expression of genes that are important for osteoclast formation and function. CYP27B1 mRNA expression increased in response to macrophage-colony stimulating factor (M-CSF)/RANKL-induced differentiation of osteoclasts, suggesting that 25(OH)D metabolism plays a role in osteoclast differentiation. 25(OH)D, in the presence of RANKL and M-CSF, time-dependently increased the expression of mRNA encoding key regulators of osteoclastogenesis, NFATc1 and c-Fos. In addition, the expression of osteoclast marker genes which relate to osteoclast function were increased in the presence of 25(OH)D. Conversely, calcitriol suppressed osteoclast differentiation. The divergent effects of calcitriol and 25(OH)D might be explained, at least in part, by their differential ability to stimulate the expression of the catabolic enzyme, CYP24. Calcitriol at 1 nM strongly induced the expression of CYP24 mRNA, whereas 25(OH)D at the physiological concentration of 50 nM induced CYP24 mRNA to an approximately 100-fold lesser extent. These results imply that 25(OH)D metabolism optimizes osteoclast differentiation through the enhancement of osteoclast marker gene expression, without down-regulation of intracellular levels of calcitriol by CYP24.

Bidirectional signaling mediated by ephrinB2 on mature osteoclasts and EphB4 on osteoblast precursors facilitates the transition from bone resorption to formation (Matsuo & Irie, 2008). Kogawa *et al.* (2010b) showed that ephrinB2 mRNA expression increased at a late stage of osteoclastogenesis in a 25(OH)D-dependent manner. This suggests that, through the coupling of bone resorption to formation, mature osteoclasts derived in the presence of adequate levels of 25(OH)D may better promote ensuing osteoblast activity during remodeling.

16.15.7 Osteoclastogenesis

Osteoclastogenesis, the formation of osteoclasts, involves a complex interaction of osteoclast precursors, osteoblasts, and bone marrow-derived stromal cells (Suda *et al.*, 1992).

During osteoclastogenesis, osteoclast precursors interact closely with osteoblasts and/or stromal cells (Takahashi *et al.*, 1988). Calcitriol facilitates this interaction by inducing the expression of the intercellular adhesion molecule, ICAM-1, on osteoblasts (Okada *et al.*, 2002). Calcitriol also up-regulates expression of the integrin $\alpha_V\beta_3$ on the surface of osteoclast precursors (Medhora *et al.*, 1993). This adhesion molecule binds to the bone matrix proteins osteopontin and bone sialoprotein. Expression of $\alpha_V\beta_3$ coincides with osteoclast differentiation and is essential to the resorptive process.

Calcitriol increases osteoclastogenesis by both up-regulating RANKL and down-regulating its antagonist, osteoprotegerin (Horwood *et al.*, 1998).

Stromal cells of the bone marrow control osteoclastogenesis through the production of cytokines capable of promoting the proliferation and differentiation of osteoclast precursors. One particularly potent cytokine, IL-11, can be stimulated by calcitriol (Romas *et al.*, 1996). The addition of anti-IL-11 antibody to bone marrow cell cultures suppresses the ability of calcitriol to induce osteoclast development (Girasole *et al.*, 1994), suggesting that IL-11 is an essential factor for the osteoclastogenic effect of calcitriol. Another cytokine, M-CSF, is essential for osteoclast differentiation from progenitors, but its production by osteoblasts/stromal cells does not appear to be regulated by calcitriol (Suda *et al.*, 1992).

Sato *et al.* (1991) reported that mouse bone-derived stromal cells and osteoblasts produce the third component of complement (C3) in response to calcitriol. Bone C3 is also produced tissue-specifically by calcitriol *in vivo* (Jin *et al.*, 1992). The addition of anti-C3 antibody to mouse bone marrow cultures completely inhibits the calcitriol-induced formation of osteoclast-like cells, suggesting that the C3 produced by stromal cells in response to calcitriol is somehow involved in osteoclast formation (Sato *et al.*, 1991).

16.16 24,25-DIHYDROXYVITAMIN D_3

16.16.1 Introduction

Historically, $24,25(OH)_2D_3$ has been considered by many to possess little or no intrinsic activity, its formation serving only to divert the metabolism of $25(OH)D_3$ away from calcitriol. However, several reports indicate that $24,25(OH)_2D_3$ is a functionally independent hormone and plays a crucial role in intramembranous and endochondral bone formation and in the repair of bone fractures. The importance of $24,25(OH)_2D_3$ in endochondral bone formation is indicated by its accumulation in growth plate cartilage when normal rats are injected with $[^3H]25(OH)D_3$ (Seo *et al.*, 1996).

16.16.2 Effect on Calcium Current in Osteosarcoma Cells

Li *et al.* (1996) demonstrated a dual nongenomic effect of $24,25(OH)_2D_3$ on the L-type calcium channel current in osteosarcoma cells. At a low physiological concentration (1×10^{-8} M), $24,25(OH)_2D_3$ activated the PKA signal pathway leading to an increase in current amplitude, whereas a higher concentration (1×10^{-5} M) reduced the current amplitude via the PKC signal pathway. In comparison, a high concentration of calcitriol (1×10^{-6} M) increased the current amplitude.

16.16.3 Regulation of Chondrocyte Maturation and Differentiation

Boyan *et al.* (2001) reviewed the results of their many experiments which examined the biochemical effects of calcitriol and $24,25(OH)_2D_3$ on cultured chondrocytes at two distinct stages of cell development. Cells were selected from the cartilaginous growth plates at the costochondral junction of the ribs of adolescent rats. Resting zone cells are fully committed cartilage cells which undergo proliferation when suitably stimulated. Growth zone cells taken from the prehypertropic/upper hypertropic zones are capable of producing matrix vesicles. Table 16.2 shows differential responses to calcitriol and $24,25(OH)_2D_3$ in the two cell types. In the case of all four biochemical parameters, growth zone chondrocytes respond exclusively to calcitriol, whereas resting zone chondrocytes respond exclusively to $24R,25(OH)_2D_3$.

The differential effects of the two vitamin D hormones on phospholipase A_2 activity are reflected in prostaglandin production, since stimulation of phospholipase A_2 releases arachidonic acid (the chief precursor of prostaglandins) from membrane phospholipids. Accordingly, calcitriol stimulates prostaglandin E_2 production in growth zone cells, whereas $24,25(OH)_2D_3$ inhibits production by resting zone cells. In the case of PKC activation, there is a marked difference in the time course of the two responses. Calcitriol exerted its effect within 9 minutes, whereas the maximal effect of $24,25(OH)_2D_3$ was noted at 90 minutes. Use of actinomycin D and cycloheximide demonstrated that the calcitriol effect is nongenomic; in contrast, the $24,25(OH)_2D_3$ effect is genomic, requiring both new gene transcription and new protein synthesis. The effect of $24,25(OH)_2D_3$ in stimulating PKC activity is stereospecific since $24R,25(OH)_2D_3$ is able

TABLE 16.2
Biochemical Responses of Chondrocytes to $1\alpha,25(OH)_2D_3$ and $24R,25(OH)_2D_3$

	Growth Zone Cells		Resting Zone Cells	
Parameter	**1α,25**	**24*R*,25**	**1α,25**	**24*R*,25**
Alkaline phosphatase activity	Increase	No response	No response	Increase
PLA_2 activity	Increase	No response	No response	Decrease
PKC activity	Increase	No response	No response	Increase
Fluidity of plasma membrane	Increase	No response	No response	Increase

to elicit the response, but $24S,25(OH)_2D_3$ is not able. The effects of hormone on phospholipid metabolism account for the changes in membrane fluidity, which affect membrane transport processes and the activities of membrane-bound enzymes. In addition, calcitriol stimulates Ca^{2+} efflux from growth zone chondrocytes, whereas $24,25(OH)_2D_3$ inhibits Ca^{2+} efflux from resting zone cells. Overall, $24,25(OH)_2D_3$ targets less mature cells and calcitriol targets more mature cells. Thus membrane receptors for these metabolites are differentially expressed as a function of cell maturation.

The rapid, membrane-mediated effects in response to $24,25(OH)_2D_3$, but not calcitriol, seen in resting zone chondrocytes suggests that a membrane receptor(s) exists which is specific for $24,25(OH)_2D_3$. Pedrozo *et al.* (1999) reported that a hybrid analog of calcitriol with 0.1% affinity for the nVDR stimulated PKC activity in plasma membranes from resting zone chondrocytes, and that this effect was not blocked by antibody generated to the calcitriol membrane receptor (Ab99). This provided definitive evidence of a membrane receptor for $24,25(OH)_2D_3$ in the resting zone chondrocytes.

16.16.4 Regulation of Matrix Vesicles

From their studies of matrix vesicles isolated from cultured chondrocytes, Boyan *et al.* (1994) proposed that calcitriol and $24,25(OH)_2D_3$ synthesized by chondrocytes diffuse into the extracellular matrix and interact directly with the vesicle membrane receptor. This initiates a cascade of biochemical events which lead to maturation of the matrix vesicle, hydroxyapatite crystal formation, loss of integrity of the vesicle membrane, and eventual release of active proteases. The proteases then degrade proteoglycan aggregates in the vicinity of the matrix vesicles, facilitating extracellular matrix calcification. Boyan *et al.* (1994) discovered that the chondrocyte plasma membrane and matrix vesicle membrane possess different isoforms of PKC – α and ζ (zeta), respectively. This allows chondrocytes to regulate events in matrix vesicles without comparable effect on their plasma membranes.

16.16.5 Bone Fracture Repair

There is substantial evidence that $24,25(OH)_2D_3$ at physiological concentration is essential for bone fracture repair. Both the renal 24-hydroxylase activity and the circulating $24,25(OH)_2D_3$ levels were highly elevated 10–12 days after chick tibial fracture (Seo & Norman, 1997). Oral administration of $24R,25(OH)_2D_3$ in combination with calcitriol was effective in healing tibial fracture in chicks, whereas a combination of $24S,25(OH)_2D_3$ and calcitriol or calcitriol alone resulted in poor healing (Seo *et al.*, 1997). The ability of the bone-healing system to distinguish between $24R,25(OH)_2D_3$ and $24S,25(OH)_2D_3$ is further evidence for the presence of a receptor. Accordingly, Kato *et al.* (1998) separated fracture-healing callus tissue into nuclear, cytosol, and membrane fractions and demonstrated the presence of a specific receptor for $24R,25(OH)_2D_3$ in the membrane fraction. Local injection of $24R,25(OH)_2D_3$ into the proximal epiphyseal growth plate of the tibiae of vitamin D-deficient chicks resulted in disappearance of the rachitic lesions in the same leg; there were no signs of healing in response to local injection of calcitriol (Lidor *et al.*, 1987). These observations suggest that $24R,25(OH)_2D_3$ may exert effects both locally via autocrine or paracrine mechanisms, or systemically via endocrine mechanisms.

16.16.6 Regulation of Intramembranous Bone Formation

St-Arnaud's group (St-Arnaud, 1999) engineered a strain of mice deficient for the 24-hydroxylase enzyme (24-hydroxylase-ablated or knockout mice) and showed that homozygotes for this mutation cannot effectively clear calcitriol from their circulation. This leads to perinatal lethality, most probably secondary to hypercalcemia caused by sustained elevated levels of calcitriol. The role of $24,25(OH)_2D_3$ during development was addressed by breeding mutant animals that had survived to adulthood. When fertile mutant homozygous females are mated to heterozygous males, litters comprise an equal proportion of mutant homozygotes and control heterozygous littermates. Because of the impaired 24-hydroxylase activity of the female, homozygous embryos are completely deprived of vitamin D metabolites hydroxylated at position 24. Heterozygous littermates can synthesize those metabolites because they carry one functional allele of the 24-hydroxylase gene. Histological examination of the bones of homozygous mutants born of homozygous females revealed an accumulation of unmineralized osteoid at sites of intramembranous ossification. Control heterozygous littermates showed normal bone structure. Homozygous mutants born from heterozygous females also showed normal bone histology, presumably because the embryo can receive maternal $24,25(OH)_2D_3$ via the placenta. These results demonstrate a key role for $24,25(OH)_2D_3$ in the developmental regulation of intramembranous bone formation. The impairment of matrix calcification in the knockout mice may be due to elevated calcitriol levels resulting from altered metabolism rather than a direct action of $24,25(OH)_2D_3$ (St-Arnaud *et al.*, 2000).

16.17 BONE-RELATED VITAMIN D DEFICIENCY DISEASES

16.17.1 Overview

Vitamin D deficiency can arise from lack of sunlight exposure, lack of dietary vitamin D intake, or impaired metabolism of the vitamin to its active form, calcitriol. At the onset of deficiency, there is a decreased efficiency of intestinal calcium absorption and a consequent fall in the plasma calcium level. In response to the hypocalcemia, the parathyroids are stimulated to secrete PTH. This hormone, acting with whatever calcitriol is still present at the onset of deficiency, elicits the mobilization of Ca^{2+} and Pi from bone. In addition, PTH by itself causes an increase in the renal reabsorption of calcium and the excretion of phosphate. The effect of PTH on phosphate excretion is usually great enough to override increased phosphate mobilization from the bone. As a result of these events, the plasma Ca^{2+} concentration is restored to a normal fasting level, but the Pi concentration falls below the normal level.

During prolonged vitamin D deficiency, the increase in PTH secretion necessary to maintain calcium homeostasis causes extreme osteoclastic resorption of bone. This in turn causes the bone to become progressively weaker and imposes marked physical stress on the bone, resulting in rapid osteoblastic activity, The osteoblasts lay down large quantities of osteoid but, because of insufficient Ca^{2+} and Pi, calcification does not occur. The failure to calcify newly formed bone matrix leads ultimately to rickets in children or osteomalacia in adults.

16.17.2 Rickets

Rickets, a word from the Anglo-Saxon *wrikken* (to twist), is the classic vitamin D disease in children. The disease is characterized by bow legs or knock knees, curvature of the spine, and pelvic and thoracic bone deformities. These deformities result from the mechanical stresses of body weight and muscular activity applied to the soft uncalcified bone.

Without vitamin D, the cartilaginous growth plate of the growing child fails to calcify. With this defect, the cartilage cannot be replaced by bone on the diaphyseal side and the growth plate becomes progressively thicker. This results in enlargement of the joints in the knees, wrists, and ankles.

The prevalence of rickets among city-dwelling children during the Industrial Revolution was attributable to a limited exposure to sunlight and a lack of sufficient vitamin D in the diet. The narrow streets and alleys in which the children lived and the smoke-polluted atmosphere were responsible for the lack of sunlight. During the 1930s, the practice of adding provitamin D to milk followed by UV irradiation drastically reduced the incidence of rickets in the United States and some European countries. Later, the commercial production of crystalline vitamin D_2 led to its use in the fortification of foods. Nowadays, rickets is a rare disease among the indigenous populations of the United States and Europe, but it is still evident among the children of immigrants, particularly Asians in Europe.

16.17.3 Osteomalacia

In adults, when the skeleton is fully developed, vitamin D is still necessary for the continuous remodeling of bone. During prolonged vitamin D deficiency, the newly formed, uncalcified bone gradually replaces the older bone tissue and the weakened bone structure is easily prone to fracture. This condition, osteomalacia, should not be confused with osteoporosis in which the ratio of mineral to osteoid is unchanged. In osteomalacia, the epiphyses do not swell, as they do in rickets, because the epiphyseal growth plates no longer exist. Patients with osteomalacia frequently suffer from muscle weakness and bone tenderness or pain in the spine, shoulder, ribs, or pelvis. Pelvis deformation can occur, causing potential problems with childbirth. Women with low vitamin D status may develop osteomalacia after several pregnancies because they are unable to replace the calcium lost from their bone reserves to the fetus *in utero* and in lactation.

16.17.4 Vitamin D-Dependent Rickets

Vitamin D-dependent rickets is a rare inherited disorder in which clinical and biochemical features of rickets are evident despite an adequate intake of vitamin D. This disorder is classified into type I and type II disease states, both of which appear to follow an autosomal recessive pattern of inheritance (Brown *et al.*, 2000).

16.17.4.1 Type I

Type I vitamin D-dependent rickets arises from impaired renal synthesis of calcitriol, which may be due to a mutation in the gene encoding 1α-hydroxylase. The disease is diagnosed by normal blood levels of 25(OH)D and profoundly decreased levels of calcitriol. At birth, affected children appear healthy, but during the first year or two of life severe hypocalcemia with tetany becomes evident. The hypocalcemia leads to secondary hyperparathyroidism with elevated PTH levels and hypophosphatemia. The calcium and phosphate deficiencies result in impaired mineralization of newly forming bone, producing the classical symptoms of rickets. The treatment of type I vitamin D-dependent rickets is long-term administration of physiological doses of calcitriol.

16.17.4.2 Type II

Type II vitamin D-dependent rickets, now more commonly called hereditary vitamin D-resistant rickets (HVDRR), arises from a lack of responsiveness of target tissues to calcitriol. In almost all cases this is due to a mutation in the gene encoding the nuclear VDR. Some mutations lead to defective ligand binding, while others lead to defective binding of the hormone–receptor complex to the DNA. Hewison *et al.* (1993) described an exceptional case attributable not to a mutation of the VDR gene, but to a defect in VDR translocation to the nucleus. Whereas the type I disease state is characterized by depressed levels of calcitriol, this metabolite is elevated in the type II state. Impaired hormonal function at the intestine and bone causes deficiencies in calcium and phosphate, leading to rickets within months of birth. Afflicted children are often growth retarded and suffer convulsions due to tetany. Some children have total scalp and body alopecia, including eyebrows and, in some cases, eyelashes. The treatment of type II vitamin D-deficient rickets is supra-physiological doses of calcitriol (Malloy *et al.*, 1999).

16.17.5 Vitamin D-Resistant Rickets

Vitamin D-resistant rickets is a group of hereditable abnormalities of renal phosphate transport, the most common of which is X-linked hypophosphatemia. The rickets cannot be explained solely by the severe hypophosphatemia that is present and the undefined pathological mechanism may involve both abnormal phosphate transport and renal 1-hydroxylase function. Treatment entails a combination of oral phosphate and calcitriol (Brown *et al.*, 2000).

16.18 DIETARY CALCIUM AND SERUM 25-HYDROXYVITAMIN D STATUS IN RELATION TO BONE MINERAL DENSITY

A study was conducted to determine the relative importance of dietary calcium intake and serum 25(OH)D status in regard to hip bone mineral density (BMD) (Bischoff-Ferrari *et al.*, 2009). The data were obtained from a large sample taken from the US NHANES III population-based survey that included hip BMD measurement, serum concentration of 25(OH)D, and dietary calcium intake. The sample comprised 4958 community-dwelling women and 5003 men ≥20 years of age. Calcium supplement users and individuals with a prior radius or hip fracture were excluded. Results were presented for six subgroups determined by gender and three concentrations of 25(OH)D (<50, 50–74, 75+ nM). Among both sexes, BMD increased stepwise and significantly with increasing 25(OH)D concentrations. The association between calcium intake and total hip BMD was significantly modified by gender and serum 25(OH) D concentration, but not by age, physical activity, or race/ethnicity. For women, a higher calcium intake (>566 mg/day) was significantly associated with higher BMD only when serum 25(OH)D concentration was less than 50 nM. That is, a calcium intake beyond 566 mg/day is sufficient for bone mineralization, provided that the 25(OH)D concentration is greater than 50 nM. This may be explained by a calcium-sparing effect of higher 25(OH)D concentrations (Heaney *et al.*, 2003). Among men, there was no significant association between a higher calcium intake (626 mg/day) and BMD within any of the three concentrations of 25(OH) D. In conclusion, only women with serum 25(OH)D concentrations of less than 50 nM seem to benefit from a higher calcium intake.

16.19 CLINICAL STUDIES INVESTIGATING THE EFFECT OF HIGH-DOSE VITAMIN D SUPPLEMENTATION ON POSTMENOPAUSAL BONE LOSS

Postmenopausal women, whose estrogen levels naturally decline, are at the highest risk of developing osteoporosis. The results of meta-analyses examining the relationship between vitamin D supplementation and bone fracture reduction in elderly people have been inconsistent. An international group of experts carried out a pooled analysis of double-blind, randomized, controlled trials of vitamin D supplementation that involved persons aged 65 years or older (Bischoff-Ferrari *et al.*, 2012). The analysis was designed to estimate the effects of vitamin D supplementation on bone fracture prevention according to the actual intake of each participant, rather than simply the dose to which the participant was randomly assigned. The source data were obtained for 30,011 participants (mean age, 76 years; 91% women) from 11 trials. The results showed a significant 30% reduction in the incidence of hip fracture and a 14% reduction in the risk of any nonvertebral fracture at the highest actual-intake level of vitamin D (median, 800 IU daily; range 792 to 2,000) in treated participants, as compared with controls. Notably, there was no reduction in the risk of hip fracture at any actual-intake level lower than 792 IU per day. Analysis of baseline levels of serum 25(OH) D produced a dose-response relationship between vitamin D and fracture risk. A 61 nM concentration of 25(OH)D reduced the risk of hip fracture by 37% compared with a concentration of 30 nM. The conclusion that a daily vitamin D supplement of 800 IU reduces the risk of bone fracture in elderly women supports the recommendation from the Institute of Medicine (2011) that men and women aged 71 years or older receive 800 IU of vitamin D per day, superseding the 1997 recommendation of 600 IU per day.

Continuing with Bischoff-Ferrari *et al.* (2012), the pooled analysis of eight trials that used combined vitamin D and calcium supplementation showed that the risk of fracture is reduced only at the highest actual-intake level of vitamin D. At this level of vitamin D (≥800 IU per day), a smaller amount of calcium (<1,000 mg per day) was more beneficial in reducing the risk of fracture than a larger amount. Pooled results from a meta-analysis of cohort studies and clinical trials showed that taking calcium supplements without vitamin D did not reduce hip fracture risk in men or women and could possibly increase the risk in women (Bischoff-Ferrari *et al.*, 2007). Thus, studies of the prevention of bone fracture should not consider calcium supplementation alone, but, rather, should focus on the optimal combination of calcium plus vitamin D.

A 1-year double-blind, randomized, controlled clinical trial was carried out among 297 postmenopausal women aged 50 to 80 years to compare high-dose vitamin D (6500 IU/day) with the standard dose of 800 IU/day for the treatment of osteoporosis. The results showed that the high dose was not better than the standard dose in improving bone mineral density in the hip and spine and was less efficient in reducing bone turnover (Grimnes *et al.*, 2012).

16.20 IMMUNOREGULATORY PROPERTIES

16.20.1 Beneficial Effects of Vitamin D in the Human Immune System

A beneficial role for vitamin D in the human immune system is suggested by the increase in susceptibility to infection of vitamin D-deficient children (El-Radi *et al.*, 1982). There are many laboratory reports that vitamin D deficiency impairs an animal's ability to mount an acute inflammatory response. Macrophages from vitamin D-deficient mice exhibited decreased migration and phagocytosis, which were both corrected when vitamin D was restored to the diet (Bar-Shavit *et al.*, 1981). The impaired phagocytic response could be corrected by incubating the macrophages with calcitriol, but other vitamin D metabolites were without effect. Macrophages from vitamin D-deficient mice also had a defective capacity to produce cytokines (Kankova *et al.*, 1991). Yang *et al.* (1993b) showed that vitamin D deficiency suppresses cell-mediated immunity *in vivo*. The provision of a vitamin D-sufficient diet for 8 weeks (but not 3 weeks) corrected the impaired immunity in mice.

Immunosuppressive effects of pharmacological doses of calcitriol and related analogs have been documented *in vivo* in animal models of immune-mediated disease such as experimental autoimmune encephalomyelitis (EAE) (Lemire & Archer, 1991; Cantorna *et al.*, 1996; Mattner *et al.*, 2000), autoimmune thyroiditis (Fournier *et al.*, 1990), and autoimmune diabetes (Gregori *et al.*, 2002). In addition, *in vivo* treatment with calcitriol prolongs graft survival (Hullett *et al.*, 1998; Gregori *et al.*, 2001). Daily injections of the hormone markedly suppressed immunoglobulin production and thymic lymphocyte proliferation in mice (Yang *et al.*, 1993a). The immunosuppressive properties of calcitriol have extended the therapeutic potential of vitamin D analogs in the treatment of autoimmune diseases and graft rejection. Calcitriol itself cannot be used therapeutically at above physiological concentrations because of its toxicity (hypercalcemic activity). Various analogs display reduced hypercalcemic activity compared to the hormone, but their immunosuppressive activity is enhanced 10- to 100-fold. In EAE (a mouse model of human multiple sclerosis), intraperitoneal injection of calcitriol prevents the disease entirely or arrests its development (Lemire, 1992). In another murine model of autoimmunity, the experimental autoimmune thyroiditis, Fournier *et al.* (1990) demonstrated a beneficial effect of combined treatment with calcitriol and cyclosporin A. Johnsson *et al.* (1995) reported enhanced cardiac allograft survival with a combination of a vitamin D analog (MC 1288) and cyclosporin A.

Lymphocytes from rheumatoid arthritis patients, but not from normal subjects, spontaneously express the VDR (Manolagas *et al.*, 1989). Exposure of T lymphocytes from rheumatoid arthritis patients to a combination of cyclosporin A and calcitriol resulted in a synergistic inhibition of both IL-2 production and the proliferation of cells (Gepner *et al.*, 1989). This *in vitro* synergism occurred at a concentration of cyclosporin A (1 ng/ml) that was 100 times lower than the optimal dose *per se*. The potentiating concentration of calcitriol was in the physiological, nontoxic range. Boissier *et al.* (1992) found that a combination of cyclosporin A and calcitriol was additively beneficial in the treatment of adjuvant arthritis, an animal model of rheumatoid arthritis.

16.20.2 Presence of the VDR in Cells of the Immune System

The VDR has been identified in almost all nucleated cell types in the body, including malignant and non-malignant cells of hematopoietic origin. Peripheral blood monocytes express the VDR constitutively. Resting T and B lymphocytes do not express the VDR: only when activated by mitogens and antigens do these cells express the receptor. Interaction of T cell receptors with the peptide–MHC complex of antigen-presenting tissue cells and cells of the immune system leads to T cell activation and subsequent gene expression events (Manolagas *et al.*, 1990).

16.20.3 Induction of Antimicrobial Peptides in Various Cell Types

The promoters of the genes encoding the human antimicrobial peptides cathelicidin (hCAP-18) and defensin-β2 (hBD-2) contain VDREs that mediate calcitriol-dependent gene expression (Wang *et al.*, 2004). Liu *et al.* (2006) reported that TLR1 and TLR2 activation of human monocytes and macrophages up-regulated expression of the VDR and the 1α-hydroxylase genes, triggering calcitriol-mediated secretion of hCAP-18 and killing of intracellular *Mycobacterium tuberculosis*. Liu *et al.* (2006) also observed that sera from African-American individuals, known to have increased susceptibility to tuberculosis, had low vitamin D status and were inefficient at producing cathelicidin.

16.20.4 *In vitro* Effects of Calcitriol on Monocytes/Macrophages

Calcitriol inhibits proliferation and promotes differentiation of bone marrow-derived macrophage precursors (Clohisy *et al.*, 1987). The relatively quiescent resident macrophages are activated by immunogenic stimuli to become activated macrophages, now with a greatly enhanced ability to phagocytose and destroy pathogens.

Calcitriol appears to be essential for macrophage activation. Gavison and Bar-Shavit (1989) reported that macrophages from vitamin D-deficient mice injected with activating or eliciting agents had defective anti-tumor activity and an impaired respiratory burst (low production of hydrogen peroxide and superoxide). Activity of the lysosomal enzyme acid phosphatase was unaffected by vitamin D deficiency. Incubation of the vitamin D-deficient macrophages with calcitriol markedly enhanced their anti-tumor activity, but did not affect the cells' capacity to produce hydrogen peroxide and superoxide, or acid phosphatase. Abe *et al.* (1984) reported that calcitriol increases the number of Fc receptors on the surface of macrophages and induces cytotoxicity. Pretreatment of normal human monocytes with calcitriol enhances their responsiveness to various chemoattractants (Girasole *et al.*, 1990). Augmented monocyte chemotaxis to FMLP is associated with an increased number of high-affinity binding sites for this chemoattractant. The hormone is also able to stimulate the migratory capacity of monocytes obtained from patients with acquired immune deficiency syndrome (AIDS), a condition associated with impaired monocyte chemotaxis. Calcitriol increased production of IL-1 by human peripheral blood monocytes (Bhalla *et al.*, 1986) and also enhanced these cells' capacity for lipopolysaccharide-triggered release of tumor necrosis factor (Rook *et al.*, 1987). Incubation of monocytes with calcitriol at 45°C led to an increased synthesis of heat shock proteins accompanied by a relative preservation of total protein synthesis (Polla *et al.*, 1986).

Normal human macrophages activated by IFN-γ (Koeffler *et al.*, 1985) or lipopolysaccharide (LPS) (Reichel *et al.*, 1987a) can independently induce the conversion of 25(OH)D_3 to calcitriol: activated lymphocytes and neutrophils do not induce this conversion. The levels of calcitriol produced may be sufficient to influence the function of lymphocytes in a paracrine manner. Overbergh *et al.* (2000) confirmed the presence of 25(OH)D_3-1α-hydroxylase in murine macrophages and showed that its expression is up-regulated by immune stimuli, particularly IFN-γ. Unlike the renal 1α-hydroxylase, no clear negative feedback by calcitriol on its own production could be observed. This induction of calcitriol required all three of the Jak–STAT, p38 MAP kinase, and NF-κB signal transduction pathways (Stoffels *et al.*, 2006). The following scenario can be envisaged in which calcitriol acts in a paracrine manner to down-regulate the immune response locally at sites of inflammation. In the early phase of inflammation, when only macrophages are involved, 1α-hydroxylase activity in macrophages is absent or low. The macrophages are activated to secrete cytokines that recruit T cells and natural killer cells. These latter cells provide the IFN-γ that up-regulates 1α-hydroxylase expression, The calcitriol produced can now perform its immunosuppressive role and prevent unrestricted immune responses.

Overbergh *et al.* (2000) further showed that non-obese diabetic (NOD) mice, which spontaneously develop type 1 diabetes, have a defect in their macrophage 1α-hydroxylase up-regulation upon stimulation. The consequent defect in calcitriol secretion by stimulated macrophages would appear therefore to contribute to autoimmunity. Supporting this possibility is the demonstration that treating NOD mice with high doses of calcitriol can prevent insulitis (Mathieu *et al.*, 1992) and diabetes (Mathieu *et al.*, 1994).

It has been postulated that vitamin D may have a protective role in tuberculosis infection (Davies, 1985). The bacterium responsible for tuberculosis is a bacillus, *Mycobacterium tuberculosis*. Monocytes, although phagocytic, have only a limited capacity to kill *M. tuberculosis* and can themselves become infected with the bacillus *in vitro*. However, any bacilli released by dying monocytes are rapidly phagocytosed by other cells, especially macrophages, and destroyed. Rook *et al.* (1986) reported that calcitriol inhibited the growth of *M. tuberculosis* in cultured human monocytes and IFNγ enhanced this inhibition. They also showed that incubation of monocytes with IFNγ led to increased ability to metabolize 25(OH)D_3 to calcitriol. Based on these observations, a possible scenario is that infected monocytes produce IFN-γ which induces these cells to synthesize calcitriol from circulating 25(OH) D_3. The locally produced hormone stimulates monocytes to differentiate into macrophages, which are better equipped to deal with the infection.

16.20.5 *In vitro* Effects of Calcitriol on Natural Killer Cells

Merino *et al.* (1989) showed that calcitriol inhibits the generation of cytotoxic activity from cultured natural killer (NK) cells. The hormone was, however, unable to interfere with the cytotoxic function of cells already established, placing the inhibition at the level of NK cell activation.

16.20.6 *In vitro* Effects of Calcitriol on Dendritic Cells

Adorini *et al.* (2003) reviewed several *in vitro* studies, performed either on monocyte-derived dendritic cells (DCs) from human peripheral blood or on bone marrow-derived mouse DCs, that showed that calcitriol and its analogs inhibit the differentiation and maturation of DCs, which results in a marked decrease in their capacity to activate naïve T cells. The calcitriol treatment alters the normal DC phenotype to a tolerogenic phenotype, characterized by decreased expression of CD40, CD80, and CD86 co-stimulatory molecules, and a diminished secretion of pro-inflammatory cytokines such as 1L-12 and IL-23. Moreover, the expression of immunoglobulin-like transcript 3 (ILT3) and the programed death-ligand 1 (PD-L1), receptors that negatively influence T cell activation, is up-regulated, and the production of tolerogenic cytokines such as IL-10 and TNF-α is enhanced in calcitriol-induced DCs. Importantly, the induction of a tolerogenic phenotype by calcitriol is stable after removal of the hormone. Calcitriol-induced tolerogenic DCs allow a shift from T helper 1 (Th1)- and Th17-mediated inflammatory responses toward Th2- and Treg-mediated regulatory responses (Vanherwegen *et al.*, 2019).

Human peripheral blood monocytes can be differentiated into immature DCs by culture with granulocyte-macrophage colony-stimulating factor and IL-4. Incubation of these monocyte-derived immature DCs with lipopolysaccharide induces them to mature. Addition of calcitriol prevents LPS-induced maturation and maintains the DCs at the immature stage, characterized by high mannose receptor expression. The DCs do not progress to a mature antigen-presenting phenotype upon withdrawal of calcitriol and are resistant to subsequent maturational stimuli. The arrest of maturation results in the decreased expression of maturation markers CD40, CD80, and CD86. The DCs are tolerogenic, as shown by impaired production of IL-12 and enhanced production of IL-10 (Penna & Adorini, 2000).

Tolerogenic DC induction by calcitriol results from VDR signaling events that promote the rapid transcriptional activation of multiple metabolic pathways involved in glucose metabolism, which have an influence on the immune phenotype (Ferreira *et al.*, 2015). Vanherwegen *et al.* (2019) showed that calcitriol rapidly induces expression of the glycolytic enzyme 6-phospho-fructo-2-kinase/fructose-2,6-biphosphatase 4 (PFKFB4), leading to a metabolic reprograming toward aerobic glycolysis and glucose oxidation in human monocyte-derived DCs. Using an inhibitor of PFKFB4, it was demonstrated that PFKFB4 contributes to the tolerogenic profile induced by calcitriol by mediating the up-regulation of tolerogenic markers and regulatory cytokines. Thus, calcitriol imprints DCs with tolerogenic properties by reprograming their glucose metabolism.

Penna *et al.* (2007) examined the effects of calcitriol on tolerogenic properties of conventional DCs (cDCs) and plasmacytoid DCs (pDCs). The above-mentioned inhibition of Th1 cells and enhancement of Tregs that results when DCs are treated with calcitriol was shown to apply only to cDCs. The intrinsically high capacity of pDCs to induce Tregs was unaffected by calcitriol treatment. No observable skewing to the Th2 phenotype was induced by calcitriol treatment in cultures containing either cDCs or pDCs. Calcitriol has a differential capacity to modulate chemokine production in the two DC subsets. CCL17 and CCL22, which are constitutively produced in both DC subsets, are able to recruit Tregs through binding to the CCR4 receptor on these cells (Iellum *et al.*, 2001). The effect of calcitriol on the constitutive production of these chemokines in the cDCs was to inhibit CCL17 and up-regulate CCL22 in a dose-dependent manner. In contrast, calcitriol had no effect on the constitutive production of either chemokine in the pDCs.

Penna *et al.* (2007) also compared the effect of calcitriol on CCL17 and CCL22 production by DC subsets unstimulated or stimulated by CD40 ligation. Stimulation up-regulated CCL17 production in cDCs, but not in pDCs. Calcitriol inhibited CCL17 production by unstimulated and stimulated cDCs equally well, with negligible effects on CCL17 production by pDCs. CCL22 production was up-regulated by CD40 ligation both in cDCs and pDCs, and was further enhanced in unstimulated (but not stimulated) cDCs by calcitriol. In CD40-stimulated pDCs, calcitriol had no substantial effect on CCL22 production.

Penna *et al.* (2007) further reported that production of IL-12 (the cDC signature cytokine) and expression of MHC class II and CD40, CD80, and CD86 co-stimulatory molecules are markedly inhibited by calcitriol in cDCs, while the inhibitory ILT3 receptor is strongly up-regulated. Conversely, neither IFN-α (the pDC signature cytokine) nor MHC class II, co-stimulatory molecules, or ILT3 are affected by calcitriol in pDCs. All these molecules are controlled by signal transduction pathways involving the activation of NF-κB as the terminal step. The different responses of cDCs and pDCs to calcitriol may be due to different effects of this hormone on such pathways. Calcitriol was shown to reduce both the phosphorylation of the p65 (RelA) subunit of NF-κB and nuclear translocation of NF-κB (p50/p65 dimer) in cDCs but not in pDCs.

Hewison *et al.* (2003) were the first to show that monocyte-derived DCs (moDCs), like macrophages, are able to synthesize calcitriol *in vitro* by up-regulating 1α-hydroxylase expression. Synthesis of calcitriol increased markedly during DC differentiation from cultured monocyte precursors. However, and in direct contrast to monocytes, VDR expression was reduced in immature DCs. Thus, as moDCs differentiate in culture, they develop an increasing capacity to synthesize calcitriol, but in parallel a reduced VDR-mediated response. The results are consistent with a model in which differential regulation of ligand and receptor contributes to a relative resistance of immature and mature DCs to the inhibitory effect of ligand as compared with their monocyte precursors.

Finally, Ferreira *et al.* (2014) showed that tolerogenic DCs derived from calcitriol-treated NOD mice efficiently restrained T cell proliferation and promptly migrated to the pancreas and liver of adult NOD mice. This study highlights the potential of calcitriol-induced tolerogenic DCs to restore immune tolerance in an autoimmune setting.

16.20.7 Polarization of Naïve CD4⁺ T Cells

Upon antigen stimulation, naïve undifferentiated $CD4^+$ T cells (Th0 cells) differentiate into the Th1 and Th2 subsets, initiated by the cytokines produced during the immune response. *In vitro*, Th1 and Th2 cells can be generated by activating Th0 cells in the presence of IL-12 or IL-4, respectively. IL-12 activates the STAT4 signaling pathway to induce IFN-γ transcription and initiate Th1 development, while IL-4 triggers the STAT6 signaling pathway, resulting in IL-4 transcription and Th2 development. Once established, the two subsets have distinct cytokine secretion profiles. IFN-γ is the Th1 signature cytokine, while IL-4 is the defining Th2 cytokine. Th1 cells secrete IFN-γ and TNF-β (lymphotoxin) and direct cell-mediated responses against intracellular microbes, while Th2 cells secrete IL-4, IL-5, and IL-6 and mediate humoral responses against extracellular parasites.

Calcitriol inhibits the secretion of IL-12 by macrophages and B cells (Lemire *et al.*, 1995), thereby preventing the differentiation of Th0 cells to Th1 cells. IL-12 is composed of two disulfide-linked subunits, p35 and p40, encoded by two different genes and regulated primarily at the transcriptional level. Both subunits contribute to the biological activity of IL-12, although p40 is more inducible than p35. D'Ambrosio *et al.* (1998) showed that calcitriol inhibits IL-12 production by macrophages and DCs by suppressing transcriptional activation of the p35 and p40 genes. Transcriptional repression of the p40 gene was dependent on the expression of the VDR–RXR heterodimer and its binding to the VDRE. The repressive effect mapped to a region in the p40 promoter containing a binding site for NF-κB (p40–κB). Activation of monocytic cells in the presence of calcitriol resulted in reduced binding of NF-κB to the p40–κB site. Thus, calcitriol may negatively regulate IL-12 production by interfering with the activation and/or nuclear translocation of NF-κB.

Calcitriol inhibits IFN-γ synthesis by peripheral blood lymphocytes independently of IL-2 (Reichel *et al.*, 1987b). Mattner *et al.* (2000) showed that calcitriol and a nonhypercalcemic analog are selective and potent inhibitors of Th1 development *in vitro* and *in vivo* without inducing a deviation to the Th2 phenotype. These results demonstrate that calcitriol has the potential to inhibit Th1 development and direct CD4$^+$ T cell polarization toward the Th2 phenotype.

Boonstra *et al.* (2001) demonstrated that calcitriol acts directly on Th0 cells and can, in the absence of antigen-presenting cells, enhance the development of Th2 cells. Calcitriol inhibited Th1 and augmented Th2 development by inhibiting IFN-γ production and enhancing IL-4, IL-5, and IL-10 production. An increased expression of the Th2-specific transcription factors GATA-3 and c-*maf* correlated with the increased production of Th2 cytokines after calcitriol treatment. The calcitriol-induced effects were largely mediated by IL-4, because neutralization of IL-4 almost completely abrogated the augmented Th2 cell development after calcitriol treatment.

Staeva-Vieira and Freedman (2002) showed that calcitriol inhibits not only the Th1 cytokine IFN-γ, but also the Th2 cytokine IL-4 in Th0 cells during their *in vitro* polarization. This effect was most dramatic when calcitriol was present from the onset of the differentiation process; if the hormone was added after the polarization had ensued, the inhibition was significantly diminished. In activated CD4$^+$ cells, calcitriol was still able to inhibit IFN-γ, but had no effect on IL-4 production. The results indicated that inhibition of these two cytokines in naïve cells by calcitriol is due to a repression of activated transcription of the cytokine genes.

16.20.8 Regulation of T Cell Proliferation

In contrast to the stimulatory effects of calcitriol on the innate arm of the immune response, the principal action of the hormone on the acquired immune response, mediated by lymphocytes, is immunosuppression. The cellular receptors for calcitriol are expressed only in activated lymphocytes (Bhalla *et al.*, 1983), implying that calcitriol has an effect on these cells only during an immune response. Calcitriol inhibits T-lymphocyte proliferation after these cells have been activated. The hormone prevents entry of the cells into the S phase of the cell cycle by blocking the transition from early G_1 to late G_1 (Rigby *et al.*, 1985).

Calcitriol inhibits the growth-promoting cytokine IL-2 (Tsoukas *et al.*, 1984). Alroy *et al.* (1995) demonstrated that calcitriol represses IL-2 gene transcription by a direct, VDR-dependent effect. In the absence of intracellular calcitriol, the IL-2 gene in T lymphocytes is activated by the binding of a T cell-specific transcription factor, NFATp, to an NF-AT-1 element and subsequent recruitment of the ubiquitous transcription factors Jun and Fos (AP-1) to form an NFATp–AP-1 complex. VDR–RXR heterodimers, which would form in response to the intracellular presence of calcitriol, directly inhibit the interaction between AP-1 and NFATp. Moreover, the stable binding of VDR–RXR to the NF-AT-1 element blocks the binding of any NFATp–AP-1 that may subsequently be formed; however, prebound NFATp–AP-1 cannot be destabilized by VDR–RXR.

Calcitriol administration to mice increases expression of TGF-β1, a potent inhibitor of T and B cell proliferation and differentiation (Cantorna *et al.* (1998). The suppressive effect of calcitriol on lymphocyte proliferation is countered indirectly by IL-1 produced by monocytes in response to stimulation by calcitriol (Bhalla *et al.*, 1986). IL-1 potentiates the release of IL-2 from activated T cells and the IL-2 stimulates lymphocyte proliferation. These differential effects of calcitriol provide a finely tuned mechanism for regulating T cell proliferation.

Jeffery *et al.* (2009) showed that stimulation of CD4$^+$ T cells in the presence of calcitriol down-regulated production of pro-inflammatory cytokines IFN-γ, IL-17, and IL-21, hallmark cytokines of Th1, Th17, and TFH effector responses, respectively. This was a direct effect of calcitriol, occurring in the absence of antigen-presenting cells. Despite these decreases in cytokine production, calcitriol had no obvious effect on T cell proliferation, with ~90% of cells entering cell division. Treatment of stimulated T cells with a combination of calcitriol and IL-2 resulted in a robust population of CTLA-4$^+$Foxp3$^+$ cells that could suppress proliferation of resting CD4$^+$ T cells. Foxp3 is a marker of Tregs and CTLA-4 is a receptor that plays an important role in the suppressive activity of Tregs (Wing *et al.*, 2008). Thus, calcitriol and IL-2 have direct synergistic effects on activated T cells, acting as potent anti-inflammatory agents and inducers of a Treg phenotype.

16.20.9 Homing of T Cells and B Cells to the Skin

Migration of effector or memory T cells into the skin depends on their expression of chemokine receptors CCR4 and CCR10. The ligand for CCR4 is CCL17, which is presented on the endothelium of dermal blood vessels (Campbell *et al.*, 1999). CCR10 binds CCL27 (CTACK, for cutaneous T cell-attracting chemokine), which is expressed by epidermal keratinocytes and preferentially attracts skin-homing memory T cells (Morales *et al.*, 1999). Another ligand for CCR10 is CCL28 (MEC, for mucosae-associated epithelial chemokine), which is expressed by the epithelial cells of the gut (small and large intestine) and other mucosal tissues (Wang *et al.*, 2000; Pan *et al.*, 2000). Skin-tropic T cells also require the expression of cutaneous lymphocyte antigen (CLA), a carbohydrate ligand which binds to E-selectin expressed on venular endothelial cells in the dermis. Imprinting of skin tropism occurs during activation of naïve T cells by antigens presented by DCs in skin-draining lymph nodes. Signals from the DCs not only induce proliferation of naïve T cells and differentiation into effector cells, but also induce expression of CLA, CCR4, and CCR10.

Calcitriol derived from vitamin D_3 synthesized in the skin during exposure to sunlight plays an important role in T cell migration to the epidermis in humans (Sigmundsdóttir *et al.*, 2007). Mature cutaneous DCs travel through the afferent lymphatics to the draining lymph nodes, where they present skin antigens to naïve T cells and help induce skin-homing properties. When cultured with vitamin D_3, DCs enriched from the afferent lymph readily convert vitamin D_3 to calcitriol. Within the skin-draining lymph nodes, IL-12 produced by resident DCs enhances the calcitriol-induced up-regulation of CCR10 in T cells that have arrived from the blood. This enables homing of the T cells to the epidermis through attraction to the epidermal chemokine CCL27 and retention mediated by several adhesion receptors. Calcitriol enforces the selective cutaneous trafficking of T cells by inhibiting the spontaneous up-regulation of the gut-homing receptor $\alpha_4\beta_7$ and preventing the *de novo* induction of the small intestinal T cell chemokine receptor CCR9 by retinoic acid. Thus, T cells exposed to calcitriol in skin-draining lymph nodes will preferentially home to the skin.

Kunkel *et al.* (2003) demonstrated that IgA-secreting plasma cells present in various mucosal tissues commonly express CCR10. Shirakawa *et al.* (2008) showed that calcitriol efficiently induces CCR10 in terminally differentiating human B cells *in vitro* and identified the VDRE in the human CCR10 promoter. Surprisingly, the major regulatory elements in the human VDRE were totally missing in the mouse CCR10 promoter and calcitriol was unable to induce CCR10 expression in terminally differentiating mouse B cells. The calcitriol-induced expression of CCR10 in human B cells might enable them to migrate to various mucosal tissues via CCL28.

16.20.10 Summary

In its capacity as an immunoregulator, calcitriol functions as a paracrine hormone. Activation of macrophages by IFNγ at the site of infection or tissue inflammation induces 1α-hydroxylase activity and localized generation of calcitriol from $25(OH)D_3$. The calcitriol thus produced enhances the immune response in many ways. As discussed above, it activates quiescent resident macrophages; promotes differentiation of bone marrow-derived macrophage precursors; and recruits monocytes to the affected site by chemotaxis and induction of adhesion molecules in capillary endothelial cells. These stimulatory effects of calcitriol are countered by its suppressive effect on Th1 cell proliferation and Th1 cell-induced antibody production. Tolerogenic cDCs induced by a short treatment with calcitriol are probably responsible for the capacity of this hormone to induce Treg cells that are able to mediate transplantation tolerance and arrest the development of type 1 diabetes.

16.21 VITAMIN D AND DIABETES

Type 1 diabetes is due to autoimmune destruction of pancreatic β cells leading to absolute insulin deficiency. Type 2 diabetes is characterized by two primary defects: insulin resistance (diminished tissue sensitivity to insulin) and impaired β cell function (delayed or inadequate insulin release).

16.21.1 Effects of Vitamin D on Insulin Secretion

Calcitriol is considered to be a modulator of insulin secretion because vitamin D deficiency in rats is associated with marked impairment of insulin secretion (Chertow *et al.*, 1983). Insulin secretion is a process dependent on changes in intracellular calcium concentration. Experimentally diabetic rats have calcium deficiencies that can be corrected with calcitriol or insulin. The effects of vitamin D on β cells may be due to its regulation of extracellular calcium and calcium flux through the β cell. Vitamin D or calcium deficiency may alter the balance between intracellular and extracellular calcium in β cells, thereby interfering with insulin secretion and possibly synthesis (Danescu *et al.*, 2009).

Administration of calcitriol to vitamin D-deficient rats was shown to increase circulating insulin levels (Clark *et al.*, 1981). These authors proposed a positive feedback relationship between kidney proximal tubule cells (the site of 1α-hydroxylase action) and pancreatic β cells in which insulin and calcitriol act to maintain the release or production of the other. In β cells, calcitriol acts by increasing mitochondrial calcium levels through increased synthesis of calcium-binding protein, thereby increasing calcium required for insulin release. Insulin, in turn, sensitizes the proximal tubule cells to respond to parathyroid hormone (PTH), which activates the 1α-hydroxylase. The calcitriol produced has a direct action on intestine, bone, and kidney to increase blood calcium. This feedback is limited by the activity of the 1α-hydroxylase, which is inhibited by rising calcium, phosphorus, and calcitriol levels.

16.21.2 Type 1 Diabetes

Full activation of naïve T cells involves signaling through the T cell receptor and complementary signals initiated by the binding of B7 (CD80/CD86) to the CD28 receptor. Salomonen *et al.* (2000) showed that B7/CD28 co-stimulation is essential for the development and homeostasis of the $CD4^+CD25^+$ regulatory T cells that control spontaneous autoimmune diabetes in the non-obese diabetic (NOD) mouse.

Pancreatic β cells can be directly destroyed by macrophages and $CD8^+$ cytotoxic T cells, which are activated, respectively, by IFN-γ and IL-2 produced by $CD4^+$ Th1 cells (Zella & DeLuca, 2003). IL-12 administration induces Th1 cells and accelerates type 1 diabetes in NOD mice (Trembleau *et al.*, 1995). Since calcitriol inhibits the production of IFN-γ, IL-2, and IL-12 (Lemire, 1995), this hormone may be able to disrupt both the initiation and the progression of the Th1-mediated pathogenesis of type 1 diabetes. Short-term treatment of NOD mice with nonhypercalcemic doses of a calcitriol analog prevented the clinical onset of the disease (Gregori *et al.*, 2002). The treatment inhibited IL-12 production and pancreatic infiltration of Th1 cells, while increasing the frequency of nTregs in pancreatic lymph nodes. Mathieu *et al.* (1994) showed that calcitriol can prevent type 1 diabetes in NOD mice, probably through the correction of their defective suppressive mechanisms.

Zella *et al.* (2003) demonstrated that a vitamin D-sufficient status alone confers partial protection against the development of type 1 diabetes in the NOD mouse. Vitamin D deficiency resulted in an earlier onset and greater incidence of the disease in both male and female mice when compared to vitamin D-sufficient animals. These results support the observations in humans that vitamin D status may be an environmental factor in determining type 1 diabetes risk. VDR gene polymorphisms affect the onset pattern of type 1 diabetes (Motohashi *et al.*, 2003).

16.21.3 Type 2 Diabetes

Calcium (Ca^{2+}) is essential for insulin-mediated intracellular processes in insulin-responsive tissues, such as skeletal muscle and adipose tissue, with a very narrow concentration range needed for optimal insulin-mediated functions. Calcitriol acts indirectly to benefit insulin action by regulating extracellular calcium and ensuring normal calcium influx through cell membranes and an adequate pool of cytosolic calcium (Sergeev, 2016). Calcitriol also acts directly by stimulating the expression of insulin receptor on muscle cells and fat cells, thereby enhancing insulin responsiveness for glucose transport (Maestro *et al.*, 2000).

16.22 VITAMIN D AND CARDIOVASCULAR DISEASE

16.22.1 Vitamin D as a Negative Regulator of the Renin–Angiotensin System

In vitro studies (Sigmund, 2002) and experiments with VDR knockout mice have shown that vitamin D is a potent negative regulator of the renin–angiotensin system. VDR knockout mice had more than a threefold increase in renin mRNA expression and more than a 2.5-fold increase in plasma angiotensin II levels compared with wild-type mice (Li *et al.*, 2002). Augmentation of angiotensin II in the knockout mice led to the development of hypertension and cardiac hypertrophy. This study showed that vitamin D directly regulates renin gene expression independently of calcium status. Data reported by Xiang *et al.* (2005) suggested that the cardiac hypertrophy seen in VDR knockout mice is a consequence of both the systemic and cardiac renin–angiotensin system and supports the notion that calcitriol regulates cardiac functions, at least in part, through the renin–angiotensin system.

16.22.2 Relationship between Vitamin D Deficiency and Cardiovascular Disease

The expression of the VDR in vascular endothelial cells (Merke *et al.*, 1989) suggests potential roles of vitamin D in anti-thrombotic functions. Evidence that vitamin D is important for normal cardiovascular function was the observation by Weishaar and Simpson (1987a) that rats maintained on a low vitamin D_3 diet displayed heightened contractile responses of isolated cardiac and vascular smooth muscle, transient increases in systolic blood pressure, and elevation of serum creatine kinase (CK). Because most of the CK in the body normally exists in muscle, a rise in serum CK is indicative of muscle damage. This applies to cardiomyocytes, which release CK when they are damaged as a result of a heart attack. The changes in cardiac muscle contractility that accompanied vitamin D_3 deficiency could not be prevented by prevention of hypocalcemia, whereas the changes in vascular muscle contractility could be so prevented and also reversed by restoration of serum calcium to normal after an initial period of vitamin D_3 deficiency (Weishaar & Simpson, 1987b). This suggests that dietary vitamin D_3 has a direct effect on cardiac muscle contractility and an indirect effect on vascular smooth muscle contractility. In the latter case, the vascular muscle responds to the hypocalcemia that accompanies vitamin D_3 depletion. Aihara *et al.* (2004) compared VDR knockout mice with their wild-type littermates to demonstrate that VDR activation elicits antithrombotic effects, such as suppression of adenosine diphosphate-induced platelet aggregation and reduced multi-organ thrombus formation after lipopolysaccharide injection.

A number of observational studies in humans have linked vitamin D deficiency to cardiovascular disease (Nemerovski *et al.*, 2009). The first evidence that vitamin D deficiency could lead to human cardiovascular disease came from patients with end-stage renal disease (ESRD). The damaged kidneys fail to convert 25(OH)D to calcitriol and the lack of the active metabolite leads to secondary hyperparathyroidism and elevated levels of circulating PTH (Zitterman *et al.*, 2006). The excessive PTH levels cause an increase in blood pressure and cardiac contractility (Rostand & Drüeke, 1999). The sustained stress on the heart leads to cardiac hypertrophy, myocardial fibrosis, and heart failure. Gluba-Brzózka *et al.* (2018) reviewed the impact of vitamin D therapy on the cardiovascular system in patients with ESRD. In a meta-analysis of 20 observational studies, the administration of various vitamin D analogs was associated with lower mortality compared to no treatment. However, many studies indicated serious adverse effects of vitamin D therapy, including hypercalcemia, which would overwhelm protective effects.

16.23 VITAMIN D AND CANCER

16.23.1 Introduction

Independent studies have demonstrated that vitamin D_3 metabolites or analogs prevent the development of tumors in experimental models of colonic carcinogenesis (Pence & Buddingh, 1988; Belli *et al.*, 1992; Wali *et al.*, 1995). Calcitriol inhibits proliferation while inducing apoptosis and differentiation of cancer cells; however, the supraphysiological doses ($>10^{-9}$ M) of this hormone needed to elicit these effects cause hypercalcemia (Pols *et al.*, 1990). For treatment of cancer, analogs of vitamin D which have growth suppression activity but do not cause hypercalcemia have potential benefit.

This section discusses some of the mechanisms governing the anti-cancer effects of calcitriol and focuses mainly on prostate cancer.

16.23.2 Human Prostate Cancer Cell Line Models

Three commonly used human prostate cancer cell lines representing progressive transformation are LNCaP (least transformed), PC-3 (intermediate transformation), and DU-145 (most transformed). These epithelial cell lines were derived from different metastases of prostate carcinoma: LNCaP from a lymph node metastasis, PC-3 from a bone metastasis, and DU-145 from a brain metastasis. All three of these cancer cell lines contain the VDR, which mediates the anti-proliferative and differentiating actions of calcitriol. The levels of VDR are approximately two- to threefold higher in PC-3 cells than the amount in the LNCaP and DU-145 cells (Zhao *et al.*, 1997). LNCaP cells, but not PC-3 or DU-145 cells, also contain the androgen receptor (AR), which is required for calcitriol action. The more recently established cell lines MDAPCa 2a and MDAPCa 2b were, like PC-3 cells, derived from a bone metastasis and are models for advanced prostate cancer. Like LNCaP cells, MDA cells express both the VDR and AR.

The AR in LNCaP cells contains a single T877A mutation in the ligand-binding domain which alters ligand specificity. The mutated AR can be activated not only by androgens but also by progestins and estrogens as well as some anti-androgens. The anti-androgen Casodex does not activate the mutated AR, but acts as an androgen antagonist by inhibiting both dissociation of the heteromeric complex of the AR with heat-shock proteins and the subsequent high-affinity binding of the receptor to the nucleus (Veldscholte *et al.*, 1992a,b).

16.23.3 Control of the Cell Cycle

Calcitriol can inhibit cell proliferation by blocking the G_1 to S transition of the cell cycle and arresting cells in G_0/G_1. The growth arrest is due to a decrease in cyclin-dependent kinase 2 (CDK2) activity and to an increase in the expression of cyclin-dependent kinase inhibitors p21 and p27, which negatively regulate cell cycle progression. In addition, calcitriol can dephosphorylate the pocket protein retinoblastoma (Rb). This protein is a major gatekeeper of the restriction point at the end of the G_1 phase. When dephosphorylated, Rb binds and inactivates E2F transcription factors that are crucial for cell cycle progression (Ylikomi *et al.*, 2002).

16.23.4 Inhibition of Cancer Cell Proliferation

Many of the functions of the VDR are the result of the activity of a VDR–RXR heterodimer, in which the RXR partner is a specific receptor for 9-*cis*-retinoic acid. Blutt *et al.* (1997) examined the response of LNCaP human prostate cancer cells to calcitriol, 9-*cis*-retinoic acid, and combination treatments. The results showed that a combination of calcitriol and 9-*cis*-retinoic acid was more potent than individual treatments in inhibiting cell proliferation, presumably because these hormones bind to their respective receptors in the VDR–RXR heterodimer. The synthetic calcitriol analog EB1089 likewise synergizes with 9-*cis*-retinoic. These treatments caused LNCaP cells to accumulate in the quiescent (G_0) and post-mitotic interphase (G_1) of the cell cycle, suggesting that calcitriol can regulate one or more factors critical for cell entry to the DNA synthesis phase (S phase). EB1089 inhibited the proliferation of NCI-H929 myeloma cells by arresting the cells in G_1 (Park *et al.*, 2000). Regulation of cellular progression from G_1 to S phase is the step most frequently altered in cancers (Sherr, 1996).

Calcitriol concentrations of 10–100 nM are required to cause accumulation of LNCaP cells in the G_0/G_1 phase of the cell cycle; however, such concentrations will induce hypercalcemia and so cannot be used clinically. By contrast, EB1089 can produce comparable growth inhibitory responses at the much lower concentrations of 0.1^{-1} nM and without inducing hypercalcemia (Blutt *et al.*, 1997).

Consistent with accumulation of cells in G_1/G_0, prolonged calcitriol treatment of LNCaP cells resulted in decreased retinoblastoma (Rb) protein phosphorylation, repressed E2F transcriptional activity, increased protein levels of p21 by translational or post-translational regulation, and decreased CDK2 activity. This effect persisted for 72 hours after calcitriol removal (Zhuang & Burnstein, 1998). The following scenario can be envisaged. The increased production of p21 decreases the kinase activity of CDK2, so Rb is less phosphorylated. In the unphosphorylated state, Rb binds and inactivates E2F transcription factors that are crucial for cell cycle progression.

Different human prostate cancer cell lines exhibit variable sensitivity to calcitriol. In PC-3 cells stably expressing VDR at levels comparable to LNCaP cells, calcitriol treatment resulted in only moderate growth inhibition compared to LNCaP cells (Zhuang *et al.*, 1997). Thus, to attain maximal growth inhibition of prostate cancer cells, an adequate expression of the VDR is required as well as the induction of proteins such as p21 that stall the progression from G_1 to S phase of the cell cycle.

16.23.5 Induction of Apoptosis

Blutt *et al.* (2000) showed that calcitriol and EB1089 induce apoptosis of LNCaP cells and down-regulate anti-apoptotic Bcl-2 and Bcl-xL proteins. The pro-apoptotic proteins, Bax, Mcl-1, and Bcl-xS, were unaffected by calcitriol treatment. Overexpression of Bcl-2 blocked calcitriol-induced apoptosis and reduced, but did not eliminate, calcitriol-induced growth inhibition. The authors concluded that both regulation of cell cycle and the apoptotic pathways are involved in calcitriol action in prostate cancer cells. Park *et al.* (2000) reported similar findings in EB1089-treated NCI-H929 myeloma cells and showed that caspase-3 and p38 MAP kinase were activated during apoptosis. Chen *et al.* (2001) showed that calcitriol increases apoptosis in Caco-2 cells.

16.23.6 Inhibition of Angiogenesis

Angiogenesis, the formation of new blood vessels from an existing vascular bed, is a prerequisite for the entry of detached tumor cells into the circulation during metastasis. Majewski *et al.* (1996) compared the anti-angiogenic effects of retinoids (all-*trans*, 9-*cis*, and 13-*cis* retinoic acid), calcitriol, and IL-12, a cytokine with strong antitumor activity *in vivo* (Brunda *et al.*, 1993). Each of these compounds, when injected intradermally into immunosuppressed BALB/c mice, significantly decreased angiogenesis induced by human tumor cell lines. Angiogenesis inhibition was dose-dependent and comparable for all compounds tested. Combination of calcitriol and retinoids resulted in a synergistic inhibition of angiogenesis.

Mantell *et al.* (2000) evaluated the *in vitro* effect of calcitriol on bovine aortic endothelial cells treated with vascular endothelial growth factor (VEGF) to induce angiogenesis. The hormone markedly inhibited endothelial cell sprouting and morphogenesis and had a smaller inhibitory effect on proliferation. In tumor xenografts induced to overexpress VEGF, calcitrol inhibited angiogenesis at a dose that does not induce hypercalcemia in mice.

16.23.7 Actions of Vitamin D_3 Analogs on Human Prostate Cancer Cell Lines

Skowronski *et al.* (1995) compared the biological actions of calcitriol to those of other vitamin D metabolites and several synthetic analogs of vitamin D_3 on LNCaP and PC-3 human prostate cancer cells *in vitro*. The synthetic analogs tested were known to exhibit less hypercalcemic activity *in vivo* and included EB1089 (Figure 16.7).

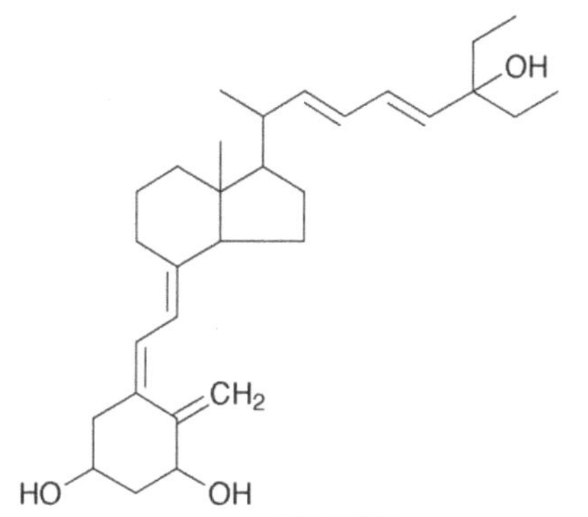

FIGURE 16.7 Structure of EB 1089.

The following properties were examined: (1) binding affinity for the VDR, (2) inhibition of cell proliferation, (3) stimulation of prostate-specific antigen (PSA) secretion by LNCaP cells, and (4) induction of $25(OH)D_3$-24*R*-hydroxylase mRNA. PSA direct a differentiation marker for prostatic epithelial cells and its enhanced secretion indicates that calcitriol may initiate differentiation and thus decrease cellular proliferation. The experiments are described in the following paragraphs.

The binding affinity of calcitriol and analogs for the VDR of LNCaP cells was assessed in competition experiments with [^{3}H]calcitriol. Calcitriol had the highest affinity of the natural metabolites, competing for 50% of the sites. $1\alpha,24,25(OH)_3D_3$ and $25(OH)D_3$ were about three- and sixfold less potent than calcitriol, respectively, whereas $24,25(OH)_2D_3$ had barely any affinity. EB 1089 was about twofold more potent than calcitriol as a competitor for the VDR.

In cell proliferation experiments, calcitriol had the greatest potency of the natural metabolites, inhibiting 50% growth of LNCaP cells at ~1 nM. The growth inhibition exhibited by the other metabolites was roughly proportional to their ability to compete for the VDR. EB1089 exhibited threefold greater anti-proliferative activity than calcitriol.

Treatment of LNCaP cells with calcitriol resulted in a dose-dependent stimulation of PSA, signifying an effect on cell differentiation. $1\alpha,24R,25(OH)_3D_3$ exhibited about sevenfold lower potency than calcitriol, whereas $25(OH)D_3$ and $24,25(OH)_2D_3$ were relatively inactive. EB1089 stimulated the production of PSA.

PC-3 cells respond to calcitriol by the induction of 24-hydroxylase mRNA. Of the other natural metabolites tested, only $1\alpha,24,25(OH)_3D_3$ was able to induce this enzyme. This metabolite and EB1089 had approximately equal potency to calcitriol.

To interpret these findings, it is important to state that the binding assay was performed in cytosolic extracts, whereas the functional response assays were performed with intact cells in a medium supplemented with 5% fetal bovine serum containing the vitamin D-binding protein (DBP). The binding affinity of EB1089 and other analogs to DBP was much less than that of calcitriol. Therefore, a higher proportion of the free active form of the analogs in the medium may partially explain their higher biological activity. Finally, the strong anti-proliferative effects of several synthetic analogs known to exhibit less calcemic activity than calcitriol suggest that these compounds may be useful as an additional therapeutic option for the treatment of prostate cancer.

16.23.8 Vitamin D and Androgen Interactions in Prostate Cancer

Androgens play a crucial role in the maintenance of the prostate gland and are involved in the development and progression of prostate cancer. The major prostatic androgen is 5α-dihydrotestosterone (DHT), whose hormonal action is mediated by the androgen receptor (AR). Patients with metastatic prostate cancer who have failed the primary therapy (surgical removal of the prostate or radiation therapy) receive drugs that block the production of androgens (androgen deprivation therapy, ADT). Isbarn *et al.* (2009) have critically weighed the benefits of ADT against its possible side effects. Although ADT controls tumor progression for perhaps 2–3 years, the cancer eventually becomes androgen independent (AIPC) with limited treatment options (Moreno *et al.*, 2005).

Calcitriol exerts anti-proliferative actions in several prostate cancer cell lines, with LNCaP exhibiting the greatest sensitivity (Miller *et al.*, 1995). LNCaP cells express the AR and respond to androgen stimulation. However, the response depends on the concentration of androgen added to the culture medium. At low concentrations (10^{-12}–10^{-10} M) of DHT, LNCaP cells underwent active proliferation, but produced very little PSA. At high concentrations (10^{-7} M), proliferation was arrested and production of PSA was increased (Lee *et al.*, 1995).

Zhao *et al.* (1997) investigated the possible interactions between calcitriol and DHT in the regulation of LNCaP cellular function. They used two different culture media: one (FBS medium) was supplemented with serum that contained endogenous androgens; the other (CSS medium) was supplemented with charcoal-stripped serum wherein endogenous androgens are depleted. LNCaP cell proliferation was dose-dependently inhibited by calcitriol when cells were cultured in FBS medium. Calcitriol-treated cells in the FBS medium showed a fivefold increase in PSA secretion, similar to the increase seen in DHT-treated cells. The anti-androgen Casodex completely blocked calcitriol anti-proliferative and PSA stimulation activities when cells were cultured in FBS medium. In combination, calcitriol and DHT synergistically enhanced PSA secretion 22-fold in FBS medium. This synergistic effect was even greater (up to 50-fold) when cells were cultured in CSS medium, but in this medium, calcitriol alone no longer inhibited cell proliferation or induced PSA secretion. The synergism between calcitriol and DHT appeared to be due, at least in part, to the up-regulation of AR by calcitriol. Levels of PSA in PC-3 and DU-145 cells, which lack the AR, did not change with exposure to calcitriol; these cells respond to calcitriol by the induction of 24-hydroxylase. In conclusion, the anti-proliferative and PSA stimulation activities of calcitriol in LNCaP cells are androgen dependent.

Because cellular responsiveness to androgen depends on AR abundance, Zhao *et al.* (1999) analyzed the ability of calcitriol and 9-*cis* retinoic acid to regulate the level of AR gene expression in LNCaP cells. Calcitriol and 9-*cis* retinoic acid were previously shown to act synergistically to inhibit the proliferation of LNCaP cells (Blutt *et al.*, 1997). Calcitriol up-regulated AR expression at both mRNA and protein levels. Induction of AR mRNA by calcitriol was completely abolished by cycloheximide, indicating that this induction required new protein synthesis. Similar to calcitriol, 9-*cis* retinoic acid also induced AR mRNA expression, and this effect of both hormones was additive. Calcitriol inhibited LNCaP cell proliferation 50%, whereas 9-*cis* retinoic acid only reduced proliferation by 10%. However, combination treatments with both hormones caused 80% growth inhibition. In the presence of Casodex, neither calcitriol nor 9-*cis* retinoic acid inhibited cell proliferation individually or in combination, showing a dependency on AR for the antiproliferative response of the two hormones. In contrast dibutyryl cAMP, although a well-known up-regulator of AR in LNCaP cells (Mizokami *et al.*, 1994), inhibited LNCaP cell proliferation by an AR-independent mechanism. These results showed that growth inhibition of LNCaP cells by calcitriol and 9-*cis* retinoic acid is mediated by AR-dependent mechanisms and is preceded by the induction of AR gene expression. Calcitriol inhibited the growth of the AR-negative PC-3 cell line, as well as primary cultures of human prostate cells, by mechanisms other than androgen signaling.

The data indicate that the calcitriol signaling pathway in LNCaP cells begins with the binding of calcitriol to the VDR and the formation of a heterodimer with RXR. The VDR–RXR heterodimer binds to the VDRE in the promoter region of target genes that encode protein mediators of the up-regulation of AR mRNA in response to calcitriol. The increased AR mRNA causes an increase in AR protein levels, which mediates the inhibitory action of androgens on LNCaP cell growth.

Zhao *et al.* (2000) examined the actions and interactions of calcitriol and DHT on the MDAPCa 2a and 2b cell lines and compared the results to those in LNCaP cells. Both MDA cell lines were shown to express high-affinity VDRs and low-affinity ARs with a binding capacity for DHT 50-fold lower than that of the AR in LNCaP cells. Calcitriol significantly up-regulated AR mRNA in all three cell lines. Calcitriol caused significant growth inhibition in MDA cells, greater (for 2b cells) or lesser (for 2a cells) than that in LNCaP cells. The growth inhibitory effect of calcitriol on LNCaP cells was blocked by Casodex, but this effect of calcitriol was not blocked in MDA cells. Thus, growth inhibition of MDA cells by calcitriol is androgen independent, while that of LNCaP cells is androgen dependent. Whereas calcitriol increased PSA secretion by MDA cells in FBS-containing medium, this effect by DHT was minimal. Zhao *et al.* (1997) previously showed that calcitriol increases PSA by AR signaling in LNCaP cells. It therefore seems likely that calcitriol regulates PSA in MDA cells by some other mechanism(s). For example, calcitriol may induce a more differentiated MDA phenotype that secretes more PSA per cell. The finding that calcitriol inhibits the growth of MDA cells independently of androgens suggests that calcitriol may have clinical usefulness in the treatment of advanced prostate cancer after its progression into an androgen-independent state.

16.23.9 Role of Insulin-Like Growth Factor Binding Protein-3 in Calcitriol-Induced Growth Inhibition of Human Prostate Cancer Cells

Treatment with either calcitriol (Sprenger *et al.*, 2001) or the EB1089 analog (Nickerson & Huynh, 1999) induces IGFBP-3 expression in cultures of normal prostate epithelial cells and rat ventral prostate, respectively. Stewart and Weigel (2005) showed that calcitriol decreases IGFBP-2 mRNA levels and increases IGFBP-3 mRNA in LNCaP, PC-3, and C4-2 human prostate cancer cell lines. Calcitriol also produced a time-dependent increase in the levels of secreted IGFBP-3 in LNCaP, C4-2, and LAP-4 cells, but not in calcitriol-resistant DU-145 cells.

Boyle *et al.* (2001) produced strong evidence that calcitriol up-regulates IGFBP-3 in LNCaP cells and that IGFBP-3 is a crucial mediator of the anti-proliferative action of calcitriol in these cells. Significant up-regulation of IGFBP-3 at the mRNA and protein levels was observed at calcitriol concentrations as low as 1 nM, reaching a maximal plateau of threefold over control levels at higher doses. Treatment of LNCaP cells with IGFBP-3 caused cell growth inhibition comparable to treatment with calcitriol. Blocking IGFBP-3 activity with immuno-neutralizing antibody or blocking its production with antisense oligonucleotide abolished the growth inhibitory action of calcitriol, indicating that this effect depends on the presence of IGFBP-3 in LNCaP cells. In the IGFBP-3 antibody and antisense experiments, a growth medium (OPTI-MEM-GM) was used which supports LNCaP cell growth in the absence of insulin-like growth factors and insulin-like growth factor binding proteins. Calcitriol and IGFBP-3 independently up-regulated the cell cycle inhibitory protein p21 in LNCaP cells. It has been established that the induction of p21 is a mechanism by which calcitriol acts to inhibit cell growth (M. Liu *et al.*, 1996). Blocking IGFBP-3 activity completely prevented the calcitriol-induced up-regulation of p21, demonstrating a dependency on IGFBP-3 for calcitriol-induced growth inhibition in LNCaP cells through p21. Peng *et al.* (2004) identified a functional DR3-type VDRE in the IGFBP-3 gene promoter in LNCaP cells and demonstrated that calcitriol is able to directly activate IGFBP-3 expression at the transcriptional level through binding of the hormone-bound VDR–RXR heterodimer to the VDRE. The experiments reported by Boyle *et al.* (2001) were performed in serum-free media conditions that require cells to rely on endogenous production of growth factors. Stewart and Weigel (2005) showed that calcitriol-induced up-regulation of IGFBP-3 was not required for the growth inhibitory effects of calcitriol in prostate cancer cells grown in the presence of serum, which contains steroids and several growth factors.

16.23.10 Sensitization of Colon Cancer Cells to TGF-β1 Growth Inhibitory Effects by Calcitriol

Signaling pathways activated by calcitriol and TGF-β have synergistic inhibitory effects on cell growth. TGF-β is secreted by cells as an inactive precursor called latent TGF-β, in which TGF-β is non-covalently bound to a peptide–protein complex that prevents binding of the cytokine to its receptor. Activation of latent TGF-β is effected by the type II insulin-like growth factor receptor (IGF-IIR), which has two distinct binding sites (MacDonald *et al.*, 1988). One site binds to IGF-II, leading to its degradation, while the second site binds a variety of proteins bearing mannose-6-phosphate residues, including latent TGF-β. Although the growth of premalignant human colonic cells was inhibited by TGF-β, several colon cancer cell lines grown under identical conditions were resistant to TGF-β (Manning *et al.*, 1991). Chen *et al.* (2002) found that calcitriol increased the abundance of active TGF-β1 by stimulating the expression of IGF-IIR in Caco-2 cells. The increased active TGF-β1 mediated, at least in part, the calcitriol-induced inhibition of cell growth. Although TGF-β1 alone had no effect on growth of Caco-2 cells, the combination of calcitriol and TGF-β1 caused significantly more growth inhibition than calcitriol alone. Thus calcitriol sensitized these cells to the growth inhibitory effects of TGF-β1. This sensitization was shown to be independent of the IGF-IIR. Calcitriol increased the expression of the type I (but not type II) TGF-β1 receptor and this receptor was shown to contribute to the calcitriol-induced inhibition of cell growth.

16.23.11 TGF-β1 as an Autocrine Mediator of U937 Leukemia Cell Growth Arrest and Differentiation Induced by Calcitriol and Retinoids

Calcitriol and 9-*cis* retinoic acid act synergistically in combination to inhibit the proliferation and induce the differentiation of U937 cells in cases where each agent alone is relatively ineffective (Marti *et al.*, 1997). U937 cells are human promonocytic leukemia cells that differentiate into cells with macrophage characteristics. At first glance, this synergy seems easy to explain: conversely to all-*trans* retinoic acid, which binds and activates only RAR, 9-*cis*-retinoic acid is an activating ligand for both RXR and RAR. Moreover, RXR forms functional heterodimers with RAR or VDR. Therefore, by associating 9-*cis* retinoic acid and calcitriol, one could expect to induce a higher transcriptional activity of both RXR–RAR and RXR–VDR heterodimers, owing to the occupancy of all possible binding sites. However, many data suggest that the actual mechanism could be more complex.

It is well established that TGF-β induces G_1 cell cycle arrest by regulating the activity of the G_1 cyclin-CDK complexes (Polyak, 1996). TGF-β1 also dose-dependently induces differentiation of U937 promonocytes into macrophages (Kamijo *et al.*, 1990). Defacque *et al.* (1999) showed that TGF-β1 is an endogenous mediator of the synergistic effects of calcitriol and retinoids on U937 cell growth inhibition and differentiation. Latent TGF-β1 accumulated in cultures of U937 cells treated with a combination of all-*trans* retinol and calcitriol. Calcitriol, all-*trans*-retinoic acid and recombinant TGF-β1 alone had only a slight effect on cell differentiation, as assessed by phagocytic activity. A combination of TGF-β1 and calcitriol induced a high level of phagocytic activity, similar to that induced by a combination of all-*trans*-retinoic acid and calcitriol. Addition of TGF-β to the latter combination did not increase phagocytic activity. TGF-β1 alone did not affect proliferation, but arrested growth when combined with calcitriol. A combination of all-*trans*-retinoic acid and calcitriol also arrested growth and addition of TGF-β1 enhanced this effect. Neutralizing antibodies and antisense oligonucleotide were used to inhibit the effects of TGF-β1. Cell growth inhibition induced by calcitriol plus all-*trans*-retinoic acid was completely reversed, and differentiation was reduced by about 75% when both inhibitors were associated.

16.23.12 Regulation of Prostaglandin Metabolism

Prostaglandin (PG) synthesis begins with the intracellular release of arachidonic acid from the plasma membrane via the action of phospholipase A_2. Cyclooxygenase-2 (COX-2) is the rate-limiting enzyme that catalyzes the conversion of arachidonic acid to PGs; 15-prostaglandin dehydrogenase (15-PGDH) is involved in PG inactivation. Increased expression of COX-2 promotes angiogenesis (Fujita *et al.*, 2002) and invasion (Nithipatikom *et al.*, 2002) of prostate cancer cells. Calcitriol down-regulates the gene encoding COX-2 and up-regulates the gene encoding 15-PGDH, thereby decreasing the PG-mediated proliferative stimulus (Moreno *et al.*, 2005).

16.23.13 Induction of Mitogen-Activated Protein Kinase Phosphatase 5 (MKP5) in Prostate Cancer

Inflammation is a critical component of tumor progression (Coussens & Werb, 2002) and chronic or recurrent prostate inflammation can initiate and promote prostate cancer development (Nelson *et al.*, 2004). Elevated serum levels of the pro-inflammatory cytokines TNF-α and IL-6 are associated with aggressive pathology and decreased survival of prostate cancer patients (Michalaki *et al.*, 2004). TNF-α released by inflammatory cells binds cell-surface receptors, which signal through multiple pathways, including p38 MAP kinase, to increase production of IL-6 and other cytokines (Brinkman *et al.*, 1999). p38 is also activated by various types of stress, such as heat shock, osmotic shock, UV irradiation, and hypoxia (Ono & Han, 2000). The primary mechanism for increased IL-6 production following p38 activation is stabilization of IL-6 mRNA and consequent increased translation. The 3′-untranslated region of IL-6 mRNA contains AU-rich elements (AREs) composed of repeats of the AUUUA motif, which are known to mediate RNA degradation. Signaling through p38 counteracts ARE-based mRNA degradation via its downstream effector molecule MAP kinase-activated protein kinase 2 (MAPKAPK-2) (Winzen *et al.*, 1999). Mitogen-activated protein kinase phosphatase 5 (MKP5), also known as dual-specificity phosphatase 10 (DUSP10), dephosphorylates and inactivates p38, thus destabilizing IL-6 mRNA and reducing IL-6 production (Theodosiou *et al.*, 1999). Nonn *et al.* (2006) showed that treatment of cultured prostatic epithelial cells with calcitriol up-regulates MKP5, thus inhibiting p38 kinase activity and deceasing production of pro-carcinogenic inflammatory cytokines like IL-6. The observed dependency of MKP5 up-regulation on the VDR and the presence of a VDRE in the MKP5 gene promoter region suggests that MKP5 is directly regulated by calcitriol at the transcriptional level. This study has implications for prostate cancer prevention by vitamin D.

16.23.14 Induction of E-Cadherin and Inhibition of B-Catenin Signaling in Colon Cancer Cells

E-cadherin is a transmembrane linker protein of the zonula adherens that forms part of the junctional complex joining a cell with its neighbor. β-catenin binds directly to the cytoplasmic tail of E-cadherin and indirectly to actin filaments of the cytoskeleton. Loss of E-cadherin is a common event during the transition from adenoma to carcinoma. When cell–cell adhesion is decreased, cells can become invasive. β-Catenin not only mediates cytoskeletal attachment, but is the central player in the Wnt signaling pathway. In the absence of a Wnt signal, β-catenin is phosphorylated by a complex containing glycogen synthase kinase (GSK). Phosphorylated β-catenin is then ubiquitinated and degraded in proteosomes. In the presence of Wnt, interaction of Wnt ligands with their membrane receptors blocks GSK, leading to the accumulation of unphosphorylated β-catenin in the cytosol. After translocation to the nucleus, β-catenin associates with T cell transcription factor (TCF) to activate genes involved in the control of cell proliferation and invasiveness, such as c-*myc*, c-*jun*, cyclin D1, and peroxisome proliferator-activated receptor (PPAR). β-Catenin also controls the regulated export of TCF from the nucleus.

In colon tumors, the regulated nuclear export of β-catenin is reduced and the consequent accumulation of β-catenin in the nucleus leads to persistent activation of the target genes. Pálmer *et al.* (2001) used the SW480 human colorectal cancer cell line to examine the mechanisms of action of calcitriol and several nonhypercalcemic analogs in colon cancer cells. These compounds (1) induced the nuclear export of β-catenin and its relocalization to the plasma membrane; (2) promoted a direct VDR–β-catenin interaction, thereby decreasing TCF–β-catenin complexes by competition; and (3) induced expression of E-cadherin, causing cancer cells to become more adhesive and hence less invasive. These results suggest that liganded VDR may hinder the loss of differentiation and the increase in proliferation that occurs at the early stages of carcinogenesis.

16.23.15 Induction of BRCA1

About 10% of cases of human breast cancer are associated with inherited defects in two breast cancer susceptibility genes, *BRCA1* and *BRCA2*. The encoded proteins interact with a wide range of other proteins involved in transcription, chromosome maintenance, DNA repair, and control of the cell cycle. BRCA1 is a nuclear tumor suppressor protein that helps suppress human breast cancer. Mutations in these genes can predispose to errors in DNA replication and cancers of the breast, ovaries, and prostate gland (Somasundarum, 2003).

Campbell *et al.* (2000) presented evidence that calcitriol, acting through the VDR, has an anti-proliferative effect on breast and prostate cancer cell lines and this effect is associated with induced expression of BRCA1 mRNA and protein. Estrogens, acting through the estrogen receptor (ER), also mediated the induction of BRCA1 mRNA and protein. Cross-talk between VDR and ER signaling pathways may reflect the binding of transcription factors that are induced by both calcitriol and estrogens to the promoter of the BRCA1 gene. In the most aggressive breast and prostate cancer cell lines, neither calcitriol nor estrogen was able to transactivate the BRCA1 promoter.

16.24 SYSTEMIC LUPUS ERYTHEMATOSUS

Systemic lupus erythematosus (referred to here as lupus) is an autoimmune disease in which the immune system mistakenly attacks healthy tissues in many parts of the body. B cells from lupus patients have an intrinsic tendency to overreact to immunological stimulation during antigenic challenge (Grammer & Lipsky, 2003). The various symptoms of lupus include vasculitis, kidney diseases, arthritis, and central nervous system abnormalities (Lisnevskaia *et al.*, 2014). The disease is visible as a 'butterfly' skin rash over the cheeks and nose. The overactive B cells produce excessive amounts of autoantibodies, especially those against nuclear components such as double-stranded DNA. Lupus is strongly associated with defects in the clearance of apoptotic cells and the accumulation of apoptotic debris. Failure of phagocytes to remove apoptotic material leads to nuclear fragments being captured by antigen-presenting cells – a situation that can lead to the development of antinuclear antibodies. Persons with genetic defects in early components of the classical pathway of complement activation tend to develop lupus. The first component, Clq, binds to apoptotic cells and aids their clearance, therefore a deficiency of Clq could at least partly explain why lupus patients fail to clear apoptotic cells efficiently. Interestingly, decreased calcitriol serum concentrations have been reported in patients with lupus (Huisman *et al.*, 2001).

Given that B cell hyperactivity is critical in the pathogenesis of lupus, it seems rational to direct therapies toward the depletion of B cells. Looney *et al.* (2004) demonstrated the effectiveness of therapeutic B cell depletion in lupus patients using rituximab, which is a chimeric monoclonal antibody specific for human CD20 (a B cell marker). Rituximab is known to deplete B cells by antibody-dependent cell-mediated cytotoxicity, complement-mediated cell lysis, and/or apoptosis (Reff *et al.*, 1994). CD20 is lost from B cells during their terminal differentiation into plasma cells. Thus, pathogenic plasma cells may not be directly removed and may continue to produce antibodies (Gorman *et al.*, 2004).

Chen *et al.* (2007) examined the effects of calcitriol on human B cells in culture to test their hypothesis that calcitriol is involved in maintaining B cell homeostasis. The hormone did not affect the survival of unstimulated cells, whereas it inhibited the ongoing proliferation of stimulated cells and induced their apoptosis after a while. The generation of antibody-secreting plasma cells and class-switched memory B cells was significantly inhibited by calcitriol. The hormone decreased mRNA levels of cyclin D, CDK4, and CDK6 which, in combination, facilitate progress through the G_1 phase of the cell cycle. It also increased the expression of p27, the protein that inhibits cell cycle progression at the G_1/S transition; neither p21 nor p18 expression were affected. Chen *et al.* (2007) also found that lupus patients had significantly diminished levels of 25(OH)D and calcitriol, and the latter was related to disease activity and production of antinuclear antibodies. These results show that calcitriol is indeed involved in maintaining B cell homeostasis and that deficiency of calcitriol contributes to B cell hyperactivity in lupus. Thus correction of vitamin D deficiency may be a useful alternative to B cell depletion for the treatment of lupus and other B cell-mediated autoimmune disorders.

16.25 THERAPEUTIC APPLICATIONS OF VITAMIN D ANALOGS

Therapy with vitamin D itself is severely limited because of its hypercalcemic effects at supraphysiological doses. Vitamin D analogs that are efficacious but do not induce hypercalcemia have potential therapeutic use (Nagpal *et al.*, 2001). Vitamin D analogs that have an additional side chain, known as Gemini analogs, have been found to be 100 to 1,000 times more potent than calcitriol in their *in vitro* anti-proliferative activity. Studies in mice suggest that they may be useful in the treatment of some cancers, including colon cancer (Spina *et al.*, 2005).

There have been several trials to assess the efficacy of vitamin D compounds in the treatment of postmenopausal osteoporosis. The most critical parameter for successful treatment, a decrease in fracture rate, was observed in some, but not all, studies (Brown *et al.*, 2000). In one Japanese study (Shiraki *et al.*, 1996), new fracture occurrence in the group treated with 1α-hydroxyvitamin D_3 was around one-third of that in the placebo group. The ideal vitamin D analog would be one which promotes bone formation and slow resorption, yet has less tendency than calcitriol to produce hypercalcemia.

When cultured human epidermal keratinocytes are exposed to physiological concentrations of calcitriol, the cells cease to proliferate and start to differentiate (Smith *et al.*, 1986). The inhibition of proliferation has been utilized in the treatment of hyperproliferative diseases of the skin. Psoriasis, for example, can be effectively treated by topical application of the vitamin D analog calcipotriol, which is about 200 times less potent than calcitriol in its effects on calcium metabolism, although similar in receptor affinity (Kragballe, 1992).

16.26 TOXICITY

An excessive chronic intake of vitamin D can result in toxicity with a fatal outcome. As in vitamin A toxicity, hypervitaminosis D results from the excessive consumption of vitamin D supplements, and not from the consumption of usual diets. Toxic concentrations of vitamin D have not resulted from unlimited exposure to sunshine. Vitamin D intoxication can be a concern in patients with specific diseases being treated with unusual amounts of vitamin D or analogs of the vitamin. Several reviews conclude that vitamin D toxicity is not seen before serum 25(OH)D exceeds 500–750 nmol/l (Vieth, 2006; Jones, 2008; Heaney, 2008).

In Great Britain, during the 1940s and early 1950s, an epidemic of 'idiopathic hypercalcemia' broke out in newborn infants, who failed to thrive and exhibited symptoms of toxicity. This epidemic was eventually traced to over-supplementation of commercial infant milk formulas with vitamin D. The government policy was to supplement milk with up to 2,000 IU (50 μg) of vitamin D to compensate for nutritional deprivation that British children had suffered during World War II. To allow for anticipated degradation of vitamin D during processing and storage, some manufacturers put 1.5 to 2 times the correct amount of vitamin D into the preprocessed milk.

Vitamin D toxicity is due primarily to the hypercalcemia caused by the increased intestinal absorption of calcium, together with increased resorption of bone. The cause of the hypercalcemia is therefore a drastic exaggeration of the normal physiological action of vitamin D. The increased serum calcium level can lead to a variety of non-specific symptoms, such as anorexia, nausea, vomiting, muscle weakness, and constipation. Polyuria and polydipsia result from the failure of the kidney to concentrate the urine. The hypercalciuria that accompanies hypercalcemia encourages the formation of kidney stones in the renal tubules. Chronic hypercalcemia results in irreversible calcification of the kidneys (nephrocalcinosis), causing permanent damage to the glomeruli and renal tubules. Calcium salts may be deposited in other extraskeletal tissues as well, such as the heart, blood vessels, and lungs. The renal damage results in a decrease in the glomerular filtration rate and severe hypertension. In long-term hypervitaminosis D, the excessive bone resorption results in part of the bone being replaced by fibrous tissue.

High amounts of 25(OH)D can promote calcium translocation in intestine and bone *in vitro*, suggesting that overwhelming concentrations of 25(OH)D can displace calcitriol from the VDR and directly elicit the biological responses. Brumbaugh and Haussler (1973) predicted from their data that 25(OH)D_3 must be present in 150 times the concentration of calcitriol to displace the physiological hormone. Hypervitaminosis D patients typically exhibit a 15-fold increase in plasma 25(OH)D concentrations compared to normal individuals, but their calcitriol levels are not substantially altered (Hughes *et al.*, 1976). These observations have led to the general conclusion that 25(OH)D, rather than calcitriol, is responsible for vitamin D toxicity.

An alternative hypothesis, presented by Vieth (1990), is that calcitriol is, in fact, the agent causing toxicity. This hypothesis is based on the differential binding affinities of the various vitamin D metabolites for the DBP in the plasma. The 25(OH)D metabolite binds much more tightly to the DBP than does calcitriol. Therefore, when the plasma concentration of 25(OH)D increases manyfold, a certain fraction of the circulating calcitriol will be displaced from the DBP by 25(OH)D, thereby increasing the concentration of free calcitriol. The liberated hormone is now able to interact with a greater than normal number of VDRs in target cells and elicit an exaggerated response.

Hypercalcemia resulting from excessive intake of the parent vitamin D can persist for weeks or months after intake has ceased, because of the accumulation of this vitamin in adipose tissue and its gradual release into the circulation. Treatment must therefore be continued for a long time to counteract the hypercalcemic response. The treatment includes drugs to enhance urinary excretion of calcium and drugs to diminish the calcium efflux from bone and absorption of calcium from the intestine.

FURTHER READING

Anderson, P.H. & Atkins, G.J. (2008) The skeleton as an intracrine organ for vitamin D metabolism. *Molecular Aspects of Medicine*, **29**, 397–406.

Christakos, S., Raval-Pandya, M., Wernyj, R.P. & Yang, W. (1996) Genomic mechanisms involved in the pleiotropic actions of 1,25-dihydroxyvitamin D_3. *Biochemical Journal*, **316**, 361–71.

Haussler, M.R., Whitfield, G.K., Haussler, C.A., Hsieh, J.-C., Thompson, P.D., Selznick, S.H., Dominguez, C.E. & Jurutka, P.W. (1998) The nuclear vitamin D receptor: Biological and molecular regulatory properties revealed. *Journal of Bone and Mineral Research*, **13**, 325–49.

Hayes, C.E., Nashold, F.E., Spach, K.M. & Pedersen, L.B. (2003) The immunological functions of the vitamin D endocrine system. *Cellular and Molecular Biology*, **49**, 277–300.

Vanherwegen, A.-S., Gysemans, C. & Mathieu, C. (2017) Regulation of immune function by vitamin D and its use in diseases of immunity. *Endocrinology & Metabolism Clinics of North America*, **46**, 1061–94.

van Leeuwen, J.P.T.M., van Driel, M., van den Bemd, G.J.C.M. & Pols, H.A.P. (2001) Vitamin D control of osteoblast function and bone extracellular matrix mineralization. *Critical Reviews in Eukaryotic Gene Expression*, **11**, 199–226.

Ylikomi, T., Laaksi, I., Lou, Y.-R., Martikainen, P., Miettinen, S., Pennanen, P., Purmonen, S., Syvälä, H., Vienonen, A. & Tuohimaa, P. (2002) Antiproliferative action of vitamin D. *Vitamins and Hormones*, **64**, 357–406.

Zittermann, A., Schleifthoff, S.S. & Koerfer, R. (2005) Putting cardiovascular disease and vitamin D insufficiency into perspective. *British Journal of Nutrition*, **94**, 483–92.

REFERENCES

Abe, E., Shiina, Y., Miyaura, C., Tanaka, H., Hayashi, T., Kanegasaki, S., Saito, M., Nishii, Y., DeLuca, H.F. & Suda, T. (1984) Activation and fusion induced by 1α,25-dihydroxyvitamin D_3 and their relation in alveolar macrophages. *Proceedings of the National Academy of Sciences of the U.S.A.*, **81**, 7112–6.

Adams, J.S., Clemens, T.L., Parrish, J.A. & Holick, M.F. (1982) Vitamin-D synthesis and metabolism after ultraviolet irradiation of normal and vitamin D-deficient subjects. *New England Journal of Medicine*, **306**, 722–5.

ADHR Consortium. (2000) Autosomal dominant hypophosphataemic rickets is associated with mutations in *FGF23*. *Nature Genetics*, **26**, 345–8.

Adorini, L., Penna, G., Giarratana, N. & Uskokovic, M. (2003) Tolerogenic dendritic cells induced by vitamin D receptor ligands enhance regulatiory T cells inhibiting allograft rejection and autoimmune diseases. *Journal of Cellular Biochemistry*, **88**, 227–33.

Aihara, K., Azuma, H., Akaike, M., Ikeda, Y., Yamashita, M., Sudo, T., Hayashi, H., Yamada, Y., Endoh, F., Fujimura, M., Yoshida, T., Yamaguchi, H., Hashizume, S., Kato, M., Yoshimura, K., Yamamoto, Y., Kato, S. & Matsumoto, T. (2004) Disruption of nuclear vitamin D receptor gene causes enhanced thrombogenicity in mice. *Journal of Biological Chemistry*, **279**(34), 35798–802.

Alevizaki, C.C., Ikkos, D.G. & Singhelakis, P. (1973) Progressive decrease of true intestinal calcium absorption with age in normal man. *Journal of Nuclear Medicine*, **14**, 760–2.

Alroy, I., Towers, T.L. & Freedman, L.P. (1995) Transcriptional repression of the interleukin-2 gene by vitamin D_3: Direct inhibition of NFATp/AP-1 complex formation by a nuclear hormone receptor. *Molecular and Cellular Biology*, **15**, 5789–99.

Anderson, P.H., O'Loughlin, P.D., May, B.K. & Morris, H.A. (2005) Modulation of CYP27B1 and CYP24 mRNA expression in bone is independent of circulating $1,25(OH)_2D_3$ levels. *Bone*, **36**, 654–62.

Anderson, P.H., Sawyer, R.K., May, B.K., O'Loughlin, P.D. & Morris, H.A. (2007) 25-Hydroxyvitamin D requirement for maintaining skeletal health utilising a Sprague-Dawley rat model. *Journal of Steroid Biochemistry and Molecular Biology*, **103**, 592–5.

Anderson, P.H., Sawyer, R.K., Moore, A.J., May, B.K., O'Loughlin, P.D. & Morris, H.A. (2008) Vitamin D depletion induces RANKL-mediated osteoclastogenesis and bone loss in a rodent model. *Journal of Bone and Mineral Research*, **23**, 1789–97.

Armbrecht, H.J., Zenser, T.V., Bruns, M.E.H. & Davis, B.B. (1979) Effect of age on intestinal calcium absorption and adaptation to dietary calcium. *American Journal of Physiology*, **236**, E769–74.

Armbrecht, H.J., Zenser, T.V. & Davis, B.B. (1980) Effect of vitamin D metabolites on intestinal calcium absorption and calcium-binding protein in young and adult rats. *Endocrinology*, **106**, 469–75.

Armbrecht, H.J., Boltz, M.A. & Wongsurawat, N. (1992) Expression of plasma membrane calcium pump mRNA in rat intestine — Effect of age and 1,25-dihydroxyvitamin D. *Journal of Bone and Mineral Research*, **7**, S166.

Armbrecht, H.J., Chen, M.L., Hodam, T.L. & Boltz, M.A. (1997) Induction of 24-hydroxylase cytochrome P450 mRNA by 1,25-dihydroxyvitamin D and phorbol esters in normal rat kidney (NRK-52E) cells. *Journal of Endocinology*, **153**, 199–205.

Armbrecht, H.J., Hodam, T.L., Boltz, M.A., Partridge, N.C., Brown, A.J. & Kumar, V.B. (1998) Induction of the vitamin D 24-hydroxylase (CYP24) by 1,25-dihydroxyvitamin D_3 is regulated by parathyroid hormone in UMR106 osteoblastic cells. *Endocrinology*, **139**, 3375–81.

Armbrecht, H.J., Boltz, M.A. & Kumar, V.B. (1999) Intestinal plasma membrane calcium pump protein and its induction by $1,25(OH)_2D_3$ decrease with age. *American Journal of Physiology*, **277**, G41–7.

Arnaud, S.B., Goldsmith, R.S., Lambert, P.W. & Go, V.L.W. (1975) 25-Hydroxyvitamin D_3: Evidence of an enterohepatic circulation in man. *Proceedings of the Society for Experimental Biology and Medicine*, **149**, 570–2.

Arriagada, G., Paredes, R., Olate, J., van Wijnen, A., Lian, J.B., Stein, G.S., Stein, J.L., Onate, S. & Montecino, M. (2007) Phosphorylation at serine 208 of the 1α,25-dihydroxy vitamin D_3 receptor modulates the interaction with transcriptional coactivators. *Journal of Steroid Biochemistry & Molecular Biology*, **103**, 425–9.

Atkins, G.J., Andersoin, P.H., Findlay, D.M., Welldon, K.J., Vincent, C., Zannettino, A.C.W., O'Loughlin, P.D. & Morris, H.A. (2007) Metabolism of vitamin D_3 in human osteoblasts; Evidence for autocrine and paracrine activities of 1α,25-dihydroxyvitamin D_3. *Bone*, **40**, 1517–28.

Baran, D.T. & Sorensen, A.M. (1994) Rapid actions of 1α-25-dihydroxyvitamin D_3 physiologic role. *Proceedings of the Society of Experimental Biology and Medicine*, **207**, 175–9.

Baran, D.T., Sorensen, A.M. & Honeyman, T.W. (1988) Rapid action of 1,25-dihydroxyvitamin D_3 on hepatocyte phospholipids. *Journal of Bone and Mineral Research*, **3**, 593–600.

Baran, D.T., Sorensen, A.M., Shalhoub, V., Owen, T., Oberdorf, A., Stein, G. & Lian, J. (1991) 1α,25-Dihydroxyvitamin D_3 rapidly increases cytosolic calcium in clonal rat osteosarcoma cells lacking the vitamin D receptor. *Journal of Bone and Mineral Research*, **6**, 1269–75.

Baran, D.T., Sorensen, A.M., Shalhoub, V., Owen, T., Stein, G. & Lian, J. (1992) The rapid nongenomic actions of 1α,25-dihydroxyvitamin D_3 modulate the hormone-induced increments in osteocalcin gene transcription in osteoblast-like cells. *Journal of Cellular Biochemistry*, **50**, 124–9.

Baran, D.T., Quail, J.M., Ray, R., Leszyk, J. & Honeyman, T. (2000) Annexin II is the membrane receptor that mediates the rapid actions of 1α,25-dihydroxyvitamin D_3. *Journal of Cellular Biochemistry*, **78**, 34–46.

Bar-Shavit, Z., Noff, D., Edelstein, S., Meyer, M., Shibolet, S. & Goldman, R. (1981) 1,25-Dihydroxyvitamin D_3 and the regulation of macrophage function. *Calcified Tissue International*, **33**, 673–6.

Bedo, G., Sanstisteban, P. & Aranda, A. (1989) Retinoic acid regulates growth hormone gene expression. *Nature*, **339**, 231–4.

Belleli, A., Shany, S., Levy, J., Guberman, R. & Lamprecht, S.A. (1992) Aprotective role of 1–25-dihydroxyvitamin D_3 in chemically induced rat colonic carcinogenesis. *Carcinogenesis*, **13**, 2293–8.

Bergwitz, C. & Jüppner, H. (2010) Regulation of phosphate homeostasis by PTH, vitamin D, and FGF23. *Annual Review of Medicine*, **61**, 91–104.

Bhalla, A.K., Amento, E.P., Clemens, T.L., Holick, M.F. & Krane, S.M. (1983) Specific high-affinity receptors for 1,25-dihydroxyvitamin D_3 in human peripheral blood mononuclear cells: Presence in monocytes and induction in T lymphocytes following activation. *Journal of Clinical Endocrinology and Metabolism*, **57**, 1308–10.

Bhalla, A.K., Amento, E.P. & Krane, S.M. (1986) Differential effects of 1,25-dihydroxyvitamin D_3 on human lymphocytes and monocyte/macrophages: Inhibition of interleukin-2 and augmentation of interleukin-1 production. *Cellular Immunology*, **98**, 311–22.

Bischoff-Ferrari, H.A., Giovannucci, E., Willett, W.C., Dietrich, T. & Dawson-Hughes, B. (2006) Estimation of optimal serum concentrations of 25-hydroxyvitamin D for multiple health outcomes. *American Journal of Clinical Nutrition*, **84**, 18–28.

Bischoff-Ferrari, H.A., Dawson-Hughes, B., Baron, J.A., Burckhardt, P., Li, R., Spiegelman, D., Specker, B., Orav, J.E., Wong, J.B., Staehelin, H.B., O'Reilly, E., Kiel, D.P. & Willett, W.C. (2007) Calcium intake and hip fracture risk in men and women: A meta-analysis of prospective cohort studies and randomized controlled trials. *American Journal of Clinical Nutrition*, **86**, 1780–90.

Bischoff-Ferrari, H.A., Kiel, D.P., Dawson-Hughes, B., Orav, J.E., Li, R., Spiegelman, D., Dietrich, T. & Willett, W.C. (2009) Dietary calcium and serum 25-hydroxyvitamin D status in relation to BMD among U.S. adults. *Journal of Bone and Mineral Research*, **24**, 935–42.

Bischoff-Ferrari, H.A., Willett, W.C., Oray, E.J., Lips, P., Meunier, P.J., Lyons, R.A., Flicker, L., Wark, J., Jackson, R.D., Cauley, J.A., Meyer, H.E., Pfeifer, M., Sanders, K.M., Stähelin, H.B., Theiler, R. & Dawson-Hughes, B. (2012) A pooled analysis of vitamin D dose requirements for fracture prevention. *New England Journal of Medicine*, **367**, 40–9.

Bissonnette, M., Tien, X.-Y., Niedziela, S.M., Hartmann, S.C., Frawley, B.P. Jr., Roy, H.K., Sitrin, M.D., Perlman, R.L. & Brasitus, T.A. (1994) $1,25(OH)_2$ vitamin D_3 activates PKC-α in Caco-2 cells: A mechanism to limit secosteroid-induced rise in $[Ca^{2+}]$. *American Journal of Physiology*, **267**, G465–75.

Blanco, J.C.G., Wang, I.-M., Tsai, S.Y., Tsai, M.-J., O'Malley, B.W., Jurutka, P.W., Haussler, M.R. & Ozato, K. (1995) Transcription factor TFIIB and the vitamin D receptor cooperatively activate ligand-dependent transcription. *Proceedings of the National Academy of Sciences of the U.S.A.*, **92**, 1535–9.

Blutt, S.E., Allegretto, E.A., Pike, J.W. & Weigel, N.L. (1997) 1,25-Dihydroxyvitamin D_3 and 9-*cis*-retinoic acid act synergistically to inhibit the growth of LNCaP prostate cells and cause accumulation of cells in G1. *Endocrinology*, **138**, 1491–7.

Blutt, S.E., McDonnell, T.J., Polek, T.C. & Weigel, N.L. (2000) Calcitriol-induced apoptosis in LNCaP cells is blocked by overexpression of Bcl-2. *Endocrinology*, **141**, 10–17.

Boissier, M.-C., Chiocchia, G. & Fournier, C. (1992) Combination of cyclosporin A and calcitriol in the treatment of adjuvant arthritis. *Journal of Rheumatology*, **19**, 754–7.

Boonstra, A., Barrat, F.J., Crain, C., Heath, V.L., Savelkoul, H.F.J. & O'Garra, A. (2001) 1α,25-Dihydroxyvitamin D_3 has a direct effect on naive $CD4^+$ T cells to enhance the development of Th2 cells. *Journal of Immunology*, **167**, 4974–80.

Borowitz, S.M. & Ghishan, F.K. (1989) Phosphate transport in human jejunal brush-border membrane vesicles. *Gastroenterology*, **96**, 4–10.

Bourdeau, A., Atmani, F., Grosse, B. & Lieberherr, M. (1990) Rapid effects of 1,25-dihydroxyvitamin D_3 and extracellular Ca^{2+} on phospholipid metabolism in dispersed porcine parathyroid cells. *Endocrinology*, **127**, 2738–43.

Boyan, B.D., Dean, D.D., Sylvia, V.L. & Schwartz, Z. (1994) Nongenomic regulation of extracellular matrix events by vitamin D metabolites. *Journal of Cellular Biochemistry*, **56**, 331–9.

Boyan, B.D., Sylvia, V.L., Dean, D.D. & Schwartz, Z. (2001) $24,25\text{-}(OH)_2D_3$ regulates cartilage and bone via autocrine and endocrine mechanisms. *Steroids*, **66**, 363–74.

Boyle, B.J., Zhao, X.-Y., Cohen, P. & Feldman, D. (2001) Insulin-like growth factor binding protein-3 mediates 1α,25-dihydroxyvitamin D_3 growth inhibition in the LNCaP prostate cancer cell line through p21/waf1. *Journal of Urology*, **165**, 1319–24.

Brasitus, T.A., Dudeja, P.K., Eby, B. & Lau, K. (1986) Correction by 1,25-dihydroxycholecalciferol of the abnormal fluidity and lipid composition of enterocyte brush border membranes in vitamin D-deprived rats. *Journal of Biological Chemistry*, **261**(35), 16404–9.

Brinkman, B.M.N., Telliez, J.-P., Schievella, A.R., Lin, L.-L. & Goldfield, A.E. (1999) Engagement of tumor necrosis factor (TNF) receptor 1 leads to ATF-2 and p38 mitogen-activated protein kinase-dependent TNF-α gene expression. *Journal of Biological Chemistry*, **274**(43), 30882–6.

Bronner, F. (1990) Intestinal calcium transport: The cellular pathway. *Mineral and Electrolyte Metabolism*, **16**, 94–100.

Bronner, F., Pansu, D. & Stein, W.D. (1986) An analysis of intestinal calcium transport across the rat intestine. *American Journal of Physiology*, **250**, G561–9.

Brown, A.J., Dusso, A. & Slatopolsky, E. (1999) Vitamin D. *American Journal of Physiology*, **277**, F157–75.

Brown, A., Dusso, A. & Slatopolsky, E. (2000) Vitamin D. In: *The Kidney*, 3rd edn, Vol. 1. (Ed. by D.W. Seldin & G. Giebisch), pp. 1047–90. Lippincott Williams & Wilkins, Philadelphia.

Brown, T.A. & DeLuca, H.F. (1990) Phosphorylation of the 1,25-dihydroxyvitamin D_3 receptor. A primary event in 1,25-dihydroxyvitamin D_3 action. *Journal of Biological Chemistry*, **265**(17), 10025–9.

Brumbaugh, P.F. & Haussler, M.R. (1973) 1α,25-Dihydroxyvitamin D_3 receptor: Competitive binding of vitamin D analogs. *Life Sciences*, **13**, 1737–46.

Brunda, M.J., Luistro, L., Warrier, R.R., Wright, R.B., Hubbard, B.R., Murphy, M., Wolf, S.F. & Gately, M.K. (1993) Antitumor and antimetastatic activity of interleukin-12 against murine tumors. *Journal of Experimental Medicine*, **178**, 1223–30.

Bullamore, J.E., Wilkinson, R., Gallagher, J.C., Nordin, B.E.C. & Marshall, D.H. (1970) Effects of age on calcium absorption. *Lancet*, **II**, 535–7.

Caffrey, J.M. & Farach-Carson, M.C. (1989) Vitamin D_3 metabolites modulate dihydropyridine-sensitive calcium currents in clonal rat osteosarcoma cells. *Journal of Biological Chemistry*, **264**(34), 20265–74.

Cai, Q., Chandler, J.S., Wasserman, R.H., Kumar, R. & Penniston, J.T. (1993) Vitamin D and adaptation to dietary calcium and phosphate deficiencies increase intestinal plasma membrane calcium pump gene expression. *Proceedings of the National Academy of Sciences of the U.S.A.*, **90**, 1345–9.

Campbell, J., Haraldsen, G., Pan, J., Rottman, J., Qin, S., Ponath, P., Andrew, D.P., Warnke, R., Ruffing, N., Kassam, N., Wu, L. & Butcher, E.C. (1999) The chemokine receptor CCR4 in vascular recognition by cutaneous but not intestinal memory T cells. *Nature*, **400**, 776–80.

Campbell, M.J., Gombart, A.F., Kwok, S.H., Park, S. & Koeffler, H.P. (2000) The anti-proliferative effects of 1α,25$(OH)_2D_3$ on breast and prostate cancer cells are associated with induction of *BRCA1* gene expression. *Oncogene*, **19**, 5091–7.

Canaff, L. & Hendy, G.N. (2002) Human calcium sensing receptor gene. *Journal of Biological Chemistry*, **277**(33), 30337–50.

Cantley, L.K., Russell, J., Lettieri, D. & Sherwood, L.M. (1985) 1,25-Dihydroxyvitamin D_3 suppresses parathyroid hormone secretion from bovine parathyroid cells in tissue culture. *Endocrinology*, **117**, 2114–9.

Cantorna, M.T., Hayes, C.E. & DeLuca, H.F. (1996) 1,25-Dihydroxyvitamin D_3 reversibly blocks the progression of relapsing encephalomyelitis, a model of multiple sclerosis. *Proceedings of the National Academy of Sciences of the U.S.A.*, **93**, 7861–4.

Cantorna, M.T., Woodward, W.D., Hayes, C.E. & DeLuca, H.F. (1998) 1,25-Dihydroxyvitamin D_3 is a positive regulator for the two anti-encephalitogenic cytokines TGF-β1 and IL-4. *Journal of Immunology*, **160**, 5314–19.

Cao, X., Ross, P., Zhang, L., MacDonald, P.N., Chappel, J. & Teitelbaum, S.L. (1993) Cloning of the promoter for the avian integrin β3 subunit gene and its regulation by 1,25-dihydroxyvitamin D_3. *Journal of Biological Chemistry*, **268**(36), 27371–80.

Chandra, S., Fullmer, C.S., Smith, C.A., Wasserman, R.H. & Morrison, G.H. (1990) Ion microscopic imaging of calcium transport in the intestinal tissue of vitamin D-deficient and vitamin D-replete chickens: A ^{44}Ca stable isotope study. *Proceedings of the National Academy of Sciences of the U.S.A.*, **87**, 5715–9.

Chen, A., Davis, B.H., Bissonnette, M., Scaglione-Sewell, B. & Brasitus, T.A. (1999) 1,25-Dihydroxyvitamin D_3 stimulates activator protein-1-dependent Caco-2 cell differentiation. *Journal of Biological Chemistry*, **274**(50), 35505–13.

Chen, A., Bissonnette, M., Sitrin, M.D. & Brasitus, T.A. (2001) ERK and its downstream target, Ets-1, are involved in the 1,25$(OH)_2D_3$-induced inhibition of cell growth and stimulation of apoptosis in Caco-2 cells. *Gastroenterology*, **120**, A665 (abstract).

Chen, A., Davis, B.H., Sitrin, M.D., Brasitus, T.A. & Bissonnette, M. (2002) Transforming growth factor-β1 signaling contributes to Caco-2 cell growth inhibition induced by 1,25$(OH)_2D_3$. *American Journal of Physiology*, **283**, G864–74.

Chen, K.-S. & DeLuca, H.F. (1995) Cloning of the human 1α,25-dihydroxyvitamin D_3 24-hydroxylase gene promoter and identification of two vitamin D-responsive elements. *Biochimica et Biophysica Acta*, **1263**, 1–9.

Chen, M.L., Boltz, M.A. & Armbrecht, H.J. (1993) Effects of 1,25-dihydroxyvitamin D_3 and phorbol ester on 25-hydroxyvitamin D_3 24-hydroxylase cytochrome P450 messenger ribonucleic acid levels in primary cultures of rat renal cells. *Endocrinology*, **132**, 1782–8.

Chen, S., Sims, G.P., Chen, X.X., Gu, Y.Y., Chen, S. & Lipsky, P.E. (2007) Modulatory effects of 1,25-dihydroxyvitamin D_3 on human B cell differentiation. *Journal of Immunology*, **179**, 1634–47.

Chen, Y., Zhang, J., Ge, X., Du, J., Deb, D.K. & Li, Y.C. (2013) Vitamin D receptor inhibits nuclear factor κB by interacting with Iκ β kinase β protein. *Journal of Biological Chemistry*, **288**(27), 19450–8.

Chertow, B.S., Sivitz, W.I., Baranetsky, N.G., Clark, S.A., Waite, A. & DeLuca, H.F. (1983) Cellular mechanisms of insulin release: The effects of vitamin D deficiency and repletion on rat insulin secretion. *Endocrinology*, **113**, 1511–8.

Christensen, E.I. & Birn, H. (2001) Megalin and cubilin; synergistic endocytic receptors in renal proximal tubule. *Americam Journal of Physiology*, **280**, F562–73.

Civitelli, R., Kim, Y.S., Gunsten, S.L., Fujimori, A., Huskey, M., Avioli, L.V. & Hruska, K.A. (1990) Nongenomic activation of the calcium message system by vitamin D metabolites in osteoblast-like cells. *Endocrinology*, **127**, 2253–62.

Clark, S.A., Stumpf, W.E. & Sar, M. (1981) Effect of 1,25 dihydroxyvitamin D_3 on insulin secretion. *Diabetes*, **30**, 382–8.

Clemens, T.L., Henderson, S.L., Adams, J.S. & Holick, M.F. (1982) Increased skin pigment reduces the capacity of skin to synthesize vitamin D_3. *Lancet*, **I**(8263), 74–6.

Clements, M.R., Chalmers, T.M. & Fraser, D.R. (1984) Enterohepatic circulation of vitamin D: A reappraisal of the hypothesis. *Lancet*, **I**(8391), 1376–9.

Clohisy, D.R., Bar-Shavit, Z., Chappel, J.C. & Teitelbaum, S.L. (1987) 1,25-Dihydroxyvitamin D_3 modulates bone marrow macrophage precursor proliferation and differentiation. Upregulation of the mannose receptor. *Journal of Biological Chemistry*, **262**(33), 15922–9.

Coussens, L.M. & Werb, Z. (2002) Inflammation and cancer. *Nature*, **420**, 860–7.

D'Ambrosio, D., Cippitelli, M., Cocciolo, M.G., Mazzeo, D., Di Lucia, P., Lang, R., Sinigaglia, F. & Panina-Bordignon, P. (1998) Inhibition of IL-2 production by 1,25-dihydroxyvitamin D_3. *Journal of Clinical Investigation*, **101**, 252–62.

Danescu, L.G., Levy, S. & Levy, J. (2009) Vitamin D and diabetes mellitus. *Endocrine*, **35**, 11–17.

Danisi, G. & Murer, H. (1991) Inorganic phosphate absorption in small intestine. In: *Handbook of Physiology, Section 6: The Gastrointestinal System, Vol. IV, Intestinal Absorption and Secretion*. (Ed. by S.G. Schultz), pp. 323–36. American Physiological Society, Bethesda, MD.

Darwish, H.M. & DeLuca, H.F. (1992) Identification of a 1,25-dihydroxyvitamin D_3-response element in the 5′-flanking region of the rat calbindin D-9k gene. *Proceedings of the National Academy of Sciences of the U.S.A.*, **89**, 603–7.

Davies, P.D.O. (1985) A possible link between vitamin D deficiency and impaired host defence to *Mycobacterium tuberculosis*. *Tubercle*, **66**, 301–6.

de Boland, A.R. & Norman, A. (1990) Evidence for involvement of protein kinase C and cyclic adenosine 3′,5′-monophosphate-dependent protein kinase in the 1,25-dihydroxy-vitamin D_3-mediated rapid stimulation of intestinal calcium transport, (transcaltachia). *Endocrinology*, **127**, 39–45.

de Boland, A.R., Nemere, I. & Norman, A.W. (1990) Ca^{2+}-channel agonist BAY K8644 mimics 1,25$(OH)_2$-vitamin D_3 rapid enhancement of Ca^{2+} transport in chick perfused duodenum.

Biochemical and Biophysical Research Communications, **166**, 217–22.

Defacque, H., Piquemal, D., Basset, A., Marti, J. & Commes, T. (1999) Transforming growth factor-β1 is an autocrine mediator of U937 cell growth arrest and differentiation induced by vitamin D_3 and retinoids. *Journal of Cellular Physiology*, **178**, 109–119.

Demay, M.B., Kiernan, M.S., DeLuca, H.F. & Kronenberg, H.M. (1992a) Characterization of 1,25-dihydroxyvitamin D_3 receptor interactions with target sequences in the rat osteocalcin gene. *Molecular Endocrinology*, **6**, 557–62.

Demay, M.B., Kiernan, M.S., DeLuca, H.F. & Kronenberg, H.M. (1992b) Sequences in the human parathyroid hormone gene that bind the 1,25-dihydroxyvitamin D_3 receptor and mediate transcriptional repression in response to 1,25-dihydroxyvitamin D_3. *Proceedings of the National Academy of Sciences of the U.S.A.*, **89**, 8097–8101.

Desai, R.K., van Wijnen, A.J., Stein, J.L., Stein, G.S. & Lian, J.B. (1995) Control of 1,25-dihydroxyvitamin D_3 receptor-mediated enhancement of osteocalcin gene transcription: Effects of perturbing phosphorylation pathways by okadaic acid and staurosporine. *Endocrinology*, **136**, 5685–93.

Ducy, P., Desbois, C., Boyce, B., Pinero, G., Story, B., Dunstan, C., Smith, E., Bonadio, J., Goldstein, S., Gundberg, C., Bradley, A. & Karsenty, G. (1996) Increased bone formation in osteocalcin-deficient mice. *Nature*, **382**, 448–52.

Dusso, A.S., Brown, A.J. & Slatopolsky, E. (2005) Vitamin D. *American Journal of Physiology (Renal Physiology)*, **289**, F8–28.

Eastwood, J.B., Stamp, T.C., De Wardener, H.E., Bordier, P.J. & Arnaud, C.D. (1977) The effect of 25-hydroxy vitamin D_3 in the osteomalacia of chronic renal failure. *Clinical Science and Molecular Medicine*, **52**, 499–508.

El-Radhi, A.S., Mansor, N., Majeed, M. & Ibrahim, M. (1982) High incidence of rickets in children with wheezy bronchitis in a developing country. *Journal of the Royal Society of Medicine*, **75**, 884–7.

Falzon, M. (1996) DNA sequences in the rat parathyroid hormone-related peptide gene responsible for 1,25-dihydroxyvitamin D_3-mediated transcriptional repression. *Molecular Endocrinology*, **10**, 672–81.

Farach-Carson, M.C., Sergeev, I. & Norman, A.W. (1991) Nongenomic actions of 1,25-dihydroxyvitamin D_3 in rat osteosarcoma cells: Structure-function studies using ligand analogs. *Endocrinology*, **129**, 1876–84.

Favus, M.J. & Tembe, V. (1992) The use of pharmacologic agents to study mechanisms of intestinal calcium transport. *Journal of Nutrition*, **122**, 683–6.

Favus, M.J., Tembe, V., Ambrosic, K.A. & Nellans, H.N. (1989) Effects of $1,25(OH)_2D_3$ on enterocyte basolateral membrane calcium transport in rats. *American Journal of Physiology*, **256**, G613–7.

Ferreira, G.B., Gysemans, C.A., Demengeot, J., da Cunha, J.P.M.C.M., Vanherwegen, A.-S., Overburgh, L., Van Belle, T.L., Pauwels, F., Verstuyf, A., Korf, H. & Mathieu, C. (2014) 1,25-Dihydroxyvitamin D_3 promotes tolerogenic dendritic cells with functional migratory properties in NOD mice. *Journal of Immunology*, **192**, 4210–20.

Ferreira, G.B., Vanherwegen, A.-S., Eelen, G., Gutiérrez, A.C.F., Van Lommel, L., Marschal, K., Verlinden, L., Verstuyf, A., Nogueira, T., Georgiadou, M., Schuit, F., Eizirik, D., Carmeliet, P., Overbergh, L. & Mathieu, C. (2015) Vitamin D_3 induces tolerance in human dendritic cells by activation of intracellular metabolic pathways. *Cell Reports*, **10**, 711–25.

Fleet, J.C. & Wood, R.J. (1999) Specific $1,25(OH)_2D_3$-mediated regulation of transcellular calcium transport in Caco-2 cells. *American Journal of Physiology*, **276**, G958–64.

Fournier, C., Gepner, P., Sadouk, M. & Charreire, J. (1990) *In vivo* beneficial effects of cyclosporin A and 1,25-dihydroxyvitamin D_3 on the induction of experimental autoimmune thyroiditis. *Clinical Immunology and Immunopathology*, **54**, 53–63.

Franceschi, R.T. (1999) The developmental control of osteoblast-specific gene expression: Role of specific transcription factors and the extracellular matrix environment. *Critical Revues in Oral Biology & Medicine*, **10**, 40–57.

Fraser, D.R. (1983) The physiological economy of vitamin D. *Lancet*, **I**(8331), 969–72.

Friedman, P.A. & Gesek, F.A. (1993) Vitamin D_3 accelerates PTH-dependent calcium transport in distal convoluted tubule cells. *American Journal of Physiology*, **265**, F300–8.

Fujita, H., Koshida, K., Keller, E.T., Takahashi, Y., Yoshimoto, T., Namiki, M. & Mizokami, A. (2002) Cyclooxygenase-2 promotes prostate cancer progression. *Prostate*, **53**, 232–40.

Galitzer, H., Ben-Dov, I., Lavi-Moshayoff, V., Naveh-Many, T. & Silver, J. (2008) Fibroblast growth factor 23 acts on the parathyroid to decrease parathyroid hormone secretion. *Current Opinion in Nephrology and Hypertension*, **17**, 363–7.

García-Villalba, P., Au-Fliegner, M., Samuels, H.H. & Aranda, A. (1993) Interaction of thyroid hormone and retinoic acid receptors on the regulation of the rat growth hormone gene promoter. *Biochemical and Biophysical Research Communications*, **191**, 580–6.

García-Villalba, P., Jimenez-Lara, A.M. & Aranda, A. (1996) Vitamin D interferes with transactivation of the growth hormone gene by thyroid hormone and retinoic acid. *Molecular and Cellular Biology*, **16**, 318–27.

Gardiner, E.M., Baldock, P.A., Thomas, G.P., Sims, N.A., Henderson, N.K., Hollis, B., White, C.P., Sunn, K.L., Morrison, N.A., Walsh, W.R. & Eisman, J.A. (2000) Increased formation and decreased resorption of bone in mice with elevated vitamin D receptor in mature cells of the osteoblastic lineage. *FASEB Journal*, **14**, 1908–16.

Gavison, R. & Bar-Shavit, Z. (1989) Impaired macrophage activation in vitamin D_3 deficiency: Differential in vitro effects of 1,25-dihydroxyvitamin D_3 on mouse peritoneal macrophage functions. *Journal of Immunology*, **143**, 388–90.

Gepner, P., Amor, B. & Fournier, C. (1989) 1,25-Dihydroxyvitamin D_3 potentiates the in vitro inhibitory effects of cyclosporin A on T cells from rheumatoid arthritis patients. *Arthritis and Rheumatism*, **32**, 31–6.

Ghijsen, W.E.J.M. & van Os, C.H. (1982) 1α,25-Dihydroxyvitamin D_3 regulates ATP-dependent calcium transport in basolateral plasma membranes of rat enterocytes. *Biochimica et Biophysica Acta*, **689**, 170–2.

Ghijsen, W.E.J.M., De Jong, M.D. & van Os, C.H. (1983) Kinetic properties of Na^+/Ca^{2+} exchange in basolateral plasma membranes of rat small intestine. *Biochimica et Biophysica Acta*, **730**, 85–94.

Gill, R.K. & Christakos, S. (1993) Identification of sequence elements in mouse calbindin-D28k gene that confer 1,25-dihydroxyvitamin D_3-and butyrate-inducible responses. *Proceedings of the National Academy of Sciences of the U.S.A.*, **90**, 2984–8.

Girasole, G., Wang, J.M., Pedrazzoni, M., Pioli, G., Balotta, C., Passeri, M., Lazzarin, A., Ridolfo, A. & Mantovani, A.

(1990) Augmentation of monocyte chemotaxis by 1α,25-dihydroxyvitamin D_3. Stimulation of defective migration of AIDS patients. *Journal of Immunology*, **145**, 2459–64.

Girasole, G., Passeri, G., Jilka, R.L. & Manolagas, S.C. (1994) Interleukin-11: A new cytokine critical for osteoclast development. *Journal of Clinical Investigation*, **93**, 1516–24.

Gluba-Brzózka, A., Franczyk, B., Cialkowska-Rysz, A., Olszewski, R. & Rysz, J.N. (2018) Impact of vitamin D on the cardiovascular system in advanced chronic kidney disease (CKD) and dialysis patients. *Nutrients*, **10**, 709. DOI: 10.3390/nu10060709

Gorman, C., Leandro, M. & Isenberg, D. (2004) Does B cell depletion have a role to play in the treatment of systemic lupus erythematosus? *Lupus*, **13**, 312–6.

Grammer, A.C. & Lipsky, P.E. (2003) B cell abnormalities in systemic lupus erythematosus. *Arthritis Researh & Therapy*, **5**, S22–7.

Gregori, S., Casorati, M., Amuchastegui, S., Smiroldo, S., Davalli, A.M. & Adorini, L. (2001) Regulatory T cells induced by 1α,25-dihydroxyvitamin D_3 and mycophenolate mofetil treatment mediate transplantation tolerance. *Journal of Immunology*, **167**, 1945–53.

Gregori, S., Giarratana, N., Smiroldo, S., Uskokovic, M. & Adorini, L. (2002) A1α,25-dihydroxyvitamin D_3 analog enhances regulatory T-cells and arrests autoimmune diabetes in NOD mice. *Diabetes*, **51**, 1367–74.

Grimnes, G., Joakimsen, R., Figenschau, Y., Torjesen, P.A., Almås, B. & Jorde, R. (2012) The effect of high-dose vitamin D on bone mineral density and bone turnover markers in postmenopausal women with low bone mass—A randomized controlled 1-year trial. *Osteoporosis International*, **23**, 201–11.

Gross, M. & Kumar, R. (1990) Physiology and biochemistry of vitamin D-dependent calcium binding proteins. *American Journal of Physiology*, **259**, F195–209.

Grosse, B., Bourdeau, A. & Lieberherr, M. (1993) Oscillations in inositol 1,4,5-triphosphate and diacylglycerol induced by vitamin D_3 metabolites in confluent mouse osteoblasts. *Journal of Bone and Mineral Research*, **8**, 1059–69.

Guggino, S.E., Lajeunesse, D., Wagner, J.A. & Snyder, S.H. (1989) Bone remodelling signaled by a dihydropyridine- and phenylalkylamine-sensitive calcium channel. *Proceedings of the National Academy of Sciences of the U.S.A.*, **86**, 2957–60.

Guo, B., Aslam, F., van Wijnen, A.J., Roberts, S.G.E., Frenkel, B., Green, M.R., DeLuca, H.F., Lian, J.B., Stein, G.S. & Stein, J.L. (1997) YY1 regulates vitamin D receptor/retinoid X receptor mediated transactivation of the vitamin D responsive osteocalcin gene. *Proceedings of the National Academy of Sciences of the U.S.A.*, **94**, 121–6.

Gurlek, A. & Kumar, R. (2001) Regulation of osteoblast growth by interactions between transforming growth factor-β and 1α,25-dihydroxyvitamin D_3. *Critical Reviews in Eukaryotic Gene Expression*, **11**, 299–317.

Halline, A.G., Davidson, N.O., Skarosi, S.F., Sitrin, M.D., Tietze, C., Alpers, D.H. & Brasitus, T.A. (1994) Effects of 1,25-dihydroxyvitamin D_3 on proliferation and differentiation of Caco-2 cells. *Endocrinology*, **134**, 1710–17.

Hattenhauer, O., Traebert, M., Murer, H. & Biber, J. (1999) Regulation of small intestinal Na-Pi type IIb cotransporter by dietary phosphate intake. *American Journal of Physiology*, **277**, G756–62.

Haussler, M.R., Whitfield, G.K., Haussler, C.A., Hsieh, J.-C., Thompson, P.D., Selznik, S.H., Dominguez, C.E. & Jurutka, P.W. (1998) The nuclear vitamin D receptor: Biological and molecular regulatory properties revealed. *Journal of Bone and Mineral Research*, **13**, 325–49.

Heaney, R.P. (2008) Vitamin D—Criteria for safety and efficacy. *Nutrition Revues*, **66**, S178–81.

Heaney, R.P., Dowell, M.S., Hale, C.A. & Bendich, A. (2003) Calcium absorption varies within the reference range for serum 25-hydroxyvitamin D. *Journal of the American College of Nutrition*, **22**, 142–6.

Hewison, M., Rut, A.R., Kristjansson, K., Walker, R.E., Dillon, M.J., Hughes, M.R. & O'Riordan, J.L.H. (1993) Tissue resistance to 1,25-dihydroxyvitamin D without a mutation of the vitamin D receptor gene. *Clinical Endocrinology*, **39**, 663–70.

Hewison, M., Freeman, L., Hughes, S.V., Evans, K.N., Bland, R., Eliopoulos, A.G., Kilby, M.D., Moss, P.A. & Chakraverty, R. (2003) Differential regulation of vitamin D receptor and its ligand in human monocyte-derived dendritic cells. *Journal of Immunology*, **170**, 5382–90.

Hoenderop, J.G.J., van der Kemp, A.W.C.M., Hartog, A., van de Graaf, S.F.J., van Os, C.H., Willems, P.H.G.M. & Bindels, R.J.M. (1999a) Molecular identification of the apical Ca^{2+} channel in 1,25-dihydroxyvitamin D_3-responsive epithelia. *Journal of Biological Chemistry*, **274**(13), 8375–8.

Hoenderop, J.G.J., van der Kemp, A.W.C.M., Hartog, A., van Os, C.H., Willems, P.H.G.M. & Bindels, R.J.M. (1999b) The epithelial calcium channel, ECaC, is activated by hyperpolarization and regulated by cytosolic calcium. *Biochemical and Biophysical Research Communications*, **261**, 488–92.

Holick, M.F. (1987) Photosynthesis of vitamin D in the skin: Effect of environmental and life-style variables. *Federation Proceedings*, **46**, 1876–82.

Holick, M.F., MacLaughlin, J.A. & Doppelt, S.H. (1981) Regulation of cutaneous previtamin D_3 photosynthesis in man: Skin pigment is not an essential regulator. *Science*, **211**, 590–3.

Holick, M.F., Matsuoka, L.Y. & Wortsman, J. (1989) Age, vitamin D, and solar ultraviolet. *Lancet*, **II**(8671), 1104–5.

Holick, M.F., Biancuzzo, R.M., Chen, T.C., Klein, E.K., Young, A., Bibuld, D., Reitz, R., Salameh, W., Ameri, A. & Tannenbaum, A.D. (2008) Vitamin D_2 is as effective as vitamin D_3 in maintaining circulating concentrations of 25-hydroxyvitamin D. *Journal of Clinical Endocrinology & Metabolism*, **93**, 677–81.

Homaiden, F.R., Donowitz, M., Weiland, G.A. & Sharp, G.W.G. (1989) Two calcium channels in basolateral membranes of rabbit ileal epithelial cells. *American Journal of Physiology*, **257**, G86–93.

Hörlein, A.J., Näär, A.M., Heinzel, T., Torchia, J., Gloss, B., Kurokawa, R., Ryan, A., Kamei, Y., Söderström, M., Glass, C.K. & Rosenfeld, M.G. (1995) Ligand-independent repression by the thyroid hormone receptor mediated by a nuclear receptor co-repressor. *Nature*, **377**, 397–404.

Horwood, N.J., Elliott, J., Martin, T.J. & Gillespie, M.T. (1998) Osteotropic agents regulate the expression of osteoclast differentiation factor and osteoprotegerin in osteoblastic stromal cells. *Endocrinology*, **139**, 4743–6.

Hsieh, J.-C., Jurutka, P.W., Galligan, M.A., Terpening, C.M., Haussler, C.A., Samuels, D.S., Shimizu, Y., Shimizu, N. & Haussler, M.R. (1991) Human vitamin D receptor is selectively phosphorylated by protein kinase C on serine 51, a residue crucial to its trans-activation function. *Proceedings of the National Academy of Sciences of the U.S.A.*, **88**, 9315–9.

Hsieh, J.-C., Jurutka, P.W., Nakajima, S., Galligan, M.A., Haussler, C.A., Shimizu, Y., Shimizu, N., Whitfield, G.K.

& Haussler, M.R. (1993) Phosphorylation of the human vitamin D receptor by protein kinase C. Biochemical and functional evaluation of the serine 51 recognition site. *Journal of Biological Chemistry*, **268**(20), 15118–26.

Hughes, M.R., Baylink, D.J., Jones, P.G. & Haussler, M.R. (1976) Radioligand receptor assay for 25-hydroxyvitamin D_2/D_3 and 1α,25-dihydroxyvitamin D_2/D_3. *Journal of Clinical Investigation*, **58**, 61–70.

Huisman, A.M., White, K.P., Algra, A., Harth, M., Vieth, R., Jacobs, J.W., Bijlsma, J.W. & Bell, D.A. (2001) Vitamin D levels in women with sytemic lupus erythematosus and fibromyalgia. *Journal of Rheumatology*, **28**, 2535–9.

Hullett, D.A., Cantorna, M.T., Redaelli, C., Humpal-Winter, J., Hayes, C.E., Sollinger, H.W. & DeLuca, H.F. (1998) Prolongation of allograft survival by 1,25-dihydroxyvitamin D_3. *Transplantation*, **66**, 824–8.

Iellem, A., Mariani, M., Lang, R., Recalde, H., Panina-Bordignon, P., Sinigaglia, F. & D'Ambrosia, D. (2001) Unique chemotactic response profile and specific expression of chemokine receptors CCR4 and CCR8 by CD4+CD25+ regulatory T cells. *Journal of Experimental Medicine*, **194**, 847–53.

Ingram, R.T., Bonde, S.K., Riggs, B.L. & Fitzpatrick, L.A. (1994) Effects of transforming growth factor beta (TGFβ and 1,25-dihydroxyvitamin D_3 on the function, cytochemistry and morphology of normal osteoblast-like cells. *Differentiation*, **55**, 153–63.

Institute of Medicine. (2011) *Dietary Reference Intakes for Calcium and Vitamin D.* National Academy Press, Washington, DC.

Isbarn, H., Boccon-Gibod, L., Carroll, P.R., Montorsi, F., Schulman, C., Smith, M.R., Sternberg, C.N. & Studer, U.E. (2009) Androgen deprivation therapy for the treatment of prostate cancer: Consider both benefits and risks. *European Urology*, **55**, 62–75.

Jeffery, L.E., Burke, F., Mura, M., Zheng, Y., Qureshi, O.S., Hewison, M., Walker, L.S.K., Lammas, D.A., Raza, K. & Sansom, D.M. (2009) 1,25-Dihydroxyvitamin D_3 and IL-2 combine to inhibit T cell production of inflammatory cytokines and promote development of regulatory T cells expressing CTLA-4 and FoxP3. *Journal of Immunology*, **183**, 5458–67.

Jenis, L.G., Lian, J.B., Stein, G.S. & Baran, D.T. (1993) 1α,25-Dihydroxyvitamin D-induced changes in intracellular pH in osteoblast-like cells modulate gene expression. *Journal of Cellular Biochemistry*, **53**, 234–9.

Jin, C.H., Shinki, T., Hong, M.H., Sato, T., Yamaguchi, A., Ikeda, T., Yoshiki, S., Abe, E. & Suda, T. (1992) 1α,25-Dihydroxyvitamin D_3 regulates *in vivo* production of the third component of complement (C3) in bone. *Endocrinology*, **131**, 2468–75.

Johansson, S. & Melhus, H. (2001) Vitamin Aantagonizes calcium response to vitamin D in man. *Journal of Bone and Mineral Research*, **16**, 1899–905.

Johnsson, C., Binderup, L. & Tufveson, G. (1995) The effects of combined treatment with the novel vitamin D analogue MC 1288 and cyclosporine A on cardiac allograft survival. *Transplant Immunology*, **3**, 245–50.

Jones, G. (2008) Pharmacokinetics of vitamin D toxicity. *American Journal of Clinical Nutrition*, **88**, 582S–6S.

Jones, G., Strugnell, S.A. & DeLuca, H.F. (1998) Current understanding of the molecular actions of vitamin D. *Physiological Reviews*, **78**, 1193–231.

Jurutka, P.W., Hsieh, J.-C., Nakajima, S., Haussler, C.A., Whitfield, G.K. & Haussler, M.R. (1996) Human vitamin D receptor phosphorylation by casein kinase II at ser-208 potentiates transcripional activation. *Proceedings of the National Academy of Sciences of the U.S.A.*, **93**, 3519–24.

Kahlen, J.-P. & Carlberg, C. (1996) Functional characterization of a 1,25-dihydroxyvitamin D_3 receptor binding site found in the rat atrial natriuretic factor promoter. *Biochemical and Biophysical Research Communications*, **218**, 882–6.

Kamijo, R., Takeda, K., Nagumo, M. & Konno, K. (1990) Effects of combinations of transforming growth factor-β1 and tumor necrosis factor on induction of differentiation of human myelogenous leukemic cell lines. *Journal of Immunology*, **144**, 1311–16.

Kamimura, S., Gallieni, M., Zhong, M., Beron, W., Slatopolsky, E. & Dusso, A. (1995) Microtubules mediate cellular 25-hydroxyvitamin D_3 trafficking and the genomic response to 1,25-dihydroxyvitamin D_3 in normal human monocytes. *Journal of Biological Chemistry*, **270**(38), 22160–6.

Kankova, M., Luini, W., Pedrazzoni, M., Riganti, F., Sironi, M., Bottazzi, B., Mantovani, A. & Vecchi, A. (1991) Impairment of cytokine production in mice fed a vitamin D_3-deficient diet. *Immunology*, **73**, 466–71.

Kato, A., Seo, E.-G., Einhorn, T.A., Bishop, J.E. & Norman, A.W. (1998) Studies on 24R,25-dihydroxyvitamin D_3: Evidence for a nonnuclear membrane receptor in the chick tibial fracture-healing callus. *Bone*, **23**, 141–6.

Kerry, D.M., Dwivedi, P.P., Hahn, C.N., Morris, H.A., Omdahl, J.L. & May, B.K. (1996) Transcriptional synergism between vitamin D-responsive elements in the rat 25-hydroxyvitamin D_3 34-hydroxylase (*CYP24*) promoter. *Journal of Biological Chemistry*, **271**(47), 29715–21.

Khare, S., Bissonnette, M., Scaglione-Sewell, B., Wali, R.K., Sitrin, M.D. & Brasitus, T.A. (1999) 1,25-Dihydroxyvitamin D_3 and TPA activate phospholipase D in Caco-2 cells: Role of PKC-α. *American Journal of Physiology*, **276**, G993–1004.

Khoury, R., Ridall, A.L., Norman, A.W. & Farach-Carson, M.C. (1994) Target gene activation by 1,25-dihydroxyvitamin D_3 in osteosarcoma cells is independent of calcium influx. *Endocrinology*, **135**, 2446–53.

Khundmiri, S.J., Rane, M.J. & Lederer, E.D. (2003) Parathyroid hormone regulation of type II sodium-phosphate cotransporters is dependent on a kinase anchorimg protein. *Journal of Biological Chemistry*, **278**(12), 10134–41.

Kikuchi, K. & Ghishan, F.K. (1987) Phosphate transport by basolateral plasma membranes of human small intestine. *Gastroenterology*, **93**, 106–13.

Kim, R.H., Li, J.J., Ogata, Y., Yamuchi, M., Freedman, L.P. & Sodek, J. (1996) Identification of a vitamin D_3-response element that overlaps a unique inverted TATA box in the rat bone sialoprotein gene. *Biochemical Journal*, **318**, 219–26.

Koeffler, H.P., Reichel, H., Bishop, J.E. & Norman, A.W. (1985) γ-Interferon stimulates production of 1,25-dihydoxyvitamin D_3 by normal human macrophages. *Biochemical and Biophysical Research Communications*, **127**, 596–603.

Kogawa, M., Anderson, P.H., Findlay, D.M., Morris, H.A. & Atkins, G.J. (2010a) The metabolism of 25-(OH)vitamin D_3 by osteoclasts and their precursors regulates the differentiation of osteoclasts. *Journal of Steroid Biochemistry & Molecular Biology*, **121**, 277–80.

Kogawa, M., Findlay, D.M., Anderson, P.H., Ormsby, R., Vincent, C., Morris, H.A. & Atkins, G.J. (2010b) Osteoclastic metabolism of 25(OH)-vitamin D_3: A potential mechanism for optimization of bone resorption. *Endocrinology*, **151**, 4613–25.

Kolek, O.I., Hines, E.R., Jones, M.D., LeSueur, L.K., Lipko, M.A., Kiela, P.R., Collins, J.F., Haussler, M.R. & Ghishan, F.K. (2005) 1α,25-Dihydroxyvitamin D_3 upregulates FGF23 gene expression in bone: The final link in a renal-gastrointestinal-skeletal axis that controls phosphate transport. *American Journal of Physiology (Liver Physiology)*, **289**, G1036–42.

Komori, T., Yagi, H., Nomura, S., Yamaguchi, A., Sasaki, K., Deguchi, K., Shimizu, Y., Bronson, R.T., Gao, Y.H., Inada, M., Sato, M., Okamoto, R., Kitamura, Y., Yoshiki, S. & Kishimoto, T. (1997) Targeted disruption of Cbfal results in complete lack of bone formation owing to maturational arrrest of osteoblasts. *Cell*, **89**, 755–64.

Kragballe, K. (1992) Vitamin D analogues in the treatment of psoriasis. *Journal of Cellular Biochemistry*, **49**, 46–52.

Kremer, R., Sebag, M., Champigny, C., Meerovitch, K., Hendy, G.N., White, J. & Goltzman, D. (1996) Identification and characterization of 1,25-dihydroxyvitamin D_3-responsive repressor sequences in the rat parathyroid hormone-related peptide gene. *Journal of Biological Chemistry*, **271**(27), 16310–16.

Krishnan, A.V. & Feldman, D. (1991) Activation of protein kinase-C inhibits vitamin D receptor gene expression. *Molecular Endocrinology*, **5**, 605–12.

Krishnan, A.V. & Feldman, D. (1992) Cyclic adenosine 3′,5′-monophosphate up-regulates 1,25-dihydroxyvitamin D_3 receptor gene expression and enhances hormone action. *Molecular Endocrinology*, **6**, 198–206.

Krishnan, A.V., Cramer, S.D., Bringhurst, F.R. & Feldman, D. (1995) Regulation of 1,25-dihydroxyvitamin D_3 receptors by parathyroid hormone in osteoblastic cells: Role of second messenger pathways. *Endocrinology*, **136**, 705–12.

Kunkel, E.J., Kim, C.H., Lazarus, N.H., Vierra, M.A., Soler, D., Bowman, E.P. & Butcher, E.C. (2003) CCR10 expression is a common feature of circulating and mucosal epithelial tissue IgAAb-secreting cells. *Journal of Clinical Investigation*, **111**, 1001–10.

Lawson, E. (1985) Vitamin D. In: *Fat-Soluble Vitamins. Their Biochemistry and Applications.* (Ed. by A.T. Diplock), pp. 76–153. Heinemann, London.

Lee, C., Sutkowski, D.M., Sensibar, J.A., Zelner, D., Kim, I., Amsel, I., Shaw, N., Prins, G.S. & Kozlowski, J.M. (1995) Regulation of proliferation and production of prostate-specific antigen in androgen-sensitive prostatic cancer cells, LNCaP, by dihydrotestosterone. *Endocrinology*, **136**, 796–803.

Lehenkari, P., Hentunen, T.A., Laitala-Leinonen, T., Tuukkanen, J. & Väänänen, H.K. (1998) Carbonic anhydrase II plays a major role in osteoclast differentiation and bone resorption by effecting the steady state intracellular pH and Ca^{2+}. *Experimental Cell Research*, **242**, 128–37.

Lemire, J.M. (1992) Immunomodulatory role of 1,25-dihydroxyvitamin D_3. *Journal of Cellular Biochemistry*, **49**, 26–31.

Lemire, J.M. & Archer, D.C. (1991) 1,25-Dihydroxyvitamin D_3 prevents the in vivo induction of murine experimental autoimmune encephalomyelitis. *Journal of Clinical Investigation*, **87**, 1103–7.

Lemire, J.M., Archer, D.C., Beck, L. & Spiegelberg, H.L. (1995) Immunosuppressive actions of 1,25-dihydroxyvitamin D_3: Preferential inhibition of Th1 functions. *Journal of Nutrition*, **125**, 1704S–8S.

Li, B., Chik, C.L., Taniguchi, N., Ho, A.K. & Karpinski, E. (1996) 24,25(OH)2 Vitamin D_3 modulates the L-type Ca^{2+} channel current in UMR 106 cells: Involvement of protein kinase A and protein kinase C. *Cell Calcium*, **19**, 193–200.

Li, Y.C., Kong, J., Wei, M., Chen, Z.-F., Liu, S.Q. & Cao, L.-P. (2002) 1,25-dihydroxyvitamin D_3 is a negative endocrine regulator of the renin-angiotensin system. *Journal of Clinical Investigation*, **110**, 229–38.

Lian, J.B. & Stein, G.S. (1992) Transcriptional control of vitamin D-regulated proteins. *Journal of Cellular Biochemistry*, **49**, 37–45.

Lidor, C., Atkin, I., Ornoy, A., Dekel, S. & Edelstein, S. (1987) Healing of rachitic lesions in chicks by 24R,25-dihydroxycholecalciferol administered locally into bone. *Journal of Bone and Mineral Research*, **2**, 91–8.

Lieberherr, M., Grosse, B., Duchambon, P. & Drücke, T. (1989) A functional cell surface type receptor is required for the early action of 1,25-dihydroxyvitamin D_3 on the phosphoinositide metabolism in rat enterocytes. *Journal of Biological Chemistry*, **264**(34), 20403–6.

Lisnevskaia, L., Murphy, G. & Isenberg, D. (2014) Systemic lupus erythematosus. *Lancet*, **384**, 1878–88.

Liu, M., Lee, M.-H., Cohen, M., Bommakanti, M. & Freedman, L.P. (1996) Transcriptional activation of the Cdk inhibitor p21 by vitamin D_3 leads to the induced differentiation of the myelomonocytic cell line U937. *Genes & Development*, **10**, 142–53.

Liu, P.T., Stenger, S., Li, H., Wenzel, L., Tan, B.H., Krutzik, S.R., Ochoa, M.T., Schauber, J., Wu, K., Meinken, C., Kamen, D.L., Wagner, M., Bals, R., Steinmeyer, A., Zügel, U., Gallo, R.L., Eisenberg, D., Hewison, M., Hollis, B.W., Adams, J.S., Bloom, B.R. & Modlin, R.L. (2006) Toll-like receptor triggering of a vitamin D-mediated human antimicrobial response. *Science*, **311**, 1770–3.

Liu, S., Gupta, A. & Quarles, L.D. (2007) Emerging role of fibroblast growth factor 23 in a bone—Kidney axis regulating systemic phosphate homeostasis and extracellular matrix mineralization. *Current Opinion in Nephrology and Hypertension*, **16**, 329–35.

Liu, S.M., Koszewski, N., Lupez, M., Malluche, H.H., Olivera, A. & Russell, J. (1996) Characterization of a response element in the 5′-flanking region of the avian (chicken) PTH gene that mediates negative regulation of gene transcription by 1,25-dihydroxyvitamin D_3 and binds the vitamin D_3 receptor. *Molecular Endocrinology*, **10**, 206–15.

Looney, R.J., Anolik, J.H., Campbell, D., Felgar, R.E., Young, F., Arend, L.J., Sloand, J.A., Rosenblatt, J. & Sanz, I. (2004) B cell depletion as a novel treatment for systemic lupus erythematosus: A phase I/II dose-escalation trial of rituximab. *Arthritis & Rheumatism*, **50**, 2580–9.

Lötscher, M., Wilson, P., Nguyen, S., Kaissling, B., Biber, J., Murer, H. & Levi, M. (1996) New aspects of adaptation of rat renal Na-Pi cotransporter to alterations in dietary phosphate. *Kidney International*, **49**, 1012–18.

Lowe, K.E., Maiyar, A.C. & Norman, A.W. (1992) Vitamin D-mediated gene expression. *Critical Reviews in Eukaryotic Gene Expression*, **2**, 65–109.

MacDonald, P.N., Dowd, D.R., Nakajima, S., Galligan, M.A., Reeder, M.C., Haussler, C.A., Ozato, K. & Haussler, M.R. (1993) Retinoid X receptors stimulate and 9-*cis* retinoic acid inhibits 1,25-dihydroxyvitamin D_3-activated expression of the rat osteocalcin gene. *Molecular and Cellular Biology*, **13**, 5907–17.

MacDonald, P.N., Sherman, D.R., Dowd, D.R., Jefcoat, S.C. Jr. & DeLisle, R.K. (1995) The vitamin D receptor interacts with general transcription factor IIB. *Journal of Biological Chemistry*, **270**(9), 4748–52.

MacDonald, R.G., Pfeffer, S.R., Coussens, L., Tepper, M.A., Brocklebank, C.M., Mole, J.E., Anderson, J.K., Chen, E., Czech, M.P. & Ullrich, A. (1988) A single receptor binds both insulin-like growth factor II and manose-6-phosphate. *Science*, **239**, 1134–7.

MacLaughlin, J. & Holick, M.F. (1985) Aging decreases the capacity of human skin to produce vitamin D_3. *Journal of Clinical Investigation*, **76**, 1536–8.

MacLaughlin, J.A., Anderson, R.R. & Holick, M.F. (1982) Spectral character of sunlight modulates photosynthesis of previtamin D_3 and its photoisomers in human skin. *Science*, **216**, 1001–3.

MacLaughlin, J.A., Cantley, L.C. & Holick, M.F. (1990) $1,25(OH)_2D_3$ increases calcium and phosphatidylinositol metabolism in differentiating cultured human keratinocytes. *Journal of Nutritional Biochemistry*, **1**, 81–7.

Maehata, Y., Takamizawa, S., Ozawa, S., Kato, Y., Sato, S., Kubota, E. & Hata, R.-I. (2006) Both direct and collagen-mediated signals are required for active vitamin D_3-elicited differentiation of human osteoblastic cells: Roles of osterix, an osteoblast-related transcription factor. *Matrix Biology*, **25**, 47–58.

Maestro, B., Campión, J., Dávila, N. & Calle, C. (2000) Stimulation by 1,25-dihydroxyvitamin D_3 of insulin receptor expression and insulin responsiveness for glucose transport in U-937 human promonocytic cells. *Endocrine Journal*, **47**, 383–91.

Majewski, S., Skopinska, M., Marczak, M., Szmurlo, A., Bollag, W. & Jablonska, S. (1996) Vitamin D_3 is a potent inhibitor of tumor cell-induced angiogenesis. *Journal of Investigative Dermatology Symposium Proceedings*, **1**, 97–101.

Malloy, P.J., Pike, J.W. & Feldman, D. (1999) The vitamin D receptor and the syndrome of hereditary 1,25-dihydroxyvitamin D-resistant rickets. *Endocrine Reviews*, **20**, 156–88.

Mangalam, H.J., Albert, V.R., Ingraham, H.A., Kapiloff, M., Wilson, L., Nelson, C., Elsholtz, H. & Rosenfeld, M.G. (1989) Apituitary POU domain protein, Pit-1, activates both growth hormone and prolactin promoters transcriptionally. *Genes & Development*, **3**, 946–58.

Manning, A.M., Williams, A.C., Game, S.M. & Paraskeva, C. (1991) Differential sensitivity of human colonic adenoma and carcinoma cells to transforming growth factor β (TGF-β): Conversion of an adenoma cell line to a tumorigenic phenotype is accompanied by a reduced response to the inhibitory effects of TGF-β. *Oncogene*, **6**, 1471–6.

Manolagas, S.C., Burton, D.W. & Deftos, L.J. (1981) 1,25-Dihydroxyvitamin D_3 stimulates the alkaline phosphatase activity of osteoblast-like cells. *Journal of Biological Chemistry*, **256**(14), 7115–17.

Manolagas, S.C., Hustmyer, F.G. & Yu, X.-P. (1989) 1,25-Dihydroxyvitamin D_3 and the immune system. *Proceedings of the Society of Experimental Biology and Medicine*, **191**, 238–45.

Manolagas, S.C., Hustmyer, F.G. & Yu, X.-P. (1990) Immunomodulating properties of 1,25-dihydroxyvitamin D_3. *Kidney International*, **38**, S9–S16.

Mantell, D.J., Owens, P.E., Bundred, N.J., Mawer, E.B. & Canfield, A.E. (2000) 1α,25-Dihydroxyvitamin D_3 inhibits angiogenesis in vitro and in vivo. *Circulation Research*, **87**, 214–20.

Martell, R.F., Simpson, R.U. & Taylor, J.M. (1987) 1,25-Dihydroxyvitamin D_3 regulation of phorbol ester receptors in HL-60 leukemia cells. *Journal of Biological Chemistry*, **262**(12), 5570–5.

Marti, J., Commes, T., Defacque, H. & Sevilla, C. (1997) Vitamin D and 9-cis retinoic acid: An efficient partnership for the induction of myelomonocytic cell growth inhibition and differentiation. *Leukemia Research*, **21**, 173–6.

Mathieu, C., Laureys, J., Sobis, H., Vandeputte, M., Waer, M. & Bouillon, R. (1992) 1,25-Dihydroxyvitamin D_3 prevents insulitis in NOD mice. *Diabetes*, **41**, 1491–5.

Mathieu, C., Waer, M., Laureys, J., Rutgeerts, O. & Bouillon, R. (1994) Prevention of autoimmune diabetes in NOD mice by 1,25 dihydroxyvitamin D_3. *Diabetologia*, **37**, 552–8.

Matsuo, K. & Irie, N. (2008) Osteoclast—Osteoblast communication. *Archives of Biochemistry and Biophysics*, **473**, 201–9.

Matsuoka, L.Y., Ide, L., Wortsman, J., MacLaughlin, J.A. & Holick, M.F. (1987) Sunscreens suppress cutaneous vitamin D_3 synthesis. *Journal of Clinical Endocrinology and Metabolism*, **64**, 1165–8.

Mattner, F., Smiroldo, S., Galbiati, F., Muller, M., Di Lucia, P., Poliani, P.L., Martino, G., Panina-Bordignon, P. & Adorini, L. (2000) Inhibition of Th1 development and treatment of chronic-relapsing experimental allergic encephalomyelitis by a non-hypercalcemic analogue of 1,25-dihydroxyvitamin D_3. *European Journal of Immunology*, **30**, 498–508.

Mawer, E.B., Backhouse, J., Holman, C.A., Lumb, G.A. & Stanbury, S.W. (1972) The distribution and storage of vitamin D and its metabolites in human tissues. *Clinical Science*, **43**, 413–31.

Medhora, M.M., Teitelbaum, S., Chappel, J., Alvarez, J., Mimura, H., Ross, F.P. & Hruska, K. (1993) 1α,25-Dihydroxyvitamin D_3 up-regulates expression of the osteoclast integrin αvβ3. *Journal of Biological Chemistry*, **268**(2), 1456–61.

Merino, F., Alvarez-Mon, M., de la Hera, A., Alés, J.E., Bonilla, F. & Durantez, A. (1989) Regulation of natural killer cytotoxicity by 1,25-dihydroxyvitamin D_3. *Cellular Immunology*, **118**, 328–36.

Merke, J., Milde, P., Lewicka, S., Hügel, U., Klaus, G., Mangelsdorf, D.J., Haussler, M.R., Rauterberg, E.W. & Ritz, E. (1989) Identification and regulation of 1,25-dihydroxyvitamin D_3 receptor activity and biosynthesis of 1,25-dihydroxyvitamin D_3. Studies in cultured bovine aortic endothelial cells and human dermal capillaries. *Journal of Clinical Investigation*, **83**, 1903–15.

Michalaki, V., Odontiadis, M., Syrigos, K., Charles, P. & Waxman, J. (2004) Serum levels of IL-6 and TNF-α correlate with clinicopathological features and patient survival in patients with prostate cancer. *British Journal of Cancer*, **90**, 2312–16.

Miller, G.J., Stapleton, G.E., Hedlund, T.E. & Moffat, K.A. (1995) Vitamin D receptor expression, 24-hydroxylase activity, and inhibition of growth by 1α,25-dihydroxyvitamin D_3 in seven human prostate carcinoma cell lines. *Clinical Cancer Research*, **1**, 997–1003.

Mizokami, A., Yeh, S.-Y. & Chang, C. (1994) Identification of 3′,5′-cyclic adenosine monophosphate response element and other *cis*-acting elements in the human androgen receptor gene promoter. *Molecular Endocrinology*, **8**, 77–88.

Montagna, W. & Carlisle, K. (1979) Structural changes in aging human skin. *Journal of Investigative Dermatology*, **73**, 47–53.

Morales, J., Homey, B., Vicari, A.P., Hudak, S., Oldham, E., Hedrick, J., Orozco, R., Copeland, N.G., Jenkins, N.A., McEvoy, L.M. & Zlotnik, A. (1999) CTACK, a skin-associated chemokine that preferentially attracts skin-homing

memory T cells. *Proceedings of the National Academy of Sciences of the U.S.A.*, **96**(25), 14470–5.
Morelli, S., de Boland, A.R. & Boland, R.L. (1993) Generation of inositol phosphates, diacylglycerol and calcium fluxes in myoblasts treated with 1,25-dihydroxyvitamin D_3. *Biochemical Journal*, **289**, 675–9.
Moreno, J., Krishnan, A.V. & Feldman, D. (2005) Molecular mechanisms mediating the anti-proliferative effects of vitamin D in prostate cancer. *Journal of Steroid Biochemistry & Molecular Biology*, **97**, 31–6.
Motohashi, Y., Yamada, S., Yanagawa, T., Maruyama, T., Suzuki, R., Niino, M., Fukazawa, T., Kasuga, A., Hirose, H., Matsubara, K., Shimada, A. & Saruta, T. (2003) Vitamin D receptor gene polymorphism affects onset pattern of type 1 diabetes. *Journal of Clinical Endocrinology & Metabolism*, **88**, 3137–40.
Murayama, A., Takeyama, K., Kitanaka, S., Kodera, Y., Kawaguchi, Y., Hosoya, T. & Kato, S. (1999) Positive and negative regulations of the renal 25-hydroxyvitamin D_3 1α-hydroxylase gene by parathyroid hormone, calcitonin, and 1α,25$(OH)_2D_3$ in intact animals. *Endocrinology*, **140**, 2224–31.
Murer, H., Forster, I., Hernando, N., Lambert, G., Traebert, M. & Biber, J. (1999) Posttranscriptional regulation of the proximal tubule NaPi-II transporter in response to PTH and dietary P1. *American Journal of Physiology (Renal Physiology)*, **277**, F676–84.
Murer, H., Hernando, N., Forster, I. & Biber, J. (2000) Proximal tubular phosphate reabsorption: Molecular mechanisms. *Physiological Reviews*, **80**, 1373–409.
Murer, H., Hernando, N., Forster, I. & Biber, J. (2001) Molecular aspects in the regulation of renal inorganic phosphate reabsorption: The type IIa sodium/inorganic phosphate co-transporter as the key player. *Current Opinion in Nephrology and Hypertension*, **10**, 555–61.
Murer, H., Köhler, K., Lambert, G., Stange, G., Biber, J. & Forster, I. (2002) The renal type II Na/Pi cotransporter. *Cell Biochemistry and Biophysics*, **36**, 215–20.
Nagpal, S., Lu, J. & Boehm, M.F. (2001) Vitamin D analogs: Mechanism of action and therapeutic applications. *Current Medicinal Chemistry*, **8**, 1661–79.
Need, A.G., Morris, H.A., Horowitz, M. & Nordin, B.E.C. (1993) Effects of skin thickness, age, body fat, and sunlight on serum 25-hydroxyvitamin D. *American Journal of Clinical Nutrition*, **58**, 882–5.
Nelson, W.G., De Marzo, A.M., DeWeese, T.L. & Isaacs, W.B. (2004) The role of inflammation in the pathogenesis of prostate cancer. *Journal of Urology*, **172**, S6–S11.
Nemere, I. (1995) Nongenomic effects of 1,25-dihydroxyvitamin D_3: Potential relation of a plasmalemmal receptor to the acute enhancement of intestinal calcium transport in chick. *Journal of Nutrition*, **125**, 1695S–8S.
Nemere, I. (1996a) Apparent nonnuclear regulation of intestinal phosphate transport: Effects of 1,25-dihydroxyvitamin D_3, 24,25-dihydroxyvitamin D_3, and 25-hydroxyvitamin D_3. *Endocrinology*, **137**, 2254–61.
Nemere, I. (1996b) Parathyroid hormone rapidly stimulates phosphate transport in perfused duodenal loops of chicks: Lack of modulation by vitamin D metabolites. *Endocrinology*, **137**, 3750–5.
Nemere, I. (1999) 24,25-Dihydroxyvitamin D_3 suppresses the rapid actions of 1,25-dihydroxyvitamin D_3 and parathyroid hormone on calcium transport in chick intestine. *Journal of Bone and Mineral Research*, **14**, 1543–9.
Nemere, I. & Farach-Carson, M.C. (1998) Membrane receptors for steroid hormones: A case for specific cell surface binding sites for vitamin D metabolites and estrogens. *Biochemical and Biophysical Research Communications*, **248**, 443–449.
Nemere, I., Yoshimoto, Y. & Norman, A.W. (1984) Calcium transport in perfused duodena from normal chicks: Enhancement within fourteen minutes of exposure to 1,25-dihydroxyvitamin D_3. *Endocrinology*, **115**, 1476–83.
Nemere, I., Leathers, V. & Norman, A.W. (1986) 1,25-Dihydroxyvitamin D_3-mediated intestinal calcium transport. *Journal of Biological Chemistry*, **261**(34), 16106–14.
Nemere, I., Theofan, G. & Norman, A.W. (1987) 1,25-Dihydroxyvitamin D_3 regulates tubulin expression in chick intestine. *Biochemical and Biophysical Research Communications*, **148**, 1270–6.
Nemere, I., Dormanen, M.C., Hammond, M.W., Okamura, W.H. & Norman, A.W. (1994) Identification of a specific binding protein for 1α,25-dihydroxyvitamin D_3 in basal-lateral membranes of chick intestinal epithelium and relationship to transcaltachia. *Journal of Biological Chemistry*, **269**(38), 23750–6.
Nemere, I., Schwartz, Z., Pedrozo, H., Sylvia, V.L., Dean, D.D. & Boyan, B.D. (1998) Identification of a membrane receptor for 1,25-dihydroxyvitamin D_3 which mediates rapid activation of protein kinase C. *Journal of Bone and Mineral Research*, **13**, 1353–9.
Nemere, I., Ray, R. & McManus, W. (2000) Immunochemical studies on the putative plasmalemmal receptor for 1,25$(OH)_2D_3$. I. Chick intestine. *American Journal of Physiology*, **278**, E1104–14.
Nemerovski, C.W., Dorsch, M.P., Simpson, R.U., Bone, H.G., Aaronson, K.D. & Bleske, B.E. (2009) Vitamin D and cardiovascular disease. *Pharmacotherapy*, **29**, 691–708.
Nickerson, T. & Huynh, H. (1999) Vitamin D analogue EB1089-induced prostate regression is associated with increased gene expression of insulin-like factor binding proteins. *Journal of Endocrinology*, **160**, 223–9.
Nishizuka, Y. (1992) Intracellular signaling by hydrolysis of phospholipids and activation of protein kinase C. *Science*, **258**, 607–14.
Nithipatikom, K., Isbell, M.A., Lindholm, P.F., Kajdacsy-Balla, A., Kaul, S. & Campell, W.B. (2002) Requirement of cyclo-oxygenase-2 expression and prostaglandins for human prostate cancer cell invasion. *Clinical and Experimental Metastasis*, **19**, 593–601.
Noda, M. & Denhardt, D.T. (2002) Osteopontin. In: *Principles of Bone Biology*, 2nd edn, Vol. I. (Ed. by J.P. Bilezikien, L.G. RA directz & G.A. Rodan), pp. 239–50. Academic Press, San Diego.
Noda, M., Vogel, R.L., Craig, A.M., Prahl, J., DeLuca, H.F. & Denhardt, D.T. (1990) Identification of a DNA sequence responsible for binding of the 1,25-dihydroxyvitamin D_3 receptor and 1,25-dihydroxyvitamin D_3 enhancement of mouse secreted phosphoprotein 1 (*Spp-1* or osteopontin) gene expression. *Proceedings of the National Academy of Sciences of the U.S.A.*, **87**, 9995–9.
Nonn, L., Peng, L., Feldman, D. & Peehl, D.M. (2006) Inhibition of p38 by vitamin D reduces interleukin-6 production in normal prostate cells via mitogen-activated protein kinase phosphatase 5: Implications for prostate cancer prevention by vitamin D. *Cancer Research*, **66**, 4516–24.
Norman, A.W. (1998) Receptors for 1α,25$(OH)_2D_3$: Past, present, and future. *Journal of Bone and Mineral Research*, **13**, 1360–9.

Okada, Y., Morimoto, I., Ura, K., Watanabe, K., Eto, S., Kumegawa, M., Raisz, L., Pilbeam, C. & Tanaka, Y. (2002) Cell-to-cell adhesion via intercellular adhesion molecule-1 and leukocyte function-associated antigen-1 pathway is involved in $1\alpha,25(OH)_2D_3$, PTH and IL-1α-induced osteoclast differentiation and bone resorption. *Endocrine Journal*, **49**, 483–95.

Oñate, S.A., Tsai, S.Y., Tsai, M.-J. & O'Malley, B.W. (1995) Sequence and characterization of a coactivator for the steroid hormone receptor superfamily. *Science*, **270**, 1354–7.

Ono, K. & Han, J. (2000) The p38 signal transduction pathway. Activation and function. *Cellular Signaling*, **12**, 1–13.

Ott, S.M. & Elder, G. (2013) Osteoporosis associated with chronic kidney disease. In: *Osteoporosis*, 4th edn, vol. 2 (Ed. by R. Marcus, D. Feldman, D.W. Dempster, M. Luckey & J.A. Cauley), pp. 1387–1424. Elsevier, Waltham, MA.

Overbergh, L., Decallonne, B., Valckx, D., Verstuyf, A., Depovere, J., Laureys, J., Rutgeerts, O., Saint-Arnaud, R., Bouillon, R. & Mathieu, C. (2000) Identification and immune regulation of 25-hydroxyvitamin D-1-α-hydroxylase in murine macrophages. *Clinical and Experimental Immunology*, **120**, 139–46.

Owen, T.A., Aronow, A.M., Shalhoub, V., Barone, L.M., Wilming, L., Tassinari, M.S., Kennedy, M.B., Pockwinse, S., Lian, J.B. & Stein, G.S. (1990) Progressive development of the rat osteoblast phenotype in vitro: Reciprocal relationships in expression of genes associated with osteoblast proliferation and differentiation during formation of the bone extracellular matrix. *Journal of Cell Physiology*, **143**, 420–30.

Owen, T.A., Aronow, M.S., Barone, L.M., Bettencourt, B., Stein, G.S. & Lian, J.B. (1991) Pleiotropic effects of vitamin D on osteoblast gene expression are related to the proliferative and differentiated state of the bone cell phenotype: Dependency upon basal levels of gene expression, duration of exposure, and bone matrix competency in normal rat osteoblast cultures. *Endocrinology*, **128**, 1496–504.

Ozono, K., Liao, J., Kerner, S.A., Scott, R.A. & Pike, J.W. (1990) The vitamin D-responsive element in the human osteocalcin gene. Association with a nuclear proto-oncogene enhancer. *Journal of Biological Chemistry*, **265**(35), 21881–8.

Pálmer, H.G., González-Sancho, J.M., Espada, J., Berciano, M.T., Puig, I., Baulida, J., Quintanilla, M., Cano, A., de Herreros, A.G., Lafarga, M. & Muñoz, A. (2001) Vitamin D_3 promotes the differentiation of colon carcinoma cells by the induction of E-cadherin and the inhibition of β-catenin signaling. *Journal of Cell Biology*, **154**, 369–87.

Pan, J., Kunkel, E.J., Gosslar, U., Lazarus, N., Langdon, P., Broadwell, K., Vierra, M.A., Genovese, M.C., Butcher, E.C. & Soler, D. (2000) Cutting edge: A novel chemokine ligand for CCR10 and CCR3 expressed by epithelial cells in mucosal tissues. *Journal of Immunology*, **165**, 2943–9.

Pansu, D., Bellaton, C. & Bronner, F. (1981) Effect of Ca intake on saturable and nonsaturable components of duodenal Ca transport. *American Journal of Physiology*, **240**, G32–7.

Pansu, D., Bellaton, C., Roche, C. & Bronner, F. (1983) Duodenal and ileal calcium absorption in the rat and effects of vitamin D. *American Journal of Physiology*, **244**, G695–700.

Paredes, R., Arriagada, G., Cruzat, F., Villagra, A., Olate, J., Zaidi, K., van Wijnen, A., Lian, J.B., Stein, G.S., Stein, J.L. & Montecino, M. (2004) Bone-specific transcription factor Runx2 interacts with the 1α,25-dihydroxyvitamin D_3 receptor to up-regulate rat osteocalcin gene expression in osteoblastic cells. *Molecular and Cellular Biology*, **24**, 8847–61.

Park, W.H., Seol, J.G., Kim, E.S., Hyen, J.M., Jung, C.W., Lee, C.C., Binderup, L., Koeffler, H.P., Kim, B.K. & Lee, Y.Y. (2000) Induction of apoptosis by vitamin D analogue EB1089 in NCI-H929 myeloma cells via activation of caspase 3 and p38 MAP kinase. *British Journal of Haematology*, **109**, 576–83.

Pedrozo, H.A., Schwartz, Z., Rimes, S., Sylvia, V.L., Nemere, I., Posner, G.H., Dean, D.D. & Boyan, B.D. (1999) Physiological importance of the $1,25(OH)_2D_3$ membrane receptor and evidence for a membrane receptor specific for $24,25(OH)_2D_3$. *Journal of Bone and Mineral Research*, **14**, 856–67.

Pence, B.C. & Buddingh, F. (1988) Inhibition of dietary fat-promoted colon carcinogenesis in rats by supplemental calcium or vitamin D_3. *Carcinogenesis*, **9**, 187–90.

Peng, J.-B., Chen, X.-Z., Berger, U.V., Vassilev, P.M., Tsukaguchi, H., Brown, E.M. & Hediger, M.A. (1999) Molecular cloning and characterization of a channel-like transporter mediating intestinal calcium absorption. *Journal of Biological Chemistry*, **274**(32), 22739–46.

Peng, J.-B., Chen, X.-Z., Berger, U.V., Vassilev, P.M., Brown, E.M. & Hediger, M.A. (2000) A rat kidney-specific calcium transporter in the distal nephron. *Journal of Biological Chemistry*, **275**(36), 28186–94.

Peng, L., Malloy, P.J. & Feldman, D. (2004) Identification of a functional vitamin D response element in the human insulin-like growth factor binding protein-3 promoter. *Molecular Endocrinology*, **18**, 1109–19.

Penna, G. & Adorini, L. (2000) 1α,25-Dihydroxyvitamin D_3 inhibits differentiation, maturation, activation, and survival of dendritic cells leading to impaired alloreactive T cell activation. *Journal of Immunology*, **164**, 2405–11.

Penna, G., Amuchastegui, S., Giarratana, N., Daniel, K.C., Vulcano, M., Sozzani, S. & Adorini, L. (2007) 1,25-Dihydroxyvitamin D_3 selectively modulates tolerogenic properties in myeloid but not plasmacytoid dendritic cells. *Journal of Immunology*, **178**, 145–53.

Polla, B.S., Healy, A.M., Amento, E.P. & Krane, S.M. (1986) 1,25-Dihydroxyvitamin D_3 maintains adherence of human monocytes and protects them from thermal injury. *Journal of Clinical Investigation*, **77**, 1332–9.

Polly, P., Carlberg, C., Eisman, J.A. & Morrison, N.A. (1996) Identification of a vitamin D response element in the fibronectin gene that is bound by a vitamin D_3 receptor homodimer. *Journal of Cellular Biochemistry*, **60**, 322–33.

Pols, H.A.P., Schilte, J.P., Herrmann-Erlee, M.P.M., Visser, T.J. & Birkenhäger, J.C. (1986) The effects of 1,25-dihydroxyvitamin D_3 on growth, alkaline phosphatase and adenylate cyclase of rat osteoblast-like cells. *Bone and Mineral*, **1**, 397–405.

Pols, H.A.P., Birkenhäger, J.C., Foekens, J.A. & van Leeuwen, J.P.T.M. (1990) Vitamin D: A modulator of cell proliferation and differentiation. *Journal of Steroid Biochemistry and Molecular Biology*, **37**, 873–6.

Polyak, K. (1996) Negative regulation of cell growth by TGF-β. *Biochimica et Biophysica Acta*, **1242**, 185–99.

Price, P.A. (1985) Vitamin K-dependent formation of bone Gla protein (osteocalcin) and its function. *Vitamins and Hormones*, **42**, 65–108.

Prosser, D.E. & Jones, G. (2004) Enzymes involved in the activation and inactivation of vitamin D. *Trends in Biochemical Sciences*, **29**, 664–73.

Quamme, G.A. (1985) Phosphate transport in intestinal brush-border membrane vesicles: Effect of pH and dietary phosphate. *American Journal of Physiology*, **249**, G168–76.

Quamme, G.A. & Wong, N.L.M. (1984) Phosphate transport in the proximal convoluted tubule: Effect of intraluminal pH. *American Journal of Physiology*, **246**, F323–33.

Quélo, I., Kahlen, J.-P., Rascle, A., Jurdic, P. & Carlberg, C. (1994) Identification and characterization of a vitamin D_3 response element of chicken carbonic anhydrase II. *DNA and Cell Biology*, **13**, 1181–7.

Rachez, C., Suldan, Z., Ward, J., Chang, C.-P.B., Burakov, D., Erdjument-Bromage, H., Tempst, P. & Freedman, L.P. (1998) A novel protein complex that interacts with the vitamin D_3 receptor in a ligand-dependent manner and enhances VDR transactivation in a cell-free system. *Genes & Development*, **12**, 1787–800.

Reff, M.E., Carner, K., Chambers, K.S., Chinn, P.C., Leonard, J.E., Raab, R., Newman, R.A., Hanna, N. & Anderson, D.R. (1994) Depletion of B cells in vivo by a chimeric mouse human monoclonal antibody to CD20. *Blood*, **83**, 435–45.

Reichel, H., Koeffler, H.P., Bishop, J.E. & Norman, A.W. (1987a) 25-Hydroxyvitamin D_3 metabolism by lipopolysaccharide-stimulated normal human macrophages. *Journal of Clinical Endocrinology and Metabolism*, **64**, 1–9.

Reichel, H., Koeffler, H.P., Tobler, A. & Norman, A.W. (1987b) 1α,25-Dihydroxyvitamin D_3 inhibits γ-interferon synthesis by normal human peripheral blood lymphocytes. *Proceedings of the National Academy of Sciences of the U.S.A.*, **84**, 3385–9.

Reinhardt, T.A. & Horst, R.L. (1990) Parathyroid hormone down-regulates 1,25-dihydroxyvitamin D receptors (VDR) and VDR messenger ribonucleic acid *in vitro* and blocks homologous up-regulation of VDR *in vivo*. *Endocrinology*, **127**, 942–8.

Rhodes, S.J., Chen, R., DiMattia, G.E., Scully, K.M., Kalla, K.A., Lin, S.-C., Yu, V.C. & Rosenfeld, M.G. (1993) A tissue-specific enhancer confers Pit-1-dependent morphogen inducibility and autoregulation on the *pi-1* gene. *Genes & Development*, **7**, 913–32.

Rigby, W.F.C., Noelle, R.J., Krause, K. & Fanger, M.W. (1985) The effects of 1,25-dihydroxyvitamin D_3 on human T lymphocyte activation and proliferation: A cell cycle analysis. *Journal of Immunology*, **135**, 2279–86.

Roche, C., Bellaton, C., Pansu, D., Miller, A. III. & Bronner, F. (1986) Localization of vitamin D-dependent active Ca^{2+} transport in rat duodenum and relation to CaBP. *American Journal of Physiology*, **251**, G314–20.

Rohde, C.M., Manatt, M., Clagett-Dame, M. & DeLuca, H.F. (1999) Vitamin Aantagonizes the action of vitamin D in rats. *Journal of Nutrition*, **129**, 2246–50.

Romas, E., Udagawa, N., Zhou, H., Tamura, T., Saito, M., Taga, T., Hilton, D.J., Suda, T., Ng, K.W. & Martin, T.J. (1996) The role of gp130-mediated signals in osteoclast development: Regulation of interleukin 11 production by osteoblasts and distribution of its receptor in bone marrow cultures. *Journal of Experimental Medicine*, **183**, 2581–91.

Rook, G.A.W., Steele, J., Fraher, L., Barker, S., Karmali, R., O'Riordan, J. & Stanford, J. (1986) Vitamin D_3, gamma interferon, and control of proliferation of *Mycobacterium tuberculosis* by human monocytes. *Immunology*, **57**, 159–63.

Rook, G.A.W., Taverne, J., Leveton, C. & Steele, J. (1987) The role of gamma-interferon, vitamin D_3 metabolites and tumour necrosis factor in the pathogenesis of tuberculosis. *Immunology*, **62**, 229–34.

Rostand, S.G. & Drüeke, T.B. (1999) Parathyroid hormone, vitamin D, and cardiovascular disease in chronic renal failure. *Kidney International*, **56**, 383–92.

Rowe, D.W. & Kream, B.E. (1982) Regulation of collagen synthesis in fetal rat calvaria by 1,25-dihydroxyvitamin D_3. *Journal of Biological Chemistry*, **257**(14), 8009–15.

Salomon, B., Lenschow, D.J., Rhee, L., Ashourian, N., Singh, B., Sharpe, A. & Bluestone, J.A. (2000) B7/CD28 costimulation is essential for the homeostasis of the $CD4^{+}CD25^{+}$ immunoregulatory T cells that control autoimmune diabetes. *Immunity*, **12**, 431–40.

Sampson, H.W., Matthews, J.L., Martin, J.H. & Kunin, A.S. (1970) An electron microscopic localization of calcium in the small intestine of normal, rachitic, and vitamin D-treated rats. *Calcified Tissue Research*, **5**, 305–16.

Sato, T., Hong, M.H., Jin, C.H., Ishimi, Y., Udagawa, N., Shinki, T., Abe, E. & Suda, T. (1991) The specific production of the third component of complement by osteoblastic cells treated with 1α,25-dihydroxyvitamin D_3. *FEBS Letters*, **285**, 21–4.

Scaglione-Sewell, B., Abraham, C., Bissonnette, M., Skarosi, S.F., Hart, J., Davidson, N.O., Wali, R.K., Davis, B.H., Sitrin, M. & Brasitus, T.A. (1998) Decreased PKC-α expression increases cellular proliferation, decreases differentiation, and enhances the transformed phenotype of Caco-2 cells. *Cancer Research*, **58**, 1074–81.

Schräder, M., Müller, K.M., Nayeri, S., Kahlen, J.-P. & Carlberg, C. (1994) Vitamin D_3—Thyroid hormone receptor heterodimer polarity directs ligand sensitivity of transactivation. *Nature*, **370**, 382–6.

Schräder, M., Nayeri, S., Kahlen, J.-P., Müller, K.M. & Carlberg, C. (1995) Natural vitamin D_3 response elements formed by inverted palindromes: Polarity-directed ligand sensitivity of vitamin D_3 receptor-retinoid X receptor heterodimer-mediated transactivation. *Molecular and Cellular Biology*, **15**, 1154–61.

Schräder, M., Kahlen, J.-P. & Carlberg, C. (1997) Functional characterization of a novel type of 1α,25-dihydroxyvitamin D_3 response element identified in the mouse c-*fos* promoter. *Biochemical and Biophysical Research Communications*, **230**, 646–51.

Schrivastava, A. & Calame, K. (1994) An analysis of genes regulated by the multi-functional transcriptional regulator Yin Yang-1. *Nucleic Acids Research*, **22**, 5151–5.

Schüle, R., Umesono, K., Mangelsdorf, D.J., Bolado, J., Pike, J.W. & Evans, R.M. (1990) Jun—Fos and receptors for vitamins A and D recognize a common response element in the human osteocalcin gene. *Cell*, **61**, 497–504.

Schwab, S.J., Klahr, S. & Hammerman, M.R. (1984) Na^{+} gradient-dependent Pi uptake in basolateral membrane vesicles from dog kidney. *American Journal of Physiology*, **246**, F663–9.

Seo, E.-G. & Norman, A.W. (1997) Three-fold induction of renal 25-hydroxyvitamin D_3-24-hydroxylase activity and increased serum 24,25-dihydroxyvitamin D_3 levels are correlated with the healing process after chick tibial fracture. *Journal of Bone and Mineral Research*, **12**, 598–606.

Seo, E.-G., Schwartz, Z., Dean, D.D., Norman, A.W. & Boyan, B.D. (1996) Preferential accumulation *in vivo* of 24R,25-dihydroxyvitamin D_3 in growth plate cartilage of rats. *Endocrine*, **5**, 147–55.

Seo, E.-G., Einhorn, T.A. & Norman, A.W. (1997) 24R,25-Dihydroxyvitamin D_3: An essential vitamin D_3 metabolite for both normal bone integrity and healing of tibial fracture in chicks. *Endocrinology*, **138**, 3864–72.

Sergeev, I.N. (2016) 1,25-dihydroxyvitamin D_3 and type 2 diabetes: Ca^{2+}-dependent molecular mechanisms and the role of vitamin D status. *Hormone Molecular Biology and Clinical Investigation*, **26**, 61–5.

Sherr, C.J. (1996) Cancer cell cycles. *Science*, **274**, 1672–7.

Shimada, T., Hasegawa, H., Yamazaki, Y., Muto, T., Hino, R., Takeuchi, Y., Fujita, T., Nakahara, K., Fukumoto, S. & Yamashita, T. (2004) FGF-23 is a potent regulator of vitamin D metabolism and phosphate homeostasis. *Journal of Bone and Mineral Research*, **19**, 429–35.

Shinki, T., Jin, C.H., Nishimura, A., Nagai, Y., Ohyama, Y., Noshiro, M., Okuda, K. & Suda, T. (1992) Parathyroid hormone inhibits 25-hydroxyvitamin D_3-24-hydroxylase mRNA expression stimulated by 1α,25-dihydroxyvitamin D_3 in rat kidney but not in intestine. *Journal of Biological Chemistry*, **267**(19), 13757–62.

Shirakawa, A.-K., Nagakubo, D., Hieshima, K., Nakayama, T., Jin, Z. & Yoshie, O. (2008) I,25-Dihydroxyvitamin D_3 induces CCR10 expression in terminally differentiating human B cells. *Journal of Immunology*, **180**, 2786–95.

Shiraki, M., Kushida, K., Yamazaki, K., Nagai, T., Inoue, T. & Orimo, H. (1996) Effects of 2 year's treatment of osteoporosis with 1α-hydroxy vitamin D_3 on bone mineral density and incidence of fracture: A placebo-controlled, double-blind prospective study. *Endocrine Journal*, **43**, 211–20.

Sierra, J., Villagra, A., Paredes, R., Cruzat, F., Gutierrez, S., Javed, A., Arriagda, G., Olate, J., Imschenetzky, M., van Wijnen, A.J., Lian, J.B., Stein, G.S., Stein, J.L. & Montecino, M. (2003) Regulation of the bone-specific osteocalcin gene by p300 requires Runx2/Cbfal and the vitamin D_3 receptor but not p300 intrinsic histone acetyltransferase activity. *Molecular and Cellular Biology*, **23**, 3339–51.

Sigmund, C.D. (2002) Regulation of renin expression and blood pressure by vitamin D_3. *Journal of Clinical Investigation*, **110**, 155–6.

Sigmundsdóttir, H., Pan, J., Debes, G.F., Alt, C., Habtezion, A., Soler, D. & Butcher, E.C. (2007) DCs metabolize sunlight-induced vitamin D_3 to 'program' T cell attraction to the epidermal chemokine CCL27. *Nature Immunology*, **8**, 285–93.

Silver, J., Russell, J. & Sherwood, L.M. (1985) Regulation by vitamin D metabolites of messenger ribonucleic acid for preproparathyroid hormone in isolated bovine parathyroid cells. *Proceedings of the National Academy of Sciences of the U.S.A.*, **82**, 4270–3.

Silver, J., Naveh-Many, T., Mayer, H., Schmeizer, H.J. & Popvtzer, M.M. (1986) Regulation by vitamin D metabolites of parathyroid hormone gene transcription in vivo in the rat. *Journal of Clinical Investigation*, **78**, 1296–301.

Silver, J., Kilav, R. & Naveh-Many, T. (2002) Mechanisms of secondary hyperparathyroidism. *American Journal of Physiology (Renal Physiology)*, **283**, F367–76.

Simboli-Campbell, M., Gagnon, A.M., Franks, D.J. & Welsh, J.E. (1994) 1,25-Dihydroxyvitamin D_3 translocates protein kinase Cβ to nucleus and enhances plasma membrane association of protein kinase Cα in renal epithelial cells. *Journal of Biological Chemistry*, **269**(5), 3257–64.

Skowronski, R.J., Peehl, D.M. & Feldman, D. (1995) Actions of vitamin D_3 analogs on human prostate cancer cell lines: Comparison with 1,25-dihydroxyvitamin D_3. *Endocrinology*, **136**, 20–6.

Smith, E.L., Walworth, N.C. & Holick, M.F. (1986) Effect of 1α,25-dihydroxyvitamin D_3 on the morphologic and biochemical differentiation of cultured human epidermal keratinocytes grown in serum-free conditions. *Journal of Investigative Dermatology*, **86**, 709–14.

Sneddon, W.B., Bogado, C.E., Kiernan, M.S. & Demay, M.B. (1997) DNA sequences downstream from the vitamin D response element of the rat osteocalcin gene are required for ligand-dependent transactivation. *Molecular Endocrinology*, **11**, 210–17.

Somasundaram, K. (2003) Breast cancer gene 1 (BRCA1): Role in cell cycle regulation and DNArepair—Perhaps through transcription. *Journal of Cellular Biochemistry*, **88**, 1084–91.

Sorensen, A.M. & Baran, D.T. (1993) 1α,25-Dihydroxyvitamin D_3 rapidly alters phospholipid metabolism in the nuclear envelope of osteoblasts. *Journal of Bone and Mineral Research*, **8**, S127.

Sorensen, A.M., Bowman, D. & Baran, D.T. (1993) 1α,25-Dihydroxyvitamin D_3 rapidly increases nuclear calcium levels in rat osteosarcoma cells. *Journal of Cellular Biochemistry*, **52**, 237–42.

Spina, C., Tangpricha, V., Yao, M., Zhou, W., Wolfe, M.M., Maehr, H., Uskokovic, M., Adorini, L. & Holick, M.F. (2005) Colon cancer and solar ultraviolet B radiation and prevention and treatment of colon cancer in mice with vitamin D and its Gemini analogs. *Journal of Steroid Biochemistry & Molecular Biology*, **97**, 111–20.

Sprenger, C.C., Peterson, A., Lance, R., Ware, J.L., Drivdahl, R.H. & Plymate, S.R. (2001) Regulation of proliferation of prostate epithelial cells by 1,25-dihydroxyvitamin D_3 is accompanied by an increase in insulin-like growth factor binding protein-3. *Journal of Endocrinology*, **170**, 609–18.

Staeva-Vieira, T.P. & Freedman, L.P. (2002) 1,25-Dihydroxyvitamin D_3 inhibits IFN-γ and IL-4 levels during in vitro polarization of primary murine $CD4^+$ T cells. *Journal of Immunology*, **168**, 1181–9.

St-Arnaud, R. (1999) Targeted inactivation of vitamin D hydroxylases in mice. *Bone*, **25**, 127–9.

St-Arnaud, R., Arabian, A., Travers, R., Barletta, F., Raval-Pandya, M., Chapin, K., Depovere, J., Mathieu, C., Christakos, S., Demay, M.B. & Glorieux, F.H. (2000) Deficient mineralization of intramembranous bone in vitamin D-24-hydroxylase-ablated mice is due to elevated 1,25-dihydroxyvitamin D and not to the absence of 24,25-dihydroxyvitamin D. *Endocrinology*, **141**, 2658–66.

Stein, G.S. & Lian, J.B. (1993) Molecular mechanisms mediating proliferation/differentiation interrelationships during progressive development of the osteoblast phenotype. *Endocrine Reviews*, **14**, 424–42.

Stein, W.D. (1992) Facilitated diffusion of calcium across the rat intestinal epithelial cell. *Journal of Nutrition*, **122**, 651–6.

Stewart, L.V. & Weigel, N.L. (2005) Role of insulin-like growth factor binding proteins in 1α,25-dihydroxyvitamin D_3-induced growth inhibition of human prostate cancer cells. *Prostate*, **64**, 9–19.

Stoffels, K., Overbergh, L., Giulietti, A., Verlinden, L., Bouillon, R. & Mathieu, C. (2006) Immune regulation of 25-hydroxyvitamin-D_3-1α-hydroxylase in human monocytes. *Journal of Bone and Mineral Research*, **21**, 37–47.

Subramaniam, N., Leong, G.M., Cock, T.-E., Flanagan, J.L., Fong, C., Eisman, J.A. & Kouzmenko, A.P. (2001) Cross-talk between 1,25-dihydroxyvitamin D_3 and transforming growth factor-β signaling requires binding of VDR and Smad3 proteins to their cognate DNA recognition elements. *Journal of Biological Chemistry*, **276**(19), 15741–6.

Suda, T., Takahashi, N. & Abe, E. (1992) Role of vitamin D in bone resorption. *Journal of Cellular Biochemistry*, **49**, 53–8.

Sun, J., Kong, J., Duan, Y., Szeto, F.L., Liao, A., Madara, J.L. & Li, Y.C. (2006) Increased NF-κB activity in fibroblasts lacking the vitamin D receptor. *American Journal of Physiology*, **291**, E315–22.

Takahashi, N., Yamana, H., Yoshiki, S., Roodman, G.D., Mundy, G.M., Jones, S.J., Boyde, A. & Suda, T. (1988) Osteoclast-like cell formation and its regulation by osteotropic hormones in mouse bone marrow cultures. *Endocrinology*, **122**, 1373–82.

Taketani, Y., Miyamoto, K., Tanaka, K., Katai, K., Chikamori, M., Tatsumi, S., Segawa, H., Yamamoto, H., Morita, K. & Takeda, E. (1997) Gene structure and functional analysis of the human Na^+/phosphate co-transporter. *Biochemical Journal*, **324**, 927–34.

Takito, J., Shinki, T., Sasaki, T. & Suda, T. (1990) Calcium uptake by brush-border and basolateral membrane vesicles in chick duodenum. *American Journal of Physiology*, **258**, G16–23.

Theodosiou, A., Smith, A., Gillieron, C., Arkinstall, S. & Ashworth, A. (1999) MKP5, a new member of the MAP kinase phosphatase family, which selectively dephosphorylates stress-activated kinases. *Oncogene*, **18**, 6981–8.

Thompson, P.D., Jurutka, P.W., Haussler, C.A., Whitfield, G.K. & Haussler, M.R. (1998) Heterodimeric DNA binding by the vitamin D receptor and retinoid X receptors is enhanced by 1,25-dihydroxyvitamin D_3 and inhibited by 9-*cis*-retinoic acid. Evidence for allosteric receptor interactions. *Journal of Biological Chemistry*, **273**(14), 8483–91.

Thompson, P.D., Hsieh, J.-C., Whitfield, G.K., Haussler, C.A., Jurutka, P.W., Galligan, M.A., Tillman, J.B., Spindler, S.R. & Haussler, M.R. (1999) Vitamin D receptor displays DNA binding and transactivation as a heterodimer with the retinoid X receptor, but not with the thyroid hormone receptor. *Journal of Cellular Biochemistry*, **75**, 462–80.

Tian, X.Q., Chen, T.C., Matsuoka, L.Y., Wortsman, J. & Holick, M.F. (1993) Kinetic and thermodynamic studies of the conversion of previtamin D_3 to vitamin D_3 in human skin. *Journal of Biological Chemistry*, **268**(20), 14888–92.

Tornquist, K. & Tashjian, A.H. Jr. (1989) Dual actions of 1,25-dihydroxycholecalciferol on intracellular Ca^{2+} in GH4C1 cells: Evidence for effects on voltage-operated Ca^{2+} channels and Na^+/Ca^{2+} exchange. *Endocrinology*, **124**, 2765–76.

Trembleau, S., Penna, G., Bosi, E., Mortara, A., Gately, M.K. & Adorini, L. (1995) Interleukin 12 administration induces T helper type 1 cells and accelerates autoimmune diabetes in NOD mice. *Journal of Experimental Medicine*, **181**, 817–21.

Tsoukas, C.D., Provvedini, D.M. & Manolagas, S.C. (1984) 1,25-Dihydroxyvitamin D_3: A novel immunoregulatory hormone. *Science*, **224**, 1438–40.

van Driel, M., Koedam, M., Buurman, C.J., Hewison, M., Chiba, H., Uitterlinden, A.G., Pols, H.A.P. & van Leeuwen, J.P.T.M. (2006) Evidence for auto/paracrine actions of vitamin D in bone: 1α-hydroxylase expression and activity in human bone cells. *FASEB Journal*, **20**, E1811–19.

Vanherwegen, A.-S., Eelen, G., Ferreira, G.B., Ghesquière, B., Cook, D.P., Nikolic, T., Roep, B., Carmeliet, P., Telang, S., Mathieu, C. & Gysemans, C. (2019) Vitamin D controls the capacity of human dendritic cells to induce functional regulatory T cells by regulation of glucose metabolism. *Journal of Steroid Biochemistry and Molecular Biology*, **187**, 134–45.

Veldscholte, J., Berrevoets, C.A., Brinkmann, A.O., Grootegoed, J.A. & Mulder, E. (1992a) Anti-androgens and the mutated androgen receptor of LNCaP cells: Differential effects on binding affinity, heat-shock protein interaction, and transcription activation. *Biochemistry*, **31**, 2393–9.

Veldscholte, J., Berrevoets, C.A., Ris-Stalpers, C., Kuiper, G.G.J.M., Jenster, G., Trapman, J., Brinkmann, A.O. & Mulder, E. (1992b) The androgen receptor in LNCaP cells contains a mutation in the ligand binding domain which affects steroid binding characteristics and response to anti-androgens. *Journal of Steroid Biochemistry and Molecular Biology*, **41**, 665–9.

Vennekens, R., Hoenderop, J.G.J., Prenen, J., Stuiver, M., Willems, P.H.G.M., Droogmans, G., Nilius, B. & Bindels, R.J.M. (2000) Permeation and gating properties of the novel epithelial Ca^{2+} channel. *Journal of Biological Chemistry*, **275**(6), 3963–9.

Vieth, R. (1990) The mechanisms of vitamin D toxicity. *Bone and Mineral*, **11**, 267–72.

Vieth, R. (2006) Critique of the considerations for establishing the tolerable upper intake level for vitamin D: Critical need for revision upwards. *Journal of Nutrition*, **136**, 1117–22.

Wali, R.K., Baum, C.L., Sitrin, M.D. & Brasitus, T.A. (1990) 1,25(OH)2 vitamin D_3 stimulates membrane phosphoinositide turnover, activates protein kinase C, and increases cytosolic calcium in rat colonic epithelium. *Journal of Clinical Investigation*, **85**, 1296–303.

Wali, R.K., Bissonnette, M., Khare, S., Hart, J., Sitrin, M.D. & Brasitus, T.A. (1995) 1α,25-Dihydroxy-16-ene-23-yne-26,27-hexafluorocholecalciferol, a noncalcemic analogue of 1α,25-dihydroxyvitamin D_3, inhibits azoxymethane-induced colonic tumorigenesis. *Cancer Research*, **55**, 3050–4.

Wali, R.K., Kong, J., Sitrin, M.D., Bissonnette, M. & Li, Y.C. (2003) Vitamin D receptor is not required for the rapid actions of 1,25-dihydroxyvitamin D_3 to increase intracellular calcium and activate protein kinase C in mouse osteoblasts. *Journal of Cellular Biochemistry*, **88**, 794–801.

Walker, J.J., Yan, T.S. & Quamme, G.A. (1987) Presence of multiple sodium-dependent phosphate transport processes in proximal brush-border membranes. *American Journal of Physiology*, **252**, F226–31.

Walters, J.R.F. (1989) Calbindin-D9k stimulates the calcium pump in rat enterocyte basolateral membranes. *American Journal of Physiology*, **256**, G124–8.

Walters, J.R.F. & Weiser, M.M. (1987) Calcium transport by rat duodenal villus and crypt basolateral membranes. *American Journal of Physiology*, **252**, G170–7.

Walters, S.N., Reinhardt, T.A., Dominick, M.A., Horst, R.L. & Littledike, E.T. (1986) Intracellular location of unoccupied 1,25-dihydroxyvitamin D receptors: A nuclear—Cytoplasmic equilibrium. *Archives of Biochemistry and Biophysics*, **246**, 366–73.

Wang, T.-T., Nestel, F.P., Bourdeau, V., Nagai, Y., Wang, Q., Liao, J., Tavera-Mendoza, L., Lin, R., Hanrahan, J.H., Mader, S. & White, J.H. (2004) Cutting edge: 1,25-dihydroxyvitamin

D_3 is a direct inducer of antimicrobial peptide gene expression. *Journal of Immunology*, **173**, 2909–12.

Wang, W., Soto, H., Oldham, E.R., Buchanan, M.E., Homey, B., Catron, D., Jenkins, N., Copeland, N.G., Gilbert, D.J., Nguyen, N., Abrams, J., Kershenovich, D., Smith, K., McClanahan, T., Vicari, A.P. & Zlotnik, A. (2000) Identification of a novel chemokine (CCL28), which binds CCR10 (GPR2). *Journal of Biological Chemistry*, **275**(29), 22313–23.

Wasserman, R.H. & Fullmer, C.S. (1995) Vitamin D and intestinal calcium transport: Facts, speculations and hypotheses. *Journal of Nutrition*, **125**, 1971S–79S.

Wasserman, R.H., Smith, C.A., Brindak, M.E., De Talamoni, N., Fullmer, C.S., Penniston, J.T. & Kumar, R. (1992) Vitamin D and mineral deficiencies increase the plasma membrane calcium pump of chicken intestine. *Gastroenterology*, **102**, 886–94.

Webb, A.R., Kline, L. & Holick, M.F. (1988) Influence of season and latitude on the cutaneous synthesis of vitamin D_3: Exposure to winter sunlight in Boston and Edmonton will not promote vitamin D_3 synthesis in human skin. *Journal of Clinical Endocrinology and Metabolism*, **67**, 373–8.

Webb, A.R., deCosta, B.R. & Holick, M.F. (1989) Sunlight regulates the cutaneous production of vitamin D_3 by causing its photodegradation. *Journal of Clinical Endocrinology and Metabolism*, **68**, 882–7.

Weishaar, R.E. & Simpson, R.U. (1987a) Vitamin D_3 and cardiovascular function in rats. *Journal of Clinical Imvestigation*, **79**, 1706–12.

Weishaar, R.E. & Simpson, R.U. (1987b) Involvement of vitamin D_3 with cardiovascular function. II. Direct and indirect effects. *American Journal of Physiology (Endocrinology and Metabolism)*, **253**, E675–83.

Whitmarsh, A.J. & Davis, R.J. (1996) Transcription factor AP-1 regulation by mitogen-activated protein kinase signal transduction pathways. *Journal of Molecular Medicine*, **74**, 589–607.

Wilson, H.D., Schedl, H.P. & Christensen, K. (1989) Calcium uptake by brush-border membrane vesicles from the rat intestine. *American Journal of Physiology*, **257**, F446–53.

Wing, K., Onishi, Y., Prieto-Martin, P., Yamaguchi, T., Miyara, M., Fehervari, Z., Nomura, T. & Sakaguchi, S. (2008) CTLA-4 control over Foxp3$^+$ regulatory T cell function. *Science*, **322**, 271–5.

Winzen, R., Kracht, M., Ritter, B., Wilhelm, A., Chen, C.-Y.A., Shyu, A.-B., Müller, M., Gaestel, M., Resch, K. & Holtmann, H. (1999) The p38 MAP kinase pathway signals for cytokine-induced mRNAstabilization via MAP kinase-activated protein kinase 2 and an AU-rich region-targeted mechanism. *EMBO Journal*, **18**, 1969–80.

Wood, R.J., Fleet, J.C., Cashman, K., Bruns, M.E. & DeLuca, H.F. (1998) Intestinal calcium absorption in the aged rat: Evidence of intestinal resistance to $1,25(OH)_2$ vitamin D. *Endocrinology*, **139**, 3843–8.

Xiang, W., Kong, J., Chen, S., Cao, L.-P., Qiao, G., Zheng, W., Liu, W., Li, X., Gardner, D.G. & Li, Y.C. (2005) Cardiac hypertrophy in vitamin D receptor knockout mice: Role of the systemic and cardiac renin-angiotensin systems. *American Journal of Physiology (Endocrinology and Metabolism)*, **288**, E125–32.

Xie, Z. & Bikle, D.D. (1997) Cloning of the human phospholipase C-γ1 promoter and identification of a DR6-type vitamin D-responsive element. *Journal of Biological Chemistry*, **272**(10), 6573–7.

Xiong, Y., Hannon, G.J., Zhang, H., Casso, D., Kobayashi, R. & Beach, D. (1993) p21 is a universal inhibitor of cyclin kinases. *Nature*, **366**, 701–4.

Yan, T., Walker, J. & Quamme, G. (1988) Sodium-independent phosphate transport in brush-border membrane vesicles prepared from the outer medulla of pig kidneys. *Progress in Clinical and Biological Research*, **252**, 81–6.

Yanagi, Y., Suzawa, M., Kawabata, M., Miyazono, K., Yanagisawa, J. & Kato, S. (1999) Positive and negative modulation of vitamin D receptor function by transforming growth factor-β signaling through Smad proteins. *Journal of Biological Chemistry*, **274**(19), 12971–4.

Yanagisawa, J., Yanagi, Y., Masuhiro, Y., Suzawa, M., Watanabe, M., Kashiwagi, K., Toriyabe, T., Kawabata, M., Miyazono, K. & Kato, S. (1999) Convergence of transforming growth factor-β and vitamin D signaling pathways on SMAD transcriptional coactivators. *Science*, **283**, 1317–21.

Yang, S., Smith, C. & DeLuca, H.F. (1993a) 1α,25-Dihydroxyvitamin D_3 and 19-nor-1α,25-dihydroxyvitamin D_2 suppress immunoglobulin production and thymic lymphocyte proliferation in vivo. *Biochimica et Biophysica Acta*, **1158**, 279–86.

Yang, S., Smith, C., Prahl, J.M., Luo, X. & DeLuca, H.F. (1993b) Vitamin D deficiency suppresses cell-mediated immunity *in vivo*. *Archives of Biochemistry and Biophysics*, **303**, 98–106.

Ylikomi, T., Laaksi, I., Lou, Y.-R., Martikainen, P., Miettinen, S., Pennanen, P., Purmonen, S., Syvälä, H., Vienonen, A. & Tuohimaa, P. (2002) Antiproliferative action of vitamin D. *Vitamins and Hormones*, **64**, 357–406.

Zelinski, J.M., Sykes, D.E. & Weiser, M.M. (1991) The effect of vitamin D on rat intestinal plasma membrane Ca-pump mRNA. *Biochemical and Biophysical Research Communications*, **179**, 749–55.

Zella, J.B. & DeLuca, H.F. (2003) Vitamin D and autoimmune diabetes. *Journal of Cellular Biochemistry*, **88**, 216–22.

Zella, J.B., McCary, L.C. & DeLuca, H.F. (2003) Oral administration of 1,25-dihydroxyvitamin D_3 completely protects NOD mice from insulin-dependent diabetes mellitus. *Archives of Biochemistry and Biophysics*, **417**, 77–80.

Zhang, Q., Wrana, J.L. & Sodek, J. (1992) Characterizaton of the promoter region of the porcine *opn* (osteopontin, secreted phosphoprotein 1) gene. Identificaton of positive and negative regulatory elements and a 'silent' second promoter. *European Journal of Biochemistry*, **207**, 649–59.

Zhao, X.-Y., Ly, L.H., Peehl, D.M. & Feldman, D. (1997) 1α,25-Dihydroxyvitamin D_3 actions in LNCaP human prostate cancer cells are androgen-dependent. *Endocrinology*, **138**, 3290–8.

Zhao, X.-Y., Ly, L.H., Peehl, D.M. & Feldman, D. (1999) Induction of androgen receptor by 1α,25-dihydroxyvitamin D_3 and 9-cis retinoic acid in LNCaP human prostate cancer cells. *Endocrinology*, **140**, 1205–12.

Zhao, X.-Y., Peehl, D.M., Navone, N.M. & Feldman, D. (2000) 1α,25-Dihydroxyvitamin D_3 inhibits prostate cancer cell growth by androgen-dependent and androgen-independent mechanisms. *Endocrinology*, **141**, 2548–56.

Zhuang, S.-H. & Burnstein, K.L. (1998) Antiproliferative effect of 1α,25-dihydroxyvitamin D_3 in human prostate cancer cell line LNCaP involves reduction of cyclin-dependent kinase 2 activity and persistent G1 accumulation. *Endocrinology*, **139**, 1197–207.

Zhuang, S.-H., Schwartz, G.G., Cameron, D. & Burnstein, K.L. (1997) Vitamin D receptor content and transcriptional activity do not fully predict antiproliferative effects of vitamin D in human prostate cancer cell lines. *Molecular and Cellular Endocrinology*, **126**, 83–90.

Zierold, C., Darwish, H.M. & DeLuca, H.F. (1995) Two vitamin D response elements function in the rat 1,25-dihydroxyvitamin D 24-hydroxylase promoter. *Journal of Biological Chemistry*, **270**(4), 1675–8.

Zierold, C., Mings, J.A. & DeLuca, H.F. (2003) Regulation of 25-hydroxyvitamin D_3-24 hydroxylase mRNAby 1,25-dihydroxyvitamin D_3 and parathyroid hormone. *Journal of Cellular Biochemistry*, **88**, 234–7.

Zittermann, A., Schleithoff, S.S. & Koerfer, R. (2006) Vitamin D insufficiency in congestive heart failure: Why and what to do about it? *Heart Failure Reviews*, **11**, 25–33.

17 Vitamin E

17.1 CHEMISTRY

Vitamin E is represented by a family of structurally related compounds (vitamers) which show pronounced quantitative differences in biological activity. Eight of these vitamers are known to occur in nature, having been isolated from vegetable oils and other plant materials. The vitamers comprise four tocopherols and four tocotrienols whose parent structures are tocol and tocotrienol, respectively. They are designated as alpha- (α), beta- (β), gamma- (γ), and delta- (δ) according to the number and position of substituent methyl groups on the chromanol ring of tocol or tocotrienol (Table 17.1).

Tocotrienol differs from tocol in having three double bonds in the phytyl side chain. The tocol molecule exhibits optical isomerism attributable to the three asymmetric carbon atoms at positions 2, 4′, and 8′, thus the tocopherols can exist as one of eight possible diastereoisomers. The tocotrienols possess only one center of asymmetry at position 2, in addition to sites of geometrical isomerism at positions 3′ and 7′ in the unsaturated side chain. The *RS* system of asymmetric configuration is used to specify the chirality of vitamin E compounds. In nature, the tocopherols exist exclusively as their 2*R*, 4′ *R*, 8′ *R* stereoisomers (*RRR*-forms), while the tocotrienols exist exclusively in the 2*R*, 3′ *trans*, 7′ *trans* configuration. The stereochemical structures of tocol and tocotrienol are depicted in Figure 17.1.

Totally synthetic α-tocopherol is produced by the condensation of trimethylhydroquinone with synthetic phytol or isophytol. This method of synthesis results in all-racemic 2*RS*, 4′ *RS*, 8′ *RS*-α-tocopherol (all-*rac*-α-tocopherol), which is a mixture of all eight possible diastereoisomers in virtually equal proportions. The four enantiomeric pairs are *RRR*/*SSS*, *RRS*/*SSR*, *RSS*/*SRR*, and *RSR*/*SRS*. In this chapter, the term 'vitamin E' is used as the generic descriptor for all tocol and tocotrienol derivatives that exhibit qualitatively the biological activity of *a*-tocopherol. The term 'tocopherol' refers specifically to the methyl-substituted derivatives of tocol and is not synonymous with the term 'vitamin E.' The tocopherols and tocotrienols may be referred to collectively as tocochromanols.

17.2 STABILITY

Vitamin E vitamers are slowly oxidized by atmospheric oxygen to form mainly biologically inactive quinones; the oxidation is accelerated by light, heat, alkalinity, and certain trace metals. The vitamers can withstand heating in acid or alkaline solutions provided that oxygen and UV radiation are excluded. The presence of ascorbic acid completely prevents the catalytic effect of iron(III) and copper(II) on vitamin E oxidation by maintaining these metals in their lower oxidation states (Cort *et al.*, 1978). The tocotrienols, by virtue of their unsaturated side chains, are more susceptible to destruction than the tocopherols.

Vitamin E is available commercially for use as a dietary supplement in both its natural form, *RRR*-α-tocopherol, and synthetic form, all-*rac*-α-tocopherol. The bioavailability of all-*rac*-α-tocopherol in humans is roughly half that of *RRR*-α-tocopherol (Burton *et al.*, 1998). These forms of α-tocopherol are usually sold as acetate esters or, less frequently, as succinate esters. Because tocopherol esters lack the reactive hydroxyl group, air and light have practically no destructive effect. In the food industry, α-tocopherol is frequently used as an antioxidant to stabilize animal fats: tocopheryl esters cannot function as antioxidants.

17.3 BIOPOTENCY AND INTERCONVERSION OF VITAMIN E UNITS

The relative percentage biopotencies of tocopherols and tocotrienols evaluated by the rat resorption–gestation assay are listed in Table 17.2.

Manufacturers of vitamin E-fortified foods and supplements continue to label their products with the international unit (IU) of vitamin E activity, which is defined as 1 mg of all-*rac*-α-tocopheryl acetate or 0.67 mg *RR*R-α-tocopherol. Vitamin E activity is also reported as α-tocopherol equivalents (α-TE). Factors for the conversion of the tocopherols and tocotrienols to α-TE units are based on the biological activity of the various forms as determined using the rat fetal resorption assay. α-TEs are defined as:

α-TE = α-tocopherol, mg × 1.0; β-tocopherol, mg × 0.5; γ-tocopherol, mg × 0.1; δ-tocopherol, mg × 0.03; α-tocotrienol, mg × 0.3; β-tocotrienol, mg × 0.05. The biological activities of γ- and δ-tocotrienol were below detection (Institute of Medicine, 2000).

It is now known that only the naturally occurring *RRR*-α-tocopherol and the 2*R*-stereoisomeric forms of α-tocopherol (*RRR*-, *RRS*-, *RSS*-, and *RSR*-α-tocopherol) obtained from synthetic all-*rac*-α-tocopherol ester are maintained in plasma and delivered to tissues. This is because of the discriminating ability of the hepatic α-tocopherol transfer protein and the fact that the vitamin E vitamers are not interconvertible in the human body. Therefore, when establishing recommended intakes, the definition of vitamin E is limited to the 2*R*-stereoisomeric forms of α-tocopherol. Based on this definition, all-*rac*-α-tocopherol has one-half the activity of *RRR*-α-tocopherol (Institute of Medicine, 2000). This 2:1 activity ratio for natural and synthetic

DOI: 10.1201/9781003410300-20

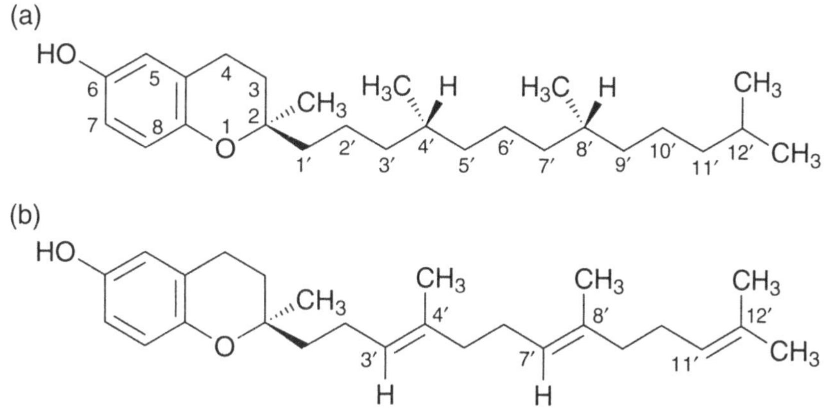

FIGURE 17.1 Stereochemical structures of tocol and tocotrienol. (a) *RRR*-tocol, (b) 2*R*, 3′-*trans*, 7′-*trans*-tocotrienol.

TABLE 17.1
Designation of Tocopherols and Tocotrienols

Methyl Substitution of Tocol or Tocotrienol Molecule	Tocopherol (T)	Tocotrienol (T3)
5,7,8-Trimethyl	α-T	α-T3
5,8-Dimethyl	β-T	β-T
7,8-Dimethyl	γ-T	γ-T3
8-Methyl	δ-T	δ-T3

TABLE 17.2
Relative Percentage Biopotencies of Vitamin E Compounds by the Rat Fetal Resorption Assay

Compound[a]	Relative Biopotency (%)	Reference
α-Tocopherol	100	Joffe and Harris (1943)
β-Tocopherol	40	Joffe and Harris (1943)
γ-Tocopherol	8	Joffe and Harris (1943)
δ-Tocopherol	1	Stern *et al.* (1947)
α-Tocotrienol	21	Bunyan *et al.* (1961)
β-Tocotrienol	4	Bunyan *et al.* (1961)

[a] *RRR* forms.

vitamin E has been demonstrated in human studies (Burton *et al.*, 1998). In food analysis, the individual E vitamers can be separated and measured by high-performance liquid chromatography.

17.4 FOOD SOURCES

The major naturally occurring sources of vitamin E are cereal grain products, nuts, beans, and seeds. Vegetable oils are the richest dietary sources of vitamin E. The distribution of tocopherols and tocotrienols varies greatly among vegetable oils, with different vitamers predominating (Table 17.3) (Syväoja *et al.*, 1986). Fish, meat, eggs, dairy products, and green leafy vegetables also provide significant amounts. Major food sources of vitamin E in the United States include margarine, mayonnaise and salad dressings, fortified breakfast cereals, vegetable shortenings and cooking oils, peanut butter, eggs, potato crisps, whole milk, tomato products, and apples (Sheppard *et al.*, 1993). In meat and meat products, most of the vitamin E is in the form of α-tocopherol, but in foods of plant origin, α-tocopherol is frequently accompanied by smaller amounts of other vitamers, particularly γ-tocopherol. In wheat flour, the predominant E vitamer is β-tocotrienol.

17.5 ABSORPTION, TRANSPORT, AND DELIVERY TO TISSUES

17.5.1 Stable Isotope Labeling

Biokinetic studies of vitamin E have been radically improved by the application of stable isotope labeling and gas chromatography–mass spectrometry (GC–MS) to the measurement of absorption, transport, uptake, and retention of tocopherols in humans and laboratory animals (Burton & Traber, 1990). Deuterated tocopherols containing three or six atoms of deuterium (^2H) per molecule may be ingested safely because deuterium is a stable isotope and has no deleterious effects. The deuterated tocopherols do not undergo any measurable exchange at the positions of substitution. Stable isotope labeling makes it possible to evaluate, within the same subject, the simultaneous competitive uptake of two (or even three) different forms of α-tocopherol labeled with different amounts of deuterium. This approach has the great advantage of eliminating the effects of variations in solute uptake between individual subjects. Furthermore, the deuterated tocopherol taken with a meal is easily distinguishable from the nondeuterated tocopherol already present in the subject.

17.5.2 Intestinal Absorption

Absorption of vitamin E depends on the simultaneous digestion and absorption of dietary fat. Individuals with impaired fat absorption invariably exhibit low vitamin E

TABLE 17.3
Distribution of Tocopherols and Tocotrienols in Selected Vegetable Oils

Oil	α-T	β-T	γ-T	δ-T	α-T3	β-T3	γ-T3	δ-T3
Corn (maize)	25.69	0.95	75.23	3.25	1.50	–	2.03	–
Olive	11.91	–	1.34	–	–	–	–	–
Palm	6.05	–	tr	–	5.70	0.82	11.34	3.33
Peanut	8.86	0.38	3.50	0.85	–	–	–	–
Rapeseed	18.88	–	48.59	1.20	–	–	–	–
Safflower	44.92	1.20	2.56	0.65	–	–	–	–
Soyabean	9.53	1.31	69.86	23.87	–	–	–	–
Sunflower	62.20	2.26	2.67	–	–	–	–	–

Source: Syväoja, E.-L. *et al.* (1986) *J. Am. Oil Chem. Soc.*, **63**, 328. With permission.
Values are mean values (n = 6–10 determinations) in mg/100g of each oil purchased from three to five different manufacturers.
–, not detected; tr, trace.

utilization. Before absorption, any ingested supplemental α-tocopheryl acetate or succinate is hydrolyzed in the lumen of the small intestine by pancreatic esterase, which requires bile salts as co-factors. The free α-tocopherol thus formed, together with vitamin E of natural origin, is solubilized within mixed micelles for passage across the unstirred water layer of the intestinal lumen. The micelles dissociate and the liberated vitamin E is absorbed by simple diffusion. The absorption efficiency of vitamin E in the rat is greater when the vitamin is solubilized in micelles containing medium-chain triglycerides as compared with long-chain triglycerides (Gallo-Torres *et al.*, 1971). Within the enterocyte, vitamin E is incorporated into chylomicrons and the chylomicrons are secreted into the bloodstream via the lymphatic route.

Absorption studies have focused mainly on α-tocopherol, yet the dietary intake of γ-tocopherol is about twice that of α-tocopherol in the United States and γ-tocopherol contributes an estimated 20% of the total vitamin E activity in a typical United States (Bieri & Poukka Evarts, 1974). Despite this, the concentration of γ-tocopherol averages only 10–15% of the concentration of α-tocopherol in adult human plasma (Drevon, 1991). Studies in thoracic duct-cannulated rats have shown no significant difference in the rate of absorption of α- and γ-tocopherol and even a 50-fold excess of α-tocopherol did not affect the absorption of γ-tocopherol (Traber *et al.*, 1986).

The few reported studies in normal human subjects have shown that the absorption of vitamin E is incomplete. Estimates of the absorption of physiological oral doses of tritiated *all-rac*-α-tocopherol based on the measurement of unabsorbed fecal radioactivity gave efficiencies ranging from 51% to 86% (mean 72%) (Kelleher & Losowsky, 1970) and 55% to 79% (mean 69%) (MacMahon & Neale, 1970). Pharmacological doses of vitamin E are absorbed with progressively less efficiency as the dose is increased until eventually a limit is reached. Even if dosages exceed the daily requirement 100-fold, the plasma concentrations do not increase beyond two- to fourfold (Kayden & Traber, 1993). There is no compensatory increase in absorption in the presence of vitamin E deficiency (Losowsky *et al.*, 1972).

There have been many conflicting reports concerning the relative absorption efficiency of α-tocopherol and its acetate ester. The situation has been clarified by Burton *et al.* (1988) using deuterium-substitution and GC–MS in studies with rats and humans. In the human study, a capsule containing an equimolar mixture of *RRR*-α-tocopherol and *RRR*-α-tocopheryl acetate was taken with an evening meal on successive days. The amount of α-tocopherol from the free phenol form was found to be equal to that from the acetate in plasma and erythrocytes, signifying an equal efficiency of absorption. These two forms should therefore be accorded equal potency on a molar basis.

17.5.3 Transport and Delivery to Tissues

Major pathways of vitamin E transport are shown in Figure 17.2. After absorption, vitamin E enters the circulation incorporated into the outer layer of chylomicrons. The chylomicrons are attacked by lipoprotein lipase located on the capillary walls of most tissues. During chylomicron lipolysis, some of the vitamin E, along with other surface components, is transferred to circulating high-density lipoproteins (HDL) and some is transferred to extrahepatic tissues. Interchange of vitamin E takes place between HDL and low-density lipoproteins (LDL), which are the principal carriers of the vitamin in human plasma. Hydrolysis of the core triglycerides results in the formation of chylomicron remnants, which hepatic parenchymal cells take up by receptor-mediated endocytosis. The liver selectively secretes α-tocopherol into the bloodstream in very low-density lipoproteins (VLDL). Non-alpha E vitamers and excess α-tocopherol are returned to the intestine in the bile and excreted in the feces. Lipolysis of the VLDL transforms these particles into intermediate-density lipoproteins (IDL).

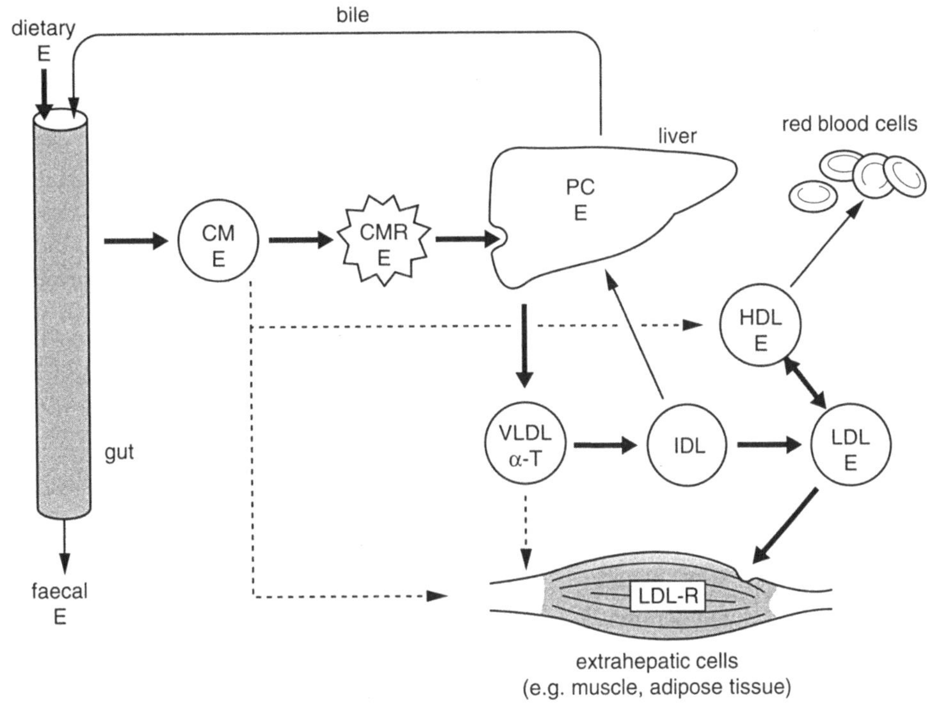

FIGURE 17.2 Major pathways of vitamin E transport. See text for details. •Reproduced with permission of Taylor & Francis from Drevon (1991), *Free Radical Research*, Vol. 14, pp. 229–46 (http://www.tandf.co.uk/journals/titles/10715762.html) Figure 3 on page 234.

Some IDL particles are taken up by the liver; the remainder are converted to LDL. Cells expressing the LDL receptor (LDL-R) take up the vitamin E in LDL by receptor-mediated endocytosis. Erythrocytes and other cells take up the vitamin E in HDL.

The liver secretes α-tocopherol in the VLDL in preference to γ-tocopherol, the latter being preferentially excreted in the bile (Traber & Kayden, 1989). This discrimination explains why α-tocopherol is the predominant circulating form of vitamin E, despite γ-tocopherol being the predominant dietary form. The liver further discriminates between stereoisomers of α-tocopherol, secreting *RRR*-α-tocopherol in preference to *SRR*-α-tocopherol (Traber *et al.*, 1990). This discriminatory ability has been attributed to a hepatic tocopherol-binding protein known as the α-tocopherol transfer protein that is present exclusively in hepatocytes (Yoshida *et al.*, 1992). The protein seems to function by stimulating the secretion of cytosolic α-tocopherol into the sinusoidal space where it becomes associated with nascent VLDL (Arita *et al.*, 1997).

17.5.4 Excretion

The major route of excretion of ingested vitamin E is fecal elimination. The non-*RRR*-α-tocopherol forms of vitamin E that are not preferentially secreted by the liver in VLDL, together with any excess α-tocopherol, are probably excreted unchanged in bile. Increasing doses of supplemental vitamin E in humans result in increasing urinary excretion of 2,5,7,8-tetramethyl-2-(2′-carboxyethyl)-6-hydroxychroman (α-CEHC), a metabolite that results from degradation of the phytyl tail. Urinary excretion of this metabolite may indicate excess vitamin E intake (Institute of Medicine, 2000). Sebaceous glands in the skin may secrete vitamin E to provide antioxidant protection for cutaneous lipids.

17.5.5 Deposition in Tissues and Turnover

No organ functions as a storage site for α-tocopherol, releasing it on demand. More than 90% of the human body pool of α-tocopherol is located in fat droplets within adipose tissue. The accumulation of vitamin E in adipose tissue is due simply to its lipid solubility. Some tissues, such as erythrocytes, liver, and spleen, have rapid turnover times, quickly replacing 'old' with 'new' α-tocopherol. The brain has the slowest α-tocopherol turnover time by far. In general, the vitamin E content of the nervous system is spared during vitamin E depletion. Within the nervous system, the peripheral nerves are the most responsive to vitamin E concentrations in the diet (Traber, 1999).

17.6 ANTIOXIDANT ROLE

Vitamin E is the major antioxidant in the lipid environment of cellular and subcellular membranes, and also in plasma lipoproteins, protecting the vital phospholipids from peroxidation. γ-Tocopherol inhibits lipid peroxidation more efficiently than α-tocopherol *in vitro* (Wolf, 1997). The relative susceptibility of microsomes from various tissues to lipid peroxidation correlates with vitamin E content. Notably, highly oxygenated heart and lung tissues have a high content of membranous vitamin E (Kornbrust & Mavis, 1980). The vitamin, chiefly α-tocopherol, is specifically and actively incorporated within biological membranes, where it stabilizes the lipoprotein structure. The tocopherol molecule is anchored in the phospholipid bilayer by means of its phytyl side chain with its polar head group located very close to the lipid-water interface. Protection by vitamin E against endothelial cell injury by hydroperoxides has been demonstrated (Hennig *et al.*, 1987).

The order of *in vitro* antioxidant activities of the tocopherols conforms to their oxidation potentials and parallels their biological activities, that is, $\alpha > \beta > \gamma > \delta$. The relative antioxidant activities (%) *in vitro* are $\alpha = 100$, $\beta = 71$, $\gamma = 68$, and $\delta = 28$ (Burton & Ingold, 1981). There is no significant difference between the β and γ positional isomers.

By scavenging peroxyl radicals that otherwise would propagate the chain reaction of lipid peroxidation, vitamin E acts as a chain-breaking antioxidant. It does this by donating a hydrogen atom from its phenolic hydroxyl group to a lipid peroxyl radical (LOO•), forming the tocopheroxyl radical (TO•) and a lipid hydroperoxide (LOOH) (reaction 17.1):

$$\text{LOO}\bullet + \text{T-OH} \longrightarrow \text{LOOH} + \text{T-O}\bullet \quad (17.1)$$

The tocopheroxyl radical, being relatively stable and unreactive, is unable to continue the chain reaction by attacking other species. Instead, it may combine with another peroxyl radical to form inactive molecular products (reaction 17.2) or it may be reduced to tocopherol by reduced coenzyme Q (reaction 17.3) or by ascorbate (reaction 17.4):

$$\text{T-O}\bullet + \text{LOO}\bullet \longrightarrow \text{molecular products} \quad (17.2)$$

$$\text{T-O}\bullet + \text{CoQH}_2 \longrightarrow \text{T-OH} + \text{CoQH}\bullet + \text{H}^+ \quad (17.3)$$

$$\text{T-O}\bullet + \text{AH}^- \longrightarrow \text{T-OH} + \text{A}^{-}\bullet \quad (17.4)$$

The semiquinone radical (CoQH•) produced in reaction 17.3 can be converted back to reduced coenzyme Q ($CoQH_2$) through the electron transport chain in the mitochondria. The ascorbyl radical (A^-•) produced in reaction 17.4 can be converted back to ascorbate (AH^-) by semidehydroascorbate reductase.

Whatever the fate of the tocopheroxyl radical, the result of vitamin E intervention is termination of the lipid peroxidation chain reaction.

The diverse vitamin E deficiency signs observed in animals are attributable to secondary effects of the widespread damage caused to the membranes of muscle and nerve cells by lipid peroxidation. The prevention of encephalomalacia in chicks by adding low concentrations of synthetic antioxidants to vitamin E-deficient diets strongly suggests that, in preventing this particular disease, vitamin E acts solely as an antioxidant and that such action is non-specific. Vitamin E and selenium can act independently in preventing exudative diathesis and liver necrosis: synthetic antioxidants have little or no effect upon these diseases.

Selenium is implicated in certain vitamin E deficiency diseases because of its role as an essential co-factor for glutathione peroxidase, the enzyme which converts potentially harmful lipid peroxides to harmless hydroxy acids. A constant supply of dietary sodium selenate is essential to maintain the enzyme's activity.

The sulfur-containing amino acids, methionine and cystine, exert a sparing effect on both selenium and vitamin E as shown by their ability to delay the onset of necrotizing myopathy in the chick. This effect is presumably due to an increased *de novo* synthesis of glutathione.

17.7 VITAMIN E AND ATHEROSCLEROSIS

17.7.1 Introduction

The results of several clinical studies have suggested that α-tocopherol supplementation has potential beneficial effects with regard to coronary artery disease (CAD) (Singh *et al.*, 2005). Biological and clinical aspects of vitamin E on atherosclerosis prevention have been reviewed (Neuzil *et al.*, 2001; Munteanu & Zingg, 2007). Vitamin E counteracts many of the atherogenic effects of oxidized LDL (listed in Section 11.6.1) by mechanisms that may be independent of its antioxidant properties. The inhibitory effect of vitamin E on vascular smooth muscle cell proliferation is attributable to its inhibitory effect on protein kinase C (PKC) activity. Attraction and adhesion of immune cells to the endothelial cell lining of vessel walls and subsequent transendothelial migration are important inflammatory events associated with the formation of fatty streaks and development of atherosclerosis. By inhibiting monocyte–endothelial cell adhesion and preventing monocyte transendothelial migration, vitamin E helps to prevent the development of atherosclerosis. Vitamin E preserves arterial smooth muscle relaxation by enhancing the production of nitric oxide. Platelet aggregation is abnormally increased in patients with CAD (Fuster *et al.*, 1992). Vitamin E inhibits platelet aggregation and delays arterial thrombus formation. This may be the major mechanism by which vitamin E decreases the frequency of acute cardiac events in patients with pre-existing CAD.

17.7.2 Maintenance of Arterial Endothelial Cell Integrity

The arterial endothelium is a selectively permeable barrier, one cell thick, that lies between the lumen and the vascular smooth muscle. It is able to respond to physical and chemical stimuli by producing substances that control coagulation, fibrinolysis, arterial tone, and vascular growth (Celermajer, 1997). Disruption of endothelial integrity increases the uptake of LDL into the arterial wall and initiates the development of atherosclerosis. Among the factors that contribute to endothelial cell injury are free radical-mediated reactions such as lipid peroxidation. Endothelial cells are protected by antioxidant defense systems that include superoxidase dismutase and vitamin E, which scavenge free radicals (Hennig & Chow, 1988).

17.7.3 Protection of Low-Density and High-Density Lipoproteins from Oxidation

LDL oxidation has a major role in the development of atherosclerosis. The release of superoxide by phagocytic monocytes during the respiratory burst induces oxidation of LDL and renders it toxic to proliferating cells (Cathcart *et al.*, 1989). Monocytes from healthy human subjects taking oral vitamin E supplements (1200 IU per day) showed lower superoxide production and a reduced capacity to oxidize LDL (Devaraj *et al.*, 1996). This effect of vitamin E appeared to be mediated via inhibition of PKC. Vitamin E may therefore protect circulating LDL from oxidation induced by activated phagocytes. The enzyme responsible for superoxide production in phagocytes is NADPH-oxidase. Activation of this enzyme, elicited by appropriate stimulation of the phagocytic cell, requires translocation of several cytosolic enzyme components to the membrane. PKC is involved in the activation of NADPH-oxidase and can phosphorylate one of its cytosolic components, $p47^{phox}$. Cachia *et al.* (1998a) studied the effect of vitamin E on NADPH-oxidase activation elicited by 12-*O*-tetradecanoylphorbol-13-acetate (TPA) in human monocytes. They found that α-tocopherol inhibited translocation and phosphorylation of $p47^{phox}$. The results suggested that the attenuating effect of α-tocopherol on the respiratory burst is due to inhibition of PKC activity.

The lysolecithin that accumulates in oxidized LDL increases production of superoxide anion in the walls of blood vessels, which may further enhance LDL oxidation (Ohara *et al.*, 1994). When human monocytes were stimulated by phorbol ester to produce superoxide *in vitro*, the addition of native LDL inhibited superoxide production in a manner highly sensitive to the increasing α-tocopherol content; the free form of α-tocopherol produced lower inhibition compared with the lipoprotein-associated form (Cachia *et al.*, 1998b). It was suggested that a vitamin E-induced decrease in monocyte superoxide production could lead to a decrease in lysolecithin production in LDL. Lysolecithin is responsible for many of the atherogenic properties of oxidized LDL and any means of reducing its production would promote an anti-atherogenic status of vessels.

Several reports indicate protection of LDL and HDL from oxidation following supplementation of human diets with α-tocopherol (Dieber-Rotheneder *et al.*, 1991; Jialal & Grundy, 1992; Princen *et al.*, 1992; Reaven *et al.*, 1993; Jialal *et al.*, 1995; Suzukawa *et al.*, 1995). The protective effect of vitamin E against oxidative modification of HDL helps to prevent atherosclerosis because oxidized HDL loses its ability to stimulate efflux of cholesterol from foam cells (Nagano *et al.*, 1991). The results of these reports suggest that supplementation with vitamin E would be beneficial for ameliorating atherosclerosis.

17.7.4 Down-Regulation of Scavenger Receptor Expression and Activity in Macrophages

Two reports showed that α-tocopherol reduces foam cell formation by decreasing the expression and down-regulating the activity of scavenger receptors in human monocyte-derived macrophages *in vitro*. Two scavenger receptors were studied: the class A scavenger receptor (SR-A) expressed on acLDL and the CD36 member of class B scavenger receptors expressed on oxLDL.

Teupser *et al.* (1999) observed dose-dependent down-regulation of SR-A activity and a decrease of SR-A mRNA expression at α-tocopherol concentrations comparable to those normally found in human plasma (1, 5, 10, and 50 μM). γ-Tocopherol had no significant effect on SR-A activity. Transcriptional activation of the scavenger receptor gene in macrophages was found to be mediated by activator protein-1 (AP-1) and co-operating transcription factors (Moulton *et al.*, 1994). α-Tocopherol reduced the activation of AP-1 during the adherence of freshly isolated rabbit macrophages (Teupser *et al.*, 1999). These observations led the latter authors to hypothesize that α-tocopherol may inhibit or delay AP-1-mediated transcriptional activity in macrophages. β-Tocopherol had an effect on SR-A expression comparable to that of α-tocopherol, but did not affect AP-1 activity. This observation indicates that, in addition to AP-1, other transcription factors may be involved in the tocopherol-mediated reduction of SR-A activity

Devaraj *et al.* (2001) reported that α-tocopherol at concentrations of 25, 50, and 100 μM decreased expression of CD36 protein in macrophages. CD36 mRNA expression was unaffected, demonstrating that α-tocopherol down-regulates CD36 at the post-translational level. α-Tocopherol significantly inhibited cholesteryl ester accumulation after incubation of macrophages with acLDL and oxLDL. Neither β- nor γ-tocopherol had these effects, and neither PKC nor PPARγ were implicated.

17.7.5 Inhibition of Vascular Smooth Muscle Cell Proliferation

Vascular smooth muscle cell (VSMC) proliferation is an important event in the formation of the fibrous atherosclerotic plaque. α-Tocopherol at physiological concentrations specifically inhibits mitogen-induced proliferation of VSMCs and certain other cell types in a parallel manner to the inhibition of PKC activity (Boscoboinik *et al.*, 1991a, b). α-Tocopherol inhibits PKC only at the cellular level; addition of α-tocopherol to recombinant PKC *in vitro* does not result in inhibition. The inhibitory effect of vitamin E on VSMC proliferation and PKC activity is specific to α-tocopherol; the other tocopherols and tocotrienols do not inhibit PKC (Chatelain *et al.*, 1993). β-Tocopherol prevents the inhibitory effect of α-tocopherol on both processes (Azzi *et al.*, 2001). Since α-tocopherol and β-tocopherol have very similar radical-scavenging abilities, this property is clearly not involved in the inhibitory ability of α-tocopherol. Although various isoforms of PKC (α, β, δ, ε, ζ, and μ) have been shown to be present in rat aortic smooth muscle cells, only PKCα is inhibited by α-tocopherol (Ricciarelli *et al.*, 1998). The inhibition is indirect and not attributable to a decreased synthesis of the enzyme. α-Tocopherol induces the activity of protein phosphatase type 2A (Ricciarelli *et al.*, 1998), an enzyme which dephosphorylates PKCα (Hansra *et al.*, 1996). Azzi *et al.* (1997) proposed that the specific activation of protein phosphatase type 2A by α-tocopherol results in inactivation of PKCα, followed by inhibition of smooth muscle cell proliferation.

The degree of inhibition of cell proliferation by α-tocopherol depends on the mitogen responsible for stimulating growth. Proliferation induced by platelet-derived growth factor (PDGF), endothelin, and unmodified LDL was almost completely inhibited by α-tocopherol, whereas proliferation produced by other mitogens, such as bombesin and lysophosphatidic acid, was only moderately or slightly inhibited (Azzi *et al.*, 1993). The pro-inflammatory cytokine, interleukin-1β (IL-1β) is also pro-atherogenic, as it stimulates smooth muscle cell proliferation induced by PDGF (Libby & Hansson, 1991). α-Tocopherol decreases the release of IL-1β from lipopolysaccharide-activated human monocytes by inhibition of 5-lipoxygenase (Devaraj & Jialal, 1999). In cultured smooth muscle cells, α-tocopherol activated the cellular release of the growth-inhibiting cytokine, transforming growth factor-β (TGF-β) (Özer *et al.*, 1995).

17.7.6 Inhibition of Monocyte–Endothelial Cell Adhesion

Devaraj *et al.* (1996) reported that monocytes isolated from healthy human subjects supplemented with 1200 IU (804 mg) per day of α-tocopherol over 8 weeks were less able, when activated, to adhere to human umbilical vein endothelial cells compared with monocytes isolated from placebo controls. The vitamin E-enriched monocytes also showed a 90% decrease in the release of interleukin 1β (IL-1β) when activated. IL-1β is a pro-atherogenic, pro-inflammatory cytokine that promotes monocyte–endothelial cell adhesion. In patients with primary hypercholesterolemia, vitamin E supplementation at 400 IU (268 mg) per day of α-tocopherol for six weeks did not affect monocyte adhesion to endothelial cells *ex vivo* (Williams *et al.*, 1997).

Wu *et al.* (1999) showed that pre-treatment of cultured human aortic endothelial cells (HAECs) and U937 monocytic cells with α-tocopherol dose-dependently reduced cytokine-stimulated adhesion of these cells to each other. α-Tocopherol also dose-dependently reduced IL-1β-mediated HAEC production of chemokines (IL-8 and MCP-1) and adhesion molecules (ICAM-1, VCAM-1, and E-selectin) that are responsible for the adhesion of monocytes to endothelium. In addition, α-tocopherol enrichment of HAECs suppressed their IL-1β-stimulated production of IL-6, which contributed to the reduced monocyte–endothelial adhesion. Pre-treatment of U937 cells with α-tocopherol also reduced lipopolysaccharide-induced adhesion of these cells to human umbilical vein endothelial cells mediated by a concomitant reduction in the production of the adhesion molecules MAC-1 and VLA-4 (Islam *et al.*, 1998).

17.7.7 Prevention of Monocyte Transendothelial Migration

Incubation of co-cultures of human aortic endothelial and smooth muscle cells with LDL in the presence of human serum resulted in an increased synthesis of monocyte chemotactic protein 1 (MCP-1) mRNA and protein. This was accompanied by an increase in the adhesion of monocytes (but not neutrophil-like cells) to the endothelial monolayer and an increased transmigration of monocytes into the subendothelial space. The increase in monocyte migration was most likely due to the increased levels of MCP-1, since it was completely blocked by a specific antibody to MCP-1. Pretreatment of the co-cultures with α-tocopherol before the addition of LDL prevented the LDL-induced monocyte transmigration (Navab *et al.*, 1991).

17.7.8 Preservation of Nitric Oxide-Mediated Arterial Smooth Muscle Relaxation

Dysregulation of arterial relaxation is a common complication associated with atherosclerosis. Nitric oxide (NO), an important relaxant of arterial smooth muscle, is released from vascular endothelial cells in response to a large number of agents which activate endothelial surface receptors (Furchgott & Zawadzki, 1980). α-Tocopherol was shown to increase the activity of endothelial NO synthase (eNOS) and enhance NO production by activating a protein kinase which phosphorylates eNOS at serine 1177. Co-incubation with ascorbic acid amplified this effect of α-tocopherol (Heller *et al.*, 2004). Exposure of the vascular wall to oxidized LDL (oxLDL) inhibits the release of NO, thereby preventing relaxation (Kugiyama *et al.*, 1990). This dysfunction has, in part, been attributed to PKC stimulation (Ohgushi *et al.*, 1993).

Keaney *et al.* (1996) fed rabbits diets deficient in or supplemented with α-tocopherol and examined the effects of the vitamin on oxLDL-induced endothelial dysfunction. Exposure of thoracic aorta segments from vitamin E-deficient animals to ox-LDL produced dose-dependent inhibition of acetylcholine-mediated relaxation, while similarly treated segments from animals consuming α-tocopherol showed no such inhibition. Vessel resistance to endothelial dysfunction in the vitamin-replete animals was strongly correlated with vascular content of α-tocopherol. Incorporation of α-tocopherol into the vasculature limited endothelial dysfunction induced by phorbol ester, a direct activator of PKC. Using cultured human aortic endothelial cells, Keaney *et al.* (1996) confirmed that oxLDL stimulates endothelial PKC and that cellular incorporation of α-tocopherol inhibits this stimulation. Desrumaux *et al.* (1999) showed that plasma phospholipid transfer protein, by supplying vascular endothelial cells with α-tocopherol, plays a distinct role in the prevention of endothelial dysfunction.

17.7.9 Protection of Prostacyclin Generation in Arteries

Prostacyclin is a strong vasodilator and an inhibitor of platelet aggregation. The decrease in prostacyclin production by arterial endothelial cells at the site of plaque formation (Gryglewski *et al.*, 1978) may be caused by oxidative stress.

Prostacyclin synthesis in cultured aortic endothelial cells was depressed by the presence of high glucose concentration in the medium and synthesis could be restored by the simultaneous addition of vitamin E (Kunisaki *et al.*, 1992). Vitamin E probably acted as an antioxidant by preventing the build-up of lipid hydroperoxides which are inhibitory to prostacyclin synthesis. The results suggested that vitamin E may restore depressed prostacyclin production by the vascular wall in hyperglycemic conditions such as those seen in patients with diabetes mellitus. Szczeklik *et al.* (1985) reported that feeding an atherogenic diet to rabbits for a week resulted in elevation of plasma lipid hydroperoxides and a 90% decrease in arterial generation of prostacyclin. Enrichment of the atherogenic diet with 100 mg of vitamin E daily prevented the increase in lipid hydroperoxides and protected the prostacyclin-generating system in arteries.

Chan & Leith (1981) showed that prostacyclin synthesis in dissected rabbit aorta is decreased when the tissue is depleted of vitamin E. Enrichment of cultured human endothelial cells with vitamin E caused an increase in arachidonic acid release and spontaneous prostacyclin synthesis (Tran & Chan, 1988, 1990). A potentiating effect of vitamin E on arachidonic acid release in megakaryocytes was attributed to an up-regulation of phospholipase A_2 (Chan *et al.*, 1998).

The above findings suggest that a defect in the local production of prostacyclin may be an underlying factor in the pathogenesis of atherosclerosis in terms of platelet aggregation and monocyte adhesion to endothelial cells. An increased dietary intake of vitamin E can overcome this defect.

17.7.10 Inhibition of Platelet Aggregation

Platelet aggregation plays an important role in thrombosis and cardiovascular events (Handin, 1996) and α-tocopherol has long been known to inhibit platelet aggregation *in vitro*. In patients with primary hypercholesterolemia, vitamin E supplementation (400 IU per day for six weeks) caused a significant decrease in thrombin-induced platelet aggregation *ex vivo* (Williams *et al.*, 1997). Mabile *et al.* (1999) studied the effects of a range of oral doses of natural α-tocopherol on platelet aggregation in healthy volunteers. Daily doses of 75 IU (50 mg) for two weeks resulted in a significant decrease of aggregation. Subsequent intake of 200 IU (134 mg) followed by 400 IU (268 mg), each for two weeks, gradually increased the inhibitory effect. For dose comparison, the Recommended Dietary Allowance (RDA) of vitamin E for adults is 15 mg/day of α-tocopherol (22 IU).

Freedman and Keaney (2001) examined the effects of *RRR*-α-tocopherol, *RRR*-α-tocopheryl acetate, and all-*rac*-α-tocopherol on platelet aggregation induced by arachidonic acid. Aggregation was inhibited by 57% and 52% with *RRR*-α-tocopherol and *RRR*-α-tocopheryl acetate, respectively, whereas all-*rac*-α-tocopherol produced no inhibition. As α-tocopheryl acetate is not an antioxidant, the inhibitory action of vitamin E on platelet aggregation cannot be entirely an antioxidation effect.

Nitric oxide (NO) is a potent smooth muscle relaxant and platelet aggregation inhibitor. It is formed naturally within arterial endothelial cells by the oxidative deamination of L-arginine, catalyzed by constitutive nitric oxide synthase (cNOS). Endothelial dysfunction caused by physical or chemical damage to the endothelium (Celermajer, 1997) allows LDL to enter the subendothelial space of the arterial intima. Here, the LDL is mildly oxidized to minimally oxidized LDL, which is chemotactic for circulating monocytes. The monocytes differentiate into macrophages, which produce oxygen-derived free radicals that further modify minimally oxidized LDL to completely oxidized LDL (oxLDL). Among the free radicals produced by macrophages is superoxide, which reacts with endothelial-derived NO to produce peroxynitrite. Peroxynitrite can directly inactivate endothelial superoxide dismutase (SOD) through formation of nitrotyrosine (Yamakura *et al.*, 1998), thereby preventing the enzymatic conversion of superoxide to hydrogen peroxide and oxygen. The harmful effect of peroxynitrite is best counteracted by γ-tocopherol and to a lesser extent by α-tocopherol (Wolf, 1997).

Two studies from the same laboratory (Li *et al.*, 1999; Saldeen *et al.*, 1999) examined the effects of feeding α- and γ-tocopherol on endothelial NO synthase (eNOS) and SOD activity and protein expression in rats. ADP-induced platelet aggregation was measured and the patterns of arterial thrombosis were studied. Plasma α-tocopherol levels were significantly increased, but γ-tocopherol levels fell in rats fed α-tocopherol. This phenomenon was also reported by Handleman *et al.* (1985). Plasma γ-tocopherol levels were increased and α-tocopherol levels did not change in rats fed γ-tocopherol. Both α- and γ-tocopherol decreased platelet aggregation and delayed arterial thrombus formation. Both tocopherols increased SOD activity in plasma and aortic homogenates as well as SOD protein expression in aortic homogenates. γ-Tocopherol was significantly more potent than α-tocopherol in all these effects. Both tocopherols increased NO generation and eNOS activity; however, only γ-tocopherol increased eNOS protein expression in aortic homogenates. These results provided a possible mechanism for the inhibitory action of vitamin E on platelet aggregation and suggested that γ-tocopherol is more potent than α-tocopherol in inhibiting platelet-dependent thrombus formation. A third study (Li *et al.*, 2001) revealed that α-, γ-, and δ-tocopherol have similar inhibitory effects on human platelet aggregation and that a combination of these three vitamers, as found in natural foods, has a greater inhibitory effect than each individual vitamer.

17.7.11 Protection against oxLDL-Mediated Cytotoxicity

Marchant *et al.* (1995) incubated LDL with copper ions for varying periods in the presence or absence of α-tocopherol. They then incubated these LDL preparations with human monocyte–macrophages and measured the toxicity produced in the cells. Toxicity increased with increasing duration of copper-catalyzed oxidation in the absence of α-tocopherol. The presence of α-tocopherol protected the LDL from oxidation and the cells from toxicity. Martin *et al.* (1998) reported that enrichment of human aortic endothelial cells with α-tocopherol *in vitro* dose-dependently increased their resistance to cytotoxic injury from oxLDL.

17.7.12 Effects of Vitamin E Supplementation on Experimentally Induced Atherosclerosis

Verlangieri and Bush (1992) studied the effects of α-tocopherol supplementation on experimentally induced atherosclerosis in monkeys using duplex ultrasound scanning and B-mode imaging to monitor lesion formation and progression. The animals were randomly assigned to one of four groups and atherosclerosis was monitored over a 36-month period. One group was fed a basal diet, while three other groups consumed an atherogenic diet (basal diet plus 0.4% cholesterol, w/w). Two of the latter groups also received α-tocopherol, one at the onset of the study (prevention group) and the other after atherosclerosis was established (regression group). The dosage of α-tocopherol was approximately eight times the daily requirement for primates. A steady initial rise in percent stenosis was detected in the three groups fed an atherogenic diet. However, stenosis in the unsupplemented animals progressed more rapidly and to greater extent than stenosis in the prevention group. In the regression group, stenosis reached a plateau at 4 months post-supplementation, while that of the unsupplemented group continued to rise. The data indicated that, while α-tocopherol does not totally prevent atherosclerosis, it appears to lessen the severity and reduce the rate of disease progression. Moreover, supplemental vitamin E may regress well-established lesions.

17.7.13 Epidemiological Studies and Clinical Trials

Epidemiological studies have shown increased protection against coronary artery disease in subjects consuming vitamin E in daily doses >100 mg taken for more than two years (Rimm *et al.*, 1993; Stampfer *et al.*, 1993). In an epidemiological study of 11,178 elderly persons (Losonczy *et al.*, 1996), use of vitamin E supplements was associated with reduced risk of death from cardiovascular disease compared with non-use. A 7-year study of 34,486 postmenopausal women (Kushi *et al.*, 1996) revealed that vitamin E supplementation (all as α-tocopherol) had no effect on the risk of cardiovascular disease mortality, whereas the intake of vitamin E from food (mainly as γ-tocopherol) is inversely associated with such risk.

Two clinical trials (HOPE Study Investigators [2000] and GISSI-Prevenzione Investigators 1999]) failed to show any beneficial effect of vitamin E supplements on cardiovascular events and cardiac death. In these trials, the vitamin E preparations contained α-tocopherol alone, so do not represent the spectrum of vitamers found in foods. γ-Tocopherol is significantly more potent than α-tocopherol in decreasing platelet aggregation and delaying arterial thrombosis (Saldeen *et al.*, 1999). Also, γ-tocopherol, but not α-tocopherol, is reduced in patients with coronary heart disease (Öhrvall *et al.*, 1996). These findings strongly suggest that vitamin E supplements should be formulated as a combination of α- and γ-tocopherols, rather than α-tocopherol alone.

17.8 EFFECTS OF VITAMIN E ON THE IMMUNE SYSTEM

Vitamin E supplementation has a stimulatory effect on both humoral and cellular immunity. Its effect is most pronounced in infectious diseases where phagocytosis is the main defense mechanism; it is least effective where cell-mediated immunity is the main defense. Vitamin E deficiency predictably leads to impairment of immune function and reduced disease resistance. The effective dietary dose for optimal immune protection is about four to six times higher than vitamin E levels found in normal diets (Tengerdy, 1989).

The mechanism for the immune-enhancing effect of vitamin E has not yet been established. One hypothesis is that vitamin E, through its antioxidant function, could decrease the production of immunosuppressive PGE_2 by activated macrophages. Meydani *et al.* (1986) showed that increasing the vitamin E intake of aged (24-month-old) mice to almost 20 times their recommended daily allowances for six weeks enhanced the mitogenic response of splenocytes and improved the response to a delayed-type hypersensitivity (DTH) skin test. The vitamin E-supplemented mice had significantly less PGE_2 than non-supplemented controls, suggesting that the immunostimulatory effect of the vitamin might have been due to its inhibition of PGE_2 synthesis.

Beharka *et al.* (1997) conducted experiments in which combinations of purified macrophages and T cells from young or old mice were cultured together. Key observations were: (1) macrophages from old mice suppressed proliferation and IL-2 secretion by T cells from young mice; (2) addition of vitamin E increased proliferation and IL-2 production by co-cultures containing cells from old mice;

(3) addition of vitamin E to co-cultures containing cells from old mice significantly reduced PGE_2 production by mitogen-stimulated macrophages; there was no such effect in co-cultures containing cells from young mice; (4) addition of indomethacin produced similar effects to vitamin E; and (5) a synthetic fat-soluble antioxidant, DPPD, did not mimic the effects of vitamin E. These results demonstrate that increased production of PGE_2 by macrophages contributes to the age-associated decrease in T cell function. The results support the idea that the immunostimulatory effect of vitamin E is attributable, at least partly, to inhibition of PGE_2 synthesis by macrophages.

Meydani *et al.* (1990) conducted a double-blind, placebo-controlled study in healthy elderly adults residing in a metabolic research unit to investigate the effect of vitamin E supplementation (800 mg *dl*-α-tocopheryl acetate/day for 30 days) on immune responses. The following significant changes were observed in the vitamin-supplemented group and not in the placebo group: (1) improved delayed-type hypersensitivity (DHT) reaction; (2) enhanced mitogen-induced T cell proliferation and IL-2 production, the increase in IL-2 concentration being positively correlated with changes in plasma α-tocopherol concentration; and (3) reduction in PGE_2 production by peripheral blood mononuclear cells. The results closely paralleled the early study in aged mice (Meydani *et al.*, 1986) and demonstrated that short-term supplementation of healthy elderly subjects with 800 mg *dl*-α-tocopheryl acetate improved immune responsiveness.

Activated T cells may undergo activation-induced cell death (AICD), that is, apoptosis using the Fas/FasL death-receptor pathway (Section 8.9.6). AICD is a means of maintaining peripheral lymphocyte homeostasis. Li-Weber *et al.* (2002) showed that vitamin E suppresses the activity of the transcription factors NF-κB and AP-1, thereby blocking expression of FasL and protecting T cells from AICD. Since AICD is a major cause of T cell depletion in AIDS, these authors examined 35 HIV-1-positive individuals and found that their T cells are more susceptible to AICD than are T cells isolated from healthy controls. Diet supplementation with vitamin E was shown to protect HIV-1-infected individuals from T cell depletion by AICD.

17.9 VITAMIN E AND CANCER

There is much stronger evidence for α-tocopheryl succinate and γ-tocopherol than for α-tocopherol in the prevention and treatment of cancer (Neuzil, 2003). The effects of α-tocopheryl succinate against cancer cells include the induction of apoptosis (Schneider, 2005), while those of γ-tocopherol include inhibition of cell cycle progression via down-regulation of cyclins D1 and E (Gysin *et al.*, 2002). The evidence for a protective role of vitamin E is most convincing in carcinomas of the prostate and gastrointestinal tract (Sung *et al.*, 2003). In the Women's Health Study (2005) with 39,876 healthy US women aged 45 years or older, the administration of 600 IU of natural-source vitamin E (α-tocopherol) on alternate days did not significantly affect the incidence of colon, lung, or total cancers. Using a population-based cohort, Slatore *et al.* (2008) found no evidence that use of supplemental vitamin E is associated with a lower incidence of lung cancer. In contrast to the often assumed benefits or at least lack of harm, supplemental vitamin E was associated with a small increased risk of lung cancer that was most prominent in current smokers and those with non-small cell lung cancer.

17.10 VITAMIN E DEFICIENCY IN ANIMALS

17.10.1 Introduction

Vitamin E deficiency in animals is readily demonstrable and results in a variety of pathological conditions that affect the muscular, cardiovascular, reproductive, and central nervous systems as well as the liver, kidney, brain, and erythrocytes. There is a marked difference between animal species in their susceptibility to different deficiency disorders. A complex biochemical interrelationship exists between vitamin E and the trace element selenium. Unsaturated fat, sulfur-containing amino acids, and synthetic fat-soluble antioxidants are also implicated in some disorders. Consequently, in order to experimentally induce a particular deficiency syndrome in a given species, it is usually necessary to adjust the balance of these nutrients in the diet. The most extensively studied deficiency syndromes are listed in Table 17.4.

17.10.2 Fetal Resorption

In female rats deprived of vitamin E, all reproductive events are normal up to implantation of the fertilized ova. Several days later, however, the developing fetus shows abnormalities followed by intra-uterine death, rapid autolysis, and resorption. A defect in the fetal blood vessels may be the primary event leading to death of the fetus (Nelson, 1980). This disease can be prevented by administering an adequate dose of vitamin E as late as the tenth day of pregnancy. The synthetic antioxidant DPPD is at least as effective as α-tocopherol in preventing fetal resportion, but ethoxyquin, which readily prevents encephalomalacia in chicks, is inactive (Draper *et al.*, 1964). Selenium compounds have no effect on fetal resorption in rats.

17.10.3 Erythrocyte Hemolysis

Erythrocyte plasma membranes are particularly vulnerable to lipid peroxidation because of their direct exposure to molecular oxygen and the presence of hemoproteins which are catalysts of peroxidation. Erythrocytes isolated from blood samples of vitamin E-depleted rats exhibit spontaneous hemolysis when added to dilute solutions of dialuric acid, whereas erythrocytes of rats receiving vitamin E are resistant to this hemolysis. This early manifestation of vitamin E deficiency can be prevented by certain synthetic antioxidants administered to the animal or added to the cell

TABLE 17.4
Some Vitamin E Deficiency Diseases

Disease	Experimental Animal	Tissue Affected	Severity Dependent on Dietary PUFA	Prevented by Vitamin E	Se	Synthetic Antioxidants	Sulfur Amino Acids
Fetal resorption	Female rat	Vascular system of fetus	Yes	Yes	No	Yes	No
Erythrocyte hemolysis	Rat, chick, man (premature infant)	Erythrocytes	Yes	Yes	No	Yes	No
Encephalo-malacia	Chick	Cerebellum	Yes	Yes	No	Yes	No
Exudative diathesis	Chick	Vascular system	No	Yes	Yes	No	No
Liver necrosis	Rat, pig	Liver	No	Yes	Yes	No	No
Testicular atrophy	Male rat	Testes	No	Yes	No	No	No
Necrotizing myopathy	Chick	Skeletal muscle	No	Yes	No	?	No
No myopathy	Chick	Skeletal muscle	–[a]	Yes	No	No	Yes

[a] A low level (0.5%) of linoleic acid was necessary to produce dystrophy; higher levels did not increase the amount of vitamin E required for prevention.
–, not detected.
Source: Scott (1980).

suspension *in vitro* as well as by vitamin E. Selenium compounds have no effect on erythrocyte hemolysis.

17.10.4 Encephalomalacia

This nutritional disorder occurs in growing chicks fed vitamin E-deficient diets containing adequate amounts of selenium for the prevention of exudative diathesis and sufficient methionine or cystine for the prevention of necrotizing myopathy. Encephalomalacia is manifested by lesions of the cerebellum, the part of the brain concerned with co-ordination of movement. The cerebellum is softened, swollen, and edematous with minute hemorrhages on the surface and greenish-yellow necrotic areas. The necrosis may be the result of thrombosis in the capillaries. Once established, the lesions are irreversible. The main symptoms are ataxia of gait and stance, backward or downward retraction of the head, tremors, spasms of the limb muscles, and eventually prostration, stupor, and death within a few hours. The incidence and severity of the disease are markedly increased with increasing levels of linoleic acid in the diet. Low concentrations of synthetic antioxidants such as DPPD and ethoxyquin in the diet readily prevent encephalomalacia, but selenium has no effect.

17.10.5 Exudative Diathesis

This is a vascular disease of chicks which develops as a result of feeding diets that are low in both vitamin E and selenium. The disease can be induced for experimental purposes by feeding diets based on Torula yeast, which is low in micronutrients but contains substantial amounts of unsaturated fatty acids. The most obvious manifestation is a massive accumulation of a greenish fluid under the skin of the breast and abdomen. Internally, the edema extends to the muscles and many organs, including the heart and lungs. The edema is the result of a leakage of plasma from the capillaries caused by an increased permeability of the capillary walls. The disease can be prevented by administration of either vitamin E or selenium, provided that the selenium deficiency is not too severe (Thompson & Scott, 1969). A severe deficiency of selenium causes degeneration of the exocrine component of the pancreas and consequent impairment of dietary lipid absorption, which will affect the absorption of vitamin E (Thompson & Scott, 1970). In this event, extremely high doses of vitamin E are required to prevent exudative diathesis. Some synthetic antioxidants, including DPPD and ethoxyquin, are also effective, but only at concentrations distinctly greater than those required to prevent encephalomalacia.

17.10.6 Liver Necrosis

Necrotic liver degeneration develops in weanling rats after commencement of a diet based on Torula yeast, which is deficient in both vitamin E and selenium and low in sulfur-containing amino acids. Necrosis is preceded by degeneration of the sinusoidal cellular plasma membrane and lipid peroxidation has been detected late in the progress of the disease. The onset of necrosis is delayed by cystine, which appears to have a sparing action on the amount of vitamin E or selenium required to prevent the disease.

17.10.7 Testicular Atrophy

In male rats depleted of vitamin E from early life there is no testicular injury until the onset of sexual maturity, when a progressive degeneration of the germinal epithelium of the seminiferous tubules occurs and the testes atrophy. The resultant sterility does not respond to vitamin E and is truly permanent.

17.10.8 Necrotizing Myopathy

This disease is manifested as a progressive muscular weakness which affects the skeletal muscles of many vertebrate species. It was originally called nutritional muscular dystrophy, but this term suggests an etiological relationship between the myopathy of vitamin E deficiency and human muscular dystrophy. Although many of the pathological lesions are similar in these two diseases, human muscular dystrophy is genetically determined and does not respond to vitamin E treatment.

Necrotizing myopathy is characterized histologically by marked variation in the cross-sectional diameter of the muscle fibers, segmental fragmentation with interstitial edema and necrosis, and, in the later stages, extensive replacement of muscle tissue by connective tissue. The disease can be detected in its early stages by an increased excretion of creatine in the urine (creatinuria), which is the result of a loss of creatine from the affected muscles. Creatine excretion is often expressed as the creatine:creatinine ratio, the excretion of creatinine being relatively constant on a body weight basis.

Necrotizing myopathy in rabbits, guinea pigs, rats, and monkeys responds primarily to vitamin E. Selenium is not capable of completely replacing vitamin E in these species, although it does reduce the vitamin requirement. The myopathy, as studied in the chick, does not respond to dietary synthetic antioxidants at levels several times those needed to prevent encephalomalacia. The disease is induced in the chick when the dietary vitamin E is accompanied by a deficiency in the sulfur-containing amino acids, methionine and cystine. Approximately 0.5% of dietary linoleic acid (but not linolenic acid) is necessary to produce myopathy. Concentrations above 0.5% do not increase the amount of vitamin E required for prevention. The chick appears to be unique in that the myopathy can be prevented in the absence of vitamin E by supplementing the diet with cystine or methionine. Cystine is about twice as effective as methionine on an equal sulfur basis (Scott, 1970).

17.11 VITAMIN E DEFICIENCY IN HUMANS

Apart from hemolytic anemia in premature infants, vitamin E, in the context of human nutrition, has long been considered 'a vitamin looking for a disease.' It is now recognized that vitamin E is responsible for the neurological abnormalities that had been described in patients with long-term disorders of fat absorption.

17.11.1 Hemolytic Anemia

Newborn infants generally have low serum vitamin E levels because of the vitamin's limited transfer through the placenta. A hemolytic anemia associated with vitamin E deficiency in premature infants six to ten weeks after birth was first reported by Oski and Barness (1967). This deficiency syndrome was further investigated in infants fed commercial milk formulas that were high in PUFA and relatively low in vitamin E (Hassan *et al.*, 1966; Ritchie *et al.*, 1968). Control infants were fed identical formulas supplemented with vitamin E. The syndrome consisted of hemolytic anemia, edema, and skin lesions. The erythrocytes lysed when treated *in vitro* with dilute hydrogen peroxide (i.e. the cell contents leaked out of the damaged cell membrane) and the blood film showed abnormal red cell morphology, such as spiky and fragmented cells. Erythrocyte survival was shortened and an increase in the number of reticulocytes (erythrocyte precursors newly arrived in the blood from the bone marrow) indicated a response to increased erythrocyte destruction. Erythroid hyperplasia was observed in the bone marrow and an increased platelet count was indicative of a general increase in bone-marrow activity. The infants were restless, breathing was noisy, and there was a watery nasal discharge. Edema appeared and slowly progressed until it involved the entire face, lower limbs, and genitalia. The edema is analogous to the exudative diathesis observed in vitamin E-deficient chicks. The skin lesions began on the sides of the face extending into the neck and adjacent parts of the scalp. All of the symptoms were associated with low serum vitamin E levels; the symptoms were not observed in the controls, which had higher serum vitamin E levels. The symptoms disappeared in response to oral vitamin E therapy; there was no response to iron or vitamin B_{12}. The lengthening of erythrocyte survival coincident with the rise in serum vitamin E was direct *in vivo* evidence that vitamin E prevented hemolytic anemia. The therapeutic effect of vitamin E in these experiments is presumably attributable to its ability to protect the vital phospholipids in cell membranes from peroxidative degeneration. Nowadays, infant milk formulas contain added vitamin E and an adequate ratio of vitamin E to PUFA; this has almost completely eradicated hemolytic anemia.

It is well documented that a diet rich in polyunsaturated fat, but which does not contain a correspondingly high amount of vitamin E, induces deficiency signs in animals. This also applies to humans as shown by the above experiments with premature infants. In a long-term human study (the Elgin project), adult male volunteers received a diet in which about half of the fat content was composed of vitamin E-stripped lard. After 30 months, this fraction of the fat content was replaced by stripped corn oil and 9 months later the amount of stripped corn oil was doubled. No manifestations of anemia were observed and it was not until the 72nd month that a well-controlled study of erythrocyte survival was performed. The data obtained showed that the erythrocytes of the vitamin E-depleted subjects were being destroyed at a rate about 8–10% faster than in the subjects in the control groups. The experiment was terminated soon after these observations, but it is logical to assume that if the diet had been made more deficient, the pathology would have been more severe (Horwitt, 1976).

17.11.2 Fat Malabsorption

Because of the intimate association between intestinal absorption of dietary fat and vitamin E, any condition causing the prolonged malabsorption of fat (steatorrhea) will lead to a secondary deficiency of vitamin E. Thus, patients with a variety of chronic fat malabsorption conditions exhibit low plasma vitamin E concentrations. The major non-genetic causes of steatorrhea associated with a symptomatic vitamin E deficiency state are chronic cholestatic hepatobiliary disorders, cystic fibrosis, and short bowel syndrome. Abetalipoproteinemia and homozygous hypobetalipoproteinemia are genetic causes of steatorrhea.

17.11.2.1 Chronic Cholestatic Hepatobiliary Disorders

These disorders include diseases of the liver and of the intrahepatic and extrahepatic bile ducts. The impaired bile flow leads to an insufficient concentration of bile constituents in the intestinal lumen and a consequent failure to produce micelles. The result is malabsorption of dietary fat-soluble substances. Because of their low vitamin E body stores, infants with cholestatic liver disease show symptoms of neuropathy as early as the second year of life, the neurological damage becoming irreversible if the vitamin E deficiency is not corrected. Correction of the deficiency by oral administration requires very high doses of vitamin E (100–200 IU per kg per day) or the use of a water-soluble form (α-tocopheryl polyethylene glycol-1000 succinate) which forms micelles. Alternatively, vitamin E can be administered by intramuscular injection.

17.11.2.2 Cystic Fibrosis

In cystic fibrosis, increased viscosity of pancreatic secretions causes obstruction of pancreatic ducts leading ultimately to destruction and fibrosis of the exocrine pancreas. The resultant failure to secrete pancreatic digestive enzymes causes steatorrhea and vitamin E deficiency. Despite the common observation of neuroaxonal lesions in the posterior column of the spinal cord at autopsy, overt neurological dysfunction is rare in vitamin E-deficient cystic fibrosis patients. Most patients who do exhibit neurological dysfunction also have fibrotic livers.

17.11.2.3 Short Bowel Syndrome

Short bowel syndrome is a collection of signs and symptoms used to describe the nutritional consequences of major surgical resections of the small intestine. Resections are carried out for treatment of Crohn's disease and mesenteric vascular thrombosis, among other disorders. The causes of vitamin E deficiency in these conditions are a reduced intestinal absorptive surface area and excessive fecal bile acid losses. Although low plasma vitamin E concentrations may be present within several years of surgical resection, ten to twenty years of severe malabsorption are generally required before the manifestation of neurological symptoms. This is because of the prior accumulation of vitamin E in most tissues and its relatively slow release from nervous tissues.

17.11.2.4 Abetalipoproteinemia

Chylomicrons contain apoB-48, among other apoproteins, while VLDL and LDL contain apoB-100. These two apoB proteins are encoded by the same gene, apoB-48 being synthesized in the intestinal mucosa and apoB-100 in the liver. Abetalipoproteinemia is a rare inborn error of lipoprotein production and transport characterized by undetectable or very small amounts of apoB-containing lipoproteins (chylomicrons, VLDL, and LDL) in the circulation. The underlying genetic defect in abetalipoproteinemia is a mutation in the gene coding for the microsomal triglyceride-transfer protein. This protein is essential for lipoprotein assembly in the Golgi apparatus; without it the lipoproteins are not secreted by the intestine or liver. Abetalipoproteinaemic patients become vitamin E deficient because the steatorrhea caused by the absence of chylomicrons severely impairs absorption of the vitamin. Furthermore, the lack of VLDL secretion by the liver means that no LDL can be formed, and so any vitamin E that might have been absorbed cannot be transported in the usual manner. The treatment of abetalipoproteinemic patients with massive oral doses of vitamin E (100 IU/kg/day) allows a small proportion to be absorbed, resulting in detectable plasma levels and correction of *in vitro* erythrocyte hemolysis (Traber *et al.*, 1993). Normal plasma levels are rarely, if ever, attained. Interestingly, the enterocytes of abetalipoproteinemic patients are able to synthesize HDL, which do not require apoB for their formation (Deckelbaum *et al.*, 1982). It is possible that this abnormally produced enteric HDL facilitates the intestinal secretion and plasma transport of vitamin E in the absence of the apoB-containing lipoproteins. The principal clinical features of abetalipoproteinemia are steatorrhea and spiky erythrocytes (both congenital), pigmented retinopathy, and a chronic progressive neurological disorder. The characteristic neurological and retinal symptoms manifest in the first decade of life, evolving into a crippling ataxia with visual impairment by the second or third decades (Sokol, 1989).

17.11.2.5 Homozygous Hypobetalipoproteinemia

Patients with this condition have a defect in the apoB gene and secrete lipoproteins containing truncated forms of apoB. These defective lipoproteins can transport minor amounts of vitamin E but they have a rapid turnover and constitute only a tiny fraction of the circulating liporoteins in these patients.

17.11.3 Ataxia with Vitamin E Deficiency (AVED)

Ataxia with vitamin E deficiency (AVED) uniquely represents a primary vitamin E deficient state. Originally called 'isolated vitamin E deficiency syndrome,' and later 'familial isolated vitamin E' (FIVE) deficiency, AVED is the result of a mutation in the gene for α-tocopherol transfer protein (α-TTP) on chromosome 8. Infants born with this syndrome have normal gastrointestinal function and yet their plasma vitamin E levels are only 1% of normal. There is either a complete absence of α-TTP or a defect in the α-tocopherol-binding region of the protein (Traber, 1994); in either case, there is impaired hepatic secretion of α-tocopherol in VLDL. The dramatic fall in the plasma level of vitamin E level is due to the rapid removal of α-tocopherol from the plasma to the liver and excretion in the bile, with no α-TTP to salvage it. The ataxia and other neurological symptoms appear between the ages of 4 and 18 years. They are manifestations of the neurological damage that arises from the impaired delivery of vitamin E to the nervous tissues, which are especially sensitive to variations in plasma vitamin E. When given vitamin E supplements (about 1 g per day), patients maintain normal plasma α-tocopherol concentrations and progression of the neurological damage is halted. If patients stop taking the supplements, their plasma concentrations fall to deficiency levels within days and the damage progresses.

17.11.4 Clinical Features and Histopathology of Vitamin E Deficiency

Sokol (1988) compared the clinical features found in abetalipoproteinemia, chronic childhood cholestasis, other fat malabsorption disorders, and isolated vitamin E deficiency (now known as AVED). The most common findings include loss of deep tendon reflexes, truncal and limb ataxia, loss of positional and vibratory sensation, muscle weakness, and dysarthria. Ophthalmoplegia (impairment of eye movements) and pigmented retinopathy are common features in abetalipoproteinemia, cholestasis and other fat malabsorption disorders, but they are not seen in AVED. A possible explanation is the fact that the first three disorders represent secondary deficiency states, malabsorption being the primary cause. In contrast, AVED, with no evidence of fat malabsorption or of other nutritional deficiencies, represents a primary deficiency state. A concomitant deficiency of vitamin A is probably required to produce the ocular symptoms present in cases of secondary vitamin E deficiency, the two vitamins acting synergistically.

Sokol (1988) also described the histopathology of vitamin E deficiency in humans. Axonal degeneration and demyelination of large-caliber neurons are universal in both primary and secondary advanced deficiency. Disturbance in function of the posterior columns of the spinal cord, sensory nerves, and spinocerebellar tracts account for the loss of vibratory and positional sensation and truncal and limb ataxia. The nerve degeneration presumably originates from peroxidation of constituent phospholipids. Peroxidative injury and the formation of lipopigments is the cause of pigmented retinopathy commonly seen in older patients with abetalipoproteinemia.

17.12 VITAMIN E TOXICITY

The tolerance and safety of vitamin E has been reviewed (Kappus & Diplock, 1992) and a Hazard Identification compiled (Institute of Medicine, 2000). There is no evidence of adverse effects from the consumption of vitamin E naturally occurring in foods. The question of toxicity arises when vitamin E is used in pharmacological amounts for 'therapeutic' purposes. Animal studies have shown that α-tocopherol is not mutagenic, carcinogenic, or teratogenic. In human studies with double-blind protocols and in large population studies, oral vitamin E supplementation resulted in few side effects even at doses as high as 3,200 IU per day (Bendich & Machlin, 1988). However, most human studies have been conducted over periods of a few weeks to a few months, so the possible chronic effects of multiyear ingestion of high doses of α-tocopherol are unknown. In view of the known inhibitory effect of high doses of vitamin E on platelet aggregation and two reported cases of postsurgery bleeding associated with intakes of 800–1200 IU of vitamin E per day, it is advisable to discourage vitamin E supplementation two weeks before and following surgery (Bendich & Machlin, 1993). In addition, vitamin supplements should not be taken by individuals receiving anticoagulant therapy.

FURTHER READING

Azzi, A. & Stocker, A. (2000) Vitamin E: Non-antioxidant roles. *Progress in Lipid Research*, **39**, 231–55.

Devaraj, S. & Jialal, I. (1998) The effects of alpha-tocopherol on critical cells in atherogenesis. *Current Opinion in Lipidology*, **9**, 11–15.

Keaney, J.F. Jr., Simon, D.I. & Freedman, J.E. (1999) Vitamin E and vascular homeostasis: Implications for atherosclerosis. *FASEB Journal*, **13**, 965–76.

Moriguchi, S. & Muraga, M. (2000) Vitamin E and immunity. *Vitamins and Hormones*, **59**, 305–36.

Steiner, M. (1991) Influence of vitamin E on platelet function in humans. *Journal of the American College of Nutrition*, **10**, 466–73.

REFERENCES

Arita, M., Nomura, K., Arai, H. & Inoue, K. (1997) α-Tocopherol transfer protein stimulates the secretion of α-tocopherol from a cultured liver cell line through a brefeldin A-insensitive pathway. *Proceedings of the National Academy of Sciences of the U.S.A.*, **94**, 12437–41.

Azzi, A., Boscoboinik, D., Chatelain, E., Özer, N.K. & Stäuble, B. (1993) d-α-Tocopherol control of cell proliferation. *Molecular Aspects of Medicine*, **14**, 265–71.

Azzi, A., Boscoboinik, D., Clément, S., Özer, N.K., Ricciarelli, R., Stocker, A., Tasinato, A. & Şirikçi, Ö. (1997) Signalling functions of α-tocopherol in smooth muscle cells. *International Journal for Vitamin and Nutrition Research*, **67**, 343–9.

Azzi, A., Breyer, I., Feher, M., Ricciarelli, R., Stocker, A., Zimmer, S. & Zingg, J.-M. (2001) Nonantioxidant functions of α-tocopherol in smooth muscle cells. *Journal of Nutrition*, **131**, 378S–81S.

Beharka, A.A., Wu, D., Han, S.N. & Meydani, S.N. (1997) Macrophage prostaglandin production contributes to the age-associated decrease in T cell function which is reversed by the dietary antioxidant vitamin E. *Mechanisms of Ageing and Development*, **93**, 59–77.

Bendich, A. & Machlin, L.J. (1988) Safety of oral intake of vitamin E. *American Journal of Clinical Nutrition*, **48**, 612–19.

Bendich, A. & Machlin, L.J. (1993) The safety of oral intake of vitamin E: Data from clinical studies from 1986 to 1991. In: *Vitamin E in Health and Disease.* (Ed. L. Packer & J. Fuchs), pp. 411–16. Marcel Dekker, Inc., New York.

Bieri, J.G. & Poukka Evarts, R. (1974) Gamma tocopherol: Metabolism, biological activity and significance in human vitamin E nutrition. *American Journal of Clinical Nutrition*, **27**, 980–6.

Boscoboinik, D., Szewczyk, A. & Azzi, A. (1991a) α-Tocopherol (vitamin E) regulates vascular smooth muscle cell proliferation and protein kinase C activity. *Archives of Biochemistry and Biophysics*, **286**, 264–9.

Boscoboinik, D., Szewczyk, A., Hensey, C. & Azzi, A. (1991b) Inhibition of cell proliferation by α-tocopherol. Role of protein kinase C. *Journal of Biological Chemistry*, **266**(10), 6188–94.

Bunyan, J., McHale, D., Green, J. & Marcinkiewicz, S. (1961) Biological potencies of ε- and ζ_1-tocopherol and 5-methyltocol. *British Journal of Nutrition*, **15**, 253–7.

Burton, G.W. & Ingold, K.U. (1981) Autoxidation of biological molecules. I. The antioxidant activity of vitamin E and related chain-breaking phenolic antioxidants *in vitro*. *Journal of the American Chemical Society*, **103**, 6472–7.

Burton, G.W. & Traber, M.G. (1990) Vitamin E: Antioxidant activity, biokinetics, and bioavailability. *Annual Review of Nutrition*, **10**, 357–82.

Burton, G.W., Traber, M.G., Acuff, R.V., Walters, D.N., Kayden, H., Hughes, L. & Ingold, K.U. (1998) Human plasma and tissue α-tocopherol concentrations in response to supplementation with deuterated natural and synthetic vitamin E. *American Journal of Clinical Nutrition*, **67**, 669–84.

Cachia, O., El Benna, J., Pedruzzi, E., Descomps, B., Gougerot-Pocidalo, M.-A. & Leger, C.-L. (1998a) α-Tocopherol inhibits the respiratory burst in human monocytes. *Journal of Biological Chemistry*, **273**(49), 32801–5.

Cachia, O., Léger, C.L. & Descomps, B. (1998b) Monocyte superoxide production is inversely related to normal content of α-tocopherol in low-density lipoprotein. *Atherosclerosis*, **138**, 263–9.

Cathcart, M.K., McNally, A.K., Morel, D.W. & Chisolm, G.M. III (1989) Superoxide anion participation in human monocyte-mediated oxidation of low-density lipoprotein and conversion of low-density lipoprotein to a cytotoxin. *Journal of Immunology*, **142**, 1963–9.

Celermajer, D.S. (1997) Endothelial dysfunction? Does it matter? Is it reversible? *Journal of the American College of Cardiology*, **30**, 325–33.

Chan, A.C. & Leith, M.K. (1981) Decreased prostacyclin synthesis in vitamin E-deficient rabbit aorta. *American Journal of Clinical Nutrition*, **34**, 2341–7.

Chan, A.C., Wagner, M., Kennedy, C., Chen, E., Lanuville, O., Mezl, V.A., Tran, K. & Choy, P.C. (1998) Vitamin E up-regulates arachidonic acid release and phospholipase A_2 in megakaryocytes. *Molecular and Cellular Biochemistry*, **189**, 153–9.

Chatelain, E., Boscoboinik, D.O., Bartoli, G.-M., Kagan, V.E., Gey, F.K., Packer, L. & Azzi, A. (1993) Inhibition of smooth muscle cell proliferation and protein kinase C activity by tocopherols and tocotrienols. *Biochimica et Biophysica Acta*, **1176**, 83–9.

Cort, W.M., Mergens, W. & Greene, A. (1978) Stability of alpha- and gamma-tocopherol: Fe^{3+} and Cu^{2+} interactions. *Journal of Food Science*, **43**, 797–8.

Deckelbaum, R.J., Eisenberg, S., Oschry, Y., Cooper, M. & Blum, C. (1982) Abnormal high density lipoproteins of abetalipoproteinemia: Relevance to normal HDL metabolism. *Journal of Lipid Research*, **23**, 1274–82.

Desrumaux, C., Deckert, V., Athias, A., Masson, D., Lizard, G., Palleau, V., Gambert, P. & Lagrost, L. (1999) Plasma phospholipid transfer protein prevents vascular endothelium dysfunction by delivering α-tocopherol to endothelial cells. *FASEB Journal*, **13**, 883–92.

Devaraj, S., Li, D. & Jialal, I. (1996) The effects of alpha tocopherol supplementation on monocyte function. Decreased lipid oxidation, interleukin-1β secretion, and monocyte adhesion to endothelium. *Journal of Clinical Investigation*, **98**, 756–63.

Devaraj, S. & Jialal, I. (1999) α-Tocopherol decreases interleukin-1β release from activated human monocytes by inhibition of 5-lipoxygenase. *Arteriosclerosis, Thrombosis, and Vascular Biology*, **19**, 1125–33.

Devaraj, S., Hugou, I. & Jialal, I. (2001) α-Tocopherol decreases CD36 expression in human monocyte-derived macrophages. *Journal of Lipid Research*, **42**, 521–7.

Dieber-Rotheneder, M., Puhl, H., Waeg, G., Striegl, G. & Esterbauer, H. (1991) Effect of oral supplementation with D-α-tocopherol on the vitamin E content of human low density lipoproteins and resistance to oxidation. *Journal of Lipid Research*, **32**, 1325–32.

Draper, H.H., Bergan, J.G., Chiu, M., Csallany, A.S. & Boaro, A.V. (1964) A further study of the specificity of the vitamin E requirement for reproduction. *Journal of Nutrition*, **84**, 395–400.

Drevon, C.A. (1991) Absorption, transport and metabolism of vitamin E. *Free Radical Research Communications*, **14**, 229–46.

Freedman, J.E. & Keaney, J.F. Jr. (2001) Vitamin E inhibition of Platelet aggregation is independent of antioxidents activity. *Journal of Nutrition,* **131**, 374S–7S.

Furchgott, R.F. & Zawadzki, J.V. (1980) The obligatory role of endothelial cells in the relaxation of arterial smooth muscle by acetylcholine. *Nature*, **288**, 373–6.

Fuster, V., Badimon, L., Badimon, J.J. & Chesebro, J.H. (1992) The pathogenesis of coronary artery disease and the acute coronary syndrome. *New England Journal of Medicine*, **326**, 310–18.

Gallo-Torres, H.E., Weber, F. & Wiss, O. (1971) The effect of different dietary lipids on the lymphatic appearance of vitamin E. *International Journal for Vitamin and Nutrition Research*, **41**, 504–15.

GISSI-Prevenzione Investigators. (1999) Dietary supplementation with n-3 polyunsaturated fatty acids and vitamin E after myocardial infarction: Results of the GSSI-Prevenzione trial. *Lancet*, **354**, 447–55.

Gryglewski, R.J., Dembinska-Kiec, A., Chytkowski, A. & Gryglewska, T. (1978) Prostacyclin and thromboxane biosynthetic capacities of the heart, arteries and platelets at various stages of experimental atherosclerosis in rabbits. *Atherosclerosis*, **31**, 385–94.

Gysin, R., Azzi, A. & Visarius, T. (2002) γ-Tocopherol inhibits human cancer cell cycle progression and cell proliferation by downregulation of cyclins. *FASEB Journal*, **16**, 1952–4.

Handin, R.I. (1996) Platelets and coronary heart disease. *New England Journal of Medicine*, **334**, 1126–7.

Handleman, G.J., Machlin, L.J., Fitch, K., Weiter, J.J. & Dratz, E.A. (1985) Oral α-tocopherol supplements decrease plasma γ-tocopherol in humans. *Journal of Nutrition*, **115**, 807–13.

Hansra, G., Bornancin, F., Whelan, R., Hemmings, B.A. & Parker, P.J. (1996) 12-*O*-tetradecanoylphorbol-13-acetate-induced dephosphorylation of protein kinase Cα correlates with the presence of a membrane-associated protein phosphatase 2A heterotrimer. *Journal of Biological Chemistry*, **271**(51), 32785–8.

Hassan, H., Hashim, S.A., Van Itallie, T.B. & Sebrell, W.H. (1966) Syndrome in premature infants associated with low plasma vitamin E levels and high polyunsaturated fatty acid diet. *American Journal of Clinical Nutrition*, **19**, 147–57.

Heller, R., Werner-Felmayer, G. & Werner, E.R. (2004) α-Tocopherol and endothelial nitric oxide synthesis. *Annals of the New York Academy of Sciences*, **1031**, 74–85.

Hennig, B. & Chow, C.K. (1988) Lipid peroxidation and endothelial cell injury: Implications in atherosclerosis. *Free Radical Biology & Medicine*, **4**, 99–106.

Hennig, B., Enoch, C. & Chow, C.K. (1987) Protection by vitamin E against endothelial cell injury by linoleic acid hydroperoxides. *Nutrition Research*, **7**, 1253–9.

HOPE (Heart Outcomes Prevention Evaluation) Study Investigators. (2000) Vitamin E supplementation and cardiovascular events in high-risk patients. *New England Journal of Medicine*, **342**, 154–60.

Horwitt, M.K. (1976) Vitamin E; a reexamination. *American Journal of Clinical Nutrition*, **29**, 569–78.

Institute of Medicine. (2000) Vitamin E. In: *Dietary Reference Intakes for Vitamin C, Vitamin E, Selenium, and Carotenoids*, pp. 186–283. National Academy Press, Washington, DC.

Islam, K.N., Devaraj, S. & Jialal, I. (1998) α-Tocopherol enrichment of monocytes decreases agonist-induced adhesion to human endothelial cells. *Circulation*, **98**, 2255–61.

Jialal, I. & Grundy, S.M. (1992) Effect of dietary supplementation with alpha-tocopherol on the oxidative modification of low density lipoprotein. *Journal of Lipid Research*, **33**, 899–906.

Jialal, I., Fuller, C.J. & Huet, B.A. (1995) The effect of α-tocopherol supplementation on LDL oxidation. A dose-response study. *Arteriosclerosis, Thrombosis, and Vascular Biology*, **15**, 190–8.

Joffe, M. & Harris, P.L. (1943) The biological potency of the natural tocopherols and certain derivatives. *Journal of the American Chemical Society*, **65**, 925–7.

Kappus, H. & Diplock, A.T. (1992) Tolerance and safety of vitamin E: A toxicological position report. *Free Radical Biology & Medicine*, **13**, 55–74.

Kayden, H.J. & Traber, M.G. (1993) Absorption, lipoprotein transport, and regulation of plasma concentrations of vitamin E in humans. *Journal of Lipid Research*, **34**, 343–58.

Keaney, J.F. Jr., Guo, Y., Cunningham, D., Shwaery, G.T., Xu, A. & Vita, J.A. (1996) Vascular incorporation of α-tocopherol prevents endothelial dysfunction due to oxidized LDL by inhibiting protein kinase C stimulation. *Journal of Clinical Investigation*, **98**, 386–94.

Kelleher, J. & Losowsky, M.S. (1970) The absorption of α-tocopherol in man. *British Journal of Nutrition*, **24**, 1033–47.

Kornbrust, D.J. & Mavis, R.D. (1980) Relative susceptibility of microsomes from lung, heart, liver, kidney, brain and testes to lipid peroxidation: Correlation with vitamin E content. *Lipids*, **15**, 315–22.

Kugiyama, K., Kerns, S.A., Morrisett, J.D., Roberts, R. & Henry, P.D. (1990) Impairment of endothelium-dependent arterial relaxation by lysolecithin in modified low-density lipoproteins. *Nature*, **344**, 160–2.

Kunisaki, M., Umeda, F., Inoguchi, T. & Nawata, H. (1992) Vitamin E restores reduced prostacyclin synthesis in aortic endothelial cells cultured with a high concentration of glucose. *Metabolism*, **41**, 613–21.

Kushi, L.H., Folsom, A.R., Prineas, R.J., Mink, P.J., Wu, Y. & Bostick, R.M. (1996) Dietary antioxidant vitamins and death from coronary heart disease in postmenopausal women. *New England Journal of Medicine*, **334**, 1156–62.

Li, D., Saldeen, T., Romeo, F. & Mehta, J.L. (1999) Relative effects of α- and γ-tocopherol on low-density lipoprotein oxidation and superoxide dismutase and nitric oxide synthase activity and protein expression in rats. *Journal of Cardiovascular and Pharmacological Therapeutics*, **4**, 219–26.

Li, D., Saldeen, T., Romeo, F. & Mehta, J.L. (2001) Different isoforms of tocopherols enhance nitric oxide synthase phosphorylation and inhibit human platelet aggregation and lipid peroxidation: Implications in therapy with vitamin E. *Journal of Cardiovascular and Pharmacological Therapeutics*, **6**, 155–61.

Libby, P. & Hansson, G.K. (1991) Involvement of the immune system in human atherogenesis: Current knowledge and unanswered questions. *Laboratory Investigation*, **64**, 5–15.

Li-Weber, M., Weigand, M.A., Giaisi, M., Süss, D., Treiber, M.K., Baumann, S., Ritsou, E., Breitkreutz, R. & Krammer, P.H. (2002) Vitamin E inhibits CD95 ligand expression and protects T cells from activation-induced cell death. *Journal of Clinical Investigation*, **110**, 681–90.

Losonczy, K.G., Harris, T.B. & Havlik, R.J. (1996) Vitamin E and vitamin C supplement use and risk of all-cause and coronary heart disease mortality in older persons: The Established Populations for Epidemiologic Studies of the Elderly. *American Journal of Clinical Nutrition*, **64**, 190–6.

Losowsky, M.S., Kelleher, J., Walker, B.E., Davies, T. & Smith, C.L. (1972) Intake and absorption of tocopherol. *Annals of the New York Academy of Sciences*, **203**, 212–22.

Mabile, L., Bruckdorfer, K.R. & Rice-Evans, C. (1999) Moderate supplementation with natural α-tocopherol decreases platelet aggregation and low-density lipoprotein oxidation. *Atherosclerosis*, **147**, 177–85.

MacMahon, M.T. & Neale, G. (1970) The absorption of α-tocopherol in control subjects and in patients with intestinal malabsorption. *Clinical Science*, **38**, 197–210.

Marchant, C.E., Law, N.S., van der Veen, C., Hardwick, S.J., Carpenter, K.L.H. & Mitchinson, M.J. (1995) Oxidized low-density lipoprotein is cytotoxic to human monocyte-macrophages: Protection with lipophilic antioxidants. *FEBS Letters*, **358**, 175–8.

Martin, A., Wu, D., Meydani, S.N., Blumberg, J.B. & Meydani, M. (1998) Vitamin E protects human aortic endothelial cells from cytotoxic injury induced by oxidized LDL in vitro. *Journal of Nutritional Biochemistry*, **9**, 201–8.

Meydani, S.N., Meydani, M., Verdon, C.P., Shapiro, A.A., Blumberg, J.B. & Hayes, K.C. (1986) Vitamin E supplementation suppresses prostaglandin E_2 synthesis and enhances the immune response of aged mice. *Mechanisms of Ageing and Development*, **34**, 191–201.

Meydani, S.N., Barklund, M.P., Liu, S., Meydani, M., Miller, R.A., Cannon, J.G., Morrow, F.D., Rocklin, R. & Blumberg, J.B.

(1990) Vitamin E supplementation enhances cell-mediated immunity in healthy elderly subjects. *American Journal of Clinical Nutrition*, **52**, 557–63.

Moulton, K.S., Semple, K., Wu, H. & Glass, C.K. (1994) Cell-specific expression of the macrophage scavenger receptor gene is dependent on PU.1 and a composite AP-1/ets motif. *Molecular and Cellular Biology*, **14**, 4408–18.

Munteanu, A. & Zingg, J.-M. (2007) Cellular, molecular and clinical aspects of vitamin E on atherosclerosis prevention. *Molecular Aspects of Medicine*, **28**, 538–90.

Nagano, Y., Arai, H. & Kita, T. (1991) High density lipoprotein loses its effect to stimulate efflux of cholesterol from foam cells after oxidative modification. *Proceedings of the National Academy of Sciences of the U.S.A.*, **88**(15), 6457–61.

Navab, M., Imes, S.S., Hama, S.Y., Hough, G.P., Ross, L.A., Bork, R.W., Valente, A.J., Berliner, J.A., Drinkwater, D.C., Laks, H. & Fogelman, A.M. (1991) Monocyte transmigration induced by modification of low density lipoprotein in cocultures of human aortic wall cells is due to induction of monocyte chemotactic protein 1 synthesis and is abolished by high density lipoprotein. *Journal of Clinical Investigation*, **88**, 2039–46.

Nelson, J.S. (1980) Pathology of vitamin E deficiency. In: *Vitamin E. A Comprehensive Treatise.* (Ed. by L.J. Machlin), pp. 397–428. Marcel Dekker, Inc., New York.

Neuzil, J. (2003) Vitamin E succinate and cancer treatment: A vitamin E prototype for selective antitumour activity. *British Journal of Cancer*, **89**, 1822–6.

Neuzil, J., Weber, C. & Kontush, A. (2001) The role of vitamin E in atherosclerosis: Linking the chemical, biological and clinical aspects of the disease. *Atherosclerosis*, **157**, 257–83.

Ohara, Y., Peterson, T.E., Zheng, B., Kuo, J.F. & Harrison, D.G. (1994) Lysophosphatidylcholine increases vascular superoxide anion production via protein kinase C activation. *Arteriosclerosis and Thrombosis*, **14**, 1007–13.

Ohgushi, M., Kugiyama, K., Fukunaga, K., Murohara, T., Sugiyama, S., Miyamoto, E. & Yasue, H. (1993) Protein kinase C inhibitors prevent impairment of endothelium-dependent relaxation by oxidatively modified LDL. *Arteriosclerosis and Thrombosis*, **13**, 1525–32.

Öhrvall, M., Sundlöf, G. & Vessby, B. (1996) Gamma, but not alpha, tocopherol levels in serum are reduced in coronary heart disease patients. *Journal of Internal Medicine*, **239**, 111–7.

Oski, F.A. & Barness, L.A. (1967) Vitamin E deficiency: A previously unrecognized cause of hemolytic anemia in the premature infant. *Journal of Pediatrics*, **70**, 211–20.

Özer, N.K., Boscoboinik, D. & Azzi, A. (1995) New roles of low density lipoproteins and vitamin E in the pathogenesis of atherosclerosis. *Biochemistry and Molecular Biology International*, **35**, 117–24.

Princen, H.M.G., van Poppel, G., Vogelezang, C., Buytenhek, R. & Kok, F.J. (1992) Supplementation with vitamin E but not with β-carotene in vivo protects low density lipoprotein from lipid peroxidation *in vitro.* Effect of cigarette smoking. *Arteriosclerosis and Thrombosis*, **12**, 554–62.

Reaven, P.D., Khouw, A., Beltz, W.F., Parthasarathy, S.P. & Witztum, J.L. (1993) Effect of dietary antioxidant combinations in humans. Protection of LDL by vitamin E but not by β-carotene. *Arteriosclerosis and Thrombosis*, **13**, 590–600.

Ricciarelli, R., Tasinato, A., Clément, S., Özer, N.K., Boscoboinik, D. & Azzi, A. (1998) α-Tocopherol specifically inactivates cellular protein kinase C α by changing its phosphorylation state. *Biochemical Journal*, **334**, 243–9.

Rimm, E.B., Stampfer, M.J., Ascherio, A., Giovannucci, E., Colditz, G.A. & Willett, W.C. (1993) Vitamin E consumption and the risk of coronary heart disease in men. *New England Journal of Medicine*, **328**, 1450–6.

Ritchie, J.H., Fish, M.B., McMasters, V. & Grossman, M. (1968) Edema and hemolytic anemia in premature infants. A vitamin E deficiency syndrome. *New England Journal of Medicine*, **279**, 1185–90.

Saldeen, T., Li, D. & Mehta, J.L. (1999) Differential effects of α- and γ-tocopherol on low-density lipoprotein oxidation, superoxide activity, platelet aggregation and arterial thrombogenesis. *Journal of the American College of Cardiology*, **34**, 1208–15.

Schneider, C. (2005) Chemistry and biology of vitamin E. *Molecular Nutrition & Food Research*, **49**, 7–30.

Scott, M.L. (1970) Studies on vitamin E and related factors in nutrition and metabolism. In: *The Fat-Soluble Vitamins.* (Ed. by H.F. DeLuca & J.W. Suttie), pp. 355–68. The University of Wisconsin Press, Madison.

Scott, M.L. (1980) Advances in our understanding of vitamin E. *Federation Proceedings*, **39**, 2736–9.

Sheppard, A.J., Pennington, J.A.T. & Weihrauch, J.L. (1993) Analysis and distribution of vitamin E in vegetable oils and foods. In: *Vitamin E in Health and Disease.* (Ed. by L. Packer & J. Fuchs), pp. 9–31. Marcel Dekker, Inc., New York.

Singh, U., Devaraj, S. & Jialal, I. (2005) Vitamin E, oxidative stress, and inflammation. *Annual Review of Nutrition*, **25**, 151–74.

Slatore, C.G., Littman, A.J., Au, D.H., Satia, J.A. & White, E. (2008) Long-term use of supplemental multivitamins, vitamin C, vitamin E, and folate does not reduce the risk of lung cancer. *American Journal of Respiratory and Critical Care Medicine*, **177**, 524–30.

Sokol, R.J. (1988) Vitamin E deficiency and neurologic disease. *Annual Review of Nutrition*, **8**, 351–73.

Sokol, R.J. (1989) Vitamin E and neurologic function in man. *Free Radical Biology and Medicine*, **6**, 189–207.

Stampfer, M.J., Hennekens, C.H., Manson, J.E., Colditz, G.A., Rosner, B. & Willett, W.C. (1993) Vitamin E consumption and the risk of coronary disease in women. *New England Journal of Medicine*, **328**, 1444–9.

Stern, M.H., Robeson, C.D., Weisler, L. & Baxter, J.G. (1947) δ-Tocopherol. I. Isolation from soybean oil and properties. *Journal of the American Chemical Society*, **69**, 869–74.

Sung, L., Greenberg, M.L., Koren, G., Tomlinson, G.A., Tong, A., Malkin, D. & Feldman, B.M. (2003) Vitamin E: The evidence for multiple roles in cancer. *Nutrition and Cancer*, **46**, 1–4.

Suzukawa, M., Ishikawa, T., Yoshida, H. & Nakamura, H. (1995) Effect of in-vivo supplementation with low-dose vitamin E on susceptibility of low-density lipoprotein and high-density lipoprotein to oxidative modification. *Journal of the American College of Nutrition*, **14**, 46–52.

Syväoja, E.-L., Piironen, V., Varo, P., Koivistoinen, P. & Salminen, K. (1986) Tocopherols and tocotrienols in Finnish foods: Oils and fats. *Journal of the Oil Chemists Society*, **63**, 328–9.

Szczeklik, A., Gryglewski, R.J., Domagala, B., Dworski, R. & Basista, M. (1985) Dietary supplementation with vitamin E in hyperlipoproteinemias: Effects on plasma lipid peroxides, antioxidant activity, prostacyclin generation and platelet aggregability. *Thrombosis and Haemostasis*, **54**, 425–30.

Tengerdy, R.P. (1989) Vitamin E, immmune response, and disease resistance. *Annals of the New York Academy of Sciences*, **570**, 335–44.

Teupser, D., Thiery, J. & Seidel, D. (1999) α-Tocopherol downregulates scavenger receptor activity in macrophages. *Atherosclerosis*, **144**, 109–15.

Thompson, J.N. & Scott, M.L. (1969) Role of selenium in the nutrition of the chick. *Journal of Nutrition*, **97**, 335–42.

Thompson, J.N. & Scott, M.L. (1970) Impaired lipid and vitamin E absorption related to atrophy of the pancreas in selenium-deficient chicks. *Journal of Nutrition*, **100**, 797–809.

Traber, M.G. (1994) Determinants of plasma vitamin E concentrations. *Free Radical Biology & Medicine*, **16**, 229–39.

Traber, M.G. (1999) Vitamin E. In: *Modern Nutrition in Health and Disease*, 9th edn. (Ed. by M.E. Shils, J.A. Olson, M. Shike & A.C. Ross), pp. 347–62. Lippincott Williams & Wilkins, Philadelphia.

Traber, M.G. & Kayden, H.J. (1989) Preferential incorporation of α-tocopherol vs γ-tocopherol in human lipoproteins. *American Journal of Clinical Nutrition*, **49**, 517–26.

Traber, M.G., Kayden, H.J., Green, J.B. & Green, M.H. (1986) Absorption of water-miscible forms of vitamin E in a patient with cholestasis and in thoracic duct-cannulated rats. *American Journal of Clinical Nutrition*, **44**, 914–23.

Traber, M.G., Burton, G.W., Ingold, K.U. & Kayden, H.J. (1990) *RRR*- and *SRR*-γ-tocopherols are secreted without discrimination in human chylomicrons, but *RRR*-α-tocopherol is preferentially secreted in very low density lipoproteins. *Journal of Lipid Research*, **31**, 675–85.

Traber, M.G., Cohn, W. & Muller, D.P.R. (1993) Absorption, transport and delivery to tissues. In: *Vitamin E in Health and Disease*. (Ed. by L. Packer & J. Fuchs), pp. 35–51. Marcel Dekker, Inc., New York.

Tran, K. & Chan, A.C. (1988) Effect of vitamin E enrichment on arachidonic acid release and cellular phospholipids in cultured human endothelial cells. *Biochimica et Biophysica Acta*, **963**, 468–75.

Tran, K. & Chan, A.C. (1990) *R,R,R*-α-tocopherol potentiates prostacyclin release in human endothelial cells. Evidence for structural specificity of the tocopherol molecule. *Biochimica et Biophysica Acta*, **1043**, 189–97.

Verlangieri, A.J. & Bush, M.J. (1992) Effects of d-α-tocopherol supplementation on experimentally induced primate atherosclerosis. *Journal of the American College of Nutrition*, **11**, 131–8.

Williams, J.C., Forster, L.A., Tull, S.P., Wong, M., Bevan, R.J. & Ferns, G.A.A. (1997) Dietary vitamin E supplementation inhibits thrombin-induced platelet aggregation, but not monocyte adhesiveness, in patients with hypercholesterolaemia. *International Journal of Experimental Pathology*, **78**, 259–66.

Wolf, G. (1997) γ-Tocopherol: An efficient protector of lipids against nitric oxide-initiated peroxidative damage. *Nutrition Reviews*, **55**, 376–8.

Women's Health Study. (2005) Vitamin E in the primary prevention of cardiovascular disease and cancer: The Women's Health Study: A randomized controlled trial. *Journal of the American Medical Association*, **294**, 56–65.

Wu, D., Koga, T., Martin, K.R. & Meydani, M. (1999) Effect of vitamin E on human aortic endothelial cell production of chemokines and adhesion to monocytes. *Atherosclerosis*, **147**, 297–307.

Yamakura, F., Taka, H., Fujimura, T. & Murayama, K. (1998) Inactivation of human manganese-superoxide dismutase by peroxynitrite is caused by exclusive nitration of tyrosine 34 to 3-nitrotyrosine. *Journal of Biological Chemistry*, **273**(23), 14085–9.

Yoshida, H., Yusin, M., Ren, I., Kuhlenkamp, J., Hirano, T., Stolz, A. & Kaplowitz, N. (1992) Identification, purification, and immunochemical characterization of a tocopherol-binding protein in rat liver cytosol. *Journal of Lipid Research*, **33**, 343–50.

18 Vitamin K

18.1 CHEMISTRY

The term 'vitamin K' is a generic term used for all compounds possessing co-factor activity for γ-glutamylcarboxylase. Two forms of vitamin K exist in nature: vitamin K_1 and vitamin K_2.

Vitamin K_1, known as phylloquinone, is synthesized by green plants and possesses unequivocal biological activity in humans. Phylloquinone is a component of the photosynthetic electron transport system and it occurs exclusively in the thylakoid membranes of the chloroplasts. The phylloquinone molecule comprises a methyl-substituted naphthoquinone ring structure attached at C-3 to a phytyl side chain composed of three saturated and one unsaturated isoprene units, each unit having five carbon atoms. The structure of phylloquinone, together with that of phytol, is shown in Figure 18.1. Phylloquinone, having a 20-carbon side chain, may be designated $K_{1(20)}$ to distinguish it from structural analogs such as $K_{1(25)}$, a synthetic compound with an extra isoprene unit. The natural form of phylloquinone is the *trans* isomer, whose stereochemistry is 2′-*trans*,7′ *R*,11′ *R*. Synthetic phylloquinone usually contains both the *trans* and the *cis* isomers, but only the *trans* isomer is essentially responsible for the vitamin's co-factor activity.

Vitamin K_2 refers to a family of structural analogs called menaquinones that are synthesized exclusively by bacteria. There is no certain evidence for utilization of menaquinones by humans, but they are generally assumed to contribute to vitamin K nutriture. Menaquinones have a common ring structure that is identical to that of phylloquinone and side chains composed of a number of unsaturated isoprene units. They are designated MK_n (where n is the number of isoprene units), e.g. menaquinone-4 (MK-4). The predominant menaquinones in nature range from MK-4 to MK-13; those with more than 8 isoprene units (i.e. MK-9 and upward) are referred to arbitrarily as long-chain menaquinones. The geranylgeranyl side chain of MK-4 confers unique properties among menaquinones with regard to bone metabolism. The structure of MK-4 is shown in Figure 18.1 together with the parent of its side chain, geranylgeraniol.

Menadione (vitamin K_3), the naphthoquinone ring without the side chain, is a metabolite of oral vitamin K that lacks intrinsic biological activity, but is converted in the body to MK-4 (Thijssen *et al.*, 2006).

18.2 FOOD SOURCES

The highest concentrations of vitamin K (in the form of phylloquinone) are found in green leafy vegetables, e.g. cabbage, broccoli, Brussels sprout, and spinach. Such vegetables are the top contributors to vitamin K intake in the American diet. Other types of vegetables (roots, bulbs, and tubers), cereal grains and their milled products, and fruits and fruit juices are poor sources of vitamin K. Animal products (meat, fish, milk products, and eggs) contain low concentrations of phylloquinone, but appreciable amounts of menaquinones are present in liver.

Some vegetable oils, including canola (rapeseed), soybean, and olive oil, are rich sources of phylloquinone, whereas peanut and corn (maize) are not. Soybean oil is the most commonly consumed vegetable oil in the American diet. The addition of phylloquinone-rich vegetable oils in the processing and cooking of foods that are otherwise poor sources of vitamin K makes them potentially important dietary sources of the vitamin. This is particularly evident, for example, when chicken, eggs, and potatoes are fried in certain vegetable oils. Those margarines, mayonnaises, and regular-calorie salad dressings that are derived from phylloquinone-rich vegetable oils are second to green leafy vegetables in their phylloquinone content. The addition of these fats and oils to mixed dishes and desserts has an important impact on the amount of vitamin K in the American diet.

Various menaquinones have been found in fermented foods (Sakano *et al.*, 1988), salmon, shellfish, beef, pork, chicken, egg yolk, cheese, and butter (Hirauchi *et al.*, 1989b) but the amounts may not be nutritionally significant in some of these foods. Livers of ruminant species (e.g. cow) contain significant concentrations (10–20 µg/100 g) of some menaquinones (Hirauchi *et al.*, 1989a), while cheese contains significant quantities of MK-8 (5–10 µg/100g) and MK-9 (10–20 µg/100 g) (Shearer *et al.*, 1996).

18.3 ABSORPTION OF DIETARY VITAMIN K

Phylloquinone, the major form of vitamin K in the diet (>90%), is absorbed in the jejunum and less efficiently in the ileum, in a process that is dependent on the normal flow of bile and pancreatic juice (Shearer *et al.*, 1974). Both long- and short-chain menaquinones are readily absorbed by rats after oral ingestion (Groenen-van Dooren *et al.*, 1995) and therefore dietary menaquinones are likely to be absorbed along with phylloquinone.

Dietary vitamin K is solubilized within mixed micelles in the duodenum. The low pH microclimate of the unstirred layer lining the luminal surface of the jejunum reduces fatty acid solubility in the mixed micelles, allowing release of the fatty acids together with other lipid constituents. Phylloquinone can then diffuse across the brush-border membrane of the enterocyte. Within the enterocyte, phylloquinone is incorporated into chylomicrons. After

DOI: 10.1201/9781003410300-21

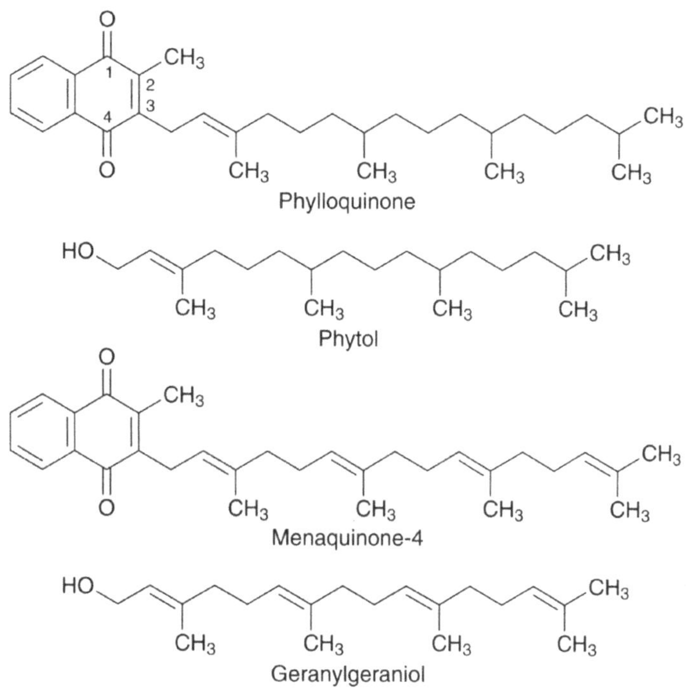

FIGURE 18.1 Chemical structures of phylloquinone, phytol, menaquinone-4, and geranylgeraniol.

processing through the Golgi apparatus, the chylomicrons are discharged from the enterocyte by exocytosis across the basolateral membrane and enter the central lacteal of the villus. From there, they pass into the larger lymphatic vessels draining the intestine and enter the blood circulation via the thoracic duct.

The efficiency of vitamin K absorption varies widely depending on the source of the vitamin and the amount of fat in the meal. Pure phylloquinone is absorbed with an efficiency of 80% (Shearer *et al.*, 1974). The phylloquinone present in cooked spinach was only 4% as bioavailable as that from a commercial detergent suspension of phylloquinone. Adding butter to the spinach increased this to 13% (Gijsbers *et al.*, 1996). The absorption of phylloquinone was six times higher after the ingestion of a 500-µg phylloquinone tablet than after the ingestion of 150 g of raw spinach containing 495 µg phylloquinone (Garber *et al.*, 1999). The phylloquinone from a phylloquinone-fortified oil was absorbed better than that from an equivalent amount from cooked broccoli, regardless of adjustment to triglyceride concentrations (Booth *et al.*, 2002). The tight binding of phylloquinone to the thylakoid membranes of chloroplasts explains the poor bioavailability of the vitamin in green plants. The free phylloquinone in vegetable oils, margarines, and dairy products is well absorbed owing to the stimulating effect of fat.

18.4 BACTERIALLY SYNTHESIZED MENAQUINONES AS A POSSIBLE ENDOGENOUS SOURCE OF VITAMIN K

The large intestine of healthy adult humans contains a microflora of bacteria, many species of which synthesize menaquinones ranging mainly from MK-6 to MK-11. The menaquinones are incorporated into the bacterium's cytoplasmic membrane where they function under reduced (anaerobic) conditions as redox compounds in bacterial respiration. Among the major forms, MK-10 and MK-11 are produced by *Bacteroides* species, MK-8 by Enterobacteria species, MK-7 by *Veillonella* species, and MK-6 by *Eubacterium lentum* (Shearer, 1993).

Conly and Stein (1992a) reported quantitative and qualitative measurements of phylloquinone and menaquinones at different sites within the human intestinal tract. Overall, long-chain menaquinones (MK-9, -10, and -11) predominated. Menaquinones were found mostly in the distal colon (10 fecal samples) and totaled 19.85 ± 0.36 µg/g dry weight. Menaquinones in two samples of terminal ileal contents taken during appendectomy totaled 8.85 µg/g dry weight. Little menaquinone was found in samples of proximal jejunal contents collected by means of a nasojejunal tube (totaling 0.03 µg/g dry weight).

The extent to which the menaquinones produced by the enteric bacteria can be absorbed and utilized by the human host as a source of vitamin K is a continuing subject of debate. Two factual observations imply that bacterially produced menaquinones could be absorbed. First, about 90% of liver stores of vitamin K are in the form of menaquinones (Shearer, 1992) despite phylloquinone predominating in the diet. Second, the various menaquinones found in liver are remarkably consistent with the menaquinone profile of human intestinal content. However, it has not been possible to prove that the hepatic menaquinones do not originate from the diet. Patients with marginal or no vitamin K intake develop bleeding episodes associated with a prolonged prothrombin time only when they are given broad-spectrum antibiotics (Allison *et al.*, 1987). The obvious inference from this observation is that the prolongation of prothrombin time is caused ultimately by destruction of the gut microflora and consequent vitamin K deficiency. However, Lipsky (1988) pointed out that *N*-methylthiotetrazole-containing antibiotics suppress vitamin K-dependent biosynthesis of prothrombin. Collectively, these observations suggest, but do not prove, that enteric menaquinones can be absorbed and utilized to some extent.

A greater wealth of evidence suggests that the menaquinones produced by colonic bacteria cannot be absorbed in amounts sufficient to prevent vitamin K deficiency under conditions of dietary deprivation. For example, the rat, which is notoriously resistant to dietary deprivation of vitamin K, displays signs of deficiency when coprophagy is prevented (Barnes & Fiala, 1959). This implies that the rat can only utilize enterically synthesized menaquinones through coprophagy and subsequent absorption from the jejunum. Another study showed that rats fed a vitamin K-deficient diet developed signs of vitamin K deficiency, despite the fact that the amounts of menaquinones in the colon had actually increased compared with a control group of rats fed a normal diet (Uchida & Komeno, 1988). The importance of sufficient dietary vitamin K consumption in rats was confirmed by Groenen-van Dooren *et al.* (1993). In this study, rats receiving a low-fiber, vitamin K-deficient diet developed hypoprothrombinemia when coprophagy was prevented.

Ichihashi *et al.* (1992) examined the colonic absorption of MK-9 and MK-4 using the rat *in situ* loop method. MK-9 is a typical bacterially produced menaquinone, whereas MK-4 is a minor bacterial product, forming less than 0.5% of the total pool of intestinal menaquinones (Groenen-van Dooren *et al.*, 1993). [^{14}C]MK-4 was shown to be absorbed via the portal bloodstream, even when bile was not present, although the absorption from the colon was slower than that from the jejunum. There was no absorption of MK-4 from the colon by the lymphatic route, even when bile was present. In contrast, when [^{14}C]MK-9 was introduced into the colonal loop without bile, most of the radioactivity was recovered from the loop, and no transfer of MK-9 into the lymph or blood was observed. The data from these experiments suggest that menaquinones of short chain length, which constitute only a small and perhaps negligible part of the bacterially produced menaquinones, are absorbed from the colon via the portal pathway. Absorption rates decrease markedly with an increase in chain length and the bulk of the bacterially produced menaquinones is not absorbed from the colon by any route. The extremely poor colonic absorption of MK-4 and MK-9 in rats was also reported by Groenen-van Dooren *et al.* (1995).

The menaquinones incorporated into membranes of viable bacteria are not available for absorption. However, Conly and Stein (1992b) described *in vitro* experiments showing that significant amounts of biologically active menaquinones can be secreted or liberated from bacteria. Being strongly lipophilic, the bacterially synthesized menaquinones require the presence of bile salts and the formation of mixed micelles for absorption to take place. Because bile salts are reabsorbed in the distal ileum and the amounts remaining are degraded by colonic bacteria, there is no opportunity for absorption of menaquinones to take place in the colon. Indeed, colonic absorption of MK-9 in rats is extremely poor, as shown by Groenen-van Dooren *et al.* (1995). However, bearing in mind the appreciable amounts of menaquinones found in the terminal ileum of two subjects (Conly & Stein, 1992a), and considering the possibility that contents from the cecum (where large amounts of bacteria reside) may backwash past the ileocecal valve into the ileum, one can envisage some degree of bile salt-mediated absorption taking place in this region. In addition, Hollander *et al.* (1977) demonstrated ileal absorption of MK-9 in the conscious rat and showed that the absorption rate increased with increasing bile salt concentration.

In a human study (Udall, 1965), ten patients receiving daily doses of warfarin (an antagonist of vitamin K in the blood-clotting process) were given large (500 mg) doses of phylloquinone directly into the cecum via a nasopharyngeal tube. No significant change in prothrombin time was observed in any patient, indicating that the vitamin was not absorbed. In contrast, one additional patient given an equal dose of phylloquinone by mouth showed the expected response, i.e. a fall in the prothrombin time. Ferland *et al.* (1993) reported that subclinical vitamin K deficiency can be induced in healthy adults by dietary deprivation of the vitamin, suggesting that absorption of bacterially synthesized menaquinones may not be sufficient to sustain adequate vitamin K status.

Conly *et al.* (1994) demonstrated that menaquinones can be absorbed directly from the human ileum and be functionally active. Their study consisted of an experimental phase followed by a control phase, using the same four volunteers. The volunteers were started on a vitamin K-deficient diet and then given adjusted doses of warfarin to maintain a stable elevated prothrombin time. A 1.5-mg dose of mixed menaquinones (MK-4–MK-9) extracted from harvested *Staphylococcus aureus* was then placed directly into the ileum by means of a nasoileal tube after an overnight fast. Within 24 h of menaquinone administration, the prothrombin time decreased significantly and the factor VII level increased significantly, indicating that the menaquinones had been absorbed and utilized. The results of this study provide an explanation as to why starvation or a complete lack of dietary intake of vitamin K alone cannot induce a clinically manifest vitamin K deficiency.

In summary, the published data do not support the tenuous assumption that menaquinones synthesized by the intestinal microflora are directly absorbed in the colon. The normal bile salt-mediated mechanism of absorption is precluded because bile salts are reabsorbed in the distal ileum. Given that long-chain colonic menaquinones are integral components of the lipophilic bacterial cytoplasmic membrane, and even in the free form are totally insoluble in water, it is difficult to envisage how significant amounts can be solubilized for absorption in a region where bile salts are lacking. However, the possibility exists that some bacterially synthesized menaquinones may be absorbed from the distal ileum by the bile salt-mediated lymphatic pathway. Arguing against this possibility is a report by Shearer *et al.* (1992) that virtually all the menaquinones present in human ileal juice could be removed by simple centrifugation or by passage through a 0.2-μm filter. This suggests that the menaquinones in ileal juice are still membrane-bound and not in the micellar form which would favor the lymphatic absorption process.

18.5 PLASMA TRANSPORT OF ABSORBED PHYLLOQUINONE AND UPTAKE BY LIVER

Like vitamin E, vitamin K has no specific plasma carrier protein and is transported in plasma by lipoproteins. Circulating chylomicrons pick up apoC-II and apoE from HDL in the blood. Lipoprotein lipase attached to the luminal surface of capillary endothelia in extrahepatic tissues hydrolyzes the triglycerides in the chylomicrons, releasing free fatty acids and diglycerides, which the tissue cells take up by pinocytosis. The lipase is activated by the apoC-II that the chylomicrons have just acquired. As a result of lipolysis, apoA and apoC-II are lost, but apoE is retained. The vitamin K-loaded chylomicron remnants are carried in the blood to the liver, where they are reduced in size by the lipolytic action of hepatic lipase. Endocytosis of the particles is mediated by the LDL receptor-related protein (LRP) located on hepatocyte sinusoidal surfaces.

After an overnight fast, some 50–70% of the circulating phylloquinone is associated with a triglyceride-rich lipoprotein (TRL) fraction that comprises chylomicrons and chylomicron remnants of intestinal origin and VLDL secreted by the liver. The remainder is approximately equally distributed between LDL and HDL (Shearer & Newman, 2008). Liver and bone can take up apoE-enriched TRL through avid binding to the glycocalyx on the surface of hepatocytes and osteoblasts (Newman *et al.*, 2002).

Lamon-Fava *et al.* (1998) examined the plasma transport of phylloquinone by plasma lipoproteins in healthy subjects after taking a pharmacological dose (50 μg/kg body weight) of phylloquinone with a fat-rich meal. Comparisons between the 3 h and 12 h postprandial timepoints showed that the proportion of phylloquinone carried by TRL decreased from 91% to 70%. Conversely, the proportions carried by LDL and HDL increased progressively from 3% to 14% for LDL and from 4% to 11% for HDL. Although the constituents of the TRL fraction could not be resolved, it was conjectured that some phylloquinone taken up by the liver re-emerges later incorporated into VLDL, which, after further lipolysis, become LDL particles still retaining their phylloquinone cargo. Another possibility is that some of the phylloquinone carried by LDL and HDL derives from the transfer of phylloquinone from intestinally derived chylomicrons and chylomicron remnants.

Schurgers and Vermeer (2002) examined the plasma transport of phylloquinone and two menaquinones (MK-4 and MK-9) in healthy subjects after ingesting a meal of toast and marmalade spread with 2 μmol of each vitamer dissolved in 30 g of corn oil. Serum and lipoprotein profiles of each vitamer were measured for up to 48 h. During the first 8 h following intake, the serum contained much higher concentrations of phylloquinone than MK-4 or MK-9. Poor absorption of the two menaquinones does not explain this observation, as another study in humans has shown that the absorption of MK-7 is far more complete than that of phylloquinone (Koivu-Tikkanen *et al.*, 2000). The explanation might be different rates of tissue uptake: maximal concentrations in serum were reached at 2.5, 4, and 5 h postprandially for MK-4, phylloquinone, and MK-9, respectively. These differences between vitamers could be attributed to their relative hydrophobicities, which would determine their positions in lipoprotein particles and their availability for transference. MK-4, being the least hydrophobic, may be loosely associated with the polar surface of the particle; MK-9, the most hydrophobic, may be present in the lipid core; and phylloquinone may occupy an intermediate position.

With regard to the incorporation of phylloquinone, MK-4, and MK-9 into lipoproteins (TRL, LDL and HDL), phylloquinone was found mainly in the TRL fraction, reaching a concentration of ~30 nmol/L at about 5 h and dropping to ~5 nmol/L at about 10 h. Phylloquinone concentrations in the LDL and HDL fractions reached ~5 nmol/L at about 5 h. Lower concentrations (~3.5–4.5 nmol/L) of MK-4 were distributed almost equally over the TRL, LDL, and HDL fractions up to 4 h. After 8 h, the proportion of MK-4 in LDL increased sharply until some 80% of that remaining was present in LDL. MK-9 was initially present only in the TRL fraction, reaching ~4.5 nmol/L at about 5 h. MK-9 appeared in the LDL fraction only at 8 h after intake and remained detectable (~1 nmol/L) until at least 48 h. This infers an initial uptake of MK-9 by the liver, from where it is released slowly via LDL and remains available for all kinds of tissues possessing LDL receptors. At no time point was MK-9 found in the HDL fraction.

18.6 MATERNAL TO FETAL TRANSFER OF PHYLLOQUINONE

The transfer of phylloquinone from the maternal to fetal circulation is poor. Despite a 500-fold increase in maternal plasma phylloquinone concentration following the intravenous administration of 1 mg of phylloquinone to pregnant women at term, the corresponding increase in cord plasma was only about fivefold. The levels attained in cord plasma (0.10–0.14 ng/ml) after injection were at most near the lower end of the normal fasting adult range (0.10–0.66 ng/ml) (Shearer *et al.*, 1982). When pregnant women were given daily oral doses of 20 mg phylloquinone for at least 3 days, cord plasma levels of phylloquinone were boosted 30-fold at mid-trimester and 60-fold at term. Again, these levels were substantially lower than corresponding supplemented maternal levels (Mandelbrot *et al.*, 1988). The large concentration gradient of phylloquinone between maternal and neonatal plasma suggests that phylloquinone does not cross the placenta readily. Alternatively, uptake by fetal plasma is low, perhaps because of low levels of transporting lipoproteins.

The cord plasma of premature infants increased by an average of 2.3-fold after their mothers received 5 mg of phylloquinone intramuscularly several hours to 35 min before delivery (Yang *et al.*, 1989). Thus, supplemental phylloquinone given to the mother antenatally can be transferred to premature infants, but to a lesser degree than to term babies.

18.7 STORAGE AND CATABOLISM OF VITAMIN K IN THE LIVER

18.7.1 Storage

The liver has a limited capacity for long-term storage of vitamin K compared to vitamin A. Surprisingly, phylloquinone comprises only about 10% of the total liver stores of vitamin K. Menaquinones ranging from MK-4 to MK-13 make up the bulk of stores with the long-chain forms (MK-9–MK-13) constituting 73% of total vitamin K (Usui *et al.*, 1990). Unlike phylloquinone, which undergoes rapid turnover, the hepatic turnover of long-chain menaquinones is low presumably because of their high affinity for membranes (Shearer, 1992). The contrasting turnovers of phylloquinone and menaquinones may account for the predominance of the latter in liver. Whether the menaquinones originate from the diet or from bacterial synthesis, their strong retention relative to phylloquinone would enable concentrations to gradually build up while phylloquinone is being constantly utilized and metabolized. In support of this concept, hepatic stores of phylloquinone are rapidly depleted during dietary restriction of vitamin K, but hepatic stores of menaquinones are not (Usui *et al.*, 1990). Also, the common hepatic menaquinones (MK-9–MK-13) are not detectable in plasma, suggesting that they are not easily mobilized. It appears, therefore, that the large hepatic pool of menaquinones does not contribute significantly to vitamin K nutriture, but represents a very slow turnover of the extremely lipophilic long-chain menaquinones. Further work is needed to establish the origin of hepatic menaquinones and their nutritional relevance.

In the liver of the human fetus, phylloquinone is detectable as early as 10 weeks gestation, and at term the concentration is about one-fifth the value in adults. Hepatic concentrations of menaquinones are usually undetectable at birth and in the first week of life. The gradual build-up of hepatic stores of menaquinones is consistent with the colonization of the neonatal gut by enteric bacteria.

18.7.2 Catabolism and Excretion

Phylloquinone is extensively catabolized by the liver which explains the rapid turnover of hepatic reserves and depletion during dietary deprivation. The catabolic products are excreted partly in the urine and partly in the feces via the bile. Based on tracer experiments with labeled phylloquinone, it appears that 60–70% of the amounts of phylloquinone absorbed from each meal will ultimately be excreted (Shearer *et al.*, 1996). Humans excrete phylloquinone and menaquinones by a common degradative pathway whereby the polyisoprenoid side chain is shortened and the metabolites are conjugated, mainly with glucuronic acid, and excreted (Shearer *et al.*, 2012).

18.8 CONVERSION OF PHYLLOQUINONE TO MENAQUINONE-4

A remarkable observation in vitamin K metabolism is the conversion of phylloquinone to MK-4. The conversion appears to be localized mainly in extrahepatic tissues. In the rat, the addition of phylloquinone to a vitamin K-free diet leads to low liver and plasma MK-4 concentrations, but in brain, pancreas, salivary gland and sternum, the concentration of MK-4 exceeds that of phylloquinone (Thijssen & Drittij-Reijnders, 1994). The absence of MK-3, MK-5, and MK-6 in the rat experiments indicates that the MK-4 synthesis is specific (Thijssen *et al.*, 1996). Humans show tissue-specific accumulation of phylloquinone and MK-4 with distribution patterns comparable to those in the rat (Thijssen & Drittij-Reijnders, 1996).

It had been generally assumed that phylloquinone is converted enzymatically to MK-4 in the extrahepatic tissues. Thijssen *et al.* (2006) were unable to demonstrate any such conversion in any of the various cell lines that they tested; however, all of the cells were able to synthesize MK-4 in the presence of menadione. The rapid appearance of menadione in urine after oral (but not subcutaneous) administration of vitamin K suggested that side-chain cleavage of phylloquinone occurs during intestinal absorption and that circulating menadione is converted to MK-4 in the tissues. Hirota *et al.* (2013) confirmed that this is indeed the case. The conversion of dietary phylloquinone to menadione is evidently not dependent on the gut flora, as it can take place in gnotobiotic (germ-free) rats and in aseptic mammalian cell cultures (Davidson *et al.*, 1998; Ronden *et al.*, 1998).

18.9 BIOCHEMICAL FUNCTIONS

18.9.1 The Vitamin K-Dependent Glutamate γ-Carboxylation Reaction

The only known biochemical function of vitamin K is as a co-factor for γ-glutamyl carboxylase, an enzyme integral to the endoplasmic reticulum membrane that catalyzes a unique post-translational conversion of specific glutamate (Glu) residues into γ-carboxyglutamate (Gla) residues in certain proteins, referred to as Gla proteins. The reaction adds another carboxyl group to the γ-carbon of Glu (Figure 18.2). Because Gla is a much stronger Ca^{2+} chelator than Glu, the conversion increases the Ca^{2+}-binding capacity of a Gla protein (Vermeer, 1990). The binding of Ca^{2+} to a γ-carboxylated Gla protein leads to the conformational change that is required for its biological activity. In the absence of vitamin K or in the presence of a vitamin K antagonist, such as warfarin, the synthesis of biologically active Gla proteins is impaired.

18.9.2 The Vitamin K Epoxide Cycle

In the liver, γ-glutamyl carboxylation is linked to the vitamin K epoxide cycle – a metabolic cycle which serves to conserve the pool of vitamin K available to the carboxylase (Figure 18.3). The vitamin K must be in the reduced (hydroquinone) form for carboxylation to take place. The carboxylation reaction is catalyzed by γ-glutamyl carboxylase (E_1) coupled to a vitamin K epoxidase (E_2) which simultaneously converts vitamin K hydroquinone to vitamin K 2,3-epoxide. The epoxide is reduced to the quinone by vitamin K epoxide reductase (E_3). The cycle is completed by the reduction of the recycled quinone by a vitamin K reductase (E_4). The reductase enzymes E_3 and E_4 are dithiol-dependent and are inhibited by vitamin K antagonists such as warfarin and dicoumarol. It is probable that E_3 and E_4 are one and the same enzyme. Exogenous vitamin K from the diet may enter the cycle via an NAD(P)H-dependent vitamin K reductase (E_5), which is not inhibited by warfarin (Shearer, 1992).

18.10 ROLE OF VITAMIN K IN BLOOD COAGULATION

The blood coagulation system is described in Section 7.10.5.

The liver synthesizes a group of seven Gla proteins that have a regulatory function in blood coagulation: factor II (prothrombin) and factors VII, IX, and X have a coagulant function, while proteins C, S, and Z have an anticoagulant function.

The chick bioassay for vitamin K is based upon the degree of lowering of elevated blood clotting times in vitamin K-depleted chicks. Blood clotting measurements (actually prothrombin times) are rapidly determined following the addition of a clotting agent (thromboplastin) and calcium chloride solution to oxalated or citrated blood. The chick is the animal of choice because its vitamin K requirement is fivefold that of the rat, it is readily depleted of vitamin K, and coprophagy (fecal recycling) is easier to control. The chick's higher requirement for vitamin K compared with the rat is at least partly attributable to the short length of its colon and rapid transit time.

Matschiner and Doisy (1966) determined the molar activities of several forms of vitamin K using the chick

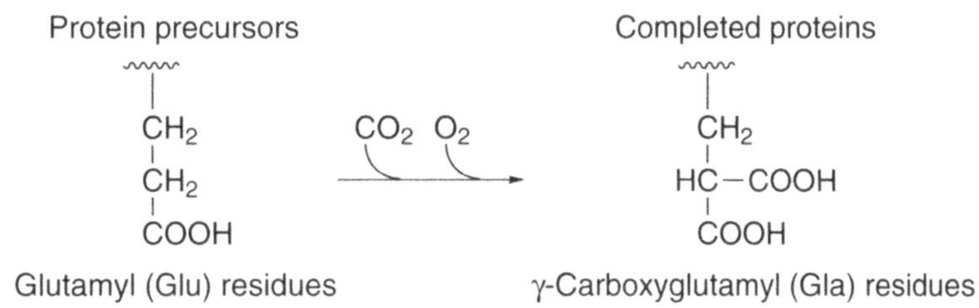

FIGURE 18.2 The vitamin K-dependent glutamate γ-carboxylation reaction.

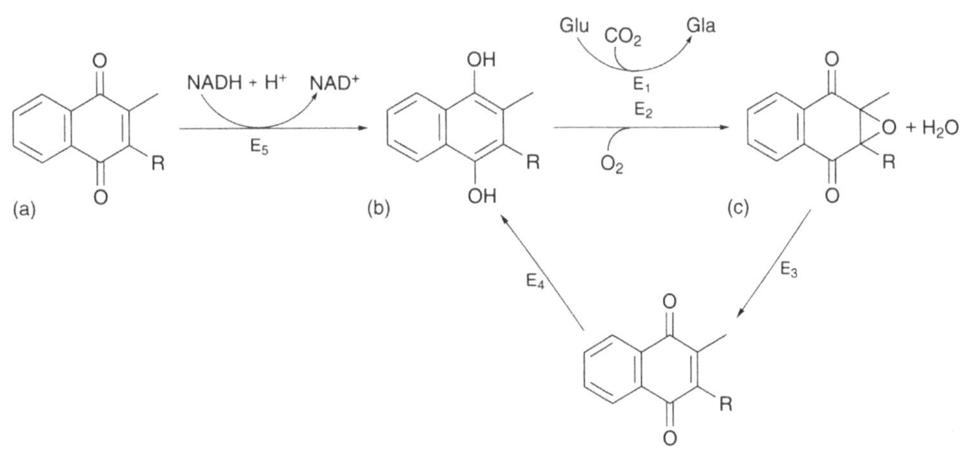

FIGURE 18.3 The vitamin K epoxide cycle.

bioassay. Compared to phylloquinone, which was arbitrarily assigned an activity of 100, MK-4 had the highest activity (156) followed by MK-7 (122) and MK-5 (116).

18.11 ROLE OF VITAMIN K IN BONE METABOLISM

18.11.1 Gla Proteins Occurring in Bone

The most abundant Gla protein in bone is osteocalcin (also known as bone Gla protein), which comprises about 15 to 20% of noncollagen protein in bone. Human osteocalcin is a relatively small molecule (6.5 kDa) and contains three Gla residues, which confer high binding affinity for calcium. In mice, osteocalcin is encoded by three genes: *Bglap* and *Bglap2* are specifically expressed in osteoblasts in bone, whereas *Bglap3* is expressed in non-ostoid tissues, including the kidneys and lungs. Just one osteocalcin gene (*BGLAP*) has been identified in humans and rats. Approximately 20% of osteocalcin newly synthesized by osteoblasts in the bone is set free in the bloodstream (Vermeer *et al.*, 1995). Osteocalcin is up-regulated by $1\alpha,25(OH)_2D_3$. the active metabolite of vitamin D.

Osteocalcin is involved in the homeostatic regulation of bone mass and bone remodeling. For example, it inhibits hydroxyapatite crystal growth through high-affinity binding to the mineral (Romberg *et al.*, 1986). Osteocalcin is chemotactic for osteoclast precursors (Malone *et al.*, 1982) and stimulates osteoclast differentiation (Glowacki *et al.*, 1991), suggesting an involvement in bone remodeling.

The role of osteocalcin in bone has been investigated using osteocalcin knockout (*Ocn*$^{-/-}$) mice in which *Bglap* and *Bglap2* were deleted from embryonic stem cells. However, the results from two research groups using this approach have been conflicting. Ducy *et al.* (1996) showed that bone formation rates were significantly increased in the long bones and vertebrae of the *Ocn*$^{-/-}$ mice. This acceleration occurred without a comparable increase in the number of osteoblasts, indicating that each osteoblast was producing more bone-forming proteins. The increase in bone mass and strength occurred without impairing bone resorption. These findings indicate that osteocalcin is a negative regulator of bone formation. Moriishi *et al.* (2020) were unable to demonstrate a negative regulatory effect. They found instead that osteocalcin is essential for correct alignment of apatite crystallites on collagen fibers. In wild-type mice, the collagen fibers were parallel to the longitudinal direction of long bone and apatite crystallites were parallel to the collagen fibers. In *Ocn*$^{-/-}$ mice, the orientation of collagen fibers remained unchanged, but the alignment of apatite crystallites was severely disrupted and the bone strength was reduced in the longitudinal direction. These findings indicate that osteocalcin is important for bone strength. This discrepancy between the results of two similar experiments has not been explained, but could be due to differences in the genetic background of the *Ocn*$^{-/-}$ mouse lines (Komari, 2020).

Another bone-associated protein, perostin, is secreted by mesenchymal stromal cells –the cells of the bone marrow stroma that differentiate into many mesenchymal cell lineages, including osteoblasts and chondrocytes. A subset of this secreted protein is γ-carboxylated in a vitamin K-dependent manner. Carboxylated periostin is abundantly found in mineralized bone nodules *in vitro*, suggesting a role in extracellular matrix mineralization (Coutu *et al.*, 2008). Periostin is up-regulated during fracture healing, which indicates its importance in bone maintenance and regeneration (Nakazawa *et al.*, 2004).

Besides its role as an anticoagulant, protein S is implicated in bone homeostasis. Children with inherited protein S deficiency not only suffer from recurrent thrombosis, but also have severely reduced bone mass (osteopenia) (Pan *et al.*, 1990).

18.11.2 Vitamin K Status and Osteoporosis

Individuals carrying the E4 allele of apoE experience a higher incidence of bone fractures during their lifetimes than do individuals without the E4 allele (Kohlmeier *et al.*, 1998). The increased risk of hip and wrist fracture in women with the apoE4 allele is not explained by bone density, impaired cognitive function, or falling (Cauley *et al.*, 1999). This predisposition toward bone fracture is consistent with E3/4 and E4/4 phenotypes having lower plasma phylloquinone levels than normal.

Vitamin K sufficiency of the bone is related to the degree of γ-carboxylation of osteocalcin and this in turn is related to the plasma phylloquinone concentration. As vitamin K intake decreases, circulating osteocalcin seems to be the first Gla protein to occur in an undercarboxylated form (Vermeer *et al.*, 1995). Circulating osteocalcin is about 92% γ-carboxylated in healthy young adults on a normal diet. Daily supplementation with 250 μg of phylloquinone increases osteocalcin carboxylation to 96%, while 1,000-μg supplements are required to achieve 100% carboxylation (Binkley *et al.*, 2002). These observations reveal that a diet sufficient to maintain normal clotting would not be able to maximize γ-carboxylation of osteocalcin and probably other vitamin K-dependent proteins. It remains unknown whether maximal osteocalcin carboxylation is necessary for optimal bone health.

Most studies have shown that the circulating levels of total osteocalcin increase with aging in normal women, especially after the menopause. This increase is likely to reflect an increase in bone turnover, which is associated with low bone mass in all skeletal regions (Ravn *et al.*, 1996). The γ-carboxylation of circulating osteocalcin is significantly impaired in women over 80 years of age (Plantalech *et al.*, 1991). Also, in elderly women, high concentrations of circulating undercarboxylated osteocalcin are associated with low hip bone mineral density (BMD) (Szulc *et al.*, 1994) and increased risk of hip fracture (Szulc *et al.*, 1993, 1996; Vergnaud *et al.*, 1997). Plasma levels of phylloquinone and of the menaquinones MK-7 and MK-8 are depressed in elderly women within a few hours of hip fracture, suggesting that vitamin K is sequestered from the circulation for use at the fracture site (Hodges *et al.*, 1993). Vitamin K_1 supplementation (1,000 μg/day) corrected undercarboxylation of osteocalcin in postmenopausal women (Knapen *et al.*, 1989; Douglas *et al.*, 1995) and decreased two markers of bone resorption, urinary calcium and hydroxyproline excretion (Knapen *et al.*, 1989, 1993). Booth *et al.* (1999) reported that 15 days of dietary vitamin K depletion led to increased bone turnover as measured by serum osteocalcin and urinary NTx (*N*-telopeptides of type I collagen) concentration. These markers were subsequently normalized by 10 days of phylloquinone repletion (200 μg/day). As elevated bone turnover is associated with rapid bone loss, vitamin K insufficiency would be expected to contribute to the development of osteoporosis. However, associations do not necessarily imply causation and no direct evidence for the participation of decreased plasma vitamin K in osteopenia in the elderly has been reported.

In a prospective study involving 72,327 women (Feskanich *et al.*, 1999), dietary vitamin K intakes less than 109 μg/day were associated with an increased risk of hip fracture. Booth *et al.* (2003) assessed dietary vitamin K intake with a food-frequency questionnaire in 1112 men and 1479 women (mean ± SD age: 59 ± 9 years) and measured BMD of the hip and spine. Women in the lowest quartile of vitamin K intake (mean: 70.2 μg/day) had significantly lower BMD at at the hip and spine than did those in the highest quartile of intake (mean: 309 μg/day). No significant association was found between dietary vitamin K intake and BMD in men.

Tamatani *et al.* (1998) evaluated the possible participation of circulating levels of testosterone, vitamin D metabolites, and vitamin K in osteopenia in elderly men. No significant correlation between plasma testosterone and BMD was observed, despite the age-related decrease in plasma testosterone. However, elderly men with decreased BMD showed significant decreases in the circulating levels of 25-hydroxyvitamin D, phylloquinone, and MK-7 compared with elderly men with normal BMD.

18.11.3 Effects of Menaquinone-4 on Bone Metabolism

Akedo *et al.* (1992) reported that MK-4 suppresses the proliferation of osteoblastic cells *in vitro*. Warfarin reversed this effect, implicating the γ-carboxylation system in the modulation of proliferation. Koshihara *et al.* (1996) reported that MK-4 enhanced 1,25-dihydroxyvitamin D_3-induced mineralization by human osteoblasts *in vitro*. This was due to enhanced γ-carboxylation of the osteocalcin induced by 1,25-dihydroxyvitamin D_3, and accumulation of carboxylated osteocalcin in the extracellular matrix, causing mineralization (Koshihara & Hoshi, 1997). Hara *et al.* (1993) reported that MK-4 inhibited the bone resorption induced by interleukin-1α, prostaglandin E_2, parathyroid hormone (PTH), and 1,25-dihydroxyvitamin D_3 in a dose-dependent manner *in vitro*. MK-4 also inhibited the prostaglandin E_2 production stimulated by interleukin-1α. Koshihara *et al.* (1993) showed that MK-4-induced inhibition of prostaglandin synthesis in cultured human osteoblast-like periosteal cells was reduced by cycloheximide, indicating that newly synthesized protein participates in the inhibitory effect.

Akiyama *et al.* (1994) examined the effects of MK-4 on osteoclast-like multinucleated cell formation in bone marrow cell cultures. MK-4 showed the most potent inhibitory effect on cell formation when present in cultures during the last three days, suggesting that the vitamin blocks cell differentiation and/or cell fusion. MK-4 did not influence 1,25-dihydroxyvitamin D_3-induced osteoclast-like cell formation when present in the culture during the first four days, indicating that it does not affect proliferation of osteoclast precursor cells. MK-4 did not affect the proliferation of many other cell types in the bone marrow culture, suggesting that the observed inhibitory effect of MK-4 on osteoclast-like cells was not a result of cytotoxicity.

Hara *et al.* (1995) compared the effects of phylloquinone and MK-4 on bone resorption *in vitro*. Calcium concentration in the medium was used as a parameter of bone resorption. MK-4 inhibited the calcium release from mouse calvaria organ cultures induced by 1,25-dihydroxyvitamin D_3 or prostaglandin E_2, and it also inhibited osteoclast-like cell formation induced by 1,25-dihydroxyvitamin D_3 in coculture of spleen cells and stromal cells at the same concentrations. In contrast, the same doses of phylloquinone had no effects on bone resorption and osteoclast-like cell formation in these *in vitro* systems. The inhibitory effect of MK-4 on the calcium release from calvaria was not affected by the addition of warfarin, suggesting that the effect of MK-4 is not due to γ-carboxylation coupling with the vitamin K epoxide cycle. The structures of MK-4 and phylloquinone differ only in their side chains (Figure 18.1), therefore whether the difference in their effects is related to the differences in side chain structure was evaluated in the coculture system. Geranylgeraniol inhibited osteoclast-like cell formation to almost the same degree as MK-4, whereas the effect of phytol was weak. Moreover, multi-isoprenyl alcohols of two to seven units, except the four-unit geranylgeraniol, did not affect osteoclast-like cell formation. Thus the specific inhibitory effect of MK-4 is attributable to the geranylgeranyl side chain.

Kameda *et al.* (1996) demonstrated that MK-4, but not phylloquinone, inhibits bone resorption by targeting osteoclasts to undergo programed cell death (apoptosis). MK-4 did not induce apoptosis in other cell types in unfractionated bone cells. Calcitonin, which strongly inhibits osteoclastic bone resorption via calcitonin receptors, did not cause osteoclast apoptosis. MK-4 might be an appropriate therapeutic drug against bone diseases with excess bone resorption, because of its selective and direct induction of osteoclast apoptosis.

18.11.4 Clinical Use of Menaquinone-4 in Osteoporosis

A number of Japanese studies have claimed beneficial results using synthetic MK-4 (menatetranone) in the treatment of osteoporosis. The rationale includes the possibility that MK-4 may have different effects on bone metabolism than phylloquinone. The dosage currently used (45 mg/day) is far in excess of daily vitamin K requirements and any effect must be regarded as pharmacological rather than a dietary correction of a nutritional deficiency. MK-4 was shown to be effective in increasing bone mineral density of cortical bone in osteoporotic patients (Orimo *et al.*, 1998) as well as preventing the occurrence of new fractures and sustaining lumbar bone mineral density (Shiraki *et al.*, 2000). In the latter study, MK-4 treatment enhanced γ-carboxylation of the osteocalcin molecule. There were no significant changes in bone resorption markers, therefore the prevention of bone fractures by MK-4 may not be caused entirely by inhibition of bone resorption.

18.12 ROLE OF VITAMIN K IN VASCULAR CALCIFICATION

18.12.1 Introduction

Vascular calcification is a common feature of aging and is associated with several disease states, including atherosclerosis, osteoporosis, chronic kidney disease, and diabetes. It is an actively regulated, cell-mediated process, and therefore has potential as a therapeutic target. Two vitamin K-dependent proteins, matrix Gla protein (MGP) and Gla-rich protein (GRP), are natural inhibitors of soft tissue calcification. Osteocalcin knockout mice have normal blood vessels, indicating that osteocalcin is not involved in the inhibition of vascular calcification (Shanahan *et al.*, 1998).

18.12.2 Matrix Gla Protein (MGP)

MGP is a 14-kDa protein that contains five Gla residues (Wallin *et al.*, 2000). It was discovered as a component in demineralized bone extracts (Price *et al.*, 1983), but has since been found to be also present in a wide variety of soft tissues such as cartilage, heart, lung, kidney, and arterial vessel wall (Schurgers *et al.*, 2001). In the arterial intima, MGP is synthesized by macrophages and to a lesser extent by vascular smooth muscle cells directly abutting calcified regions (Shanahan *et al.*, 2000). MGP binds tightly and selectively to hydroxyapatite crystal nuclei and prevents their growth and ability to seed daughter crystals (Price *et al.*, 1998). MGP also inhibits the osteogenic effect of bone morphogenetic protein-2 by forming a BMF-2–MGP complex (Sweatt *et al.*, 2003). This protein–protein interaction requires the conformational change in MGP induced by calcium binding to the Gla residues.

Murshed *et al.* (2004) demonstrated that extracellular matrix mineralization is regulated locally and not systemically by MGP. To be effective, MGP needs to be γ-carboxylated, because mutating the Gla residues into aspartic acid residues leads to the synthesis of non-functional MGP in mice and to the death of the animal (Murshed *et al.*, 2004). MGP is up-regulated by vitamin D, the major regulator of calcium in the body (Cancela *et al.*, 1990). Because vitamin K can enhance γ-carboxylation, it might be effective in increasing the efficacy of MGP.

Luo *et al.* (1997) reported that deletion of the *MGP* gene in mice causes rapid calcification of the elastic lamellae in the tunica media of the arteries, but not of the arterioles, capillaries, or veins. The entire media is replaced by chondrocytes, producing a typical cartilage that starts to progressively calcify at birth. By three to six weeks of age, calcification is so extensive that the arteries become rigid tubes and, within eight weeks of age, death occurs due to rupture of the thoracic or abdominal aorta. There is also inappropriate calcification of proliferating chondrocytes at the epiphyseal plate of growing long bones, resulting in stunted bone growth and osteopenia. The vascular phenotype of the MGP-deficient mouse suggests that MGP is an essential inhibitor of arterial calcification. Furthermore, it indicates that vascular calcification occurs spontaneously if not actively inhibited. Murshed *et al.* (2004) demonstrated that reintroducing the MGP gene into the vascular smooth muscle cells of MGP-deficient mice could prevent the development of vascular calcification. In humans, mutations in the MGP gene are responsible for Keutel syndrome, a rare inherited disease characterized by multiple peripheral pulmonary stenoses, neural hearing loss, short terminal phalanges, mid-facial hypoplasia, and abnormal calcification of the cartilage of the auricles, nose, larynx, trachea, and ribs (Munroe *et al.*, 1999).

Sweatt *et al.* (2003) showed that in aged rats fed a normal vitamin K-containing diet, MGP accumulates in calcified lesions within the aorta. However, the MGP is poorly γ-carboxylated and cannot acquire the Ca^{2+}-induced conformation that is required for BMP-2 binding. Thus, BMP-2 produced in the lesions will not be neutralized and is free to exert its activity. As the aging rats were fed a normal diet, production of under-γ-carboxylated MGP in the lesions must be an aging phenomenon.

Jie *et al.* (1995) reported that postmenopausal women with calcified atherosclerotic lesions had higher levels of undercarboxylated osteocalcin and a lower dietary vitamin K intake than women without calcifications. This study demonstrated that aortic calcification is associated with a reduced vitamin K status. Furthermore, the presence of atherosclerotic calcifications was associated with a lower bone mass (Jie *et al.*, 1996). On the basis that MGP is produced by the arterial wall as a defense mechanism against calcification, an insufficiency of vitamin K could lead to the production of non-functional MGP, and hence inappropriate calcification.

Price *et al.* (1998) showed that feeding rats with the vitamin K antagonist warfarin induces aortic calcification with a pathology similar to that seen in the MGP null mouse (Luo *et al.*, 1997). The warfarin treatment markedly increased the levels of MGP mRNA and protein in calcifying arteries and decreased the level of MGP in serum. Schurgers *et al.* (2007) investigated this phenomenon and showed that a high vitamin K intake (200 times the daily requirement of the liver) could prevent and even reverse warfarin-induced arterial calcification and the associated decreased arterial distensibility in rats. They demonstrated that most of the MGP that had been synthesized in response to warfarin treatment was under-γ-carboxylated and hence inactive. In healthy, non-treated animals, a relatively low vitamin K intake was sufficient for complete MGP carboxylation and for preventing arterial calcification. Plasma levels of MGP were decreased in warfarin-fed animals, but after feeding the high vitamin K diet for 6 weeks the plasma MGP levels increased significantly. The logical explanation for this increase is that the decreased vascular calcification provides fewer binding sites for MGP and the protein is free to circulate. Schurgers *et al.* (2007) postulated that arterial calcium deposits are removed by phagocytosis carried out by smooth muscle cells under conditions of calcification inhibition. The transition of smooth muscle cells from a contractile to a migratory, phagocytic phenotype has been demonstrated (Sandison *et al.*, 2016).

18.12.3 Gla-Rich Protein (GRP)

Viegas *et al.* (2008) isolated a novel 10.2-kDa protein from the calcified cartilage of Adriatic sturgeon and named it Gla-rich protein (GRP). This particular protein contained 16 Gla residues in its 72 amino acid sequence, which is the highest Gla percent (22%) among the vitamin K-dependent proteins so far identified. GRP was also found in rat cartilage and bone. Human GRP contains 15 Gla residues (Cancela *et al.*, 2012). Viegas *et al.* (2009) showed that human GRP is a circulating protein that is also expressed in soft tissues of rats and humans, including the skin and the walls of small and medium blood vessels. When tissues are affected by pathological calcifications, GRP accumulates at sites of mineral deposition, where it shows a high mineral binding affinity. Viegas *et al.* (2015) showed that GRP expression and accumulation are up-regulated with calcification simultaneously with osteocalcin and matrix Gla protein. They also demonstrated the association of a GRP–MGP–fetuin-A complex with calcification sites. Fetuin-A is a circulating vitamin K-independent glycoprotein that can inhibit vascular calcification by a variety of mechanisms (Reynolds *et al.*, 2005). Viegas *et al.* (2014) investigated the specific accumulation pattern of γ-carboxylated (cGRP) and undercarboxylated (ucGRP) protein forms in healthy and cancerous tissue. Although both forms have mineral-binding affinity *in vitro* and are present in healthy tissue, ucGRP is the predominant form in cancerous tissue. Viegas *et al.* (2015) showed a predominance of ucGRP relative to cGRP in calcified aortic valve disease. The use of an *ex vivo* model of vascular calcification confirmed that cGMP but not ucGMP inhibits calcification. The accumulation of GRP at sites of pathological calcifications and its high calcium-binding capacity through its multiple Gla residues suggest that GRP may regulate calcium availability in the extracellular environment.

18.12.4 The Rotterdam Study

Geleijnse *et al.* (2004) examined the association of dietary intake of phylloquinone (from dark-green leafy vegetables and vegetable oils) and menaquinone (mainly MK-4 from eggs and meat) with the incidence of coronary heart disease (CHD), all-cause mortality, and aortic calcification in the Rotterdam Study – a prospective, population-based study to assess the occurrence of diseases of the elderly and to clarify their determinants. The results indicated that an adequate intake of menaquinone is associated with a reduced risk of coronary heart disease. Intake of phylloquinone was not significantly associated with this risk.

18.13 ROLE OF VITAMIN K IN THE DEVELOPING NERVOUS SYSTEM

A more recently discovered Gla protein encoded by a growth arrest gene and known as Gas6 has a wide tissue distribution, including the nervous system. Gas6 is a ligand for a class of tyrosine kinase receptors and as such is involved in cell cycle regulation and cell–cell adhesion. In the nervous system, Gas6 is a growth factor for Schwann cells and is implicated in neuronal survival. This lesser-known aspect of vitamin K physiology has been reviewed by Tsaioun (1999).

18.14 VITAMIN K DEFICIENCY

18.14.1 Deficiency in Adults

In adult humans, clinical vitamin K deficiency manifests as bleeding at various sites in the body. Abnormal blood coagulation is more likely to arise from secondary causes such as malabsorption syndromes or biliary obstruction than from a dietary inadequacy of vitamin K. However, subclinical deficiency, manifested as decreased urinary γ-carboxyglutamic acid excretion, has been induced in healthy adults by dietary deprivation of the vitamin (Ferland *et al.*, 1993).

In a placebo-controlled study involving healthy young and elderly adults, Binkley *et al.* (2000) reported that vitamin K supplementation (1,000 μg of synthetic phylloquinone/day) resulted in a 10-fold increase in serum phylloquinone concentration. The mean percentage undercarboxylated osteocalcin decreased from 7.6% to 3.4% without significant differences by age or sex. The results showed that the usual dietary practices in the population studied did not provide adequate vitamin K for maximal osteocalcin carboxylation. Further research is needed to establish whether maximal osteocalcin γ-carboxylation is important for optimum bone mass density and whether submaximal osteocalcin γ-carboxylation should be used as a marker of vitamin K nutritional status.

18.14.2 Deficiency in Infants

Plasma concentrations of the Gla-containing blood clotting factors (factors II, VII, IX, and X) in normal newborns range between 30 and 60% of adult values (Vermeer & Hamulyák, 1991). These relatively low values are not due to vitamin K deficiency as raising cord blood levels of phylloquinone to the endogenous maternal range by maternal oral supplementation does not improve coagulation in the fetus or neonate (Mandelbrot *et al.*, 1988). Also, there is no detectable difference in coagulation between breastfed and formula-fed infants in the first month of life, despite the marked differences in serum phylloquinone concentrations (Pietersma-de Bruyn *et al.*, 1990). The likely explanation for the low neonatal concentrations of vitamin K-dependent clotting factors is reduced synthesis of their precursor proteins. In the mouse, gestational factor IX mRNA levels are <5% of adult levels up to two days before birth, when levels begin to rise steeply, reaching 43% of adult levels at birth. This is followed by a gradual increase until adult levels are reached at about 24 days of age (Yao *et al.*, 1991).

About 30% of full-term infants have low vitamin K status as indicated by the presence of des-γ-carboxyprothrombin (also known as PIVKA-II) in their plasma during the first week of life (Motohara *et al.*, 1985). Des-γ-carboxyprothrombin represents undercarboxylated prothrombin and is a sensitive hemostatic marker of subclinical vitamin K deficiency. The low vitamin K status, coupled with the low concentrations of vitamin K-dependent clotting factor precursor proteins, makes infants at birth and in early life susceptible to a syndrome referred to nowadays as vitamin K deficiency bleeding (VKDB) of early infancy. This disease, formerly known as hemorrhagic disease of the newborn, has a reported incidence of between two and ten cases per 100,000 births (Shearer, 1995a). Three syndromes have been identified according to their time of presentation: early, classic, and late VKDB. Early VKDB presents within 24 hours of birth and is commonly manifested as bleeding within the gut and around the genitalia. Classic VKDB presents one to seven days after birth and the bleeding is usually gastrointestinal, dermal, nasal, or from circumcision. Late VKDB, which presents 2 to 12 weeks after birth, is the most serious syndrome and is frequently associated with some abnormality of liver function. It has a 50% incidence of intracranial hemorrhage, resulting in death or severe and permanent brain damage (Shearer, 1995b).

Owing to limited placental transfer of maternal phylloquinone to the fetus, babies are born with low liver reserves of vitamin K. After birth, it takes several weeks before the liver stores of menaquinones attain adult levels. The absence of an intestinal microflora during the first few days of life may be significant in this regard. The newborn is entirely dependent on milk for its supply of vitamin K and hence any delay in the establishment of lactation may be a risk factor for classic VKDB. The vitamin K content of mature human milk ranges from 0.85 to 9.2 μg L^{-1} with a mean concentration of 2.5 μg L^{-1}, but can be increased by maternal intakes of pharmacological doses of the vitamin. By comparison, bovine milk contains 5 μg L^{-1} and infant formulas contain 50–100 μg L^{-1} (Institute of Medicine, 2001). Two major risk factors for VKDB are exclusive breastfeeding and not giving vitamin K prophylaxis at birth. Premature babies now routinely receive intramuscular or (less effectively) oral doses of vitamin K as a prophylactic measure against VKDB.

18.15 TOXICITY OF MENADIONE

There is no evidence of toxicity associated with the oral intake of either phylloquinone or menaquinones (Institute of Medicine, 2001). Menadione and its water-soluble derivatives are toxic when administered to the newborn, so this practice is prohibited. The unsubstituted 3 position of menadione can link up with the sulfhydryl group of reduced glutathione (GSH), preventing its oxidation to GSSG. The oxidation of GSH to GSSG, catalyzed by glutathione peroxidase, is coupled to the reduction of harmful peroxidized fatty acids to harmless hydroxy fatty acids, therefore menadione, by depleting cells of GSH, can cause oxidative membrane damage and hemolytic anemia. Chiou and Tzeng (2000) observed that menadione-resistant K300 cells are characterized by a reduction of reactive oxygen species as compared to the parental P19 cells during menadione's action.

FURTHER READING

Shearer, M.J. (2000) Role of vitamin K and Gla proteins in the pathophysiology of osteoporosis and vascular calcification. *Current Opinion in Clinical Nutrition and Metabolic Care*, **3**, 433–8.

Shearer, M.J. (2009) Vitamin K in parenteral nutrition. *Gastroenterology*, **137**, S105–18.

Shearer, M.J. & Newman, P. (2008) Metabolism and cell biology of vitamin K. *Thrombosis and Haemostasis*, **100**, 530–47.

Vermeer, C., Gijsbers, B.L.M.G., Craciun, A.M., Groenen-van Dooren, M.C.L. & Knapen, M.H.J. (1996) Effects of vitamin K on bone mass and bone metabolism. *Journal of Nutrition*, **126**, 1187S–91S.

Weber, P. (1997) Management of osteoporosis: Is there a role for vitamin K? *International Journal for Vitamin and Nutrition Research*, **67**, 350–6.

REFERENCES

Akedo, Y., Hosoi, T., Inoue, S., Ikegami, A., Mizuno, Y., Kaneki, M., Nakamura, T., Ouchi, Y. & Orimo, H. (1992) Vitamin K_2 modulates proliferation and function of osteoblastic cells *in vitro*. *Biochemical and Biophysical Research Communications*, **187**, 814–20.

Akiyama, Y., Hara, K., Tajima, T., Murota, S. & Morita, I. (1994) Effect of vitamin K_2 (menatetrenone) on osteoclast-like cell formation in mouse bone marrow cultures. *European Journal of Pharmacology*, **263**, 181–5.

Allison, P.M., Mummah-Schendel, L.L., Kindberg, C.G., Harms, C.S., Bang, N.U. & Suttie, J.W. (1987) Effects of a vitamin K-deficient diet and antibiotics in normal human volunteers. *Journal of Laboratory and Clinical Medicine*, **110**, 180–8.

Barnes, R.H. & Fiala, G. (1959) Effects of the prevention of coprophagy in the rat. VI. Vitamin K. *Journal of Nutrition*, **68**, 603–14.

Binkley, N.C., Krueger, D.C., Engelke, J.A., Foley, A.L. & Suttie, J.W. (2000) Vitamin K supplementation reduces serum concentrations of under-γ-carboxylated osteocalcin in healthy young and elderly adults. *American Journal of Clinical Nutrition*, **72**, 1523–8.

Binkley, N.C., Krueger, D.C., Kawahara, T.N., Engelke, J.A., Chappell, R.J. & Suttie, J.W. (2002) A high phylloquinone intake is required to achieve maximal osteocalcin γ-carboxylation. *American Journal of Clinical Nutrition*, **76**, 1055–60.

Booth, S.L., Gundberg, C.M., McKeown, N.M., Morse, M.O. & Wood, R.J. (1999) Vitamin K depletion increases bone turnover. *Journal of Bone and Mineral Research*, **14**, S393 (abstr).

Booth, S.L., Lichtenstein, A.H. & Dallal, G.E. (2002) Phylloquinone absorption from phylloquinone-fortified oil is greater than from a vegetable in younger and older men and women. *Journal of Nutrition*, **132**, 2609–12.

Booth, S.L., Broe, K.E., Gagnon, D.R., Tucker, K.L., Hannan, M.T., McClean, R.R., Dawson-Hughes, B., Wilson, P.W.F., Cupples, L.A. & Kiel, D.P. (2003) Vitamin K intake and bone mineral density in women and men. *American Journal of Clinical Nutrition*, **77**, 512–16.

Cancela, L., Hsieh, C.-L., Francke, U. & Price, P.A. (1990) Molecular structure, chromosome assignment, and promoter organization of the human matrix Gla protein gene. *Journal of Biological Chemistry*, **265**(25), 15040–8.

Cancela, M.L., Conceição, N. & Laizé, V. (2012) Gla-rich protein, a new player in tissue calcification. *Advances in Nutrition*, **3**, 174–81.

Cauley, J.A., Zmuda, J.M., Yaffe, K., Kuller, L.H., Ferrell, R.E., Wisniewski, S.R. & Cummings, S.R. (1999) Apolipoprotein E polymorphism: A new genetic marker of hip fracture risk – The study of osteoporotic fractures. *Journal of Bone and Mineral Research*, **14**, 1175–81.

Chiou, T.-J. & Tzeng, W.-F. (2000) The roles of glutathione and antioxidant enzymes in menadione-induced oxidative stress. *Toxicology*, **154**, 75–84.

Conly, J.M. & Stein, K. (1992a) Quantitative and qualitative measurements of K vitamins in human intestinal contents. *American Journal of Gastroenterology*, **87**, 311–6.

Conly, J.M. & Stein, K. (1992b) The production of menaquinones (vitamin K_2) by intestinal bacteria and their role in maintaining coagulation homeostasis. *Progress in Food and Nutrition Science*, **16**, 307–43.

Conly, J.M., Stein, K., Worobetz, L. & Rutledge-Harding, S. (1994) The contribution of vitamin K_2 (menaquinones) produced by the intestinal microflora to human nutritional requirements for vitamin K. *American Journal of Gastroenterology*, **89**, 915–23.

Coutu, D.L., Wu, J.H., Monette, A., Rivard, G.-É., Blostein, M.D. & Galipeau, J. (2008) Periostin, a member of a novel family of vitamin K-dependent proteins, is expressed by mesenchymal stromal cells. *Journal of Biological Chemistry*, **283**(26), 17991–8001.

Davidson, R.T., Foley, A.L., Engelke, J.A. & Suttie, J.W. (1998) Conversion of dietary phylloquinone to tissue menaquinone-4 in rats is not dependent on gut bacteria. *Journal of Nutrition*, **128**, 220–3.

Douglas, A.S., Robins, S.P., Hutchison, J.D., Porter, R.W., Stewart, A. & Reid, D.M. (1995) Carboxylation of osteocalcin in post-menopausal osteoporotic women following vitamin K and D supplementation. *Bone*, **17**, 15–20.

Ducy, P., Desbois, C., Boyce, B., Pinero, G., Story, B., Dunstan, C., Smith, E., Bonadio, J., Goldstein, S., Gundberg, C., Bradley, A. & Karsenty, G. (1996) Increased bone formation in osteocalcin-deficient mice. *Nature*, **382**, 448–52.

Ferland, G., Sadowski, J.A. & O'Brien, M.E. (1993) Dietary induced subclinical vitamin K deficiency in normal human subjects. *Journal of Clinical Investigation*, **91**, 1761–8.

Feskanich, D., Weber, P., Willett, W.C., Rockett, H., Booth, S.L. & Colditz, G.A. (1999) Vitamin K intake and hip fractures in women: A prospective study. *American Journal of Clinical Nutrition*, **69**, 74–9.

Garber, A.K., Binkley, N.C., Krueger, D.C. & Suttie, J.W. (1999) Comparison of phylloquinone bioavailability from food sources or a supplement in human subjects. *Journal of Nutrition*, **129**, 1201–3.

Geleijnse, J.M., Vermeer, C., Grobbee, D.E., Schurgers, L.J., Knapen, M.H.J., van der Meer, I.M., Hofman, A. & Witteman, J.C.M. (2004) Dietary intake of menaquinone is associated with a reduced risk of coronary heart disease: The Rotterdam Study. *Journal of Nutrition*, **134**, 3100–5.

Gijsbers, B.L.M.G., Jie, K.-S.G. & Vermeer, C. (1996) Effect of food composition on vitamin K absorption in human volunteers. *British Journal of Nutrition*, **76**, 223–9.

Glowacki, J., Rey, C., Glimcher, M.J., Cox, K.A. & Lian, J. (1991) A role for osteocalcin in osteoclast differentiation. *Journal of Cellular Biochemistry*, **45**, 292–302.

Groenen-van Dooren, M.M.C.L., Soute, B.A., Jie, K.S.-G., Thijssen, H.H.W. & Vermeer, C. (1993) The relative effects of phylloquinone and menaquinone-4 on the blood coagulation factor synthesis in vitamin K-deficient rats. *Biochemical Pharmacology*, **45**, 433–7.

Groenen-van Dooren, M.M.C.L., Ronden, J.E., Soute, B.A.M. & Vermeer, C. (1995) Bioavailability of phylloquinone and menaquinones after oral and colorectal administration in vitamin K-deficient rats. *Biochemical Pharmacology*, **50**, 797–801.

Hara, K., Akiyama, Y., Tajima, T. & Shiraki, M. (1993) Menatetrenone inhibits bone resorption partly through inhibition of PGE_2 synthesis in vitro. *Journal of Bone and Mineral Research*, **8**, 535–42.

Hara, K., Akiyama, Y., Nakamura, T., Murota, S. & Morita, I. (1995) The inhibitory effect of vitamin K_2 (menatetrenone) on bone resorption may be related to its side chain. *Bone*, **16**, 179–84.

Hirauchi, K., Sakano, T., Notsumoto, S., Nagaoka, T., Morimoto, A., Fujimoto, K., Masuda, S. & Suzuki, Y. (1989a) Measurement of K vitamins in animal tissues by high-performance liquid chromatography with fluorimetric detection. *Journal of Chromatography, Biomedical Applications*, **497**, 131–7.

Hirauchi, K., Sakano, T., Notsumoto, S., Nagaoka, T., Morimoto, A., Fujimoto, K., Masuda, S. & Suzuki, Y. (1989b) Measurement of K vitamins in foods by high-performance liquid chromatography with fluorometric detection. *Vitamins (Japan)*, **63**, 147–51 (in Japanese).

Hirota, Y., Tsugawa, N., Nakagawa, K., Suhara, Y., Tanaka, K., Uchino, Y. *et al.* (2013) Menadione (vitamin K_3) is a catabolic product of oral phylloquinone (vitamin K_1) in the intestine and a circulating precursor of tissue menaquinone-4 (vitamin K_2) in rats. *Journal of Biological Chemistry*, **288**(46), 33071–80.

Hodges, S.J., Akesson, K., Vergnaud, P., Obrant, K. & Delmas, P.D. (1993) Circulating levels of vitamin K_1 and K_2 decreased in elderly women with hip fractures. *Journal of Bone and Mineral Research*, **8**, 1241–5.

Hollander, D., Rim, E. & Ruble, P.E. Jr. (1977) Vitamin K_2 colonic and ileal in vivo absorption: Bile, fatty acids, and pH effects on transport. *American Journal of Physiology*, **233**, E124–9.

Ichihashi, T., Takagishi, Y., Uchida, K. & Yamada, H. (1992) Colonic absorption of menaquinone-4 and menaquinone-9 in rats. *Journal of Nutrition*, **122**, 506–12.

Institute of Medicine. (2001) Vitamin K. In: *Dietary Reference Intakes for Vitamin A, Vitamin K, Arsenic, Boron, Chromium, Copper, Iodine, Iron, Manganese, Molybdenum, Nickel, Silicon, Vanadium, and Zinc*, pp. 162–96. National Academy Press, Washington, DC.

Jie, K.-S.G., Bots, M.L., Vermeer, C., Witteman, J.C.M. & Grobbee, D.E. (1995) Vitamin K intake and osteocalcin levels in women with and without aortic atherosclerosis: A population-based study. *Atherosclerosis*, **116**, 117–23.

Jie, K.-S.G., Bots, M.L., Vermeer, C., Witteman, J.C.M. & Grobbee, D.E. (1996) Vitamin K status and bone mass in women with and without aortic atherosclerosis: A population-based study. *Calcified Tissue International*, **59**, 352–6.

Kameda, T., Miyazawa, K., Mori, Y., Yuasa, T., Shiokawa, M., Nakamaru, Y., Mano, H., Hakeda, Y., Kameda, A. & Kumegawa, M. (1996) Vitamin K_2 inhibits osteoclastic bone resorption by inducing osteoclast apoptosis. *Biochemical and Biophysical Research Communications*, **220**, 515–9.

Knapen, M.H.J., Hamulyák, K. & Vermeer, C. (1989) The effect of vitamin K supplementation on circulating osteocalcin (bone Gla protein) and urinary calcium excretion. *Annals of Internal Medicine*, **111**, 1001–5.

Knapen, M.H.J., Jie, K.-S.G., Hamulyák, K. & Vermeer, C. (1993) Vitamin K-induced changes in markers for osteoblast activity and urinary calcium loss. *Calcified Tissue International*, **53**, 81–5.

Kohlmeier, M., Saupe, J., Schaefer, K. & Asmus, G. (1998) Bone fracture history and prospective bone fracture risk of hemodialysis patients are related to apolipoprotein E genotype. *Calcified Tissue International*, **62**, 278–81.

Koivu-Tikkanen, T.J., Schurgers, L.J., Thijssen, H.H. & Vermeer, C. (2000) Intestinal, hepatic, and circulating vitamin K levels at low and high intakes of vitamin K in rats. *British Journal of Nutrition*, **83**, 185–90.

Komori, T. (2020) Functions of osteocalcin in bone, pancreas, testis, and muscle. *International Journal of Molecular Sciences*, **21**, 7513. DOI: 10.3390/ijms21207513

Koshihara, Y. & Hoshi, K. (1997) Vitamin K_2 enhances osteocalcin accumulation in the extracellular matrix of human osteoblasts in vitro. *Journal of Bone and Mineral Research*, **12**, 431–8.

Koshihara, Y., Hoshi, K. & Shiraki, M. (1993) Vitamin K_2 (menatetrenone) inhibits prostaglandin synthesis in cultured human osteoblast-like periosteal cells by inhibiting prostaglandin H synthase activity. *Biochemical Pharmacology*, **46**, 1355–62.

Koshihara, Y., Hoshi, K., Ishibashi, H. & Shiraki, M. (1996) Vitamin K2 promotes 1α,25(OH)2vitamin D3-induced mineralization in human periosteal osteoblasts. *Calcified Tissue International*, **59**, 466–73.

Lamon-Fava, S., Sadowski, J.A., Davidson, K.W., O'Brien, M.E., McNamara, J.R. & Schaefer, E.J. (1998) Plasma lipoproteins as carriers of phylloquinone (vitamin K_1) in humans. *American Journal of Clinical Nutrition*, **67**, 1226–31.

Lipsky, J.J. (1988) Antibiotic-associated hypoprothrombinaemia. *Journal of Antimicrobial Chemotherapy*, **21**, 281–300.

Luo, G., Ducy, P., McKee, M.D., Pinero, G.J., Loyer, E., Behringer, R.R. & Karsenty, G. (1997) Spontaneous calcification of arteries and cartilage in mice lacking matrix GLA protein. *Nature*, **386**, 78–81.

Malone, J.D., Teitelbaun, S.L., Griffin, G.L., Senior, R.M. & Kahn, A.J. (1982) Recruitment of osteoclast precursors by purified bone matrix constituents. *Journal of Cell Biology*, **92**, 227–30.

Mandelbrot, L., Guillaumont, M., Leclercq, M., Lefrère, J.J., Gozin, D., Daffos, F. & Forestier, F. (1988) Placental transfer of vitamin K_1 and its implications in fetal hemostasis. *Thrombosis and Haemostasis*, **60**, 39–43.

Matschiner, J.T. & Doisy, E.A. Jr. (1966) Bioassay of vitamin K in chicks. *Journal of Nutrition*, **90**, 97–100.

Moriishi, T., Ozasa, R., Ishimoto, T., Nakano, T., Hasegawa, T., Miyazaki, T., Liu, W., Fukuyama, R., Wang, Y., Komori, H., Qin, X., Amizuka, N. & Komori, T. (2020) Osteocalcin is necessary for the alignment of apatite crystallites, but not glucose metabolism, testosterone synthesis, or muscle mass. *PloS Genetics*, **16**(5), e1008586. DOI: 10.1371/journal.pgen.1008586

Motohara, K., Endo, F. & Matsuda, J. (1985) Effect of vitamin K administration on acarboxy prothrombin (PIVKA-II) levels in newborns. *Lancet*, **II**(8449), 242–4.

Munroe, P.B., Olgunturk, R.O., Fryns, J.-P., Van Maldergem, L., Ziereisen, F., Yuksel, B., Gardiner, R.M. & Chung, E. (1999) Mutations in the gene encoding the human matrix Gla protein cause Keutel syndrome. *Nature Genetics*, **21**, 142–4.

Murshed, M., Schinke, T., McKee, M.D. & Karsenty, G. (2004) Extracellular matrix mineralization is regulated locally; different roles of two gla-containing proteins. *Journal of Cell Biology*, **165**, 625–30.

Nakazawa, T., Nakajima, A., Seki, N., Okawa, A., Kato, M., Moriya, H., Amizuka, N., Einhorn, T.A. & Yamazaki, M. (2004) Gene expression of periostin in the early stage of fracture healing detected by cDNA microarray analysis. *Journal of Orthopaedic Research*, **22**, 52–5.

Newman, P., Bonello, F., Wierzbicki, A.S., Lumb, P., Savidge, G.F. & Shearer, M.J. (2002) The uptake of lipoprotein-borne phylloquinone (vitamin K_1) by osteoblasts and osteoblast-like cells: Role of heparan sulfate proteoglycans and apolipoprotein E. *Journal of Bone and Mineral Research*, **17**, 426–33.

Orimo, H., Shiraki, M., Tomita, A., Morii, H., Fujita, T. & Ohata, M. (1998) Effects of menatetrenone on the bone and calcium metabolism in osteoporosis: A double-blind placebo-controlled study. *Journal of Bone and Mineral Metabolism*, **16**, 106–12.

Pan, E.Y., Gomperts, E.D., Millen, R. & Gilsanz, V. (1990) Bone mineral density and its association with inherited protein S deficiency. *Thrombosis Research*, **58**, 221–31.

Pietersma-de Bruyn, A.L.J.M., van Haard, P.M.M., Beunis, M.H., Hamulyák, K. & Kuijpers, J.C. (1990) Vitamin K_1 levels and coagulation factors in healthy term newborns till 4 weeks after birth. *Haemostasis*, **20**, 8–14.

Plantalech, L., Guillaumont, M., Vergnaud, P., Leclercq, M. & Delmas, P.D. (1991) Impairment of gamma carboxylation of circulating osteocalcin (bone gla protein) in elderly women. *Journal of Bone and Mineral Research*, **6**, 1211–6.

Price, P.A., Urist, M.R. & Otawara, Y. (1983) Matrix Gla protein, a new γ-carboxyglutamiic acid-containing protein which is associated with the organic matrix of bone. *Biochemical and Biophysical Research Communications*, **117**, 765–71.

Price, P.A., Faus, S.A. & Williamson, M.K. (1998) Warfarin causes rapid calcification of the elastic lamellae in rat arteries and heart valves. *Arteriosclerosis, Thrombosis, and Vascular Biology*, **18**, 1400–7.

Ravn, P., Fledelius, C., Rosenquist, C., Overgaard, K. & Christiansen, C. (1996) High bone turnover is associated with low bone mass in both pre- and postmenopausal women. *Bone*, **19**, 291–8.

Reynolds, J.L., Skepper, J.N., McNair, R., Kasama, T., Gupta, K., Weissberg, P.L., Jahnen-Dechent, W. & Shanahan, C.M. (2005) Multifunctional roles for serum protein fetuin-A in inhibition of human vascular smooth muscle cell calcification. *Journal of the American Society of Nephrology*, **16**, 2920–30.

Romberg, R.W., Werness, P.G., Riggs, B.L. & Mann, K.G. (1986) Inhibition of hydroxyapatite crystal growth by bone-specific and other calcium-binding proteins. *Biochemistry*, **25**, 1176–80.

Ronden, J.F., Drittij-Reijnders, M.J., Vermeer, C. & Thijssen, H.H.W. (1998) Intestinal flora is not an intermediate in the phylloquinone–menaquinone-4 conversion in the rat. *Biochimica et Biophysica Acta*, **1379**, 69–75.

Sakano, T., Notsumoto, S., Nagaoka, T., Morimoto, A., Fujimoto, K., Masuda, S., Suzuki, Y. & Hirauchi, K. (1988) Measurement of K vitamins in food by high-performance liquid chromatography with fluorometric detection. *Vitamins (Japan)*, **62**, 393–8 (in Japanese).

Sandison, M.E., Dempster, J. & McCarron, J.G. (2016) The transition of smooth muscle cells from a contractile to a migratory, phagocytic phenotype: A direct demonstration of phenotypic modulation. *Journal of Physiology*, **594**, 6189–209.

Schurgers, L.J. & Vermeer, C. (2002) Differential lipoprotein transport pathways of K-vitamins in healthy subjects. *Biochimica et Biophysca Acta*, **1570**, 27–32.

Schurgers, L.J., Dissel, P.E.P., Spronk, H.M.H., Soute, B.A.M., Dhore, C.R., Cleutjens, J.P.M. & Vermeer, C. (2001) Role of vitamin K and vitamin K-dependent proteins in vascular calcification. *Zeitschrift für Kardiologie*, **90**, 57–63. (In English)

Schurgers, L.J., Spronk, H.M.H., Skepper, J.N., Hackeng, T.M., Shanahan, C.M., Vermeer, C., Weissberg, P.L. & Proudfoot, D. (2007) Post-translational modifications regulate matrix Gla protein function: Importance for inhibition of vascular smooth muscle cell calcification. *Journal of Thrombosis and Haemostasis*, **5**, 2503–11.

Shanahan, C.M., Proudfoot, D., Farzaneh-Far, A. & Weissberg, P.L. (1998) The role of Gla proteins in vascular calcification. *Critical Reviews in Eukaryotic Gene Expression*, **8**, 357–75.

Shanahan, C.M., Proudfoot, D., Tyson, K.L., Cary, N.R.B., Edmonds, M. & Weissberg, P.L. (2000) Expression of mineralization-regulating proteins in association with human vascular calcification. *Zeitschrift für Kardiologie*, **89**, II/63–8.

Shearer, M.J. (1992) Vitamin K metabolism and nutriture. *Blood Reviews*, **6**, 92–104.

Shearer, M.J. (1993) Vitamin K. In: *Encyclopaedia of Food Science, Food Technology and Nutrition*, Vol. 7. (Ed. by R. Macrea, R.K. Robinson & M.J. Sadler), pp. 4804–16. Academic Press, London.

Shearer, M.J. (1995a) Vitamin K: Its physiological role and the assessment of clinical and subclinical states. *Journal of the International Federation of Clinical Chemistry*, **7**, 88–95.

Shearer, M.J. (1995b) Vitamin K. *Lancet*, **345**, 229–34.

Shearer, M.J. & Newman, P. (2008) Metabolism and cell biology of vitamin K. *Thrombosis and Haemostasis*, **100**, 530–47.

Shearer, M.J., McBurney, A. & Barkhan, P. (1974) Studies on the absorption and metabolism of phylloquinone (vitamin K_1) in man. *Vitamins and Hormones*, **32**, 513–42.

Shearer, M.J., Rahim, S., Barkhan, P. & Stimmler, L. (1982) Plasma vitamin K_1 in mothers and their newborn babies. *Lancet*, **II**(8296), 460–3.

Shearer, M.J., Kries, R.V. & Saupe, J. (1992) Comparative aspects of human vitamin K metabolism and nutriture. *Journal of Nutritional Science and Vitaminology*, Special edn, S-13-3, 413–6.

Shearer, M.J., Bach, A. & Kohlmeier, M. (1996) Chemistry, nutritional sources, tissue distribution and metabolism of vitamin K with special reference to bone health. *Journal of Nutrition*, **126**, 1181S–6S.

Shearer, M.J., Fu, X. & Booth, S.L. (2012) Vitamin K nutrition, metabolism, and requirements: Current concepts and future research. *Advances in Nutrition*, **3**, 182–95.

Shiraki, M., Shiraki, Y., Aoki, C. & Miura, M. (2000) Vitamin K_2 (menatetrenone) effectively prevents fractures and sustains lumbar bone mineral density in osteoporosis. *Journal of Bone and Mineral Research*, **15**, 515–21.

Sweatt, A., Sane, D.C., Hutson, S.M. & Wallin, R. (2003) Matrix Gla protein (MGP) and bone morphogenetic protein-2 in aortic calcified lesions of aging rats. *Journal of Thrombosis and Haemostasis*, **1**, 178–85.

Szulc, P., Chapuy, M.-C., Meunier, P.J. & Delmas, P.D. (1993) Serum undercarboxylated osteocalcin is a marker of the risk of hip fracture in elderly women. *Journal of Clinical Investigation*, **91**, 1769–74.

Szulc, P., Arlot, M., Chapuy, M.-C., Duboeuf, F., Meunier, P.J. & Delmas, P.D. (1994) Serum undercarboxylated osteocalcin correlates with hip bone mineral density in elderly women. *Journal of Bone and Mineral Research*, **9**, 1591–5.

Szulc, P., Chapuy, M.-C., Meunier, P.J. & Delmas, P.D. (1996) Serum undercarboxylated osteocalcin is a marker of the risk of hip fracture: A three year follow-up study. *Bone*, **18**, 487–8.

Tamatani, M., Morimoto, S., Nakajima, M., Fukuo, K., Onishi, T., Kitano, S., Niinobu, T. & Ogihara, T. (1998) Decreased circulating levels of vitamin K and 25-hydroxyvitamin D in osteopenic elderly men. *Metabolism*, **47**, 195–9.

Thijssen, H.H.W. & Drittij-Reijnders, M.J. (1994) Vitamin K distribution in rat tissues: Dietary phylloquinone is a source of tissue menaquinone-4. *British Journal of Nutrition*, **72**, 415–25.

Thijssen, H.H.W. & Drittij-Reijnders, M.J. (1996) Vitamin K status in human tissues: Tissue-specific accumulation of phylloquinone and menaquinone-4. *British Journal of Nutrition*, **75**, 121–7.

Thijssen, H.H.W., Drittij-Reijnders, M.J. & Fischer, M.A.J.G. (1996) Phylloquinone and menaquinone-4 distribution in rats: Synthesis rather than uptake determines menaquinone-4 organ concentrations. *Journal of Nutrition*, **126**, 537–43.

Thijssen, H.H.W., Vervoort, L.M.T., Schurgers, L.J. & Shearer, M.J. (2006) Menadione is a metabolite of oral vitamin K. *British Journal of Nutrition*, **95**, 260–6.

Tsaioun, K.I. (1999) Vitamin K-dependent proteins in the developing and aging nervous system. *Nutrition Reviews*, **57**, 231–40.

Uchida, K. & Komeno, T. (1988) Relationships between dietary and intestinal vitamin K, clottimg factor levels, plasma vitamin K, and urinary Gla. In: *Current Advances in Vitamin K*

Research. (Ed. by J.W. Suttie), pp. 477–92. Elsevier Applied Science, New York.

Udall, J.A. (1965) Human sources and absorption of vitamin K in relation to anticoagulation activity. *Journal of the American Medical Association*, **194**, 127–9.

Usui, Y., Tanimura, H., Nishimura, N., Kobayashi, N., Okanoue, T. & Ozawa, K. (1990) Vitamin K concentrations in the plasma and liver of surgical patients. *American Journal of Clinical Nutrition*, **51**, 846–52.

Vergnaud, P., Garnero, P., Meunier, P.J., Bréart, G., Kamihagi, K. & Delmas, P.D. (1997) Undercarboxylated osteocalcin measured with a specific immunoassay predicts hip fracture in elderly women: The EPIDOS study. *Journal of Clinical Endocrinology and Metabolism*, **82**, 719–24.

Vermeer, C. (1990) γ-Carboxyglutamate-containing proteins and the vitamin K-dependent carboxylase. *Biochemical Journal*, **266**, 625–36.

Vermeer, C. & Hamulyák, K. (1991) Pathophysiology of vitamin K deficiency and oral anticoagulants. *Thrombosis and Haemostasis*, **66**, 153–9.

Vermeer, C., Jie, K.-S.G. & Knapen, M.J.H. (1995) Role of vitamin K in bone metabolism. *Annual Review of Nutrition*, **15**, 1–22.

Viegas, C.S.B., Simes, D.C., Laizé, V., Williamson, M.K., Price, P.A. & Cancela, M.L. (2008) Gla-rich protein (GRP), a new vitamin K-dependent protein identified from sturgeon cartilage and highly conserved in vertebrates. *Journal of Biological Chemistry*, **283**(52), 36655–64.

Viegas, C.S.B., Cavaco, S., Neves, P.L., Ferreira, A., João, A., Williamson, M.K., Price, P.A., Cancela, M.L. & Simes, D.C. (2009) Gla-rich protein is a novel vitamin K-dependent protein present in serum that accumulates at sites of pathological calcifications. *American Journal of Pathology*, **175**, 2288–98.

Viegas, C.S.B., Herfs, M., Rafael, M.S., Enriquez, J.L., Teixeira, A., Luis, I.M., van't Hoofd, C.M.R., João, A., Maria, V.L., Cavaco, S., Ferreira, A., Serra, M., Theuwissen, E., Vermeer, C. & Simes, D.C. (2014) Gla-rich protein is a potential new vitamin K target in cancer: Evidence for a direct GRP-mineral interaction. *BioMed Research International*, **2014**, Article ID 340216, 14 pages. DOI: 10.1155/2014/340216

Viegas, C.S.B., Rafael, M.S., Enriquez, J.L., Teixeira, A., Vitorino, R., Luis, I.M., Costa, R.M., Santos, S., Cavaco, S., Neves, J., Macedo, A.L., Willems, B.A.G., Vermeer, C. & Simes, D.C. (2015) Gla-rich protein acts as a calcification inhibitor in the human cardiovascular system. *Arteriosclerosis, Thrombosis, and Vascular Biology*, **35**, 399–408.

Wallin, R., Cain, D., Hutson, S.M., Sane, D.C. & Loeser, R. (2000) Modulation of the binding of matrix Gla protein (MGP) to bone morphogenetic protein-2 (BMP-2). *Thrombosis and Haemostasis*, **84**, 1039–44.

Yang, Y.-M., Simon, N., Maertens, P., Brigham, S. & Liu, P. (1989) Maternal-fetal transport of vitamin K_1 and its effects on coagulation in premature infants. *Journal of Pediatrics*, **115**, 1009–13.

Yao, S.-N., DeSilva, A.H., Kurachi, S., Samuelson, L.C. & Kurachi, K. (1991) Characterization of a mouse factor IX cDNA and developmental regulation of the factor IX gene expression in liver. *Thrombosis and Haemostasis*, **65**, 52–8.

Part IV

The Water-Soluble Vitamins

19 Thiamin (Vitamin B_1)

19.1 CHEMISTRY AND BIOLOGICAL ACTIVITY

Although the term vitamin B_1 is in common use, thiamin is the preferred generic term for this vitamin and will be used in this text. 'Thiamin,' rather than 'thiamine,' is used in accordance with the nomenclature policy of the International Union of Nutritional Sciences Committee on Nomenclature.

The thiamin molecule comprises substituted pyrimidine and thiazole moieties linked by a methylene bridge (Figure 22.1). It is a quaternary amine, which exists as a monovalent or divalent cation depending on the pH of the solution. Three phosphorylated forms of thiamin occur in nature. In living tissues the predominant form is the diphosphate ester, usually referred to as thiamin pyrophosphate (TPP) (Figure 19.1), which serves as a coenzyme in several metabolic pathways. Small amounts of the monophosphate and triphosphate esters also occur in animal tissues. Thiamin triphosphate has no coenzyme function, but it has a role (not yet completely understood) in nerve transmission. Thiamin monophosphate appears to be biologically inactive.

19.2 FOOD SOURCES AND BIOAVAILABILITY

19.2.1 Food Sources

All plant and animal tissues contain thiamin, and so the vitamin is present in all natural unprocessed foods. Rich sources of thiamin include yeast and yeast extract, wheat bran, oatmeal, wholegrain cereals, pulses, nuts, lean pork, heart, kidney, and liver. Beef, lamb, chicken, eggs, vegetables, and fruits contain intermediate amounts, while milk contains a relatively low amount. The milling of cereals removes most of the thiamin, so white flour, breakfast cereals, and, in certain countries, polished rice are enriched by addition of the vitamin.

The rice grain (Figure 19.2) comprises the inedible hull and the edible portion (caryopsis), which is enveloped by a fibrous pericarp. The mass distribution of the caryopsis is pericarp (1–2 %), seed coat, nucellus, and aleurone layer (4–6%), embryo (2–3%), and starchy endosperm (89–94%) (Zhou *et al*., 2002). Removal of the hull during milling produces brown rice; further milling produces a bran fraction and white rice. The bran fraction comprises the multi-layered skin (pericarp, seed coat, nucellus, and aleurone layer) and the embryo; it contains most of the grain's oil, protein, vitamin, and mineral content. White rice comprises the endosperm, which contains most of the grain's starch content.

Thiamin and the other water-soluble vitamins are concentrated in the aleurone layer and embryo of the rice grain. The vitamins that would be lost from the grain during milling can be retained to some extent by parboiling before milling, as this allows the vitamins to diffuse into the endosperm. Parboiling increases the thiamin content of milled rice despite the vitamin's thermal lability.

In most animal tissues, over 90% of the thiamin is phosphorylated, with TPP predominating. Exceptions are pig skeletal muscle (Egi *et al*., 1986) and chicken skeletal white muscle (Miyoshi *et al*., 1990), in which the triphosphate constitutes 70–80% of the total thiamin present. The natural vitamin B_1 content of most cereals and cereal products, including white flour made from wheat, is present almost entirely in the form of nonphosphorylated thiamin

19.2.2 Bioavailability

Few studies have been conducted on the bioavailability of thiamin present naturally in foods, but it is generally considered to be high.

19.3 DIGESTION AND ABSORPTION OF DIETARY THIAMIN

The human body cannot synthesize thiamin and so must obtain this vitamin from the diet. Dietary sources of phosphorylated thiamin, mainly TPP, are completely hydrolyzed to free thiamin in the small intestinal lumen by different phosphatases, including alkaline phosphatase. Absorption of thiamin then takes place in the duodenum and proximal jejunum by a dual mechanism that is predominantly saturable at low, physiological concentrations of the vitamin (<2.5 μmol/L) and largely non-saturable (diffusive) at higher concentrations (Laforenza *et al*., 1997). A significant amount of the thiamin that is taken up by the enterocyte is phosphorylated, mainly to TPP, via the action of the cytoplasmic thiamin pyrophosphokinase (Ricci & Rindi, 1992). Intracellular TPP is dephosphorylated by microsomal phosphatases to free thiamin before exit from the cell (Rindi, 1984).

Said *et al*. (1999) examined the uptake of tritiated thiamin ([^{3}H]thiamin) by monolayers of human-derived intestinal epithelial Caco-2 cells. Thiamin uptake was found to be temperature- and energy-dependent and occurred with minimal metabolic alterations to the transported substrate. Decreasing incubation buffer pH from 8 to 5 led to a progressive inhibition of thiamin uptake. This sensitivity to pH could not be attributed to changes in thiamin's ionic state, since the molecule exists as a monovalent cation over the pH range examined. The uptake process was independent of Na^+, as indicated by the lack of significant

DOI: 10.1201/9781003410300-23

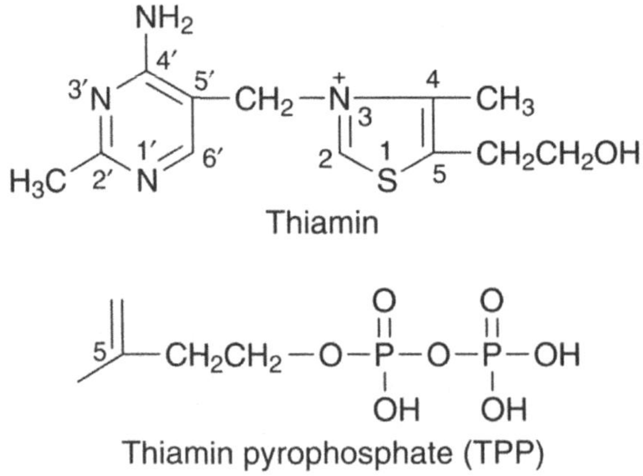

FIGURE 19.1 Structures of thiamin and thiamin pyrophosphate.

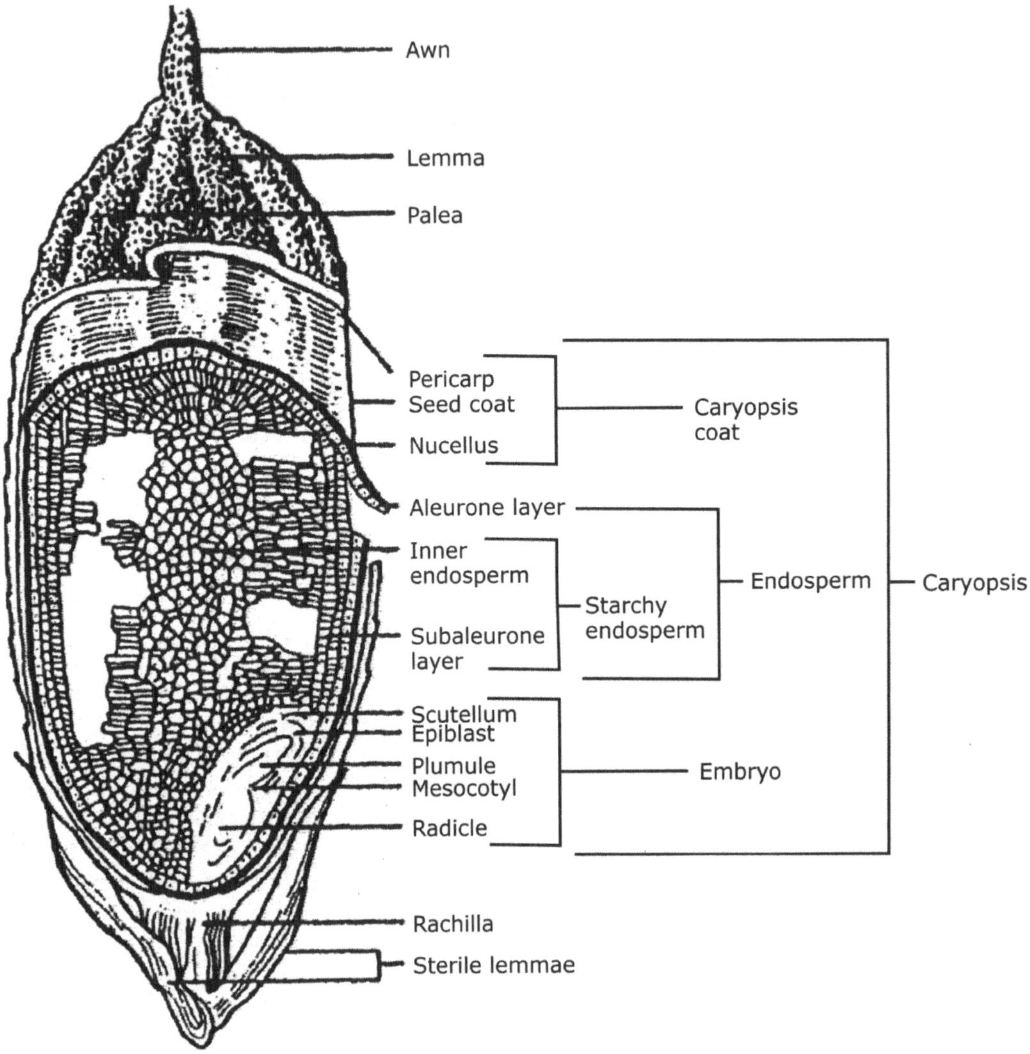

FIGURE 19.2 Structure of the rice grain in longitudinal section. The rice grain consists of the outer husk or hull and the edible portion (caryopsis). Dehulling reveals the pericarp or silverskin, which is composed of several layers of crushed cells covered by a thin cuticle. The underlying seed coat or tegmen is a single layer of crushed cells lined on the inner side by a thick cuticle. Adjacent to the seed coat cuticle is another thick cuticle that covers a crushed cell layer known as the nucellus. The aleurone layer is composed of up to seven layers of cuboidal cells. The starchy endosperm comprises the subaleurone layer and the inner endosperm, which consists of thin-walled parenchyma cells filled with starch granules. •From Zhou *et al.* (2002) Composition and functional properties of rice. *International Journal of Food Science and Technology*, **37**, 849–68. Figure 1 on page 851. ©2002 Blackwell Science Ltd.

inhibition of uptake upon removal of Na^+, as well as by the lack of effect when the cells were pretreated with the Na^+/K^+ ATPase (sodium pump) inhibitor ouabain. This conclusion was in agreement with the findings of Laforenza *et al.* (1997), using human intestinal biopsy specimens. Evidence for a carrier-mediated process was the saturation of thiamin uptake as a function of concentration and the competitive inhibition of uptake by structural analogs of thiamin. Inhibition of the intracellular Ca^{2+}/calmodulin second messenger pathway resulted in a significant inhibition of thiamin uptake, implicating this pathway in the regulation of thiamin absorption. The effect of one such inhibitor, trifluoperazine, was to decrease V_{max} (maximal transport rate), but not the apparent K_m (half-saturation concentration) of the thiamin uptake process. This indicates that the inhibitory effect of trifluoperazine is mediated via down-regulation of thiamin carriers at the brush-border membrane, with no effect on carrier affinity.

The use of purified intestinal membrane vesicles permits transport mechanisms in brush-border and basolateral membranes to be studied separately, with no interference from intracellular metabolism. Dudeja *et al.* (2001) investigated the mechanism for [^{3}H]thiamin uptake by brush-border membrane vesicles (BBMVs) from human jejunum. Uptake was saturable, inhibited by structural analogs of thiamin (but not by unrelated organic cations), and not influenced by changes in membrane electrical potential – features indicative of a carrier-mediated, electroneutral process. Uptake was independent of Na^+ and ATP. An outwardly directed downhill H^+ gradient (pH 5.5_{in}/pH 7.5_{out}) significantly stimulated thiamin uptake with a distinct overshoot phenomenon being observed during the initial phase of uptake. This could be explained by a possible involvement of a thiamin$^+/H^+$ antiport system with a 1:1 stoichiometry. Dudeja *et al.* (2003) demonstrated the presence of an ATP-independent thiamin$^+/H^+$ antiport system in human jejunal basolateral membrane vesicles (BLMVs).

Thiamin uptake experiments using rat proximal small intestinal BBMVs (Laforenza *et al.*, 1998) produced similar findings to corresponding experiments using human BBMVs (Dudeja *et al.*, 2001). However, while thiamin transport in human BLMVs was ATP-independent (Dudeja *et al.*, 2003), thiamin transport in rat BLMVs was directly coupled to ATP hydrolysis by Na^+/K^+ ATPase (Laforenza *et al.*, 1993).

The molecular identity of the thiamin carrier protein in the human intestine emerged after the cloning of human thiamin transporters hTHTR-1 and hTHTR-2 (the products of the *SLC19A2* and *SLC19A3* genes, respectively) from a number of human tissues, including the placenta (Dutta *et al.*, 1999; Eudy *et al.*, 2000). These two transporters are differentially localized in polarized cells of various tissues, including the small and large intestine, kidney, liver, brain, and placenta. Such polarization establishes a pathway for vectorial transport of thiamin across epithelia and allows for independent regulation of thiamin import and export pathways in cells. hTHTR-2 targets to the apical membrane and hTHTR-1 is biased more, but not exclusively, to the basolateral membrane. Thus, hTHTR-2 is positioned to play a crucial role in the uptake of thiamin by the epithelial cell, whereas hTHTR-1 is strategically positioned to contribute to thiamin exit from the cell (Boulware *et al.*, 2003). Thiamin transport mediated by THTR-1 and THTR-2 is pH-dependent. For THTR-1, influx is highest in the pH range 8.0 to 8.5; influx at 8.0 is threefold higher than at 6.0 (Dutta *et al.*, 1999). For THTR-2 influx is optimal at around neutral pH; influx is 3.5-fold higher at pH 7.4 than at 5.0 (Rajgopal *et al.*, 2001). Said *et al.* (2004) showed that hTHTR-2 is expressed along the length of the human gastrointestinal tract, with the highest level being in the duodenum and jejunum. This finding coincides with the higher ability of the proximal part of small intestine to absorb dietary thiamin. They also showed that hTHTR-2 is localized to the brush-border (apical) membrane of Caco-2 cells, whereas hTHTR-1 is localized to both apical and basolateral membranes. hTHTR-2-mediated uptake of thiamin by Caco-2 cells was shown to be saturable in the nanomolar range in contrast to hTHTR-1, which mediates uptake in the micromolar range (Dutta *et al.*, 1999).

Dietary thiamin deficiency in the mouse leads to an increase in intestinal thiamin uptake and an up-regulation of hTHTR-2 (but not hTHTR-1) expression (Reidling & Said, 2005). Conversely, THTR-2 deficiency in mice leads to an impairment in intestinal thiamin uptake and a decrease in blood thiamin levels. In contrast, THTR-1 deficiency in mice does not impair intestinal thiamin uptake because there is a compensatory up-regulation of TRHTR-2 expression (Reidling *et al.*, 2010).

In summary, thiamin absorption by the human small intestine is a carrier-mediated mechanism that utilizes the energy supplied by an outwardly directed H^+ gradient to power the uphill movement of luminal thiamin into the enterocyte via the brush-border membrane. A similar ATP-independent thiamin$^+/H^+$exchange mechanism facilitates exit of thiamin from the enterocyte to the serosa via the basolateral membrane. Thiamin absorption is adaptively regulated according to dietary intake of the vitamin.

19.4 ABSORPTION OF BACTERIALLY SYNTHESIZED THIAMIN IN THE LARGE INTESTINE

It has long been recognized that the normal microflora of the large intestine synthesizes a considerable amount of thiamin and that this thiamin exists in the free and phosphorylated forms. *In vitro* experiments using human-derived colonic epithelial NCM460 cells demonstrated that human colonocytes possess a pH- and Na^+-independent, high-affinity, carrier-mediated system for TTP uptake that involves thiamin$^+/H^+$ exchange and is under the regulation of the Ca^{2+}/calmodulin second messenger pathway (Said *et al.*, 2001). This mechanism has all of the features described for uptake of thiamin by Caco-2 cells (Said *et al.*, 1999).

Nabokina *et al.* (2014) identified the human colonic TPP transporter, hTPPT, as the product of the *SLC44A4* gene. hTPPT was predominantly expressed at the apical membrane of colonocytes, with negligible expression in the small intestine. The transport system was highly specific for TPP, being unaffected by free thiamin or TMP. Considerable metabolism of TPP to TMP and free thiamin occurred in the colonocytes following uptake. The authors proposed that part of the TPP taken up is used for the metabolic needs of the coloncytes, and the rest being metabolized. Free thiamin could then exit the cell via THTR-1 expressed at the basolateral membrane.

19.5 POST-ABSORPTIVE METABOLISM

Following absorption, thiamin is carried by the portal blood to the liver. Both nonphosphorylated thiamin and thiamin monophosphate circulate in the bloodstream, the former bound to plasma proteins. In normal adults, 20–30% of total thiamin in the plasma is protein-bound (Davis *et al.*, 1984).

Said *et al.* (2002) used cultured HepG2 cells as a model to study thiamin uptake by human liver cells. The results showed evidence of a thiamin$^+$/H$^+$ antiport mechanism, as described previously for thiamin uptake by small intestinal Caco-2 cells. In addition, the HepG2 cells were shown to express both THTR-1 and THTR-2, with the former being predominant. A thiamin$^+$/H$^+$ antiport mechanism was also shown to be present in basolateral (sinusoidal) membrane vesicles prepared from rat liver (Moseley *et al.*, 1992).

Thiamin is converted to its coenzyme form, TPP, by the catalytic action of thiamin pyrophosphokinase, which is present in all mammalian nucleated cells.

$$\text{Thiamin} + \text{ATP} \rightarrow \text{TPP} + \text{AMP} \quad (19.1)$$

A mitochondrial membrane carrier encoded by the *SLC25A19* gene mediates TPP transport across the inner mitochondrial membrane into the mitochondrion for utilization in carbohydrate metabolism and energy production (Lindhurst *et al.*, 2006).

In the brain and other nervous tissue, some of the TPP is converted to thiamin triphosphate (TTP) by thiamin pyrophosphate-ATP phosphoryltransferase:

$$\text{TPP} + \text{ATP} \rightarrow \text{TTP} + \text{ADP} \quad (19.2)$$

Nervous tissue also contains thiamin pyrophosphatase which converts small amounts of TPP to thiamin monophosphate (TMP) and inorganic phosphate:

$$\text{TPP} \rightarrow \text{TMP} + \text{PPi} \quad (19.3)$$

Also in nervous tissue, TTP may be hydrolyzed by thiamin triphosphatase to yield TPP and inorganic phosphate.

$$\text{TTP} \rightarrow \text{TPP} + \text{PPi} \quad (19.4)$$

Similarly, in nervous tissue, TMP may hydrolyzed to thiamin by thiamin monophosphatase.

$$\text{TMP} \rightarrow \text{Thiamin} + \text{PPi} \quad (19.5)$$

The human body contains approximately 30 mg of total thiamin of which about 80% is TPP, 10% is thiamin triphosphate, and the remainder is thiamin monophosphate and free thiamin. About half of the body content of total thiamin is found in skeletal muscles, the remainder being distributed mainly in the liver, heart, kidney, and brain. The biological half-life of [^{14}C]thiamin in the body is 9 to 18 days. Because of this relatively high turnover rate and low storage capacity, a continuous dietary intake of vitamin B_1 is necessary. During deprivation, there is a rather rapid loss of the vitamin from all tissues except the brain.

19.6 RENAL REABSORPTION

Circulating thiamin is filtered in the renal glomeruli and then salvaged via reabsorption by the epithelial cells of the proximal convoluted tubule to prevent its loss in the urine. Gastaldi *et al.* (2000) used BBMVs from rat kidney proximal tubules to study thiamin transport in renal epithelial cells. The presence of a H$^+$ gradient across the vesicle membrane enhanced [^{3}H]thiamin uptake only when the direction of the gradient was from inside to outside ($pH_{in} < pH_{out}$), suggesting that the uptake mechanism in the rat is thiamin$^+$/H$^+$ antiport.

Ashokkumar *et al.* (2006) studied the mechanism and regulation of renal thiamin uptake using the human-derived renal proximal tubular epithelial HEK-293 cell line. Their findings were similar to those reported for thiamin uptake in human intestinal and hepatic epithelial cells, where the carrier-mediated process was suggested to involve a thiamin$^+$/H$^+$exchange mechanism. The results showed that thiamin uptake is pH-dependent, Na$^+$-independent, and involves the activity of both hTHTR-1 and THTR-2.

With regard to regulation of renal thiamin uptake, maintaining the HEK-293 cells in a thiamin-deficient growth medium led to a significant increase in initial rate of [^{3}H]thiamin uptake. This increase in uptake was associated with a substantial increase in endogenous hTHTR-1 and hTHTR-2 mRNA and protein levels, as well as an increased activity of the *SLC19A2* and *SLC19A3* promoters. These findings implicate a transcriptional mechanism in the observed adaptive regulation of renal thiamin uptake (Ashokkumar *et al.*, 2006). Furthermore, thiamin uptake was shown to be regulated by an intracellular Ca^{2+}/calmodulin-mediated pathway. Experiments using transgenic mice carrying the human *SLC19A2* and *SLC19A3* promoters showed that dietary thiamin deficiency is associated with an increase in the transcriptional activity of both *SLC19A2* and *SLC19A3* promoters *in vivo* in the kidney (Reidling & Said, 2005).

19.7 THIAMIN UPTAKE BY PANCREATIC ACINAR CELLS AND BETA CELLS

The pancreas is a large gland with both exocrine and endocrine functions. The major (~82%) exocrine portion is composed of polarized acinar cells that secrete digestive enzymes. The endocrine portion comprises the islets of Langerhans, which are scattered throughout the pancreas. The islets contain beta cells, which produce insulin. A deficiency of thiamin leads to a severe reduction in the production of digestive enzymes (Singh, 1982) and a marked impairment in insulin synthesis and secretion (Rathanaswami & Sundaresan, 1991; Rathanaswami *et al.*, 1991).

Subramanya *et al.* (2011) used freshly isolated rat primary pancreatic acinar cells and cultured rat-derived AR42J pancreatic acinar cells as models to study the mechanism of thiamin uptake by pancreatic acinar cells. Thiamin uptake by both cell models was shown to be carrier-mediated and associated with the expression of both THTR-1 and THTR-2, with THTR-1 being predominant. Subramanian *et al.* (2012) used a gene-specific small interfering RNA (siRNA) silencing approach and *SLC19A2* and *SLC19A3* knockout mouse models to show that these transporters are involved in carrier-mediated thiamin uptake by pancreatic acinar cells.

Mee *et al.* (2009) showed that thiamin uptake by cultured mouse-derived pancreatic beta-TC-6 cells is carrier-mediated and involves both THTR-1 and THTR-2, with THTR-1 being predominant. Involvement of a carrier-mediated process was confirmed in mouse and human pancreatic islets. Uptake was pH-dependent and Na^+-independent, suggesting the involvement of a thiamin$^+$/H^+exchange mechanism. Maintaining the beta-TC-6 cells at a high concentration of thiamin led to a significant decrease in thiamin uptake. This was associated with a significant down-regulation of hTHTR-1 and hTHTR-2 expression, as well as a decreased activity of the *SLC19A2* and *SLC19A3* promoters. The response of THTR-1 to the changes in thiamin concentration was more pronounced than that of THTR-2, demonstrating differential regulation of these two transporters in pancreatic beta cells. In addition, the process appeared to be under the regulation of Ca^{2+}/calmodulin- and protein-tyrosine kinase-mediated pathways.

19.8 TRANSPORT IN THE BRAIN

Brain cells need a continuous supply of thiamin to replace that used in cerebral metabolism. The turnover of cerebrospinal fluid (CSF) also creates a need for continuing transfer of thiamin from blood to brain. Experimental depletion of thiamin from the brain causes neuromuscular disturbances, especially ataxia and convulsions.

19.8.1 Transport across the Blood–Brain Barrier

Greenwood *et al.* (1982) used an *in vivo* technique to study thiamin transport across the blood–brain barrier of rats and found that the normal rate of thiamin entry into the brain was of a similar order of magnitude to the rate of thiamin turnover in different brain regions. This comparison shows that, although blood-brain transport is normally adequate to meet the needs of the brain for thiamin and to replace the obligatory losses, there is little margin of spare capacity. The transport of thiamin was evidently carrier-mediated and could be saturated by increasing the concentration of vitamin in the bloodstream. Less than 10% of the total flux was provided by a non-saturable process.

Thiamin monophosphate, which accompanies thiamin in plasma and CSF, is also transported across the blood–brain barrier by a saturable mechanism. The half-saturation concentration (K_m) ranged between 2.6 and 5.0 μM, depending on the cerebral region (Patrini *et al.*, 1988). This compares with K_m values of 1.9–2.6 μM for saturable thiamin transport across the blood–brain barrier (Reggiani *et al.*, 1984).

19.8.2 Transport Across the Choroid Plexus–CSF Barrier

The entry of thiamin from blood into CSF via the choroid plexus and from the extracellular space of brain into brain cells themselves are both saturable processes (Spector, 1976). In brain slices from rabbits and rats, the uptake system or thiamin is half-saturated at ~0.3 μM, which is similar to the normal concentration of total thiamin in the CSF of these species (Spector, 1982). This means that the entry of excessive amounts of the vitamin into the brain is prohibited. Indeed, it has been shown that after the intravenous injection of large doses of thiamin, CSF and brain total thiamin levels are little changed.

Reidling and Said (2005) showed that dietary thiamin deficiency in the mouse leads to an increase in thiamin uptake by the brain and an up-regulation of hTHTR-1 (but not hTHTR-2) expression. They used transgenic mice carrying the human *SLC19A2* and *SLC19A3* promoters to demonstrate that thiamin deficiency is associated with an increase in the transcriptional activity of the *SLC19A2* (but not the *SLC19A3*) promoter *in vivo* in the brain.

19.9 PLACENTAL TRANSPORT

Schenker *et al.* (1990) studied the transfer of physiological concentrations of thiamin across human placenta using an *in vitro* perfusion system which eliminated the confounding variables of maternal or fetal metabolism. A unidirectional maternal-to-fetal transfer of thiamin took place against a concentration gradient, implying active transport. There was no evidence of phosphorylation of thiamin during its transfer. Saturation kinetics and inhibition of transport by structural analogs of thiamin were indicative of a carrier-mediated process. The results of this study indicate the existence of a carrier-mediated, active transport mechanism for thiamin transport across the microvillous membrane of the syncytiotrophoblast. Grassl (1998) demonstrated the existence of a thiamin$^+$/H^+ antiport mechanism in human placental BBMVs, and Dutta *et al.* (1999) cloned the human thiamin transporter hTHTR-1 from human placenta.

19.10 BIOCHEMICAL FUNCTIONS

In mammalian tissues, TPP is a coenzyme for three mitochondrial enzymes involved in the oxidative decarboxylation of α-keto acids; these enzymes are the pyruvate dehydrogenase complex (EC 1.2.4.1), α-ketoglutarate dehydrogenase (EC 1.2.4.2), and branched-chain α-keto acid dehydrogenase (EC 1.2.4.4). The involvement of these dehydrogenases in the tricarboxylic acid cycle is shown in Figure 19.3. In addition, TPP is a coenzyme for transketolase (EC 2.2.1.1), which is found in the cytosol. Low levels of thiamin in cells lead to impairment in energy metabolism and oxidative stress.

19.10.1 Oxidative Decarboxylation of a-Keto Acids

Pyruvate dehydrogenase is a multienzyme complex which catalyzes the conversion of pyruvate to acetyl-CoA at a central point between glycolysis, lipid metabolism, and the tricarboxylic acid cycle. The enzyme complex is composed of three subunit proteins that require TPP, Mg^{2+}, NAD^+, FAD, lipoic acid, and coenzyme A (CoA–SH) as co-factors. The overall reaction (Figure 19.4) is the sum of the individual reactions.

α-Ketoglutarate dehydrogenase, which is also composed of three subunits, catalyzes the conversion of α-ketoglutarate to succinyl-CoA in the tricarboxylic acid cycle. The overall reaction is shown in Figure 19.5. The composition of the multi-enzyme complex, the co-factor requirements, and the steps in the formation of succinyl-CoA are similar to the oxidative decarboxylation of pyruvate.

Branched-chain α-keto acid dehydrogenase catalyzes the oxidative decarboxylation of the branched-chain α-keto acids, α-ketoisocaproic acid, α-keto-β-methylvaleric acid, and α-ketoisovaleric acid, which are formed after deamination of leucine, isoleucine, and valine, respectively (Figure 19.6). The oxidative decarboxylation reactions are similar to those for pyruvate and α-ketoglutarate and require the same co-factors. The decarboxylation products, isovaleryl-CoA, α-methylbutyryl-CoA and isobutyryl-CoA, are indirectly involved with the tricarboxylic acid cycle, as shown in Figure 19.3.

19.10.2 Transketolation

In the pentose phosphate pathway, TPP is the coenzyme for two reversible transketolase reactions involving the transfer of a two-carbon ketol unit from a donor molecule to an acceptor molecule (Figure 19.7). In the first reaction, the ketol group from xylulose-5-phosphate is donated to ribose-5-phosphate to form sedoheptulose-7-phosphate and glyceraldehyde-3-phosphate. In the second reaction, the ketol group from xylose-5-phosphate is donated to

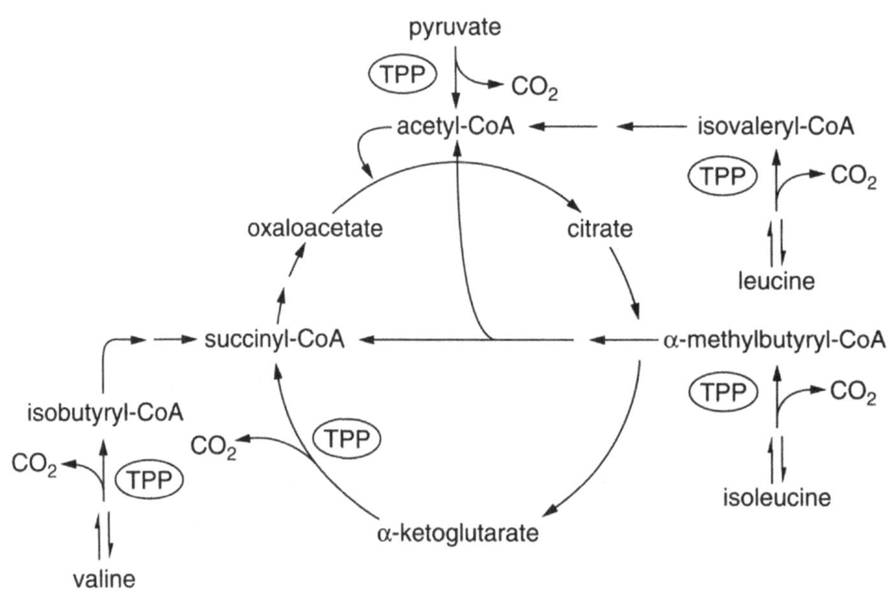

FIGURE 19.3 Involvement of thiamin pyrophosphate (TPP) in the tricarboxylic acid cycle.

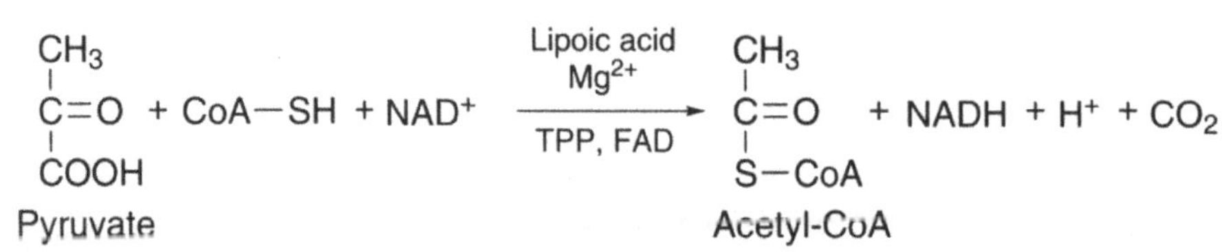

FIGURE 19.4 Conversion of pyruvate to acetyl-CoA.

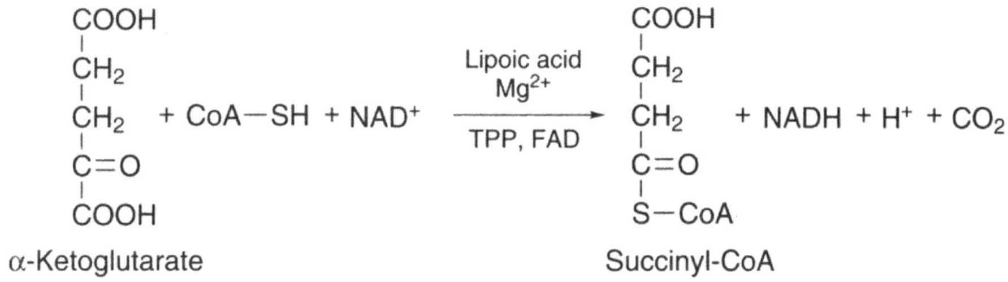

FIGURE 19.5 Conversion of α-ketoglutarate to succinyl-CoA.

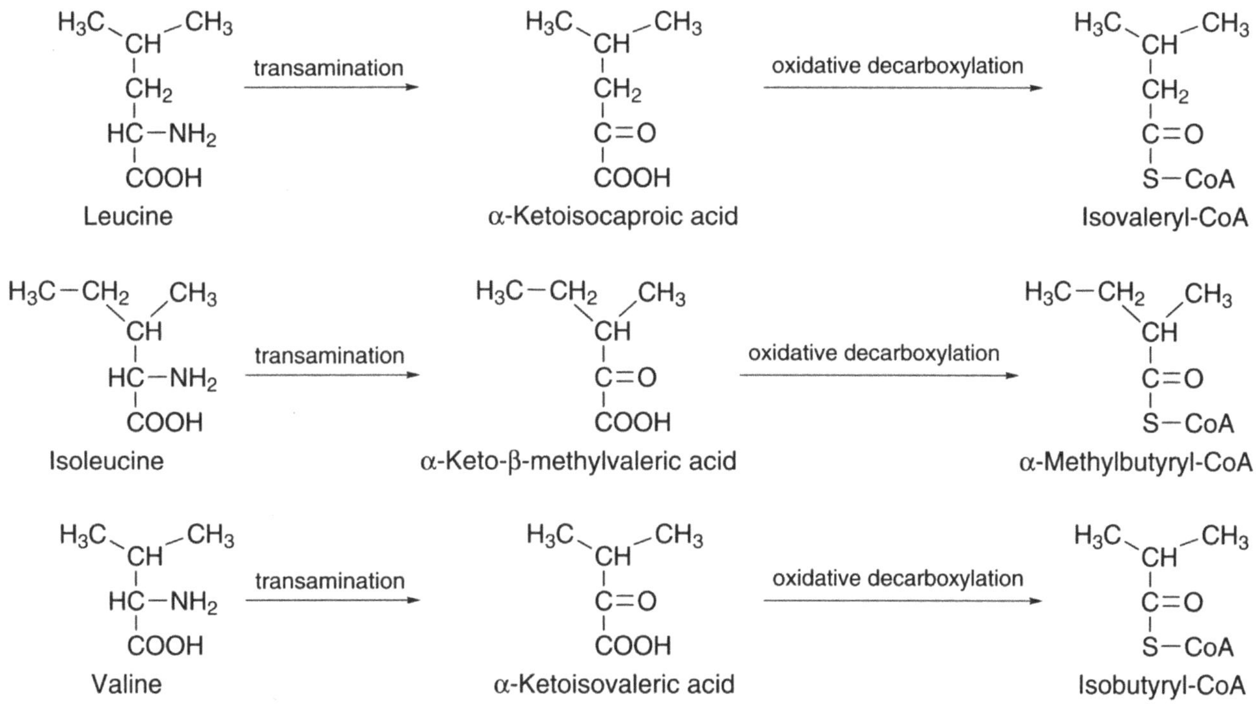

FIGURE 19.6 Oxidative decarboxylation of branched-chain α-keto acids.

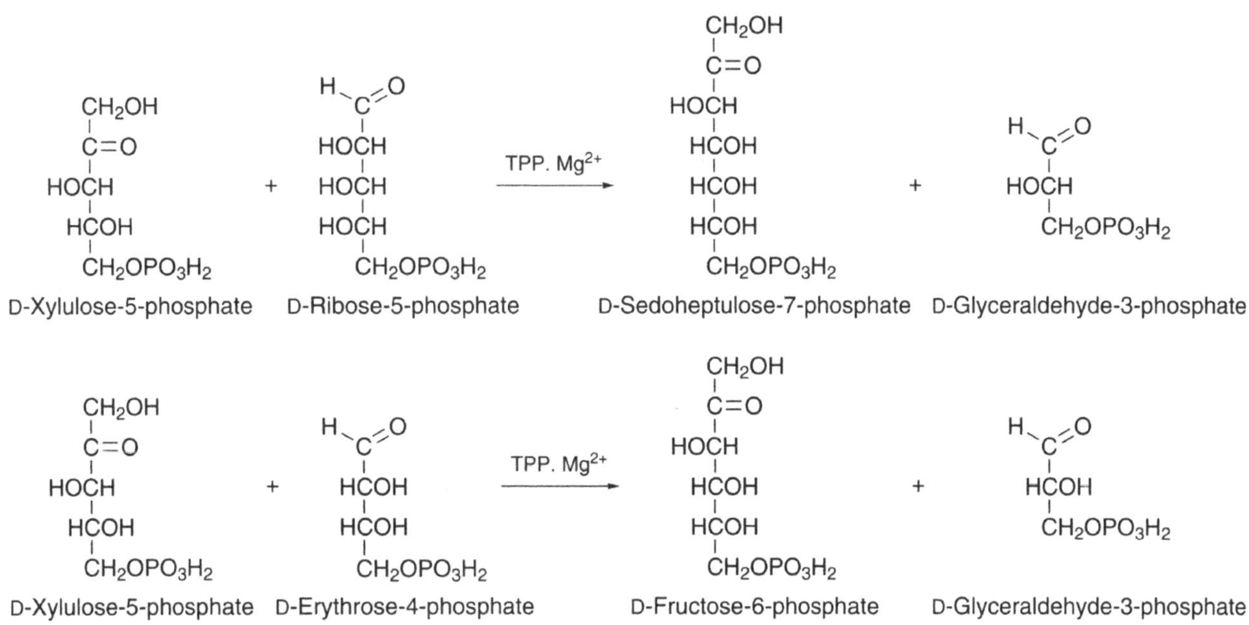

FIGURE 19.7 Two transketolation reactions in the pentose phosphate pathway.

erythrose-4-phosphate to form fructose-6-phosphate and glyceraldehyde-3-phosphate.

19.11 METABOLIC CONSEQUENCES OF THIAMIN DEFICIENCY

Proper functioning of the nervous and cardiovascular systems relies upon normal carbohydrate metabolism, and this in turn depends upon a dietary supply of thiamin for conversion to TPP. The pyruvate dehydrogenase complex and α-ketoglutarate dehydrogenase, which require TPP as a coenzyme, are important in the main energy-yielding pathway of carbohydrate metabolism, the tricarboxylic acid cycle. The activities of these enzymes are decreased in selective regions of the brain that are reversibly damaged as a result of thiamin deprivation (Butterworth *et al.*, 1985, 1986). The decreased enzyme activities are due not only to a decrease in the level of coenzyme, but also of apoenzyme. It is logical to suppose that impairment of these dehydrogenases due to thiamin deficiency will lead to loss of energy production in the form of ATP. Actually, this does not happen – at least not in the brain. The brain finds alternative metabolic pathways to bypass the TPP-dependent steps, thereby maintaining and even increasing energy production. One possible pathway is the GABA shunt, which bypasses α-ketoglutarate dehydrogenase (Page *et al.*, 1989) (Figure 19.8). Thus, energy deprivation is not believed to be the direct cause of brain lesions resulting from thiamin deficiency, although the compensatory metabolic changes necessary for maintaining energy may affect biosynthetic processes.

Preventing the synthesis of acetylcholine through lack of conversion of pyruvate to acetyl-CoA will impair nerve impulse transmission at synapses, and blocking the pentose phosphate pathway though impaired transketolase activity will lead to a reduced synthesis of DNA through lack of ribose. Dreyfus and Hauser (1965) showed that transketolase is more severely affected by thiamin deficiency than is pyruvate dehydrogenase, and the reduction of transketolase activity is anatomically correlated to brain lesions.

19.12 NEUROPHYSIOLOGICAL FUNCTIONS

19.12.1 Introduction

There is increasing evidence that thiamin, specifically thiamin triphosphate, is somehow involved in nerve membrane function. This property appears to be independent of the known coenzyme role of TPP. The evidence is substantiated by the finding that thiamin triphosphate, which accounts for 1% of total thiamin in rat brain, makes up 90% of total thiamin in the electric organ of the electric eel (Bettendorff *et al.*, 1987). In the lamb, vitamin B_1 deprivation for four weeks led to a 20% depletion of total thiamin in the brain, with a similar percentage loss of free thiamin, thiamin monophosphate, and TPP. There was, however, no appreciable fall in thiamin triphosphate (Thornber *et al.*, 1980).

Most of the thiamin present in the brain and peripheral nerves is in the coenzyme form, TPP. The 1% or so of thiamin triphosphate present in whole brain is largely concentrated in the membrane fraction (Matsuda & Cooper, 1981). Fluorescence microscopy shows that the vitamin is localized in the membranes of peripheral nerves rather than in the axoplasm (Tanaka & Cooper, 1968). A complete set of enzymes catalyzing the interconversion of thiamin and its phosphate esters has been isolated and purified from nervous tissue (Fox & Duppel, 1975).

19.12.2 Role of Thiamin Triphosphate in Nerve Conduction and Synaptic Transmission

Thiamin triphosphate may play a direct role in nerve conduction or it may be implicated in synaptic transmission.

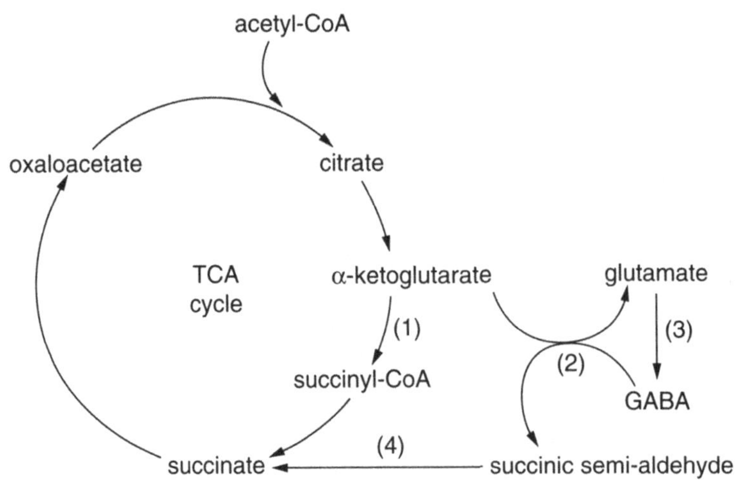

FIGURE 19.8 The γ-aminobutyric acid (GABA) shunt and tricarboxylic acid cycle. α-Ketoglutarate is aminated to glutamate, which is decarboxylated to GABA. GABA is inactivated by transamination to succinic semi-aldehyde, which is oxidized to succinate, an intermediate in the tricarboxylic acid cycle. Enzymes: (1) α-Ketoglutarate dehydrogenase, (2) GABA aminotransferase and other aminotransferases, (3) glutamate decarboxylase, and (4) succinic semi-aldehyde dehydrogenase.

19.12.2.1 Nerve Conduction

When Eichenbaum and Cooper (1971) UV-irradiated electrically stimulated vagus nerves dissected from a rabbit, action potentials were completely abolished after 2 h and never spontaneously reappeared when the irradiated nerve was immersed in physiological solution. The irradiation almost completely destroyed endogenous thiamin in the nerve, as expected. Immersion of the irradiated nerve in physiological solution containing 1 mM thiamin restored the action potentials after about 1 h. Thiamin is known to be rapidly taken up by the vagus nerve, so the delay may be explained by the two enzymatic steps required to convert thiamin to its triphosphate ester.

Sasa *et al.* (1976) studied the effects of thiamin, thiamin triphosphate, and pyrithiamin on the excitability of the perfused giant axon of the crayfish. A stimulating current was delivered to the axon every 2 s. The following treatments were performed sequentially. Addition of thiamin to the perfusate produced an increase in the rising rate (*dV/dt*) of the action potential as early as 5 min; there was no accompanying effect on either the resting membrane potential or the threshold potential. Perfusing the axon with physiological solution restored *dV/dt* to control rates in 40 min. Addition of pyrithiamin elicited a decrease of *dV/dt* and prolongation of the duration of the action potential in 60 to 90 min; again the resting membrane potential and threshold potential were unaffected. Perfusing with physiological solution did not restore the action potential within 30 min, but addition of thiamin did so in 30 min. Addition of thiamin triphosphate produced a similar effect to thiamin. The additional finding that protein-binding thiamin was reduced in axons treated with pyrithiamin indicates a replacement of thiamin located in the membrane. Since *dV/dt* of the action potential reflects the sodium conductance, the finding that thiamin and thiamin triphosphate increased *dV/dt* could be construed as an increase in sodium conductance. Conversely, the finding that pyrithiamin decreased *dV/dt* could mean an impairment of sodium conductance. It was inferred from these experiments that thiamin plays an essential role in the membrane excitability of the crayfish giant axon.

In a series of experiments summarized by Itokawa (1977), treatment of perfused spinal cords or sciatic nerves with a variety of neuroactive agents released free thiamin and thiamin monophosphate. The nerve preparations were obtained from either bullfrogs or rats made deficient in thiamin by dietary restriction and injected with radioactive thiamin. The neuroactive agents also released free thiamin and thiamin monophosphate from the membrane-myelin fraction obtained from homogenized rat brain, spinal cord, and sciatic nerves. As the bulk of thiamin in the membrane fraction consisted of TPP and thiamin triphosphate, the release phenomenon presumably involved dephosphorylation. One of the neuroactive agents, tetrodotoxin, acts as a nerve poison by inhibiting the early inward current of sodium. Itokawa (1976) postulated that the propagation of an action potential involves the shift of TPP or thiamin triphosphate through dephosphorylation from a specific site in the sodium channel to allow the early inward current of sodium. Tetrodotoxin, by displacing these phosphate esters, would occupy the site and prevent the flow of current.

Goldberg *et al.* (1975), using voltage-clamped squid giant axons, showed that the magnitudes of both sodium and potassium conductance were decreased by treatment with thiamin antimetabolites. However, there was no appreciable change in the kinetics of the conductance changes, hinting that the the mechanism of ion channel operation was unaffected. Most probably, the antimetabolites prevented the operation of a certain percentage of channels: those channels that remained open during perfusion with the antimetabolites functioned normally.

Fox and Duppel (1975) showed that B_1 vitamers (thiamin triphosphate > TPP > thiamin) applied internally to the cut internodes of frog sciatic nerve preparations prevented the exponential decline of sodium and potassium currents in the node of Ranvier. Neither thiamin triphosphate nor TPP was active when applied to the node externally. Tetrodotoxin did not alter this property of the thiamin compounds, implying that the tetrodotoxin-induced release of thiamin from nerve membranes (shown by other investigators) is not related to the mechanism by which tetrodotoxin blocks the sodium channels. Fox and Duppel reasoned that the thiamin dephosphorylation and rephosphorylation process is probably not directly coupled to the excitation process. They suggested that the thiamin phosphates control the number of functioning voltage-gated ion channels by stabilizing the density of negative surface charges at the inner side of the nerve membrane.

Further evidence for a role of thiamin triphosphate as a membrane electrical field stabilizer was provided by Bettendorff *et al.* (1990). These authors reported that thiamin triphosphate (1 μM) increased the uptake of radioactive chloride ($^{36}Cl^-$) by rat brain membrane vesicles, while TPP, thiamin monophosphate, and thiamin had no significant effect. The opening of chloride channels to allow an influx of negatively charged chloride ions into the postsynaptic neuron contributes to neuronal inhibition. Thus, if thiamin triphosphate controls the opening of chloride channels, it could act as a membrane stabilizer.

19.12.2.2 Synaptic Transmission

One of the characteristics of thiamin-deficiency encephalopathy is its predilection for specific brain structures with sparing of neighboring ones. This selective vulnerability of certain brain regions to thiamin deficiency has been suggested to have a metabolic basis.

Butterworth (1982) reviewed evidence that the neurological signs of thiamin deficiency may involve a defect in synaptic transmission. Changes in the concentration or metabolism of catecholamines (noradrenaline and adrenaline), serotonin (5-hydroxytryptamine), γ-aminobutyric acid (GABA), and acetylcholine were reported in certain regions of the brain of pyrithiamin-treated animals. In addition to these recognized neurotransmitters, changes were also observed in the dicarboxylic amino acids glutamic acid and aspartic acid. These amino acids act like neurotransmitters in that they open sodium and potassium ionic channels and cause a rapid, powerful, excitatory response.

The synthesis of GABA, acetylcholine, glutamic acid, and aspartic acid is directly associated with the metabolism of glucose in the brain. There is evidence to suggest that a decreased activity of TPP-dependent pyruvate and α-ketoglutarate dehydrogenase and transketolase due to vitamin B_1 deficiency may be ultimately responsible for a block in the synthesis of one or more of these neurotransmitters.

19.12.3 Subacute Necrotizing Encephalomyelopathy

Subacute necrotizing encephalomyelopathy (Leigh's disease) is a rare recessively inherited degenerative disease of the central nervous system which generally becomes symptomatic in the first year of life and is fatal. Diagnosis is complicated by the wide variety of symptoms, which include swallowing difficulties, abnormal respiration, ataxia, ophthalmoplegia, hypotonia, convulsions, and progressive mental deterioration. The neuropathology at autopsy shows characteristic lesions of the brain stem and spinal cord, notable features being capillary infiltration and demyelination of axons. Dietary vitamin B_1 deficiency is not a factor in the etiology of Leigh's disease. Early reports describing encouraging responses to large doses of thiamin or thiamin derivatives have not been borne out by further investigations (Blass, 1981).

Infants with Leigh's disease exhibit a deficiency of thiamin triphosphate in the brain. A substance, possibly a lipoprotein, that inhibits the synthesis of thiamin triphosphate from TPP via the thiamin pyrophosphate-ATP phosphotransferase has been detected in the blood, urine, and cerebrospinal fluid of these patients (Cooper & Pincus, 1979). This substance only inhibits the brain enzyme; the phosphotransferase that catalyzes the formation of thiamin triphosphate in the liver is unaffected. Leigh's disease appears therefore to be the result of disordered brain metabolism, possibly caused by the genetic lack of some enzyme. The resultant deficiency of thiamin triphosphate explains the neurological symptoms and neuropathological changes.

19.13 EXPERIMENTAL THIAMIN DEFICIENCY IN ANIMALS

19.13.1 Deficiency in the Chick

The symptoms of thiamin deficiency in chicks were studied by Gries and Scott (1972). One-day-old chicks were kept on a feed consisting of isolated soybean protein, glucose, cellulose, and maize oil, with added amino acids, vitamins (except thiamin), and minerals. Thiamin was added in stepped dosages from 0.1 to 0.8 mg per kg of feed. Chicks in the lower thiamin dosage groups (up to 0.4 mg/kg feed) survived for only seven to ten days, apparently only a few days after the supply of thiamin in the yolk sac was exhausted. Some of the chicks developed nervous disorders, apathy, and tremor as early as the third or fourth day of life. These symptoms increased in severity up to ataxia, inability to stand, and twisting of the neck. The severity of the spasms were enhanceable by startling the chicks. Chicks which showed these high-grade nervous disorders died within a few hours. The chicks in the 0.6 and 0.8 mg/kg dosage groups developed the same symptoms somewhat later, but nevertheless died within about 14 days. Every thiamin-deficient chick had severe dilation of the duodenal crypts of Lieberkühn and moderate to severe vacuolation of pancreatic acinar cells. The control chicks, kept on a diet containing 15 mg thiamin per kg feed, developed normally.

19.13.2 Deficiency in the Rat

When rats are given a thiamin-deficient diet, they show abnormalities of posture and equilibrium after several weeks. These neurological signs can be promptly reversed by the administration of 5 to 10 μg of thiamin. If no more thiamin is given, severe signs reappear in three to six days. Pathological examination of the nervous systems of rats subjected to several bouts of deficiency reveals anatomical lesions of certain parts of the brain. The lesions consist of areas of tissue destruction and a marked proliferation of glial cells. Within the center of lesions, all myelinated nerve fibers are destroyed (Dreyfus & Victor, 1961). Experimental thiamin deficiency is produced more rapidly (12 to 16 days) and more effectively by the use of pyrithiamin, a direct antagonist of TPP that also crosses the blood–brain barrier.

When [^{14}C]glutamate was injected into the brains of thiamin-deficient rats, the specific radioactivity of GABA in the brains rose by 45–50 %, suggesting a considerable increase in GABA shunt activity (Page *et al.*, 1989). This increase in brain GABA level may explain the anorexia observed in vitamin B_1-deficient rats. GABA aminotransferase, the enzyme principally responsible for GABA catabolism, is selectively inhibited by 1-(*n*-decyl)-3-pyrazolidinone. When this inhibitor was administered to rats, brain GABA levels were increased threefold and anorexia was observed in the absence of other symptoms (White *et al.*, 1982). This finding is consistent with speculations that GABA is implicated in appetite-controlling mechanisms (Kimura & Kuriyama, 1975).

Thiamin deficiency in rats leads to a marked impairment in insulin synthesis and secretion by pancreatic beta cells (Rathanaswami & Sundaresan, 1991; Rathanaswami *et al.*, 1991).

19.14 THIAMIN DEFICIENCY IN HUMANS

19.14.1 Causes and Effects

A deficiency of thiamin may occur in situations of poor diet, chronic alcoholism, excessive diarrhea or vomiting, malabsorption, and genetic metabolic defects. Overloading the tissues with glucose without adequate thiamin coverage can precipitate deficiency, as can the use of diuretics. Diseases in which the metabolic rate is elevated (e.g. hyperthyroidism) can also lead to deficiency. Some researchers have demonstrated that secondary deficiency of a particular B group vitamin can be induced by excessive dosing with another vitamin of this group. On the other hand, deficiencies in vitamins B_6 and B_{12} induced vitamin B_1 deficiency in rats, even when dietary thiamin levels were normal (Nishino & Itokawa, 1977). Howard *et al.* (1974) confirmed that folate deficiency in rats impairs thiamin absorption.

The combined results from several human studies have shown that inducement of thiamin deficiency in adults produces a wide range of disorders involving the gastrointestinal tract, central and peripheral nervous systems, and cardiovascular system. Anorexia and associated weight loss are constant findings. Indigestion results from hypochlorhydria. Gastric atony results in severe constipation through lack of gut motility. Thus, when a barium meal is given, there is incomplete emptying of the stomach and pooling of the barium in segments of the small intestine, producing the so-called step-ladder pattern. There are paresthesias with a stocking- or glove-type distribution, and impairment in perception of light, touch, pinprick, temperature, and vibratory sensation. Subjects experience difficulty in rising from a squatting position, owing to weakness of the calf muscles. The deep tendon reflexes (patella and Achilles) disappear after a while. Electrocardiograms show irregularities, and subjects complain of shortness of breath, consciousness of the heartbeat, irregularities in heart rhythm, and discomfort in the chest after exertion. Mental changes develop, such as loss of emotional control, paranoid trends, manic or depressive episodes, and confusion.

The classic disease resulting from a gross deficiency of vitamin B_1 in humans is beriberi, which is prevalent in Far Eastern populations where polished rice is the staple diet. Not only is this diet deficient in vitamin B_1, but the high carbohydrate intake increases the requirement for the vitamin, and the consumption of anti-thiamin factors will exacerbate the deficiency. In other parts of the world where such a diet is not consumed, chronic alcoholism gives rise to the Wernicke-Korsakoff syndrome, a form of beriberi that affects the brain.

19.14.2 Beriberi

The development of beriberi, its symptoms, and its pathology are extremely variable making it difficult to describe a clinical picture or sequence of development. Many of the early writers described three forms of beriberi in adult humans: dry (wasting) and wet (edematous) beriberi, which are chronic forms, and acute, fulminating (cardiac) beriberi. Which of these forms predominates depends on the circumstances. Thiamin deprivation accompanied by malnutrition and low physical activity tends to favor beriberi presenting in the dry form, whereas high carbohydrate intake and high physical activity during thiamin deprivation predispose to wet beriberi. It should be emphasized that any one of these forms may merge into another.

Dry beriberi is a disease of the peripheral nervous system involving bilateral impairment of sensory, motor, and reflex functions. The pathological findings are segmental thinning of myelin in peripheral nerves, progressing to degeneration of fiber tracts. The neuropathy begins in the feet and legs and then extends up the body. Early signs of dry beriberi often include sensations of pins-and-needles and numbness in the feet. The legs, especially the calves, feel heavy and weak so that walking becomes uncomfortable. As the disease progresses, there is a marked wasting of the leg muscles and even slight pressure applied to the calves elicits severe pain. The characteristic foot and wrist drop develop and there may be complete flaccid paralysis of the lower, and occasionally upper, extremities.

In wet beriberi, thiamin deficiency affects the cardiovascular system by causing arteriolar dilation throughout the circulatory system and by weakening the heart muscle. The vasodilation causes a twofold increase in the venous return of blood to the heart. Physical signs of wet beriberi are indicative of high-output cardiac failure; they include tachycardia, rapid circulation time, elevated peripheral venous pressure, and widespread edema. Pathological changes are an enlarged heart, swollen liver, and the presence of fluid in the pericardial, pleural, and abdominal cavities. Microscopically, the cardiac muscle fibers show fragmentation and hydropic degeneration with interstitial edema.

The acute form of beriberi, known in Japan as 'shoshin,' usually results in sudden death from heart failure. The sufferer experiences severe dyspnea, violent palpitation of the heart, and intense precordial pain. The heart and liver become enlarged and there is neck vein distension, cyanosis, and extreme tachycardia. Edema is variable in this form of beriberi.

Infantile beriberi occurs in breastfed babies between the second and fifth months of life. The mother may display no signs of beriberi although obviously her milk must be of low vitamin B_1 content through poor diet. As in adults, several clinical syndromes may occur and the condition may be chronic or acute. Vomiting is one of the most important early signs of infantile beriberi. In severe cases the child appears to be crying but no sound is heard or only a thin whine. This characteristic feature, aphonia, is due either to paralysis of the laryngeal nerve or to edema of the vocal cords. A chronic, pseudomeningeal form of beriberi which affects the central nervous system and may cause convulsions occurs in older infants aged seven to nine months.

Beriberi is treated with a proprietary thiamin preparation, the dosage and route depending on the patient's condition. To prevent further recurrences, a nutritious diet containing all of the B group vitamins should be instituted. Severe cardiac (shoshin) beriberi constitutes a medical emergency requiring immediate treatment with thiamin, given intravenously.

19.14.3 Wernicke-Korsakoff Syndrome

The Wernicke-Korsakoff syndrome is a thiamin deficiency disease particularly associated with chronic alcoholics who derive more than half their daily calories from ethanol every day. Binge drinkers who eat normally between bouts and who are in a reasonable state of health are not prone to the disease. Alcoholic beverages contain water, ethanol, variable amounts of carbohydrate, and little else of nutritive value. Apart from beers, protein and vitamin content of these beverages is extremely low. Vitamin B_1 deficiency in the alcoholic is a combination of several factors. The main factor is reduced food intake due to depressed consciousness during inebriation and hangover. Suppression of appetite by alcohol has been postulated, but not studied much. Gastritis and diarrhea caused by alcohol will impair digestion and absorption, and alcoholic pancreatitis results in decreased secretion of digestive enzymes. Impairment of thiamin absorption has been reported in severely alcoholic patients (Tomasulo *et al.*, 1968). In cases of alcoholic cirrhosis of the liver, there may be a decreased conversion of thiamin to TPP and a decreased capacity of the liver to store vitamin B_1 (Leevy & Baker, 1968). The mortality of the Wernicke-Korsakoff syndrome is 90% without therapy, heart failure being the usual cause of death.

The Wernicke-Korsakoff syndrome has two components, Wernicke's encephalopathy (or Wernicke's disease) and Korsakoff's psychosis. The former is specific to vitamin B_1 deficiency, whereas the latter may be seen in association with other disorders of the nervous system. Treatment of Wernicke-Korsakoff syndrome with intravenous injections of thiamin will eradicate the symptoms of encephalopathy (with abstinence from alcohol), but the psychosis is irreversible.

Wernicke's disease is characterized by a triad of clinical signs: eye abnormalities, lack of co-ordination, and altered state of consciousness. They occur simultaneously, or one may precede the other by days or weeks. The eye signs are caused by paralysis of one or more of the eye muscles (ophthalmoplegia). Typical signs are photophobia, nystagmus (oscillation of the eyeballs), strabismus (crossed eyes), and diplopia (double vision). In advanced cases, there may be complete loss of eye movement and the pupils may become non-reactive. The disorder of co-ordination is seen as a broad-based stance and ataxia (staggering gait). Common mental signs are listlessness, inattentiveness, apathy, and confusion. There is sometimes delirium and, in extreme cases, stupor and coma.

The most notable feature of Korsakoff's psychosis is a form of amnesia in which events of ordinary daily life are forgotten as quickly as they occur but events in the distant past are well remembered. The patient is alert and can converse, think, and solve problems, but is unable to memorize new information. The symptom of confabulation (storytelling) is an attempt by the patient to hide the amnesia.

Neuropathological changes in Wernicke's encephalopathy are seen as brain lesions distributed in a bilaterally symmetrical fashion in the mammillary bodies, superior vermis of the cerebellum, hypothalamic nuclei, and other diencephalic structures (Reuler *et al.*, 1985). Histology shows necrosis of neurons and glial cells. There is also capillary damage with endothelial proliferation and pinpoint hemorrhages. Damage to specific regions of the brain can account for the clinical features of Wernicke-Korsakoff syndrome. Thus, for example, the nystagmus is due to damage of the sixth cranial nerve; the ataxia is related to loss of neurons in the superior vermis of the cerebellum; and the amnesia is associated with atrophy of the mammillary bodies.

Although the Wernicke-Korsakoff syndrome results from a lack of dietary thiamin, two clinical observations suggest that genetic factors are important in its pathogenesis: it develops in only a minority of alcoholics and other chronically malnourished persons, and it occurs much more frequently among Europeans than among Asians. Blass and Gibson (1977) found that the transketolase apoenzyme in tissue-cultured cells from patients with Wernicke-Korsakoff syndrome had an affinity for its coenzyme (TPP) that was only one-tenth that of the normal apoenzyme. The abnormality persisted through more than 20 generations of culture in medium containing excess thiamin and no ethanol, and therefore appeared to be genetic rather than dietary. Thus the abnormality was a mutation of the gene for transketolase. The abnormality appeared to be specific for transketolase, as pyruvate dehydrogenase and α-ketoglutarate dehydrogenase were unaffected. Persistent aberrations had previously been found in erythrocyte transketolase from the patients, even after they had been treated with thiamin for months while in hospital. Nixon *et al.* (1984) demonstrated a highly significant association between a particular variant of erythrocyte transketolase and the Wernicke-Korsakoff syndrome, supporting the concept that the syndrome has a genetic as well as a dietary origin.

In individuals with an unmutated transketolase gene, a moderate deficiency in thiamin will have little effect on transketolase activity. However, in individuals with a mutant transketolase gene, the same level of thiamin deficiency will reduce the level of TPP below that needed to saturate the apoenzyme, thereby reducing enzyme activity and exacerbating the Wernicke-Korsakoff syndrome. This type of genetic abnormality is an example of an inborn predisposition to metabolic disorders. Unlike inborn errors of metabolism, inborn predispositions are likely to be clinically silent unless the person with the predisposition faces an appropriate stress. In cases of Wernicke-Korsakoff syndrome, the stress is a deficiency of thiamin.

19.15 THIAMIN AND DIABETES

Diabetic nephropathy develops progressively over 5–40 years of diabetes and is associated with a high risk of cardiovascular disease and mortality (Parving *et al.*, 2001). The initial stage of development of nephropathy, incipient nephropathy, is characterized by the presence of serum albumin in the urine, known as microalbuminuria. Hyperglycemia is a risk factor for the development of incipient nephropathy in both type 1 and type 2 diabetes. Biochemical dysfunction in hyperglycemia is triggered by accumulation of triose phosphates arising from high cytosolic glucose concentrations. Babaei-Jadidi *et al.* (2003) reported that stimulation of the reductive pentose phosphate pathway by high-dose therapy with thiamin and the thiamin monophosphate derivative benfotiamine inhibited the development of incipient nephropathy in experimental diabetes. The thiamin therapy countered the accumulation of triose phosphates and strongly inhibited the development of microalbuminuria. These effects were achieved without change in elevated plasma glucose concentration. Experimental diabetes was associated with thiamin deficiency characterized by a marked decrease in plasma thiamin concentration and decreased activity and expression of the thiamin-dependent enzyme transketolase in renal glomeruli (Babaei-Jadidi *et al.*, 2004). Patients with type 1 or type 2 diabetes exhibit low plasma concentrations and high renal clearance of thiamin (Thornalley *et al.*, 2007).

Rabbani *et al.* (2009) conducted a randomized, double-blind, placebo-controlled pilot study in 40 type 2 diabetic patients with microalbuminuria to assess whether oral supplements of thiamin could reverse microalbuminuria. The 40 patients were given 3 × 100 mg capsules of thiamin or placebo per day for three months with a two-month follow-up washout period. The results showed a decreased urinary albumin excretion in the thiamin-treated patients with no significant decrease in patients receiving placebo. There were no adverse effects of therapy and no effect on glycemic control. It was concluded that thiamin supplements at high dose may provide improved therapy for incipient diabetic nephropathy. Whether thiamin therapy prevents the progression from incipient to overt nephropathy remains to be investigated.

Incubation of monolayers of human proximal tubular epithelial cells (hPTECs) in a medium containing a high concentration of glucose led to decreased expression and abundance of THTR-1 and THTR-2, together with a concomitant decrease in net thiamin transport from the apical to the basolateral sides of the monolayers. Furthermore, intensive insulin therapy attenuates, but does not correct, the increase in fractional excretion of thiamin in streptozotocin-induced diabetic rats (Larkin *et al.*, 2012). These findings imply that diabetic patients are prone to thiamin deficiency because their high blood glucose levels cause a decreased expression of thiamin transporters in the renal tubular epithelium and consequent reduced renal reabsorption of the vitamin.

19.16 TOXICITY

Thiamin is nontoxic by the oral route because excess amounts of ingested thiamin are rapidly excreted in the urine. Large parenteral doses of thiamin administered over a long period have been reported to produce clinical manifestations and, in some cases, even death (Cumming *et al.*, 1981).

FURTHER READING

Sauberlich, H.E. (1967) Biochemical alterations in thiamine deficiency – Their interpretation. *American Journal of Clinical Nutrition*, **20**, 528–42.

REFERENCES

Ashokkumar, B., Vaziri, N.D. & Said, H.M. (2006) Thiamin uptake by the human-derived renal epithelial (HEK-293) cells: Cellular and molecular mechanisms. *American Journal of Physiology (Renal Physiology)*, **291**, F796–805.

Babaei-Jadidi, R., Karachalias, N., Ahmed, N., Battah, S. & Thornalley, P.J. (2003) Prevention of incipient diabetic nephropathy by high-dose thiamine and benfotiamine. *Diabetes*, **52**, 2110–20.

Babaei-Jadidi, R., Karachalias, N., Kupich, C., Ahmed, N. & Thornalley, P.J. (2004) High dose thiamine therapy counters dyslipidaemia in streptozotocin-induced diabetic rats. *Diabetologia*, **47**, 2235–46.

Bettendorff, L., Michel-Cahay, C., Grandfils, C., De Rycker, C. & Schoffeniels, E. (1987) Thiamine triphosphate and membrane-associated thiamine triphosphatases in the electric organ of *Electrophorus electricus*. *Journal of Neurochemistry*, **49**, 495–502.

Bettendorff, L., Wins, P. & Schoffeniels, E. (1990) Regulation of ion uptake in membrane vesicles from rat brain by thiamine compounds. *Biochemical and Biophysical Research Communications*, **171**, 1137–44.

Blass, J.P. (1981) Thiamin and the Wernicke-Korsakoff syndrome. In: *Vitamins in Human Biology and Medicine*. (Ed. by M.H. Briggs), pp. 107–35. CRC Press, Inc., Boca Raton, FL.

Blass, J.P. & Gibson, G.E. (1977) Abnormality of a thiamine-requiring enzyme in patients with Wernicke-Korsakoff syndrome. *New England Journal of Medicine*, **297**, 1367–70.

Boulware, M.J., Subramanian, V.S., Said, H.M. & Marchant, J.S. (2003) Polarized expression of members of the solute carrier *SLC19A* gene family of water-soluble multi-vitamin transporters: Implications for physiological function. *Biochemical Journal*, **376**, 43–8.

Butterworth, R.F. (1982) Neurotransmitter function in thiamine-deficiency encephalopathy. *Neurochemistry International*, **4**, 449–64.

Butterworth, R.F., Giguère, J.-F. & Besnard, A.-M. (1985) Activities of thiamine-dependent enzymes in two experimental models of thiamine-deficiency encephalopathy. 1. The pyruvate dehydrogenase complex. *Neurochemical Research*, **10**, 1417–28.

Butterworth, R.F., Giguère, J.-F. & Besnard, A.-M. (1986) Activities of thiamine-dependent enzymes in two experimental models of thiamine-deficiency encephalopathy. 2. α-Ketoglutarate dehydrogenase. *Neurochemical Research*, **11**, 567–77.

Cooper, J.R. & Pincus, J.H. (1979) The role of thiamine in nervous tissue. *Neurochemical Research*, **4**, 223–39.

Cumming, F., Briggs, M. & Briggs, M. (1981) Clinical toxicology of vitamin supplements. In *Vitamins in Human Biology and Medicine*. (Ed. by M.H. Briggs), pp. 187–243. CRC Press, Inc., Boca Raton, FL.

Davis, R.E., Icke, G.C., Thom, J. & Riley, W.J. (1984) Intestinal absorption of thiamin in man compared with folate and pyridoxal and its subsequent urinary excretion. *Journal of Nutritional Science and Vitaminology*, **30**, 475–82.

Dreyfus, P.M. & Hauser, G. (1965) The effect of thiamine deficiency on the pyruvate decarboxylase system of the central nervous system. *Biochimica et Biophysica Acta*, **104**, 7–84.

Dreyfus, P.M. & Victor, M. (1961) Effects of thiamine deficiency on the central nervous system. *American Journal of Clinical Nutrition*, **9**, 414–25.

Dudeja, P.K., Tyagi, S., Kavilaveettil, R.J., Gill, R. & Said, H.M. (2001) Mechanism of thiamine uptake by human jejunal brush-border membrane vesicles. *American Journal of Physiology (Cell Physiology)*, **281**, C786–92.

Dudeja, P.K., Tyagi, S., Gill, R. & Said, H.M. (2003) Evidence for a carrier-mediated mechanism for thiamine transport to human jejunal basolateral membrane vesicles. *Digestive Diseases and Sciences*, **48**, 109–15.

Dutta, B., Huang, W., Molero, M., Kekuda, R., Leibach, F.H., Devoe, L.D., Ganapathy, V. & Prasad, P.D. (1999) Cloning of the human thiamine transporter, a member of the folate transporter family. *Journal of Biological Chemistry*, **274**(45), 31925–9.

Egi, Y., Koyama, S., Shikata, H., Yamada, K. & Kawasaki, T. (1986) Content of thiamin phosphate esters in mammalian tissues – An extremely high concentration of thiamin triphosphate in pig skeletal muscle. *Biochemistry International*, **12**, 385–90.

Eichenbaum, J.W. & Cooper, J.R. (1971) Restoration by thiamine of the action potential in ultraviolet irradiated nerves. *Brain Research*, **32**, 258–60.

Eudy, J.D., Spiegelstein, O., Barber, R.C., Windarczyk, B.J., Talbot, J. & Finnell, R.H. (2000) Identification and characterization of human and mouse SLC19A3 gene: A novel member of the reduced folate family of micronutrient transporter genes. *Molecular Genetics and Metabolism*, **71**, 581–90.

Fox, J.M. & Duppel, W. (1975) The action of thiamine and its di- and triphosphates on the slow exponential decline of the ionic currents in the node of Ranvier. *Brain Research*, **89**, 287–302.

Gastaldi, G., Cova, E., Verri, A., Laforenza, U., Faelli, A. & Rindi, G. (2000) Transport of thiamin in rat renal brush border membrane vesicles. *Kidney International*, **57**, 2043–54.

Goldberg, D.J., Begenisich, T.B. & Cooper, J.R. (1975) Effects of thiamine antagonists on nerve conduction. II. Voltage clamp experiments with antimetabolites. *Journal of Neurobiology*, **6**, 453–62.

Grassl, S.M. (1998) Thiamine transport in human placental brush border membrane vesicles. *Biochimica et Biophysica Acta*, **1371**, 213–22.

Gries, C.L. & Scott, M.L. (1972) The pathology of thiamin, riboflavin, pantothenic acid and niacin deficiencies in the chick. *Journal of Nutrition*, **102**, 1269–86.

Greenwood, J., Love, E.R. & Pratt, O.E. (1982) Kinetics of thiamine transport across the blood-brain barrier in the rat. *Journal of Physiology*, **327**, 95–103.

Howard, L., Wagner, C. & Schenker, S. (1974) Malabsorption of thiamin in folate-deficient rats. *Journal of Nutrition*, **104**, 1024–32.

Itokawa, Y. (1976) Thiamine and nerve membrane. *Journal of Nutritional Science and Vitaminology*, **22**, 17–19.

Itokawa, Y. (1977) Thiamine (vitamin B_1) and diseases related to vitamin B_1 deficiencies. *Journal of Applied Nutrition*, **29**, 5–16.

Kimura, H. & Kuriyama, K. (1975) Distribution of gamma-aminobutyric acid (GABA) in the rat hypothalamus: Functional correlates of GABA with activities of appetite controlling mechanisms. *Journal of Neurochemistry*, **24**, 903–7.

Laforenza, U., Gastaldi, G. & Rindi, G. (1993) Thiamine outflow from the enterocyte: A study using basolateral membrane vesicles from rat small intestine. *Journal of Physiology*, **468**, 401–12.

Laforenza, U., Patrini, C., Alvisi, C., Faelli, A., Licandro, A. & Rindi, G. (1997) Thiamine uptake in human intestinal biopsy specimens, including observations from a patient with acute thiamine deficiency. *American Journal of Clinical Nutrition*, **66**, 320–6.

Laforenza, U., Orsenigo, M.N. & Rindi, G. (1998) A thiamine/H^+ antiport mechanism for thiamine entry into brush border membrane vesicles from rat small intestine. *Journal of Membrane Biology*, **161**, 151–61.

Larkin, J.R., Zhang, F., Godfrey, L., Molostvov, G., Zehnder, D., Rabbani, N. & Thornalley, P.J. (2012) Glucose-induced down regulation of thiamine transporters in the kidney proximal tubular epithelium produces thiamine insufficiency in diabetes. *Plos One*, **7**(12), e53175. DOI: 10.1371/journal.pone.0053175

Leevy, C.H. & Baker, H. (1968) Vitamins and alcoholism. Introduction. *American Journal of Clinical Nutrition*, **21**, 1325–8.

Lindhurst, M.J., Fiermonte, G., Song, S., Struys, E., De Leonardis, F., Schwartzberg, P.L., Chen, A., Castegna, A., Verhoeven, N., Mathews, C.K., Palmieri, F. & Biesecker, L.G. (2006) Knockout of *Slc25a19* causes mitochondrial thiamine pyrophosphate depletion, embryonic lethality, CNS malformations, and anemia. *Proceedings of the National Academy of Sciences of the U.S.A.*, **103**(43), 15927–32.

Matsuda, T. & Cooper, J.R. (1981) The separation and determination of thiamin and its phosphate esters in brain. *Analytical Biochemistry*, **117**, 203–7.

Mee, L., Nabokina, S.M., Sekar, V.T., Subramanian, V.S., Maedler, K. & Said, H.M. (2009) Pancreatic beta cells and islets take up thiamin by a regulated carrier-mediated process: Studies using mice and human pancreatic preparations. *American Journal of Physiology (Gastrointestinal and Liver Physiology)*, **297**, G197–206.

Miyoshi, K., Egi, Y., Shioda, T. & Kawasaki, T. (1990) Evidence for *in vivo* synthesis of thiamin triphosphate by cytosolic adenylate kinase in chicken skeletal muscle. *Journal of Biochemistry*, **108**, 267–70.

Moseley, R.H., Vashi, P.G., Jarose, S.M., Dickinson, C.J. & Permoad, P.A. (1992) Thiamine transport by basolateral rat liver plasma membrane vesicles. *Gastroenterology*, **103**, 1056–65.

Nabokina, S.M., Inoue, K., Subramanian, V.S., Valle, J.E., Yuasa, H. & Said, H.M. (2014) Molecular identification and functional characterization of the human colonic thiamine pyrophosphate transporter. *Journal of Biological Chemistry*, **289**(7), 4405–16.

Nishino, K. & Itokawa, Y. (1977) Thiamin metabolism in vitamin B_6 or vitamin B_{12} deficient rats. *Journal of Nutrition*, **107**, 775–82.

Nixon, P.F., Kaczmarek, M.J., Tate, J., Kerr, R.A. & Price, J. (1984) An erythrocyte transketolase isoenzyme pattern associated with the Wernicke-Korsakoff syndrome. *European Journal of Clinical Investigation*, **14**, 278–81.

Page, M.G., Ankoma-Sey, V., Coulson, W.F. & Bender, D.A. (1989) Brain glutamate and γ-aminobutyrate (GABA) metabolism in thiamin-deficient rats. *British Journal of Nutrition*, **62**, 245–53.

Parving, H.-H., Harrington, J.T., Lewis, J., Boner, G., Lewis, E.J., Viberti, G., Gilbert, R., Raz, I., Cooper, M. & Rossing, K. (2001) Diabetic nephropathy: Prevention and treatment. *Kidney International*, **60**, 2041–55.

Patrini, C., Reggiani, C., Laforenza, U. & Rindi, G. (1988) Blood–brain transport of thiamine monophosphate in the rat: A kinetic study in vivo. *Journal of Neurochemistry*, **50**, 90–3.

Rabbani, N., Alam, S.S., Riaz, S., Larkin, J.R., Akhtar, M.W., Shafi, T. & Thornalley, P.J. (2009) High-dose thiamine therapy for patients with type 2 diabetes and microalbuminuria: A randomised, double-blind placebo-controlled pilot study. *Diabetologia*, **52**, 208–12.

Rajgopal, A., Edmondson, A., Goldman, I.D., Zhao, R. (2001) SLC19A3 encodes a second thiamine transporter ThTr2. *Biochimica et Bioephysica Acta*, 1537, 175--8.

Rathanaswami, P. & Sundaresan, R. (1991) Effects of thiamine deficiency on the biosynthesis of insulin in rats. *Biochemistry International*, **24**, 1057–62.

Rathanaswami, P., Pouranyl, A. & Sundaresan, R. (1991) Effects of thiamine deficiency on the secretion of insulin and the metabolism of glucose in isolated rat pancreatic islets. *Biochemistry International*, **25**, 577–83.

Reggiani, C., Patrini, C. & Rindi, G. (1984) Nervous tissue thiamine metabolism in vivo. I. Transport of thiamine and thiamine monophosphate from plasma to different brain regions of the rat. *Brain Research*, **293**, 319–27.

Reidling, J.C. & Said, H.M. (2005) Adaptive regulation of intestinal thiamin uptake: Molecular mechanism using wild-type and transgenic mice carrying hTHTR-1 and −2 promoters. *American Journal of Physiology (Gastrointestinal and Liver Physiology)*, **288**, G1127–34.

Reidling, J.C., Lambrecht, N., Kassir, M. & Said, H.M. (2010) Impaired intestinal vitamin B_1 (thiamin) uptake in thiamin transporter-2-deficient mice. *Gastroenterology*, **138**, 1802–9.

Reuler, J.B., Girard, D.E. & Cooney, T.G. (1985) Wernicke's encephalopathy. *New England Journal of Medicine*, **312**, 1035–9.

Ricci, V. & Rindi, G. (1992) Thiamin uptake by rat isolated enterocytes: Relationship between transport and phosphorylation. *Archives Internationales de Physiologie, de Biochimie et de Biophysique*, **100**, 275–9.

Rindi, G. (1984) Thiamin absorption by small intestine. *Acta Vitaminologica et Enzymologica*, **6**, 47–55.

Said, H.M., Ortiz, A., Kumar, C.K., Chatterjee, N., Dudeja, P.K. & Rubin, S. (1999) Transport of thiamine in human intestine: Mechanism and regulation in intestinal epithelial cell model Caco-2. *American Journal of Physiology: Cell Physiology*, **277**, C645–51.

Said, H.M., Ortiz, A., Subramanian, V.S., Neufeld, E.J., Moyer, M.P. & Dudeja, P.K. (2001) Mechanism of thiamine uptake by human colonocytes: Studies with cultured colonic epithelial cell line NCM460. *American Journal of Physiology (Gastrointestinal and Liver Physiology)*, **281**, G144–50.

Said, H.M., Reidling, J.C. & Ortiz, A. (2002) Cellular and molecular aspects of thiamin uptake by human liver cells: Studies with cultured HepG2 cells. *Biochimica et Biophysica Acta*, **1567**, 106–12.

Said, H.M., Balamurugan, K., Subramanian, V.S. & Marchant, J.S. (2004) Expression and functional contribution of hTHTR-2 in thiamin absorption in human intestine. *American Journal of Physiology (Gastrointestinal and Liver Physiology)*, **286**, G491–8.

Sasa, M., Takemoto, I., Nishino, K. & Itokawa, Y. (1976) The role of thiamine on excitable membrane of crayfish giant axon. *Journal of Nutritional Science and Vitaminology*, **22**, 21–46.

Schenker, S., Johnson, R.F., Hoyumpa, A.M. & Henderson, G.I. (1990) Thiamine-transfer by human placenta: Normal transport and effects of ethanol. *Journal of Laboratory and Clinical Medicine*, **116**, 106–15.

Singh, M. (1982) Effect of thiamin deficiency on pancreatic acinar cell function. *American Journal of Clinical Nutrition*, **36**, 500–4.

Spector, R. (1976) Thiamine transport in the central nervous system. *American Journal of Physiology*, **230**, 1101–7.

Spector, R. (1982) Thiamin homeostasis in the central nervous system. *Annals of the New York Academy of Sciences*, **378**, 344–54.

Subramanian, V.S., Subramanya, S.B. & Said, H.M. (2012) Relative contribution of THTR-1 and THTR-2 in thiamin uptake by pancreatic acinar cells: Studies utilizing *Slc19a2* and *Slc19a3* knockout mouse models. *American Journal of Physiology (Gastrointestinal and Liver Physiology)*, **302**, G572–8.

Subramanya, S.B., Subramanian, V.S., Sekar, V.T. & Said, H.M. (2011) Thiamin uptake by pancreatic acinar cells: Effect of chronic alcohol feeding/exposure. *American Journal of Physiology (Gastrointestinal and Liver Physiology)*, **301**, G896–904.

Tanaka, C. & Cooper, J.R. (1968) The fluorescent microscopic localization of thiamine in nervous tissue. *Journal of Histochemistry and Cytochemistry*, **16**, 362–5.

Thornalley, P.J., Babaei-Jadidi, R., Al Ali, H., Rabbani, N., Antonysunil, A., Larkin, J., Ahmed, A., Rayman, G. & Bodmer, C.W. (2007) High prevalence of low plasma thiamine concentration in diabetes linked to a marker of vascular disease. *Diabetologia*, **50**, 2164–70.

Thornber, E.J., Dunlop, R.H. & Gawthorne, J.M. (1980) Thiamin deficiency in the lamb: Changes in thiamin phosphate esters in the brain. *Journal of Neurochemistry*, **35**, 713–7.

Tomasulo, P.A., Kater, R.M.H. & Iber, F.L. (1968) Impairment of thiamine absorption in alcoholism. *American Journal of Clinical Nutrition*, **21**, 1341–4.

White, H.L., Howard, J.L., Cooper, B.R., Soroko, F.E., McDermed, J.D., Ingold, K.J. & Maxwell, R.A. (1982) A novel inhibitor of gamma-aminobutyrate aminotransferase with anorectic activity. *Journal of Neurochemistry*, **39**, 271–3.

Zhou, Z., Robards, K., Helliwell, S. & Blanchard, C. (2002) Composition and functional properties of rice. *International Journal of Food Science and Technology*, **37**, 849–68.

20 Flavins
Riboflavin, FMN, and FAD (Vitamin B_2)

20.1 CHEMISTRY

The principal vitamin B_2-active flavins found in nature are riboflavin, riboflavin-5′-phosphate (flavin mononucleotide, FMN), and riboflavin-5′-adenosyldiphosphate (flavin adenine dinucleotide, FAD). The structures of these compounds are depicted in Figure 20.1. The parent riboflavin molecule comprises a dimethyl-isoalloxazine fused ring structure attached to the sugar ribitol. FMN and FAD are commonly referred to as flavin nucleotides.

In biological tissues, FAD and, to a lesser extent, FMN occur almost entirely as prosthetic groups for a large variety of flavin enzymes (flavoproteins). In most flavoproteins the flavins are bound tightly but non-covalently to the apoenzyme. In mammalian tissues less than 10% of the FAD is covalently attached to specific amino acid residues of four important apoenzymes. These are found within succinate and sarcosine dehydrogenases, monoamine oxidase, and gulonolactone oxidase in which FAD is peptide-linked to an *N*-histidyl or *S*-cysteinyl residue via the 8-methyl group.

The term 'riboflavin' is ambiguous: it is either the generic descriptor for all biologically active flavins or it refers specifically to the parent riboflavin molecule. The term riboflavin is unambiguous when the specific riboflavin molecule is used and measured experimentally. Otherwise, it is up to the reader to discern the appropriate alternative according to the context in which the term is used. In this text, flavins or vitamin B_2 are used as generic terms.

20.2 FOOD SOURCES AND BIOAVAILABILITY

20.2.1 Food Sources

Living cells require FMN and FAD as the prosthetic groups of a variety of enzymes, and hence the flavins are found, at least in small amounts, in all natural unprocessed foods. In most foods the predominant form of vitamin B_2 is protein-bound FAD. Yeast extract is exceptionally rich in vitamin B_2, and liver and kidney are also rich sources. Wheat bran, eggs, meat, milk, and cheese are important sources in diets containing these foods. Cereal grains contain relatively low concentrations of flavins, but are important sources in those parts of the world where cereals constitute the staple diet. The milling of cereals results in considerable loss (up to 60%) of vitamin B_2, so white flour is enriched by addition of the vitamin. The enrichment of bread and breakfast cereals contributes significantly to the dietary supply of vitamin B_2. Polished rice is not usually enriched, since the yellow color of the vitamin would make the rice visually unacceptable. However, most of the flavin content of the whole brown rice is retained if the rice is steamed prior to milling. This process drives the water-soluble vitamins in the germ and aleurone layers into the endosperm (Cooperman & Lopez, 1991).

20.2.2 Bioavailability

The bioavailability of vitamin B_2 appears to be related to the digestibility of the food, being generally high for meat and dairy products and lower for plant products. Bioavailability is impaired by the excessive consumption of alcohol, which appears to inhibit the intestinal FMN phosphatase and FAD pyrophosphatase necessary to release riboflavin for absorption (Pinto *et al.*, 1984).

20.3 DIGESTION AND ABSORPTION OF DIETARY VITAMIN B_2

The absorption of riboflavin is efficient along the length of the small intestine (i.e. jejunum and ileum) in the guinea pig (Hegazy & Schwenk, 1983) and the rabbit (Said *et al.*, 1993a). The FMN and FAD present in the ingested food are released from non-covalent binding to flavoproteins as a consequence of acidification in the stomach and gastric and intestinal proteolysis. Riboflavin is similarly released from its association with binding proteins (Merrill *et al.*, 1981).

The flavin coenzymes are hydrolyzed in the upper small intestine to free riboflavin, which is then absorbed. Hydrolysis of both FMN and FAD is effected by alkaline phosphatase (EC 3.1.3.1), which has a broad specificity and is located on the brush-border membrane of the enterocyte (Daniel *et al.*, 1983a). Two additional brush-border enzymes, FMN phosphatase and FAD pyrophosphatase, participate in the degradation of the flavin coenzymes (Akiyama *et al.*, 1982). The considerably smaller amounts of covalently bound flavins are released as 8α-(peptidyl) riboflavins, which are absorbed along with the free riboflavin (Chia *et al.*, 1978).

In vitro studies using rat everted jejunal sacs have shown that absorption of riboflavin takes place by a saturable, energy-dependent process at physiologically relevant concentrations and by simple diffusion at higher concentrations (Daniel *et al.*, 1983b; Said *et al.*, 1985; Middleton,

DOI: 10.1201/9781003410300-24

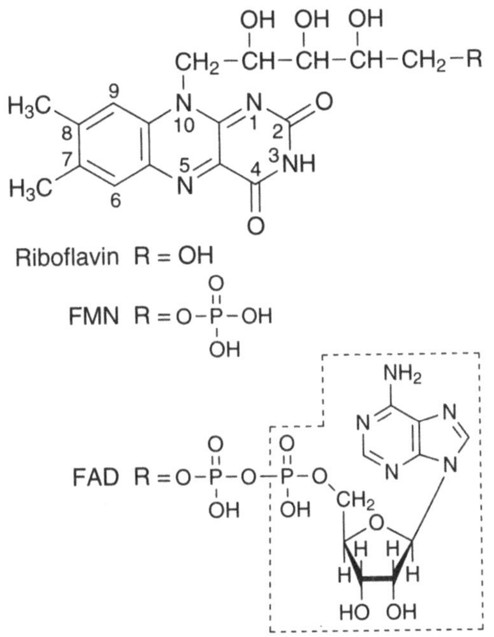

FIGURE 20.1 Structures of riboflavin, FMN, and FAD.

1990). This dual process of absorption has been confirmed under *in vivo* conditions (Feder *et al.*, 1991). Absorption is enhanced by the presence of bile salts in the intestinal lumen (Mayersohn *et al.*, 1969). A role of bile salts is also indicated by impaired absorption in cases of biliary obstruction in children (Jusko *et al.*, 1971). Bile salts may increase the absorption of riboflavin in two ways: (1) the detergent action of bile salts enhances the solubility of riboflavin in the intestinal lumen and increases the permeability of the brush-border membrane to the vitamin; and (2) bile salts inhibit gastric emptying and proximal intestinal transit, resulting in an increased residence time of riboflavin at the absorption sites.

Transport of riboflavin across brush-border and basolateral membrane vesicles prepared from rabbit small intestine was found to be carrier-mediated, Na^+-independent, and electroneutral (Said *et al.*, 1993a, b). Using human-derived Caco-2 intestinal epithelial cells, Said and Ma (1994) confirmed the involvement of a carrier-mediated process in the initial phase of riboflavin uptake (a three-minute incubation time). Riboflavin uptake was Na^+- and pH-independent and the initial phase occurred without metabolic alteration of the transported riboflavin. Inhibitors of anion transport did not produce inhibition of riboflavin uptake by Caco-2 cells, thus riboflavin does not appear to act as an anion with regard to its intestinal transport.

Three human riboflavin transporters (hRFTs) have been cloned and characterized, namely, hRFT-1 (Yonezawa *et al.*, 2008), hRFT-2 (Yamamoto *et al.*, 2009) and hRFT-3 (Yao *et al.*, 2010). All three transporters are expressed in the small intestine, with hRFT-2 expression being significantly higher than that of the other two. Subramanian *et al.* (2011) showed that hRFT-2 is exclusively expressed at the apical membrane, where it facilitates riboflavin entry into the absorptive epithelium; hRFT-1 is mainly expressed at the basolateral membrane; while hRFT-3 is mostly localized inside intracellular vesicular structures, with some expression at the basolateral membrane. Expression of RFT-1 and to some extent RFT-3 at the basolateral membrane suggests that these transporters are responsible for regulating riboflavin exit out of absorptive epithelia, and possibly for riboflavin uptake from blood in the absence of luminal riboflavin.

Some of the absorbed riboflavin is phosphorylated to FMN within the cytosol of the enterocyte by flavokinase (ATP:riboflavin 5′-phosphotransferase, EC 2.7.1.26) and most of the FMN is further converted to FAD by FAD synthetase (ATP:FMN adenylyltransferase, EC 2.7.7.2). Both of these metabolic steps are energy-dependent, requiring ATP. Gastaldi *et al.* (1999) investigated the energy dependency of the riboflavin uptake process in isolated rat enterocytes by comparing de-energized cells (cells treated with rotenone) with normal cells. Short (3 min) and long (20 min) incubation times were selected as these times represent membrane events and intracellular metabolic events, respectively. The results showed that in the initial (3 min) phase, the saturable uptake of [^{3}H]riboflavin is mainly an energy-independent process with high affinity and low capacity, whereas in the later (20 min) phase the saturable uptake is strictly energy-dependent and has an increased capacity. The results further showed that transport in the initial phase is due solely to riboflavin binding to carrier proteins on the brush-border membrane. Saturable uptake in the later phase is due to high-affinity binding of riboflavin to the cytosolic enzymes, flavokinase and FAD synthetase. The enzymatic conversion of riboflavin to FMN and FAD is ATP-dependent and this accounts for the energy dependency of the later phase. Intracellular phosphorylation of riboflavin was confirmed as the crucial and limiting step in riboflavin transport by isolated rat enterocytes (Gastaldi *et al.*, 2000).

To summarize, the small intestine is well adapted to completely extracting the small amounts of riboflavin that are largely bound within the ingested flavin coenzymes. The coenzymes are dephosphorylated in the lumen and the liberated riboflavin is extracted very efficiently by a high-affinity, carrier-mediated transport system, which is distributed along the entire length of the small intestine. The uptake mechanism is Na^+-independent and electroneutral in nature. After uptake, some of the riboflavin is metabolically trapped within the enterocyte as FMN. The energy used in riboflavin absorption is not required for membrane uptake, but rather for riboflavin metabolism within the enterocyte. Thus, intracellular metabolism is probably the driving force behind the internalization of riboflavin. The vitamin is dephosphorylated to permit its exit across the basolateral membrane; this also takes place by a carrier-mediated, Na^+-independent and electroneutral mechanism.

20.3.1 Adaptive Regulation of Riboflavin Absorption

Feeding rats with a vitamin B_2-deficient diet caused a significant and specific up-regulation in riboflavin uptake by rat intestinal brush-border membrane vesicles compared with controls. Conversely, over-supplementing rats with riboflavin caused a down-regulation in the vitamin's uptake. Both up- and down-regulation were mediated through changes in the number of the functional riboflavin uptake carriers and/or their activity (decreased or increased V_{max}) with no effect on the affinity of the transport system (unchanged K_m) (Said & Mohammadkhani, 1993). Similarly. growing Caco-2 monolayers in a riboflavin-deficient medium caused significant enhancement of riboflavin uptake, whereas an over-supplemented medium suppressed riboflavin uptake (Said & Ma, 1994). This adaptive up- or down-regulation of riboflavin uptake was further studied by Said *et al.* (1994), who established a role for the 'second messenger' cyclic AMP (cAMP) in the regulation mechanism. They found that compounds that increased intracellular cAMP concentration through different mechanisms caused a significant and concentration-dependent inhibition (down-regulation) in riboflavin uptake. This inhibition of riboflavin uptake was in contrast to other findings of stimulation of intestinal uptake of D-glucose by these compounds (Sharp & Debnam, 1994), indicating that the effect of cAMP-stimulating compounds on riboflavin intestinal uptake is not generalized in nature.

20.4 ABSORPTION OF BACTERIALLY SYNTHESIZED VITAMIN B_2 IN THE LARGE INTESTINE

The normal microflora of the large intestine synthesize considerable amounts of vitamin B_2, a significant portion of which exists as free riboflavin. The amount of vitamin B_2 synthesized depends on the diet, being significantly higher following consumption of a plant-based diet compared with a meat-based diet (Iiuma, 1955). In a study with human subjects, Sorrell *et al.* (1971) showed that riboflavin instilled directly into the lumen of the mid-transverse colon was absorbed, as judged by an increase in plasma riboflavin concentrations. Colonic absorption of FMN sodium has also been demonstrated in the rat (Kasper, 1970).

Said *et al.* (2000) demonstrated the existence of a high-affinity, carrier-mediated transport system in the large intestine using cultured human colonic epithelial cells (NCM460 cells). Saturable uptake of riboflavin by these cells was energy-dependent and Na^+-independent. Transport was regulated by the Ca^{2+}/calmodulin cell signaling pathway but not by the protein kinase C pathway. An adaptive up- and down-regulation of riboflavin uptake took place when NCM460 cells were grown in a riboflavin-deficient or over-supplemented medium, respectively. These findings are similar to those in the small intestine and suggest that the same mechanism may be operating in the large intestine to absorb the bacterially synthesized riboflavin.

20.5 POST-ABSORPTIVE METABOLISM

Following absorption, B_2 vitamers are carried by the portal blood to the liver. About 50% of circulating flavins is riboflavin, with somewhat less FAD and less than 10% FMN. The concentration of riboflavin in human plasma is about 0.03 μM on average (McCormick, 1989). A proportion of the circulating flavins is bound loosely to albumin and tightly to some immunoglobulins. The extent to which flavins are bound to plasma proteins is not believed to be crucial in regulating tissue availability of the vitamin (White & Merrill, 1986). Erythrocytes contain four to five times more flavin than plasma. There is a relatively slow equilibration of free riboflavin between the plasma and erythrocytes, hence flavin levels in erythrocytes are less subject to recent dietary intake. For this reason, the activity of an erythrocyte enzyme, glutathione reductase, is used as an indicator of vitamin B_2 status.

Uptake of riboflavin by human-derived cultured liver cells is by means of a carrier-mediated, energy-dependent, Na^+-independent system which appears to be regulated by an intracellular Ca^{2+}/calmodulin-mediated transduction pathway and by substrate level in the growth medium (Said *et al.*, 1998). The liver is the major storage site of the vitamin and contains about one-third of the total body flavins, 70–90% of which is in the form of FAD. Free riboflavin constitutes less than 5% of the stored flavins. Other storage sites are the spleen, kidney, and cardiac muscle. These depots maintain significant amounts of the vitamin even in severe deficiency states.

Cellular entry of riboflavin in all tissues is followed by phosphorylation to FMN in an irreversible reaction (reaction 20.1) catalyzed by flavokinase.

$$\text{Riboflavin} + \text{ATP} \xrightarrow{\text{Flavokinase}} \text{FMN} + \text{ADP} \quad (20.1)$$

Some FMN is utilized by certain flavoenzymes; the remainder undergoes a reversible reaction (reaction 20.2) with ATP to give FAD and inorganic pyrophosphate. FAD contributes 70–90% of total flavins in most tissues.

$$\text{FMN} + \text{ATP} \overset{\text{FAD synthetase}}{=\!=\!=\!=} \text{FAD} + \text{PPi} \quad (20.2)$$

FAD is inhibitory to the synthetase and thereby confers regulation upon its own synthesis (Yamada *et al.*, 1990).

Free FAD that is not bound to apoenzyme (enzyme protein) is rapidly hydrolyzed to FMN and AMP (reaction 20.3) and free FMN is rapidly hydrolyzed to riboflavin and inorganic phosphate (reaction 23.4).

$$\text{FAD} \xrightarrow{\text{Nucleotide pyrophosphatase}} \text{FMN} + \text{AMP} \quad (20.3)$$

$$\text{FMN} \xrightarrow{\text{Non-specific phosphatases}} \text{Riboflavin} + \text{Pi} \quad (20.4)$$

The activities of enzymes involved in the specific biosynthesis of FMN and FAD are depressed by riboflavin deficiency, while those involved in the non-specific degradation of the flavocoenzymes are not affected by riboflavin status (Lee & McCormick, 1983).

20.6 RENAL REABSORPTION AND SECRETION

In the kidney, riboflavin that has been filtered through the glomerulus can undergo either reabsorption or secretion in the proximal tubule before it is finally excreted in the urine. Reabsorption is the process by which riboflavin in the tubular lumen enters the epithelial cell via the brush-border membrane, exits the cell at the basolateral membrane, and diffuses into the peritubular capillary. Secretion is the process by which riboflavin in the peritubular capillary is transported across the epithelial cell and into the tubular lumen in the opposite direction to reabsorption. As at least 40% of plasma riboflavin is thought to be unbound in humans (Rose, 1988), renal reabsorption is an important conservation mechanism. Renal secretion removes excess riboflavin from the body at a rate several times higher than the glomerular filtration rate. The kidney, together with the intestine, therefore plays an important role in maintaining riboflavin homeostasis in the body.

Kumar *et al.* (1998) studied riboflavin uptake by cultures of human-derived renal proximal tubule epithelial cells. They demonstrated uptake via an energy-dependent, Na^+-independent, carrier-mediated system that adapted according to the concentration of riboflavin in the growth medium. The adaptive regulatory effect of riboflavin was mediated via changes in the number and/or activity as well as affinity of the riboflavin uptake carriers. Using specific modulators of intracellular signal transduction pathways, it was shown that protein kinase A, protein kinase C, and protein tyrosine kinase were not involved in regulating riboflavin uptake. In contrast, inhibition of the Ca^{2+}/calmodulin signal transduction pathway resulted in a significant inhibition of riboflavin uptake, implicating this system in the regulation of riboflavin transport. The effect of one inhibitor, calmidazolium, appeared to be mediated through decreases in both the number/activity and affinity of the riboflavin uptake carriers.

20.7 TRANSPORT IN THE BRAIN

The concentration of total riboflavin in brain, unlike in liver and kidney, is maintained relatively constant even in the face of severe vitamin B_2 deficiency or after massive doses of intravenous riboflavin. Because CSF is constantly leaving the central nervous system, riboflavin must be continually supplied to the newly formed fluid. As in other tissues, riboflavin that enters the brain is enzymatically phosphorylated to FMN, which can then be converted to FAD. In rat brain, more than 90% of the total riboflavin is present as FAD and FMN. About 10% of the total riboflavin in the brain of normal rats turns over per hour.

Riboflavin, but not FMN or FAD, enters the central nervous system principally through the blood–brain barrier (Spector, 1980a). Uptake of [^{14}C]riboflavin by rabbit brain slices *in vitro* took place by a saturable system that depended on the conversion of accumulated riboflavin to FMN and FAD (Spector, 1980b). These enzymatic conversions require ATP. Energy dependency was shown by the inhibition of uptake by dinitrophenol and low-temperature (1°C) incubation. The system was one-half saturated at the normal plasma concentration of riboflavin (~0.03 μM). This means that the entry of excessive amounts of riboflavin from blood to brain is prohibited.

Spector and Boose (1979) studied the capacity of the isolated choroid plexus, the anatomical locus of the blood–CSF barrier, to transport [^{14}C]riboflavin. With concentrations of [^{14}C]riboflavin of 0.7 μM or greater in the incubation medium, the choroid plexus accumulated [^{14}C]riboflavin against a large concentration gradient, thus demonstrating active transport. Uptake did not depend on intracellular binding or phosphorylation of the vitamin. The half-saturation concentration (K_m) was 78 μM (cf. normal plasma concentration of riboflavin of ~0.03 μM), which means that the system has the potential to transport high concentrations of riboflavin before it becomes saturated. Studies using the isolated choroid plexus do not show direction of transport.

Spector (1980a) injected [^{14}C]riboflavin directly into the ventricular CSF of anesthetized rabbits. Some of the riboflavin entered the brain by a saturable accumulation system that depended in part on phosphorylation of the riboflavin, but the majority of the labeled vitamin left the CSF extremely rapidly and was not found in the brain. The disappearance of the injected riboflavin means that the choroid plexus must have actively transported the vitamin from the CSF into the bloodstream. Direct pictorial evidence of this property of the choroid plexus was provided by fluorescence microscopy (Spector, 1980c).

In conclusion, the controlled entry and exit of riboflavin from the central nervous system provide the means for maintaining total riboflavin homeostasis in brain cells. The main entry is via the blood–brain barrier followed by high-affinity saturable uptake by brain cells and conversion of riboflavin to FMN, which can then be converted to FAD. The choroid plexus actively transports riboflavin from blood into the CSF, but perhaps more importantly it has the capacity to transport excess riboflavin in the reverse direction.

20.8 PLACENTAL TRANSPORT

During pregnancy, a riboflavin carrier protein (RCP) synthesized by the maternal liver is required for transporting riboflavin in maternal blood across the placental chorionic villus (Adiga, 1994). A specific membrane receptor for RCP mediates endocytosis of the RCP–riboflavin complex and the riboflavin is released into the fetal blood. Sustained production of RCP is dictated by the concentration of circulating estrogen, which binds to specific receptors on the hepatocytes. Estrogen is produced by the ovary in response to stimulation by follicle-stimulating hormone (FSH) and luteinizing hormone (LH). These gonadotrophic hormones are secreted from the anterior pituitary gland in response to hypothalamic releasing hormones, FSH-RH and LH-RH. The placenta also synthesizes and secretes RCP under the influence of estrogen secreted directly by the placenta. This placental RCP carries the endocytosed riboflavin to the fetus. The placental estrogen reaches the fetal circulation and induces RCP synthesis in the fetal liver. This fetal RCP conserves the riboflavin in the fetal circulation for efficient utilization by the rapidly growing fetus.

20.9 BIOCHEMICAL FUNCTIONS

In mammalian tissues, the flavins FAD and FMN function as coenzymes for a wide variety of enzymes known as flavoproteins, which are concerned with oxidoreductive processes. Most of these enzymes require FAD rather than FMN as the coenzyme. The flavins are perhaps better classed as prosthetic groups because they remain tightly bound to the apoenzyme during the catalytic reactions. In most flavoproteins the binding is non-covalent. Less than 10% of the FAD is covalently attached to the protein (apoenzyme) of four important enzymes: succinate and sarcosine dehydrogenases, monoamine oxidase, and L-gluconolactone oxidase. The means of attachment is a peptide link to an *N*-histidyl or *S*-cysteinyl residue via the 8-methyl group of FAD.

The flavoproteins play a wide variety of roles in intermediary metabolism. For example, in the oxidative decarboxylation of pyruvate to acetyl-CoA and α-ketoglutarate to succinyl-CoA in the tricarboxylic acid cycle, FAD serves as an intermediate electron carrier, passing the electrons to NAD and forming NADH. The NADH is reoxidized by donating its electrons to FMN incorporated within the electron transport chain.

In the following reactions, there is no involvement of NAD, $FADH_2$ being the final reduced product. In the β-oxidation of fatty acids FAD is a coenzyme for the transferase which catalyzes the oxidation of saturated fatty acyl CoA to α,β-unsaturated fatty acyl-CoA (Figure 20.2). FAD is also a coenzyme for succinate dehydrogenase which catalyzes the oxidation of succinate to fumarate in the tricarboxylic acid cycle. The $FADH_2$ produced from these reactions is reoxidized by passing its electrons to coenzyme Q in the electron-transport chain.

As a coenzyme for an oxidase such as xanthine oxidase, FAD transfers electrons directly to oxygen with the formation of hydrogen peroxide. Xanthine oxidase catalyzes two steps in the formation of uric acid.

In addition, flavin coenzymes are necessary for reactions involving other vitamins. For example, an FMN-dependent oxidase – pyridoxine phosphate oxidase – catalyzes the conversion of pyridoxine phosphate and pyridoxamine phosphate to pyridoxal phosphate, the primary vitamin B_6 coenzyme. An FAD-dependent hydroxylase is involved in the conversion of kynurenine to 3-hydroxykynurenine in the pathway of NAD synthesis from tryptophan. In folate metabolism an FAD-dependent dehydrogenase reduces 5,10-methylenetetrahydrofolate to 5-methyltetrahydrofolate, which interfaces with the vitamin B_{12}-dependent formation of methionine from homocysteine.

20.10 REGULATION OF RIBOFLAVIN CONVERSION TO FMN AND FAD BY THYROID HORMONE

The tight binding of FMN or FAD to their respective apoenzymes confers stability to the holoenzyme and an insufficiency of coenzyme leads to loss of enzyme activity through proteolysis. There is evidence (Rivlin & Langdon, 1966; Rivlin, 1970; Lee & McCormick, 1985) that thyroid hormone regulates the conversion of riboflavin to its functional coenzyme forms. The mechanism of regulation is on biosynthetic rather than degradative steps. Hepatic flavokinase activity was diminished in hypothyroid rats, causing decreased FMN and FAD synthesis and consequent decreases in the activities of a number of FMN- and FAD-dependent enzymes. FAD synthetase activity was also diminished, but to a lesser degree than flavokinase. Furthermore, riboflavin levels were reduced in the livers of hypothyroid rats. Treatment of the hypothyroid rats with thyroid hormone restored coenzyme levels and enzyme activities to normal. In the case of hyperthyroid rats, hepatic levels of FMN and FAD were not increased above normal despite a twofold increase in the activity of flavokinase. Even when supplemental riboflavin was administered, coenzyme levels were not increased above normal.

Cimino *et al.* (1987) reported that in hypothyroid adult humans, the activity of the FAD-containing enzyme erythrocyte glutathione reductase was reduced to levels observed during riboflavin deficiency. After two weeks of therapy with thyroxine and without supplementation with riboflavin, the enzyme activity reverted to normal.

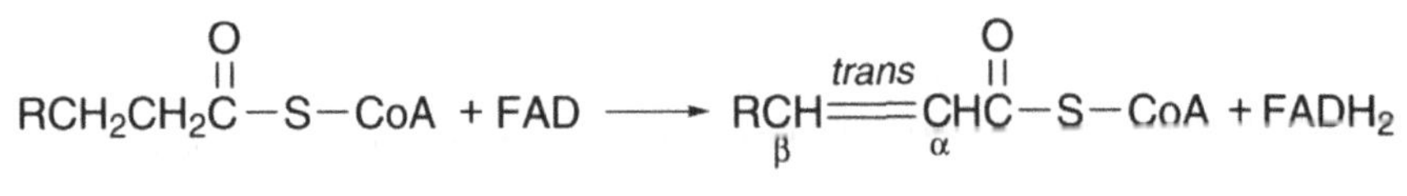

FIGURE 20.2 Oxidation of saturated fatty acyl-CoA to α,β-unsaturated fatty acyl-CoA.

It is well established that thyroid hormone increases the synthesis of protein at the transcriptional and translational levels. Yet the restoration of flavoprotein enzyme activity produced by thyroid hormone treatment in hypothyroidism was not prevented by inhibiting protein synthesis with actinomycin-D (Rivlin & Langdon, 1966). Apparently, therefore, thyroid hormone does not induce the synthesis of flavokinase apoenzyme; rather it appears to stimulate the conversion of an inactive precursor form of flavokinase to the active form or, alternatively, decrease the proteolytic conversion of active to inactive form (Lee & McCormick (1985). The reduced levels of substrate (riboflavin) could explain the reduced activity of flavokinase in hypothyroidism.

Rivlin and Langdon (1966) offered a plausible explanation for the apparent upper limit in the hepatic concentration of FMN and FAD in hyperthyroidism. The concentration of FMN/FAD remaining in the liver cells is restricted by the quantity of apoenzyme to which it can be stably bound. Excess free coenzyme would not be stored but would be destroyed enzymatically. Thus the elevated flavokinase activity seen in hyperthyroidism is not accompanied by abnormally high levels of coenzyme. From these data it appears that thyroid hormone regulates FMN and FAD synthesis by altering the activity of flavokinase. Excessive concentrations of FMN/FAD are prevented by the enzymatic destruction of coenzyme that is not stably bound to the limited amounts of apoenzyme.

20.11 EXPERIMENTAL VITAMIN B_2 DEFICIENCY IN ANIMALS

Vitamin B_2 deficiency diseases in animals have been demonstrated by experimentation with the feeding of purified diets of known composition. Due to the essential role of vitamin B_2 in the release of energy and the utilization of nutrients, a vitamin B_2 deficiency can be manifested in a great number of symptoms which vary with the animal species under study. The following deficiency signs have been observed.

- **Growth.** Reduced appetite and slow or halted growth.
- **Skin and mucous membranes.** Dry, scaly dermatitis, loss of hair, and rough coat in the pig, calf, and lamb. Seborrheic inflammation of buccal and nasal mucosae, corner of the mouth, and eyelids. Inflammation of mucous membranes of the digestive tract causes vomiting and diarrhea. Severe dilation of the duodenal crypts of Lieberkühn and slight vacuolation of pancreatic acinar cells in chicks.
- **Eye.** Sensitivity to light. Vascularization of the cornea and cataract formation in the rat.
- **Nervous system.** Spasms, flaccid paralysis, and ataxia due to myelin degeneration of peripheral nerve bundles and cranial nerves. In the chick, paralysis of the extremities is characterized by curled toes and sitting with outstretched legs.
- **Reproductive system.** In poultry, there is a fall in egg production and a lowered hatchability. Pigs exhibit sterility, fetal resorption, premature births, and stillbirths.

20.12 VITAMIN B_2 DEFICIENCY IN HUMANS

No pathologically severe symptoms attributed to vitamin B_2 deficiency have been observed in humans. Deficiency symptoms have been induced experimentally in volunteers whose diets were lacking only in the vitamin or who were fed vitamin B_2 antagonists. Symptoms usually include lesions of the lips (cheilosis) and angles of the mouth (angular stomatitis), a fissured and magenta-colored tongue (glossitis), seborrheic follicular keratosis of the nose and forehead, and dermatitis of the anogenital region. Ophthalmic symptoms are a superficial vascularization of the cornea accompanied by intense photophobia.

20.13 TOXICITY

The low solubility of riboflavin and the limited capacity of intestinal absorption mechanisms probably account for the lack of toxicity following oral doses of up to 20 mg of riboflavin daily (Cumming *et al.*, 1981).

REFERENCES

Adiga, P.R. (1994) Riboflavin carrier protein in reproduction. In: *Vitamin Receptors. Vitamins as Ligands in Cell Communication.* (Ed. by K. Dakshinamurti), pp. 137–76. Cambridge University Press, Cambridge.

Akiyama, T., Selhub, J. & Rosenberg, I.H. (1982) FMN phosphatase and FAD pyrophosphatase in rat intestinal brush borders: Role in intestinal absorption of dietary riboflavin. *Journal of Nutrition*, **112**, 263–8.

Chia, C.P., Addison, R. & McCormick, D.B. (1978) Absorption, metabolism, and excretion of 8α- (amino acid) riboflavins in the rat. *Journal of Nutrition*, **108**, 373–81.

Cimino, J.A., Jhangiani, S., Schwartz, E. & Cooperman, J.M. (1987) Riboflavin metabolism in the hypothyroid adult. *Proceedings of the Society for Experimental Biology and Medicine*, **184**, 151–3.

Cooperman, J.M. & Lopez, R. (1991) Riboflavin. In: *Handbook of Vitamins*, 2nd edn. (Ed. by L.J. Machlin), pp. 283–310. Marcel Dekker Inc., New York.

Cumming, F., Briggs, M. & Briggs, M. (1981) Clinical toxicology of vitamin supplements. In: *Vitamins in Human Biology and Medicine.* (Ed. by M.H. Briggs), pp. 187–243. CRC Press, Boca Raton, FL.

Daniel, H., Binninger, E. & Rehner, G. (1983a) Hydrolysis of FMN and FAD by alkaline phosphatase of the intestinal brush-border membrane. *International Journal of Vitamin and Nutrition Research*, **53**, 109–14.

Daniel, H., Wille, U. & Rehner, G. (1983b) In vitro kinetics of the intestinal transport of riboflavin in rats. *Journal of Nutrition*, **113**, 636–43.

Feder, S., Daniel, H. & Rehner, G. (1991) In vivo kinetics of intestinal absorption of riboflavin in rats. *Journal of Nutrition*, **121**, 72–9.

Gastaldi, G., Laforenza, U., Casirola, D., Ferrari, G., Tosco, M. & Rindi, G. (1999) Energy depletion differently affects membrane transport and intracellular metabolism of riboflavin taken up by isolated rat enterocytes. *Journal of Nutrition*, **129**, 406–9.

Gastaldi, G., Ferrari, G., Verri, A., Casirola, D., Novella Orsenigo, M. & Laforenza, U. (2000) Riboflavin phosphorylation is the crucial event in riboflavin transport by isolated rat enterocytes. *Journal of Nutrition*, **130**, 2556–61.

Hegazy, E. & Schwenk, M. (1983) Riboflavin uptake by isolated enterocytes of guinea pigs. *Journal of Nutrition*, **113**, 1702–7.

Iinuma, S. (1955) Synthesis of riboflavin by intestinal bacteria. *Journal of Vitaminology*, **1**(2), 6–13.

Jusko, W.J., Levy, G., Yaffe, S.J. & Allen, J.E. (1971) Riboflavin absorption in children with biliary obstruction. *American Journal of Diseases of Children*, **121**, 48–52.

Kasper, H. (1970) Vitamin absorption in the colon. *American Journal of Proctology*, **21**, 341–5.

Kumar, C.K., Yanagawa, N., Ortiz, A. & Said, H.M. (1998) Mechanism and regulation of riboflavin uptake by human renal proximal tubule epithelial cell line HK-2. *American Journal of Physiology*, **274**, F104–10.

Lee, S.-S. & McCormick, D.B. (1983) Effect of riboflavin status on hepatic activities of flavin-metabolizing enzymes in rats. *Journal of Nutrition*, **113**, 2274–9.

Lee, S.-S. & McCormick, D.B. (1985) Thyroid hormone regulation of flavocoenzyme biosynthesis. *Archives of Biochemistry and Biophysics*, **237**, 197–201.

McCormick, D.B. (1989) Two interconnected B vitamins: Riboflavin and pyridoxine. *Physiological Reviews*, **69**, 1170–98.

Mayersohn, M., Feldman, S. & Gibaldi, M. (1969) Bile salt enhancement of riboflavin and flavin mononucleotide absorption in man. *Journal of Nutrition*, **98**, 288–96.

Merrill, A.H. Jr., Lambeth, J.D., Edmondson, D.E. & McCormick, D.B. (1981) Formation and mode of action of flavoproteins. *Annual Review of Nutrition*, **1**, 281–317.

Middleton, H.M. III. (1990) Uptake of riboflavin by rat intestinal mucosa in vitro. *Journal of Nutrition*, **120**, 588–93.

Pinto, J., Huang, Y. & Rivlin, R. (1984) Selective effects of ethanol and acetaldehyde upon intestinal enzymes metabolizing riboflavin: Mechanism of reduced flavin bioavailability due to ethanol. *American Journal of Clinical Nutrition*, **39**, 685 (abstr).

Rivlin, R.S. (1970) Regulation of flavoprotein enzymes in hypothyroidism and in riboflavin deficiency. *Advances in Enzyme Regulation*, **8**, 239–50.

Rivlin, R.S. & Langdon, R.G. (1966) Regulation of hepatic FAD levels by thyroid hormone. *Advances in Enzyme Regulation*, **4**, 45–58.

Rose, R.C. (1988) Transport of ascorbic acid and other water-soluble vitamins. *Biochimica et Biophysica Acta*, **947**, 335–66.

Said, H.M. & Ma, T.Y. (1994) Mechanism of riboflavine uptake by Caco-2 human intestinal epithelial cells. *American Journal of Physiology*, **266**, G15–21.

Said, H.M. & Mohammadkhani, R. (1993) Uptake of riboflavin across the brush border membrane of rat intestine: Regulation by dietary vitamin levels. *Gastroenterology*, **105**, 1294–8.

Said, H.M., Hollander, D. & Duong, Y. (1985) A dual, concentration-dependent transport system for riboflavin in rat intestine *in vitro*. *Nutrition Research*, **5**, 1269–79.

Said, H.M., Mohammadkhani, R. & McCloud, E. (1993a) Mechanism of transport of riboflavin in rabbit intestinal brush border membrane vesicles. *Proceedings of the Society for Experimental Biology and Medicine*, **202**, 428–34.

Said, H.M., Hollander, D. & Mohammadkhani, R. (1993b) Uptake of riboflavin by intestinal basolateral membrane vesicles: A specialized carrier-mediated process. *Biochimica et Biophysica Acta*, **1148**, 263–8.

Said, H.M., Ma, T.Y. & Grant, K. (1994) Regulation of riboflavin intestinal uptake by protein kinase A: Studies with Caco-2 cells. *American Journal of Physiology*, **267**, G955–9.

Said, H.M., Ortiz, A., Ma, T.Y. & McCloud, E. (1998) Riboflavin uptake by the human-derived liver cells Hep G2: Mechanism and regulation. *Journal of Cellular Physiology*, **176**, 588–94.

Said, H.M., Ortiz, A., Moyer, M.P. & Yanagawa, N. (2000) Riboflavin uptake by human-derived colonic epithelial NCM460 cells. *American Journal of Physiology*, **278**, C270–6.

Sharp, P.A. & Debnam, E.S. (1994) The role of cyclic AMP in the control of sugar transport across the brush border and basolateral membranes of rat jejunal enterocytes. *Experimental Physiology*, **79**, 203–14.

Sorrell, M.F., Frank, O., Thomson, A.D., Aquino, H. & Baker, H. (1971) Absorption of vitamins from the large intestine *in vivo*. *Nutrition Report International*, **3**, 143–8.

Spector, R. (1980a) Riboflavin homeostasis in the central nervous system. *Journal of Neurochemistry*, **35**, 202–9.

Spector, R. (1980b) Riboflavin accumulation by rabbit brain slices *in vitro*. *Journal of Neurochemistry*, **34**, 1768–71.

Spector, R. (1980c) Riboflavin transport in the central nervous system. Characterization and effects of drugs. *Journal of Clinical Investigation*, **66**, 821–31.

Spector, R. & Boose, B. (1979) Active transport of riboflavin by the isolated choroid plexus *in vitro*. *Journal of Biological Chemistry*, **254**(20), 10286–9.

Subramanian, V.S., Subramanya, S.B., Rapp, L., Marchant, J.S., Ma, T.Y. & Said, H.M. (2011) Differential expression of human riboflavin transporters-1, -2, and -3 in polarized epithelia: A key role for hRFT-2 in intestinal riboflavin uptake. *Biochimica et Biophysica Acta*, **1808**, 3016–21.

White, H.B. & Merrill, A.H. Jr. (1986) Riboflavin-binding proteins. *Annual Review of Nutrition*, **8**, 279–99.

Yamada, Y., Merrill, A.H. Jr. & McCormick, D.B. (1990) Probable reaction mechanisms of flavokinase and FAD synthetase from rat liver. *Archives of Biochemistry and Biophysics*, **278**, 125–30.

Yamamoto, S., Inoue, K., Ohta, K., Fukatsu, R., Maeda, J., Yoshida, Y. & Yuasa, H. (2009) Identification and functional characterization of rat riboflavin transporter 2. *Journal of Biochemistry*, **145**, 437–43.

Yao, Y., Yonezawa, A., Yoshimatsu, H., Matsuda, S., Katsuda, T. & Inui, K. (2010) Identification and comparative functional characterization of a new human riboflavin transporter hRFT3 expressed in the brain. *Journal of Nutrition*, **140**, 1220–28.

Yonezaura, A., Masuda, S., Katsura, T. & Inui, K. (2008) Identification and functional characterization of a novel human and rat riboflavin transporter, RFT1. *American Journal of Physiology*, **295**, C632–41.

21 Niacin
Nicotinic Acid and Nicotinamide

21.1 CHEMISTRY

Throughout this text, niacin is the generic descriptor for two vitamers, nicotinic acid (pyridine-3-carboxylic acid) and its amide, nicotinamide (pyridine-3-carboxamide) (Figure 21.1). *Note.* In American terminology, the term niacin denotes specifically nicotinic acid, and niacinamide denotes the amide.

In living tissues nicotinamide is the reactive moiety of the nicotinamide nucleotide coenzymes NAD (nicotinamide adenine dinucleotide) and NADP (nicotinamide adenine dinucleotide phosphate). The notation NAD(P) is used to mean either NAD or NADP. The structure of NAD can be envisaged as the adenosine diphosphate-ribosyl moiety, hereafter abbreviated as ADP-ribose, attached covalently to nicotinamide through a β-*N*-glycosidic linkage (Figure 21.1). NAD glycohydrolases hydrolyze the *N*-glycosidic linkage of NAD, yielding free ADP-ribose, nicotinamide, and a proton. Most cellular NAD and NADP is stored in the cytoplasm bound to protein.

In mature cereal grains (those examined included maize, wheat, rice, barley, and sorghum), 85–90% of the total niacin content exists in chemically bound forms of nicotinic acid, the remainder being present as free nicotinic acid and nicotinamide (Ghosh *et al.*, 1963). Investigations by Mason *et al.* (1973) revealed that the bound nicotinic acid of wheat bran is incorporated in a number of macromolecules that are both polysaccharide and glycopeptide in character. Mason and Kodicek (1973) subjected preparations of bound nicotinic acid of wheat bran to partial acid hydrolysis and identified a subunit as nicotinoyl glucose (Figure 21.2). The nicotinoyl ester bond linking the nicotinic acid and glucose moieties most probably exists in the polysaccharide and glycopeptide fractions isolated from wheat bran, and is so far the only established linkage in bound nicotinic acid.

21.2 FOOD SOURCES AND BIOAVAILABILITY

21.2.1 Food Sources

Niacin can be synthesized in the human body from the essential amino acid L-tryptophan. Approximately 60 mg of L-tryptophan yield 1 mg of niacin; therefore, to calculate the niacin equivalent (NE) of a diet, one adds one-sixtieth of the weight of tryptophan present to the weight of preformed niacin.

Because of the contribution of tryptophan, foods containing balanced protein are important contributors to niacin intake. Lean red meat, poultry, and liver contain high levels of both niacin and tryptophan and, together with legumes, are important sources of the vitamin. Peanut butter is an excellent source of niacin. Cheese and eggs are relatively poor sources of preformed niacin, but these high-protein foods contain ample amounts of tryptophan and therefore have a high niacin equivalent. Fruits and vegetables provide useful amounts, depending upon the dietary intake. Other useful sources are wholegrain cereals, bread, tea, and coffee.

In mature cereal grains, most of the niacin is present as bound nicotinic acid and is concentrated in the aleurone and germ layers. Milling to produce white flour removes most of the vitamin with the bran. In the UK it is compulsory by law to add niacin to white flour (mostly 70% extraction rate) at 16 mg kg^{-1}. All flour other than wholemeal (100% extraction) must be enriched.

As discussed later, some plant-derived foods contain niacin in chemically bound forms that result in their bioavailabilities being low. Most food composition tables give total niacin, and are compiled from the results of analyses in which nicotinic acid is liberated from unavailable bound forms by hydrolysis with acid or alkali. Therefore, tabulated niacin contents for many plant foods, particularly mature cereals, overestimate their value in providing biologically available niacin.

21.2.2 Bioavailability

The majority of the bound nicotinic acid in mature cereal grains is biologically unavailable after conventional cooking (Wall & Carpenter, 1988). Nicotinoyl glucose itself is readily utilized, so why should this compound be unavailable when present in plant tissues? Mason and Kodicek (1973) suggested that its incorporation within indigestible celluloses and hemicelluloses prevents access of the gastrointestinal esterases to the nicotinoyl ester bonds. Alternatively, esterase activity may be poor: the methyl ester of nicotinic acid was only 15% as effective as the free acid in supporting the growth of rats (Wall & Carpenter, 1988). About 10% of the total niacin was released as free nicotinic acid after extraction of sorghum meal with 0.1 N HCl (Magboul & Bender, 1982). This suggests that a small proportion of bound nicotinic acid can be hydrolyzed by gastric juice and made available.

Pellagra, the disease caused by a deficiency of both niacin and tryptophan, has commonly been found in population groups having maize as their staple food. The generally

DOI: 10.1201/9781003410300-25

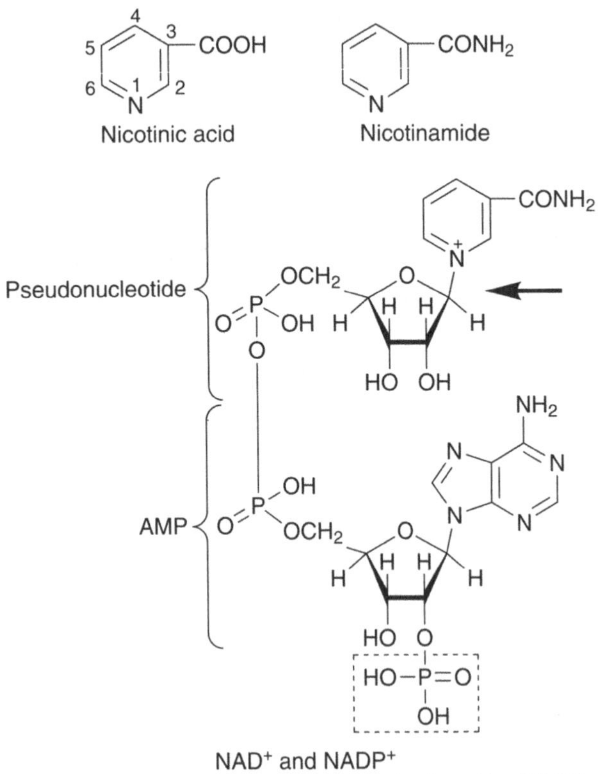

FIGURE 21.1 Structures of niacin compounds and the nicotinamide nucleotides. The nitrogen in the pyridine ring of nicotinamide is positively charged when it is a component of NAD(P). Arrow shows the β-*N*-glycosidic linkage separating the ADP-ribose and nicotinamide moieties.

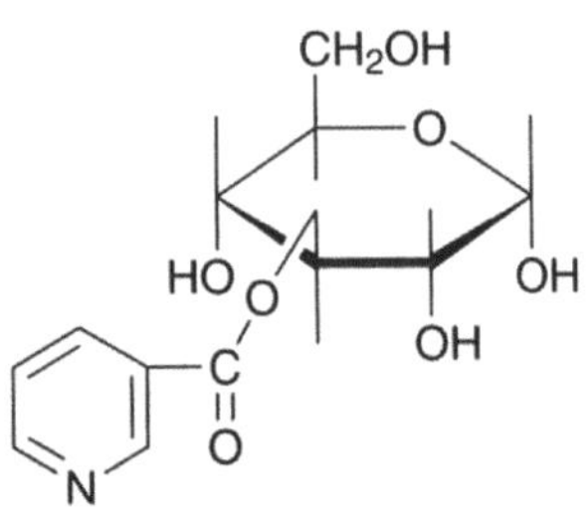

FIGURE 21.2 Structure of β-3-*O*-nicotinoyl-D-glucose.

accepted explanation for this association is the unavailability of niacin in maize, coupled with a very low proportion of tryptophan in zein (the major protein in maize). Mexican and Central American peasants, and also Hopi Indians in Arizona, rely upon maize as a staple food and yet do not experience pellagra. The explanation for this paradox lies in the way in which these people prepare the maize for bread-making. In the traditional preparation of Mexican tortillas (Cravioto *et al.*, 1945), the maize is soaked at alkaline pH in lime-water before baking and this process releases the nicotinic acid from its bound forms. In the making of piki bread the Hopi Indians use wood ash, which is alkaline and also results in the liberation of nicotinic acid. The availability of nicotinic acid in tortillas baked from maize treated with lime-water has been demonstrated in pigs by Kodicek *et al.* (1959).

21.3 DIGESTION AND ABSORPTION OF DIETARY NIACIN

21.3.1 Digestion

Much of the available niacin in the diet will be in the form of the nicotinamide nucleotides NAD and NADP, which are not transportable across cell membranes. Nicotinamide is the major (if not the only) form of niacin absorbed when nicotinamide nucleotides are ingested at the usual level of intake. Nicotinamide is absorbed with little hydrolysis to nicotinic acid and is the main circulating vitamer (Henderson & Gross, 1979). NAD(P) in animal tissues is rapidly hydrolyzed *post mortem*, so it is likely that much of the niacin in meat will be free nicotinamide (Bender & Bender, 1986).

Studies using vascularly perfused rat intestine and live animals (Gross & Henderson, 1983) have led to the following pathway of NAD(P) degradation to nicotinamide (Figure 21.3). The initial step is an attack at the phosphodiester bond of NAD to form nicotinamide mononucleotide and adenosine monophosphate (AMP). The enzyme responsible, NAD pyrophosphatase (EC 3.6.1.22), is present in intestinal secretions and to a much lesser extent in pancreatic juice. Nicotinamide mononucleotide is rapidly hydrolyzed to nicotinamide riboside and inorganic phosphate. In the final and rate-limiting step, nicotinamide riboside reacts slowly with inorganic phosphate to form nicotinamide and ribose-1-phosphate. The last two reactions require the presence of intestinal cells, indicating that the enzymes are membrane-bound or intracellular.

21.3.2 Absorption

Early studies of intestinal uptake of nicotinic acid and nicotinamide did not demonstrate any specialized transport mechanism for these substrates. The transport mechanism was attributed to simple diffusion aided by metabolic trapping. According to this theory, nicotinic acid and nicotinamide diffuse unaided into the enterocyte where they are rapidly converted to NAD. The NAD thus formed is not transportable across cell membranes. The lowering of the intracellular concentrations of nicotinic acid and nicotinamide due to the accumulation of NAD creates a concentration gradient across the brush-border membrane that increases the diffusion rates of the vitamers (Schuette & Rose, 1983).

The human sodium-coupled monocarboxylate transporter 1 (SMCT1) has a suggested role in intestinal uptake of nicotinic acid, but its affinity for substrate (apparent K_m = 0.23–0.3 mM) is low considering that nicotinic acid concentrations are in the micromolar range (Said, 2011).

Nicotinic acid has a pK_a of 4.8, which means that it will be fully ionized in the neutral pH environment in the

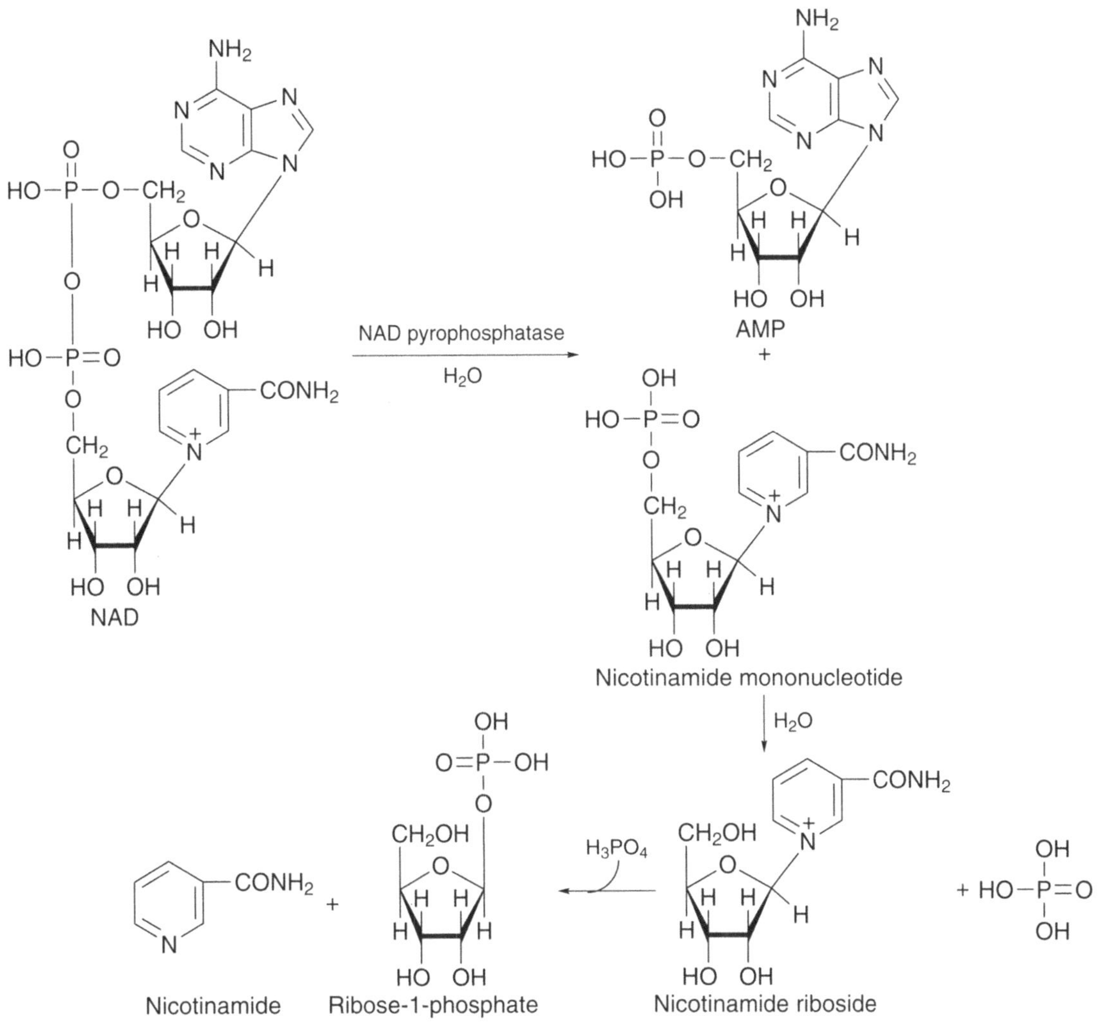

FIGURE 21.3 Pathway of NAD degradation to nicotinamide in rat intestine.

intestinal lumen and absorbed in the form of the nicotinate anion. Ionized compounds diffuse poorly through cell membranes, so simple diffusion as a significant mechanism for nicotinate absorption would not be expected. Nabokina *et al.* (2005) discovered the existence of a high-affinity, carrier-mediated mechanism for the uptake of physiological concentrations of nicotinic acid at the apical membrane of human enterocytes. *In vitro* studies with human-derived intestinal epithelial Caco-2 cells showed uptake of [^{3}H]nicotinic acid to be both temperature- and energy-dependent, and to occur with minor metabolic alterations in the transported substrate. Also, the uptake process was Na^+-independent as indicated by the inability of the Na^+–K^+-ATPase inhibitor ouabain to affect the substrate uptake. Uptake was highly dependent on the pH of the incubation buffer – decreasing the buffer pH from 8.0 to 5.0 led to a marked (~fivefold) increase in nicotinate uptake. This dependency on an acidic pH suggested the possible involvement of a nicotinate$^+$/H^+ co-transport system (symport), which was supported by the finding of significant inhibition of nicotinate uptake after pretreatment of the Caco-2 cells with a protonophore. The specificity of the uptake process for nicotinic acid was based on the observations that nicotinamide and structural analogs of nicotinic acid had no effect on substrate uptake. Saturation was observed in nicotinate uptake over the micromolar concentration range. The apparent Michaelis constant (K_m) of the saturable component was 0.53 ± 0 .08 μM, indicating a high affinity of the carrier for its substrate, and the maximal velocity (V_{max}) was 13.32 ± 0.58 pmol of substrate per mg protein per 3 min. These kinetic parameters suggested that the transport system could be responsible for the absorption of dietary nicotinic acid (estimated to be in the micromolar range). With regard to the regulation of nicotinic acid uptake by Caco-2 cells, no role for PKA-, PKC-, or Ca^{2+}/calmodulin-mediated pathways was evident; however, involvement of a protein tyrosine kinase-mediated intracellular signal transduction pathway was apparent. The relevance of these findings to the native human intestine was established by demonstrating significant inhibition of ^{3}H-nicotinic acid uptake by unlabeled niacin in purified jejunal brush-border membrane vesicles. The studies with the membrane vesicles also confirmed the dependency on an acidic pH for nicotinic acid uptake. No reports have been found concerning the mechanism involved in niacin exit out of the enterocytes across the basolateral membrane.

21.4 ABSORPTION OF BACTERIALLY SYNTHESIZED NIACIN IN THE LARGE INTESTINE

Niacin synthesized by the normal microflora of the large intestine has been largely ignored as a potential exogenous source of the vitamin due to the long-standing belief that the large intestine is incapable of absorbing it. Kumar *et al.* (2013) tested the ability of mammalian colonocytes to take up physiological concentrations of niacin. They used nicotinic acid as a substrate and non-transformed human colonic epithelial NCM460 cells, as well as native human colonic apical membrane vesicles obtained from organ donors, and intact mouse colonic loops *in vivo*. Uptake of ^{3}H-nicotinic acid showed energy and temperature dependence with little metabolism of the substrate during transport. The process was independent of sodium and dependent on an acidic pH. Pretreatment of the NCM460 cells with a protonophore inhibited nicotinate uptake, which suggested the possible involvement of a nicotinate/H^+ co-transport system. There was a significant inhibition of ^{3}H-nicotinic acid uptake by unlabeled nicotinic acid and nicotinamide, but not by other structural analogs, demonstrating an uptake specificity for nicotinic acid and nicotinamide. Uptake included a saturable component of which the apparent K_m was 2.5 ± 0.8 μM and the V_{max} was 27.4 ± 3.74 pmol substrate per mg protein per 5 min. Although no regulatory role for the intracellular PKA- and PKC-mediated pathways could be demonstrated, a role for the intracellular protein tyrosine kinase and Ca^{2+}/calmodulin-mediated pathways was observed. The uptake process was also found to be adaptively regulated by the extracellular substrate level, with higher uptake by cells maintained in the presence of low compared with high nicotinic acid concentrations. The results showed, for the first time, the existence of an efficient, specific, and regulated carrier-mediated mechanism for the uptake of nicotinic acid and nicotinamide by human and mouse colonocytes. These findings raise the possibility that the bacterially synthesized niacin in the large intestine may contribute to the overall niacin body status and especially toward the nutritional needs of the local colonocytes. This is only a possibility as the microflora of the large intestine tend to retain their niacin.

21.5 POST-ABSORPTIVE METABOLISM

The liver is the major site of NAD synthesis from either of the niacin vitamers or from tryptophan. The liver receives ingested nicotinamide, nicotinic acid, and tryptophan via the portal vein, as well as nicotinamide released into the bloodstream from extrahepatic tissues. Said *et al.* (2007) identified a high-affinity (apparent K_m of 0.73 ± 0.16 μM) carrier-mediated and regulated system for nicotinic acid transport in human hepatocytes. In the liver, the concentrations of niacin vitamers are normally high enough to saturate the enzymes involved in converting them to NAD, hence utilization of the vitamers for NAD synthesis is limited to normal hepatic concentrations (Bender *et al.*, 1982). Some of the NAD synthesized in the liver is utilized by the liver itself. Surplus NAD is hydrolyzed to nicotinamide, which is then released into the general circulation, accompanied by any nicotinic acid that was not metabolized. On reaching the tissues, the niacin vitamers are used for the intracellular synthesis of NAD and NADP. There is a continuous turnover of NAD and NADP in the body and very little storage.

21.6 EXCRETION OF NIACIN METABOLITES

Niacin vitamers in excess of requirements for NAD synthesis are converted into various metabolites, which are excreted in the urine. The metabolites excreted by humans are shown in Figure 21.4. The principal metabolite of nicotinamide is N^1-methylnicotinamide, which is produced in most tissues by the action of nicotinamide *N*-methyltransferase. N^1-methylnicotinamide can be further metabolized to N^1-methyl-2-pyridone-5-carboxamide and N^1-methyl-4-pyridone-3-carboxamide in a ratio of 5.4:1 in a normal subject (Mrochek *et al.*, 1976). Nicotinamide-N^1-oxide is normally a minor metabolite, unless large (3 g) doses of nicotinamide are ingested. Nicotinic acid is excreted mainly as nicotinuric acid by conjugation with glycine and as trigonelline (N^1-methylnicotinic acid) by methylation. Nicotinic acid can give rise to N^1-methylnicotinamide and the pyridones as a result of its incorporation into NAD(P) and subsequent hydrolysis to release nicotinamide. Similarly, nicotinamide can give rise to nicotinic acid, and hence nicotinuric acid and trigonelline, as a result of deamidation (Bender, 2003a).

21.7 RENAL REABSORPTION OF NIACIN

Both nicotinamide and nicotinic acid are actively reabsorbed from the glomerular filtrate, so at normal plasma concentrations there is little or no urinary excretion of either vitamer. A sodium-coupled transporter for nicotinic acid was demonstrated in the kidney by Schuette and Rose (1986). The molecular identity of this transporter remained elusive until Gopal *et al.* (2005) discovered that the sodium-coupled monocarboxylate transporter (SMCT) is a transporter for nicotinate and a variety of other monocarboxylates. SMCT is expressed abundantly in the colon, ileum, kidney, and thyroid gland. SMCT transports nicotinate with a Na^+/nicotinate stoichiometry of 2:1 and an apparent K_m of 296 ± 88 μM. The electrogenic nature of the transport process renders it highly active, making SMCT a candidate for the renal reabsorption of nicotinic acid. The transporter does not recognize nicotinamide as a substrate, indicating the essential nature of the free carboxylate group for interaction with the substrate-binding site.

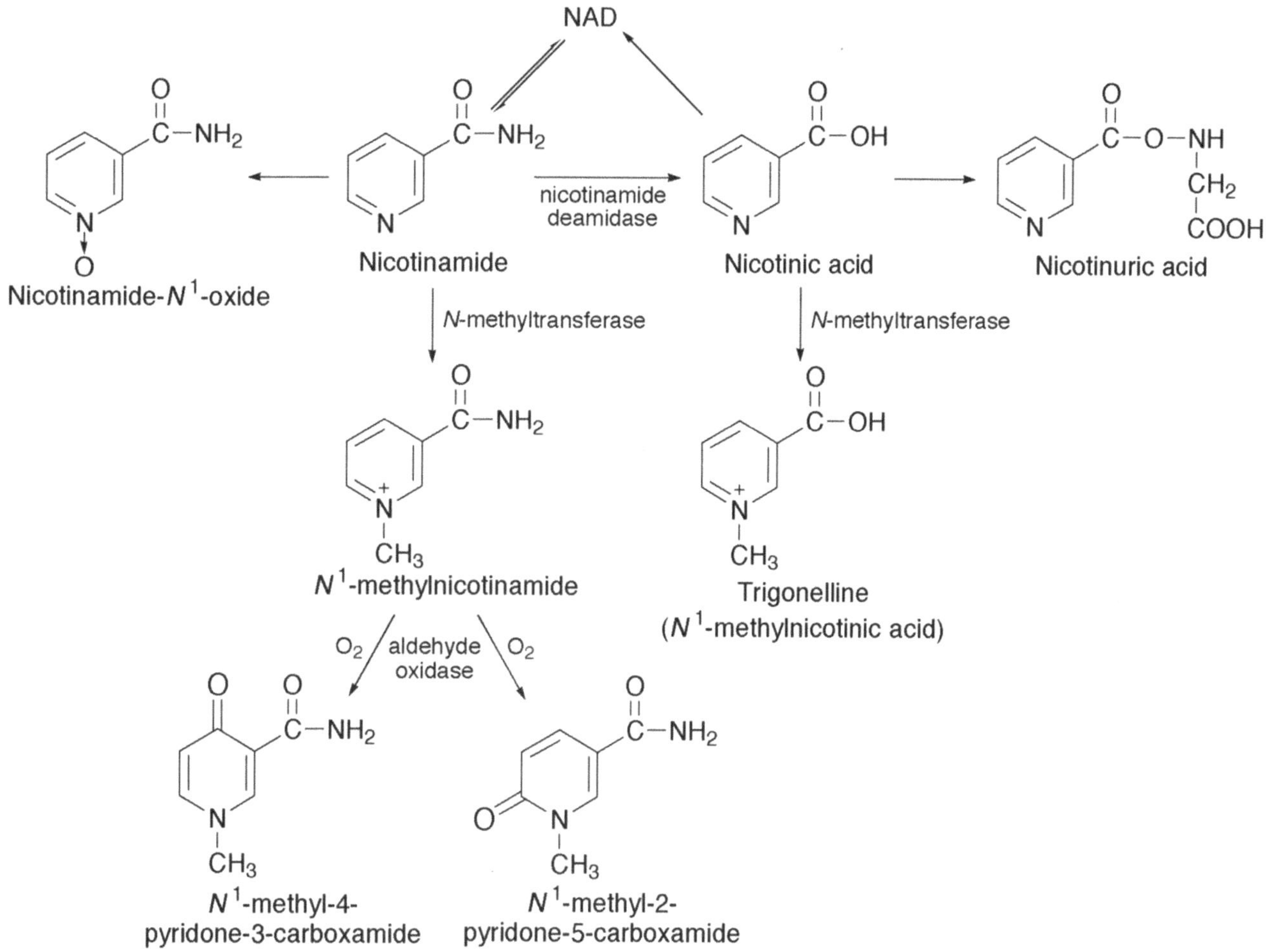

FIGURE 21.4 Metabolites of nicotinamide and nicotinic acid.

21.8 PATHWAYS OF NAD SYNTHESIS IN MAMMALS

With reference to Figure 21.5, NAD can be synthesized *de novo* from dietary sources of tryptophan (reactions 1 to 9). Under normal conditions, almost all the dietary intake of tryptophan, apart from the small amount that is used for protein synthesis, is metabolized by this pathway. Tryptophan-2,3-dioxygenase, the first enzyme in the oxidative pathway of tryptophan metabolism,, is rate-limiting under normal conditions. The synthesis of this enzyme is induced by glucocorticoids (cortisol in humans and corticosterone in the rat) and glucagon. The combined effects of these hormones in starvation, together with the tryptophan released from muscle proteins, can lead to a very considerable increase in the rate of tryptophan oxidation (Bender & Bender, 1986).

NAD can be cleaved by NAD-consuming enzymes (reaction 10) and the liberated nicotinamide can be converted to nicotinic acid (reaction 11). The Preiss-Handler pathway synthesizes NAD by recycling nicotinamide and nicotinic acid (reactions 11, 12, 8, and 9).

There are also two pathways to NAD from nicotinamide riboside, both leading to nicotinamide mononucleotide (Chi & Sauve, 2013). In one pathway, exogenous nicotinamide riboside is phosphorylated to nicotinamide mononucleotide (reaction 13); in the second pathway, nicotinamide riboside is degraded to nicotinamide (reaction 14), which is then converted to nicotinamide mononucleotide (reaction 15). The nicotinamide mononucleotide produced by both pathways is converted to NAD by nicotinamide mononucleotide adenylyl transferase (reaction 16).

21.9 TRANSPORT IN THE BRAIN

Niacin and NAD levels in brain are homeostatically regulated. In plasma, nicotinamide is normally the predominant (if not the only) form of niacin present. Transport studies using rabbit brain slices (Spector & Kelley, 1979) showed that uptake of [^{14}C]nicotinamide by brain cells *in vitro* was saturable and dependent on the production of intracellular energy. Spector (1979) reported that when [^{14}C]nicotinamide was injected into the ventricle of the brain of conscious rabbits, some of the radioactivity was incorporated into intracellular NAD and some left the brain and cerebrospinal fluid (CSF) extremely rapidly by a non-saturable system. This rapid equilibration of nicotinamide between CSF and plasma suggests that there is no control of the concentration of nicotinamide in the CSF and extracellular

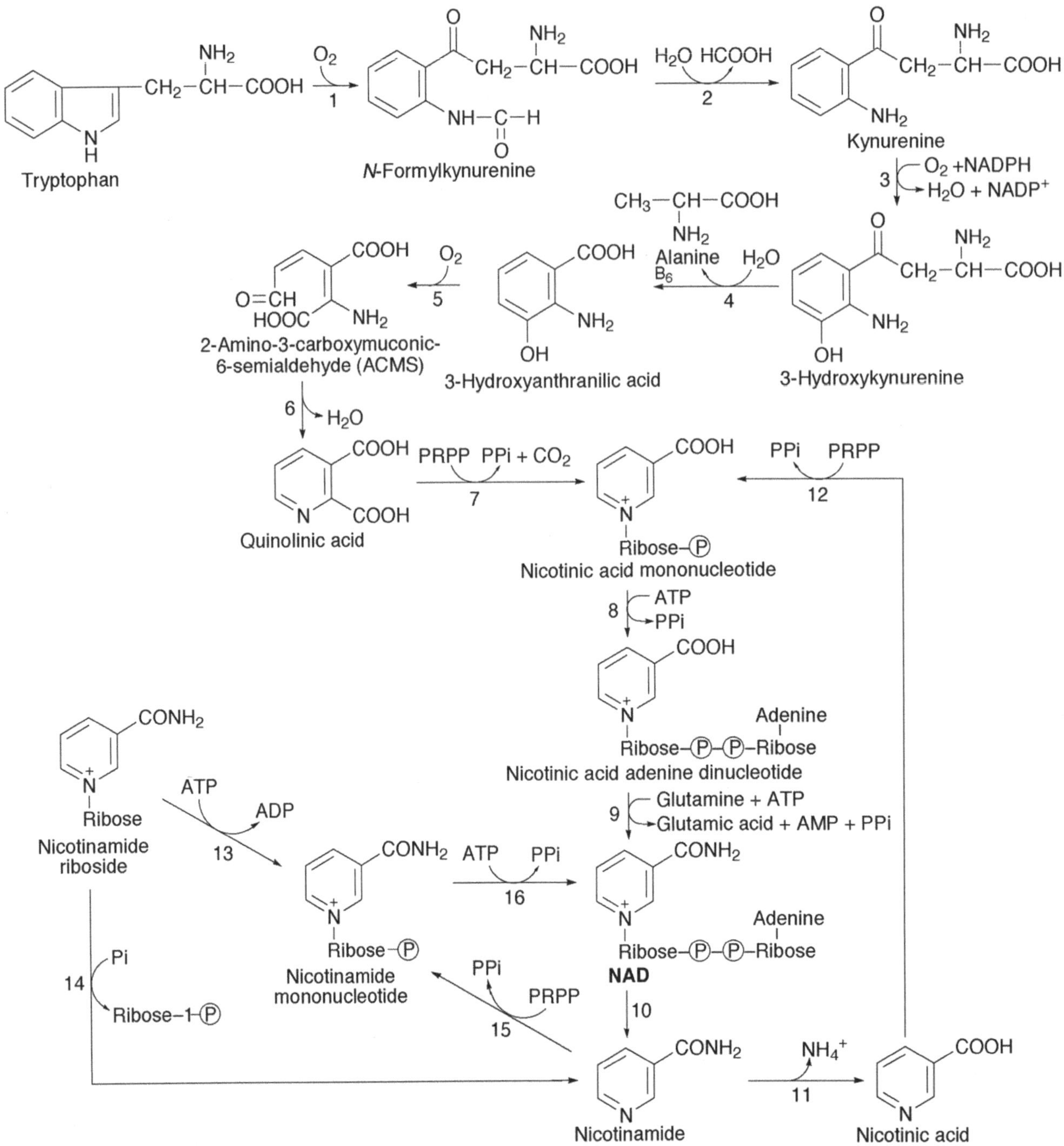

FIGURE 21.5 Pathways of NAD synthesis in mammals. NAD can be synthesized *de novo* from tryptophan and from exogenous nicotinamide riboside. There are also salvage pathways in which nicotinamide is recycled. The following enzymes correspond to the numbered reactions: 1, tryptophan-2,3- dioxygenase; 2, formamidase; 3, kynurenine-3-monooxygenase; 4, kynureninase; 5, 3-hydroxy-anthranilate-3,4-dioxygenase; 6, spontaneous non-enzymatic reaction; 7, quinolinate phosphoribosyltransferase; 8, nicotinic acid mononucleotide adenylyltransferase; 9, NAD synthetase; 10, NAD^+-consuming enzymes; 11, nicotinamide deamidase; 12, nicotinic acid phosphoribosyltransferase; 13, nicotinamide riboside kinase; 14, purine nucleoside phosphorylase; 15, nicotinamide phosphoribosyl transferase; 16, nicotinamide mononucleotide adenylyl transferase. PPRP, 5-phosphoribosyl-α-pyrophosphate; PPi, pyrophosphate.

space of brain: nicotinamide concentrations in these compartments reflect concentrations in plasma. Once within the extracellular space, nicotinamide enters brain cells by a concentration-dependent, saturable accumulation system. On entry into the brain cells, much of the nicotinamide is incorporated into NAD.

Spector (1987) measured the unidirectional influx of [^{14}C]nicotinamide across cerebral capillaries (the anatomical locus of the blood-brain barrier) using an *in situ* rat brain perfusion technique. Transport of nicotinamide was much faster than could be explained by simple diffusion alone and was not saturable with 10 mM nicotinamide in the perfusate. However, with periods of infusion longer than 30 seconds, there was substantial backflow of [^{14}C]nicotinamide into the perfusate. At a concentration of 1.7 μM, nicotinamide transport was not inhibited by 3-acetylpyridine. The non-saturability and lack of inhibition in the presence of a structural analog indicate that nicotinamide transport is not carrier-mediated. The data suggested that most, if not all, of the nicotinamide that enters brain from blood gains access to the extracellular space of brain directly via the blood–brain barrier.

From the above findings the following inferences can be made. The saturability seen in the brain slices and *in vivo* studies is due not to saturation of carrier (probably no carrier exists) but to saturation of the enzymes involved in the intracellular conversion of nicotinamide to NAD. This metabolic conversion requires ATP and so accounts for the observed energy dependency in these experiments. Niacin levels in brain are controlled by the saturable system of nicotinamide uptake by brain cells. Since CSF has a nicotinamide concentration of ~0.7 μM, the uptake system (with half-saturation concentration of 0.8 μM) is normally approximately half-saturated. This means that the entry of excessive amounts of nicotinamide into the brain is prohibited. As to the initial uptake mechanism, it can be speculated that nicotinamide that has entered the extracellular space of brain mainly via the blood–brain barrier enters brain cells by diffusion, accelerated by the favorable concentration gradient created by metabolic trapping.

21.10 ROLE OF NICOTINAMIDE COENZYMES IN OXIDATION–REDUCTION REACTIONS

The nicotinamide nucleotides, NAD and NADP, serve as coenzymes for hundreds of dehydrogenases that catalyze a wide variety of oxidation-reduction reactions. More precisely, these compounds act as co-substrates as they leave the catalytic site of the enzyme at the end of the reaction. Examples of typical reactions are listed in Table 21.1.

The nicotinamide coenzymes are not tightly bound to their apoenzymes and are not, therefore, prosthetic groups. There is, however, an important exception: glyceraldehyde-3-phosphate dehydrogenase in mammalian muscle contains a binding site for NAD in each of its four subunits and the coenzyme is found in crystalline preparations of the enzyme. This enzyme has the distinction of catalyzing the first reaction in the glycolytic pathway in which a high-energy phosphate compound is formed where none previously existed.

Although NAD and NADP undergo similar reversible reduction at the molecular level, their metabolic functions are entirely different. The NADH produced during the tricarboxylic acid cycle is reoxidized by transferring its electrons to the electron transport chain, thereby providing ATP by the coupled process of oxidative phosphorylation. NADH required for gluconeogenesis is generated via the oxidation of fatty acids and deamination of amino acids. NADP, on the other hand, participates in the pentose phosphate pathway which is not primarily a provider of energy. The resulting NADPH is used as a reducing agent to synthesize, among other materials, fatty acids, glutamate, cholesterol (precursor of steroid hormones), and deoxyribonucleotides (precursors of DNA).

21.11 ROLE OF NAD IN ADP-RIBOSYLATION

NAD functions in ADP-ribosylation, a reversible post-translational modification of proteins in which the ADP-ribose moiety of NAD is transferred to acceptor proteins, thereby altering their function. ADP-ribosylation reactions are classified into two major groups: mono(ADP-ribosyl)ation and poly(ADP-ribosyl)ation.

TABLE 21.1
Reactions Catalyzed by Nicotinamide Nucleotide Coenzymes within the Glycolytic Pathway/Tricarboxylic Acid Cycle and Pentose Phosphate Pathway of Glucose Metabolism

Enzyme	Substrate	Product	Coenzyme
Glycolysis/TCA cycle			
Glyceraldehyde-3-phosphate dehydrogenase	Glyceraldehyde-3-phosphate	1,3-Diphosphoglycerate	NAD^+
Lactate dehydrogenase	Lactate	Pyruvate	NAD^+
Pyruvate dehydrogenase	Pyruvate	Acetyl CoA + CO_2	NAD^+
Isocitrate dehydrogenase	Isocitrate	α-Ketoglutarate +CO_2	NAD^+
α-Ketoglutarate dehydrogenase	α-Ketoglutarate	Succinyl CoA + CO_2	NAD^+
Malate dehydrogenase	Malate	Oxaloacetate	NAD^+
Pentose phosphate pathway			
Glucose-6-phosphate dehydrogenase	Glucose-6-phosphate	6-Phosphogluconic acid[a]	$NADP^+$
6-Phosphogluconic acid dehydrogenase	6-Phosphogluconic acid	Ribulose-5-phosphate + CO_2	$NADP^+$

[a] The δ-lactone of phosphogluconic acid is formed initially.

21.11.1 Mono(ADP-Ribosyl)ation

In mono(ADP-ribosyl)ation, the ADP-ribose moiety is split off from an NAD molecule by hydrolysis of the high-energy β-*N*-glycosidic bond and enzymatically transferred to a specific amino acid of an acceptor protein with the simultaneous release of nicotinamide (Figure 21.6). The mono-ADP-ribosyltransferases are either ectoenzymes that modify cell surface proteins, such as integrins and defensins, or endoenzymes that act on intracellular proteins involved in cell signaling and metabolism.

21.11.2 Role of Poly(ADP-Ribosyl)ation in the Repair of DNA Strand Breaks

Poly(ADP-ribosyl)ation facilitates the repair of DNA strand breaks, including those produced by enzymatic excision of damaged DNA. The repair process is initiated by poly(ADP-ribose) polymerase-1 (PARP-1), which binds to and is activated by single- and double-strand DNA breaks. PARP-1 utilizes NAD for the synthesis of ADP-ribose polymers, which are covalently attached to several nuclear proteins, including PARP-1 itself (automodification). DNA damage, particularly DNA strand breaks, stimulates NAD biosynthesis. Since NAD is required for the synthesis of poly(ADP-ribose), a deficiency of dietary niacin will impair PARP-dependent DNA repair. PARP-1 is inhibited by nicotinamide and its analogs (Hageman & Stierum, 2001).

Poly(ADP-ribosyl)ation begins with the transient binding of catalytically inactive PARP-1 to strand breaks in the DNA, which immediately activates the enzyme and sterically prevents access of DNA repair enzymes. It has been suggested that the binding of PARP-1 to the breakage site might possibly constitute a signal that switches off DNA synthesis to ensure that lesions are not replicated before repair (Sato & Lindahl, 1992). The activated PARP-1 splits off ADP-ribose units from NAD molecules, liberating the nicotinamide component of NAD. The ADP-ribose units join together to form linear or branched poly(ADP-riboses), which are covalently attached to acceptor proteins in the nucleus and to PARP-1 itself. In the automodification reaction, up to 28 polymers of different sizes and structural complexities extend from the PARP-1 protein. The first ADP-ribose unit to be split off forms an *O*-glycosidic bond with a γ-carboxy group of a glutamic acid residue on the acceptor protein. The polymer is assembled by the addition of successive ADP-ribose units to the initial unit. The ADP-ribose units in the polymer are linked via glycosidic ribose–ribose 1′′→2′ bonds. The chain length of poly(ADP-riboses) is heterogeneous and can reach 200 units *in vitro*. Long polymers are branched with a frequency of one branch per linear section of 20 to 50 ADP-ribose units (D'Amours *et al.*, 1999).

The automodified PARP-1 has reduced affinity for DNA due to electrostatic repulsion of the negatively charged polymer. This leads to release of the enzyme and allows access of the DNA repair machinery to the breakage site. The DNA repair machinery is recruited as a complex containing at least XRCC1, DNA polymerase-β, and DNA ligase II. The polymerase repairs the damaged DNA and the ligase rejoins the DNA strands. XRCC1 acts as a molecular scaffold by interacting with all components of this complex and facilitates recruitment by interacting with the BRCT motif in the automodification domain of PARP-1. Ligase activity is strongly inhibited by histones, but the inhibition is reversed by poly(ADP-riboses) formed on PARP-1 in response to DNA breakage (Ohashi *et al.*, 1983). Histones are well-known targets for poly(ADP-ribosyl)ation and a histone shuttle mechanism has been demonstrated in which a histone–poly(ADP-ribose) complex is temporarily displaced from the DNA at the breakage site, causing local decondensation of chromatin and exposing naked DNA to the repair proteins (Realini & Althaus, 1992).

After completion of DNA repair, the poly(ADP-ribose) is rapidly degraded by the action of two distinct enzymes: pADPr glycohydrolase (PARG) and ADP-ribosyl protein lyase. PARG cleaves the glycosidic ribose–ribose bonds linking the ADP-ribose units and the lyase removes the remaining proximal ADP-ribose unit attached to the acceptor protein (Davidovic *et al.*, 2001). Degradation begins immediately upon the initiation of polymer synthesis and can be completed within minutes after cessation of synthesis (D'Amours *et al.*, 1999). The liberated ADP-ribose units are catabolized to AMP and ribose 5-phosphate by ADP pyrophosphatases (D'Amours *et al.*, 1999). As a result of polymer degradation, the modified state of PARP-1 is reversed and nucleosomal order is restored.

One other member of the PARP family, PARP-2, has a catalytic domain with 69% similarity to that of PARP-1. Despite its low capacity to synthesize poly(ADP-ribose), PARP-2 is an active player in DNA repair. PARP-1 and PARP-2 homo- and heterodimerize and poly(ADP-ribosyl)ate each other. In addition, PARP-2 interacts with XRCC1, DNA polymerase β, and DNA ligase III, as does PARP-1 (Schreiber *et al.*, 2002).

21.12 THE SIR2-CATALYZED DEACETYLATION REACTION

Sir2 (silent information regulator 2) is the founding member of an evolutionarily conserved family of class III histone/

ADP-ribose-nicotinamide + protein ⟶ ADP-ribose-protein + nicotinamide + H^+
(NAD)

FIGURE 21.6 Mono(ADP-ribosyl)ation. A single unit of ADP-ribose is transferred to the target protein with the concomitant release of nicotinamide and a proton.

protein deacetylases that requires NAD as a co-substrate. These enzymes, collectively called sirtuins, include five homologs in yeast (ySir2 and Hst1–4) and seven in humans (SIRT1–7). Yeast Sir2 (and also mouse Sir2) deacetylates lysines 9 and 14 of histone H3 and specifically lysine 16 of H4 (Imai *et al.*, 2000). SIRT1, SIRT6, and SIRT7 are nuclear proteins which show different subnuclear localizations; SIRT3, SIRT4, and SIRT5 are localized in mitochondria; and SIRT2 is localized in the cytoplasm (Michishita *et al.*, 2005). Given that the mitochondrion is an organelle involved in both aging and energy metabolism (Merry, 2004), an interesting possibility is that the mitochondrial SIRT proteins may function in linking metabolic and aging processes.

In the Sir2-catalyzed deacetylation reaction (Figure 21.7), NAD acts as a co-substrate. The first step is the cleavage of the high-energy glycosidic bond in NAD that joins the ADP-ribose moiety of nicotinamide. The acetyl group on the protein's lysine ε-amino group is then transferred to the 2′-OH of the ribose moiety of ADP-ribose, producing the deacetylated protein, nicotinamide, and the unique metabolite 2′-*O*-acetyl-ADP ribose. The reaction is initiated by formation of a long-lived *O*-alkylamidate intermediate and the release of nicotinamide. The nicotinamide produced functions as a non-competitive inhibitor of the deacetylation reaction through a base-exchange mechanism in which nicotinamide reacts with the enzyme-bound intermediate before deacetylation. Depletion of the intermediate reverses the first half of the reaction and regenerates the reaction products (Denu, 2005).

Nicotinamide inhibition of Sir2 is relieved *in vitro* by isonicotinamide, which causes an increase in Sir2 deacetylation activity (Sauve *et al.*, 2005). Pnc1, the deamidase that converts nicotinamide to nicotinic acid in the NAD salvage pathway, has been proposed as a regulator of Sir2 function (Anderson *et al.*, 2002). By lowering the concentration of nicotinamide formed as a product of Sir2 action and in pathways of NAD metabolism, Pnc1 would reduce inhibition on Sir2, while simultaneously increasing the availability of NAD for Sir2 by increasing the flux through the salvage pathway. Expression of the *PNC1* gene is significantly induced during various stress conditions (Anderson *et al.*, 2003). This raises the possibility that stress or nutrient limitation can increase Sir2-dependent gene silencing through Pnc1 production.

Although sirtuins were identified originally as histone lysine deacetylases, regulating gene expression through chromatin modifications, numerous non-histone substrates have been identified, expanding the potential role of this class of enzymes to virtually all known cellular processes. For example, SIRT1 reduces the transcriptional activity of p53 by deacetylation of lysine 382, thereby inhibiting the p53-dependent apoptotic response and promoting cell survival under stress (Vaziri *et al.*, 2001). SIRT1 also reduces

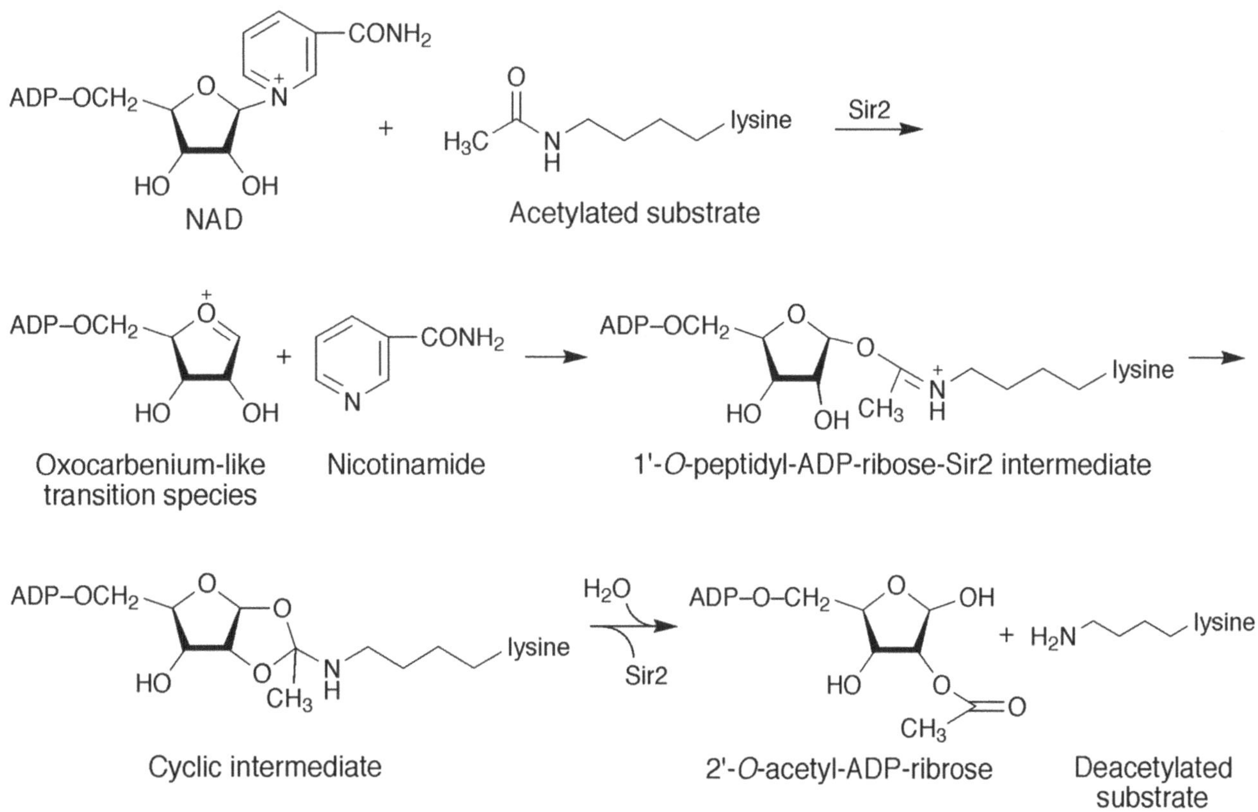

FIGURE 21.7 Abbreviated scheme for Sir2 deacetylation reactions. Reactions with protein substrates produce an equilibrium mixture of 2′- and 3′-*O*-acetyl-ADP-ribose (only the 2′-acetyl isomer is shown), nicotinamide, and deacetylated lysine side chains.

the transcriptional activity of NF-κB by deacetylating the RelA/p65 subunit at lysine 310 (Yeung *et al.*, 2004). In this context, SIRT1 augments apoptosis because pro-survival gene products of NF-κB transcription that are responsible for overcoming TNF-α-induced apoptosis are no longer produced. SIRT1 has a dual effect on the FOXO3 transcription factor: SRT1 increases FOXO3's ability to induce cell cycle arrest and resistance to oxidative stress, but inhibits FOXO3's ability to induce apoptosis (Brunet *et al.*, 2004). By deacetylating FOXO transcription factors, SIRT1 might tip FOXO-dependent responses away from cell death and toward stress resistance. SIRT1 represses the transcriptional activity of p300 by deacetylating lysines 1020 and 1024 within the CRD1 (cell cycle regulatory domain 1) transcriptional repression domain. These residues also serve as acceptor lysines for modification by the ubiquitin-like SUMO protein. The SUMO-specific protease SSP3 relieves SIRT1 repression of p300 (Bouras *et al.*, 2005). SIRT1 functions in glucose homeostasis as a modulator of the transcriptional co-activator PGC-1α (Rodgers *et al.*, 2005). A nutrient signaling response that is mediated by pyruvate induces SIRT1 in the liver during fasting. Once induced, SIRT1 activates PGC-1α by deacetylation of specific lysines to co-ordinate the increase in expression of gluconeogenic genes with the repression of glycolytic genes.

21.13 EXPERIMENTAL NIACIN DEFICIENCY IN ANIMALS

In a deficiency of niacin, the activities of the niacin coenzymes NAD and NADP are decreased. The involvement of the niacin coenzymes in energy production makes it likely that a deficiency will have an effect on the utilization of nutrients. A deficiency of niacin is characterized by severe metabolic disorders affecting particularly the skin, mucous membranes, and the digestive organs. The following deficiency signs have been observed in animals deprived of niacin.

- **Growth.** Loss of appetite, reduced growth, and loss of body weight in pigs and poultry..
- **Skin and mucous membranes.** Squamous dermatitis in pigs and poultry. Poor feathering in poultry. Inflamed mucous membranes of the digestive tract and digestive disorders. Dilation of the duodenal crypts of Lieberkühn and vacuolation of pancreatic acinar cells in chicks. Characteristic 'black tongue' in dogs.
- **Nervous system.** Disordered reflexes, ataxia, paralysis, and epileptic attacks in dogs.
- **Bones.** Deformation of the femur in poultry.
- **Blood.** Normocytic anemia in pigs and fur-bearing animals.
- **Reproductive system.** Reduced egg production and hatchability in laying hens.

21.14 PELLAGRA AS A NUTRITIONAL DISEASE IN HUMANS

21.14.1 Etiology of Pellagra

Pellagra, the classical niacin deficiency disease, is known to affect human populations subsisting on maize diets. The association of pellagra and maize consumption has been attributed to the low tryptophan content of maize and to the poor availability of its nicotinic acid content. The disease is also common in regions of India where the cereal sorghum is the dietary staple. The tryptophan content of sorghum is nearly as high as that of rice, a cereal that is not associated with pellagra. However, sorghum and maize share a common feature – a high content of leucine. Leucine causes growth retardation in rats when added at a 1% level to a low-protein diet, but growth is normal when the protein content is increased. The results of many investigations strongly suggest that a relative excess of dietary leucine causes pellagra by inhibiting the synthesis of NAD from tryptophan (Gopalan, 1968; Magboul & Bender, 1983). In particular, Bender (1983) reported that a high dietary intake of leucine both reduces the activity of kynureninase and increases the activity of ACMS decarboxylase (picolinate carboxylase) in rats. These two effects will result in a considerable reduction in the synthesis of NAD from tryptophan.

21.14.2 Symptoms of Pellagra

Typically, pellagra is an adult disease and rarely occurs during infancy. Untreated pellagra results in death from multiorgan failure. The classical features of endemic pellagra are dermatitis, inflammation of the mucous membranes, diarrhea, and psychiatric disturbances (Bender, 2003b; Hegyi *et al.*, 2004). The dermatitis often appears after exposure to sunlight and resembles sunburn. The skin becomes red and blistered and frequently peels off in large areas. In chronic cases, the skin becomes rough and thickened with a brown pigmentation. In acute pellagra, the mucous membranes of the gastrointestinal and genitourinary tracts are severely inflamed. The mouth becomes extremely sore and the tongue is swollen and scarlet in color. Chewing and swallowing are painful and even liquids may be refused. Inflammation of the small and large intestine is manifested by diarrhea, abdominal pain, and soreness of the rectum. Hypermotility of the gastrointestinal tract and the loss of appetite lead to profound loss of weight. Inflammation of the lower urinary tract causes urethritis, with increased micturition accompanied by a burning sensation. In the female, severe vaginitis is observed and amenorrhea is common. Bender (1984) vividly described neurological and neuropsychiatric signs. Early signs include tremor, irritability, anxiety and depression, with delirium and dementia sometimes occurring in severe and chronic cases. Unless the disease is treated, the inevitable outcome is death. Fortunately, the response to nicotinamide therapy is rapid and dramatic.

21.14.3 Effect of Estrogens in the Etiology of Pellagra

A conspicuous feature of pellagra is that twice as many women as men have succumbed to it historically (Miller, 1978). This phenomenon has prompted investigations into estrogens as possible factors in the etiology of the disease. Conjugated estrogens have been shown to inhibit two enzymes in the pathway of NAD synthesis from tryptophan, namely kynurenine-3-monooxygenase and kynureninase. Feeding ovariectomized rats with a diet containing estrone sulfate led to significant inhibition of kynurenine-3-monooxygenase (Bender & Totoe, 1984). This inhibitory effect is probably due to repression of the synthesis of kynurenine-3-monooxygenase, as no such effect took place *in vitro*. Bender and Wynick (1981) prepared apo-kynureninase (enzyme protein without the pyridoxal phosphate coenzyme) to demonstrate that estrone sulfate inhibited kynureninase non-competitively with respect to the coenzyme and competitively with respect to the substrate (kynurenine). Estrone glucuronide, which is a human metabolite, inhibited kynureninase in a similar manner, but non-conjugated estrone was not inhibitory. This result disproved a long-standing hypothesis that conjugated estrogens inhibit kynureninase by competing with the coenzyme. It is considered likely that, at the high-estrogen phases of the menstrual cycle, there would be significant inhibition of kynureninase in normal women (Bender & Bender, 1986).

21.15 INFLUENCE OF NIACIN DEFICIENCY ON APOPTOSIS AND CANCER

21.15.1 Apoptosis

Besides its function in DNA repair, PARP-1 mediates cell death after subjection of cells to physiological damage. Apoptosis was implicated in PARP-1-mediated cell death when Yu *et al.* (2002) showed that PARP-1 activation is required for translocation of apoptosis-inducing factor (AIF) from the mitochondria to the nucleus during chemically induced cell death, and that AIF is necessary for PARP-1-dependent cell death. Their data support a model in which activation of PARP-1 signals the release of AIF from mitochondria, resulting in a caspase-independent pathway of apoptosis. Since PARP-1 activation depends on a supply of NAD, the potential exists for an increased fraction of damaged cells evading apoptosis in the niacin-deficient state.

21.15.2 Cancer

Tissues that have high NAD levels, such as the liver and lung, are not prone to DNA damage during dietary niacin deficiency. In contrast, the bone marrow is very sensitive to changes in dietary niacin levels (Kirkland, 2003). Niacin deficiency delays DNA excision repair and increases spontaneous and nitrosourea-induced chromosomal instability in rat bone marrow (Kostecki *et al.*, 2007).

Depletion of nicotinamide and the resulting decreased NAD concentration in cultured cells derived from human breast, skin, and lung tissues is accompanied by decreased expression of p53, demonstrating that niacin modulates p53 expression in human cells (Jacobson *et al.*, 1999). This has important implications, since diminished p53 function is strongly associated with tumorigenesis in the above tissues (Donehower *et al.*, 1992).

Boyonoski *et al.* (2002a) showed that niacin deficiency in rats decreased the latency of cancer development and dramatically reduced bone marrow poly(ADP-ribose) levels following 12 doses of ethylnitrosourea (ENU). Conversely, niacin supplementation with nicotinic acid or nicotinamide increased the latency of cancer development and bone marrow poly(ADP-ribose) levels (Boyonoski *et al.*, 2002b).

21.16 NICOTINAMIDE AND TYPE 1 DIABETES PREVENTION

Type 1 diabetes is characterized by the slow destruction of the β cells of the islets of Langerhans, which is accompanied by an intra-islet chronic inflammatory process. The β cell toxins alloxan and streptozotocin are used experimentally to produce diabetes in rodents. These chemicals induce DNA strand breaks in the nuclei of pancreatic islet cells. Alloxan generates oxygen radicals, especially the highly reactive hydroxyl radical, that break islet DNA. The streptozotocin-induced DNA breaks are mainly associated with the alkylating activity of the agent itself (Okamoto & Takasawa, 2003). These breaks in the DNA induce DNA repair, which involves the activation of PARP and consequent depletion of its substrate, NAD. The fall in cellular NAD decreases ATP and inhibits cellular functions including insulin synthesis and secretion. Thus the β cell ultimately dies. The NAD depletion and the decrease in β cell functions are prevented by radical scavengers such as superoxide dismutase and catalase, and by the PARP inhibitors, nicotinamide and 3-aminobenzamide. Although nicotinamide is not a potent antioxidant, it is able to scavenge oxygen radicals, but not nitric oxide (Kolb & Burkart, 1999). Theoretically, therefore, nicotinamide could prevent the development of diabetes by preventing NAD depletion through PARP inhibition and protecting islets from damage by free radicals.

Pretreatment with high doses of nicotinamide prevents the decrease in proinsulin synthesis in islets from rats treated with alloxan or streptozotocin (Uchigata *et al.*, 1983). Prophylactic treatment of non-obese diabetic (NOD) female mice with high-dose injections of nicotinamide prevents spontaneous diabetes, but is less effective in reversing the disease once it is established (Yamada *et al.*, 1982).

To assess whether high-dose nicotinamide prevents or delays clinical onset of diabetes in humans, a large-scale intervention study was organized by the European Nicotinamide Diabetes Intervention Trial (ENDIT) Group

(2004). In the randomized, double-blind, placebo-controlled trial, 552 participants recruited from 18 European countries, Canada, and the USA were randomly allocated oral modified-release nicotinamide (1.2 g/m^2) or placebo for five years. The participants were first-degree relatives (parent, sibling, child) of people with type 1 diabetes. The proportion of participants who developed diabetes within five years was almost identical in those treated with nicotinamide and those on placebo. Thus, unlike the results with rodents, pharmacological doses of nicotinamide are ineffective in the prevention of type 1 diabetes in humans. The failure to confirm findings in animal studies might be related to the lower dose used in humans, or to the fact that nicotinamide was given at a much later stage of the disease process in humans than in rodents. Another possible explanation relates to species-related differences in the islets themselves.

21.17 TOXICITY

There is no evidence of adverse effects associated with high intake of naturally occurring niacin from food sources. Therapeutic doses of 30 to 100 mg/day of nicotinic acid cause flushing, a term which covers reddening of the skin and accompanying sensation of burning and itching. Massive doses (3 to 6 g/day) of nicotinic acid used as a drug for the long-term treatment of hyperlipidemia produce pathological signs of liver toxicity (Baggenstoss *et al.*, 1967). Sustained release preparations of nicotinic acid are much more toxic than the plain formulation, because they permit more prolonged maintenance of blood and tissue concentrations of the drug (DiPalma & Thayer, 1991). Nicotinamide does not cause vasodilatation, but is otherwise two to three times as toxic as the acid (Miller & Hayes, 1982; Alhadeff *et al.*, 1984).

REFERENCES

Alhadeff, L., Gualtieri, T. & Lipton, M. (1984) Toxic effects of water-soluble vitamins. *Nutrition Reviews*, **42**, 33–40.

Anderson, R.M., Bitterman, K.J., Wood, J.G., Medvedik, O., Cohen, H., Lin, S.S., Manchester, J.K., Gordon, J.I. & Sinclair, D.A. (2002) Manipulation of a nuclear NAD$^+$ salvage pathway delays aging without altering steady-state NAD$^+$ levels. *Journal of Biological Chemistry*, **277**(21), 18881–90.

Anderson, R.M., Bitterman, K.J., Wood, J.G., Medvedik, O. & Sinclair, D.A. (2003) Nicotinamide and *PNC1* govern lifespan extension by calorie restriction in *Saccharomyces cervisiae*. *Nature*, **423**, 181–5.

Baggenstoss, A.H., Christensen, N.A., Berge, K.G., Baldus, W.P., Spieckerman, R.E. & Ellefson, R.D. (1967) Fine structural changes in the liver in hypercholesterolemic patients receiving long-term nicotinic acid therapy. *Mayo Clinic Proceedings*, **42**, 385–99.

Bender, D.A. (1983) Effects of a dietary excess of leucine on the metabolism of tryptophan in the rat: A mechanism for the pellagragenic action of leucine. *British Journal of Nutrition*, **50**, 25–32.

Bender, D.A. (1984) B vitamins in the nervous system. *Neurochemistry International*, **6**, 297–321.

Bender, D.A. (2003a) Niacin. Physiology. In: *Encyclopedia of Food Science and Nutrition*, 2nd edn, Vol. 7. (Ed. by B. Caballero, L. Trugo & P. Finglas), pp. 4119–28. Academic Press, London.

Bender, D.A. (2003b) Pellagra. In: *Encyclopedia of Food Science and Nutrition*, 2nd edn, Vol. 7. (Ed. by B. Caballero, L. Trugo & P. Finglas), pp. 4456–60. Academic Press, London.

Bender, D.A. & Bender, A.E. (1986) Niacin and tryptophan metabolism: The biochemical basis of niacin requirements and recommendations. *Nutrition Abstracts and Reviews (Series A)*, **56**(10), 695–719.

Bender, D.A. & Totoe, L. (1984) Inhibition of tryptophan metabolism by oestrogens in the rat: A factor in the aetiology of pellagra. *British Journal of Nutrition*, **51**, 219–24.

Bender, D.A. & Wynick, D. (1981) Inhibition of kynureninase (L-kynurenine hydrolase, EC 3.7.1.3) by oestrone sulphate: An alternative explanation for abnormal results of trypyophan load tests in women receiving oestrogenic steroids. *British Journal of Nutrition*, **45**, 269–75.

Bender, D.A., Magboul, B.I. & Wynick, D. (1982) Probable mechanisms of regulation of the utilization of dietary tryptophan, nicotinamide and nicotinic acid as precursors of nicotinamide nucleotides in the rat. *British Journal of Nutrition*, **48**, 119–27.

Bouras, T., Fu, M., Sauve, A.A., Wang, F., Quong, A.A., Perkins, N.D., Hay, R.T., Gu, W. & Pestell, R.G. (2005) SIRT1 deacetylation and repression of p300 involves lysine residues 1020/1024 within the cell cycle regulatory domain 1. *Journal of Biological Chemistry*, **280**(11), 10264–76.

Boyonoski, A.C., Spronck, J.C., Gallacher, L.M., Jacobs, R.M., Shah, G.M., Poirier, G.G. & Kirkland, J.B. (2002a) Niacin deficiency decreases bone marrow poly(ADP-ribose) and the latency of ethylnitrosourea-induced carcinogenesis in rats. *Journal of Nutrition*, **132**, 108–14.

Boyonoski, A.C., Spronck, J.C., Jacobs, R.M., Shah, G.M., Poirier, G.G. & Kirkland, J.B. (2002b) Pharmacological intakes of niacin increase bone marrow poly(ADP-ribose) and the latency of ethylnitrosourea-induced carcinogenesis in rats. *Journal of Nutrition*, **132**, 115–20.

Brunet, A., Sweeney, L.B., Sturgill, J.F., Chua, K.F., Greer, P.L., Lin, Y., Tran, H., Ross, S.E., Mostoslavsky, R., Cohen, H.Y., Hu, L.S., Cheng, H.-L., Jedrychowski, M.P., Gygi, S.P., Sinclair, D.A., Alt, F.W. & Greenberg, M.E. (2004) Stress-dependent regulation of FOXO transcription factors by the SIRT1 deacetylase. *Science*, **303**, 2011–15.

Chi, Y. & Sauve, A.A. (2013) Nicotinamide riboside, a trace nutrient in foods, is a vitamin B3 with effects in energy metabolism and neuroprotection. *Current Opinion in Clinical Nutrition and Metabolic Care*, **16**, 657–61.

Cravioto, R.O., Anderson, R.K., Lockhart, E.E., Miranda, F.P. & Harris, R.S. (1945) Nutritive value of the Mexican tortilla. *Science*, **102**, 91–3.

D'Amours, D., Desnoyers, S., D'Silva, I. & Poirier, G.G. (1999) Poly(ADP-ribosyl)ation reactions in the regulation of nuclear functions. *Biochemical Journal*, **342**, 249–68.

Davidovic, L., Vodenicharov, M., Affar, E.B. & Poirier, G.G. (2001) Importance of poly(ADP-ribose) glycohydrolase in the control of poly(ADP-ribose) metabolism. *Experimental Cell Research*, **268**, 7–13.

Denu, J.M. (2005) Vitamin B_3 and sirtuin function. *Trends in Biochemical Sciences*, **30**, 479–83.

DiPalma, J.R. & Thayer, W.S. (1991) Use of niacin as a drug. *Annual Review of Nutrition*, **11**, 169–87.

Donehower, L.A., Harvey, M., Slagle, B.L., McArthur, M.J., Montgomery, C.A. Jr., Butel, J.S. & Bradley, A. (1992) Mice deficient for p53 are developmentally normal but susceptible to spontaneous tumors. *Nature*, **356**, 215–21.

European Nicotinamide Diabetes Intervention Trial (ENDIT) Group. (2004) European nicotinamide diabetes intervention trial (ENDIT): a randomised controlled trial of intervention before the onset of type 1 diabetes. *Lancet*, **363**, 925–31.

Ghosh, H.P., Sarkar, P.K. & Guha, B.C. (1963) Distribution of the bound form of nicotinic acid in natural materials. *Journal of Nutrition*, **79**, 451–3.

Gopal, E., Fei, Y.-J., Miyauchi, S., Zhuang, L., Prasad, P.D. & Ganapathy, V. (2005) Sodium-coupled and electrogenic transport of B-complex vitamin nicotinic acid by sic5a8, a member of the Na/glucose co-transporter gene family. *Biochemical Journal*, **388**, 309–16.

Gopalan, C. (1968) Leucine and pellagra. *Nutrition Reviews*, **11**, 323–6.

Gross, C.J. & Henderson, L.M. (1983) Digestion and absorption of NAD by the small intestine of the rat. *Journal of Nutrition*, **113**, 412–20.

Hageman, G.J. & Stierum, R.H. (2001) Niacin, poly(ADP-ribose) polymerase-1 and genomic stability. *Mutation Research*, **475**, 45–56.

Hegyi, J., Schwartz, R.A. & Hegyi, V. (2004) Pellagra: Dermatitis, dementia, and diarrhea. *International Journal of Dermatology*, **43**, 1–5.

Henderson, L.M. & Gross, C.J. (1979) Metabolism of niacin and nicotinamide in perfused rat intestine. *Journal of Nutrition*, **109**, 654–62.

Imai, S.-I., Armstrong, C.M., Kaeberlein, M. & Guarente, L. (2000) Transcriptional silencing and longevity protein Sir2 is an NAD-dependent histone deacetylase. *Nature*, **403**, 795–800.

Jacobson, E.L., Shieh, W.M. & Huang, A.C. (1999) Mapping the role of NAD metabolism in prevention and treatment of carcinogenesis. *Molecular and Cellular Biochemistry*, **193**, 69–74.

Kirkland, J.B. (2003) Niacin and carcinogenesis. *Nutrition and Cancer*, **46**, 110–118.

Kodicek, E., Braude, R., Kon, S.K. & Mitchell, K.G. (1959) The availability to pigs of nicotinic acid in *tortill*a baked from maize treated with lime-water. *British Journal of Nutrition*, **13**, 363–84.

Kolb, H. & Burkart, V. (1999) Nicotinamide in type 1 diabetes. *Diabetes Care*, **22**, B16–20.

Kostecki, L.M., Thomas, M., Linford, G., Lizotte, M., Toxopeus, L., Bartleman, A.P. & Kirkland, J.B. (2007) Niacin deficiency delays DNA excision repair and increases spontaneous and nitrosourea-induced chromosomal instability in rat bone marrow. *Mutation Research*, **625**, 50–61.

Kumar, J.S., Subramanian, V.S., Kapadia, R., Kashyap, M.L. & Said, H.M. (2013) Mammalian colonocytes possess a carrier-mediated mechanism for uptake of vitamin B3 (niacin): Studies utilizing human and mouse colonic preparations. *American Journal of Physiology (Gastrointestinal and Liver Physiology)*, **305**, G207–13.

Magboul, B.I. & Bender, D.A. (1982) The nature of niacin in sorghum. *Proceedings of the Nutrition Society*, **41**, 50A.

Magboul, B.I. & Bender, D.A. (1983) The effects of a dietary excess of leucine on the synthesis of nicotinamide nucleotides in the rat. *British Journal of Nutrition*, **49**, 321–9.

Mason, J.B. & Kodicek, E. (1973) The chemical nature of the bound nicotinic acid of wheat bran: Studies of partial hydrolysis products. *Cereal Chemistry*, **50**, 637–46.

Mason, J.B., Gibson, N. & Kodicek, E. (1973) The chemical nature of the bound nicotinic acid of wheat bran: Studies of nicotinic acid-containing macromolecules. *British Journal of Nutrition*, **30**, 297–311.

Merry, B.J. (2004) Oxidative stress and mitochondrial function with aging – The effects of calorie restriction. *Aging Cell*, **3**, 7–12.

Michishita, E., Park, J.Y., Burneskis, J.M., Barrett, J.C. & Horikawa, I. (2005) Evolutionarily conserved and nonconserved cellular localizations and functions of human SIRT proteins. *Molecular Biology of the Cell*, **16**, 4623–35.

Miller, D.F. (1978) Pellagra deaths in the United States. *American Journal of Clinical Nutrition*, **31**, 558–9.

Miller, D.R. & Hayes, K.C. (1982) Vitamin excess and toxicity. In: *Nutritional Toxicolog*, Vol. 1. (Ed. by J.N. Hathcock), pp. 81–133. Academic Press, New York.

Mrochek, J.E., Jolley, R.L., Young, D.S. & Turner, W.J. (1976) Metabolic response of humans to nicotinic acid and nicotinamide. *Clinical Chemistry*, **22**, 1821–7.

Nabokina, S.M., Kashyap, M.L. & Said, H.M. (2005) Mechanism and regulation of human intestinal niacin uptake. *American Journal of Physiology (Cell Physiology)*, **289**, C97–103.

Ohashi, Y., Ueda, K., Kawaichi, M. & Hayaishi, O. (1983) Activation of DNA ligase by poly(ADP-ribose) in chromatin. *Proceedings of the National Academy of Sciences of the U.S.A.*, **80**, 3604–7.

Okamoto, H. & Takasawa, S. (2003) Recent advances in physiological and pathological significance of NAD^+ metabolites: Roles of poly(ADP-ribose) and cyclic ADP-ribose in insulin secretion anf diabetogenesis. *Nutrition Research Reviews*, **16**, 253–66.

Realini, C.A. & Althaus, F.R. (1992) Histone shuttling by poly(ADP ribosylation). *Journal of Biological Chemistry*, **267**(26), 18858–65.

Rodgers, J.T., Lerin, C., Haas, W., Gygi, S.P., Spiegelman, B.M. & Puigserver, P. (2005) Nutrient control of glucose homeostasis through a complex of PGC-1α and SIRT1. *Nature*, **434**, 113–8.

Said, H.M. (2011) Intestinal absorption of water-soluble vitamins in health and disease. *Biochemical Journal*, **437**, 357–72.

Said, H.M., Nabokina, S.M., Balamurugan, K., Mohammed, Z.M., Urbina, C. & Kashyap, M.L. (2007) Mechanism of nicotinic acid transport in human liver cells: Experiments with HepG2 cells and primary hepatocytes. *American Journal of Physiology (Cell Physiology)*, **293**, C1773–8.

Sato, M.S. & Lindahl, T. (1992) Role of poly(ADP-ribose) formation in DNA repair. *Nature*, **356**, 356–8.

Sauve, A.A., Moir, R.D., Schramm, V.L. & Willis, I.M. (2005) Chemical activation of Sir2-dependent silencing by relief of nicotinamide inhibition. *Molecular Cell*, **17**, 595–601.

Schreiber, V., Amé, J.-C., Dollé, P., Schultz, I., Rinaldi, B., Fraulob, V., Ménissier-de Murcia, J. & de Murcia, G. (2002) Poly(ADP-ribose) polymerase-2 (PARP-2) is required for efficient base excision DNA repair in association with PARP-1 amf XRCC1. *Journal of Biological Chemistry*, **277**(25), 23028–36.

Schuette, S.A. & Rose, R.C. (1983) Nicotinamide uptake and metabolism by chick intestine. *American Journal of Physiology*, **245**, G531–8.

Schuette, S.A. & Rose, R.C. (1986) Renal transport and metabolism of nicotinic acid. *American Journal of Physiology*, **250**, C694–703.

Spector, R. (1979) Niacin and niacinamide transport in the central nervous system. In vivo studies. *Journal of Neurochemistry*, **33**, 895–904.

Spector, R. (1987) Niacinamide transport through the blood-brain barrier. *Neurochemical Research*, **12**, 27–31.

Spector, R. & Kelley, P. (1979) Niacin and niacinamide accumulation by rabbit brain slices and choroid plexus *in vitro*. *Journal of Neurochemistry*, **33**, 291–8.

Uchigata, Y., Yamamoto, H., Nagai, H. & Okamoto, H. (1983) Effect of poly(ADP-ribose) synthetase inhibitor administration to rats before and after injection of alloxan and streptozotocin on islet proinsulin synthesis. *Diabetes*, **32**, 316–318.

Vaziri, H., Dessain, S.K., Ng Eaton, E., Imai, S.-I., Frye, R.A., Pandita, T.K., Guarente, L. & Weinberg, R.A. (2001) *hSIR-2^{SIRT1}* functions as an NAD-dependent p53 deacetylase. *Cell*, **107**, 149–59.

Wall, J.S. & Carpenter, K.J. (1988) Variation in availability of niacin in grain products. *Food Technology*, **42**(10), 198–202, 204.

Yamada, K., Nonaka, K., Hanafusa, T., Miyazaki, A., Toyoshima, H. & Tarui, S. (1982) Prevention and therapeutic effects of large-dose nicotinamide injections on diabetes associated with insulitis. *Diabetes*, **31**, 749–53.

Yeung, F., Hoberg, J.E., Ramsey, C.S., Keller, M.D., Jones. D.R., Frye, R.A & Mayo, M.W. (2004) Modulation of NF-κB-dependent transcription and cell survival by the SIRT1 deacetylase. *EMBO Journal*, **23**, 2369–80.

Yu, S.-W., Wang, H., Poitras, M.F., Coombs, C., Bowers, W.J., Federoff, H.J., Poirier, G.G., Dawson, T.M. & Dawson, V.L. (2002) Mediation of poly(ADP-ribose) polymerase-1-dependent cell death by apoptosis-inducing factor. *Science*, **297**, 259–63.

22 Vitamin B_6

22.1 CHEMISTRY AND BIOLOGICAL ACTIVITY

Vitamin B_6 is the generic descriptor for a group of six vitamers (Figure 22.1) that exhibits qualitatively in rats the biological activity of pyridoxine (2-methyl-3-hydroxy-4,5-bis(hydroxymethyl)pyridine). Three nonphosphorylated B_6 vitamers differ in the substituent (R) at the fourth position of the pyridine ring: R is an alcohol in pyridoxine (PN), an aldehyde in pyridoxal (PL), and an amine in pyridoxamine (PM). The other three B_6 vitamers are 5′-phosphate esters of PN, PL, and PM, namely pyridoxine 5′-phosphate (PNP), pyridoxal 5′-phosphate (PLP), and pyridoxamine 5′-phosphate (PMP). The six B_6 vitamers are considered to have approximately equivalent biopotency in humans on the basis of their ultimate conversion to PLP. Synthetic 4-deoxypyridoxine is a potent vitamin B_6 antagonist used to study the effects of vitamin B_6 deficiency.

PLP represents the biologically active form of vitamin B_6 and serves as a coenzyme for a variety of enzymes. As a coenzyme, PLP is bound tightly to the apoenzyme by a Schiff base (aldimine) linkage (–N=CH–) formed through condensation of the 4-aldehyde group with the primary amino group, most commonly the ε-amino group of specific lysine residues (Figure 22.2). This binding protects the PLP in cells from alkaline phosphatase hydrolysis and helps to maintain normal intracellular PLP levels. PMP is also biologically active, as PLP and PMP are interconverted during the actions of the aminotransferase family of enzymes. Although PNP is not a coenzyme, it is an important intermediate in the conversion of dietary PN to PLP.

A ubiquitous bound form of pyridoxine that occurs in plant tissues is a glucoside conjugate, pyridoxine-5′-β-D-glucoside (Figure 22.3), abbreviated to PN-glucoside. A more complex derivative of PN-glucoside containing cellobiose and 5-hydroxydioxindole-3-acetic acid moieties has been identified as a major form of vitamin B_6 in rice bran and a minor form in wheat bran and legumes (Tadera & Orite, 1991).

22.2 FOOD SOURCES AND BIOAVAILABILITY

22.2.1 Food Sources

Vitamin B_6 is present in all natural unprocessed foods, with yeast extract, wheat bran, and liver containing particularly high concentrations. Other important sources include wholegrain cereals, nuts, pulses, lean meat, fish, kidney, potatoes, and other vegetables. Milk, eggs, and fruits contain relatively low concentrations of the vitamin.

The dietary B_6 vitamers of natural origin are principally PLP, PMP, PN, and glycosylated forms of PN. PN, glycosylated PN, and PMP are the predominant vitamers in plant-derived foods, while PLP is mostly present in foods of animal origin. PNP, being a quantitatively minor metabolite in vitamin B_6 metabolism, is scarcely present in foods. Synthetic pyridoxine hydrochloride (PN.HCl) is used in the fortification of foods on account of its chemical stability.

No generalizations can be made as to one group of plant foods consistently having a high PN-glucoside content. Typical sources of PN-glucoside (expressed as a percentage of the total vitamin B_6 present) are bananas (5%), raw broccoli (35%), raw green beans (58%), raw carrots (70%), and orange juice (69%) (Gregory & Ink, 1987). PN-glucoside accounted for 10–15% of the total vitamin B_6 in typical American mixed diets (Gregory *et al.*, 1991), but would be proportionally higher in vegetarian diets.

22.2.2 Bioavailability

The bioavailability of vitamin B_6 in foods is highly variable, owing largely to the presence of poorly utilizable PN-glucoside in plant tissues. As expected, vitamin B_6 generally has a lower availability from plant-derived foods than from animal tissues (Nguyen & Gregory, 1983). Based on plasma PLP levels in male human subjects, the bioavailability of the vitamin in an average American diet ranged from 61% to 81%, with a mean of 71% (Tarr *et al.*, 1981). The bioavailability of PN-glucoside relative to that of PN was found to be 50.0 ± 7.0% (Nakano *et al.*, 1997).

22.3 DIGESTION AND ABSORPTION OF NON-GLYCOSYLATED VITAMIN B_6

Humans cannot synthesize vitamin B_6 and a dietary source of the vitamin is necessary to achieve adequate intake. Vitamin B_6 is present in foods mainly as the PN, PLP, and PMP vitamers. In many fruits and vegetables, 30% or more of the total vitamin B_6 is present as PN-glucoside. As formation of a Schiff base is reversible in an acid medium, the PLP bound to protein dissociates from the protein under normal gastric acid conditions. The release of PLP from its association with protein is an important step in the subsequent absorption of vitamin B_6, as binding to protein inhibits the next step, hydrolysis of PLP by alkaline phosphatase (Middleton, 1986). The polar nature of phosphorylated B_6 vitamers prevents them from readily crossing cell membranes, therefore dephosphorylation is required before they can be absorbed. Middleton (1985) used the *in vivo* luminally perfused segment of rat small intestine to study PN uptake over a 30-minute period. Their results showed (1) a gradient of decreasing rates of uptake from proximal

DOI: 10.1201/9781003410300-26

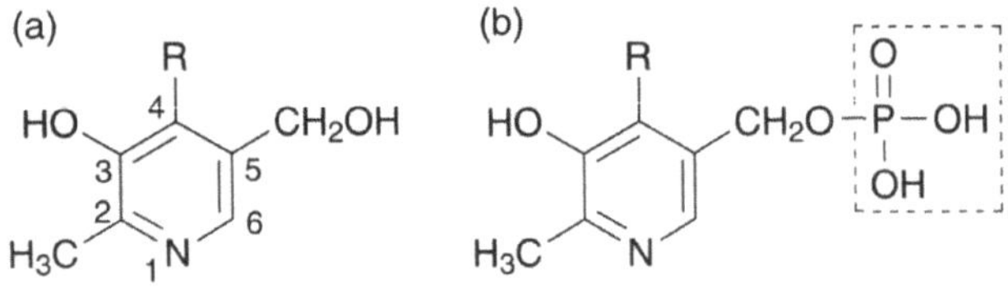

R	Vitamers
CH_2OH	PN, PNP
CHO	PL, PLP
CH_2NH_2	PM, PMP

FIGURE 22.1 Structures of vitamin B_6 compounds showing (a) nonphosphorylated and (b) phosphorylated forms.

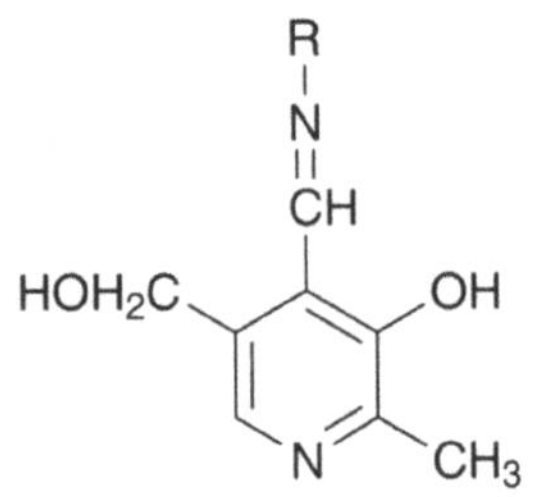

FIGURE 22.2 Attachment of PLP to the apoenzyme by a Schiff base (aldimine) linkage.

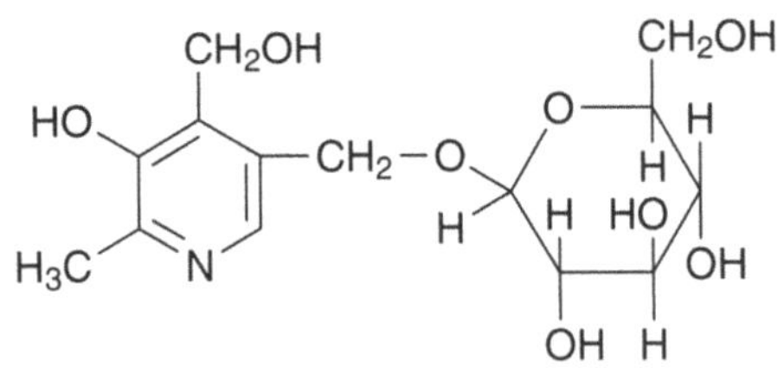

FIGURE 22.3 Structure of pyridoxine-5′-β-D-glucoside.

to distal small intestine, and (2) a saturable component of uptake, which was most evident in duodenum, less so in proximal jejunum, and absent in distal ileum.

Dietary PLP and PMP are dephosphorylated by membrane-bound alkaline phosphatase in the intestinal lumen, and the free PL and PM diffuse into the enterocytes (Mehanso *et al.*, 1979; Hamm *et al.*, 1979). PL and PM are converted to their corresponding phosphates by the catalytic action of cytoplasmic pyridoxal kinase, and transaminases interconvert PLP and PMP. It is conjectured that the lowering of the intracellular concentrations of PL and PM due to the accumulation of their phosphates creates a concentration gradient across the brush-border membrane that increases the diffusion rate of these vitamers (Middleton, 1985). Biochemists call this accelerated passive uptake mechanism 'metabolic trapping.' The intracellular phosphorylated vitamers are largely dephosphorylated by nonspecific phosphatases, thus permitting their easy diffusion across the basolateral membrane and release to the portal circulation. When supplemental PN.HCl is absorbed, it too is phosphorylated and dephosphorylated before release from the enterocyte (Middleton, 1978).

Said *et al.* (2003) examined the nature of PN uptake by human enterocytes using the human-derived intestinal epithelial Caco-2 cell line as an *in vitro* model and [^{3}H] PN as the uptake substrate. Uptake was linear with time for up to 10 min of incubation and a three-minute incubation time was chosen to represent the initial rate of uptake. Three observations provided conclusive evidence of a carrier-mediated uptake mechanism: (1) the initial rate of PN uptake was saturable as a function of concentration in the pH 5.5-buffered incubation medium; (2) uptake was inhibited by structural analogs of PN; and (3) efflux of [^{3}H]PN from Caco-2 cells pre-loaded with the labeled vitamer was stimulated when the cells were incubated in the presence of unlabeled PN. The initial rate of PN uptake occurred with minimal metabolic alterations in the transported substrate and was unaffected by theophylline and penicillamine. These two compounds are known inhibitors of pyridoxine kinase, the intracellular enzyme responsible for phosphorylation of B_6 vitamers, thus it can be inferred from their lack of interference that metabolic trapping has no significant role in the initial uptake process. The transport process was found to be temperature- and energy-dependent, but Na^+-independent. PN uptake was pH-dependent, with higher uptake at pH 5 compared with pH 8, therefore the acidic microclimate at the luminal surface of the small intestine would be favorable for uptake. Said *et al.* (2003) further showed that modulators of the cAMP-dependent protein kinase (PKA)-mediated signaling pathway caused significant reduction in PN uptake, thus implicating this intracellular pathway in the regulation of uptake. The effect of this reduction was to decrease V_{max}, but not the apparent K_m, indicating a decrease in the number of PN carriers, with no effect on carrier affinity. Modulators of PKC-, Ca^{2+}/calmodulin-, and nitric oxide-mediated pathways had no effect on PN uptake.

No specific transporter for vitamin B_6 has yet been found. Yamashiro *et al.* (2020) used Caco-2 cells to show that the thiamin transporters, THTR-1 and THTR-2, can also transport PN on the basis of the positive charge on thiamin and PN molecules. The specific uptake by THTR-1 was extensively inhibited by PM and, to a lesser extent, by 4-deoxpyridoxine; however, PL, PLP and 4-pyridoxic acid were not inhibitory. THTR-2 exhibited almost the same characteristics, although PL exhibited a slight inhibitory effect as a minor difference. Among water-soluble vitamins, only thiamin was a potent inhibitor for both transporters. There were marked differences in the pH at which transport of thiamin and PN is optimal. Whereas thiamin uptake favored neutral to basic conditions. PN uptake favored acidic conditions, with an optimum of pH 5. Another difference was that inhibition of PN transport by thiamin was competitive, whereas inhibition of thiamin transport by PN was non-competitive. This suggested that the two substrates may bind to different sites of the transporter proteins. These results suggest a possible role for THTR-1 and THTR-2 in PN transport in the acidic surface microclimate of the small intestine.

Varying dietary vitamin B_6 supply had no effect on intestinal absorption of the vitamin (Roth-Maier *et al.*, 1982), demonstrating that absorption is not adaptively regulated by the level of dietary intake.

22.4 ABSORPTION AND INTESTINAL PROCESSING OF PN-GLUCOSIDE

Three possible events could explain the intestinal absorption and processing of dietary PN-glucoside. (1) A small part of the ingested PN-glucoside may be unabsorbed and eliminated with the feces. (2) PN-glucoside may be partially hydrolyzed by a brush-border membrane enzyme and the free PN and glucose absorbed. (3) PN-glucoside may be absorbed intact by simple diffusion and partially hydrolyzed by PN-glucoside hydrolase in the cytosol of enterocytes.

The only known mammalian brush-border β-glucosidase is lactase-phlorizin hydrolase (LPH) (Mackey *et al.*, 2002). Although this enzyme is primarily responsible for the hydrolysis of dietary lactose, it can also hydrolyze PN-glucoside. LPH has two distinct active sites – a lactase active site for the hydrolysis of lactose and a phlorizin hydrolase active site for the hydrolysis of phlorizin. PN-glucoside is hydrolyzed at the lactase active site and therefore lactose competitively inhibits LPH-catalyzed PN-glucoside hydrolysis. In the context of human nutrition, the inhibitory effect of lactose would depend on the composition of the meal consumed. Only if a dairy product and a plant-derived food were consumed together would lactose affect vitamin B_6 bioavailability.

Intestinal PN-glucoside hydrolysis was once thought to be catalyzed by a broad specificity β-glucosidase (BSβG) (EC. 3.2.1.21), a cytosolic enzyme found in the intestine and liver (Trumbo *et al.*, 1990). McMahon *et al.* (1997) purified a novel cytosolic β-glucosidase, designated pyridoxine-β-D-glucoside hydrolase (PN-glucoside hydrolase), from porcine jejunal mucosa. They showed that this enzyme, not the broad specificity β-glucosidase, is responsible for the cytosolic hydrolysis of PN-glucoside. The hydrolase was also capable of hydrolyzing lactose, but not sucrose.

22.5 ABSORPTION OF BACTERIALLY SYNTHESIZED VITAMIN B_6 IN THE LARGE INTESTINE

Evidence that intestinally produced vitamin B_6 is available to the host comes from balance studies showing that the amount of vitamin excreted in the urine exceeds the amount ingested. In patients with mid-transverse colostomies, absorption of vitamin B_6 takes place equally well whether the vitamin is given orally or instilled directly into the lumen of the large intestine (Sorrell *et al.*, 1971).

Said *et al.* (2008) investigated the mechanism involved in PN uptake by mammalian colonic epithelial cells (colonocytes) using the young adult mouse colonic epithelial (YAMC) cell line as an *in vitro* model and [^{3}H]PN as the uptake substrate. Uptake was linear with time for up to 10 min of incubation and a three-minute incubation time was chosen to represent the initial rate of uptake. Evidence of a carrier-mediated uptake mechanism were the observations that (1) the initial phase of PN uptake was saturable; (2) uptake was inhibited by structural analogs of PN; and (3) efflux of [^{3}H]PN from pre-loaded YAMC cells was stimulated when unlabeled PN was present in the incubation medium. The initial rate of PN uptake occurred with no metabolic alterations and was unaffected by inhibitors of pyridoxine kinase (theophylline, penicillamine, and isoniazid). Thus intracellular metabolic trapping is not a significant uptake mechanism. The transport process was found to be temperature- and energy-dependent, Na^+-independent, and sensitive to pH, with higher uptake at pH ≥6.5 compared with lower pH. Said *et al.* (2008) also studied uptake of [^{3}H]PN using purified human colonic apical membrane vesicles (AMV). Evidence of a carrier-mediated mechanism in this model was based on the observation of a significant inhibition in [^{3}H]PN uptake by unlabeled PN.

Said *et al.* (2008) further showed that the uptake of PN by YAMC cells is regulated by substrate availability. Switching the maintenance condition of the cells from high to low PN levels resulted in increased uptake, which appeared to be mediated via transcriptional and translational mechanisms. Uptake was inhibited by inhibitors of the intracellular Ca^{2+}/calmodulin-mediated pathway, but not by inhibitors of the PKA-, PKC-, and nitric oxide-mediated pathways. These observations indicated that colonic uptake of PN is regulated by extracellular vitamin B_6 concentrations and by an intracellular Ca^{2+}/calmodulin-mediated pathway.

22.6 POST-ABSORPTIVE METABOLISM

22.6.1 Metabolic Processes

An overall metabolic scheme for the absorption, metabolism, and transport of vitamin B_6 is depicted in Figure 22.4.

The liver is the primary site of vitamin B_6 metabolism and contains all the enzymes required for the interconversion of the various B_6 vitamers. The major vitamin B_6-metabolizing reactions that occur in human liver are shown in Figure 22.5. A single enzyme, pyridoxal kinase (EC 2.7.1.35), catalyzes the phosphorylation of PN, PL, and PM. The PNP and PMP are subsequently oxidized to PLP by PNP (PMP) oxidase (EC 1.4.3.5), an FMN-dependent enzyme. The PLP can be reversibly converted to PMP through aminotransferases and phosphorylated vitamers can be dephosphorylated by phosphatases. An FAD-dependent aldehyde oxidase oxidizes PL irreversibly to 4-pyridoxic acid (Merrill *et al.*, 1984). Pyridoxal kinase has been detected in all mammalian tissues investigated (McCormick *et al.*, 1961), whereas PNP (PMP) oxidase is confined to only a few mammalian tissues and cannot be detected in rat small intestine or hind limb muscle (Buss *et al.*, 1980). The specific activity of PMP (PNP) oxidase is

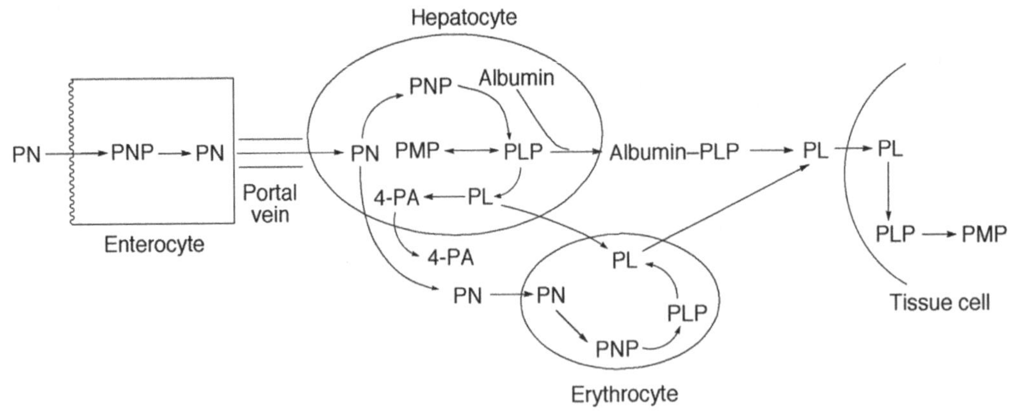

FIGURE 22.4 Overall metabolic scheme for the absorption, metabolism, and transport of vitamin B_6. The scheme shows that ingested pyridoxine is metabolized in both the liver and erythrocyte. The PLP in plasma and the PL in erythrocytes represent the major B_6 vitamers available to tissues.

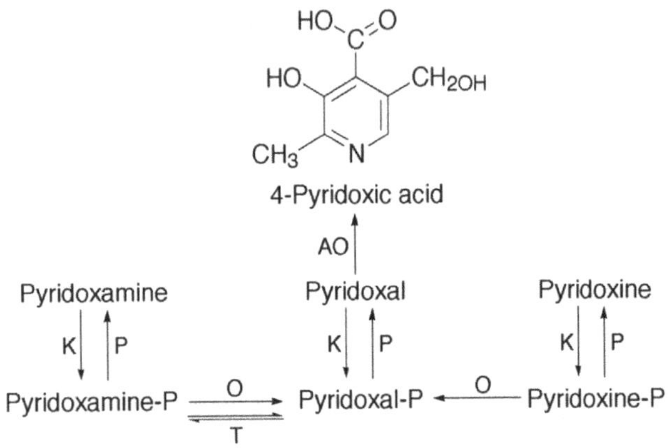

FIGURE 22.5 Major metabolic transformations of vitamin B_6 that occur in liver. K, kinase; O, pyridoxine (pyridoxamine) 5′-phosphate oxidase; P, phosphatase; AO, aldehyde oxidase; T, aminotransferases.

highest in the liver; lower activities are found in the kidney, brain, and human (but not rat) erythrocytes (Lumeng *et al.*, 1985).

The newly absorbed nonphosphorylated B_6 vitamers are conveyed to the liver in the portal blood. The vitamers diffuse into the liver cells and are rapidly converted to the phosphorylated forms. The cell plasma membrane is impermeable to the phosphorylated vitamers, which accumulate in the cell. This metabolic trapping creates a concentration gradient across the membrane, which provides the impetus for passive but rapid uptake of the vitamers (Mehansho *et al.*, 1980). In the liver, the newly synthesized PLP constitutes a rapidly mobilizable pool. Some of the newly synthesized PLP is bound by apoenzymes, some is transported to the bloodstream, and some is dephosphorylated. The PL thus formed is released from the hepatocyte or oxidized to the terminal catabolite 4-pyridoxic acid.

The principal forms of vitamin B_6 in liver are PLP and PMP. PNP is usually present in only trace quantities because of its rapid oxidation to PLP. The liver releases regulated amounts of both PLP and PL into the circulation. PLP, being a polar molecule, is unable to cross cell membranes unaided and, presumably, is released as a complex with albumin. A unique protein-mediated transport mechanism for PLP is indicated by the fact that PMP is not released from the liver. In plasma, PLP is almost totally bound to albumin, which prevents the urinary excretion of PLP and protects it against rapid hydrolysis to PL. Prior to uptake by extrahepatic tissues, PLP is dissociated from the albumin and dephosphorylated by a membrane-bound phosphatase. Cellular uptake of PL is by diffusion and metabolic trapping by phosphorylation. Interconversion between PLP and PMP occurs within the tissues. Some 80% of the total body pool of about 250 mg of vitamin B_6 is in skeletal muscle, mostly in the form of PLP bound to glycogen phosphorylase. The liver contains a further 10% of the body pool.

The PL released from the liver is taken up by erythrocytes and bonded to the α-chain of hemoglobin by Schiff base formation with the amino-terminal valine (Ink *et al.*, 1982). Erythrocytes also take up PN that escapes metabolism in the liver and converts it to PL via PNP and PLP (Anderson *et al.*, 1971). This situation could arise if a meal contains a high amount of PN or if a PN.HCl supplement is taken. The PL is released from the erythrocytes and taken up by extracellular tissues. A unique acid phosphatase that specifically dephosphorylates PLP, PNP, PMP, 4-pyridoxic acid, and 4-deoxy-PNP has been purified from human erythrocytes (Fonda, 1992). The uptake and plasma clearance of PN by erythrocytes *in vivo* has been demonstrated in a human subject given PN intravenously (Anderson *et al.*, 1989).

PLP binds avidly to proteins and at unphysiologically high concentrations is an inhibitor of numerous enzymes. It is important, therefore, to control the cellular concentration of PLP. PLP limits its rate of synthesis by competitively inhibiting PNP (PMP) oxidase (Merrill *et al.*, 1978). In liver cells, protein binding and hydrolysis of PLP function jointly to maintain PLP at physiological levels. When the PLP concentration exceeds the binding capacity of cellular proteins, excess free PLP is rapidly hydrolyzed by alkaline phosphatase to PL (Li *et al.*, 1974). This source of PL is oxidized to 4-pyridoxic acid, which is released into the circulation and excreted in the urine. As a metabolic end-product, urinary 4-pyridoxic acid excretion reflects the major portion of vitamin B_6 intake that has been absorbed and metabolized.

22.6.2 Inhibitory Effect of PN-Glucoside on the Metabolism of Co-Ingested Pyridoxine

Nakano *et al.* (1997) reported that PN-glucoside partially inhibits the utilization of co-ingested PN in humans. The inhibitory effect may be due to the competitive inhibition of hepatic uptake of PN by PN-glucoside, as demonstrated in isolated rat hepatocytes (Zhang *et al.*, 1993). Nakano and colleagues concluded that the inhibitory effect of PN-glucoside probably has little nutritional significance when typical US diets are consumed.

22.7 RENAL REABSORPTION

Bowman and McCormick (1989) studied the uptake of [^{3}H]PN by freshly isolated rat renal proximal tubular cells and demonstrated saturability, with an apparent K_m of 1.3 µM. Initial uptake was significantly decreased after preincubation with ethionine, an ATP-trapping agent, or with ouabain, an inhibitor of the sodium–potassium ATPase. Preincubation with the diuretic drug amiloride inhibited PN uptake in a concentration-related fashion. Amiloride is an inhibitor of facilitated sodium entry into epithelial cells via the apical membrane and acts by competing with sodium ions for binding to an external site on the sodium–hydrogen antiporter (Benos, 1982). The results of these experiments indicate that renal reabsorption of PN is sodium-dependent and might be uniquely mediated by Na^+/H^+ exchange. Following uptake, intracellular PN is subject to phosphorylation and conversion to PLP. Bowman and McCormick (1989) found that the majority of absorbed vitamin, about 88% after 30 minutes, is trapped metabolically with the appearance of a significant amount of PLP.

Said *et al.* (2002) used the opossum-derived renal tubule epithelial cell line, opossum kidney (OK), as an *in vitro* model system to examine the nature of PN reabsorption in the renal tubule. Their observations that uptake of PN was saturable and inhibited by structural analogs provided evidence of a carrier-mediated mechanism. Uptake of PN was specific in nature and occurred with no metabolic alterations. The transport process was found to be temperature- and energy-dependent, Na^+-independent and sensitive to pH, with a higher uptake at alkaline or neutral buffer pH compared with acidic pH.

Said *et al.* (2002) further showed that growing OK cells in a vitamin B_6-deficient medium led to a significant up-regulation of PN uptake. This up-regulation was specific for PN, reversible and inhibited by actinomycin D and cyclohexamide (inhibitors of transcription and translation, respectively). The up-regulation was mainly mediated via an increase in V_{max} with a slight change in the apparent K_m, indicating an increase in the number of PN carriers, with little effect on carrier affinity. Pretreating OK cells with different inhibitors of the Ca^{2+}/calmodulin-mediated pathway led to a significant reduction in PN uptake via a decreased V_{max}. Modulators of PKA- and nitric oxide-mediated pathways had no effect on PN uptake. These observations indicated that uptake of PN by renal tubular epithelia is regulated by extracellular vitamin B_6 concentrations and by an intracellular Ca^{2+}/calmodulin-mediated pathway.

22.8 TRANSPORT IN THE BRAIN

Vitamin B_6 levels in brain are homeostatically regulated. Although it is relatively easy to produce symptomatic vitamin B_6 deficiency in animals, levels of the vitamin in brain (and heart) are somewhat better maintained in deficiency states than they are in liver and kidney. Conversely, massive parenteral doses (daily intravenous injections of 200 mg/kg of PN for three days to rabbits) elevated the brain levels of PLP by an average of only 39% (Spector & Shikuma, 1978).

In brain, vitamin B_6 exists predominantly in the enzymatically active forms PLP and PMP at concentrations much higher than in plasma and cerebrospinal fluid (CSF). In rabbits, the concentrations of vitamin B_6 in plasma, CSF, brain, and choroid plexus were, respectively, 0.30, 0.39, 8.90, and 15.10 µmol L^{-1} or kg^{-1} (Spector, 1978a).

Spector's group studied the *in vitro* uptake and release of tritium-labeled vitamin B_6 in rabbit brain slices and isolated choroid plexuses (Spector, 1978b; Spector & Greenwald, 1978). Uptake of [^{3}H]PN by both tissues was inhibited by (1) low temperature (2°C) and dinitrophenol, demonstrating energy dependence; (2) pyridoxal azine, demonstrating dependence on the activity of intracellular pyridoxal kinase; and (3) unlabeled nonphosphorylated B_6 vitamers and, to lesser extent, phosphorylated B_6 vitamers, demonstrating saturability of the uptake system. There was no detectable metabolism of [^{3}H]PN to [^{3}H]pyridoxic acid in brain slices or choroid plexus. From 70 to 80% of the labeled vitamin B_6 in both tissues was phosphorylated after a 30-minute incubation in [^{3}H]PN. Phosphorylated B_6 vitamers were taken up much less readily than nonphosphorylated vitamers. These studies are not conclusive in separating active transport from metabolic trapping because both pyridoxal kinase and active transport require ATP, and therefore depletion of ATP could affect either process. Furthermore, dinitrophenol is known to inhibit mammalian pyridoxal kinase as well as preventing ATP synthesis by uncoupling oxidative phosphorylation from electron flow through the electron-transport chain.

The activity of pyridoxal kinase in brain is unimpaired by moderate and severe vitamin B_6 deficiency (McCormick *et al.*, 1961). Spector & Shikuma (1978) showed that pyridoxal kinase activity and vitamin B_6 accumulation by brain slices and choroid plexus are not affected by various drugs that alter the concentrations of PLP or biogenic amines in brain.

Spector (1978a) showed that, during one pass through the cerebral circulation, [^{3}H]PN was cleared from the circulation no more rapidly than mannitol. Mannitol, a molecule of similar size and shape to PN, is known to be transported by diffusion. Spector (1978a) also confirmed *in vivo*, by injection directly into the ventricular CSF of rabbits, that nonphosphorylated B_6 vitamers enter brain cells by a saturable accumulation process. Kinetic studies conducted by Spector & Greenwald (1978) revealed marked differences in the uptake of vitamin B_6 by choroid plexus and brain. The half-saturation concentrations and rate maxima for accumulation were ~0.2 μM and 1.0–2.0 μmol/kg per 30 min for brain slices and 7.0 μM and 40 μmol/kg per 30 min for isolated choroid plexus. Assuming the absence of a membrane carrier for vitamin B_6 uptake, the kinetic constants refer to the binding of substrate to pyridoxal kinase and are therefore values of K_m and V_{max}. The differences in the constants for vitamin B_6 uptake by brain cells and choroid plexus are presumably due to factors (e.g. intracellular pH) that cause variations in enzyme activity.

The addition of carrier (unlabeled) PN to the incubation medium not only inhibited uptake of [^{3}H]PN by brain slices and choroid plexus, but accelerated dephosphorylation and release of previously accumulated labeled vitamin B_6 (Spector, 1978b). These two mechanisms, acting together, would tend to keep the total intracellular vitamin B_6 concentrations constant. One striking difference in the release mechanism was that choroid plexus released predominantly phosphorylated B_6 vitamers, whereas brain slices released only nonphosphorylated B_6 vitamers (Spector, 1978b). This makes the choroid plexus, rather than brain cells, the likely source of the phosphorylated B_6 vitamers in CSF.

Vitamin B_6 is transported in the reverse direction (i.e. from brain and/or CSF into blood) more rapidly than mannitol (Spector, 1978a). This suggests that the transport mechanism for vitamin B_6 in this direction involves a mechanism other than simple diffusion.

In conclusion, Spector's data show that circulating vitamin B_6 can enter the brain via the blood–CSF barrier (choroid plexus). The finding that PN was extracted no more rapidly than mannitol during one pass through the cerebral circulation argues against significant entry of PN via the blood–brain barrier. There is a saturable transport system (K_m = 0.7 μM) within the choroid plexus that regulates the entry of free (unbound) nonphosphorylated B_6 vitamers from plasma into the CSF. The vitamers find their way into the extracellular space of brain and enter brain cells by a high-affinity (K_m =~0.2 μM) saturable accumulation system. The transport system in both choroid plexus and brain cells appears to be simple (or possibly facilitated) diffusion accelerated by the concentration gradient created by phosphorylation of the transported B_6 vitamers (metabolic trapping). PN, PL, and PM have comparable affinity for the vitamin B_6 transport systems, as they also do for pyridoxal kinase. Excessive concentrations of phosphorylated B_6 vitamers within brain cells are dephosphorylated intracellularly and transported out of the cells.

22.9 BIOCHEMICAL FUNCTIONS

22.9.1 Overview

PLP, the physiologically active B_6 vitamer, serves as a coenzyme for over 100 different enzymes involved in almost all aspects of cellular biochemistry and metabolism. Reactions catalyzed by PLP include deamination, transamination (which can also be catalyzed by PMP), decarboxylation, transsulfuration, and desulfuration. A selection of vitamin B_6-dependent enzymes is listed in Table 22.1. In amino acid metabolism, vitamin B_6 is required in a variety of enzymes

TABLE 22.1
A Selection of Vitamin B_6-Requiring Enzymes

Enzyme	Function
Aminotransferases	Transamination (transfer of an amino group from an amino acid to an α-keto acid)
Serine hydroxymethyltransferase	Removal of the β carbon from serine to form glycine.
	Tetrahydrofolate (THF) accepts the carbon atom to yield 5,10-methylene-THF.
Serine dehydratase	Degradation of serine to pyruvate
P-protein of the glycine-cleavage system	Glycine catabolism in the mitochondria
Histidine decarboxylase	Decarboxylation of histidine to form histamine
Glutamate decarboxylase	Decarboxylation of glutamate to form γ-aminobutyric acid (GABA)
Cysteine sulfinate decarboxylase	Decarboxylation of cysteine sulfonate to hypotaurine
Phosphatidylserine decarboxylase	Decarboxylation of phosphatidylserine to phosphatidylethanolamine
δ-Aminolevulinate synthase	Condensation of succinyl-CoA with glycine followed by decarboxylation to form δ-aminolevulinic acid
Cystathionine β-synthase	Condensation of homocysteine with serine to form cystathionine
3-Ketosphinganine synthase	Condensation of palmitoyl-CoA with serine to form 3-ketosphinganine
Cystathionine β-lyase	Breakdown of cystathionine to homocysteine
Cystathionine γ-lyase	Breakdown of cystathionine to α-ketobutyric acid and cysteine
Selenocysteine β-lyase	Breakdown of selenocysteine to selenide (Se^{2-}) and alanine
Glycogen phosphorylase	Glycogen phosphorolysis to release glucose-1--phosphate

concerned with the interconversion of amino acids, the synthesis of non-essential amino acids, and the metabolism of amino acids in excess of the amounts required for protein synthesis. Aminotransferases facilitate the transfer of amino groups between amino acids and keto acids, and thus represent an important link among amino acid, carbohydrate, and fat metabolism. Many neurotransmitters are formed by the PLP-dependent decarboxylation of amino acids. PLP is the coenzyme for glycogen phosphorylase, a key enzyme in the utilization of liver and muscle glycogen reserves. PLP is involved in the synthesis of sphingomyelin, which is an important constituent of the myelin sheath of nerve cell axons. Finally, PLP, together with folate and vitamin B_{12}, affects the pathways of one-carbon metabolism, which are important for the synthesis of nucleotides and methionine.

22.9.2 One-carbon Metabolism and Transsulfuration

PLP is a coenzyme for four enzymes involved in one-carbon metabolism and transsulfuration (Figure 22.6). Two of these enzymes, serine hydroxymethyltransferase and glycine decarboxylase (a component of the glycine cleavage system), transfer one-carbon units to tetrahydrofolate (THF) from serine and glycine, respectively. The 5,10-methylene-THF thus formed is reduced to 5-methyl-THF, which then donates its methyl group to homocysteine to form methionine. A considerable proportion of methionine is activated by ATP to form *S*-adenosylmethionine, which serves as a methyl donor to a variety of acceptors. In the transsulfuration pathway, cystathionine β-synthase catalyzes the condensation of homocysteine with serine to form cystathionine in a PLP-dependent irreversible reaction. The fourth PLP-dependent enzyme is cystathionine γ-lyase, which cleaves cystathionine to cysteine and α-ketobutyrate.

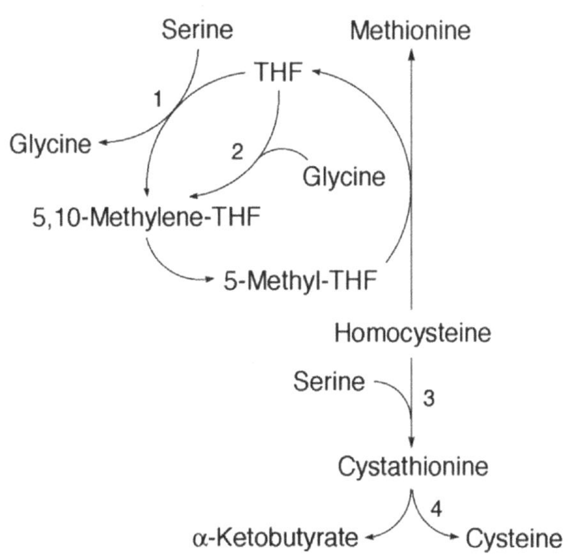

FIGURE 22.6 PLP-dependent enzymes in one-carbon metabolism and transsulfuration. Enzymes: (1) serine hydroxymethyltransferase, (2) glycine decarboxylase of the glycine cleavage system, (3) cystathionine β-synthase, (4) cystathionine γ-lyase.

22.9.3 Biosynthesis of Nucleic Acids

The 5,10-methylene-THF formed by the transfer of a carbon unit from serine or glycine to THF is used as such for the synthesis of thymidylate (deoxythymidine monophosphate) or is oxidized to formyl-THF, which is used for purine synthesis (see Section 28.15.2). Vitamin B_6-deficient rats have been shown to exhibit a decreased rate of production of one-carbon units and a decreased ability to synthesize DNA, RNA, and proteins (Trakatellis *et al.*, 1997).

The following experimental observations further indicate that a vitamin B_6 deficiency state impairs nucleic acid synthesis:

- A decreased incorporation of L-valine-1-C^{14} into proteins of liver, spleen, and serum (Trakatellis & Axelrod, 1964).
- A decreased ability to incorporate DL-leucine-1-C^{14} into ribosomes isolated from liver and spleen, which may be linked to the observed decrease in the number of polysomes per unit weight of these tissues (Montjar *et al.*, 1965).
- A decreased incorporation of labeled precursors of nucleic acids into liver and spleen DNA and RNA, particularly messenger RNA (Trakatellos & Axelrod, 1965).

22.9.4 Metabolism of Tryptophan

PLP is a coenzyme for kynureninase, the enzyme which catalyzes the conversion of 3-hydroxykynurenine to 3-hydroxyanthranilic acid in the oxidative pathway of tryptophan metabolism (Figure 21.5). Under normal conditions in the body, tryptophan dioxygenase is the rate-limiting enzyme and the pathway proceeds through 3-hydroxyanthranilic acid. However, in vitamin B_6 deficiency, kynureninase loses its activity. This situation leads to the accumulation of kynurenine and 3-hydroxykynurenine, which are then converted to kynurenic acid and xanthurenic acid, respectively, by the action of kynurenine aminotransferase (Figure 22.7). This enzyme is also PLP-dependent, but for some unknown reason it is little affected in vitamin B_6 deficiency. Kynurenic and xanthurenic acids are excreted in the urine.

22.9.5 Biosynthesis of Porphyrin

The heme proteins (e.g. cytochromes, hemoglobin, and myoglobin) have in common a cyclic tetrapyrrole structure called a porphyrin. In the biosynthesis of porphyrin (Figure 22.8) PLP is a coenzyme for δ-aminolevulinic acid synthase, which catalyzes the formation of δ-aminolevulinic acid from succinyl-CoA and glycine. In the presence of the appropriate enzyme, two molecules of δ-aminolevulinic acid condense by dehydration and porphobilinogen is

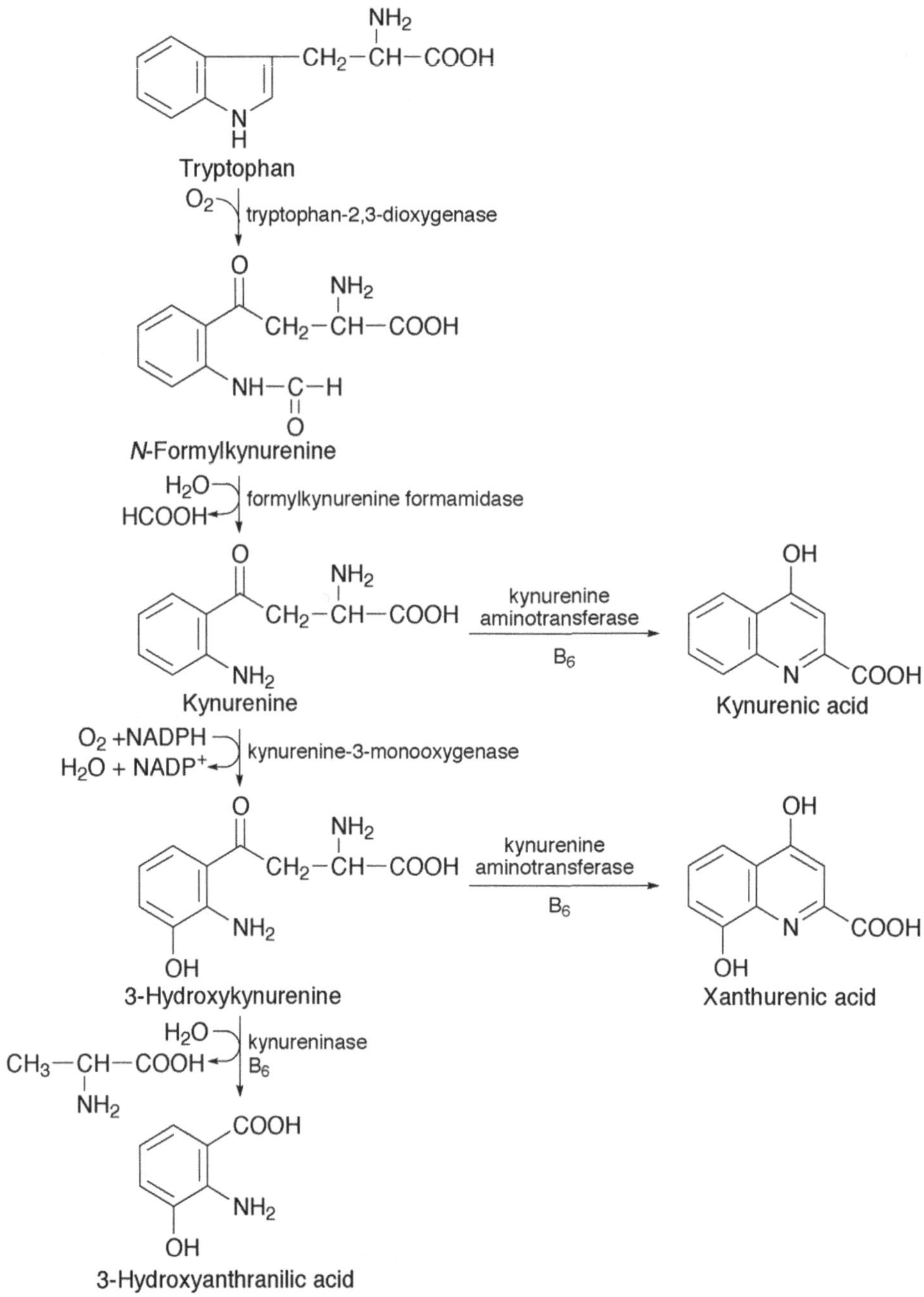

FIGURE 22.7 Effect of vitamin B_6 deficiency on the biochemical conversion of tryptophan to NAD in the liver. In vitamin B_6 deficiency the tryptophan metabolites kynurenic acid and xanthurenic acid appear in the urine.

formed. Several additional enzymes acting sequentially combine four molecules of porphobilinogen into protoporphyrin IX. Finally, to form heme, a specific ferrochetalase inserts the ferrous iron into the tetrapyrrole ring.

22.9.6 Biosynthesis of Biogenic Amines

Several physiologically important amines, collectively referred to as biogenic amines (Table 22.2 and Figure 22.9), are formed by the non-oxidative decarboxylation of amino acids (reaction 22.1). The enzymes involved, amino acid decarboxylases, all require PLP as coenzyme.

$$\begin{array}{c} R \\ | \\ H - C - NH_3^+ \\ | \\ COO- \end{array} \longrightarrow \begin{array}{c} R \\ | \\ H - C - NH_2 + CO_2 \\ | \\ H \end{array} \qquad (22.1)$$

22.9.7 Role in Glycogenolysis

PLP is a coenzyme for glycogen phosphorylase, the enzyme that splits terminal glucose units from the branches of glycogen during glycogenolysis. Rat skeletal muscle contains as

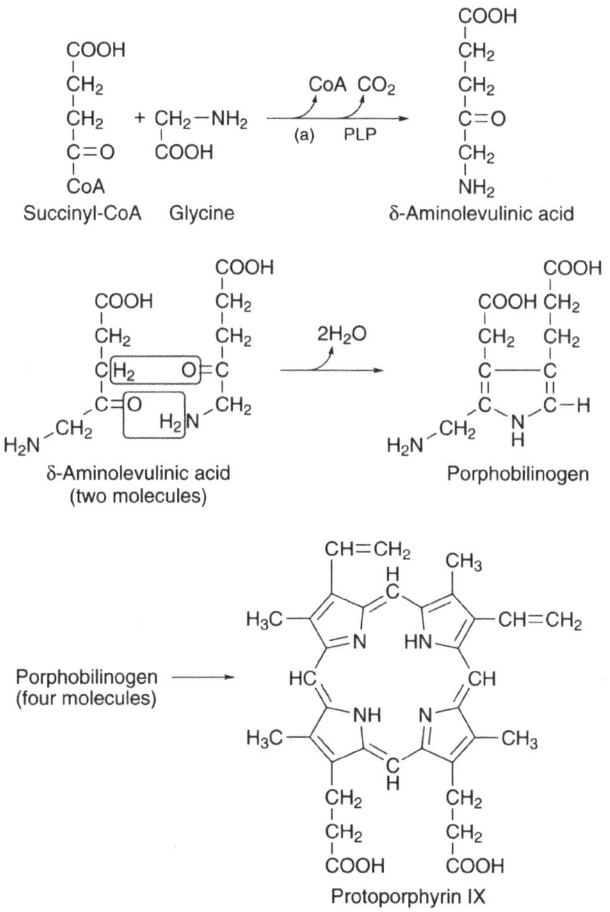

FIGURE 22.8 PLP as a coenzyme in the biosynthesis of porphyrin. Enzyme: (a) δ-aminolevulinic acid synthase.

much as 50% of the total body content of vitamin B_6, most of which is bound stoichiometrically to glycogen phosphorylase (Leklem, 1988). The glycogen phosphorylase accumulates in the muscle during abundance of vitamin B_6 in the diet (Black *et al.*, 1977), but this reservoir of enzyme does not decrease if the rats are deprived of vitamin B_6 over a period of at least eight weeks (Black *et al.*, 1978). Prolonged deficiency does ultimately lead to enzyme depletion, but only after anorexia develops and weight loss occurs. From a variety of data, Black *et al.* (1978) and other groups have postulated that the reservoir of muscle glycogen phosphorylase provides an endogenous source of PLP to meet the urgent need for glucose synthesis during starvation. The PLP released from the muscle combines with appropriate apoenzymes to boost the amounts of amino acid catabolizing enzymes. The relevant enzymes act upon glucogenic amino acids (e.g. glutamate) to produce precursors (e.g. α-ketoglutarate) for glucose synthesis by gluconeogenesis.

22.9.8 Biosynthesis of Sphingomyelin

The sphingolipids are membrane lipids composed of a polar head group and two long nonpolar tails. One of the tails is the long-chain amino alcohol sphingosine and the other is a long-chain fatty acid. Sphingomyelins are a class of sphingolipids whose fatty acid constituents are mainly palmitic (16:0) and stearic (18:0) acid and whose polar head groups are phosphocholine or phosphoethanolamine. Sphingomyelins are important constituents of the myelin sheath that surrounds and electrically insulates nerve cell axons; they account for 8.5% by weight of total lipid in human myelin.

The biosynthesis of sphingomyelin (Figure 22.10) begins with palmitoyl-coenzyme A (palmitoyl-CoA), the principal product of fatty acid synthesis. Palmitoyl-CoA condenses with serine to produce 3-ketosphinganine. The enzyme that catalyzes this reaction, 3-ketosphinganine synthase, is PLP-dependent. The subsequent steps leading to the production of ceramide involve keto group reduction, acetylation, and an FAD-dependent oxidation to create a double bond. The final step is attachment of the head group to ceramide through donation of the phosphocholine group of phosphatidylcholine or the phosphoethanolamine group of phosphatidylethanolamine.

22.10 REGULATION OF STEROID HORMONE ACTION

Steroid hormones (androgens, estrogens, progestins, and the corticosteroids) are able to enter cells, bind to

TABLE 22.2
Biogenic Amines Derived from Amino Acids or Amino Acid Derivatives by PLP-Dependent Non-Oxidative Decarboxylation

Amino Acid	Amine	Function of Amine
Glutamic acid	γ-Aminobutyric acid (GABA)	Inhibitory neurotransmitter in the brain and spinal cord.
Histidine	Histamine	Amplifies the immunoresponse by increasing capillary permeability and smooth muscle contraction. Stimulates gastric acid secretion.
Dihydroxyphenylalanine (DOPA)	Dihydroxyphenylethylamine (dopamine)	Intermediary in the biosynthesis of noradrenaline and adrenaline (neurohormonal agents). Is active itself as an inhibitory transmitter in certain parts of the brain.
5-Hydroxytryptophan	5-Hydroxytryptamine (serotonin)	Inhibition of pain signals in the spinal cord. Associated with promotion of sleep.

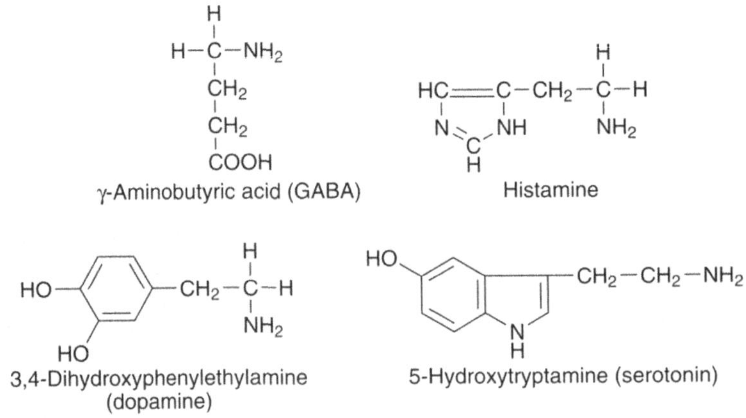

FIGURE 22.9 Structures of some biogenic amines.

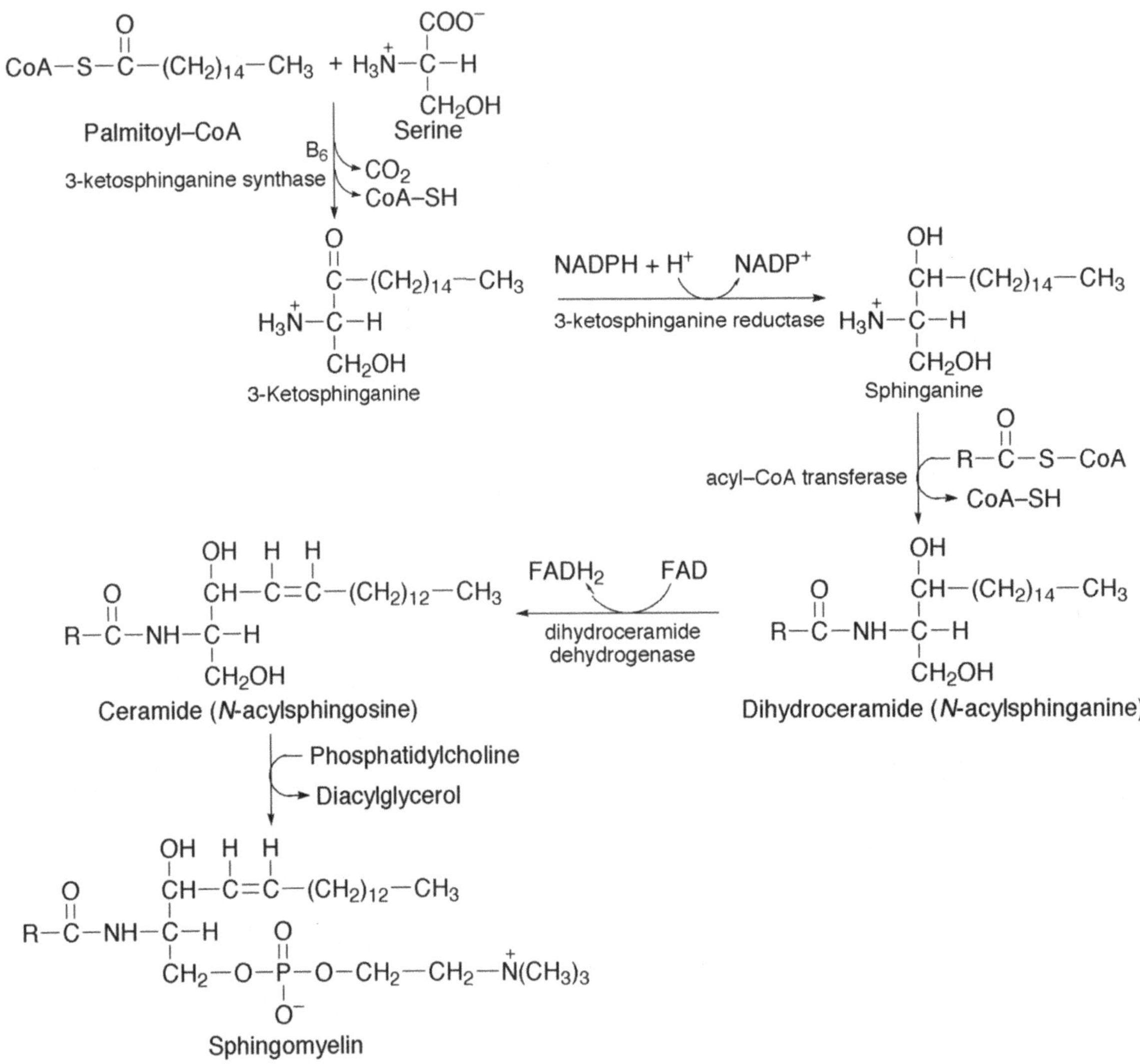

FIGURE 22.10 Biosynthesis of sphingomyelin.

receptors, and directly regulate gene transcription. The synthesized proteins carry out the ultimate effect of the hormone. Evidence accrued in the following studies suggests a physiological role for vitamin B_6 in modulating steroid hormone action.

In a number of rat studies, Bender's group produced *in vivo* evidence that vitamin B_6 may be involved in the normal physiological action of steroid hormones. In male rats, Symes *et al.* (1984) showed that the uptake and accumulation of tracer [1,2,6,7-^{3}H]testosterone in the nucleus of the prostate gland were significantly increased in vitamin B_6-deficient animals compared with vitamin B_6-adequate controls. In a corresponding study of female rats (Bowden *et al.*, 1986), the animals were segregated according to the phase of the estrous cycle to avoid the inherent variations of both plasma concentration of estrogen and the concentration of estrogen receptor in the uterus during the course of the estrous cycle. As for testosterone in the male, uptake and accumulation of tracer [2,4,6,7-^{3}H]17β-estradiol in uterine nuclei were significantly increased in vitamin B_6-deficient animals throughout the estrous cycle; there were no significant differences at anestrus.

Bender's group also found evidence of enhanced sensitivity to steroid hormone action in vitamin B_6-deficient rats of both sexes. In the male, testosterone is secreted by the interstitial cells of Leydig in the testes, but only when these cells are stimulated by luteinizing hormone (LH) released by the anterior pituitary gland in response to hypothalamic gonadotrophin-releasing hormone (Gn-RH). Circulating testosterone exerts negative feedback control at the level of the hypothalamus, switching off the supply of pituitary LH and thereby stopping testicular secretion of testosterone (Figure 22.11). Symes *et al.* (1984) found that the plasma concentration of testosterone in vitamin B_6-deficient male rats was only 25% of that in vitamin B_6-adequate controls. This unexplainable reduction in plasma testosterone was not accompanied by a reduction in the relative weight of the prostate gland as might have been expected; nor was it accompanied by a rise in plasma LH. These two observations suggest that there may be enhanced sensitivity of the hypothalamus to negative feedback by testosterone in vitamin B_6 deficiency, leading to normal (or reduced) plasma concentrations of LH and normal growth of the prostate despite considerably reduced circulating concentrations of testosterone.

In the female rat, ovarian secretion of estrogen is stimulated by LH and follicle-stimulating hormone (FSH) released from the anterior pituitary gland in response to hypothalamic Gn-RH. During most of the estrous cycle, circulating estrogen exerts negative feedback control at the level of the hypothalamus, suppressing the release of LH

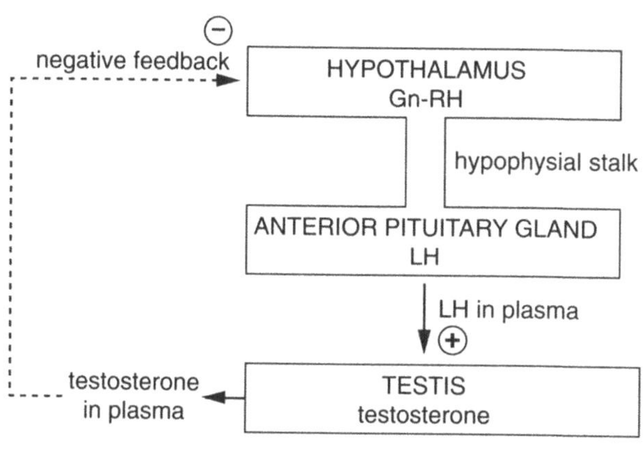

FIGURE 22.11 Negative feedback of luteinizing hormone (LH) secretion by testosterone. –, inhibition; +, stimulation. Gn-RH, gonadotrophin-releasing hormone.

and FSH. The major event of ovulation is preceded by a massive outflow of LH from the pituitary (the preovulatory surge) caused by positive feedback of estrogen upon the hypothalamus. Bowden *et al.* (1986) reported that in ovariectomized rats, doses of ethynyl-estradiol that had no effect on circulating LH in control animals (i.e. submaximal doses) lowered plasma LH levels in vitamin B_6-deficient animals. As in the male rat, this suggests that vitamin B_6 deficiency leads to enhanced sensitivity of the hypothalamus to negative feedback by steroid hormone.

22.11 MODULATION OF GENE EXPRESSION BY VITAMIN B_6

22.11.1 Expression of the Cytosolic Aspartate Aminotransferase Gene

Aspartate aminotransferase (AST; EC 2.6.1.1) is a PLP-dependent enzyme which exists as two isozymes, one cytosolic (cAST) and the other mitochondrial (mAST). cAST expression is induced by glucocorticoid hormones (mainly hydrocortisone) secreted by the adrenal cortex (Pavé-Preux *et al.*, 1988). Kondo and Okada (1985) showed that the induction of cAST by hydrocortisone in the liver of adrenalectomized vitamin B_6-deficient rats is suppressed by the simultaneous administration of pyridoxine. Oka *et al.* (1995a) provided *in vitro* evidence to suggest that PLP modulates cAST gene expression by directly interacting with the glucocorticoid receptor and preventing it from binding to the glucocorticoid-responsive element on the DNA. Other B_6 vitamers (PMP, PL, PM, and PN) did not show this inhibitory effect.

22.11.2 Expression of the Albumin Gene

Expression of the albumin gene is regulated by tissue-specific transcription factors such as HNF-1 and C/EBP (Cereghini *et al.*, 1987). Oka *et al.* (1995b) showed that PLP modulates albumin gene expression in rat liver by interacting directly with tissue-specific transcription factors and thereby preventing their binding to DNA. This inhibition of transcription factor binding to DNA was specific for PLP and involved formation of a Schiff base between its aldehyde group and primary amino groups (Oka *et al.*, 2001). Chiku *et al.* (1993) reported that depletion of amino acid supply to rats during total parenteral nutrition causes a decrease in the rate of albumin synthesis in the liver. This was shown to be due, at least in part, to an increased PLP concentration in the liver and consequent inactivation of tissue-specific transcription factors (Oka *et al.*, 1995b).

22.12 EFFECTS OF VITAMIN B_6 ON THE IMMUNE SYSTEM

22.12.1 Introduction

Through the involvement of vitamin B_6 in the biosynthesis of nucleic acids, and hence protein synthesis, it is expected that a deficiency of this vitamin would have detrimental effects on the immune system. Among reported effects are (1) changes in the structure and cellular content of lymphoid organs, (2) inhibition of humoral responsiveness to a variety of test antigens, and (3) an inhibition of various cell-mediated immune responses. As the effects of deficiency due to any one factor are difficult to elucidate in human beings, most of the experimental work has been done on laboratory animals. Reported effects of vitamin B_6 deficiency in laboratory animals are listed below.

- Atrophy of the thymus (Stoerk *et al.*, 1947).
- Impaired maturation of lymphocytes in the thymus (Willis-Carr & St. Pierre, 1978).
- Prevention of lymphocyte egress from the thymus, leading to a decrease in the number of circulating lymphocytes (lymphopenia) (Schwab *et al.*, 2005).
- Impaired proliferation of thoracic duct lymphocytes and loss of cellular immunocompetance (Robson & Schwarz, 1975).
- Reduced production and cytotoxicity of cytotoxic T cells (Sergeev *et al.*, 1978; Ha *et al.*, 1984a).
- Reduced number of antibody-producing cells in the spleen (Kumar & Axelrod, 1968).
- Impairment of circulating antibody response to antigenic stimuli (Axelrod & Hopper, 1960; Axelrod *et al.*, 1961).
- Stimulation of specific IgE antibody in response to antigen and an accompanying increase in IL-4, which is known to induce IgE production through class switching (Doke *et al.*, 1997).
- Impairment of the primary and secondary antibody response (Axelrod, 1958).
- Decreased synthesis of DNA and RNA in liver and spleen (Trakatellis & Axelrod, 1965).
- Reduced number of polysomes per unit weight of liver and spleen (Montjar *et al.*, 1965).
- Reduced delayed-type hypersensitive skin response (Axelrod *et al.*, 1963).
- Increased survival time of allogeneic skin grafts (Axelrod *et al.*, 1958).

22.12.2 The Influence of Vitamin B_6 on Lymphoid Organs

Stoerk *et al.* (1947) showed that the thymuses of vitamin B_6-deficient rats were atrophied and totally depleted of lymphocytes. Willis-Carr and St. Pierre (1978) showed that the severe depletion of functional T cells in vitamin B_6-deficient rats appears to be caused by the inability of thymic epithelial cells to effect the differentiation of T cell precursors. Thymic T cell precursors of vitamin B_6-deficient rats were undamaged, therefore the immunological defects were presumably due to a blocking of the synthesis and/or release of a humoral factor that is produced by thymic epithelial cells. A likely candidate for such a humoral factor is thymulin, a hormone of exclusively thymic origin. Chandra and Puri (1985) found significantly reduced thymulin activity in rats deprived of dietary vitamin B_6 and given the anti-vitamin deoxypyridoxine in their drinking water.

Lymphocyte egress from the thymus and secondary lymphoid organs requires a differential in sphingosine-1-phosphate (S1P) concentration at the site of transmigration and is critically dependent on expression of the sphingosine-1-phosphate receptor ($S1P_1$) by the lymphocytes. The concentration of S1P in the interstitial fluid of lymphoid organs is much lower than that in blood and lymph due to the action of S1P lyase, which cleaves S1P into phosphoethanolamine and 2-hexadecanal. S1P lyase is a vitamin B_6-dependent enzyme and therefore it might be supposed that a deficiency of vitamin B_6 would, by preventing lymphocyte egress, cause lymphocytes to accumulate in lymphoid organs. This would lead to a decrease in the number of circulating lymphocytes (lymphopenia) and eventual immune deficiency. Evidence to support this supposition is a report that treating mice with deoxypyridoxine caused a build-up of mature lymphocytes in the thymus and induced lymphopenia (Schwab *et al.*, 2005).

22.12.3 The Influence of Vitamin B_6 on Humoral Immune Responses

Vitamin B_6-deficient animals exhibit a marked impairment of circulating antibody response to various antigenic stimuli, including diphtheria toxoid (Pruzansky & Axelrod, 1955; Axelrod *et al.*, 1961) and influenza virus (Axelrod & Hopper, 1960). Vitamin B_6 is required for both the primary and secondary phases of the antibody response (Axelrod & Trakatellis, 1964a). In vitamin B_6-deficient rats immunized with sheep red blood cells, the reduction in circulating antibodies was accompanied by a decreased number of antibody-forming cells in the spleen. The cell numbers were restored to normal by the administration of pyridoxine shortly before immunization (Kumar & Axelrod, 1968).

22.12.4 The Influence of Vitamin B_6 on Cell-Mediated Immune Responses

Robson and Schwarz (1975) reported a dramatic 85–95% reduction in the number of thoracic duct lymphocytes and a significant reduction in cellular immunocompetence in vitamin B_6-deficient rats. These conclusions were based on the results of two tests: (1) the *in vitro* mixed lymphocyte reaction (MLR) and (2) the *in vivo* normal lymphocyte transfer reaction (NLT). In the MLR, lymphocytes from test Lewis strain rats (in this case, vitamin B_6-deficient and control rats) are cultured with genetically dissimilar lymphocytes taken from normally nourished F_1 hybrid rats. If the lymphocytes are immunocompetent, they will become activated and then they will proliferate and transform into the larger lymphoblasts. The extent of blastogenesis is quantitated by exposing the cultures to [^{3}H]thymidine and then measuring the incorporation of the radioactivity into DNA. In the NLT, lymphocytes from donor Lewis rats (the test rats) are injected into the ventral abdominal wall of F_1 hybrid rats. Immunologically competent donor cells produce a graft-versus-host reaction in the skin of the F_1 rat. The impaired proliferation of lymphocytes and loss of cellular immunocompetence may perhaps be attributed to a cessation of T lymphocyte development within the thymus of the vitamin B_6-deficient animal.

Ha *et al.* (1984a) reported significantly reduced primary and secondary T cell-mediated cytotoxicity in vitamin B_6-deficient mice. The level of vitamin B_6 intake had no effect on complement-dependent antibody-mediated cytotoxicity, phagocytic responses of macrophages, or natural killer (NK) cell activity. Trakatellis *et al.* (1992) showed that vitamin B_6 deficiency impairs the synthesis of IL-2, which is required for T cell and B cell proliferation.

22.12.5 Effect of Vitamin B_6 Deficiency on the Induction of Immune Tolerance

Axelrod *et al.* (1963) demonstrated that vitamin B_6-deficient guinea pigs immunized with BCG (attenuated *Mycobacterium tuberculosis*) exhibited a depressed delayed-type hypersensitivity skin response to purified protein derivative (PPD), an antigen derived from *M. tuberculosis*. The animals rapidly acquired reactivity when supplied with vitamin B_6. However, results of *in vitro* tests of cellular sensitivity and passive transfer experiments indicated that the sensitization mechanism in vitamin B_6-deficient animals had not been inhibited at the cellular level. Having ruled out skin tissue changes and inanition (caloric restriction) as causative factors, the authors explained the altered skin sensitivity by a decreased ability of sensitized cells to react with PPD. They speculated that pyridoxine or its coenzyme (PLP) is an essential component in the sequence of reactions between the sensitized cell and antigen. These experimental results suggest that vitamin B_6 deficiency states, induced by dietary means and/or the administration of deoxypyridoxine, may be utilized to ameliorate the undesirable consequences of hypersensitivity reactions.

Delayed-type hypersensitivity reactions are known to be a major cause of tissue rejection. Since vitamin B_6 deficiency is associated with a depressed delayed-type hypersensitivity reaction, it is logical to suppose that the deficiency would also prolong the life of a skin graft. Indeed, vitamin B_6-deficient rats of certain strains rejected allografts more slowly than did animals supplied with the vitamin (Axelrod *et al.*, 1958; Fisher *et al.*, 1958). This finding indicates that vitamin B_6 deficiency can inhibit the immune response to foreign antigens, thereby prolonging the survival of an allograft. However, the effects of vitamin B_6 deficiency are non-specific and may also increase susceptibility to infection in transplant recipients. It therefore becomes necessary to induce a specific state of tolerance to donor antigens. It is well documented that tolerance to tissue allografts can be induced in adult mice by injecting viable splenic cells genetically identical with those of the skin donor (Brent & Gowland, 1962).

Axelrod and Trakatellis (1964b) demonstrated that a dose of splenic cells ineffective in inducing tolerance to skin grafts in adult mice can be rendered effective if the recipients are deficient in vitamin B_6 at the time of splenic cell administration. This study used male mice of the CBA/J strain as recipients and male mice of the C3H/HeJ strain as splenic cell and skin donors. The recipient mice were fed a purified complete (control) diet during 0 to 14 days of the experiment, after which they received the purified diet with pyridoxine omitted and were given daily injections of deoxypyridoxine during the next 14 to 35 days. The C3H/HeJ donor mice were fed the control diet throughout the experiment. Splenic cells from the C3H/HeJ mice were administered on day 28. From day 35, deoxypyridoxine treatment was discontinued and the recipient mice resumed consumption of the control diet for the remainder of the experiment. Growth response was immediate and on day 47 the mice received a skin graft from a C3H/HeJ donor. Since these allografts survived for longer periods of time than controls, the mice were judged to be tolerant of the allogeneic tissue.

22.12.6 Mode of Action of Vitamin B_6 in Immune Responses

In vitamin B_6-deficient rats, the spleens show a diminished synthesis of DNA and RNA (Trakatellis & Axelrod, 1965) and reduced polysome formation (Montjar *et al.*, 1965). *In vitro* incorporation of tritiated thymidine into splenic cells of vitamin B_6-deficient rats is dramatically impaired, demonstrating the diminished synthesis of DNA (Axelrod, 1971). Thus, the inhibitory effects of vitamin B_6 deficiency on the immune response could be related to the reduced cell proliferation resulting from an inadequate supply of DNA, as well as to a decreased production of specific mRNAs essential for the synthesis of antibodies. It is known that administration of an antigen stimulates an intensive multiplication of host cells in the spleen and lymph nodes, which increases the requirements for DNA. Vitamin B_6 is required for DNA and RNA synthesis because of its involvement in the generation of the necessary one-carbon units from serine, therefore its absence would impair antibody synthesis.

22.13 VITAMIN B_6 AND PLATELET AGGREGATION

Kobzar *et al.* (2009) studied the effect of different combinations of vitamin B_6 vitamers on platelet aggregation *in vitro*. When tested separately, the vitamers only weakly inhibited aggregation. The IC_{50} values for PL, PN, and PM did not differ significantly from each other and only PLP was significantly more effective ($P<0.05$) than other vitamers (Table 22.3). The inhibition of ADP-induced platelet aggregation by pairwise combinations of the vitamers was significantly stronger than that of single vitamers. The combinations of three and four vitamers were yet more effective, inhibiting aggregation at the concentration of 4

TABLE 22.3
IC_{50} Values for the Inhibition of ADP-Induced Platelet Aggregation with the Vitamers of Vitamin B_6

B_6 Vitamer	IC_{50} (mM)
Pyridoxal 5′-phosphate (PLP)	0.14 ± 0.04
Pyridoxal (PL)	1.3 ± 0.3
Pyridoxine (PN)	1.5 ± 0.2
Pyridoxamine (PM)	1.6 ± 0.4

The data are mean ± SEM for the blood samples of five patients.
IC_{50} is a vitamer concentration causing a 50% inhibition of ADP-induced platelet aggregation.
Source: from Kobzar *et al.* (2009).

μM. This concentration is equal to that found in the blood plasma of humans after supplementation of 100 mg vitamin B_6 daily (Ubbink *et al.*, 1987). Evidently, the blood concentration of vitamin B_6 found after vitamin B_6 supplementation can inhibit platelet aggregation. Indeed, treatment of male volunteers with a single daily oral dose of 5 mg/kg vitamin B_6 for four weeks inhibited ADP- or adrenaline-induced platelet aggregation by 48% and 41% ($P<0.001$), respectively, whereas there was no change with placebo treatment (Sermet *et al.*, 1995).

Kobzar *et al.* (2009) showed that the inhibition of platelet aggregation by the mixture of vitamin B_6 vitamers is mediated through IP receptors that can be activated by prostacyclin as well as by PGE_1. Prostacyclin, which is the main product of arachidonic acid metabolism in the endothelial cells of blood vessels and is not synthesized in platelets, can inhibit platelets through IP receptors only *in vivo.* PGE_1, on the other hand, can be synthesized in platelets from dihomo-γ-linolenic acid and can mediate the inhibition of platelet aggregation also *in vitro.* It may therefore be suggested that the combined B_6 vitamers activate PGE_1 synthesis in platelets, leading to the inhibition of aggregation through IP receptors. Platelet aggregation can also be inhibited through DP receptors by PGD_2 and by nitric oxide produced in platelets by nitric oxide synthase (NOS) (Schwarz *et al.*, 2001). However selective inhibition of DP receptors and NOS did not prevent the inhibitory effect of the vitamers, indicating that PGD_2 and NO do not participate in the inhibition of platelet aggregation by vitamin B_6.

22.14 VITAMIN B_6 STATUS AND CORONARY ARTERY DISEASE

A borderline vitamin B_6 deficiency (plasma PLP concentration <30 nmol/L) is strongly associated with the risk of coronary artery disease (CAD) in humans (Lin *et al.*, 2006). It is well established that moderate elevation in plasma homocysteine concentration is a clear risk factor for CAD (Section 25.15.2). Because the catabolism of homocysteine through cystathionine β-synthase requires PLP, the association between low plasma PLP and CAD has been presumed to be due to an elevation of plasma homocysteine. Measurements of total plasma homocysteine were not available in a number of the early studies, therefore the association of low PLP and CAD was not adjusted for this amino acid. Miller *et al.* (1992) showed that fasting plasma homocysteine concentrations are not elevated in vitamin B_6 deficiency. This is probably because of the ability of *S*-adenosylmethionine to activate the homocysteine catabolic enzyme cystathionine β-synthase, despite a decrease in the availability of this enzyme's co-factor, PLP. It is now accepted that the association between vitamin B_6 and the risk of CAD is independent of the homocysteine pathway, at least for a significant part (Friso *et al.*, 2012).

Atherosclerosis is a chronic inflammatory response that has become excessive. Among markers of inflammation, acute-phase proteins are proven to be strong and independent risk factors for CAD (Ridker *et al.*, 2000). In a case-control study with 742 participants, Friso *et al.* (2004) observed an inverse and homocysteine-independent relation between plasma concentrations of PLP and two major markers of inflammation – high-sensitivity C-reactive protein (hs-CRP) and fibrinogen. This finding in a group of CAD subjects substantiated and extended a previous observation in a population-based study (Friso *et al.*, 2001). The CAD risk as a result of low PLP was enhanced by an increased ratio of LDL to HDL as well as by an elevated hs-CRP concentration.

In a nested case-control study based on the Nurses' Health Study cohort and including 144 cases of myocardial infarction (of which 21 were fatal), women in the highest quartile of blood PLP levels (≥70 nanomoles/liter) had a 79% lower risk of myocardial infarction compared to those in the lowest quartile (<27.9 nanomoles/L (Page *et al.*, 2009). A prospective study that followed a Japanese cohort of over 40,000 middle-aged individuals for 11.5 years reported a 48% lower risk of myocardial infarction in those of the highest quartile (mean, 1.6 mg/day) versus the lowest quartile (mean, 1.3 mg/day) vitamin B_6 intake in non-supplement users (Ishihara *et al.*, 2008).

22.15 VITAMIN B_6 STATUS AND DIABETES

Vitamin B_6 deficiency in rats impairs the pancreatic secretion of insulin and glucagon and lowers circulating insulin levels (Toyota *et al.*, 1981). Depressed insulin levels may be the direct result of the deficiency of vitamin B_6, which is known to function in amino acid and protein metabolism, or an adaptive mechanism to the deficiency. Adaptive mechanisms could include the effects of hormonal systems, such as growth hormone (GH). Malaisse *et al.* (1968) showed that GH had no apparent immediate effect on insulin secretory activity of the pancreatic β cell *in vitro.* Pancreatic tissue removed from hypophysectomized rats secreted an abnormally low amount of insulin in response to stimulation by glucose. Such secretion rose toward the normal after treatment of the rats with ovine GH for four days. Prolonged administration of GH, in the form of implanted GH-secreting tumors, to normal rats increased rate of growth, the store of pancreatic insulin, and the rate of insulin secretion induced by glucose *in vitro*. These findings support the concept that GH, by some chronic process, can increase the sensitivity of the β cells to glucose and probably stimulates synthesis of insulin. As serum and pituitary GH levels are decreased in vitamin B_6-deficient rats (Makris & Gershoff, 1973), a deficiency in vitamin B_6 could decrease insulin synthesis, at least in part, through a deficiency in GH.

Kuno (1960) showed degenerative changes in β cells in the islets of Langerhans of rats fed a vitamin B_6-deficient diet. A significant proportion of people with either type 1 or type 2 diabetes mellitus have been shown to have low plasma levels of PLP (Hollenbeck *et al.*, 1983; Hamfelt & Soderhjelm, 1983). Related to this is the observation that acute ingestion of glucose results in a significant decrease in plasma PLP concentration for at least five hours (Leklem & Hollenbeck, 1990). In vitamin B_6 deficiency, a large amount of xanthurenic acid is produced because the main pathway of tryptophan breakdown is blocked. Xanthurenic acid forms a complex with insulin and inhibits glucose uptake into fat cells (Kotake *et al.*, 1975).

22.16 VITAMIN B_6 STATUS AND COLORECTAL CANCER

22.16.1 Introduction

Vitamin B_6 is a coenzyme for serine hydroxymethyltransferase, which catalyzes the transfer of the one-carbon β-hydroxymethyl group of serine to tetrahydrofolate (THF), forming glycine and 5,10-methylene-THF. The one-carbon moiety of 5,10-methylene-THF can be directed into pathways that lead to the synthesis of methionine and DNA base constituents. A deficiency of vitamin B_6 will impair DNA synthesis, repair, and methylation, thereby increasing the risk of cancer development (Ames, 2001). Several animal studies have shown a protective role for vitamin B_6 against colorectal carcinogenesis (Matsubara *et al.*, 2003). Vitamin B_6 may suppress colorectal cancer by reducing cell proliferation, oxidative stress, nitric oxide production, and the recruitment of new capillary blood vessels by tumor cells (angiogenesis) (Matsubara *et al.*, 2003). Although many studies focus on colorectal cancer (colon cancer and rectal cancer combined), the risk factors for the two types of cancer may differ (Iacopetta, 2002; Wei *et al.*, 2004).

22.16.2 Laboratory Studies of Colorectal Carcinogenesis

DiSorbo and Litwack (1981) showed that the culturing of FAZA rat hepatoma cells in medium supplemented with high concentrations (5 mM) of pyridoxine significantly elevated the intracellular concentration of PLP. DiSorbo and Litwack (1982) then showed that culturing Fu5-5 rat hepatoma cells, rat glial cells, and human kidney cells in the 5 mM pyridoxine-supplemented medium retarded cell growth, and raising the pyridoxine level to 10 mM killed the cells. The preliminary data suggested that pyridoxine causes cell death through conversion to PLP, which blocks DNA synthesis by inhibiting DNA polymerase. Pyridoxine is converted to PLP by pyridoxine (pyridoxamine) 5′-phosphate oxidase, which has high activity in liver, brain, and kidney. The pyridoxine-supplemented medium had little effect on the growth of the human mammary cancer cell line MCF-7, due probably to the low level of the oxidase in these cells.

DiSorbo and Nathanson (1983) studied the growth characteristics of a human malignant melanoma cell line (designated NEL) cultured in pyridoxal-supplemented medium. At the end of a 72-hour growth period, a 0.5 mM concentration of pyridoxal inhibited cell growth by 90%, and a 0.1 mM concentration by 35%. Both pyridoxal and PLP were found to be necessary to produce the inhibitory effect.

DiSorbo *et al.* (1985) demonstrated that vitamin B_6 can inhibit the growth of B16 melanoma cells both *in vitro* and *in vivo*. In the *in vitro* studies, melanoma cells grown for three days in medium supplemented with 5.0 mM pyridoxine or 0.5 mM pyridoxal showed an 80% reduction in cell proliferation compared with control culture. Culturing the cells for more than 48 hours in 2.5 mM pyridoxal-supplemented medium resulted in cell death. Co-incubation of deoxypyridoxine with pyridoxal decreased the growth inhibition of the melanoma cells. These findings indicate that the B16 melanoma cells lack the pyridoxine (pyridoxamine) 5′-phosphate oxidase required to convert pyridoxine to PLP. The conversion of pyridoxal to PLP is rapid because all cells contain pyridoxal kinase. In the *in vivo* studies, mice pretreated with pyridoxal (0.5 g/kg body weight) for two weeks and then injected with melanoma cells had a 62% reduction in tumor weights compared with controls three weeks after injection. If tumors were first established in mice and then treated with pyridoxal (0.5 g/kg body weight) for six days, a 39% reduction in tumor growth was observed. There were no differences in body weights or liver weights in any of the animal groups.

Molina *et al.* (1997) found that the growth of human hepatoma HepG2 cells was completely inhibited in medium supplemented with PN in the millimolar range. The growth inhibition of HepG2 cells was accompanied by a marked inhibition of gene expression, protein synthesis, and secretion of serum proteins, particularly albumin. Pyridoxal was as effective as PN, but pyridoxamine showed no inhibitory action. Morphological features of the PN-treated HepG2 cells were compatible with depressed transcriptional activity in these cells.

With regard to the mechanism by which PLP inhibits cell growth, PLP has been shown to inhibit various enzymes through Schiff base formation between the aldehyde group of PLP and the free amino group(s) present at the enzyme's active site. For example, PLP inhibits various glycolytic and pentose phosphate shunt enzymes (Srivastava & Beutler, 1972), RNA polymerases (Martial *et al.*, 1975), and DNA polymerases (Modak, 1976). PLP has also been shown to induce a conformational change of the transfer RNA molecule through specific binding to guanine at position 20 of the dihydrouridine loop, resulting in the inhibition of amino acid acceptance and ribosomal binding (Kopelovich & Wolfe, 1977). It could be postulated that PLP inhibits cell growth by reducing the cell's energy source and/or limiting the synthesis of macromolecules needed for cell proliferation.

Gridley *et al.* (1987) examined the effects of dietary vitamin B_6 at intake levels ranging from deficient to megadose levels on the development of herpes simplex virus type 2-transformed (H238) cell-induced tumors. Mice (n = 120) were fed 20% casein diets containing pyridoxine at 0.2, 1.2 (control), 7.7, or 74.3 mg/kg diet for four to 11 weeks. After four weeks of dietary treatment, 120 of the mice received an injection of H238 cells; mice without H238 injection served as controls. Strikingly, the control mice fed 1.2 mg pyridoxine had the largest primary tumor volume, the highest incidence of lung metastases, and the greatest number of metastatic nodules per animal at seven weeks post-injection. Mice fed 7.7 and 74.3 mg pyridoxine had tumor volumes that were 14 and 32% less, respectively, than those of the control mice. Mice fed 0.2 mg pyridoxine (the deficient level) had the lowest tumor volume. The results of this study suggest that high dietary intake of vitamin B_6 may have suppressed tumor development by enhancing immune function, whereas low dietary intake may have inhibited tumor development by suppressing important metabolic processes related to cell growth. In contradiction to the results of this study, Ha *et al.* (1984b) found that a comparable vitamin B_6 deficiency *enhanced* growth of tumors induced by the Maloney sarcoma virus. These investigators suggested that vitamin B_6 deficiency may have impaired the immune system of the mice, allowing greater tumor growth.

Komatsu *et al.* (2001) studied the effect of supplemental dietary vitamin B_6 on azoxymethane-induced colon tumorigenesis in mice. Supplementation at the 7, 14, and 35 mg/kg diet levels significantly suppressed the incidence of tumors (Table 22.4) and the number of tumors per mouse compared with the 1 mg/kg level. The tumor volumes did not differ significantly among all four groups of mice. The majority of the tumors were identified as adenomas and were located mainly in the distal colon. Proliferation of cells in all of the colon epithelium correlated with the number of tumors and was significantly reduced by supplemental vitamin B_6 at the 7, 14, and 35 mg/kg levels compared with the 1 mg/kg level. Cell proliferation in the distal colon did not differ significantly among the 7, 14, and 35 mg/kg vitamin B_6-supplemented groups. The 7, 14, and 35 mg PN.HCl/kg diets also significantly suppressed expressions of the protein products of the c-*myc* and c-*fos* proto-oncogenes in the colonic crypts. Overall, the greatest suppression of tumorigenesis by dietary vitamin B_6 was observed in the mice fed 14 and 35 mg PN.HCl/kg diets. Vitamin B_6 had no significant effect on colonic epithelium apoptosis. In the mice fed the 35 mg PN.HCl/kg diet, no toxicity symptoms were observed throughout the entire feeding period. The results suggest that supplementing diets with vitamin B_6 suppresses colon tumorigenesis by reducing cell proliferation.

TABLE 22.4
Effect of Dietary Vitamin B_6 on the Incidence of Colon Tumors in Azoxymethane-Treated Mice

Pyridoxine HCl (mg/kg Diet)	Number of Mice with Colon Tumors		
	Total	Adenoma	Adenocarcinoma
1	13[a]	10[a]	4
7	5[b]	2[b]	3
14	2[b]	2[b]	0
35	1[b]	1[b]	0

Thirty-four mice per diet group were examined for tumorigenesis.
Values in a column not sharing a superscript letter are significantly different (χ^2 test, $P<0.05$).
Source: from Komatsu *et al.* (2001).

22.16.3 Epidemiological Studies of Colorectal Cancer Risk

The overall conclusion from epidemiological studies is that low vitamin B_6 intake and low blood PLP concentrations are associated with the risk of developing colon cancer, but not rectal cancer. Epidemiological studies that examine vitamin B_6 intake and colorectal cancer risk are confounded by factors associated with health. Although some of these factors have been taken into account in the statistical analysis, there is always the possibility that the inverse associations that have been reported might be simply due to correlations between vitamin B_6 intake and healthy behaviors (Zhang *et al.*, 2013).

22.17 VITAMIN B_6 DEFICIENCY DURING BRAIN DEVELOPMENT

Vitamin B_6 is critical at certain stages of brain development (Kirksey *et al.*, 1990). A deficiency of vitamin B_6 leads to incomplete myelination of nerve fibers by impairing the synthesis of sphingomyelin and very-long-chain fatty acids (Dakshinamurti, 1982). Progeny of rats fed a vitamin B_6-deficient diet during the last week of gestation displayed neurological abnormalities that culminated in convulsive seizures accompanied by high-pitched cries and frequently ended in death (Dakshinamurti *et al.*, 1990). Electroencephalograms (EEGs) obtained from vitamin B_6-deficient rats exhibited high-voltage spikes and wave discharges in contrast to slow-wave activity in control rats. Levels of the inhibitory neurotransmitter γ-aminobutyric acid (GABA) in the cerebellum of these animals were significantly decreased when compared with appropriate controls (Stephens *et al.*, 1971). Wasynczuk *et al.* (1983) observed decreased GABA levels in the extrapyramidal motor system of vitamin B_6-deficient 15-day-old rats. Injection of these rats with PN.HCl resulted in increased GABA concentration in the substantia nigra after three hours. In this study, numbers of synapses per millimeter square were significantly less in the caudate nucleus and putamen of the vitamin B_6-deficient pups compared to controls. Fewer of these synapses in conjunction with decreased GABA are conducive to ineffective nerve inhibition in vitamin B_6 deficiency.

22.18 EXPERIMENTAL VITAMIN B_6 DEFICIENCY IN ANIMALS

Domesticated animals can be rendered vitamin B_6-deficient by formulating a diet that lacks the vitamin. The following deficiency signs have been observed:

- **Growth.** The retention of nitrogen is impaired and growth is retarded. Appetite is diminished and the feed is poorly utilized.
- **Skin.** Lesions of the skin are a sign of vitamin B_6 deficiency in many species. Acrodynia in the rat is characterized by severe inflammation and encrustation of eyes, nose, paws, and tail; flaking and thickening of the skin of the ears; depletion of the subcutaneous layer of fat; and atrophy of the hair follicles and sebaceous glands. The higher the level of protein in the feed, the more severe is the acrodynia.
- **Nervous system.** Neurological disorders occur as ataxia, increased irritability, and convulsions; these disorders are probably the result of a reduction in the activity of vitamin B_6-dependent enzymes in the brain – enzymes such as γ-aminobutyric acid transaminase. Demyelination of peripheral nerves has been demonstrated histologically in pigs.
- **Blood.** A common blood disorder is microcytic hypochromic anemia, the condition in which the number and size of the erythrocytes are considerably reduced and the concentration of hemoglobin rapidly falls. The impaired synthesis of porphyrin explains the lowered concentration of hemoglobin. A reduced leukocyte count is a consequence of the atrophy of lymphatic tissue. There is also a change in the ratio of certain proteins in the blood serum. The level of *total* protein in the serum remains constant, but the proportions of the albumins and γ-globulins are reduced, whilst the α-globulin is proportionately increased. These changes adversely affect the resistance to disease because the γ-globulins function as carriers of the antibodies.
- **Immune system.** Impairment of humoral and cell-mediated immune responses.
- **Reproductive system.** Reproduction is seriously impaired because of the steroid hormonal disorders which are induced by a deficiency of vitamin B_6. Spermatogenesis is reduced in male animals and the breeding female experiences metabolic disorders that impair the growth of the fetus. In poultry, there is a fall in egg production and a lowered hatchability.

22.19 VITAMIN B_6 DEFICIENCY IN HUMANS

Vitamin B_6 deficiency in humans is defined as a plasma PLP concentration of less than 20 nmol/L (Institute of Medicine, 1998).

22.19.1 Deficiency Symptoms

A vitamin B_6 deficiency state in humans has not been recognized under natural conditions, nor have clinical manifestations of deficiency been induced in adult subjects by dietary restrictions alone. Vilter *et al.* (1953) reported the clinical effects of vitamin B_6 deficiency induced by deoxypyridoxine in fifty consenting patients with a wide variety of chronic diseases. Thirty of these subjects were given diets deficient in vitamins of the B complex, while twenty were on regular hospital diets. Both groups of patients developed clinical signs that are symptomatic of other B vitamin deficiency states. Cheilosis, angular stomatitis, and glossitis, which occur in vitamin B_2 deficiency, did not respond to riboflavin supplementation, but disappeared when pyridoxine was administered. The pellagra-like dermatitis that characterizes niacin deficiency did not respond to nicotinamide, but did so to pyridoxine. Neuropathy, a sign of vitamin B_1 deficiency, did not respond to thiamin, but did so to pyridoxine.

In a well-controlled study conducted by Hodges *et al.* (1962), six healthy male volunteers were divided into pairs and given a basic formulated diet administered by nasogastric tube twice daily. The first pair of men received a complete formula, including pyridoxine; the second pair received the same diet but without pyridoxine; and the third pair were given deoxypyridoxine in addition to the pyridoxine-free diet. The men receiving pyridoxine-free diets, but not those receiving complete diets, developed adverse symptoms and signs of illness, which were more severe in the men given the anti-vitamin. The most obvious symptoms were gastrointestinal disturbances and epithelial changes. Both men in the anti-vitamin group had scaling of the skin, foul breath, severe gingivitis, soreness and discoloration of the tongue, and dry cracked lips. No objective neurological changes could be demonstrated. After vitamin B_6 was restored to the diet and the anti-vitamin discontinued, one man recovered promptly and the other recovered gradually.

Elderly people are more likely to exhibit vitamin B_6 deficiency than young people. This is demonstrated by an increased urinary excretion of xanthuronic acid in response to an oral load of tryptophan. indicating a reduced capacity to complete the normal process of tryptophan metabolism (Ranke *et al.*, 1960).

22.19.2 Immunological Signs of Vitamin B_6 Deficiency

Subjects consuming vitamin B_6-free diets or given deoxypyridoxine exhibited lymphopenia (Vilter *et al.*, 1953; Cheslock & McCully, 1960) and low antibody responses to tetanus and typhoid (Hodges *et al.*, 1962). Elderly subjects with biochemical evidence of vitamin B_6 deficiency experience a significant decrease in the responsiveness of their peripheral blood lymphocytes to the mitogens concanavalin A and PHA. Increasing the intake of vitamin B_6 improves the responsiveness and eventually normalizes vitamin B_6 status (Merrill & Henderson, 1987).

22.19.3 Effects of Vitamin B_6 Supplementation and Deprivation on Vitamin Status

Rose *et al.* (1976) studied the effect of age on vitamin B_6 metabolism in 617 healthy men aged 18 to 90 using plasma PLP as the primary criterion of vitamin B_6 status. About one-third of the subjects were taking supplementary vitamins on their own initiative. The average plasma PLP of the men not taking a supplement ($n = 414$) was 12.3 ± 0.3 ng/ml, with 25% of the values below 7.5 ng/ml and 7% below 5 ng/ml. There was a statistically significant decrease in plasma PLP with age of 0.9 ng/ml per decade. For those taking a supplement, the average plasma PLP was 20.5 ± 1.0 ng/ml, with only 8% of the values below 7.5 ng/ml and none below 5 ng/ml.

Talbott *et al.* (1987) studied the effect of vitamin B_6 supplementation on lymphocyte responses in fifteen healthy women aged 65–81 years. Eleven subjects received 50 mg/day PN.HCl and the other four subjects (controls) received a placebo. Lymphocyte proliferation to T and B cell mitogens, lymphocyte subpopulations, and plasma PLP were measured before and after one and two months of supplementation. *In vitro* lymphocyte proliferation from both T and B cell mitogens increased as a result of supplementation, particularly in individuals with initially low plasma PLP concentrations. After two months of supplementation there was an increase in the percentage of $CD4^+$ T helper cells and no change in the percentage of $CD8^+$ cytotoxic T cells. This study suggests that cell-mediated immune function in the elderly may be improved by vitamin B_6 supplementation. However, the dose of pyridoxine provided was more than 25 times the recommended dietary allowance (RDA) for this age group and therefore the risk of chronic toxicity must be considered.

Meydani *et al.* (1991) studied the effect of vitamin B_6 deficiency on immune response in eight healthy adults (four males and four females) aged 61–67 years. The protocol consisted of the following steps: a five-day baseline period during which the subjects ate self-selected diets; a vitamin B_6-depletion period of up to 20 days during which the subjects were provided with a formulated diet devoid of vitamin B_6; a vitamin B_6 repletion period of three stages, each lasting 21 days, in which the amount of vitamin B_6 ingested was increased to 2.88 ± 0.17 mg/day for males and 1.90 ± 0.18 mg/day for females in the third stage; and a four-day final period during which the subjects were given 50 mg/day of PN.HCl. During vitamin B_6 depletion, all subjects showed significant decreases in the percentage and total number of lymphocytes, and also IL-2 production, that correlated with a decreased plasma PLP and an increase in post-tryptophan urinary xanthurenic acid excretion. Significant decreases in the proliferative responses of peripheral blood lymphocytes to T and B cell mitogens were also observed. These indices of immune response returned to baseline levels after the three 21-day stages of vitamin B_6 repletion. These findings suggest that for maintenance of normal immune function, elderly people may require amounts of vitamin B_6 higher than those currently recommended.

22.19.4 Vitamin B_6 Deficiency in Infants

In the 1950s, a widespread occurrence of convulsions in infants was traced to an unfortified liquid milk-based canned formula that had undergone heat-sterilization in manufacture, rendering it deficient in vitamin B_6. The convulsions ceased when the formula was changed to evaporated milk (Coursin, 1954; Molony & Parmelee, 1954). It was subsequently discovered that the thermal processing of milk products and other animal-derived foods promotes the chemical reduction of protein-bound Schiff base forms of PL and PLP to peptide-linked ε-pyridoxyllysine or its 5′-phosphate derivative (Gregory & Kirk, 1977; Gregory *et al.*, 1986). ε-Pyridoxyllysine has 50% vitamin B_6 activity relative to PN and exhibits antivitamin B_6 activity when fed to rats at low levels in vitamin B_6-deficient diets (Gregory, 1980). The antagonistic effect of ε-pyridoxyllysine would exacerbate the loss of naturally occurring vitamin B_6 during heat sterilization. PN-HCl, the form of vitamin B_6 used in the fortification of foods, is stable during retort processing (Gregory & Hiner, 1983).

Investigations into the neurochemical basis of the infantile convulsions seen in vitamin B_6 deficiency point to an imbalance between excitatory and inhibitory neurotransmitters. The various neurotransmitters are synthesized from amino acids by PLP-dependent decarboxylases. In the vitamin B_6-deficient rat, there is a decrease in the brain content of inhibitory GABA and a concomitant increase in excitatory glutamate from which GABA is formed (Dakshinamurti *et al.*, 1990). Notably, convulsions have not been observed in vitamin B_6-deficient adult rats despite significant reductions in brain GABA levels (Bayoumi *et al.*, 1972). Therefore, it is possible that additional factors are necessary for producing convulsions in vitamin B_6-deficient neonatal rats. One such factor may be the tryptophan metabolite 3-hydroxykynurenine, which accumulates during vitamin B_6 deficiency due to low activity of PLP-dependent kynureninase. 3-Hydroxykynurenine induces convulsions when injected into brain ventricles of adult mice (Lapin, 1978a) or intraperitoneally in immature rats (Lapin, 1978b). Guilarte and Wagner (1987) found that endogenously produced 3-hydroxykynurenine accumulates to extremely high concentrations in the brains of vitamin B_6-deficient neonatal (but not adult) rats.

22.20 VITAMIN B_6 DEPENDENCY

Vitamin B_6 dependency is a rare, genetically determined disorder that is distinct from vitamin B_6 deficiency. Whereas the dietary requirement of vitamin B_6 for a normal adult is 1.3–1.7 mg/day, some vitamin B_6-dependent conditions require as much as 200–600 mg daily of PN.HCl to effect a maximal clinical response (Frimpter *et al.*, 1969). At least five vitamin B_6-dependent syndromes have been described: infantile convulsions, hypersideremic anemia, xanthurenic aciduria, cystathioninuria, and homocystinuria.

22.20.1 Infantile Convulsions

Hunt *et al.* (1954) reported that seizures occurring in two infant siblings within two weeks of birth could not be controlled by anticonvulsants, but could be controlled by daily intramuscular injection of large doses of PN.HCl. If not treated with pyridoxine, the infants developed severe mental retardation. Since this first report, approximately 100 similar cases in neonates and infants have appeared in the literature (Baxter, 1999). Lott *et al.* (1978) described the neuropathological and biochemical changes in the brain of a child who had died from clinically proven vitamin B_6-dependent seizures. This syndrome is an example of an autosomal recessive inborn error of metabolism.

There is considerable evidence that underproduction of GABA is primarily responsible for the vitamin B_6-dependent seizures (Wood, 1975). GABA is formed from glutamic acid through the action of glutamic acid decarboxylase (GAD), which requires PLP as a coenzyme. It has been proposed that a mutation of the *GAD* gene alters the PLP-binding site of GAD and impairs its affinity for the coenzyme, thus reducing enzyme activity. The resultant lowering of cerebral GABA levels and decrease in nerve inhibition accounts for the seizures. Intravenous administration of pyridoxine results in sufficient binding of PLP to the apoenzyme, despite the poor affinity of the binding site (Dakshinamurti, 1982).

22.20.2 Enzymatic B_6 Dependency Syndromes

The following syndromes are attributable to genetic defects in vitamin B_6-dependent enzymes and are examples of an inborn error of metabolism. Hypersideremic anemia results from failure of iron utilization for hemoglobin synthesis. In this syndrome, the defective enzyme is δ-aminolevulinate synthase, which catalyzes the condensation of glycine with succinyl-CoA to form δ-aminolevulinic acid. In xanthurenic aciduria, cystathioninuria, and homocystinuria, there is excessive urinary excretion of xanthurenic acid, cystathionine, and homocystine, respectively. The respective enzymes that are affected are δ-aminolevulinate synthase, kynureninase, and cystathionine synthase (Frimpter *et al.*, 1969).

22.21 NEUROTOXICITY

Many animal and human studies have proved conclusively that chronic megadosing with pyridoxine causes neurotoxicity. The pathogenesis and biochemical basis of the toxicity is unknown, but various theories have been proposed.

22.21.1 Animal Studies of Pyridoxine Neurotoxicity

Krinke *et al.* (1981) treated six dogs with daily oral doses of 300 mg/kg of PN.HCl and used two dogs as controls. The treated animals developed limb ataxia within nine days and, by day 30, several animals were unable to walk. However, the animals did not lose muscle strength. Microscopic examination of nervous tissue sampled throughout the nervous system revealed selective damage to the cell bodies of neurons in the dorsal root ganglia and trigeminal ganglia. Membrane-bound vacuoles in the cytoplasm of large neurons fused to form large cavities that eventually reduced the cytoplasm to a thin rim. Some neurons were necrotic, but many smaller diameter neurons appeared unaffected, even after prolonged intoxication. Wallerian-like degeneration was present throughout the length of sensory nerve fibers in the dorsal roots, dorsal funiculi of the spinal cord, descending tract of the trigeminal nerve in the brain stem, and in scattered fibers of all peripheral nerves sampled. The changes were especially prominent in the sural nerve (a sensory nerve in the hind limb), but even in this nerve only selected fibers were damaged. No changes were detected in the other sampled areas of the central or peripheral nervous systems. It appeared that axonal degeneration occurred secondary to deterioration of the parent cell body. This pattern of nervous system degeneration may be called a primary sensory neuronopathy, understood as a disruption of nerve cell metabolism. This disruption can cause death of the cells, rendering them unable to maintain their axons, which rapidly degenerate along their entire length. Schaeppi and Krinke (1982) showed that, in dogs, damage of afferent nerves from muscle spindles and from mechanoreceptors in the skin accounted for the observed functional lesions, such as loss of the patellar reflex and impairment of the tactile placing response. Small and slowly conducting nerve fibers associated with pain perception retained normal structure and function.

Xu *et al.* (1989) administered PN.HCl to rats by intraperitoneal injection using two dose regimens: short-term, high-dose and long-term, low-dose. Animals receiving the highest dose in the short term (1200 mg/kg for 10 days) developed severe limb and truncal ataxia. There was severe necrosis of dorsal root ganglion nerve cell bodies, accompanied by degeneration of their central and peripheral axons. None of the animals in the long-term, low-dose group (150 and 300 mg/kg/day for 12 weeks) developed ataxia. Abnormalities of dorsal root ganglion nerve cell bodies were minor, but degenerative changes were found in portions of their peripheral and central sensory axons. The neurons most affected by pyridoxine intoxication were lumbar dorsal root neurons. By virtue of their large axoplasmic volume and high physiological activity, these neurons presumably have the greatest metabolic requirements, rendering them particularly susceptible to toxins.

Windebank *et al.* (1985) reported that rats injected intraperitoneally with 200 mg/kg/day of PN.HCl for up to 60 days developed ataxia. There was a highly selective loss of sensory axons from dorsal root ganglion neurons, with relative sparing of neuron cell bodies. The animals recovered in 12 weeks after cessation of pyridoxine treatment and the axons regenerated. The results suggested that pyridoxine alters neuronal metabolism so that the neuron cannot support the axon and degeneration occurs. Providing the poisoning is not too severe, regeneration can occur after removal of the toxin. In this case, the neuron can provide axonal energy metabolism, axonal transport, and synthesis of proteins required for axonal integrity. In severe poisoning, the neuron will die and regeneration will not occur.

In order to investigate the underlying cause of pyridoxine neuropathy, Krinke *et al.* (1985) injected rats intraperitoneally with 600 mg/kg of PN.HCl twice daily for 1, 2, 3, 4, 8, and 14 days. The rationale for using this massive dose was to induce an early and obvious response and to determine the target site for the toxicity. Ataxia was first noticed after three days of treatment. By the eighth day, the rats were unable to walk, despite the preservation of muscle force. The primary histological change, observed in dorsal root neurons 24 hours after treatment, was a swollen membranous profile and focal clearing of the cytoplasmic matrix in both the axon hillock and the first axonal segment. A slight accumulation of neurofilaments in some cell bodies was also observed, indicating an inability to transport them into the axon. These initial changes proved to be reversible. After two days of treatment, the dorsal root ganglia contained many cell bodies that showed vacuolated cytoplasm, dispersion of the Nissl bodies, and peripheral displacement of the nucleus. Secondary degeneration of the peripheral processes was first observed on the day following the third day of treatment. After prolonged periods of treatment (up to 14 days) there was a widespread damage to the sensory nerve fibers in the dorsal spinal columns. Corticospinal motor fibers and motor neurons in the posterior horns and ventral spinal roots remained intact. The authors presumed that a critical amount of pyridoxine can damage the metabolism of the dorsal root neuron in the axon hillock area and thereby impair axonal maintenance by the cell body.

In summary, animal studies have shown that, in megadoses, pyridoxine causes a primary injury to the cell bodies of first-order sensory neurons residing in dorsal root and trigeminal ganglia, with secondary degeneration of their central and peripheral axons. When given in lower doses, however, pyridoxine produces Wallerian-like axonal degeneration, with relative sparing of the cell body.

22.21.2 Human Studies of Pyridoxine Neurotoxicity

Clinical studies that reported pyridoxine-induced neurotoxicity have been summarized by the Scientific Committee on Food (2000). In the following selected studies, elevated serum PLP levels after megadosing indicated high pyridoxine ingestion and absorption.

Schaumburg *et al.* (1983) reported seven adult cases of vitamin B_6 abuse in which daily oral doses of pyridoxine ranging from 2 to 6 g taken over periods of 2 to 40 months produced signs of chronic neurotoxicity. The patients were unable to perceive joint position, cold, vibration, light touch, and pinpricks, indicating a sensory peripheral neuropathy. Sensory-nerve conduction was slowed, but motor-nerve conduction was normal. There was a progressive numbness of the feet, an unsteady gait, and an inability to walk unaided. Three patients had the sensation of electric shocks shooting down the spine and into the legs and soles of the feet upon neck flexion (Lhermitte's sign), which signified an impairment of central conduction. Nerve biopsies in two patients demonstrated widespread non-specific axonal degeneration. Most patients improved gradually after discontinuation, but some patients still had residual symptoms. In a series of 40 cases, the time of onset and severity of toxic symptoms were closely related to pyridoxine dose levels. Individuals taking 500 mg/day did not experience the onset of symptoms for several years, while those taking 7–10 g/day experienced symptoms within two months (Schaumburg & Berger, 1988).

Albin *et al.* (1987) reported two unusual cases of acute sensory neuropathy in association with high-dose parenteral pyridoxine therapy. A husband and wife were admitted to hospital after ingesting large amounts of the poisonous mushroom *Gyromitra esculenta*. Pyridoxine was administered intravenously over a three-day period, with the woman receiving a total dose of 132 g and the man 183 g. The rationale is that the active toxin, monomethylhydrazine, inhibits a pyridoxine-dependent step in the synthesis of the neurotransmitter γ-aminobutyric acid. Both patients developed a loss of sensation over the entire body and ataxia. Twelve months after onset, neither patient was able to walk and neither recovered their joint position sense, making them unable to stand unaided.

In a study conducted by Berger *et al.* (1992), five healthy volunteers were given oral doses of pyridoxine in neurotoxic amounts. Three low-dose subjects (nos. 1, 2, and 3) took 1 g/day and two high-dose subjects (nos. 4 and 5) took 3 g/day. Clinical neurological examinations included the testing of deep tendon reflexes, appreciation of pinprick, cold, and vibration, motor strength, and gait. Physiological evaluations included quantitative measurement of sural nerve conduction velocity and sensory potential amplitude and the monitoring of vibration and cold thresholds. Within one day of symptom onset, pyridoxine was discontinued and serum PLP measurements were taken. Serum PLP levels in subjects 1, 2, and 4 were 650, 275, and 750 nM, respectively.

PLP levels were not available for subject 5. The surprisingly low level of 80 nM in subject 3 indicated poor absorption of the megadose. In subjects 1, 2 and 4, serum PLP levels returned to normal at 4, 2 and 6 weeks, respectively, after dosing was stopped. Two of the low-dose subjects (nos. 1 and 2) showed symptoms after 7 and 4.5 months of treatment, respectively. The other low-dose subject (no. 3) did not experience symptoms and treatment was discontinued after 14 months when sensory threshold results became abnormal. The high-dose subjects (nos. 4 and 5) showed symptoms after 3.5 and 1.5 months, respectively. The initial clinical symptoms in subjects 1 and 2 were numbness and pins-and-needles sensation in the toes. Subject 2 experienced mild gait imbalance and Lhermitte's sign concurrent with the sensations in the toes. Subjects 4 and 5 initially developed numbness and tingling in the toes concurrent with similar, although less intense, sensations in the fingertips. Mild gait imbalance was reported by subject 4 as an early symptom. All subjects (except no. 3) reported an increase in symptom intensity for two to three weeks after stopping pyridoxine administration ('coasting'). Sensory symptoms gradually resolved – in the low-dose group over four to six weeks and in the high-dose group over six to eight weeks. Elevated cold thresholds in the toes of low-dose subjects preceded lesser elevations of vibration thresholds. High-dose subjects had concurrent and relatively equal elevations of cold and vibration thresholds, which continued to increase after stopping pyridoxine administration. Sural nerve conduction velocity and sensory potential amplitude were evaluated in subjects 1, 3, and 4. The only change was a reduction in potential amplitude observed in subject 1.

Kulkantrakorn (2014) studied three patients who had taken a combination tablet containing 100 mg of vitamin B_1, 200 mg of vitamin B_6, and 200 μg of vitamin B_{12}, one tablet three times a day. Two patients took this dosage (600 mg of pyridoxine per day) over three years, and the third over ten years. These patients, aged 80, 83, and 83, presented with numbness and ataxia for 3–8 months. Blood levels of PLP were markedly elevated at 105, 81, and 67 times the upper normal limits (8.70–27.20 μg/l). Examination in all three patients showed normal motor power and absence of deep tendon reflexes in all four limbs. Impaired touch, vibratory, and proprioception sensation in length-dependent and symmetric patterns were found in both lower limbs, while pin-prick sensation and thermal sensation were preserved. The degree of numbness and sensory loss was moderate to severe. They were present from toes to mid-calf in two patients and to ankle in the other patient. Wide-based gait, a positive Romberg's test, and normal plantar reflex were noted. Lhermitte's sign was absent. All findings suggested sensory ataxia. Two years after vitamin discontinuation, all three patients showed no significant improvement in the neuropathy and gait.

Finally, Gdynia *et al.* (2008) reported a unique case of a 75-year-old male who had self-medicated a daily dosage of 9.6 g of pyridoxine for one year. This was the highest dosage of long-term pyridoxine administration ever reported in the medical literature. The patient's blood level of pyridoxine measured 1850 μg/l (normal level 40–120 μg/l). Neurological examination of the wheelchair-bound patient demonstrated a symmetric tetraparesis with distally pronounced muscle weakness and positive Romberg's test. Deep-tendon reflexes were reduced or absent. He showed loss of touch, temperature, pinprick, vibration, and joint position in all distal limbs. In addition, his skin color was remarkably yellowish-brown. Electrophysiological studies revealed a sensorimotor mixed axonal-demyelinating polyneuropathy. Upon stopping pyridoxine administration, the patient's condition improved clinically and electrophysiologically, and his skin lost the yellowish color.

22.21.3 Mechanism of Neurotoxicity

The mechanism of pyridoxine neurotoxicity is not known, but several factors can lead to conjecture. Although a saturable, carrier-mediated mechanism of absorption has been demonstrated for physiological amounts of vitamin B_6, megadoses of pyridoxine would be absorbed by a non-saturable passive diffusion mechanism, resulting in very high concentrations of pyridoxine circulating in the bloodstream. Pyridoxine needs to be converted to the active form of the vitamin, PLP, by the sequential action of pyridoxal kinase and pyridoxine phosphate oxidase. The massive circulating dose of pyridoxine might temporarily saturate one or both of these enzymes. The biologically inactive pyridoxine that accumulates might then occupy enzyme binding sites that are normally occupied by the PLP coenzyme, thereby acting as a competitive inhibitor for PLP. For example, binding to amino acid decarboxylases would inhibit the synthesis of neurotransmitters. Thus, pyridoxine megadoses might create in the peripheral nervous system a deficiency state for PLP, resulting in neurotoxicity. The presence of the blood–brain barrier limits the amount of pyridoxine entering the central nervous system. Cell bodies of the motor neurons are located in nuclei within the spinal cord, whereas the cell bodies of peripheral sensory neurons are located in ganglia outside the central nervous system. The sensory neurons are therefore not protected by the blood–brain barrier, which could explain the selective toxicity of pyridoxine for the sensory pathways.

22.22 TOXICITY

There are no reports of adverse effects associated with high intake of naturally occurring vitamin B_6 from food sources. However, therapeutic doses of pyridoxine taken over long periods can cause neurotoxicity.

The acute toxicity of pyridoxine, expressed as the single oral dose that kills 50% of the experimental animals (the LD_{50}), is 4 g/kg body weight for the free base and 6 g/kg for the hydrochloride in the rat (Unna & Antopol, 1940). Considering this high LD_{50}, vitamin B_6 toxicity would appear to be of little practical concern. Indeed, prolonged

TABLE 22.5
Selected Clinical Studies on Neuropathy Associated with Pyridoxine

Dose (mg/day)	Duration (Months)	Total Dose (g)	Results	Reference
800	17	408	No adverse effects reported	Gill and Rose (1986)
800	24	576	Sensory neuropathy	Berger and Schaumburg (1984)
900	12	324	No adverse effects reported	Palareti *et al.* (1986)
1,000	12	365	Sensory neuropathy	Waterston and Gilligan (1987)

Source: from Bendich and Cohen (1990).

megadosing with pyridoxine has been prescribed by medical practitioners as a remedy for various conditions, including migraine headaches, carpal tunnel syndrome, schizophrenia, infantile convulsions, and childhood autism (Bendich & Cohen, 1990). There have also been cases of individuals self-prescribing high doses of the vitamin for various reasons, and tablets of up to 1,000 mg PN.HCl are available over the counter. In adult humans, daily oral doses of vitamin B_6 in excess of 200 mg per kg body weight taken for long periods can produce a range of non-specific systemic symptoms, including nausea, vomiting, diarrhea, tachypnoea, and rash. As previously discussed, chronic megadosing with pyridoxine causes neurotoxicity. The neuropathy is reversible following withdrawal of the pyridoxine supplement and no reports have been found of permanent damage to the nervous system following chronic ingestion of large doses of pyridoxine (Dakshinamurti & Dakshinamurti, 2007).

Bendich & Cohen (1990) reviewed preclinical and clinical data on pyridoxine neurotoxicity. Table 22.5 summarizes selected cases from clinical data. Doses of less than 500 mg/day of pyridoxine for up to two years have not been associated with neuropathy and are considered to be safe. Daily doses exceeding 500 mg/day should be avoided unless there are sound reasons for such treatment. In such cases, neurological testing on a regular basis should be part of the treatment program.

FURTHER READING

Allgood, V.E. & Cidlowski, J.A. (1991) Novel role for vitamin B_6 in steroid hormone action: A link between nutrition and the endocrine system. *Journal of Nutritional Biochemistry*, **2**, 523–34.

REFERENCES

Albin, R.L., Albers, J.W., Greenberg, H.S., Townsend, J.B., Lynn, R.B., Burke, J.M. Jr. & Alessi, A.G. (1987) Acute sensory neuropathy-neuronopathy from pyridoxine overdose. *Neurology*, **37**, 1729–32.

Ames, B.N. (2001) DNA damage from micronutrient deficiencies is likely to be a major cause of cancer. *Mutation Research*, **475**, 7–20.

Anderson, B.B., Fulford-Jones, C.E., Child, J.A., Beard, M.E.J. & Bateman, C.J.T. (1971) Conversion of vitamin B_6 compounds to active forms in the red blood cell. *Journal of Clinical Investigation*, **50**, 1901–9.

Anderson, B.B., Perry, G.M., Clements, J.E. & Greany, M.F. (1989) Rapid uptake and clearance of pyridoxine by red blood cell in vivo. *American Journal of Clinical Nutrition*, **50**, 1059–63.

Axelrod, A.E. (1958) The role of nutritional factors in the antibody responses of the anamnestic process. *American Journal of Clinical Nutrition*, **6**, 119–25.

Axelrod, A.E. (1971) Immune processes in vitamin deficiency states. *American Journal of Clinical Nutrition*, **24**, 265–71.

Axelrod, A.E. & Hopper, S. (1960) Effects of pantothenic acid, pyridoxine and thiamine deficiencies upon antibody formation to influenza virus PR-8 in rats. *Journal of Nutrition*, **72**, 325–30.

Axelrod, A.E. & Trakatellis, A.C. (1964a) Relationship of pyridoxine to immunological phenomena. *Vitamins and Hormones*, **22**, 591–607.

Axelrod, A.E. & Trakatellis, A.C. (1964b) Induction of tolerance to skin homografts by administering splenic cells to pyridoxine-deficient rats. *Proceedings of the Society for Experimental Biology and Medicine*, **116**, 206–10.

Axelrod, A.E., Fisher, B., Fisher, E., Chiung Puh Lee, Y. & Walsh, P. (1958) Effect of a pyridoxine deficiency on skin grafts in the rat. *Science*, **127**, 1388–9.

Axelrod, A.E., Hopper, S. & Long, D.A. (1961) Effects of pyridoxine deficiency upon circulating antibody formation and skin hypersensitivity reactions to diphtheria toxoid in guinea pigs. *Journal of Nutrition*, **74**, 58–64.

Axelrod, A.E., Trakatellis, A.C., Bloch, H. & Stinebring, W.R. (1963) Effect of pyridoxine deficiency upon delayed hypersensitivity in guinea pigs. *Journal of Nutrition*, **79**, 161–7.

Baxter, P. (1999) Epidemiology of pyridoxine dependent and pyridoxine responsive seizures in the UK. *Archives of Disease in Childhood*, **81**, 431–3.

Bayoumi, R.A., Kirwan, J.R. & Smith, W.R.D. (1972) Some effects of dietary vitamin B_6 deficiency and 4-deoxypyridoxine on γ-aminobutyric acid metabolism in rat brain. *Journal of Neurochemistry*, **19**, 569–76.

Bendich, A. & Cohen, M. (1990) Vitamin B_6 safety issues. *Annals of the New York Academy of Sciences*, **585**, 320–30.

Benos, D.J. (1982) Amiloride: A molecular probe of sodium transport in tissues and cells. *American Journal of Physiology*, **242**, C131–45.

Berger, A.R. & Schaumburg, H.H. (1984) More on neuropathy from pyridoxine abuse. *New England Journal of Medicine*, **311**, 986–7.

Berger, A.R., Schaumburg, H.H., Schroeder, C., Apfel, S. & Reynolds, R. (1992) Dose response, coasting, and differential fiber vulnerability in human toxic neuropathy: A prospective study of pyridoxine neurotoxicity. *Neurology*, **42**, 1367–70.

Black, A.L., Guirard, B.M. & Snell, E.E. (1977) Increased muscle phosphorylase in rats fed high levels of vitamin B_6. *Journal of Nutrition*, **107**, 1962–8.

Black, A.L., Guirard, B.M. & Snell, E.E. (1978) The behavior of muscle phosphorylase as a reservoir for vitamin B_6 in the rat. *Journal of Nutrition*, **108**, 670–7.

Bowden, J.-F., Bender, D.A., Coulson, W.F. & Symes, E.K. (1986) Increased uterine uptake and nuclear retention of [^{3}H]oestradiol through the oestrous cycle and enhanced end-organ sensitivity to oestrogen stimulation in vitamin B_6 deficient rats. *Journal of Steroid Biochemistry*, **25**, 359–65.

Bowman, B.B. & McCormick, D.B. (1989) Pyridoxine uptake by rat renal proximal tubular cells. *Journal of Nutrition*, **119**, 745–9.

Brent, L. & Gowland, G. (1962) Induction of tolerance of skin homografts in immunologically competent mice. *Nature*, **196**, 1298–301.

Buss, D.D., Hamm, M.W., Mehanso, H. & Henderson, L.M. (1980) Transport and metabolism of pyridoxine in the perfused small intestine and the hind limb of the rat. *Journal of Nutrition*, **110**, 1655–63.

Cereghini, S., Raymondjean, M., Carranca, A.G., Herbomel, P. & Yaniv, M. (1987) Factors involved in control of tissue-specific expression of albumin gene. *Cell*, **50**, 627–38.

Chandra, R.K. & Puri, S. (1985) Vitamin B-6 modulation of immune responses and infection. In: *Vitamin B-6: Its Role in Health and Disease*. (Ed. by R.D. Reynolds & J.E. Leklem), pp. 163–75. Alan R. Liss, Inc., New York.

Cheslock, K.E. & McCully, M.T. (1960) Response of human beings to a low-vitamin B_6 diet. *Journal of Nutrition*, **70**, 507–13.

Chiku, K., Mochida, H., Yamamoto, M. & Natori, Y. (1993) Amino acids suppress intracellular protein degradation in rat liver during parenteral nutrition. *Journal of Nutrition*, **123**, 1771–6.

Coursin, D.B. (1954) Convulsive seizures in infants with pyridoxine-deficient diet. *Journal of the American Medical Association*, **154**, 406–8.

Dakshinamurti, K. (1982) Neurobiology of pyridoxine. In: *Advances in Nutritional Research*, Vol. 4. (Ed. by H.H. Draper), pp. 143–79. Plenum Press, New York.

Dakshinamurti, S. & Dakshinamurti, K. (2007) Vitamin B_6. In: *Handbook of Vitamins*, 4th edn. (Ed. by J. Zempleni, R.B. Rucker, D.B. McCormick & J.W. Suttie), pp. 315–59. CRC Press, New York.

Dakshinamurti, K., Paulose, C.S., Viswanathan, M., Siow, Y.L. & Sharma, S.K. (1990) Neurobiology of pyridoxine. *Annals of the New York Academy of Sciences*, **585**, 128–44.

DiSorbo, D.M. & Litwack, G. (1981) Changes in the intracellular levels of pyridoxal 5′-phosphate affect the induction of tyrosine aminotransferase by glucocorticoids. *Biochemical and Biophysical Research Communications*, **99**, 1203–8.

DiSorbo, D.M. & Litwack, G. (1982) Vitamin B_6 kills hepatoma cells in culture. *Nutrition and Cancer*, **3**, 216–22.

DiSorbo, D.M. & Nathanson, L. (1983) High-dose pyridoxal supplemented culture medium inhibits the growth of a human malignant melanoma cell line. *Nutrition and Cancer*, **5**, 10–15.

DiSorbo, D.M., Wagner, R. Jr. & Nathanson, L. (1985) *In vivo* and *in vitro* inhibition of B16 melanoma growth by vitamin B_6. *Nutrition and Cancer*, **7**, 43–52.

Doke, S., Inagaki, N., Hayakawa, T. & Tsuge, H. (1997) Effect of vitamin B_6 deficiency on an antibody production in mice. *Bioscience, Biotechnology and Biochemistry*, **61**, 1331–6.

Fisher, B., Axelrod, A.E., Fisher, E.R., Lee, S.H. & Calvanese, N. (1958) The favorable effect of pyridoxine deficiency on skin homograft survival. *Surgery*, **44**, 149–67.

Fonda, M.L. (1992) Purification and characterization of a vitamin B6-phosphate phosphatase from human erythrocytes. *Journal of Biological Chemistry*, **267**(22), 15978–83.

Frimpter, G.W., Andelman, R.J. & George, W.F. (1969) Vitamin B_6-dependency syndromes. *American Journal of Clinical Nutrition*, **22**, 794–805.

Friso, S., Jacques, P.F., Wilson, P.W., Rosenberg, I.H. & Selhub, J. (2001) Low circulating vitamin B_6 is associated with elevation of the inflammation marker C-reactive protein independently of plasma homocysteine levels. *Circulation*, **103**, 2788–91.

Friso, S., Girelli, D., Martinelli, N., Olivieri, O., Lotto, V., Bozzini, C., Pizzolo, F., Faccini, G., Beltrame, F. & Corrocher, R. (2004) Low plasma vitamin B-6 concentrations and modulation of coronary artery disease risk. *American Journal of Clinical Nutrition*, **79**, 992–8.

Friso, S., Lotto, V., Corrocher, R. & Choi, S.W. (2012) Vitamin B6 and cardiovascular disease. *Subcellular Biochemistry*, **56**, 265–90.

Gdynia, H.J., Müller, T., Sperfeld, A.D., Kühnlein, P., Otto, M., Kassubek, J. & Ludolph, A.C. (2008) Severe sensorimotor neuropathy after intake of highest dosages of vitamin B6. *Neuromuscular Disorders*, **18**, 156–8.

Gill, H.S. & Rose, G.A. (1986) Mild metabolic hyperoxaluria and its response to pyridoxine. *Urologia Internationalis*, **41**, 393–6.

Gregory, J.F. III (1980) Effects of ε-pyridoxyllysine bound to dietary protein on the vitamin B-6 status of rats. *Journal of Nutrition*, **110**, 995–1005.

Gregory, J.F. III & Hiner, M.E. (1983) Thermal stability of vitamin B6 compounds in liquid model food systems. *Journal of Food Science*, **48**, 1323–7, 1339.

Gregory, J.F. III & Ink, S.L. (1987) Identification and quantification of pyridoxine-β-glucoside as a major form of vitamin B6 in plant-derived foods. *Journal of Agricultural and Food Chemistry*, **35**, 76–82.

Gregory, J.F. III & Kirk, J.R. (1977) Interaction of pyridoxal and pyridoxal phosphate with peptides in a model food system during thermal processing. *Journal of Food Science*, **42**, 1554–7, 1561.

Gregory, J.F. III, Ink, S.L. & Sartain, D.B. (1986) Degradation and binding to food proteins of vitamin B-6 compounds during thermal processing. *Journal of Food Science*, **51**, 1345–51.

Gregory, J.F. III, Trumbo, P.R., Bailey, L.B., Toth, J.P., Baumgartner, T.G. & Cerda, J.J. (1991) Bioavailability of pyridoxine-5′-β-D-glucoside determined in humans by stable-isotopic methods. *Journal of Nutrition*, **121**, 177–86.

Gridley, D.S., Stickney, D.R., Nutter, R.L., Slater, J.M. & Shultz, T.D. (1987) Suppression of tumor growth and enhancement of immune status with high levels of dietary vitamin $B6_6$ in BALB/c mice. *Journal of the National Cancer Institute*, **78**, 951–9.

Guilarte, T.R. & Wagner, H.N. Jr. (1987) Increased concentrations of 3-hydroxykynurenine in vitamin B_6 deficient neonatal rat brain. *Journal of Neurochemistry*, **49**, 1918–26.

Ha, C., Miller, L.T. & Kerkvliet, N.I. (1984a) The effect of vitamin B_6 deficiency on cytotoxic immune responses of T cells, antibodies, and natural killer cells, and phagocytosis by macrophages. *Cellular Immunology*, **85**, 318–29.

Ha, C., Kerkvliet, N.I. & Miller, L.T. (1984b) The effect of vitamin B-6 deficiency on host susceptibility to moloney sarcoma virus-induced tumor growth in mice. *Journal of Nutrition*, **114**, 938–45.

Hamfelt, A. & Soderhjelm, L. (1983) Plasma pyridoxal phosphate in diabetes. *American Journal of Clinical Nutrition*, **38**, 841–2.

Hamm, M.W., Mehanso, H. & Henderson, L.M. (1979) Transport and metabolism of pyridoxamine and pyridoxamine phosphate in the small intestine of the rat. *Journal of Nutrition*, **109**, 1552–9.

Hodges, R.E., Bean, W.B., Ohlson, M.A. & Bleiler, R.E. (1962) Factors affecting human antibody responses. IV. Pyridoxine deficiency. *American Journal of Clinical Nutrition*, **11**, 180–6.

Hollenbeck, C.B., Leklem, J.E., Riddle, M.C. & Connor, W.E. (1983) The composition and nutritional adequacy of subject-selected high carbohydrate, low fat diets in insulin-dependent diabetes mellitus. *American Journal of Clinical Nutrition*, **38**, 41–51.

Hunt, A.D. Jr., Stokes, J. Jr., McCrory, W.W. & Stroud, H.H. (1954) Pyridoxine dependency: Report of a case of intractable convulsions in an infant controlled by pyridoxine. *Pediatrics*, **13**, 140–5.

Iacopetta, B. (2002) Are there two sides to colorectal cancer? *International Journal of Cancer*, **101**, 403–8.

Ink, S.L., Mehansho, H. & Henderson, L.M. (1982) The binding of pyridoxal to hemoglobin. *Journal of Biological Chemistry*, **257**(9), 4753–7.

Institute of Medicine. (1998) *Dietary Reference Intakes for Thiamin, Riboflavin, Niacin, Vitamin B6, Folate, Vitamin B12, Pantothenic Acid, Biotin, and Choline*. National Academy Press, Washington, DC.

Ishihara, J., Hiroyasu, I., Manai, I., Motoki, I., Katsutoshi, O., Yoshikuni, K., Akira, O. & Shoichiro, T. (2008) Intake of folate, vitamin B_6 and vitamin B_{12} and the risk of CHD: The Japan Public Health Center-based Prospective Study Cohort I. *Journal of the American College of Nutrition*, **27**, 127–36.

Kirksey, A., Morré, D.M. & Wasynczuk, A.Z. (1990) Neuronal development in vitamin B_6 deficiency. *Annals of the New York Academy of Sciences*, **585**, 202–18.

Kobzar, G., Mardla, V., Rätsep, I. & Samel, N. (2009) Effect of vitamin B_6 vitamers on platelet aggregation. *Platelets*, **20**, 120–4.

Komatsu, S., Watanabe, H., Oka, T., Tsuge, H., Nii, H. & Kato, N. (2001) Vitamin B-6-supplemented diets compared with a low vitamin B-6 diet suppress azoxymethane-induced colon tumorigenesis in mice by reducing cell proliferation. *Journal of Nutrition*, **131**, 2204–7.

Kondo, T. & Okada, M. (1985) Effect of pyridoxine administration on the induction of cytosolic aspartate aminotransferase in the liver of rats treated with hydrocortisone. *Journal of Nutritional Science and Vitaminology*, **31**, 509–17.

Kopelovich, L. & Wolfe, G. 0(1977) Modification of ribonucleic acid by vitamin B_6. I. Specific interaction of pyridoxal 5′-phosphate with transfer ribonucleic acid. *Biochemistry*, **16**, 3721–6.

Kotake, T., Ueda, T., Mori, T., Murakami, E. & Hattori, M. (1975) The physiological significasnce of the xanthurenic acid-insulin complex. *Journal of Biochemistry*, **77**, 685–7.

Krinke, G., Schaumburg, H.H., Spencer, P.S., Suter, J., Thomann, P. & Hess, R. (1981) Pyridoxine megavitaminosis produces degeneration of peripheral sensory neuroms (sensory neuronopathy) in the dog. *Neurotoxicology*, **2**, 13–24.

Krinke, G., Naylor, D.C. & Skorpil, V. (1985) Pyridoxine megavitaminosis: An analysis of the early changes induced with massive doses of vitamin B_6 in rat primary sensory neurons. *Journal of Neuropathology and Experimental Neurology*, **44**, 117–29.

Kulkantrakorn, K. (2014) Pyridoxine-induced sensory ataxic neuronopathy and neuropathy: Revisited. *Neurological Sciences*, **35**, 1827–30.

Kumar, M. & Axelrod, A.E. (1968) Cellular antibody synthesis in vitamin B_6-deficient rats. *Journal of Nutrition*, **96**, 53–9.

Kuno, M. (1960) Changes of pancreas and parotid gland, especially of endocrine part, in vitamin B_6 deficient rats on a view point of diabetic development. *Vitamins* (Japan Vitamin Society), **19**(1), 140–54. (In Japanese)

Lapin, I.P. (1978a) Stimulant and convulsive effects of kynurenines injected into brain ventricles in mice. *Journal of Neural Transmission*, **42**, 37–43.

Lapin, I.P. (1987b) Convulsions and tremor in immature rats after intraperitonal injection of kynurenine and its metabolites. *Pharmacological Research Communications*, **10**, 81–4.

Leklem, J.E. (1988) Vitamin B_6: Of reservoirs, receptors and requirements. *Nutrition Today*, **23**, 4–10.

Leklem, J.E. & Hollenbeck, C.B. (1990) Acute ingestion of glucose decreases plasma pyridoxal 5′-phosphate and total vitamin B-6 concentration. *American Journal of Clinical Nutrition*, **51**, 832–6.

Li, T.-K., Lumeng, L. & Veitch, R.L. (1974) Regulation of pyridoxal 5′-phosphate metabolism in liver. *Biochemical and Biophysical Research Communications*, **61**, 677–84.

Lin, P.-T., Cheng, C.-H., Liaw, Y.-P., Lee, B.-J., Lee, T.-W. & Huang, Y.-C. (2006) Low pyridoxal 5′-phosphate is associated with increased risk of coronary artery disease. *Nutrition*, **22**, 1146–51.

Lott, I.T., Coulombe, T., DiPaolo, R.V., Richardson, E.P. Jr. & Levy, H.L. (1978) Vitamin B_6-dependent seizures: Pathology and chemical findings in brain. *Neurology*, **28**, 47–54.

Lumeng, L., Li, T.-K., Lui, A. & Roudebush, R.L. (1985) The interorgan transport and metabolism of vitamin B-6. In: *Vitamin B-6: Its Role in Health and Disease*. (Ed. by R.D. Reynolds & J.E. Leklem), pp. 35–54. Alan R. Liss, Inc., New York.

Mackey, A.D., Henderson, G.N. & Gregory, J.F. III (2002) Enzymatic hydrolysis of pyridoxine-5′-β-D-glucoside is catalysed by intestinal lactase-phlorizin hydrolase. *Journal of Biological Chemistry*, **277**(30), 26858–64.

Makris, A. & Gershoff, S.N. (1973) Growth hormone levels in vitamin B_6-deficient rats. *Hormone and Metabolic Research*, **5**, 457–61.

Malaisse, W.J., Malaisse-Lagae, F., King, S. & Wright, P.H. (1968) Effect of growth hormone on insulin secretion. *American Journal of Physiology*, **215**, 423–8.

Martial, J., Zaldivar, J., Bull, P., Venegas, A. & Valenzuela, P. (1975) Inactivation of rat liver RNA polymerases I and II and yeast RNA polymerase I by pyridoxal 5′-phosphate. Evidence for the participation of lysyl residues at the active site. *Biochemistry*, **14**, 4907–11.

Matsubara, K., Komatsu, S., Oka, T. & Kato, N. (2003) Vitamin B_6-mediated suppression of colon tumorigenesis, cell proliferation, and angiogenesis (review). *Journal of Nutritional Biochemistry*, **14**, 246–50.

McCormick, D.B., Gregory, M.E. & Snell, E.E. (1961) Pyridoxal phosphokinases. I. Assay, distribution, purification, and properties. *Journal of Biological Chemistry*, **236**(7), 2076–84.

McMahon, L.G., Nakano, H., Levy, M.-D. & Gregory, J.F. III (1997) Cytosolic pyridoxine-β-D-glucoside hydrolase from porcine jejunal mucosa. *Journal of Biological Chemistry*, **272**(51), 32025–33.

Mehansho, H., Hamm, M.W. & Henderson, L.M. (1979) Transport and metabolism of pyridoxal and pyridoxal phosphate in the small intestine of the rat. *Journal of Nutrition*, **109**, 1542–51.

Mehansho, H., Buss, D.D., Hamm, M.W. & Henderson, L.M. (1980) Transport and metabolism of pyridoxine in rat liver. *Biochimica et Biophysica Acta*, **631**, 112–23.

Merrill, A.H. Jr. & Henderson, J.M. (1987) Diseases associated with defects in vitamin B_6 metabolism or utilization. *Annual Review of Nutrition*, **7**, 137–56.

Merrill, A.H., Horiike, K. & McCormick, D.B. (1978) Evidence for the regulation of pyridoxal 5′-phosphate formation in liver by pyridoxamine (pyridoxine) 5′-phosphate oxidase. *Biochemical and Biophysical Research Communications*, **83**, 984–90.

Merrill, A.H. Jr., Henderson, J.M., Wang, E., McDonald, B.W. & Millikan, W.J. (1984) Metabolism of vitamin B-6 by human liver. *Journal of Nutrition*, **114**, 1664–74.

Meydani, S.N., Ribaya-Mercado, J.D., Russell, R.M., Sahyoun, N., Morrow, F.D. & Gershoff, S.N. (1991) Vitamin B-6 deficiency impairs interleukin 2 production and lymphocyte proliferation in elderly adults. *American Journal of Clinical Nutrition*, **53**, 1275–80.

Middleton, H.M. III (1978) Jejunal phosphorylation and dephosphorylation of absorbed pyridoxine.HCl in vitro. *American Journal of Physiology*, **235**, E272–8.

Middleton, H.M. III (1985) Uptake of pyridoxine by in vivo perfused segments of rat small intestine: A possible role for intracellular vitamin metabolism. *Journal of Nutrition*, **115**, 1079–88.

Middleton, H.M. III (1986) Intestinal hydrolysis of pyridoxal 5′-phosphate in vitro and in vivo in the rat. Effect of protein binding and pH. *Gastroenterology*, **91**, 343–50.

Miller, J.W., Ribaya-Mercado, J.D., Russell, R.M., Shepard, D.C., Morrow, F.D., Cochary, E.F., Sadowski, J.A., Gershoff, S.N. & Selhub, J. (1992) Effect of vitamin B-6 deficiency on fasting plasma homocysteine concentrations. *American Journal of Clinical Nutrition*, **55**, 1154–60.

Modak, M.J. (1976) Observations on the pyridoxal 5′-phosphate inhibition of DNA polymerases. *Biochemistry*, **15**, 3620–6.

Molina, A., Oka, T., Muñoz, S., Chikamori-Aoyama, M., Kuwahata, M. & Natori, Y. (1997) Vitamin B_6 suppresses growth and expression of albumin gene in a human hepatoma cell line HepG2. *Nutrition and Cancer*, **28**, 206–211.

Molony, C.J. & Parmelee, A.H. (1954) Convulsions in young infants as a result of pyridoxine (vitamin B_6) deficiency. *Journal of the American Medical Association*, **154**, 405–406.

Montjar, M., Axelrod, A.E. & Trakatellis, A.C. (1965) Effect of pyridoxine deficiency upon polysomes and messenger RNA of rat tissues. *Journal of Nutrition*, **85**, 45–51.

Nakano, H., McMahon, L.G. & Gregory, J.F. III (1997) Pyridoxine-5′-β-D-glucoside exhibits incomplete bioavailability as a source of vitamin B-6 and partially inhibits the utilization of co-ingested pyridoxine in humans. *Journal of Nutrition*, **127**, 1508–13.

Nguyen, L.B. & Gregory, J.F. III (1983) Effects of food composition on the bioavailability of vitamin B-6 in the rat. *Journal of Nutrition*, **113**, 1550–60.

Oka, T., Komori, N., Kuwahata, M., Hiroi, Y., Shimoda, T., Okada, M. & Natori, Y. (1995a) Pyridoxal 5′-phosphate modulates expression of cytosolic aspartate aminotransferase gene by inactivation of glucocorticoid receptor. *Journal of Nutritional Science and Vitaminology*, **41**, 363–75.

Oka, T., Komori, N., Kuwahata, M., Okada, M. & Natori, Y. (1995b) Vitamin B_6 modulates expression of albumin gene by inactivating tissue-specific DNA-binding protein in rat liver. *Biochemical Journal*, **309**, 243–8.

Oka, T., Sugitatsu, H., Nordin, H., Thakur, M.K., Aoyama, M., Sasagawa, T., Suzuki, I. & Tsuji, H. (2001) Pyridoxal 5′-phosphate inhibits DNA binding of HNF1. *Biochimica et Biophysica Acta*, **1568**, 189–96.

Page, J.H., Ma, J., Chiuve, S.E., Stampfer, M.J., Selhub, J., Mason, J.E. & Rimm, E.B. (2009) Plasma vitamin B_6 and risk of myocardial infarction in women. *Circulation*, **120**, 649–55.

Palareti, G., Salardi, S., Piazzi, S., Legnani, C., Poggi, M., Grauso, F., Caniato, A., Coccheri, S. & Cacciari, E. (1986) Blood coagulation changes in homocystinuria: Effects of pyridoxine and other specific therapy. *Journal of Pediatrics*, **109**, 1001–6.

Pavé-Preux, M., Ferry, N., Bouguet, J., Hanoune, J. & Barouki, R. (1988) Nucleotide sequence and glucocorticoid regulation of the mRNA for the isoenzymes of rat aspartate aminotransferase. *Journal of Biological Chemistry*, **263**(25), 17459–66.

Pruzansky, J. & Axelrod, A.E. (1955) Antibody production to diphtheria toxoid in vitamin deficiency states. *Proceedings of the Society for Experimental Biology and Medicine*, **89**, 323–5.

Ranke, E., Tauber, S.A., Horonick, A., Ranke, B., Goodhart, R.S. & Chow, B.F. (1960) Vitamin B_6 deficiency in the aged. *Journal of Gerontology*, **15**, 41–4.

Ridker, P.M., Hennekens, C.H., Buring, J.E. & Rifai, N. (2000) C-reactive protein and other markers of inflammation in the prediction of cardiovascular disease in women. *New England Journal of Medicine*, **342**, 836–43.

Robson, L.C. & Schwarz, M.R. (1975) Vitamin B_6 deficiency and the lymphoid system. 1. Effects on cellular immunity and *in vitro* incorporation of ^{3}H-uridine by small lymphocytes. *Cellular Immunology*, **16**, 135–44.

Rose, C.S., György, P., Butler, M., Andres, R., Norris, A.H., Shock, N.W., Tobin, J., Brin, M. & Spiegel, H. (1976) Age differences in vitamin B_6 status of 617 men. *American Journal of Clinical Nutrition*, **29**, 847–53.

Roth-Maier, D.A., Zinner, P.M. & Krichgessner, M. (1982) Effect of varying dietary vitamin B_6 supply on intestinal absorption on B_6. *International Journal for Vitamin and Nutriction Research*, **52**, 272–9.

Said, H.M., Ortiz, A. & Vaziri, N.D. (2002) Mechanism and regulation of vitamin B_6 uptake by renal tubular epithelia: Studies with cultured OK cells. *American Journal of Physiology*, **282**, F465–71.

Said, H.M., Ortiz, A. & Ma, T.Y. (2003) A carrier-mediated mechanism for pyridoxine uptake by human intestinal epithelial Caco-2 cells: Regulation by a PKA-mediated pathway. *American Journal of Physiology*, **285**, C1219–25.

Said, Z.M., Subramanian, V.S., Vaziri, N.D. & Said, H.M. (2008) Pyridoxine uptake by colonocytes: A specific and regulated carrier-mediated process. *American Journal of Physiology*, **294**, C1192–7.

Schaeppi, U. & Krinke, G. (1982) Pyridoxine neuropathy: Correlation of functional tests and neuropathology in beagle dogs treated with large doses of vitamin B_6. *Agents and Actions*, **12**, 575–82.

Schaumburg, H.H. & Berger, A. (1988) Pyridoxine neurotoxicity. *Current Topics in Nutrition and Disease*, **19**, 403–13.

Schaumburg, H., Kaplan, J., Windebank, A., Vick, N., Rasmus, S., Pleasure, D. & Brown, M.J. (1983) Sensory neuropathy from pyridoxine abuse. *New England Journal of Medicine*, **309**, 445–8.

Schwab, S.R., Pereira, J.P., Matloubian, M., Xu, Y., Huang, Y. & Cyster, J.G. (2005) Lymphocyte sequestration through S1P lyase inhibition and disruption of S1P gradients. *Science*, **309**, 1735–9.

Schwarz, U.R., Walter, U. & Eigenthaler, M. (2001) Taming platelets with cyclic nucleotides. *Biochemical Pharmacology*, **62**, 1153–61.

Scientific Committee on Food. (2000) Opinion of the Scientific Committee on food on the tolerable upper intake level of vitamin B_6. SFC/CS/NUT/UPPLEV/16. Final 28 November 2000. European Commission, Brussels.

Sergeev, A.V., Bykovskaja, S.N., Luchanskaja, L.M. & Rauschenbach, M.O. (1978) Pyridoxine deficiency and cytotoxicity of T lymphocytes *in vitro*. *Cellular Immunology*, **38**, 187–92.

Sermet, A., Aybak, M., Ulak, G., Güzel, C. & Denli, O. (1995) Effect of oral pyridoxine hydrochloride supplementation on *in vitro* plazelet sensitivity to different agonists. *Arzneimittel-Forschung*, **45**, 19–21.

Sorrell, M.F., Frank, O., Thomson, A.D., Aquino, H. & Baker, H. (1971) Absorption of vitamins from the large intestine *in vivo*. *Nutrition Report International*, **3**, 143–8.

Spector, R. (1978a) Vitamin B_6 transport in the central nervous system: *In vivo* studies. *Journal of Neurochemistry*, **30**, 881–7.

Spector, R. (1978b) Vitamin B_6 transport in the central nervous system: *In vitro* studies. *Journal of Neurochemistry*, **30**, 889–97.

Spector, R. & Greenwald, L.L. (1978) Transport and metabolism of vitamin B_6 in rabbit brain and choroid plexus. *Journal of Biological Chemistry*, **253**(7), 2373–9.

Spector, R. & Shikuma, S.N. (1978) The stability of vitamin B_6 accumulation and pyridoxal kinase activity in rabbit brain and choroid plexus. *Journal of Neurochemistry*, **31**, 1403–10.

Srivastava, S.K. & Beutler, E. (1972) The effect of normal red cell constituents on the activities of red cell enzymes. *Archives of Biochemistry and Biophysics*, **148**, 249–55.

Stephens, M.C., Havlicek, V. & Dakshinamurti, K. (1971) Pyridoxine deficiency and development of the central nervous system. *Journal of Neurochemistry*, **18**, 2407–16.

Stoerk, H.C., Elsen, H.N. & John, H.M. (1947) Impairment of antibody response in pyridoxine-deficient rats. *Journal of Experimental Medicine*, **85**, 365–71.

Symes, E.K., Bender, D.A., Bowden, J.-F. & Coulson, W.F. (1984) Increased target tissue uptake of, and sensitivity to, testosterone in the vitamin B_6 deficient rat. *Journal of Steroid Biochemistry*, **20**, 1089–93.

Tadera, K. & Orite, K. (1991) Isolation and structure of a new vitamin B_6 conjugate in rice bran. *Journal of Food Science*, **56**, 268–9.

Talbott, M.C., Miller, L.T. & Kerkvliet, N.I. (1987) Pyridoxine supplementation: Effect on lymphocyte responses in elderly persons. *American Journal of Clinical Nutrition*, **46**, 659–64.

Tarr, J.B., Tamura, T. & Stokstad, E.L.R. (1981) Availability of vitamin B-6 and pantothenate in an average American diet in man. *American Journal of Clinical Nutrition*, **34**, 1328–37.

Toyota, T., Kai, Y., Kakizaki, M., Ohtsuka, H., Shibata, Y. & Goto, Y. (1981) The endocrine pancreas in pyridoxine deficient rats. *Tohoku Journal of Experimental Medicine*, **134**, 331–6.

Trakatellis, A.C. & Axelrod, A.E. (1964) Effect of pyridoxine deficiency upon valine incorporation into tissue proteins of the rat. *Journal of Nutrition*, **82**, 483–8.

Trakatellis, A.C. & Axelrod, A.E. (1965) Effect of pyridoxine deficiency on nucleic acid metabolism in the rat. *Biochemical Journal*, **95**, 344–9.

Trakatellis, A., Dimitriadou, A., Exindari, M., Scountzou, J., Koliakos, G., Christodoulou, D., Malissiovas, N., Antoniadis, A. & Polyzoni, T. (1992) Effect of pyridoxine deficiency on immunological phenomena. *Postgraduate Medical Journal*, **68**, S70–7.

Trakatellis, A., Dimitriadou, A. & Trakatelli, M. (1997) Pyridoxine deficiency: New approaches in immunosuppression and chemotherapy. *Postgraduate Medical Journal*, **73**, 617–22.

Trumbo, P.R., Banks, M.A. & Gregory, J.F. III (1990) Hydrolysis of pyridoxine-5′-β-D-glucoside from mammalian tissues. *Proceedings of the Society for Experimental Biology and Medicine*, **195**, 240–6.

Ubbink, J.B., Serfontein, W.J., Becker, P.J. & de Villiers, L.S. (1987) Effect of different levels of oral pyridoxine supplementation on plasma pyridoxal-5′-phosphate and pyridoxal levels and urinary vitamin B-6 excretion. *American Journal of Clinical Nutrition*, **46**, 78–85.

Unna, K. & Antopol, W. (1940) Toxicity of vitamin B_6. *Proceedings of the Society for Experimental Biology and Medicine*, **43**, 116–18.

Vilter, R.W., Mueller J.F., Glazer, H.S., Jarrold, T., Abraham, J., Thompson, C. & Hawkins, V.R. (1953) The effect of vitamin B_6 deficiency induced by desoxypyridoxine in human beings. *Journal of Laboratory and Clinical Medicine*, **42**, 335–57.

Wasynczuk, A., Kirksey, A. & Morré, D.M. (1983) Effects of maternal vitamin B-6 deficiency on specific regions of developing rat brain: The extrapyramidal motor system. *Journal of Nutrition*, **113**, 746–54.

Waterston, J.A. & Gilligan, B.S. (1987) Pyridoxine neuropathy. *Medical Journal of Austalia*, **146**, 640–2.

Wei, E.K., Giovannucci, E., Wu, K., Rosner, B., Fuchs, C.S., Willett, W.C. & Colditz, G.A. (2004) Comparison of risk factors for colon and rectal cancer. *International Journal of Cancer*, **108**, 433–42.

Willis-Carr, J.I. & St. Pierre, R.L. (1978) Effects of vitamin B6 deficiency on thymic epithelial cells and T lymphocyte differentiation. *Journal of Immunology*, **120**, 1153–9.

Windebank, A.J., Low, P.A., Blexrud, M.D., Schmelzer, J.D. & Schaumburg, H.H. (1985) Pyridoxine neuropathy in rats: Specific degeneration of sensory axons. *Neurology*, **35**, 1617–22.

Wood, J.D. (1975) The role of γ-aminobutyric acid in the mechanism of seizures. *Progress in Neurobiology*, **5**, 77–95.

Xu, Y., Sladky, J.T. & Brown, M.J. (1989) Dose-dependent expression of neuronopathy after experimental pyridoxine intoxication. *Neurology*, **39**, 1077–63.

Yamashiro, T., Yasujima, T., Said, H.M. & Yuasa, H. (2020) pH dependent pyridoxine transport by SLC19A2 and SLC19A3: Implications for absorption in acidic microclimates. *Journal of Biological Chemistry*, **295**(50), 16998–17008.

Zhang, X., Ma, J., Smith-Warner, S.A., Lee, J.E. & Giovannucci, E. (2013) Vitamin B6 and colorectal cancer: Current evidence and future directions. *World Journal of Gastroenterology*, **19**, 1005–10.

Zhang, Z., Gregory, J.F. III & McCormick, D.B. (1993) Pyridoxine-5′-β-D-glucoside competitively inhibits uptake of vitamin B-6 into isolated rat liver cells. *Journal of Nutrition*, **123**, 85–9.

23 Pantothenic Acid

23.1 CHEMISTRY

Structures of pantothenic acid and compounds containing a pantothenate moiety are shown in Figure 23.1. Pantothenic acid comprises a derivative of butyric acid (pantoic acid) joined by a peptide linkage to the amino acid β-alanine. In nature, pantothenic acid occurs only rarely in the free state, but it is very widely distributed as an integral part of the structures of coenzyme A (CoA) and 4-phosphopantetheine. The latter serves as a covalently attached prosthetic group of acyl carrier protein. The steric configuration of the pantothenic acid moiety is important for enzymatic recognition in biochemical reactions involving CoA and acyl carrier protein.

23.2 FOOD SOURCES AND BIOAVAILABILITY

23.2.1 Food Sources

Coenzyme A is the major pantothenic acid-containing compound present in foods of both animal and plant origin, accompanied by small amounts of other bound forms (phosphopantothenic acid, pantetheine, and phosphopantetheine). Notable exceptions are human and bovine milk in which free (unbound) pantothenic acid constitutes around 90% of the total pantothenate content.

Pantothenic acid is widely distributed in foods. It is particularly abundant in animal organs (liver, kidney, heart, brain) and also in egg yolk, peanuts, and broad beans. Lean meat, milk, potatoes, and green leafy vegetables contain lesser amounts, but these will be important food sources if consumed in sufficient quantity. Highly refined foods such as sugar, fats and oils, and corn starch are totally devoid of the vitamin.

23.2.2 Bioavailability

Little information is at hand regarding the nutritional availability to the human of pantothenic acid in food commodities. Based on the urinary excretion of pantothenic acid, the availability for male human subjects ingesting the 'average American diet' ranged from 40% to 61% with a mean of 50% (Tarr *et al.*, 1981).

23.3 DIGESTION AND ABSORPTION OF DIETARY PANTOTHENIC ACID

Humans and other mammals cannot synthesize pantothenic acid and therefore they rely on dietary sources of the vitamin. Ingested CoA, the major dietary form of pantothenic acid, is hydrolyzed in the intestinal lumen to pantetheine by the non-specific action of pyrophosphatases and phosphatase. Pantetheine is then split into pantothenic acid and β-mercaptoethylamine by the action of pantetheinase secreted from the intestinal mucosa into the lumen (Shibata *et al.*, 1983). Absorption of the liberated pantothenic acid takes place mainly in the jejunum, although there is no significant difference in the rate of uptake in the upper, middle, or lower intestine. Within the alkaline medium of the intestinal chyme, the vitamin exists primarily as the pantothenate anion.

A carrier protein, designated sodium-dependent multivitamin transporter (SMVT), has been cloned from rat (Prasad *et al.*, 1998) and human (Wang *et al.*, 1999) placenta and from rabbit intestine (Prasad *et al.*, 1999). This transporter mediates placental and intestinal uptake of pantothenate, biotin, certain biotin analogs, and the essential metabolite lipoate. Transfection of the cloned intestinal cDNA into COS-7 cells has confirmed that the functional characteristics of the cloned transporter are similar to those observed in native intestinal membranes regarding substrate specificity, kinetics, and inhibitor profiles (Chatterjee *et al.*, 1999). The transporter appears to interact primarily, though not exclusively, with the long side-chain of the substrate containing the carboxylate group, which is present in pantothenate, biotin, and lipoate. mRNA transcripts of SMVT were shown to be present in all of the tissues that were tested (intestine, liver, kidney, heart, lung, skeletal muscle, brain, and placenta), suggesting that this carrier protein may be involved in the uptake of pantothenate, biotin, and lipoate by all cell types (Prasad *et al.*, 1998). Chatterjee *et al.* (1999) identified four distinct variants (I–IV) of the SMVT and determined their tissue distribution. Variants II, III, and IV were present in the small intestine, with II predominating; variant I was present in the placenta, but not in the small intestine.

The number of Na^+ ions interacting with SMVT was calculated to be 1.6 for the uptake of pantothenate and 2.0 for the uptake of biotin (Prasad *et al.*, 1999). This indicates that, for every pantothenate or biotin molecule transported, two Na^+ ions are co-transported, i.e. a Na^+:vitamin stoichiometry of 2:1. Since both pantothenic acid and biotin are monovalent anions at physiological pH, the transport process involving this carrier is electrogenic. Thus, both the Na^+ gradient and the potential difference across the brush-border membrane drive the transport process.

Unlike the other water-soluble vitamins which are absorbed by specific carrier-mediated systems (ascorbic acid, biotin, and thiamin), the absorption of pantothenic

DOI: 10.1201/9781003410300-27

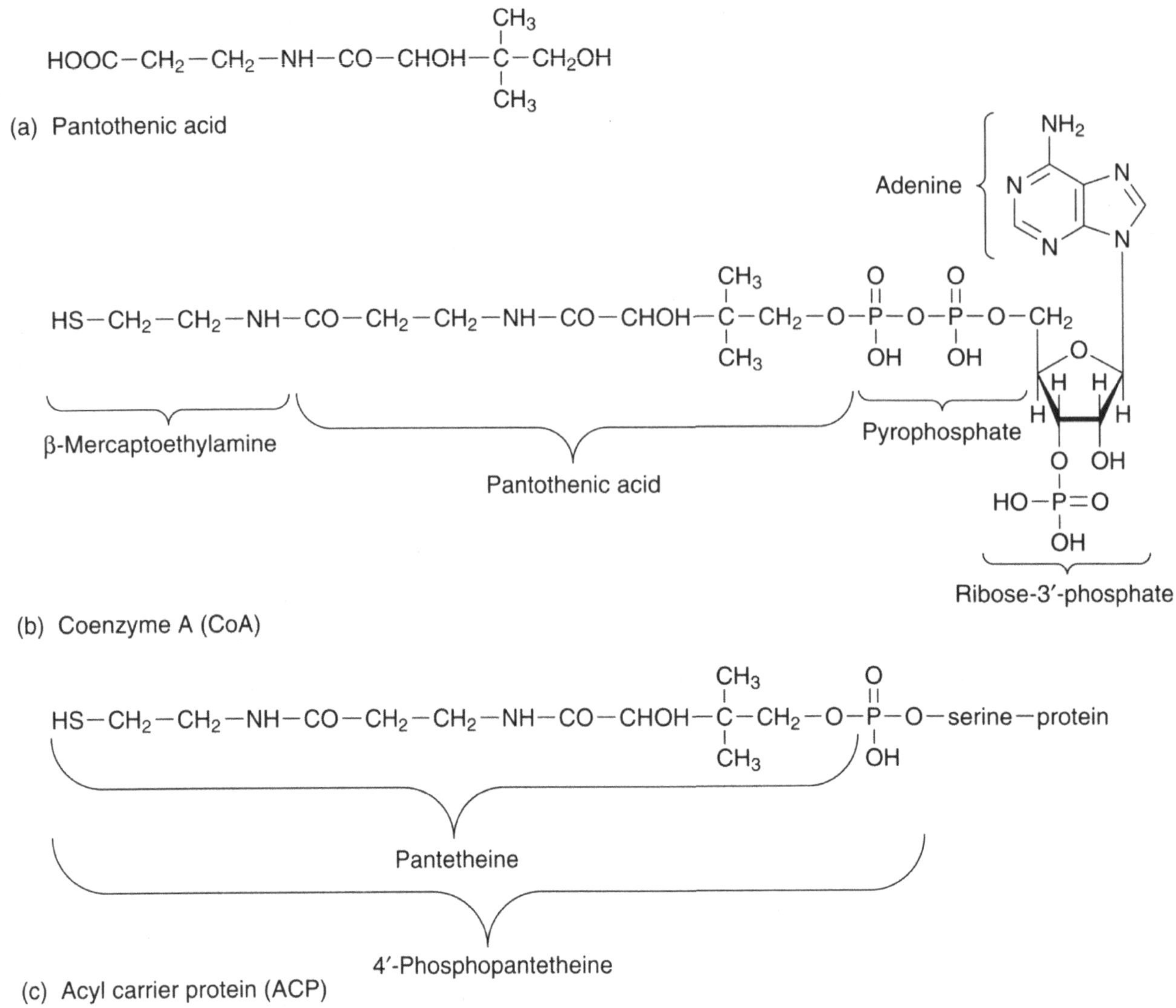

FIGURE 23.1 Structures of (a) pantothenic acid, (b) coenzyme A, and (c) acyl carrier protein.

acid is not adaptively regulated by its level of dietary intake (Stein & Diamond, 1989). The absence of clear-cut deficiency symptoms in humans and the lack of toxicity at high doses could explain why an adaptively regulated absorption mechanism has not evolved for pantothenic acid.

23.4 ABSORPTION OF BACTERIALLY SYNTHESIZED PANTOTHENIC ACID IN THE LARGE INTESTINE

The normal microflora of the large intestine synthesizes pantothenic acid, but it is not known how much, if any, of this endogenous vitamin is available to the host tissues. In human subjects, absorption of pantothenic acid takes place equally well whether the vitamin is given orally or instilled directly into the lumen of the mid-transverse colon (Sorrell *et al.*, 1971). Said *et al.* (1998) showed that uptake of pantothenic acid by human colonic epithelial NCM460 cells is mediated by the SMVT.

23.5 POST-ABSORPTIVE METABOLISM

After absorption, free pantothenic acid is conveyed in the portal circulation to the liver where the majority is taken up by sodium-coupled, secondary active transport (Smith & Milner, 1985). The heart takes up pantothenic acid by a similar mechanism (Lopaschuk *et al.*, 1987). Beinlich *et al.* (1990) observed that pantothenic acid inhibited biotin accumulation in the perfused rat heart, suggesting that these two vitamins share a common uptake mechanism in the heart. Uptake of pantothenic acid by red blood cells, unlike other tissues and cell types, takes place by simple diffusion (Annous & Song, 1995).

In mammalian tissues (but not in red blood cells) CoA is synthesized from pantothenic acid in five enzymatic steps. Three substrates are needed to synthesize CoA: pantothenic acid, ATP, and cysteine. The rate-controlling step in the synthesis is the conversion of pantothenic acid to 4′-phosphopantothenic acid by pantothenate kinase. Tissue levels of CoA are kept in check by feedback inhibition of pantothenate kinase by CoA, acetyl-CoA, or a related metabolite (Robishaw & Neely, 1985).

In the event of a drastically reduced intake of pantothenic acid, such as would occur during food deprivation, the liver, and possibly other tissues, is able to maintain nearly constant CoA levels for some considerable time. In fasting rats, pantothenic acid uptake by the liver is stimulated by the natural rise in glucagon, and incorporation of pantothenic acid into CoA is stimulated by glucagon and cortisol (Smith & Savage, 1980). In contrast to the liver, uptake of pantothenic acid by heart and skeletal muscle of fasting rats is reduced, and yet the rate of pantothenic acid conversion to CoA is increased (Reibel *et al.*, 1981). Evidently, myocardial and muscle CoA synthesis is not governed by the availability of pantothenic acid to these tissues, but rather is controlled intracellularly by regulation of enzymes involved in the CoA synthetic and/or degradative pathways. Reibel *et al.* (1981) postulated that the large amounts of pantothenic acid stored in muscle can be shifted to the liver where endogenous concentrations of the vitamin are normally low.

23.6 RENAL REABSORPTION AND EXCRETION

Pantothenic acid acid derived from the degradation of CoA is excreted intact in urine. Karnitz *et al.* (1984) showed that transport of pantothenic acid in rat kidney brush-border membrane vesicles from the proximal convoluted tubule is a carrier-mediated, Na^+-dependent active transport system that has a K_m of 7.30 μM and a V_{max} of 23.8 pmol/mg protein per min. Isolated perfused rat kidneys were employed to study excretion of pantothenic acid at various concentrations in the perfusate. At physiological concentrations, the filtered pantothenic acid was largely reabsorbed, whereas at higher concentrations pantothenic acid was secreted. Thus pantothenic acid excretion in the rat kidney appears to be regulated by the processes of tubular reabsorption and tubular secretion. Barbarat and Podevin (1986) showed that the renal transport system is electrogenic with a 2:1 Na^+:pantothenate$^-$ stoichiometry.

The only labeled compound excreted after rats were injected with [^{14}C]pantothenic acid was free pantothenic acid (Karnitz *et al.*, 1984). The amount excreted varies proportionally with dietary intake over a wide range of intake values (Tahiliani & Beinlich, 1991). Both fasting and diabetes result in decreased excretion (Reibel *et al.*, 1981), thus conserving whole-body pantothenic acid under these conditions.

23.7 PLACENTAL TRANSPORT

Evidence of an electrogenic Na^+-pantothenate (2:1) cotransport mechanism was reported in human placental microvillous membrane vesicles (Grassl, 1992) and two human placental cell lines (Prasad *et al.*, 1997). The system was inhibited by biotin and lipoate. Prasad *et al.* (1999) deduced from amino acid sequencing that the SMVT involved in Na^+-coupled transport of pantothenate and biotin in the intestine is also shared by these vitamins in placental transport.

23.8 TRANSPORT IN THE BRAIN

Spector *et al.* (1986) measured the unidirectional influx of [^{3}H]pantothenic acid across cerebral capillaries (the anatomical locus of the blood–brain barrier) using an *in situ* rat brain perfusion technique. Transport took place by a low-capacity, saturable system that was inhibited by biotin and medium-chain fatty acids. The data suggested that most, if not all, of the pantothenic acid in brain gains entry through the blood–brain barrier. The half-saturation concentration for the transport process was 19 μM, which is an order of magnitude greater than the plasma concentration of 2 μM. Normally, the concentration of plasma pantothenic acid is regulated by the kidneys, but if one were to increase plasma levels by parenteral means, unnaturally high concentrations would gain access to the extracellular space of brain. Thus transport across the blood–brain barrier plays no role in the regulation of brain CoA levels; rather, subsequent enzymatic steps in the brain cells regulate the conversion of pantothenic acid to CoA. This conversion is very slow, unlike that in other tissues such as the liver and heart, and could at least partly explain why the brain is the organ most difficult to deplete of pantothenic acid and CoA. The inhibition of pantothenic acid transport by biotin provided further evidence that these two vitamins share the SMVT characterized by Prasad *et al.* (1998). The inhibitory effect of medium-chain fatty acids has clinical significance, as in several disease states (e.g. liver failure) there are enormous increases in plasma concentrations of medium-chain and long-chain fatty acids.

The entry of pantothenic acid from blood into cerebrospinal fluid via the choroid plexus and from the extracellular space of brain into brain cells themselves are both saturable transport processes (Spector, 1986).

23.9 BIOCHEMICAL FUNCTIONS OF COENZYME A AND ACYL CARRIER PROTEIN IN CELLULAR METABOLISM

23.9.1 Coenzyme A

The synthesis of CoA from pantothenic acid is shown in Figure 23.2. In the initial and rate-limiting reaction, ATP provides the phosphate group for the phosphorylation of pantothenic acid to 4′-phosphopantothenate, which is then coupled to cysteine in another ATP-requiring reaction. The 4′-phosphopantothenoylcysteine thus formed is decarboxylated to generate 4′-phosphopantetheine, which is then coupled to adenosine monophosphate (AMP) in a pyrophosphate linkage to form dephosphocoenzyme A. This penultimate product is phosphorylated at its adenosine 3′–OH group to form CoA. Synthesis of CoA is inhibited by acetyl-CoA, malonyl-CoA, and propionyl-CoA, as well as by other longer-chain acyl CoAs.

CoA is represented in biochemical equations by HS–CoA to show the reactive thiol group. CoA forms energy-rich thioesters with weakly reactive carboxylic acids, so

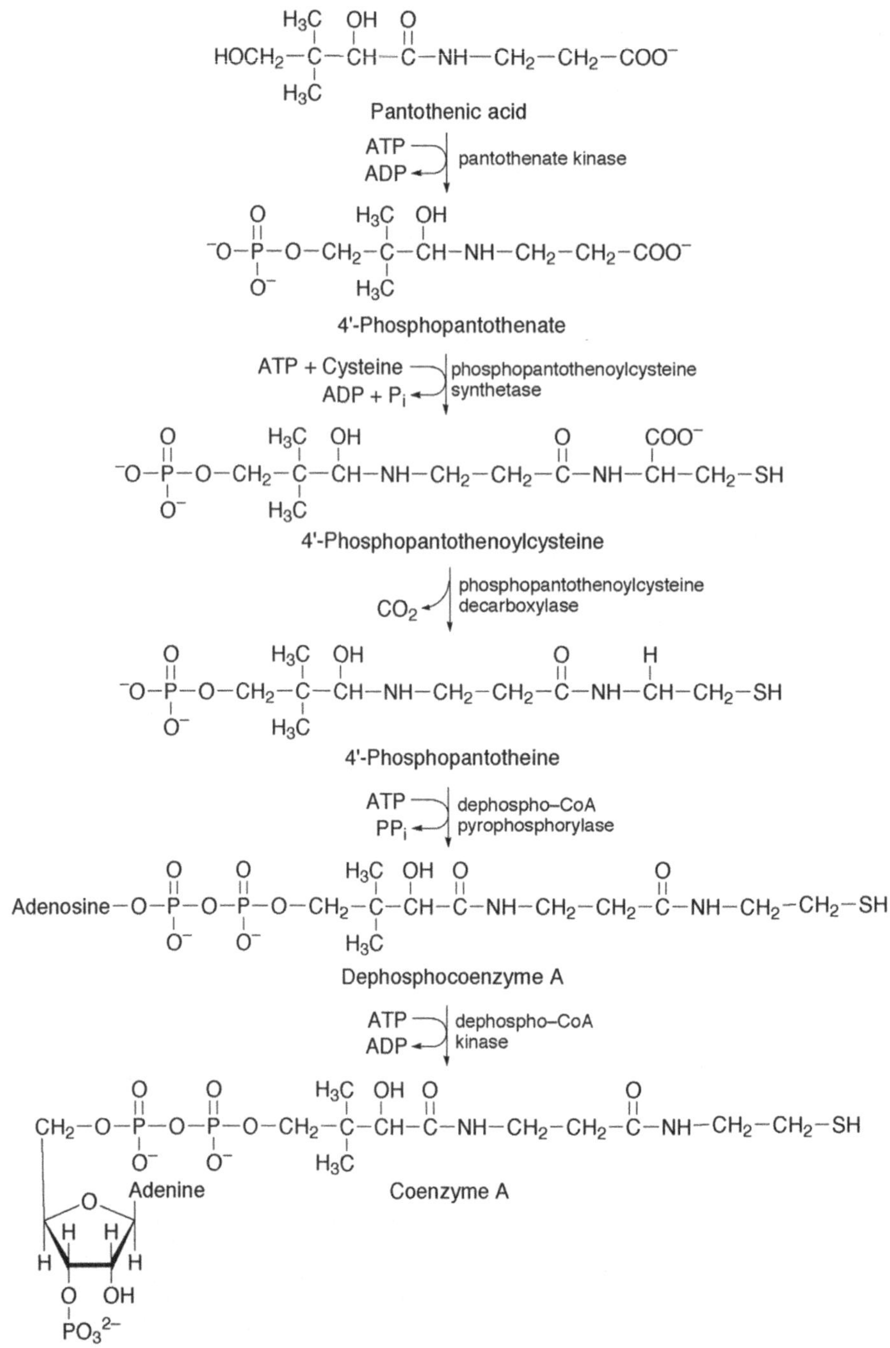

FIGURE 23.2 Synthesis of coenzyme A from pantothenic acid.

that the acyl (R–CO–) groups can participate in numerous biochemical reactions. Some of these reactions are involved in the release of energy from carbohydrates, fats, and amino acids. Eukaryotic cells contain sequestered pools of CoA and the availability of intracellular CoA is limited by the membranes surrounding the mitochondria and peroxisomes (Leonardi *et al.*, 2005).

- In carbohydrate metabolism, the pyruvate produced by glycolysis must undergo an oxidative decarboxylation reaction with CoA to form acetyl-CoA (CH_3–CO–S–CoA in biochemical reactions) before the acetyl moiety can react with oxaloacetate to produce citrate in the tricarboxylic acid cycle.
- Another tricarboxylic acid cycle intermediate, α-ketoglutarate, undergoes an oxidative decarboxylation with CoA to form succinyl-CoA. Succinyl-CoA reacts with glycine to form (via an intermediate) δ-aminolevulinic acid, which is a precursor of the porphyrin ring system of hemoglobin and cytochromes.
- CoA is required at two steps in each cycle of the β-oxidation of fatty acids, in which two carbon

units are removed per cycle to yield ultimately acetyl-CoA.

- In amino acid metabolism, leucine is deaminated to form the keto acid isovalerylformic acid, which reacts with CoA to yield a series of intermediates, ultimately giving rise to acetoacetic acid and acetyl-CoA.
- Acetyl-CoA is required for the acetylation of choline to form the neurotransmitter acetylcholine.
- The amino sugars D-glucosamine and D-galactosamine react with acetyl-CoA to form acetylated products, which are structural components of various mucopolysaccharides. For example, hyaluronic acid, which is found in connective tissue, the vitreous humor of the eye, and Wharton's jelly in the umbilical cord, consists of alternating residues of *N*-acetyl-D-glucosamine and D-glucuronic acid.
- The biosynthesis of cholesterol begins with the condensation of two molecules of acetyl-CoA to form acetoacetyl-CoA. The latter reacts with a third molecule of acetyl-CoA to form β-hydroxy-β-methylglutaryl-CoA (HMG-CoA), which, in turn, is reduced to mevalonic acid.

23.9.2 Role of Acyl Carrier Protein in the Biosynthesis of Fatty Acids

Acyl carrier protein, as an integral part of fatty acid synthase, plays an essential role in the biosynthesis of saturated fatty acids by sequestering the reaction intermediates as unique 4′-phosphopantetheine thioesters.

23.10 PHYSIOLOGICAL ROLES OF COENZYME A IN THE MODIFICATION OF PROTEINS

23.10.1 Protein Modification

Many diverse cellular proteins are modified by the covalent attachment of lipids donated by CoA or requiring CoA for their synthesis. The modifications fall into three main categories: acetylation, acylation, and isoprenylation (Magee, 1990; Casey, 1994). The alterations in protein structure may be relevant to the association of proteins with the plasma membrane or with subcellular membranes, protein–protein binding, or the targeting of proteins to specific intracellular locations. In some cases the modifications are co-translational, i.e. they take place on the growing polypeptide chain associated with the ribosome during protein synthesis; in other cases they are post-translational.

Most soluble proteins are acetylated at their amino termini as a means of altering their binding affinity for receptors or other proteins. Internal acetylation of nuclear histones weakens their association with DNA. The two long-chain fatty acids most commonly attached to proteins are myristic acid (14:0) and palmitic acid (16:0). The enzyme linking myristate to amino-terminal glycine residues by an amide bond is *N*-myristoyl transferase, which has strict sequence requirements in the protein substrate. Palmitoyl transferases link palmitate to the side chains of cysteine residues by a thioester bond. The cysteine residues can reside at any point in the primary structure of the protein; there is little evidence for any specific sequence requirements. Unlike the highly stable amide linkages to myristate, modifications of proteins by palmitate occur in thioester or oxyester linkages that are subject to hydrolysis by esterases. Cycles of palmitoylation and depalmitoylation allow the modified protein to have a regulating function. Addition of an isoprenoid chain to the cysteine residue of the primary motif CAAX (C, cysteine; AA, an aliphatic amino acid; X, the carboxyterminal amino acid) is the first step in the modification of proteins bearing this C-terminal motif. Either the 15-carbon farnesyl or the 20-carbon geranylgeranyl chain is added, depending on the sequence of the CAAX motif.

Myristoylated proteins include G protein α subunits (signal transduction), ADP-ribosylation factors (vesicular transport), myristoylated alanine-rich C kinase substrate protein (cytoskeletal rearrangements), recoverin (vision), proteins of the immune system, and several enzymes. Palmitoylated proteins include G protein α subunits, many plasma membrane-anchored receptors, cytoskeletal proteins, gap junction proteins, neuronal proteins, and the enzymes acetylcholinesterase and glutamic acid decarboxylase. Palmitate modification is also a prerequisite for the budding of transport vesicles from Golgi cisternae. Isoprenylated proteins include Ras proteins (signal transduction), Rab proteins (vesicular transport), nuclear lamins A and B (assembly and stabilization of the nuclear envelope), G protein γ subunits, and the enzymes phosphorylase kinase and rhodopsin kinase.

23.10.2 Physiological Implications of Protein Modification

The physiological implications of protein modifications involving coenzyme A are illustrated in the following examples.

23.10.2.1 Acetylation of β-Endorphin

Amino-terminal acetylation plays an important role in regulating the biological activity of the brain neurotransmitter β-endorphin. This peptide has morphine-like analgesic activity and also affects sexual behavior and learning. Acetylation deactivates β-endorphin by rendering it unable to bind to specific receptors. The modification is post-translational and occurs before or during the packaging of the peptide into the secretory granules of neurotransmitter neurons in the pituitary gland (Glembotski, 1982; Chappell *et al.*, 1986).

23.10.2.2 Histone Acetylation

In the control of gene transcription, one important chromatin remodeling system involves the enzyme-catalyzed acetylation/deacetylation of core histones. Histone acetylation, which results in activation of transcription, requires acetyl-CoA as the acetyl donor.

23.10.2.3 α-Tubulin Acetylation

Microtubules are constituents of the cytoskeleton and are composed of polymerized α- and β-tubulin dimers. A subset of the α-tubulin is modified, like the histones, by post-translational acetylation of the ε-amino group of specific lysine residues. In contrast to histone acetylation, the acetylation of α-tubulin stabilizes the polymeric structure of the microtubule; deacetylation is coupled to depolymerization (Plesofsky-Vig & Brambl, 1988).

23.10.2.4 Acylation of G Proteins

The biological activity of peptide hormones is mediated by second messengers such as cyclic AMP whose formation is triggered by the action of a G protein upon an effector enzyme. The reversible translocation of the G protein α-subunit between the plasma membrane and the cytoplasm is facilitated by the detachment and re-attachment of a palmitate group. Coenzyme A is required as the palmitate donor when Gα is palmitoylated by palmitoyl transferase.

23.10.2.5 Farnesylation of Rhodopsin Kinase

Rhodopsin kinase, the enzyme which phosphorylates the photon-stimulated receptor rhodopsin and desensitizes the visual signal, is translocated from the cytosol to the rod outer segment membrane upon light exposure. This light-induced translocation is facilitated by farnesylation of the enzyme's CAAX motif (Inglese *et al.*, 1992). The synthesis of farnesol (an isoprenyl compound) requires CoA.

23.11 IMMUNE FUNCTION

Axelrod and Hopper (1960) showed that antibody formation in response to influenza virus infection in the rat is impaired by pantothenic acid deficiency. Pantothenic acid-deficient rats were reported to exhibit depressed concentrations of circulating antibodies measured five days after immunization with human erythrocytes (Ludovici *et al.*, 1949). The use of paired-weight animals as controls showed that this depressed response was not due to inanition. Serum antibody concentration represents an equilibrium determined by rate of synthesis and release of the antibody and by its catabolism. Therefore, a decreased concentration of serum antibody may reflect: (1) an impairment of antibody synthesis, (2) a failure to release antibodies from their site of formation into the bloodstream, or (3) an accelerated rate of antibody catabolism.

Ludovici *et al.* (1951) reported that pantothenic acid-deficient rats injected intraperitoneally with human erythrocytes while receiving a complete diet maintained high serum antibody levels during a protracted period in which pantothenic acid was removed from the diet. This finding suggests that pantothenic acid deficiency does not result in an accelerated rate of antibody catabolism. Administration of massive doses of pantothenic acid to rats immunized while in a pantothenic acid-deficient state produced only a very gradual increase in serum antibody levels. Relatively normal levels were reached approximately two months after pantothenic acid repletion.

Lederer *et al.* (1975) demonstrated a marked reduction in the number of antibody-forming cells in the spleens of pantothenic acid-deficient rats injected with sheep erythrocytes. The defect was alleviated by administration of pantothenic acid nine days prior to injection of antigen. It therefore appeared that a deficiency of pantothenic acid leads to metabolic alterations that prevent the normal sequence of events leading to antibody-forming cell production.

A deficiency of pantothenic acid has been shown to impair the secretion of newly synthesized proteins out of hepatocytes (Axelrod, 1971). The evidence was derived from the analysis of ^{14}C-labelled protein radioactivity associated with the smooth and rough endoplasmic reticulum after varying pulse periods with labeled amino acids. The results indicated that in the hepatocytes of control rats, the secretory proteins move to the smooth endoplasmic reticulum after synthesis on the rough reticulum and are finally secreted out of the cell. In the hepatocytes of pantothenic acid-deficient rats, however, there appears to be a block in the release of the newly synthesized proteins, resulting in their accumulation in the smooth endoplasmic reticulum. A comparable defect in the intracellular transport of antibody by splenic cells could (if found) contribute to the impairment of antibody response in pantothenic acid deficiency.

23.12 PANTOTHENIC ACID DEFICIENCY IN ANIMALS

Smith and Song (1996) have reviewed the comparative nutrition of pantothenic acid. Natural pantothenic acid deficiency in domesticated animals has only been encountered in pigs. Pantothenic acid deficiency has been induced experimentally in many species of animals and birds by feeding diets containing low levels of the vitamin. The following deficiency signs have been observed.

- **Growth.** A decrease in appetite leads to reduced food intake and reduced growth rate or weight loss. There is also poor feed utilization.
- **Skin.** Dermatitis in pigs. Loss of hair pigment (graying) followed by hair loss in rats and mice. Faded and rough plumage in poultry. Formation of crusts in the beak area and between the toes and sealing of the eyelids by a viscous substance in chicks. Infrequent and slight dilation of the duodenal crypts of Lieberkühn and vacuolation of pancreatic acinar cells.
- **Nervous system.** 'Goose-stepping' of the hind legs in pigs and ataxia in chicks are associated with demyelination of the motor neurons.
- **Immune system.** Necrotic lymphocytes in lymphoid organs. Inhibition of antibody formation.

- **Adrenals.** Rodents are particularly prone to hypertrophy, necrosis, and hemorrhage of the adrenal glands with consequent impairment of adrenal endocrine function.
- **Reproductive system.** Degenerative changes in the female genital tract and loss of ovarian function in pigs. Decrease in spermatogenesis with atrophic seminiferous tubules and reduced number of seminal epithelia in rats. Reduced hatchability in poultry.

23.13 PANTOTHENIC ACID DEFICIENCY IN HUMANS

Because of the widespread occurrence of pantothenic acid in food, human deficiency does not normally occur. Human pantothenic acid deficiency has been studied in healthy male volunteers given an emulsified artificial diet by stomach tube. In one study (Hodges *et al.*, 1958), two subjects received the basic diet devoid of pantothenic acid, a second pair received the same diet with added antagonist (omega-methyl pantothenic acid), and a third pair (the controls) received the diet supplemented with pantothenic acid. After about four weeks, subjects receiving the antagonist and those in the deficient group began to show similar symptoms of illness. Clinical observations were irritability, restlessness, drowsiness, insomnia, impaired motor co-ordination, and neurological manifestations such as numbness and 'burning feet' syndrome. The most persistent and troublesome symptoms were fatigue, headache, and the sensation of weakness. Among the laboratory tests, the loss of eosinopenic response to adrenocorticotropic hormone indicated adrenocortical insufficiency.

23.14 TOXICITY

There are no known toxic effects of oral pantothenic acid in humans or animals.

REFERENCES

Annous, K.F. & Song, W.O. (1995) Pantothenic acid uptake and metabolism by red blood cells of rats. *Journal of Nutrition*, **125**, 2586–93.

Axelrod, A.E. (1971) Immune processes in vitamin deficiency states. *American Journal of Clinical Nutrition*, **24**, 265–71.

Axelrod, A.E. & Hopper, S. (1960) Effects of pantothenic acid, pyridoxine and thiamine deficiencies upon antibody formation to influenza virus PR-8 in rats. *Journal of Nutrition*, **72**, 325–30.

Barbarat, B. & Podevin, R.-A. (1986) Pantothenate-sodium cotransport in renal brush-border membranes. *Journal of Biological Chemistry*, **261**(31), 14455–60.

Beinlich, C.J., Naumovitz, R.D., Song, W.O. & Neely, J.R. (1990) Myocardial metabolism of pantothenic acid in chronically diabetic rats. *Journal of Molecular and Cellular Cardiology*, **22**, 323–32.

Casey, P.J. (1994) Lipid modifications of G proteins. *Current Opinion in Cell Biology*, **6**, 219–25.

Chappell, M.C., O'Donohue, T.L., Millington, W.M. & Kempner, E.S. (1986) The size of enzymes acetylating α-melanocyte-stimulating hormone and β-endorphin. *Journal of Biological Chemistry*, **261**(3), 1088–90.

Chatterjee, N.S., Kumar, C.K., Ortiz, A., Rubin, S.A. & Said, H.M. (1999) Molecular mechanism of the intestinal biotin transport process. *American Journal of Physiology*, **277**, C605–13.

Glembotski, C.C. (1982) Characterization of the peptide acetyltransferase activity in bovine and rat intermediate pituitaries responsible for the acetylation of β-endorphin and α-melanotropin. *Journal of Biological Chemistry*, **257**(17), 10501–9.

Grassl, S.M. (1992) Human placental brush-border membrane Na^+-pantothenate cotransport. *Journal of Biological Chemistry*, **267**(32), 22902–6.

Hodges, R.E., Ohlson, M.A. & Bean, W.B. (1958) Pantothenic acid deficiency in man. *Journal of Clinical Investigation*, **37**, 1642–57.

Inglese, J., Glickman, J.F., Lorenz, W., Caron, M.G. & Lefkowitz, R.J. (1992) Isoprenylation of a protein kinase. Requirement of farnesylation/α-carboxyl methylation for full enzymatic activity of rhodopsin kinase. *Journal of Biological Chemistry*, **267**(3), 1422–5.

Karnitz, L.M., Gross, C.J. & Henderson, L.M. (1984) Transport and metabolism of pantothenic acid by rat kidney. *Biochimica et Biophysica Acta*, **769**, 486–92.

Lederer, W.H., Kumar, M. & Axelrod, A.E. (1975) Effects of pantothenic acid deficiency on cellular antibody synthesis in rats. *Journal of Nutrition*, **105**, 17–25.

Leonardi, R., Zhang, Y.-M., Rock, C.O. & Jackowski, S. (2005) Coenzyme A: Back in action. *Progress in Lipid Research*, **44**, 125–53.

Lopaschuk, G.D., Michalak, M. & Tsang, H. (1987) Regulation of pantothenic acid transport in the heart. Involvement of a Na^+-cotransport system. *Journal of Biological Chemistry*, **262**(8), 3615–9.

Ludovici, P.P., Axelrod, A.E. & Carter, B.B. (1949) Circulating antibodies in vitamin deficiency states. Pantothenic acid deficiency. *Proceedings of the Society for Experimental Biology and Medicine*, **72**, 81–3.

Ludovici, P.P., Axelrod, A.E. & Carter, B.B. (1951) Circulating antibodies in vitamin deficiency states. Pantothenic acid and pyridoxine deficiencies. *Proceedings of the Society for Experimental Biology and Medicine*, **76**, 665–70.

Magee, A.I. (1990) Lipid modification of proteins and it relevance to protein targeting. *Journal of Cell Science*, **97**, 581–4.

Plesofsky-Vig, N. & Brambl, R. (1988) Pantothenic acid and coenzyme A in cellular modification of proteins. *Annual Review of Nutrition*, **8**, 461–82.

Prasad, P.D., Ramamoorthy, S., Leibach, F.H. & Ganapathy, V. (1997) Characterization of a sodium-dependent vitamin transporter mediating the uptake of pantothenate, biotin and lipoate in human placental choriocarcinoma cells. *Placenta*, **18**, 527–33.

Prasad, P.D., Wang, H., Kekuda, R., Fujita, T., Fei, Y.-J., Devoe, L.D., Leibach, F.H. & Ganapathy, V. (1998) Cloning and functional expression of a cDNA encoding a mammalian sodium-dependent vitamin transporter mediating the uptake of pantothenate, biotin, and lipoate. *Journal of Biological Chemistry*, **273**(13), 7501–6.

Prasad, P.D., Wang, H., Huang, W., Fei, Y.-J., Leibach, F.H., Devoe, L.D. & Ganapathy, V. (1999) Molecular and functional characterization of the intestinal Na^+-dependent multivitamin transporter. *Archives of Biochemistry and Biophysics*, **366**, 95–106.

Reibel, D.K., Wyse, B.W., Berkich, D.A. & Palko, W.M. (1981) Effects of diabetes and fasting on pantothenic acid metabolism in rats. *American Journal of Physiology*, **240**, E597–601.

Robishaw, J.D. & Neely, J.R. (1985) Coenzyme A metabolism. *American Journal of Physiology*, **248**, E1–9.

Said, H.M., Ortiz, A., McCloud, E., Dyer, D., Moyer, M.P. & Rubin, S. (1998) Biotin uptake by human colonic epithelial NCM460 cells: A carrier-mediated process shared with pantothenic acid. *American Journal of Physiology*, **275**, C1365–71.

Shibata, K., Gross, C.J. & Henderson, L.M. (1983) Hydrolysis and absorption of pantothenate and its coenzymes in the rat small intestine. *Journal of Nutrition*, **113**, 2107–15.

Smith, C.M. & Milner, R.E. (1985) The mechanism of pantothenate transport by rat liver parenchymal cells in primary culture. *Journal of Biological Chemistry*, **260**(8), 4823–31.

Smith, C.M. & Savage, C.R. Jr. (1980) Regulation of Coenzyme A biosynthesis by glucagon and glucocorticoid in adult rat liver parenchymal cells. *Biochemical Journal*, **188**, 175–84.

Smith, C.M. & Song, W.O. (1996) Comparative nutrition of pantothenic acid. *Journal of Nutritional Biochemistry*, **7**, 312–21.

Sorrell, M.F., Frank, O., Thomson, A.D., Aquino, H. & Baker, H. (1971) Absorption of vitamins from the large intestine in vivo. *Nutrition Reports International*, **3**, 143–8.

Spector, R. (1986) Pantothenic acid transport and metabolism in the central nervous system. *American Journal of Physiology*, **250**, R292–7.

Spector, R., Sivesind, C. & Kinzenbaw, D. (1986) Pantothenic acid transport through the blood-brain barrier. *Journal of Neurochemistry*, **47**, 966–71.

Stein, E.D. & Diamond, J.M. (1989) Do dietary levels of pantothenic acid regulate its intestinal uptake in mice? *Journal of Nutrition*, **119**, 1973–83.

Tahiliani, A.G. & Beinlich, C.J. (1991) Pantothenic acid in health and disease. *Vitamins and Hormones*, **46**, 165–228.

Tarr, J.B., Tamura, T. & Stokstad, E.L.R. (1981) Availability of vitamin B-6 and pantothenate in an average American diet in man. *American Journal of Clinical Nutrition*, **34**, 1328–37.

Wang, H., Huang, W., Fei, Y.-J., Xia, H., Yang-Feng, T.L., Leibach, F.H., Devoe, L.D., Ganapathy, V. & Prasad, P.D. (1999) Human placental Na^+-dependent multivitamin transporter. Cloning, functional expression, gene structure, and chromosomal localization. *Journal of Biological Chemistry*, **274**(21), 14875–83.

24 Biotin

24.1 CHEMISTRY

The biotin molecule (Figure 24.1) is a fusion of an imidazolidone ring with a tetrahydrothiophene ring bearing a pentanoic acid side chain. The molecule contains three asymmetric carbon atoms, and hence eight stereoisomers are possible. Of these, only the dextrorotatory *d*-(+)-biotin occurs in nature and possesses vitamin activity. Biotin synthesized industrially by the Hoffmann-La-Roche process is also in the *d*-form (Achuta Murthy & Mistry, 1977).

In animal and plant tissues only a small proportion of the biotin present occurs in the free state. The majority is covalently bound to the protein structure (apoenzyme) of biotin-dependent enzymes via an amide bond between the carboxyl group of biotin and the ε-amino group of a lysine residue. Proteolysis of such enzymes liberates a natural water-soluble fragment called biocytin (biotin-ε-lysine) (Figure 24.2), which is biologically active.

24.2 FOOD SOURCES AND BIOAVAILABILITY

24.2.1 Food Sources

Biotin is present in all natural foodstuffs, but the content of even the richest sources is very low when compared with the content of most other water-soluble vitamins. Biotin is not commonly used in fortified foods, apart from infant formulas. Typical values of some rich natural sources of biotin include ox liver (33 μg per 100 g), whole eggs (20 μg per 100 g), dried soya beans (65 μg per 100 g), and peanuts (72 μg per 100 g) (Holland *et al.*, 1991). Other good sources include yeast, wheat bran, oatmeal, and some vegetables. Muscle meats, fish, dairy products, and cereals contain smaller amounts, but are important contributors to the dietary intake. Most of the biotin content of animal products, nuts, cereals, and yeast is in a protein-bound form. A higher percentage of free, water-extractable biotin occurs in vegetables, green plants, fruit, milk, and rice bran (Lampen *et al.*, 1942).

24.2.2 Bioavailability

The bioavailability of biotin in various foods and the effect of protein binding on bioavailability have not been studied directly in human subjects. Studies of animal feed ingredients using growth assays have shown that biotin in various cereals (but not in maize) is largely unavailable to the chicken (Frigg, 1984), turkey (Misir & Blair, 1988a), and weanling pig (Misir & Blair, 1988b). Most, if not all, of the biotin content of human milk is in the free form and thus is likely to be completely bioavailable to the infant (Heard *et al.*, 1987).

24.3 DIGESTION AND ABSORPTION OF DIETARY BIOTIN

Humans and other mammals cannot synthesize biotin and therefore must obtain the vitamin from exogenous sources via intestinal absorption. Biotin in foods exists as the free vitamin and as protein-bound forms in variable proportions. Because the amide bond linking biotin and lysine residues is not hydrolyzed by gastrointestinal proteases, proteolytic digestion of protein-bound biotin releases not biotin, but biocytin and biotin-containing short peptides (biotinyl peptides). Biotinidase (biotin-amidohydrolase, EC 3.5.1.12), which is present in pancreatic juice and in intestinal mucosa, is uniquely capable of hydrolyzing biocytin to yield free biotin and lysine. The optimal pH for biotinidase hydrolysis of biocytin is pH 4.5–5.5 (Chauhan & Dakshinamurti, 1986), which is appropriate for releasing biotin in the acidic milieu within the lumen of the duodenum and proximal jejunum (pH range 3.5–7) (McMahon, 2002). There is a precipitous decrease in hydrolase activity above pH 7 (the pH of serum is 7.4).

Avidin, the glycoprotein in egg white, uniquely binds biotin, both in the free form and as the prosthetic group of enzymes. The complex is stable over a wide pH range and is not dissociated in the gastrointestinal tract. When eggs are cooked, the avidin is denatured and the biotin is liberated.

Intestinal absorption of biotin in the rat proceeds largely by a saturable (carrier-mediated) process at concentrations less than 5 μM, whereas at concentrations above 25 μM, non-saturable uptake (simple diffusion) predominates (Bowman *et al.*, 1986). The rate of biotin transport in human intestine was shown to increase with a decrease in pH of the incubation medium from 8 to 5.5 (Said *et al.*, 1988a). In adult humans, biotin absorption is significantly greater in the duodenum and jejunum than in the ileum, owing to a decrease in V_{max} (the number of biotin transport carriers) distally with no changes in the apparent K_m (affinity of the carrier toward its substrate) (Said *et al.*, 1988a). The sodium-dependent multivitamin transporter (SMVT), which handles pantothenate, biotin, and lipoate, mediates an electrogenic process with a Na^+:biotin$^-$ stoichiometry of 2:1 (Prasad *et al.*, 2000). The SMVT is the major (if not the only) biotin uptake system that operates in intestinal and liver epithelial cells (Balamurugan *et al.*, 2003).

Said and Mohammedkhan (1988) reported developmental changes in the transepithelial transport of biotin in intact rat intestinal tissue preparations. Further investigation by

DOI: 10.1201/9781003410300-28

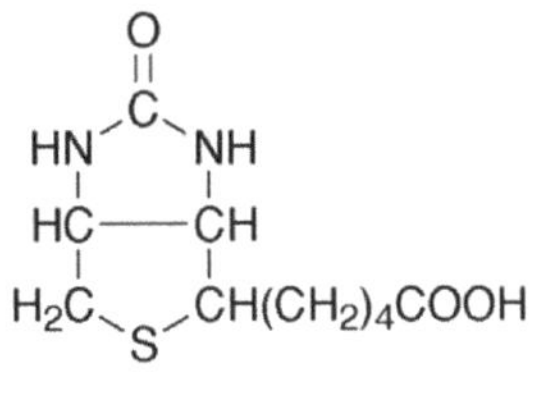

FIGURE 24.1 Structure of *d*-biotin.

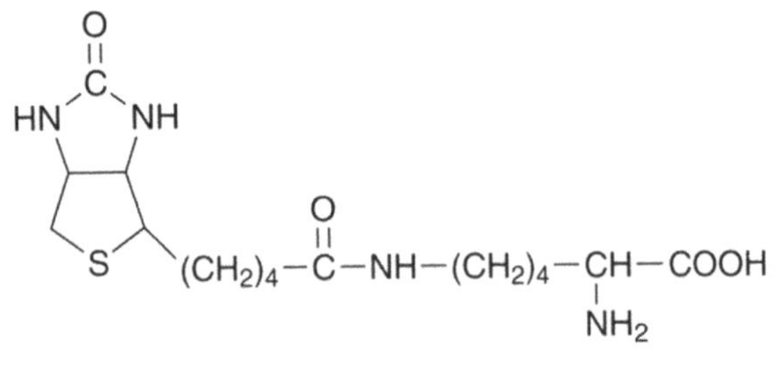

FIGURE 24.2 Structure of biocytin.

Nabokina *et al.* (2003) showed that changes in the kinetics of intestinal biotin uptake during maturation were associated with an increase in the level of the SMVT protein from suckling to adult rats. Also observed was a higher transcription rate for SMVT in adult compared with suckling rats, which supported the suggestion that the developmental changes involve regulation by transcriptional mechanism(s).

Experiments using specific activators and inhibitors have shown that biotin absorption in human-derived Caco-2 cells is regulated by the protein kinase C (PKC) and Ca^{2+}/calmodulin cell signaling pathways, but not by the protein kinase A pathway (Said, 1999). Activation of PKC by a phorbol ester (phorbol 12-myristate 13-acetate) inhibited biotin uptake; conversely, inhibition of PKC with staurosporine and chelerythrin stimulated biotin uptake. The inhibition was due to a decrease in the number of functional biotin uptake carriers (decreased V_{max}) with no effect upon their affinity (apparent K_m unchanged). In contrast, inhibition of the Ca^{2+}/calmodulin pathway by calmidazolium and trifluoroperazine inhibited biotin uptake, also through a decrease in the number of functional carriers. Simultaneous addition of phorbol ester and calmidazolium to the cultured cells produced a greater degree of inhibition than addition of either compound alone, which suggested that these compounds use different mechanisms to exert their effects.

The exit of biotin from the human enterocyte, i.e transport across the basolateral membrane, is a carrier-mediated process which has a higher affinity for the substrate than the transport system at the brush-border membrane. The basolateral transport process is independent of sodium, electrogenic in nature, and not capable of accumulating biotin against a concentration gradient (Said *et al.*, 1988b).

Studies using rat intestinal brush-border membrane vesicles (Said *et al.*, 1989) and human Caco-2 cells (Ma *et al.*, 1994) have shown that the intestinal biotin uptake process is adaptively up-regulated during biotin deficiency, due to an increase in SMVT protein and mRNA levels. This up-regulation is attributable to transcriptional regulatory mechanism(s) and not to changes in mRNA stability (Reidling *et al.*, 2007). There are conflicting data from rat studies regarding the intestinal transport of biocytin. On the one hand, Dakshinamurti *et al.* (1987) concluded that the transport is carrier-mediated and more efficient than the uptake of free biotin. Said *et al.* (1993), on the other hand, found that the transport of biocytin occurs by a non-mediated process that is independent of sodium, pH, energy, and temperature, and is significantly less efficient than that of free biotin. The mechanism for the transport of biotinyl peptides is unknown; it might involve a non-specific pathway for peptide absorption or possibly a specific biotin transporter (Mock, 1990a).

24.4 ABSORPTION OF BACTERIALLY SYNTHESIZED BIOTIN IN THE LARGE INTESTINE

The normal microflora of the large intestine (mainly in the cecum) can synthesize large amounts of biotin, much of which exists as the unbound molecule. This is inferred by the fecal content of biotin exceeding dietary intake three- to sixfold. Balance studies in humans have shown that urinary content often exceeds dietary intake, suggesting that enterically synthesized biotin can be absorbed and utilized to some extent by the host. In a study with human subjects (Sorrell *et al.*, 1971), when biotin was instilled directly into the lumen of the mid-transverse colon, the concentration of plasma biotin increased, although not as much as when the same dose of vitamin was given orally. Bowman and Rosenberg (1987) examined biotin absorption using the rat intestinal loop, an *in vivo* preparation in which the intestinal blood supply and lymphatics remain intact. Their data showed that the large intestine of the rat is capable of absorbing nutritionally significant amounts of biotin.

Said *et al.* (1998) studied the mechanism and regulation of biotin transport in the large intestine using the human colonic epithelial cell line NCM460 as an *in vitro* model system. Uptake of biotin by these cells involved a Na^+-dependent, carrier-mediated transport system that was competitively inhibited by pantothenic acid and appeared to be under the regulation of an intracellular PKC-mediated pathway. These findings are similar to those in the small intestine and suggest that a common mechanism mediated by the SMVT may be involved in the transport of biotin and pantothenic acid in the small and large intestinal epithelia.

24.5 BIOTIN UPTAKE BY THE LIVER

The liver extracts the majority of the newly absorbed biotin from the portal blood and is the major site of biotin utilization and metabolism. Pacheco-Alvarez *et al.* (2005) showed that biotin availability regulates expression of the

SMVT and the rate of biotin uptake in human liver cells. Their results showed that HepG2 cells grown in a biotin-deficient medium have a lower rate of biotin uptake than normal cells. The impairment in biotin uptake is associated with a reduction in the amount of both SMVT protein mass and mRNA levels. SMVT expression is restored by biotin supplementation, while lipoic acid and pantothenic acid, the other substrates of SMVT, do not have a significant effect. As discussed in Section 24.11.5, this negative effect of biotin deficiency on biotin uptake by HepG2 cells spares biotin depletion of the brain.

24.6 BIOTINIDASE AS A BIOTIN CARRIER IN PLASMA

Approximately 81% of the total biotin in plasma is free (unbound), 7% is reversibly bound to plasma proteins, and 12% is covalently bound (Mock & Malik, 1992). Biocytin is generated in the breakdown of biotin-dependent carboxylases. In plasma, at pH 7.4, biotinidase cleaves biocytin, but instead of releasing the biotin, a complex, biotinyl-biotinidase (Figure 24.3), is formed through the sulfhydryl group of a cysteine moiety in the protein (Hymes & Wolf, 1999). The stability of the thioester bond at >pH 7 allows biotinyl-biotinidase to act as a carrier protein for biotin in plasma. Either low pH or a nucleophile, such as hydroxylamine, can remove the biotin from the enzyme, thereby permitting biotinidase to interact with another biocytin molecule.

24.7 RENAL REABSORPTION AND EXCRETION

The renal reabsorption of biotin plays an essential role in biotin homeostasis and conservation. Concentrative uptake of free biotin into brush-border membrane vesicles (BBMVs) isolated from human kidney cortex was shown to be carrier mediated and Na^+ dependent (Baur & Baumgartner, 1993). Intracellular biotin is thought to diffuse passively across the basolateral membrane into the peritubular fluid to complete the process of renal reabsorption (Podevin & Barbarat, 1986).

Having shown that intestinal transport of biotin undergoes clear developmental changes during the early stages of life (Said & Mohammedkhani, 1988), Nabokina *et al.* (2003) compared the effect of development on renal and intestinal transport of biotin using purified renal and intestinal BBMVs isolated from suckling and adult rats. Their results showed that, unlike intestinal uptake of biotin, renal uptake is not developmentally regulated; that is, renal uptake is similar in suckling and adult rats.

Balamurugan *et al.* (2005) used human-derived proximal tubular epithelial HK-2 cells as a model to study biotin uptake by human renal epithelial cells. In common with biotin uptake by human intestinal epithelial cells, biotin uptake by renal epithelial cells occurs via the SMVT system and is regulated by the intracellular PKC-mediated pathway. However, whereas activation of the PKC pathway inhibits biotin uptake in the small intestine (Said, 1999), it stimulates biotin uptake in the kidney. This difference may be attributed to the above-mentioned different responses to ontogenic regulation, which suggests that the renal and intestinal biotin uptake processes are under different regulation during maturation. Biotin uptake by HK-2 cells is also stimulated by the Ca^{2+}/calmodulin-mediated pathway. In addition, the uptake process is adaptively regulated by extracellular biotin level, which involves transcriptional regulatory mechanism(s).

Biotin is excreted in the urine with about half of it unchanged and the balance as biotin metabolites (Mock *et al.*, 1993). The biliary excretion of biotin and metabolites is only a minor route in the overall elimination of biotin from the body (Zemplini *et al.*, 1997).

24.8 THE BIOTIN CYCLE

Much of the biotin in foods exists as the prosthetic group of carboxylase enzymes and is not readily available for absorption. In order to conserve what little biotin is absorbed and to maintain requirements, mammals are able to recycle the vitamin. The biocytin and biotinyl peptides released by proteolysis in the small intestine are hydrolyzed by biotinidase to yield free biotin, which is absorbed into the bloodstream. Once inside the cell, biotin is covalently attached to a specific lysine residue on the apocarboxylase by the enzyme holocarboxylase synthetase (HCS), forming the active holoenzyme (see next paragraph). Turnover of carboxylases involves proteolysis to biotinyl peptides and cleavage of the peptides by cytoplasmic or plasma biotinidase. The biotin produced is used in biotinylation of new carboxylases, thus completing the cycle (Pacheco-Alvarez *et al.*, 2002).

The attachment of biotin to the apoenzyme is an adenosine triphosphate (ATP)-dependent, two-step reaction (Chapman-Smith & Cronan, 1999). In the first half reaction, HCS reacts with ATP to form biotinyl-5′-adenosine monophosphate (AMP) plus pyrophosphate (PPi). In the second half reaction, an amide bond is formed between the mixed anhydride carbon atom of biotinyl-5′-AMP and the ε-amino group of the specific lysine residue in the apoenzyme; this reaction releases AMP. Once the amide bond is formed, the biotin moiety remains attached throughout the lifetime of the protein molecule.

Pacheco-Alvarez *et al.* (2004) investigated the regulation of the biotin cycle in cultured cells and rat liver under

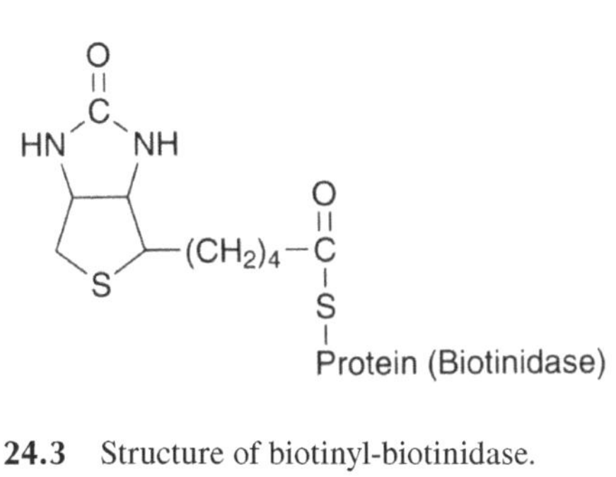

FIGURE 24.3 Structure of biotinyl-biotinidase.

conditions of biotin starvation and replenishment. They found that SMVT and HCS, proteins critical to the utilization of exogenous or dietary biotin, respond to biotin deprivation by reduction in the steady-state level of their mRNAs. In contrast, biotin reutilization via biotinidase, the enzyme responsible for biotin recycling during carboxylase turnover, is unaffected by biotin status.

24.9 TRANSPORT IN THE BRAIN

Spector and Mock (1987) measured the unidirectional influx of ^{3}[H]biotin across cerebral capillaries (the anatomical locus of the blood–brain barrier) using an *in situ* rat brain perfusion technique. Transport took place by a low-capacity, saturable system that was inhibited by pantothenic acid. This inhibition suggests that the SMVT characterized by Prasad *et al.* (1998) is responsible for the shared transport of biotin and pantothenic acid into the brain. The concentration of plasma biotin necessary to one-half saturate the entry of tracer ^{3}[H]biotin into brain, 100 μM, is several orders of magnitude higher than the normal plasma concentration of biotin (0.02 μM). This means that blood–brain transport plays no part in brain homeostasis of total biotin. If one were to increase plasma biotin levels by parenteral means, unnaturally high concentrations would enter the extracellular space of brain. Normally, the brain relies on the kidneys to maintain constant plasma levels of biotin. Biotin is similar to pantothenic acid in this respect, but differs from thiamin, folate, and vitamin C whose blood–brain transport systems play an important homeostatic role.

The isolated rabbit choroid plexus was unable to accumulate ^{3}[H]biotin from a medium concentration of 1 nM (Spector & Mock, 1987). This implies that there is no active transport process for transferring biotin between blood and cerebrospinal fluid: the route of biotin entry into the brain is mainly, if not exclusively, via the blood–brain barrier.

Spector and Mock (1988) investigated the observed resistance of brain to depletion of total biotin in deficiency states. They concluded that the resistance is due to the minimal incorporation of biotin into the appropriate apoenzymes and hence slow turnover of the holoenzymes in brain; it is not due to the extremely slow turnover of biotin in the cerebrospinal fluid and brain.

24.10 PLACENTAL TRANSPORT

Evidence of an electrogenic Na^{+}-biotin (2:1) co-transport system was reported in human placental microvillous membrane vesicles (Grassl, 1992) and two human placental cell lines (Prasad *et al.*, 1997). The transport system was inhibited by pantothenic acid and lipoate. Prasad *et al.* (1999) proved that the same transporter, the SMVT, is responsible for the uptake of biotin and pantothenate in placenta and intestine.

In human choriocarcinoma cells (JAr cells), biotin concentrations in culture media correlated negatively with expression of SMVT, as judged by cellular transport rates of biotin, abundance of SMVT protein, and transcriptional activity of SMVT reporter-gene constructs (Crisp *et al.*, 2004). This regulation of SMVT gene expression by the supply of biotin is the primary mechanism by which biotin homeostasis is maintained in the placenta.

24.11 BIOCHEMICAL FUNCTIONS

24.11.1 Biotin as a Coenzyme for Carboxylases

Biotin is a covalently bound prosthetic group in four carboxylases: pyruvate carboxylase (PC; EC 6.4.1.1), propionyl-CoA carboxylase (PCC; EC 6.4.1.3), β-(or 3-) methylcrotonyl-CoA carboxylase (β-MCC; 6.4.1.4), and acetyl-CoA carboxylase (ACC; 6.4.1.2), with the latter enzyme existing as two genetically distinct isoforms, ACC-1 and ACC-2. All are mitochondrial enzymes, except for ACC-1, which is cytosolic. The enzymes participate in gluconeogenesis, fatty acid synthesis, and amino acid metabolism. An overall scheme of biotin-dependent carboxylase reactions is shown in Figure 24.4.

In biotin deficiency, the four carboxylases are present as catalytically inactive apoenzymes. As soon as biotin is provided, holocarboxylase synthetase mediates its incorporation into the apoenzymes to create the active holoenzymes (Achuta Murthy & Mistry, 1977). The biotin prosthetic group of the holoenzyme functions in carboxylation

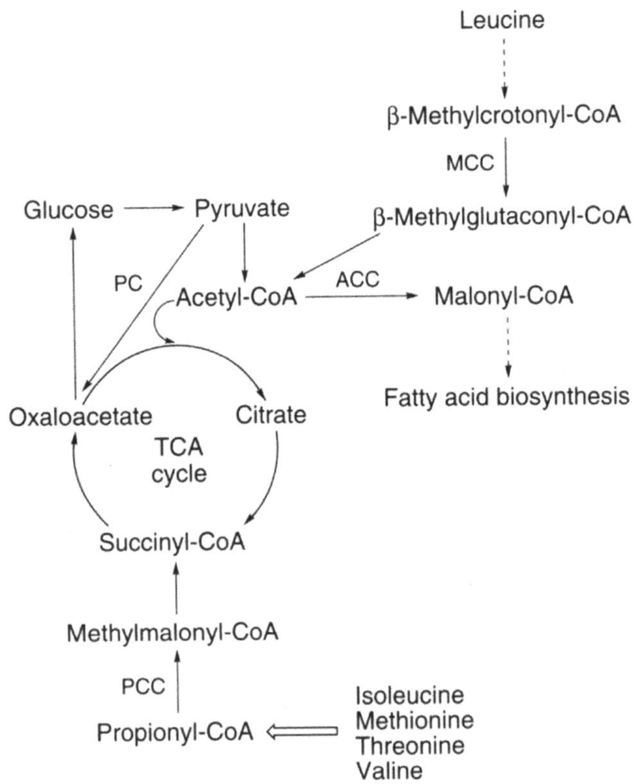

FIGURE 24.4 Overall scheme of biotin-dependent carboxylase reactions. MCC, β-methylcrotonyl-CoA carboxylase; PC, pyruvate carboxylase; ACC, acetyl-CoA carboxylase; PCC, propionyl-CoA carboxylase.

reactions as a specialized carrier of a carboxyl group (Seki, 2006). The biotin-catalyzed carboxyl group transfer reaction consists of two partial reactions involving two different catalytic sites. This is exemplified by the conversion of pyruvate to oxaloacetate, which is catalyzed by PC.

$$\text{Enz} - \text{biotin} - \text{HCO}_3 + \text{ATP} \xrightarrow{\text{Mg}^{2+}} \text{Enz} - \text{biotin} - \text{CO}_2 + \text{ADP} + \text{Pi}$$

$$\text{Enz} - \text{biotin} - \text{CO}_2 + \text{pyruvate} \rightarrow \text{Enz} - \text{biotin} + \text{oxaloacetate}$$

At catalytic site 1, an ATP-activated carboxyl group derived from bicarbonate (HCO_3^-) is attached to the biotin ring structure at the N-1 position. In effect, the HCO_3^- is dehydrated by its reaction with ATP, and the CO_2 reacts with biotin to form the carboxybiotinyl-enzyme depicted in Figure 24.5. The long, flexible chain connecting biotin and the apoenzyme allows the biotin to transfer the carboxyl group to catalytic site 2, where it reacts with the substrate. The conversion of pyruvate to oxaloacetate is an important anaplerotic reaction – a reaction that replenishes an intermediate in the tricarboxylic acid (TCA) cycle (Achuta Murthy & Mistry, 1977).

24.11.2 Gluconeogenesis

In glycolysis, the conversion of phosphoenolpyruvate to pyruvate, catalyzed by pyruvate kinase, is irreversible and there is no alternative enzyme that can reverse this reaction. In gluconeogenesis (the reversal of glycolysis) the pyruvate kinase system is circumvented by a bypass pathway, which involves PC activity in the mitochondrion.

Gluconeogenesis is impaired in biotin deficiency. The reduction in gluconeogenesis is initiated by a decrease in PC activity and subsequently enhanced by a shortage of cytoplasmic reducing equivalents, as indicated by a reduction in the ratio of lactate to pyruvate (Arinze & Mistry, 1971).

24.11.3 Lipogenesis

Two biotin-dependent carboxylases, ACC and PCC, are linked directly to lipogenesis. ACC catalyzes the conversion of acetyl-CoA to malonyl-CoA – the first and rate-limiting step in the synthesis of long-chain fatty acids (Figure 24.6). If malonyl-CoA formation is compromised during biotin deficiency, a decrease in the production of long-chain fatty acids would be expected. Skin from biotin-deficient rats contained 70% less total fatty acid compared to normal skin (Proud *et al.*, 1990). The reduction was due mainly to changes in 16:0, 16:1, 18:0, 18:1, 18:2, and 20:1.

Propionyl-CoA, produced by catabolism of isoleucine, methionine, threonine, and valine is converted by PCC to D-methylmalonyl-CoA. The latter compound is enzymatically epimerized to its L stereoisomer, which in turn undergoes an intramolecular rearrangement to form succinyl-CoA. The rearrangement is catalyzed by a mutase, which requires coenzyme B_{12} for its activity (Figure 24.7). The succinyl-CoA produced enters the TCA cycle. In the absence of biotin, the conversion of propionyl-CoA to D-methylmalonyl-CoA is impaired, causing an accumulation of propionyl-CoA.

Kramer *et al.* (1984) demonstrated that biotin deficiency in rats caused significantly increased concentrations of 15:0, 17:0, and total odd-chain fatty acids in liver phospholipids. Increases of these two odd-chain fatty acids were also demonstrated in plasma and heart phospholipids, suggesting a generalized abnormality in phospholipid composition (Mock *et al.*, 1988b). These results can be explained by the increased levels of propionyl-CoA, which is a primer for odd-chain fatty acid synthesis. In three patients who developed biotin deficiency during parenteral alimentation, the percentage of odd-chain fatty acids (15:0, 17:0) in serum increased for each of the four major lipid classes: cholesterol esters, phospholipids, triglycerides, and free fatty acids (Mock *et al.*, 1988a). Taken together, these findings

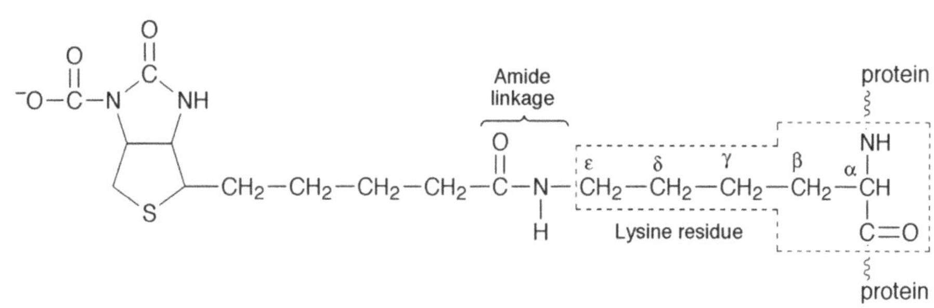

FIGURE 24.5 Structure of carboxybiotinyl-enzyme.

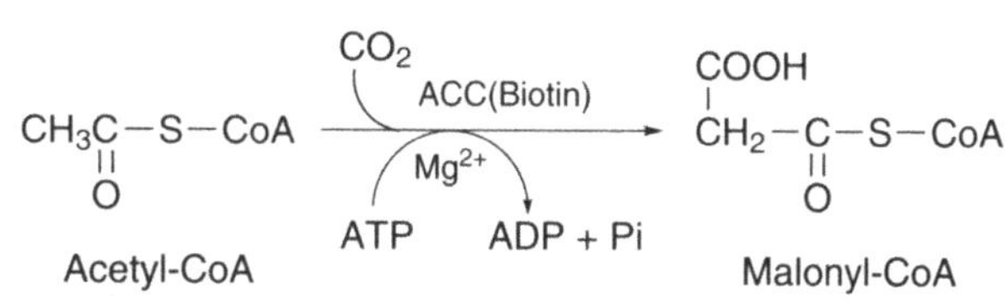

FIGURE 24.6 Carboxylation of acetyl-CoA. ACC, acetyl-CoA carboxylase.

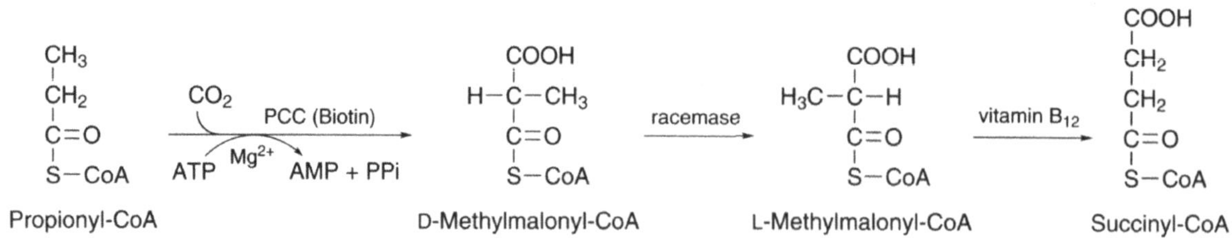

FIGURE 24.7 Oxidation of propionyl-CoA. PCC, propionyl-CoA carboxylase.

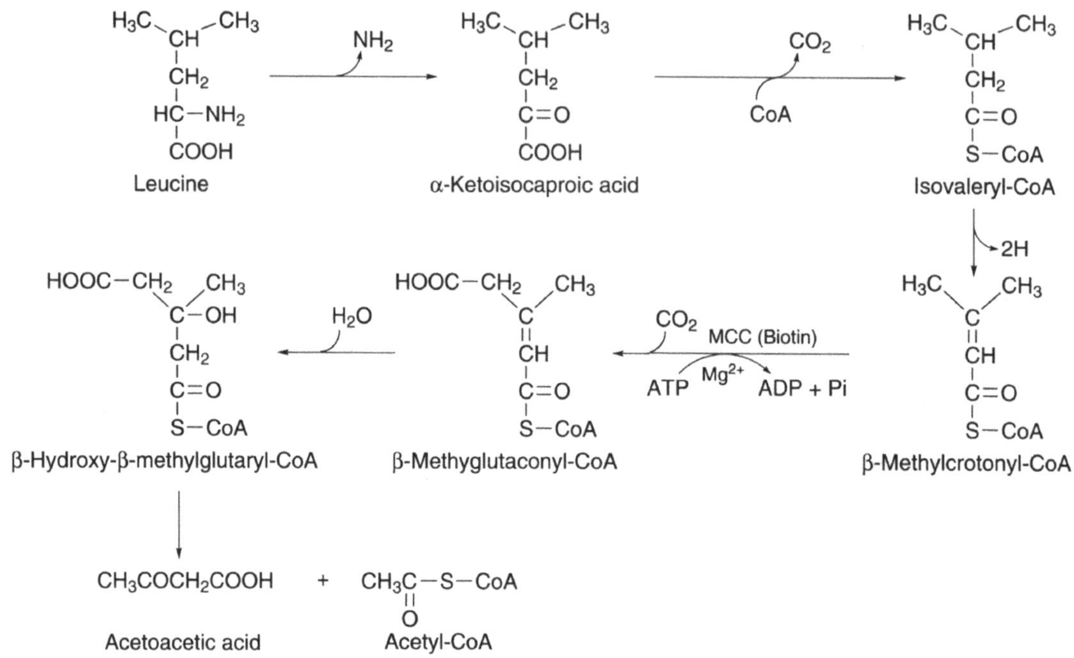

FIGURE 24.8 Pathway for the catabolism of leucine. MCC, β-methylcrotonyl-CoA carboxylase.

suggest that odd-chain fatty acid accumulation could be a marker for reduced PCC activity in biotin deficiency.

PC plays an indirect role in fatty acid synthesis. This process takes place in the cytosol, but the starting metabolite, acetyl-CoA, is generated by the TCA cycle within the mitochondria. Since the mitochondrial membrane is impermeable to acetyl-CoA, this metabolite has to reach the cytosol by an indirect means. It does so by donating its acetyl group to oxaloacetate, forming citrate which can cross the mitochondrial membrane. In the cytosol, the citrate is cleaved to re-form oxaloacetate and acetyl-CoA. The constant removal of oxaloacetate from the mitochondria necessitates its constant replenishment by the PC-catalyzed carboxylation of pyruvate (Achuta Murthy & Mistry, 1977).

PC is also implicated in the generation of NADPH by the pyruvate/malate pathway (Sugden & Holnesss, 2011). In this pathway, the oxaloacetate formed in the mitochondrion by the PC-catalyzed decarboxylation of pyruvate is converted to malate by malate dehydrogenase. Malate can be transported into the cytosol by the dicarboxylate carrier and converted to pyruvate in a reaction that generates NADPH. The pyruvate can then be transported back to the mitochondrion by the mitochondrial pyruvate carrier and converted to oxaloacetate. The significance of the pyruvate/malate pathway in lipogenesis is that NADPH is required for fatty acid synthesis. Normally, the demand for NADPH is met by the pentose phosphate pathway, but in circumstances of enhanced lipogenesis, this pathway may not be able to supply adequate amounts of reducing equivalents (Achuta Murthy & Mistry, 1977).

β-Methylcrotonyl-CoA arises from the catabolism of leucine. The fourth biotin-dependent enzyme, β-MCC catalyzes the conversion of β-methylcrotonyl-CoA to β-methylglutaconyl-CoA. This reaction product is hydroxylated to a compound which is cleaved to form acetyl-CoA and acetoacetate (Figure 24.8). The acetyl-CoA can enter the TCA cycle directly or be used for fatty acid synthesis.

24.11.4 Effects of Biotin Deficiency on the Activities of Biotin-Dependent Carboxylases

Two studies in the same laboratory, using rats and chicks fed an egg-white diet, showed a fall in biotin-dependent carboxylase activities in various tissues during the progress of biotin deficiency. In the first study, Arinze and Mistry

(1971) measured the activities of PC, PCC, and ACC in the livers of rats and chicks, and also ACC activity in rat adipose tissue. Hepatic PC and PCC activities were rapidly reduced in both species, with less than 20% of the control (biotin-injected animals) activities remaining after 4 weeks of biotin deprivation. ACC activity in rat adipose tissue also fell markedly, but the activity of this enzyme in rat and chick liver was less dramatically reduced. Even in severely deficient animals (7–8-week-old rats and over 4-week-old chicks on the egg-white diet) more than 50% of the control activities were still present. The high residual activity of ACC in liver of the biotin-deficient animal may be related to the rate of degradation of the enzyme in liver in comparison to its degradation rate in adipose tissue. Hepatic enzyme activity is rapidly restored in deficient rats after biotin administration. Within 2 hours after a single injection of biotin (200 μg), PC and PCC activities increased over 20 times, representing about 40% and 30% of the control activity, respectively. A two- to threefold increase in liver ACC occurred within 2 hours, representing about 80% of the control activity (Arinze & Mistry, 1971).

In the second study, Chiang and Mistry (1974) measured the activities of PC and PCC in the liver, kidney, heart, and brain of biotin-deficient rats during the progress of deprivation. Both enzyme activities in liver, kidney, and brain decreased at similar rates, with hepatic enzymes being the most adversely affected. After 20 days of deprivation, hepatic PC and PCC activities were 10% and 5%, respectively, of the control, falling to 2% and 5%, respectively after 40 days. Both PC and PCC activities in the kidney were 25% of control after 40 days. In the brain, PC and PCC activities were 65% and 60% of control after 40 days. PCC activity in the heart fell to about 15% of control after 30 days; PC activity in the heart of the deficient animal was negligible.

In the same study (Chiang & Mistry, 1974), rats that were fed a 20% egg-white diet for 35 days were given a single injection of 200 μg biotin and the tissues were analyzed for PC and PCC activities at intervals after injection. The results showed a rapid restoration of enzyme activities in all the tissues examined. After 12 hours, the enzyme activities in liver, kidney, and brain were on a par with those of control animals. The kidney responded more rapidly than the other tissues, with enzyme activities reaching near control values within 2 hours of biotin injection. In contrast, PCC of the heart attained only 30% of the control value in 2 hours and even at the end of 12 hours less than 50% of the activity was restored.

24.11.5 Paradoxical Regulation of Biotin Utilization in Brain and Liver

Rodríguez-Meléndez *et al.* (2001) studied the effects of biotin deficiency on the genetic expression of HCS, PC, and PCC in selected organs of rats fed a biotin-deficient diet. In the liver, kidney, and muscle, the mRNA level of HCS and the protein levels (but not mRNA levels) and activities of the carboxylases were reduced. By contrast, PC and PCC were unaffected in the brains of the biotin-deficient rats. When biotin was injected into the deficient rats, the HCS mRNA levels in liver increased to near normal levels; however, this was a delayed effect, apparent at 24 hours after administration. Pacheco-Alvarez *et al.* (2004) demonstrated that mRNA levels of enzymes involved in biotin transport and utilization, including HCS, are down-regulated during biotin deficiency in liver, while remaining constitutively expressed in brain. This pattern of regulation during biotin deficiency aims to maintain the essential biotin requirement of the brain at the expense of other organs such as the liver and kidneys.

24.12 REGULATION OF GENE EXPRESSION BY BIOTIN

Experiments using rats have shown that biotin affects the transcription of genes encoding two critical enzymes of glucose metabolism: hepatic and pancreatic glucokinase (a key enzyme in glycolysis) and hepatic (but not renal) phosphoenolpyruvate carboxykinase (PEPCK – a key enzyme in gluconeogenesis). Biotin stimulates expression of hepatic glucokinase (Chauhan & Dakshinamurti, 1991) and suppresses expression of hepatic PEPCK (Dakshinamurti & Li, 1994). In addition, biotin stimulates expression of pancreatic glucokinase and the hormone insulin (Romero-Navarro *et al.*, 1999).

Holocarboxylase synthetase is required to mediate the biotin-dependent increase in mRNA levels of HCS, ACC-1 (but not ACC-2), and the α subunit of PCC in HepG2 cells (Solórzano-Vargas *et al.*, 2002); PC and SMVT are under similar control (Pacheco-Alvarez *et al.*, 2004). The signal transduction cascade shown in Figure 24.9 (Rodríguez-Meléndez & Zemplini, 2003) demonstrates how biotin can activate transcription of these genes. Biotinyl-AMP produced in the first half reaction of the biotin cycle activates soluble guanylyl cyclase (sGC). This enzyme catalyzes the conversion of guanosine triphosphate to cyclic guanosine monophosphate, which in turn activates protein kinase G (PKG). The latter enzyme of this HCS–sGC–PKG pathway phosphorylates and activates proteins that stimulate gene transcription.

24.13 POST-TRANSLATIONAL HISTONE MODIFICATIONS

Post-translational histone modifications affect gene expression by causing structural alterations to chromatin. The most notable modification mediated by biotin is histone biotinylation. Biotinylation of histones increases in response to cell proliferation in human peripheral blood mononuclear cells (Stanley *et al.*, 2001). Hymes *et al.* (1995) reported that histones are biotinylated when incubated with biotinidase and biocytin at pH 7 to 9. This demonstrated that the stable biotinyl-biotinidase complex formed during hydrolysis of biocytin at pH ≥7 can transfer the biotin moiety to histone

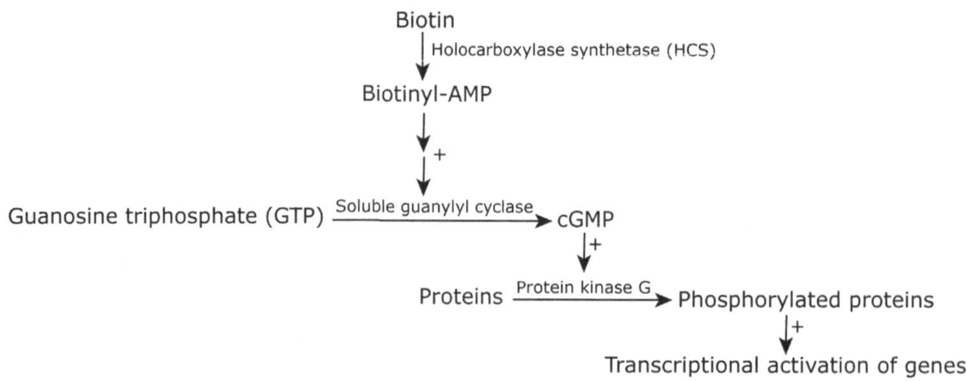

FIGURE 24.9 Regulation of gene expression by biotinyl-AMP. Activation of metabolic pathways is denoted by '+.'

lysine residues, at least *in vitro*. The biocytin required for biotinylation of histones is generated in the breakdown of biotin-dependent carboxylases.

There is evidence that biotinylation of histones plays a role in cell proliferation, gene silencing, and cellular response to DNA damage (Kothapalli *et al.*, 2005). Ballard *et al.* (2002) showed that biotinidase is able to catalyze both biotinylation and debiotinylation of histones *in vitro*. They suggested that covalent modification of biotinidase might be one mechanism to favor either biotinylation or debiotinylation of histones. Glycosylation has been identified as a post-translational modification of biotinidase (Cole *et al.*, 1994). The confirmation by Stanley *et al.* (2001) that biotin attaches to histones *in vivo* suggested that the above findings are physiologically relevant; however, there are several concerns that challenge the biological relevance of biotinidase in histone biotinylation. The biocytin substrate required for biotinylation is a product of proteolysis and biotinidase is not present in the nucleus. Moreover, neither the biotinidase-catalyzed addition nor removal of biotin from histones has been demonstrated *in vivo*.

Narang *et al.* (2004) made the surprising discovery that HCS, besides catalyzing the covalent attachment of biotin to apocarboxylases in the cytoplasm, targets to the nucleus and catalyzes the attachment of biotin to histones. All histone classes (H1, H2A, H2B, H3, and H4) could be biotinylated in the presence of ATP. It was further shown that multiple carboxylase deficiency patients, who are deficient in HCS, exhibit dramatically reduced histone biotinylation.

Biotin may play a role in the packaging of DNA into chromatin, because histones have been shown to dissociate from the DNA in biotin-deficient rats (Petrelli *et al.*, 1976). Biotin deficiency is associated with decreased phosphorylation and methylation, and increased acetylation of histones (Petrelli *et al.*, 1978). These phenomena could be reversed within days of readministration of biotin to normal levels.

24.14 IMMUNE FUNCTION

In this section, Rabin (1983) refers to the afferent and efferent limbs of adaptive immunity. The afferent pathway comprises the capture of antigen, antigen presentation, and the activation of helper and cytotoxic T cells. The efferent pathway comprises activated T cells homing to the site of infection or tumor and attempting to eradicate their targets (Dunn & Okada, 2015).

Wiedmann *et al.* (2003) describe the effects of biotin supplementation on gene expression of cytokines and biotin-dependent carboxylases in human peripheral blood mononuclear cells (PBMCs). PBMCs refer to blood cells having a round nucleus and comprise lymphocytes (T cells, B cells, and natural killer cells) and monocytes. The gradient centrifugation procedure of Zemplini and Mock (1998) was used to isolate PBMCs. In this procedure, the monocytes are removed, so the cell preparation essentially contains only lymphocytes.

Experimental autoimmune (or allergic) encephalomyelitis (EAE) produced by immunization with myelin basic protein (MBP) lends itself to studies in which nutritional effects on the immune system can be evaluated. EAE has a characteristic clinical course of hind-leg paralysis followed by recovery. The recovery phase is related to activity of the suppressor lymphocyte population. Thus, the role of different lymphocyte populations (inducer and suppressor) can be evaluated in this autoimmune disease. Rabin (1983) investigated the effect of dietary biotin deficiency on the immune system and the development of EAE. Rats were pair-fed biotin-deficient and biotin-adequate diets. Total T cells, T helper cells, and T suppressor cells were quantitatively the same in the spleens of rats fed biotin-adequate or deficient diets. Biotin deficiency produced a marked reduction in the size and cellularity of the thymus in animals that were not immunized, reflecting a reduced number of intrathymic T cells. Biotin deficiency produced a depressed immune response to immunization with sheep erythrocytes, an antigen that does not induce an autoimmune disease. Following immunization with guinea pig MBP, biotin-deficient animals showed no signs of EAE in contrast to biotin-sufficient animals, which had pronounced hind-leg weakness or paralysis. An adoptive lymphocyte transfer procedure was employed to determine whether the afferent or efferent limb of adaptive immunity was altered by biotin deficiency. Spleen lymphocytes isolated from rats immunized with guinea pig MBP were injected intraperitoneally into non-immunized animals and the recipients were evaluated daily for clinical signs of EAE. The lymphocytes isolated from immunized biotin-sufficient animals were capable of producing EAE in both biotin-sufficient and biotin-deficient recipients. The lymphocytes from immunized biotin-deficient animals failed to produce any signs of disease in biotin-deficient recipients and could only produce tail flaccidity and an occasional perivascular lymphocyte infiltrate in biotin-sufficient recipients. These results indicate that the afferent limb of adaptive immunity is impaired in biotin deficiency.

Biotin-deficient rats produced a reduced number of splenic antibody-forming cells in response to immunization with sheep erythrocytes. This decreased antibody response was partially restored to normal by administration of biotin shortly before immunization (Kumar & Axelrod, 1978). In biotin-deficient guinea pigs, the number of circulating neutrophils was increased, whereas the number of B and T cells was decreased. Incubation of lymphocytes from the deficient guinea pigs with biotin increased the number of B and T cells (Petrelli *et al.*, 1981).

Báez-Saldaña *et al.* (1998) investigated the mechanism by which biotin deficiency affects the immune status of mice during 20 weeks of experimentation. Mice were divided into deficiency, supplemental, and control groups. The deficiency group was fed a biotin-free diet with 30% dried egg white as a protein source. The supplemental group received the same diet as the deficient group, but with biotin added in excess of the amount potentially bound by the egg white. The control group was fed a commercial, non-purified diet. From weeks 7 to 20, there was a progressive weight loss in the deficient group. The specific activities of two biotin-dependent enzymes, PC and PCC, decreased in the livers and spleen lymphocytes of the deficient animals. With respect to the effects of biotin efficiency on the immune system, the spleens of deficient mice were smaller than those of the supplemented and control mice; this was reflected in the lower number of cells obtained from them. The percentage of B cells in the spleen was significantly lower in the deficient group than in the control and supplemental groups, with a concomitant increase in the percentage of T cells, mainly of the helper ($CD4^+$) phenotype.

Wiedmann *et al.* (2003) quantified gene expression of cytokines in human lymphocytes isolated from healthy adults before and after supplementation with 8.8 μmol biotin/day for 21 days. Cells were cultured *ex vivo* with concanavalin A for 21 hours to simulate stimulation with antigens. Biotin supplementation increased the abundance of mRNA encoding IL-1β and IFN-γ, and decreased the abundance of mRNA encoding IL-4. The authors pointed out that an increased abundance of mRNA encoding cytokines such as IL-1β and IFN-γ is not necessarily paralleled by increased levels of these cytokines in extracellular fluids. Previous studies have shown that biotin supplementation of healthy adults (Zemplini *et al.*, 2001) and Jurkat cells (a human T cell line) in culture (Manthey *et al.*, 2002) may cause decreased net secretion of cytokines by PBMCs and Jurkat cells despite increased abundance of mRNA encoding these cytokines. Binding of cytokines to receptors triggers both intracellular signaling cascades and internalization and degradation of cytokines. Biotin supplementation causes increased synthesis of cytokine receptors, leading to increased endocytosis and, thus, decreased extracellular accumulation of cytokines.

Biotin is known to be required for the biosynthesis of palmitic acid from acetyl-CoA and hence for the synthesis of its derivatives, such as stearic and oleic acids. The attendant synthesis of structural lipids of appropriate fatty acid composition plays a prominent role in the generation of cytotoxic T cells. This explains why the observed requirement for biotin for the *in vitro* generation of cytotoxic T cells can be largely replaced by the addition of fatty acids to the medium. The binding of biotin to avidin inhibits cytotoxic T cell generation, but inhibition is completely reversed by the addition of excess biotin (Kung *et al.*, 1979).

24.15 EXPERIMENTAL BIOTIN DEFICIENCY IN ANIMALS

24.15.1 Deficiency Signs

Biotin deficiency states can be induced in animals by dietary deprivation, feeding a biotin antagonist such as raw egg white or destroying the biotin-producing gut flora by antibiotic treatment. The following deficiency signs have been observed.

- **General condition.** Reduced growth, reduced appetite, and poor feed conversion. Reduced resistance to infections.
- **Skin.** In poultry there is brittle feathering and dermatitis with encrustation and cracking of the beak, eyelids, comb, and feet. In pigs there is hair loss, dry scaly skin, and cracking of the hooves. In mink there is depigmentation and severe loss of fur.
- **Locomotory system.** Perosis (leg deformity) in chickens and turkeys. Pigs become stiff-legged and develop an abnormal gait. In calves there is paralysis of the hind legs and an inability to stand.
- **Nervous disorders.** Ataxia in newly hatched chicks.
- **Reproductive system.** Poor hatchability and embryonic malformations in chickens.
- **Pregnancy.** Biotin deficiency in pregnant mice and hamsters is very teratogenic. The predominant malformations are skeletal, such as micrognathia, cleft palate, and micromelia (Watanabe, 1983, 1993; Watanabe & Endo, 1984).

24.15.2 Effects of Biotin Deficiency on Skin Morphology and Fatty Acid Composition in Rats

Proud *et al.* (1990) investigated the effects of biotin deficiency on the morphology, carboxylase activities, and fatty acid composition of the skin. Rats were rendered biotin deficient by being fed a semipurified diet containing egg-white solids as a source of the biotin-binding protein, avidin. The basal diet contained 170 g/kg of a mixture of cotton-seed and cod liver oils (5:1). Cotton-seed oil contains 52.89% and 0.17%, respectively, of the essential fatty acids, linoleic acid (18:2) and linolenic acid (18:3). Skin changes in the deficient rats included 'spectacle eye' (circumocular alopecia with inflammation), erythema, scaling dermatitis, and site-specific alopecia. Skin biopsies from haired and alopecic sites had less subcutaneous fat and showed lipophilic follicular plugging, vascular engorgement, epidermal hyperplasia, and abnormal keratinization. Mean activities of the mitochondrial biotin-dependent carboxylases (PC, PCC, and MCC) in the skin of biotin-deficient animals were reduced to 3–18% of control, whereas the cytosolic enzyme (ACC) was reduced only to 38–61%. The total fatty acid content of haired and alopecic skin of deficient rats was 30% of those in the corresponding skin sites of control animals. Skin from deficient rats contained less of several long-chain fatty acids (16:0, 16:1, 18:0, 18:1, and 18:2) and more very-long-chain fatty acids, with 24:1 and 26:1 being significantly higher than controls. These adverse effects of biotin deficiency on the fatty acid composition of the skin occurred despite adequate intake of the essential fatty acids (linoleic and linolenic acid). In conclusion, reduced activities of the biotin-dependent carboxylases lead to altered fatty acid composition and morphological abnormalities of the skin of rats.

When Liposyn (a fat emulsion containing 77% linoleic acid) was administered parenterally to biotin-deficient rats, the rats did not develop any of the skin changes associated with biotin deficiency (Mock, 1990b). This suggests that a deficiency in linoleic acid caused by some abnormality of polyunsaturated fatty acid metabolism plays a pathogenic role in the cutaneous manifestations of biotin deficiency. The decreases in long-chain and very-long-chain fatty acids observed by Proud *et al.* (1990) suggest that the skin changes may be due to the impairment of fatty acid elongation as a result of reduced ACC activity.

24.16 DIETARY DEFICIENCY OF BIOTIN IN HUMANS

Biotin is so widely distributed in foods that it is doubtful whether a true dietary deficiency of the vitamin has ever occurred in human adults capable of utilizing it. Biotin deficiency will occur under the unusual circumstance of prolonged biotin-devoid parenteral alimentation (Mock *et al.*, 1981). Of concern is a report that approximately one-third of children with severe protein-energy malnutrition are biotin-deficient (Velázquez *et al.*, 1995). Artificial biotin deficiency states have been induced in healthy volunteers by being fed low-biotin diets containing a high proportion of raw egg white. An initial dry scaly dermatitis was followed by non-specific symptoms that included anorexia and extreme lassitude. All of the symptoms responded to injections of 150–300 μg of biotin per day.

A unique opportunity to study dietary biotin deficiency was presented to Baugh *et al.* (1968) by a 62-year-old female patient who had consumed six raw eggs and 4 pints of skimmed milk daily for 18 months. This diet had been recommended by a physician (ill-advisedly as we now know) as a dietary supplement to provide a high intake of essential amino acids to aid liver regeneration following a diagnosis of liver cirrhosis. During this dietary period, the patient took vitamin supplements, which included biotin, and she also received 100 μg of vitamin B_{12} by injection monthly. Thus the stage was set, unintentionally, for the development of biotin deficiency due to the avidin content of raw egg whites, uncomplicated by deficiencies of other vitamins or common nutrients. Clinical manifestations included anorexia, nausea, vomiting, glossitis, pallor, depression, lassitude, substernal pain, scaly dermatitis, and desquamation of the lips. All symptoms cleared or improved markedly after 2 to 5 days of parenteral (by injection) vitamin therapy providing 200 μg of biotin daily, while the patient continued her pretreatment diet. In contrast to other case reports, the patient did not exhibit anemia, muscle pains, hypercholesterolemia, or electrocardiographic abnormalities.

Seborrheic dermatitis of the scalp and a more generalized dermatitis known as Leiner's disease have been reported in breast-fed infants when the mother is malnourished. These symptoms are relieved when biotin is administered to the mother.

The scaly dermatitis seen in human biotin deficiency is also a common symptom of essential fatty acid (EFA) deficiency. One might therefore suppose that biotin deficiency somehow produces fat malabsorption, and the consequent EFA deficiency can be alleviated by parenteral administration of EFA. However, the abnormalities in fatty acid composition observed in the biotin-deficient rats (Kramer *et al.*, 1984; Mock *et al.*, 1988b) are not typical of EFA deficiency *per se*. Moreover, patients who developed the cutaneous symptoms of biotin deficiency during parenteral nutrition had been given generous amounts of EFA parenterally, and fatty acid composition of plasma phospholipid was not compatible with simple EFA deficiency (Mock *et al.*, 1988a).

Studies of biotin status during pregnancy provide evidence that marginal biotin deficiency develops in a substantial proportion of women during normal pregnancy. Biotin deficiency is teratogenic in several animal species at degrees of deficiency that produce no obvious signs in the pregnant animal. There is concern that biotin deficiency may be teratogenic in humans, with abnormal lipid metabolism being a potential causative factor (Zemplini & Mock, 2000).

24.17 MULTIPLE CARBOXYLASE DEFICIENCY

Multiple carboxylase deficiency (MCD) is a rare autosomal recessive disorder of biotin utilization associated with deficient activity of all four biotin-dependent carboxylases. This is a classic example of an inborn error of metabolism. Because of the vital role of these enzymes in protein, carbohydrate, and lipid metabolism, their deficiency leads to severe life-threatening disease. MCD may be caused by a primary deficiency of HCS, required for the biotin-dependent activation of apocarboxylases, or of biotinidase, which functions in the salvage of biotin from degraded biotin-dependent carboxylases. The two forms of MCD differ biochemically in that the HCS-deficient patients have normal plasma biotin concentrations but decreased carboxylase activities, whereas patients with biotinidase deficiency have subnormal plasma biotin concentrations and normal carboxylase activities. MCD arising from either inherited defect causes a block in the biotin-dependent metabolic pathways with characteristic accumulation and urinary excretion of organic acids such as lactate, 3-hydroxyisovalerate, 3-methylcrotonylglycine, and methylcitrate.

Solórzano-Vargas *et al.* (2002) experimented on cultured fibroblasts from MCD patients and healthy persons. When deprived of biotin, HCS mRNA levels in both mutant and normal fibroblasts were reduced to ~35% of control values after 14 days. The mutant cells required 100 times more biotin than control cells to restore HCS mRNA to normal level. Because the mutated HCS is unable to bind biotin, biotinyl-5′-AMP, the substrate for HCS-mediated transfer of biotin to apocarboxylases, cannot be formed.

The two forms of MCD usually become symptomatic in early infancy or childhood. The incidence of biotinidase deficiency is about one in 60,000 and that of HCS deficiency seems to be even lower. Dupuis *et al.* (1999) identified six HCS mutations in MCD patients that were responsible for their reduced HCS activity. Four mutations were located in the biotin-binding domain and were associated with an impaired affinity for biotin. The other mutations, located outside the biotin-binding domain, were associated with a more limited biotin responsiveness that could be explained by the degree of residual enzyme activity present.

The clinical presentation and age of onset of MCD are extremely variable. HCS deficiency may present in the first days of life, while biotinidase deficiency usually becomes manifest between the second and fifth months of age, depending on the amount of free biotin in the diet. However, onset of HCS deficiency is delayed (2–21 months) in some cases and therefore classifying the two disorders as neonatal or early-onset and infantile or late-onset MCD should be discouraged. Clinical symptoms common to both disorders include neurological abnormalities (hypotonia, seizures, ataxia) and cutaneous changes (skin rash, alopecia). In healthy persons receiving an adequate diet, biotinidase activity in the brain is relatively very low (Suchy *et al.*, 1985) and so the brain relies largely on biotin that is transferred across the blood–brain barrier. This feature of the brain could explain the rapid onset of neurological symptoms observed in biotinidase deficiency.

Munnich *et al.* (1980) reported the case of a 12-month-old boy with MCD who presented with total alopecia, skin eruptions, and other symptoms of biotin deficiency. Since the patient's plasma linoleic acid (C18:2) concentration was lower than normal, the patient was treated with a mixture of polyunsaturated fatty acids (PUFAs) orally and topically before starting biotin therapy. The PUFA mixture was composed of 11% C18:1, 71% C18:2, 8% C18:3, and 0.3% C20:4. This treatment resulted in an increase in plasma linoleic acid, a dramatic improvement of the dermatological condition, and hair growth on the whole scalp. The authors speculated that the deficiency of linoleic acid in the patient might arise secondary to the impaired elongation of long-chain fatty acids as a consequence of deficient malonyl-CoA synthesis by the biotin-dependent ACC.

Fortunately, virtually all MCD patients with HCS or biotinidase deficiency can be successfully treated with pharmacological doses of biotin. If left untreated, severe forms of MCD can lead to coma and death.

24.18 TOXICITY

Oral pharmacological doses of biotin are considered safe. For example, lifelong treatment of biotinidase deficiency patients with biotin doses that exceed the normal dietary intake by 300 times does not produce overt signs of toxicity. Likewise, no signs of toxicity were reported after acute oral and intravenous administration of doses that exceeded the dietary biotin intake by up to 600 times (Zemplini *et al.*, 2009).

FURTHER READING

Zemplini, J., Wijeratne, S.S.K. & Hassan, Y.I. (2009) Biotin. *BioFactors*, **35**, 36–45.

REFERENCES

Achuta Murthy, P.N. & Mistry, S.P. (1977) Biotin. *Progress in Food and Nutrition Science*, **2**, 405–55.

Arinze, J.C. & Mistry, S.P. (1971) Activities of some biotin enzymes and certain aspects of gluconeogenesis during biotin deficiency. *Comparative Biochemistry and Physiology*, **38B**, 285–94.

Báez-Saldaña, A., Diaz, G., Espinoza, B. & Ortega, E. (1998) Biotin deficiency induces changes in subpopulations of spleen lymphocytes in mice. *American Journal of Clinical Nutrition*, **67**, 431–7.

Balamurugan, K., Ortiz, A. & Said, H.M. (2003) Biotin uptake by human intestinal and liver epithelial cells: Role of the SMVT system. *American Journal of Physiology (Gastrointestinal amd Liver Physiology)*, **285**, G73–7.

Balamurugan, K., Vaziri, N.D. & Said, H.M. (2005) Biotin uptake by human proximal tubular epithelial cells: Cellular and molecular aspects. *American Journal of Physiology (Renal Physiology)*, **288**, F823–31.

Ballard, T.D., Wolff, J., Griffin, J.B., Stanley, J.S., van Calcar, S. & Zemplini, J. (2002) Biotinidase catalyzes debiotinylation of histones. *European Journal of Nutrition*, **41**, 78–84.

Baugh, C.M., Malone, J.H. & Butterworth, C.E. Jr. (1968) Human biotin deficiency. A case history of biotin deficiency induced by raw egg consumption in a cirrhotic patient. *American Journal of Clinical Nutrition*, **21**, 173–82.

Baur, B. & Baumgartner, E.R. (1993) Na^+-dependent biotin transport into brush-border membrane vesicles from human kidney cortex. *European Journal of Physiology*, **422**, 499–505.

Bowman, B.B. & Rosenberg, I.H. (1987) Biotin absorption by distal rat intestine. *Journal of Nutrition*, **117**, 2121–6.

Bowman, B.B., Selhub, J. & Rosenberg, I.H. (1986) Intestinal absorption of biotin in the rat. *Journal of Nutrition*, **116**, 1266–71.

Chapman-Smith, A. & Cronan, J.E. Jr. (1999) Molecular biology of biotin attachment to proteins. *Journal of Nutrition*, **129**, 477S–84S.

Chauhan, J. & Dakshinamurti, K. (1986) Purification and characterization of human serum biotinidase. *Journal of Biological Chemistry*, **261**(9), 4288–75.

Chauhan, J. & Dakshinamurti, K. (1991) Transcriptional regulation of the glucokinase gene by biotin in starved rats. *Journal of Biological Chemistry*, **266**(16), 10035–8.

Chiang, G.S. & Mistry, S.P. (1974) Activities of pyruvate carboxylase and propionyl CoA carboxylase in rat tissues during biotin deficiency and restoration of the activities after biotin administration. *Proceedings of the Society for Experimental Biology and Medicine*, **146**, 21–4.

Cole, H., Reynolds, T.R., Lockyer, J.M., Buck, G.A., Denson, T., Spence, J.E., Hymes, J. & Wolf, B. (1994) Human serum biotinidase cDNA cloning, sequence, and characterization. *Journal of Biological Chemistry*, **269**(9), 6566–70.

Crisp, S.E.R.H., Griffin, J.B., White, B.R., Toombs, C.F., Camporeale, G., Said, H.M. & Zemplini, J. (2004) Biotin supply affects rates of cell proliferation, biotinylation of carboxylases and histones, and expression of the gene encoding the sodium-dependent multivitamin transporter in JAr choriocarcinoma cells. *European Journal of Nutrition*, **43**, 23–31.

Dakshinamurti, K. & Li, W. (1994) Transcriptional regulation of liver phosphoenolpyruvate carboxykinase by biotin in diabetic rats. *Molecular and Cellular Biochemistry*, **132**, 127–32.

Dakshinamurti, K., Chauhan, J. & Ebrahim, H. (1987) Intestinal absorption of biotin and biocytin in the rat. *Bioscience Reports*, **7**, 667–73.

Dunn, G.P. & Okada, H. (2015) Principles of immunology and its nuances in the central nervous system. *Neuro-Oncology*, **17**, vii3–vii8.

Dupuis, L., Campeau, E., Leclere, D. & Gravel, R.A. (1999) Mechanism of biotin responsiveness in biotin-responsive multiple carboxylase deficiency. *Molecular Genetics and Metabolism*, **66**, 80–90.

Frigg, M. (1984) Available biotin content of various feed ingredients. *Poultry Science*, **63**, 750–3.

Grassl, S.M. (1992) Human placental brush-border membrane Na^+-biotin cotransport. *Journal of Biological Chemistry*, **267**(25), 17760–5.

Heard, G.S., Redmond, J.B. & Wolf, B. (1987) Distribution and bioavailability of biotin in human milk. *Federation Proceedings*, **46**, 897 (abstr).

Holland, B., Welch, A.A., Unwin, I.D., Buss, D.H., Paul, A.A. & Southgate, D.A.T. (1991) *McCance and Widdowson's The Composition of Foods*, 5th edn, Royal Society of Chemistry and Ministry of Agriculture, Fisheries and Food.

Hymes, J. & Wolf, B. (1999) Human biotinidase isn't just for recycling biotin. *Journal of Nutrition*, **129**, 485S–9S.

Hymes, J., Fleischhauer, K. & Wolf, B. (1995) Biotinylation of histones by human serum biotinidase: Assessment of biotinyl-transferase activity in sera from normal individuals and chidren with biotinidase deficiency. *Biochemical and Molecular Medicine*, **56**, 76–83.

Kothapalli, N., Camporeale, G., Kueh, A., Chew, Y.C., Oommen, A.M., Griffin, J.B. & Zemplini, J. (2005) Biological functions of biotinylated histones. *Journal of Nutritional Biochemistry*, **16**, 446–8.

Kramer, T.R., Briske-Anderson, M., Johnson, S.B. & Holman, R.T. (1984) Effects of biotin deficiency on polyunsaturated fatty acid metabolism in rats. *Journal of Nutrition*, **114**, 2047–52.

Kumar, M. & Axelrod, A.E. (1978) Cellular antibody synthesis in thiamin, riboflavin, biotin and folic acid-deficient rats. *Proceedings of the Society for Experimental Biology and Medicine*, **157**, 421–3.

Kung, J.T., Mackenzie, C.G. & Talmage, D.W. (1979) The requirement for biotin and fatty acids in the cytotoxic T-cell response. *Cellular Immunology*, **48**, 100–10.

Lampen, J.O., Bahler, G.P. & Peterson, W.H. (1942) The occurrence of free and bound biotin. *Journal of Nutrition*, **23**, 11–21.

Ma, T.Y., Dyer, D.L. & Said, H.M. (1994) Human intestinal cell line Caco-2: A useful model for studying cellular and molecular regulation of biotin uptake. *Biochimica et Biophysica Acta*, **1189**, 81–8.

Manthey, K.C., Griffin, J.B. & Zemplini, J. (2002) Biotin supply affects expression of biotin transporters, biotinylation of carboxylases, and metabolism of interleukin-2 in Jurkat cells. *Journal of Nutrition*, **132**, 887–92.

McMahon, R.J. (2002) Biotin in metabolism and molecular biology. *Annual Review of Nutrition*, **22**, 221–39.

Misir, R. & Blair, R. (1988a) Biotin bioavailability of protein supplements and cereal grains for starting turkey poults. *Poultry Science*, **67**, 1274–80.

Misir, R. & Blair, R. (1988b) Biotin bioavailability from protein supplements and cereal grains for weanling pigs. *Canadian Journal of Animal Science*, **68**, 523–32.

Mock, D.M. (1990a) Biotin. In: *Present Knowledge in Nutrition*, 6th edn. (Ed. by M.L. Brown), pp. 189–207. International Life Sciences Institute, Nutrition Foundation, Washington, DC.

Mock, D.M. (1990b) Evidence for a pathogenic role of ω6 polyunsaturated fatty acid in the cutaneous manifestations of biotin deficiency. *Journal of Pediatric Gastroenterology and Nutrition*, **10**, 222–9.

Mock, D.M. & Malik, M.I. (1992) Distribution of biotin in human plasma: Most of the biotin is not bound to protein. *American Journal of Clinical Nutrition*, **56**, 427–32.

Mock, D.M., DeLorimeer, A.A., Liebman, W.M., Sweetman, L. & Baker, H. (1981) Biotin deficiency: An unusual complication of parenteral alimentation. *New England Journal of Medicine*, **304**, 820–3.

Mock, D.M., Johnson, S.B. & Holman, R.T. (1988a) Effects of biotin deficiency on serum fatty acid composition: Evidence for abnormaties in humans. *Journal of Nutrition*, **118**, 342–8.

Mock, D.M., Mock, N.I., Johnson, S.B. & Holman, R.T. (1988b) Effects of biotin deficiency on plasma and tissue fatty acid composition: Evidence for abnormalities in rats. *Pediatric Research*, **24**, 396–402.

Mock, D.M., Lankford, G.L. & Cazin, J. Jr. (1993) Biotin and biotin analogs in human urine: Biotin accounts for only half of the total. *Journal of Nutrition*, **123**, 1844–51.

Munnich, A., Saudubray, J.M., Coude, F.X., Charpentier, C., Saurat, J.H. & Frezal, J. (1980) Fatty-acid-responsive alopecia in multiple carboxylase deficiency. *Lancet*, **I**(8177), 1080–1.

Nabokina, S.M., Subramanian, V.S. & Said, H.M. (2003) Comparative analysis of ontogenic changes in renal and intestinal biotin transport in the rat. *American Journal of Physiology (Renal Physiology)*, **284**, F737–42.

Narang, M.A., Dumas, R., Ayer, L.M. & Gravel, R.A. (2004) Reduced histone biotinylation in multiple carboxylase deficiency patients: A nuclear role for holocarboxylase synthetase. *Human Molecular Genetics*, **13**, 15–23.

Pacheco-Alvarez, D., Solórzano-Vargas, R.S. & León-Del-Rio, A. (2002) Biotin in metabolism and its relationship to human disease. *Archives of Medical Research*, **33**, 439–47.

Pacheco-Alvarez, D., Solórzano-Vargas, R.S., Gravel, R.A., Cervantes-Roldán, R., Velázquez, A. & León-Del-Rio, A. (2004) Paradoxical regulation of biotin utilization in brain and liver and implications for inherited multiple carboxylase deficiency. *Journal of Biological Chemistry*, **279**(50), 522312–18.

Pacheco-Alvarez, D., Solórzano-Vargas, R.S., González-Noriega, A., Michalak, C., Zempleni, J. & León-Del-Rio, A. (2005) Biotin availability regulates expression of the sodium-dependent multivitamin transporter and the rate of biotin uptake in HepG2 cells. *Molecular Genetics and Metabolism*, **85**, 301–7.

Petrelli, F., Marsili, G. & Moretti, P. (1976) RNA, DNA, histones and interactions between histone proteins and DNA in the liver of biotin deficient rats. *Biochemistry and Experimental Biology*, **12**, 461–5.

Petrelli, F., Coderoni, S., Moretti, P. & Paparelli, M. (1978) Effect of biotin on phosphorylation, acetylation, methylation of rat liver histones. *Molecular Biology Reports*, **4**, 87–92.

Petrelli, F., Moretti, P. & Campanati, G. (1981) Studies on the relationships between biotin and the behaviour of B and T lymphocytes in the guinea pig. *Experientia*, **37**, 1204–6.

Podevin, R.-A. & Barbarat, B. (1986) Biotin uptake mechanisms in brush-border and basolateral membrane vesicles isolated from rabbit kidney cortex. *Biochimica et Biophysica Acta*, **856**, 471–81.

Prasad, P.D., Ramamoorthy, S., Leibach, F.H. & Ganapathy, V. (1997) Characterization of a sodium-dependent vitamin transporter mediating the uptake of pantothenate, biotin and lipoate in human choriocarcinoma cells. *Placenta*, **18**, 527–33,

Prasad, P.D., Wang, H., Kekuda, R., Fujita, T., Fei, Y.-J., Devoe, L.D., Leibach, F.H. & Ganapathy, V. (1998) Cloning and functional expression of a cDNA encoding a mammalian sodium-dependent vitamin transporter mediating the uptake of pantothenate, biotin, and lipoate. *Journal of Biological Chemistry*, **273**(13), 7501–6.

Prasad, P.D., Wang, H., Huang, W., Fei, Y.-J., Leibach, F.H., Devoe, L.D. & Ganapathy, V. (1999) Molecular and functional characterization of the intestinal Na^+-dependent multivitamin transporter. *Archives of Biochemistry and Biophysics*, **366**, 95–106.

Prasad, P.D., Srinivas, S.R., Wang, H., Leibach, F.H., Devoe, L.D. & Ganapathy, V. (2000) Electrogenic nature of rat sodium-dependent multivitamin transport. *Biochemical and Biophysical Research Communications*, **270**, 836–40.

Proud, V.K., Rizzo, W.B., Patterson, J.W., Heard, G.S. & Wolf, B. (1990) Fatty acid alterations and carboxylase deficiencies in the skin of biotin-deficient rats. *American Journal of Clinical Nutrition*, **51**. 853–8.

Rabin, B.S. (1983) Inhibition of experimentally induced autoimmunity in rats by biotin deficiency. *Journal of Nutrition*, **113**, 2316–22.

Reidling, J.C., Nabokina, S.M. & Said, H.M. (2007) Molecular mechanisms involved in the adaptive regulation of human intestinal biotin uptake: A study of the hSMVT system. *American Journal of Physiology (Gastrointestinal and Liver Physiology)*, **292**, G275–G281.

Rodriguez-Meléndez, R. & Zemplini, J. (2003) Regulation of gene expression by biotin (Review). *Journal of Nutritional Biochemistry*, **14**, 680–90.

Rodríguez-Meléndez, R., Cano, S., Méndez, S.T. & Velázquez, A. (2001) Biotin regulates the genetic expression of holocarboxylase synthetase and mitochondial carboxylases in rats. *Journal of Nutrition*, **131**, 1909–13.

Romero-Navarro, G., Cabrera-Valladares, G., German, M.S., Matshinsky, F.M., Velázquez, A., Wang, J. & Fernández-Mejia, C. (1999) Biotin regulation of pancreatic glucokinase and insulin in primary cultured rat islets and in biotin-deficient rats. *Endocrinology*, **140**, 4595–600.

Said, H.M. (1999) Cellular uptake of biotin: Mechanisms and regulation. *Journal of Nutrition*, **129**, 490S–3S.

Said, H.M. & Mohammedkhani, R. (1988) Ontogenesis of the intestinal transport of biotin in the rat. *Gastroenterology*, **94**, 68–79.

Said, H.M., Redha, R. & Nylander, W. (1988a) Biotin transport in the human intestine: Site of maximum transport and effect of pH. *Gastroenterology*, **95**, 1312–7.

Said, H.M., Redha, R. & Nylander, W. (1988b) Biotin transport in basolateral membrane vesicles of human intestine. *Gastroenterology*, **94**, 1157–63.

Said, H.M., Mock, D.M. & Collins, J.C. (1989) Regulation of intestinal biotin transport in the rat: Effect of biotin deficiency and supplementation. *American Journal of Physiology (Gastrointestinal and Liver Physiology)*, **256**, G306–11.

Said, H.M., Thuy, L.P., Sweetman, L. & Schatzman, B. (1993) Transport of the biotin dietary derivative (*N*-biotinyl-L-lysine) in rat small intestine. *Gastroenterology*, **104**, 75–80.

Said, H.M., Ortiz, A., McCloud, E., Dyer, D., Moyer, M.P. & Rubin, S. (1998) Biotin uptake by human colonic epithelial NCM460 cells: A carrier-mediated process shared with pantothenic acid. *American Journal of Physiology (Cell Physiology)*, **275**, C1365–71.

Seki, M. (2006) Biological significance and development of practical synthesis of biotin. *Medicinal Research Reviews*, **26**, 434–82.

Solórzano-Vargas, R.S., Pacheco-Alvarez, D. & León-Del-Rio, A. (2002) Holocarboxylase synthetase is an obligate participant in biotin-mediated regulation of its own expression and of biotin-dependent carboxylases mRNA levels in human cells, *Proceedings of the National Academy of Sciences of the U,S.A.*, **99**(8), 5325–30.

Sorrell, M.F., Frank, O., Thomson, A.D., Aquino, H. & Baker, H. (1971) Absorption of vitamins from the large intestine in vivo. *Nutrition Report International*, **3**, 143–8.

Spector, R. & Mock, D. (1987) Biotin transport through the blood-brain barrier. *Journal of Neurochemistry*, **48**, 400–4.

Spector, R. & Mock, D.M. (1988) Biotin transport and metabolism in the central nervous system. *Neurochemical Research*, **13**, 213–9.

Stanley, J.S., Griffin, J.B. & Zemplini, J. (2001) Biotinylaton of histones in human cells. Effects of cell proliferation. *European Journal of Biochemistry*, **268**, 5424–9.

Suchy, S.F., McVoy, J.S. & Wolf, B. (1985) Neurologic symptoms of biotinidase deficiency: Possible explanation. *Neurology*, **35**, 1510–11.

Sugden, M.C. & Holness, M.J. (2011) The pyruvate carboxylase-pyruvate dehydrogenase axis in islet pyruvate metabolism. Going round in circles? *Islets*, **3**, 302–19.

Velázquez, A., Terán, M., Báez, A., Gutiérrez, J. & Rodríguez, R. (1995) Biotin supplementation affects lymphocyte carboxylases and plasma biotin in severe protein-energy malnutrition. *American Journal of Clinical Nutrition*, **61**, 385–91.

Watanabe, T. (1983) Teratogenic effects of biotin deficiency in mice. *Journal of Nutrition*, **113**, 574–81.

Watanabe, T. (1993) Dietary biotin deficiency affects reproductive function and prenatal development in hamsters. *Journal of Nutrition*, **123**, 2101–8.

Watanabe, T. & Endo, A. (1984) Teratogenic effects of avidin-induced biotin deficiency in mice. *Teratology*, **30**, 91–4.

Wiedmann, S., Eudy, J.D. & Zemplini, J. (2003) Biotin supplementation causes increased expression of genes encoding interferon-γ, interleukin-1β, and 3-methylcrotonyl-CoA carboxylase, and causes decreased expression of the gene encoding interleukin-4 in human peripheral blood mononuclear cells. *Journal of Nutrition*, **133**, 716–9.

Zemplini, J. & Mock, D.M. (1998) Uptake and metabolism of biotin by human peripheral blood mononuclear cells. *American Journal of Physiology (Cell Physiology)*, **275**, C382–8.

Zemplini, J. & Mock, D.M. (2000) Marginal biotin deficiency is teratogenic. *Proceedings of the Society for Experimental Biology and Medicine*, **223**, 14–21.

Zemplini, J., Green, G.M., Spannagel, A.W. & Mock, D.M. (1997) Biliary excretion of biotin and biotin metabolites is quantitatively minor in rats and pigs. *Journal of Nutrition*, **127**, 1496–500.

Zemplini, J., Helm, R.M. & Mock, D.M. (2001) In vivo biotin supplementation at a pharmacologic dose decreases proliferation rates of human peripheral blood mononuclear cells and cytokine release. *Journal of Nutrition*, **131**, 1479–84.

Zemplini, J., Wijeratne, S.S.K. & Hassan, Y.I. (2009) Biotin. *BioFactors*, **35**, 36–45.

25 Folate

25.1 CHEMISTRY

The term 'folate' is used as the generic descriptor for all derivatives of pteroic acid that demonstrate vitamin activity in humans. The structure of the parent folate compound, folic acid, comprises a pterin ring joined by a methylene bridge to *p*-aminobenzoic acid, which in turn is coupled via an α-peptide bond to a single molecule of L-glutamic acid (Figure 25.1). The pterin and methylene moieties combined constitute a pteridine moiety.

(Note: In the present context, the term 'folic acid' refers specifically to pteroylmonoglutamic acid which, with reference to the pteroic acid and glutamate moieties, can be abbreviated to PteGlu. 'Folate' is a non-specific term referring to any folate compound with vitamin activity. 'Folacin' is a non-approved term synonymous with 'folate.')

Folic acid is not a common natural physiological form of the vitamin. In most natural foods, the pteridine ring is reduced to give either the 7,8-dihydrofolate (DHF) or 5,6,7,8-tetrahydrofolate (THF) (Figure 25.1). These reduced forms can be substituted with a covalently bonded one-carbon adduct attached to nitrogen positions 5 or 10 or bridged across both positions. The following substituted forms of THF are important intermediates in folate metabolism: 10-formyl-THF, 5-methyl-THF, 5-formimino-THF, 5,10-methylene-THF, and 5,10-methenyl-THF (Figure 25.1). Methotrexate (4-amino-10-methylfolic acid; Figure 25.2) is a folate antagonist which is used as an anti-cancer drug.

An important structural aspect of the 5,6,7,8-tetrahydrofolates is the stereochemical orientation at the C-6 asymmetric carbon of the pteridine ring. Of the two diastereomers, 6*S* and 6*R* (formerly called 6*l* and 6*d*), only the 6*S* is biologically active and occurs in nature. Methods of chemical synthesis of tetrahydrofolates, whether by catalytic hydrogenation or chemical reduction, yield a racemic product (i.e. a mixture of both diastereomers).

All folate compounds exist predominantly as polyglutamates containing typically from five to seven glutamate residues in γ-peptide linkage. The γ-peptide bond is unique in mammalian biochemistry. Folate conjugates are abbreviated to $PteGlu_n$ derivatives, where n is the number of glutamate residues; for example, $5\text{-}CH_3\text{-}H_4PteGlu_3$ refers to triglutamyl-5-methyltetrahydrofolic acid. Assuming that the polyglutamyl side chain extends to no more than seven residues, the theoretical number of folates approaches 150 (Baugh & Krumdieck, 1971).

Fundamentally, folate monoglutamates are the circulatory and membrane-transportable forms of the vitamin, whereas polyglutamates are the intracellular biochemical and storage forms.

25.2 FOOD SOURCES AND BIOAVAILABILITY

25.2.1 Food Sources

Polyglutamyl folate is an essential biochemical constituent of living cells, and most foods contribute some folate. The folates generally exist in nature bound to proteins (Baugh & Krumdieck, 1971) and they are also bound to storage polysaccharides (various types of starch and glycogen) in foods (Černá & Káš, 1983). In the United States, dried beans, eggs, greens, orange juice, sweet corn, peas, and peanut products are good natural sources of folate that are inexpensive and available all the year round. Folic acid fortification of white wheat flour and grain products contributes to the dietary intake. Although liver, all types of fortified breakfast cereals, cooked dried beans, asparagus, spinach, broccoli, and avocados provide the highest amount of folate per average serving, several of these foods do not rank highly in terms of actual dietary folate intake in the United States because of their low rate of consumption; in fact, orange juice ranks the highest (Subar *et al.*, 1989).

25.2.2 Bioavailability

The bioavailability of folate depends, among other factors, on whether the folate is in the polyglutamyl form (i.e. food folate in natural sources) or in the monoglutamyl form (i.e. synthetic folic acid in supplements and fortified foods). Polyglutamyl folates, unlike folic acid, must undergo enzymatic deconjugation in the small intestine before they can be absorbed. Organic acids present in many foods can inhibit conjugase activity (Wei & Gregory, 1998) and this may partly explain why the polyglutamyl folate present in natural food sources is less bioavailable than the folic acid present in fortified foods and supplements (Cuskelly *et al.*, 1996). Physical entrapment of folates in the cellular structure of plant materials can adversely affect the rate of folate absorption (van het Hof *et al.*, 1999; Castenmiller *et al.*, 2000). Gregory *et al.* (1991) reported a bioavailability of around 50% for $PteGlu_6$ relative to PteGlu as judged by urinary excretion when stable isotopes of these compounds were administered orally to human subjects. In a 92-day metabolic study in women, dietary folate appeared to be no more than 50% bioavailable when compared with folic acid in a formulated diet (Sauberlich *et al.*, 1987). Pfeiffer *et al.* (1997) reported that folic acid added to cereal is highly bioavailable and the matrix of cereal-grain foods has little inhibitory effect on intestinal absorption of the added folate.

There is little information regarding the effects of food on the bioavailability of folic acid in supplements. Absorption

DOI: 10.1201/9781003410300-29

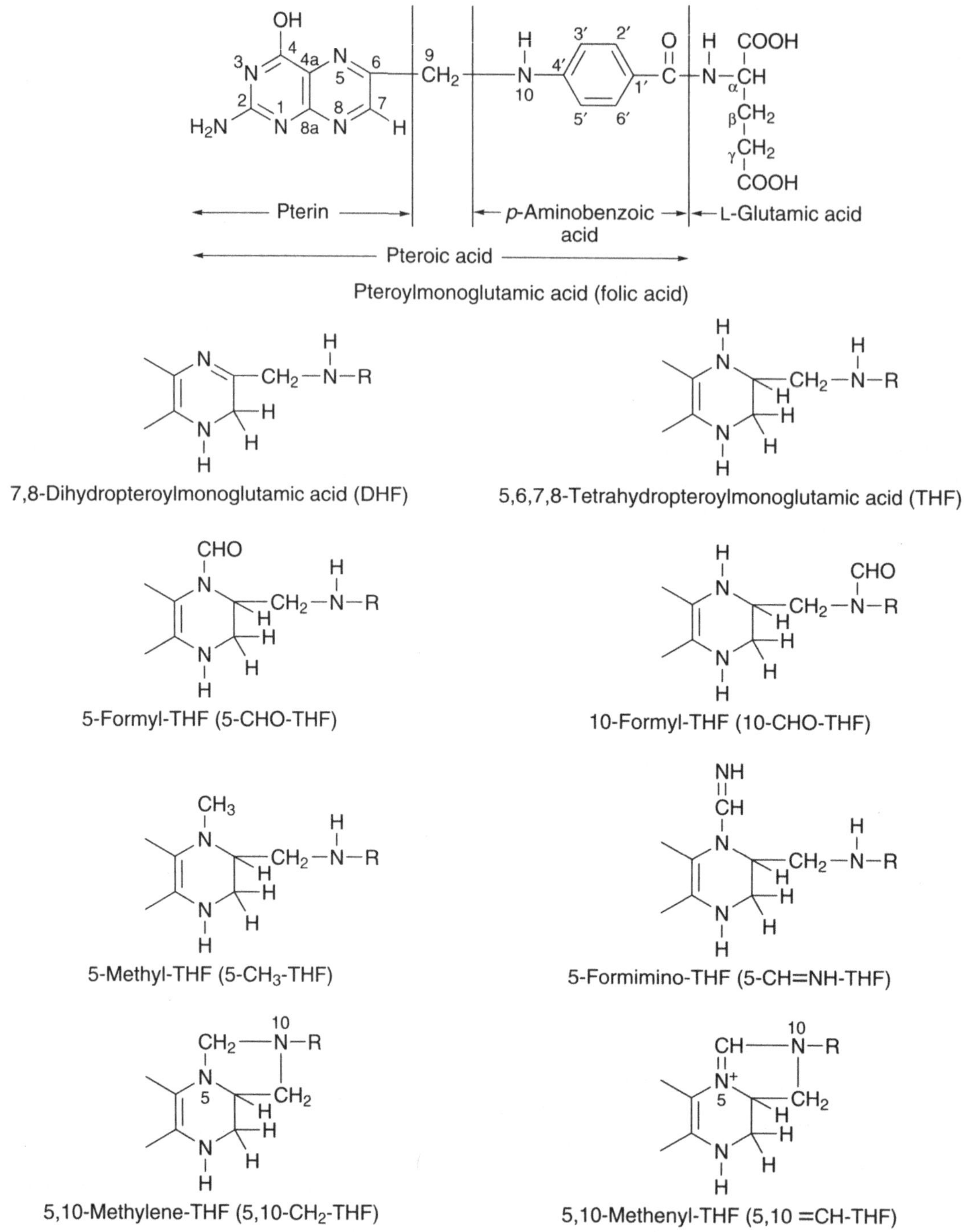

FIGURE 25.1 Structures of folate compounds. For the reduced derivatives of folic acid, only the relevant part of the pteridine ring is shown.

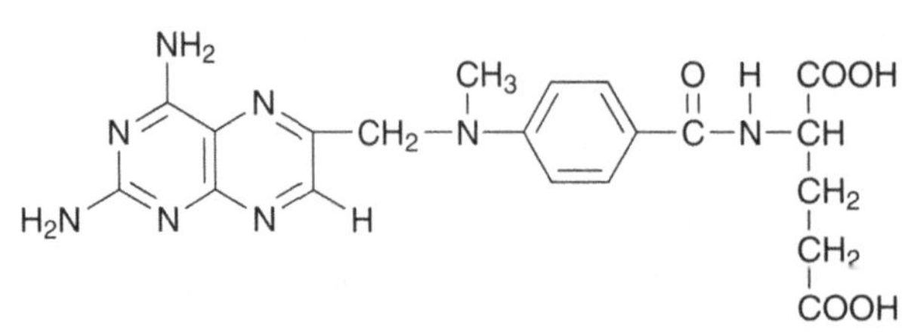

FIGURE 25.2 Structure of methotrexate.

of a dose of [$^{13}C_5$]folic acid was reduced by an insignificant 15% when taken immediately after a light meal (Pfeiffer *et al.*, 1997), providing an estimated 85% for the bioavailability of folic acid supplement consumed with food.

25.3 THE FOLATE TRANSPORTERS

25.3.1 Introduction

Transepithelial transport of folates is mediated by a combination of influx and efflux transporters at specific membrane locations. Transporters mediating transit in one direction are opposed by others mediating transit in the opposite direction, providing a transport system that maintains folate homeostasis in the epithelial cell.

Folates exist primarily as anions at physiological pH due to the fully ionized state of the two glutamate carboxyl groups, and transport is markedly influenced by changes in pH. Cellular uptake of folate is mediated specifically by three functionally different membrane transport proteins: (1) the reduced folate carrier (RFC), (2) the proton-coupled folate transporter (PCFT), and (3) the folate receptors (FRs). Neither PCFT nor RFC transports the polyglutamylated derivatives of folates or antifolates (Zhao *et al.*, 2017). The folate transporters and receptors show different localization patterns in the polarized epithelial cells in intestine, liver, kidney, choroid plexus, and placenta. They have significantly different affinities for folate and function at different pH optima. In rats, dietary folate deficiency up-regulates RFC and PCFT expression at the intestinal brush-border membrane (Thakur *et al.*, 2015). In contrast, oversupplementation of rats with folate leads to decreased RFC and PCFT protein expression in intestinal cells (Dev *et al.*, 2011).

25.3.2 The Reduced Folate Carrier

The RFC, a product of the *SLC19A1* gene, is the major transporter that delivers folates to systemic tissues at physiologically neutral pH. The carrier, cloned by Dixon *et al.* (1994), is ubiquitously expressed, notably on the brush-border membrane of enterocytes and the basolateral membrane of renal epithelial cells. RFC is an organic anion antiporter that exchanges folates with intracellular organic phosphates. Organic phosphates are synthesized in adenosine triphosphate (ATP)-dependent reactions and are highly concentrated in cells. RFC has a relatively high affinity (K_m 1–10 μM) for methotrexate and reduced folates (5-methyl-THF and 5-formyl-THF) but a very low affinity (K_m 200–400 μM) for folic acid (Rijnboutt *et al.*, 1996). RFC-mediated folate transport is optimum at pH 7.4; activity decreases with decreasing pH and there is little activity below pH 6.5. In contrast to PCFT, the affinity of RFC for folates does not change appreciably at low pH and the decrease in function is due to a decrease in influx V_{max} (Zhao *et al.*, 2009a).

Hou *et al.* (2014) found that increasing extracellular folate caused progressively elevated human RFC (hRFC) transcripts (newly transcribed RNA) and total hRFC proteins due to enhanced hRFC transcript stability. Despite this up-regulation of hRFC, cell-surface hRFC and folate transport activity decreased with increasing exogenous folate. This impairment of intracellular trafficking and plasma membrane targeting was attributed to increased retention of the newly synthesized carrier in the endoplasmic reticulum. Thus, post-transcriptional regulation of hRFC is an adaptive mechanism in response to folate excess or deficiency.

Expression of RFC mRNA and protein was up-regulated in cultured human blood–brain barrier endothelial cells (hCMEC/D3 cells) and also in isolated mouse brain capillaries after exposure to 1,25-dihydroxyvitamin D_3. Transport assays with methotrexate showed an increase in RFC functional activity in the 1,25-dihydroxyvitamin D_3-treated cells compared to controls. RFC protein expression in hCMEC/D3 cells could be down-regulated by VDR-targeting siRNA, confirming the role of the VDR in the direct regulation of RFC (Alam *et al.*, 2017).

25.3.3 The Proton-Coupled Folate Transporter

In humans and rodents, the PCFT is functionally expressed at high levels in the epithelia of the small intestine, kidney, placenta, liver, and spleen, where it accounts for the low-pH folate transport activity. Within the small intestine, the highest PCFT levels are found in the brush-border membrane of the duodenum and proximal jejunum. Although PCFT is present at the basolateral membrane of the choroid epithelial cell and the luminal face of the blood–brain barrier endothelial cell, its low pH optimum makes it ineffective at the neutral pH at these interfaces. Qiu *et al.* (2006) identified *SLC46A1* as the human gene that encodes the PCFT and showed that a loss-of-function mutation in this gene is the cause of hereditary folate malabsorption.

The PCFT functions as a folate$^-$/H$^+$ symporter. Folate transport is electrogenic, indicating that there is a net translocation of positive charges as each folate molecule is transported. As folates are bivalent anions, more than two protons must be co-transported with each folate molecule to account for the positive charge of the PCFT–folate–proton complex. Folate uptake by PCFT conforms to Michaelis-Menten kinetics and is pH-dependent. Affinity of folic acid for the transporter was high at pH 5.5 (K_m 1.3 ± 0.1 μM) and low to negligible at pH 7.5 (K_m 56.2 ± 5.6 μM). Intermediate K_m values for folic acid were 1.5 ± 0.1 μM at pH 6.0, 2.7 ± 0.2 μM at pH 6.5, and 6.0 ± 2.6 μM at pH 7.0 (Qiu *et al.*, 2006). PCFT also transports the antifolate drug methotrexate (Yuasa *et al.*, 2009).

Little is known about the regulation of *PCFT* gene expression. Levels of mRNA encoding PCFT protein increased ~13-fold in the proximal small intestine of mice fed a folate-deficient diet, consistent with the critical role that PCFT plays in intestinal folate absorption (Qiu *et al.*, 2007).

25.3.4 The Folate Receptors

Two isoforms of the human folate receptor (FRα and FRβ) expressed at the cell surface of systemic tissues have high affinities for folic acid (K_m 0.1–1 nM) and reduced folates (K_m 1–10 nM), but relatively low affinities for methotrexate (K_m >100 nM) (Rijnboutt *et al.*, 1996). FRα and FRβ are differentially expressed in different tissues. FRα, which is encoded by the *FOLR1* gene, is expressed on the membrane of some epithelial tissues, in particular the apical brush-border membrane of proximal renal tubular cells and the basolateral membrane of choroid epithelial cells. FRα mRNA was not detected in human blood–brain barrier endothelial cells, but was found in isolated rodent brain capillaries, suggesting species-specific differences in the expression of this receptor (Alam *et al.*, 2017). FRβ is expressed in placenta, spleen, and thymus (Zhao *et al.*, 2009a). Neither isoform is expressed in enterocytes or hepatocytes.

The cellular expression of FRs is regulated inversely by the extracellular folate concentration. Up-regulation of the receptor in response to low extracellular folate can result from increased rates of gene transcription and/or increased stability of receptor mRNA. In addition, translational or post-translational changes may be involved (Antony, 1996). It is postulated that FRs function as folate scavengers when folate supply is low or when rapid cell growth requires elevated uptake of folate for DNA biosynthesis (O'Shannessy *et al.*, 2011).

Folates can be transported into cells by a specific pathway of receptor-mediated endocytosis. The FRs (α and β) are anchored to the plasma membrane by a glycosylphosphoinositol (GPI) moiety. On binding of folate to the receptor, the liganded receptor is internalized by an inward folding (invagination) and pinching off of the membrane to form an endosome. The endosome migrates along microtubules to a perinuclear endosomal compartment, where its interior is acidified to a pH of ~6.0–6.5 by the action of a membrane proton pump. Acidification causes reduced folates to dissociate from the receptor; the affinity of folic acid for the FR is so high that it does not significantly dissociate from the receptor. The PCFT is expressed on the endosome membrane and mediates folate export from the acidified endosome into the cytoplasm (Zhao *et al.*, 2009b). The receptors are recycled to the cell surface to bind new ligand.

Folate uptake by certain cells takes place by receptor-mediated potocytosis in which a GPI-anchored membrane protein within a caveola is a specific receptor for 5-methyl-THF. On binding of the folate molecule to the receptor, the caveola closes and acidification of the caveola space by the action of a membrane proton pump causes the folate to dissociate from the receptor. The high concentration of folate in the caveola space creates a gradient that favors movement of folate across the membrane by facilitated diffusion. When the folate reaches the cytoplasm, it is covalently modified by the addition of multiple glutamic acid residues. The polyglutamyl folate is unable to diffuse out of the cell and is therefore trapped within the cytoplasm. The caveola then reopens, thereby exposing the receptors for the next cycle (Anderson *et al.*, 1992).

25.4 ABSORPTION OF DIETARY FOLATE

Humans and other mammals cannot synthesize folate in their tissues and thus they must obtain the vitamin from exogenous sources via intestinal absorption. The intestine is exposed to two sources of folate: (1) dietary folate and (2) folate synthesized by bacteria in the large intestine. The latter source is available to the host tissues through direct absorption in the colon.

25.4.1 Deconjugation of Polyglutamyl Folate

The folates naturally present in foods exist largely in protein-bound form, the predominant vitamers being polyglutamyl forms of THF, 5-methyl-THF and 10-formyl-THF (Gregory, 1984). Polyglutamyl folates, being large and strongly electronegative molecules, are not transportable into cells and, before they can be absorbed, they must be hydrolyzed to monoglutamate forms. None of the known proteases in saliva, gastric juice, or pancreatic secretions are capable of splitting the γ-peptide bonds in the polyglutamyl side chain. Polyglutamyl folate can, however, be hydrolyzed by folate conjugase, which is a trivial name for pteroylpolyglutamate hydrolase (EC 3.4.12.10, also known as folylpoly-γ-glutamyl carboxypeptidase). As much as 50–75% of dietary polyglutamyl folate can be absorbed after deconjugation to monoglutamyl folate (Butterworth *et al.*, 1969). The presence of conjugase activity in many raw foods of both plant and animal origin results in a high proportion of the dietary folate being already monoglutamyl when presented to the intestinal mucosa (Gregory, 1989).

Two folate conjugases have been found in human jejunal tissue fractions. One, a brush-border exopeptidase, has a pH optimum of 6.7–7.0 and is activated by Zn^{2+}. The other, an intracellular endopeptidase of mainly lysosomal origin, has a pH optimum of 4.5 and no defined metal requirement. The brush-border conjugase splits off terminal glutamate residues one at a time and is thought to be the principal enzyme in the hydrolysis of polyglutamyl folate. Brush-border conjugase from the jejunum of the human and pig possess similar enzymatic properties (Gregory *et al.*, 1987) and thus the porcine enzyme can be used to study folate bioavailability in humans. Interestingly, the human and the pig are the only species in which intestinal brush-border conjugase activity has been demonstrated. Bhandari and Gregory (1990) showed that extracts from certain foods (e.g. legumes, tomatoes, and orange juice) can inhibit brush-border conjugase activity from human and porcine intestine *in vitro*. Organic acids may be responsible for this inhibition (Wei & Gregory, 1998). Such inhibition may be a factor affecting the bioavailability of polyglutamyl folates in diets containing these foods. The intracellular conjugase may play no role in the digestion of dietary folate, being, instead, concerned with folate metabolism within the enterocyte (Halsted, 1990).

Significant conjugase activity has also been reported in the pancreatic juice of pigs and humans (Gregory, 1995). Bhandari *et al.* (1990) found the porcine pancreatic enzyme to be Zn^{2+}-dependent with maximum activity at pH 4.0–4.5. Feeding stimulated secretion of pancreatic juice, including conjugase activity. Chandler *et al.* (1991) calculated that conjugase activity in porcine pancreatic juice was minor relative to the activity of the jejunal brush-border conjugase.

25.4.2 Transport Mechanisms

In the human, the entire small intestine is capable of absorbing monoglutamyl folate. Absorption is somewhat greater in the proximal than in the distal jejunum which, in turn, is much greater than in the ileum. Folate transport across the brush-border membrane of the enterocyte proceeds by two parallel processes (Selhub & Rosenberg, 1981). At physiological concentrations (<5 μM) of luminal folate, transport occurs primarily by a saturable process with a pH optimum of 5.0–5.5 (Alam *et al.*, 2020). At higher concentrations, folate transport occurs by a non-saturable process with characteristics of simple diffusion. The saturable component is discussed in the following with no further mention of unsaturable transport.

Influx of luminal folate across the brush-border membrane of the enterocyte is mediated specifically by the PCFT, which operates by an electrogenic folate$^-$/H^+ co-transport mechanism at an optimal pH range of 5.0–5.5. Because of the acidic microclimate at the luminal surface (pH 5.8–6.0), an inwardly directed proton gradient exists and the energy generated by the downhill movement of H^+ into the cell powers the movement of folate in the same direction (Said, 2011). Until the discovery of the PCFT by Qiu *et al.* (2006), this transport route was attributed to the RFC. Despite being abundantly expressed along the entire intestinal brush-border membrane, the RFC has no significant function as a brush-border membrane folate transporter since it has negligible activity below pH 6.5 (Zhao *et al.*, 2009a).

Folate exit from the enterocyte across the basolateral membrane into the lamina propria of the villus is carrier-mediated and sensitive to the effect of anion exchange inhibition. In addition, the exit mechanism is electroneutral, Na^+-independent, and has a higher affinity for the substrate than has the system at the brush-border membrane (Said & Redha, 1987). Neither PCFT nor RFC is expressed at the basolateral membrane (Zhao *et al.*, 2009a).

During intestinal transport, some of the folate is converted within the enterocyte to 5-methyl-THF in a pH-dependent manner (Strum, 1979). This conversion is extensive at pH 6.0 and negligible at pH 7.5, presumably because dihydrofolate reductase, the rate-limiting enzyme in the reduction and methylation process, has an acidic pH optimum. The percent conversion is reduced by increasing the concentration of folate in the mucosal medium, thus indicating saturation of the process. Since at higher concentrations most transported folate remains unmodified, intestinal conversion of absorbed folic acid is not obligatory for transport into the circulation.

25.4.3 Influence of Folate-Binding Protein on the Absorption of Folate from Milk

Human milk and the milk of other mammals contain soluble proteins that avidly and specifically bind the folate present in the milk. These proteins are referred to as folate-binding proteins (FBPs). At pH values above 6.0, cow's milk FBP exists as an aggregating system in which a polymer composed of at least 16 monomers is reversibly formed (Svendsen *et al.*, 1979; Pedersen *et al.*, 1980). Concentration and pH effects on the equilibrium constant of this reversible system explain the variations in the molecular weight of FBP reported in the literature. FBP has both physicochemical and immunological species-specificity among mammals (Iwai *et al.*, 1983). The binding of folate to FBP in milk is tight, but not covalent (Ford *et al.*, 1969), and the molar stoichiometry is 1:1 (Salter *et al.*, 1981). FBP is present in excessive amounts, so that the milk has the capacity to bind added folate (Ford, 1974). The folate in bovine milk occurs predominantly as 5-methyl-THF, of which about 60% is the monoglutamate (Wagner, 1985). In unprocessed and pasteurized milk, the folate is bound to FBP, but in UHT-processed milk and yogurt the folate exists in the free (unbound) form (Wigertz *et al.*, 1996).

Little is known about the manner of absorption of bound milk folate. Tani *et al.* (1983) showed that, *in vitro*, folic acid was entirely dissociated from FBP at pH 4.5 and below, but at pH 7.4 dissociation was negligible. Dissociation was completely reversible and pepsin did not impair the binding activity of the protein. These observations indicate that, *in vitro*, the bovine milk FBP is not inactivated under the acidic conditions in the stomach.

Tani *et al.* (1983) studied the influence of FBP from bovine milk on the absorption of folate in the gastrointestinal tract of the rat. When bound ^{14}C-labeled folic acid was administered intragastrically to rats by stomach tube, a considerable amount of folic acid was released from FBP in the stomach and it recombined with FBP in the jejunum. It appears that, when the ingested material is transferred from the stomach to the duodenum, it is not yet fully neutralized, allowing a small part of the unbound folic acid to be absorbed before being recombined with FBP as the material is neutralized in the jejunum. Compared with free folic acid, bound folic acid was more gradually absorbed in the small intestine, but finally the total amount of bound folic acid absorbed was almost the same as that of free folic acid. Free folic acid was rapidly absorbed in the jejunum, whereas bound folic acid was only slightly absorbed in this region. In the ileum, however, the absorption rates of the two forms of folic acid were almost equal. These results suggest that folic acid bound to FBP is absorbed by a manner different from that of free folic acid in the adult rat. A slower rate of folate absorption was also demonstrated in adult rats by Said *et al.* (1986) using human milk FBP and 5-methyl-THF as the substrate.

Milk has evolved for the nutritional support of the neonate, so absorption studies in neonates might shed more light on the role of FBP in milk than studies in adults. Mason and Selhub (1988) demonstrated that, with respect to preferential jejunal uptake and sulfasalazine inhibition, free folic acid absorption in the neonatal rat is similar to that in the adult. However, the characteristics of FBP-bound folic acid absorption are very different from those of free folic acid. The uptake of folate bound to FBP from rat milk occurs preferentially in the ileum of the suckling rat as opposed to the jejunum. The absorption rate was slower than that for free folic acid and sulfasalazine was ineffective. In the suckling animal, a variety of pancreatic and intestinal absorptive functions are not fully developed, a situation which could allow the FBP to reach the ileum in an active form. This situation was demonstrated by Salter and Mowlem (1983), who showed that a proportion of goat's milk FBP administered orally to neonatal goats survived along the length of the small intestine. Protease inhibitors inherent to colostrum may assist the passage of bound folate into the small intestine (Laskowski & Laskowski, 1951). Additional evidence for the role of FBP in the absorption of milk folate by neonates was provided by Salter and Blakeborough (1988), who reported that the addition of goat's milk FBP enhanced the transport of 5-methyl-THF in brush-border membrane vesicles isolated from the small intestine of neonatal goats.

Tani and Iwai (1984) reported that the incorporation of radiolabeled folic acid into folate-requiring intestinal bacteria was considerably diminished when it was bound to FBP. Also, the presence of goat's colostrum considerably reduced the uptake of folic acid by folate-requiring bacteria (Ford, 1974). These observations suggest that milk FBP prevents the uptake of bound folate by intestinal bacteria, thereby indirectly promoting folate absorption by the infant.

Mason and Selhub (1988) postulated that, in neonates, the absorption of bound folate occurs by endocytosis (or similar process) of the intact folate–protein complex. Subsequent proteolytic digestion inside the enterocyte would then release the bound folate for export into the portal circulation. Suckling animals are known to absorb proteins by a process that decreases markedly at the time of weaning and is more active in the ileum than in the jejunum (Henderson, 1990).

25.4.4 Adaptive Response to Folate Deprivation

Said *et al.* (2000) induced folate deficiency in rats by feeding a folate-deficient diet that contained an antibiotic to decrease the bacterial synthesis of folate in the intestine. Using everted intestinal sacs and brush-border membrane vesicles, they showed that folate deprivation causes a specific up-regulation in the transport of physiological concentrations of folic acid across the apical membrane of both the small and large intestines. The effect in the small intestine took place not only in the jejunum, but also in the ileum, a region that does not usually absorb folate. In addition to the up-regulation of transepithelial folate transport, folate deficiency was associated with a tenfold increase in the activity of brush-border membrane conjugase. The intestine is therefore able to maximize its ability to extract the limited amount of folate ingested during periods of deprivation.

25.4.5 Differentiation-Dependent Up-Regulation of Intestinal Folate Uptake

The differentiation of intestinal epithelial cells as they move from the crypt region to the villus tip is accompanied by alterations in levels of expression of many genes, including those involved in nutrient uptake. Subramanian *et al.* (2008) examined the effects of differentiation of intestinal epithelial cells on the intestinal folate uptake process. The results from studies with Caco-2 cells showed a progressive increase in carrier-mediated folic acid uptake with differentiation. They also showed a progressive increase in the level of expression of hRFC and hPCFT protein and mRNA with differentiation, supported by a significant increase in activity of the *hRFC* and *hPCFT* promoters. Studies with native mouse intestine showed that folate uptake in the differentiated cells of the villus tip was higher than that in the undifferentiated cells of the crypt, and this was again associated with a higher level of expression of the mRFC and mPCFT protein and mRNA. These findings suggest the possible involvement of a transcriptional mechanism(s) in the differentiation-dependent regulation of the intestinal folate uptake process.

25.5 ABSORPTION OF BACTERIALLY SYNTHESIZED FOLATE IN THE SMALL AND LARGE INTESTINE

The bulk of the gut flora is concentrated in the large intestine, but a significant population of bacteria resides in the small intestine. Many of the bacterial species that inhabit the human large and small intestine are capable of synthesizing folate from a precursor pterin, 6-hydroxymethyl-7.8-dihydropterin (HMDP). The folates synthesized are typically tri-, tetra-, and pentaglutamyl forms. The enzyme 6-hydroxymethyl-7,8-dihydropterin pyrophosphokinase catalyzes the transfer of pyrophosphate from ATP to HMDP to form 6-hydroxymethyl-7,8-dihydropterin pyrophosphate (DHPPP). The next enzyme in the pathway, 7,8-dihydropteroate synthase, catalyzes the condensation of DHPPP with *para*-aminobenzoic acid to form 7,8-dihydropteroate and pyrophosphate (Bermingham *et al.*, 2000). A third enzyme, 7,8-dihydropteroate synthase, adds a glutamate molecule to 7,8-dihydropteroate to form 7,8-dihydropteroylmonoglutamic acid (DHF). The addition of second and third glutamic residues can only occur after DHF has been reduced to THF (Stokstad & Koch, 1967).

Hydrolysis of polyglutamyl folate to monoglutamyl folate by bacterial conjugase is a necessary step for folate absorption. In the rat (Rong *et al.*, 1991) and piglet (Asrar & O'Connor, 2005), a portion of bacterially synthesized monoglutamyl folate is absorbed by the colon and incorporated into the tissues. The folate synthesized in the large intestine of human infants was shown to be predominantly 5-methyl-THF of which about 50% was monoglutamated; the quantities produced were sufficient to potentially affect the folate status of the host (Kim *et al.*, 2004). Camilo *et al.* (1996) established that folate synthesized by bacteria in the human small intestine is assimilated by the host, but the contribution to folate nutriture from this source is uncertain. An increase in fiber intake may promote increased folate biosynthesis by intestinal microflora (Houghton *et al.*, 1997).

Two papers from the same group provide the first direct *in vivo* evidence for folate absorption across the intact human colon. In the first paper (Aufreiter *et al.*, 2009), a physiological dose of ^{13}C-labeled 5-formyl-THF was infused directly into the cecum of healthy adults during screening colonoscopy. Three or more weeks later, each subject received an intravenous injection of the labeled 5-formyl-THF. Blood samples were collected before and after each treatment. Plasma analysis of labeled and unlabeled 5-formyl-THF and its metabolite 5-methyl-THF was performed by liquid chromatography coupled to tandem mass spectrometry (LC-MS/MS). Both labeled 5-formyl-THF and 5-methyl-THF were detected. There was also a significant rise in unlabeled mean 5-formyl-THF, which suggests the displacement of endogenous folate from the colonocyte and/or liver. This observation serves as a caution in interpreting folate studies using unlabeled folates, because the portion of the plasma response due to the test dose cannot be distinguished from that due to displacement of endogenous folate. The apparent rate of folate absorption across the colon was 0.6 ± 0.2 nmol/h (mean $\pm$ SEM) based on the appearance of labeled 5-methyl-THF in plasma after cecal infusion of labeled 5-formyl-THF. In comparison, the rate of appearance of labeled 5-methyl-THF after intravenous injection of labeled 5-formyl-THF was 7 ± 1.2 nmol/h.

The above report shows that preformed folate can be absorbed across the colon during colonoscopy, that is, after bowel cleansing. In the second paper, Lakoff *et al.* (2014) evaluated folate absorption across the human colon with an intact microbiome by using specially designed enteric-coated caplets to deliver ^{13}C-labeled 5-formyl-THF past the ileocecal junction. About 4 weeks after receiving the caplets, each subject received an intravenous injection of the labeled 5-formyl-THF. Labeled 5-formyl-THF was not detected in the plasma of any subject after consumption of test caplets; however, its metabolite 5-methyl-THF was detected in seven out of nine subjects who completed the study. The rate of appearance of labeled 5-methyl-THF in plasma was 0.33 ± 0.09 (caplet) and 5.8 ± 1.2 (intravenous) nmol/h. The mean apparent absorption across the colon was 46%.

Studies using a human-derived colonic epithelial cell line NCM460 (Kumar *et al.*, 1997) and purified apical membrane vesicles (AMVs) isolated from the colon of human organ donors (Dudeja *et al.*, 1997) demonstrated the involvement of a carrier-mediated, pH-dependent system in folate uptake across the luminal membrane of the colonocyte. The PCFT that facilitates folate absorption in the acidic (pH 5.8–6.0) environment of the small intestine is also present in the colon (Visentin *et al.*, 2014), but it is uncertain whether the pH at the colonic absorptive surface is in the range that would allow for appreciable transport activity. The uptake system in the colon was found to be similar in most respects to that of the small intestine. Energy dependency was demonstrated using different metabolic inhibitors. The K_m for the apical membrane transport system of the human colon was 8.2 ± 2.4 μM (Dudeja *et al.*, 1997) compared with 1.69 μM for its small intestinal counterpart (Said *et al.*, 1987), indicating a relatively low affinity of the colonic carrier for folate. Conversely, the V_{max} of the apical membrane transport system of the colon (19.8 ± 2.9 pmol per mg protein per 10 s) was three to four times higher than that of the small intestine, indicating a greater uptake capacity of the carrier.

Dudeja *et al.* (2001) used human colonic basolateral membrane vesicles (BLMVs) to investigate the mechanism of folate exit out of the colonocyte. The transport characteristics of folic acid across the colonic basolateral membrane were similar to those across the apical membrane with regard to pH dependence, K_m, DIDS sensitivity, and electroneutrality. The two processes, however, showed some differences with regard to uptake capacity of the carrier system involved. This was indicated by the difference in the V_{max} of the two uptake processes, which were reported at 8.10 ± 0.36 and 19.8 ± 2.9 pmol per mg protein per 10 s for human colonic BLMV and AMV, respectively.

The apparent rate of folate absorption across the colon reported by Aufreiter *et al.* (2009) is considerably lower than that reported for the small intestine (Wright *et al.*, 2003, 2005). However, the transit time in the colon (24–72 hours) is considerably longer than that in the small intestine (3 ± 1 hours), which allows for a greater opportunity for absorption to occur. Furthermore, whereas folate absorption across the small intestine relies on oral intake, and hence is intermittent, the availability of microbial folate for absorption across the colon is continuous.

25.6 PLASMA TRANSPORT

Following the ingestion of foods containing natural amounts of folate, the vitamin circulates in the plasma as reduced monoglutamates, primarily 5-methyl-THF, in concentrations ranging from 10 to 35 nM. High oral doses of folic acid bypass the normal folate absorption mechanism, resulting in both folic acid and 5-methyl-THF appearing in the bloodstream (Kelly *et al.*, 1997). Folate in plasma seems to be distributed in three fractions: free folate and that loosely bound to plasma proteins exist in similar proportions; less than 5% is bound to high-affinity binders (Herbert, 1999). Folate is not taken up by circulating erythrocytes during their 120-day lifespan; it is only taken up by the developing erythrocytes in the bone marrow.

25.7 FOLATE DISPOSITION AND TRANSPORT IN THE LIVER

Approximately 10–20% of the 5-methyl-THF in the portal blood is taken up by the liver during the first pass through this organ. Some of the folate that enters the liver is converted to polyglutamate derivatives and stored. The majority of folate arriving at the liver is secreted into the bile and returned to the intestine for reabsorption in the enterohepatic cycle. Mobilization of folates from the liver requires their hydrolysis to monoglutamyl forms, which is catalyzed by a lysosomal conjugase (Zhao *et al.*, 2017). Monoglutamyl 5-methyl-THF formed in the liver, as well as that delivered directly from the portal vein, can reach the systemic circulation via the hepatic vein and enter the extrahepatic tissues. In the tissues, 5-methyl-THF is demethylated by the action of vitamin B_{12}-dependent methionine synthetase (Steinberg, 1984). The resultant THF can now have glutamate residues (usually five to eight) added to it. The resultant polyglutamated THF is the functional coenzyme, accepting and transferring one-carbon units in intermediary metabolism.

Monoglutamyl folates in the portal blood are delivered to the hepatic sinusoids where they are transported across the hepatocyte sinusoidal membrane into the cytoplasm. The PCFT is highly expressed at the sinusoidal membrane and mediates folate uptake by a folate$^-$/H^+ co-transport mechanism. A Na^+/H^+ antiporter at the sinusoidal membrane may generate a local acidic region to facilitate PCFT-mediated uptake (Zhao *et al.*, 2009a). Although RFC is detected at the sinusoidal membrane, its level of activity is low in the acidic environment (Zhao *et al.*, 2009a).

25.8 INTRACELLULAR METABOLISM

The glutamate moieties of polyglutamyl folates are linked covalently by γ-peptide bonds forming an oligo-γ-glutamyl chain. The chain length changes in response to folate, methionine, and vitamin B_{12} status and to changes in growth rates. The enzyme responsible for the stepwise addition of glutamate residues to intracellular folates is folate polyglutamyl synthetase (FPGS). This enzyme is located mainly in the cytosol, although some activity is found in the mitochondrial fraction. The enzyme requires ATP as a source of energy and has optimal activity at pH 8.5–9.0.

$$\underset{\text{(THF)}}{H_4PteGlu_n} + ATP + glutamate \rightarrow H_4PteGlu_{n+1} + ADP + Pi \quad (25.1)$$

FPGS activity is particular high in the liver. The polyglutamyl folates in rat liver are mainly pentaglutamates with smaller amounts of hexaglutamates and trace amounts of heptaglutamates. When rats are dosed with radioactive folate, pentaglutamates account for most of the label in the liver after about 6 hours. Labeled hexaglutamates appear in the liver after 2 to 3 days and heptaglutamates after about 4 weeks (Shane, 1989).

The preferred substrate for FPGS is THF; DHF and 10-formyl-THF are also effective substrates, but 5-methyl-THF, 5,10-methylene-THF, and folic acid are very poor substrates. When THF is incubated with purified FPGS, the hexaglutamate and smaller amounts of the heptaglutamate accumulate. The substrate specificity determines which polyglutamyl folates are predominant in a particular tissue.

Polyglutamylation fulfills three important roles in folate metabolism. Firstly, the addition of at least three glutamate residues to monoglutamyl folate is necessary in order for folates to be retained by cells after uptake. Secondly, polyglutamyl folates have greater affinities for enzymes than their monoglutamyl counterparts; for most (but not all) folate-dependent enzymes, hexaglutamates are the most effective substrates. Thirdly, polyglutamylation facilitates the channeling of substrates between different enzymes that are organized into clusters. Channeling is the process whereby the reaction product of one enzyme is transferred directly to the next enzyme without achieving equilibrium with the cell medium.

25.9 RECYCLING OF FOLATE VIA THE ENTEROHEPATIC PATHWAY

The majority of 5-methyl-THF arriving at the liver from the intestine and taken up is not demethylated and converted to polyglutamate; instead it is quickly released for distribution to extrahepatic tissues. The initial route for this distribution is the enterohepatic circulation, whereby the folate is discharged into the bile and subsequently reabsorbed by the small intestine before re-entering the systemic circulation. Accompanying 5-methyl-THF in the bile are larger amounts of non-methylated tetrahydrofolates which represent folates salvaged from dying cells such as senescent erythrocytes and hepatocytes (Shin *et al.*, 1995). Any folic acid that might have been absorbed and released into the portal circulation without modification is exclusively taken up by the liver and either converted into one-carbon derivatives of THF prior to rapid release into bile or polyglutamated and incorporated into the hepatic folate pool (Steinberg, 1984). Hepatic reduction and derivatization of folic acid provides another source of non-methylated tetrahydrofolates present in bile (Shin *et al.*, 1995).

The recycling of folate via the enterohepatic pathway may account for as much as 50% of the folate that ultimately reaches the extrahepatic tissues. Disruption of the enterohepatic cycle by bile drainage results in a fall of the serum folate level to 30–40% of normal within 6 hours – a much more dramatic drop than that seen with a folate-deficient diet. Eventually, the serum folate level stabilizes, despite continuing losses in the bile. This suggests a net flux of folate into the plasma compartment from tissue pools.

Release of stored folate from cells of any tissue requires hydrolysis of the polyglutamates to monoglutamates by intracellular conjugase.

The maintenance of a normal level of plasma folate depends on regular increments of exogenous folate from the diet. The enterohepatic circulation of folate evens out the intermittent intake of dietary folate. The liver plays a major role in maintaining folate homeostasis because of its capacity to store about 50% of the total body folate, its relatively rapid folate turnover, and the large folate flux through the enterohepatic circulation (Steinberg, 1984). In situations of dietary folate deficiency, the liver does not respond by releasing its folate stores. Rather, the nonproliferating, less metabolically active tissues mobilize their folate stores and return monoglutamyl folate to the liver. This folate is released by the liver via the enterohepatic cycle and distributed to the tissues that most require it – in particular, those with actively proliferating cells. The kidney plays its part in conserving body folate by actively reabsorbing folate from the glomerular filtrate. In addition, a pathway exists that is capable of salvaging folate released from senescent erythrocytes.

25.10 RENAL REABSORPTION AND SECRETION OF FOLATE

Circulating free folate, but not the protein-bound fraction, is filtered at the glomerulus and then completely reabsorbed in the proximal tubules against a concentration gradient, indicative of active transport. The reabsorption system becomes saturated at plasma folate concentrations above 0.2 mM. There is strong evidence that FRα mediates tubular transport of folate in the luminal-to-serosal direction. FRα is highly expressed at the apical membrane of the proximal tubule epithelial cell, where its function is to capture folate molecules in the glomerular filtrate. Treatment of cultured proximal tubule cells with colchicine, which disrupts the vesicular trafficking that takes place during endocytosis, inhibits folate transport across the apical membrane, but not across the basolateral membrane (Morshed *et al.*, 1997). This suggests that folate uptake by the apical membrane involves internalization of the FR–folate complex by some form of endocytosis. On the basis of elaborate experiments, Birn *et al.* (1997) postulated that folate in the glomerular filtrate binds to FRα and is internalized by receptor-mediated endocytosis. After dissociation of the FR–folate complex in the endosomes, folate is transported across the basolateral membrane for release into the renal vascular circulation, while the receptor migrates back into the apical membrane through dense apical tubules.

25.11 CATABOLISM AND EXCRETION

The initial step in folate catabolism is cleavage of intracellular polyglutamyl folates at the C9–N10 bond, catalyzed by carboxypeptidase G (EC 3.4.17.11). The resulting *p*-aminobenzoylglutamates are hydrolyzed to the monoglutamates, which are *N*-acetylated before excretion; the pterin moiety is excreted either unchanged or as biologically inactive derivatives.

Except in cases of malabsorption, the body is very adept at conserving folate of dietary origin. Folate secreted into the bile is reabsorbed in the small intestine. Much of the circulating folate binds to plasma proteins and so cannot be filtered in the kidney. The free circulating folate that is filtered is captured by the folate receptors in the renal tubules and reabsorbed into the bloodstream. The result is very little folate loss by fecal or urinary excretion.

25.12 FOLATE TRANSPORT IN THE BRAIN

25.12.1 Introduction

Folate transport in the brain is critical for the development, function, and repair of the central nervous system and cerebral folate deficiency causes a number of neurological disorders. Regulation of cerebral folate transport is essential to prevent excessive amounts of folate from entering the brain. Toxicity can be demonstrated by the seizures induced by injecting folate directly into the brain. Folate homeostasis in brain is implied by the fact that it is much more difficult to cause folate depletion in brain than in other organs, such as the liver and bone marrow. On the other hand, it is difficult to raise the concentration of 5-methyl-THF in brain by raising the plasma levels of this metabolite. Although the brain is able to take up folic acid, it lacks the enzyme (dihydrofolate reductase) that converts it to functional reduced forms (Zhao *et al.*, 2017).

There are two potential routes of folate entry into the brain: the blood–cerebrospinal fluid (CSF) barrier at the choroid plexus and the blood–brain barrier at the cerebral vascular endothelium. The primary route is across the choroid plexus mediated by FRα and PCFT. Clinical reports have shown that patients with FRα or PCFT mutations can sometimes present with mild neurological deficits, indicating compensatory transport of folate across the blood–brain barrier despite suboptimal CSF folate levels. Folate homeostasis in the brain is maintained by the specific localization of folate-importing and -exporting transporters at opposing membranes of the choroid plexus epithelium and the blood–brain barrier endothelium.

25.12.2 Transport across the Blood–CSF Barrier

The concentration of folate in the CSF is normally two to three times higher than in blood (Zhao *et al.*, 2009a), consistent with active transport from blood to CSF across the choroid plexus. The mechanism of folate transport is specific for reduced folates, saturable and energy-dependent, with 5-methyl-THF being the preferred substrate (Spector & Lorenzo, 1975a,b). The entry of 5-methyl-THF from blood into CSF is approximately one-half saturated at the normal plasma concentration of ~0.02 μM. Thus, if plasma folate levels are increased to much above normal levels by intravenous injection of folate, the uptake mechanism becomes fully saturated and the brain is protected from excessive amounts of folate. In order to enter the CSF from the blood, the folate molecule must be transported across the choroid plexus epithelial cell, first crossing the basolateral membrane facing the blood and then crossing the apical membrane facing the CSF. Diffusion of folate across the choroid plexus is prevented by the tight junctions of adjacent choroid epithelial cells. Once within the CSF, folate molecules can easily diffuse through the leaky ependymal layer into the interstitial fluid of the brain.

Transport of folate from blood to CSF across the blood–CSF barrier is primarily mediated by FRα. FRα has a polarized distribution in the choroid epithelial cell, with high density at the apical membrane facing the CSF and low density at the basolateral membrane facing the blood. RFC is selectively expressed at the apical membrane. PCFT is expressed at the basolateral membrane as well as the limiting membranes of intracellular vesicles. Loss-of-function mutations in the *FOLR1* and *SLC6A1* genes result in a loss of FRα and PCFT, respectively, and an impaired delivery of folate to the brain via the choroid plexus. A hereditary FRα deficiency results in a childhood onset neurodegenerative disease known as cerebral folate deficiency (Grapp *et al.*, 2012). The equivalent disease caused by a loss of PTFC is hereditary folate malabsorption (Qiu *et al.*, 2006).

Grapp *et al.* (2013) proposed the following mechanism of choroid plexus transcytosis and exosome shuttling for the delivery of folate into the human brain parenchyma. At physiological plasma concentrations, 5-methyl-THF binds to the FRα at the basolateral membrane of the choroid plexus. The 5-methyl-THF–FRα complex is internalized by receptor-mediated endocytosis, translocated into GPI-anchored protein-enriched early endosomal compartments, and further transferred to late endosomal compartments called multivesicular endosomes (or bodies). Intraluminal vesicles of multivesicular endosomes containing FRα are generated by inward budding of the limiting membrane. The multivesicular endosomes fuse with the apical membrane of the choroid epithelial cell and the intralumenal vesicles are released as exosomes into the CSF. The FRα-containing exosomes diffuse passively through the ependymal layer of the ventricles and are taken up by astrocytes before distribution in the brain parenchyma.

Zhao *et al.* (2009b) showed that FRα in PCFT- and RFC-deficient HeLa R1-11 cells was able to bind and trap folates in vesicles, but with minimal export into the cytoplasm. They demonstrated a role for PCFT in FRα-mediated endocytosis by serving as a route of export of folates from acidified endosomes. In this process, the binding of folate to FRα anchored to the cell surface triggers the internalization of the receptor–ligand complex in endocytic vesicles that localize PCFT and are distinct from FR-containing intraluminal vesicles. Within the cytoplasm, the vesicles fuse with one another to form endosomes, whose contents are acidified by a proton pump. Acidification results in the dissociation of folate from the receptor and the transvesicular proton gradient provides the driving force that facilitates PCFT-mediated folate export from the endosome. RFC is expressed at the apical membrane of choroid plexus epithelial cells and could serve as a route for folate efflux into the CSF (Wang *et al.*, 2001).

25.12.3 Transport across the Blood–Brain Barrier

A number of *in vitro* and *in vivo* studies have demonstrated transport of folates and methotrexate across the blood–brain barrier, with 5-methyl-THF appearing to be the preferred substrate (Alam *et al.*, 2020). Wu and Pardridge (1999) examined the transport of [^{3}H]5-methyl-THF at this barrier *in vivo* using both intravenous injection/pharmacokinetic and internal carotid artery perfusion methods. Parallel *in vitro* studies were also performed with isolated human brain capillaries. The uptake of labeled 5-methyl-THF by either rat brain or human brain capillaries was saturable and could be equally inhibited by 10 μM concentrations of either unlabeled 5-methyl-THF or unlabeled folic acid. The competition by 10 μM folic acid indicated that folic acid is also transported by this uptake system.

RFC and PCFT are present at the luminal blood-facing membrane of blood–brain barrier endothelial cells, where the pH should be 7.4, but FRα is absent. The RFC has a neutral pH optimum, but a very low affinity for folic acid. By contrast, the PCFT is a high-affinity folate transporter, but has a low pH optimum. Thus, neither of these transporters would be expected to contribute significantly to the transport of folates to the brain via the blood–brain barrier.

In FRα knockout mice, injection of 1α,25-dihydroxyvitamin D_3 (calcitriol) up-regulated the RFC expression at the blood–brain barrier, thereby compensating for its inherent low affinity for folate and the absence of FRα-mediated folate uptake at the choroid plexus (Alam *et al.*, 2019). This finding suggests that an RFC-augmenting drug could be a therapy for the treatment of neurodegenerative disorders arising from failure of folate transport across the choroid plexus.

25.13 PLACENTAL TRANSPORT OF FOLATE

Total folate concentrations in fetal blood are generally five- to sixfold higher than those in the mother, with the placental concentration being several-fold higher. This implies the existence of active transport of folate across the placenta from mother to fetus.

Prasad *et al.* (1994a,b) demonstrated the selective expression of FRα in placental trophoblast cells. Prasad *et al.* (1995) accomplished the molecular cloning of the human placental RFC. The presence of a V-type H^+-ATPase (a type of proton pump) has been demonstrated in brush-border membrane vesicles (BBMVs) isolated from human placentas (Simon *et al.*, 1992). Blocking the V-type proton pump, or abolishing the transmembrane proton gradient, inhibits 5-methyl-THF transport into JAR human placental choriocarcinoma cells. This suggests that the transport of 5-methyl-THF into the cytoplasm of placental trophoblast cells is dependent on a transmembrane H^+ gradient generated by a V-type proton pump.

Yasuda *et al.* (2008) investigated the folate transport mechanism using human term placental BBMVs. They also investigated the uptake of folate by the placenta and fetus during the course of pregnancy in the rat. Based on the effects of specific inhibitors, it was apparent that the three folate transporters, FRα, RFC, and PCFT, are involved in the folate transport mechanism in the human placenta. Both FRα and RFC were expressed on the apical (maternal) side of the placental villi, but the precise location of PCFT could not be determined. Accumulation of folate after intravenous injection increased with the progress of gestation in the rat placenta and the fetus. Furthermore, increases in the expression levels of mRNA of FRα, RFC, and PCFT in the rat placenta during pregnancy were observed. These findings suggest that transport of folate in the placenta increases in response to increased need for folate during fetal development.

According to Alam *et al.* (2020), FRα, FRβ, and RFC mediate the uptake of folate from maternal blood into the syncytiotrophoblast across the apical membrane, while PCFT and RFC mediate the efflux of folate into the fetal circulation across the basolateral membrane.

25.14 BIOCHEMICAL FUNCTIONS

Folate plays an essential role in the mobilization of one-carbon units in intermediary metabolism. Such reactions are essential for the synthesis of the purines (adenine and guanine) and one of the pyrimidine nucleotides (deoxythymidine monophosphate), which are required for DNA synthesis. Folate is also directly involved in the metabolism of certain amino acids. The role of folate in the metabolism of homocysteine is discussed in Section 25.15.1.

Folic acid, the parent compound, is not biochemically active, but becomes so after reduction to THF and substitution of one-carbon units on the N-5 and/or the N-10 positions. THF accepts one-carbon groups from various degradative reactions in amino acid metabolism. The THF derivatives thus formed are coenzymes that serve as donors of one-carbon units in a variety of synthetic reactions. The folate coenzymes are interconvertible by virtue of reversible reactions. There is, however, one important exception: 5-methyl-THF cannot be converted back to 5,10-methylene-THF under physiological conditions.

25.14.1 Reduction of Folic Acid to THF

Folic acid and subsequently DHF are both reduced by dihydrofolate reductase (EC 1.5.1.3) to generate THF.

$$\underset{\text{(folic acid)}}{\text{PteGlu}} + \text{NADPH} + \text{H}^+ \rightarrow \underset{\text{(DHF)}}{\text{H}_2\text{PteGlu}} + \text{NADP}^+ \quad (25.2)$$

$$\underset{\text{(DHF)}}{\text{H}_2\text{PteGlu}} + \text{NADPH} + \text{H}^+ \rightarrow \underset{\text{(THF)}}{\text{H}_4\text{PteGlu}} + \text{NADP}^+ \quad (25.3)$$

Reaction 25.2 is essential for utilizing the synthetic form of the vitamin, folic acid. The enzyme is the target of the anti-cancer drug methotrexate (Figure 25.2), which binds extremely tightly to the active site. By inhibiting the enzyme, methotrexate produces a partial depletion of intracellular-reduced folates, thereby impeding folate-dependent one-carbon reactions and conferring cytotoxicity. Methotrexate treatment also results in the accumulation of DHF and 10-formyl-DHF. The latter compound, which is not found in normal tissues, may be responsible for the drug-related cytotoxicity, since it inhibits both glycinamide ribotide transformylase and thymidylate synthase (Baram *et al.*, 1988). Chemotherapy consists of alternating periods of administration of methotrexate and 5-formyl-THF (leucovorin) in order to replete the normal tissues.

25.14.2 Folate Coenzymes in One-Carbon Transfer Reactions

25.14.2.1 Interconversion of Serine and Glycine

Reaction 25.4 (Figure 25.3) is the major pathway for serine catabolism under normal conditions and the β carbon of serine is the major source of one-carbon units for folate metabolism. The reversible reaction between serine and glycine is catalyzed by serine hydroxymethyltransferase (EC 2.1.2.1) which requires pyridoxal phosphate (vitamin B_6) as coenzyme. Mammalian cells contain two isozymes of the hydroxymethyltransferase, one cytosolic and one mitochondrial, encoded by separate genes. The cytosolic isozyme appears to be the predominant one in liver. The formation of 5,10-methylene-THF plays a central role in one-carbon metabolism as its one-carbon moiety can be directed into pathways for the synthesis of methionine, purine, and pyrimidine nucleotide.

$$\text{Serine} + \text{THF} \xrightarrow{B_6} \text{glycine} + 5,10-\text{methylene}-\text{THF} + \text{H}_2\text{O} \quad (25.4)$$

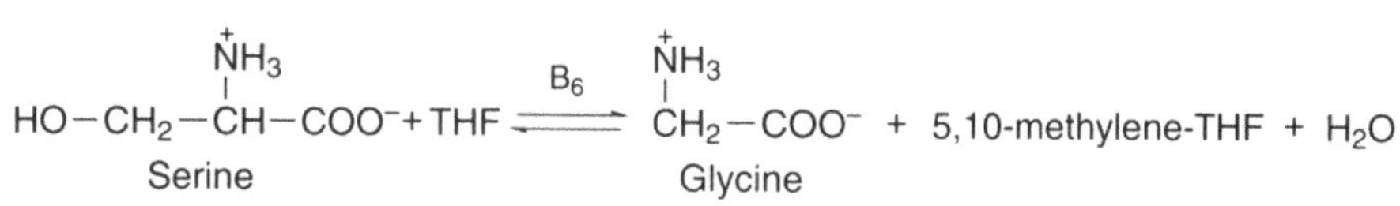

FIGURE 25.3 Interconversion of serine to glycine.

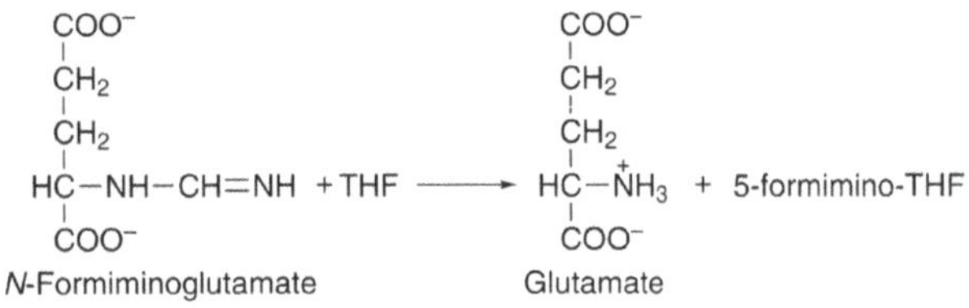

FIGURE 25.4 Formimination of THF.

5-Methyl-THF is a potent inhibitor of serine hydroxymethyltransferase, so when there is an adequate concentration of substituted folates from other sources, serine can be spared for energy-yielding metabolism or gluconeogenesis.

25.14.2.2 Synthesis of 5,10-Methenyl-THF from a Product of Histidine Catabolism

Formiminoglutamate, a product of histidine catabolism, provides an alternative source of one-carbon units to the principal source, serine. Transfer of the formimino group of formiminoglutamate to THF (reaction 25.5) (Figure 25.4) and removal of the =NH group by the bifunctional enzyme 5-formimino-THF cyclodeaminase/transferase yields 5,10-methenyl-THF (reaction 25.6).

$$N\text{-formiminoglutamate} + \text{THF} \rightarrow \text{glutamate} + \text{5-formimino-THF} \quad (25.5)$$

$$\text{5-Formimino-THF} + 2H^+ \rightarrow \text{5,10-methenyl-THF} + NH_4^+ \quad (25.6)$$

25.14.2.3 Synthesis of Deoxythymidine Monophosphate

Deoxythymidine monophosphate (dTMP) is the nucleotide of the pyrimidine thymine and is one of the four nucleotide constituents of DNA. (Note: dTMP is commonly referred to as thymidylate.) The transfer of a one-carbon unit from deoxyuridine monophosphate (dUMP) to dTMP (reaction 25.7) (Figure 25.5) is required for DNA synthesis. The enzyme responsible for the methylation, thymidylate synthase (EC 2.1.1.45), is rate limiting in the synthesis of DNA by the action of DNA polymerase. The transferred methylene group (CH_2) is reduced to a methyl group (CH_3), the reducing component being supplied by the reduced pteridine ring of 5,10-methylene-THF, which is converted to DHF. The DHF is then rapidly reduced to THF by dihydrofolate reductase, after which 5,10-methylene-THF can be regenerated through reaction 25.4. The pathway for the synthesis of dTMP is shown in Figure 25.6.

$$\text{dUMP} + \text{5,10-methylene-THF} \rightarrow \text{dTMP} + \text{DHF} \quad (25.7)$$

dTMP is not an immediate precursor for DNA synthesis. The nucleotide must first be converted to deoxythymidine triphosphate (dTTP) by two consecutive reactions with ATP. The high-energy bonds of the β and γ terminal phosphates of dTTP and other nucleotide triphosphates provide the energy required to form each phosphodiester bond of the growing strand of complementary DNA. The loss of the pyrophosphate results in the nucleotide monophosphates being incorporated into the new polynucleotide chain.

25.14.2.4 Purine Biosynthesis

Adenine and guanine are the two common purine base constituents of DNA and RNA. In the second step of purine biosynthesis, glycine reacts with 5-phosphoribosyl-1-amine to produce glycinamide ribonucleotide (GAR). GAR then reacts with 10-formyl-THF to produce formylglycinamide ribonucleotide (FGAR). The transfer of the formyl group from the folate coenzyme to GAR is catalyzed by GAR transformylase (EC 2.1.2.2). This reaction (reaction 25.8) (Figure 25.7) gives rise to the C-8 position of the eventual purine ring.

$$\text{glycinamide ribonucleotide (GAR)} + \text{10-formyl-THF} \rightarrow \text{formylglycinamide ribonucleotide (FGAR)} + \text{THF} \quad (25.8)$$

A similar folate-dependent reaction (reaction 25.9) (Figure 25.8) is involved in the penultimate (ninth) step of purine biosynthesis, giving rise to the C-2 position of the nearly formed purine ring. In this reaction, the transference of formate to 5-amino-4-imidazole carboxamide ribonucleotide (AICAR) to produce FAICAR is catalyzed by AICAR transformylase (EC 2.1.2.3).

$$\text{5-amino-4-imidazole carboxamide ribonucleotide (AICAR)} + \text{10-formyl-THF} \rightarrow \text{formyl-5-amino-4-imidazole carboxamide ribonucleotide (FAICAR)} + \text{THF} \quad (25.9)$$

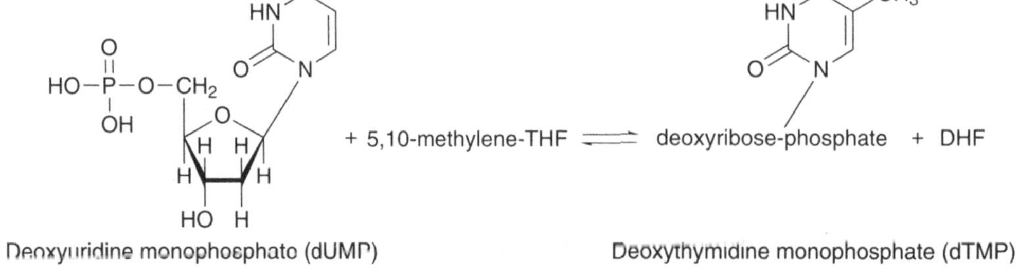

FIGURE 25.5 Methylation of deoxyuridine monophosphate (dUMP) to deoxythymidine monophosphate (dTMP).

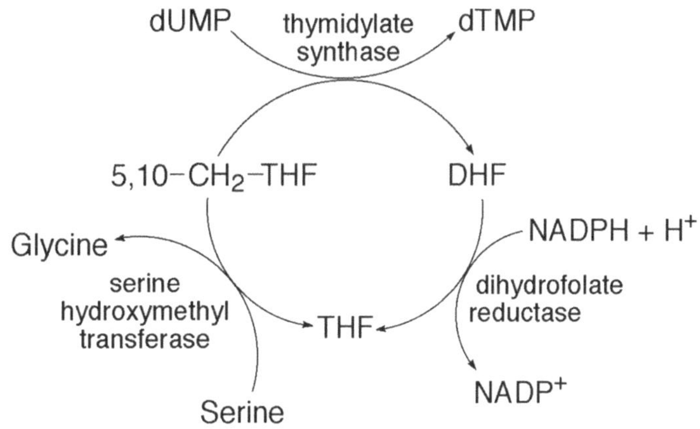

FIGURE 25.6 Metabolic pathway for the synthesis of deoxythymidine monophosphate.

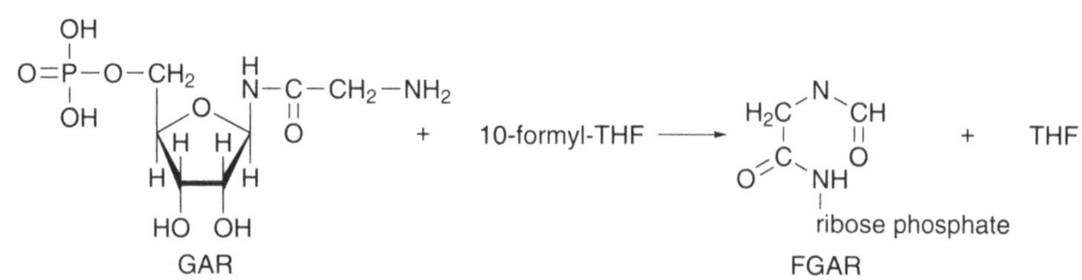

FIGURE 25.7 Conversion of glycinamide ribonucleotide (GAR) to formylglycinamide ribonucleotride (FGAR).

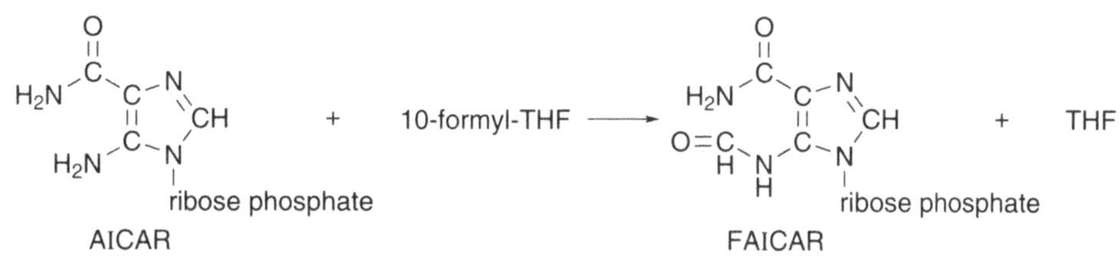

FIGURE 25.8 Conversion of AICAR to FAICAR.

The 10-formyl-THF used in the biosynthesis of purines can be produced by reactions catalyzed by a trifunctional enzyme known as C_1-THF synthase (Shane, 1989). This enzyme has two domains in its protein structure, each containing a folate-binding site. One domain contains two enzyme activities: (1) 5,10-methylene-THF dehydrogenase (EC 1.5.1.5), which converts 5,10-methylene-THF to 5,10-methenyl-THF (reaction 25.10) and (2) 5,10-methenyl-THF cyclohydrolase (EC 3.5.4.9), which converts 5-10-methenyl-THF to 10-formyl-THF (reaction 25.11). The second domain contains the third enzyme activity, namely 10-formyl-THF synthetase (EC 6.3.4.3) which catalyzes the conversion of formate to 10-formyl-THF (reaction 25.12).

$$\text{5,10-methylene-THF} + NADP^+ \rightarrow \text{5,10-methenyl-THF} + NADPH \quad (25.10)$$

$$\text{5,10-methenyl-THF} + H_2O \rightarrow \text{10-formyl-THF} + H^+ \quad (25.11)$$

$$THF + HCOOH + ATP \rightarrow \text{10-formyl-THF} + ADP + Pi \quad (25.12)$$

Formate is derived metabolically from the indole ring of tryptophan. Under normal physiological conditions, formate constitutes only a minor source of one-carbon units for folate metabolism in mammalian tissues.

25.14.2.5 Disposal of Excess One-Carbon Units

Excess one-carbon units formed during the catabolism of histidine, tryptophan, glycine, serine, or formate can be eliminated by reaction 25.13, which is catalyzed by 10-formyl-THF dehydrogenase (EC 1.5.1.6). This enzyme is inhibited by the product THF and regulated by the ratio of 10-formyl-THF: free THF in the tissue.

$$\text{10-formyl-THF} + NADP^+ + H_2O \rightarrow THF + NADPH + H^+ + CO_2 \quad (25.13)$$

25.14.2.6 Choline Catabolism

Choline catabolism takes place exclusively in liver mitochondria (Wagner, 1996). Choline first undergoes a two-step oxidation to betaine (trimethylglycine), followed by the transfer of one of the methyl groups of betaine to homocysteine to generate methionine and dimethylglycine. The dimethylglycine undergoes an oxidative demethylation to sarcosine (*N*-methylglycine) which in turn is similarly converted to glycine. Reactions 25.14 and 25.15 (Figure 25.9) are catalyzed respectively by the flavoproteins dimethylglycine dehydrogenase (EC 1.5.99.2) and sarcosine dehydrogenase (EC 1.5.99.1). These enzymes can function at a maximal rate in the absence of folate. However, THF effectively traps the formaldehyde produced and thus minimizes the accumulation of this potentially toxic compound in the cell (Brody, 1991).

$$\text{N,N-dimethylglycine} + \text{THF} + \text{FAD} \rightarrow \text{sarcosine} + \text{5,10-methylene-THF} + \text{FADH}_2 \quad (25.14)$$

$$\text{sarcosine} + \text{THF} + \text{FAD} \rightarrow \text{glycine} + \text{5,10-methylene-THF} + \text{FADH}_2 \quad (25.15)$$

Cytoplasmic sarcosine, if transported into the mitochondria, is an additional potential source of one-carbon units for mitochondrial metabolism. Cytoplasmic sarcosine is formed by methylation of glycine by *S*-adenosylmethionine (SAM) in a reaction catalyzed by glycine *N*-methyltransferase (Shane, 1989).

25.14.2.7 Serine and Glycine Metabolism

Mitochondria contain a serine hydroxymethyltransferase isozyme for the reversible interconversion of serine and glycine (see reaction 25.4). The major pathway of glycine metabolism is via the oxidative decarboxylation of glycine catalyzed by a multisubunit enzyme complex known as the glycine cleavage system (Okamura-Ikeda *et al.*, 1987). The result of glycine catabolism is that the α-carbon of glycine is converted to a one-carbon compound which enters the intracellular folate pool.

The glycine cleavage system comprises three enzymes, designated P, T, and L proteins, and a carrier protein, known as the H-protein. The P-protein, a vitamin B_6-dependent enzyme, is commonly referred to as glycine decarboxylase. The H-protein contains covalently bound lipoic acid, which exists in either the oxidized or reduced form (Figure 25.10). The four proteins participate in reactions 25.16 to 25.18 in the following manner. In reaction 25.16, glycine forms a Schiff base with pyridoxal phosphate (vitamin B_6), the coenzyme of P-protein. The disulfide (oxidized) form of lipoate on the H-protein then combines with the Schiff base, liberating CO_2 from the glycine carboxyl group. In reaction 25.17, the H-protein-intermediate complex serves as a substrate for T-protein which, in the presence of THF, catalyzes the decomposition of the complex to H-protein, ammonia, and formaldehyde. The formaldehyde is transferred to THF forming 5,10-methylene-THF. In reaction 25.18, the disulfhydryl lipoate on H-protein is reoxidized by the NAD$^+$-dependent L-protein.

$$\text{CH}_2(\text{NH}_2)\text{COOH} + \text{H}-\text{protein}-\text{lipoylS}_2 \xrightarrow{\text{P-protein (B}_6)} \text{H}-\text{protein}-\text{lipoyl}(\text{SH})\text{S}-\text{CH}_2\text{NH}_2 + \text{CO}_2 \quad (25.16)$$

$$\text{H}-\text{protein}-\text{lipoyl}(\text{SH})\text{S}-\text{CH}_2\text{NH}_2 + \text{THF} \xrightarrow{\text{T-protein}} \text{5,10}-\text{methylene}-\text{THF} + \text{H}-\text{protein}-\text{lipoyl}(\text{SH})_2 + \text{NH}_3 \quad (25.17)$$

$$\text{H}-\text{protein}-\text{lipoyl}(\text{SH})_2 + \text{NAD}^+ \xrightarrow{\text{L-protein}} \text{H}-\text{protein}-\text{lipoylS}_2 + \text{NADH} + \text{H}^+ \quad (25.18)$$

The overall reaction is:

$$\text{glycine} + \text{THF} + \text{NAD}^+ \rightarrow \text{5,10-methylene-THF} + \text{NH}_3 + \text{CO}_2 + \text{NADH} + \text{H}^+ \quad (25.19)$$

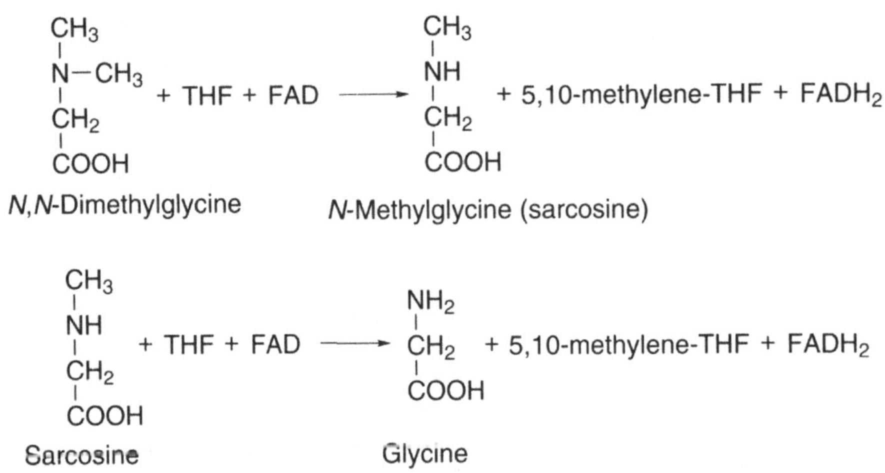

FIGURE 25.9 Two-step conversion of dimethylglycine to glycine.

The glycine cleavage system, although reversible *in vitro*, is not utilized for glycine synthesis *in vivo*. However, coupling of the mitochondrial serine hydroxymethyltransferase and glycine cleavage system provides a mechanism for the synthesis of serine from glycine. One molecule of serine can arise by reversal of the transferase-catalyzed reaction. The 5,10-methylene-THF required for the reversed reaction arises from oxidative decarboxylation of an additional molecule of glycine via the cleavage system (Shane, 1989).

25.15 HOMOCYSTEINE-RELATED OCCLUSIVE ARTERIAL AND THROMBOTIC DISEASES

25.15.1 Cellular Metabolism of Homocysteine

Homocysteine is a sulfur-containing amino acid formed as an intermediate during the conversion of methionine to cysteine. It is not a constituent of protein and hence is not present in the diet. If the balance between synthesis and turnover of homocysteine is interrupted, homocysteine and its derivatives begin to accumulate in cells. The cell is able to export potentially toxic homocysteine, leading to the appearance of excess homocysteine in the blood (hyperhomocysteinemia).

In humans, the metabolism of homocysteine involves two pathways in which homocysteine undergoes (1) remethylation and (2) transsulfuration (Selhub & Miller, 1992). The overall metabolism of homocysteine is shown in Figure 25.11. Individual reactions are shown in Figure 25.12.

In remethylation, homocysteine acquires a methyl group from 5-methyl-THF or from betaine to form methionine. Remethylation by either route is essential for maintaining adequate intracellular pools of methionine as well as for maintaining concentrations of homocysteine at nontoxic concentrations.

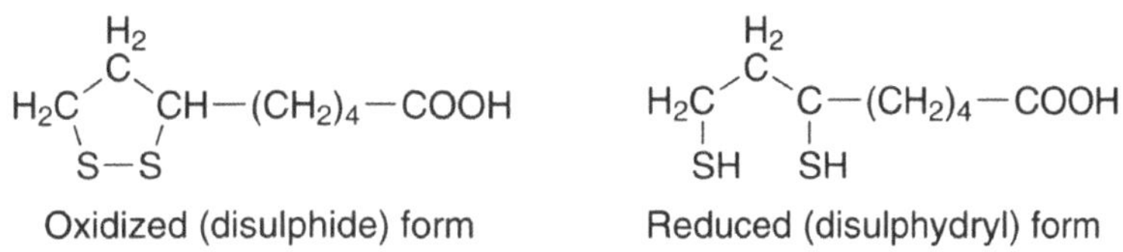

FIGURE 25.10 Structures of oxidized and reduced lipoic acid.

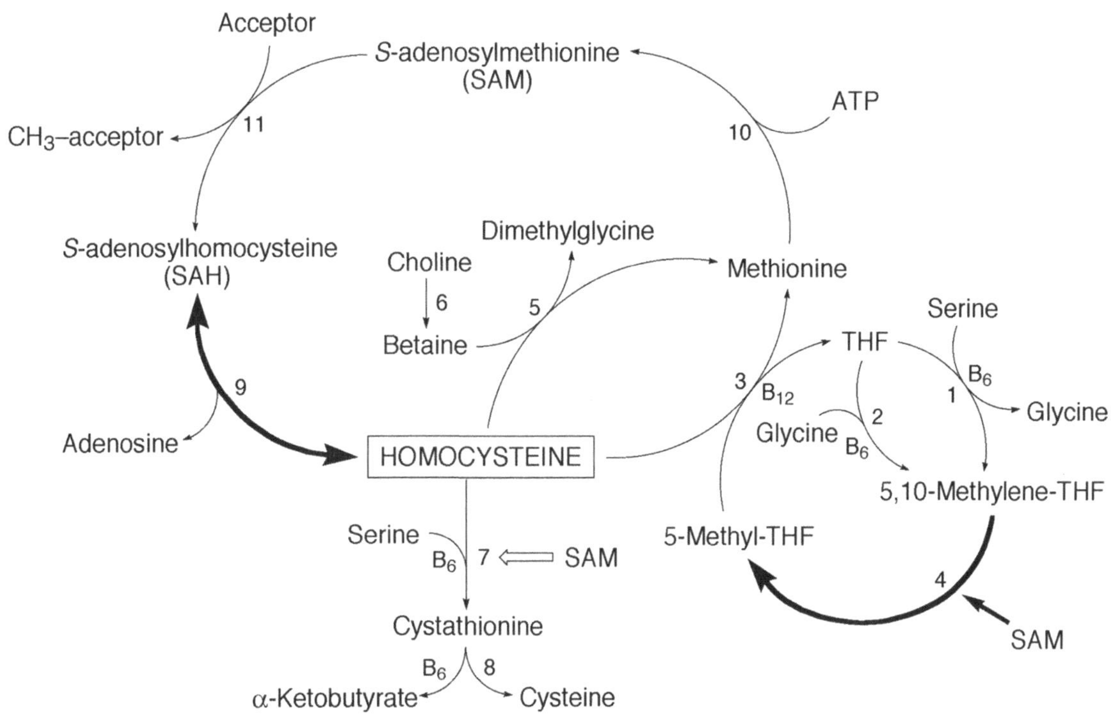

FIGURE 25.11 Pathways of homocysteine metabolism showing involvement of vitamins B_6 and B_{12} as coenzymes. Enzyme reactions that are regulated by *S*-adenosylmethionine (SAM) are indicated by large arrows; closed arrow indicates inhibition, open arrow indicates activation. Enzymes: (1) serine hydroxymethyltransferase, (2) glycine decarboxylase (P-protein of the glycine cleavage system), (3) methionine synthase, (4) 5,10-methylene-THF reductase (irreversible), (5) betaine:homocysteine methyltransferase, (6) choline dehydrogenase, (7) cystathionine β-synthase, (8) cystathione γ-lyase, (9) *S*-adenosylhomocysteine hydrolase (reversible) (10) methionine-adenosyl transferase, (11) methyltransferases. Hydrolysis of *S*-adenosylhomocysteine (SAH) is readily reversible, but proceeds in the hydolysis direction upon efficient removal of the hydrolysis products. The reduction of 5,10-methylene-THF is irreversible under physiological conditions.

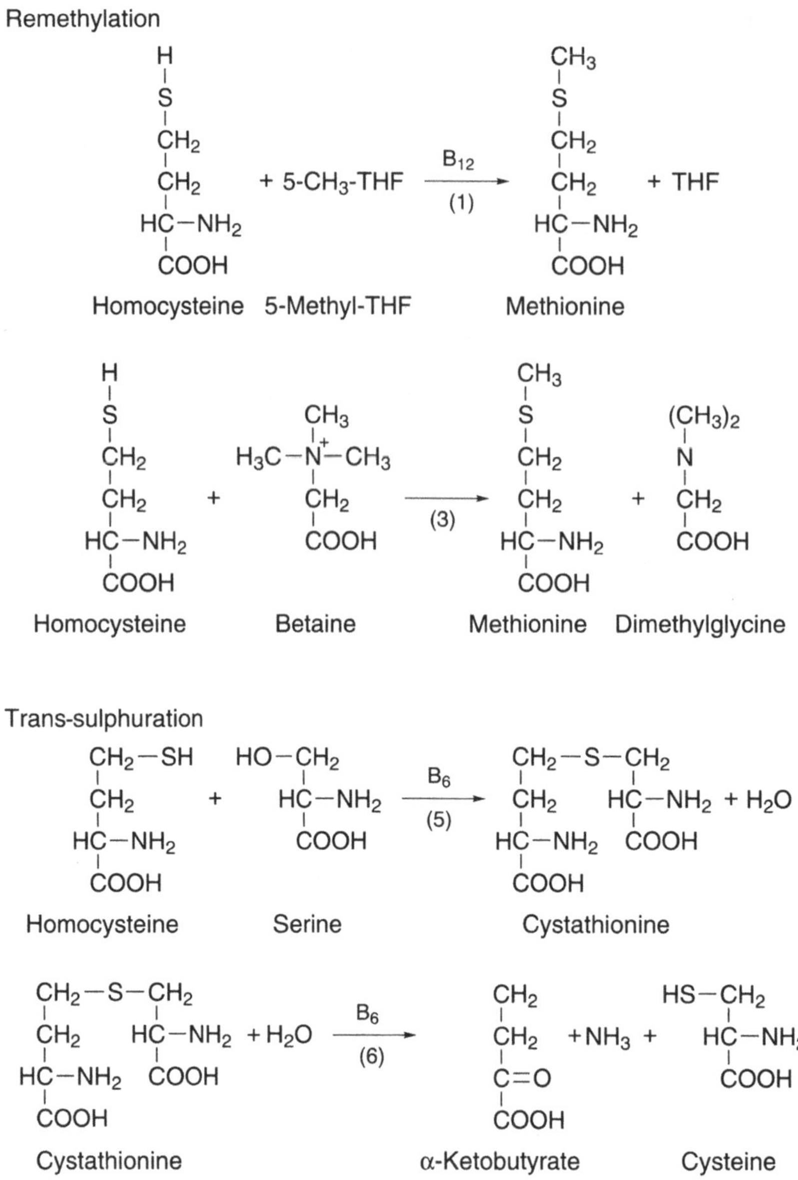

FIGURE 25.12 Individual reactions involved in homocysteine metabolism. Numbers in parentheses correspond to the enzymes numbered in Figure 25.10.

A considerable proportion of methionine is activated by ATP to form *S*-adenosylmethionine. SAM serves primarily as a methyl donor to a variety of acceptors, including DNA, neurotransmitters, phospholipids, and hormones. After transfer of the methyl group, SAM is converted to *S*-adenosylhomocysteine (SAH) by the action of various methyltransferases. Hydrolysis of SAH regenerates homocysteine, which can be utilized to start a new cycle of methyl-group transfer. Hydrolysis of SAH is readily reversible with equilibrium dynamics that strongly favor SAH synthesis rather than hydrolysis. In fact, the only reason that this reaction proceeds in the hydrolytic direction is efficient removal of the hydrolysis products. Normally, multiple routes for removal of adenosine and homocysteine ensure that SAH does not accumulate (Yi *et al.*, 2000).

Yi *et al.* (2000) reported that plasma SAH levels in healthy young women increased linearly with mild elevations in homocysteine levels. The increase in plasma SAH was also associated with a progressive increase in lymphocyte DNA hypomethylation. These data support an indirect mechanism for homocysteine pathogenicity secondary to SAH-mediated inhibition of the DNA methyltransferase.

The remethylation of homocysteine to methionine is catalyzed by methionine synthase (EC 2.1.1.13) which has an enzyme-bound methylcobalamin (vitamin B_{12}) co-factor. The co-factor acts as an intermediate methyl-group carrier between 5-methyl-THF and homocysteine. Over time, the essential methylcob(I)alamin co-factor becomes oxidized to methylcob(II)alamin, rendering methyl synthase inactive. An auxiliary enzyme, methionine synthase reductase, catalyzes the regeneration of functional cobalamin co-factor by reductive methylation via a reaction in which SAM is employed as a methyl donor.

The methyl group of 5-methyl-THF is synthesized *de novo* when a carbon unit is transferred from a carbon source, such as serine, to THF. The product, 5,10-methylene-THF, is subsequently reduced to 5-methyl-THF, which is a poor substrate for polyglutamylation, Since methionine synthase is unique in converting 5-methyl-THF to THF, which is readily polyglutamated, the enzyme is rate-limiting for the cellular accumulation of polyglutamyl folates.

The reaction with betaine uses preformed methyl groups because betaine is derived from choline, which in part is obtained from the diet and in part is synthesized through successive methylations of phosphatidylethanolamine.

The reduction of 5,10-methylene-THF to 5-methyl-THF (reaction 25.20) is catalyzed by 5,10-methylenetetrahydrofolate reductase (MTHFR) (EC 1.1.1.68) and is irreversible under physiological conditions. A non-covalently bound flavin adenine nucleotide (FAD) coenzyme accepts reducing equivalents from NADPH and transfers them to 5,10-methylene-THF.

$$\text{5,10-methylene-THF} + FADH_2 + NADPH + H^+ \rightarrow \text{5-methyl-THF} + FAD + NADP^+ \quad (25.20)$$

Reaction 25.20 is the only reaction capable of producing 5-methyl-THF in the body. The 5-methyl-THF is used exclusively for the regeneration of methionine from homocysteine.

In the transsulfuration pathway, homocysteine condenses with serine to form cystathionine in a vitamin B_6-dependent irreversible reaction. Cystathionine is then hydrolyzed by a second vitamin B_6-dependent enzyme to form cysteine and α-ketobutyrate. Excess cysteine is oxidized to taurine and eventually to inorganic sulfates. Thus, in addition to the synthesis of cysteine, this transsulfuration pathway effectively catabolizes potentially toxic homocysteine, which is not required for methyl group transfer.

SAM is both an allosteric inhibitor of MTHFR and an activator of cystathionine β-synthase. The ability of SAM to act as an enzyme effector provides a means by which remethylation and transsulfuration can be coordinated. When cellular SAM concentration is low, the synthesis of 5-methyl-THF will proceed uninhibited, whereas cystathionine synthesis will be suppressed. This will result in the conservation of homocysteine for methionine synthesis. Conversely, when SAM concentration is high, inhibition of 5-methyl-THF synthesis is accompanied by the diversion of homocysteine through the transsulfuration pathway because of stimulated cystathionine synthesis.

Increased synthesis of deoxythymidine monophosphate (dTMP) in actively proliferating cells elevates DHF levels (see reaction 25.7). Inhibition of the allosteric MTHFR by DHF provides a regulatory mechanism for ensuring that priority is given to DNA synthesis over methionine formation in proliferating tissues (Matthews & Baugh, 1980).

25.15.2 Hyperhomocysteinemia

In plasma, 70–80% of homocysteine is bound to plasma proteins, chiefly albumin; only about 1% circulates as free homocysteine. The remaining 20–30% circulates as homocysteine disulfide (homocystine) or as the mixed disulfide, homocysteine–cysteine. Plasma homocysteine assays measure total homocysteine, which is the sum of the homocysteine moieties present in all of the above forms. Because variable changes in plasma homocysteine concentration have been observed postprandially, it is customary to obtain measurements in the fasting state. Normal levels of fasting plasma homocysteine are considered to be between 5 and 15 μmol/L. Higher fasting values are classified arbitrarily as moderate (16–30), intermediate (31–100), and severe (>100 μmol L^{-1}) hyperhomocysteinemia. The methionine loading test has been used to accentuate abnormalities of the homocysteine metabolic pathways. The test measures fasting plasma homocysteine before and 2 hours after an oral dose of methionine (100 mg per kg body weight). An elevated post-loading homocysteine level indicates an abnormality.

Hyperhomocysteinemia can result from inherited defects in enzymes necessary for either transsulfuration or remethylation and from acquired deficiencies in vitamin coenzymes. Renal insufficiency can also lead to hyperhomocysteinemia. Subclinical folate deficiency is commonly associated with hyperhomocysteinemia, presumably because of decreased remethylation of homocysteine. Kang *et al.* (1987) found elevated total homocysteine levels in 84% of subjects with subnormal folate levels. The mean homocysteine level in the low-folate subjects was about fourfold greater than the mean level in the control subjects.

An association between mild hyperhomocysteinemia and increased risk of occlusive vascular disease in the coronary, cerebral, and peripheral arteries has been demonstrated in case-control (Selhub *et al.*, 1995; European Concerted Action Project, 1997) and prospective (Stampfer *et al.*, 1992; Arnesen *et al.*, 1995; Perry *et al.*, 1995) studies. Plasma homocysteine concentration is a strong predictor of mortality in patients with angiographically confirmed coronary artery disease (Nygard *et al.*, 1997). Whether hyperhomocysteinemia is a causal risk factor for the disease or simply a marker of another prothrombotic risk factor(s) is debatable (Kuller & Evans, 1998).

Up to 30% of patients with coronary artery disease had homocysteine elevations that were 10–50% greater than the level observed among normal subjects (Clarke *et al.*, 1991). Subjects with hyperhomocysteinemia have a two- to threefold increase in risk of developing cardiovascular disease or venous thrombosis (den Heijer *et al.*, 1998). *In vitro* studies have shown that high concentrations of homocysteine can promote a prothrombotic state at the luminal surface of the blood vessel (Lentz, 1998). An association between impaired endothelium-dependent vasodilation and hyperhomocysteinemia was demonstrated in children with homozygous homocystinuria (Celermajer *et al.*, 1993), in monkeys fed a methionine-enriched diet (Lentz *et al.*,

1996), in methionine-loaded healthy humans (Bellamy *et al.*, 1998; Chambers *et al.*, 1998), and in non-induced hyperhomocysteinemic healthy middle-aged (Woo *et al.*, 1997) and elderly (Tawakol *et al.*, 1997) humans. In healthy human subjects, even physiological increments in plasma homocysteine following oral administration of methionine or an animal protein meal impaired endothelium-dependent vasodilation (Chambers *et al.*, 1999b). Plasma homocysteine concentration can be decreased by dietary supplementation with folic acid, which suggests that hyperhomocysteinemia may be a treatable risk factor for vascular disease.

25.15.3 Effects of Homocysteine on Vascular Function

The following *in vivo* observations provide insight into potential mechanisms of vascular dysfunction in moderate hyperhomocysteinemia. Infusion of homocystine caused patchy desquamation of arterial endothelial cells, proliferation of arterial smooth muscle cells, and intimal thickening in baboons (Harker *et al.*, 1976). The endothelial injury, by exposing the subcellular matrix, could lead to platelet aggregation. In rats, acute methionine-induced hyperhomocysteinemia enhanced platelet aggregation in response to thrombin and ADP as well as the thrombin-induced platelet synthesis of thromboxane. It also stimulated the basal and lipopolysaccharide-induced tissue factor activity of macrophages (Durand *et al.*, 1997). In healthy humans, coagulation and circulating adhesion molecules significantly increased after methionine loading, but not after placebo or methionine ingestion with antioxidant vitamins C and E (Nappo *et al.*, 1999). Hyperhomocysteinemia in minipigs caused deterioration of elastic material of the arterial wall (Charpiot *et al.*, 1998).

Homocysteine can be viewed as an early risk factor for atherosclerosis. When added to plasma it undergoes autooxidation, forming homocystine, homocysteine-mixed disulfides, and homocysteine thiolactone. During oxidation of the sulfhydryl group, superoxide and hydrogen peroxide are generated. Both of these oxygen-derived molecules give rise to the harmful hydroxyl radical, the former by reacting with nitric oxide. The hydroxyl radical initiates lipid peroxidation at the endothelial cell surface, thus accounting for the cytotoxic effects of superoxide and hydrogen peroxide. Upchurch *et al.* (1997) showed that, compared with control cells, cultured aortic endothelial cells treated with homocysteine showed a significant reduction in glutathione peroxidase activity. Thus, in addition to increasing the generation of hydrogen peroxide, homocysteine may selectively impair the endothelial cell's ability to detoxify hydrogen peroxide.

Acute methionine-induced hyperhomocysteinemia impaired vascular responses to L-arginine, the natural precursor of nitric oxide, in healthy humans (Nappo *et al.*, 1999). The impairment of endothelial function was prevented by a 1 g (Chambers *et al.*, 1999a) or 2 g (Kanani *et al.*, 1999) daily supplement of ascorbic acid, suggesting that the effects of hyperhomocysteinemia are produced through an increase in oxidative stress. By scavenging reactive oxygen species, ascorbic acid may prevent the formation of the cytotoxic hydroxyl radical. *In vivo* evidence for oxidative stress during hyperhomocysteinemia was the demonstration of increased lipid peroxidation after methionine loading in humans (Domagala *et al.*, 1997) and rats (Durand *et al.*, 1997).

25.16 FOLATE STATUS AND NEURAL TUBE DEFECTS

25.16.1 Introduction

Neural tube defects (NTDs) comprise a range of congenital malformations resulting from a failure of the neural tube to close during embryonic development, late in the fourth week after conception. NTDs include spina bifida, in which the developing neural tube fails to close posteriorly, causing the herniation of meninges (meningocele), neural tissue (myelocele), or both (myelomeningocele). Spina bifida causes varying degrees of motor paralysis as well as impaired bowel, bladder, and sexual function. Failure of the anterior cranial portion of the neural tube to close leads to anencephaly, a fatal condition in which the brain is exposed and incompletely developed, and encephalocele, in which meninges and neural tissue herniate through a defect in the skull.

NTDs are among the most common of severe birth defects, affecting about 4,000 pregnancies in the United States each year. Of these pregnancies, approximately 2,500 result in babies born with spina bifida and anencephaly; the remaining NTD-affected pregnancies are electively terminated (Mulinare, 1995). In many countries the prevalence of NTD is approximately 1 in 1,000 births. Over 95% of NTD-affected pregnancies occur in women who have not previously had an affected pregnancy. Women who have had a previous NTD-affected pregnancy are 10–20 times more likely to have a subsequent affected pregnancy than those who have not. Some rare cases of NTD (e.g. Meckel-Gruber syndrome) have a purely genetic basis. However, the majority of cases have an underlying genetic basis that can be influenced by a variety of environmental (nutritional, geographical, and socioeconomic) factors (Scott *et al.*, 1990). Polymorphisms involved in folate and homocysteine metabolism are genetic determinants of NTD.

25.16.2 Nutritional Studies

An association exists between low maternal folate status, hyperhomocysteinemia, and increased incidence of NTDs (Eskes, 1998). Because closure of the embryonic neural tube takes place very early in pregnancy, at a time when many women will be unaware that they are pregnant, any intervention to prevent NTDs must be initiated around the periconceptional period – an 8-week period 4 weeks before and after conception (Mulinare *et al.*, 1988). Beginning in the 1980s, numerous studies have investigated the effect of periconceptional folic acid supplementation on recurrence or occurrence risk for NTDs (Institute of Medicine, 1998). Two representative studies are those reported by the Vitamin Study Research Group of the Medical Research Council in the United Kingdom (MRC, 1991) and the National Institute of Hygiene, a World Health Organization Collaborating Centre for the Community Control of Hereditary Diseases in Hungary (Czeizel & Dudás, 1992).

The two representative studies confirmed that folic acid reduces both the occurrence and recurrence of NTDs. This important finding led to a number of recommendations regarding folic acid supplementation and the prevention of NTDs (summarized by Picciano *et al.*, 1994). The 1998 US recommended dietary allowance (RDA) for folate during pregnancy is 600 μg per day of dietary folate equivalents (Institute of Medicine, 1998). In order to achieve this recommended intake, all women of childbearing age are advised to consume 400 μg of folic acid daily from fortified foods, supplements, or both in addition to consuming food folate from a varied diet. To boost the contribution of food folate to total folate intake, the Food and Drug Administration (FDA) in the United States decreed in March 1996 that folic acid must be added to all enriched cereal grains (e.g. enriched bread, pasta, flour, breakfast cereals, and rice) at 1.4 mg kg^{-1} of grain, with compliance mandatory from January 1, 1998. Cereal products were chosen for folate fortification because they are widely consumed in the United States and because they are suitable vehicles for this purpose. It was estimated that this public health measure would increase the folate intake of most US women by about 80 to 100 μg per day. This estimate was based on the assumption that the fortified products would actually contain the amount of folic acid required by the FDA's regulation. However, it is standard manufacturing practice to add more than the stipulated amount of vitamin to allow for potential vitamin loss during processing and throughout the shelf-life. In practice, the increased intake of folate resulting from fortification is approximately twice as high as projected (Choumenkovitch *et al.*, 2002). Honein *et al.* (2001) reported a 19% decrease in the birth prevalence of NTDs following folic acid fortification of the US food supply.

In the United Kingdom there is no current policy for mandatory folate fortification. Recommendations by the Department of Health (1992) suggest that the daily intake of an extra 400 μg of folate can be achieved through the ample consumption of naturally folate-rich foods. However, Cuskelly *et al.* (1996) reported that, compared with supplements and fortified food, consumption of extra folate as natural food folate is ineffective at increasing folate status. This is because the polyglutamyl folate present in natural food sources is less bioavailable than the monoglutamyl folic acid present in supplements and fortified foods. Even though protection can be achieved through the use of supplements, there is no guarantee that women will actually take the supplements. An audit of compliance with the UK Department of Health recommendation for routine folate prophylaxis indicated that, at least six months after its issue, very few women had complied (Clark & Fisk, 1994).

Despite the beneficial effects of supplemental folate in reducing the incidence of NTDs, no direct link has been found to connect maternal folate deficiency with NTDs. A study of nonpregnant women who had had two NTD-affected pregnancies did not exhibit folate deficiency (erythrocyte folate levels were not depressed). However, unlike control women, there was no correlation between dietary intake of folate and both serum and erythrocyte folate levels (Wild *et al.*, 1994). It appeared that women at increased risk for NTD may have defective folate metabolism and an increased requirement for folate.

Exactly how folate therapy reduces NTD risk is not known. The teratogenic effect of folate deficiency could result from an insufficient supply of nucleic acid precursors to the rapidly dividing embryonic cells. However, the highly specific and predictable nature of the congenital defects caused by folate deficiency argues against a general limitation of nucleic acid availability. Rosenquist *et al.* (1996) showed that treatment of avian embryos with physiological concentrations of exogenous homocysteine induces NTDs that are typical of folate deficiency. This implies that the congenital defects that arise from maternal folate deficiency are a result of hyperhomocysteinemia and, conversely, that folate supplementation protects against dysmorphogenesis by lowering maternal plasma homocysteine levels. Bearing in mind that homocysteine promotes proliferation of cultured vascular smooth muscle cells (Tsai *et al.*, 1994), Rosenquist *et al.* (1996) postulated that the teratogenic effect of homocysteine is based on a growth factor-like effect, altering with this effect the expression of genes critical to the process of neural tube closure/neural crest migration.

25.17 FOLATE STATUS AND CANCER

Following an observation that folic acid causes regression of mammary tumors in mice (Leuchtenberger *et al.*, 1945), Farber *et al.* (1947) administered folate to children with leukemia: unexpectedly, the folate treatment accelerated rather than slowed progression of the cancer. Reasoning that a deficiency of folate in cancerous tissues would reduce cell proliferation by slowing DNA synthesis, Farber *et al.* (1948) subsequently showed that a folic acid antagonist, 4-aminopteroyl glutamic acid (aminopterin), produced complete remissions in children with acute leukemia. This discovery paved the way for cancer chemotherapy using a number of antifolate agents, such as methotrexate and 5-fluorouracil (Kim, 1999).

Contrary to the inhibitory effect of antifolates on established tumors, a large body of epidemiological, clinical, and experimental evidence suggests that folate deficiency in normal tissues may be a factor that predisposes them to neoplastic transformation, and that tumor development can be suppressed by folate supplementation. Experimental evidence includes reports by Cravo *et al.* (1992) and Mason and Levesque (1996). Epidemiological studies collectively suggest an association between reduced folate status (measured by either folate intake or folate blood levels) and the risk of several malignancies, including cancer of the colorectum, oropharynx, esophagus, stomach, pancreas, lungs, cervix, ovary, and breast, and neuroblastoma and leukemia.

Although severe folate deficiency in rats impairs natural killer (NK) cell-mediated cytotoxicity, a moderate deficiency of the vitamin sufficient to enhance colorectal cancer (CRC) has no such effect on NK cells. Furthermore, a modest level of folate supplementation above the basal dietary requirement (BDR) does not enhance NK-mediated cytotoxicity. Thus, NK-mediated cytotoxicity is not a likely mechanism by which folate status modulates carcinogenesis (Kim *et al.*, 2002).

Folate is essential for DNA synthesis and repair through its role in the synthesis of the purine ring and the pyrimidine nucleotide, thymidylate. Consequently, a deficiency of folate in normal tissues with rapidly replicating cells results in ineffective DNA synthesis and repair (Duthie *et al.*, 2002). Division of cells with unrepaired or misrepaired DNA damage leads to mutations. If these relate to critical genes, such as proto-oncogenes or tumor suppressor genes, cancer may result. The role of folate in the synthesis of *S*-adenosylmethionine is also relevant to cancer. SAM donates its methyl group to DNA, among other acceptors, and a deficiency of folate can lead to hypomethylation of DNA (Davis & Uthus, 2004). As DNA methylation is a mechanism for silencing transcription (Ng & Bird, 1999), hypomethylation of DNA has the potential to alter the normal control of gene expression. Hypomethylation also alters chromatin conformation, thereby allowing access of DNA-damaging agents and endonucleases, which destabilize the DNA and make it more susceptible to strand breaks (Kim *et al.*, 1997). Imbalanced DNA methylation is a common occurrence in carcinogenesis (Laird & Jaenisch, 1994).

Folate, in the form of 5,10-methylene-THF, acts as methyl donor for the enzyme thymidylate synthase, which converts deoxyuridine monophosphate to deoxythymidine monophosphate. During folate deficiency, a block in the methylation of dUMP to dTMP leads to an increase in the cellular levels of deoxyuridine triphosphate (dUTP) and misincorporation of uracil into the DNA molecule in place of thymine. Normally, deoxyuridine triphosphatase (dUTPase) hydrolyzes excess dUTP to pyrophosphate and dUMP, thereby providing more substrate for thymidylate synthase. Misincorporated uracil residues are removed from the DNA molecule by uracil-DNA glycosylase. Subsequent removal of the base-free sugar from the DNA backbone creates a single-strand break that can be sealed by DNA ligase activity. However, if dTMP is continually limited under conditions of folate deficiency, uracil misincorporation and repair may occur repeatedly in what has been termed a 'futile' repair cycle (Duthie, 1999). Lymphocytes cultured under conditions of folate deficiency have a two- to threefold increase in the level of misincorporated uracil and an increased frequency of DNA strand breaks (Duthie & Hawdon, 1998). Kim *et al.* (1997) showed that, in rats, dietary folate depletion is capable of producing DNA strand breaks and hypomethylation within a highly conserved region of the p53 tumor suppressor gene. Excessive DNA uracil content as well as increased numbers of chromosomal breaks are observed in folate-deficient humans, and both defects are reversed by folate administration (Blount *et al.*, 1997). Simultaneous removal of two closely spaced uracil residues on opposite DNA strands can result in a double-strand DNA break, which is difficult to repair. Unrepaired double-stranded DNA breaks contribute to the increased risk of cancer. Blount and colleagues estimated the incidence of two closely spaced uracils on opposite strands to be 50 times higher in folate-deficient individuals than in normal individuals.

Kim (2007) reviewed evidence of dual modulatory effects of folate on CRC development and progression in rats, depending on the timing and dose of folate intervention. Folate deficiency has an inhibitory effect on the progression of established colorectal neoplasms, whereas folic acid supplementation has a promoting effect. In contrast, folate deficiency in normal colorectal mucosa seems to predispose it to neoplastic transformation and modest levels of folic acid supplementation (four to ten times above the BDR) suppress the development of CRC. A very high supplemental dose of folic acid (20 times the BDR) did not confer additional protection and, in some animals, promoted rather than suppressed the progression of existing microscopic colonic neoplasia to macroscopic neoplasia.

The above observations in rats confirm that the timing of folate intervention is critical in providing an effective and safe chemopreventive effect on CRC. Moderate folate intake may decrease risk of early lesions of colorectal neoplasia, whereas high intakes of folate after the neoplastic lesion develops may increase risk of CRC. The latter effect has raised concern that fortification of flour and uncooked cereal grains with synthetic folic acid may have been wholly or partly responsible for the observed increase in CRC in the mid-1990s (Mason *et al.*, 2007). From the analysis of data from the Nurses' Health Study and Health Professionals Follow-Up Study, Lee *et al.* (2011) did not find any clear evidence that intake of folate, including folic acid added to foods or used as supplements, increased the risk of developing CRC. Their findings suggested that a benefit of folate supplementation is limited to the early preadenoma stages of CRC.

25.17.1 Chemotherapeutic Potential of Methotrexate

The folate antagonist methotrexate has a chemotherapeutic potential in the treatment of malignant disease. Methotrexate binds to and inhibits the cytosolic enzyme dihydrofolate reductase, thereby interfering with the synthesis of reduced folate coenzymes that are necessary for the synthesis of purines and a pyrimidine. Unbound intracellular methotrexate concentrations in excess of the maximum binding by dihydrofolate reductase is necessary to completely inhibit the enzyme activity. Methotrexate is transported by the folate receptor, but its affinity for the receptor is relatively low. However, malignant cells express very high levels of folate receptor compared to the corresponding normal tissue (Ross *et al.*, 1994), and this overexpression allows methotrexate to enter the cells and inhibit their proliferation (Kane *et al.*, 1986).

25.18 EXPERIMENTAL FOLATE DEFICIENCY IN ANIMALS

Folate deficiency has been produced experimentally in many animal species. The various species react differently to a folate deficiency, hence symptoms of a different nature may develop. The following deficiency signs have been observed:

- **General condition and growth.** General weakness, diminished appetite, and retarded growth.
- **Digestive tract.** Gastro-intestinal disturbances leading to diarrhea.
- **Skin.** Rough plumage, dermatitis, and feather depigmentation in chicks.
- **Bones.** Perosis (leg deformity) in chicks.
- **Nervous system.** Cervical paralysis in turkeys.
- **Blood and vascular system.** Macrocytic anemia, leukopenia, and thrombopenia (abnormally low platelet count).
- **Reproductive system.** In poultry, there is reduced egg production and hatchability of the eggs, embryonic mortality, and malformation of the surviving embryos.

25.19 FOLATE DEFICIENCY IN HUMANS

25.19.1 General Features of Folate Deficiency

A deficiency of folate leads primarily to metabolic failures in systems in which folate is involved as a coenzyme. Consequently, the compounds concerned (amino acids, and purine and pyrimidine bases) cannot be produced in sufficient concentrations. It is therefore the metabolic deficiency of these compounds that causes the subclinical or clinical symptoms of folate deficiency. The tissues affected are those with a rapid growth or regeneration rate of the cells, and with a high protein turnover rate, such as epithelia, epidermis, bone marrow, and the tissues of organs.

Folate deficiency can result, in the absence of disease, from reduced ingestion or absorption, or from increased utilization. Dietary folate deficiency is common among people who, for various reasons, eat little fruit or fresh vegetables. Absorption is impaired in alcoholics. There is an increased utilization of folate during pregnancy owing to the need to transfer an extra 100–300 μg of folate per day to the fetus (Beatty & Wickramasinghe, 1993). Experimental folate deficiency is difficult to produce under normal circumstances, but the study of patients suffering from malabsorption problems such as tropical sprue (atrophy of the jejenum) or the use of folate antagonists has yielded useful clinical information.

Folate deficiency affects DNA synthesis and repair, resulting in increases in DNA strand breakage, chromosomal aberrations, and micronuclei formation (an indicator of chromosomal damage). Everson *et al.* (1988) reported an exceptionally high frequency of micronucleated red blood cells in a patient with mild folate deficiency (1.9 ng/ml plasma folate). Supplementation with folic acid reduced the level of micronuclei to normal after only 8 days.

Folate deficiency is associated with increased risk of occlusive vascular disease, neural tube defects, and cancer. The classic manifestation of folate deficiency is megaloblastic anemia.

25.19.2 Megaloblastic Anemia

A deficiency of folate gives rise to a type of anemia known as megaloblastic anemia. A clinically indistinguishable anemia is also produced by vitamin B_{12} deficiency but, because it is accompanied by neurological damage, it is referred to as pernicious anemia. Both types of anemia are the result of abnormal nuclear maturation caused by impaired DNA synthesis. The impaired DNA synthesis is presumed to be attributable to reduced intracellular levels of polyglutamyl 5,10-methylene-THF. As shown in reaction 25.7, this important folate is involved in the formation of deoxythymidine monophosphate, one of the two pyrimidine nucleotide constituents of DNA. The defect in DNA synthesis leads to a variety of secondary disturbances which result in the premature death of many hemopoietic cells in the bone marrow.

Megaloblastic anemia caused by folate deficiency manifests as megaloblastosis of the bone marrow and macrocytosis of the circulating erythrocytes. Examination of the bone marrow is of great diagnostic importance. The erythrocyte precursor cells (erythroblasts) in the bone marrow fail to proliferate rapidly and exist as gigantic cells called megaloblasts at all stages of maturation. It is the existence of such cells that gives rise to the term 'megaloblastic anemia.' The increased size of megaloblasts is apparent both in the cytoplasm and in the nucleus. The nuclei contain smaller quantities of condensed chromatin than the nuclei of normoblasts of similar maturity and thus have an open, sieve-like appearance.

The circulating erythrocytes which are derived from mature megaloblasts are also abnormally large and are referred to as macrocytes. The mean corpuscular volume of macrocytes ranges from 100 to 160 μm^3 compared with 90 to 95 μm^3 for normal erythrocytes. The macrocytes are generally oval in shape (erythrocytes are biconcave discs) but some fragmented and irregularly shaped cells are also present. There is a reduction in red cell count and sometimes incredibly low values are found. The hemoglobin content of individual macrocytes is increased owing to their larger size, but there is little change in the hemoglobin concentration of whole blood.

White blood cells and platelets are also produced in the bone marrow. In the differentiating granulocyte series of white blood cells (neutrophils, eosinophils, and basophils) giant, abnormally shaped metamyelocytes are found. Megakaryocytes (precursors of platelets) may also be larger than usual. In advanced folate/B_{12} deficiency, the total white blood cell count and platelet count may be low. Circulating neutrophils are characterized by an increased number of nuclear segments. The presence of hypersegmented neutrophils is a valuable clue in diagnosing folate/B_{12} deficiency when red cell changes are masked by a co-existent iron deficiency or the anemia of chronic disease.

As in all cases of anemia, the body adjusts its cardiopulmonary system to compensate for the diminished oxygen-carrying capacity of the blood, so in mild anemia the subject may not be aware of any problems. Eventually, however, the progressing anemia leads to symptoms of weakness, fatigue, shortness of breath, and palpitations. The sufferer may also experience headache, irritability, and an inability to concentrate. Visible signs of megaloblastic anemia in white-skinned people are a marked pallor and a slight jaundice, giving the skin a distinctive lemon-yellow tinge.

Megaloblastosis is not confined to developing cells in the bone marrow – all other rapidly dividing cell types will be affected, including epithelial cells lining the gastrointestinal, respiratory, and urinogenital tracts. A notable feature is glossitis where the tongue is sore at the edges, bright red in color, and smooth in texture. Gastrointestinal disturbances caused by defective gut epithelia have adverse consequences upon overall nutritional status. Male infertility results from impaired spermatogenesis.

25.20 HUMAN REQUIREMENT

The results of many studies have led to the universal conclusion that an intake of folate adequate for the prevention of clinical deficiency (megaloblastic anemia) may be suboptimal for long-term health. Epidemiological studies have shown that dietary folate deficiency is associated with an increased risk of certain types of cancer. It will be interesting to see what impact the US 1998 cereal-food fortification policy (also adopted in Canada) has on the health and life expectancy of American and Canadian people.

In the United States, fortification of grain products with folic acid is mandatory and all women of childbearing age are advised to take a daily 400 μg supplement of folic acid. This level of intake, together with the consumption of supplemented processed foods, creates a state of folate oversupplementation in a significant segment of the population (Ulrich & Potter, 2006). Ashokkumar *et al.* (2007) examined the effect of long-term oversupplementation with folic acid on folate intestinal absorption and renal reabsorption using as models the human-derived Caco-2 and HK-2 cell lines. Maintaining the intestinal Caco-2 cells and the renal HK-2 cells in a folic acid-oversupplemented growth medium buffered at pH 5.5 led to a significant down-regulation of folate uptake in both cell types compared with uptake by cells maintained under a folate-sufficient condition. No such down-regulation was observed in a medium buffered at pH 7.4. The observed down-regulation of folate uptake was associated with significant reductions of FR, PCFT, and RFC expression.

25.21 TOXICITY

Folate is generally considered to have a low acute and chronic toxicity for humans. No adverse effects have been associated with the consumption of folate normally found in fortified foods (Butterworth & Tamura, 1989). Safety is concerned primarily with ingestion of folate supplements and is related to the occasional presence of metabolically unaltered folic acid in the circulation (Kelly *et al.*, 1997).

People with undiagnosed vitamin B_{12} deficiency are particularly at risk if they take folic acid supplements without concomitant vitamin B_{12} supplementation because the obvious symptoms of megaloblastic anemia will be cured by folic acid, leaving the hidden neurological symptoms of B_{12} deficiency to progress. This applies particularly to the elderly because of their propensity to vitamin B_{12} malabsorption. It is important to make sure that folic acid or multivitamin supplements contain sufficient vitamin B_{12} to provide adequate absorption even in the total absence of intrinsic factor. The Recommended Dietary Allowance (RDA) for vitamin B_{12} is 2.4 μg per day. Assuming that 1% of ingested vitamin B_{12} is absorbed passively, a supplemental level of 500 μg would provide twice the RDA in the event of pernicious anemia.

It is possible that anticonvulsant drugs become less effective if the patient is taking high doses of folic acid, giving rise to increased fit frequency in epileptics (Butterworth & Tamura, 1989). Folic acid supplements should therefore be used with caution in patients with epilepsy. Reports of an association between supplemental folate intake and impaired intestinal zinc absorption have not been substantiated in the more recent literature (Institute of Medicine, 1998).

FURTHER READING

Bailey, L.B. & Gregory, J.F. III (1999) Polymorphisms of methylenetetrahydrofolate reductase and other enzymes: Metabolic significance, risks and impact on folate requirement. *Journal of Nutrition*, **129**, 919–22.

Lucock, M. (2000) Folic acid: Nutritional biochemistry, molecular biology, and role in disease processes. *Molecular Genetics and Metabolism*, **71**, 121–38.

Stehouwer, C.D.A. & Jacobs, C. (1998) Abnormalities of vascular function in hyperhomocysteinaemia: Relationship to atherothrombotic disease. *European Journal of Pediatrics*, **157**, S107–11.

Steinberg, S.E. (1984) Mechanisms of folate homeostasis. *American Journal of Physiology*, **246**, G319–24.

Welch, G.N. & Loscalzo, J. (1998) Homocysteine and atherothrombosis. *New England Journal of Medicine*, **338**, 1042–50.

REFERENCES

Alam, C., Hoque, M.T., Finnell, R.H., Goldman, I.D. & Bendayan, R. (2017) Regulation of reduced folate carrier (RFC) by vitamin D receptor at the blood-brain barrier. *Molecular Pharmaceutics*, **14**, 3848–58.

Alam, C., Aufreiter, S., Georgiou, C.J., Hoque, M.T., Finnell, R.H., O'Connor, D.L., Goldman, I.D. & Bendaya, R. (2019) Upregulation of reduced folate carrier by vitamin D enhances brain folate uptake in mice lacking folate receptor alpha. *Proceedings of the National Academy of Sciences of the U.S.A.*, **116**(35), 17531–40.

Alam, C., Kondo, M., O'Connor, D.L. & Bendayan, R. (2020) Clinical implications of folate transport in the central nervous system. *Trends in Pharmacological Sciences*, **41**, 349–61.

Anderson, R.G.W., Kamen, B.A., Rothberg, K.G. & Lacey, S.W. (1992) Potocytosis: Sequestration and transport of small molecules by caveolae. *Science*, **255**, 410–11.

Antony, A.C. (1996) Folate receptors. *Annual Review of Nutrition*, **16**, 501–21.

Arnesen, E., Refsum, H., Bønaa, K.H., Ueland, P.M., Førde, O.H. & Nordrehaug, J.E. (1995) Serum total homocysteine and coronary heart disease. *International Journal of Epidemiology*, **24**, 704–9.

Ashokkumar, B., Mohammed, Z.M., Vaziri, N.D. & Said, H.M. (2007) Effect of folate oversupplementation on folate uptake by human intestinal and renal epithelial cells. *American Journal of Clinical Nutrition*, **86**, 159–66.

Asrar, F.M. & O'Connor, D.L. (2005) Bacterially synthesized folate and supplemental folic acid are absorbed across the large intestine of piglets. *Journal of Nutritional Biochemistry*, **16**, 587–93.

Aufreiter, S., Gregory, J.F. III, Pfeiffer, C.M., Fazili, Z., Kim, Y.-I., Marcon, N., Kamalaporn, P., Pencharz, P.B. & O'Connor, D.L. (2009) Folate is absorbed across the colon of adults: Evidence from cecal infusion of ^{13}C-labeled [6S]-5-formyltetrahydrofolic acid. *American Journal of Clinical Nutrition*, **90**, 116–23.

Baram, J., Chabner, B.A., Drake, J.C., Fitzhugh, A.L., Sholar, P.W. & Allegra, C.J. (1988) Identification and biochemical properties of 10-formyldihydrofolate, a novel folate found in methotrexate-treated cells. *Journal of Biological Chemistry*, **263**(15), 7105–11.

Baugh, C.M. & Krumdieck, C.L. (1971) Naturally occurring folates. *Annals of the New York Academy of Sciences*, **186**, 7–28.

Beatty, C. & Wickramasinghe, S.N. (1993) Megaloblastic anaemias. In: *Encyclopaedia of Food Science, Food Technology and Nutrition*, Vol. 1. (Ed. by R. Macrea, R.K. Robinson & M.J. Sadler), pp. 171–7. Academic Press, London.

Bellamy, M.F., McDowell, I.W.F., Ramsey, M.W., Brownlee, M., Bones, C., Newcombe, R.G. & Lewis, M.J. (1998) Hyperhomocysteinemia after an oral methionine load acutely impairs endothelial function in healthy adults. *Circulation*, **98**, 1848–52.

Bermingham, A., Bottomley, J.R., Primrose, W.U. & Derrick, J.P. (2000) Equilibrium and kinetic studies of substrate binding to 6-hydroxymethyl-7,8-dihydropterin pyrophosphokinase from *Escherichia coli*. *Joural of Biological Chemistry*, **275**(24), 17962–7.

Bhandari, S.D. & Gregory, J.F. III (1990) Inhibition by selected food components of human and porcine intestinal pteroylpolyglutamate hydrolase activity. *American Journal of Clinical Nutrition*, **51**, 87–94.

Bhandari, S.D., Gregory, J.F. III, Renuart, D.R. & Merritt, A.M. (1990) Properties of pteroypolyglutamate hydrolase in pancreatic juice of the pig. *Journal of Nutrition*, **120**, 467–75.

Birn, H., Nielsen, S. & Christensen, E.I. (1997) Internalization and apical-to-basolateral transport of folate in rat kidney proximal tubule. *American Journal of Physiology*, **272**, F70–8.

Blount, B.C., Mack, M.M., Wehr, C.M., MacGregor, J.T., Hiatt, R.A., Wang, G., Wickramasinghe, S.N., Everson, R.B. & Ames, B.N. (1997) Folate deficiency causes uracil misincorporation into human DNA and chromosome breakage: Implications for cancer and neuronal damage. *Proceedings of the National Academy of Sciences of the U.S.A.*, **94**, 3290–5.

Brody, T. (1991) Folic acid. In: *Handbook of Vitamins*, 2nd edn. (Ed. by L.J. Machlin), pp. 453–89. Marcel Dekker, Inc., New York.

Butterworth, C.E. Jr. & Tamura, T. (1989) Folic acid safety and toxicity: A brief review. *American Journal of Clinical Nutrition*, **50**, 353–8.

Butterworth, C.E. Jr., Baugh, C.M. & Krumdieck, C. (1969) A study of folate absorption and metabolism in man utilizing carbon-14-labeled polyglutamates synthesized by the solid phase method. *Journal of Clinican Investigation*, **48**, 1131–42.

Camilo, E., Zimmerman, J., Mason, J.B., Golner, B., Russell, R., Selhub, J. & Rosenberg, I.H. (1996) Folate synthesized by bacteria in the human upper small intestine is assimilated by the host. *Gastroenterology*, **110**, 991–8.

Castenmiller, J.J.M., van den Poll, C.J., West, C.E., Brouwer, I.A., Thomas, C.M.G. & van Dusseldorp, M. (2000) Bioavailability of folate from processed spinach in humans. *Annals of Nutrition and Metabolism*, **44**, 163–9.

Celermajer, D.S., Sorensen, K., Ryalls, M., Robinson, J., Thomas, O., Leonard, J.V. & Deanfield, J.E. (1993) Impaired endothelial function occurs in the systemic arteries of children with homozygous homocystinuria but not in their heterozygous parents. *Journal of the American College of Cardiology*, **22**, 854–8.

Černá, J. & Káš, J. (1983) New conception of folacin assay in starch or glycogen containing food samples. *Nahrung*, **27**, 957–64.

Chambers, J.C., McGregor, A., Jean-Marie, J. & Kooner, J.S. (1998) Acute hyperhomocysteinaemia and endothelial dysfunction. *Lancet*, **351**, 36–7.

Chambers, J.C., McGregor, A., Jean-Marie, J., Obeid, O.A. & Kooner, J.S. (1999a) Demonstration of rapid onset vascular endothelial dysfunction after hyperhomocysteinemia. An effect reversible with vitamin C therapy. *Circulation*, **99**, 1156–60.

Chambers, J.C., Obeid, O.A. & Kooner, J.S. (1999b) Physiological increments in plasma homocysteine induce vascular endothelial dysfunction in normal human subjects. *Arteriosclerosis and Thrombosis*, **19**, 2922–7.

Chandler, C.J., Harrison, D.A., Buffington, C.A., Santiago, N.A. & Halsted, C.H. (1991) Functional specificity of jejunal brush-border pteroylpolyglutamate hydrolase in pig. *American Journal of Physiology*, **260**, G865–72.

Charpiot, P., Bescond, A., Augier, T., Chareyre, C., Fraterno, M., Rolland, P.-H. & Garcon, D. (1998) Hyperhomocysteinemia induces elastolysis in minipig arteries: Structural consequences, arterial site specificity and effect of captopril-hydrochlorothiazide. *Matrix Biology*, **17**, 559–74.

Choumenkovitch, S.F., Selhub, J., Wilson, P.W.F., Rader, J.I., Rosenberg, I.H. & Jacques, P.F. (2002) Folic acid intake from fortification in United States exceeds predictions. *Journal of Nutrition*, **132**, 2792–8.

Clark, N.A.C. & Fisk, N.M. (1994) Minimal compliance with the Department of Health recommendation for routine folate prophylaxis to prevent fetal neural tube defects. *British Journal of Obstetrics and Gynaecology*, **101**, 709–10.

Clarke, R., Daly, L., Robinson, K., Naughten, E., Cahalane, S., Fowler, B. & Graham, I. (1991) Hyperhomocysteinemia: An independent risk factor for vascular disease. *New England Journal of Medicine*, **324**, 1149–55.

Cravo, M.L., Mason, J.B., Dayal, Y., Hutchinson, M., Smith, D., Selhub, J. & Rosenberg, I.H. (1992) Folate deficiency enhances the development of colonic neoplasia in dimethylhydrazine-treated rats. *Cancer Research*, **52**, 5002–6.

Cuskelly, G.J., McNulty, H. & Scott, J.M. (1996) Effect of increasing dietary folate on red-cell folate: Implications for prevention of neural tube defects. *Lancet*, **347**, 657–9.

Czeizel, A.E. & Dudás, I. (1992) Prevention of the first occurrence of neural-tube defects by periconceptional vitamin supplementation. *New England Journal of Medicine*, **327**, 1832–5.

Davis, C.D. & Uthus, E.O. (2004) DNA methylation, cancer susceptibility, and nutrient interactions. *Experimental Biology and Medicine*, **229**, 988–95.

den Heijer, M., Brouwer, I.A., Bos, G.M.J., Blom, H.J., van der Put, N.M.J., Spaans, A.P., Rosendaal, F.R., Thomas, C.M.G., Haak, H.L., Wijermans, P.W. & Gerrits, W.B.J. (1998) Vitamin supplementation reduces blood homocysteine levels. A controlled trial in patients with venous thrombosis and healthy volunteers. *Arteriosclerosis, Thrombosis, and Vascular Biology*, **18**, 356–61.

Department of Health. (1992) Folic acid and the prevention of neural tube defects. Report from an expert advisory group. Department of Health, London.

Dev, S., Wani, N.A. & Kaur, J. (2011) Regulatory mechanisms of intestinal folate uptake in a rat model of folate oversupplementation. *British Journal of Nutrition*, **105**, 827–35.

Dixon, K.H., Lanpher, B.C., Chiu, J., Kelley, K. & Cowan, K.H. (1994) A novel cDNA restores reduced folate carrier activity and methotrexate sensitivity to transport deficient cells. *Journal of Biological Chemistry*, **269**(1), 17–20.

Domagala, T.B., Libura, M. & Szczeklik, A. (1997) Hyperhomocysteinemia following oral methionine load is associated with increased lipid peroxidation. *Thrombosis Research*, **87**, 411–16.

Dudeja, P.K., Torania, S.A. & Said, H.M. (1997) Evidence for the existence of a carrier-mediated folate uptake mechanism in human colonic luminal membranes. *American Journal of Physiology*, **272**, G1408–15.

Dudeja, P.K., Kode, A., Alnounou, M., Tyagi, S., Torania, S., Subramanian, V.S. & Said, H.M. (2001) Mechanism of folate transport across the human colonic basolateral membrane. *American Journal of Physiology (Gastrointestinal and Liver Physiology)*, **281**, G54–60.

Durand, P., Lussier-Cacan, S. & Blache, D. (1997) Acute methionine load-induced hyperhomocysteinemia enhances platelet aggregation, thromboxane biosynthesis, and macrophage-derived tissue factor activity in rats. *FASEB Journal*, **11**, 1157–68.

Duthie , S.J. (1999) Folic acid deficiency and cancer: Mechanisms of DNA instability. *British Medical Bulletin*, **55**, 578–92.

Duthie, S.J. & Hawdon, A. (1998) DNA instability (strand breakage, uracil misincorporation, and defective repair) is increased by folic acid depletion in human lymphocytes *in vitro. FASEB Journal*, **12**, 1491–7.

Duthie, S.J., Narayanan, S., Brand, G.M., Pirie, L. & Grant, G. (2002) Impact of folate deficiency on DNA stability. *Journal of Nutrition*, **132**, 2444S–9S.

Eskes, T.K.A.B. (1998) Neural tube defects, vitamins and homocysteine. *European Journal of Pediatrics*, **157**, S139–41.

European Concerted Action Project. (1997) Plasma homocysteine as a risk factor for vascular disease. *Journal of the American Medical Association*, **277**, 1775–81.

Everson, R.B., Wehr, C.M., Erexson, G.L. & MacGregor, J.T. (1988) Association of marginal folate depletion with increased human chromosomal damage in vivo: Demonstration by analysis of micronucleated erythrocytes. *Journal of the National Cancer Institute*, **80**, 525–9.

Farber, S., Cutler, E.C., Hawkins, J.W., Harrison, J.H., Peirce, E.C. & Lenx, G.G. (1947) The action of pteroylglutamic conjugates on man. *Science*, **106**, 619–21.

Farber, S., Diamond, L.K., Mercer, R.D., Sylvester, R.F. & Wolff, V.A. (1948) Temporary remission of acute leukemia in children produced by folic acid antagonist, 4-aminopteroyl glutamic acid (aminopterin). *New England Journal of Medicine*, **238**, 787–93.

Ford, J.E. (1974) Some observations on the possible nutritional significance of vitamin B_{12}- and folate-binding proteins in milk. *British Journal of Nutrition*, **31**, 243–57.

Ford, J.E., Salter, D.N. & Scott, K.J. (1969) The folate-binding protein in milk. *Journal of Dairy Research*, **36**, 435–46.

Grapp, M., Just, I.A., Linnankivi, T., Wolf, P., Lücke, T., Häusler, M., Gärtner, J. & Steinfeld, R. (2012) Molecular characterization of folate receptor 1 mutations delineates cerebral folate transport deficiency. *Brain*, **135**, 2022–31.

Grapp, M., Wrede, A., Schweizer, M., Hüwel, S., Galla, H.-J., Snaidero, N., Simons, M., Bückers, J., Low, P.S., Urlab, H., Gärtner, J. & Steinfeld, R. (2013) Choroid plexus transcytosis and exosome shuttling deliver folate into brain parenchyma. *Nature Communications*, **4**, 2123. DOI: 10.1038/ncomms3123

Gregory, J.F. III (1984) Determination of folacin in foods and other biological materials. *Journal of the Association of Official Analytical Chemists*, **67**, 1015–19.

Gregory, J.F. III (1989) Chemical and nutritional aspects of folate research: Analytical procedures, methods of folate synthesis, stability, and bioavailability of dietary folates. *Advances in Food and Nutrition Research*, **33**, 1–101.

Gregory, J.F. III (1995) The bioavailability of folate. In: *Folate in Health and Disease*. (Ed. by L.B. Bailey), pp. 195–235. Marcel Dekker, Inc., New York.

Gregory, J.F. III, Ink, S.L. & Cerda, J.J. (1987) Comparison of pteroylpolyglutamate hydrolase (folate conjugase) from porcine and human intestinal brush border membrane. *Comparative Biochemistry and Physiology*, **88B**, 1135–41.

Gregory, J.F. III, Bhandari, S.D., Bailey, L.B., Toth, J.P., Baumgartner, T.G. & Cerda, J.J. (1991) Relative bioavailability of deuterium-labeled monoglutamyl and hexaglutamyl folates in human subjects. *American Journal of Clinical Nutrition*, **53**, 736–40.

Halsted, C.H. (1990) Intestinal absorption of dietary folates. In: *Folic Acid Metabolism in Health and Disease*. (Ed. by M.F. Picciano, E.L.R. Stokstad & J.F. Gregory III), pp. 23–45. Wiley-Liss, Inc., New York.

Harker, L.A., Ross, R., Slichter, S.J. & Scott, C.R. (1976) Homocystine-induced arteriosclerosis.. The role of endothelial cell injury and platelet response in its genesis. *Journal of Clinical Investigation*, **58**, 731–41.

Henderson, G.B. (1990) Folate-binding proteins. *Annual Review of Nutrition*, **10**, 319–35.

Herbert, V. (1999) Folic acid. In: *Modern Nutrition in Health and Disease*, 9th edn. (Ed. by M.E. Shils, J.A. Olson, M. Shike & A.C. Ross), pp. 433–46. Lippincott Williams & Wilkins, Philadelphia.

Honein, M.A., Paulozzi, L.J., Mathews, T.J., Erickson, J.D. & Wong, L.-Y.C. (2001) Impact of folic acid fortification of the US food supply on the occurrence of neural tube defects. *Journal of the American Medical Association*, **285**, 2981–6.

Hou, Z., Orr, S. & Matherly, L.H. (2014) Post-transcriptional regulation of the human reduced folate carrier as a novel adaptive mechanism in response to folate excess or deficiency. *Bioscience Reports*, **34**, 457–68.

Houghton, L.A., Green, T.J., Donovan, U.M., Gibson, R.S., Stephen, A.M. & O'Connor, D.L. (1997) Association between dietary fiber intake and the folate status of a group of female adolescents. *American Journal of Clinical Nutrition*, **66**, 1414–21.

Institute of Medicine. (1998) *Dietary Reference Intakes for Thiamin, Riboflavin, Niacin, Vitamin B_6, Folate, Vitamin B_{12}, Pantothenic Acid, Biotin, and Choline.* National Academy Press, Washington, DC.

Iwai, K., Tani, M. & Fushiki, T. (1983) Electrophoretic and immunological properties of folate-binding protein isolated from bovine milk. *Agricultural and Biological Chemistry*, **47**, 1523–30.

Kanani, P.M., Sinkey, C.A., Browning, R.L., Allaman, M., Knapp, H.R. & Haynes, W.G. (1999) Role of oxidant stress in endothelial dysfunction produced by experimental hyperhomocyst(e)inemia in humans. *Circulation*, **100**, 1161–8.

Kane, M.A., Portillo, R.M., Elwood, P.C., Antony, A.C. & Kolhouse, J.F. (1986) The influence of extracellular folate concentration on methotrexate uptake by human KB cells. Partial characterization of a membrane-associated methotrexate binding protein. *Journal of Biological Chemistry*, **261**(1), 44–9.

Kang, S.-S., Wong, P.W.K. & Norusis, M. (1987) Homocysteinemia due to folate deficiency. *Metabolism*, **36**, 458–62.

Kelly, P., McPartlin, J., Goggins, M., Weir, D.G. & Scott, J.M. (1997) Unmetabolized folic acid in serum: Acute studies in subjects consuming fortified food and supplements. *American Journal of Clinical Nutrition*, **65**, 1790–5.

Kim, T.H., Yang, J., Darling, P.B. & O'Connor, D.L. (2004) A large pool of available folate exists in the large intestine of human infants and piglets. *Journal of Nutrition*, **134**, 1389–94.

Kim, Y.-I. (1999) Folate and carcinogenesis: Evidence, mechanisms, and implications. *Journal of Nutritional Biochemistry*, **10**, 66–88.

Kim, Y.-I. (2007) Folate and colorectal cancer: An evidence-based critical review. *Molecular Nutrition and Food Research*, **51**, 267–92.

Kim, Y.-I., Pogribny, I.P., Basnakian, A.G., Miller, J.W., Selhub, J., James, S.J. & Mason, J.B. (1997) Folate deficiency in rats induces DNA strand breaks and hypomethylation within the p53 tumor suppressor gene. *American Journal of Clinical Nutrition*, **72**, 46–52.

Kim, Y.-I., Hayek, M., Mason, J.B. & Meydani, S.N. (2002) Severe folate deficiency impairs natural killer cell-mediated cytotoxicity in rats. *Journal of Nutrition*, **132**, 1361–7.

Kuller, L.H. & Evans, R.W. (1998) Homocysteine, vitamins, and cardiovascular disease. *Circulation*, **98**, 196–9.

Kumar, C.K., Moyer, M.P., Dudeja, P.K. & Said, H.M. (1997) A protein-tyrosine kinase-regulated, pH-dependent, carrier-mediated uptake system for folate in human normal colonic epithelial cell line NCM460. *Journal of Biological Chemistry*, **272**(10), 6226–31.

Laird, P.W. & Jaenisch, R. (1994) DNA methylation and cancer. *Human Molecular Genetics*, **3**, 1487–95.

Lakoff, A., Fazili, Z., Aufreiter, S., Pfeiffer, C.M., Connolly, B., Gregory, J.F. III, Pencharz, P.B. & O'Connor, D.L. (2014) Folate is absorbed across the human colon: Evidence by using enteric-coated caplets containing ^{13}C-labeled [6S]-5-formyltetrahydrofolate. *American Journal of Clinical Nutrition*, **100**, 1278–86.

Laskowski, M. & Laskowski, M. (1951) Crystalline trypsin inhibitor from colostrum. *Journal of Biological Chemistry*, **190**, 563–73.

Lee, J.E., Willett, W.C., Fuchs, C.S., Smith-Warner, S.A., Wu, K., Ma, J. & Giovannucci, E. (2011) Folate intake and risk of colorectal cancer and adenoma: Modification by time. *American Journal of Clinical Nutrition*, **93**, 817–25.

Lentz, S.R. (1998) Mechanisms of thrombosis in hyperhomocysteinemia. *Current Opinion in Hematology*, **5**, 343–9.

Lentz, S.R., Sobey, C.G., Piegors, D.J., Bhopatkar, M.Y., Faraci, F.M., Malinow, M.R. & Heistad, D.D. (1996) Vascular dysfunction in monkeys with diet-induced hyperhomocyst(e) inemia. *Journal of Clinical Investigation*, **98**, 24–9.

Leuchtenberger, R., Leuchtenberger, C., Laszlo, D. & Lewisohn, R. (1945) The influence of "folic acid" on spontaneous breast cancers in mice. *Science*, **101**, 46.

Mason, J.B. & Levesque, T. (1996) Folate: Effects on carcinogenesis and the potential for cancer chemoprevention. *Oncology*, **10**, 1727–43.

Mason, J.B. & Selhub, J. (1988) Folate-binding protein and the absorption of folic acid in the small intestine of the suckling rat. *American Journal of Clinical Nutrition*, **48**, 620–5.

Mason, J.B., Dickstein, A., Jacques, P.F., Haggarty, P., Selhub, J., Dallal, G. & Rosenberg, I.H. (2007) A temporal association between folic acid fortification and an increase in colorectal cancer rates may be illuminating important biological principles: A hypothesis. *Cancer Epidemiology, Biomarkers & Prevention*, **16**, 1325–9.

Matthews, R.G. & Baugh, C.M. (1980) Interactions of pig liver methylenetetrahydrofolate reductase with methylenetetrahydropteroylpolyglutamate substrates and with dihydropteroylpolyglutamate inhibitors. *Biochemistry*, **19**, 2040–5.

Morshed, K.M., Ross, D.M. & McMartin, K.E. (1997) Folate transport proteins mediate the bidirectional transport of 5-methyltetrahydrofolate in cultured human proximal tubule cells. *Journal of Nutrition*, **127**, 1137–47.

MRC Vitamin Study Research Group. (1991) Prevention of neural tube defects: Results of the Medical Research Council Vitamin Study. *Lancet*, **338**, 131–7.

Mulinare, J. (1995) Public health perspectives on folic acid and neural tube defects. *Cereal Foods World*, **40**, 58–61.

Mulinare, J., Cordero, J.F., Erickson, J.D. & Berry, R.J. (1988) Periconceptional use of multivitamins and the occurrence of neural tube defects. *Journal of the American Medical Association*, **260**, 3141–5.

Nappo, F., De Rosa, N., Marfella, R., De Lucia, D., Ingrosso, D., Perna, A.F., Farzati, B. & Giugliano, D. (1999) Impairment of endothelial functions by acute hyperhomocysteinemia and reversal by antioxidant vitamins. *Journal of the American Medical Association*, **281**, 2113–8.

Ng, H.-H. & Bird, A. (1999) DNA methylation and chromatin modification. *Current Opinion in Genetics & Development*, **9**, 158–63.

Nygård, O., Nordrehaug, J.E., Refsum, H., Ueland, P.M., Farstad, M. & Vollset, S.E. (1997) Plasma homocysteine levels and mortality in patients with coronary artery disease. *New England Journal of Medicine*, **337**, 230–6.

Okamura-Ikeda, K., Fujiwara, K. & Motokawa, Y. (1987) Mechanism of the glycine cleavage reaction. Properties of the reverse reaction catalyzed by T-protein. *Journal of Biological Chemistry*, **262**(14), 6746–9.

O'Shannessy, D.J., Somers, E.B., Albone, E., Cheng, X., Park, Y.C., Tomkowicz, B.E., Hamuro, Y., Kohl, T.O., Forsyth, T.M., Smale, R., Fu, Y.-S. & Nicolaides, N.C. (2011) Characterization of the human folate receptor alpha via novel antibody-based probes. *Oncotarget*, **2**, 1227–43.

Pedersen, T.G., Svendsen, I., Hanson, S.I., Holm, J. & Lyngbye, J. (1980) Aggregation of a folate-binding protein from cow's milk. *Carlsberg Research Communications*, **45**, 161–6.

Perry, I.J., Refsum, H., Morris, R.W., Ebrahim, S.B., Ueland, P.M. & Shaper, A.G. (1995) Prospective study of serum total homocysteine concentration and risk of stroke in middle-aged British men. *Lancet*, **346**, 1395–8.

Pfeiffer, C.M., Rogers, L.M., Bailey, L.B. & Gregory, J.F. III (1997) Absorption of folate from fortified cereal-grain products and of supplemental folate consumed with or without food determined by using a dual-label stable-isotope protocol. *American Journal of Clinical Nutrition*, **66**, 1388–97.

Picciano, M.F., Green, T. & O'Connor, D.L. (1994) The folate status of women and health. *Nutrition Today*, **29**, 20–9.

Prasad, P.D., Mahesh, V.B., Leibach, F.H. & Ganapathy, V. (1994a) Functional coupling between a bafilomycin A_1-sensitive proton pump and a probenecid-sensitive folate transporter in human placental choriocarcinoma cells. *Biochimica et Biophysica Acta*, **1222**, 309–14.

Prasad, P.D., Ramamoorthy, S., Moe, A.J., Smith, C.H., Leibach, F.H. & Ganapathy, V. (1994b) Selective expression of the high-affinity isoform of the folate receptor (FR-α) in the human placental syncytiotrophoblast and choriocarcinoma cells. *Biochimica et Biophysica Acta*, **1223**, 71–5.

Prasad, P.D., Ramamoorthy, S., Leibach, F.H. & Ganapathy, V. (1995) Molecular cloning of the human placental folate transporter. *Biochemical and Biophysical Research Communications*, **206**, 681–7.

Qiu, A., Jansen, M., Sakaris, A., Min, S.H., Chattopadhyay, S., Tsai, E., Sandoval, C., Zhao, R., Akabas, M.H. & Goldman, I.D. (2006) Identification of an intestinal folate transporter and the molecular basis for hereditary folate malabsorption. *Cell*, **127**, 917–28.

Qiu, A., Min, S.H., Jansen, M., Malhotra, U., Tsai, E., Cabelof, D.C., Matherly, L.H., Zhao, R., Akahas, M.H. & Goldman, I.D. (2007) Rodent intestinal folate transporters (SLC46A1): Secondary structure, functional properties, and responses to dietary folate restriction. *American Journal of Physiology*, **293**, C1669–78.

Rijnboutt, S., Jansen, G., Posthuma, G., Hynes, J.B., Schornagel, J.H. & Strous, G.J. (1996) Endocytosis of GPI-linked membrane folate receptor-α. *Journal of Cell Biology*, **132**, 35–47.

Rong, N., Selhub, J., Goldin, B.R. & Rosenberg, I.H. (1991) Bacterially synthesized folate in rat large intestine is incorporated into host tissue folyl polyglutamates. *Journal of Nutrition*, **121**, 1955–9.

Rosenquist, T.H., Ratashak, S.A. & Selhub, J. (1996) Homocysteine induces congenital defects of the heart and neural tube: Effect of folic acid. *Proceedings of the National Academy of Sciences of the U.SA.*, **93**, 15227–32.

Ross, J.F., Chaudhuri, P.K. & Ratnam, M. (1994) Differential regulation of folate receptor isoforms in normal and malignant tissues in vivo and in established cell lines. *Cancer*, **73**, 2432–43.

Said, H.M. (2011) Intestinal absorption of water-soluble vitamins in health and disease. *Biochemical Journal*, **437**, 357–72.

Said, H.M. & Redha, R. (1987) A carrier-mediated transport for folate in basolateral membrane vesicles of rat small intestine. *Biochemical Journal*, **247**, 141–6.

Said, H.M., Ghishan, F.K. & Redha, R. (1987) Folate transport by human intestinal brush-border membrane vesicles. *American Journal of Physiology*, **252**, G229–36.

Said, H.M., Horne D.W. & Wagner, C. (1986) Effect of human milk folate-binding protein on folate intestinal transport. *Archives of Biochemistry and Biophysics*, **251**, 114–20.

Said, H.M., Ma, T.Y., Ortiz, A., Tapia, A. & Valerio, C.K. (1997) Intracellular regulation of intestinal folate uptake: Studies with cultured IEC-6 epithelial cells. *American Journal of Physiology*, **272**, C729–36.

Said, H.M., Chatterjee, N., ul Haq, R., Subramanian, V.S., Ortiz, A., Matherly, L.H., Sirotnak, F.M., Halsted, C. & Rubin, S.A. (2000) Adaptive regulation of intestinal folate uptake: Effect of dietary folate deficiency. *American Journal of Physiology*, **279**, C1889–95.

Salter, D.N. & Blakeborough, P. (1988) Influence of goat's-milk folate-binding protein on transport of 5-methyltetrahydrofolate in neonatal-goat small intestinal brush-border-membrane vesicles. *British Journal of Nutrition*, **59**, 497–507.

Salter, D.N. & Mowlem, A. (1983) Neonatal role of milk folate-binding protein: Studies on the course of digestion of goat's milk folate binder in the 6-d-old kid. *British Journal of Nutrition*, **50**, 589–96.

Salter, D.N., Scott, K.J., Slade, H. & Andrew, P. (1981) The preparation and properties of folate-binding protein from cow's milk. *Biohemical Journal*, **193**, 469–76.

Sauberlich, H.E., Kretch, M.J., Skala, J.H., Johnson, H.L. & Taylor, P.C. (1987) Folate requirement and metabolism in nonpregnant women. *American Journal of Clinical Nutrition*, **46**, 1016–28.

Scott, J.M., Kirke, P.N. & Weir, D.G. (1990) The role of nutrition in neural tube defects. *Annual Review of Nutrition*, **10**, 277–95.

Selhub, J. & Miller, J.W. (1992) The pathogenesis of homocysteinemia: Interruption of the coordinate regulation by *S*-adenosylmethionine of the remethylation and transsulfuration of homocysteine. *American Journal of Clinical Nutrition*, **55**, 131–8.

Selhub, J. & Rosenberg, I.H. (1981) Folate transport in isolated brush border membrane vesicles from rat intestine. *Journal of Biological Chemistry*, **256**(9), 4489–93.

Selhub, J., Jacques, P.F., Bostom, A.G., D'Agostino, R.B., Wilson, P.W.F., Belanger, A.J., O'Leary, D.H., Wolf, P.A., Schaefer, E.J. & Rosenberg, I.H. (1995) Association between plasma homocysteine concentrations and extracranial carotid-artery stenosis. *New England Journal of Medicine*, **332**, 286–91.

Shane, B. (1989) Folylpolyglutamate synthesis and role in the regulation of one-carbon metabolism. *Vitamins and Hormones*, **45**, 263–335.

Shin, H.-C., Takakuwa, F., Shimoda, M. & Kokue, E. (1995) Enterohepatic circulation kinetics of bile-active folate derivatives and folate homeostasis in rats. *American Journal of Physiology*, **269**, R421–5.

Simon, B.J., Kulanthaivel, P., Burckhardt, G., Ramamoorthy, S., Leibach, F.H. & Ganapathy, V. (1992) Characterization of an ATP-driven H^+ pump in human placental brush-border membrane vesicles. *Biochemical Journal*, **287**, 423–30.

Spector, R. & Lorenzo, A.V. (1975a) Folate transport by the choroid plexus in vitro. *Science*, **187**, 540–2.

Spector, R. & Lorenzo, A.V. (1975b) Folate transport in the central nervous system. *American Journal of Physiology*, **229**, 777–82.

Stampfer, M.J., Malinow, M.R., Willett, W.C., Newcomer, L.M., Upson, B., Ullmann, D., Tishler, P.V. & Hennekens, C.H. (1992) A prospective study of plasma homocyst(e)ine and risk of myocardial infarction in US physicians. *Journal of the American Medical Association*, **268**, 877–81.

Steinberg, S.E. (1984) Mechanisms of folate homeostasis. *American Journal of Physiology*, **246**, G319–24.

Stokstad, E.L.R. & Koch, J. (1967) Folic acid metabolism. *Physiological Reviews*, **47**, 83–116.

Strum, W.B. (1979) Enzymatic reduction and methylation of folate following pH-dependent, carrier-mediated transport in rat jejunum. *Biochimica et Biophysica Acta*, **554**, 249–57.

Subar, A.F., Block, G. & James, L.D. (1989) Folate intake and food sources in the US population. *American Journal of Clinical Nutrition*, **50**, 508–16.

Subramanian, V.S., Reidling, J.C. & Said, H.M. (2008) Differentiation-dependent regulation of the intestinal folate uptake process: Studies with Caco-2 cells and native mouse intestine. *American Journal of Physiology*, **295**, C828–35.

Svendsen, I., Martin, B., Pedersen, T.G., Hansen, S.I., Holm, J. & Lyngbye, J. (1979) Isolation and characterization of the folate-binding protein from cow's milk. *Carlsberg Research Communications*, **44**, 89–99.

Tani, M. & Iwai, K. (1984) Some nutritional effects of folate-binding protein in bovine milk on the bioavailability of folate to rats. *Journals of Nutrition*, **114**, 778–85.

Tani, M., Fushiki, T. & Iwai, K. (1983) Influence of folate-binding protein from bovine milk on the absorption of folate in gastrointestinal tract of rat. *Biochimica et Biophysica Acta*, **757**, 274–81.

Tawakol, A., Omland, T., Gerhard, M., Wu, J.T. & Creager, M.A. (1997) Hyperhomocyst(e)inemia is associated with impaired endothelium-dependent vasodilation in humans. *Circulation*, **95**, 1119–21.

Thakur, S., Rahat, B., Hamid, A., Najar, R.A. & Kaur, J. (2015) Identification of regulatory mechanisms of intestinal folate transport in condition of folate deficiency. *Journal of Nutritional Biochemistry*, **26**, 1084–94.

Tsai, J.-C., Perrella, M.A., Yoshizumi, M., Hsieh, C.-M., Haber, E., Schlegel, R. & Lee, M.-E. (1994) Promotion of vascular smooth muscle cell growth by homocysteine: A link to atherosclerosis. *Proceedings of the National Academy of Sciences of the U.SA.*, **91**, 6369–73.

Ulrich, C.M. & Potter, J.D. (2006) Folate supplementation: Too much of a good thing? *Cancer Epidemiology, Biomarkers & Prevention*, **15**, 189–93.

Upchurch, G.R. Jr., Welch, G.N., Fabian, A.J., Pigazzi, A., Keaney, J.F. Jr. & Loscalzo, J. (1997) Stimulation of endothelial nitric oxide production by homocyst(e)ine. *Atherosclerosis*, **132**, 177–85.

van het Hof, K.H., Tijburg, L.B.M., Pietrzik, K. & Weststrate, J.A. (1999) Influence of feeding different vegetables on plasma levels of carotenoids, folate and vitamin C. Effect of disruption of the vegetable matrix. *British Journal of Nutrition*, **82**, 203–12.

Visentin, M., Diop-Bove, N., Zhao, R. & Goldman, I.D. (2014) The intestinal absorption of folates. *Annual Review of Physiology*, **76**, 251–74.

Wagner, C. (1985) Folate-binding proteins. *Nutrition Research*, **43**, 293–9.

Wagner, C. (1996) Symposium on the subcellular compartmentation of folate metabolism. *Journal of Nutrition*, **126**, 1228S–34S.

Wang, Y., Zhao, R., Russell, R.G. & Goldman, I.D. (2001) Localization of the murine reduced folate carrier as assessed by immunohistochemical analysis. *Biochimica et Biophysica Acta*, **1513**, 49–54.

Wei, M.-M. & Gregory, J.F. III (1998) Organic acids in selected foods inhibit intestinal brush border pteroylpolyglutamate hydrolase in vitro: Potential mechanism affecting the bioavailability of dietary polyglutamyl folate. *Journal of Agricultural and Food Chemistry*, **46**, 211–9.

Wigertz, K., Hansen, I., Høier-Madsen, M., Holm, J. & Jägerstad, M. (1996) Effect of milk processing on the concentration of folate-binding protein (FBP), folate-binding capacity and retention of 5-methyltetrahydrofolate. *International Journal of Food Sciences and Nutrition*, **47**, 315–22.

Wild, J., Schorah, C.J. & Smithells, R.W. (1994) Investigation of folate intake and metabolism in women who have had two pregnancies complicated by neural tube defects. *British Journal of Obstetrics and Gynaecology*, **101**, 197–202.

Woo, K.S., Chook, P., Lolin, Y.I., Cheung, A.S.P., Chan, L.T., Sun, Y.Y., Sanderson, J.E., Metreweli, C. & Celermajer, D.S. (1997) Hyperhomocyst(e)inemia is a risk factor for arterial endothelial dysfunction in humans. *Circulation*, **96**, 2542–4.

Wright, A.J.A., Finglas, P.M., Dainty, J.R., Hart, D.J., Wolfe, C.A., Southon, S. & Gregory, J.F. (2003) Single oral doses of ^{13}C forms of pteroylmonoglutamic acid and 5-formyltetrahydrofolic acid elicit differences in short-term kinetics of labelled and unlabelled folates in plasma: Potential problems in interpretation of folate bioavailability studies. *British Journal of Nutrition*, **90**, 363–71.

Wright, A.J.A., Finglas, P.M., Dainty, J.R., Wolfe, C.A., Hart, D.J., Wright, D.M. & Gregory, J.F. (2005) Differential kinetic behavior and distribution for pteroylglutamic acid and reduced folates: A revised hypothesis of the primary site of PteGlu metabolism in humans. *Journal of Nutrition*, **135**, 619–23.

Wu, D. & Pardridge, W.M. (1999) Blood-brain barrier transport of reduced folic acid. *Pharmaceutical Research*, **16**, 415–19.

Yasuda, S., Hasui S., Yamamoto, C., Yoshioka, C., Kobayashi, M. Itagaki, S., Hirano, T. & Iseki, K. (2008) Placental folate transport during pregnancy. *Bioscience, Biotechnology and Biochemistry*, **72**, 80112–18.

Yi, P., Melnyk, S., Pogribna, M., Pogribny, I.P., Hine, R.J. & James, S.J. (2000) Increase in plasma homocysteine associated with parallel increases in plasma *S*-adenosylhomocysteine and lymphocyte DNA hypomethylation. *Journal of Biological Chemistry*, **275**(38), 29318–23.

Yuasa, H., Inoue, K. & Hayashi, Y. (2009) Molecular and functional characteristics of protein-coupled folate transporter. *Journal of Pharmaceutical Sciences*, **98**, 1608–16.

Zhao, R., Matherly, L.H. & Goldman, I.D. (2009a) Membrane transporters and folate homeostasis: Intestinal absorption and transport into systemic compartments and tissues. *Expert Reviews in Molecular Medicine*, **11**, e4. DOI: 10.1017/S1462399409000969

Zhao, R., Min S.H., Wang, Y., Campanella, E., Low, P.S. & Goldman, I.D. (2009b) A role for the proton-coupled folate transporter (PCFT-SLC46A1) in folate receptor-mediated endocytosis. *Journal of Biological Chemistry*, **284**(7), 4267–74.

Zhao, R., Aluri, S. & Goldman, I.D. (2017) The proton-coupled folate transporter (PCFT-SLC46A1) and the syndrome of systemic and cerebral folate deficiency of infancy: Hereditary folate malabsorption. *Molecular Aspects of Medicine*, **53**, 57–72.

26 Vitamin B_{12}

26.1 CHEMISTRY

In accordance with the literature on nutrition and pharmacology, the term 'vitamin B_{12}' is used in this text as the generic descriptor for all cobalamins that exhibit antipernicious anemia activity. Individual cobalamins will be referred to by their specific names (e.g. cyanocobalamin).

The cobalamin molecule is a six-co-ordination cobalt complex containing a corrin ring system substituted with numerous methyl, acetamide, and propionamide radicals (Figure 26.1). Methylene bridges link the pyrrole rings A to B, B to C, and C to D, but not A to D, which are linked directly. The cobalt atom, which may assume an oxidation state of (I), (II), or (III), is linked by four equatorial co-ordinate bonds to the four pyrrole nitrogens, and by an axial co-ordinate bond to a 5,6-dimethylbenzimidazol (DMB) moiety, which extends in α-glycosidic linkage to ribose-3-phosphate. The phosphate group is linked to the D ring of the corrin structure via a substituted propionamide chain. The pseudonucleotide (DMB base plus sugar phosphate) is orientated perpendicularly to the corrin structure. The remaining axial co-ordinate bond at the X position links the cobalt atom to a cyano (–CN) group in the case of cyanocobalamin ($C_{63}H_{88}O_{14}N_{14}PCo$, MW = 1355.4) or, depending on the chemical environment, to some other group (e.g. –OH in hydroxocobalamin and –HSO_3 in sulfitocobalamin.

There are two vitamin B_{12} coenzymes with known metabolic activity in humans, namely methylcobalamin and 5′-deoxyadenosylcobalamin (frequently abbreviated to adenosylcobalamin and also known as coenzyme B_{12}). The methyl or adenosyl ligands of the coenzymes occupy the X position in the corrin structure. The coenzymes are bound intracellularly to their protein apoenzymes through a covalent peptide link, or in milk and plasma to specific transport proteins. The enzyme-bound cobalamins exist as cob(I)alamins.

Cyanocobalamin is the most stable of the vitamin B_{12}-active cobalamins and is the one mostly used in pharmaceutical preparations and food supplementation. Aqueous solutions of cyanocobalamin are stable in air at room temperature if protected from light. On exposure to light, the cyano group dissociates from cyanocobalamin and hydroxocobalamin is formed. In neutral and acid solution hydroxocobalamin exists in the form of aquocobalamin (Gräsbeck & Salonen, 1976). This photolytic reaction does not cause a loss of activity.

26.2 FOOD SOURCES

Naturally occurring vitamin B_{12} originates solely from synthesis by bacteria and other microorganisms growing in soil or water, in sewage, and in the rumen and intestinal tract of animals. Any traces of the vitamin that may be detected in plants are due to microbial contamination from the soil or manure or, in the case of certain legumes, to bacterial synthesis in the root nodules.

Vitamin B_{12} is ubiquitous in foods of animal origin and is derived from the animal's ingestion of cobalamin-containing animal tissue or microbiologically contaminated plant material, in addition to vitamin absorbed from the animal's own digestive tract. Liver is the outstanding dietary source of the vitamin, followed by kidney and heart. Muscle meats, fish, eggs, cheese, and milk are other important food sources. Vitamin B_{12} activity has been reported in yeast, but this has since been attributed to the presence of noncobalamin corrinoids or vitamin B_{12} originating from the enriching medium (Herbert, 1988). About 5 to 30% of the reported vitamin B_{12} in foods may be microbiologically active noncobalamin corrinoids rather than true B_{12} (National Research Council, 1989).

Vitamin B_{12} in foods exists in several forms as reported by Farquharson and Adams (1976). Meat and fish contain mostly adenosyl- and hydroxocobalamins; these compounds, accompanied by methylcobalamin, also occur in dairy products, with hydroxocobalamin predominating in milk. Sulfitocobalamin is found in canned meats and fish. Cyancobalamin was only detected in small amounts in egg white, cheeses, and boiled haddock.

26.3 DIGESTION AND ABSORPTION OF DIETARY VITAMIN B_{12}

Humans appear to be entirely dependent on a dietary intake of vitamin B_{12}. Although microbial synthesis of the vitamin occurs in the human colon, it is apparently not absorbed. Strict vegetarians may obtain limited amounts of vitamin B_{12} through ingestion of the vitamin-containing root nodules of certain legumes and from plant material contaminated with microorganisms.

The absorption and transport of the vitamin B_{12} naturally present in foods takes place by specialized mechanisms that accumulate the vitamin and deliver it to cells that require it. The high efficiency of these mechanisms enables 50–90% of the minute amount of B_{12} present in a typical omnivorous diet (approximately 10 μg) to be absorbed and delivered to cells. The specificity is such that all natural forms of vitamin B_{12} are absorbed and transported in the same way; structurally similar but biologically inactive analogs, which are metabolically useless and possibly harmful, bypass the transport mechanisms and are eliminated from the body.

DOI: 10.1201/9781003410300-30

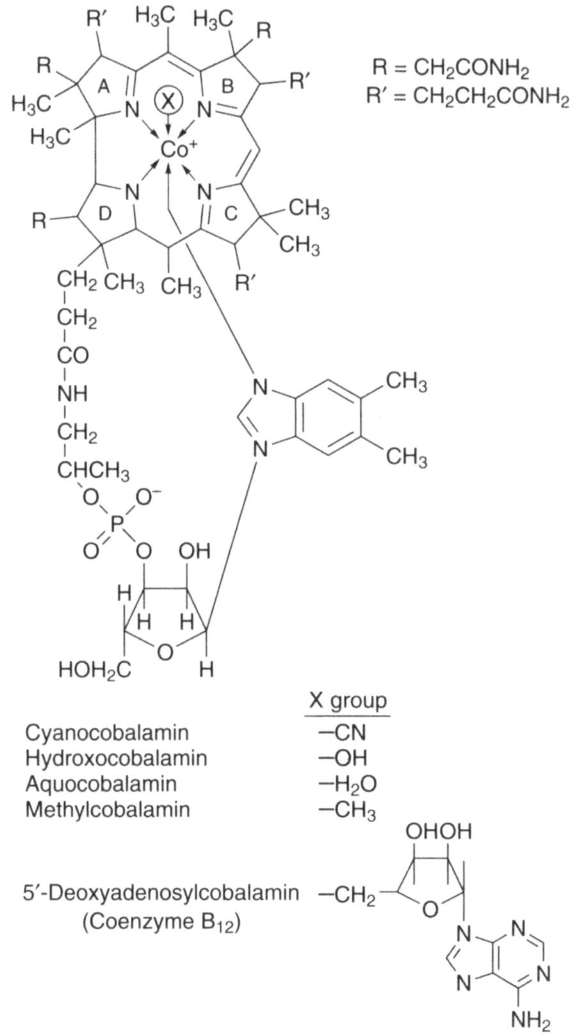

FIGURE 26.1 Structures of vitamin B_{12} compounds.

Ingested protein-bound cobalamins are released by the combined action of hydrochloric acid and pepsin in the stomach. Gastric juice also contains two functionally distinct cobalamin-binding proteins: (1) haptocorrin (there are actually several haptocorrins, which are also known as R binders or cobalophilin) and (2) intrinsic factor. Haptocorrin binds a wide variety of cobalamin analogs in addition to vitamin B_{12}, whereas intrinsic factor binds B_{12} vitamers with high specificity and equal affinity. The 5,6-dimethylbenzimidazole moiety is essential for recognition by intrinsic factor (Seetharam & Alpers, 1991). The haptocorrin originates in saliva while intrinsic factor is synthesized and secreted directly into gastric juice by the parietal cells of the stomach. At the acid pH of the stomach, cobalamins have a greater affinity for haptocorrin than for intrinsic factor. Therefore cobalamins leave the stomach and enter the duodenum bound to haptocorrin and accompanied by free intrinsic factor. In the mildly alkaline environment of the jejunum, pancreatic proteases, particularly trypsin, partially degrade both free haptocorrin and haptocorrin complexed with cobalamins. Intrinsic factor, which is resistant to proteolysis by pancreatic enzymes, then binds avidly to the released B_{12} vitamers.

The intrinsic factor–B_{12} complex is carried down to the ileum where it binds avidly to specific receptors on the brush border of the ileal enterocyte. The presence of calcium ions and a pH above 5.5 are necessary to induce the appropriate configuration of the receptor for binding (Donaldson, 1987). The intrinsic factor–B_{12} complex is absorbed intact (Seetharam *et al.*, 1985), but the precise mechanism of absorption and subsequent events within the enterocyte have yet to be elucidated. It is possible that ileal absorption of the intrinsic factor–B_{12} complex is accomplished by receptor-mediated endocytosis (Donaldson, 1985), but clathrin-coated pits or vesicles have not been found. The B_{12} is subsequently released at an intracellular site, possibly either in lysosomes or prelysosomal vesicles (endosomes), both of which are acidic (Seetharam, 1994). The intrinsic factor appears to be degraded by proteolysis after releasing its bound B_{12}, there being no apparent recycling of intrinsic factor to the brush-border membrane.

The intrinsic factor-mediated system is capable of handling between 1.5 and 3.0 μg of vitamin B_{12}. The limited capacity of the ileum to absorb B_{12} can be explained by the limited number of receptor sites, there being only about one receptor per microvillus. Saturation of the system at one meal does not preclude absorption of normal amounts of the vitamin some hours later. The entire absorptive process, from ingestion of the vitamin to its appearance in the portal vein, takes 8–12 hours.

Absorption can also occur by simple diffusion across the entire small intestine. This process probably accounts for the absorption of only 1 to 3% of the vitamin consumed in ordinary diets, but can provide a physiologically significant source of the vitamin when it is administered as free cobalamin in pharmacological doses of 30 μg or more.

With regard to efficiency of vitamin B_{12} absorption in humans, the percentage of ingested vitamin B_{12} that is absorbed decreases as the actual amount in the diet increases. At intakes of 0.5 μg or less, about 70% of the available vitamin B_{12} is absorbed. At an intake of 5.0 μg, a mean of 28% is absorbed (range, 2–50%) while at a 10-μg intake the mean absorption is 16% (range, 0–34%) (Herbert, 1987). When 100 μg or more of crystalline vitamin B_{12} is taken, the absorption efficiency drops to 1%, and the excess vitamin is excreted in the urine.

In normal human subjects, the vitamin B_{12} in lean mutton (Heyssel *et al.*, 1966), chicken meat (Doscherholmen *et al.*, 1978), and fish (Doscherholmen *et al.*, 1981) is absorbed as efficiently as a comparable amount of crystalline cyanocobalamin administered orally in aqueous solution. In contrast, the vitamin B_{12} in eggs is poorly absorbed (Doscherholmen *et al.*, 1975) owing to the presence of distinct vitamin B_{12} binding proteins in egg white and egg yolk (Levine & Doscherholmen, 1983).

26.4 DELIVERY OF VITAMIN B_{12} TO TISSUES AND CELLULAR UPTAKE

An estimated 80% of the circulating vitamin B_{12} (200–300 pg mL^{-1}) is bound to transcobalamin I. This protein, a plasma haptocorrin, serves no apparent physiological role in cellular uptake of B_{12}; however, it does prevent circulating B_{12} from being filtered by the kidney, thus obviating the need for renal reabsorption. Vitamin B_{12} destined for tissue uptake binds to a plasma β-globulin, transcobalamin II, with the exclusion of most inactive cobalamin analogs that may have gained access to the circulation. The binding of B_{12} to transcobalamin II takes place either within the ileal enterocyte or at its serosal surface. The transcobalamin II–B_{12} complex is conveyed in the portal blood to the liver which takes up approximately 50% of the vitamin. The remainder is conveyed to other tissues, where cellular uptake of the complex is accomplished by receptor-mediated endocytosis. Following lysosomal degradation of transcobalamin II, the vitamin B_{12} is released into the cytoplasm in the form of hydroxocobalamin. This compound is either converted directly to methylcobalamin in the cytoplasm or eventually to adenosylcobalamin in the mitochondria.

26.5 CONSERVATION OF VITAMIN B_{12}

The body is extremely efficient at conserving vitamin B_{12}. Unlike the other water-soluble vitamins, vitamin B_{12} is stored in the liver, primarily in the form of adenosylcobalamin. In an adult man, the total body store of vitamin B_{12} is estimated to be 2,000 to 5,000 μg, of which about 80% is in the liver. The amount stored in the liver increases with age, more than doubling between the ages of 20 and 60. The remainder of the stored vitamin is located in muscle, skin, and blood plasma. Only 2 to 5 μg of vitamin B_{12} are lost daily through metabolic turnover, regardless of the amount stored in the body.

Approximately 0.5–5 μg of cobalamin is secreted into the jejunum daily, mainly in the bile. Of this, at least 65–75% is reabsorbed in the ileum along with dietary sources of the vitamin (Ellenbogen & Cooper, 1991). Reabsorption of biliary vitamin B_{12} requires intrinsic factor: in the absence of intrinsic factor all the cobalamin is excreted with the feces rather than recirculated. The enterohepatic cycle allows the excretion of unwanted nonvitamin cobalamin analogs, which constitute about 60% of the corrinoids secreted in bile, and returns vitamin B_{12} relatively free of analogs.

The binding of vitamin B_{12} with transcobalamins I and II prevents the vitamin molecule from being excreted in the urine as it passes through the kidney. Only if the circulating B_{12} exceeds the vitamin-binding capacity of the blood is the excess excreted; this typically occurs only after injection of cobalamin. The binding with plasma proteins, negligible urinary loss, and an efficient enterohepatic circulation, together with the slow rate of turnover, explains why strict vegetarians, whose absorptive capacity is intact, take 20 years or more to develop signs of deficiency. People with absorptive malfunction develop deficiency signs within 2–3 years.

26.6 BIOCHEMICAL FUNCTIONS

In humans, only two enzymes require a B_{12} coenzyme: methionine synthase (EC 2.1.1.13) and L-methylmalonyl-coenzyme A (CoA) mutase (EC 5.4.99.2).

26.6.1 Methionine Synthase

Methionine synthase catalyzes the conversion of homocysteine to methionine and regenerates THF. A deficiency of vitamin B_{12} lowers intracellular levels of 5,10-methylene-THF, which is required for DNA synthesis. The following two pathways suggest how this might happen.

26.6.1.1 The Methyl Trap Hypothesis

It can be seen from Figure 25.11 that a lack of methionine synthase activity resulting from vitamin B_{12} deficiency leads to reduced synthesis of methionine and THF and accumulation of homocysteine and 5-methyl-THF. The reduction in THF leads to reduced availability of 5,10-methylene-THF, which is needed to convert deoxyuridine monophosphate (dUMP) to deoxythymidine monophosphate (dTMP) for DNA synthesis. 5-Methyl-THF does not directly participate in DNA synthesis. The deficiency in methionine means that less *S*-adenosylmethionine (SAM) is produced. SAM normally suppresses 5,10-methylene-THF reductase, so an impaired production of SAM will reduce this suppression and promote the conversion of 5,10-methylene-THF to 5-methyl-THF. The latter compound becomes metabolically trapped because (1) the reductase reaction is irreversible *in vivo* and (2) methionine synthase is rendered inactive by the lack of its coenzyme, methylcobalamin. This series of events is known as the methyl trap hypothesis.

A potential way of circumventing the methyl trap is provided by the following reactions in which BH_4 is tetrahydrobiopterin (Figure 26.2), BH_3OH is 4a-hydroxytetrahydrobiopterin, and qBH_2 is quinonoid dihydrobiopterin.

$$\text{Phenylalanine} + O_2 + BH_4 \rightarrow \text{Tyrosine} + BH_3OH \quad (26.1)$$

$$BH_3OH \rightarrow qBH_2 + H_2O \quad (26.2)$$

$$qBH_2 + NAD{+}\ H^+ \rightarrow BH_4 + NAD^+ \quad (26.3)$$

$$qBH_2 + 5\text{-}CH_3\text{-THF} \rightarrow BH_4 + 5{,}10\text{-}CH_2\text{-THF} \quad (26.4)$$

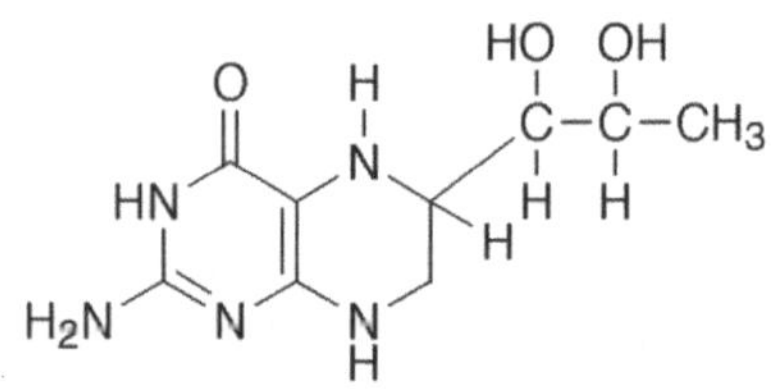

FIGURE 26.2 Structure of tetrahydrobiopterin.

It is evident that if adequate amounts of qBH_2 are present, reaction 26.4 can make 5,10-methylene-THF available for one-carbon metabolism. Whether this escape reaction from the methyl trap actually occurs will depend on the presence of the various enzymes and substrates required to sustain the above reactions (Kaufman, 1991).

26.6.1.2 The Formate Starvation Hypothesis

According to this hypothesis, the lack of methionine caused by vitamin B_{12} deficiency leads to impaired synthesis of 'active' formate (Figure 26.3) and consequent reduced synthesis of 10-formyl-THF, a precursor of 5,10-methylene-THF. In support of this hypothesis, impaired deoxythymidine synthesis is either not improved or only marginally improved when THF is added to vitamin B_{12}-deficient bone marrow cells, but is corrected completely with formyl-THF (Chanarin *et al.*, 1989).

26.6.2 L-Methylmalonyl-Coenzyme A Mutase

L-Methylmalonyl-CoA mutase requires adenosylcobalamin for the conversion of L-methylmalonyl-CoA to succinyl-CoA, an important intermediary in the tricarboxylic acid cycle.

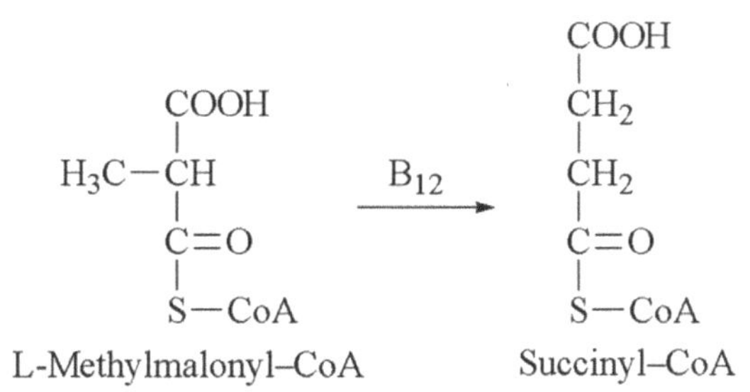

(26.5)

L-Methylmalonyl-CoA is generated from propionyl-CoA, which in turn arises from (1) the degradation of carbon skeletons of methionine, isoleucine, threonine, and valine and (2) the beta-oxidation of fatty acids with odd-numbered chains. Propionyl-CoA is converted in a biotin-dependent reaction to D-methylmalonyl-CoA, which then undergoes racemization to L-methylmalonyl-CoA.

26.7 VITAMIN B_{12} DEFICIENCY IN ANIMALS

In the vitamin B_{12} deficiency state, the activity of its coenzymes is reduced and the respective physiological functions are impaired. Any clinical or subclinical symptom occurring in vitamin B_{12}-deficient animals must be understood as a consequence of such metabolic failures. The following deficiency signs have been observed:

- **General condition.** Loss of appetite, variable feed intake, reduced growth.
- **Skin.** Rough skin and dermatitis in pigs. Defective feathering in growing chicks and turkeys.
- **Nervous system.** Increased irritability, uncoordinated movements, and painful hindquarters in pigs. Leg weakness and perosis in chickens and turkeys. Ataxia in calves fed vitamin B_{12}-free milk-replacer diets.
- **Reproductive system.** Reduced egg production and impaired hatchability in chickens and turkeys. High mortality in rats after feeding a diet free of vitamin B_{12} for several generations.

26.8 VITAMIN B_{12} DEFICIENCY IN HUMANS

The vitamin B_{12} deficiency disease pernicious anemia manifests as a megaloblastic anemia and a neuropathy. The megaloblastic anemia is clinically indistinguishable from that induced by folate deficiency. About 12% of patients with pernicious anemia present with neuropathy alone (Lindenbaum *et al.*, 1988). Unless treated, pernicious anemia is fatal in 1 to 3 years following diagnosis.

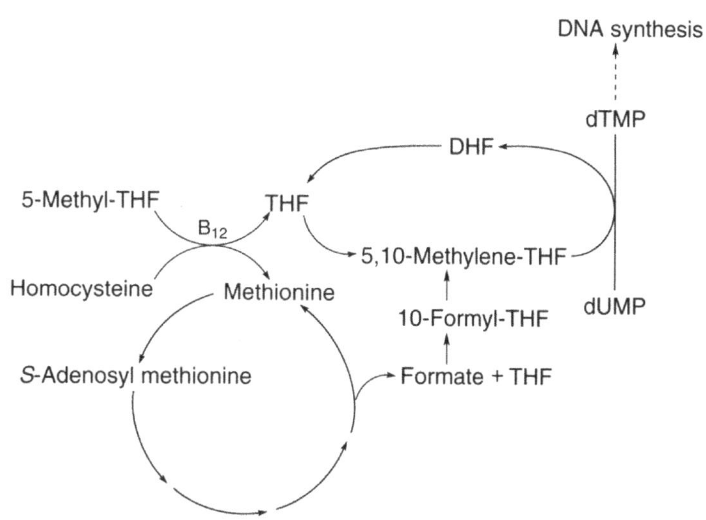

FIGURE 26.3 Pathways showing how a lack of methionine can result in reduced synthesis of 5,10-methylene-THF.

26.8.1 Causes of Vitamin B_{12} Malabsorption

Vitamin B_{12} deficiency results from a failure of intestinal absorption or subsequent transport to the tissues; it is rarely, if ever, caused by a lack of B_{12} in the diet. Disorders of vitamin B_{12} absorption and transport have been discussed by Kapadia and Donaldson (1985) and just a few examples of malabsorption are mentioned here.

- Elderly people are prone to atrophic gastritis, a condition in which the gastric oxyntic mucosa atrophies to such an extent that virtually no hydrochloric acid or intrinsic factor is secreted.
- In patients with diverticula, strictures, and fistulas of the small intestine, stagnant regions of the lumen may become contaminated with colonic bacteria which can take up much of the dietary vitamin B_{12} passing by. Bacteria can take up vitamin B_{12} bound to intrinsic factor, although not as avidly as they can take up the free vitamin. Intrinsic factor and bacteria have a similar affinity for B_{12}, so it is possible that bacterial uptake could take place following the vitamin's release from haptocorrin but before its transfer to intrinsic factor.
- The fish tapeworm *Diphyllobothrium latum* competes with the host for B_{12}, making it less available for absorption.
- A common inherited disorder is an autoimmune reaction with formation of antibodies against intrinsic factor. Such cases involve two types of antibody: type I prevents intrinsic factor from binding to cobalamin, and type II blocks the binding of the intrinsic factor–cobalamin complex to the ileal receptor.

26.8.2 Megaloblastic Anemia

Incoming dietary folate is converted mainly to monoglutamyl 5-methyl-THF which circulates in the plasma. Normally, the 5-methyl-THF is demethylated by the action of methionine synthase, and glutamate residues are attached to the resultant THF to facilitate cellular uptake. However, in the absence of vitamin B_{12}, demethylation cannot take place and the monoglutamyl 5-methyl-THF, being a poor substrate for polyglutamyl synthetase, is poorly incorporated into the B_{12}-deficient cell. Thus a deficiency in vitamin B_{12} results in an inability of cells to utilize folate for DNA synthesis and this leads to megaloblastic anemia. This interrelationship between folate and vitamin B_{12} explains why dietary deficiencies of vitamin B_{12} and folate result in morphologically identical anemias.

The direct cause of the megaloblastic anemia that results from vitamin B_{12} deficiency is a defective production of erythrocytes in the bone marrow due to impaired DNA synthesis. This impairment is a consequence of reduced intracellular levels of 5,10-methylene-THF, which is needed to synthesize deoxythymidine monophosphate, a precursor for DNA synthesis. The hematological changes and clinical signs of megaloblastic anemia are described in Section 25.19.2.

26.8.3 Neuropathy

The neurological changes associated with the neuropathy reflect the inability to manufacture the lipid component of myelin, which results in a generalized demyelination of nerve tissue. The myelin sheath insulates axons, enabling them to conduct nerve impulses at very rapid rates. Neuropathy begins in the peripheral nerves and progresses to the spinal cord and brain. The earliest neurological symptoms are often a tingling, numbness, and a feeling of muscle weakness in the limbs. There is a loss of vibratory and position sense in the feet and a feeling of walking on cotton wool. Motor disturbances are shown by inco-ordination of the legs and a loss of fine co-ordination of the fingers. Impaired bowel and bladder control are manifested by a sense of urinary urgency and constipation. An uncommon symptom, found mainly in carriers of the fish tapeworm, is impaired bilateral vision due to demyelination of the optic nerves. Neuropsychiatric symptoms appear when demyelination progresses to the deep white matter of the cerebral hemispheres. These symptoms may include irritability, memory disturbances, mild depression, and apathy. Uncommonly, more serious mental problems such as psychosis and dementia develop. The response to treatment with vitamin B_{12} depends on the duration of symptoms prior to diagnosis. Treatment will halt progression of neurological impairment, but much of the damage present will remain. If untreated, the patient will die.

The metabolic basis for the neuropathy of vitamin B_{12} deficiency has been the subject of much debate. One plausible explanation proposed by Scott (1992) and Metz (1992) is based on the impaired synthesis of methionine that leads to a deficiency of *S*-adenosylmethionine, a general donor of methyl groups in transmethylation reactions. Baldwin and Carnegie (1971) demonstrated the specific methylation of an arginine residue in myelin basic protein, and so a deficiency of SAM could explain the demyelination of neural tissue. As shown in Figure 25.11, methionine can be synthesized independently of vitamin B_{12} by the reaction between betaine (a product of choline catabolism) and homocysteine that has accumulated as a result of impaired methionine synthase activity. This reaction takes place in the liver but in neural tissue of human, pig, and rat there is a complete absence of betaine:homocysteine methyltransferase in the three species examined (McKeever *et al.*, 1991). Thus, when vitamin B_{12} becomes deficient, neural tissue becomes deprived of a vital substrate, SAM, required for myelination.

There is some evidence to support the above hypothesis that impairment of methionine synthase activity due to lack of vitamin B_{12} leads to blocking of SAM production and consequent demyelination. The evidence is based on the inactivation of methionine synthase by the anesthetic gas nitrous oxide (N_2O). Treatment of monkeys (Scott *et al.*, 1981) and pigs (Weir *et al.*, 1988) with nitrous oxide causes a neuropathy similar to that caused by vitamin B_{12} deficiency; the neuropathy is alleviated by dietary supplementation with methionine. Further evidence comes from children with a deficiency of 5,10-methylenetetrahydrofolate reductase – an inborn error of metabolism associated with reduced concentrations of methionine and SAM in the cerebrospinal fluid. These patients exhibit demyelination in the brain and degeneration of the spinal cord, but there is no megaloblastic bone marrows or anemia (Hyland *et al.*, 1988).

26.9 TOXICITY

Vitamin B_{12} is nontoxic, even in doses that exceed the minimal daily adult human requirement by 10,000 times (Herbert & Das, 1994). Vitamin B_{12} is not broken down within the body. Ingested amounts that exceed the limited binding capacity in plasma and tissues are excreted unchanged in the urine and feces.

FURTHER READING

Glusker, J.P. (1995) Vitamin B_{12} and the B_{12} coenzymes. *Vitamins and Hormones*, **50**, 1–76.

REFERENCES

Baldwin, G.S. & Carnegie, P.R. (1971) Specific enzymatic methylation of an arginine in the experimental allergic encephalomyelitis protein from human myelin. *Science*, **171**, 579–81.

Chanarin, I., Deacon, R., Lumb, M. & Perry, J. (1989) Cobalamin-folate interrelations. *Blood Reviews*, **3**, 211–5.

Donaldson, R.M. (1985) How does cobalamin (vitamin B_{12}) enter and traverse the ileal cell? *Gastroenterology*, **88**, 1069–71.

Donaldson, R.M. (1987) Intrinsic factor and the transport of cobalamins. In: *Physiology of the Gastrointestinal Tract*, 2nd edn. (Ed. by L.R. Johnson), pp. 959–73. Raven Press, New York.

Doscherholmen, A., McMahon, J. & Ripley, D. (1975) Vitamin B_{12} absorption from eggs. *Proceedings of the Society for Experimental Biology and Medicine*, **149**, 987–90.

Doscherholmen, A., McMahon, J. & Ripley, D. (1978) Vitamin B_{12} assimilation from chicken meat. *American Journal of Clinical Nutrition*, **31**, 825–30.

Doscherholmen, A., McMahon, J. & Economon, P. (1981) Vitamin B_{12} absorption from fish. *Proceedings of the Society for Experimental Biology and Medicine*, **167**, 480–4.

Ellenbogen, L. & Cooper, B.A. (1991) Vitamin B_{12}. In: *Handbook of Vitamins*, 2nd edn. (Ed. by L.J. Machlin), pp. 491–536. Marcel Dekker, New York.

Farquharson, J. & Adams, J.F. (1976) The forms of vitamin B-12 in foods. *British Journal of Nutrition*, **36**, 127–36.

Gräsbeck, R. & Salonen, E.-M. (1976) Vitamin B_{12}. *Progress in Food and Nutrition Science*, **2**, 193–231.

Herbert, V. (1987) Recommended dietary intakes (RDI) of vitamin B-12 in humans. *American Journal of Clinical Nutrition*, **45**, 671–8.

Herbert, V. (1988) Vitamin B-12: Plant sources, requirements, and assays. *American Journal of Clinical Nutrition*, **48**, 852–8.

Herbert, V. & Das, K.C. (1994) Folic acid and vitamin B_{12}. In: *Modern Nutrition in Health and Disease*, 8th edn, Vol. 1. (Ed. by M.E. Shils, J.A. Olson & M. Shike), pp. 402–25. Lea & Febiger, Philadelphia.

Heyssel, R.M., Bozian, R.C., Darby, W.J. & Bell, M.C. (1966) Vitamin B_{12} turnover in man. The assimulation of vitamin B_{12} from natural foodstuff by man and estimates of minimal daily dietary requirements. *American Journal of Clinical Nutrition*, **18**, 176–84.

Hyland, K., Smith, I., Bottiglieri, T., Perry, J., Wendel, U., Clayton, P.T. & Leonard, J.V. (1988) Demyelination and decreased *S*- adenosylmethionine in 5,10-methylenetetrahydrofolate reductase deficiency. *Neurology*, **38**, 459–62.

Kapadia, C.R. & Donaldson, R.M. Jr. (1985) Disorders of cobalamin (vitamin B_{12}) absorption and transport. *Annual Review of Medicine*, **36**, 93–110.

Kaufman, S. (1991) Some metabolic relationships between biopterin and folate: Implications for the "methyl trap hypothesis". *Neurochemical Research*, **16**, 1031–6.

Levine, A.S. & Doscherholmen, A. (1983) Vitamin B_{12} bioavailability from egg yolk and egg white: Relationship to binding proteins. *American Journal of Clinical Nutrition*, **38**, 436–9.

Lindenbaum, J., Healton, E.B., Savage, D.G., Brust, J.C.M., Garrett, T.J., Podell, E.R., Marcell, P.D., Stabler, S.P. & Allen, R.H. (1988) Neuropsychiatric disorders caused by cobalamin deficiency in the absence of anemia or macrocytosis. *New England Journal of Medicine*, **318**, 1720–8.

McKeever, M.P., Weir, D.G., Molloy, A. & Scott, J.M. (1991) Betaine–homocysteine methyltransferase: Organ distribution in man, pig and rat and subcellular distribution in the rat. *Clinical Science*, **81**, 551–6.

Metz, J. (1992) Cobalamin deficiency and the pathogenesis of nervous system disease. *Annual Review of Nutrition*, **12**, 59–79.

National Research Council. (1989) Water-soluble vitamins. In: *Recommended Dietary Allowances*, 10th edn, pp. 115–73. National Academy Press, Washington, DC.

Scott, J.M., Dinn, J.J., Wilson, P. & Weir, D.G. (1981) Pathogenesis of subacute combined degeneration: A result of methyl group deficiency. *Lancet*, **II**(8242), 334–7.

Scott, J.M. (1992) Folate–vitamin B_{12} interrelationships in the central nervous system. *Proceedings of the Nutrition Society*, **51**, 219–24.

Seetharam, B. (1994) Gastrointestinal absorption and transport of cobalamin (vitamin B_{12}). In: *Physiology of the Gastrointestinal Tract*, 3rd edn. (Ed. by L.R. Johnson), pp. 1997–2026. Raven Press, New York.

Seetharam, B. & Alpers, D.H. (1991) Gastric intrinsic factor and cobalamin absorption. In: *Handbook of Physiology, Section 6: The Gastrointestinal System, Vol. IV. Intestinal Absorption and Secretion*. Volume (Ed. by M. Field & R.A. Frizzell), pp. 437–61. American Physiological Society, Bethesda, MD.

Seetharam, B., Presti, M., Frank, B., Tiruppathi, C. & Alpers, D.H. (1985) Intestinal uptake and release of cobalamin complexed with rat intrinsic factor. *American Journal of Physiology*, **248**, G326–331.

Weir, D.G., Keating, S., Molloy, A., McPartlin, J., Kennedy, S., Blanchflower, J., Kennedy, D.G., Rice, D. & Scott, J.M. (1988) Methylation deficiency causes vitamin B_{12}-associated neuropathy in the pig. *Journal of Neurochemistry*, **51**, 1949–52.

27 Vitamin C

27.1 CHEMISTRY

The term 'vitamin C' refers to both ascorbic acid and dehydroascorbic acid, since the latter oxidation product is reduced back to ascorbic acid in the body. The principal natural compound with vitamin C activity is L-ascorbic acid. There are two enantiomeric pairs (mirror images) of the 2-hexenono-1,4-lactone structure; namely, L- and D-ascorbic acid and L- and D-isoascorbic acid (Figure 27.1). D-Ascorbic acid and L-isoascorbic acid are devoid of vitamin C activity and do not occur in nature. D-Isoascorbic acid (commonly known as erythorbic acid) is an epimer of L-ascorbic acid, the structural difference being the orientation of the hydrogen and hydroxyl group at the fifth carbon atom. D-Isoascorbic acid is also not found in natural products, apart from its occurrence in certain microorganisms. It possesses similar reductive properties to L-ascorbic acid, but exhibits only 5% of the antiscorbutic activity of L-ascorbic acid in guinea pigs (Pelletier & Godin, 1969).

At around neutral pH, ascorbic acid exists as the ascorbate anion due to facile ionization of the hydroxyl group on C-3. Ascorbate is easily and reversibly oxidized to dehydro-L-ascorbic acid, forming the ascorbyl radical (also known as semidehydroascorbate) as an intermediate (Figure 27.2). The delocalized nature of the unpaired electron in the ascorbyl radical makes it a relatively unreactive free radical, and two ascorbyl radicals can react together non-enzymatically to produce ascorbate and dehydroascorbic acid. In the body, enzymes are available to reduce the ascorbyl radical and dehydroascorbic acid back to ascorbate. Dehydroascorbic acid is not a true organic acid as it contains no readily ionizable protons. In aqueous solution, dehydroascorbic acid exists not as the 2,3-diketo compound, but as the bicyclic hemiketal hydrate. In buffered solution at neutral or alkaline pH, dehydroascorbic acid undergoes a nonreversible oxidation in which the two rings open to give 2,3-diketogulonic acid in a straight-chain structure.

Humans and other primates, guinea pigs, fruit-eating bats, some exotic birds (e.g. the red-vented bulbul), and fish such as Coho salmon, rainbow trout, and carp, lack the enzyme L-gulono-γ-lactone oxidase, which catalyzes the final step in the biosynthesis of ascorbic acid, and thus rely on their diet to provide vitamin C. Other animal species can synthesize ascorbic acid from glucose and have no need for dietary vitamin C.

27.2 FOOD SOURCES AND BIOAVAILABILITY

27.2.1 Food Sources

Fresh fruits (especially citrus fruits and blackcurrants) and green vegetables constitute rich sources of vitamin C. Potatoes contain moderate amounts but, because of their high consumption, represent the most important source of the vitamin in the British diet. Liver (containing 10–40 mg per 100 g), kidney, and heart are good sources, but muscle meats and cereal grains do not contain the vitamin in measurable amounts. Human milk provides enough ascorbic acid to prevent scurvy in breast-fed infants, but preparations of cow's milk are a poor source owing to oxidative losses incurred during processing. A typical Western diet containing fruits and vegetables provides upwards of 200 mg vitamin C per day, a level at which tissues will probably be saturated (Levine *et al.*, 1999).

27.2.2 Bioavailability

In a human study involving 68 adult male nonsmokers (Mangels *et al.*, 1993), ascorbic acid bioavailability from oranges, orange juice, and cooked broccoli was not significantly different to that from synthetic ascorbic acid, indicating that the bioavailability of vitamin C in these natural food sources is high.

27.3 VITAMIN C TRANSPORT PROTEINS

Two isoforms of the Na^+-dependent L-ascorbic acid transporter, SVCT1 and SVCT2, have been cloned from rat kidney and rat brain, respectively, and characterized (Tsukaguchi *et al.*, 1999). Despite their close sequence homology and similar function, the two isoforms are discretely distributed (Table 27.1). SVCT1 is mainly confined to bulk-transporting epithelia (intestine, kidney, liver), whereas SVCT2 serves a host of metabolically active cells and specialized tissues in the brain, eye, and other organs. Daruwala *et al.* (1999) cloned and characterized the human SVCT1 and SVCT2 transporters, which are products of the *SLC23A1* and *SLC23A2* genes, respectively. Human SVCT2 has a higher affinity for ascorbic acid than SVCT1 (Liang *et al.*, 2001). The optimum pH for ascorbic acid transport by both SVCT-1 and SVCT-2 is pH 7.5: at pH 5.5, transport is reduced by ~90%. Both transporter isoforms are expressed in the intestine: SVCT1 is expressed at the apical membrane and SVCT2 is localized basolaterally (Boyer *et al.*, 2005). The transporters are differentially expressed along the intestinal tract, with expression levels lower in the colon than in the jejunum (Subramanian *et al.*, 2017).

Dehydroascorbic acid can be transported by the sodium-independent facilitative glucose transporter isoforms GLUT1 and GLUT3 with an affinity similar to or lower than that for glucose (Rumsey *et al.*, 1997). GLUT1 has wide tissue distribution, while GLUT3 is primarily expressed

DOI: 10.1201/9781003410300-31

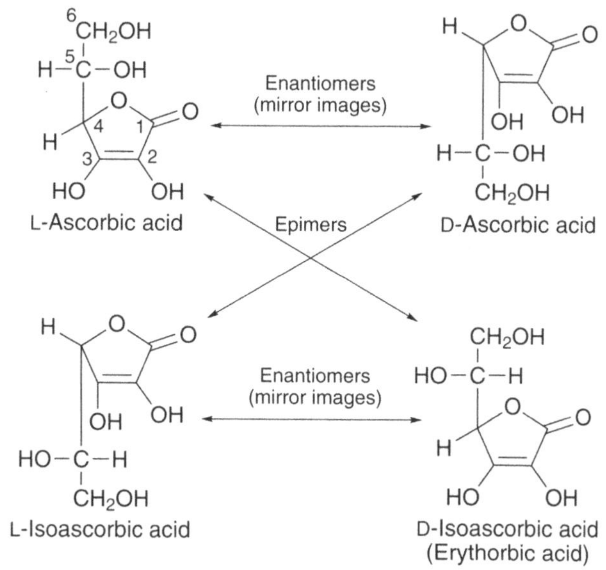

FIGURE 27.1 Stereoisomers of 2-hexenono-1,4-lactone.

in brain, placenta, testis, and platelets. Dehydroascorbic acid uptake has been demonstrated in many human cells and tissues, including neutrophils, fibroblasts, erythrocytes, platelets, the blood–brain barrier, and placenta, all of which express relatively high levels of either GLUT1 or GLUT3. In these cells and tissues, dehydroascorbic acid is transported across the cell membrane and accumulated in the reduced form, ascorbic acid, which is not transportable through the bidirectional glucose transporters (Vera *et al.*, 1995).

27.4 ABSORPTION OF DIETARY VITAMIN C

27.4.1 General Principles of Absorption

Approximately 80–90% of the vitamin C content of a given foodstuff exists in the reduced form, ascorbic acid; the remainder is in the oxidized form, dehydroascorbic acid. Ascorbic acid and dehydroascorbic acid are absorbed by separate transport mechanisms in animal species that depend upon dietary vitamin C (Figure 27.3). The absorption mechanism for ascorbic acid is sodium-coupled, secondary active transport. Inside the absorptive cell (enterocyte) of the intestinal epithelium, dehydroascorbic acid is enzymatically reduced, and the accumulated ascorbic acid is transported across the basolateral membrane to the bloodstream. In addition to uptake at the brush-border membrane, dehydroascorbic acid from the bloodstream can be taken up at the basolateral membrane, reduced within the cell, and returned to the circulation in the form of the useful and nontoxic ascorbic acid. The serosal uptake of dehydroascorbic acid from the bloodstream and intracellular reduction to ascorbic acid take place in animal species that do not have a dietary vitamin C requirement, as well as those species that do.

The ability of the enterocyte to absorb dehydroascorbic acid efficiently is important because, apart from the indigenous dehydroascorbic acid content of the diet, additional

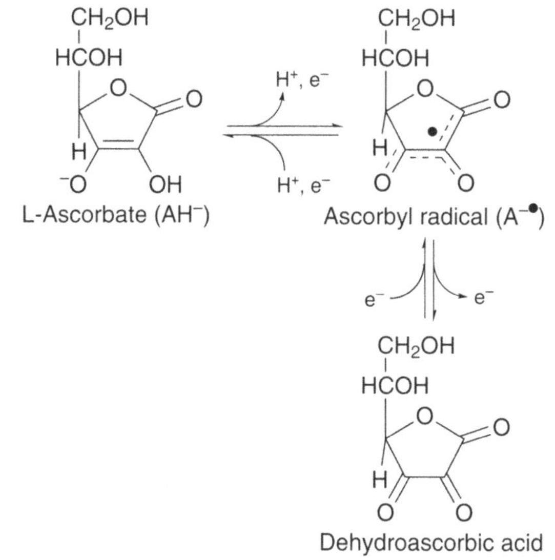

FIGURE 27.2 Oxidation of ascorbate. Note the delocalized unpaired electron in the ascorbyl radical.

oxidation of ascorbic acid occurs in the gastrointestinal tract as the vitamin functions to maintain other nutrients such as iron in the reduced state. The intestinal uptake and reduction of dehydroascorbic acid explains why this compound, orally administered, maintains plasma concentrations of ascorbic acid and prevents scurvy. The overall system of intestinal transport and metabolism is designed to maximize the conservation of vitamin C and also to maintain the vitamin in its nontoxic reduced state, whether it is derived from the diet or from the circulation.

27.4.2 Transport Mechanisms

Ascorbic acid and dehydroascorbic acid are transported into human cells by separate mechanisms (Welch *et al.*, 1995).

TABLE 27.1
Distribution of Na^+-Dependent L-Ascorbic Acid Transporters (SVCT1 and SVCT2) in Rat Tissues

Tissue	SVCT1	SVCT2
Intestine (enterocytes)	+	+
Liver (hepatocytes)	+	–
Kidney (proximal tubule)	+	–
Neurons throughout the central nervous system	–	+
Brain (meninges, choroid plexus)	–	+
Osteoblasts	–	+
Retina (inner nuclear layer)	–	+
Endocrine glands (various)	–	+
Stomach (gastric glands)	–	+
Immune system organs (spleen, thymus)	–	+
Testis (interstitial cells, spermatocytes)	–	+

+ indicates abundant expression; – indicates lack of expression or weak expression.
Source: Tsukaguchi *et al.* (1999).

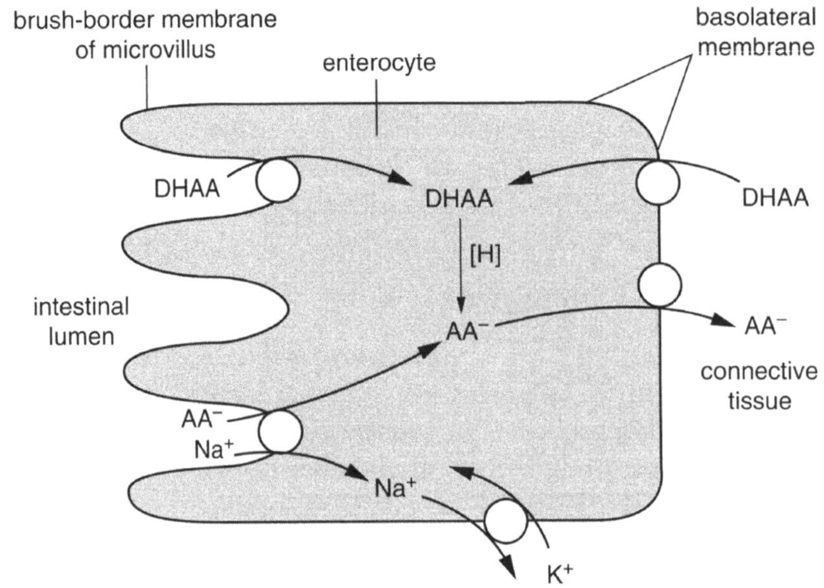

FIGURE 27.3 Model of intestinal transport of the L-ascorbate anion (AA^-) and uncharged dehydroascorbic acid (DHAA) in vitamin C-dependent animals. Thick arrowed lines indicate directional pathways; [H] signifies enzymatic reduction.

27.4.2.1 Ascorbic Acid

Ascorbic acid is absorbed by both small and large intestine. Uptake is lower in the colon than in the small intestine, commensurate with the differential expression of the transporters along the intestinal tract. Absorption of physiological intakes of ascorbic acid by guinea pigs takes place mainly in the ileum and occurs as a result of specific carrier-mediated mechanisms in the brush-border and basolateral membranes of enterocytes (Mellors *et al.*, 1977; Patterson *et al.*, 1982; Bianchi *et al.*, 1986). Ascorbic acid is 99.9% ionized within the pH range of intestinal chyme, and therefore it is the ascorbate anion (specifically L-ascorbate$^-$) that is transported across the brush-border membrane.

The mechanism of ascorbate absorption is sodium-coupled, secondary active transport with a Na^+:ascorbate stoichiometry of 2:1, making the process electrogenic (Liang *et al.*, 2001). The immediate energy for the sodium-coupled transport of ascorbate is provided by the inward sodium concentration gradient, which in turn is created and maintained by the sodium pump at the basolateral membrane. Phloridzin, a well-known fully competitive inhibitor of D-glucose transport across the small intestinal brush-border membrane, does not inhibit L-ascorbate uptake, demonstrating that D-glucose and L-ascorbate do not share the same carrier. Uptake of L-ascorbate is, however, competitively inhibited by D-isoascorbic acid, making the latter a potential antivitamin C.

Ascorbate leaves the enterocyte by sodium-independent, facilitated diffusion at the basolateral membrane (Bianchi *et al.*, 1986). Although the high intracellular concentration of ascorbate and the negative membrane potential are energetically favorable toward the exit of ascorbate, a carrier protein is required to facilitate transport of the anion across the lipid bilayer.

27.4.2.2 Dehydroascorbic Acid

Dehydroascorbic acid, lacking the dissociable hydrogens at carbon atoms 2 and 3, does not ionize and is therefore unable to be co-transported with sodium. In the guinea pig intestine, intraluminal dehydroascorbic acid is transported across the brush-border membrane by facilitated diffusion, driven by the steep concentration gradient maintained by its intracellular reduction to ascorbate. In contrast, rat small intestine takes up dehydroascorbic acid from the lumen only very slowly. Dehydroascorbic acid is taken up from the blood across the basolateral membrane by facilitated diffusion in both guinea pig and rat (Rose *et al.*, 1988).

27.4.3 Efficiency of Ascorbate Absorption in Humans

The usual dietary intake of vitamin C ranges from 30 to 180 mg per day and over this range the efficiency of absorption is 70–90% (Institute of Medicine, 2000). Brush-border uptake by the sodium-coupled, secondary active transport mechanism reaches its maximum rate at a relatively low luminal concentration. Beyond physiological intakes, absorption becomes progressively less efficient, falling from 75% of a single 1-g dose to 16% of a single 12-g dose (Table 27.2). This fall-off in efficiency occurs because absorption of high luminal concentrations of vitamin C takes place mainly by simple diffusion, and this passive movement proceeds at a very low rate.

The ingestion of eight 0.125-g doses of ascorbate spaced throughout the day produced a 72% increase in absorption compared to a single 1-g dose (Yung *et al.*, 1981). The absorption efficiency of a single dose can be improved if the ascorbate is ingested in the form of a sustained-release

TABLE 27.2
Absorption of Large, Single, Oral Doses of Ascorbic Acid in Humans

Dose Ingested (g)	Absorption Efficiency (%)	
	Expt 1[a]	Expt 2[b]
1	75	—
1.5	—	50
2	44	—
3	39	36
4	28	—
5	20	—
6	—	26
12	—	16

[a]Hornig *et al.* (1980).
[b]Kübler and Gehler (1970).

capsule (Sacharin *et al.*, 1976). The ingestion of 1 g of ascorbate immediately after a fatty meal produced a 69% increase in absorption compared to the same dose given on an empty stomach (Yung *et al.*, 1981). The divided dose effect is consistent with a saturable absorption mechanism, while the after-meal effect indicates a slowing of gastric emptying.

27.4.4 Adaptive Regulation of Ascorbate Absorption

In the guinea pig, intestinal absorption of ascorbate is adaptively regulated in a transient and reversible manner by the level of dietary ascorbate (Karasov *et al.*, 1991). The mechanism of regulation is an increase or decrease in the number of carriers at both brush-border and basolateral membranes of enterocytes in response to low or high concentrations of ascorbate in the blood. The rationale for adaptive regulation is that carriers are most needed at low dietary ascorbate levels; at excessive levels the required amount of ascorbate can be absorbed by fewer carriers, aided by passive diffusion. As ascorbate does not provide metabolizable energy, there is nothing to gain from the cost of synthesizing and maintaining carriers when the vitamin supply is in excess.

MacDonald *et al.* (2002) showed that incubating human intestinal Caco-2 cells with a high concentration of ascorbic acid (4.5 mg/ml) for 24 hours decreased ascorbic acid uptake and reduced expression of the SVCT1 transporter. This was evidence that the concentration of vitamin C in the intestinal lumen regulates absorption of the vitamin by reducing expression of the predominant transporter SVCT1.

Amano *et al.* (2010) used a strain of senescence marker protein-30 (SMP30)/gluconolactonase (GNL) knockout mice, which cannot synthesize ascorbic acid, to examine whether ascorbic acid and dehydroascorbic acid regulate their respective transporters. They found that SVCT1 and SVCT2 expression levels and ascorbate uptake ability were significantly enhanced in the hepatocytes of vitamin C-depleted mice, indicating that ascorbic acid regulates SVCT1 and SCVT2 in the liver.

27.5 PLASMA TRANSPORT AND TISSUE DISTRIBUTION

Vitamin C in plasma occurs as free (nonprotein-bound) ascorbate with concentrations ranging from 5 to 90 μM; dehydroascorbic acid is present at <2% of ascorbate and is often undetectable (Rumsey & Levine, 1998). Ascorbic acid accumulates at mM concentrations in neutrophils, lymphocytes, monocytes, and platelets. Uptake of ascorbate by human leukocytes is a stereospecific, sodium-dependent, active transport process that shows different transport kinetics for different cell types (Moser, 1987). Red cell levels of ascorbate are probably less than plasma levels (Rumsey & Levine, 1998). Nearly all studies show that smokers have decreased plasma and leukocyte ascorbate concentrations compared to nonsmokers. Smoking increases the metabolic turnover of vitamin C, apparently due to increased oxidative stress (Kallner *et al.*, 1981).

Vitamin C is widely distributed to the cells of the body, all of which take up and utilize the vitamin. Tissues with the greatest concentration of ascorbate include adrenal and pituitary glands with 30–50 mg per 100 g, followed by liver, spleen, eye lens, pancreas, kidney, and brain with between 10 and 30 mg per 100 g (Rumsey & Levine, 1998). Liver has the greatest store by virtue of its size.

27.6 RENAL REABSORPTION

27.6.1 General Principles

The kidney actively reabsorbs ascorbate present in the glomerular filtrate, thereby maximizing vitamin C conservation in the body and helping the intestine to maintain the circulating vitamin in its useful, reduced state. The kidneys of all mammals handle vitamin C in a similar manner. Renal reabsorption of vitamin C is an essential process for humans as, without it, urinary loss would far exceed the average daily intake of the vitamin. Although species that have the ability to synthesize ascorbic acid might be able to replace that lost in the urine, the metabolic costs would be high.

27.6.2 Transport Mechanisms

27.6.2.1 Ascorbic Acid

Renal uptake of the L-ascorbate anion at the brush-border membrane of the absorptive cell of the proximal convoluted tubule is, like intestinal uptake in the human or guinea pig, a sodium-coupled, secondary active transport system (Rose, 1986; Bowers-Komro & McCormick, 1991). Unlike the corresponding intestinal transport system, however, the renal system is electrogenic, indicating a Na^+/ascorbate$^-$ coupling ratio of 2:1 (Toggenburger *et al.*, 1981). As the loaded carrier bears a net positive charge, its transport is accelerated by the negative membrane potential. Rapid renal reabsorption of ascorbate is essential considering that the transit time in the proximal tubule is only about 10 seconds. Ascorbate is transported across the basolateral membrane by sodium-independent facilitated diffusion (Bianchi & Rose, 1985a).

27.6.2.2 Dehydroascorbic Acid

The mechanism of dehydroascorbic acid transport in renal brush-border (Bianchi & Rose, 1985b) and basolateral (Bianchi & Rose, 1985a) membrane vesicles appears to be facilitated diffusion. A favorable gradient for continued renal reabsorption is maintained by intracellular reduction of recently reabsorbed dehydroascorbic acid (Rose, 1989). Dehydroascorbic acid is taken up also from the blood across the basolateral cell membrane and subsequently reduced to ascorbate, which is then returned to the circulation (Rose, 1989). The kidney participates with the intestine and blood components in promoting reduction of dehydroascorbic acid derived from the blood.

27.7 CATABOLISM AND EXCRETION

In humans, the first step in the catabolism of ascorbic acid is reversible oxidation to dehydroascorbic acid by a variety of non-specific enzymatic and non-enzymatic reactions. Dehydroascorbic acid is hydrolyzed enzymatically and irreversibly to 2,3-diketogulonic acid, which in turn is decomposed to a variety of compounds, including oxalic and threonic acids, L-xylose, and ascorbate-2-sulfate. The major route for the elimination of catabolic products is urinary excretion; negligible amounts of ascorbic acid or its catabolites are excreted in feces. Vitamin C intakes greater than the urinary threshold provide little or no increase in the ascorbate body pool (Kallner *et al.*, 1979). When large amounts of vitamin C are ingested, appreciable amounts are broken down in the intestinal lumen to compounds that are degraded to carbon dioxide, which is eliminated by exhalation (Kallner *et al.*, 1985). This presystemic formation of carbon dioxide may result from microbiological or chemical degradation of ascorbate.

27.8 TRANSPORT IN THE BRAIN

In vitamin C-deprived guinea pigs, the brain retains substantial amounts of ascorbic acid when other tissues are severely depleted. On the other hand, raising the plasma level of ascorbic acid does not markedly increase the concentration of ascorbate in the cerebrospinal fluid (CSF) or brain. The importance of ascorbic acid homeostasis in the brain lies in its functions as a neuromodulator and neuroprotective agent (Rebec & Pierce, 1994). The vitamin also enables Schwann cells to assemble a basal lamina, which is required for myelination to proceed (Eldridge *et al.*, 1987, 1989).

Ascorbic acid concentrations in CSF and brain are maintained within a narrow range. The concentration of ascorbic acid in the brain exceeds that of plasma by more than tenfold. Because ascorbic acid is not synthesized in the central nervous system or bound in the CSF or brain, there must be active (uphill) transport mechanisms for ascorbic acid at the blood–brain and/or blood–CSF barriers, and by brain cells themselves. Ascorbic acid is not able to pass through the blood–brain barrier, but its immediate oxidation product, dehydroascorbic acid, can do so by means of GLUT1 (Agus *et al.*, 1997). Ascorbic acid enters the CSF by active transport via the choroid plexus and then diffuses into the extracellular space of the brain. It is then actively transported across the brain cell membrane and concentrated in the brain cell. Saturable ascorbic acid uptake by cultured rat cerebral astrocytes (a type of glial cell) was shown to be sodium dependent and electrogenic with a Na^+/ascorbate$^-$ stoichiometry of 2:1 (Wilson, 1989; Wilson *et al.*, 1991). Entry of ascorbic acid into CSF is half-saturated at the normal plasma level, whereas entry into brain is almost completely saturated at the normal CSF level (Spector, 1977). Thus the transport systems are ideally set for homeostasis.

27.9 PLACENTAL TRANSPORT

Uptake of vitamin C in its oxidized form, dehydroascorbic acid, has been demonstrated in the isolated guinea pig placenta (Leichtweiss *et al.*, 1987), the isolated human near-term placenta (Rybakowski *et al.*, 1995), human placental membrane vesicles (Ingermann *et al.*, 1986), and human placental tissue fragments (Choi & Rose, 1989). Placental uptake of dehydroascorbic acid proceeded much more rapidly than uptake of ascorbic acid. This can be explained by the high frequency of facilitative glucose transporters on the microvillous membrane of the syncytiotrophoblaste coupled to the effective intracellular reduction of the transported dehydroascorbic acid to ascorbic acid. The finding of avid cellular uptake and reduction of dehydroascorbic acid supports the concept that the placenta clears this toxic compound from the maternal circulation and delivers the useful ascorbic acid to the fetus.

Prasad *et al.* (1998) measured the uptake of radioactive vitamin C by the human placental choriocarcinoma cell line, JAR, under conditions in which the vitamin existed either in the oxidized or reduced form. Both forms were transported into the cells, but by different mechanisms. Ascorbate was transported by a high-affinity/low-capacity active carrier with a binding constant (K_m) of 22 μM. Dehydroascorbic acid was transported by a low-affinity/high-capacity facilitative carrier that was sodium independent and was inhibited by D-glucose. As both GLUT1 and GLUT3 are expressed in JAR cells, these glucose transporters are presumably responsible for the uptake of dehydroascorbic acid.

Rajan *et al.* (1999) isolated the human sodium-dependent vitamin C transporter (SVCT2) from placenta and found a Na^+:ascorbate$^-$ stoichiometry of 2:1. The functional characteristics of the induced transport activity were similar to those described for ascorbate transport in JAR cells.

27.10 BIOCHEMICAL AND NEUROCHEMICAL FUNCTIONS

Ascorbic acid acts as a co-factor for eight mammalian enzymes (Table 27.3) by keeping enzyme-bound iron or copper ions in the necessary reduced state. Other reducing agents are far less effective as co-factors *in vitro* and so these enzymes are considered to be ascorbic acid-dependent. The roles of such enzymes in four biochemical processes are discussed below (Sections 27.10.1–27.10.4). Ascorbic acid also enhances the synthesis of at least one of the prostaglandins and is implicated in the release of insulin from pancreatic islets.

Ascorbic acid modulates neurotransmitter systems by regulating neural receptor synthesis and facilitating the release of transmitters.

27.10.1 Biosynthesis of Collagen

Collagen is the major macromolecule of most connective tissues. It is composed of three α chain subunits that are wound together to form a triple helix. Cross linking gives the molecule a rigid and inextensible structure. There are over 25 different α chains that associate to yield 15 different types of collagen. Type I collagen, which is found in large quantities in skin and bone, comprises two α1(I)-chains and one α2(I)-chain. The amino acid composition of collagen is unusual among animal proteins in that it has an abundance of proline and 4-hydroxyproline and a few residues of 3-hydroxyproline and hydroxylysine. The hydroxyproline residues are necessary for proper structural conformation and stability; hydroxylysine residues take part in cross-linking and facilitate subsequent glycosylation and phosphorylation.

Collagen α chains are synthesized in a precursor form known as proα chains, which have additional noncollagenous amino acid sequences (propeptides) at both amino and carboxyl termini. The presence of hydroxyproline and hydroxylysine arises through the post-translational hydroxylation of particular proline and lysine residues in the polypeptide chain. Within the cisternae of the rough endoplasmic reticulum, the newly synthesized proα chains encounter three hydroxylating enzymes. Two of these enzymes, prolyl-4-hydroxylase and prolyl-3-hydroxylase, convert proline residues to 4-hydroxyproline or 3-hydroxyproline respectively, and the third, lysyl hydroxylase, converts lysine residues to hydroxylysine. Following amino acid modification, the propeptides at the carboxyl termini of two proα1 and one proα2 chains associate and bond through disulfide bridges. Triple helix formation then takes place as the protein passes through the endoplasmic reticulum. Following attachment of carbohydrate moieties to the carboxy terminal propeptides, the procollagen molecules are transported to the cell surface within secretory granules. Enzymatic removal of the propeptides during the process of extrusion allows the collagen molecules to spontaneously assemble into fibrils. These are then cross-linked by a series of covalent bonds and deposited into the extracellular matrix.

Ascorbic acid stimulates collagen synthesis through increased transcription of procollagen genes (Hitomi & Tsukagoshi, 1996). Also, ascorbic acid is an essential cofactor for the post-translational hydroxylation of proline and lysine residues in the polypeptide chain. Each of the enzymes concerned contains an iron ion (maintained in the ferrous state by ascorbate) and requires molecular oxygen and α-ketoglutarate as co-substrates (Prockop *et al.*, 1979) (Figure 27.4). The absence of wound healing is one of the features of scurvy that can be attributed to impaired collagen synthesis arising from lack of vitamin C.

The pathway of collagen synthesis is tightly coupled through feedback regulation (Schwarz *et al.*, 1987). Proline hydroxylation stabilizes the triple helical conformation of the procollagen. This conformation increases the secretion rates by sixfold and this in turn leads to an increase in translational efficiency. Therefore ascorbate levels, solely by controlling the activity of the proline hydroxylation step, can control the chain of events through the whole pathway.

27.10.2 Biosynthesis of Carnitine

Carnitine (3-hydroxy-4-*N*-trimethylaminobutyric acid; Figure 27.5) is essential for the β-oxidation of long-chain

TABLE 27.3
Ascorbic Acid-Dependent Mammalian Enzymes

Enzyme	Function
Prolyl-4-hydroxylase (EC 1.14.11.2)	Collagen synthesis
Prolyl-3-hydroxylase (EC 1.14.11.7)	Collagen synthesis
Lysyl hydroxylase (EC 1.14.11.4)	Collagen synthesis
ε-*N*-Trimethyllysine hydroxylase (EC 1.14.11.8)	Carnitine synthesis
γ-Butyrobetaine hydroxylase (EC 1.14.11.1)	Carnitine synthesis
Dopamine β-hydroxylase (EC 1.14.17.1)	Noradrenaline synthesis
Peptidylglycine α-amidating monooxygenase (EC 1.14.17.3)	Peptide amidation
4-Hydroxyphenylpyruvate hydroxylase (EC 1.13.11.27)	Tyrosine metabolism

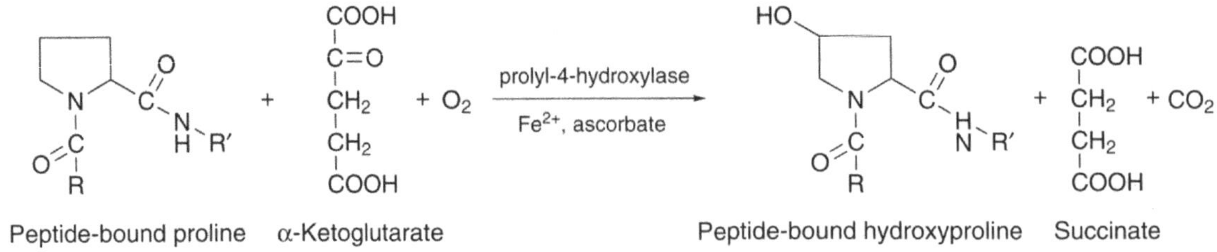

FIGURE 27.4 Hydroxylation of proline.

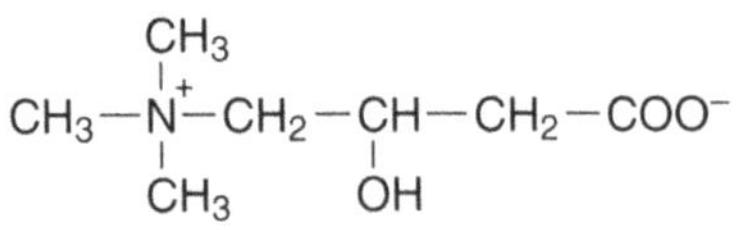

FIGURE 27.5 L-Carnitine.

fatty acids, producing energy in the mitochondria. Neither free fatty acids nor fatty acyl coenzymes can penetrate the inner membranes of mitochondria, but acylcarnitine can readily do so. The translocation of fatty acids from the cytosol to the β-oxidation site in the matrix of the mitochondrion is therefore dependent upon carnitine.

The synthesis of carnitine commences with the conversion of specific peptide-linked lysine residues to trimethyllysine within certain proteins. Proteins in which trimethyllysine has been found include histones, myosin, calmodulin, and cytochrome c. Free trimethyllysine released by proteolytic cleavage is converted to carnitine in a number of steps which include two hydroxylations. The enzymes responsible for these hydroxylations are ε-*N*-trimethyllysine hydroxylase and γ-butyrobetaine hydroxylase. Like the hydroxylases involved in collagen biosynthesis, both enzymes contain ferrous iron and require molecular oxygen and α-ketoglutarate as co-substrates. Ascorbate is required as a co-factor to maintain the iron in the ferrous state (Rebouche, 1991).

Vitamin C deficiency results in a variable decrease in carnitine levels of skeletal muscle, heart muscle, liver, and kidney. Although it has never been systematically tested, investigators have postulated that the fatigue and muscle weakness observed in human scurvy is attributable to the lack of fatty acid-based energy production, resulting from depleted carnitine levels in skeletal muscle. Carnitine status depends on dietary carnitine intake, a modest rate of carnitine biosynthesis, and efficient conservation of carnitine by renal reabsorption (Rebouche, 1988). The rate of carnitine synthesis in scorbutic guinea pigs was shown to be much higher than the rate in normal guinea pigs (Rebouche, 1995a), thus impaired synthesis of carnitine does not appear to be responsible for carnitine depletion in scurvy. Further investigation (Rebouche, 1995b) showed that the factor responsible is increased urinary excretion of carnitine resulting from inefficient renal reabsorption. In support of the guinea pig studies, Jacob and Pianalto (1997) showed that urinary carnitine excretion increased in healthy men during experimental vitamin C depletion. However, the increased excretion had no substantial effect on carnitine status (plasma carnitine and serum triglycerides) over the nine-week period of vitamin C depletion. The mechanism by which vitamin C deficiency increases the rate of carnitine excretion is not known.

27.10.3 Biosynthesis of Noradrenaline

As a neurotransmitter, noradrenaline (norepinephrine) is synthesized in the nerve terminals of postganglionic sympathetic neurones, referred to as adrenergic neurones. As a hormone, noradrenaline, along with adrenaline, is synthesized in the chromaffin cells of the adrenal medulla. The final and probably rate-limiting step in the biosynthetic pathway – the conversion of dopamine to noradrenaline (Figure 27.6) – is catalyzed by dopamine-β-hydroxylase in the presence of molecular oxygen. Dopamine-β-hydroxylase (EC 1.14.17.1) is a tetramer, each monomer containing two copper ions maintained in the cuprous (Cu^+) state by ascorbate. The enzyme is localized within storage vesicles in both adrenergic neurones and adrenomedullary chromaffin cells. Intravesicular ascorbate is maintained in the reduced state by electron transport across the vesicle membrane, cytosolic ascorbate being the most likely electron donor. Cytosolic semidehydroascorbate generated in the extra- to intravesicular electron transfer reaction may be reduced by the outer mitochondrial membrane enzyme semidehydroascorbate reductase (EC 1.6.5.4) to complete the ascorbate regeneration cycle (Menniti *et al.*, 1986). In this manner, a stable rate of noradrenaline biosynthesis can be maintained in the absence of exogenous ascorbate.

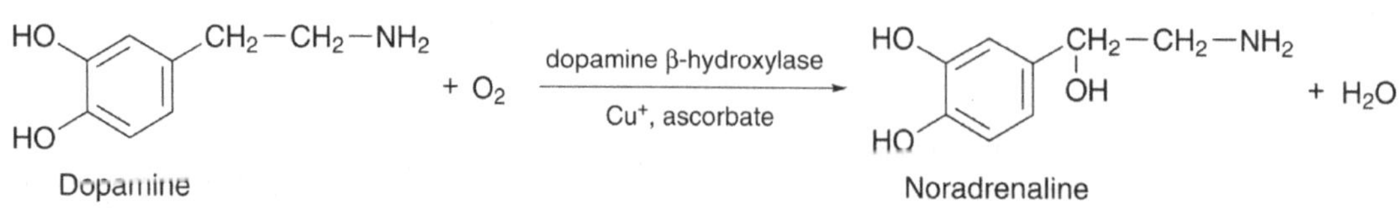

FIGURE 27.6 Hydroxylation of dopamine.

27.10.4 Amidation of Peptides to Hormones and Hormone-Releasing Factors

A large number of peptides act as hormones and hormone-releasing factors. The vast majority of these peptides are initially synthesized as larger, inactive precursor molecules, which are converted to their active forms through a series of post-translational modifications. One such final step of activation is carboxy-terminal α-amidation, which is essential for biological activity. Examples of α-amidated peptides include α- and γ-melanotropins, calcitonin, pro-ACTH, vasopressin, oxytocin, cholecystokinin, gastrin, gastrin-releasing peptide, and releasing factors for growth hormone, corticotropin and thyotropin. Direct precursors to α-amidated peptides have glycine as the carboxy terminal amino acid. The enzyme responsible for the amidation is the bifunctional peptidylglycine α-amidating monooxygenase (PAM), which is found in secretory granules in many neuroendocrine tissues. PAM is actually a precursor protein which undergoes endoproteolytic cleavage to generate two enzymes, peptidylglycine α-hydroxylating monooxygenase (PHM) and peptidyl-α-hydroxyglycine α-amidating lyase (PAL). The two-step nature of the amidation is shown in Figure 27.7. Step 1 requires molecular oxygen, cuprous ion, and ascorbate as the reductant (Eipper & Mains, 1991).

27.10.5 Enhancement of Prostaglandin Synthesis

The prostaglandins (PG) are a group of hormone-like lipids, which are formed in the body from derivatives of essential fatty acids, particularly arachidonic acid (20:4*n*-6). Prostaglandins have a vast range of effects and modulate cardiovascular, pulmonary, immune, and reproductive functions. PGE_1, for example, is required for T lymphocyte formation, regulation of collagen and cholesterol metabolism, and regulation of responsiveness to insulin; it is also an inhibitor of platelet aggregation (Horrobin *et al.*, 1979).

Physiological concentrations of vitamin C stimulate the formation of PGE_1 (but not PGE_2) from dihomo-γ-linolenic acid in human platelets (Manku *et al.*, 1979). In many vitamin C deficiency states, PGE_1 has an action similar to that of vitamin C supplementation. It is therefore possible that ascorbate stimulation of PGE_1 biosynthesis could contribute to many well-known actions of vitamin C.

27.10.6 Release of Insulin

It is well established that insulin function is impaired in scorbutic guinea pigs and that the symptoms are alleviated by restoring vitamin C to the guinea pigs' diet. Wells *et al.* (1995) showed that isolated pancreatic islets from scorbutic guinea pigs failed to rapidly release insulin in response to elevated D-glucose. When ascorbic acid (5 mM of the stable 2-phosphate derivative) was added to the perfusion medium concurrently with elevation of medium D-glucose, islets from scorbutic guinea pigs released insulin as rapidly as control guinea pig islets and to a somewhat greater extent. Ascorbic acid 2-phosphate without elevated D-glucose had no effect on insulin release by islets from normal or scorbutic guinea pigs. The pancreas from scorbutic guinea pigs contained 2.4 times more insulin than that from control animals, suggesting that the decreased insulin output from the scorbutic islets was not due to decreased insulin synthesis but to impaired insulin secretion.

27.10.7 Modulation of Neurotransmitter Systems

Of the subcellular fractions of rat brain, the synaptosomal cytoplasm has the highest content of ascorbic acid (Kuo *et al.*, 1979). This fraction contains the nerve terminals, where synthesis, storage, and release of transmitters take place.

27.10.7.1 Regulation of Acetylcholine Receptor Synthesis

Acetylcholine is the neurotransmitter at synapses between motor neurons and muscle cells, often called the neuromuscular junction. Arrival of an action potential at the terminus of a presynaptic motor neuron induces opening of voltage-gated Ca^{2+} channels and subsequent release of acetylcholine, which diffuses across the synaptic cleft and triggers opening of the ligand-gated acetylcholine receptors in the muscle cell plasma membrane. The resulting influx of Na^+ causes opening of voltage-gated Na^+ channels and generation of an action potential that leads to the release of stored

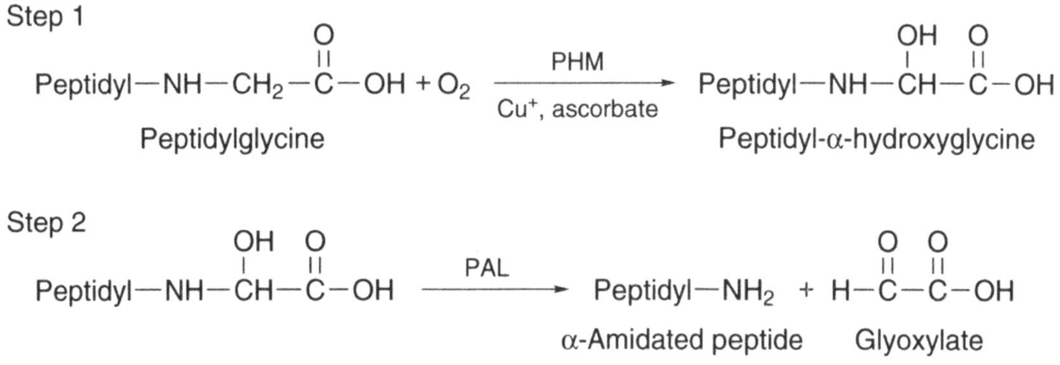

FIGURE 27.7 Amidation of peptidylglycine.

Ca^{2+} into the cytosol. The resulting rise in cytosolic Ca^{2+} causes muscle contraction.

The acetylcholine receptor is a pentameric protein composed of two α subunits and one of each β, γ, and δ subunits arranged around a central pore. Binding of acetylcholine triggers conformational changes in the receptor subunits that can cause channel opening, allowing the passage of both Na^+ and K^+ ions with their shell of bound water molecules. Ascorbic acid increases the number of acetylcholine receptors in cultures of the L_5 muscle cell line. This effect is specific for ascorbic acid and is not a general property of reducing agents (Knaak *et al.*, 1986). Ascorbic acid increases the mRNA for the α-subunit, but not the mRNAs for the other subunits (Horovitz *et al.*, 1989). Thus increased expression of acetylcholine receptors in response to ascorbic acid is controlled by the availability of the α-subunit mRNA.

27.10.7.2 Regulation of Neurotransmitter Release

Ascorbic acid stimulates the release of acetylcholine and noradrenaline from isolated synaptic vesicles (Kuo *et al.*, 1979). Release of each transmitter requires the presence of adenosine triphosphate, Mg^{2+}, and Ca^{2+}. Ascorbic acid also enhances the neurotransmitter-evoked release of luteinizing hormone-releasing hormone (LH-RH) from the mediobasal hypothalmus *in vitro*, probably by enhancing the activity of endogenous noradrenaline (Miller & Cicero, 1986). The neuropeptide atrial natriuretic factor (ANF) acts in the brain to alter salt appetite and drinking behavior among other functions. Ascorbic acid acts alone or in synergism with forskolin to stimulate the production and secretion of ANF in rat hypothalamic neurons in culture (Huang *et al.*, 1993). The synergistic effect operates at the genomic level, with enhanced production of pro-ANF mRNA, and is mediated, at least in part, through the adenylyl cyclase-cyclic adenosine monophosphate signaling system. The role of ascorbic acid in facilitating hypothalamic LH-RH and ANF release appears to be independent of its antioxidant properties.

27.11 ROLE OF ASCORBIC ACID IN MESENCHYMAL DIFFERENTIATION

Ascorbate is involved in the development of connective tissue cells, such as myoblasts, adipocytes, chondrocytes, and osteoblasts, which all arise from a common mesenchymal precursor cell.

27.11.1 Myogenesis and Adipogenesis

The formation of skeletal muscle (myogenesis) entails, among other events, the differentiation of myoblasts and their fusion to produce multinucleated myotubes. Fat is produced and stored in adipocytes. Collagen synthesis, and hence the presence of ascorbate, is an essential requirement for myogenesis and adipogenesis. This was demonstrated *in vitro* by the inhibition of myoblast differentiation and preadipocyte cell differentiation when collagen synthesis was inhibited by ethyl-3,4-dihydroxybenzoate; addition of ascorbate at high concentration partially reversed the drug's effect (Nandan *et al.*, 1990; Ibrahimi *et al.*, 1992). Ascorbate likely promotes myoblast differentiation through the increased expression of a myogenic regulatory protein, myogenin (Mitsumoto *et al.*, 1994).

27.11.2 Chondrogenesis

Treatment of cultured prehypertrophic chondrocytes with ascorbate induces the production of type X collagen mRNA after 24 hours (Leboy *et al.*, 1989). The same treatment also increases the activity of alkaline phosphatase. The delayed response implies that ascorbate does not activate transcription directly; rather it activates a pathway which leads to enhanced transcription. In control cultures grown in the absence of ascorbate, type X collagen mRNA production is markedly lower, alkaline phosphatase activity is insignificant, and there is no evidence of calcium deposition. The ability of ascorbate to modulate chondrocyte gene expression, unlike the osteoblast system, does not depend on production of a collagen-rich matrix (Sullivan *et al.*, 1994).

27.11.3 Osteogenesis

The importance of ascorbic acid for osteogenesis is evident from the defective bone extracellular matrix seen in vitamin C deficiency (scurvy) and the presence of a specific transport system for ascorbate in osteoblasts (Wilson & Dixon, 1989). *In vivo*, administered [^{14}C]ascorbate accumulates preferentially in sites of active osteogenesis (Hammarström, 1966). Preosteoblast culture systems require the addition of ascorbic acid in order to induce the expression of osteoblast marker proteins (alkaline phosphatase and osteocalcin) and to form a mineralized extracellular matrix (Franceschi & Iyer, 1992).

Franceschi and Iyer (1992) investigated the role of collagen synthesis in osteoblast differentiation using the MC3T3-E1 cell line as a model system. These cells display a time-dependent and sequential expression of osteoblast characteristics analogous to *in vivo* bone formation (Quarles *et al.*, 1992). Within 24 hours of ascorbic acid addition to cell cultures, cells had increased levels of type I collagen mRNA and had begun synthesizing a collagen-containing extracellular matrix. After two to three days, the cells had increased alkaline phosphatase mRNA and enzyme activity, followed by osteocalcin mRNA after four to six days. Mineralization of the collagen matrix began after six to eight days. Ascorbate also stimulated cell proliferation, the cells becoming incorporated into the collagen matrix to form multilayered structures with a discrete zone of mineral. Blocking collagen synthesis with specific inhibitors prevented ascorbic acid-dependent cell proliferation and expression of osteoblast marker genes without affecting noncollagen protein synthesis or basal proliferation rates.

In a discussion of the above work, Franceschi (1992) considered that once vitamin C has formed some critical amount of collagen matrix, an extracellular environment is established that is permissive for subsequent expression of osteoblast marker genes. They speculated that collagen or some other extracellular matrix protein may interact with cell surface adhesion receptors, thereby generating a signal that stimulates proliferation and marker gene expression. Harada *et al.* (1991) reported that peptides containing the sequence Arg-Gly-Asp completely blocked the proliferative activity of ascorbate. This finding suggests that integrins may be the signaling receptors. Integrins are known to interact with proteins via a Arg-Gly-Asp sequence (among other sequences) and can act as true signaling receptors in a variety of cell types (Hynes, 1992). The Arg-Gly-Asp sequence is present in type I collagen and also in fibronectin (Dedhar *et al.*, 1987).

27.12 ANTIOXIDANT ROLE

Ascorbate is an effective scavenger of all aggressive reactive oxygen species within the aqueous environment of the cytosol and extracellular fluids. These species include hydroxyl, superoxide anion, and nonlipid peroxyl radicals together with the non-radicals singlet oxygen and hydrogen peroxide (Sies & Stahl, 1995). Ascorbate reacts with free radicals to produce the ascorbyl radical and detoxified product through a single-electron transfer. Figure 27.8 shows a possible scheme in which ascorbate can be recycled during the scavenging process.

Ascorbate is not the only antioxidant in aqueous systems: other water-soluble antioxidants such as protein thiols and urate are also present. However, ascorbate is the only endogenous antioxidant that effectively protects the lipids in blood plasma (and also low-density lipoprotein) against oxidative damage initiated by nonlipid peroxyl radicals generated in the aqueous phase. This is observed as a complete cessation of lipid peroxidation when ascorbate is added to plasma; other endogenous antioxidants, including α-tocopherol, do not have this effect (Frei, 1991). Apparently, ascorbate traps virtually all peroxyl radicals generated in the aqueous phase before they can diffuse into the lipid phase. Thus, ascorbate acts as the first and major line of antioxidant defense in the protection of lipoidal plasma constituents and low-density lipoprotein. In this action, ascorbate spares vitamin E, the chain-breaking antioxidant in the lipid phase (Doba *et al.*, 1985).

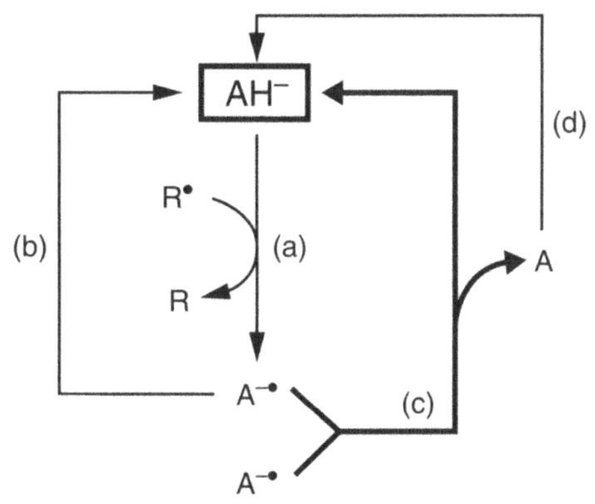

FIGURE 27.8 Possible scheme for recycling of ascorbate during the process of free radical scavenging. Ascorbate (AH^-) donates an electron to the free radical (R•), forming detoxified product (R) and is itself oxidized to the ascorbyl radical (A^-•), also known as semidehydroascorbate (reaction a). Ascorbyl radical is converted to ascorbate by semidehydroascorbate reductase (reaction b). In addition, pairs of ascorbyl radicals disproportionate nonenzymatically to form one molecule each of dehydroascorbic acid (A) and ascorbate (reaction c). The rapid rate of disproportionation of the ascorbyl radical prevents substantial interactions with other substances. Dehydroascorbic acid can be reduced to ascorbate by either a glutathione-dependent or an NADPH-dependent reductase (reaction d).

In its role as a lipid-soluble, chain-breaking antioxidant in biomembranes and lipoproteins, vitamin E (tocopherol, T-OH) scavenges lipid peroxyl free radicals and itself is converted to the tocopheroxyl radical (T-O•) (Section 17.6, reaction 17.1). Lipid peroxyl radicals, because of their location in lipid environments, cannot be scavenged by ascorbate anion. However, *in vitro* studies using phospholipid liposomes as model biomembranes have shown that ascorbate (AH^-) restores the antioxidant activity of vitamin E by converting the tocopheroxyl radical back to the phenolic tocopherol (reaction 27.1). Ascorbate works at the lipid-water interface of membranes, very close to the polar head groups of tocopherol.

$$\text{T-O}\bullet + AH^- \rightarrow \text{T-OH} + A^-\bullet \quad (27.1)$$

Whether vitamin C regenerates vitamin E *in vivo* is debatable. Burton *et al.* (1990) found no evidence for an interaction between the two vitamins in guinea pigs and concluded that any such interaction must be negligible in comparison with the normal turnover of vitamin E.

As discussed above, ascorbate is an excellent antioxidant but, paradoxically, it could theoretically behave as a pro-oxidant at lower concentrations (Buettner & Jurkiewicz, 1996). This crossover effect from pro-oxidant to antioxidant is dependent on the ability of transition metals in their reduced forms (e.g. Fe^{2+} and Cu^+) to catalyze the generation of free radicals. Ascorbate, being a powerful reducing agent, maintains transition metals in their catalytic reduced forms. At a high concentration of ascorbate, the length of free radical chain reactions will be small owing to ascorbate's free radical scavenging action. As the concentration of ascorbate is lowered, there will come a point where its antioxidant action is negligible but its capacity to reduce catalytic metals is still sufficient. At this crossover point ascorbate switches from being an antioxidant to a pro-oxidant. Whether this situation could occur *in vivo* is debatable.

The antioxidant action of vitamin C has a wide variety of protective roles in the body. For example:

- The DNA in human sperm is protected from free radical damage (Fraga *et al.*, 1991).

- Lung tissue is protected from free radical damage resulting from inhalation of tobacco smoke, pollutants, and ozone.
- Ocular tissue is protected from photo-oxidative damage that can ultimately result in cataract formation.
- The high concentrations of ascorbate in neutrophils and macrophages and its release on stimulation protect these phagocytes and host tissue during the respiratory burst in which reactive oxygen species are produced to kill phagocytosed pathogens.

27.13 IMMUNE FUNCTION

27.13.1 Introduction

There is a large body of evidence that vitamin C plays an important role in the biochemistry of the human immune system, particularly in the stimulation of phagocytosis. Leukocytes accumulate ascorbic acid after uptake from the plasma by active transport (Moser, 1987), suggesting an involvement of the vitamin in the normal function of these cells. The concentration of ascorbate in monocytes, for example, is over 80 times higher than that in plasma (Evans *et al.*, 1982) and macrophages contain about twice as much ascorbate as neutrophils and monocytes (Schmidt & Moser, 1985). Vitamin C accumulation in activated human neutrophils is increased as much as tenfold above the concentrations present in resting neutrophils as a result of a novel vitamin recycling mechanism. Extracellular ascorbate is oxidized to dehydroascorbic acid by oxidants generated by the activated neutrophil. The dehydroascorbic acid is preferentially taken up by the neutrophil and reduced intracellularly to ascorbate within minutes (Washko *et al.*, 1993). Ascorbate, as an antioxidant, protects phagocytes from self-destruction by reactive oxidants (Muggli, 1993). It also neutralizes reactive oxidants released extracellularly by activated phagocytes, thereby preventing damage to surrounding host tissue (Anderson & Lukey, 1987).

27.13.2 Effects of Ascorbic Acid on Metabolic Pathways in Phagocytes

Cooper *et al.* (1971) reported that the *in vitro* addition of ascorbic acid markedly stimulated the hexose monophosphate shunt (HMS) in human neutrophils, rabbit alveolar macrophages, and in neutrophils from a patient with chronic granulomatous disease. DeChatelet *et al.* (1972) suggested the following sequence of reactions to explain the ascorbate-induced HMS activity.

$$\text{Ascorbate} + O_2 \xrightarrow{Cu^{2+}} \text{Dehydroascorbic acid} + H_2O_2 \quad (27.2)$$

$$\text{Ascorbate} + H_2O_2 + 2H^+ \rightarrow \text{Dehydroascorbic acid} + 2H_2O \quad (27.3)$$

$$\text{Dehydroascorbic acid} + 2\text{GSH} \rightarrow \text{Ascorbate} + \text{GSSG} \quad (27.4)$$

$$\text{GSSG} + \text{NADPH} + H^+ \rightarrow 2\text{GSH} + \text{NADP} \quad (27.5)$$

$$\text{Glucose-6-phosphate} + \text{NADP} \rightarrow \text{Initiation of HMS} \quad (27.6)$$

Antimicrobial activity may be apparent during both reactions 27.2 and 27.3 in which ascorbate is oxidized to dehydroascorbic acid. The oxidation of ascorbic acid by a peroxide plus copper ions is accompanied by a strong bactericidal, fungicidal, and virucidal effect (Ericsson & Lundbeck, 1955). Miller (1969) demonstrated that a mixture of ascorbic acid and hydrogen peroxide is bactericidal *in vitro* and renders gram-negative bacteria sensitive to lysis by lysozyme. DeChatelet *et al.* (1972) also observed bactericidal activity of ascorbic acid and hydrogen peroxide *in vitro*. The remaining reactions (reactions 27.4 to 27.6) serve to explain the HMS stimulation by ascorbate. Dehydroascorbic acid is reduced by reduced glutathione (GSH) (reaction 27.4) and the oxidized glutathione (GSSG) thus formed is reduced by NADPH to produce NADP (reaction 27.5). The NADP in turn reacts with glucose-6-phosphate to initiate the HMS (reaction 27.6).

Ascorbic acid was shown to inhibit the myeloperoxidase-H_2O_2-halide reaction, which has been implicated in the bactericidal activity of neutrophils, but, despite this inhibition, the neutrophils were still able to kill bacteria (McCall *et al.*, 1971). This finding supports the concept that the capacity of the neutrophil to destroy bacteria does not depend solely on a single mechanism.

Shilotri (1977) assessed the effects of vitamin C on phagocytosis by studying metabolic changes in neutrophils isolated from ascorbic acid-deficient guinea pigs, using *Escherichia coli* as the test organism. Activities of the glycolytic and HMS pathways were decreased during phagocytosis and bactericidal activity was low compared with these activities in control neutrophils. These findings showed that bacterial engulfment by neutrophils (as indicated by glycolytic activity) and also bacterial destruction (as indicated by HMS activity) are impaired in vitamin C activity.

27.13.3 Neutralization of Harmful Extracellular Oxidants and Proteases

27.13.3.1 Reactive Oxidants

Phagocytes are susceptible to attack by the reactive oxidants that they produce and their ability to accumulate ascorbate affords them some protection. There are notable differences in the degree of protection afforded to neutrophils and macrophages. Neutrophils die after phagocytosis and so they only need to be protected from oxidation in processes leading up to this event. Their ascorbate content is low relative to that of macrophages and their intracellular ascorbate pool does not show a decline during phagocytosis

(Oberritter *et al.*, 1986). Macrophages, on the other hand, can perform repeated phagocytosis and are thus better protected against self-destruction. Three independent observations provide evidence for the role of ascorbate as an antioxidant in protecting macrophages.

- The ascorbate content of macrophages is twice that of neutrophils and monocytes (Schmidt & Moser, 1985).
- Macrophages increase their consumption of ascorbate during phagocytosis (Oberritter *et al.*, 1986); the intracellular pool of ascorbate can be replenished afterwards by uptake from plasma.
- Ascorbate offers partial protection to macrophages against phagocytosis-induced destruction *in vitro* (McGee & Myrvik, 1979).

27.13.3.2 Elastase

The lysosomes of neutrophils contain large quantities of elastase, an enzyme capable of degrading almost all components of the host's extracellular matrix. Plasma and interstitial fluids contain a series of antiproteinases, including α_1-proteinase, that can limit the destructive effects of elastase on extracellular substrates by forming an enzyme–inhibitor complex. Hypochlorous acid released by neutrophils during phagocytosis inactivates the protective antiproteinases, allowing unregulated elastase and other proteolytic enzymes to attack and degrade host tissues (Weiss, 1989).

Ascorbic acid at physiological concentrations can scavenge the myeloperoxidase-derived hypochlorous acid at rates sufficient to protect α_1-proteinase against inactivation by this oxidant (Halliwell *et al.*, 1987). Lunec and Blake (1985) reported a decreased level of ascorbate and increased level of dehydroascorbate in the synovial fluid and sera of patients with rheumatoid arthritis compared with normal fluid and sera. This presumably reflects the increased antioxidant activity of vitamin C at inflammatory sites. This depletion of ascorbic acid may give hypochlorous acid free rein to destroy the antiproteinases that protect inflamed tissues from attack by proteases; in short, ascorbic acid depletion may allow proteolytic damage.

27.13.4 Antihistamine Activity

Histamine stimulates the early phase inflammatory response by causing localized vasodilation and increased capillary permeability, thereby enhancing the flow of immune factors to the site of inflammation. After a while, the increasing histamine concentration reaches a level at which the amine begins to suppress the immune response.

Histamine is degraded non-enzymatically by ascorbic acid *in vitro* (Uchida *et al.*, 1989) and an inverse relationship between blood histamine and vitamin C status has been demonstrated (Clemetson, 1980; Johnston *et al.*, 1996). In view of this antihistamine activity, an elevated consumption of vitamin C-rich foods may be particularly valuable in allergy-prone individuals.

27.13.5 Enhancement of Microtubule Assembly in Neutrophils

Boxer *et al.* (1979) reported that exposure of thin sections of human neutrophils to ascorbic acid resulted in an increase in assembled microtubules. This response was diminished by prior incubation of the neutrophils to colchicine, a compound known to promote disassembly of microtubules. In addition, ascorbic acid directly enhanced the assembly of microtubules from purified tubulin *in vitro*. Microtubules may be involved not only in the internal organization of cells, but also in phagocyte chemotaxis and degranulation (Schmidt & Moser, 1985).

27.13.6 Stimulation of Neutrophil Chemotaxis

The incubation of human leukocytes with ascorbic acid at concentrations achievable in normal tissues increased chemotactic migration of the cells: the enhanced mobility correlated with the cells' ability to assemble microtubules (Boxer *et al.*, 1979). Stimulation of neutrophil chemotaxis *in vivo* has been reported in normal adult volunteers following the daily ingestion of 1 g (Boxer *et al.*, 1976) or 2–3 g of vitamin C (Anderson *et al.*, 1980). Ganguly *et al.* (1976) reported a reduced migration of macrophages from vitamin C-deficient guinea pigs.

Johnston *et al.* (1992) examined the effect of supplemental vitamin C on blood histamine levels and neutrophil chemotaxis in healthy men and women. The data indicated that chronic oral administration of the vitamin (2 g ascorbic acid per day for 2 weeks) significantly lowered blood histamine levels, and that this fall in histamine was directly related to a rise in neutrophil chemotaxis. There was no *direct* effect of ascorbate on chemotaxis. Blood histamine and neutrophil chemotaxis did not change 4 hours following a single 2-g dose of ascorbic acid, although plasma ascorbate rose 150%. Thus oral administration of ascorbate reduced blood histamine levels only when taken chronically. Johnston *et al.* (1996) reported that blood histamine levels were not significantly affected until mean plasma vitamin C concentrations rose over 20 µmol/L, a level threefold greater than that associated with overt scurvy.

27.13.7 Natural Killer Cell Activity

An *in vivo* effect of ascorbic acid on enhancement of human natural killer cell activity has been reported at a dosage of 60 mg per kg body weight (Vojdani & Ghoneum, 1993).

27.13.8 Regulation of the Complement Component C1q

When guinea pigs were fed tissue-saturating amounts of vitamin C, plasma C1q concentrations were significantly higher than in those animals fed only enough ascorbate for adequate growth and for the prevention of scurvy (Haskell & Johnston, 1991). When healthy men and women were given 500 mg ascorbate three times daily with meals for four weeks, their plasma C1q levels were not significantly altered (Johnston, 1991). Hence, significantly enhanced C1q production may occur only during activation of the immune system, and not in healthy, non-infected individuals.

27.13.9 Enhancement of Lymphocyte Blastogenesis

Using cultured spleen cells from an inbred strain of rat that does not synthesize vitamin C, Oh and Nakano (1988) observed that ascorbic acid enhanced lymphocyte blastogenesis through inhibition of the biosynthesis of immunosuppressive histamine.

27.13.10 T Cell Function

Anthony *et al.* (1979) reported that vitamin C deficiency in guinea pigs diminishes cytotoxic T cell activity, implying that the vitamin is important for proper function of the $CD8^{+}T$ cell. In a study of five healthy volunteers hospitalized in a metabolic unit, Kay *et al.* (1982) found that experimental vitamin C deficiency at the scorbutic level does not alter T cell numbers or impair *in vitro* T cell function. The same applied to a patient admitted with spontaneous scurvy.

Campbell *et al.* (1999) showed that three T cell death pathways were inhibited when mouse T cells were incubated with ascorbic acid. Both activated and resting T cells were resistant to death induced by growth factor withdrawal and to spontaneous apoptosis when treated with ascorbic acid. Resting T cells, but not activated T cells, were resistant to steroid-induced death when treated with ascorbic acid. The authors hypothesized that the various death pathways are funneled into a common pathway and it is at this point that vitamin C can block the progression of death. They also found that effector T cells were more likely to enter the DNA synthesis phase (S phase) of the cell cycle if treated with ascorbic acid. These results suggest that vitamin C may boost immunity by increasing survival of resting and activated T cells.

27.13.11 Antibody Production

Anthony *et al.* (1979) examined the effect of ascorbic acid deficiency on the primary humoral response in guinea pigs immunized with chicken erythrocytes. The guinea pigs were divided into three dietary groups: ascorbic acid-deficient, pair-fed, and *ad libitum*-fed control. No difference in antibody titers was observed among the three dietary groups, indicating that vitamin C is not involved in antibody production.

27.13.12 Enhancement of Interferon Synthesis

The participation of vitamin C in protection against some viral infections may be in the enhancement of interferon biosynthesis as demonstrated *in vivo* and *in vitro*. The level of circulating interferon induced in mice by inoculation with leukemia virus was enhanced by the addition of ascorbate to the drinking water (Siegel, 1974). Ascorbate also enhanced the interferon levels produced by cultured human embryo fibroblasts in response to Newcastle Disease virus (Dahl & Degré, 1976; Karpińska *et al.*, 1982).

TABLE 27.4
Effect of Ascorbic Acid (8 μg ml⁻¹) on Interferon-γ Production and Lymphocyte Proliferation

Culture	Culture Condition	Counts Per Min (Mean Triplicate)	Interferon-γ (Units Per ml)
1	Cells alone	230	0
2	Influenza A	2300	58
3	Influenza A + ascorbic acid	4100	59
4	Interferon-γ	3800	—
5	Interferon-γ + ascorbic acid	6800	—
6	Ascorbic acid	200	0

Source: Cunningham-Rundles *et al.* (1993).

27.13.13 Regulation of Cytokines

Vitamin C has an indirect effect on lymphocyte proliferation through its action on cytokines, as shown *in vitro* by Cunningham-Rundles *et al.* (1993). Ascorbic acid suppressed proliferation response to interleukin-2, suggesting a basis for the vitamin's inhibitory effect on mitogen-induced lymphocyte proliferation. In contrast, ascorbic acid enhanced the proliferative response to interferon-γ, without inhibiting the production of interferon-γ that accompanied the response to influenza A (Table 27.4).

27.14 VITAMIN C AND CARDIOVASCULAR DISEASE

Ascorbic acid through its numerous metabolic and antioxidant effects may inhibit some of the steps involved in atherosclerosis and thrombosis, thus reducing the risk of cardiovascular disease.

27.14.1 Cholesterol Metabolism

Studies of animals that either synthesize (rat, rabbit) or do not synthesize (guinea pig, monkey) vitamin C have shown that vitamin C is intimately involved in cholesterol metabolism. Guinea pigs subjected to chronic vitamin C deficiency exhibit increased cholesterol levels in blood plasma and liver due to slower conversion of cholesterol to bile acids (Ginter *et al.*, 1971; Ginter, 1973). The impaired conversion results from a decreased activity of the rate-limiting liver enzyme cholesterol 7α-hydroxylase (Horio *et al.*, 1989). When guinea pigs, rats, and rabbits are rendered hypercholesterolemic by being fed a high-cholesterol diet, vitamin C supplementation lowers their blood cholesterol levels.

27.14.2 Lipoprotein Profile

Diets low in vitamin C lead to a redistribution of cholesterol among the various plasma lipoproteins. Vitamin C deficiency in ODS rats (rats with a hereditary inability to synthesize ascorbic acid) leads to an increase in potentially pro-atherogenic low-density lipoprotein (LDL) cholesterol and a decrease in high-density lipoprotein cholesterol, resulting in hypercholesterolemia (Uchida *et al.*, 1990).

27.14.3 Protection of LDL against Peroxidative Modification

Physiological concentrations of ascorbic acid protect LDL against copper-catalyzed peroxidative modification *in vitro*, maintaining the ability of LDL to be recognized by appropriate LDL receptors and not by the scavenger receptor of macrophages (Sakuma *et al.*, 2001). This protective action preserves LDL's indigenous lipid-soluble antioxidants, except for ubiquinol, the reduced form of coenzyme Q (Retsky & Frei, 1995). Ascorbic acid spares, rather than regenerates, LDL-associated α-tocopherol, i.e. prevents α-tocopherol oxidation in the first place. The dilemma of whether ascorbate acts as a pro-oxidant or as an antioxidant when interacting with LDL has been addressed by Lynch *et al.* (1996). Ascorbate protects native or mildly oxidized LDL against further metal ion-dependent oxidation; only if LDL becomes extensively oxidized does ascorbate acts as a pro-oxidant.

27.14.4 Effects on Nitric Oxide-Mediated Arterial Relaxation

Several studies have shown that an acute application of ascorbic acid enhanced endothelium-dependent vasodilation in patients with diabetes, coronary artery disease, hypertension, hypercholesterolemia, or chronic heart failure, and in cigarette smokers (Ting *et al.*, 1996; Levine *et al.*, 1996; Solzbach *et al.*, 1997; Ting *et al.*, 1997; Hornig *et al.*, 1998; Heitzer *et al.*, 1996). Long-term ascorbic acid treatment (500 mg/day) produced a sustained improvement in endothelium-dependent vasodilation in patients with coronary artery disease (Gokce *et al.*, 1999). Kanani *et al.* (1999) demonstrated that administration of ascorbic acid prevents induction of endothelial dysfunction by homocysteine. These findings may be attributable to the scavenging of superoxide by ascorbate, thus preventing the reaction between superoxide and nitric oxide to form hydroxyl radicals and nitrogen dioxide, both of which can initiate lipid peroxidation.

Heller *et al.* (1999) demonstrated that preincubation of cultured endothelial cells with ascorbic acid led to a threefold increase of the cellular production of nitric oxide after stimulation with ionomycin or thrombin. Ascorbate did not induce the expression of nitric oxide synthase and appeared to act through an effect on the availability or affinity of the enzyme co-factor tetrahydrobiopterin. The findings suggest that saturation of the vascular tissue with ascorbate provides the optimum reaction conditions for adequate nitric oxide synthesis and that a decrease in intracellular ascorbate leads to endothelial dysfunction.

27.14.5 Enhancement of Prostacyclin Formation

The formation of prostacyclin (PGI_2), a member of the prostaglandin family which protects the arterial wall against deposition of platelets, is inhibited by hydroperoxides of unsaturated fatty acids. *In vitro* studies have shown that physiological concentrations of ascorbic acid enhance the formation of prostacyclin by aortic rings by protecting the cyclo-oxygenase and PGI-synthase (Beetens & Herman, 1983).

27.14.6 Effects of Vitamin C Supplementation on LDL Oxidation

Rifici and Khachadurian (1993) administered vitamin C (1 g/day) and vitamin E (800 IU/ day), both separately and in combination, to healthy female and male subjects and examined oxidation of lipoproteins *in vitro.* Vitamin E administration alone produced a 52% inhibition and vitamin C alone a 15% inhibition of copper-catalyzed thiobarbituric-acid reactive substances production; the combination of vitamins produced a 63% inhibition. Harats *et al.* (1998) reported that in young healthy male subjects consuming a diet high in saturated fats, supplementation with citrus fruits containing an estimated 500 mg per day of vitamin C reduced the *in vitro* susceptibility of LDL to oxidation. Mosca *et al.* (1997) reported that antioxidant supplementation with a combination of 800 IU of vitamin E, 1,000 mg of vitamin C, and 24 mg of β-carotene significantly reduced the susceptibility of LDL to oxidation in patients with coronary artery disease. The response produced by a similar combination containing half the amounts of each antioxidant was nonsignificant.

27.14.7 Epidemiological Studies

Current evidence from epidemiological studies on the role of vitamin C in the prevention of cardiovascular disease is inconclusive, with some studies showing a very strong correlation between vitamin C intake and incidence of cardiovascular events and other studies showing no correlation at all (Lynch *et al.*, 1996; Institute of Medicine, 2000).

27.15 VITAMIN C DEFICIENCY IN ANIMALS

The scorbutic guinea pig exhibits capillary fragility, impaired wound healing, muscular atrophy of the feet, and diminished activity. Apart from guinea pigs and monkeys, experimental vitamin C deficiency is difficult to produce in animals, due to their ability to synthesize the vitamin in the body organs. Under normal conditions, synthesis of vitamin C covers the individual requirements of the healthy animal.

Exceptions are calves and piglets, which seem unable to synthesize adequate amounts of the vitamin during the first 10 days of life. Further, some fully grown domestic animals show clear signs of reduced synthesis or increased requirement for vitamin C, induced by various stress factors (heat, cold, infections and parasitic diseases, poor diet). In poultry, these stress factors can lead to deterioration in growth, egg production, and quality of egg shells.

27.16 VITAMIN C DEFICIENCY IN HUMANS

The classic vitamin C deficiency disease in humans is scurvy. Fully developed scurvy is rarely seen nowadays, but clinical signs of mild scurvy are found quite frequently in alcoholics and drug addicts. The symptoms described below have been observed in patients with scurvy (Chazan & Mistilis, 1963) and in experimentally induced scurvy (Hodges *et al.*, 1969, 1971).

27.16.1 Signs and Symptoms of Scurvy

Early signs of scurvy in adults are weakness, easy fatigue, and listlessness, followed by shortness of breath and aching bones, joints, and muscles. Progressive changes in the skin then appear after about 4 months of complete vitamin C deprivation. A horny material piles up around the openings of hair follicles, and the hair becomes fragmented and coiled. Red spots of pinpoint to pinhead size (petechiae) caused by the rupture of small blood vessels appear first on the feet and ankles, and then spread upwards. Thereafter, bruises appear over large areas of skin, particularly on the legs. Bruising is the manifestation of hemorrhages in subcutaneous tissue, beneath the periosteum of bones, and in the synovia of joints. The gums become swollen and bleeding, especially where there is advanced dental caries. Hemorrhages are caused by rupture of capillaries, which are fragile because of impaired ascorbic acid-dependent synthesis of vascular basement membrane (Priest, 1970). Wounds fail to heal and old wounds reopen. The sufferer is visibly anemic due in part to hemolysis caused by peroxidative damage to the erythrocyte plasma membrane (Goldberg, 1963). Vitamin C deficiency in adults may cause osteoporosis due to a diminished production of organic matrix in bones. The corresponding symptoms in infantile scurvy are impaired ossification and bone growth.

Kinsman and Hood (1971) studied the psychological aspects of vitamin C deficiency in healthy volunteers. They measured four behavioral areas: physical fitness (strength, co-ordination, and balance), mental functions (memory, vigilance, and problem solving), psychomotor performance tasks (reaction time, manipulative skills, and hand-arm steadiness), and personality. Three areas of change associated with vitamin C deficiency were found: physical fitness involving bending or twisting of the legs, psychomotor tasks, and measures of personality. The changes in the physical fitness could be accounted for by the pronounced joint pain in the legs that occurred during the deficiency period. The decrements in psychomotor performance were attributed to a reduced motivational level. The personality changes corresponded to the classical 'neurotic triad' of the Minnesota Multiphasic Personality Inventory, i.e. hypochondriasis, depression, and hysteria. Elevation of this triad is also found in prolonged semistarvation and deficiencies of B-complex vitamins.

In fully developed scurvy, as witnessed and recorded at sea by James Lind in 1752, the body is covered with spots and bruises, and the skin overlying the joints becomes discolored from the hemolyzed blood in and around them. There may be bleeding into the peritoneal cavity and pericardial sac as well as into joints. The gums become swollen, spongy, and of a livid blue-red color. The swelling can develop to such an extent that the gum tissue completely encases and hides the teeth. The spongy gums bleed on the slightest touch and become secondarily infected, leading to loosening of the teeth and gangrene. Death preceded by dyspnea, cyanosis, and convulsions is inevitable in the continuing absence of vitamin C.

27.16.2 Metabolism of Vitamin C in Experimental Human Scurvy

Experimental scurvy was investigated in five healthy adult males who were maintained on a synthetic liquid diet devoid of vitamin C. This work was presented in two consecutive papers (Hodges *et al.*, 1971; Baker *et al.*, 1971). The metabolism of the vitamin was followed through the use of isotopically labeled ascorbic acid during the depletion and repletion stages of the experiment. A direct relationship was demonstrated between clinical signs of scurvy and the body pool size of ascorbic acid. The five subjects had an average body pool of 1500 mg ascorbic acid at the beginning of the study. They developed clinical signs of scurvy in 84 to 97 days, the signs appearing when the body pool of ascorbic acid had been depleted to 300 mg. Upon repletion of the ascorbate pool beyond a level of 300 mg, clinical symptoms of scurvy began to recede. In a previous study (Baker *et al.*, 1969), an intake of 6.5 mg per day of ascorbic acid was sufficient to alleviate and cure the clinical signs of scurvy in one of the subjects.

27.17 TOXICITY

Vitamin C is considered to be harmless in single large oral doses, but the regular intake of excessive amounts may lead to gastrointestinal disturbances, such as nausea, abdominal cramps, and diarrhea (Hoffer, 1971). These effects are attributed to the osmotic effect of unabsorbed ascorbate passing through the intestine. The fact that oxalic acid is a normal product of ascorbic acid catabolism prompts concern about the possible formation of calcium oxalate stones in the kidneys. Excess ascorbic acid is mostly excreted in the urine unchanged, and the amount metabolized to oxalate is normally limited, regardless of vitamin C intake. So for most people, high intakes of vitamin C do not pose a risk for kidney stone formation. Individuals who are prone to kidney stones have some metabolic abnormality and excrete unusually high amounts of oxalate. These individuals and patients with kidney disorders are advised not to take pharmacological doses of vitamin C (Sauberlich, 1994).

Dehydroascorbic acid is nontoxic when given by mouth because it is reduced to ascorbic acid in the tissues of the gastrointestinal tract. However, dehydroascorbic acid itself disrupts membrane structure and, if given intravenously, causes a diabetic condition by increasing the permeability of pancreatic B cells, resulting in loss of insulin (Bianchi & Rose, 1986).

FURTHER READING

Duarte, T.L. & Lunec, J. (2005) Review: When is an antioxidant not an antioxidant? A review of novel actions and reactions of vitamin C. *Free Radical Research*, **39**, 671–86.

Leibovitz, B. & Siegel, B.V. (1981) Ascorbic acid and the immune system. *Advances in Experimental Medicine and Biology*, **135**, 1–25.

REFERENCES

Agus, D.B., Gambhir, S.S., Pardridge, W.M., Spielholz, C., Baselga, J., Vera, J.C. & Golde, D.W. (1997) Vitamin C crosses the blood-brain barrier in the oxidized form through the glucose transporters. *Journal of Clinical Investigation*, **100**, 2842–8.

Amano, A., Aigaki, T., Maruyama, N. & Ishigami, A. (2010) Ascorbic acid depletion enhances expression of the sodium-dependent vitamin C transporters, SVCT1 and SVCT2, and uptake of ascorbic acid in livers of SMP30/GNL knockout mice. *Archives of Biochemistry and Biophysics*, **496**, 38–44.

Anderson, R. & Lukey, P.T. (1987) A biological role for ascorbate in the selective neutralization of extracellular phagocyte-derived oxidants. *Annals of the New York Academy of Sciences*, **498**, 229–47.

Anderson, R., Oosthuizen, R., Maritz, R., Theron, A. & Van Rensburg, A.J. (1980) The effects of increasing weekly doses of ascorbate on certain cellular and humoral immune functions in normal volunteers. *American Journal of Clinical Nutrition*, **33**, 71–6.

Anthony, L.E., Kurahara, C.G. & Taylor, K.B. (1979) Cell-mediated cytotoxicity and humoral immune response in ascorbic acid-deficient guinea pigs. *American Journal of Clinical Nutrition*, **32**, 1691–8.

Baker, E.M., Hodges, R.E., Hood, J., Sauberlich, H.E. & March, S.C. (1969) Metabolism of ascorbic-1-^{14}C acid in experimental human scurvy. *American Journal of Clinical Nutrition*, **22**, 549–58.

Baker, E.M., Hodges, R.E., Hood, J., Sauberlich, H.E., March, S.C. & Canham, J.E. (1971) Metabolism of ^{14}C- and ^{3}H-labeled L-ascorbic acid in human scurvy. *American Journal of Clinical Nutrition*, **24**, 444–54.

Beetens, J.R. & Herman, A.G. (1983) Vitamin C increases the formation of prostacyclin by aortic rings from various species and neutralizes the inhibitory effect of 15-hydroperoxy-arachidonic acid. *British Journal of Pharmacology*, **80**, 249–54.

Bianchi, J. & Rose, R.C. (1985a) Transport of L-ascorbic acid and dehydro-L-ascorbic acid across renal cortical basolateral membrane vesicles. *Biochimica et Biophysica Acta*, **820**, 265–73.

Bianchi, J. & Rose, R.C. (1985b) Na^+-indepenent dehydro-L-ascorbic acid uptake in renal brush-border membrane vesicles. *Biochimica et Biophysica Acta*, **819**, 75–82.

Bianchi, J. & Rose, R.C. (1986) Dehydroascorbic acid and cell membranes: Possible disruptive effects. *Toxicology*, **40**, 75–82.

Bianchi, J., Wilson, F.A. & Rose, R.C. (1986) Dehydroascorbic acid and ascorbic acid transport systems in the guinea pig ileum. *American Journal of Physiology*, **250**, G461–8.

Bowers-Komro, D.M. & McCormick, D.B. (1991) Characterization of ascorbic acid uptake by isolated rat kidney cells. *Journal of Nutrition*, **121**, 57–64.

Boxer, L.A., Watanabe, A.M., Rister, M., Besch, H.R. Jr., Allen, J. & Baehner, R.L. (1976) Correction of leukocyte function in Chediak-Higashi syndrome by ascorbate. *New England Journal of Medicine*, **295**, 1041–5.

Boxer, L.A., Vanderbilt, B., Bonsib, S., Jersild, R., Yang, H.-H. & Baehner, R.L. (1979) Enhancement of chemotactic responses and microtubule assembly in human leukocytes by ascorbic acid. *Journal of Cellular Physiology*, **100**, 119–26.

Boyer, J.C., Campbell, C.E., Sigurdson, W.J. & Kuo, S.M. (2005) Polarized localization of vitamin C transporters, SVCT1 and SVCT2, in epithelial cells. *Biochemical and Biophysical Research Communications*, **334**, 150–6.

Buettner, G.R. & Jurkiewicz, B.A. (1996) Catalytic metals, ascorbate and free radicals: Combinations to avoid. *Radiation Research*, **145**, 532–41.

Burton, G.W., Wronska, U., Stone, L., Foster, D.O. & Ingold, K.U. (1990) Biokinetics of dietary *RRR*-α-tocopherol in the male guinea pig at three dietary levels of vitamin C and two levels of vitamin E. Evidence that vitamin C does not "spare" vitamin E *in vivo*. *Lipids*, **25**, 199–210.

Campbell, J.D., Cole, M., Bunditrutavorn, B. & Vella, A.T. (1999) Ascorbic acid is a potent inhibitor of various forms of T cell apoptosis. *Cellular Immunology*, 194, 1–5.

Chazan, J.A. & Mistilis, S.P. (1963) The pathophysiology of scurvy. A report of seven cases. *American Journal of Medicine*, **34**, 350–8.

Choi, J.-L. & Rose, R.C. (1989) Transport and metabolism of ascorbic acid in human placenta. *American Journal of Physiology*, **257**, C110–13.

Clemetson, C.A.B. (1980) Histamine and ascorbic acid in human blood. *Journal of Nutrition*, **110**, 662–8.

Cooper, M.R., McCall, C.E. & DeChatelet, L.R. (1971) Stimulation of leukocyte hexose monophosphate shunt activity by ascorbic acid. *Infection and Immunity*, **3**, 851–3.

Cunningham-Rundles, W.F., Berner, Y. & Cunningham-Rundles, S. (1993) Interaction of vitamin C in lymphocyte activation: Current status and possible mechanisms of action. In: *Nutrient Modulation of the Immune Response*. (Ed. S. Cunningham-Rundles), pp. 91–103. Marcel Dekker, Inc., New York.

Dahl, H. & Degré, M. (1976) The effect of ascorbic acid on production of human interferon and the antiviral activity *in vitro*. Acta Pathologica et Microbiologica Scandinavica, Section B, **84**, 280–4.

Daruwala, R., Song, J., Koh, W.S., Rumsey, S.C. & Levine, M. (1999) Cloning and functional characterization of the human sodium-dependent vitamin C transporters hSVCT1 and hSVCT2, *FEBS Letters*, **460**, 480–4.

DeChatelet, L.R., Cooper, M.R. & McCall, C.E. (1972) Stimulation of the hexose monophosphate shunt in human neutrophils by ascorbic acid: Mechanism of action. *Antimicrobial Agents and Chemotherapy*, **1**, 12–16.

Dedhar, S., Ruoslahti, E. & Pierschbacher, M.D. (1987) A cell surface receptor complex for collagen type I recognizes the Arg-Gly-Asp sequence. *Journal of Cell Biology*, **104**, 585–93.

Doba, T., Burton, G.W. & Ingold, K.U. (1985) Antioxidant and co-antioxidant activity of vitamin C. The effect of vitamin C, either alone or in the presence of vitamin E or a water-soluble vitamin E analogue, upon the peroxidation of aqueous multilamellar phospholipid liposomes. *Biochimica et Biophysica Acta*, **835**, 298–303.

Eipper, B.A. & Mains, R.E. (1991) The role of ascorbate in the biosynthesis of neuroendocrine peptides. *American Journal of Clinical Nutrition*, **54**, 1153S–56S.

Eldridge, C.F., Bunge, M.B., Bunge, R.P. & Wood, P.M. (1987) Differentiation of axon-related Schwann cells in vitro. I. Ascorbic acid regulates basal lamina assembly and myelin formation. *Journal of Cell Biology*, **105**, 1023–34.

Eldridge, C.F., Bunge, M.B. & Bunge, R.P. (1989) Differentiation of axon-related Schwann cells *in vitro*. II. Control of myelin formation by basal lamina. *Journal of Neuroscience*, **9**, 625–38.

Ericsson, Y. & Lundbeck, H. (1955) Antimicrobial effect in vitro of the ascorbic acid oxidation. I. Effect on bacteria, fungi and viruses in pure cultures. *Acta Pathologica et Microbiologica Scandinavica*, **37**, 493–506.

Evans, R.M., Currie, L. & Campbell, A. (1982) The distribution of ascorbic acid between various cellular components of blood, in normal individuals, and its relation to the plasma concentration. *British Journal of Nutrition*, **47**, 473–82.

Fraga, C.G., Motchnik, P.A., Shigenaga, M.K., Helbock, H.J., Jacob, R. A. & Ames, B.N. (1991) Ascorbic acid protects against endogenous oxidative DNA damage in human sperm. *Proceedings of the National Academy of Sciences of the U.S.A.*, **88**, 11003–6.

Franceschi, R.T. (1992) The role of ascorbic acid in mesenchymal differentiation. *Nutrition Reviews*, **50**, 65–70.

Franceschi, R.T. & Iyer, B.S. (1992) Relationship between collagen synthesis and expression of the osteoblast phenotype in MC3T3-E1 cells. *Journal of Bone and Mineral Research*, **7**, 235–46.

Frei, B. (1991) Ascorbic acid protects lipids in human plasma and low-density lipoprotein against oxidative damage. *American Journal of Clinical Nutrition*, **54**, 1113S–18S.

Ganguly, R., Durieux, M.-F. & Waldman, R.H. (1976) Macrophage function in vitamin C-deficient guinea pigs. *American Journal of Clinical Nutrition*, **29**, 762–5.

Ginter, E. (1973) Cholesterol: Vitamin C controls its transformation to bile acids. *Science*, **179**, 702–4.

Ginter, E., Červe, J., Nemec, R. & Mikuš, L. (1971) Lowered cholesterol catabolism in guinea pigs with chronic ascorbic acid deficiency. *American Journal of Clinical Nutrition*, **24**, 1238–45.

Gokce, N., Keaney, J.F. Jr., Frei, B., Holbrook, M., Olesiak, M., Zachariah, B.J., Leeuwenburgh, C., Heinecke, J.W. & Vita, J.A. (1999) Long-term ascorbic acid administration reverses endothelial vasomotor dysfunction in patients with coronary artery diseases. *Circulation*, **99**, 3234–40.

Goldberg, A. (1963) The anaemia of scurvy. *Quarterly Journal of Medicine*, **32**, 51–64.

Halliwell, B., Wasil, M. & Grootveld, M. (1987) Biologically significant scavenging of the myeloperoxidase-derived oxidant hypochlorous acid by ascorbic acid. Implications for antioxidant protection in the inflamed rheumatoid joint. *FEBS Letters*, **213**, 15–17.

Hammarström, L. (1966) Autoradiographic studies on the distribution of C^{14}-labelled ascorbic acid and dehydroascorbic acid. *Acta Physiologica Scandinavica*, **70** (Supplementum 289), 1–84.

Harada, S., Matsumoto, T. & Ogata, E. (1991) Role of ascorbic acid in the regulation of proliferation in osteoblast-like MC3T3-E1 cells. *Journal of Bone and Mineral Research*, **6**, 903–8.

Harats, D., Chevion, S., Nahir, M., Norman, Y., Sagee, O. & Berry, E.M. (1998) Citrus fruit supplementation reduces lipoprotein oxidation in young men ingesting a diet high in saturated fat: Presumptive evidence for an interaction between vitamins C and E in vivo. *American Journal of Clinical Nutrition*, **67**, 240–5.

Haskell, B.E. & Johnston, C.S. (1991) Complement component C1q activity and ascorbic acid nutriture in guinea pigs. *American Journal of Clinical Nutrition*, **54**, 1228S–30S.

Heitzer, T., Just, H. & Münzel, T. (1996) Antioxidant vitamin C improves endothelial dysfunction in chronic smokers. *Circulation*, **94**, 6–9.

Heller, R., Münscher-Paulig, F., Gräbner, R. & Till, U. (1999) L-Ascorbic acid potentiates nitric oxide synthesis in endothelial cells. *Journal of Biological Chemistry*, **274**(12), 8254–60.

Hitomi, K. & Tsukagoshi, N. (1996) Role of ascorbic acid in modulation of gene expression. In: *Subcellular Biochemistry, Vol. 25. Ascorbic Acid: Biochemistry and Biomedical Cell Biology.* (Ed. J. R. Harris), pp. 41–56. Plenum Press, New York.

Hodges, R.E., Baker, E.M., Hood, J., Sauberlich, H.E. & March, S.C. (1969) Experimental scurvy in man. *American Journal of Clinical Nutrition*, **22**, 535–48.

Hodges, R.E., Hood, J., Canham, J.E., Sauberlich, H.E. & Baker, E.M. (1971) Clinical manifestations of ascorbic acid deficiency in man. *American Journal of Clinical Nutrition*, **24**, 432–43.

Hoffer, A. (1971) Ascorbic acid and toxicity. *New England Journal of Medicine*, **285**, 635–6.

Horio, F., Ozaki, K., Oda, H., Makino, S., Hayashi, Y. & Yoshida, A. (1989) Effect of dietary ascorbic acid, cholesterol and PCB on cholesterol and bile acid metabolism in a rat mutant unable to synthesize ascorbic acid. *Journal of Nutrition*, **119**, 409–15.

Hornig, B., Arakawa, N., Kohler, C. & Drexler, H. (1998) Vitamin C improves endothelial function of conduit arteries in patients with chronic heart failure. *Circulation*, **97**, 363–8.

Hornig, D., Vuilleumier, J.-P. & Hartmann, D. (1980) Absorption of large, single, oral intakes of ascorbic acid. *International Journal for Vitamin and Nutrition Research*, **50**, 309–14.

Horovitz, O., Knaack, D., Podleski, T.R. & Salpeter, M.M. (1989) Acetylcholine receptor α-subunit mRNA is increased by ascorbic acid in cloned L5 muscle cells; Northern blot analysis and in situ hybridization. *Journal of Cell Biology*, **108**, 1823–32.

Horrobin, D.F., Oka, M. & Manku, M.S. (1979) The regulation of prostaglandin E1 formation: A candidate for one of the fundamental mechanisms involved in the actions of vitamin C. *Medical Hypotheses*, **5**, 849–58.

Huang, W., Yang, Z., Lee, D., Copolov, D.L. & Lim, A.T. (1993) Ascorbic acid enhances forskolin-induced cyclic AMP production and pro-ANF mRNA expression of hypothalamic neurons in culture. *Endocrinology*, **132**, 2271–3.

Hynes, R.O. (1992) Integrins: Versatility, modulation, and signaling in cell adhesion. *Cell*, **69**, 11–25.

Ibrahimi, A., Bonino, F., Bardon, S., Ailhaud, G. & Dani, C. (1992) Essential role of collagens for terminal differentiation of preadipocytes. *Biochemical and Biophysical Research Communications*, **187**, 1314–22.

Ingermann, R.L, Stankova, L. & Bigley, R.H. (1986) Role of monosaccharide transporter in vitamin C uptake by placental membrane vesicles. *American Journal of Physiology*, **250**, C637–41.

Institute of Medicine (2000) *Dietary Reference Imtakes for Vitamin C, Vitamin E, Selenium, and Carotenoids*. National Academy Press, Washington, D. C.

Jacob, R.A. & Pianalto, F.S. (1997) Urinary carnitine excretion increases during experimental vitamin C depletion of healthy men. *Journal of Nutritional Biochemistry*, **8**, 265–9.

Johnston, C.S. (1991) Complement component C1q unaltered by ascorbate supplementation in healthy men and women. *Journal of Nutritional Biochemistry*, **2**, 499–501.

Johnston, C.S., Martin, L.J. & Cai, X. (1992) Antihistamine effect of supplemental ascorbic acid and neutrophil chemotaxis. *Journal of the American College of Nutrition*, **11**, 172–6.

Johnston, C.S., Solomon, R.E. & Corte, C. (1996) Vitamin C depletion is associated with alterations in blood histamine and plasma free carnitine in adults. *Journal of the American College of Nutrition*, **15**, 586–91.

Kallner, A., Hartmann, D. & Hornig, D. (1979) Steady-state turnover and body pool of ascorbic acid in man. *American Journal of Clinical Nutrition*, **32**, 530–9.

Kallner, A.B., Hartmann, D. & Hornig, D.H. (1981) On the requirements of ascorbic acid in man: Steady-state turnover and body pool in smokers. *American Journal of Clinical Nutrition*, **34**, 1347–55.

Kallner, A., Hornig, D. & Pellikka, R. (1985) Formation of carbon dioxide from ascorbate in man. *American Journal of Clinical Nutrition*, **41**, 609–13.

Kanani, P.M., Sinkey, C.A., Browning, R.L., Allaman, M., Knapp, H.R. & Haynes, W.G. (1999) Role of oxidant stress in endothelial dysfunction produced by experimental hyperhomocyst(e)inemia in humans. *Circulation*, **100**, 1161–8.

Karasov, W.H., Darken, B.W. & Bottum, M.C. (1991) Dietary regulation of intestinal ascorbate uptake in guinea pigs. *American Journal of Physiology*, **260**, G108–18.

Karpińska, T., Kawecki, Z. & Kandefer-Szerszeń, M. (1982) The influence of ultraviolet irradiation, L-ascorbic acid and calcium chloride on the induction of interferon in human embryo fibroblasts. *Archivum Immunologiae et Therapiae Experimentalis*, **30**, 33–7.

Kay, N.E., Holloway, D.E., Hutton, S.W., Bone, N.D. & Duane, W.C. (1982) Human T-cell function in experimental ascorbic acid deficiency and spontaneous scurvy. *American Journal of Clinical Nutrition*, **36**, 127–30.

Kinsman, R.A. & Hood, J. (1971) Some behavioral effects of ascorbic acid deficiency. *American Journal of Clinical Nutrition*, **24**, 455–64.

Knaak, D., Shen, I., Salpeter, M.M. & Podleski, T.R. (1986) Selective effects of ascorbic acid on acetylcholine receptor number and distribution. *Journal of Cell Biology*, **102**, 795–802.

Kübler, W. & Gehler, J. (1970) On the kinetics of the intestinal absorption of ascorbic acid: A contribution to the calculation of an absorption process that is not proportional to the dose. *International Journal for Vitamin and Nutrition Research*, **40**, 442–53. (In German)

Kuo, C.-H., Hata, F., Yoshida, H., Yamatodani, A. & Wada, H. (1979) Effect of ascorbic acid on release of acetylcholine from synaptic vesicles prepared from different species of animals and release of noradrenaline from synaptic vesicles of rat brain. *Life Sciences*, **24**, 911–16.

Leboy, P.S., Vaias, L., Uschmann, B., Golub, E., Adams, S.L. & Pacifici, M. (1989) Ascorbic acid induces alkaline phosphatase, type X collagen, and calcium deposition in cultured chick chondrocytes. *Journal of Biological Chemistry*, **264**(29), 17281–6.

Leichtweiss, H.-P., Lisbôa, B. & Steinborn, C. (1987) Uptake of ascorbic acid and its oxidized products in the isolated guinea pig placenta. *Trophoblast Research*, **2**, 501–14.

Levine, G.N., Frei, B., Koulouris, S.N., Gerhard, M.D., Keaney, J.F. Jr. & Vita, J.A. (1996) Ascorbic acid reverses endothelial vasomotor dysfunction in patients with coronary artery disease. *Circulation*, **93**, 1107–13.

Levine, M., Rumsey, S.C., Daruwala, R., Park, J.B. & Wang, Y. (1999) Criteria and recommendations for vitamin C intake. *Journal of the American Medical Association*, **281**, 1415–23.

Liang, W.-J., Johnson, D. & Jarvis, S.M. (2001) Vitamin C transport systems of mammalian cells. *Molecular Membrane Biology*, **18**, 87–95.

Lunec, J. & Blake, D.R. (1985) The determination of dehydroacorbic acid and ascorbic acid in the serum and synovial fluid of patients with rheumatoid arthritis. *Free Radical Research Communications*, **1**, 31–9.

Lynch, S.M., Gaziano, J.M. & Frei, B. (1996) Ascorbic acid and atherosclerotic cardiovacular disease. In: *Subcellular Biochemistry, Vol. 25. Ascorbic Acid: Biochemistry and Biomedical Cell Biology*. (Ed. J.R. Harris), pp. 331–67. Plenum Press, New York.

MacDonald, L., Thumser, A.E. & Sharp, P. (2002) Decreased expression of the vitamin C transporter SVCT1 by ascorbic acid in a human intestinal epithelial cell line. *British Journal of Nutrition*, **87**, 97–100.

Mangels, A.R. Block, G., Frey, C.M., Patterson, B.H., Taylor, P.R., Norkus, E.P. & Levander, O.A. (1993) The bioavailability to humans of ascorbic acid from oranges, orange juice and cooked broccoli is similar to that of synthetic ascorbic acid. *Journal of Nutrition*, **123**, 1054–61.

Manku, M.S., Oka, M. & Horrobin, D.F. (1979) Differential regulation of the formation of prostaglandins and related substances from arachidonic acid and from dihomogammalinolenic acid. II. Effects of vitamin C. *Prostaglandins and Medicine*, **3**, 129–37.

McCall, C.E., DeChatelet, L.R., Cooper, M.R. & Ashburn, P. (1971) The effects of ascorbic acid on bactericidal mechanisms of neutrophils. *Journal of Infectious Diseases*, **124**, 194–8.

McGee, M.P. & Myrvik, Q.N. (1979) Phagocytosis-induced injury of normal and activated alveolar macrophages. *Infection and Immunity*, **26**, 910–15.

Mellors, A.J., Nahrwold, D.L. & Rose, R.C. (1977) Ascorbic acid flux across mucosal border of guinea pig and human ileum. *American Journal of Physiology*, **233**, E374–9.

Menniti, F.S., Knoth, J. & Diliberto, E.J. Jr. (1986) Role of ascorbic acid in dopamine β-hydroxylation. *Journal of Biological Chemistry*, **261**(36), 16901–8.

Miller, T.E. (1969) Killing and lysis of gram-negative bacteria through the synergistic effect of hydrogen peroxide, ascorbic acid, and lysozyme. *Journal of Bacteriology*, **98**, 949–55.

Miller, B.T. & Cicero, T.J. (1986) Ascorbic acid enhances the release of luteinizing hormone-releasing hormone from the mediobasal hypothalamus *in vitro*. *Life Sciences*, **39**, 2447–54.

Mitsumoto, Y., Liu, Z. & Klip, A. (1994) A long-lasting vitamin C derivative, ascorbic acid 2-phosphate, increases myogenin gene expression and promotes differentiation in L6 muscle cells. *Biochemical and Biophysical Research Communications*, **199**, 394–402.

Mosca, L., Rubenfire, M., Mandel. C., Rock, C., Tarshis, T., Tsai, A. & Pearson, T. (1997) Antioxidant nutrient supplementation reduces the susceptibility of low density lipoprotein to oxidation in patients with coronary artery disease. *Journal of the American College of Cardiology*, **30**, 392–9.

Moser, U. (1987) Uptake of ascorbic acid by leukocytes. *Annals of the New York Academy of Sciences*, **498**, 200–15.

Muggli, R. (1993) Vitamin C and phagocytes. In: *Nutrient Modulation of the Immune Response*. (Ed. S. Cunningham-Rundles), pp. 75–90. Marcel Dekker, Inc., New York.

Nandan, D., Clarke, E.P., Ball, E.H. & Sanwal, B.D. (1990) Ethyl-3,4-dihydroxybenzoate inhibits myoblast differentiation: Evidence for an essential role of collagen. *Journal of Cell Biology*, **110**, 1673–9.

Oberritter, H., Glatthaar, B., Moser, U. & Schmidt, K.H. (1986) Effect of functional stimulation on ascorbate content in phagocytes under physiological and pathological conditions. *International Archives of Allergy and Applied Immunology*, **81**, 46–50.

Oh, C. & Nakano, K. (1988) Reversal by ascorbic acid of supresssion by endogenous histamine of rat lymphocyte blastogenesis. *Journal of Nutrition*, **118**, 639–44.

Patterson, L.T., Nahrwold, D.L. & Rose, R.C. (1982) Ascorbic acid uptake in guinea pig intestinal mucosa. *Life Sciences*, **31**, 2783–91.

Pelletier, O. & Godin, C. (1969) Vitamin C activity of D-isoascorbic acid for the guinea pig. *Canadian Journal of Physiology and Pharmacology*, **47**, 985–91.

Prasad, P.D., Huang, W., Wang, H., Leibach, F.H. & Ganapathy, V. (1998) Transport mechanisms for vitamin C in the JAR human placental choriocarcinoma cell line. *Biochimica et Biophysica Acta*, **1369**, 141–51.

Priest, R.E. (1970) Formation of epithelial basement membrane is restricted by scurvy *in vitro* and is stimulated by vitamin C. *Nature*, **225**, 744–5.

Prockop, D.J., Kivirikko, K.I., Tuderman, L. & Guzman, N.A. (1979) The biosynthesis of collagen and its disorders. *New England Journal of Medicine*, **301**, 13–23.

Quarles, L.D., Yohay, D.A., Lever, L.W., Caton, R. & Wenstrup, R.J. (1992) Distinct proliferative and differentiated stages of murine MC3T3-E1 cells in culture: An in vitro model of osteoblast development. *Journal of Bone and Mineral Research*, **7**, 683–92.

Rajan, D.P., Huang, W., Dutta, B., Devoe, L.D., Leibach, F.H., Ganapathy, V. & Prasad, P.D. (1999) Human placental sodium-dependent vitamin C transporter (SVCT2): Molecular cloning and transport function. *Biochemical and Biophysical Research Communications*, **262**, 762–8.

Rebec, G.V. & Pierce, R.C. (1994) A vitamin as neuromodulator: Ascorbate release into the extracellular fluid of the brain regulates dopaminergic and glutamatergic transmission. *Progress in Neurobiology*, **43**, 537–65.

Rebouche, C.J. (1988) Carnitine metabolism and human nutrition. *Journal of Applied Nutrition*, **40**, 99–111.

Rebouche, C.J. (1991) Ascorbic acid and carnitine biosynthesis. *American Journal of Clinical Nutrition*, **54**, 1147S–52S.

Rebouche, C.J. (1995a) The ability of guinea pigs to synthesize carnitine at a normal rate from ε-*N*-trimethyllysine or γ-butyrobetaine in vivo is not compromised by experimental vitamin C deficiency. *Metabolism. Clinical and Experimental*, **44**, 624–9.

Rebouche, C.J. (1995b) Renal handling of carnitine in experimental vitamin C deficiency. *Metabolism. Clinical and Experimental*, **44**, 1639–43.

Retsky, K.L. & Frei, B. (1995) Vitamin C prevents metal ion-dependent initiation and propagation of lipid peroxidation n human low-density lipoprotein. *Biochimica et Biophysica Acta*, **1257**, 279–87.

Rifici, V.A. & Khachadurian, A.K. (1993) Dietary supplementation with vitamins C and E inhibits in vitro oxidation of lipoproteins. *Journal of the American College of Nutrition*, **12**, 631–7.

Rose, R.C. (1986) Ascorbic acid transport in mammalian kidney. *American Journal of Physiology*, **250**, F627–32.

Rose, R.C. (1989) Renal metabolism of the oxidized form of ascorbic acid (dehydro-L-ascorbic acid). *American Journal of Physiology*, **256**, F52–6.

Rose, R.C., Choi, J.-L. & Koch, M.J. (1988) Intestinal transport and metabolism of oxidized ascorbic acid (dehydroascorbic acid). *American Journal of Physiology*, **254**, G824–8.

Rumsey, S.C. & Levine, M. (1998) Absorption, transport, and disposition of ascorbic acid in humans. *Journal of Nutritional Biochemistry*, **9**, 116–30.

Rumsey, S.C., Kwon, O., Xu, G.W., Burant, C.F., Simpson, I. & Levine, M. (1997) Glucose transporter isoforms GLUT1 and GLUT3 transport dehydroascorbic acid. *Journal of Biological Chemistry*, **272**(30), 18982–9.

Rybakowski, C., Mohar, B., Wohlers, S., Leichtweiss, H.-P. & Schröder, H. (1995) The transport of vitamin C in the isolated near-term placenta. *European Journal of Obstetrics & Gynecology and Reproductive Biology*, **62**, 107–114.

Sacharin, R., Taylor, T. & Chasseaud, L.F. (1976) Blood levels and bioavailability of ascorbic acid after administration of a sustained-release formulation to humans. *International Journal for Vitamin and Nutrition Research*, **47**, 68–74.

Sakuma, N., Yoshikawa, M., Hibino, T., Sato, T., Kamiya, Y., Ohte, N., Tamai, N., Kunimatsu, M., Kimura, G. & Inoue, M. (2001) Ascorbic acid protects against peroxidative modification of low-density lipoprotein, maintaining its recognition by LDL receptors. *Journal of Nutritional Science and Vitaminology*, **47**, 28–31.

Sauberlich, H.E. (1994) Pharmacology of vitamin C. *Annual Review of Nutrition*, **14**, 371–91.

Schmidt, K. & Moser, U. (1985) Vitamin C – A modulator of host defense mechanisms. International Journal for Vitamin and Nutrition Research, Supplement No. 27, 363–79.

Schwarz, R.I., Kleinman, P. & Owens, N. (1987) Ascorbate can act as an inducer of the collagen pathway because most steps are tightly coupled. *Annals of the New York Academy of Sciences*, **498**, 172–85.

Shilotri, P.G. (1977) Glycolytic, hexose monophosphate shunt and bactericidal activities of leukocytes in ascorbic acid deficient guinea pigs. *Journal of Nutrition*, **107**, 1507–12.

Siegel, B.V. (1974) Enhanced interferon response to murine leukemia virus by ascorbic acid. *Infection and Immunity*, **10**, 409–10.

Sies, H. & Stahl, W. (1995) Vitamins E and C, β-carotene, and other carotenoids as antioxidants. *American Journal of Clinical Nutrition*, **62**, 1315S–21S.

Solzbach, U., Hornig, B., Jeserich, M. & Just, H. (1997) Vitamin C improves endothelial dysfunction of epicardial coronary arteries in hypertensive patients. *Circulation*, **96**, 1513–19.

Spector, R. (1977) Vitamin homeostasis in the central nervous system. *New England Journal of Medicine*, **296**, 1393–8.

Subramanian, V.S., Srinivasan, P., Wildman, A.J., Marchant, J.S. & Said, H.M. (2017) Molecular mechanism(s) involved in differential expression of vitamin C transporters along the intestinal tract. *American Journal of Physiology (Gastrointestinal and Liver Physiology)*, **312**, G340–G347.

Sullivan, T.A., Uschmann, B., Hough, R. & Leboy, P.S. (1994) Ascorbate modulation of chondrocyte gene expression is independent of its role in collagen secretion. *Journal of Biological Chemistry*, **269**(36), 22500–6.

Ting, H.H., Timimi, F.K., Boles, K.S., Creager, S.J., Ganz, P. & Creager, M.A. (1996) Vitamin C improves endothelium-dependent vasodilation in patients with non-insulin-dependent diabetes mellitus. *Journal of Clinical Investigation*, **97**, 22–8.

Ting, H.H., Timimi, F.K., Haley, E.A., Roddy, M.-A., Ganz, P. & Creager, M.A. (1997) Vitamin C improves endothelium-dependent vasodilation in forearm resistance vessels of humans with hypercholesterolemia. *Circulation*, **95**, 2617–22.

Toggenburger, G., Häusermann, M., Mütsch, B., Genoni, G., Kessler, M., Weber, F., Hornig, D., O'Neill, B. & Semenza, G. (1981) Na^+-dependent, potential-sensitive L-ascorbate transport across brush border membrane vesicles from kidney cortex. *Biochimica et Biophysica Acta*, **646**, 433–43.

Tsukaguchi, H., Tokui, T., Mackenzie, B., Berger, U. V., Chen, X.-Z., Wang, Y., Brubaker, R.F. & Hediger, M.A. (1999) A family of mammalian Na^+-dependent L-ascorbic acid transporters. *Nature*, **399**, 70–75.

Uchida, K., Mitsui, M. & Kawakishi, S. (1989) Monooxygenation of *N*-acetylhistamine mediated by L-ascorbic acid. *Biochimica et Biophysica Acta*, **991**, 377–9.

Uchida, K., Nomura, Y., Takase, H., Tasaki, T., Seo, S., Hayashi, Y. & Takeuchi, N. (1990) Effect of vitamin C depletion on serum cholesterol and lipoprotein levels in ODS (*od*/*od*) rats unable to synthesize ascorbic acid. *Journal of Nutrition*, **120**, 1140–7.

Vera, J.C., Rivas, C.I., Velásquez, F.V., Zhang, R.H., Concha, I.I. & Golde, D.W. (1995) Resolution of the facilitated transport of dehydroascorbic acid from its intracellular accumulation as ascorbic acid. *Journal of Biological Chemistry*, **270**(40), 23706–12.

Vojdani, A. & Ghoneum, M. (1993) In vivo effect of ascorbic acid on enhancement of human natural killer cell activity. *Nutrition Research*, **13**, 753–64.

Washko, P.W., Wang, Y. & Levine, M. (1993) Ascorbic acid recycling in human neutrophils. *Journal of Biological Chemistry*, **268**(21), 15531–5.

Weiss, S.J. (1989) Tissue destruction by neutrophils. *New England Journal of Medicine*, **320**, 365–76.

Welch, R.W., Wang, Y., Crossman, A. Jr., Park, J.B., Kirk, K.L. & Levine, M. (1995) Accumulation of vitamin C (ascorbate) and its oxidized metabolite dehydroascorbic acid occurs by separate mechanisms. *Journal of Biological Chemistry*, **270**(21), 12584–92.

Wells, W.W., Dou, C.-Z., Dybas, L.N., Jung, C.-H., Kalbach, H.L. & Xu, D.P. (1995) Ascorbic acid is essential for the release of insulin from scorbutic guinea pig pancreatic islets. *Proceedings of the National Academy of Sciences of the U.S.A.*, **92**, 11869–73.

Wilson, J.X. (1989) Ascorbic acid uptake by a high-affinity sodium-dependent mechanism in cultured astrocytes. *Journal of Neurochemistry*, **53**, 1064–71.

Wilson, J.X. & Dixon, S.J. (1989) High-affinity sodium-dependent uptake of ascorbic acid by rat osteoblasts. *Journal of Membrane Biology*, **111**, 83–91.

Wilson, J.X., Jaworski, E.M. & Dixon, S.J. (1991) Evidence for electrogenic sodium-dependent ascorbate transport in rat astroglia. *Neurochemical Research*, **16**, 73–8.

Yung, S., Mayersohn, M. & Robinson, J.B. (1981) Ascorbic acid absorption in man: Influence of divided dose and food. *Life Sciences*, **28**, 2505–11.

Index

Page numbers in *italic* indicate figures and **bold** indicates tables.